T0301006

Numerical Methods for Scientists and Engineers

Numerical Methods for Scientists and Engineers: With Pseudocodes is designed as a primary textbook for a one-semester course on Numerical Methods for sophomore or junior-level students. It covers the fundamental numerical methods required for scientists and engineers, as well as some advanced topics which are left to the discretion of instructors.

The objective of the text is to provide readers with a strong theoretical background on numerical methods encountered in science and engineering, and to explain how to apply these methods to practical, real-world problems. Readers will also learn how to convert numerical algorithms into running computer codes.

Features:

- Numerous pedagogic features including exercises, "pros and cons" boxes for each method discussed, and rigorous highlighting of key topics and ideas
- Suitable as a primary text for undergraduate courses in numerical methods, but also as a reference to working engineers
- A Pseudocode approach that makes the book accessible to those with different (or no) coding backgrounds, which does not tie instructors to one particular language over another
- A dedicated website featuring additional code examples, quizzes, exercises, discussions, and more: https://github.com/zaltac/NumMethodsWPseudoCodes
- A complete Solution Manual and PowerPoint Presentations are available (free of charge) to instructors at www.routledge.com/9781032754741

Numerical Analysis and Scientific Computing Series

Series Editors:

Frederic Magoules, Choi-Hong Lai

About the Series

This series, comprising of a diverse collection of textbooks, references, and handbooks, brings together a wide range of topics across numerical analysis and scientific computing. The books contained in this series will appeal to an academic audience, both in mathematics and computer science, and naturally find applications in engineering and the physical sciences.

Modelling with Ordinary Differential Equations
A Comprehensive Approach
Alfio Borzì

Numerical Methods for Unsteady Compressible Flow Problems
Philipp Birken

A Gentle Introduction to Scientific Computing
Dan Stanescu, Long Lee

Introduction to Computational Engineering with MATLAB
Timothy Bower

An Introduction to Numerical Methods
A MATLAB® Approach, Fifth Edition
Abdelwahab Kharab, Ronald Guenther

The Sequential Quadratic Hamiltonian Method
Solving Optimal Control Problems
Alfio Borzì

Advances in Theoretical and Computational Fluid Mechanics
Existence, Blow-up, and Discrete Exterior Calculus Algorithms
Terry Moschandreou, Keith Afas, Khoa Nguyen

Stochastic Methods in Scientific Computing
From Foundations to Advanced Techniques
Massimo D'Elia, Kurt Langfeld, Biagio Lucini

For more information about this series please visit: https://www.crcpress.com/Chapman--HallCRC-Numerical-Analysis-and-Scientific-Computing-Series/book-series/CHNUANSCCOM

Numerical Methods for Scientists and Engineers

With Pseudocodes

Zekeriya Altaç

CRC Press
Taylor & Francis Group
Boca Raton London New York

CRC Press is an imprint of the
Taylor & Francis Group, an **informa** business

A CHAPMAN & HALL BOOK

Designed cover image: peterschreiber.media/Shutterstock

First edition published 2025
by CRC Press
2385 NW Executive Center Drive, Suite 320, Boca Raton FL 33431

and by CRC Press
4 Park Square, Milton Park, Abingdon, Oxon, OX14 4RN

CRC Press is an imprint of Taylor & Francis Group, LLC

© 2025 Zekeriya Altaç

Library of Congress Cataloging-in-Publication Data
Names: Altaç, Zekeriya, author.
Title: Numerical methods for scientists and engineers : with pseudocodes /
Zekeriya Altaç.
Description: First edition. | Boca Raton : C&H/CRC Press, 2025. | Series:
Chapman and Hall/CRC numerical analysis and scientific computing series
| Includes bibliographical references and index.
Identifiers: LCCN 2024020497 (print) | LCCN 2024020498 (ebook) | ISBN
9781032754741 (hardback) | ISBN 9781032756424 (paperback) | ISBN
9781003474944 (ebook)
Subjects: LCSH: Numerical analysis--Textbooks. | Numerical analysis--Data
processing--Textbooks.
Classification: LCC QA297 .A538 2025 (print) | LCC QA297 (ebook) | DDC
518--dc23/eng/20240522
LC record available at https://lccn.loc.gov/2024020497
LC ebook record available at https://lccn.loc.gov/2024020498

ISBN: 978-1-032-75474-1 (hbk)
ISBN: 978-1-032-75642-4 (pbk)
ISBN: 978-1-003-47494-4 (ebk)

DOI: 10.1201/9781003474944

Typeset in Latin Modern font
by KnowledgeWorks Global Ltd.

Publisher's note: This book has been prepared from camera-ready copy provided by the authors.

Access the Instructor Resources: www.routledge.com/9781032754741

I dedicate this book to my beloved uncle Aydın BİNGÖL, whom I have always looked up to in all my endeavors. Without his support, I would not have been able to get my master's and doctoral degrees in the USA and to be academically where I am today. I will always be grateful to him.

Contents

List of Figures

List of Tables

Preface

This book is intended primarily as a core textbook for courses taught under the general title of "Numerical Methods." It is also suitable as a main or supplementary textbook for upper-level undergraduate or graduate courses, as well as a reference book for practicing engineers.

The objective of the text can be summarized as follows: (i) Provide a strong theoretical background on numerical methods encountered in science and engineering education and beyond; (ii) Understand and apply the methods to practical problems; (iii) Develop the ability to choose a suitable method for a given problem; (iv) Gain the skill of converting numerical algorithms into running computer codes, emphasized *via* pseudocodes. I explain in detail the strategy implemented in the text to achieve the aforementioned objectives under the following main headings:

A. LEARNING OBJECTIVES

Each chapter begins with the "Learning Objectives," which briefly describe educational goals and competencies to be achieved. The theoretical and applied objectives make up the broader aims of the course. The learning objectives inform the students what they should master by the end of each chapter.

B. CONTENT

The text content is designed for a one-semester course. It consists of 11 chapters covering the fundamental methods as well as some advanced topics with significant reference values, which are left to the discretion of instructors. In the arrangement of the chapters, the contents of the subsequent chapters have been taken into account so that a student new to a subject can digest the course material chapter by chapter. In cases where there is little interdependence between the chapters, a sufficiently brief introduction to the pertinent topic is provided before it is covered in greater detail in the following chapter(s).

My thoughts on how to teach and design the course and its contents have been shaped over the past three decades while teaching engineering students. The S&E students want to see the relevance of this math course and its material in their professional education. Accordingly, in each chapter, the importance and application areas of the topics in S&E are briefly stated. Every method is kept as simple as possible, and proofs have only been included where they help students understand the underlying theory of the method better.

The S&E students learn best when they are motivated by solved examples and end-of-chapter exercises. Besides, this is a field in which problem solving by hand is essential to understanding the material, so a problem-solving perspective is adopted in this text. Immediately after a method is presented, its application is reinforced with example problems that comprehensively and thoroughly illustrate and explain the complexities and potential difficulties of the method. Most examples and exercise problems consist of problems with known exact solutions, allowing a clear and concise assessment of the performance of the method. There are a total of 108 solved example problems, in which the solution is explained step-by-step and through error assessments. Furthermore, maximum effort has been put into preparing examples and end-of-chapter exercise problems from various disciplines, not only to demonstrate the relevance of the methods in S&E but also to illustrate the strengths and weaknesses of the presented methods. Each chapter ends with a **Closure** that summarizes the key features of the numerical methods presented.

C. ATTENTION BOXES

In order to grab the attention of students, key points, good practices, pitfalls, or warnings in connection with the application of a numerical method are highlighted throughout the text using the "attention boxes" with a hand pointing to information provided.

D. "PROS AND CONS" BOXES

Numerical methods devised for a specific mathematical problem differ from each other in objective, scope, degree of accuracy, amount of computation, and so on. In this regard, no single numerical method is suitable for handling every problem in its class. Consequently, there are usually several methods to solve a given problem numerically. For this reason, the basics of a numerical method and associated errors are presented in the theoretical part, and a clear and concise idea of when the method performs well or poorly is provided as well. This kind of knowledge is critical and necessary, even if one uses *modules* or *procedures* from available software packages. As a result, the ability to select a suitable method for a numerical task is crucial for the accurate and successful implementation of any numerical method. In this context, after a numerical method is presented, its advantages and disadvantages are presented in a "Pros and Cons Box." With this approach, by the end of the course, the students are expected to have gained the ability to question and understand the purpose and limitations of the methods or modules before using them.

E. USE OF PSEUDOCODES

All numerical algorithms in this text are presented as self-contained pseudomodules that can be converted and used in any program. The modules are written as simply as possible (with very few commands) so that translation to other programming platforms is as easy as possible. This approach blends the algorithms and methods with sufficient mathematical theory and supports easy implementation of the algorithms.

Why pseudocode? Today's engineers and scientists are expected to become proficient in one or more programming languages (C/C++, Python, Visual Basic, Fortran 90/95, etc.), as well as a number of software packages (MatLab, Mathematica, Maple, etc.) known as the *Computer* or *Symbolic Algebra System* (CAS or SAS). To be more general and inclusive, the text is not tied to any specific programming language or to any specific CAS software. This approach aims to teach numerical methods that students will encounter in their future courses or careers while enabling them to gain the ability to prepare and/or use well-structured programs in a variety of languages.

There are basically three main reasons why not to adopt popular "programming languages" or CAS software:

(1) Experience has shown over the last several decades that programming languages and/or software are highly transient in nature. They can become partially or completely

popular in one field (or course) or unpopular in another, or they can be replaced very quickly by other language or software in certain disciplines.

(2) Teaching numerical methods together with a particular programming language, or CAS, runs the risk of blinding the student to the underlying principles of the methods. Such practices inadvertently encourage students to simply copy, compile, and run the given source codes (or simply use or type the specified built-in *function, procedure,* or *module*) rather than attempt to understand the methods in depth, including their limitations or advantages. Furthermore, this attitude generally leads students to believe that any method can be applied to similar problems by utilizing readily available codes or software without having any knowledge or sufficient understanding of the relevant details.

(3) A majority of problems that scientists and engineers face today cannot be tackled by employing a single method (or ready-to-use modules). At some point, programming proficiency is required to implement a series of numerical methods within a broader computer program to achieve a more global computational objective. In this respect, the best way to learn numerical methods is to reinforce the application skills with computer programming while taking the "Numerical Methods" course.

The motivation for adopting a pseudocode approach is that the texts providing brief algorithms target an audience with some programming experience. A beginner-level undergraduate student without sufficient hands-on programming skills, in general, struggles to convert such algorithms into running computer programs. The pseudocodes adopted in the text mimic the full features of a real program, including input/output statements, structured coding, and line-by-line explanatory comments and annotations, so that translation to other programming languages is as simple as possible. By incorporating the pseudocode approach adopted here as a teaching tool, a student with minimum programming skills should be able to not only grasp the logic of the numerical method but also gain insight on how to implement it in a programming language of his or her choice.

The text contains a total of 96 pseudocodes. Every numerical algorithm is presented as a completely independent pseudomodule that can be converted and used with other programs. Some pseudocodes use one or more of the pseudomodules presented in prior sections or chapters. This text also puts forth a convention for writing pseudocodes since there is no common standard. However, basic structured programming statements and commands of common programming languages are adopted in this pseudocode convention, which is presented in the Appendix section under the heading "How to Read and Write Pseudocodes." With this as a reference, a student with an elementary programming background can easily follow, write, and convert the pseudocodes to any programming language s/he desires.

F. END-OF-CHAPTER "EXERCISE" PROBLEMS:

The text includes more than 1500 end-of-chapter exercise problems presented at the end of each chapter, organized according to sections. These problems emphasize the mechanics of the method and are designed to give practice in applying the algorithms presented. The majority of the exercises are made up of problems with true (exact) solutions, allowing for an accurate quantitative performance assessment of the methods. Some problems that are purely mathematical in nature are designed not only to reinforce methods through practice but also to demonstrate the strengths and weaknesses of a method. Additional applied or realistic problems are chosen from S&E disciplines to help reinforce understanding of methods and demonstrate their relevance in applied problems.

The exercises are designed to be assigned to students as self-study or homework problems. Some exercises do require computer programming to carry out calculations. Even in such cases, it is important for the students to solve most of the exercise problems or, at least, the first few steps by hand or by using spreadsheets to ensure full comprehension of the details of an algorithm before attempting to program it. Such exercise problems can also be used to validate the results of a computer program or to compare them with the results of other available software.

G. END-OF-CHAPTER "COMPUTER ASSIGNMENTS"

It is important for students to be able to not only select and use suitable modules (methods) from the pertinent software but also program the numerical algorithms in a language of their (or the instructor's) choice. To develop programming and parametric analysis skills, additional exercise problems 115 are provided as "Computer Assignments" at the end of each chapter, which require using and/or writing and running computer code (of the instructor's choice) and analyzing the results. The instructors may use or modify some exercises or computer assignment problems to allow methodological analysis and comparisons. By the end of the course, the student should realize and appreciate the importance of this skill in his profession.

A NOTE TO THE INSTRUCTOR

This author's experience and thoughts on teaching numerical methods were shaped as follows: applying the course contents effectively to practical problems requires theoretical knowledge as well as computational experience. The knowledge of how well or poorly a method will perform is extremely important and necessary. The student should be able to understand the purpose and limitations of a method or module to know whether it applies to his problem or not. Most problems that today's professionals deal with in practice require the use of several numerical methods and cannot be simply solved by adopting a standard module or subprogram. For such problems, students should be able to find solutions by putting together a number of suitable numerical methods and developing skills in adopting numerical methods for a new problem.

The course instructor has a significant influence on the direction of the course and what students can gain from it. In this context, this text empowers the instructors to utilize Excel and/or MatLab, Mathematica, MathCad, and similar CAS software. Besides implementing readily available modules or procedures specific to a particular algorithm, students can be directed to develop simple, well-structured programs to extend the base capabilities of such software environments. Alternatively, this strategy can be employed in programming languages such as Python, C/C++, Visual Basic, Fortran 90, and so on. Combined with realistic problems that require analysis, this approach allows students to be able to identify, select, and employ algorithms (or modules) for a specific task, allowing them to reinforce the numerical methods by experimenting with competing methods and helping them to explore their advantages and limitations.

I will be maintaining the webpage, set up specifically for this text to aid the students as well as instructors, which can be accessed at

https://github.com/zaltac/NumMethodsWPseudoCodes

It will be dynamic, and I will be posting modules in C/C++, Python, Visual Basic, Fortran, Mathematica and MatLab, PowerPoint presentations, solutions to additional exercises, errata, and more. I encourage the instructors who adopt this text as a coursebook to provide me with feedback on how I can improve the materials that I will be sharing on this website.

A NOTE TO THE STUDENT

Numerical methods is a very different area of mathematics and certainly different from your previous math courses. In math courses, the tasks are clearly defined, with a very concrete notion of "the right answer." Here, we are concerned with computing approximations, and this involves a slightly different kind of thinking. We have to understand what we are approximating well enough to construct a reasonable approximation, and we have to be able to think clearly and logically enough to analyze the accuracy and performance of that approximation.

The pseudocodes are not intended to be programs with all possible routes or advanced programming techniques. They are designed in a simple way that will help you understand and implement the algorithms. Writing your own programs will help you understand how the method works. In this regard, the computational effort that you put in is an important part of learning numerical methods.

ACKNOWLEDGMENTS

I am indebted to a number of people for their encouragement and assistance during the preparation of this and previous texts. I would like to express my gratitude to all of my students and colleagues for their support and encouragement. Particular thanks are due to Necati MAHİR, Hasan Hüseyin ERKAYA, Nevzat KIRAÇ, Mesut TEKKALMAZ, Nihal UĞURLUBİLEK, and Zerrin SERT for proofreading the Turkish or English versions of the manuscript. I would like to extend a special thanks to Gökçe Mehmet AY for reintroducing me to the LaTeX world. I would also like to thank Callum Fraser of CNC Press for his continued interest and advice in making this project possible.

I thank my mother Muzaffer, my siblings Mithat, Afitap, and Asuman, my daughter Delilah, and my son-in-law Mahmut for their unconditional love and support, as well as my grandchildren Efe and Meryem, who fill my heart with joy each and every day. Finally, I thank my wife Leslie for her love, enduring patience, and unconditional support, without which the book would probably not have been completed.

Zekeriya ALTAÇ
April, 2024.
Eskişehir, Turkiye

e-mail: altacz@gmail.com
https://github.com/zaltac/NumMethodsWPseudoCodes

Author Bio

Zekeriya Altaç is currently a professor of mechanical engineering at Eskişehir Osmangazi University and he received his B.Sc. degree in 1983 from Istanbul Technical University in Turkey. He received a scholarship from the Ministry of Education of Turkey to study nuclear engineering in the USA. He attended Iowa State University and earned his M.Sc. (1986) and Ph.D. (1989) in Nuclear Engineering. Upon returning to Turkey, he worked as an assistant professor at the Mechanical Engineering Department of Anadolu University between 1989 and 1992 in Eskişehir, Turkey. Later, he joined the Department of Mechanical Engineering at Eskişehir Osmangazi University as an associate professor in 1993 and received a tenure-track position in 1998, where he is currently working. His areas of interest and expertise are computational nuclear reactor physics, computational fluid mechanics and heat transfer, and numerical (computational) methods in general. He is the author of over 100 journal articles, conference and research papers, books, and book translations. The author has books in Turkish on computer programming (BASIC, FORTRAN) and the application of CFD software (ANSYS, FLUENT). Most of his research deals with developing efficient numerical methods and/or employing computational techniques in practical scientific and engineering problems. He served as an advisor at the Von Karman Institute Technical Advisory Committee (VKI-TAC) in Brussels, Belgium, between 2009 and 2011. He has also served as Eskişehir Osmangazi University's vice president (2007-2011), as head of the Mechanical Engineering Department (2012-2021), and held positions on numerous conferences, symposiums, and academic committees, such as the engineering college board and university senate.

Numerical Algorithms and Errors

NUMERICAL algorithms were first developed for hand calculations in the 19th and early 20th centuries to tackle mathematical problems such as solving small systems of linear equations, inverting matrices, approximating integrals, and solving ordinary differential equations. Early numerical solutions were obtained manually until the introduction of the computer after World War II. Rapid technological advances led to the development of the high-speed computers we use today. As a result, making use of the computer's capability to perform long sequences of calculations has brought about advances and further technological developments in many fields, and computers and mathematical sciences naturally became inseparable.

Numerical analysis, a whole new subject born out of the advent of computers, is a branch of mathematics that deals with the theory, development, and analysis of approximation methods for solving a wide range of problems in applied mathematics. Having said that, *numerical methods* are concerned with numerical algorithms and their applications to solve science and engineering problems that do not have analytical solutions or are very difficult to solve. In this respect, most numerical methods only give solutions that are approximations of the true solutions. Thus, the computed results are only approximations (estimates), inevitably containing errors that are for the most part specific to the numerical method. For instance, many problems require an infinite sequence of operations to obtain approximate solutions, but in practice, only a finite number of these are processed. In such cases, the level of accuracy of a solution depends on how many sequences of operations the analyst is willing to perform or execute. Hence, it is important for an analyst to understand and estimate the errors incurred and, preferably, the lower and upper bounds of the numerical method implemented.

Over the last several decades, computers have become indispensable in engineering design and scientific research due to their speed and computational power. Nowadays, computers are widely used to design programs or software to numerically simulate various physical events or to design engineering systems that often require the use of multiple programming languages. As a result of the developments in computer technologies in parallel with technological developments, the functions demanded from programming languages have both increased and diversified. Over time, some programming languages can become obsolete as they lose their ability to meet the needs and requirements of programmers, and outdated languages are replaced by emerging, more functional languages. In practice, today, a single programming language is often not enough to meet the needs of engineers or scientists, who eventually have to be proficient in more than one programming language to achieve their goals. The main motivation for presenting algorithms in the form of pseudocodes in this book is to embrace the common programming languages of not only our time but also the near future by not adhering to a specific programming language.

This chapter introduces the general concept of numerical algorithms as a tool for solving numerical problems. The implementation of the pseudocode convention adopted in this text is presented with elementary examples illustrating control and conditional constructions, accumulators, loops, and modular programming. After briefly covering various error types and error sources encountered in mathematical modeling and numerical simulation of physical problems, errors arising due to floating-point representation of numbers and floating-point arithmetic during the storage and processing of numbers by computers are also examined. The general concept of error propagation and its application in error analysis are also presented. Different strategies for minimizing computational errors are also given. Finally, the concepts of truncation error, convergence, and stability are illustrated with commonly encountered problems.

1.1   NUMERICAL ALGORITHMS IN PSEUDOCODES

A numerical *algorithm* is a set of instructions outlining how to carry out the numerical solution of a certain mathematical problem. Although most algorithms for short and/or simple calculations can also be carried out by hand, the term *numerical algorithm* almost always implies a procedure for computer implementation to obtain an approximate solution to a mathematical problem.

The task of selecting and implementing a numerical algorithm for a particular problem is not always easy. In most cases, there are numerous algorithms devised to solve a

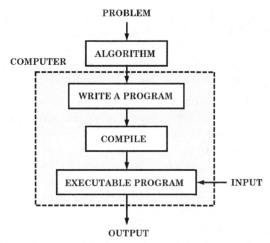

FIGURE 1.1: Process of finding a numerical solution of a problem.

class of mathematical problems. These algorithms usually differ from each other in accuracy, efficiency, reliability, and/or stability, which results in varying performances. Some algorithms yield results very quickly, while others either diverge or converge slowly due to their unfavorable convergence properties. In brief, analysts need to seek and implement algorithms that are not only accurate and reliable but also efficient and robust, which is easier said than done. Selecting a suitable algorithm for a problem is just the beginning of the problem-solving process.

The numerical solution to a problem often requires putting together a collection of numerical algorithms, involving, for instance, integration, interpolation, the solution of matrix equations, and so on. An analyst has to select and implement the most suitable algorithms in order to obtain the best possible numerical solutions for his problems. Once the overall numerical algorithm for a problem is established, a computer program can then be devised and converted into an executable program.

A typical process flowchart for finding the numerical solution to a problem *via* computers is presented in Fig. 1.1. Once a suitable algorithm has been designed, the rest of the process is carried out on the *computer*, a device capable of carrying out a set of instructions communicated by a *computer program*. A computer program is basically a tool that directs the computer to do the tasks that we want it to do. In this context, there are two kinds of computer programs: (1) human-readable programs and (2) machine-readable programs. A *computer programmer* uses a suitable *programming language* to convert the algorithm into a human-readable (instructions) program, which is referred to as *source code*. No matter what the programming language is, the source code is also an artificial language that still needs to be translated into machine-readable (binary) instructions so that it can be executed by computers. Translating instructions into machine language is carried out by a *compiler*– software that creates a machine-readable executable program. An executable program can then be run as many times as needed to find approximate solutions for various scenarios.

An algorithm eventually has to be programmed using a programming language to create an executable program. However, there is not just one programming language out there; there are many! Computers are used almost everywhere, from household appliances to cell phones, from artificial intelligence to manufacturing, and so on. That is why many computer programming languages have been developed to date, as each language has its own limitations. Web developers utilize HTML, PHP, and JavaScript to create web applications, while Kotlin and Java are used for mobile applications. Scientists and engineers make use of C, C++, Python, Visual Basic, Pascal, Fortran, and so on, along with symbolic processing software such as MatLab, Maple, Mathematica, etc. In short, professionals in different fields use different programming tools because different projects have different needs and goals, and no single language usually meets the needs of the users.

A computer is a tool that provides an environment for running programs; it merely executes the instructions prepared by its programmer. Sometimes a program may contain error(s), flaw(s), or fault(s) that produce an incorrect or unexpected result or behave in unexpected ways. We cannot blame computers for any one of these undesirable results because a computer does exactly what it is instructed to do. So the question is, *Who is to blame?* The blame may lie in the choice or misapplication of the algorithm(s) or outright programmer error. If a suitable numerical algorithm is chosen and implemented correctly with sufficient clarity, then the programmer's errors can be justifiably questioned. Programmers should check the coding of the algorithm, test it module by module and as a whole to determine whether it works properly, and make necessary corrections. In this respect, a well-prepared algorithm before writing a computer program saves the programmer from many troubles (writing, checking, debugging, etc.). A successful algorithm is one that can be understood by a programmer (who may not even be a numerical analyst) and can be turned into a computer program that generates the correct results.

1.1.1 ELEMENTS OF A PSEUDOCODE

In this text, numerical algorithms are presented in pseudocodes. A pseudocode is simply a text-based fake code that specifies a detailed description of an algorithm. Presenting an algorithm in pseudocode has the following benefits:

- It mimics structured programming languages, describing the step-by-step implementation of an algorithm;

- It is intended for human rather than compiler interpretation, so it is a lot easier for people to follow, relate to, and understand;

- It is a language-independent, efficient way of characterizing the key features of an algorithm;

- Well-thought-out and well-crafted pseudocodes contribute to an easier and faster translation into an actual programming language and save programmers a lot of trouble when debugging;

- Pseudocodes address all programmers, regardless of their programming background or knowledge.

For the reasons stated above, numerical algorithms in this textbook are presented in pseudocodes. There is, however, no generally agreed-upon convention or accepted standard norm for writing pseudocodes nor is there a standard, well-defined pseudocode syntax. Moreover, pseudocodes are not compiled and executed by computers; they are processed by humans. Thus, the correctness of a pseudocode cannot be verified until it is programmed, compiled, and run. For this reason, it is important that pseudocodes are clear and concise. In this context, the pseudocode style adopted in this textbook omits many of the details of a real programming language yet contains roughly the same level of detail, so that the pseudocode is complete enough to allow line-by-line translation into any source code.

In order for a pseudocode to achieve its intended objective, the following characteristics should be adhered to when designing a pseudocode:

- *Definiteness:* A pseudocode must specify every step and should follow a sequence of steps (or instructions) in the right order. It must be clear and unambiguous, and it should be detailed enough to include error handling;

- *Finiteness:* A pseudocode must terminate after a finite number of steps with either an output or error message, and it should not go into an infinite loop or terminate without any clue;

- *Effectiveness:* A pseudocode must be free of unnecessary or redundant steps or instructions.

Pseudocodes, like real programs, have three basic elements:

1. *Input(s).* A pseudocode requires at least one or more inputs, which include the form, amount, and type of data;

2. *Statements.* It is the body of the algorithm; a sequence of expressions, commands, and operations (looping, branching, conditionals, etc.) is embedded here;

3. *Output(s).* It should produce at least one (or more) output (form, type, amount, etc.) that is the intended result.

A short guide on how to read and/or write pseudocodes (a summary of pseudocode syntax, statements, structures, etc.) is presented in Appendix A. The body (*statements* block) of the pseudocode must describe precisely the order of instructions, statements, modules, and so on in the program.

A pseudocode, just as in normal programming languages, contains three basic constructs: sequence, selection, and iteration (or repetition).

Sequence constructs: The statements are processed and executed in the same sequential order they are written in the program.

Selection (branching) constructs: These constructs (If-Then and If-Then-Else) decide which of the two sets of expression blocks will be processed. The selection constructs start by testing a condition (a Boolean result: True or False). Depending on the result, the program is directed down in one of the two available paths.

Iteration (repetition or looping) constructs: There are two kinds of repetition structures: count-controlled iteration (For-construct) and condition-controlled iteration (While- and Repeat-Until). A block of instructions is repeated a number of times until a condition is met.

Think of a pseudocode (i.e., *algorithm*) very much like a cake recipe. *Inputs* are the ingredients (with measurements and amounts): 1 cup of butter, 3 cups of flour, 4 eggs, and so on. *Statements*, the main body of the cake recipe, contain step-by-step instructions for making a cake, e.g., (i) mix cream, butter, and sugar until fluffy; (ii) add eggs and beat well; (iii) sift flour, baking powder, and salt into a bowl, etc., and so on, and finally (ix) bake in a preheated oven at 190°C for 25 minutes. When one follows these steps exactly with the correct input (ingredients and measurements), the *output* (i.e., the final product) is a baked cake. In other words, the success of an end result (the accuracy of the output) is determined by how well and accurately a process is described in the *statements* section. Notice that all instructions in this recipe are not only specified without skipping any critical steps but also itemized in the correct order (*definiteness*). The process of making a cake requires a finite number of steps (*finiteness*). The recipe does not contain any instructions such as *add ground beef to the pot* or *add a tea spoon of cumin to the sauce*, which would be irrelevant since none of these instructions are part of making a cake (*effectiveness*).

1.1.2 SEQUENTIAL CONSTRUCTS

The instructions in sequential constructs are processed one after another in a sequence. Even though each instruction in a code is stated in a specific order, some of the instructions may be interchangeable. For example, consider estimating the cost of the paint for a large room ($width \times length \times height$). A pseudocode segment may be given as follows:

Read: $width, length, height$	\ Read in the room dimensions
Read: $paint$	\ Read the cost of paint per unit area
$A1 \leftarrow 2 * width * height$	\ Find area of front & back walls
$A2 \leftarrow 2 * length * height$	\ Find area of left & right walls
$Atop \leftarrow width * length$	\ Find area of ceiling
$A \leftarrow A1 + A2 + Atop$	\ Find total surface area
$cost \leftarrow A * paint$	\ Calculate cost of paint
Write: "Cost of paint is", $cost$	\ Print out cost of paint

In the first and second lines, the room dimensions and paint cost (per unit area) are supplied into the code, respectively, using **Read** statements, the order of which can be changed. Next, the total room surface area should be determined to estimate the cost of the paint. Following **Read** statements, the calculation of front-back, left-right, and ceiling wall areas is carried out with three statements. Note that these three statements are also interchangeable. The following (fourth) statement gives the total surface area. The fifth statement computes the cost of paint by multiplying the total area by the cost of the paint per unit area. Finally, the cost is printed out as output information. It should be pointed out, however, that the last three statements are not interchangeable.

1.1.3 CONDITIONAL CONSTRUCTS

Selection of the part of a code to be executed is done with conditional (**If-Then** or **If-Then-Else**) constructs. An **If** construct is a logical (boolean) expression that takes the value **True** or **False**. Recall that logical expressions are built with relational operators ($>$, $\geq$, $<$, $\leq$, $=$, $\neq$) and may be combined with logical operators (**And, Or, Not**) to form more complex logical expressions.

Consider the following pseudocode segment:

Read: $name, grade$	\ Read in $name$ and $grade$
Write: $name,$ "Your grade is", $grade$	\ Print out $name$ and $grade$
If $[grade < 50]$ **Then**	\ If-1, Case of $grade < 50$
Write: "NP"	\ Print NP (Not Passed) as letter grade
Else	\ Case of $grade \geq 50$
Write: "P"	\ Print P (Passed) as letter grade
End If	
If $[grade = 100]$ **Then**	\ If-2, Case of $grade = 100$
Write: "CONGRATULATIONS", $name$	\ Print "congratulation"
End If	

First, the student's $name$ and $grade$ are read into the code and printed out in the next statement. Here, both **If-Then-Else** and **If-Then**-constructs are used in this code segment. The **If-Then-Else**-construct 1, which is based on $grade < 50$ (a logical expression), is used for printing out the student's letter grade as NP (Not Passed) or P (Passed) for $grade \geq 50$. Finally, in the **If-2**, the message "CONGRATULATIONS $name$" is printed for the student who received 100.

More complicated If-structures are constructed by nesting If-Then-Else-constructs. Consider the following step-wise varying function:

$$f(x) = \begin{cases} 2 - x, & x < -1 \\ x^2 + 2, & -1 \leqslant x \leqslant 3 \\ 3x + 2 & 3 < x \leqslant 5 \end{cases}$$

A pseudocode segment of this function is established by nesting three If-Then-Else-constructs as follows:

Read: x \ Read x from file or console
If $\begin{bmatrix} x < -1 \end{bmatrix}$ **Then** \ If-1, case of $x < -1$
  $f \leftarrow 2 - x$
Else \ case of $x \geqslant -1$
  If $\begin{bmatrix} x \leqslant 3 \end{bmatrix}$ **Then** \ If-2, case of $-1 \leqslant x \leqslant 3$
    $f \leftarrow 2 + x^2$
  Else \ case of $x > 3$
    If $\begin{bmatrix} x \leqslant 5 \end{bmatrix}$ **Then** \ If-3, case of $3 < x \leqslant 5$
      $f \leftarrow 3x + 2$
    Else \ case of $x > 5$
      Write: "x is Out-of-Range." \ Print out a warning
    End If \ End of If-3
  End If \ End of If-2
End If \ End of If-1

Note that three nested If-Then-Else-constructs are used to define the function with respect to step intervals. If $x < -1$ (If-1), then $f(x) = 2 - x$ is processed; otherwise, x corresponds to an $x \geqslant -1$ interval that requires further refining. In If-2, $f(x) = 2 + x^2$ is executed if $x \leqslant 3$, which falls within $-1 \leqslant x \leqslant 3$. If x is not in $-1 \leqslant x \leqslant 3$ interval, then it must be greater than 3. Next, If-3 narrows down the interval to $3 < x \leqslant 5$, for which case $f(x) = 3x + 2$ is evaluated; otherwise, to ensure correct data processing, an "x is Out-of-Range" message is printed for $x > 5$ since $f(x)$ is not defined for this interval.

1.1.4 LOOPS AND ACCUMULATORS

Most programming languages are furnished with *sequential* and *conditional* loop constructs. These loops are represented by the For-(definite counting loop), While-, and Repeat-Until (indefinite counting loops) constructs.

A common task in most algorithms is to find the sum (or product) of a series of values. This is accomplished through the use of an accumulator. An *accumulator* is a variable that a program uses to perform the sum (or product) of successive values in a loop construct. For example, the following pseudocode segment illustrates the sum of integer numbers from 1 to 10 using a For-construct (a counting loop).

$S \leftarrow 0$ \ Initialize accumulator by setting $S = 0$
For $\begin{bmatrix} k = 1, 10 \end{bmatrix}$ \ Iterate block statement for $k = 1$ to 10
  $S \leftarrow S + k$ \ Add present value of k to S; the accumulator
End For \ Increment counter variable, $k + 1$

To visualize how this code works, the process of a series sum or a product using an accumulator is illustrated in Fig. 1.2. In Fig. 1.2a, the accumulator is initialized by zero

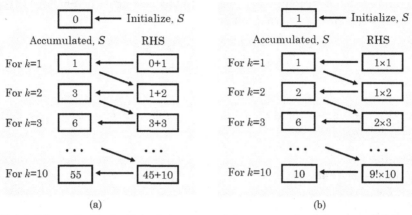

(a)   (b)

FIGURE 1.2: Schematic illustration of the accumulator process for (a) sum and (b) product.

$(S \leftarrow 0)$, which is moved into the memory location of S. The loop index (or counter) is k, and the loop (iteration) block consists of only one expression: $S \leftarrow S + k$. Starting with $k = 1$, the rhs of this expression is processed first. The value on the rhs $(0 + 1 = 1)$ is moved to the memory location of the accumulator as denoted by the $\leftarrow$ symbol, where its prior "0" value is overwritten by "1" $(S \leftarrow 1)$. Note that in Fig. 1.2 the memory location of S is denoted with boxes. For $k = 2$, the most recent value of the accumulator is used to calculate the rhs $(S \leftarrow 1 + 2)$, and the sum is moved to the lhs (the memory location of the accumulator), overwriting on its previous value, i.e., $S \leftarrow 3$. The process is repeated as long as the $k \leqslant 10$ condition remains True, which is executed for $k = 1$ to 10 with $\Delta k = 1$ increments (*see* flowchart in Appendix A). The procedure gives the sum $\sum_{k=1}^{n} k$ for any n. This construction can be easily extended to any $\sum_{k=1}^{n} a_k$ by simply replacing "k" with the general term of a series, i.e., $a_k f(k)$.

The procedure outlined above can also be thought of as generating a sequence using the $S_k = S_{k-1} + a_k$ recursion formula for $k = 1, 2, \ldots, n$ with $S_0 = 0$. For every k, the corresponding S_k is the partial sum from 1 up to k: i.e., $S_k = \sum_{i=1}^{k} a_i$.

The computation of 10!, which requires a series of multiplications, is illustrated using an accumulator in Fig. 1.2b. The accumulator is initialized as $S \leftarrow 1$, which is stored in the memory location of S. The loop index remains k, and the loop block is replaced by $S \leftarrow S \times k$. Starting with $k = 1$, the rhs becomes "1" $(1 \times 1 = 1)$, and it is carried to the memory location of S. For $k = 2$, the most recent value of the accumulator is used to calculate the rhs as $S \leftarrow 1 \times 2$. This product is moved to the memory location of S, overwriting the accumulator's previous value of "1." This process is repeated in the same way until the $k \leqslant 10$ condition becomes False (i.e., $k > 10$). This process gives $\prod_{k=1}^{n} k$ for any n. This construction can be extended to any $\prod_{k=1}^{n} a_k$ by replacing "k" with a general term, a_k. This process can also be represented as a recursive procedure by setting $S_0 = 1$. Then, a sequence of values is generated according to the $S_k = S_{k-1} \times k$ recursion formula for $k = 1, 2, \ldots, n$, which eventually yields $S_k = k!$ for an arbitrary k.

 Always indent the body of the statement(s) block in If-Then-, If-Then-Else, or loop constructs by two or more spaces to highlight the blocks and improve the readability of the code.

Using a While-construct, we consider evaluating the same summation. A pseudocode segment may be given as

Initialize: n \ Initialize number of terms
$S \leftarrow 0$ \ Initialize accumulator
$k \leftarrow 0$ \ Initialize counter
While $\left[k \leqslant n\right]$ \ Execute block so long as $k \leqslant n$
$\quad k \leftarrow k + 1$ \ Increment counter by $+1$
$\quad S \leftarrow S + k$ \ Add present value of k to S; the accumulator
End While   \ Go to top of the loop

First, the number of terms, n (a user-supplied value), is provided. Next, the accumulator is initialized by assigning $S \leftarrow 0$. The While-loop cycles an unknown number of times until it stops when a criterion that does not necessarily involve the loop control variable is met. However, in this pseudocode segment, a stopping criterion ($k \leqslant n$) is tied to the iteration counter (k). Hence, the iteration counter k is also initialized as $k \leftarrow 0$. This example requires two accumulators: one for the counter and the other for the summation. However, the counter is placed before the accumulator so that the loop can run from $k = 1$ to 10. The loop condition ($k \leqslant n$) is evaluated before the iteration block. The exchange of data in the memory location of the accumulator and the index variable works in the same way. While counting the number of terms (k), the value of the accumulator in the memory location is continuously updated, leading to a recursive operation. The block statements ($k \leftarrow k + 1$ and $S \leftarrow S + k$) are processed n-times until the counter limit ($k > n$) is reached.

Next, we code the same example by using a Repeat-Until construct. The pseudocode segment is given below:

Initialize: n \ Initialize number of terms
$S \leftarrow 0$ \ Initialize accumulator
$k \leftarrow 0$ \ Initialize counter
Repeat   \ Execute block statements
$\quad k \leftarrow k + 1$ \ Increment counter by $+1$
$\quad S \leftarrow S + k$ \ Add present value of k to accumulator
Until $\left[k = n\right]$ \ Exit loop when $k = n$, else go to top of the loop

The procedure works pretty much the same as the While-construct; however, the loop stopping (termination) condition ($k > n$) is evaluated at the end of the loop (*see* Appendix A). After initializing the counter and accumulator ($k \leftarrow 0$ and $S \leftarrow 0$), the iteration block is executed, resulting in $k \leftarrow 1$ and $S \leftarrow 1$. The logical expression becomes $1 < n$, which is True for $n = 10$. The second iteration yields $k \leftarrow 2$, $S \leftarrow 3$, and $2 < n$ (True). The iteration block is executed in the same way for $k > 2$ until the loop *condition* becomes False (i.e., $k > n$). Note that a counter variable is not required for neither While- nor Repeat-Until construct, only the stopping criterion must be satisfied.

 In programming, the letters i, j, k, m, n, etc. are generally adopted as *counter variables* because they are also used as index variables in mathematics (e.g., $\sum_{i=1}^{n} a_i$ or $\prod_{j=1}^{n} b_j$). Declare a counter variable as Integer because it is operationally easier and faster to increase the value of an integer-type variable.

Modular Programming

- A module can be reused as many times as needed; thus, modular programming reduces program complexity by eliminating repetition of tasks and results in shorter codes;
- It makes reading and understanding a code simpler and easier;
- Maintaining or debugging codes is easier since errors are localized in a module that can be independently modified and tested;
- Software development is easier since each team focuses on a smaller part of the entire code;
- A module can also be used in other projects that require the processing of the same task;
- Establishing a library of modules that work reduces (coding, debugging, etc.) the project completion time.

- Modular programming may require extra time and/or budget for the program development of a project;
- It is a critical, yet cumbersome, task to prepare "complete and detailed documentation" for each module, which is communicated between the design teams;
- Preparing a collection of modules can be a challenging task.

1.1.5 PSEUDOCODES, FUNCTION MODULES AND MODULES

A program or module in any programming language is organized as follows: name, list of variables, input statements, processing/computing blocks (involving sequential and/or conditional constructs), and output statements. Every program or module has a name and a list of internal (specific to the module) and/or external (common with other modules) variables. In this respect, pseudocodes are organized in the same way.

Modular programming is the name of the general concept of dividing a complicated computer program into smaller yet separate modules (i.e., procedures or sub-programs). Each module (sub-program or procedure) comprises everything that is needed to carry out only one aspect of the functionality. Modules can be utilized for a range of applications and functions in conjunction with other modules in a general program. A module may also make use of other module(s). Many programming languages allow two types of modules (procedures, subroutines, sub-programs, etc.) and functions (built-in and user-defined). Hence, it is wise to adopt a modular coding approach to write efficient programs. The advantages of modular programming outweigh its disadvantages; thus, in this text, every method has been presented with a pseudocode module.

Pseudocode 1.1 is an example of a pseudo-program containing various elements of the pseudocode convention presented in this text. All comments and annotations are distinguished from the actual code statements with light fonts. The pseudo-program is delimited by the **Program-End Program** statement. It contains a header block with a brief description of the code. Also following the **USES** statement, any user-defined **Module** and/or **Function Module** used in the code (i.e., program dependencies) are specified. In this example, two user-defined modules, **QUADRATIC_EQ** and **FUNC**, are identified in the header block. The square root function, commonly named **SQRT**, is available as a built-in function in most programming languages. Throughout this text, built-in functions such as square root, absolute value, and so on have been used in their symbolic forms but are explicitly declared in

---

Pseudocode 1.1

Program EXAMPLE   \ First executable statement: define program name
\ DESCRIPTION: A pseudocode illustrating general features of a pseudocode.
\ USES:
\   FUNC :: User-defined function, Pseudocode 1.2;
\   QUADRATIC_EQ :: A pseudomodule finding the roots of a quadratic equation.
Declare: a_n, b_n, c_n, re_2, im_2 \ Declaring array variables
Read: u , v \ Read in u and v from console or a file
Read: a, b   \ Read in arrays **a** and **b** of length n from a file
$u \leftarrow u + 2 * v$ \ Arithmetic expression-1 is assigned to u
$v \leftarrow v + u * u$ \ Arithmetic expression-2 is assigned to v
$w \leftarrow$ **FUNC**(u, v) \ Set function FUNC value to w
c $\leftarrow 2 * $**a**$ + 3 * $**b**   \ Computes $c_i = 2a_i + 3b_i$ for all i
Write: "c=",**c**   \ Print c_i for $i = 1, 2, \ldots, n$
c $\leftarrow$ **a** $*$ **b**   \ Compute $c_i = a_i b_i$ for $i = 1, 2, \ldots, n$
Write: "c=",**c**   \ Print c_i for $i = 1, 2, \ldots, n$
$d \leftarrow c_1 * c_n$ \ Arithmetic expression-3 is assigned to d
If $[d > 0]$ **Then**   \ Case of $d > 0$
    QUADRATIC_EQ$(u, d,$ **re, im**)   \ Invoke Module QUADRATIC_EQ
Else   \ Case of $d \leqslant 0$
    QUADRATIC_EQ$(u, -d,$ **re, im**)   \ Invoke Module QUADRATIC_EQ
End If
Write: "Root 1=", re_1, im_1 \ Print out $x_1 = re_1 + i * im_1$
Write: "Root 2=", re_2, im_2 \ Print out $x_2 = re_2 + i * im_2$
End Program EXAMPLE   \ End of Program

---

the USES statement with their full names. The array variables (a_n, b_n, c_n) of length n and arrays (re_2, im_2) of length 2 are declared with the **Declare** statement. If the types of the other (non-array) variables are not obvious from the context of the pseudocode, their type declarations (**Real, Integer, Logical**, etc.) should be provided to leave no room for doubt.

Reading the initial values of u and v from an input file into the program is indicated by the "**Read:** u, v" statement. Similarly, the "**Read: a, b**" statement indicates that two arrays (**a** and **b**) of length n are read into the code (likely from an input file) as whole-arrays. In the next eight lines, each line is processed in the order in which it is written. The w is computed by invoking the **FUNC** function. Since **a** and **b** are arrays (vectors) of the same length, whole-array arithmetic (*see* Appendix A) is incorporated as **c=2*a+3*b** and **c=a*b**. The **Write** statements are used anywhere in the code to report any intermediate or final value of variable(s). The pseudomodule **QUADRATIC_EQ** is invoked within conditional statements with different argument lists. The real re_2 and imaginary im_2 parts of the roots are then printed out by using the **Write** statements.

A typical structure of a pseudo-**Function Module** is illustrated in Pseudocode 1.2. A function of two independent variables is given as $func(x, y) =$ **MOD**$(x, y)\sqrt{x^2 + y^2}$. The module makes use of two built-in functions (**MOD** and **SQRT**), which are also identified with the USES statement. (In this text, an effort has been made to use built-in function names common to most programming languages.) This module checks the validity of the input data to determine a course of action to achieve effectiveness, finiteness, and definiteness properties. The case of $y > 0$ and $|x| \geqslant 10$ (**If-1** condition becomes **True**) indicates that the arguments are within the validity range. Then the mathematical operation is carried out,

---

Pseudocode 1.2

Function Module FUNC (x, y)
\ DESCRIPTION: A pseudo-function module defined as $func(x,y)=$MOD(x,y)
\ $\times \sqrt{x^2 + y^2}$ for $y > 0$ and $|x| \geqslant 10$, where x and y are integers.
\ USES:
\ MOD:: Built-in function returning the remainder of x divided by y;
\ SQRT:: A built-in function computing the square-root of a value.
Declare: x, y \ Type declarations
If $\left[y > 0 \textbf{ And } |x| \geqslant 10 \right]$ **Then** \ If-1, arguments within range?
    $F \leftarrow \textbf{MOD}(x, y)$ \ Built-in function is invoked
    FUNC $\leftarrow F * \sqrt{x^2 + y^2}$ \ Set rhs as FUNC
Else \ Identify argument not in range
    Write: "Error: Illegal Input" \ Print out an error message
    FUNC$\leftarrow 10^{30}$ \ Set a large value for FUNC
    If $\left[y \leqslant 0 \right]$ **Then** \ If-2, y has the illegal input value
        Write: "Input y is", y \ Print y to indicate $y \leqslant 0$
    End If
    If $\left[|x| < 10 \right]$ **Then** \ If-3, Case of x with an illegal input value
        Write: "Input x is", x \ Print x to indicate $|x| < 10$
    End If
End If
End Function Module FUNC

---

and the result is assigned to **FUNC** (i.e., the function name) before exiting the module. If x or y or both fall outside the validity ranges, the If-1 condition becomes **False**, for which the user is warned with the "Error: Illegal Input" message, and **FUNC** is a large set to value. If-2 and -3 are used to check if x or y are out of range, respectively, and print the input value(s).

The module **QUADRATIC_EQ** presented in Pseudocode 1.3 is an example to illustrate the general structure and features of a pseudomodule. The module is intended to give the roots of a quadratic equation of the form $x^2 + px + q = 0$ under all conditions. Note that the sub-program is delimited by **Module-End Module** statements. The arguments of the module are p, q, **re**, and **im**. Of these, p and q are the *scalar input variables* denoting the coefficients of the quadratic equation, **re** and **im**, are the *output array variables* of length 2 spared for the real and imaginary parts of the roots. Annotations can be inserted into steps (or lines) deemed necessary to explain any intended actions. The discriminant is computed first. All real and imaginary roots are determined using an If-Then-Else construct with $d < 0$ as the condition. Since we only need $\sqrt{d}$ from this point forward, we may assign $d \leftarrow \sqrt{d}$ (or $d \leftarrow \sqrt{-d}$ in case of $d < 0$) to save from memory space. The real and imaginary parts of the roots are evaluated and saved on re_k and im_k for $k = 1, 2$.

An example of a pseudo-recursive function module is presented in Pseudocode 1.4. The module **FACTORIAL** with a single argument n is intended to give $n!$ An If-Then-Else construct (If-1) is used to determine the paths for $n < 0$ and $n \geqslant 0$. For $n < 0$, the code echoes the value entered with an error message and suggests a remedy. The If-2 provides paths for $n \leqslant 1$ and $n \geqslant 2$ conditions. The **FACTORIAL** is set to "1" for $n = 0$ or 1. For $n \geqslant 2$, the number is passed to the function, where n is multiplied by $(n-1)!$ This number is passed again to the function, where it is multiplied with the factorial of $n - 2$, and so on. This procedure is repeated until n reaches 1.

Pseudocode 1.3

Module QUADRATIC_EQ (p, q, **re, im**) \ Name & variable list of the module
\ DESCRIPTION: A pseudo module to find the roots of a quadratic equation
\ of the form $x^2 + px + q = 0$.
\ USES:
\ SQRT:: A built-in function computing the square-root of a value.
Declare: Real p, q, re_2, im_2 \ Type declaration of arguments
$d \leftarrow p * p - 4 * q$ \ Compute discriminant, Δ
If $[d < 0]$ **Then** \ Case of imaginary roots, $\Delta < 0$
$\quad d \leftarrow \sqrt{-d}$ \ Evaluate $d = \sqrt{-\Delta}$
$\quad re_1 \leftarrow -p/2$; $re_2 \leftarrow re_2$ \ Real parts are the same
$\quad im_1 \leftarrow -d/2$; $im_2 \leftarrow -im_1$ \ Imaginary parts have opposite signs
Else \ Case of real roots, $\Delta \geqslant 0$
$\quad d \leftarrow \sqrt{d}$ \ Evaluate $d = \sqrt{\Delta}$
$\quad re_1 \leftarrow (-p - d)/2$; $re_2 \leftarrow (-p + d)/2$ \ Real parts of the roots
$\quad im_1 \leftarrow 0$; $im_2 \leftarrow 0$ \ No imaginary parts
End If
End Module QUADRATIC_EQ \ End of the Module

Pseudocode 1.4

Recursive Function Module FACTORIAL (n)
\ DESCRIPTION: A pseudo-function (recursive) module for computing $n!$
Declare: Integer n, FACTORIAL \ Type declarations
If $[n < 0]$ **Then** \ If-1, validity range check ($n < 0$?)
$\quad$ **Write:** "Error: Illegal Input" \ Issue an error message
$\quad$ **Write:** "You entered",n," Enter $n \geqslant 0$" \ Echo input and guide user
$\quad$ **Exit** \ Exit the module
Else \ Case of $n \geqslant 0$, compute $n!$
$\quad$ **If** $[n \leqslant 1]$ **Then** \ If-2, Case of $n = 0$ and 1
$\quad\quad$ FACTORIAL$\leftarrow 1$ \ Set FACTORIAL=1 since $0! = 1! = 1$
$\quad$ **Else** \ Case of $n > 1$
$\quad\quad$ FACTORIAL$\leftarrow n*$FACTORIAL$(n - 1)$ \ Compute $n! = n * (n - 1)!$
$\quad$ **End If** \ End of If-2
End If \ End of If-1
End Recursive Function Module FACTORIAL

1.1.6 TIPS FOR EFFICIENT PROGRAMMING

Regardless of your computer programming background, you can improve your programming skills, reduce programming and debugging time, and increase the efficiency of your programs by adhering to the following recommendations:

1. Take your time. To be able to see the "big picture," make a flow chart of the algorithm with all possible scenarios (including bounds-checking for inputs, providing error paths, etc.). Make a list of the variables, functions, and sub-programs that you will need.

2. Establish a "*list of variables*" for each program or module. First of all, use meaningful names for the variables, constants, modules, and so on to avoid confusion. A list of variables, containing a full list of input and output variables used in a program or module, should be included in the header of any module or program. Here, we will skip identifying the input and output variables in the header block to save space. However, explanations of the variables are presented in the code description paragraph. It is also a good practice to define (by commenting) the internal variables within the code. The *list* should include the *description* as well as any *units*, if applicable. Although this practice may seem like an unnecessary burden or a waste of time, it is nonetheless invaluable when you or someone else needs to modify the program at a later time.

3. Echo all input variables. Displaying any data immediately after it is read into a program is called *echoing*. Especially when a code is being developed, it should contain several temporary or permanent Write statements that echo input and display even some of the intermediate results. Failure to do so may lead to erroneous results due to incorrectly typing the input data, of which the user would be unaware. Also, echoing any input data supplied to a code ensures that the number of data, their type, and their values are not only correctly supplied but also correctly received by the code. This trick eliminates some of the compiler and runtime errors such as *"missing input data," "type mismatch," "undefined variables,"* and so on.

4. Split large programs' tasks into smaller sub-programs. Any specific task that is performed a number of times in a program or subprogram can be defined as a separate subprogram (module, procedure, or function). Creating a library of modules or procedures reduces the complexity of a program by eliminating repetitions and increasing the readability of the code, giving it a modular structure. Apart from being independent from the rest (and possibly undesirable elements) of the program, the procedures offer the benefits of easy testing and reusability. Implementing any built-in functions or procedures available in compiler software reduces coding and debugging time, thereby reducing the overall program development time.

5. Design programs to reduce memory requirements. Avoid using arrays (or arrays larger than actually required) unless they are absolutely needed because large arrays use up a lot of memory, leading to a large program that is much larger than it should be. Also, if the size of an executable file is too large, it may not run on a particular computer. Keep in mind that the use of an array is not justified unless a problem requires a list of input data (such as vectors or matrices) to be retained in memory at the same time.

6. Design programs to reduce cpu-time. The cost of arithmetic and logical operations depends on the central processing unit. In the programming of numerical algorithms, a significant portion of the cpu-time is devoted to floating-point arithmetic and logic operations. By defining the cost of an addition, subtraction, or comparison operation as one-cpu unit, the cost of other operations (albeit machine dependent) can be roughly established. To give you an idea, the cost of various operations is given with the relative cpu cost (in parenthesis) as follows: absolute value ($\sim$2), multiplication ($\sim$4), division and modulus ($\sim$10), exponentiation, square root ($\sim$50), power x^y ($\sim$100).

One way to reduce cpu-time is to replace an expression that is executed numerous times with a different expression that yields the same value but is cheaper to compute. For instance, consider calculating the real roots of a quadratic equation. A naive programmer would most likely express the computation of the roots as

$$x1 \leftarrow (-b + \sqrt{b^2 - 4*a*c})/(2*a) \qquad \backslash \text{ Compute } x1$$
$$x2 \leftarrow (-b - \sqrt{b^2 - 4*a*c})/(2*a) \qquad \backslash \text{ Compute } x2$$

Note that the computation of each root requires 2 addition or subtraction, 3 multiplications, 1 power, 1 division, and 1 square root operations, corresponding to 348 cpu-units. However, the expression leading the roots may be rearranged as follows:

$$d \leftarrow 2 * a \qquad \qquad \text{\textbackslash Find denominator}$$
$$\Delta \leftarrow \sqrt{b * b - 4 * a * c} \qquad \text{\textbackslash Evaluate } \Delta$$
$$x1 \leftarrow (-b + \Delta)/d \qquad \qquad \text{\textbackslash Find } x1$$
$$x2 \leftarrow (-b - \Delta)/d \qquad \qquad \text{\textbackslash Find } x2$$

Now, the set of expressions requires 3 additions or subtractions, 4 multiplications, 2 divisions, and 1 square root, totaling 89 cpu-units. Notice that by restating the mathematical expressions differently, a cpu-time saving of about 75% is achieved.

Some other examples of rearranging mathematical expressions are given below:

$$u \leftarrow 0.5 * b - 0.3 * c \qquad \qquad \text{\textbackslash Find } u = (b - 3 * c/5)/2$$
$$v \leftarrow d * d \qquad \qquad \qquad \text{\textbackslash Compute } v = d^2$$
$$w \leftarrow v * v \qquad \qquad \qquad \text{\textbackslash Compute } w = d^4$$
$$y \leftarrow x * (x * (a * x + b) + c) + d \qquad \text{\textbackslash Find } y = ax^3 + bx^2 + cx + d$$

Note that in the first expression, division operations are expressed as multiplications after removing the parentheses, i.e., $1/2 = 0.5$ and $3/10 = 0.3$. The second and third expressions yield d^2 and d^4 with only two multiplications, respectively. In the last expression, a polynomial is recast in nested form, requiring only additions and multiplications.

Another example of saving cpu-time is given by array applications. If a particular element of an array is to be accessed numerous times, access the value once, assign its value to a temporary scalar variable, and use the temporary variable.

For $\left[k = 1, n \right]$
$\qquad tmp \leftarrow a_k \qquad \qquad \qquad$ \ Get a_k and set $tmp = a_k$
$\qquad v \leftarrow tmp * (2 * tmp - 3) \qquad$ \ Find $v = a_k(2a_k - 3)$
$\qquad w \leftarrow a_{k+1}/tmp \qquad \qquad$ \ Find $w = a_{k+1}/a_k$
End For

 Minimize the number of conditional constructs and indefinite-loop constructs due to the cpu-time devoted to evaluating logical expressions, which require larger cpu-time. In this regard, **For**-constructs are faster than **While**- or **Repeat-Until** constructs.

1.2 BASIC DEFINITIONS AND SOURCES OF ERROR

Sources of errors in simulating scientific and engineering problems can be divided into three categories: modeling, experimental, and numerical errors.

Modelling Error. Scientists and engineers deal with the problems of the physical world, and they frequently resort to computer simulation. *Computer simulation* is the prediction of the behavior of a physical phenomenon in nature or the outcome of a physical system under consideration. The first step in numerical simulation is to derive precise and detailed mathematical models. Any mathematical model of a process or physical system may consist of one or more mathematical sub-models. The complexity of a model depends on the complexity of the physical phenomenon under consideration. *Mathematical* *modeling* includes the complete specification of all pertinent differential equations, boundary conditions, and initial conditions for the system.

Mathematical models help scientists and engineers develop an understanding of systems or processes using mathematical equations. Measuring exactly every variable or including the effects of all pertinent variables in a mathematical model may be impossible and/or impractical. Furthermore, a complete mathematical model may become way too expensive to be used in analysis with little benefit. As a result of these factors, mathematical models are revised and simplified after perhaps many simplifying assumptions, resulting in less accurate models. Hence, simplified mathematical models may not be 'exact' as compared to the underlying physical process. If a mathematical model is derived erroneously or with false or gross assumptions, the numerical simulation will not produce the results corresponding to the physical event, even if the governing mathematical equations are solved exactly by any analytical means. In other words, the simulation of a simplified physical event will deviate to some extent from its true behavior. This small or large deviation due to modeling error depends on the validity of the assumptions made.

Modeling errors may be the result of simplifying geometry (reducing it to one- or two-dimensional geometry, etc.), material description (assuming homogeneity when it is not, inadequate representation of material properties), prescribing boundary conditions, which may be difficult to do in some cases, etc. Oversimplifying a model surely requires less computational effort and yields simpler solutions, but it may also result in inaccurate solutions that cannot be accepted or tolerated. For instance, neglecting air resistance in the motion of a pendulum or the friction between the tires of an automobile and the road, etc., will give results that are somewhat inaccurate. Hence, it is not surprising that analysts end up using different mathematical models to take into account the effects of fewer or more variables. It should be kept in mind that the accuracy of the prediction of a mathematical model depends more on its ability to correctly identify the dominant parameters and their effects than on its complexity. In this regard, a simple model may sometimes be more useful than a more complex model. Therefore, there is a need to find a balance between the level of accuracy required for simulations and the complexity level of a mathematical model.

Once the numerical solution of a physical event is obtained, additional information, generally computed using numerical methods such as differentiation, integration, interpolation, etc., may also be required. These computed quantities also approximate and contribute to the numerical errors in the final results.

Experimental Error. When we speak of experimental errors, we do not refer to human errors that result from mistakes, blunders, or numerical miscalculations. Although these errors are definitely important, they can be eliminated by correctly repeating experiments and/or calculations. Yet experimental errors are inherent in any measurement, and they cannot be eliminated no matter how carefully experiments are conducted.

Experimental errors are due to observations and/or instruments or devices used in measurements. Data or empirical errors arise when a mathematical model acquires experimentally derived data as input. Such errors are mostly the result of limitations and errors in instrumentation. The true value of any measurement is therefore unknown; that is, we are not certain if a measured value is smaller or larger than the actual value. All experiments include systematic and random errors to some degree. In some cases, it is possible to reduce the magnitude of an experimental error.

Additionally, instruments may be affected by the frequency of use, humidity, temperature, pressure, magnetic field of a physical environment, etc. Instruments yield deviations in their measurement readings as a combined result of manufactured settings, human handling, and environmental factors. For this reason, all instruments need to be periodically calibrated with an instrument of known accuracy. In order to reduce modeling errors, it is important to improve the accuracy of experimental data, probably more than the precision of arithmetic operations.

Numerical error. It is introduced as a result of carrying out an arithmetic operation using either a calculator or a digital computer, or numerical methods or approximations. At this point, we exclude user errors resulting from, for instance, accidentally entering incorrect data into a numerical process.

To understand the sources of numerical errors in computations, it is important that we understand how numbers are manipulated in digital computers. When a floating-point number is converted to its binary form, on which computers base their arithmetic operations, most numbers cannot be represented in their exact forms due to the finite length of the floating-point mantissa. The digits that are not kept in the number result in chopping or rounding errors. As a result, a digital computer retains only a limited number of digits in a floating-point representation that is mostly inexact. These errors, which arise due to the inability of computers to store the exact value of the numbers, are called *machine errors*. Arithmetic operations with the stored or computed numbers yield *round-off* errors. On the other hand, implementation of a numerical method that, in reality, requires approximation(s) to a mathematical procedure results in *truncation errors*. Thus, the overall numerical error is a combination of truncation and round-off errors. Since these errors are specific to digital computers and numerical methods, they are covered in the subsequent sections.

1.2.1 DEFINITION OF ERROR

Determining and reporting the accuracy of a computed quantity is one of the most important tasks in numerical analysis. An approximation error can be the result of computed or measured data that is not precise or of approximations used instead of the real data. There are basically two ways to express the magnitude of error in a computed quantity: *absolute error* and *relative error*.

Let Q be a physical quantity, and Q and $\tilde{Q}$ denote its true and approximate values, respectively. Then error is defined as the difference between the *true* and *approximate* values; that is,

$$\text{Error,} \quad \Delta Q = Q - \tilde{Q}$$

Error and *absolute error* are often used interchangeably; however, *absolute error* is merely the absolute value of the error:

$$\text{Absolute error,} \quad |\Delta Q| = |Q - \tilde{Q}| \tag{1.1}$$

On the other hand, *relative error* is a measure of "computed" or "measured" quantity relative to the magnitude of its true value:

$$\text{Relative error,} \quad \left|\frac{\Delta Q}{Q}\right| = \left|\frac{Q - \tilde{Q}}{Q}\right| \tag{1.2}$$

where $Q \neq 0$; otherwise, the relative error is undefined.

Relative error is independent of the magnitude of the value since it is defined as a ratio of the absolute value of error to the true value. Hence, it is also a dimensionless quantity and more meaningful than absolute error. These errors are sometimes expressed in percentages (%).

$$\text{Absolute error,} \quad |Q - \tilde{Q}| \times 100\% \tag{1.3}$$

$$\text{RelativeError,} \quad \left| \frac{Q - \tilde{Q}}{Q} \right| \times 100\% \tag{1.4}$$

Absolute and/or relative errors are often used in iterative algorithms to determine the convergence level of the iterated quantity, as follows:

$$\left| Q^{(p+1)} - Q^{(p)} \right| < \varepsilon_1 \quad \text{or} \quad \left| \frac{Q^{(p+1)} - Q^{(p)}}{Q^{(p+1)}} \right| < \varepsilon_2 \tag{1.5}$$

where ε_1 and ε_2 are the desired convergence tolerances, and $Q^{(p)}$ denotes a computed numerical quantity at any p^{th} iteration. When the true value of a computed quantity is zero ($Q = 0$), its relative error becomes undefined since $Q^{(p)} \to Q$ as $p \to \infty$. For this reason, it is customary to add a very small number to the denominator to prevent "*division by zero*" or "*overflow*" errors:

$$\left| \frac{Q^{(p+1)} - Q^{(p)}}{Q^{(p+1)} + \varepsilon} \right| < \varepsilon_2 \tag{1.6}$$

where ϵ is a small real positive number.

EXAMPLE 1.1: Evaluating absolute and relative errors

Consider the following approximation for the inverse tangent function:

$$\tan^{-1} x \cong \frac{x}{1.012 + 0.267x^2} \quad \text{for } 0 \leqslant x \leqslant 1$$

Write a pseudocode segment that calculates and prints the true and approximate values of $\tan^{-1} x$ along with its absolute and relative errors from $x = 0$ to 1 with 0.1 increments.

SOLUTION:

A pseudocode segment to compute $\tan^{-1} x$, its approximation, and the pertinent errors are presented below:

```
x ← 0;  Δx ← 0.1                    \ Initialize x and Δx (increment)
While [x ≤ 1]                        \ While x ≤ 1 execute block statements
    true ← tan⁻¹ x                   \ Find true value using built-in function
    appx ← x/(1.012 + 0.267x²)          \ Find approximate value
    Ea ← |true − appx|                  \ Calculate absolute error
    Er ← Ea/true                        \ Calculate relative error
    Write: x, true, appx, Ea, Er     \ Print out computed results for x
    x ← x + Δx                          \ Find next x as x + Δx
End While                            \ Return to the top of the loop
```

In this code segment, first x and Δx are initialized. Then a **While**-construct is used to compute the desired quantities for x values from 0 to 1 with 0.1 increments. The computed results for each x are printed out within the loop. Note that an accumulator is used to determine the value of the next x.

Discussion: The same result can also be accomplished more efficiently by using a **For**-construct. Although $\tan^{-1} x$ is computed using its built-in function, its true value and resulting calculated error depend on the precision (defined as *single* or *double precision*) adopted in the actual program.

1.2.2   MEASUREMENT ERROR, ACCURACY, AND PRECISION

It is impossible to measure the true value of any physical quantity. Even if measurements are repeated multiple times, each time a different value will be recorded, and none of these measurements can be favored over the others. In other words, there will always be an error in any measurement, and from this point of view, experimentally collected data are not perfect. Therefore, *measurement error* (also referred to as *observational error*) is defined as the difference between a measured quantity and its true value or between two measured values of a physical quantity.

Scientists and engineers are aware of the presence of measurement errors in experiments. It is thus desired to minimize experimental errors as much as possible to ensure that measurements represent true value as accurately as possible. There are two kinds of errors: random and systematic errors.

Random errors (also referred to as *precision errors*), which are caused by unpredictable changes in the environment, can be easily detected and analyzed by statistical methods. We know that multiple measurements of a physical quantity always exhibit fluctuations in the measured quantity around a mean value. The statistical method for finding a quantity (Q), its uncertainty (dQ), or standard deviation (σ_Q) requires repeating the measurements and computing its average, *average deviation*, or the *standard deviation*. Thus, we hope to reduce the adverse effects of random errors by conducting a sufficient number of experiments.

Systematic errors are the result of how an experiment is conducted and can be identified by leading to results that are too high or too low. These errors cannot be mitigated simply by averaging the measured values. Systematic errors may be more difficult to detect, but their effects can be reduced by changing the way experiments (using a different method or technique) are conducted. Systematic errors are the result of a wide range of factors that arise due to poorly designed experiments, faulty calibration of measuring instruments, poorly maintained instruments, and/or faulty readings by the operator. These errors do not include human errors such as blunders and arithmetic errors, which can be eliminated once noticed. To minimize the effects of experimental errors, the analyst should know how to measure experimental errors, analyze them, and report the measurements and their uncertainties clearly and accurately.

Experimental error is best measured by the *accuracy* and *precision* of any measured quantity. *Accuracy* is a measure of the degree of closeness of a measured (or computed) value to its true value. However, since the true value of a physical quantity is generally unknown, it is seldom possible to determine the accuracy of a measurement. *Precision* is a measure of how closely multiple measurements agree with each other. Precision is also referred to as *reproducibility* or *repeatability*. *Bias*, which is a result of systematic errors, is a measure of how far a measured value is from its true value. In other words, the difference between the mean value of the measurements and their true value is the bias. The source of bias error is attributed to measuring instrument calibration errors.

Fig. 1.3 illustrates accuracy, precision, and bias by using the analogy of grouping darts in a target. A close grouping of darts (measured values) is always considered a good thing, as it indicates consistency in the action of throwing. In Fig. 1.3(a), the accuracy and precision are low, and the grouping is on the right side of the target value (i.e., bias exists). In Fig. 1.3(b), the accuracy is high with no bias, but the precision is low (i.e., the darts are far apart from each other). In Fig. 1.3(c), accuracy and bias are poor (dart group on the south-west side of the target), but precision is good (i.e., darts are close to each other). In Fig. 1.3(d), the accuracy and precision are high with no bias.

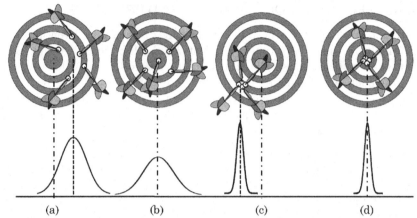

FIGURE 1.3: (a) Accuracy-low, Precision-low, Bias present (on the east of the target); (b) Accuracy-high, precision-low (No bias), (c) Accuracy-low, Precision-high, Bias present (on the SW of the target), (d) Accuracy-high, Precision-high (No Bias)

1.3 PROPAGATION OF ERROR

Since no measurement is exact, we seek the best possible values for the experimentally determined quantities. Thus, the experimental results are reported as a range of possible true values (e.g., $Q \pm dQ$) based on the limited number of measurements. In most cases, computer numbers are not exact, either due to rounding or truncation errors. The quantities derived as a result of either experiment with uncertainties or numerical computations with rounded or chopped numbers are generally put through functional manipulations (addition, subtraction, multiplication, division, etc.) to compute the desired quantity. The errors associated with each independent variable propagate through the functional manipulations, which affect the corresponding compound error in the derived quantity.

The *propagation of error* (or *uncertainty*) is defined as the effects of variable uncertainties on a function. The concept of error propagation analysis is developed to determine the effects of uncertainties (specifically random or bias errors) on a computed quantity.

Consider the calculated quantity to be $F = F(x, y)$, where x and y are independent (measured) variables. If x and y have uncertainties dx and dy, then the measured quantities are reported as $x \pm dx$ and $y \pm dy$. These average errors contribute to the error of the computed quantity F, i.e., the final result should be reported as $F \pm dF$. Assuming the average errors are small, the compound error can be estimated using the total differential formula as follows:

$$dF = \frac{\partial F}{\partial x} dx + \frac{\partial F}{\partial y} dy \qquad (1.7)$$

where dx and dy differentials are average (or absolute) errors in the x and y variables, respectively, and dF is the compound error.

Dividing both sides of Eq. (1.7) by F gives the relative error:

$$\left| \frac{dF}{F} \right| = \left| \frac{x}{F} \frac{\partial F}{\partial x} \left(\frac{dx}{x} \right) \right| + \left| \frac{y}{F} \frac{\partial F}{\partial y} \left(\frac{dy}{y} \right) \right| \qquad (1.8)$$

or multiplying both sides of Eq. (1.8) by 100, we get

$$F\% = \left| \frac{x}{F} \frac{\partial F}{\partial x} \right| (x\%) + \left| \frac{y}{F} \frac{\partial F}{\partial y} \right| (y\%) + \left| \frac{z}{F} \frac{\partial f}{\partial z} \right| (z\%) \qquad (1.9)$$

TABLE 1.1: Error propagation rules.

Expression	Standard Deviation
$F = Cx$	$\sigma_F = |C|\,\sigma_x \quad \leftrightarrow \quad \dfrac{\sigma_F}{F} = \dfrac{\sigma_x}{x}, \quad C = \text{constant}$
$F = x \pm y$	$\sigma_F = \sqrt{\sigma_x^2 + \sigma_y^2}$
$F = \dfrac{x}{y}$	$\dfrac{\sigma_F}{F} = \sqrt{\left(\dfrac{\sigma_x}{x}\right)^2 + \left(\dfrac{\sigma_y}{y}\right)^2}$
$F = x^n y^m$	$\dfrac{\sigma_F}{F} = \sqrt{\left(\dfrac{n\,\sigma_x}{x}\right)^2 + \left(\dfrac{m\,\sigma_y}{y}\right)^2}$

where $F\%$ is the percent relative error. The guiding principle in all cases is to consider the *most pessimistic* situation.

Random errors in measured x and y similarly propagate to the computed quantity F as well. In the following analysis, we assume that (a) the random errors of each measured value are independent of each other; (b) measurements follow a Gaussian distribution; and (c) there is negligible or no covariance between the errors. Under these assumptions, the differentials may take $(+/-)$ signs since these errors fluctuate about a mean value. Thus, instead of dF, we examine $(dF)^2$. Taking the square of Eq. (1.7) yields

$$
(dF)^2 = \left(\frac{\partial F}{\partial x}dx + \frac{\partial F}{\partial y}dy\right)^2 \tag{1.10}
$$

$$
= \left(\frac{\partial F}{\partial x}\right)^2 (dx)^2 + \left(\frac{\partial F}{\partial y}\right)^2 (dy)^2 + 2\left(\frac{\partial F}{\partial x}\right)\left(\frac{\partial F}{\partial y}\right)(dx)(dy)
$$

where x and y are independent variables. Since dx and dy can have positive or negative signs, the square terms are always positive. However, the cross terms may cancel each other out, and as a result, the mean value of the cross terms will be very small or zero. In this case, we may write

$$
dF = \sqrt{\left(\frac{\partial F}{\partial x}\right)^2 (dx)^2 + \left(\frac{\partial F}{\partial y}\right)^2 (dy)^2} \tag{1.11}
$$

For random errors, it is customary to replace the dx and dy differentials with the standard deviations to give

$$
\sigma_F = \sqrt{\left(\frac{\partial F}{\partial x}\right)^2 \sigma_x^2 + \left(\frac{\partial F}{\partial y}\right)^2 \sigma_y^2} \tag{1.12}
$$

Equation (1.11) or (1.12) is the *error propagation formula* for a quantity of two independent variables. Using Eq. (1.12), the standard deviation of the computed quantity F associated with common mathematical operations is tabulated in Table 1.1.

EXAMPLE 1.2: Calculate average and standard error

The density of a solid cylindrical object is to be determined. Estimate the computed error of the density if the object's circumference, height, and mass are measured with (a) the average error of ±1% and (b) the standard deviation of ±1%.

SOLUTION:

The density of a cylindrical object is determined by $\rho = m/\pi r^2 h$, where m, h, and r are the mass, height, and radius, respectively. However, the error for the radius is not given, but it can be estimated by using the perimeter ($S = 2\pi r$). Replacing the radius with the circumference leads to $\rho = 4\pi\, m/hS^2$.

(a) The measured quantities have average errors ($\pm dm$, $\pm dS$, $\pm dh$), which gives rise to $\pm d\rho$. Therefore, the measured quantities (m, h, and S) are taken as the independent variables, i.e., $\rho = f(m, h, S)$.

Considering the most pessimistic scenario, the maximum possible error in each variable will contribute to the maximum possible error in the density. In this case, the total differential formula for the density is expressed as

$$d\rho = \left|\frac{\partial \rho}{\partial m} dm\right| + \left|\frac{\partial \rho}{\partial h} dh\right| + \left|\frac{\partial \rho}{\partial S} dS\right|$$

where dm, dS, and dh are differentials that denote absolute errors. Replacing the partial derivatives by $\partial\rho/\partial m = 4\pi/hS^2$, $\partial\rho/\partial S = -8\pi m/hS^3$, and $\partial\rho/\partial h = -4\pi m/h^2 S^2$ and dividing both sides with ρ, we get

$$\frac{d\rho}{\rho} = \left|\frac{dm}{m}\right| + \left|-\frac{dh}{h}\right| + 2\left|-\frac{dS}{S}\right|$$

or multiplying both sides by 100 leads to

$$\rho\% = m\% + h\% + 2(S\%)$$

Since m, h, and S are measured with relative errors of ±1%, the relative error of the estimated density is found to be ±4%.

(b) Replacing the differentials with the standard deviations (σ_m, σ_S, σ_h) implies that the errors in m, r, and h are independent of each other. By dividing both sides with ρ, the propagation of error formula becomes

$$\frac{\sigma_\rho}{\rho} = \sqrt{\left(\frac{\sigma_m}{m}\right)^2 + \left(\frac{\sigma_h}{h}\right)^2 + \left(2\frac{\sigma_S}{S}\right)^2}$$

or

$$\rho\% = \sqrt{(m\%)^2 + (h\%)^2 + 4\,(S\%)^2}$$

In this case the relative error is obtained as ±2.45%.

Discussion: This example illustrates the two types of uncertainty estimations. The average error estimate in the density is larger than that of the random error. It should be noted that the standard deviation describes the spread of the data. On the other hand, standard error is not about the spread of our data but the accuracy of the average. Any experimental calculation should be accompanied by a proper uncertainty analysis.

1.4 NUMERICAL COMPUTATIONS

A major source of numerical errors in digital computers arises from the way numbers are stored and managed in computers. Before diving into examining numerical errors, it is necessary to briefly look into how numbers are handled on digital computers.

1.4.1 NUMBER SYSTEMS

In daily life, the *decimal system* (a 10-based numbering system) with 10 digits (from 0 to 9) is used to express integer, rational, and irrational numbers. Several numbering systems with different bases are incorporated into digital computer hardware to interpret and communicate data between its components. Binary (base 2), ternary (base 3), octal (base 8), and hexadecimal (base 16) numbering systems are the most commonly used.

The input supplied to digital computers is in decimal format, but the digital computers process any type of numeric data, letters, and special symbols in discrete form as pulses sent by electronic components. For instance, the state of an electrical pulse is interpreted as 1 for a high voltage (*on*) or 0 for a low voltage (*off*). This concept can be thought of as switching an electric circuit on and off. Hence, digital computers are designed to read and interpret data in *binary* form as a sequence of *on*'s (1's) and *off*'s (0's). In other words, devices based on binary (*on/off* logic) can be constructed, and all forms of data can be expressed in a binary system.

Computers use *bits* to represent information. A *bit* is the basic unit of storage in a computer, and it is defined as a binary, which is 0 or 1. A *bayt* is a group of eight bits used to represent a character, and it is the basic unit for measuring memory size in computers. Two or more bits are referred to as a *word*.

In the decimal number system, an integer is expressed as a polynomial having one of the ten digits ranging from 0 to 9 as coefficients.

$$N = (a_n a_{n-1} a_{n-2} \cdots a_1 a_0)_{10}$$

$$= a_n \times 10^n + a_{n-1} \times 10^{n-1} + \cdots + a_1 \times 10^1 + a_0 \times 10^0 = \sum_{k=0}^{n} a_k \times 10^k$$

For instance, 125 is expressed as $125 = 1 \times 10^2 + 2 \times 10^1 + 5 \times 10^0$.

A positive real number has an integer (N) and a fractional part (F). The fractional part is written as a decimal fraction.

$$F = (b_1 b_2 b_3 \cdots b_n \cdots)_{10}$$

$$= b_1 \times 10^{-1} + b_2 \times 10^{-2} + \cdots + b_n \times 10^{-n} + \cdots = \sum_{k=1}^{\infty} b_k \times 10^{-k}$$

where each b_n takes one of the values from 0 to 9. For example,

$$(0.125)_{10} = 1 \times 10^{-1} + 2 \times 10^{-2} + 5 \times 10^{-3}$$
$$(0.111 \cdots)_{10} = 1 \times 10^{-1} + 1 \times 10^{-2} + 1 \times 10^{-3} + \cdots$$

A floating-point representation of a decimal number in base β is expressed as

$$(a_n \ldots a_2 a_1 a_0 . b_1 b_2 b_3 b_4 \ldots)_\beta = \sum_{k=0}^{n} a_k \beta^k + \sum_{k=1}^{\infty} b_k \beta^{-k} \qquad (1.13)$$

Note that a_n's and b_n's take the values of $0, 1, 2, \ldots, \beta - 1$.

In the binary number system ($\beta = 2$), for instance, 125 is found as

$$(125)_{10} = (1111101)_2 = 1 \times 2^6 + 1 \times 2^5 + 1 \times 2^4 + 1 \times 2^3 + 1 \times 2^2 + 0 \times 2^1 + 1 \times 2^0$$

The numbers with fractional parts are expressed as

$$(0.125)_{10} = (0.001)_2 = 0 \times 2^{-1} + 0 \times 2^{-2} + 1 \times 2^{-3}$$

$$(0.111\cdots)_{10} = (0.\overline{000111})_2$$

$$= 0 \times 2^{-1} + 0 \times 2^{-2} + 0 \times 2^{-3} + 1 \times 2^{-4} + 1 \times 2^{-5} + 1 \times 2^{-6} + \cdots$$

Expressing data in binary format uses too many bits, so the binary system can be impractical due to the difficulty of reading large numbers in binary form. In order to overcome this problem, binary numbers in groups of four bits are merged to form the *hexadecimal number system*. This system, also known as "hex," has a base of 16 ($\beta = 16$) and consists of *16 symbols*: 10 numbers of the decimal system (digits ranging from 0 to 9) and six letters from A to F with letters representing 10 to 15, respectively. The hex system allows information to be represented with fewer bits, making it more useful and suitable for digital hardware. For example, 125 and 0.21875 in the hex system are expressed as

$$(125)_{10} = (7D)_{16} = 7 \times 16^1 + 13 \times 16^0$$

$$(0.21875)_{10} = (0.38)_{16} = 3 \times 16^{-1} + 8 \times 16^{-2}$$

Octal system ($\beta = 8$) which is also used to make any binary information more compact, is not as popular or common as hex or binary systems. An octal number is made up of digits ranging from 0 to 7. The octal system groups binary numbers into triplets instead of quartets. For instance, 125 in the octal system is expressed as

$$(125)_{10} = (175)_8 = 1 \times 8^2 + 7 \times 8^1 + 5 \times 8^0$$

$$(0.21875)_{10} = (0.16)_8 = 1 \times 8^{-1} + 6 \times 8^{-2}$$

1.4.2 FLOATING-POINT REPRESENTATION OF NUMBERS

In this section, computer arithmetic of floating-point numbers is presented using the standard set by the IEEE (Institute for Electrical and Electronic Engineers). In 1985, the IEEE published its first report, IEEE 754-1985 [15], that regulated standards for binary or decimal floating-point numbers, algorithms for rounding arithmetic operations, how to handle exceptions, etc. The *IEEE standard* was updated in 2019 as IEEE 754-2019 [16]. This standardization, now followed by microprocessor manufacturers, has enhanced program portability.

Real numbers on a computer are represented by a floating-point number system. We normally supply data (*numbers*) as input to a computer using the decimal system. The input is then converted to and stored in binary form. Computers have a finite word length, and for this reason, only numbers with a finite number of digits (integers or numbers with powers of 2) can be *exactly* represented. In other words, since they cannot be represented exactly on computers, most numbers are rounded. This rounding process alone can lead to errors due to the inexact storage of *machine numbers*. A machine number of x (i.e., a finite floating-point approximation) is denoted by $fl(x)$.

EXAMPLE 1.3: Converting decimals to binary, octal, and hex numbers

Convert 587, 18.75, and 1.5625 (in the decimal system) to equivalent binary, octal, and hexadecimal numbers.

SOLUTION:

Conversion of 587 (an integer) leads to

$$(587)_{10} = 1 \times 2^9 + 0 \times 2^8 + 0 \times 2^7 + 1 \times 2^6 + 0 \times 2^5 + 0 \times 2^4 + 1 \times 2^3 + 0 \times 2^2$$
$$+ 1 \times 2^1 + 1 \times 2^0 = (1001001011)_2$$
$$(587)_{10} = 1 \times 8^3 + 1 \times 8^2 + 1 \times 8^1 + 3 \times 8^0 = (1113)_8$$
$$(587)_{10} = 2 \times 16^2 + 4 \times 16^1 + B \times 16^0 = (24B)_{16}$$

where B denotes 11.

Conversion of 18.75 (a real number) gives

$$(18.75)_{10} = 1 \times 2^4 + 0 \times 2^3 + 0 \times 2^2 + 1 \times 2^1 + 0 \times 2^0 + 1 \times 2^{-1} + 1 \times 2^{-2}$$
$$= (10010.11)_2$$
$$(18.75)_{10} = 2 \times 8^1 + 2 \times 8^0 + 6 \times 8^{-1} = (22.6)_8$$
$$(18.75)_{10} = 1 \times 16^1 + 2 \times 16^0 + C \times 16^{-1} = (12.C)_{16}$$

where C denotes 12. Finally, converting 1.5625 yields

$$(1.5625)_{10} = 1 \times 2^0 + 1 \times 2^{-1} + 0 \times 2^{-2} + 0 \times 2^{-3} + 1 \times 2^{-4} = (1.1001)_2$$
$$(1.5625)_{10} = 1 \times 8^0 + 4 \times 8^{-1} + 4 \times 8^{-2} = (1.44)_8$$
$$(1.5625)_{10} = 1 \times 16^0 + 9 \times 16^{-1} = (1.9)_{16}$$

Discussion: Note that the decimals in this example are made up of the sum of positive and negative powers of 2. Hence, they are represented "exactly" in the binary, octal, and hexadecimal number systems.

A non-negative real number is represented with an integer part, a fractional part, and a decimal point, separating the integer and fractional parts, i.e., 123.4567. This representation is referred to as *decimal notation*. Any real number in the decimal system can be expressed in *normalized scientific notation*. This is accomplished by shifting the decimal point and expressing the number with a power of 10 such that all digits on the right of the decimal point are not zero; for example, 0.00012 (0.12×10^{-3}), 0.125 (0.125×10^0), and 23.4 (0.234×10^2).

Arithmetic calculations account for the major source of numerical errors. This is because computers carry out calculations in floating-point arithmetic. An n-digit *floating-point representation* of a real number in base-β has the form

$$M \times \beta^e \quad \text{or} \quad \pm (0.m_1 m_2 \cdots m_n)_\beta \times \beta^e$$

where β denotes the *base* of the number system ($\beta = 2$ for binary, $\beta = 10$ for decimal systems), M is a β-fraction called *mantissa* ($\beta^{-1} \leqslant M < 1$ or $M=0$), containing the normalized value of the number, $\pm$ denotes the sign of the number, $m_1 \neq 0$, m_2, m_3, $\ldots$, m_n are the decimal digits (0, 1, 2, $\ldots$, $\beta - 1$), and e is an integer (positive, negative, or zero) called the *exponent*, which effectively defines the offset from normalization. The total number of bits assigned to a number is fixed by computer architecture.

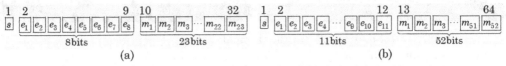

FIGURE 1.4: IEEE 754 Floating-Point Standard (a) single, (b) double precision.

EXAMPLE 1.4: Expressing numbers in Normal Scientific Notation

Express 123.4567, 0.01234567, and 123456.7 in normalized scientific notation.

SOLUTION:

Applying the normalized scientific notation rule results in

$$12.34567 \ = \ 0.1234567 \times 10^2$$
$$0.01234567 \ = \ 0.1234567 \times 10^{-1}$$
$$123456.7 \ = \ 0.1234567 \times 10^6$$

Discussion: Notice that the leading decimal in any fraction is not zero unless the number itself is zero; for instance, expressing 0.001234567 as $0.01234567 \times 10^{-1}$ or 0.001234567×10^0 can be viewed as a format option, but these expressions cannot be regarded as the normalized scientific notation.

1.4.3 PRECISION OF A NUMERICAL QUANTITY

The definition of *precision* for a numerical quantity in computers is different from that of an experimental value. The *precision* of a numerical value describes the number of digits (or bits) that are used to represent a numerical value. The precision of floating-point numbers on any computer is determined by the word length of the computer. However, computing systems generally provide floating-point numbers of different lengths. A floating-point number with a short form (32 bits or 8 bytes) is called a *single-precision* floating-point number. As depicted in Fig. 1.4(a), each number is represented by a 24-bit mantissa (including a one-bit sign) and an 8-bit exponent. The exponents lie between −126 and 127. Exponent values of −127 and 128 are reserved for floating-point exceptions such as "*infinity*" due to division by zero or NaN ("*Not a Number*"). In other words, single precision allows for the representation of numbers with an accuracy of seven decimals, ranging from $5.877 \times 10^{-39} (= 2^{-127})$ to $3.4 \times 10^{38} (= 2^{128})$.

A floating-point number represented by a long form (64 bits or 16 bytes) is called a *double-precision* floating-point number. A double-precision number uses 53 bits for the mantissa (including the sign) and 11 bits for the exponent, as shown in Fig. 1.4(b). The range of exponents becomes −1022 and 1023, which means that the numbers between $2.225 \times 10^{-308} (= 2^{-1022})$ and $8.988 \times 10^{308} (= 2^{1023})$ can be represented with an accuracy of sixteen decimals. Calculation in double precision doubles the memory storage requirements and the cpu-time as compared to single precision. Some compilers allow the use of *quadruple* and *octuplet* precision.

Regardless of its "precision," when a real number is too large or too small to be represented in the normalized scientific notation, we may encounter a representation problem since the range of floating-point numbers is finite. When an arithmetic operation results in a number larger than the maximum allowed number, it is referred to as *overflow*. Similarly, an

---

Pseudocode 1.5

Function Module MACHEPS (x)
\ DESCRIPTION: A pseudo Function module to estimate machine epsilon.
While $\left[1 + x/2 > 1 \right]$ \ x is large enough so that $1 + x/2 \neq 1$
    $x \leftarrow x/2$ \ Reduce magnitude of x by halving
End While
MACHEPS$\leftarrow x$ \ Set smallest value of x to MACHEPS
End Function Module MACHEPS

---

operation resulting in a smaller number than the minimum allowed is called an *underflow*. An error message is issued when an operation yielding overflow or underflow is encountered.

Machine epsilon, ε_m, represents the round-off error for a floating-point number with a given precision. It is defined as the smallest positive number (i.e., the gap between the number 1 and the next largest floating-point number); mathematically speaking, $\varepsilon_m \oplus 1 > 1$, where $\oplus$ denotes floating-point addition. Hence, we can compute ε_m from

$$\varepsilon_m = \left\{ \begin{array}{ll} \beta^{-t}, & \text{rounding} \\ \beta^{1-t}, & \text{chopping} \end{array} \right. \tag{1.14}$$

where β is the number base and t is the number of significant digits in the mantissa.

Machine epsilon can also be defined as an upper bound on the relative error due to rounding in floating-point arithmetic. It is in practice used to compare whether two floating-point numbers are equivalent. Its size is determined by the *precision* used, *type of rounding*, and *computer hardware*. For binary machines ($\beta = 2$), theoretically, it yields 1.19×10^{-7} and 2.22×10^{-16} for single ($t = 24$) and double precision ($t = 53$), respectively.

A pseudofunction module, **MACHEPS**, that determines the machine epsilon of a computer is presented in Pseudocode 1.5. The code accepts an arbitrary floating-point number x as input. Using a **While**-construct, x is halved successively until $1 \oplus (x/2) = 1$, where the machine can no longer distinguish the difference between $1 \oplus (x/2)$ and 1. When a bound for the machine epsilon is found, the value of x is set to **MACHEPS**. However, it should be noted that most compilers come with built-in functions for furnishing machine epsilon.

 Utilizing the machine epsilon, ϵ_m, in computer programs makes the programs portable, i.e., machine-independent. In other words, the programs become independent of the computer hardware or systems for which they were designed.

Equality of two real numbers. We have already learned that digital computers can rarely store or manage the "true" or "exact value" of a real number due to computer representations of numbers and round-offs. A fundamental step in many algorithms is to decide if and when an approximation is accurate enough. So testing for equality of two real numbers as $x = y$, $x - y = 0$, or $x/y = 1$ must be done cautiously. (Symbolic processors that do not use rounding easily overcome this problem.) Instead, a smart approach is to test for *near equality*; that is, one must determine the interval around the value of "0" or "1" that is sufficient to meet the criteria. The machine epsilon (or a *user-prescribed tolerance*) is used to test for either equality of two real numbers (x and y) or convergence of an iterative process (as a stopping criterion) as $|x - y| < \varepsilon_m$ or $|x/y - 1| < \varepsilon_m$, where ϵ_m is the machine epsilon.

In iterative processes, ϵ_m may be too strict, in which case it is replaced by a user-prescribed tolerance, ϵ.

 A more prudent approach is to base an equality test on the relative error, $|x - y|/|y| < \varepsilon_m$, rather than the absolute error, especially when dealing with very small or very large numbers.

1.4.4 ROUNDING-OFF AND CHOPPING

Most real numbers are approximated by their closest representation in the machine since they cannot be exactly represented by the floating-point representation. There are two ways of rounding: (*i*) optimal rounding and (*ii*) chopping.

Consider any real positive number (within the numerical range of a digital computer) expressed in the normalized scientific form as

$$0 . m_1 m_2 \cdots m_n m_{n+1} \cdots \times 10^e$$

where $1 \leqslant m_1 \leqslant 9$ and $0 \leqslant m_n \leqslant 9$ for $n > 1$.

In *optimal rounding*, the closest machine number is chosen. Rounding is done in such a fashion as to cause the fewest possible errors. Assuming that n is the maximum number of decimal digits to be rounded, the general rule for rounding off a number to n decimal places is given as

$$fl_r(0 . m_1 m_2 \cdots m_n m_{n+1} \cdots \times 10^e) = \begin{cases} 0 . m_1 m_2 \cdots m_n m_{n+1} \times 10^e, & 0 \leqslant m_{n+1} < 5 \\ 0 . m_1 m_2 \cdots (m_n + 1) \times 10^e, & 5 \leqslant m_{n+1} \leqslant 9 \end{cases}$$

Note that if m_{n+1} (the digit of the number to the right of m_n) is less than 5, then m_n remains unchanged and all digits after m_n (i.e., $m_{n+1} m_{n+2} \cdots$) are discarded. If $m_{n+1} \geqslant 5$, then the value of m_n is raised by 1 ($m_n \leftarrow m_n + 1$) and all digits after d_n are discarded.

In *chopping*, the digits after the nth place ($m_{n+1} m_{n+2} \cdots$) are omitted regardless of their magnitude, yielding the following approximation:

$$fl_c(0 . m_1 m_2 \cdots m_n m_{n+1} \cdots \times 10^e) = 0 . m_1 m_2 \cdots m_n \times 10^e$$

Chopping is easier than optimal rounding, but slightly less accurate. Notice that we used the subscripts "r" and "c" in the floating-point representation of rounded and chopped numbers, respectively. Many computers use "chopping" rather than "rounding" after each arithmetic operation.

Some examples of rounding and chopping to *four decimal places* are illustrated below:

Number	Rounding	Chopping
0.123456	0.1235	0.1234
−0.123456	−0.1235	−0.1234
0.123450	0.1235	0.1234
0.123750	0.1238	0.1237
3.141592	3.1416	3.1415
18.127653	18.1277	18.1276

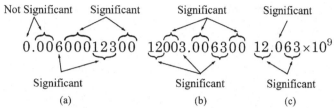

FIGURE 1.5: Depiction of significant/non-significant figures of a (a) small, (b) large, and (c) exponential number.

1.4.5 SIGNIFICANT DIGITS

The term *significant digits* (or *significant figures*) is used to indicate the accuracy of a number. Apart from its exponential part, it contains the information (i.e., digits) that contributes to the size or precision of a number.

Significant figures contain any one of the digits 1, 2, 3, ..., 9, and 0. The rules to identify significant digits are given as

- All non-zero digits 1, 2, ..., 9 are significant; e.g., 123 and 1.234 have, respectively, three and four significant digits;

- Zeros between any two non-zero digits are significant; e.g., 102 and 1.02 have three significant figures, while 12.003 has five significant digits;

- Zeros to the left of the first non-zero digit in a number are *not* significant; e.g., 0.01, 0.012, 0.00123, and 0.01234 have one, two, three, and four significant digits, respectively;

- Zeros to the right of the last non-zero digit in a number (i.e., trailing zeros) with a decimal point are significant; e.g., 1.30 and 123.000 have, respectively, three, and six significant digits;

- The significance of trailing zeros in an integer depends on the measurement; e.g., 12300 is ambiguous, but it becomes more meaningful when expressed as 1.23×10^4 (three-significant digits), 1.230×10^4 (four significant digits), or 1.2300×10^4 (five significant digits).

For example, consider that 0.00600012300 has nine significant digits (i.e., 6, 0, 1, 2, and 3). As shown in Fig. 1.5(a), zeros after the decimal point are not significant; they are used to fix the decimal point. But the zeros between 6 and 1 and the two trailing zeros are significant. The number 12003.006300 depicted in Fig. 1.5(b) has 11 significant digits. All non-zero digits are significant. The zeros between 2 and 3 and between 3 and 6, along with the two trailing zeros, are significant. The exponential number presented in Fig. 1.5(c) has five significant figures; in other words, all digits (1, 2, 0, 6, 3) are significant.

Consider the additional numbers given in Table 1.2. All numbers with non-zero digits (123456.7, 1.2345, and 0.123456) are significant. The numbers with zeros between non-zero digits (1002, 12.034, and 2.00076) are significant. The leading zeros in 0.00016 and 0.004830 vanish when expressed in the normalized scientific notation as 0.16×10^{-3} and 0.483×10^{-2}. Similarly, when 70000 and 1200 are also expressed in normalized scientific notation (0.7×10^5 and 0.12×10^4), we observe one and two significant digits, respectively. Also note that the 'trailing zeros' in 3.000 and 0.004830 are important, as these indicate the level of accuracy of the measurement as much as any other digits.

TABLE 1.2: Examples of significant digits.

Number	Significant digits	No. of significant digits
123456.7	1, 2, 3, 4, 5, 6, 7	seven
70000	7	one
1200	1, 2	two
1002	0, 1, 2	four
12.034	1, 2, 0, 3, 4	five
0.00016	1, 6	two
2.00076	2, 0, 7, 6	six
3.000	3, 0	four
0.004830	0, 3, 4, 8	four
1.2345	1, 2, 3, 4, 5	five
0.123456	1, 2, 3, 4, 5, 6	six

In Pseudocode 1.6, the pseudomodule MANTISSA_EXP which retrieves the mantissa and exponent (m, e) for an input floating-point number (fl) is presented. The module makes use of three mathematical functions—absolute value $\textsf{Abs}(x) = |x|$, $\textsf{Floor}(x)=\lfloor x \rfloor$, and $\textsf{Log}_{10}(x)$—that are incorporated in most programming languages as built-in functions; these functions are specified in the module following the USES statement. While on the subject, it should be pointed out here that, although not required, the practice of identifying module dependencies can be useful if a source code is translated into a different language or compiled and run on different computing platforms that may not share the same built-in functionality. Some computer software, including CAS, has built-in functions to extract the mantissa and exponent of a floating-point value.

The built-in function $\textsf{Floor}(x)$ returns the greatest integer less than or equal to x, while $\textsf{Log}_{10}(x)$ returns the base-10 logarithm of x. The code first determines the exponent for any $fl \neq 0$ and then the mantissa by $m = fl \times 10^{-e}$. Clearly, if $fl = 0$, then $m = e = 0$.

Pseudocode 1.6

Module MANTISSA_EXP (fl, m, e)
\ DESCRIPTION: A pseudomodule to find mantissa (m) and exponent (e)
\   of a real number (fl).
\ USES:
\   ABS :: A built-in function computing the absolute value;
\   FLOOR:: Built-in floor function;
\   LOG10:: Built-in base-10 logarithm function.
If $\left[|fl| > 0\right]$ **Then**   \ Case of $|fl| \neq 0$
    $e \leftarrow 1 + \textsf{Floor}(\textsf{Log}_{10}(|fl|))$ \ Find exponent
    $m \leftarrow fl * 10^{-e}$ \ Find mantissa
Else   \ Case of $fl = 0$
    $e \leftarrow 0; \quad m \leftarrow 0$ \ Both e and m are zero
End If
End Module MANTISSA_EXP

EXAMPLE 1.5: Chopping and rounding by significant digits

Apply *four-significant-digit* (a) chopping, (b) rounding to the following numbers:

123456789.12,   3.7841828451,   0.3183698863,   0.0881078861

SOLUTION:

These numbers are first expressed in normalized scientific notation:

$$123456789.12 = 0.12345678912 \times 10^9$$
$$3.7841828451 = 0.37841828451 \times 10^1$$
$$0.3183698863 = 0.3183698863 \times 10^0$$
$$0.0881078861 = 0.8810788610 \times 10^{-1}$$

(a) The floating-point forms using four-digit chopping are

$$123456789.12 = 0.1234 \times 10^9 = 123400000$$
$$3.7841828451 = 0.3784 \times 10^1 = 3.784,$$
$$0.3183698863 = 0.3183 \times 10^0 = 0.3183,$$
$$0.0881078861 = 0.8810 \times 10^{-1} = 0.08810$$

The absolute and relative (in parenthesis) errors are 56789.12 (0.046%), 0.000182845 (0.00483%), 0.0000698863 (0.02195%), and 0.000007886 (0.00895%) for 123456789.12, 3.7841828451, 0.3183698863, and 0.0881078861, respectively. Note that the absolute error is large for the largest number.

(b) Since the fifth digits of the first, second, and third numbers are greater than or equal to 5, the fifth digits are increased by 1, while the first number remains unchanged. So the floating-point forms using four-digit rounding become:

$$0.12345678912 \times 10^9 = 0.1235 \times 10^9 = 123500000$$
$$0.37841828451 \times 10^1 = 0.3784 \times 10^1 = 3.784,$$
$$0.3183698863 \times 10^0 = 0.3183 \times 10^0 + 0.0001 \times 10^0$$
$$= 0.3184 \times 10^0 = 0.3184,$$
$$0.881078861 \times 10^{-1} = 0.881 \times 10^{-1} + 0.0001 \times 10^{-1}$$
$$= 0.8811 \times 10^{-1} = 0.08811$$

The true absolute and relative errors turn out to be 43210.88 (0.035%), 0.000182845 (0.00483%), 0.000030114 (0.00946%), and 0.000002114 (0.0024%) for 123456789.12, 3.7841828451, 0.3183698863, and 0.0881078861, respectively. However, clearly, the round-off errors can be reduced by increasing the number of digits allowed in a representation.

Discussion: Round-off error is caused by approximating a number, which is a quantization error that results from mathematical calculations. It all boils down to is to figure out how big is the round-off error. The absolute (round-off) error for large numbers can be very large. The relative error indicating how good a number is relative to its true size should be preferred in decisions.

1.4.6 DECIMAL-PLACE AND SIGNIFICANT-DIGIT ACCURACY

In the preceding discussions, we have learned that the numbers processed by digital computers are chopped or rounded, which eventually results in numbers that differ from their true values. To terminate a numerical procedure, it is often necessary to determine the significant-digit accuracy of a computed number.

Accuracy is a measure of how close a measured or computed quantity is to its true value. A number is said to be "*accurate to n decimal places*" if the n-digits to the right of the decimal place (including the nth digit) are correct. In this case, the error is at most 10^{-n}. But the nth decimal place is where the rounding takes place, in which case the error is at most $10^{-n}/2$. Hence, for n-decimal place accuracy, it is sufficient to show that the following inequality holds:

$$|Q - \tilde{Q}| \leq \frac{1}{2} \times 10^{-n}$$

where $\tilde{Q}$ and Q denote the approximate and true values of a quantity Q.

A number is said to be "*accurate to n-significant digits*" if the n-digits to the right of the first non-zero digit are correct. If $|Q - \tilde{Q}|$ has a magnitude less than or equal to 5 in the $(n + 1)$th digit of Q, counting to the right starting from the first nonzero digit in Q, then the quantity $\tilde{Q}$ is said to have n-significant digits with respect to its true value. In other words, it indicates that you can trust a total of n digits.

The significant digit accuracy is a measure of relative error, and so the following inequality can be used to measure the number of (accurate) significant digits in $\tilde{Q}$:

$$\left| \frac{Q - \tilde{Q}}{Q} \right| \leq \frac{1}{2} \times 10^{-n} \qquad \text{or} \qquad n \cong \left\lfloor -\log_{10}\left(\left| \frac{Q - \tilde{Q}}{Q} \right| \right) \right\rfloor$$

We can then say $\tilde{Q}$ has $n-$significant digits with respect to Q. Note that when defining accurate digits, $\lfloor x \rfloor$ (*i.e.*, Floor(x) function) is used.

> Note that absolute error, $|Q - \tilde{Q}|$, gives the number of digits *after* the decimal point, while relative error, $|Q - \tilde{Q}|/|Q|$, gives the number of digits regardless of the position of the decimal point.

1.4.7 EFFECT OF ROUNDING IN ARITHMETIC OPERATIONS

Round-off errors are initially encountered when saving a real number in the computer's memory as a floating-point number. Computer arithmetic with floating-point numbers is predominantly inexact and thus yields computation errors. Although the error of a single arithmetic operation is negligibly small, the accumulation of round-off errors due to numerous arithmetic operations (in repetitive or iterative computations) can dominate calculations under certain conditions.

In this section, we briefly look at how common arithmetic operations affect round-off errors. It should be pointed out that the rounding process works basically similarly in binary, octal, and hexadecimal number systems. Hence, the normalized decimal system is used here to illustrate the effect of round-off errors on arithmetic operations. To make visualization easier, consider a hypothetical decimal computer with a 3-digit mantissa and a 1-digit exponent. When adding or subtracting two floating-point numbers, the mantissa of the number with the smaller exponent is modified to equalize the exponents of both

---

EXAMPLE 1.6: Determining significant digits

Determine the significant digit accuracy of approximations to the following numbers:

True Value	Approximation
1/7	0.1439
0.01234	0.012
1.23456	1.235
12.4	12.3462

SOLUTION:

For $Q = 1/7$ and $\tilde{Q} = 0.1439$, we find the absolute and relative errors as $|Q - \tilde{Q}| = 0.00104286$ and $|Q - \tilde{Q}|/Q = 0.0073$, respectively. Since the error is less than 5 in the third digit to the right of the first non-zero digit in Q, $\tilde{Q}$ has *two significant digits* with respect to Q. Also note that $n \cong \lfloor -\log_{10}(0.0073) \rfloor = 2$.

For $Q = 0.01234$ and $\tilde{Q} = 0.012$, similarly, the absolute and relative errors are found as $|Q - \tilde{Q}| = 0.00034$ and $|Q - \tilde{Q}|/Q = 0.0275527$. The error is less than 5 in the third digit to the right of the first non-zero digit in Q (which is 3), so $\tilde{Q}$ has *two significant digits* with respect to Q. However, the prediction yields $n \cong \lfloor -\log_{10}(0.0275527) \rfloor = 1$.

For $Q = 1.23456$ and $\tilde{Q} = 1.235$, the absolute and relative errors yield $|Q - \tilde{Q}| = 0.00044$ and $|Q - \tilde{Q}|/Q = 0.0003564$. The error is less than 5 in the fourth digit to the right of the first non-zero digit in Q so $\tilde{Q}$ has *three significant digits* with respect to Q. We also find $n \cong \lfloor -\log_{10}(0.0003564) \rfloor = 3$.

We find $|Q - \tilde{Q}| = 0.0538$ and $|Q - \tilde{Q}|/Q = 0.00433871$ for $Q = 12.4$ and $\tilde{Q} = 12.3462$. The error is less than 5 in the third digit to the right of the first non-zero digit in Q so $\tilde{Q}$ has *two significant digits* with respect to Q. Also note that $n \cong \lfloor -\log_{10}(0.00433871) \rfloor = 2$.

Discussion: In this example, determining the significant-digit accuracy of approximate numbers was explained in detail, but in practice, we need to rely on quantitative tools. The inequalities presented in this section successfully estimate the significant-digit accuracy of the given approximations.

---

numbers. This action aligns the decimal points. For instance, the normalized floating-point representations of 21×10^5 and 19×10^3 are 0.21×10^7 and 0.19×10^5, respectively. To add or subtract these numbers, the decimal point of the small number is aligned with that of the larger one as 0.0019×10^7. Then the two numbers can be added or subtracted as follows:

$$\begin{array}{cc} 0.2100 \times 10^7 & 0.2100 \times 10^7 \\ +0.0019 \times 10^7 & -0.0019 \times 10^7 \\ \hline 0.2119 \times 10^7 & 0.2081 \times 10^7 \end{array}$$

The results are then chopped to give 0.211×10^7 and 0.208×10^7, respectively. Note that the fourth digit of the arithmetic operations is lost in the computation.

The round-off errors get worse when a very large and a very small number (or two numbers that are very close to each other) are added or subtracted. Now consider adding and subtracting 2.5×10^3 and 2.3.

$$
\begin{array}{cc}
0.25000 \times 10^4 & 0.25000 \times 10^4 \\
+0.00023 \times 10^4 & -0.00023 \times 10^4 \\
\hline
0.25023 \times 10^4 & 0.24977 \times 10^4
\end{array}
$$

The addition and subtraction operations with the 3-digit mantissa result in 0.250×10^4 and 0.249×10^4, respectively. The absolute errors in the addition and subtraction operations are 2.3 and 7.7, respectively, compared to the true values of 2502.3 and 2497.7.

Since most real numbers cannot be represented in exact form, their floating-point values are used in arithmetic computations. Adopting the $fl(x)$ notation to denote the *floating-point* representation of the real number x, the difference between x and $fl(x)$ is the round-off error, which also depends on the size of x. Thus, it is measured relative to x as

$$
fl(x) = x(1 + \varepsilon) \tag{1.15}
$$

with $|\varepsilon| \leqslant 2^{-t}$ where t denotes the number of binary places in the mantissa.

The floating-point representation of finite-digit arithmetic for analogous elementary operations is given as

$$
fl(x \pm y) = (x \pm y)(1 + \varepsilon), \quad fl(x \times y) = xy(1 + \varepsilon), \quad fl(x \div y) = (x/y)(1 + \varepsilon)
$$

where x and y are floating-point numbers, and $\pm$, $\times$ and $\div$ denote binary arithmetic operations.

At this stage, without going into a detailed analysis of floating-point arithmetic, we will only mention that the computation errors propagate across multiple arithmetic operations. The reader can find a comprehensive account of errors and error propagation associated with the elementary operations in Ref. [34].

1.4.8 LOSS OF SIGNIFICANCE

The term "loss of significance" refers to the undesirable effects that arise when performing calculations with floating-points. In any computation, each floating-point operation that does not precisely represent the true arithmetic operation results in an error that may be reduced or amplified in subsequent operations. This has to do with the finite precision with which computers represent numbers. In general, it is desirable in any floating-point arithmetic operation to preserve the same number of significant digits. In practice, this is unfortunately not always possible. A common, often avoidable anomaly is observed when an operation on two numbers increases relative error significantly more than it increases absolute error.

Consider two nearly equal real numbers, $a = 0.123459526$ and $b = 0.123459876$, which have eight significant digits. Subtracting, for instance, the two floating-point numbers yields $b - a = 0.00000035 \ (= 0.35 \times 10^{-6})$, which has two significant digits. This event is referred to as *loss of significance*, which is a source of inaccuracy in most computations. In most cases, the loss of significance can be partially remedied or entirely avoided by changing the way arithmetic operations are performed. In this context, the use of rationalization, identities of trigonometric or logarithmic functions, Taylor series, increasing the precision of the computations, and so on are some of the techniques that can be exploited to guard against degradation of precision. The analyst should be alert to the possibility of loss of significance and take steps to avoid it, if possible.

EXAMPLE 1.7: Effect of rounding/chopping in arithmetic ops

Perform the following arithmetic operations: (i) exactly, using three-digit floating-point arithmetic; (ii) chopping; and (iii) rounding.

$$(a)\ \frac{7}{3} + \frac{3}{11}, \qquad (b)\ \frac{7}{3} - \frac{3}{11}, \qquad (c)\ \left(\frac{7}{3} + \frac{3}{11}\right) + \frac{1}{7}$$

SOLUTION:

(a) Using the exact and floating-point representations of the numbers, for (i), (ii), and (iii), we obtain

$$\frac{7}{3} + \frac{3}{11} = \frac{86}{33} = 0.260\overline{606} \times 10^1,$$

$$fl_c\left(\frac{7}{3} + \frac{3}{11}\right) = 0.260 \times 10^1 = 2.60$$

$$fl_r\left(\frac{7}{3} + \frac{3}{11}\right) = 0.261 \times 10^1 = 2.61$$

where subscripts c and r denote chopping and rounding operations, respectively.

(b) The operations for the subtraction case are as follows:

$$\frac{7}{3} - \frac{3}{11} = \frac{68}{33} = 0.206\overline{06} \times 10^1,$$

$$fl_c\left(\frac{7}{3} - \frac{3}{11}\right) = 0.206 \times 10^1 = 2.06,$$

$$fl_r\left(\frac{7}{3} - \frac{3}{11}\right) = 0.206 \times 10^1 = 0.206 \times 10^1 = 2.06$$

The 4th decimal place of 7/3 is 0, so it is kept as zero when rounding.

(c) The operations are as follows:

$$\left(\frac{7}{3} + \frac{3}{11}\right) + \frac{1}{7} = \frac{635}{231} = 0.2748917\cdots \times 10^1,$$

$$fl_c\left(\frac{1}{7}\right) = 0.142857\cdots \times 10^0,$$

$$fl_c\left(\frac{7}{3} + \frac{3}{11}\right) + fl_c\left(\frac{1}{7}\right) = 0.260 \times 10^1 + 0.014 \times 10^1 = 0.274 \times 10^1 = 2.74,$$

$$fl_r\left(\frac{7}{3} + \frac{3}{11}\right) + fl_r\left(\frac{1}{7}\right) = 0.261 \times 10^1 + 0.014 \times 10^1 = 0.275 \times 10^1 = 2.75$$

Discussion: This example illustrates the effect of rounding and chopping on addition and subtraction operations. In some cases, such arithmetic operations with rounded or chopped numbers may yield the same approximate result; however, the consequences of rounding or chopping a number that is very small or large can be far greater and these will be explained in the following sections. In such cases, high precision calculations should be used in arithmetic operations.

EXAMPLE 1.8: Loss of Significance

Compute the roots of $x^2 - 256x + 1 = 0$ with three significant digits.

SOLUTION:

The roots of $x^2 - px + q = 0$ are $x_{1,2} = (p \pm \sqrt{p^2 - 4q})/2$, which lead to

$$x_1 = \frac{1}{2}(256 - \sqrt{256^2 - 4}) \quad \text{and} \quad x_2 = \frac{1}{2}(256 + \sqrt{256^2 - 4})$$

or in decimal form: $x_1 = 0.00390631$, and $x_2 = 255.99609369$.

Rounding $\sqrt{256^2 - 4}$ to three significant digits results in 256, eventually resulting in $x_1 = 0$ and $x_2 = 256$ due to loss of significance. But, here, the loss of significance can be avoided by multiplying the numerator and denominator with the conjugate of the numerator to yield

$$x_1 = \frac{2q}{p + \sqrt{p^2 - 4q}} \rightarrow x_1 = \frac{2}{256 + 256} = 0.00390625 = 0.391 \times 10^{-2}$$

Discussion: Note that in this example, the way the calculation is carried out caused the loss-of-significance, not the errors in the computed data. By using an alternative arithmetic expression, the smallest root was recovered with four significant digits. Though it has been easy for this case to find an alternative way of computing the same quantity that eliminated the loss-of-significance error, this may not always be possible for every problem.

1.4.9 CONDITION AND STABILITY

Every problem requires a set of input and output data. The *condition* of a problem is concerned with the sensitivity of the problem to perturbations in the input data. If a small change in the input data yields large changes in the output, the problem is said to be *ill-conditioned*.

In modeling physical events, a model may lead to erroneous output; however, physical events with inherent instabilities are quite rare. If ill-conditioning is observed, then the problem should most likely lie in the assumptions made, in the data, or in the physical and/or numerical models or techniques being used. For instance, consider the following cubic equation, whose roots are $x = 1, 13$, and 14:

$$x^3 - 28x^2 + 209x - 182 = 0.$$

Now consider the cubic equation obtained by perturbing the coefficient of x^3 as

$$0.99x^3 - 28x^2 + 209x - 182 = 0$$

We find the roots of this equation to be $x \approx 1$, $x \approx 12.1378$, and $x \approx 15.1449$. Notice that, in comparison to a small change in the coefficient of x^3, the two large roots have significantly deviated from the true solution.

Next, consider the following perturbed cubic equation:

$$1.01x^3 - 28x^2 + 209x - 182 = 0$$

which yields a pair of complex conjugate roots with a respectable imaginary part: $x \approx$ 0.999877 and $x \approx 9.8961 \pm 1.0437\,i$. This leads us to conclude that the second and third roots are *ill-conditioned*; however, we cannot deduce anything about the condition of the first root.

Consider the following matrix $\mathbf{A}$ and a slightly perturbed matrix $\mathbf{A}'$:

$$\mathbf{A} = \begin{bmatrix} 2 & 100 & 0 \\ 0 & 2 & 100 \\ 0 & 0 & 2 \end{bmatrix} \qquad \mathbf{A}' = \begin{bmatrix} 2 & 100 & 0 \\ 0 & 20 & 100 \\ 10^{-4} & 0 & 20 \end{bmatrix}$$

which leads $|\mathbf{A}| = 8$ and $|\mathbf{A}'| = 9$. This example from matrix algebra shows that slight perturbations in numbers can lead to serious deviations in the final results.

Now, to analyze the condition of a problem, we consider a univariant function, $f(x)$, subjected to a perturbation of δx. The corresponding perturbation in f can be estimated by Taylor's formula as

$$\delta f = f(x + \delta x) - f(x) \approx \delta x\, f'(x)$$

Then, the relative error of the output can be expressed as

$$\frac{\delta f}{f} \approx \frac{x f'(x)}{f(x)} \cdot \frac{\delta x}{x}$$

which yields true equality as $\delta x \to 0$. This suggests that the *condition number* $f(x)$ be defined as

$$\kappa \cong \left| \frac{x f'(x)}{f(x)} \right|$$

Note that the *condition number* provides a measure of how large the relative perturbation in f is compared to the relative perturbation in x. A condition number of unity ($\kappa = 1$) implies that the input and output perturbations are identical. The effects of perturbation diminish for $\kappa < 1$ but amplify for $\kappa > 1$. Large condition numbers are indicative of *"ill-condition."* The *condition of a problem* is independent of floating-point number systems or numerical methods; it indicates the magnification of initial errors through exact calculations.

The *stability* of a method deals with the sensitivity of the numerical method to the propagation of round-off errors during the numerical solution. It is desirable that the cumulative effect of the round-off errors in any numerical method remain bounded, i.e., stable. The *stability of a method* refers to the influence of rounding errors as a result of inexact calculations due to finite precision arithmetic. A method that ensures accurate solutions is said to be *stable*; otherwise, the method is *unstable*.

This is the point where we will introduce the Big O notation, $\mathcal{O}(x)$, which will be extensively used throughout the text for analyzing the limiting behavior of a polynomial when x tends to either $x \to 0$ or $x \to \infty$.

For $f(x) = 2x + 3x^2 + 4x^3$, the function is expressed with its effective dominant term as $f(x) = 2x + \mathcal{O}(x^2)$, or $f(x) = 2x + 3x^2 + \mathcal{O}(x^3)$ as $x \to 0$.

Similarly, for large x, the dominant term of the polynomial is stated as $x \to \infty$ by $f(x) = 4x^3 + \mathcal{O}(x^2)$.

1.5 APPLICATION OF TAYLOR SERIES

1.5.1 TAYLOR APPROXIMATION AND TRUNCATION ERROR

Truncation errors often arise when an approximation formula is made to an infinite series by a partial (truncated) sum. A *truncation error* is defined as the difference between a true value and its truncated (approximated) value.

Consider the *Taylor series* of $f(x)$ that is infinitely differentiable at $x = a$:

$$f(x) = f(a) + \frac{f'(a)}{1!}(x-a) + \frac{f''(a)}{2!}(x-a)^2 + \cdots + \frac{f^{(n)}(a)}{n!}(x-a)^n + \cdots \qquad (1.16)$$

The Taylor series for $a = 0$ is referred to as the *Maclaurin* series.

In practice, it is impossible and impractical to sum up all the terms on the rhs of Eq. (1.16). Hence, the Taylor series is truncated and replaced with the following N-term partial sum, while higher-order terms are discarded:

$$f(x) \approx p_N(x) = f(a) + \frac{f'(a)}{1!}(x-a) + \frac{f''(a)}{2!}(x-a)^2 + \cdots + \frac{f^{(N-1)}(a)}{(N-1)!}(x-a)^{N-1} \qquad (1.17)$$

The truncated sum (approximation) is called the *Taylor polynomial*, and it results in estimates of some degree of truncation error.

Truncation error is commonly used to assess how many terms are required to get an estimate sufficiently close to a true value, and it is defined as

$$R_N(x) = f(x) - p_N(x) = \frac{f^{(N)}(\xi)}{N!}(x-a)^N, \qquad a \leqslant \xi \leqslant x \qquad (1.18)$$

which is referred to as the *remainder*.

If x is sufficiently close to a, then $(x - a)$ will be small enough so that higher-order terms in the approximation diminish. In such cases, truncating a series into a few terms generally results in an approximation that yields an estimate that is sufficiently close to the true value. In other words, if a truncation error is reasonably small, then an approximating Taylor polynomial becomes a good approximation for the function the series represents. However, Equation (1.18) has two major pitfalls: (1) ξ, which is unknown, lies somewhere between a and x; (2) the Nth derivative of $f(x)$ is required. If $f(x)$ is known, then there is no need for the Taylor series expansion in the first place! Nevertheless, Eq. (1.18) is useful to gain insight on the magnitude of the truncation error and the upper bound of the error. When assessing the truncation error, $f^{(N)}(\xi)$ is generally replaced by its maximum absolute value in $a \leqslant \xi \leqslant x$, i.e., $M = \max|f^{(N)}(\xi)|$. Then the upper bound of a truncation error can be written as

$$R_N(x) \leqslant \frac{M}{N!}(x-a)^N \qquad (1.19)$$

Now, going back to Eq. (1.19), note that the truncation error is proportional to $(x - a)^N$ with M and $N!$ being constants. This expression, however, implies nothing about the magnitude of M, yet it is still useful in assessing the comparative error of a numerical approximation based on Taylor polynomials. With $h = x - a$, the remainder is cast as

$$R_N \equiv \mathcal{O}(h^N) \qquad (1.20)$$

which implies that the truncation error is of the *order of* h^N, i.e., $|R_N| \leqslant C \cdot h^N$, where C

is a positive constant. In a way, the Big O notation provides a relative measure of accuracy as it indicates how rapidly the accuracy can be improved by reducing h.

The Maclaurin series of several commonly encountered functions is given as

$$e^x = 1 + x + \frac{x^2}{2!} + \frac{x^3}{3!} + \frac{x^4}{4!} + \ldots = \sum_{k=0}^{\infty} \frac{x^k}{k!}, \qquad (-\infty < x < \infty) \qquad (1.21)$$

$$\sin x = x - \frac{x^3}{3!} + \frac{x^5}{5!} - \frac{x^7}{7!} + \ldots = \sum_{k=0}^{\infty} \frac{x^{2k+1}}{(2k+1)!}, \qquad (-\infty < x < \infty) \qquad (1.22)$$

$$\cos x = 1 - \frac{x^2}{2!} + \frac{x^4}{4!} - \frac{x^6}{6!} + \ldots = \sum_{k=0}^{\infty} \frac{x^{2k}}{(2k)!}, \qquad (-\infty < x < \infty) \qquad (1.23)$$

$$\ln(1+x) = x - \frac{x^2}{2} + \frac{x^3}{3} - \frac{x^4}{4} + \ldots = \sum_{k=1}^{\infty} (-1)^{k+1} \frac{x^k}{k}, \qquad (-1 < x \leqslant 1) \qquad (1.24)$$

$$\frac{1}{1-x} = 1 + x + x^2 + x^3 + x^4 + \ldots = \sum_{k=0}^{\infty} x^k, \qquad (-1 < x < 1) \qquad (1.25)$$

where the specified intervals denote the convergence intervals.

Stopping (Termination) Criteria. An easy way of evaluating a continuous and differentiable function at $x = x_0$ is to truncate its Taylor (Maclaurin) series to an N-term partial sum. Evaluating Eq. (1.17) at $x = x_0$ yields

$$S_N = s_0 + s_1 + s_2 + \ldots + s_{N-1} = \sum_{k=0}^{N-1} s_k$$

where $s_k = c_k(x_0 - a)^k$ is the general term. However, the question becomes: what will be the associated error with the S_N? or is S_N good enough as an approximation?

Estimating truncation error *via* Eq. (1.18) may be impossible when $f^{(N)}(x)$ is too difficult or cumbersome to obtain. On the other hand, sometimes, even if $f^{(N)}(x)$ can be determined, finding its absolute or local maximum values to find the upper error bound (especially ξ and $f^{(N)}(\xi)$) may pose another challenging problem. That is why we need alternative quantitative tools for deciding when to terminate (truncate) an infinite series. In a computational process, this decision is usually made rather easily when increasing the degree of the polynomial (i.e., adding new terms to the series). To be more specific, to obtain an approximation that is accurate within ε tolerance, we consider in our decision-making the following relative or absolute error criterion:

For relative error: $|S_k - S_{k-1}| < \varepsilon \cdot |S_k|$ or equivalently $|s_k| < \varepsilon \cdot |S_k|$.

For absolute error: $|S_k - S_{k-1}| < \varepsilon$ or equivalently $|s_k| < \varepsilon$.

The relative error criterion is often preferred because it takes into account the relative size of the partial sum S_k and the new term s_k to be added. The absolute error criterion, which is simpler and easier to apply, requires less effort as long as both S_k and s_k are "fairly sized," i.e., S_k is not very large.

 The number of correct decimals gives us an idea of the magnitude of the absolute error. Assuming $f^{(N)}(x)$ is easily obtained, if $|R_N| < 0.5 \times 10^{-b}$, then we say that a truncated approximation is *correct to b decimal places*.

EXAMPLE 1.9: Effect of number of terms on accuracy of Taylor polynomials

Find approximating polynomials to compute the number "e" by truncating the Maclaurin series of e^x into an N-term finite sum (S_N), then (a) calculate S_N, truncation error R_N, and the true (absolute and relative) error for approximations of $N = 4$ to 10; (b) for $x = 0.01, 0.05, 0.10, 0.50$, and 1, assess the accuracy of $S_2 = 1 + x$ and $S_3 = 1 + x + x^2/2$ approximating polynomials.

SOLUTION:

(a) Using the Maclaurin series of e^x given by Eq. (1.21), an approximation is obtained by truncating the first N-terms (S_N) as

$$S_N = 1 + x + \frac{x^2}{2!} + \frac{x^3}{3!} + \frac{x^4}{4!} + \ldots + \frac{x^{N-1}}{(N-1)!}, \qquad R_N = \frac{x^N}{N!}e^\xi \quad (0 \leqslant \xi \leqslant x)$$

where R_N is the truncation error.

TABLE 1.3

N	S_N	R_N	$e - S_N$	$(e - S_N)/e$
4	2.66666667	0.11326174	0.0516152	0.0189882
5	2.70833334	0.02265235	0.0099485	0.0036598
6	2.71666667	0.00377539	0.0016152	0.0005942
7	2.71805556	0.00053934	0.0002263	0.0000832
8	2.71825397	0.00006742	0.0000279	0.0000102
9	2.71827877	7.4909×10^{-6}	3.0586×10^{-6}	1.1252×10^{-6}
10	2.71828153	7.4909×10^{-7}	3.0289×10^{-7}	1.1142×10^{-7}

To evaluate e, we set $x = 1$ in the expressions for S_N and R_N. The absolute maximum of e^ξ on $[0,1]$ occurs at $\xi = 1$. Thus, setting $M = e$, the upper bound for the truncation error is found to be $R_N = e/N!$.

In Table 1.3, the approximation (S_N), truncation error (R_N), and true (absolute and relative) errors are presented for $N = 4$ to 10 to depict the effect of N on the approximation for the number e. Notice that a four-term approximation (S_4) yields 2.66666667, and its true absolute error (0.0516152) is less than the estimated truncation error (=0.11326174). As N increases, both the true and truncation errors are reduced. It is also worth noting that the true error is always less than the estimated truncation error for all N.

TABLE 1.4

x	S_2	R_2	S_3	R_3
0.01	1.0100000	0.0001359	1.0100050	4.530×10^{-7}
0.05	1.0500000	0.0033978	1.0512500	0.0000566
0.1	1.1000000	0.0135914	1.0105000	0.0004530
0.2	1.2000000	0.0543656	1.2200000	0.0036244
0.5	1.5000000	0.3397852	1.6250000	0.0566309
1.0	2.0000000	1.3591409	2.5000000	0.4530470

The effect of x on the accuracy of two- and three-term approximations (S_2 and S_3), as well as the upper bounds for the truncation errors (R_2 and R_3), are presented

Pseudocode 1.7

Function Module EXPE (x, N)
  \ DESCRIPTION: A pseudo function module to compute e^x using an N-term
  \   Taylor polynomial approximation, i.e., $\sum_{k=0}^{N-1} x^k/k!$.
  $term \leftarrow 1$ \ Initialize $term$, $term = x^0/0! = 1$
  $sum \leftarrow 1$ \ Initialize accumulator, sum
  For $\left[k = 1, N-1 \right]$ \ Loop: series sum by accumulation
    $term \leftarrow term * x/k$ \ yields $(x/1)(x/2) \cdots (x/k) = x^k/k!$
    $sum \leftarrow sum + term$ \ Add the general $term$ to accumulator
  End For
  EXPE$\leftarrow sum$ \ Set accumulated sum to EXPE
End Function Module EXPE

in Table 1.4. Notice that $R_N = M\, x^N/N!$ is proportional to the power of x; thus, the truncation error for $x < 1$ decays with relatively few terms. However, for large $x \geqslant 1$, it also increases, resulting in large approximation errors for S_2 and S_3. Nevertheless, the magnitude of the errors for three-term approximations ($R_2 > R_3$) is naturally smaller than that of two-term approximations.

Discussion: This example illustrates that the value of S_N differs from the true value by no more than R_N. That is, the true error of an approximation obtained by a truncated sum from a series is always less than or equal to the truncation error ($E_{true} \leqslant |R_N|$). In other words, the truncation error can be confidently used to estimate the accuracy of an approximation in cases where the true error cannot be determined.

A pseudofunction, **EXPE**, that computes e^x for an arbitrary x with an n-term Taylor approximation is presented in Pseudocode 1.7. The arguments of the function are a real number (x) and the number of terms of the approximation (N). A For-construct is used to perform the finite sum since N is supplied as an input. Note that the accumulator variable $term$, which corresponds to $x^k/k!$ (i.e., the general term), is also evaluated recursively ($term_k = term_{k-1} * (x/k)$ with $term_0 = 1$) primarily to (1) reduce cpu-time by averting the direct computation of $k!$ and x^k; and (2) avoid possible *overflow* errors when calculating the numerator and denominator separately. On exit, the N-term approximation (accumulator variable sum) is set to **EXPE**. This module, however, does not provide much information about the accuracy of the calculated approximation.

A pseudofunction, **EXPA**, computing e^x for an arbitrary x with the known accuracy using the Maclaurin series is given in Pseudocode 1.8. The arguments of the module are x (a real number) and ϵ, a user-supplied tolerance that may be reasonably replaced with the machine epsilon, **MACHEPS** (*see* Pseudocode 1.5). The first three lines are the initializations: the counter with $k = 0$, the term with the first term of the series $term = a_0$, and the accumulator with $sum = term$. A While-construct with the relative error criterion ($|term| > \varepsilon \cdot |sum|$) is adopted to carry out the series approximation since the number of terms required is not known beforehand. In the block statements, the computation of $(k+1)$th term ($term = x^k/k!$) is carried out in the same manner as in Pseudocode 1.7, i.e., $sum + term$. The loop is terminated when the condition becomes **False**. Then the value of the accumulator is assigned to the function **EXPA** on exit. Notice that this code applies to a convergent series; therefore, it does not include any safeguard against the possibility of divergence.

Function Module EXPA (x, ε)
\ DESCRIPTION: A pseudo-function to compute e^x using the Maclaurin series
\   within ε tolerance.
\ USES:
\   ABS:: A built-in function computing the absolute value.
$k \leftarrow 0$ \ Initialize counter
$term \leftarrow 1$ \ Initialize $term$ with $a_0 = x^0/0!$
$sum \leftarrow term$ \ Initialize accumulator with 1st term, $S_0 = a_0$
While $\big[\,|term| > \varepsilon|sum|\,\big]$ \ Keep accumulating until $|a_k| < \varepsilon \cdot |S_k|$
$\quad k \leftarrow k + 1$ \ Count the term being added
$\quad term \leftarrow term * x/k$ \ Added $term$ is $a_k = x^k/k!$
$\quad sum \leftarrow sum + term$ \ Add $term$ to accumulator, $S_k = S_{k-1} + term$
End While   \ Go to top of the loop
EXPA$\leftarrow sum$ \ Set accumulated sum to EXPA
End Function Module EXPA

1.5.2   MEAN VALUE THEOREMS

The *Mean Value Theorem* for the derivative is obtained from the Taylor series using a one-term approximation of $f(x)$ at $x = b$ which is expressed as

$$f(b) = f(a) + (b - a)f'(\xi), \qquad a < \xi < b$$

which can be used to find an approximation for $f'(x)$ at some point ξ in (a,b); that is,

$$f'(\xi) = \frac{f(b) - f(a)}{b - a} \tag{1.26}$$

For the mean value theorem for the integrals, we consider two bounded and integrable functions ($f(x)$ and $g(x)$) where $g(x)$ does not change sign in $[a, b]$. There is a number ξ in the interval $[a, b]$ such that

$$\int_a^b f(x)g(x)dx = f(\xi)\int_a^b g(x)dx \tag{1.27}$$

For $g(x) = 1$, it yields

$$\int_a^b f(x)dx = f(\xi)(b - a) \qquad \text{or} \qquad f(\xi) = \frac{1}{b - a}\int_a^b f(x)dx$$

where $f(\xi)$ is the mean value of $f(x)$. Geometrically speaking, the area of the region under the curve from a to b is equal to that of a rectangle whose lengths are $f(\xi)$ and $(b - a)$.

The second mean value theorem for the integrals is given

(i) if $f(x)$ is bounded, decreasing, and non-negative function in $[a, b]$, and $g(x)$ is a bounded integrable function, then

$$\int_a^b f(x)g(x)dx = f(a)\int_a^\xi g(x)dx \tag{1.28}$$

or (ii) if $f(x)$ is bounded, increasing, and non-negative function in $[a, b]$, and $g(x)$ is a bounded integrable function, then

$$\int_a^b f(x)g(x)dx = f(b)\int_\eta^b g(x)dx \tag{1.29}$$

1.6   TRUNCATION ERROR OF SERIES

Positive Series. The truncation error, or error estimation, of a series that is not generated from the Taylor series is slightly different. To illustrate the truncation error, consider a convergent infinite series, $\sum a_n$, where $\{a_n\}$ is a sequence of positive terms with $a_n = f(n)$. Estimating the error of the N-term partial sum is equivalent to evaluating: indexError!truncation

$$R_N = S - S_N = a_{N+1} + a_{N+2} + a_{N+3} + \cdots$$

where $S = \sum_{n=0}^{\infty} a_n$ and $S_N = \sum_{n=0}^{N} a_n$. The integral test can be applied to find lower and upper bounds for the error.

The truncation error for positive, continuous, and decreasing $f(x)$ for all $x \geqslant N$ satisfies the following inequality:

$$\int_{N+1}^{\infty} f(x)dx + \frac{a_{N+1}}{2} \leqslant R_N \leqslant \int_{N}^{\infty} f(x)dx - \frac{a_{N+1}}{2} \tag{1.30}$$

Adding S_N on each side of Eq. (1.30) yields

$$S_N + \int_{N+1}^{\infty} f(x)dx + \frac{a_{N+1}}{2} \leqslant S \leqslant S_N + \int_{N}^{\infty} f(x)dx - \frac{a_{N+1}}{2} \tag{1.31}$$

which indicates that the true error cannot be larger than the length of the interval in this inequality. In cases where the integral test is impractical to assess the convergence of a series, the comparison test should be employed.

Now consider another convergent infinite series, $\sum b_n$, which is $a_n \leqslant b_n$ for all n. The remainders for both series can be expressed as

$$R_N = \sum_{n=N+1}^{\infty} a_n \qquad and \qquad B_N = \sum_{n=N+1}^{\infty} b_n$$

Since $a_n \leqslant b_n$, this leads to $R_N \leqslant B_N$, and generally, b_n is chosen such that it is a fairly simple expression, which permits the integral test to be employed. An upper bound for the partial sum is given as

$$R_N \leqslant B_N \leqslant \int_{N}^{\infty} g(x)dx \tag{1.32}$$

where $g(n) = b_n$. If the magnitude of T_N cannot be determined, then there is no benefit in estimating the truncation error using the comparison test.

Next, consider applying the ratio test to a convergent series $(\sum a_n)$ with $\{a_n\}$ being a positive and decreasing sequence with $\lim_{n \to \infty} |a_{n+1}/a_n| \to \rho < 1$. An estimate for the upper and lower bounds of the truncation error can then be found as follows:

If $|a_{n+1}/a_n|$ approaches the limit ρ from above, then we have

$$a_N \left(\frac{\rho}{1 - \rho} \right) \leqslant R_N \leqslant \frac{a_{N+1}}{1 - \dfrac{a_{N+1}}{a_N}} \tag{1.33}$$

If $|a_{n+1}/a_n|$ approaches the limit ρ from below, then we apply

$$\frac{a_{N+1}}{1 - \dfrac{a_{N+1}}{a_N}} \leqslant R_N \leqslant a_N \left(\frac{\rho}{1 - \rho} \right) \tag{1.34}$$

Alternating Series. Let $S = \sum_{n=1}^{\infty}(-1)^{n-1}a_n$ and let the N-term partial sum be $S_N = \sum_{n=1}^{N}(-1)^{n-1}a_n$, where $\{a_n\}$ is a positive sequence with $\lim_{n\to\infty} a_n \to 0$ and $\sum_{n=1}^{\infty} a_n$ converges. The standard (Leibniz's) error bound is given by

$$-a_{N+1} < R_N < a_{N+1} \tag{1.35}$$

EXAMPLE 1.10: Estimating truncation error of positive series

Consider the following infinite series:

$$\text{(a)} \ \sum_{n=1}^{\infty} \frac{1}{n^2+1} = \frac{1}{2}\left(\pi \coth \pi - 1\right), \qquad \text{(b)} \ \sum_{n=1}^{\infty} \frac{n^2}{4^n} = \frac{20}{27}$$

Find the 10-term approximation (partial sum), the true error, and error bounds and discuss the true error in relation to the error bounds.

SOLUTION:

(a) The 10-term partial sum is obtained as

$$S_{10} = \sum_{n=1}^{10} \frac{1}{n^2+1} = \frac{1}{1^2+1} + \frac{1}{2^2+1} + \frac{1}{3^2+1} + \cdots + \frac{1}{10^2+1} = 0.9817928223$$

and the true error is $E_{\text{true}} = \frac{1}{2}(\pi \coth \pi - 1) - 0.98179282 = 0.0948812$.

Using Eq. (1.31), the truncation error satisfies

$$\int_{11}^{\infty} f(x)dx + \frac{a_{11}}{2} \leq R_{10} \leq \int_{10}^{\infty} f(x)dx - \frac{a_{11}}{2}$$

Setting $f(x) = 1/(1+x^2)$ and noting $\int_N^{\infty} f(x)dx = \pi/2 - \tan^{-1}(N)$, we find

$$\frac{\pi}{2} - \tan^{-1}(11) + \frac{1/122}{2} < R_{10} < \frac{\pi}{2} - \tan^{-1}(10) - \frac{1/122}{2}$$

or $0.09475825 < R_{10} < 0.09557029$.

(b) The first 10-term partial sum yields

$$S_{10} = \sum_{n=1}^{10} \frac{n^2}{4^n} = \frac{1^2}{4^1} + \frac{2^2}{4^2} + \frac{3^2}{4^3} + \cdots + \frac{10^2}{4^{10}} = 0.74069977$$

which leads to $E_{\text{true}} = 20/27 - 0.74069977 = 4.0971 \times 10^{-5}$.

Setting $a_n = n^2/4^n$ and applying the ratio test, we find

$$\rho = \lim_{n\to\infty} \frac{a_{n+1}}{a_n} = \lim_{n\to\infty} \frac{(n+1)^2}{4n^2} \to \frac{1}{4}$$

also we have $a_{11}/a_{10} = (11^2/4^{11})(4^{10}/10^2) = 121/400$. Note that $|a_{n+1}/a_n|$ approaches the limit value of $\rho = 1/4$ from above as n is increased, in which case the upper and lower bounds for the truncation error are obtained from Eq. (1.33):

$$a_{10}\left(\frac{\rho}{1-\rho}\right) \leq R_{10} \leq \frac{a_{11}}{1 - \frac{a_{11}}{a_{10}}}$$

Substituting the numerical values results in

$$\frac{10^2}{4^{10}}\left(\frac{1/4}{1-1/4}\right) < R_{10} < \frac{11^2}{4^{11}}\left(1-\frac{121}{400}\right)^{-1}$$

or $3.1789 \times 10^{-5} < R_{10} < 4.136 \times 10^{-5}$.

Discussion: Note that the true errors in both cases are within the error bounds. If the true error happens to be near the middle of the bounding interval, adding the arithmetic average of the upper and lower error bounds to the approximation will bring it closer to the true value. For the series in Part (a), adding the arithmetic average of the upper and lower error bounds ($=0.095164270$) to the approximation yields 1.07695709, which reduces the true error to 2.8304×10^{-4}. Likewise, for the series in Part (b), adding the average of the upper and lower bounds ($= 3.65745 \times 10^{-5}$) to S_{10} results in 0.74073634, which is in error by 4.396×10^{-6}.

1.7 RATE AND ORDER OF CONVERGENCE

In numerical analysis, there is often more than one method to solve a particular problem. Many of the numerical methods covered in the subsequent chapters are iterative in nature. These methods always generate a sequence of estimates through successive approximations that converge toward a solution. Some sequences depict very slow and some very rapid convergence. In such cases, when determining the most suitable method for the numerical solution of a problem, the method whose sequence converges as rapidly as possible is preferred.

The *order of convergence* and the *rate of convergence* of a convergent sequence are quantities that indicate how quickly a sequence approaches its limit value. One of the ways in which numerical algorithms are compared is *via* their order and rates of convergence. These are also the most important factors to consider when choosing one method over another. The methods with a high order or rate of convergence require fewer iterations to reach a reasonable solution.

A sequence $\{x_n\}$ is said to *converge* or *be convergent* if

$$\lim_{n \to \infty} x_n \to L \quad \text{or} \quad \lim_{n \to \infty} |x_n - L| \to 0$$

where L is the *limit* of the sequence and denotes the value to which the sequence converges. The sequence is said to diverge if $\lim_{n \to \infty} x_n$ does not exist.

Let $\{x_n\}$ be a sequence that converges to a number L. If there exists a sequence $\{q_n\}$ that converges to zero and a constant $\lambda >$ independent of n, such that

$$|x_n - L| \leqslant \lambda |q_n| \tag{1.36}$$

for large values of n, then $\{x_n\}$ is said to converge to L with the *rate of convergence* $\mathcal{O}(q_n)$, which is commonly expressed in shorthand notation as $x_n = L + \mathcal{O}(q_n)$, and λ is called the *asymptotic error constant*.

The sequence $\{q_n\}$ is generally chosen as a p-sequence (i.e., $\sum(1/n^p)$ for $p > 0$) or a geometric sequence (i.e., $\sum r^n$, $r < 1$) to allow for simple comparison between different sequences. For instance, a sequence with a convergence rate of $\mathcal{O}(1/2^n)$ converges faster than the one with $\mathcal{O}(1/n^4)$, which in turn converges faster than a sequence with a rate of

TABLE 1.5: Convergence behavior of the sequences a_n, b_n, and c_n.

n	a_n	b_n	c_n
1	1.000000	1.000000	1.000000
2	1.500000	1.750000	1.875000
3	1.666667	1.888889	1.962963
4	1.750000	1.937500	1.984375
5	1.800000	1.960000	1.992000
⋮	...	...	...
10	1.900000	1.990000	1.999000
11	1.909091	1.991736	1.999249
12	1.916667	1.993056	1.999421
⋮	...	...	...
20	1.950000	1.997500	1.999875

convergence of $\mathcal{O}(1/n^2)$. For instance, consider $a_n = 2-1/n$, $b_n = 2-1/n^2$, and $c_n = 2-1/n^3$, all of which converge to "2." In Table 1.5, the terms of the sequences up to $n = 20$ are presented. From Eq. (1.36), we find that $|a_n - 2| \leqslant (1/n)$ (i.e., $\lambda = 1$ and $q_n = 1/n$) so that we may write $a_n = 2 + \mathcal{O}(1/n)$. Similarly, we obtain $b_n = 2 + \mathcal{O}(1/n^2)$ and $c_n = 2 + \mathcal{O}(1/n^3)$, which converges fastest.

EXAMPLE 1.11: Limit and rate of convergence of sequences

Find the limit and the rate of convergence of the following sequences:

(i) $a_n = \dfrac{3n^2 + 7}{4n^2 + 3n + 1}$,   (ii) $b_n = \dfrac{2n + 3}{\sqrt{9n^2 + 2}}$,   (iii) $c_n = n \sin\left(\dfrac{2}{n}\right)$

SOLUTION:

(i) The limit at which the sequence should converge is found as $\lim_{n \to \infty} a_n = 3/4$. To find the convergence rate, we express the general term of the sequence as a function of $(1/n)$. Dividing both the numerator and denominator with n^2 and simplifying yields:

$$a_n = \frac{3n^2 + 7}{4n^2 + 3n + 1} = \frac{3 + 7(1/n)^2}{4 + 3(1/n) + (1/n)^2}$$

Setting $x = 1/n$, the sequence becomes $(3 + 7x)/(x^2 + 3x + 4)$. For $n \to \infty$ (i.e., $x \to 0$), it can be expanded to the Maclaurin series as follows:

$$a_n = \frac{3}{4} - \frac{9}{16}\left(\frac{1}{n}\right) + \frac{127}{64}\left(\frac{1}{n}\right)^2 + \dots \quad \text{or} \quad = \frac{3}{4} - \frac{9}{16}x + \frac{127}{64}x^2 + \dots$$

For very large n, we arrive at $|a_n - 3/4| = (9/16)(1/n)$, from which the convergence rate and λ are found to be $\mathcal{O}(1/n)$ and $9/16$, respectively.

(ii) The limit value of the sequence is $\lim_{n \to \infty} b_n = 2/3$. To determine the convergence rate, the general term of the sequence is likewise expressed as a function of $1/n$ and expanded to a Maclaurin series to give

$$b_n = \frac{2n+3}{\sqrt{9n^2+2}} = \frac{2+3(1/n)}{\sqrt{9+2(1/n)^2}}$$

$$= \frac{2}{3} + \left(\frac{1}{n}\right) - \frac{2}{27}\left(\frac{1}{n}\right)^2 + \dots \quad \text{or} \quad = \frac{2}{3} + x - \frac{2}{27}x^2 + \dots$$

which may also be cast as $|b_n - 2/3| = (1/n)$ for very large n. This expression gives the convergence rate as $\mathcal{O}(1/n)$ and $\lambda = 1$.

(iii) The limit value of this sequence is $\lim_{n\to\infty} c_n = 2$. This general term is already in the form of $1/n$. Setting $1/n = x$ and expanding it to the Maclaurin series gives

$$c_n = n \sin\left(\frac{2}{n}\right) = \frac{\sin(2/n)}{(1/n)}$$

$$= 2 - \frac{4}{3}\left(\frac{1}{n}\right)^2 + \frac{4}{15}\left(\frac{1}{n}\right)^4 + \dots \quad \text{or} \quad = 2 - \frac{4}{3}x^2 + \frac{4}{15}x^4 - \dots$$

which leads to $|c_n - 2| = (4/3)(1/n)^2$, yielding the rate of convergence of $\mathcal{O}(1/n^2)$ and $\lambda = 4/3$.

Discussion: For a sequence whose general term is given by $a_n = g(n)$, a function $f(x)$ is obtained by substituting $n = 1/x$ in the general term. Then $f(x)$ is expanded into a Maclaurin series. For very large n, the series expansion will have $L + \lambda x^m$ form or $|a_n - L| = \lambda(1/n)^m$, where m is the order of convergence.

Let us consider again the sequence $\{x_n\}$ that converges to L. A stronger condition for the convergence rate is given by

$$\lim_{n\to\infty} \frac{|x_{n+1} - L|}{|x_n - L|^r} = \lim_{n\to\infty} \frac{|e_{n+1}|}{|e_n|^r} \to \lambda \tag{1.37}$$

where $e_n = x_n - L$ is the absolute error, λ is the asymptotic error constant, and r ($\geqslant 1$) is called the *order convergence*. For a linearly converging sequence, the order of convergence r is 1 while the sequence is said to have an order of convergence of $r = 2$ for quadratically converging sequences.

A practical way of estimating the rate of convergence needs to be established since the value of L in Eq. (1.37) is also being computationally predicted. For a sufficiently large number of iterations, we may replace Eq. (1.37) by

$$|e_{n+1}| \approx \lambda|e_n|^r \quad \text{or} \quad |e_n| \approx \lambda|e_{n-1}|^r \tag{1.38}$$

The order (rate) of convergence can then be estimated by eliminating λ and solving for r, which leads to

$$r \approx \frac{\log(|e_{n+1}|/|e_n|)}{\log(|e_n|/|e_{n-1}|)} = \frac{\log(|x_{n+1} - x_n|/|x_n - x_{n-1}|)}{\log(|x_n - x_{n-1}|/|x_{n-1} - x_{n-2}|)} \tag{1.39}$$

Note that the estimate for the order of convergence with Eq. (1.39) is good only for sufficiently large n.

EXAMPLE 1.12: Determining rate of convergence of a sequence

Determine the order of convergence of the following sequences:

$$\text{(a) } x_n = 2 - 1/n^3, \qquad \text{(b) } y_n = 2 - 1/2^{2^n}$$

SOLUTION:

(a) Several terms of the sequence, the absolute difference between two consecutive terms, the logarithm of the ratio of two consecutive absolute differences, and the estimate for the order of convergence by Eq. (1.39) are tabulated in Table 1.6.

TABLE 1.6

n	x_n	$|x_n - x_{n-1}|$	$\log(|e_n|/|e_{n-1}|)$	r
1	1.000000000000			
2	1.875000000000	0.875000000		
3	1.962962962963	0.087962963	−0.99770820	
4	1.984375000000	0.021412037	−0.61364186	0.61505144
5	1.992000000000	0.007625000	−0.44841814	0.73074893
6	1.995370370370	0.003370370	−0.35456222	0.79069554
7	1.997084548105	0.001714178	−0.29362178	0.82812483
8	1.998046875000	0.000962327	−0.25073323	0.85393266
⋮	⋮	⋮	⋮	⋮
30	1.999629629629	3.9651×10^{-6}	−0.05993833	0.96547547
↓	↓	↓	↓	↓
∞	2	0	0	1

Since $\lim_{n\to\infty} x_n = 2$, the sequence is convergent and the order of convergence is 1.

(b) In Table 1.7, the first six terms of the sequence, the absolute difference between two consecutive terms, the logarithm of the ratio of two consecutive absolute differences, and the estimate for the rate of convergence by Eq. (1.39) are listed. It is seen that the sequence converges much faster (i.e., the rate of convergence is larger than in (b)).

TABLE 1.7

n	y_n	$|y_n - y_{n-1}|$	$\log(|e_n|/|e_{n-1}|)$	r
1	1.75000000			
2	1.93750000	0.18750000		
3	1.99609375	0.05859375	−0.50514998	
4	1.99998474	0.00389099	−1.17779104	2.33156705
5	2	1.52593×10^{-5}	−2.40654681	2.04327144
6	2	2.3283×10^{-10}	−4.81647330	2.00140437
↓	↓	↓	↓	↓
∞	2	0	−∞	2

Discussion: Note that the sequence x_n converges rather slowly. The absolute error is reduced to the order of 10^{-6} for $n \approx 25$-30. As $r \to 1$, the sequence depicts linearly convergent behavior for increasing n. On the other hand, the sequence y_n converges to 2 much faster, in $n \approx 5$-6. The absolute error $(|y_n - y_{n-1}|)$ quickly goes to zero and r approaches 2 even for small n, indicating a quadratic convergence rate.

1.8 CLOSURE

Nowadays, science and engineering calculations have evolved to the point where they are too complex to be carried out by hand. Numerical methods are the main source for modeling science and engineering problems on computers. Communicating with computers is achieved through software designed in various programming languages. In this regard, it is important to understand and code a numerical method to be used in digital computers.

There are many programming languages out there. This is because different tasks require different tools to solve them, and each programming language has certain features that make it suitable for a specific task. Therefore, it is inevitable that a scientist or engineer will encounter problems that require the use of more than one programming language and/or symbolic processing software throughout his career. Most programming languages provide mechanisms for writing programs using the same type of constructs common to all, even though the syntax of those languages is quite different. That is why this chapter focused on presenting and explaining the basic elements of common algorithms (sequential constructs, conditional constructs, loops, accumulators, etc.) using pseudocodes. While learning numerical methods, a special emphasis is placed on writing codes that use *modular* and *efficient* programming techniques.

Most people falsely believe that when a computer is used for a computational task, the result (answer) is unquestionably correct. But in scientific computing, the computed answer rarely represents the true answer. The inexactness of the result depends on the mathematical model, or the use of experimentally derived data, and the truncation and/or round-off errors acquired in computations. Roundoff and truncation errors are common in any numerical calculation. A roundoff error occurs as the floating-point numbers are represented with finite precision. On the other hand, truncation error occurs when we do a discrete approximation to a continuous function. In this chapter, some examples of each of these errors were given, and their significance was discussed. The definitions of error were presented along with the computation and reporting of the sources of numerical errors. The errors from several variables could propagate so much as to affect the final value. The error propagation formula was presented in this connection to estimate the propagation of error in computed or estimated quantities. The reason for the inexactness of digital computation is commonly related to how numbers are stored and used in computers. The concepts of *accuracy, precision, round-off, significant digit, significant place,* and *loss of significance* are defined and illustrated with examples. Apart from the errors caused by computer hardware operations, numerical analysts may encounter *condition* and *stability* issues depending on how a numerical procedure is carried out. These problems are illustrated, as are techniques or methods to avoid them.

Most numerical methods are almost exclusively based on approximations, especially Taylor series approximations. In this respect, estimation of a quantity by truncating the Taylor series is covered in detail, including efficient programming techniques for a series sum (or product) and estimation of the truncation errors. Additionally, the basic convergence theorems of sequences and series (including alternating series) are presented. Important criteria for iterative methods are the *order* and *rate of convergence*, which are covered with examples. Often, we also encounter problems that are insensitive or well-conditioned when there are small errors in the problem parameters. In other words, a small perturbation to a problem parameter of a well-conditioned problem yields only a small perturbation of the true solution. On the other hand, sometimes we are faced with ill-conditioned problems whose parameters (i.e., variables) are sensitive to small perturbations, which result in a significant deviation from the final result.

1.9 EXERCISES

Section 1.1 Numerical Algorithms in Pseudocodes

E1.1 The period of a pendulum is determined by $T = 2\pi\sqrt{L/g}$, where L is its length and g is the acceleration of gravity. Write a pseudocode for calculating the period of a pendulum for an arbitrary input L.

E1.2 Write a pseudocode to find the area of an arbitrary triangle with the side lengths a, b, and c, using $A = \sqrt{s(s-a)(s-b)(s-c)}$ where $s = (a+b+c)/2$.

E1.3 A pump curve is approximated by $h(Q) = 31.2 + 0.003Q - 4.29 \times 10^{-6}Q^2 - 8.6 \times 10^{-10}Q^3$, where Q is the flowrate (in liters/min) and h is the head (in meters). Write a pseudocode for estimating the head (using conformal units) for an input flowrate.

E1.4 Write a pseudocode that calculates and prints out the slope of the line passing through two arbitrary input points, $P(x_1, y_1)$ and $Q(x_2, y_2)$, and the distance between them.

E1.5 Write a pseudocode to calculate the dot product ($\mathbf{u} \cdot \mathbf{v}$) of the arbitrary input vectors $\mathbf{u} = (u1, u2, u3)$ and $\mathbf{v} = (v1, v2, v3)$ and the angle between them (in degrees).

E1.6 When a circular cone is cut parallel to the base, a frustum shown in **Fig. E1.6** is obtained. The slant length (l), volume (V), lateral surface (S_{lat}), and the total surface area (S) can be determined from the following expressions. Write a pseudo-program to calculate the volume and the surface areas for arbitrary input values of R, h, and r.

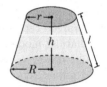

Fig. E1.6

$$l = \sqrt{h^2 + (R-r)^2}, \quad V = \frac{\pi h}{3}(R^2 + r^2 + Rr)$$
$$S_{\text{lat}} = \pi l(R+r), \quad S = \pi(l*(R+r) + R^2 + r^2)$$

E1.7 Write a pseudocode to calculate the following expression for an arbitrary x.

$$f(x) = \begin{cases} x^2 + 3x + 1, & x \leqslant 0 \\ 4x^2 - 5x + 1, & x > 0 \end{cases}$$

E1.8 Write a pseudocode to convert an input temperature value from °F to °C (*switch*=1) or from °C to °F (*switch*=2) using the conversion formula $°F = 32 + 1.8°C$. Use an integer-type *switch* variable to select a conversion formula.

E1.9 Repeat **E1.2**, extending the pseudocode to include validity checks ($a > 0$, $b > 0$, and $c > 0$) for the input variables. *Note*: When an invalid entry ($a \leqslant 0$, $b \leqslant 0$, or $c \leqslant 0$) is encountered, an error message should be prompted, asking the user for valid input. You may use a *switch* (a logical type variable) to determine if the input is valid or not.

E1.10 Consider finding the roots of a quadratic equation of the form $ax^2 + bx + c = 0$. Write a pseudocode that takes into account all possible outcomes ($a = 0$ or $b = 0$ or $c = 0$, $\Delta < 0$, $\Delta > 0$, $\Delta = 0$ and so on).

E1.11 Write a pseudocode to calculate the arithmetic, geometric, harmonic, and quadratic means of a sequence (x_i) of real numbers of length n, read sequentially from a file. *Note*: Avoid using an array variable.

$$m_a = \frac{1}{n}\sum_{i=1}^{n} x_i, \quad m_h = n\left(\sum_{i=1}^{n}\frac{1}{x_i}\right)^{-1}, \quad m_g = \left(\prod_{i=1}^{n} x_i\right)^{1/n}, \quad m_q = \left(\frac{1}{n}\sum_{i=1}^{n} x_i^2\right)^{1/2}$$

E1.12 Write a pseudocode that uses (a) **For-** and (b) **While**-constructs to find the sum and average of all odd numbers from 1 to n.

E1.13 A formula for wind chill index (WC) as a function of ambient temperature T_a (°C) and wind speed V (km/h) is given as follows:

$$\text{WC}(T_a, V) = 13 + 0.6T - 11.4V^{1/6} + 0.4TV^{1/6}, \quad T_a < 30°\text{C} \quad \text{and} \quad V > 6 \text{ km/h}$$

Write a pseudocode that accepts ambient temperature and wind speed as an input, checks the valid range of the input variables, and prompts the user to re-enter the number(s) in case *"wrong type"* or *"out-of-range"* data is detected.

E1.14 Consider the motion of an object thrown into the air. The horizontal and vertical displacements of the object relative to its initial position are described by

$$x = Vt\cos\theta, \qquad y = Vt\sin\theta - \frac{1}{2}gt^2$$

where t, V, θ, and g denote the time, launch velocity, launch angle (in degrees), and the acceleration of gravity, respectively. The velocity components, velocity magnitude, and direction of the object can then be obtained from

$$v_x = \frac{dx}{dt}, \quad v_y = \frac{dy}{dt}, \quad U = \sqrt{v_x^2 + v_y^2}, \quad \tan\alpha = \frac{v_y}{v_x}$$

Write a pseudocode to find the horizontal and vertical positions, velocity magnitude, and direction of the object until it hits the ground. *Note*: Use the **Repeat-Until** construct and assume time intervals of $\Delta t = V * \sin\theta/50$.

E1.15 Write a pseudocode to find the largest value and its position in an arbitrary array of n positive integers. *Note*: Use a **For**-construct.

E1.16 An integer array of length n consists of 1's, 2's, ..., 5's. Write a pseudocode that reads the array (**a**) and its length (n) from a file and determines the frequency of each value. *Note*: Use a **For**-construct and assume an array of maximum length 999.

E1.17 Write a pseudo-**Function Module** LOG(a, x) to compute the logarithm of a real number x to an arbitrary base a. Make use of the change of base formula.

E1.18 Write a pseudo-**Module** LINE for the problem given in **E1.4**.

E1.19 Write a pseudo-**Function Module** FX for the function given in **E1.7**.

E1.20 The double factorial function is defined as

$$n!! = \begin{cases} (2k)!! = 2 \cdot 4 \cdot 6 \cdots (2k), & n = 2k \\ (2k+1)!! = 1 \cdot 3 \cdot 5 \cdots (2k+1), & n = 2k+1 \end{cases}$$

where $0!!=1!!=1$. Write a recursive pseudo-**Function Module** DBL_FACTORIAL that evaluates a double factorial with the least computation time. *Note*: Include a validity check for n and suggest a proper action with a warning message in case of invalid input.

E1.21 The binomial coefficient, $B(n, k)$, can be efficiently computed as

$$B(n, k) = \frac{n(n-1)(n-2)\cdots(n-(k-1))}{k(k-1)(k-2)\cdots 1} = \prod_{j=1}^{k} \frac{n+1-j}{j}$$

Since it is symmetrical with regard to k and $n - k$, its computation can be further simplified as

$$B(n, k) = \begin{cases} B(n, k), & k \leqslant n/2 \\ B(n, n-k), & k > n/2 \end{cases}$$

Write a pseudo-**Function Module** B(n,k) that calculates the binomial coefficient most efficiently (with minimum arithmetic operations) for arbitrary input values of n and k.

E1.22 The Fibonacci numbers form a sequence in which each number is the sum of the two preceding ones. It is expressed algebraically by the following recurrence relation:

$$F_0 = 0, \quad F_1 = 1, \quad F_n = F_{n-1} + F_{n-2}, \quad \text{for} \quad n \geqslant 2$$

Write a pseudo-**Function Module FIBONACCI**(n) that generates the Fibonacci numbers without using arrays.

E1.23 Write a pseudocode that constructs Pascal's triangle from the first to the nth power for an arbitrary input value n. The code should print all rows, beginning with the first and ending with the last. The coefficients of the kth power are found from the following recurrence relation:

$$c_{10} = 1, \quad c_{11} = 1, \quad c_{kj} = c_{k-1,j-1} + c_{k-1,j}, \quad k > 1$$

where k is the row number. Use two one-dimensional arrays (of lengths n and $n+1$) to store two successive (prior and current) rows of the triangle. Prepare a **Module PASCALS_TRIANGLE** that finds the coefficients of Pascal's triangle for the kth row and uses arrays of a maximum length of 99. The code should also check for the bounds and validity of the input data.

Section 1.2 Basic Definitions and Sources of Error

E1.24 Describe the sources of modeling (mathematical model) errors.

E1.25 Describe a computer simulation and its purpose.

E1.26 Briefly name the sources of numerical errors.

E1.27 In your own words, define *accuracy* and *precision*.

E1.28 In your own words, define the *systematic errors* and give an example of a systematic error.

E1.29 In your own words, define *random errors* and give an example of a random error.

E1.30 Define *blunders* in coding or programming. Suggest ways or approaches to avoid blunders in programming.

E1.31 Consider the mathematical model $F_{net} = m(dv/dt) = -mg$ for a falling object. Here g is the acceleration of gravity, m is the mass, and v is the velocity of the object. Do you think that this model would be sufficient to describe the motion of a parachutist?

E1.32 Consider the mathematical model $dp/dt = a\,p(t)$ for the population growth of a species. Here a is a constant and $p(t)$ denotes the population size. Do you think that this model would be sufficient to describe the population of a species?

E1.33 Consider a cubic-shaped carbon steel specimen placed on a graphite (C) layer. The model proposed for the diffusion of carbon atoms to the bottom surface of a specimen at high temperature is given as $\partial C/\partial t = D\,\partial^2 C/\partial z^2$, where D is the diffusion constant and $C(z,t)$ is the concentration of diffusing atoms in z (vertical) direction as a function of time t. Do you think that this model adequately describes the diffusion process of carbon atoms into the specimen? Under what conditions would this mathematical model provide reasonably accurate estimates?

E1.34 The true surface area of an object is known to be $A = 1.25$ m^2. A computer model numerically estimates the surface area as 1.252 m^2. Calculate the absolute, relative, percent absolute, and relative errors of the estimate.

E1.35 Suppose the true speed of a flying object is $V = 150$ m/s. Its speed is also estimated by two measurements: 148.33 and 152.82 m/s. Calculate the absolute and relative errors as well as the percent absolute and relative errors for each velocity estimate.

E1.36 Calculate the absolute and relative errors of the following quantities with respect to the reported approximations.

(a) $Q_t = 10000, \quad Q_a = 9997,$ (b) $Q_t = 789.6573, \quad Q_a = 789.6643,$
(c) $Q_t = 1.4587, \quad Q_a = 1.3678,$ (d) $Q_t = 0.00089645, \quad Q_a = 0.00089512$

E1.37 Determine the absolute and relative errors resulting from computing $\sin(\pi/4)$ as $\sin(0.785)$.

E1.38 Find the absolute and relative errors made by calculating π as 22/7.

E1.39 Find the absolute and relative errors made by calculating e as 19/7.

E1.40 Determine the absolute and relative errors made by calculating $\sqrt{1.1}$ as 1.05.

E1.41 Determine the absolute and relative errors made by calculating $1/\sqrt{1.4}$ as 0.8.

E1.42 Determine the absolute and relative errors made by computing $e^{0.04}$ and $e^{0.1}$ using $e^x \approx 1 + x + x^2/2$.

Section 1.3 Propagation of Error

E1.43 What would be the change in volume (in %) if the height and radius of a cone differed by $+2\%$ and -2%, respectively?

E1.44 The force of attraction between the charges q_1 and q_2 at a distance r from each other is dictated by $F = kq_1q_2/r^2$, where k is a constant. Estimate the percent change in force if each charge is increased by 2% and the distance is reduced by 3%.

E1.45 The pressure of an ideal gas of volume V, held at temperature T, is determined by the ideal gas law $PV = nRT$, where n is the amount of gas and R is the universal gas constant. Estimate the percent change in pressure if the temperature of the gas is reduced by 3% and the volume is increased by 2%.

E1.46 Find the percent error in y at $x = 0.8$ if $y = 3x + 2x^4$ and the error in x is 0.01.

E1.47 Consider Newton's second law of motion, $F = ma$, for a projectile with $m = (125 \pm 2)$ kg and $a = (2 \pm 0.1)$ m^2/s. Find the maximum possible absolute error that can be made in calculating the force.

E1.48 The measured dimensions of a cone are reported with their mean uncertainty as $D = (6 \pm 0.03)$ m and $h = (4 \pm 0.02)$ m. Develop an expression for the uncertainty of volume (in percent) and estimate the volume of the cone as well as its uncertainty.

E1.49 A wire of length L_0 undergoes thermal elongation when subjected to a temperature rise $\Delta T = T - T_0$. The length of a wire as a function of temperature difference is given by $L = L_0(1 + \beta \Delta T)$, where β is the linear expansion coefficient. Develop an expression to estimate the percent uncertainty in L in terms of the uncertainties of L_0, β, and ΔT.

E1.50 The volume of a parallelepiped is given with $V = abc \sin \theta$, where $a = (8 \pm 0.01)$ cm, $b = (10 \pm 0.05)$ cm, $c = (9 \pm 0.1)$ cm, and $\theta = (\pi/6 \pm \pi/60)$ radians. Find the upper and lower bounds (i.e., uncertainty) for the estimated volume.

E1.51 The law of cosines, $a^2 = b^2 + c^2 - 2bc \cos \beta$, relates the lengths of adjacent sides of a triangle to the cosine of the angle (β) between them. If the side lengths of a triangle are given as $a = (20 \pm 0.1)$ cm, $b = (30 \pm 0.1)$ cm, and $c = (12 \pm 0.1)$ cm, estimate β and $d\beta$.

E1.52 A plot of land has the shape of a parallelogram. The lengths of adjacent sides and the angle between them are reported as $a = (15 \pm 0.2)$ m, $b = (12 \pm 0.15)$ m, and $\theta = (38 \pm 2.5)°$, respectively. Estimate the area of the plot, its average, and its standard deviations, assuming the reported errors (uncertainties) are averages or standard deviations.

E1.53 A plot of land has the shape of a trapezoid. Its bases and height are reported with their standard deviations as $a = (120 \pm 0.55)$ m, $b = (78 \pm 0.45)$ m, and $h = (53 \pm 0.38)$ m, respectively. Find the area of the land and estimate its standard deviation.

E1.54 The mass of a spherical shell (tank) can be expressed in terms of its internal radius (R) and its thickness (d) as $m = (4\pi/3)\rho((R + d)^3 - R^3)$, where $\rho = 8000$ kg/m^3. The mean values and standard deviations of radius and thickness measurements made at different points on the shell are (113 ± 1.27) cm and (2.53 ± 0.13) cm, respectively. Estimate the mass of the shell and the associated standard deviation.

E1.55 The thermal (volume) expansion of a liquid is described by $\Delta V = \beta V_0 \Delta T$, where β is the volumetric expansion coefficient, ΔT is the temperature rise, and V_0 is the initial volume of the liquid. Use the furnished data to estimate the standard deviation of ΔV. Data: $\beta = (1100 \pm 33) \times 10^{-6}$ K^{-1}, $V_0 = (0.5 \pm 0.01)$, m^3, and $\Delta T = (25 \pm 0.5)$ K.

E1.56 The acceleration of gravity can be experimentally determined by using a simple pendulum experiment. Estimate the acceleration of gravity and associated uncertainty using $T = 2\pi\sqrt{\ell/g}$ and the results of the measurements for a pendulum with $\ell = (73 \pm 0.5)$ cm and $T = (1.714 \pm 0.0096)$.

E1.57 Resistance R of a wire is related to its resistivity ρ, length L, and cross-section area A by the relationship $R = \rho L/A$. For a particular wire, the mean values and standard deviations of the measured values are $R = (0.0753 \pm 0.0005)$ Ω, $L = (122 \pm 2.3)$ cm, and $D = (1 \pm 0.04)$ mm (circular in cross section). Estimate the wire resistivity (in Ω-cm) and associated standard deviation.

E1.58 A bar, rectangular in cross section, is fixed to a wall at one end, as shown in **Fig. E1.58**. The length L, the width w, and the thickness t are reported as (25 ± 0.01) mm, (5 ± 0.01) mm, and (1 ± 0.01) mm, respectively. A relationship for the deflection of the bar upon application of force is given by $F = Ewt^3y/4L^3$. If the deflection y is measured as (2.5 ± 0.01) mm, estimate the applied force and the resulting standard deviation. The elasticity modulus of the bar is $E = 2 \times 10^5$ MPa.

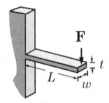

Fig. E1.58

E1.59 An Atwood machine consists of two masses coupled by a massless string over a massless pulley, as shown in **Fig. E1.59**. When the two weights are unequal, the system will move so that the heavier mass is pulled down while the lighter mass is pulled up. The acceleration of an Atwood machine is given by the relationship:

$$a = g\frac{m_1 - m_2}{m_1 + m_2}$$

(a) Estimate the system acceleration and its standard deviation for the given $m_1 = (110 \pm 1)$ g, $m_2 = (70 \pm 1)$ g, and $g = 9.81$ m/s^2. (b) Repeat (a) with $g = 9.80655$ m/s^2 and determine the effect of rounding on the result.

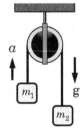

Fig. E1.59

E1.60 The thermal conductivity of an alloy is determined experimentally. A rectangular sample with a cross section of $A = (10 \pm 1)$ cm^2 and a length of $L = (20 \pm 1)$ cm was used for this experiment. All sides of the sample were perfectly insulated. When $Q = (2400 \pm 25)$ J heat was supplied to one end of the bar for a duration of $t = (120 \pm 2)$ s, the temperatures at both end points stabilized at $T_1 = (55 \pm 1)°$ C and $T_2 = (38 \pm 1)°$C. Estimate the thermal conductivity (k) of this alloy with its standard deviation, assuming that the heat transferred can be calculated by $Q/t = kA(T_1 - T_2)/L$.

Section 1.4 Numerical Computations

E1.61 Convert the following integers to their binary, octal, and hexadecimal equivalents.

(a) 36,   (b) 17,   (c) 81,   (d) 341

E1.62 Convert the following decimal numbers to their binary, octal, and hexadecimal equivalents:

(a) 0.015625,   (b) 3.4375,   (c) 38.125,   (d) 180.328125

E1.63 Convert the following numbers to their decimal equivalents:

(a) $(1110.0011)_2$, $(10111.11)_2$, $(110101.0101)_2$

(b) $(1572.432)_8$, $(30254.027)_8$, $(123.4567)_8$

(c) $(2AB1.F)_{16}$, $(DE1.A0B)_{16}$, $(234.ABC)_{16}$

E1.64 Express the following quantities in normalized scientific notation:

(a) 1.2345 (b) − 3.10203 (c) 0.00012345 (d) 123456 (e) 0.0000011, (f) 2.34×10^{-12}

E1.65 Express the following real numbers in normalized scientific notation:

(a) 54.3210, (b) 765.45, (c) 0.54325, (d) 0.00543,
(e) 0.5432×10^2, (f) 5.432×10^3, (g) 5.54×10^{-3}

E1.66 In a decimal floating-point system, the number $10/9 = 1.1111111\cdots$ does not have an exact representation with finite precision. Is there a finite floating-point system where the number can be exactly represented?

E1.67 Convert $3377/343 = (9.845481049\cdots)_{10}$ to a base-7 number system. Is the floating-point representation exact?

E1.68 Determine the number of significant digits in the following numbers:

(a) 54.3210, (b) 5.4003, (c) 0.54321, (d) 0.00543,
(e) 0.5432×10^2, (f) 5.432×10^1, (g) 0.54×10^{-3}

E1.69 Express the following approximate and true quantities in normalized scientific notation:

(a) $\tilde{Q} = 123.012$, $Q = 123.03$, (b) $\tilde{Q} = 12.3469$, $Q = 12.3458$
(c) $\tilde{Q} = 0.012313$, $Q = 0.01236$, (d) $\tilde{Q} = 999.90$, $Q = 1000$

E1.70 For the approximate $(\tilde{Q})$ and true (Q) quantities given in **E1.69**, determine the number of decimal digit accuracy of $\tilde{Q}$.

E1.71 Determine the "most accurate" one among the sets of approximate $(\tilde{Q})$ and true (Q) quantities given below.

(a) $\tilde{Q} = 785600$, $Q = 785543$, (b) $\tilde{Q} = 0.00035$, $Q = 0.0003544$
(c) $\tilde{Q} = 1.89315$, $Q = 1.8942120$, (d) $\tilde{Q} = 5.415 \times 10^{-3}$, $Q = 5.4421 \times 10^{-3}$

E1.72 Determine the number of significant digits of the following approximations for π.

(a) $\pi \cong \dfrac{22}{7}$, (b) $\pi \cong \dfrac{333}{106}$, (c) $\pi \cong \dfrac{355}{113}$

E1.73 Determine the number of significant digits in the following approximations for e.

(a) $e \cong \dfrac{19}{7}$, (b) $e \cong \dfrac{87}{32}$, (c) $e \cong \dfrac{193}{71}$

E1.74 Round the following numbers to two decimal places:

(a) 84.61485, (b) 5.3687, (c) 0.64557, (d) 0.02847

E1.75 Repeat **E1.74** for two significant (digit) figures.

E1.76 Round the following numbers to four significant digits:

(a) 54.3210, (b) 765.45, (c) 0.54325, (d) 0.00543,
(e) 0.5432×10^2, (f) 5.432×10^1, (g) 0.54×10^{-3}

E1.77 Approximate the following numbers to six decimal places by applying rounding and chopping rules.

(a) $Q_1 = 1.1234469$, (b) $Q_2 = 1.1234569$, (c) $Q_3 = 1.1234669$,
(d) $Q_4 = 2.1234569$, (e) $Q_5 = 4.1234369$

E1.78 Round the following numbers to four significant figures:

(a) 12.345678, (b) 4.120381, (c) 0.0012345, (d) 43.462180,
(e) 0.033678, (f) 0.2489741

E1.79 Round the numbers in **E1.78** to four decimal (place) figures:

E1.80 Determine the absolute and relative errors when $A = 12.4527$ is (a) chopped off and (b) rounded off to *three significant places*.

E1.81 Determine the absolute and relative errors when $A = 12.4527$ is (a) chopped off and (b) rounded off to *three decimal places*.

E1.82 Determine the absolute and relative errors when the floating-point representations of $A = 2/3$ and $B = 4/9$ are (a) chopped off and (b) rounded off to *six significant places*.

E1.83 Given $x = 5.4321$, $y = 0.5432$, and $z = 0.5433$, perform the following calculations: (i) exactly, and (ii) using *four-digit* decimal arithmetic with *chopping*.

$$\text{(a) } \frac{x}{z} - \frac{x}{y}, \qquad \text{(b) } x \times \left(\frac{1}{z} - \frac{1}{y} \right)$$

E1.84 For $x = 2/3$, $y = 3/7$, and $z = 5/9$, perform the following calculations: (i) *exactly*; (ii) using three-digit decimal arithmetic with *chopping*; (iii) using *three-digit* decimal arithmetic with *rounding*.

(a) $fl(x) + fl(y + z)$, (b) $fl(x + y) + fl(z)$, (c) $fl(x + y)$, (d) $fl(x \times y)$,
(e) $fl(x/y)$, (f) $fl(y/z)$, (g) $fl(z/x)$, (h) $fl(z/y)$

E1.85 For given $x = 1.56897463$ and $y = 1.56886938$, perform the $x - y$ operation (i) *exactly* and (ii) using *three-, four-, five-,* and *six-digit* decimal arithmetic with *chopping*.

E1.86 Repeat **E1.85** by using decimal arithmetic with *rounding*.

E1.87 Consider $p = 2.342356$ and $q = 5.134236$, which are approximated as $\tilde{p} = 2.342$ and $\tilde{q} = 5.134$. Calculate the absolute and relative error associated with the $p + q$ and $p - q$ operations.

E1.88 Consider $p = 3.1415927$ and $q = 3.1416932$, which are approximated as $\tilde{p} = 3.141$ and $\tilde{q} = 3.142$. Calculate the absolute and relative error associated with the $p + q$ and $p - q$ operations.

E1.89 Perform the following sum (i) *exactly* and (ii) using *four-digit* decimal arithmetic with rounding. Find the true error.

$$fl\left(\frac{1}{7}\right) + fl\left(\frac{1}{9}\right) + fl\left(\frac{1}{11}\right) + fl\left(\frac{1}{13}\right) + fl\left(\frac{1}{15}\right)$$

E1.90 Evaluate the following sum by applying *four-significant-digit rounding*. What is the true error?

$$fl\left(\frac{1}{\sqrt{104}}\right) + fl\left(\frac{1}{\sqrt{10.4}}\right) + fl\left(\sqrt{10.4}\right) + fl\left(\sqrt{104}\right)$$

E1.91 Evaluate $x^3 - 5.3x^2 + 6.8x + 2.71$ for $x = 1.42$ using *three-digit* decimal arithmetic with *chopping* and *rounding*.

E1.92 Repeat **E1.91** using its nested form, i.e., $(x(x - 5.3) + 6.8)x + 2.71$.

E1.93 Evaluate $1.328\,e^{3x} - 2.051\,e^{2x} + 3e^x - 1.7812$ for $x = 0.04$ using two-digit arithmetic with rounding. Find the absolute and relative errors in percent.

E1.94 Repeat **E1.93** by using the nested form of the polynomial.

E1.95 Perform the following calculations using *four-significant-places*.

(a) $0.8972 \times 10^{-8} - 0.4683 \times 10^{-8}$, (b) $\left(0.8972 \times 10^{-30}\right) \times \left(0.4683 \times 10^{-25}\right)$
(c) $\left(0.8972 \times 10^{25}\right) \times \left(0.4683 \times 10^{30}\right)$, (d) $\left(0.8972 \times 10^3\right) / \left(0.4683 \times 10^3\right)$

E1.96 Find the roots of $x^2 - 1010\,x + 2.5 = 0$ using floating-point arithmetic with *four-digit* chopping.

E1.97 Consider the mathematical equivalent expressions given below for the calculation of $(\sqrt{3} - \sqrt{2})^6$. Which is a better estimate if $\sqrt{6}$ is approximated by 2.4?

(a) $(5 - 2\sqrt{6})^3$, (b) $1/(5 + 2\sqrt{6})^3$, (c) $(485 - 198\sqrt{6})$, (d) $1/(485 + 198\sqrt{6})$

E1.98 Evaluate $\sqrt{16.42} - \sqrt{16.40}$ by applying *three-, four-*, and *five-significant-digit chopping*, and also find the true errors.

E1.99 Repeat **E1.98** after recasting the expression as $0.02/(\sqrt{16.42} + \sqrt{16.40})$.

E1.100 Calculate the following expressions for $x = 1$, 10^{-1}, $10^{-2}, \cdots$, 10^{-7}. Do you observe any problems? If yes, identify the problem and devise an alternative form to avoid it.

$$\text{(a)}\ \frac{\sec x - 1}{\tan^2 x}, \qquad \text{(b)}\ \frac{\sqrt{x^2+4}-2}{x^2}, \qquad \text{(c)}\ \frac{(x-1)^3+1}{x}$$

E1.101 Compute $x^{3/2}(\sqrt{x+1/x} - \sqrt{x-1/x})$ for $x = 10$, $10^2 \cdots$, 10^8. Do you observe any problems? If yes, identify the problem and devise an alternative form to avoid it.

E1.102 Compute $(1 - \cos 2x)/x^2$ for 10^{-1}, 10^{-2}, $\cdots$, 10^{-9}. Do you observe any problems? If yes, identify the problem and devise an alternative form to avoid it.

E1.103 Consider computing $\ln x - \ln y$ for $x \cong y$ which leads to a loss of significance. Derive an alternative form of computation to avoid this problem.

E1.104 Consider computing $\ln(\sqrt{x^2-1}-x)$ for large x. (a) Show whether $-\ln(\sqrt{x^2-1}+x)$ can be an alternative form or not; (b) which expression is more suitable for numerical computations? Explain why?

E1.105 Calculate $\sqrt{16.07} - 4$ by chopping and rounding, using *three-digit arithmetic*.

E1.106 Calculate the arithmetic operation given in **E1.105** alternatively as $0.07/(\sqrt{16.07} + 4)$.

Section 1.5 Application of Taylor Series

E1.107 Use Taylor polynomials of degree $n = 3, 4, \ldots, 7$ to find approximate values of $\sqrt{e}$, $\sqrt[3]{e}$, and $\sqrt[4]{e}$. How many terms of the series are needed to accurately estimate the approximate values to within 10^{-3}.

E1.108 Using the Maclaurin series, find a fourth-degree approximation and corresponding remainder for $f(x) = 1/(x+5)$.

E1.109 Consider estimating $\pi/4$ with (a) $\tan^{-1}(1)$ and (b) $4\tan^{-1}(1/5) - \tan^{-1}(1/239)$ (the Machnin's formula) using the Taylor polynomial given below.

$$\tan^{-1}x \cong S_N = x - \frac{x^3}{3} + \frac{x^5}{5} - \ldots + (-1)^N \frac{x^{2N+1}}{(2N+1)} = \sum_{k=0}^{N} (-1)^k \frac{x^{2k+1}}{(2k+1)!}$$

Use 10-terms (S_{10}) to calculate inverse tangents and determine the decimal-place accuracy of the estimates obtained from both approximations. *Note*: Use 10-decimal place arithmetic in calculations.

E1.110 Use the Maclaurin series to show that the following two-term approximation is fourth-order, i.e., $\mathcal{O}(x^4)$:

$$\sqrt{1+x^2} = 1 + \tfrac{1}{2}x^2$$

E1.111 Consider the Maclaurin series of $f(x) = (1+x)^{-1/3}$ on $(-1,1]$ given below: Use the three-term approximation to estimate $1.1^{-1/3}$. What can you say about the decimal-place accuracy of the estimated result?

$$(1+x)^{-1/3} = 1 - \frac{1}{3}x + \frac{1 \cdot 3}{3 \cdot 6}x^2 - \frac{1 \cdot 3 \cdot 5}{3 \cdot 6 \cdot 9}x^3 + \cdots$$

E1.112 How many terms of the series must be included to accurately predict $e^{2.75}$ to five decimal places using the series expansion of e^x? Suppose the true value of e^2 is given. How can you calculate $e^{2.75}$ with the same accuracy by reducing the number of terms (as well as cpu-time) in the series? Can you suggest a measure to further reduce the cpu-time?

E1.113 Use the following series expansion to determine a four-term approximation (S_4) and its upper error bound (R_4) to $f(x) = (2x+9)^{-1/2}$ given on $[-1/2, 1/2]$.

$$\frac{1}{\sqrt{x+1}} = 1 - \frac{1}{2}x + \frac{1\cdot 3}{2\cdot 4}x^2 + \frac{1\cdot 3\cdot 5}{2\cdot 4\cdot 6}x^3 - \cdots$$

E1.114 Use power series to find 2nd and 3rd-degree polynomial approximations to $(3+x)^{-1/2}$ for any $x \in (1/5, 9/5)$. *Hint*: Determine the most suitable point "a" for the Taylor polynomials.

E1.115 Use the power series of $(1+x)^{1/3}$ on $(-1, 1]$ to obtain a third-degree polynomial approximation to estimate $(1.92)^{1/3}$. Is the true error less than the upper error bound? Use at least seven-decimal-place accuracy in arithmetic operations.

$$(1+x)^{1/3} = 1 + \frac{x}{3} - \frac{2}{3\cdot 6}x^2 + \frac{2\cdot 5}{3\cdot 6\cdot 9}x^3 - \frac{2\cdot 5\cdot 8}{3\cdot 6\cdot 9\cdot 12}x^4 + \cdots$$

E1.116 Find a 5th-degree Maclaurin polynomial for the following functions:

$$\text{(a) } e^{\sin x}, \qquad \text{(b) } \cos(\sin x), \qquad \text{(c) } e^x \cos x$$

E1.117 Expand $f(x) = \sin x \cos x$ to the Maclaurin series. Then, use the series to estimate $\sin 36° \cos 36°$ accurate to within 10^{-3}?

E1.118 Find the range of x for which the truncation error of approximating e^x with the 9th-degree Maclaurin polynomial is bounded by 10^{-8}.

E1.119 Construct a 3rd-degree Maclaurin polynomial for $\sin 3x$ and determine the largest possible truncation error in $[0, \pi/2]$?

E1.120 Find a 3rd-degree polynomial approximation that avoids the "loss of significance" in the following expressions:

$$\text{(a) } \frac{e^{2x}-1}{x}, \qquad \text{(b) } \frac{1-e^{-x/2}}{x}, \qquad \text{(c) } \frac{(x+4)^{3/2}-3x-8}{x^2}, \qquad \text{(d) } \frac{2x-\sin 2x}{x^2}$$

E1.121 Use the Maclaurin series of $\ln(x+1)$ to construct a polynomial approximation that yields estimates accurate to within 10^{-3} in $[0, 1/2]$.

E1.122 Use the Maclaurin series of $(x+1)^{-2}$ to construct a polynomial approximation that yields estimates accurate to within 10^{-3} in $[0, 1/2]$.

E1.123 Use the Maclaurin series of $\tan^{-1} x$ to determine a four-term approximation (S_4) and its upper error bound (R_4) for any x in $[-1/2, 1/2]$.

E1.124 For $f(x) = \sin x$, construct Taylor polynomial approximations (of degree $N = 2,3,4$, and 5) about $x = \pi/4$ and find the associated upper error bounds. Use the approximations to estimate $\sin \pi/3$ and calculate the maximum possible errors.

E1.125 Use the Maclaurin series to develop Nth-degree polynomial approximation and corresponding truncation error for the following definite integrals:

$$\text{(a) } \int_0^x e^{-u^2}\, du, \qquad \text{(b) } \int_0^x \frac{\sin t^2\, dt}{t}, \qquad \text{(c) } \int_0^x \ln(1+w^4)\, dw$$

E1.126 Determine all ξ points that satisfy the mean-value theorem for $f(x) = 8x^3 + 9x^2 - 6x + 5$ on $(-2, 1)$.

E1.127 Show that $f(x) = x^5 - x^4 + x^2 - 1$ has only two points on $(-1, 1)$ that satisfy the mean value theorem.

E1.128 Determine whether the mean value theorem can be applied to $f(x) = 3 + \sqrt{x+1}$ on $[0,3]$. If so, find all possible values of ξ.

E1.129 Consider $f(x) = 2\sqrt{x}$ on $(1,9)$. Determine a point ξ whose existence is assured by the mean value theorem.

E1.130 For $f(x) = x^4 - 2x^3 - 36x^2 + 3$, determine a point ξ on $(0,3)$ that satisfies the mean value theorem for $f'(x)$.

E1.131 Consider $f(x)$ that is bounded on $(0, h)$. If $f(h) = f(0) + 4h^2$, estimate the derivative of $f(x)$ at the midpoint of the interval, i.e., $f'(h/2)$.

E1.132 Find the average value of $f(x) = 4 - x^2$ on $[0, 2]$.

E1.133 Find the average value of $f(x) = x - x^3$ on $[0,1]$.

E1.134 Find the average value of $f(x) = 8/x^3$ on $[1,2]$.

E1.135 The axial temperature distribution of a 30-cm-long circular bar is described by $T(x) = 10 + 20\cosh(2 - 3x)$, where x is the axial direction (in m) and T is the temperature (in $^\circ$C). Estimate the average temperature of the bar.

E1.136 Consider the flow of a fluid between two large horizontal parallel plates, where the bottom plate is stationary and the top plate is moving at speed U. The fluid velocity distribution, depicted in **Fig. E1.136**, is given by $v(y) = UY(1 + \Phi - \Phi Y)$, where $Y = y/b$, Φ is a constant, and b is the distance between parallel plates. Find an expression for the average fluid velocity.

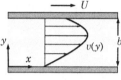

Fig. E1.136

E1.137 Find the average value of $f(x)$ in the specified intervals.

(a) On $[-3, 3]$, $\quad f(x) = \begin{cases} 1, & x < -2 \\ 3 & -2 \leqslant x < 1 \\ 2 & x \geqslant 1 \end{cases}$ (b) On $[-2, 3]$, $\quad f(x) = \begin{cases} -3x, & x < 0 \\ x^2 & 0 \leqslant x < 1 \\ 1 & x \geqslant 1 \end{cases}$

E1.138 For the given integral, find the average value for $g(x)$ on $[a, b]$.

$$\int_a^b f(x)g(x)dx = 3(a - b)^3$$

where $f(x) = x - a$ and $g(x)$ is bounded and integrable on $[a, b]$.

E1.139 Consider $f(x)$ that is bounded and integrable on $[0, a]$. Find a suitable K for some ξ that satisfies the integral in this interval.

$$\int_0^a x(a - x)f(x)dx = K \cdot f(\xi)$$

E1.140 Find ξ-values that satisfy the second mean-value theorem for

$$\int_0^1 f(x)g(x)dx$$

where $f(x) = x^2 + 1$ and $g(x) = e^{-2x}$. *Hint*: Swap $f(x)$ and $g(x)$ to find a suitable ξ for both cases.

E1.141 Consider $f(x)$ which is bounded and integrable on $[-h/2, h/2]$. What would be a suitable K for the integral to be estimated as follows?

$$\int_{-h/2}^{h/2} x^2 f(x)dx \cong K \cdot f(0)$$

Section 1.6 Truncation Error of Series

E1.142 Find the true values of the following infinite sums: *Hint*: First obtain an expression for the partial sums, S_N, and then take the limit for $N \to \infty$.

(a) $\displaystyle\sum_{k=1}^{\infty} \frac{1}{4k^2 - 1}$, (b) $\displaystyle\sum_{k=1}^{\infty} \frac{1}{4k^2 - 9}$, (c) $\displaystyle\sum_{k=1}^{\infty} \frac{2k + 1}{k^2(k + 1)^2}$, (d) $\displaystyle\sum_{k=1}^{\infty} \frac{k(4k^2 + 1)}{16(4k^2 - 1)^4}$

E1.143 For the series given in **E1.142**, calculate five-term approximations and error bounds. How many decimal places of accuracy are achieved in each series?

E1.144 Estimate the six-term partial sums, true errors, and error bounds of the following series. Compare the results with the true solution. Is the true error within the error bounds?

(a) $\displaystyle\sum_{n=1}^{\infty} \frac{1}{n^2(n^2+1)} = \frac{1}{6}(3+\pi^2-3\pi\coth\pi)$, (b) $\displaystyle\sum_{n=1}^{\infty} \frac{3^n}{n\cdot 4^n} = \ln 4$

(c) $\displaystyle\sum_{n=1}^{\infty} \frac{2n}{n^4+1} = 1.388346044$, (d) $\displaystyle\sum_{n=1}^{\infty} \frac{4^n}{(n!)^2} = I_0(4) - 1$

where $I_0(x)$ is the modified Bessel function of the first kind.

E1.145 Estimate the nine-term partial sums, true errors, and the error bounds of the following alternating series. Compare the results with the true solution.

(a) $\displaystyle\sum_{n=1}^{\infty} \frac{(-1)^{n-1}}{n^2+1} = \frac{1}{2}(1-\pi\operatorname{csch}\pi)$, (b) $\displaystyle\sum_{n=1}^{\infty} \frac{(-\pi^2)^{n+1}}{(2n)!} = 2\pi^2$

(c) $\displaystyle\sum_{n=1}^{\infty} (-1)^{n-1}\frac{n^2}{n^4+4} = \frac{1}{4}\pi\operatorname{csch}\pi$, (d) $\displaystyle\sum_{n=1}^{\infty} \frac{(-1)^n}{n(n+1)} = 1-2\ln 2$

E1.146 How many decimal places of accuracy can be expected in the worst case if the following series are replaced by 10-term approximations?

(a) $\displaystyle\sum_{n=1}^{\infty} \frac{(-1)^n}{n^2(n^2+1)}$, (b) $\displaystyle\sum_{n=1}^{\infty} \frac{(-1)^n}{n(n+1)(n+2)}$, (c) $\displaystyle\sum_{n=1}^{\infty} \frac{(-1)^n}{n^3(n+1)^3}$

Section 1.7 Rate and Order of Convergence

E1.147 Find the limits and convergence rates of the following sequences:

(a) $x_n = \dfrac{4n+1}{5n+12}$, (b) $x_n = \dfrac{n+3}{n^3+2}$, (c) $x_n = \dfrac{1}{n^2(1+2^{1/n})}$, (d) $x_n = \ln\left(\dfrac{3n+1}{3n+5}\right)$

E1.148 For three cases, the absolute errors of successive iterates x_n (e.g., $|x_n - x_{n-1}|$ for $n = 1, 2, \ldots, 7$) are given in the table below. Estimate the order of convergence for each case.

n	Case 1 (x_n)	Case 2 (y_n)	Case 3 (z_n)
1	1	2	0.9
2	0.64	1.5	0.195
3	0.262144	0.68695298	0.04200
4	0.044000	8.65×10^{-2}	0.0091500
5	1.24×10^{-3}	1.90×10^{-4}	0.0020000
6	9.85×10^{-7}	2.03×10^{-12}	0.0004350
7	6.2×10^{-13}	2.49×10^{-36}	0.0000946

E1.149 The absolute errors of successive iterates $x_{(n)}$ (i.e., $|x_n - x_{n-1}|$ for $n = 1, 2, \ldots, 7$) are given below. Estimate the order of convergence.

n	1	2	3	4	5	6	7
x_n	2	1.781	1.477	1.091	0.668	0.302	0.0836

1.10 COMPUTER ASSIGNMENTS

CA1.1 Consider the conversion of an integer number (input) to an equivalent binary number (output string). Write a pseudomodule, **Module Int2Binary** (*integer_number, binary_number*), and test your code with a computer program in the language of your choice.

CA1.2 Consider the conversion of a decimal number (input in the range $0 < x < 1$) to an equivalent binary number (23-bit string output). Write a pseudomodule, **Module Decimal2Binary** (*x, binary_number*), and test your code with a computer program in the language of your choice.

CA1.3 Write a computer program that uses *n-digit* floating-point arithmetic to perform the $\sum_{i=1}^{n} \Delta x$ operation. Perform 10 million additions with $\Delta x = 2^{-16}$ and 2×10^{-5} using both rounding and chopping to six-, seven-, and eight-digit. Compare and interpret your findings.

CA1.4 Write a pseudocode to calculate $S_N = \sum_{k=1}^{N} a_k$ and convert the code to a computer program in the language of your choice. To estimate the series sums whose true values are given below, (a) run your program for $N = 10, 20, 50$, and 100; (b) calculate the true (absolute) error for the estimates; and (c) interpret your findings.

(a) $\displaystyle\sum_{n=1}^{\infty} \frac{1}{n^2(n^2+1)} = \frac{1}{6}(3 + \pi^2 - 3\pi \coth \pi)$ (b) $\displaystyle\sum_{n=1}^{\infty} \frac{3^n}{n \cdot 4^n} = \ln 4$, (c) $\displaystyle\sum_{n=1}^{\infty} \frac{6^n}{n!} = e^6 - 1$

CA1.5 Write a computer program that finds the partial sum of the p−series, $S_N = \sum_{k=1}^{N}(1/n^p)$, using *five-* and *ten-digit* floating-point arithmetic with both *rounding* and *chopping*. In doing so, your program should be able to perform the additions (i) back-to-front ($k = N, (N-1), \cdots, 2, 1$, adding numbers from smallest to largest) and (ii) front-to-back ($k = 1, 2, \cdot, (N-1), N$, adding numbers from largest to smallest). Then investigate the effect (if any) of the direction on the addition process. Discuss your findings for $p = 1/2$, 1 and 2 with $N = 10, 50$, and 100 terms.

CA1.6 Write a program to compute $\pi/4$ using the expressions in (i), (ii), and the Machin's formulas given by (iii) and (iv). (a) Find $\pi/4$ using the Maclaurin series of $\tan^{-1}x$ with $N = 10$, 50, and 100 terms. How does accuracy improve? (b) Write another program to compute $\tan^{-1}x$ with an ε as a stopping criterion. (d) Using all three methods, how many terms of the series are necessary to obtain $\pi/4$ with 10 decimal digits? Report and discuss your findings.

(i) $\dfrac{\pi}{4} = \tan^{-1}(1)$, (ii) $\dfrac{\pi}{4} = 4\tan^{-1}\left(\dfrac{1}{5}\right) - \tan^{-1}\left(\dfrac{1}{239}\right)$,

(iii) $\dfrac{\pi}{4} = \dfrac{3}{2}\tan^{-1}\left(\dfrac{1}{\sqrt{3}}\right)$ (iv) $\dfrac{\pi}{4} = 6\tan^{-1}\left(\dfrac{1}{8}\right) + 2\tan^{-1}\left(\dfrac{1}{57}\right) + \tan^{-1}\left(\dfrac{1}{239}\right)$

CA1.7 Devise a general program to evaluate e^x for all x accurate within machine epsilon using the Maclaurin series. (a) Use your program to obtain the approximations as well as the true relative errors for negative ($x < 0$) and positive ($x > 0$) real values. (b) How does the relative error change? (c) Does a program with double-precision arithmetic eliminate the problem you observed? If not? What is the reason for this? (d) How can you overcome this problem?

CA1.8 The Taylor series expansion of the inverse sine function is defined as

$$\sin^{-1}x = x + \frac{1}{2}\frac{x^3}{3} + \frac{1 \cdot 3}{2 \cdot 4}\frac{x^5}{5} + \frac{1 \cdot 3 \cdot 5}{2 \cdot 4 \cdot 6}\frac{x^7}{7} + \cdots + \frac{1 \cdot 3 \cdots (2n-1)}{2 \cdot 4 \cdots (2n)}\frac{x^{2n+1}}{2n+1} + \cdots$$

Write a pseudo-**Function Module ARCSIN**(x) that calculates the inverse sine of any x in $[-1, 1]$ using the series expansion accurately within ε. *Note*: Make sure that the code performs a minimum number of arithmetic operations. Also, you should include warning messages, errors, and validity control statements.

CA1.9 The Bessel function of the first kind is given as follows:

$$J_n(x) = \sum_{k=0}^{\infty} (-1)^k \frac{(x/2)^{n+2k}}{k!(n+k)!} = \frac{(x/2)^n}{n!} \left\{ 1 - \frac{(x/2)^2}{(n+1)} + \frac{(x/2)^4}{(n+1)(n+2)} - \cdots \right\}$$

Write a pseudocode, **Module BESSELJ**$(n, x, \varepsilon, k, BesselJn)$, that calculates the Bessel function ($BesselJn$, real output) for any n (integer order) and x real inputs, accurate to within ε (input) tolerance. The program should also return the number of terms added (k) as output. Apply cpu-time minimization strategy. Convert your code to a program in the language of your choice and test the program for various n and x values. Finally, use the program to obtain the numerical values of $J_0(x)$, $J_1(x)$, and $J_2(x)$ on $[0, 2]$ with 0.2 increments and comparatively plot these functions. Also, you should include warning messages, errors, and validity control statements.

CA1.10 For real a and b and integer n, k, and m input variables, a computer program is to be prepared to evaluate the following definite integral:

$$d_m = \int_0^b x^n (a + bx^k)^m dx$$

The integral d_m (output) can be computed using the following recursion formula:

$$d_m = \frac{b^{n+1}(a + b^{k+1})^m}{1 + mk + n} + \frac{amk}{mk + n + 1}d_{m-1} \quad \text{and} \quad d_0 = \frac{b^{n+1}}{n+1}$$

(a) Write a pseudocode, **Module DITEG**$(a, b, m, n, k, \mathbf{d})$, that calculates the integrals up to m and returns the results in $d_0, d_1, \ldots, d_m$. Make your code as efficient as possible by keeping the number of arithmetic operations (power, multiplication, and division) to a minimum; and (b) convert your code to a computer program in the language of your choice and test the program for various input values.

CA1.11 The following expression is known as the hyperbolic sine integral function.

$$\text{Shi}(x) = \int_0^x \frac{\sinh t}{t} dt$$

(a) Obtain a polynomial approximation for $\text{Shi}(x)$ by making use of the Maclaurin series; (b) Use the approximation derived in Part (a) to write a pseudocode, **Function Module SHI**(x), that calculates an approximate value of $\text{Shi}(x)$ for an input x in $[0, 1]$ within 5×10^{-7} tolerance; and (c) convert your code to a computer program in the language of your choice and test the program for various input values. Make your code as efficient as possible by keeping the number of arithmetic operations (power, multiplication, and division) to a minimum.

CA1.12 Consider the following recursive sequences for which convergence rates are to be determined:

(a) $x_{n+1} = 2 + \dfrac{3}{x_n}$, $x_0 = 1$, (b) $x_{n+1} = \sqrt{2x_n + 15}$, $x_0 = 1$, (c) $x_{n+1} = e^{-x_n/5}$, $x_0 = 1$

(a) Write a computer program (or use a spreadsheet) to evaluate the terms of a sequence up to 11 and then tabulate $|x_n - x_{n-1}|$, $\log(|x_{n+1} - x_n|/|x_n - x_{n-1}|)$ and r; (b) Determine the limit value and the order of convergence of a sequence.

CA1.13 In the field of heat transfer, the Nusselt number is a dimensionless number that gives the ratio of convective to conductive heat transfer at a wall in a fluid. The Nusselt number for slug flow in annular ducts subjected to constant heat flux may be predicted from the following correlation [33]:

$$\text{Nu}(m) = \frac{8(m-1)(m^2-1)^2}{m^4(4\ln m - 3) + 4m^2 - 1},$$

where $m = D_2/D_1$ denotes the diameter ratio. Find simpler linear and quadratic approximations valid in the range $1 \leqslant m \leqslant 3$, and investigate the accuracy of these approximations by plotting comparatively the true and approximate values in $1 < m \leqslant 3$. Clue: Expand the Nusselt relationship to the Taylor's series about $m = 1$.

CHAPTER 2

Linear Systems: Fundamentals and Direct Methods

LEARNING OBJECTIVES

After completing this chapter, you should be able to

- refresh the basic linear algebra concepts and fundamental matrix operations;
- describe the general structure of a system of linear equations;
- realize the difference between direct and iterative methods;
- perform elementary row operations and matrix transformations;
- perform matrix inversion using the Gauss-Jordan algorithm;
- carry out forward- and back-substitution algorithms to solve upper- and lower-triangular systems of equations;
- apply Gauss elimination to solve linear systems *with* and *without* pivoting;
- understand and explain *ill-condition* and its effects on matrix algebra;
- apply Doolittle and Cholesky's decomposition techniques to solve linear systems;
- understand and implement the Gauss elimination and LU-decomposition methods to solve tridiagonal systems of equations;
- implement Gauss elimination method to solve banded linear systems;
- explain and implement LU-decomposition for banded linear systems.

M OST problems in science and engineering are reduced to problems in the field of linear algebra. A few of these applications are structural analysis, circuit analysis, optimization, least squares fitting, data analysis, network flow, eigenvalue and eigenvector problems, the solution of systems of simultaneous linear equations, the numerical solution of differential equations using finite difference or finite volume methods, etc.

Solving a linear system, inverting a matrix, and determining the eigenvalues or eigenvectors of a matrix are the most common math operations in applied sciences. The solution of a system of linear equations can be obtained by either *direct* or *iterative methods*. In the absence of round-offs or other errors, direct methods give the exact solution of the linear

DOI: 10.1201/9781003474944-2

system of equations. On the other hand, iterative methods discussed in Chapter 3 give an approximate solution of a linear system within a prescribed tolerance.

Linear algebra basics, definitions, notations, etc. are revisited to refresh the readers' memory. Comprehensive pseudocodes are provided for most of the methods presented to reinforce understanding and writing computer codes on this topic. Eigenvalues and eigenvectors are briefly discussed here, but this topic is covered in greater detail in Chapter 11.

2.1 FUNDAMENTALS OF LINEAR ALGEBRA

Systems of linear equations are conveniently expressed in matrix notation, and methods of solution for such systems can be developed very compactly using matrix algebra. Consequently, in this text, the elementary properties of matrices, vectors, and determinants are presented in this section, and shorthand matrix and vector notations are extensively used throughout the text.

2.1.1 MATRICES

A *matrix* is a rectangular arrangement of numbers in tabular form with m rows and n columns. The number of rows and columns determines the *size* or *dimension* of the matrix. An $m \times n$ *rectangular matrix* is expressed as

$$\mathbf{A} = [a_{ij}]_{m \times n} = \begin{bmatrix} a_{11} & a_{12} & a_{13} & \cdots & a_{1n} \\ a_{21} & a_{22} & a_{23} & \cdots & a_{2n} \\ a_{31} & a_{32} & a_{33} & \cdots & a_{3n} \\ \vdots & \vdots & \vdots & \ddots & \vdots \\ a_{m1} & a_{m2} & a_{m3} & \cdots & a_{mn} \end{bmatrix}_{m \times n} \tag{2.1}$$

where a_{ij} denotes the element of the ith row and the jth column. A comma is used to separate two indices when necessary to avoid confusion, e.g., $a_{n,n+1}$ instead of a_{nn+1}. Enclosing the general element within square brackets as "$[a_{ij}]_{m \times n}$" is an abbreviation for a general matrix by indicating its size. Each number in a matrix, a_{ij}, is referred to as a *matrix element* or *matrix entry*.

A matrix consisting of a single row $\mathbf{b} = [b_i]_{1 \times n}$ is referred to as a *row matrix*, while a matrix consisting of a single column $\mathbf{x} = [x_i]_{n \times 1}$ is called a *column matrix* or a *column vector*. Following are examples of such matrices:

$$\mathbf{b} = [b_i]_{1 \times n} = [\, b_1 \ \ b_2 \ \ \cdots \ \ b_n \,], \qquad \mathbf{x} = [x_i]_{n \times 1} = \begin{bmatrix} x_1 \\ x_2 \\ \vdots \\ x_n \end{bmatrix}$$

Elements of a row vector that may involve long or complicated expressions could be separated from each other with commas to avoid confusion, i.e., $\mathbf{b} = [\, a + 3, \ a - b, \ b - 2, \ \cdots, \ 2a + 3b \,]$.

 A matrix is denoted with an uppercase boldface letter ($\mathbf{A}$, $\mathbf{B}$, $\mathbf{C}$) and a vector (row or column matrices) by a lowercase boldface letter ($\mathbf{a}$, $\mathbf{b}$, $\mathbf{c}$) throughout this text.

A *square matrix*, most often encountered when solving systems of linear equations, has the same number of rows and columns: $n \times n$. The *main* (or *principal*) *diagonal* elements of a square matrix $\mathbf{A}$ are $a_{11}, a_{22}, a_{33}, \cdots, a_{nn}$.

The *trace* of a square matrix, $\mathbf{A} = [a_{ij}]_{n \times n}$, is defined as the sum of its main diagonal elements, i.e.,

$$\text{tr}(\mathbf{A}) = a_{11} + a_{22} + a_{33} + \cdots + a_{nn} = \sum_{k=1}^{n} a_{kk}$$

2.1.2 SPECIAL MATRICES

A matrix whose elements are zero is called a *zero matrix*. For example,

$$\mathbf{O} = [0]_{2\times2} = \begin{bmatrix} 0 & 0 \\ 0 & 0 \end{bmatrix}, \quad \mathbf{O} = [0]_{2\times4} = \begin{bmatrix} 0 & 0 & 0 & 0 \\ 0 & 0 & 0 & 0 \end{bmatrix}, \quad \mathbf{O} = [0]_{3\times1} = \begin{bmatrix} 0 \\ 0 \\ 0 \end{bmatrix}$$

where $\mathbf{O}$ denotes any conformable zero matrix. In cases where necessary, the notation $\mathbf{O} = [0]_{m \times n}$ may be used to refer to a zero matrix by its size.

Square matrices have additional special properties, unlike rectangular matrices. If $\mathbf{A}$ is an $n \times n$ square matrix and satisfies the $a_{ij} = a_{ji}$ condition for all i, j, then the matrix is said to be *symmetric*. For instance, the following matrices are symmetrical about the principal diagonal:

$$\begin{bmatrix} 1 & -4 \\ -4 & 5 \end{bmatrix}, \quad \begin{bmatrix} a & -x & u \\ -x & b & e \\ u & e & f \end{bmatrix}, \quad \begin{bmatrix} 3 & 2 & 1 & -3 \\ 2 & 4 & 2 & 1 \\ 1 & 2 & 7 & 3 \\ -3 & 1 & 3 & 5 \end{bmatrix}$$

A *diagonal matrix*, usually denoted by $\mathbf{D}$, is a square matrix whose off-diagonal elements are all zero ($d_{ij} = 0$ for all $i \neq j$), while *identity matrix* or *unit matrix*, $\mathbf{I}$, is a special case of a diagonal matrix whose diagonal elements are "1." For example,

$$\mathbf{D} = \begin{bmatrix} d_{11} & 0 & 0 & \cdots & 0 \\ 0 & d_{22} & 0 & \cdots & 0 \\ 0 & 0 & d_{33} & \cdots & 0 \\ \vdots & \vdots & \vdots & \ddots & \vdots \\ 0 & 0 & 0 & \cdots & d_{nn} \end{bmatrix}, \quad \mathbf{I}_n = \begin{bmatrix} 1 & 0 & 0 & \cdots & 0 \\ 0 & 1 & 0 & \cdots & 0 \\ 0 & 0 & 1 & \cdots & 0 \\ \vdots & \vdots & \vdots & \ddots & \vdots \\ 0 & 0 & 0 & \cdots & 1 \end{bmatrix} \quad (2.2)$$

where the notation $\mathbf{I}_n$ is often used to denote the identity matrix with its size ($n \times n$). The subscript n is generally omitted in cases where its size is obvious from the context.

A square matrix with zero elements below the main diagonal ($u_{ij} = 0$, $i > j$) is called an *upper triangular matrix*, and the one with zero elements above the diagonal ($\ell_{ij} = 0$, $i < j$) is called a *lower triangular matrix*.

$$\mathbf{U} = \begin{bmatrix} u_{11} & u_{12} & u_{13} & \cdots & u_{1n} \\ 0 & u_{22} & u_{23} & \cdots & u_{2n} \\ 0 & 0 & u_{33} & \cdots & u_{3n} \\ \vdots & \vdots & \vdots & \ddots & \vdots \\ 0 & 0 & 0 & \cdots & u_{nn} \end{bmatrix}, \quad \mathbf{L} = \begin{bmatrix} \ell_{11} & 0 & 0 & \cdots & 0 \\ \ell_{21} & \ell_{22} & 0 & \cdots & 0 \\ \ell_{31} & \ell_{32} & \ell_{33} & \cdots & 0 \\ \vdots & \vdots & \vdots & \ddots & \vdots \\ \ell_{n1} & \ell_{n2} & \ell_{n3} & \cdots & \ell_{nn} \end{bmatrix} \quad (2.3)$$

where $\mathbf{U}$ and $\mathbf{L}$ are generally used to denote Upper and Lower triangular matrices.

A *band matrix* or *banded matrix*, $\mathbf{B}$, is a square matrix that consists of mostly nonzero elements in a rectangular band about the main diagonal. For example, the following is a 7×7 band matrix:

$$\mathbf{B} = \begin{bmatrix} b_{11} & b_{12} & b_{13} & b_{14} & & & \\ b_{21} & b_{22} & b_{23} & b_{24} & b_{25} & & \\ b_{31} & b_{32} & b_{33} & b_{34} & b_{35} & b_{3,6} & \\ & b_{42} & b_{43} & b_{44} & b_{45} & b_{46} & b_{47} \\ & & b_{53} & b_{54} & b_{55} & b_{56} & b_{57} \\ & & & b_{64} & b_{65} & b_{66} & b_{67} \\ & & & & b_{75} & b_{76} & b_{77} \end{bmatrix} \tag{2.4}$$

where m_L and m_U are the lower and upper bandwidths, which for this example are $m_L = 2$ and $m_U = 3$. Note that $b_{ij} = 0$ if $j - i > m_U$ or $i - j > m_L$, and $b_{ij} \neq 0$ for $i - m_L \leqslant j \leqslant i + m_U$. The total bandwidth of a banded matrix is $n_b = m_L + m_U + 1$.

A *tridiagonal matrix*, $\mathbf{T}$, is a square matrix and has nonzero elements only on the main diagonal (d_i, $i = 1, 2, \ldots, n$), the first diagonal below (b_i, $i = 2, 3, \ldots, n$), and the first diagonal above it (a_i, $i = 1, 2, \ldots, n-1$). Although $t_{ij} = 0$ for all $i - j > 1$ and $j - i > 1$, a few elements of b_i or a_i may be zero. Note that it is also a special case of a banded matrix with $m_L = m_U = 1$ and $n_b = 3$.

$$\mathbf{T} = \begin{bmatrix} d_1 & a_1 & & & & \\ b_2 & d_2 & a_2 & & & \\ & b_3 & d_3 & a_3 & & \\ & & \ddots & \ddots & \ddots & \\ & & & b_{n-1} & d_{n-1} & a_{n-1} \\ & & & & b_n & d_n \end{bmatrix}$$

In most cases, it is desirable to work with matrices that are in *row-echelon* or *reduced row-echelon* forms. These matrices could be square or rectangular in shape. A matrix in row echelon form illustrated below has a pivot at its non-zero rows; in other words, all the entries to the left and below the pivot (1's here) are equal to zero.

$$\begin{bmatrix} 1 & * & * & \cdots & * & * & * \\ 0 & 1 & * & \cdots & * & * & * \\ 0 & 0 & 1 & \cdots & * & * & * \\ \vdots & \vdots & \vdots & \ddots & \vdots & \vdots & \vdots \\ 0 & 0 & 0 & \cdots & 1 & * & * \end{bmatrix}_{n \times m}$$

where * denotes a nonzero element. When the coefficient matrix of a linear system is converted into a *row echelon form*, it is easier to obtain the solution of the system using the so-called *back substitution* algorithm.

The following matrix is said to be in *reduced row echelon form*:

$$\begin{bmatrix} 1 & 0 & 0 & \cdots & 0 & * & * \\ 0 & 1 & 0 & \cdots & 0 & * & * \\ 0 & 0 & 1 & \cdots & 0 & * & * \\ \vdots & \vdots & \vdots & \ddots & \vdots & \vdots & \vdots \\ 0 & 0 & 0 & \cdots & 1 & * & * \end{bmatrix}_{n \times m}$$

Notice that the basic columns are vectors of the standard basis, i.e., vectors having all zero elements except for one equal to 1.

The *rank* of a matrix, denoted by rank($\mathbf{A}$), is the maximum number of linearly independent (unique) rows (or columns), and the number of rows (or columns) that are not obtained by a linear combination of other rows (or columns) is termed the *row rank* (or *column rank*). Consider the following matrices:

$$\mathbf{A} = \begin{bmatrix} 3 & -2 & 2 \\ 1 & 1 & 1 \\ -2 & 3 & -1 \end{bmatrix}, \quad \mathbf{B} = \begin{bmatrix} 2 & 4 & -8 \\ -1 & -2 & 4 \end{bmatrix}, \quad \mathbf{C} = \begin{bmatrix} 1 & -1 \\ 1 & 3 \\ -3 & 3 \\ -2 & 6 \end{bmatrix}$$

Note that rank($\mathbf{A}$)=2 since *row*-2 is obtained as *row*-1+*row*-3, i.e., *row*-1 and *row*-3 are the only rows that are linearly independent. On the other hand, rank($\mathbf{B}$)=1 because *row*-1= $(-2) \times$ *row*-2, i.e., there is only one linearly independent row. However, rank($\mathbf{C}$)=2 because *row*-3= $(-3) \times$ *row*-1 and *row*-4= $(-2) \times$ *row*-2, i.e., only *row*-1 and *row*-2 are the linearly independent rows.

2.1.3 OPERATIONS WITH MATRICES

2.1.3.1 *Transpose of a matrix*

A new matrix obtained by interchanging the rows and columns of the original matrix $\mathbf{A} = [a_{ij}]_{m \times n}$ is called the *transpose (matrix)* of $\mathbf{A}$, and it is expressed as

$$\mathbf{A}^T = \begin{bmatrix} a_{11} & a_{21} & a_{31} & \cdots & a_{n1} \\ a_{12} & a_{22} & a_{32} & \cdots & a_{n2} \\ a_{13} & a_{23} & a_{33} & \cdots & a_{n3} \\ \vdots & \vdots & \vdots & \ddots & \vdots \\ a_{1,m} & a_{2m} & a_{3m} & \cdots & a_{nm} \end{bmatrix}_{n \times m}$$

where the superscript T denotes the transpose operation.

Let $\mathbf{A}$ and $\mathbf{B}$ be matrices of the same (conformable) size and α be a non-zero scalar, then

- *Symmetry property:* $\mathbf{A}^T = \mathbf{A}$ if $\mathbf{A}$ is a square symmetric matrix;
- *Reflexive property:* $(\mathbf{A}^T)^T = \mathbf{A}$;
- *Addition property:* $(\mathbf{A} + \mathbf{B})^T = \mathbf{A}^T + \mathbf{B}^T = \mathbf{B}^T + \mathbf{A}^T$;
- *Exchange of order in multiplication:* $(\mathbf{AB})^T = \mathbf{B}^T \mathbf{A}^T$;
- *Scalar multiple property:* $(\alpha \mathbf{A})^T = \alpha \mathbf{A}^T$.

Also, a column vector is frequently denoted with the transpose notation to save vertical space in equations:

$$\mathbf{x} = \begin{bmatrix} x_1 & x_2 & x_3 & \cdots & x_n \end{bmatrix}^T$$

2.1.3.2 *Matrix addition/subtraction*

Addition, or *subtraction*, of the same-size matrices $\mathbf{A} = [a_{ij}]_{m \times n}$ and $\mathbf{B} = [b_{ij}]_{m \times n}$ is accomplished by adding or subtracting the corresponding elements of $\mathbf{A}$ and $\mathbf{B}$:

$$\mathbf{A} \pm \mathbf{B} = [a_{ij} \pm b_{ij}]_{m \times n}$$

```
                    Pseudocode 2.1

Module MAT_ADD (m, n, A, B, C)
\ DESCRIPTION: A pseudomodule for matrix addition, A + B = C.
Declare: a_mn, b_mn, c_mn              \ Declare matrices as array variables
For [i = 1, m]                         \ Loop i: Sweep from top row to bottom
   For [j = 1, n]                      \ Loop j: Sweep row from left to right
      c_ij ← a_ij + b_ij              \ Set a_ij + b_ij to c_ij
   End For
End For
End Module MAT_ADD
```

Let $\mathbf{A}$, $\mathbf{B}$, $\mathbf{C}$, and $\mathbf{O}$ (zero) be matrices of the same size, then

- *Commutative property:* $\mathbf{A} + \mathbf{B} = \mathbf{B} + \mathbf{A}$;
- *Associative property:* $\mathbf{A} + (\mathbf{B} + \mathbf{C}) = (\mathbf{A} + \mathbf{B}) + \mathbf{C}$;
- *Additive identity property:* $\mathbf{A} + \mathbf{O} = \mathbf{O} + \mathbf{A} = \mathbf{A}$;
- *Additive inverse property:* $\mathbf{A} + (-\mathbf{A}) = \mathbf{O}$;
- Matrix subtraction is *non-commutative* and *non-associative*, that is, $\mathbf{A} - \mathbf{B} \neq \mathbf{B} - \mathbf{A}$ and $\mathbf{A} - (\mathbf{B} - \mathbf{C}) \neq (\mathbf{A} - \mathbf{B}) - \mathbf{C}$.

A pseudomodule, MAT_ADD, for adding two matrices is presented in Pseudocode 2.1. The module requires two matrices ($\mathbf{A}$ and $\mathbf{B}$) and their size ($m \times n$) as input. The addition operation is carried out by summing up the corresponding elements of $\mathbf{A}$ and $\mathbf{B}$ ($c_{ij} \leftarrow a_{ij} + b_{ij}$) for every row and every column ($i = 1, 2, \ldots, m$, $j = 1, 2, \ldots, n$) using two For loops. The loop over j-variable sweeps all the elements in a row from left to right. The loop over the i variable sweeps rows from top to bottom. The same module can be adopted for matrix subtraction simply by replacing c_{ij} with $c_{ij} \leftarrow a_{ij} - b_{ij}$. Note that the inner or outer loops are interchangeable in this module.

2.1.3.3 Scalar multiplication

In multiplying a matrix $\mathbf{A} = [a_{ij}]_{m \times n}$ by a λ scalar number, the *scalar multiplication* $\lambda\mathbf{A}$ is obtained by multiplying each element of matrix $\mathbf{A}$ by λ as follows:

$$\lambda\mathbf{A} = \lambda[a_{ij}]_{m \times n} = [\lambda a_{ij}]_{m \times n} = \begin{bmatrix} \lambda a_{11} & \lambda a_{12} & \lambda a_{13} & \cdots & \lambda a_{1n} \\ \lambda a_{21} & \lambda a_{22} & \lambda a_{23} & \cdots & \lambda a_{2n} \\ \lambda a_{31} & \lambda a_{32} & \lambda a_{33} & \cdots & \lambda a_{3n} \\ \vdots & \vdots & \vdots & \ddots & \vdots \\ \lambda a_{m1} & \lambda a_{m2} & \lambda a_{m3} & \cdots & \lambda a_{mn} \end{bmatrix}$$

Let $\mathbf{A}$, $\mathbf{B}$, and $\mathbf{O}$ (zero) be matrices of the same size and α and β be non-zero scalar numbers, then

- *Associative property of multiplication:* $\alpha(\beta\mathbf{A}) = \beta(\alpha\mathbf{A}) = (\alpha\beta)\mathbf{A}$;
- *Distributive properties:* $\alpha(\mathbf{A} + \mathbf{B}) = \alpha\mathbf{A} + \alpha\mathbf{B}$, $(\alpha + \beta)\mathbf{A} = \alpha\mathbf{A} + \beta\mathbf{A}$;
- *Multiplicative identity property:* $1 \cdot \mathbf{A} = \mathbf{A}$;
- *Multiplicative properties of zero:* $0 \cdot \mathbf{A} = 0$ and $\alpha \cdot \mathbf{O} = 0$.

Pseudocode 2.2

Module SCALM $(m, n, \lambda, \mathbf{A}, \mathbf{B})$
\ DESCRIPTION: A pseudomodule to perform matrix addition, $\mathbf{A} + \mathbf{B} = \mathbf{C}$.
Declare: a_{mn}, b_{mn} \ Declare matrices as array variables
For $[i = 1, m]$ \ Loop i: Sweep from top row to bottom
   For $[j = 1, n]$ \ Lop j: Sweep row left to right
      $b_{ij} \leftarrow \lambda a_{ij}$ \ Assign λa_{ij} product to b_{ij}
   End For
End For
End Module SCALM

A pseudomodule, **SCALM**, performing a scalar multiplication is given in Pseudocode 2.2. As input, the module requires a matrix $(\mathbf{A})$, its size $(m \times n)$, and a real scalar number (λ). The output is the resultant matrix $\mathbf{B}$, whose elements, b_{ij}, are obtained by multiplying every element of $\mathbf{A}$ with the real-type input scalar λ; that is, $b_{ij} = \lambda a_{ij}$ for $i = 1, 2, \ldots, m$ and $j = 1, 2, \ldots, n$ for every row and column.

2.1.3.4 Inner or dot product

The *inner* or *dot product* of two vectors of the same size, $\mathbf{x} = [x_i]_{n \times 1}$ and $\mathbf{y} = [y_i]_{n \times 1}$, is denoted by $(\mathbf{x}, \mathbf{y})$ or $\mathbf{x}^T \mathbf{y}$ and is defined as

$$(\mathbf{x}, \mathbf{y}) = \mathbf{x}^T \mathbf{y} = x_1 y_1 + x_2 y_2 + \cdots + x_n y_n = \sum_{k=1}^{n} x_k y_k \qquad (2.5)$$

Let $\mathbf{x}$, $\mathbf{y}$, and $\mathbf{z}$ be real vectors and α and β be non-zero scalars, then

- *Positive-definiteness property:* $\mathbf{x} \cdot \mathbf{x} > 0$ for $\mathbf{x} \neq \mathbf{0}$ and $\mathbf{x} \cdot \mathbf{x} = 0$ *if and only* $\mathbf{x} = 0$;
- *Commutative property:* $\mathbf{x} \cdot \mathbf{y} = \mathbf{y} \cdot \mathbf{x}$;
- *Scalar multiplication property:* $(\alpha \mathbf{x}) \cdot (\beta \mathbf{y}) = (\alpha \mathbf{y}) \cdot (\beta \mathbf{x}) = \alpha\beta(\mathbf{x} \cdot \mathbf{y})$;
- *Distributive property:* $\mathbf{x} \cdot (\mathbf{y} + \mathbf{z}) = \mathbf{x} \cdot \mathbf{y} + \mathbf{x} \cdot \mathbf{z}$;
- $\mathbf{x} \cdot \mathbf{y} = 0$ when either the vectors are *orthogonal* or either $\mathbf{x} = 0$ or $\mathbf{y} = 0$.

Pseudocode 2.3

Function Module XdotY $(n, \mathbf{x}, \mathbf{y})$
\ DESCRIPTION: A pseudo-function to perform dot product, $\mathbf{x} \cdot \mathbf{y}$.
Declare: x_n, y_n \ Declare vectors as arrays of length n
$sums \leftarrow 0$ \ Initialize accumulator, sum
For $[k = 1, n]$ \ Loop k: accumulate $x_k * y_k$'s
   $sums \leftarrow sums + x_k * y_k$ \ Add $x_k * y_k$ to accumulator
End For
XdotY$\leftarrow sums$ \ Set accumulator to XdotY (dot product)
End Function Module XdotY

A pseudofunction module, XdotY, that calculates the inner product of two input vectors ($\mathbf{x}$ and $\mathbf{y}$) of the same size (n) is given in Pseudocode 2.3. The module requires a single accumulator to perform the dot product: Eq. (2.5). The accumulator variable, sum, is first initialized ($sum \leftarrow 0$), and the products of the corresponding vector elements, $x_k y_k$, are accumulated with a For-loop. The accumulated sum is then set to the real function name XdotY on exit.

2.1.3.5 Matrix multiplication

The *matrix multiplication*, $\mathbf{AB} = \mathbf{C}$, is defined only if the number of columns in $\mathbf{A}$ is equal to the number of rows in $\mathbf{B}$; that is, $\mathbf{A} = [a_{ij}]_{m \times p}$ and $\mathbf{B} = [b_{ij}]_{p \times n}$. These matrices are said to be "conformable" to multiplication. The product matrix of a multiplication of two matrices is represented by $\mathbf{C} = [c_{ij}]_{m \times n}$. An element of $\mathbf{C}$, c_{ij}, is found by multiplying corresponding elements of the ith row of $\mathbf{A}$ with the jth column of $\mathbf{B}$ and summing the p products as

$$c_{ij} = \sum_{k=1}^{p} a_{ik} b_{kj} \tag{2.6}$$

c_{ij} may also be regarded as the inner product of the ith row of $\mathbf{A}$ and the jth column of $\mathbf{B}$.

Let $\mathbf{A}$, $\mathbf{B}$, and $\mathbf{C}$ be matrices of conformable size, then

- *Non-commutative property:* $\mathbf{AB} \neq \mathbf{BA}$ even if $\mathbf{BA}$ is conformable;
- *Associative property:* $(\mathbf{AB})\mathbf{C} = \mathbf{A}(\mathbf{BC})$;
- *Distributive property:* $\mathbf{A}(\mathbf{B} + \mathbf{C}) = \mathbf{AB} + \mathbf{AC}$ or $(\mathbf{B} + \mathbf{C})\mathbf{A} = \mathbf{BA} + \mathbf{CA}$;
- *Multiplicative property of zero:* $\mathbf{A} \cdot 0 = 0 \cdot \mathbf{A} = 0$;
- *Multiplicative identity property:* $\mathbf{AI} = \mathbf{IA} = \mathbf{A}$.

A pseudomodule, MAT_MUL, performing multiplication of two conformable matrices, $\mathbf{C} = \mathbf{AB}$, is presented in Pseudocode 2.4. The module requires two matrices $\mathbf{A}$ and $\mathbf{B}$ of *conformable* sizes as input. The output is the product (resultant) matrix, $\mathbf{C} = [c_{ij}]_{m \times n}$. The inner product of the ith row of $\mathbf{A}$ (a_{i*}) and the jth column of $\mathbf{B}$ (b_{*j}) is calculated by Eq. (2.6) using an accumulator, i.e., the For-loop that runs over all k. The accumulated result for all (i, j)'s is stored on the accumulator variable c_{ij}, which represents the element of the resultant matrix at the ith row and the jth column.

2.1.3.6 Matrix-vector multiplication

Multiplication of a matrix $\mathbf{A} = [a_{ij}]_{m \times n}$ by a vector $\mathbf{x} = [x_i]_{n \times 1}$ gives a vector of length m. Multiplying a matrix by a vector may also be regarded as the "*dot product*" of $\mathbf{x}$ with every row of $\mathbf{A}$. The elements of the $\mathbf{b} = \mathbf{Ax}$ vector are shown explicitly as

$$\begin{bmatrix} a_{11} & a_{12} & a_{13} & \cdots & a_{1n} \\ a_{21} & a_{22} & a_{23} & \cdots & a_{2n} \\ a_{31} & a_{32} & a_{33} & \cdots & a_{3n} \\ \vdots & \vdots & \vdots & \ddots & \vdots \\ a_{m1} & a_{m2} & a_{m3} & \cdots & a_{mn} \end{bmatrix} \begin{bmatrix} x_1 \\ x_2 \\ x_3 \\ \vdots \\ x_n \end{bmatrix} = \begin{bmatrix} a_{11}x_1 + a_{12}x_2 + \cdots + a_{1n}x_n \\ a_{21}x_1 + a_{22}x_2 + \cdots + a_{2n}x_n \\ a_{31}x_1 + a_{32}x_2 + \cdots + a_{3n}x_n \\ \vdots \\ a_{m1}x_1 + a_{m2}x_2 + \cdots + a_{mn}x_n \end{bmatrix} = \begin{bmatrix} b_1 \\ b_2 \\ b_3 \\ \vdots \\ b_m \end{bmatrix}$$

The components of the product can be formulated as follows:

Pseudocode 2.4

Module MAT_MUL $(m, p, n, \mathbf{A}, \mathbf{B}, \mathbf{C})$
\ DESCRIPTION: A pseudomodule to perform matrix multiplication, $\mathbf{AB} = \mathbf{C}$.
Declare: a_{mp}, b_{pn}, c_{mn} \ Declare matrices as array variables
For $[i = 1, m]$ \ Loop i: Sweep from top row to bottom
  For $[j = 1, n]$ \ Loop j: Sweep row from left to right
    $c_{ij} \leftarrow 0$ \ Initialize accumulator, c_{ij}
    For $[k = 1, p]$ \ Loop k: Accumulator for $a_{ik} * b_{kj}$'s
      $c_{ij} \leftarrow c_{ij} + a_{ik} * b_{kj}$ \ Add $a_{ik} * b_{kj}$ to accumulator, c_{ij}
    End For
  End For
End For
End Module MAT_MUL

$$b_i = \sum_{k=1}^{n} a_{ik} x_k, \quad i = 1, 2, \ldots, m$$

In Pseudocode 2.5, a pseudomodule, Ax, performing $\mathbf{Ax}$ matrix-vector product is presented. The module requires a square matrix ($\mathbf{A}$), a vector ($\mathbf{x}$), and the size (n also denoting $n \times n$) as input. The output, $\mathbf{b}$, is a vector of length n. For all i's, the accumulator variable b_i is initialized and is used as an accumulator ($b_i \leftarrow b_i + a_{ik} x_k$) to compute $\sum_k a_{ik} x_k$ using the inner loop over the j index variable.

Pseudocode 2.5

Module Ax $(n, \mathbf{A}, \mathbf{x}, \mathbf{b})$
\ DESCRIPTION: A pseudomodule to perform matrix-vector multiplication, $\mathbf{Ax}$.
Declare: a_{nn}, x_n, b_n \ Declare matrix and vectors as array variables
For $[i = 1, n]$ \ Loop i: Sweep from top row to bottom
  $b_i \leftarrow 0$ \ Initialize the accumulator, b_i
  For $[k = 1, n]$ \ Loop k: Sweep row from left to right
    $b_i \leftarrow b_i + a_{ik} * x_k$ \ Add $a_{ik} * x_k$ to b_i
  End For
End For
End Module Ax

2.1.4 DETERMINANTS

A *determinant*, denoted by $\det(\mathbf{A})$ or $|\mathbf{A}|$, is a unique scalar number calculated by the elements of a square matrix $\mathbf{A}$. It provides useful information about the matrix, such as whether it is *invertible* or *not*. The determinant of a 2×2 matrix is defined as follows:

$$\begin{vmatrix} a_1 & a_2 \\ b_1 & b_2 \end{vmatrix} = a_1 b_2 - b_1 a_2,$$

Sarrus's rule is commonly used as a quick way of evaluating a 3×3 determinant. The first two columns are copied to the right side of the determinant, as shown below. Then the sum of the products of the three opposite diagonal elements (from south-west to north-east) is subtracted from the product of the main diagonal elements (from north-west to south-east).

$$\begin{vmatrix} a_1 & a_2 & a_3 \\ b_1 & b_2 & b_3 \\ c_1 & c_2 & c_3 \end{vmatrix} \begin{matrix} a_1 & a_2 \\ b_1 & b_2 \\ c_1 & c_2 \end{matrix} = a_1b_2c_3 + a_2b_3c_1 + a_3b_1c_2 - (a_3b_2c_1 + a_1b_3c_2 + a_2b_1c_3)$$

 Sarrus' rule is valid for matrices of size 2×2 and 3×3; it cannot be generalized to the determinants of matrices of size 4×4 or larger.

To generalize the determinant of a square matrix $\mathbf{A} = [a_{ij}]_{n \times n}$, we define *minor* (denoted by M_{ij}) associated with a_{ij}, which is the determinant of the $(n-1) \times (n-1)$ submatrix obtained by deleting the ith row and the jth column of matrix $\mathbf{A}$. The *cofactor* of a_{ij} is related to the minor by $C_{ij} = (-1)^{i+j} M_{ij}$. For example, for matrix $\mathbf{A} = [a_{ij}]_{3 \times 3}$, the minor and cofactor of a_{11} and a_{23} are obtained as follows:

$$M_{11} = \begin{vmatrix} a_{11} & a_{12} & a_{13} \\ a_{21} & a_{22} & a_{23} \\ a_{31} & a_{32} & a_{33} \end{vmatrix} = \begin{vmatrix} a_{22} & a_{23} \\ a_{32} & a_{33} \end{vmatrix} = a_{22}a_{33} - a_{32}a_{23}, \quad C_{11} = (-1)^{1+1} M_{11} = M_{11}$$

$$M_{23} = \begin{vmatrix} a_{11} & a_{12} & a_{13} \\ a_{21} & a_{22} & a_{23} \\ a_{31} & a_{32} & a_{33} \end{vmatrix} = \begin{vmatrix} a_{11} & a_{12} \\ a_{31} & a_{32} \end{vmatrix} = a_{11}a_{32} - a_{31}a_{12}, \quad C_{23} = (-1)^{2+3} M_{23} = -M_{23}$$

Note that the 2×2 determinant obtained after deleting the first row and first column of the matrix gives M_{11}, while in the latter case, the one obtained after deleting the second row and third column gives M_{23}.

In general, the determinant of a square matrix $\mathbf{A}$ is calculated by the *cofactor expansion method* along any row or any column as follows:

$$\det(\mathbf{A}) = a_{i1}C_{i1} + a_{i2}C_{i2} + \cdots + a_{in}C_{in} = \sum_{j=1}^{n} a_{ij}C_{ij} \quad \text{for any } i \tag{2.7}$$

or

$$\det(\mathbf{A}) = a_{1j}C_{1j} + a_{2j}C_{2j} + \cdots + a_{nj}C_{nj} = \sum_{i=1}^{n} a_{ij}C_{ij} \quad \text{for any } j \tag{2.8}$$

Properties of Determinants: Let $\mathbf{A}$ and $\mathbf{B}$ be $n \times n$ square matrices and λ be a non-zero scalar number, then

- *Reflection property:* $\det(\mathbf{A}^T) = \det(\mathbf{A})$;
- *Proportionality property:* $\det(\mathbf{A}) = 0$ if all of the elements in any row (or column) are proportional by λ to the elements in another row (or column);
- *Switching property:* If any two rows (or columns) of a determinant are interchanged, then its sign changes;
- *Scalar multiple property:* If any row (or column) of a determinant is multiplied by λ, then the value of the new determinant becomes

$\lambda \cdot \det(\mathbf{A})$; likewise, if a matrix $\mathbf{A}$ is multiplied by λ, then $\det(\lambda \mathbf{A}) = \lambda^n \det(\mathbf{A})$;

- *Invariance property:* A determinant remains unaltered if any row (or column) is replaced with a linear combination of two or more rows (or columns);
- *Triangular matrix property:* The determinant of an upper-($\mathbf{U}$) or lower-triangular ($\mathbf{L}$) or diagonal ($\mathbf{D}$) matrix is equal to the product of the main diagonal elements, i.e., $\det(\mathbf{U}) = u_{11}u_{22}\cdots u_{nn}$, $\det(\mathbf{L}) = \ell_{11}\ell_{22}\cdots\ell_{nn}$, and $\det(\mathbf{D}) = d_{11}d_{22}\cdots d_{nn}$;
- *Product property:* $\det(\mathbf{AB}) = \det(\mathbf{A})\det(\mathbf{B})$;
- *Inverse matrix property:* $\det(\mathbf{A}^{-1}) = 1/\det(\mathbf{A})$.

2.1.5 SYSTEM OF LINEAR EQUATIONS

General definition. A *system of linear equations*, or *linear system*, is a collection of m equations with n unknowns of the form:

$$\begin{aligned}
a_{11}x_1 + a_{12}x_2 + a_{13}x_3 + \cdots + a_{1n}x_n &= b_1 \\
a_{21}x_1 + a_{22}x_2 + a_{23}x_3 + \cdots + a_{2n}x_n &= b_2 \\
a_{31}x_1 + a_{32}x_2 + a_{33}x_3 + \cdots + a_{3n}x_n &= b_3 \\
\vdots \qquad\qquad \ddots \qquad\qquad \vdots \\
a_{m1}x_1 + a_{m2}x_2 + a_{m3}x_3 + \cdots + a_{mn}x_n &= b_m
\end{aligned} \tag{2.9}$$

For convenience, the system can be expressed in matrix form as

$$\mathbf{Ax} = \mathbf{b} \tag{2.10}$$

where a_{ij} and b_i are the known coefficients of matrix $\mathbf{A}$ and the rhs vector $\mathbf{b}$, respectively, and $\mathbf{x}$ is a vector of unknown coefficients expressed explicitly as

$$\mathbf{A} = \begin{bmatrix} a_{11} & a_{12} & a_{13} & \cdots & a_{1n} \\ a_{21} & a_{22} & a_{23} & \cdots & a_{2n} \\ a_{31} & a_{32} & a_{33} & \cdots & a_{3n} \\ \vdots & \vdots & \vdots & \ddots & \vdots \\ a_{m1} & a_{m2} & a_{m3} & \cdots & a_{mn} \end{bmatrix}, \quad \mathbf{x} = \begin{bmatrix} x_1 \\ x_2 \\ x_3 \\ \vdots \\ x_n \end{bmatrix}, \quad \mathbf{b} = \begin{bmatrix} b_1 \\ b_2 \\ b_3 \\ \vdots \\ b_m \end{bmatrix}$$

A linear system of n equations with n unknowns (i.e., a consistent set of equations) has a *unique solution* if it is nonsingular (i.e., $\det(\mathbf{A}) \neq 0$). A system whose number of equations is smaller than the number of unknowns ($m < n$) is said to be an *underdetermined* system. But if the number of equations is greater than that of the unknowns ($m > n$), then it is called an *over-determined* system. The solution of a linear system with m equations and n unknowns is an ordered sequence $(s_1, s_2, \ldots, s_n)$ in which each equation is satisfied for $x_1 = s_1$, $x_2 = s_2$, ..., $x_n = s_n$. The *general solution*, or the *solution set*, is the set of all possible solutions. In most practical problems, however, the number of unknowns is the same as the number of equations.

A system of linear equations is called *homogeneous* if the right-hand side of each and every equation is zero, i.e., $\mathbf{Ax} = \mathbf{0}$. When the solution of a linear system is $\mathbf{x} = \mathbf{0}$, then it is termed the *trivial solution*. A solution with at least one non-zero value is referred to as a *non-trivial solution*.

Tridiagonal systems. A system of linear equations with a tridiagonal coefficient matrix is encountered in many aspects of numerical analysis. A *tridiagonal system* is a special case of a banded system (i.e., $m_L = m_U = 1$, $n_b = 3$) with nonzero elements on the main diagonal, the first diagonal above, the first diagonal below, and the right-hand-side vector. A typical triangular n by n system is expressed as follows:

$$
\begin{bmatrix}
d_1 & a_1 & & & & \\
b_2 & d_2 & a_2 & & & \\
& b_3 & d_3 & a_3 & & \\
& & \ddots & \ddots & \ddots & \\
& & & b_{n-1} & d_{n-1} & a_{n-1} \\
& & & & b_n & d_n
\end{bmatrix}
\begin{bmatrix}
x_1 \\ x_2 \\ x_3 \\ \vdots \\ x_{n-1} \\ x_n
\end{bmatrix}
=
\begin{bmatrix}
c_1 \\ c_2 \\ c_3 \\ \vdots \\ c_{n-1} \\ c_n
\end{bmatrix}
\tag{2.11}
$$

By taking advantage of the sparsity of the tridiagonal matrix, a significant reduction in both memory and computational cost is achieved by defining the system with four arrays: b_i, d_i, a_i, and c_i for all i's.

Cramer rule. One of the direct methods that may be applied to solve a system of n linear equations with n unknowns is the *Cramer's rule*. Consider the linear system given in Eq. (2.9), where matrix $\mathbf{A}$ is invertible. The matrix $\mathbf{A}_i$ is constructed by replacing the ith column of $\mathbf{A}$ with $\mathbf{b}$. If the unique solution of a matrix equation is $\mathbf{x} = [x_1, x_2, \dots, x_n]^T$, then its solution for x_i is found from

$$
x_i = \frac{\det(\mathbf{A}_i)}{\det(\mathbf{A})}, \quad \text{for } i = 1, 2, \dots, n
$$

where

$$
\det(\mathbf{A}) =
\begin{vmatrix}
a_{11} & a_{12} & a_{13} & \cdots & a_{1n} \\
a_{21} & a_{22} & a_{23} & \cdots & a_{2n} \\
a_{31} & a_{32} & a_{33} & \cdots & a_{33} \\
\vdots & \vdots & \vdots & \ddots & \vdots \\
a_{n1} & a_{n2} & a_{n3} & \cdots & a_{nn}
\end{vmatrix}, \quad
\det(\mathbf{A}_i) =
\begin{vmatrix}
a_{11} & a_{12} & \cdots & b_1 & \cdots & a_n \\
a_{21} & a_{22} & \cdots & b_2 & \cdots & a_{2n} \\
a_{31} & a_{32} & \cdots & b_3 & \cdots & a_{3n} \\
\vdots & \vdots & \ddots & \vdots & \ddots & \vdots \\
a_{n1} & a_{n2} & \cdots & b_n & \cdots & a_{nn}
\end{vmatrix}
$$

Cramer's rule, based on computing multiple determinants, is a highly inefficient method. The memory requirement for a system of $n \times n$ linear equations is on the order of $\mathcal{O}(n^3)$, and the number of arithmetic operations is on the order of $\mathcal{O}\left((n+1)!\right)$. For this reason, it is impractical to apply to large linear systems as it requires extensive memory and cpu-time.

2.1.6 EIGENVALUES AND EIGENVECTORS

Eigenvalues, also known as characteristic values, are a set of unique scalar values associated with a set of linear equations, i.e., a matrix equation. The formal definition of a matrix eigenvalue problem is given as

$$
\mathbf{A}\mathbf{x} = \lambda\mathbf{x} \tag{2.12}
$$

where $\mathbf{A}$ is a $n \times n$ square matrix, $\mathbf{x}$ is a vector of length n, and λ is a scalar constant.

Solving an *eigenvalue problem* means finding scalar the λ values and associated $\mathbf{x}$ vectors satisfying Eq. (2.12) which may also be recast as follows:

$$
(\mathbf{A} - \lambda\mathbf{I})\mathbf{x} = \mathbf{0} \tag{2.13}
$$

In order for this homogeneous system of linear equations to have a non-trivial solution ($\mathbf{x} \neq \mathbf{0}$), the system must satisfy the following condition:

$$\det(\mathbf{A} - \lambda\mathbf{I}) = p_n(\lambda) = 0 \tag{2.14}$$

where $p_n(\lambda)$ is an nth-degree polynomial referred to as a *characteristic polynomial*, and λ denotes the set of unique scalars called the *eigenvalues*, i.e., the roots of the characteristic polynomial, λ_i for $i = 1, 2, \ldots, n$. Accordingly, it can be said that λ is an eigenvalue *if and only if* it satisfies $p_n(\lambda) = 0$. Note that the characteristic equation yields n eigenvalues (real or imaginary) that are unique to the matrix $\mathbf{A}$. A unique solution of Eq. (2.12) for a specified λ_i is denoted by $\mathbf{x}_i$ and referred to as the corresponding or *associated eigenvector*. Each eigenvalue is paired with its associated eigenvector, also frequently referred to as an *eigenpair*.

EXAMPLE 2.1: Finding eigenvalues and normalized eigenvectors

Find the eigenvalues and eigenvectors of the following matrix.

$$\mathbf{A} = \begin{bmatrix} -9 & 4 & 8 \\ -11 & 10 & 4 \\ -9 & 8 & 4 \end{bmatrix}$$

SOLUTION:

All eigenvalues and eigenvectors satisfy Eq. (2.13), so we set

$$\det(\mathbf{A} - \lambda\mathbf{I}) = \begin{vmatrix} -9 - \lambda & 4 & 8 \\ -11 & 10 - \lambda & 4 \\ -9 & 8 & 4 - \lambda \end{vmatrix} = 0$$

which results in the following characteristic polynomial:

$$p_3(\lambda) = \lambda^3 - 5\lambda^2 - 2\lambda + 24 = 0$$

All of the roots (i.e., eigenvalues) of the characteristic polynomial are real and distinct: $\lambda_1 = -2$, $\lambda_2 = 3$, and $\lambda_3 = 4$. An eigenvector ($\mathbf{x}_k$) associated with λ_k is the solution of the corresponding matrix equation $(\mathbf{A} - \lambda_k\mathbf{I})\mathbf{x}_k = \mathbf{0}$, which takes the following explicit form:

$$\begin{bmatrix} -9 - \lambda & 4 & 8 \\ -11 & 10 - \lambda & 4 \\ -9 & 8 & 4 - \lambda \end{bmatrix} \begin{bmatrix} x_1 \\ x_2 \\ x_3 \end{bmatrix} = \begin{bmatrix} 0 \\ 0 \\ 0 \end{bmatrix}$$

For $\lambda_1 = -2$, the homogeneous system becomes

$$\begin{bmatrix} -7 & 4 & 8 \\ -11 & 12 & 4 \\ -9 & 8 & 6 \end{bmatrix} \begin{bmatrix} x_1 \\ x_2 \\ x_3 \end{bmatrix} = \begin{bmatrix} 0 \\ 0 \\ 0 \end{bmatrix}$$

Depending on the choice of x_1, x_2, or x_3, there are an infinite number of solution sets that are multiples of the same solution. As a result, for instance, arbitrarily letting $x_1 = c_1$ and solving for x_2 and x_3 yields the following associated eigenvector:

$$\mathbf{x}_1 = \begin{bmatrix} x_1 \\ x_2 \\ x_3 \end{bmatrix} = c_1 \begin{bmatrix} 1 \\ 3/4 \\ 1/2 \end{bmatrix}$$

Next, for $\lambda_2 = 3$, the homogeneous system takes the form

$$\begin{bmatrix} -12 & 4 & 8 \\ -11 & 7 & 4 \\ -9 & 8 & 1 \end{bmatrix} \begin{bmatrix} x_1 \\ x_2 \\ x_3 \end{bmatrix} = \begin{bmatrix} 0 \\ 0 \\ 0 \end{bmatrix}$$

Similarly, employing the above procedure with $x_1 = c_2$, the solution of the homogeneous system for x_2 and x_3 results in the following eigenvector:

$$\mathbf{x}_2 = \begin{bmatrix} x_1 \\ x_2 \\ x_3 \end{bmatrix} = c_2 \begin{bmatrix} 1 \\ 1 \\ 1 \end{bmatrix}$$

Finally, for $\lambda_3 = 4$, we obtain

$$\begin{bmatrix} -13 & 4 & 8 \\ -11 & 6 & 4 \\ -9 & 8 & 0 \end{bmatrix} \begin{bmatrix} x_1 \\ x_2 \\ x_3 \end{bmatrix} = \begin{bmatrix} 0 \\ 0 \\ 0 \end{bmatrix}$$

Setting $x_1 = c_3$ and solving the homogeneous system for x_2 and x_3 yields its associated eigenvector:

$$\mathbf{x}_3 = \begin{bmatrix} x_1 \\ x_2 \\ x_3 \end{bmatrix} = c_3 \begin{bmatrix} 1 \\ 9/8 \\ 17/16 \end{bmatrix}$$

Eigenvectors are generally normalized to unit length (as $\mathbf{x_i}/\|\mathbf{x}_i\|$) and saved in matrix form as each eigenvector is placed in a column corresponding to the eigenvalues λ_1, λ_2, and λ_3, respectively. Consequently, the eigenvectors of this example can be expressed as follows:

$$\begin{bmatrix} \dfrac{4}{\sqrt{29}} & \dfrac{1}{\sqrt{3}} & \dfrac{16}{\sqrt{869}} \\ \dfrac{3}{\sqrt{29}} & \dfrac{1}{\sqrt{3}} & \dfrac{18}{\sqrt{869}} \\ \dfrac{2}{\sqrt{29}} & \dfrac{1}{\sqrt{3}} & \dfrac{17}{\sqrt{869}} \\ \dfrac{}{\sqrt{29}} & \dfrac{}{\sqrt{3}} & \dfrac{}{\sqrt{869}} \end{bmatrix}$$

where the first, second, and third columns correspond to the normalized associated eigenvectors of $\lambda_1 = -2$, $\lambda_2 = 3$ and $\lambda_3 = 4$, respectively.

Discussion: An $n \times n$ square matrix $\mathbf{A}$ has n independent and unique eigenvalues and eigenvectors, which are found by solving a homogeneous system: $(\mathbf{A} - \lambda\mathbf{I})\mathbf{x} = \mathbf{0}$. Normalizing an eigenvector to have a *unit length* gives the same vector regardless of the arbitrary choices of constants c_1, c_2, and so on.

Eigenpairs of a Tridiagonal *Toeplitz matrix*. In numerical analysis, the eigenvalues and eigenvectors of a tridiagonal matrix are frequently encountered. A common special form of a tridiagonal matrix is the diagonally-constant matrix (all the elements are constant along each diagonal) called the *Toeplitz matrix* (named after the German mathematician Otto Toeplitz). This kind of matrix is not confined to tridiagonal or symmetric matrices. A square Toeplitz matrix has at most $2n - 1$ unique values as opposed to n^2 in a general square matrix.

A tridiagonal Toeplitz has the following form:

$$
\mathbf{T} = \begin{bmatrix}
d & a & & & & & \\
b & d & a & & & & \\
& b & d & a & & & \\
& & \ddots & \ddots & \ddots & & \\
& & & b & d & a \\
& & & & b & d
\end{bmatrix}_{n \times n}
$$

where a, b, and c are real constants.

The exact eigenvalues of $\mathbf{T}$ are found by the following expression

$$
\lambda_k = d + 2\sqrt{ab}\cos\left(\frac{k\pi}{n+1}\right), \quad k = 1, 2, \ldots, n \tag{2.15}
$$

Additionally, the jth component of the associated eigenvector of the kth eigenvalue is obtained from

$$
v_j^{(k)} = \left(\frac{b}{a}\right)^{j/2} \sin\left(\frac{jk\pi}{n+1}\right), \quad j = k = 1, 2, \ldots, n \tag{2.16}
$$

Equations (2.15) and (2.16) are simple and very useful when determining the exact eigenpairs of some of the most common eigenvalue problems.

EXAMPLE 2.2: Application of eigenvalues and eigenvectors in physics

(1,3)-Butadiene (C_4H_6) is a conjugated diene, structurally joined together as two vinyl groups (CH_2) with a single bond. Hückel's approximation is applied to the π-orbitals of butadiene. Upon employing Schöringer's equation with simplifying assumptions, it leads to the so-called *secular equations* ($\mathbf{H\Psi} = \mathbf{E\Psi}$), where the Hamiltonian matrix ($\mathbf{H}$) and molecular orbital ($\mathbf{\Psi}$) are given as

$$
\mathbf{H} = \begin{bmatrix}
\alpha & \beta & & \\
\beta & \alpha & \beta & \\
& \beta & \alpha & \beta \\
& & \beta & \alpha
\end{bmatrix} \quad \text{and} \quad \mathbf{\Psi} = \begin{bmatrix}
\psi_1 \\
\psi_2 \\
\psi_3 \\
\psi_4
\end{bmatrix}
$$

with α and β being the Hückel parameters. Noting that every orbital energy (E_i, eigenvalue) corresponds to a molecular orbital ($\mathbf{\Psi_i}$, eigenvector), find the orbital energies and corresponding molecular orbitals for the butadiene.

SOLUTION:

The Hamiltonian matrix is a symmetric tridiagonal Toeplitz matrix and the eigensystem can be rearranged as a homogeneous system ($\mathbf{H\Psi} - \mathbf{E\Psi} = \mathbf{0}$) as

$$
\begin{bmatrix}
\alpha - E & \beta & & \\
\beta & \alpha - E & \beta & \\
& \beta & \alpha - E & \beta \\
& & \beta & \alpha - E
\end{bmatrix}
\begin{bmatrix}
\psi_1 \\
\psi_2 \\
\psi_3 \\
\psi_4
\end{bmatrix} =
\begin{bmatrix}
0 \\
0 \\
0 \\
0
\end{bmatrix}
$$

The parameter dependence (α and β) is eliminated by employing $(\alpha - E) = -x\beta$

substitution, yielding the following form:

$$
\begin{bmatrix}
0-x & 1 & & \\
1 & 0-x & 1 & \\
& 1 & 0-x & 1 \\
& & 1 & 0-x
\end{bmatrix}
\begin{bmatrix}
\psi_1 \\
\psi_2 \\
\psi_3 \\
\psi_4
\end{bmatrix}
=
\begin{bmatrix}
0 \\
0 \\
0 \\
0
\end{bmatrix}
$$

The eigenvalues of the 4×4 tridiagonal matrix are determined from Eq. (2.15) by setting $a = b = 1$ and $d = 0$. The resulting eigenvalues are obtained from $x_k = 2\cos(k\pi/5)$ for $k = 1, 2, 3$, and 4, which yield

$$
x_1 = \frac{1}{2}(1+\sqrt{5}), \quad x_2 = \frac{1}{2}(-1+\sqrt{5}), \quad x_3 = \frac{1}{2}(1-\sqrt{5}), \quad x_4 = -\frac{1}{2}(1+\sqrt{5}),
$$

Making use of the substitution $(\alpha - E) = -x\beta$, the orbital energies are found as

$$
E_{1,2} = \alpha + \frac{\beta}{2}(\pm 1 + \sqrt{5}), \qquad E_{3,4} = \alpha \pm \frac{\beta}{2}(1-\sqrt{5})
$$

For the eigenvectors, Eq. (2.16) yields $\psi_j^{(k)} = \sin(jk\pi/5)$, which for $j, k = 1, 2, 3$, and 4, rounding to four digits, results in

$$
\begin{bmatrix}
0.5877 & 0.9511 & 0.9511 & 0.5877 \\
0.9511 & 0.5877 & -0.5877 & -0.9511 \\
0.9511 & -0.5877 & -0.5877 & 0.9511 \\
0.5877 & -0.9511 & 0.9511 & -0.5877
\end{bmatrix}
$$

Discussion: The eigensystem gave four unique solutions. Columns 1 through 4 are the eigenvectors of Ψ_1 through Ψ_4, respectively.

2.1.7 VECTOR AND MATRIX NORMS

Vector norms. In any numerical computation, we are always concerned with knowing the magnitude of an approximation error. Moreover, when two vectors are frequently compared, it is more practical and efficient to measure the difference between the two vectors with a single scalar value rather than comparing the vectors element-by-element. A *vector norm*, denoted by $\|\mathbf{x}\|$, is a single positive scalar value representing the length, size, or magnitude of the vector, and it is well suited for vector comparisons.

The most commonly used vector norms belong to the family of p−norms (or ℓ_p−norms), which depend on their physical meaning. A general mathematical definition is given as

$$
\|\mathbf{x}\|_p = (x_1^p + x_2^p + \cdots + x_n^p)^{1/p} = \left(\sum_{i=1}^{n} |x_i|^p\right)^{1/p}
$$

where $\mathbf{x} = [x_1, x_2, \cdots, x_n]^T$. For $p = 1$, the ℓ_1−*norm* becomes

$$
\|\mathbf{x}\|_1 = \sum_{i=1}^{n} |x_i|
$$

which is the sum of the lengths of the vectors in a space, while for $p = 2$, the ℓ_2−*norm* (or also called the *Euclidean norm*) denotes the shortest distance between two points:

$$\|\mathbf{x}\|_2 = \left(\sum_{i=1}^{n} |x_i|^2 \right)^{1/2}$$

The $\ell_\infty-norm$ (or *infinity norm*), which gives the largest magnitude among each element of a vector (also called *maximum magnitude norm*), is mathematically expressed as

$$\|\mathbf{x}\|_\infty = \max_{1 \leqslant i \leqslant n} |x_i|$$

Properties of a vector norm: Let $\mathbf{x}$ and $\mathbf{y}$ be vectors of the same length and α be a non-zero scalar, then a vector norm satisfies the following conditions:

- *Positivity property:* $\|\mathbf{x}\| \geqslant 0$ and $\|\mathbf{x}\| = 0$ *if and only if* $\mathbf{x} = 0$;
- *Homogeneity property:* $\|\alpha\mathbf{x}\| = |\alpha|\|\mathbf{x}\|$;
- *Triangle inequality:* $\|\mathbf{x}\| + \|\mathbf{y}\| \leqslant \|\mathbf{x} - \mathbf{y}\| \leqslant \|\mathbf{x} + \mathbf{y}\|$.

Matrix norms. The *norms* of matrix $\mathbf{A}$ is denoted $\|\mathbf{A}\|$ and is similar to a vector norm in that it is a measure of the magnitude of the matrix. The two norms commonly encountered are $\ell_1- norm$ (*maximum column sum*) and $\ell_\infty-norm$ (*maximum row sum*):

$$\|\mathbf{A}\|_1 = \max_{1 \leqslant j \leqslant n} \sum_{i=1}^{n} |a_{ij}| \qquad \text{and} \qquad \|\mathbf{A}\|_\infty = \max_{1 \leqslant i \leqslant n} \sum_{j=1}^{n} |a_{ij}|$$

The ℓ_2-norm of a matrix $\mathbf{A}$ (also called *spectral norm*) is defined as

$$\|\mathbf{A}\|_2 = \sqrt{\lambda_{\max}}$$

where $\lambda_{\max}$ is the largest magnitude eigenvalue of the $\mathbf{A}^T\mathbf{A}$ product. If $\mathbf{A}$ is a symmetric matrix, $\|\mathbf{A}\|_2 = \mu_{\max}$ corresponds to the largest magnitude eigenvalue of $\mathbf{A}$. The $\|\mathbf{A}\|_2$ norm is always less than (or equal to) $\|\mathbf{A}\|_1$ and $\|\mathbf{A}\|_\infty$.

The *Frobenius norm*, or *Euclidean norm* (or ℓ_F-norm), is defined as the square root of the sum of the squares of the elements of a matrix. For a matrix $\mathbf{A} = [a_{ij}]_{n \times m}$, it is mathematically expressed as follows:

$$\|\mathbf{A}\|_F = \left(\sum_{i=1}^{n} \sum_{j=1}^{m} a_{ij}^2 \right)^{1/2}$$

Properties of matrix norms: Let $\mathbf{A}$ and $\mathbf{B}$ be matrices of the same size and α be a non-zero scalar, then matrix norms satisfy the following properties:

- *Positivity property:* $\|\mathbf{A}\| \geqslant 0$ and $\|\mathbf{A}\| = 0$ *if and only if* $\mathbf{A} = \mathbf{O}$;
- *Homogeneity property:* $\|\alpha\mathbf{A}\| = |\alpha|\|\mathbf{A}\|$;
- *Triangle inequality:* $\|\mathbf{A} + \mathbf{B}\| \leqslant \|\mathbf{A}\| + \|\mathbf{B}\|$ or $\|\mathbf{A}\| - \|\mathbf{B}\| \leqslant \|\mathbf{A} - \mathbf{B}\|$

In addition to the above-given "required" properties, some also satisfy the following (additional) properties *not* required of all matrix norms:

- *Subordinance property:* $\|\mathbf{A}\mathbf{x}\| \leqslant \|\mathbf{A}\|\|\mathbf{x}\|$;
- *Submultiplicativity property:* $\|\mathbf{A}\mathbf{B}\| \leqslant \|\mathbf{A}\|\|\mathbf{B}\|$.

EXAMPLE 2.3: Norms of a vector and a matrix

For the given matrix $\mathbf{A}$ and vector $\mathbf{x}$, calculate all possible norms.

$$\mathbf{A} = \begin{bmatrix} 2 & 1 & -2 \\ -1 & -3 & 3 \\ 3 & 4 & 2 \end{bmatrix}, \qquad \mathbf{x} = \begin{bmatrix} -2 \\ 1 \\ 3 \end{bmatrix}$$

SOLUTION:

For $\mathbf{x}$, the norms are found as

$$\|\mathbf{x}\|_1 = |-2| + |1| + |3| = 6, \quad \|\mathbf{x}\|_2 = \left((-2)^2 + 1^2 + 3^2 \right)^{1/2} = \sqrt{14}, \quad \|\mathbf{x}\|_\infty = |3| = 3$$

For the matrix $\mathbf{A}$, the norms are determined as follows:

$$\|\mathbf{A}\|_1 = \max_j \left(|2| + |-1| + |3|, \ |1| + |-3| + |4|, \ |-2| + |3| + |2| \right) = \max(6, 8, 7) = 8$$

$$\|\mathbf{A}\|_\infty = \max_i \left(|2| + |1| + |-2|, \ |-1| + |-3| + |3|, \ |3| + |4| + |2| \right) = \max(5, 7, 9) = 9$$

$$\|\mathbf{A}\|_F = \left(2^2 + (-1)^2 + 2^2 + (-1)^2 + (-3)^2 + 3^2 + 3^2 + 4^2 + 2^2 \right)^{1/2} = \sqrt{57}$$

In order to find $\|\mathbf{A}\|_2$, the maximum magnitude eigenvalue of the product $\mathbf{A}^T\mathbf{A}$ is required. The characteristic polynomial of the $\mathbf{A}^T\mathbf{A}$ is obtained as

$$\det(\mathbf{A}^T\mathbf{A} - \lambda\mathbf{I}) = \begin{vmatrix} 14-\lambda & 17 & -1 \\ 17 & 26-\lambda & -3 \\ -1 & -3 & 17-\lambda \end{vmatrix} = \lambda^3 - 57\lambda^2 + 745\lambda - 1225 = 0$$

The roots (i.e., eigenvalues) are found as $\lambda \approx 1.916$, 16.629, and 38.4552 yielding $\lambda_{max} = 38.4552$ and $\|\mathbf{A}\|_2 = \sqrt{38.4552} = 6.20122$.

Discussion: A norm gives a single unique scalar value for a vector or matrix, which is used for comparing or sorting vectors or matrices. Matrix norms are also used for sensitivity analysis and estimating the condition number of a matrix.

2.2 ELEMENTARY MATRIX OPERATIONS

Elementary matrix operations play an important role in linear algebra when inverting matrices or solving systems of linear equations. These operations are applied to augmented matrices either as row or column operations. Upon employing row (or column) operations, also referred to as *row-reduction* (or *column-reduction*), an augmented matrix is transformed into an *equivalent matrix*.

Before we proceed any further, let us introduce column and row notations, which we will extensively use in this text from this point on. The following 3×2 rectangular matrix has 3 rows and 2 columns; all rows and columns are represented as

$$\begin{bmatrix} 2 & 1 \\ 3 & -2 \\ 4 & 5 \end{bmatrix} \Rightarrow \begin{matrix} r_1 = [\ 2 \quad 1\ ] \\ r_2 = [\ 3 \quad -2\ ], \\ r_3 = [\ 4 \quad 5\ ] \end{matrix} \qquad c_1 = \begin{bmatrix} 2 \\ 3 \\ 4 \end{bmatrix}, \quad c_2 = \begin{bmatrix} 1 \\ -2 \\ 5 \end{bmatrix}$$

where r_i and c_j denote the ith row and jth column, respectively. These notations are particularly useful when developing algorithms and tracking the order of elementary operations.

As a reduction is applied to a specific row or column (r_i or c_i), it is wise to code the elementary operations with arrows ($\leftarrow$, $\uparrow$, $\leftrightarrow$) aligned with the modified row or column (r_i' or c_i').

There are basically three *elementary operations* or *transformations*: *scaling*, *pivoting*, and *elimination*. Now, we will discuss these elementary operations in greater detail.

1) **Scaling:** *The elements of a row (or a column) of a matrix can be multiplied by a nonzero real number* ($\lambda \neq 0$). For example, the notation $r_i \leftarrow \lambda r_i$ (or $c_i \leftarrow \lambda c_i$) indicates that the multiplication of the ith row (or ith column) by a scalar λ is overwritten on the ith row (or i'th column);

$$\begin{bmatrix} 2 & 1 \\ 3 & -2 \\ 4 & 5 \end{bmatrix} \sim \begin{bmatrix} \mathbf{6} & \mathbf{3} \\ 3 & -2 \\ 4 & 5 \end{bmatrix} \begin{matrix} \leftarrow 3r_1 \\ \\ \end{matrix}$$

$$\begin{bmatrix} 2 & 1 \\ 3 & -2 \\ 4 & 5 \end{bmatrix} \sim \begin{bmatrix} 2 & -2 \\ 3 & 4 \\ 4 & -10 \end{bmatrix}$$
$$\begin{matrix} \uparrow \\ -2c_2 \end{matrix}$$

Row-1 is multiplied by 3 in place; $r_1' \leftarrow 3r_1$.

Column-2 is multiplied by (-2) in place; $c_2' \leftarrow (-2)c_2$.

2) **Pivoting:** *Two rows (or two columns) of a matrix can be interchanged.* The notation $r_i \leftrightarrow r_j$ or $r_i \rightleftarrows r_j$ (or $c_i \rightleftarrows c_j$) indicates interchanging the ith row (or i'th column) with the jth row (or i'th column).

$$\begin{bmatrix} 2 & 1 \\ 3 & -2 \\ 4 & 5 \end{bmatrix} \sim \begin{bmatrix} \mathbf{4} & \mathbf{5} \\ 3 & -2 \\ \mathbf{2} & \mathbf{1} \end{bmatrix} \begin{matrix} r_1 \leftrightarrow r_3 \\ \\ \end{matrix}$$

$$\begin{bmatrix} 2 & 1 \\ 3 & -2 \\ 4 & 5 \end{bmatrix} \sim \begin{bmatrix} 1 & -2 \\ -2 & 3 \\ 5 & -4 \end{bmatrix}$$
$$\begin{matrix} c_1 \\ \updownarrow \\ c_2 \end{matrix}$$

Row-1 and Row-3 are interchanged

Col-1 and Col-2 are interchanged

3) **Elimination:** *A row (or column) of a matrix can be obtained by a linear combination of two rows (or two columns).* For given m and n, the notation $r_k \leftarrow mr_i + nr_j$ (or $c_k \leftarrow mc_i + nc_j$) denotes placing a linear combination of ith and jth rows (or columns) into the kth row (or column) or the row (or column) to which the arrow is points ($\leftarrow mr_i + nr_j$ or $\leftarrow mc_i + nc_j$).

$$\begin{bmatrix} 2 & 1 \\ 3 & -2 \\ 4 & 5 \end{bmatrix} \sim \begin{bmatrix} 2 & 1 \\ \mathbf{7} & \mathbf{0} \\ \mathbf{6} & \mathbf{0} \end{bmatrix} \begin{matrix} \\ \leftarrow r_2 + (2)r_1 \\ \leftarrow r_3 + (-5)r_1 \end{matrix}$$

$$\begin{bmatrix} 2 & 1 \\ 3 & -2 \\ 4 & 5 \end{bmatrix} \sim \begin{bmatrix} \mathbf{0} & 1 \\ \mathbf{7} & -2 \\ \mathbf{-6} & 5 \end{bmatrix}$$
$$\begin{matrix} \uparrow \\ c_1 - 2c_2 \end{matrix}$$

Row-1 times 2 is added to Row-2 in place, $r_2' \leftarrow r_2 + 2r_1$; Row-1 times (-5) is added to Row-3 in place, $r_3' \leftarrow r_3 + (-5)r_1$.

Col-2 times (-2) is added to Col-1 in place, $c_1' \leftarrow c_1 + (-2)c_1$.

When a matrix $\mathbf{A}$ is reduced to another matrix (say $\mathbf{B}$) by a series of row (or column) reductions, matrix $\mathbf{A}$ is said to be *row* (or *column*) *equivalent* to $\mathbf{B}$ and is denoted $\mathbf{A} \sim \mathbf{B}$. It is incorrect to use "=" or "$\Rightarrow$" instead of "$\sim$" for equivalent matrices.

In numerical algorithms, the elementary operations are applied to serve a specific purpose. As a divisor in the elimination process, it is important that the pivot (main diagonal) element should neither be "zero" nor have a value that is too small. A numerical procedure that requires numerous eliminations with pivot elements whose magnitudes are smaller than those of the other elements can result in significant rounding errors. *Scaling* is done mainly to select pivot elements, which helps prevent "underflow" or "overflow" of numbers and reduces susceptibility to round-off errors. On the other hand, even if the original matrix has no zeros on the principal diagonal, the elimination process can produce zeros on the diagonal. Therefore, rows (or equations) or columns (or variables in linear systems) can be modified so that the element with the largest magnitude is placed on the diagonal. This process is called "pivoting. Additionally, many algorithms require that the elements below or above the diagonal of a matrix, or both, be zero. The *elimination* procedure can be used to systematically eliminate elements above or below the main diagonal, either row by row, column by column, or both.

 How can a matrix be *equivalent* even though the original matrix is destroyed? Recall that a linear system of equations can be put into matrix form. (1) Scaling is simply multiplying both sides of an equation by the same scalar; (2) Pivoting by interchanging the two rows means changing the order of the equations; and (3) Elimination is simply a linear combination of the two equations that preserves the solution if the system has a common solution.

2.3 MATRIX INVERSION

We have already seen that matrix addition, subtraction, and multiplication are analogs to the same operations for numbers, and matrix division is not defined. To introduce the matrix analogy of division, consider the solution of $ax = b$. For $a \neq 0$, multiplying both sides of $ax = b$ by the multiplicative inverse of a (i.e., $a^{-1} = 1/a$), the solution is found as $x = a^{-1}b$ since $a^{-1}a = 1$. The concept of multiplicative inverse is applied to matrix algebra using the same reasoning.

Let $\mathbf{A}$ be an $n \times n$ square matrix. If there exists an $n \times n$ matrix $\mathbf{B}$ such that $\mathbf{AB} = \mathbf{BA} = \mathbf{I}_n$, where $\mathbf{I}_n$ is the identity matrix of order n, then the matrix $\mathbf{B}$ is the multiplicative *inverse* of $\mathbf{A}$. The inverse matrix is denoted by $\mathbf{A}^{-1}$, and left- or right-multiplying it with $\mathbf{A}$ gives the identity matrix, i.e., $\mathbf{AA}^{-1} = \mathbf{A}^{-1}\mathbf{A} = \mathbf{I}_n$. However, it should be noted that not every matrix necessarily has an inverse. Firstly, a matrix to be inverted must be a *square* matrix. Secondly, it must be *invertible* (or *non-singular*); that is, $|\mathbf{A}| \neq 0$.

The inverse of an invertible diagonal matrix $\mathbf{D} = [d_{ii}]_{n \times n}$ ($d_{ii} \neq 0$ for all i) represents the simplest case and corresponds to the reciprocal of the corresponding diagonal elements:

$$\mathbf{D}^{-1} = \begin{bmatrix} 1/d_{11} & 0 & \cdots & 0 \\ 0 & 1/d_{22} & \cdots & 0 \\ \vdots & \vdots & \ddots & \vdots \\ 0 & 0 & \cdots & 1/d_{nn} \end{bmatrix}$$

There are several ways in which a matrix can be inverted: the adjoint method, Gauss-Jordan, matrix partition, LU-decomposition, etc. In this section, much attention will be devoted to the *Gauss-Jordan Reduction* method due to its computational efficiency and low cpu cost in comparison to other existing methods.

Adjoint Method

- Fast and easy evaluation of the inverse of 2×2 or 3×3 matrices;
- Once the inverse of a matrix with variable elements is obtained, it can then be used repeatedly.

- For large n, the computational cost of matrix inversion is very expensive. The number of arithmetic operations is in the order of $(n+1)!$, and it is generally not recommended for matrices larger than 3×3.

Properties of the inverse: Let $\mathbf{A}$ and $\mathbf{B}$ (also $\mathbf{A}^T$, $\mathbf{A}^{-1}$, and $\mathbf{AB}$) be invertible (non-singular) matrices and α be a non-zero scalar, then

- *Uniqueness property:* The matrix $\mathbf{A}^{-1}$ is the unique matrix satisfying the condition $\mathbf{AA}^{-1} = \mathbf{A}^{-1}\mathbf{A} = \mathbf{I}_n$, where $\mathbf{I}_n$ is an $n \times n$ identity matrix;
- *Reflexive property:* $(\mathbf{A}^{-1})^{-1} = \mathbf{A}$;
- *Exchange of inverse/transpose operations:* $(\mathbf{A}^T)^{-1} = (\mathbf{A}^{-1})^T$;
- *Scalar multiple property:* $(\alpha\mathbf{A})^{-1} = \alpha^{-1}(\mathbf{A}^{-1})$;
- *Exchange of multiplication order:* $(\mathbf{AB})^{-1} = \mathbf{B}^{-1}\mathbf{A}^{-1}$.

2.3.1 ADJOINT METHOD

Consider an invertible matrix $\mathbf{A} = [a_{ij}]_{n \times n}$ and a matrix $\mathbf{C} = [C_{ij}]_{n \times n}$, where C_{ij} is the cofactor of a_{ij}. The *adjoint matrix* of $\mathbf{A}$ denoted $\mathrm{Adj}(\mathbf{A})$ is simply the transpose of the cofactor matrix $\mathbf{C}$, i.e., $\mathrm{Adj}(\mathbf{A}) = \mathbf{C}^T$. Then the inverse is determined by

$$\mathbf{A}^{-1} = \frac{\mathrm{Adj}(\mathbf{A})}{\det(\mathbf{A})}$$

For a 2×2 matrix, the inverse matrix yields

$$\mathbf{A}^{-1} = \begin{bmatrix} a & b \\ c & d \end{bmatrix}^{-1} = \frac{1}{\det(\mathbf{A})}\begin{bmatrix} C_{11} & C_{12} \\ C_{21} & C_{22} \end{bmatrix}^T = \frac{1}{\det(\mathbf{A})}\begin{bmatrix} d & -b \\ -c & a \end{bmatrix}$$

2.3.2 GAUSS-JORDAN METHOD

Gauss-Jordan elimination is a method that is used to find the inverse of an invertible matrix and, also, to find the solution of a linear system of equations. The method basically relies on elementary row operations. It is applied to a larger matrix called the *augmented matrix*, which (in the case of matrix inversion) is formed by adjoining matrix $\mathbf{A}$ and identity matrix $\mathbf{I}$, leading to a rectangular matrix $[\mathbf{A}|\mathbf{I}]$ of size $n \times 2n$:

$$[\mathbf{A}|\mathbf{I}] = \begin{bmatrix} a_{11} & a_{21} & \cdots & a_{1n} & 1 & 0 & \cdots & 0 \\ a_{21} & a_{22} & \cdots & a_{2n} & 0 & 1 & \cdots & 0 \\ \vdots & \vdots & \ddots & \vdots & \vdots & \vdots & \ddots & \vdots \\ a_{n1} & a_{n2} & \cdots & a_{nn} & 0 & 0 & \cdots & 1 \end{bmatrix}_{n \times 2n} \tag{2.17}$$

where the vertical line separates the two matrices.

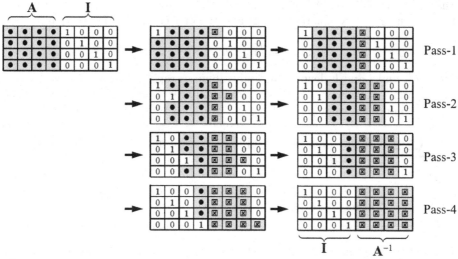

FIGURE 2.1: Illustration of the matrix inversion procedure • and ☒ denote the non-zero elements occupied in matrices **A** and **I**, respectively.

The objective is to shift the identity matrix to the left-hand side of Eq. (2.17) through a sequence of elementary row operations. The row operations, however, are carried out in a systematic manner, as illustrated in Fig. 2.1. After a series of row operations, the first pass should yield the first column of **I** in place of the first column of **A**. Note that while working on the first column of **A**, simultaneously the first column of **I** is destroyed. On moving to the next column, the element on the diagonal (the pivot) must first be normalized to "1." Then the elements above and below "1" are zeroed through proper row operations. With each pass, a column of matrix **A** is replaced sequentially from left to right by the corresponding column of the identity matrix. After n passes, **A** is transformed to **I**, and the identity matrix on the rhs is destroyed completely, leading to $[\mathbf{I}|\,\mathbf{B}]$, where **B** is now the *inverse matrix*, i.e., $\mathbf{B} = \mathbf{A}^{-1}$ and $[\mathbf{A}|\,\mathbf{I}] \to [\mathbf{I}|\,\mathbf{A}^{-1}]$.

Now we follow the step-by-step implementation of this algorithm in more detail. The first step is to *scale* (or *normalize*) the first row by multiplying with $p = 1/a_{11}$:

$$
\begin{bmatrix}
1 & a'_{12} & \cdots & a'_{1n} & b'_{11} & 0 & \cdots & 0 \\
a_{21} & a_{22} & \cdots & a_{2n} & 0 & 1 & \cdots & 0 \\
\vdots & \vdots & \ddots & \vdots & \vdots & \vdots & \ddots & \vdots \\
a_{n1} & a_{n2} & \cdots & a_{nn} & 0 & 0 & \cdots & 1
\end{bmatrix}
\begin{matrix} \leftarrow p\,r_1 \\ \\ \\ \\ \end{matrix}
\tag{2.18}
$$

where $r_i = [a_{i1} \ \dots \ a_{in} \ b_{i1} \ \dots \ b_{in}]$ denotes the ith row of the augmented matrix, b'_{11} and a'_{1j} denote the matrix element modified as a result of normalization, i.e., $b'_{11} = p$ and $a'_{1j} = p\,a_{1j}$ for $j = 1, 2, 3, \dots, n$.

Next, the elements below the first column pivot ($a_{21}, a_{31}, \dots, a_{n1}$) are eliminated by the following row operations: $r'_i \leftarrow r_i - a_{i1}r'_1$ ($i = 2, 3, \dots, n$). The first pass yields

$$
\begin{bmatrix}
1 & a'_{12} & \cdots & a'_{1n} & b'_{11} & 0 & \cdots & 0 \\
0 & a'_{22} & \cdots & a'_{2n} & b'_{21} & 1 & \cdots & 0 \\
\vdots & \vdots & \ddots & \vdots & \vdots & \vdots & \ddots & \vdots \\
0 & a'_{n2} & \cdots & a'_{nn} & b'_{n1} & 0 & \cdots & 1
\end{bmatrix}
\begin{matrix} \\ \leftarrow r_2 - a_{21}r'_1 \\ \vdots \\ \leftarrow r_n - a_{n1}r'_1 \end{matrix}
\tag{2.19}
$$

Note that the first column of $\mathbf{A}$ is now identical to $\mathbf{I}$. Meanwhile, the row operations naturally modify the elements of $\mathbf{A}$ (for rows $i \geqslant 2$) and the first column of $\mathbf{I}$ accordingly:

$$a'_{ik} = a_{ik} - s * a'_{1,k} \quad \text{and} \quad b'_{i1} = b_{i1} - s * b'_{11}, \qquad i, k = 2, 3, \ldots, n$$

where $s = a_{i1}$, and i and k denote row and column sweeps, respectively. Also, notice that, by iterating from $i = 2$ and $k = 2$, we do not actually carry out the computation that makes zero, which would be a waste of cpu-time.

> When a matrix is read into a program, make sure that the diagonal elements are not zero (or that the matrix is entered correctly) or that the matrix is invertible. In Pseudocode 2.6, the input matrix $\mathbf{A}$ is assumed to be invertible and possess non-zero diagonal elements ($a_{jj} \neq 0$).

The second pass begins by scaling the second row so that the pivot is $a'_{22} = 1$. The elimination step zeros out the elements above and below the pivot. Subsequently, with each pass, a column of $\mathbf{A}$ is replaced with the corresponding column of $\mathbf{I}$ for the rest of the columns ($j = 3, 4, \ldots, n$). In loop-j (column-loop), the normalizing multiplier ($p = 1/a_{jj}$) is prepared for scaling. The jth row is scaled with p; then all the elements in the jth row of the $\mathbf{A}$ and $\mathbf{I}$ matrices are modified using the inner loop over k variable according to

$$b_{jk}^{(j)} = p * b_{jk}^{(j-1)}, \quad \text{and} \quad a_{jk}^{(j)} = p * a_{jk}^{(j-1)}, \qquad \text{for} \quad k = 1, 2, \ldots, n$$

where we will now abandon the "prime notation" and instead use the superscript "(j)" to denote the number of modifications that have occurred, i.e., indicating that the matrix element has been overwritten j times.

Row reduction is applied with the loop-i, which sweeps the row from the pivot to the end of the row:

$$a_{ik}^{(j)} = a_{ik}^{(j-1)} - s * a_{jk}^{(j)} \quad \text{and} \quad b_{ik}^{(j)} = b_{ik}^{(j-1)} - s * b_{jk}^{(j)}, \qquad \text{for} \quad i \neq j, \ i, k = 1, 2, \ldots, n$$

where $s = a_{ij}^{(j-1)}$.

Finally, once all possible elementary operations for all columns are completed, the augmented matrix takes the form $[\mathbf{I}|\,\mathbf{B}]$:

$$\begin{bmatrix} 1 & 0 & \cdots & 0 & b_{11} & b_{12} & \cdots & b_{1n} \\ 0 & 1 & \cdots & 0 & b_{21} & b_{22} & \cdots & b_{2n} \\ \vdots & \vdots & \ddots & \vdots & \vdots & \vdots & \ddots & \vdots \\ 0 & 0 & \cdots & 1 & b_{n1} & b_{n2} & \cdots & b_{nn} \end{bmatrix} \tag{2.20}$$

where b_{ij}'s are now elements of the inverse matrix, i.e., $\mathbf{A}^{-1} = \mathbf{B}$.

> In Pseudocode 2.6, setting $p = 1/a_{jj}$ and $s = a_{ij}$ serves to speed up the computation (reduce cpu-time) by minimizing the search time to locate and retrieve an element of a matrix that would otherwise have been accessed numerous times. This is one of the important factors to take into account in programming when dealing with very large matrices.

```
                        Pseudocode 2.6

   Module INV_MAT (n, A, B)
   \ DESCRIPTION: A pseudomodule to find inverse of a matrix with Gauss-
   \   Jordan method.
   Declare: a_nn, b_nn                          \ Declare matrices as arrays
   B ← 0                                        \ Initialize B with zero matrix
   For [i = 1, n]                               \ Construct identity matrix
       b_ii ← 1                                 \ Set diagonals to "1"
   End For
   For [j = 1, n]                        \ Loop j: Sweep from top row to bottom
       p ← 1/a_jj                        \ Set normalizing multiplier, p = 1/a_jj
       For [k = 1, n]                     \ Loop k: Normalize pivot row (Row-j)
           a_jk ← p * a_jk                      \ Normalizing j^th row of A
           b_jk ← p * b_jk                      \ Normalizing j^th row of B
       End For
       For [i = 1, n]                   \ Loop i: Elimination steps on column j
           s ← a_ij                                    \ Save a_ij on s
           If [i ≠ j] Then                   \ Skip diagonal (pivot) element
               For [k = 1, n]                   \ Loop k: Apply eliminations
                   a_ik ← a_ik − s * a_jk            \ Modify i^th column of A
                   b_ik ← b_ik − s * b_jk            \ Modify i^th column of B
               End For
           End If
       End For
   End For
   \   The original matrix A is destroyed on Exit.
   End Module INV_MAT
```

The solution of a linear system of equations, Eq. (2.9), can be obtained by left-multiplying both sides of Eq. (2.10) with A^{-1} and applying $A^{-1}A = I$ identity:

$$x = A^{-1}b \tag{2.21}$$

In practice, finding the inverse matrix to solve a system of linear equations by Eq. (2.21) is not recommended. As a matter of fact, computing the inverse matrix requires far more arithmetic operations in comparison to finding the solution of the system with the Gauss elimination method (covered in Section 2.5). However, there are some cases where evaluating the inverse matrix offers some advantages.

A pseudomodule, INV_MAT, employing the Gauss-Jordan elimination method without pivoting to invert a matrix is presented in Pseudocode 2.6. As input, the module requires a matrix to be inverted (A) and its size n. On exit, the output matrix B is the inverse of A. Firstly, the identity matrix is setup: $b_{ii} = 1$ and $b_{ij} = 0$ in the code. The matrices A and B are simultaneously modified as a result of successive elementary row reductions to yield the inverse matrix. Loop$-j$ runs from 1 to n and sweeps the jth row from left (using the loop over k variable) to right of the augmented matrix after scaling by $p = 1/a_{jj}$: $r_{jk} \leftarrow p * r_{jk}$. Then the elements above and below the pivot (skipping $i = j$ with an If-construction) are eliminated inside the inner loop$-k$. Note that setting $s = a_{ij}$, which is used in eliminations, is to access the matrix element only once. During this process, the original matrix A is destroyed.

Matrix Inversion

- The method permits rescaling at each stage, which makes it less sensitive to round-off errors;
- Having an inverse matrix is useful when many sets of linear equations need to be solved with the same matrix $\mathbf{A}$ but a different $\mathbf{b}$ (i.e., $\mathbf{A}\mathbf{x}_k = \mathbf{b}_k$ for $k = 1, 2, \ldots$). Then, $\mathbf{A}^{-1}$ can be calculated once, and the solutions of the linear systems are obtained simply by matrix-vector multiplication—Eq. (2.21);
- Finding $\mathbf{A}^{-1}$ can offer invaluable information on the existence or severity of ill-conditioning (covered in Section 2.7).

- Elementary row operations as applied to the augmented matrix double the memory requirement and number of arithmetic operations;
- Arithmetic operation count is on the order of $\mathcal{O}(n^3)$; hence, for large matrices, cpu-time can get unacceptably large.

EXAMPLE 2.4: Inverse matrix by Gauss-Jordan Method

Given below are three planes that intersect each other at one point.

$$x - y - 2z + 2 = 0$$
$$2x - 3y - 5z + 5 = 0$$
$$-x + 3y + 5z - 5 = 0$$

These equations (planes) can be expressed by a system of linear equations, $\mathbf{A}\mathbf{x} = \mathbf{b}$, where

$$\mathbf{A} = \begin{bmatrix} 1 & -1 & -2 \\ 2 & -3 & -5 \\ -1 & 3 & 5 \end{bmatrix}, \qquad \mathbf{x} = \begin{bmatrix} x \\ y \\ z \end{bmatrix}, \qquad \mathbf{b} = \begin{bmatrix} -2 \\ -5 \\ 5 \end{bmatrix}$$

The intersection point can be found from $\mathbf{x} = \mathbf{A}^{-1}\mathbf{b}$. First employ the Gauss-Jordan method to obtain $\mathbf{A}^{-1}$, and then find the point of intersection, i.e., $\mathbf{x}$.

SOLUTION:

The augmented matrix is constructed as follows:

$$[\mathbf{A} \,|\, \mathbf{I}] = \begin{bmatrix} 1 & -1 & -2 & | & 1 & 0 & 0 \\ 2 & -3 & -5 & | & 0 & 1 & 0 \\ -1 & 3 & 5 & | & 0 & 0 & 1 \end{bmatrix}$$

In the first step, we skip the normalization procedure since $a_{11} = 1$ and proceed to eliminate the first column elements below a_{11}. The second and third-row elements in the first column are eliminated by the elementary row operations coded in line with the second and third rows:

$$\sim \begin{bmatrix} 1 & -1 & -2 & | & 1 & 0 & 0 \\ 0 & -1 & -1 & | & -2 & 1 & 0 \\ 0 & 2 & 3 & | & 1 & 0 & 1 \end{bmatrix} \begin{matrix} \\ \leftarrow r_2 - 2r_1 \\ \leftarrow r_3 + r_1 \end{matrix}$$

where we used "$\sim$" to indicate the matrix equivalence.

Next, the second row, r_2, is scaled to get $a_{22} = 1$ by multiplying r_2 with (-1), i.e., $r'_2 \leftarrow (-1)r_2$, where r'_2 indicates that all the elements of the second row have been modified.

$$\sim \left[\begin{array}{ccc|ccc} 1 & -1 & -2 & 1 & 0 & 0 \\ 0 & 1 & 1 & 2 & -1 & 0 \\ 0 & 2 & 3 & 1 & 0 & 1 \end{array}\right] \leftarrow (-1)r_2$$

The elements above and below the second column pivot (i.e., -1 and 2) are eliminated by adding $(-2)r_2$ to r_3 ($r'_3 \leftarrow r_3 - 2r_2$) and adding it to r_2 ($r'_3 \leftarrow r_3 + r_2$):

$$\sim \left[\begin{array}{ccc|ccc} 1 & 0 & -1 & 3 & -1 & 0 \\ 0 & 1 & 1 & 2 & -1 & 0 \\ 0 & 0 & 1 & -3 & 2 & 1 \end{array}\right] \begin{array}{l} \leftarrow r_1 + r_2 \\ \\ \leftarrow r_3 - 2r_2 \end{array}$$

In the last step, the third row need not be scaled since $a_{33} = 1$. The third column elements above the pivot (i.e., -1 and 1) are eliminated by $r'_1 \leftarrow r_1 + r_3$ (adding r_3 to r_1) and $r'_2 \leftarrow r_2 - r_3$ (adding $(-r_3)$ to r_2) row operations:

$$\sim \left[\begin{array}{ccc|ccc} 1 & 0 & 0 & 0 & 1 & 1 \\ 0 & 1 & 0 & 5 & -3 & -1 \\ 0 & 0 & 1 & -3 & 2 & 1 \end{array}\right] \begin{array}{l} \leftarrow r_1 + r_3 \\ \leftarrow r_2 - r_3 \\ \\ \end{array}$$

Notice that, at the end of the row reduction procedure, the modified matrix on the right-hand side became the inverse of $\mathbf{A}$, while the identity matrix was shifted to the left-hand side of the augmented matrix.

$$\mathbf{A}^{-1} = \left[\begin{array}{ccc} 0 & 1 & 1 \\ 5 & -3 & -1 \\ -3 & 2 & 1 \end{array}\right].$$

The intersection point can now be determined as

$$\mathbf{x} = \mathbf{A}^{-1}\mathbf{b} = \left[\begin{array}{ccc} 0 & 1 & 1 \\ 5 & -3 & -1 \\ -3 & 2 & 1 \end{array}\right]\left[\begin{array}{c} -2 \\ -5 \\ 5 \end{array}\right] = \left[\begin{array}{c} 0 \\ 0 \\ 1 \end{array}\right]$$

which means that the intersection point is (0,0,1).

Discussion: The Gauss-Jordan elimination procedure requires a series of elementary row operations to transform an original matrix into its inverse. The procedure can give true solutions if exact arithmetic can be employed. In this regard, symbolic algebra systems are able to render exact solutions. Once the inverse matrix is found, especially in hand calculations, it is always a good idea to verify whether the equality $\mathbf{A}^{-1}\mathbf{A} = \mathbf{I}$ is satisfied.

The inverse matrix method is one of the leading numerical methods used in structural engineering applications, signal processing in electrical engineering, control systems analysis, power systems, and more. Although the method is extremely useful and versatile, it has some limitations that are presented in the "Pros and Cons" box. Finding the inverse of a matrix using the adjoint method is also computationally very expensive and cpu-time intensive, especially for large matrices.

2.4 TRIANGULAR SYSTEMS OF LINEAR EQUATIONS

Systems of linear equations representing special cases in matrix algebra are expressed by $\mathbf{Lx} = \mathbf{b}$ and $\mathbf{Ux} = \mathbf{b}$, where $\mathbf{L}$ and $\mathbf{U}$ denote lower and upper triangular matrices, respectively. In such systems, one of the unknowns can be explicitly and easily found from the solution of either the first or last equation, which consists of a single unknown. The remaining unknowns are then found by marching in one direction (up or down) and substituting for the unknowns already found.

A *lower triangular system of equations*, $\mathbf{Lx} = \mathbf{b}$, has a lower triangular coefficient matrix:

$$\begin{bmatrix} \ell_{11} & 0 & 0 & \cdots & 0 \\ \ell_{21} & \ell_{22} & 0 & \cdots & 0 \\ \ell_{31} & \ell_{32} & \ell_{33} & \cdots & 0 \\ \vdots & \vdots & \vdots & \ddots & \vdots \\ \ell_{n1} & \ell_{n2} & \ell_{n3} & \cdots & \ell_{nn} \end{bmatrix} \begin{bmatrix} x_1 \\ x_2 \\ x_3 \\ \vdots \\ x_n \end{bmatrix} = \begin{bmatrix} b_1 \\ b_2 \\ b_3 \\ \vdots \\ b_n \end{bmatrix} \tag{2.22}$$

or explicitly

$$\begin{aligned} \ell_{11}x_1 &= b_1 \\ \ell_{21}x_1 + \ell_{22}x_2 &= b_2 \\ \ell_{31}x_1 + \ell_{32}x_2 + \ell_{33}x_3 &= b_3 \\ \vdots \quad \vdots \quad \vdots \quad \ddots \quad &\quad \vdots \\ \ell_{n1}x_1 + \ell_{n2}x_2 + \ell_{n3}x_3 + \cdots + \ell_{nn}x_n &= b_n \end{aligned} \tag{2.23}$$

The algorithm for finding the solution of a lower-triangular system of equations is termed *forward substitution*. Assuming $\mathbf{L}$ is not singular (i.e., $\ell_{ii} \neq 0$ for all i), the equations are visited in the forward direction from top to bottom, as shown in Fig. 2.2. First, x_1 is obtained from the first equation ($x_1 = b_1/\ell_{11}$), and it is placed in the memory space allocated for x_1. With x_1 secured, the second equation involving only x_1 and x_2 is used to find x_2: $x_2 = (b_2 - \ell_{21}x_1)/\ell_{22}$. Note that the value of x_1 is fetched from memory when executing the latter equation, i.e., mathematically speaking, the available solution of x_1 is substituted into the rhs. This procedure is repeated from the top to the bottom until all equations are reduced to and solved as a single equation with one unknown.

A general expression for x_k can be written from the kth row (or equation) as

$$\boxed{\ell_{k1}x_1 + \ell_{k2}x_2 + \cdots + \ell_{k,k-1}x_{k-1}} + \ell_{kk}x_k = b_k$$

where x_i's enclosed in the box contain all *"known solutions"* obtained in previous steps, and x_k, which is the only unknown, yields a generalized solution:

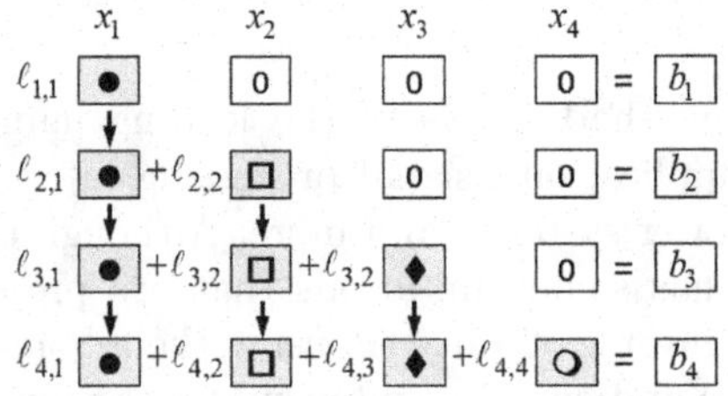

FIGURE 2.2: Illustration of the forward substitution procedure. Note: The boxes denote allocated memory for x_i's and b_i's, and symbols denote non-zero values.

```
                      Pseudocode 2.7

Module FORW_SUBSTITUTE (n,L,b, x)
\ DESCRIPTION: A pseudomodule to solve Lx = b by forward substitution.
Declare: ℓnn, xn, bn                          \ Declare array variables
x₁ ← b₁/ℓ₁₁                                    \ Solve x₁ from 1ˢᵗ equation
For [k = 2, n]                       \ Loop k: Sweep equations from top to bottom
     sums ← 0                            \ Initialize accumulator variable, sum
     For [j = 1, k − 1]          \ Loop j: Accumulate terms for j = 1, . . . , (k − 1)
          sums ← sums + ℓkj * xj               \ Add ℓkj * xj to accumulator
     End For
     xk ← (bk − sums)/ℓkk                      \ Solve xk from kᵗʰ equation
End For
End Module FORW_SUBSTITUTE
```

$$x_1 = \frac{b_1}{\ell_{11}}, \quad x_k = \frac{1}{\ell_{kk}} \left(b_k - \sum_{j=1}^{k-1} \ell_{kj} x_j \right), \quad k = 2, 3, \ldots, n \tag{2.24}$$

A pseudomodule, FORW_SUBSTITUTE, employing the forward substitution algorithm to solve a lower triangular system is given in Pseudocode 2.7. The module requires the number of equations (n), a lower triangular matrix ($\mathbf{L}$), and the rhs vector ($\mathbf{b}$) as input and places the solution on $\mathbf{x}$ as the output. Solving for x_1 is straightforward: $x_1 = b_1/\ell_{11}$. Solving x_k for $k = 2, 3, \ldots, n$ is carried out with the outer For-loop over the index variable k until all unknowns are found. The summation term in Eq. (2.24) is denoted by sum and obtained with the inner For-loop-j. Finally, the solution for the kth unknown is set to $x_k = (b_k - sum)/\ell_{kk}$.

An *upper triangular system*, $\mathbf{U}\mathbf{x} = \mathbf{b}$, has a coefficient matrix that is an upper triangular matrix:

$$\begin{bmatrix} u_{11} & u_{12} & u_{13} & \cdots & u_{1n} \\ 0 & u_{22} & u_{23} & \cdots & u_{2n} \\ 0 & 0 & u_{33} & \cdots & u_{3n} \\ \vdots & \vdots & \vdots & \ddots & \vdots \\ 0 & 0 & 0 & \cdots & u_{nn} \end{bmatrix} \begin{bmatrix} x_1 \\ x_2 \\ x_3 \\ \vdots \\ x_n \end{bmatrix} = \begin{bmatrix} b_1 \\ b_2 \\ b_3 \\ \vdots \\ b_n \end{bmatrix} \tag{2.25}$$

where $\mathbf{U}$ denotes an upper triangular matrix throughout the text.

The system is explicitly expressed as

$$u_{11}x_1 + u_{12}x_2 + u_{13}x_3 + \cdots + u_{1n}x_n = b_1$$
$$u_{22}x_2 + u_{23}x_3 + \cdots + u_{2n}x_n = b_2$$
$$u_{33}x_3 + \cdots + u_{3n}x_n = b_3 \tag{2.26}$$
$$\ddots \quad \vdots \quad \vdots$$
$$u_{nn}x_n = b_n$$

If the system is non-singular ($u_{ii} \neq 0$ for all i), the unknowns can be easily found by applying the *back substitution* algorithm. Unlike forward substitution, the algorithm starts from the last equation and progresses toward the first equation. Note that in this case, x_n is readily available from the nth equation: $x_n = b_n/u_{nn}$. Once this solution of

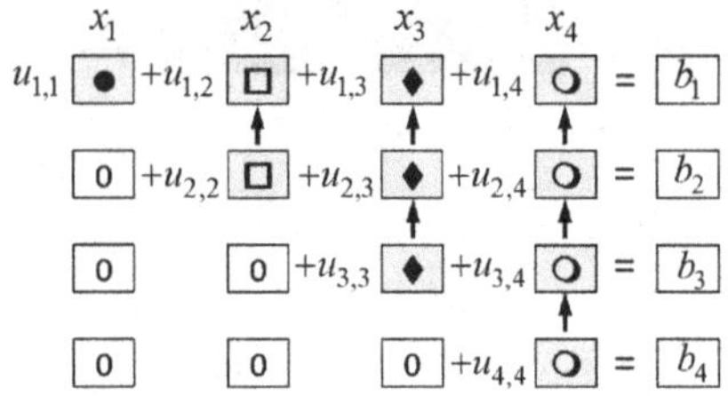

FIGURE 2.3: Illustration of the back substitution procedure. Note: The boxes denote allocated memory for x_i's and b_i's, and symbols denote non-zero values.

x_n is secured, it is substituted in the above equation to obtain x_{n-1}, yielding $x_{n-1} = (b_{n-1} - u_{n-1,n}x_n)/u_{n-1,n-1}$. The rest of the unknowns are obtained sequentially by sweeping the equations from the bottom row (equation) to the top and solving one equation with one unknown at each step, as illustrated in Fig. 2.3. The equation for the kth row is generalized as

$$u_{kk}x_k + \boxed{u_{k,k+1}x_{k+1} + u_{k,k+2}x_{k+2} + \cdots + u_{kn}x_n} = b_k$$

where the term enclosed in the box is "*known*" due to the solutions found in previous steps, and therefore the only unknown is x_k. Solving this for x_k, the generalized solution for the back substitution algorithm is obtained as

$$x_n = \frac{b_n}{u_{nn}}, \quad x_k = \frac{1}{u_{kk}}\left(b_k - \sum_{j=k+1}^{n} u_{kj}x_j\right), \quad k = (n-1), (n-2), \ldots, 2, 1 \qquad (2.27)$$

A pseudomodule, **BACK_SUBSTITUTE**, employing the back substitution algorithm to obtain the solution of an upper triangular system of linear equations is presented in Pseudocode 2.8. The module requires the number of equations (n), an upper triangular matrix (**U**), and the rhs vector (**b**) as input, and on exit, the solution is saved on **x**. x_n is easily found as $x_n = b_n/u_{nn}$. We obtain x_{n-1} after substituting x_n into the $(n-1)$th equation and solving for it. The x_k is obtained by back substituting all available x_i's ($i = n, (n-1), \ldots, k$) into the kth equation. Then the summation term in Eq. (2.27), denoted by sum, is calculated using an accumulator (**For**-loop over j). Finally, the solution for the kth unknown is obtained as $x_k = (b_k - sum)/\ell_{kk}$. This procedure is carried out sequentially using a **For** loop over k from $k = (n-1)$ until the first equation ($k = 1$) is reached.

Pseudocode 2.8

Module BACK_SUBSTITUTE (n,**U**,**b**, **x**)
\ DESCRIPTION: A pseudomodule to solve **Ux** = **b** by back substitution.
Declare: u_{nn}, x_n, b_n \ Declare array variables
$x_n \leftarrow b_n/u_{nn}$ \ Solve x_n from last equation
For $\left[k = (n-1), 1, (-1)\right]$ \ Loop k: Sweep equations from bottom to top
  $sums \leftarrow 0$ \ Initialize accumulator variable, $sums$
  For $\left[j = (k+1), n\right]$ \ Loop j: Accumulation terms for $j = (k+1), \ldots, n$
    $sums \leftarrow sums + u_{kj} * x_j$ \ Adde $u_{kj} * x_j$ to accumulator
  End For
  $x_k \leftarrow (b_k - sums)/u_{kk}$ \ Solve x_k from kth equation
End For
End Module BACK_SUBSTITUTE

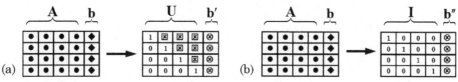

FIGURE 2.4: (a) Row echelon ($\mathbf{Ux} = \mathbf{b'}$), (b) reduced row echelon forms ($\mathbf{Ix} = \mathbf{b''}$). Symbols $\otimes$ and $\boxtimes$ denote the modified elements of $\mathbf{A}$ and $\mathbf{b}$, respectively, and any two identical symbols do not imply equality of numbers.

 Any linear system that has a unique solution can be transformed to an equivalent upper triangular system ($\mathbf{Ux} = \mathbf{b}$) through a set of row reductions. The purpose of such transformations is to make use of the simplicity and speed that the back substitution algorithm provides.

2.5 GAUSS ELIMINATION METHODS

Gauss elimination and its variants are the *direct methods* of solving a system of linear equations. The method is considered one of the most efficient and practical in this field. It requires a *row reduction* algorithm that consists of a sequence of elementary row operations applied to the *augmented matrix* formed by adjoining $\mathbf{A} = [a_{ij}]_{n \times n}$ and $\mathbf{b} = [b_i]_n$ as

$$[\mathbf{A} \,|\mathbf{b}]_{n \times (n+1)} = \begin{bmatrix} a_{11} & a_{12} & a_{13} & \cdots & a_{1n} & b_1 \\ a_{21} & a_{22} & a_{23} & \cdots & a_{2n} & b_2 \\ a_{31} & a_{32} & a_{33} & \cdots & a_{3n} & b_3 \\ \vdots & \vdots & \vdots & \ddots & \vdots & \vdots \\ a_{n1} & a_{n2} & a_{n3} & \cdots & a_{nn} & b_n \end{bmatrix}_{n \times (n+1)} \tag{2.28}$$

where the augmented matrix has n rows (each corresponding to an equation) and $(n + 1)$ columns. A vertical line separates the coefficient matrix from the rhs.

2.5.1 NAIVE GAUSS ELIMINATION

This technique consists of two phases: (1) *forward elimination*, (2) *back substitution*. The aim of the forward elimination step is to transform matrix $\mathbf{A}$ through a series of elementary row operations into an upper triangular matrix ($\mathbf{U}$). The rhs vector $\mathbf{b}$ is also simultaneously modified. This process results in an upper triangular system of equations: $\mathbf{Ux} = \mathbf{b'}$ (*see* Fig. 2.4a). Finally, the upper triangular system is solved using the back substitution algorithm (Section 2.4).

The method is identical to that of the Gauss-Jordan row reduction used to invert a matrix. The row reduction procedure is employed to create a system of equations in either $\mathbf{Ux} = \mathbf{b'}$ or $\mathbf{Ix} = \mathbf{b''}$ form (*see* Fig. 2.4). Each elementary row operation modifies the elements of both $\mathbf{A}$ and $\mathbf{b}$ as the algorithm is implemented. So, from this point forward, we will use $a^{(p)}$ superscript notation to indicate that an element has been modified p times.

The procedure begins with scaling the first row with a_{11} ($r_1^{(1)} \leftarrow r_1/a_{11}$), so all the elements in the first row are modified as $a_{1,m}^{(1)} \leftarrow val * a_{1,m}$ and $b_1^{(1)} \leftarrow val * b_1$ for $m = 1, 2, \ldots, n$, where $val = 1/a_{11}$.

$$
\begin{bmatrix}
1 & a_{12}^{(1)} & a_{13}^{(1)} & \cdots & a_{1n}^{(1)} & b_1^{(1)} \\
a_{21} & a_{22} & a_{23} & \cdots & a_{2n} & b_2 \\
a_{31} & a_{32} & a_{33} & \cdots & a_{3n} & b_3 \\
\vdots & \vdots & \vdots & \ddots & \vdots & \vdots \\
a_{n1} & a_{n2} & a_{n3} & \cdots & a_{nn} & b_n
\end{bmatrix}
\qquad r_1^{(1)} \leftarrow val * r_1
$$

Using the first row as a pivot, the elements in the first column below a_{11} (a_{i1}'s for $i = 2, 3, \ldots n$) are eliminated by $r_i^{(1)} \leftarrow r_i - s * r_1^{(1)}$ operations that yield:

$$
a_{ik}^{(1)} = a_{ik} - s * a_{1k}^{(1)} \quad \text{and} \quad b_i^{(1)} = b_i - s * b_1^{(1)}, \qquad i, k = 2, 3, \ldots, n
$$

where $s = a_{i1}$ serves to reduce cpu-time. After the first pass, the augment matrix becomes:

$$
\begin{bmatrix}
1 & a_{12}^{(1)} & a_{13}^{(1)} & \cdots & a_{1n}^{(1)} & b_1^{(1)} \\
0 & a_{22}^{(1)} & a_{23}^{(1)} & \cdots & a_{2n}^{(1)} & b_2^{(1)} \\
0 & a_{32}^{(1)} & a_{33}^{(1)} & \cdots & a_{3n}^{(1)} & b_3^{(1)} \\
\vdots & \vdots & \vdots & \ddots & \vdots & \vdots \\
0 & a_{n2}^{(1)} & a_{n3}^{(1)} & \cdots & a_{nn}^{(1)} & b_n^{(1)}
\end{bmatrix}
\qquad
\begin{array}{l}
r_2^{(1)} \leftarrow r_2 - a_{21} r_1^{(1)} \\
r_3^{(1)} \leftarrow r_3 - a_{31} r_1^{(1)} \\
\vdots \\
r_n^{(1)} \leftarrow r_n - a_{n1} r_1^{(1)}
\end{array}
$$

Note that row operations are encoded to the right of the augmented matrix, in the row where the elementary row operation is carried out.

 The scaling (normalization) procedure is *not* a requirement in Gauss elimination or Gauss-Jordan elimination methods; however, it *does* help minimize round-off errors.

The second pass also begins with scaling the second row: $r_2^{(2)} \leftarrow val * r_2^{(1)}$, where $val = 1/a_{22}^{(1)}$ (i.e., $a_{2k}^{(2)} \leftarrow val * a_{2k}^{(1)}$ and $b_2^{(2)} \leftarrow val * b_2^{(1)}$ for $k = 2, 3, \ldots, n$). Then, using the second row as the pivot, all elements in the second column below $a_{22}^{(1)}$ are eliminated by $r_i^{(2)} \leftarrow r_i^{(1)} - s * r_2^{(2)}$ for $i = 3, 4, \ldots, n$, which yield

$$
a_{ik}^{(2)} = a_{ik}^{(1)} - s * a_{2k}^{(2)} \quad \text{and} \quad b_i^{(2)} = b_i^{(1)} - s * b_2^{(2)}, \qquad \text{for} \quad i, k = 3, 4, \ldots, n
$$

where $s = a_{i2}^{(1)}$.

Upon completing the second pass, the augmented matrix becomes

$$
\begin{bmatrix}
1 & a_{12}^{(1)} & a_{13}^{(1)} & \cdots & a_{1n}^{(1)} & b_1^{(1)} \\
0 & 1 & a_{23}^{(2)} & \cdots & a_{2n}^{(2)} & b_2^{(2)} \\
0 & 0 & a_{33}^{(2)} & \cdots & a_{3n}^{(2)} & b_3^{(2)} \\
\vdots & \vdots & \vdots & \ddots & \vdots & \vdots \\
0 & 0 & a_{n3}^{(2)} & \cdots & a_{nn}^{(2)} & b_n^{(2)}
\end{bmatrix}
\qquad
\begin{array}{l}
r_2^{(2)} \leftarrow val * r_2^{(1)} \\
r_3^{(2)} \leftarrow r_3^{(1)} - a_{32}^{(1)} r_2^{(2)} \\
\vdots \\
r_n^{(2)} \leftarrow r_n^{(1)} - a_{n2}^{(1)} r_2^{(2)}
\end{array}
$$

The third, fourth, and so on passes are similarly carried out first by applying the normalization (scaling) and then elimination procedures, as illustrated earlier, which can be generalized as $r_i^{(j)} = r_i^{(j-1)} - r_k^{(j)} a_{ik}^{(j)}$ for an ith row, leading to

$$
a_{j,m}^{(j)} = val * a_{j,m}^{(j-1)}, \quad b_j^{(j)} = val * b_j^{(j-1)}, \quad val = 1/a_{jj}^{(j-1)}, \quad j, m = 1, 2, \ldots, n
$$
$$
a_{ik}^{(j)} = a_{ik}^{(j-1)} - s * a_{jk}^{(j)} \quad \text{and} \quad b_i^{(j)} = b_i^{(j-1)} - s * b_j^{(j)}, \quad s = a_{ij}^{(j-1)}, \quad i, k = (j+1), \ldots, n
$$

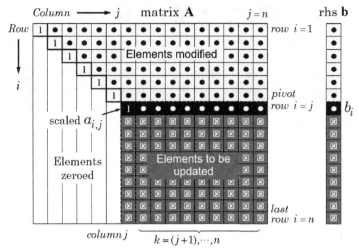

FIGURE 2.5: Illustration of the naive Gauss elimination procedure without pivoting (•
and ⊠ denote the modified (computed) and unmodified (to be modified) elements of **A**
and **b**, respectively). Any two identical symbols do not imply equality of numbers.

The forward elimination procedure is schematically illustrated in Fig. 2.5. Due to scaling, the principal diagonal elements of the processed rows consist of "1"s. The elements
below "1"s are zeroed during the forward elimination procedure, but in practice, these computations that should yield zero are not performed, which would otherwise be a waste of
cpu-time.

After n passes, the coefficient matrix is reduced to an upper triangular matrix:

$$
\left[
\begin{array}{ccccc|c}
1 & a_{12}^{(1)} & a_{13}^{(1)} & \cdots & a_{1n}^{(1)} & b_1^{(1)} \\
0 & 1 & a_{23}^{(2)} & \cdots & a_{2n}^{(2)} & b_2^{(2)} \\
0 & 0 & 1 & \cdots & a_{3n}^{(2)} & b_3^{(3)} \\
\vdots & \vdots & \vdots & \ddots & \vdots & \vdots \\
0 & 0 & 0 & \cdots & 1 & b_n^{(n)}
\end{array}
\right] \quad r_n^{(n)} \leftarrow val * r_n^{(n)}
$$

The procedure is completed when the back substitution algorithm is successfully applied
to the upper triangular matrix, as outlined in Section 2.4. Nevertheless, notice that this
algorithm assumed $|a_{ii}^{(i-1)}| > 0$, which does not give "*division by zero*" error, which is why
it is referred to as "naive Gauss Elimination." The method will clearly fail if, at least, one
of the diagonal values is zero ($|a_{ii}^{(i-1)}| = 0$). Having said that, it is possible to avoid this
problem by changing the order of the equations using a technique called pivoting, which is
also covered in Section 2.5.3.

The memory requirement and the operation count for a full system Gauss elimination
method are on the order of $\mathcal{O}(n^2)$ and $\mathcal{O}(n^3)$, respectively.

2.5.1.1 Application to underdetermined systems

A linear system is said to be underdetermined when $\mathbf{A} = [a_{ij}]_{m \times n}$ is a rectangular matrix, i.e., the number of equations is less than the number of unknowns ($m < n$). An
underdetermined linear system can have either no solution, one solution, or infinitely many
solutions. Consider, for example, a linear system consisting of one equation and two unknowns: $a + 2b = 8$, i.e., $m = 1$ and $n = 2$. This simple system has infinitely many solutions,

Gauss elimination Method

- It is a reliable method that gives the *true solution* if no round-off errors are made;
- It requires much less computation in comparison to finding $\mathbf{A}^{-1}$;
- The coefficient matrix need not be diagonally dominant;
- Pivoting is not needed if $\mathbf{A}$ is symmetric, positive-definite, or diagonally dominant;
- It is also applied to ill-conditioned systems;
- It may be used to compute the rank of a matrix;
- The determinant of a matrix can be found simultaneously while performing forward elimination step.

- This method is impractical for sparse or very large systems due to the propagation of rounding errors and excessive cpu-time;
- When no pivoting strategy is employed and a_{ii} becomes *zero* in at least one row, a "division by zero" will be encountered during the forward elimination step;
- When no pivoting strategy is employed and a_{ii} becomes very *small* ($a_{ii} \approx 0$) in at least one row, a "loss of significance" may be encountered;
- Every pass (solution step) depends on the previous pass, which makes the method is highly susceptible to round-off errors;
- The accuracy of the solution improves as the number of significant digits increases, but this does not completely eliminate the round-off errors.

which can be obtained by recasting it as $a = 8 - 2b$. We can then find a solution for a (i.e., *basic variable*) in terms of an arbitrary b (i.e., *free* or *independent variable*). In an underdetermined system, any m of the n unknowns can be selected as basic variables, but the choice of the basic variables is not usually evident. The Gauss elimination method (or preferably the Gauss-Jordan method, in Section 2.5.2) may be utilized to pick out the basic variables.

2.5.1.2 *Determining the rank of a matrix.*

If a linear system of equations is *homogeneous* (i.e., $\mathbf{Ax} = \mathbf{0}$) and also det($\mathbf{A}$)=0 (if the rank of $\mathbf{A}$ is less than n), the system $\mathbf{Ax} = \mathbf{0}$ will have r *basic* and $(n - r)$ *free variables*. In such a case, a non-trivial (non-zero) solution can be found. In other words, any set of r variables could then be expressed in terms of free variables. Like underdetermined systems, the Gauss-Jordan elimination method may be employed to obtain the solution in terms of free variables.

2.5.2 GAUSS-JORDAN ELIMINATION METHOD

The *Gauss-Jordan Elimination* method, which is similar to the Gaussian elimination, transforms the augmented matrix into the reduced row-echelon form (Fig. 2.4b). It requires, in addition to the Gauss elimination procedure, the elimination of the elements above the pivot (i.e., diagonal). The normalization of the pivot row is a requirement in this method, while in the naive Gauss elimination method it is not.

Gauss-Jordan Elimination Method

- Gauss-Jordan elimination is very stable when a pivoting technique is also employed;
- The algorithm can be applied to an augmented matrix ($[\mathbf{A}||\mathbf{I}]$) to find the inverse matrix of a matrix, $\mathbf{A}$.

- If a pivot element is very small ($a_{ii} \approx 0$), a "loss of significance" and "stability" issues associated with roundoffs may be encountered;
- A "division by zero" may be encountered, even if the system has a unique solution;
- The method requires more arithmetic operations (on the order of $\mathcal{O}(n^3)$ about 50% more) than Gauss elimination or LU decomposition methods.

In the Naive Gauss elimination algorithm presented in Section 2.5.1, the normalization of pivot elements was carried out as a standard part of the procedure. Therefore, it will be sufficient to zero out the elements above diagonal through elementary row operations. Nonetheless, there is no need to carry out the arithmetic operations with the elements above the diagonal to save cpu-time, as these will eventually become zero. The elimination procedure is only applied to the right-hand side vector, leading to

$$b_i^{(k)} = b_i^{(k-1)} - a_{ij}^{(k-1)} b_j^{(k)}, \qquad j = n, (n-2), \ldots, 2, \quad i = (j-1)(j-2), \ldots, 1 \qquad (2.29)$$

where k denotes the elimination step ($k > n$).

The final form of the system becomes

$$\begin{bmatrix} 1 & 0 & 0 & \cdots & 0 & \bigg| & b_1^{(n+k+1)} \\ 0 & 1 & 0 & \cdots & 0 & \bigg| & b_2^{(n+k)} \\ \vdots & \vdots & \ddots & \ddots & \vdots & \bigg| & b_3^{(3)} \\ 0 & 0 & \cdots & 1 & 0 & \bigg| & \vdots \\ 0 & 0 & \cdots & 0 & 1 & \bigg| & b_n^{(n)} \end{bmatrix} \qquad (2.30)$$

where the last column of the augmented matrix is the solution vector, $x_i = b_i$ for all i.

A pseudomodule, **LINEAR_SOLVE**, for solving a system of linear equations with the naive Gauss elimination ($opt = 0$) or Gauss-Jordan elimination ($opt \neq 0$) is presented in Pseudocode 2.9. The number of equations (n), coefficient matrix ($\mathbf{A}$), rhs vector ($\mathbf{b}$), and method flag (opt) are required as input. On exit, solution is stored on $\mathbf{x}$. The elimination procedure is carried out so long as $|a_{ij}| > 0$; otherwise, the procedure is terminated with a warning message. In this module, the normalization of the pivot row and elimination of elements below the pivot are carried out inside the loop$-j$ (outermost loop) in column-by-column order from left to right. The module requires a small number (ϵ), either internally defined or set to machine epsilon, to check whether the diagonal values of the matrix are zero. As a cpu saving measure, before scaling the jth row, the normalization factor is set to $val \leftarrow 1/a_{jj}$ once, and it is applied to scale the rhs ($b_j \leftarrow val * b_j$) and non-zero elements ($a_{jm} \leftarrow val * a_{jm}$ for $m = n, (n+1), \ldots, n$) of the pivot row. The loop$-i$ ranges down the jth column just below the diagonal element (from $i = j + 1$) all the way to the last row ($i = n$), where the pivot is assigned $s = a_{ij}$ and the elements are set to zero without performing the actual computations ($a_{ij} = 0$) to reduce cpu-time.

Pseudocode 2.9

Module LINEAR_SOLVE $(n,\mathbf{A}, \mathbf{b}, \mathbf{x}, opt)$
\\ DESCRIPTION: A pseudomodule to solve $\mathbf{Ax} = \mathbf{b}$ by Naive Gauss
\\   elimination or Gauss-Jordan method without *pivoting*.
\\ USES:
\\   BACK_SUBSTITUTE :: Module for back substitution, *see* Pseudocode 2.8;
\\   ABS :: A built-in function computing the absolute value.
Declare: a_{nn}, x_n, b_n \\ Declare array variables
$\varepsilon \leftarrow 10^{-12}$ \\ Define a small number or get Machine epsilon
For $\left[j = 1, n\right]$ \\ Loop j: Sweep row from left to right
    $ajj \leftarrow a_{jj}$ \\ Isolate the pivot element
    If $\left[|ajj| < \varepsilon\right]$ **Then**   \\ Check for possible "division by zero"
        Write: "Pivot row is zero ",j \\ Warning! Matrix may be singular
        Exit   \\ Exit the loop
    Else   \\ Case of non-zero pivot, $a_{ij} \neq 0$
        $val \leftarrow 1/ajj$ \\ Set normalization (scaling) factor
        $b_j \leftarrow val * b_j$ \\ Scale rhs of the pivot row
        For $\left[m = j, n\right]$ \\ Loop m: Skip zeros on the left, $m < j$
            $a_{jm} \leftarrow val * a_{jm}$ \\ Scaling pivot row of $m \geqslant j$
        End For
        For $\left[i = (j + 1), n\right]$ \\ Loop i: Eliminate elements below ith row
            $s \leftarrow a_{ij}$ \\ Save pivot on s
            $a_{ij} \leftarrow 0$ \\ Set 0 for zeroed element, no need to compute
            For $\left[k = (j + 1), n\right]$ \\ Loop k: Elimination for the ith row
                $a_{ik} \leftarrow a_{ik} - s * a_{jk}$ \\ Modify elements of kth row
            End For
            $b_i \leftarrow b_i - s * b_j$ \\ Modify the rhs of ith row
        End For
    End If
End For
If $\left[opt=0\right]$ **Then**   \\ $opt = 0$, Naive Gauss elimination solution
    BACK_SUBSTITUTE$(n,\mathbf{A}, \mathbf{b}, \mathbf{x})$ \\ Goto back substitution module
Else   \\ $opt \neq 0$, Eliminate above diagonal elements
    For $\left[j = n, 2, (-1)\right]$ \\ Loop j: Sweep rows right to left
        For $\left[i = (j - 1), 1, (-1)\right]$ \\ Loop i: Sweep from bottom to top
            $b_i \leftarrow b_i - a_{ij} * b_j$ \\ Perform operations only on rhs
        End For
        $x_j \leftarrow b_j$ \\ $\mathbf{x} = \mathbf{b}$ since $\mathbf{A}$ is now $\mathbf{I}$
    End For
    $x_1 \leftarrow b_1$ \\ Since $a_{11} = 1$ and $a_{1j} = 0$
End If
End Module LINEAR_SOLVE

The inner loop over k-index variable ranges across the ith row, starting after the pivot element all the way to the last element in the row, modifying each element appropriately $(a_{ik} \leftarrow a_{ik} - s * a_{jk})$ and the row reductions are completed with $(b_i \leftarrow b_i - s * b_j)$. At the end, after n steps, the matrix $\mathbf{A}$ is transformed to the equivalent matrix $\mathbf{A}^{(n)}$, which is in the row-echelon form:

$$
\begin{bmatrix}
1 & a_{12}^{(1)} & a_{13}^{(1)} & \cdots & a_{1n}^{(1)} & b_1^{(1)} \\
0 & 1 & a_{23}^{(2)} & \cdots & a_{2n}^{(2)} & b_2^{(2)} \\
0 & 0 & 1 & \cdots & a_{3n}^{(3)} & b_3^{(3)} \\
\vdots & \vdots & \vdots & \ddots & \vdots & \vdots \\
0 & 0 & 0 & \cdots & 1 & b_n^{(n)}
\end{bmatrix}
$$

In the case of the Gauss elimination method ($opt = 0$), the solution set is obtained by the BACK_SUBSTITUTE module, which implements the back substitution algorithm (*see* Section 2.5) since the system is in the upper triangular form.

In the case of the Gauss-Jordan elimination method ($opt \neq 0$), there is practically no need to employ row reductions to the elements above the diagonals (a_{ij}'s for $j > i$) since they are eventually eliminated. Thus, it is sufficient to carry out the required eliminations only on the rhs vector ($b_i \leftarrow b_i - a_{ij}b_j$) to save cpu-time. The row reductions are applied from the bottom row ($i = j-1$ above diagonal) to the top ($j = 1$) and from the last ($j = n$) to the second column ($j = 2$). The first column is already $x_1 = b_1$.

EXAMPLE 2.5: Application of Gauss elimination Method

Ethane (C_2H_6) burns in oxygen to form carbon dioxide (CO_2) and water (H_2O) according to the equation:

$$
x_1 C_2H_6 + x_2 O_2 \rightarrow x_3 CO_2 + x_4 H_2O
$$

Use the row reductions to balance the chemical reaction in question and demonstrate that the resulting is an underdetermined homogeneous system that has a nontrivial solution.

SOLUTION:

First, we balance the number of Carbon (C), Hydrogen (H), and Oxygen (O) atoms on each side.

$$
\begin{matrix} C :: \\ H :: \\ O :: \end{matrix} \quad
x_1 \begin{bmatrix} 2 \\ 6 \\ 0 \end{bmatrix} + x_2 \begin{bmatrix} 0 \\ 0 \\ 2 \end{bmatrix} = x_3 \begin{bmatrix} 1 \\ 0 \\ 2 \end{bmatrix} + x_4 \begin{bmatrix} 0 \\ 2 \\ 1 \end{bmatrix}
$$

Our task is to find positive integers (x_1, x_2, x_3, x_4) that balance the reaction. Expressing all equations in matrix equation form ($\mathbf{Ax} = \mathbf{0}$), we obtain

$$
\begin{bmatrix}
2 & 0 & -1 & 0 \\
6 & 0 & 0 & -2 \\
0 & 2 & -2 & -1
\end{bmatrix}
\begin{bmatrix} x_1 \\ x_2 \\ x_3 \\ x_4 \end{bmatrix}
=
\begin{bmatrix} 0 \\ 0 \\ 0 \end{bmatrix}
$$

This system is underdetermined and homogeneous as well. To reduce it into row-echelon form, the augmented matrix is established as

$$
[\mathbf{A} \,|\, \mathbf{0}]_{3 \times 5} =
\begin{bmatrix}
2 & 0 & -1 & 0 & 0 \\
6 & 0 & 0 & -2 & 0 \\
0 & 2 & -2 & -1 & 0
\end{bmatrix}
$$

Now we can begin the elimination procedure for the augmented matrix. As a first step, the first and second rows are normalized as follows:

$$\sim \begin{bmatrix} 1 & 0 & -\frac{1}{2} & 0 & \bigg| & 0 \\ 1 & 0 & 0 & -\frac{1}{3} & \bigg| & 0 \\ 0 & 2 & -2 & -1 & \bigg| & 0 \end{bmatrix} \begin{matrix} \leftarrow \frac{1}{2}r_1 \\ \leftarrow \frac{1}{6}r_2 \\ \\ \end{matrix}$$

Next, the second element of the first column is eliminated by adding $(-r_1)$ to r_2 in place of Row-2:

$$\sim \begin{bmatrix} 1 & 0 & -\frac{1}{2} & 0 & \bigg| & 0 \\ 0 & 0 & \frac{1}{2} & -\frac{1}{3} & \bigg| & 0 \\ 0 & 2 & -2 & -1 & \bigg| & 0 \end{bmatrix} \begin{matrix} \\ \leftarrow r_2 - r_1 \\ \\ \end{matrix}$$

Interchanging Row-2 with $(1/2) \times$ Row-3 results in

$$\sim \begin{bmatrix} 1 & 0 & -\frac{1}{2} & 0 & \bigg| & 0 \\ 0 & 1 & -1 & -\frac{1}{2} & \bigg| & 0 \\ 0 & 0 & \frac{1}{2} & -\frac{1}{3} & \bigg| & 0 \end{bmatrix} \begin{matrix} \\ r_2 \rightleftarrows \frac{1}{2}r_3 \\ \\ \end{matrix}$$

The third row is multiplied with "2" to obtain $a_{33} = 1$:

$$\sim \begin{bmatrix} 1 & 0 & -\frac{1}{2} & 0 & \bigg| & 0 \\ 0 & 1 & -1 & -\frac{1}{2} & \bigg| & 0 \\ 0 & 0 & 1 & -\frac{2}{3} & \bigg| & 0 \end{bmatrix} \begin{matrix} \\ \\ \leftarrow 2r_3 \end{matrix}$$

Finally, eliminating a_{13} and a_{23} yields

$$\sim \begin{bmatrix} 1 & 0 & 0 & -\frac{1}{3} & \bigg| & 0 \\ 0 & 1 & 0 & -\frac{7}{6} & \bigg| & 0 \\ 0 & 0 & 1 & -\frac{2}{3} & \bigg| & 0 \end{bmatrix} \begin{matrix} \leftarrow r_1 + \frac{1}{2}r_3 \\ \leftarrow r_2 + r_3 \\ \\ \end{matrix}$$

which is equivalent to $x_1 - x_4/3 = 0$, $x_2 - 7x_4/6 = 0$, $x_3 - 2x_4/3 = 0$. The basic (leading) variables are x_1, x_2 and x_3, and x_4 is the free (independent) variable. Setting $x_4 = u$, we obtain the general solution as $x_1 = u/3$, $x_2 = 7u/6$, $x_3 = 2u/3$. Hence, a nontrivial solution can be found for any real u. For $u = 6$, we obtain $x_1 = 2$, $x_2 = 7$, and $x_3 = 4$.

Discussion: Homogeneous underdetermined linear systems always have an infinite number of nontrivial solutions, in addition to the trivial solution ($\mathbf{x} = \mathbf{0}$). It is possible to obtain expressions for all solutions to an underdetermined system. Which one is "better" is a value judgment. In this problem, we have chosen u such that the solution is in integer form. However, it should be emphasized that even though the column vector has been augmented in this example, it is not necessary to explicitly augment the coefficient matrix with the column vector $\mathbf{b} = \mathbf{0}$ since no basic row operations will affect the zeros on the rhs.

When dealing with underdetermined systems, it may be very difficult to identify at a glance the leading and free variables of an underdetermined system; however, as in this case, the Gauss-Jordan elimination procedure is an important tool to determine the free and basic variables.

2.5.3   GAUSS-ELIMINATION WITH PIVOTING

Diagonal elements of a coefficient matrix play a vital role in implementing the Gauss elimination algorithm. The elements on the principle diagonal are called the pivots or pivot elements because of their significance in this regard. As we have discussed earlier, in order to avoid "division by zero," it is critical to know if a pivot element is or becomes zero at any step of the elimination procedure. *Pivoting* is the process of interchanging columns (order of unknowns) or rows (order of equations) to avoid having zero as the pivot element.

We have learned that the naive Gauss elimination does not employ pivoting. If any one of the pivots turns out to be zero during the row reduction process, a "division by zero" is encountered. In this case, the method is doomed to fail even if the coefficient matrix is non-singular or the linear system has a unique solution. Consider, for instance, the following coefficient matrix:

$$\begin{bmatrix} 0 & 3 \\ 2 & -2 \end{bmatrix}$$

Note that the matrix is non-singular; nonetheless, we cannot initiate the elimination procedure because $val = 1/0$ is encountered in the first pass. But this problem can be avoided simply by changing the first and second lines, as illustrated below:

$$\begin{bmatrix} 2 & -2 \\ 0 & 3 \end{bmatrix} \; r_1 \rightleftarrows r_2$$

Now the pivots of the equivalent matrix are no longer zero ($a_{11} \neq 0$ or $a_{22} \neq 0$), which allows the elimination procedure to be implemented successfully.

There are several pivoting strategies. In *complete pivoting*, both rows and columns are interchanged, whereas only the rows are interchanged in the *partial pivoting* process. The partial pivoting covered in this section is faster and least expensive. This technique requires only row interchanges to push the zero element off the main diagonal. Even though partial pivoting is more efficient and easier, it may still yield some inaccuracies in the final solution. Overall, the pivoting strategy is based on finding the largest element in the same column below the current diagonal element, which is considered the pivot, and swapping it with the current pivot row. This strategy helps prevent a division by zero or a very small number, which could lead to numerical instability. Although not required, scaling can be applied to the pivot rows selected, which leads to an algorithm that is less susceptible to round-off errors.

Gauss (-Jordan) elimination with Pivoting

- Gauss-Jordan or Gauss elimination methods with pivoting avoid possible divisions by zero;
- It minimizes round-off errors by using the largest elements as pivots;
- It does offer, in some cases, a remedy for solving ill-conditioned systems of linear equations.

- Full or partial pivoting leads to an increase in cpu-time due to searching for the largest element in a row or column, but partial pivoting requires less cpu as it needs fewer interchanges;
- The symmetry or regularity properties of the original matrix may be lost;
- It is computationally expensive for very large linear systems.

```
                    Pseudocode 2.10

  Module GAUSS_ELIMINATION_P (n,A, b, x)
  \ DESCRIPTION: A pseudomodule to solve a system of linear equations
  \    (Ax = b) by naive Gauss elimination with pivoting.
  \ USES:
  \    BACK_SUBSTITUTE :: Module back substitution (Pseudocode 2.8);
  \    ABS :: A built-in function computing the absolute value.
  Declare: a_nn, x_n, b_n                    \ Declare matrices as arrays
  For [j = 1, (n − 1)]                       \ Loop j: Sweep row from left to right
      valmax ← |a_jj|                        \ Set pivot as absolute max value
      pos ← j                                \ Find position of pivot
      For [i = (j + 1), n]          \ Loop i: Search for a greater element below a_jj
          If [|a_ij| > valmax] Then          \ If a greater value found
              valmax ← |a_ij|                \ Set a_jj as max value
              pos ← i                        \ Update position of pivot
          End If
      End For
      If [pos > j] Then                      \ A row of greater magnitude found
          For [k = j, n]                     \ Loop k: Swap rows row_pos ↔ row_j
              temp ← a_jk                    \ Use temp for swaping rows
              a_jk ← a_pos,k
              a_pos,k ← temp
          End For
          temp ← b_j                         \ Use a temp for swaping rhs
          b_j ← b_pos
          b_pos ← temp
      End If                                 \ Pivoting for jth column is complete
      For [i = (j + 1), n]                   \ Loop i: Perform forward elimination
          r ← a_ij/a_jj                      \ Set normalization factor
          a_ij ← 0                \ Set 0 for zeroed element, no need to compute
          For [k = (j + 1), n]               \ Loop k: Perform forward elimination
              a_ik ← a_ik − r ∗ a_jk         \ Row op. for kth row
          End For
          b_i ← b_i − r ∗ b_j                \ Row op. for rhs of ith row
      End For
  End For
  BACK_SUBSTITUTE(n,A,b,x)          \ Apply back substitution to get solution
  End Module GAUSS_ELIMINATION_P
```

A pseudomodule, GAUSS_ELIMINATION_P, for applying Gauss elimination with partial pivoting to solve a system of linear equations is provided in Pseudocode 2.10. The number of equations (n), the coefficient matrix (**A**), and the rhs (**b**) are required as input. On exit, the solution is stored on **x**. The basic structure of the program is the same as that of the module LINEAR_SOLVE with one fundamental difference, which requires interchanging the row with a small (or zero) pivot with the largest one of the rows below it. Afterwards, the usual forward elimination phase is performed.

The pivoting strategy is illustrated in Fig. 2.6. At the beginning of the loop over row-j, the default pivot and its position are initialized with these of the principal diagonal element and its corresponding row number ($valmax = |a_{jj}|$ and $pos = j$). With the For-loop over i,

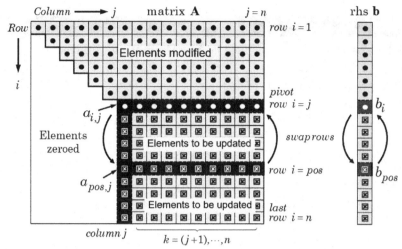

FIGURE 2.6: Illustrating the Gauss elimination procedure with pivoting (• and ⊠ denote the modified (computed) and unmodified (to be updated) elements of **A** and **b**, respectively).

the rows below the pivot (from $i = (j + 1)$ to n) are visited to find a row with the largest magnitude. When such a row is found, its magnitude and row number are set as current ones ($valmax = a_{ij}$ and $pos = i$), respectively. Whenever a different row with a larger magnitude than the current pivot is found, the current row (r_j) is swapped with the row with the largest magnitude (r_{pos}) using an If-construction with a $pos > j$ condition. Note that the contents of the jth row are $r_j = \{a_{jj}, a_{j,j+1}, \ldots, a_{jn}, b_j\}$.

The pivoting procedure is carried out column by column from left ($j = 1$) to right $j = n$ inside the outermost loop$-j$, in each row from left ($i = j + 1$) to right (n), and from top ($k = j + 1$) to bottom (n). The scaling factor is set to $r \leftarrow a_{ij}/a_{jj}$ once and is used in the forward elimination of the rows in the loop$-k$. The loop$-i$ ranges just below the pivot element (from $i = j + 1$) all the way to the last row ($i = n$), and the elements are zeroed without actually performing the calculations ($a_{ij} = 0$) to reduce the cpu-time. The inner loop over k variable ranges across the ith row, starting after the pivot element all the way to the last element in the row, modifying each element appropriately ($a_{ik} \leftarrow a_{ik} - r * a_{jk}$ for $i, k = (j + 1), \ldots, n$), and the row reductions are completed with ($b_i \leftarrow b_i - r * b_j$). At the end, the matrix **A** is transformed to an equivalent matrix $\mathbf{A}^{(n)}$, which is in row-echelon form. Once the forward elimination procedure is complete, the back substitution procedure is applied to obtain the solution.

EXAMPLE 2.6: Application of Elimination Methods

The given circuit consists of three loops and loop currents denoted by i_1, i_2 and i_3. The Kirchoff's current law for the circuit loops yields

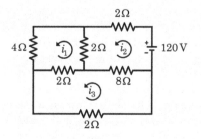

$$i_1 + 4i_2 - 6i_3 = 0$$
$$-i_1 + 6i_2 - 4i_3 = 60$$
$$4i_1 - i_2 - i_3 = 0$$

Solve the given system for the loop currents with the Gauss elimination with pivoting.

SOLUTION:

The linear system is first expressed in $\mathbf{Ax} = \mathbf{b}$ form and construct the augmented matrix:

$$\begin{bmatrix} 1 & 4 & -6 \\ -1 & 6 & -4 \\ 4 & -1 & -1 \end{bmatrix} \begin{bmatrix} i_1 \\ i_2 \\ i_3 \end{bmatrix} = \begin{bmatrix} 0 \\ 60 \\ 0 \end{bmatrix}, \qquad [\mathbf{A}\,|\,\mathbf{b}] = \left[\begin{array}{ccc|c} 1 & 4 & -6 & 0 \\ -1 & 6 & -4 & 60 \\ 4 & -1 & -1 & 0 \end{array}\right]$$

We start with the first column. The first row is interchanged with the third row, which has the largest magnitude element ($a_{31} = 4$) in this column, and then the first row is normalized as follows:

$$\left[\begin{array}{ccc|c} 4 & -1 & -1 & 0 \\ -1 & 6 & -4 & 60 \\ 1 & 4 & -6 & 0 \end{array}\right] \begin{array}{c} r_3 \leftrightarrow r_1 \\ \\ \\ \end{array} \sim \left[\begin{array}{ccc|c} 1 & -1/4 & -1/4 & 0 \\ -1 & 6 & -4 & 60 \\ 1 & 4 & -6 & 0 \end{array}\right] \begin{array}{c} \leftarrow (1/4)r_1 \\ \\ \\ \end{array}$$

Next, we proceed to eliminate the elements below a_{11} as follows:

$$\sim \left[\begin{array}{ccc|c} 1 & -1/4 & -1/4 & 0 \\ 0 & 23/4 & -17/4 & 60 \\ 0 & 17/4 & -23/4 & 0 \end{array}\right] \begin{array}{c} \\ \leftarrow r_2 + r_1 \\ \leftarrow r_3 - r_1 \end{array}$$

Note that the elementary row operations are encoded on the rhs of the augmented matrix, aligned with the modified row.

We observe that a_{22} is the largest element in the second column, so we will keep it as the pivot. Next, we normalize the second row (as $r_2 \leftarrow 4r_2/23$) and go on to eliminate the element below the pivot (i.e., a_{23}).

$$\sim \left[\begin{array}{ccc|c} 1 & -1/4 & -1/4 & 0 \\ 0 & 1 & -17/23 & 240/23 \\ 0 & 0 & -60/23 & -1020/23 \end{array}\right] \begin{array}{c} \\ \\ \leftarrow r_3 - (17/4)r_2 \end{array}$$

To complete the elimination process, the third row is scaled with a_{33}.

$$\sim \left[\begin{array}{ccc|c} 1 & -1/4 & -1/4 & 0 \\ 0 & 1 & -17/23 & 240/23 \\ 0 & 0 & 1 & 17 \end{array}\right] \equiv [\mathbf{A}\,|\,\mathbf{b}]$$

Finally, the coefficient matrix has now been transformed into an upper triangular matrix. The equivalent system of linear equations is

$$i_1 + 4i_2 - 6i_3 = 0, \qquad i_2 - i_3 = 6, \qquad i_3 = 17$$

Applying the back substitution algorithm to find the currents, we note that i_3 is readily available from the last equation: $i_3 = 17$ A. Substituting this into the second equation and solving for i_2 gives $i_2 = 23$ A. Finally, substituting both i_2 and i_3 into the first equation and solving for i_1 yields $i_1 = 10$ A.

Discussion: The Gauss elimination method with exact arithmetic gives the true solution. The Gauss-Jordan method additionally requires the elimination of the elements above the diagonals, which increases the cpu-time. The advantage of the Gauss elimination method over the Gauss-Jordan method is more apparent in large systems of equations.

2.6 COMPUTING DETERMINANTS

Using the cofactor expansion method to evaluate the determinant of an $n \times n$ matrix requires about $n \times n!$ arithmetic operations. For this reason, computing the determinant of a large matrix *via* cofactor expansion is impractical due to its immense cpu-time demand. A determinant, however, can be computed efficiently as the product of the diagonal elements once it is transformed into an upper triangular or row-echelon form.

In order to compute the determinant of a matrix quickly and efficiently, a variant of the Gauss elimination algorithm can be devised to transform the matrix to an upper triangular form. However, the elementary row operations must be used with caution. This Gauss elimination procedure works exactly as indicated above if neither pivoting nor scaling are used. But recall that both *scaling* and *pivoting* modify the original determinant in a predictable manner. In this regard, these row operations can still be applied, provided that suitable corrections are made. For instance, if a row (or column) is multiplied by a scalar λ, the determinant variable must be divided by λ as well. On the other hand, when two rows (r_i and r_j) are interchanged, the sign of the determinant is changed if $|i - j|$ is odd; however, if $|i - j|$ is even, no sign change is needed. This procedure requires the order of $\mathcal{O}(n^3)$ arithmetic operations to compute the determinant of an $n \times n$ square matrix, which is considerably less than the number of operations required by the cofactor expansion process.

The direct solution methods are not employed for solving systems of linear equations with more than a few hundred unknowns. For a system with fewer than about 100 unknowns, it is possible to obtain accurate results using single-precision arithmetic. However, in order to obtain reasonable results in systems with more than about a hundred unknowns, it is recommended to use "double precision" in arithmetic operations.

2.7 ILL-CONDITIONED MATRICES AND LINEAR SYSTEMS

A *"well-conditioned"* problem is one in which a small change in any of its numerical data causes only a small change in the solution of the problem. On the other hand, an *ill-conditioned* problem is one in which a small change in any of its numerical values causes a large change in the solution of the problem. In other words, ill-conditioned systems are extremely sensitive to round-off errors. But ill-conditioning is not a problem with the use of fractions or, in theory, infinite precision arithmetic.

Ill-condition is observed when inverting matrices and solving a system of linear equations using calculators or digital computers. In the literature, there is no clear-cut definition for "ill-condition." It is a phenomenon that manifests itself as a result of round-off errors (or small perturbations in the matrix elements) during arithmetic operations. An ill-conditioned system of linear equations converges to a solution that deviates from the true solution. The magnitudes of the deviations depend on how badly the system of equations or a matrix is conditioned. In other words, the more severe the ill-condition, the greater the magnitude of the deviations (from the true answer) will be.

When dealing with matrices or systems of linear equations, we cannot visually determine if a matrix or a linear system is ill-conditioned. However, it is important for the numerical analyst to determine whether or not an ill condition exists and how severe it is.

The *condition number*, $\kappa(\mathbf{A})$, is used as a measure of sensitivity of a matrix or system to small changes (errors or perturbations) in any one of its elements, and it is defined as

$$\kappa(\mathbf{A}) = \|\mathbf{A}\| \|\mathbf{A}^{-1}\| \tag{2.31}$$

where $\|\cdot\|$ denotes any one of the norms of a matrix. In other words, the numerical value of $\kappa(\mathbf{A})$ depends on the type of norm (indicated by a corresponding subscript such as 1, 2, F, or ∞) from which it is calculated.

The numerical accuracy of the condition number computed by Eq. (2.31) also depends on how accurately $\mathbf{A}^{-1}$ is computed. If $\mathbf{A}$ is severely ill-conditioned, estimating the condition number by evaluating $\mathbf{A}^{-1}$ not only increases the computational cost significantly but also raises questions regarding the accuracy of the result. Consequently, we should not use the numerical value of the condition number that we cannot trust.

The condition number is easiest to calculate when the matrix $\mathbf{A}$ is a diagonal matrix ($\mathbf{A} = \mathbf{D}$), and it is found from

$$\kappa(\mathbf{A}) = \kappa(\mathbf{D}) = \frac{\max_{1 \leqslant i \leqslant n} d_{ii}}{\min_{1 \leqslant i \leqslant n} d_{ii}} \tag{2.32}$$

However, determining the condition number of a matrix directly from the definition requires considerably more work than solving the actual linear system whose accuracy is to be determined. That is why, in practice, the condition number is estimated to be within an order of magnitude. A rough estimate for the condition number in terms of calculable quantities is given as $\kappa(\mathbf{A}) \approx \|\mathbf{A}\|_F / \|\det(\mathbf{A})\|$ [84]. A large value in the condition number is an indicative ill-conditioning.

Another similar but more accurate estimate for the spectral condition number was proposed by Guggenheimmer *et. al.* [10]

$$\kappa(\mathbf{A}) \approx \frac{2}{|\det(\mathbf{A})|} \left(\frac{1}{n} \sum_{i=1}^{n} \sum_{j=1}^{n} a_{ij}^2 \right)^{n/2} \gg 1 \tag{2.33}$$

where n is the order of matrix $\mathbf{A}$, and $\det(\mathbf{A})$ can be computed easily and quickly during the forward elimination procedure (*see* Section 2.6). The product of the diagonal elements resulting in an upper triangular matrix gives the determinant of $\mathbf{A}$, but the product should be calculated at every pivot row before the row is normalized (if a normalization strategy is employed).

Equation (2.33) is an upper bound for $\kappa(\mathbf{A})$, and it may result in estimates much larger than the actual condition number. Therefore, in order to minimize the round-offs, the linear system of equations is normalized. This is achieved by either dividing each row by the largest magnitude row element (i.e., $\|\mathbf{r}_i\|_\infty$) or by transforming each row of the matrix into a unit vector in Euclidean norm (i.e., $\|\mathbf{r}_i\|_2 = 1$). With this procedure, matrix $\mathbf{A}$ is transformed into a matrix $\mathbf{M}$, said to be *equilibrated* or *row-balanced*. Then Eq. (2.33) for $\mathbf{M}$ becomes

$$\kappa(\mathbf{M}) \approx \frac{2}{|\det(\mathbf{M})|} \gg 1 \tag{2.34}$$

We now have $\left(\sum \sum m_{ij}^2 / n \right)^{n/2} = 1$ for the equilibrated matrix. Ill-condition can then be based on the magnitude of $\kappa(\mathbf{M})$. For relatively small $\kappa(\mathbf{M})$ values, $\mathbf{A}$ is considered well-conditioned, whereas very large values indicate severe ill-condition.

The numerical estimate can be improved while solving the system of equations. Let us consider $\mathbf{Ax} = \mathbf{b}$ system whose solution is $\mathbf{z}$ [31]. Owing to properties of the norms, we may write

$$\|\mathbf{z}\| = \|\mathbf{A}^{-1}\mathbf{b}\| \leqslant \|\mathbf{A}^{-1}\|\|\mathbf{b}\|$$

which provides us a lower bound for $\|\mathbf{A}^{-1}\| \leqslant \|\mathbf{z}\|/\|\mathbf{b}\|$. If this bound is chosen to be as large as possible, then $\|\mathbf{A}^{-1}\|$ can be reasonably estimated. However, finding an optimum value for $\mathbf{b}$ could also be prohibitively expensive. Instead, we will adopt a heuristic approach that requires solving the $\mathbf{A}^T\mathbf{b} = \mathbf{u}$ system, where $\mathbf{u}$ is a vector promoting growth in the solution, and it is chosen such that each element is made up of (± 1)'s, i.e., $u_i = \pm 1$, $i = 1, 2, \ldots, n$. As a result, in order to estimate the condition number, the following linear systems are solved [126]:

$$\mathbf{A}^T\mathbf{b} = \mathbf{u} \quad \text{and} \quad \mathbf{Az} = \mathbf{b} \tag{2.35}$$

and use $\|\mathbf{z}\|/\|\mathbf{b}\|$ as an estimate for $\|\mathbf{A}^{-1}\|$. We then obtain

$$\kappa(\mathbf{A}) \approx \|\mathbf{A}\| \frac{\|\mathbf{z}\|}{\|\mathbf{b}\|} \tag{2.36}$$

Clearly, the condition number depends on the type of norm. Note that it is easy to compute $\|\mathbf{A}\|_1$ and $\|\mathbf{A}\|_\infty$, and $\|\mathbf{A}\|_2$ can be estimated using the Power algorithm (*see* Chapter 11).

Now, we turn our attention to the effects of ill-conditioning. To detect or assess the severity of an ill-condition, there are two simple and effective tests that require the computation of either $\mathbf{AA}^{-1} = \mathbf{I}$ or $(\mathbf{A}^{-1})^{-1} = \mathbf{A}$. Any substantial deviation from the original or expected outcome is considered evidence of an ill-conditioning. $(\mathbf{A}^{-1})^{-1} = \mathbf{A}$ is particularly important to observe since matrix inversion is performed twice, leading to noteworthy round-off accumulation. To answer the question of how badly the expected results will be distorted, we use the following "heuristic rule" for solving $\mathbf{Ax} = \mathbf{b}$: *expect to lose at least k-digits of precision, where* $k = \log \kappa(\mathbf{A})$.

The numerical treatment of ill-conditioned systems is an iterative procedure; for this reason, a remedy for an ill-conditioned system is covered in Section 3.5.

It should be noted that the systems of linear equations obtained as a result of numerical treatments of ordinary differential equations (ODEs) or partial differential equations (PDEs) are *not* ill-conditioned. The ill-conditioning is most severe in linear systems arising from least squares regressions.

EXAMPLE 2.7: Estimating Condition Number

Consider the following matrix:

$$\mathbf{A} = \begin{bmatrix} 1 & 3 & -2 \\ 2.99 & -16.11 & 4 \\ -3.01 & 1 & 2.03 \end{bmatrix}$$

(a) Use 1-, 2-, F-, and ∞-norm to estimate the norm-based condition numbers of matrix $\mathbf{A}$ and its equilibrated form $\mathbf{M}$; (b) find the spectral condition number of $\mathbf{A}$ using Eqs. (2.33) and (2.34); (c) compute $\mathbf{AA}^{-1} = \mathbf{I}$ and $(\mathbf{A}^{-1})^{-1} = \mathbf{A}$ and comment on the severity of ill-condition.

SOLUTION:

(a) We select the starting vector $\mathbf{u}$ as $\mathbf{u} = [1, -1, 1]^T$, and solve $\mathbf{A}^T\mathbf{b} = \mathbf{u}$

$$\begin{bmatrix} 1 & 2.99 & -3.01 \\ 3 & -16.11 & 1 \\ -2 & 4 & 2.03 \end{bmatrix} \begin{bmatrix} b_1 \\ b_2 \\ b_3 \end{bmatrix} = \begin{bmatrix} 1 \\ -1 \\ 1 \end{bmatrix}$$

which, with single precision, yields $\{b_1, b_2, b_3\} = \{2122.34, \ 467.881, \ 1169.54\}$. Then we solve $\mathbf{Az} = \mathbf{b}$ which gives $\{z_1, z_2, z_3\} = \{3.488 \times 10^6, \ 1.721 \times 10^6, \ 4.324 \times 10^6\}$. We can then estimate the condition number with compatible norms using Eq. (2.36).

TABLE 2.1

Norm	True, $\kappa(\mathbf{A})$	$\|\mathbf{A}\|$	$\|\mathbf{z}\|$	$\|\mathbf{b}\|$	Estimated, $\kappa(\mathbf{A})$
1	66798.6	20.110	9.533×10^6	3759.76	50989.6
2	40510.4	17.191	5.816×10^6	2468.01	40511.5
F	41668.6	17.682	5.816×10^6	2468.01	41668.6
∞	61659.5	23.101	4.324×10^6	2122.34	47063.3

Norm	True, $\kappa(\mathbf{M})$	$\|\mathbf{M}\|$	$\|\mathbf{z}\|$	$\|\mathbf{b}\|$	Estimated, $\kappa(\mathbf{M})$
1	25136.6	2.0225	2.265×10^8	20236.7	22636.7
2	15604.2	1.3585	1.382×10^8	12030.5	15604.2
F	19895.0	1.7321	1.382×10^8	12030.5	15604.2
∞	23042.5	1.6039	1.027×10^8	7941.07	20752.3

The condition numbers, $\kappa(\mathbf{A})$ and $\kappa(\mathbf{M})$, based on 1-, 2-, F-, and ∞-norms are computed using the definition, i.e., Eq. (2.31), and these results along with the estimated values from Eq. (2.36) are tabulated in Table 2.1. We notice that the estimates for both $\kappa(\mathbf{M})$ and $\kappa(\mathbf{A})$ are smaller than the "true" condition number values. However, a caution is in order here: if $\mathbf{A}$ were badly ill-conditioned, the condition numbers computed by Eq. (2.31) would be questionable.

When we analyze the condition numbers of the equilibrated matrix, it is observed that the orders of magnitude of the $\kappa(\mathbf{M})$ are about 40% of the $\kappa(\mathbf{A})$. The fact that the estimates for $\kappa(\mathbf{M})$ are also large verifies ill-conditioning. Evaluating $\log_{10} \kappa(\mathbf{A})$ suggests at least 4-5 digits of precision will be lost; on the other hand, $\log_{10} \kappa(\mathbf{M})$ predicts a loss of 3-4 digits of precision.

(b) For matrix $\mathbf{A}$, we find $|\det(\mathbf{A})| = 0.0302$, and $\sum\sum a_{ij}^2 = 312.653$. An estimate for the spectral condition number is obtained from Eq. (2.33) as

$$\kappa(\mathbf{A}) \approx \frac{2}{0.0302} \left(\frac{312.653}{3} \right)^{3/2} = 70458.8 \gg 1$$

Notice that this estimate, as an upper bound for the condition number, is larger than $\kappa(\mathbf{A}) = 40510$ value (*see* Table 2.1).

Next, we determine $\mathbf{M}$ by making each row of the matrix a unit vector in the Euclidean norm as follows:

$$\mathbf{M} = \begin{bmatrix} 1 & 3 & -2 \\ 2.99 & -16.11 & 4 \\ -3.01 & 1 & 2.03 \end{bmatrix} \begin{matrix} r_1/\|r_1\|_2 \\ r_2/\|r_2\|_2 \\ r_3/\|r_3\|_2 \end{matrix} = \begin{bmatrix} 0.267261 & 0.801784 & -0.534522 \\ 0.177276 & -0.955159 & 0.237159 \\ -0.799306 & 0.265550 & 0.539067 \end{bmatrix}$$

Similarly, for matrix $\mathbf{M}$ we find $|\det(\mathbf{M})| = 0.00012688$ and an estimate for the condition number using Eq. (2.34) leads to

$$\kappa(\mathbf{M}) \approx \frac{2}{|\det(\mathbf{M})|} = \frac{2}{0.00012688} = 15763 \gg 1$$

Note that this estimate is slightly higher than the computed true value, $\kappa(\mathbf{M}) = 15604.2$ given in Table 2.1.

(c) In order to determine the severity of "ill-conditioning," the following matrix identities are computed using double precision arithmetic: The acquired errors from the computation of $\mathbf{A}\mathbf{A}^{-1}$ and $(\mathbf{A}^{-1})^{-1}$ are

$$\mathbf{A}\mathbf{A}^{-1} - \mathbf{I} = \begin{bmatrix} 0 & 0 & 0 \\ 0.00195 & 0.00024 & 0.00049 \\ 0 & 0.00006 & 0.00024 \end{bmatrix}$$

$$(\mathbf{A}^{-1})^{-1} - \mathbf{A} = \begin{bmatrix} 0.00109 & -0.00632 & 0.00163 \\ 0.00522 & 0.00155 & -0.00483 \\ -0.00407 & 0.01084 & -0.00102 \end{bmatrix}$$

where the results above are reported correct to seven decimal places.

Discussion: This example demonstrates that estimating the condition number using the equilibrated matrix with Eq. (2.34) is closer to the true theoretical value of $\kappa(\mathbf{M})$. Since the true and estimated values of $\kappa(\mathbf{M})$ are very large, we can surely expect to observe the ill-effects of ill-condition when computing the inverse of $\mathbf{A}$ or solving linear systems with $\mathbf{A}$ as the coefficient matrix.

In general, the order of the average absolute error in the computation of $\mathbf{A}\mathbf{A}^{-1} = \mathbf{I}$ is 10^{-5}, which is about the same as the magnitude of the reciprocal of the condition number. On the other hand, the maximum errors in the computation of $\mathbf{A}^{-1}$ and $(\mathbf{A}^{-1})^{-1}$ are 0.00195 and 0.01084, respectively. The average and maximum errors of the latter are larger because of the accumulation of round-off errors owing to inverting the matrix twice. It is clear that when an ill-conditioned matrix is inverted or an ill-conditioned linear system is solved, the accumulation of round-offs will be a major problem.

2.8 DECOMPOSITION METHODS

We often encounter $\mathbf{A}\mathbf{x} = \mathbf{b}$ systems, where $\mathbf{A}$ remains fixed while $\mathbf{b}$ changes (e.g., $\mathbf{A}\mathbf{x} = \mathbf{b}_1$, $\mathbf{A}\mathbf{y} = \mathbf{b}_2$, etc.). For instance, in dynamics, matrix $\mathbf{A}$ is generally termed the "structure matrix," which depends on the structure of a mechanical or dynamic system, and vector $\mathbf{b}$ denotes the "load." Especially in structural engineering, the load denotes the applied forces, and the solution vector $\mathbf{x}$ gives the stresses at various points on the structure. Hereby, a structure is tested for various applied load scenarios.

Matrix decomposition (or factorization) methods have been developed as alternatives to Gauss elimination to reduce the computation time and effort when solving multiple linear systems consisting of the same $\mathbf{A}$. A *matrix decomposition* (or *matrix factorization*) is the process of factorizing a matrix into a product of matrices as $\mathbf{A} = \mathbf{L}\mathbf{U}$, where $\mathbf{L}$ and $\mathbf{U}$ are generally lower and upper triangular matrices, respectively. Utilizing triangular matrices in matrix computations (on the order of $\mathcal{O}(n^2)$) makes calculations easier in the process and provides advantages and efficiency in programming. However, when solving linear systems with varying rhs vectors ($\mathbf{A}\mathbf{x} = \mathbf{b}_m$) by Gauss elimination, the row reductions are applied to the augmented matrix (including the rhs vector). Therefore, the computational effort ($\mathcal{O}(n^3)$) and cpu-time are the same whether $\mathbf{A}$ is the same or not.

Many different matrix decomposition methods have been developed over the years, e.g., LU, Cholesky, QR, SVD, and so on. Each method is used for a particular class of problems. In other words, some of these methods are more useful than others, depending on the field of application. In this section, we will present the two more common methods, Doolittle-LU and Cholesky decomposition, that work with only positive definite and nonsingular diagonally dominant square matrices.

1) **Doolittle LU-Decomposition method** is particularly useful when a matrix has the form $\mathbf{A} = \mathbf{L}\mathbf{U}$, where $\mathbf{L}$ and $\mathbf{U}$ are lower and upper triangular matrices, respectively. The Doolittle decomposition requires that the diagonal elements of the lower triangular matrix $\mathbf{L}$ be "1" and an upper triangular matrix $\mathbf{U}$. An alternative method, *Crout LU-Decomposition*, factors the matrix into a lower triangular matrix $\mathbf{L}$ and an upper triangular matrix $\mathbf{U}$ whose diagonal entries are required to be "1." In Crout's decomposition, however, the errors may grow very rapidly; therefore, it is generally not recommended unless the system of linear equations is relatively small. For this reason, the Doolittle LU-decomposition is presented in this text.

2) **Cholesky Decomposition Method** is applied to matrices that are positive definite, non-singular, and symmetric. Matrix $\mathbf{A}$ can be decomposed as $\mathbf{A} = \mathbf{L}\mathbf{L}^T$, where $\mathbf{L}$ is a lower triangular matrix and $\mathbf{L}^T$ is its transpose. The method is computationally faster and more efficient since only one triangular matrix is determined.

An advantage of decomposing matrix $\mathbf{A}$ as $\mathbf{L}\mathbf{U}$ is that the system, $\mathbf{A}\mathbf{x} = \mathbf{b}$, can be rewritten as $\mathbf{L}\mathbf{U}\mathbf{x} = \mathbf{b}$. Defining $\mathbf{U}\mathbf{x} = \mathbf{y}$ leads to an upper triangular system, and $\mathbf{A}\mathbf{x} = \mathbf{b}$ eventually results in a lower triangular system, as illustrated below:

$$\mathbf{L}\mathbf{U}\mathbf{x} = \mathbf{b} \quad \{\mathbf{U}\mathbf{x} = \mathbf{y}\} \rightarrow \quad \mathbf{L}\mathbf{y} = \mathbf{b} \tag{2.37}$$

First, $\mathbf{L}\mathbf{y} = \mathbf{b}$ is solved to find an intermediate solution $\mathbf{y}$, using the *forward substitution* (Pseudocode 2.7); then $\mathbf{U}\mathbf{x} = \mathbf{y}$ is solved to obtain $\mathbf{x}$ by applying the *back substitution* (Pseudocode 2.8).

2.8.1 DOOLITTLE LU-DECOMPOSITION

The objective of this method is to find two matrices, $\mathbf{L}$ and $\mathbf{U}$, that satisfy $\mathbf{A} = \mathbf{L}\mathbf{U}$, where $\mathbf{L}$ is a lower triangular matrix with diagonal elements 1, and $\mathbf{U}$ is an upper triangular matrix.

Consider a square matrix $\mathbf{A}$, which has an LU-decomposition in the following forms:

$$\begin{bmatrix} a_{11} & a_{12} & a_{13} & \cdots & a_{1n} \\ a_{21} & a_{22} & a_{23} & \cdots & a_{2n} \\ a_{31} & a_{32} & a_{33} & \cdots & a_{3n} \\ \vdots & \vdots & \vdots & \ddots & \vdots \\ a_{n1} & a_{n2} & a_{n3} & \cdots & a_{nn} \end{bmatrix} = \begin{bmatrix} 1 & 0 & 0 & \cdots & 0 \\ \ell_{21} & 1 & 0 & \cdots & 0 \\ \ell_{31} & \ell_{32} & 1 & \cdots & 0 \\ \vdots & \vdots & \vdots & \ddots & \vdots \\ \ell_{n1} & \ell_{n2} & \ell_{n3} & \cdots & 1 \end{bmatrix} \begin{bmatrix} u_{11} & u_{12} & u_{13} & \cdots & u_{1n} \\ 0 & u_{22} & u_{23} & \cdots & u_{2n} \\ 0 & 0 & u_{33} & \cdots & u_{3n} \\ \vdots & \vdots & \vdots & \ddots & \vdots \\ 0 & 0 & 0 & \cdots & u_{nn} \end{bmatrix}$$

$$\tag{2.38}$$

Notice that the diagonal elements of $\mathbf{L}$ are set to 1, i.e., $\ell_{ii} = 1$ for $i = 1, 2, \ldots, n$.

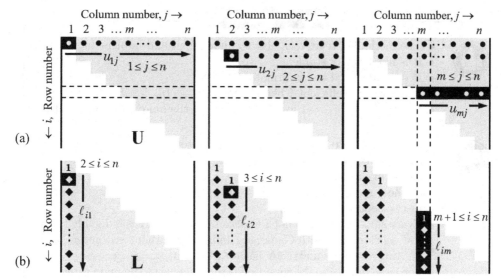

FIGURE 2.7: Depicting Doolittle LU-decomposition; (a) left to right row sweeps on $\mathbf{U}$, (b) top to bottom column sweeps on $\mathbf{L}$.

The $\mathbf{LU}$ product can be explicitly written as

$$\mathbf{A} = \begin{bmatrix} u_{11} & u_{12} & u_{13} & \cdots & u_{1n} \\ \ell_{21}u_{11} & \ell_{21}u_{12}+u_{22} & \ell_{21}u_{13}+u_{23} & \cdots & \ell_{21}u_{1n}+u_{2n} \\ \ell_{31}u_{11} & \ell_{31}u_{12}+\ell_{32}u_{22} & \ell_{31}u_{13}+\ell_{32}u_{23}+u_{33} & \cdots & \ell_{31}u_{1n}+\ell_{32}u_{2n}+u_{3n} \\ \vdots & \vdots & \vdots & \ddots & \vdots \\ \ell_{n1}u_{11} & \ell_{n1}u_{12}+\ell_{n2}u_{22} & \ell_{n1}u_{13}+\ell_{n2}u_{23}+u_{33} & \cdots & \sum_{k=1}^{n-1}\ell_{nk}u_{kn}+u_{nn} \end{bmatrix} \quad (2.39)$$

Comparing and matching the first rows of $\mathbf{A}$ and $\mathbf{LU}$ in Eq. (2.39), we find

$$u_{1j} = a_{1j}, \qquad j = 1, 2, \ldots, n \qquad (2.40)$$

which takes care of the first row of $\mathbf{U}$.

In Fig. 2.7, the steps of Doolittle decomposition on $\mathbf{U}$ and $\mathbf{L}$ matrices are schematically illustrated. Finding the first row of $\mathbf{U}$ by Eq. (2.40) does not require any calculations, so the first row of $\mathbf{A}$ is placed exactly on the first row of $\mathbf{U}$ (Fig. 2.7a). Next, comparing and matching the first columns of $\mathbf{A}$ and $\mathbf{LU}$, we obtain $a_{i1} = \ell_{i1}u_{11}$ for $i = 2, 3, \ldots, n$. Since $u_{11} = a_{11}$, for the first column (Fig. 2.7b), we deduce

$$\ell_{i1} = a_{i1}/u_{11}, \qquad i = 2, 3, \ldots, n \qquad (2.41)$$

Similarly, we compare and match the diagonal and subsequent elements of the second row, which yields $a_{2j} = \ell_{21}u_{1j} + u_{2j}$. Solving for u_{2j} gives

$$u_{2j} = a_{2j} - \ell_{21}u_{1j}, \qquad j = 2, 3, \ldots, n \qquad (2.42)$$

which takes care of the second row $\mathbf{U}$ (*see* Fig. 2.7a). The corresponding second column of $\mathbf{L}$ is likewise obtained by matching and comparing the elements below the diagonal (i.e., $\ell_{22} = 1$) as $a_{i2} = \ell_{i1}u_{12} + \ell_{i2}u_{22}$. Solving for ℓ_{i2}, we obtain

$$\ell_{i2} = \frac{a_{i2} - \ell_{i1}u_{12}}{u_{22}}, \qquad i = 3, 4, \ldots, n \qquad (2.43)$$

LU-Decomposition

- The method is relatively easy to understand and implement;
- It is numerically stable;
- It does not need the information on **b**;
- It is cpu-time efficient when solving $\mathbf{A}\mathbf{x}_m = \mathbf{b}_m$ for $m = 1, 2, \ldots$;
- Having **L** with $\ell_{ii} = 1$ for all i guarantees the uniqueness of $\mathbf{A} = \mathbf{LU}$.

- Employing it to solve a single linear system ($\mathbf{Ax} = \mathbf{b}$) is inefficient due to excessive calculations (decomposition, forward and backward substitutions);
- Memory requirement increases by twofold if **L** and **U** are kept (stored) as two separate matrices;
- Every square matrix is not LU-decomposable (if **A** is reducible to the row echelon form with Gaussian elimination without ever interchanging two rows, the matrix **A** has an LU-form);
- For large matrices, the round-off error emerges as the main drawback of the decomposition methods.

---

which are placed in **L** from top to bottom (*see* Fig. 2.7b).

Notice that the procedure is sequenced to compute one row of **U** and its corresponding column of **L** until all rows and columns of **A** are exhausted. Hence, we systematically find the elements in **L** and **U** with the help of equations obtained from matching and comparing the elements of $\mathbf{A} = \mathbf{LU}$.

The elements of the ith row of **U** and the jth row of **L** can be generalized as

$$u_{ij} = a_{ij} - \sum_{k=1}^{i-1} \ell_{ik} u_{kj}, \quad \text{for } i \leqslant j \leqslant n \quad (1 \leqslant i \leqslant n)$$

$$\ell_{ij} = \frac{1}{u_{jj}} \left(a_{ij} - \sum_{k=1}^{j-1} \ell_{ik} u_{kj} \right), \quad \text{for } i + 1 \leqslant j \leqslant n$$

(2.44)

Using two matrices (**L** and **U**) as given in the algorithm doubles the memory requirement. As depicted below, the memory requirement can be minimized by overwriting **A** with **L** and **U** and omitting the storage of 1's and 0's of **L** or **U**, as they are the givens of the method.

$$\begin{bmatrix} a_{11} & a_{12} & \cdots & a_{1n} \\ a_{21} & a_{22} & \cdots & a_{2n} \\ \vdots & \vdots & \ddots & \vdots \\ a_{n1} & a_{n2} & \cdots & a_{nn} \end{bmatrix} \leftarrow \begin{bmatrix} u_{11} & u_{12} & \cdots & u_{1n} \\ \ell_{21} & u_{22} & \cdots & u_{2n} \\ \vdots & \vdots & \ddots & \vdots \\ \ell_{n1} & \ell_{n2} & \cdots & u_{nn} \end{bmatrix}$$

A pseudomodule, LU_DECOMP, implementing the Doolittle LU-decomposition is presented in Pseudocode 2.11. The module requires the matrix **A** and its order n as input and returns **L** and **U** as output. Both triangular matrices are first initialized: $\ell_{ii} = 1$ for all i and $\ell_{ij} = 0$ for $i \neq j$, and $u_{ij} = 0$ for all i, j.

Pseudocode 2.11

Module LU_DECOMP $(n, \mathbf{A}, \mathbf{L}, \mathbf{U})$
\ DESCRIPTION: A pseudomodule to find LU-decomposition of a matrix.
Declare: $a_{nn}, u_{nn}, \ell_{nn}$ \qquad\qquad\qquad\quad \ Declare matrices as arrays
$\mathbf{L} \leftarrow 0; \ \mathbf{U} \leftarrow 0$ \qquad\qquad\qquad\quad \ Initialize $\mathbf{L}$ and $\mathbf{U}$ with zero matrices
For $[i = 1, n]$
$\quad \ell_{ii} \leftarrow 1$ \qquad\qquad\qquad\qquad\qquad\qquad\qquad \ Set $\ell_{ii} = 1$
End For
For $[i = 1, n]$
$\quad$ **For** $[j = i, n]$ \qquad\qquad\qquad\qquad \ Loop j: Find row elements of $\mathbf{U}$
$\qquad u_{ij} \leftarrow a_{ij}$ \qquad\qquad\qquad\quad \ Initialize accumulator, $u_{ij} = a_{ij}$
$\qquad$ **For** $[k = 1, (i-1)]$ \qquad\qquad \ Loop k: Accumulating loop
$\qquad\quad u_{ij} \leftarrow u_{ij} - \ell_{ik} * u_{kj}$ \qquad\qquad \ Add $-\ell_{ik} * u_{kj}$ to u_{ij}
$\qquad$ **End For**
$\quad$ **End For**
$\quad$ **For** $[j = (i+1), n]$
$\qquad \ell_{ji} \leftarrow a_{ji}$ \qquad\qquad\qquad\quad \ Initialize accumulator, $\ell_{ji} = a_{ji}$
$\qquad$ **For** $[k = 1, (i-1)]$ \qquad\qquad \ Loop k: Accumulating loop
$\qquad\quad \ell_{ji} \leftarrow \ell_{ji} - \ell_{jk} * u_{ki}$ \qquad\qquad \ Add $-\ell_{jk} * u_{ki}$ to ℓ_{ji}
$\qquad$ **End For**
$\qquad \ell_{ji} \leftarrow \ell_{ji}/u_{ii}$ \qquad\qquad\qquad\qquad\qquad \ Find u_{ii}
$\quad$ **End For**
End For
End Module LU_DECOMP

The algorithm constructs $\mathbf{U}$ and $\mathbf{L}$ by sweeping through the matrix by rows and columns, as illustrated in Fig. 2.7. The inner loop$-j$ is first used to determine the elements of $\mathbf{U}$ along the rows. u_{ij} is used as an accumulator with an initial value assigned to a_{ij}. When for $i \leqslant j$, the accumulation carried out with the loop$-k$ is completed, the accumulated result is u_{ij}. The elements of $\mathbf{L}$ are computed across columns with the second loop$-j$. Note that the matrix indices i and j in Eq. (2.44) are reversed here, since the index of the loop$-i$ also runs for the column index. A column of $\mathbf{L}$ is constructed with the second loop$-j$ from $j = i$ to n. Similarly, ℓ_{ji} is used as the accumulator, with an initial value assigned to a_{ij}. Once the accumulation is completed for $j > i$ using the loop$-k$ and dividing it by u_{ii} gives ℓ_{ij}. Note that this module requires a memory allocation of $2n^2$ for $\mathbf{L}$ and $\mathbf{U}$. Removing $\mathbf{L}$ and $\mathbf{U}$ from the Module's variable list, **Module** LU_DECOMP$(n,\mathbf{A})$, and replacing ℓ_{ij} and u_{ij}'s simply with a_{ij}'s in both the module and Eq. (2.44) reduce the memory requirement to a single matrix storage.

EXAMPLE 2.8: LU-Decomposition to solve linear systems

The smallest eigenvalue and associated eigenvector of a matrix satisfy $\mathbf{A}\mathbf{x} = \lambda^{-1}\mathbf{x}$, where $\lambda = \max|\mathbf{x}|$. A procedure for determining the smallest eigenvalue begins with a guess $\mathbf{x}^{(0)}$ vector, followed by a series of successive solutions of the matrix equation:

$$\mathbf{A}\mathbf{x}^{(p+1)} = \frac{1}{\lambda^{(p)}}\mathbf{x}^{(p)}, \quad \lambda^{(p)} = \max\left|\mathbf{x}^{(p)}\right|, \quad p = 0, 1, 2, \ldots.$$

where p denotes an estimate at the pth step. Since matrix $\mathbf{A}$ is fixed while only rhs varies, the LU-decomposition is suitable for this example. Thus, the matrix $\mathbf{A}$

is decomposed once, and then two linear systems are solved: (1) a lower-triangular $\mathbf{Ly} = (1/\lambda^{(p)})\mathbf{x}^{(p)}$ and (2) an upper-triangular system, $\mathbf{Ux}^{(p+1)} = \mathbf{y}$. Apply the Doolittle algorithm to the following matrix to find suitable $\mathbf{L}$ and $\mathbf{U}$.

$$\mathbf{A} = \begin{bmatrix} -1 & -2 & -2 & -1 \\ -2 & -6 & -7 & -4 \\ -1 & 2 & 5 & 6 \\ 1 & -4 & -3 & 3 \end{bmatrix}$$

SOLUTION:

Starting with $u_{1j} = a_{1j}$ for $j = 1, 2, 3$, and 4, using Eq. (2.40) we readily obtain

$$u_{11} = -1, \quad u_{12} = -2, \quad u_{13} = -2, \quad u_{14} = -1$$

From $\ell_{i1} = a_{i1}/u_{11}$, Eq. (2.41), for $i = 2, 3$, and 4 we find

$$\ell_{21} = a_{21}/u_{11} = -2/(-1) = 2,$$
$$\ell_{31} = a_{31}/u_{11} = (-1)/(-1) = 1,$$
$$\ell_{4,1} = a_{4,1}/u_{11} = 1/(-1) = -1$$

Using Eq. (2.42) for $j = 2, 3$, and 4 results in

$$u_{22} = a_{22} - \ell_{21}u_{12} = (-6) - (2)(-2) = -2,$$
$$u_{23} = a_{23} - \ell_{21}u_{13} = (-7) - (2)(-2) = -3,$$
$$u_{24} = a_{24} - \ell_{21}u_{14} = (-4) - (2)(-1) = -2$$

Using Eq. (2.43) for $i = 3$ and 4, we obtain

$$\ell_{32} = \frac{a_{32} - \ell_{31}u_{12}}{u_{22}} = \frac{2 - (1)(-2)}{(-2)} = -2,$$
$$\ell_{4,2} = \frac{a_{4,2} - \ell_{4,1}u_{12}}{u_{22}} = \frac{-4 - (-1)(-2)}{(-2)} = 3$$

The third row of $\mathbf{U}$ leads to $u_{3i} = a_{3i} - \ell_{31}u_{1i} - \ell_{32}u_{2i}$ which for $i = 3$ and 4 gives

$$u_{33} = a_{33} - \ell_{31}u_{13} - \ell_{32}u_{23} = 5 - (1)(-2) - (-2)(-3) = 1$$
$$u_{34} = a_{34} - \ell_{31}u_{14} - \ell_{32}u_{24} = 6 - (1)(-1) - (-2)(-2) = 3$$

Equation (2.44) gives the third column of $\mathbf{L}$:

$$\ell_{43} = \frac{1}{u_{33}}(a_{43} - \ell_{41}u_{13} - \ell_{42}u_{23}) = \frac{1}{(1)}((-3) - (-1)(-2) - (3)(-3)) = 4$$

Lastly, from Eq. (2.44), we obtain

$$u_{44} = a_{44} - \ell_{41}u_{14} - \ell_{42}u_{24} - \ell_{43}u_{34} = 3 - (-1)(-1) - (3)(-2) - (4)(3) = -4$$

Finally, collecting the elements of the lower and upper triangular matrices, we find

$$\mathbf{L} = \begin{bmatrix} 1 & 0 & 0 & 0 \\ 2 & 1 & 0 & 0 \\ 1 & -2 & 1 & 0 \\ -1 & 3 & 4 & 1 \end{bmatrix}, \quad \mathbf{U} = \begin{bmatrix} -1 & -2 & -2 & -1 \\ 0 & -2 & -3 & -2 \\ 0 & 0 & 1 & 3 \\ 0 & 0 & 0 & -4 \end{bmatrix}$$

Cholesky Decomposition

- The method requires fewer arithmetic operations;
- It outperforms the LU-decomposition by a factor of two;
- It is stable without pivoting since $\mathbf{A}$ is positive definite;
- The memory requirement is cut in half since $\mathbf{U} = \mathbf{L}^T$.

- The method can be employed to symmetric matrices only;
- Matrix $\mathbf{A}$ must be positive definite to avoid square root of negative numbers when calculating ℓ_{ii}'s.

2.8.2 CHOLESKY DECOMPOSITION

If $\mathbf{A} = [a_{ij}]_{n \times n}$ is a *positive definite* and *symmetric* matrix, the matrix $\mathbf{A}$ can be decomposed into a product of a unique lower triangular matrix $\mathbf{L}$ and its transpose: $\mathbf{A} = \mathbf{L}\mathbf{L}^T$. The decomposition matrix can be written explicitly as

$$
\begin{bmatrix}
a_{11} & a_{12} & a_{13} & \cdots & a_{1n} \\
a_{21} & a_{22} & a_{23} & \cdots & a_{2n} \\
a_{31} & a_{32} & a_{33} & \cdots & a_{3n} \\
\vdots & \vdots & \vdots & \ddots & \vdots \\
a_{n1} & a_{n2} & a_{n3} & \cdots & a_{nn}
\end{bmatrix}
=
\begin{bmatrix}
\ell_{11} & 0 & 0 & \cdots & 0 \\
\ell_{21} & \ell_{22} & 0 & \cdots & 0 \\
\ell_{31} & \ell_{32} & \ell_{33} & \cdots & 0 \\
\vdots & \vdots & \vdots & \ddots & \vdots \\
\ell_{n1} & \ell_{n2} & \ell_{n3} & \cdots & \ell_{nn}
\end{bmatrix}
\begin{bmatrix}
\ell_{11} & \ell_{21} & \ell_{31} & \cdots & \ell_{n1} \\
0 & \ell_{22} & \ell_{23} & \cdots & \ell_{2n} \\
0 & 0 & \ell_{33} & \cdots & \ell_{3n} \\
\vdots & \vdots & \vdots & \ddots & \vdots \\
0 & 0 & 0 & \cdots & \ell_{nn}
\end{bmatrix}
\tag{2.45}
$$

Upon performing $\mathbf{L}\mathbf{L}^T$, we obtain

$$
\mathbf{A} =
\begin{bmatrix}
\ell_{11}^2 & \ell_{11}\ell_{21} & \ell_{11}\ell_{31} & \cdots & \ell_{11}\ell_{n1} \\
\ell_{11}\ell_{21} & \ell_{21}^2 + \ell_{22}^2 & \ell_{21}\ell_{31} + \ell_{22}\ell_{32} & \cdots & \ell_{21}\ell_{n1} + \ell_{22}\ell_{n2} \\
\ell_{11}\ell_{31} & \ell_{21}\ell_{31} + \ell_{22}\ell_{32} & \ell_{31}^2 + \ell_{32}^2 + \ell_{33}^2 & \cdots & \ell_{31}\ell_{n1} + \ell_{32}\ell_{n2} + \ell_{33}\ell_{n3} \\
\vdots & \vdots & \vdots & \ddots & \vdots \\
\ell_{11}\ell_{n1} & \ell_{21}\ell_{n1} + \ell_{22}\ell_{n2} & \ell_{31}\ell_{n1} + \ell_{32}\ell_{n2} + \ell_{33}\ell_{n3} & \cdots & \sum_{k=1}^{n} \ell_{nk}^2
\end{bmatrix}
\tag{2.46}
$$

Inspecting and matching the first row and first column entries of $\mathbf{A}$ and $\mathbf{L}\mathbf{L}^T$, we find $\ell_{11} = \sqrt{a_{11}}$. The Cholesky decomposition is a special case of the LU-decomposition where $\mathbf{U} = \mathbf{L}^T$. So we could skip the derivation of the relations between the elements of $\mathbf{A}$ and $\mathbf{L}$ and simply replace u_{ij}'s with ℓ_{ij}'s in Eq. (2.44) to get

$$
\ell_{ij} =
\begin{cases}
\sqrt{a_{ii} - \sum_{k=1}^{i-1} \ell_{ik}^2}, & j = i \text{ and } i = 1, 2, \ldots, n \\
\dfrac{1}{\ell_{jj}} \left(a_{ij} - \sum_{k=1}^{j-1} \ell_{ik}\ell_{jk} \right), & i > j \text{ and } j = 1, 2, \ldots, (i-1)
\end{cases}
\tag{2.47}
$$

A pseudomodule, CHOLESKY_DECOMP, for decomposing a positive definite symmetric matrix using the Cholesky's method is given in Pseudocode 2.12. The module requires an input matrix $\mathbf{A}$ of size n and returns a lower triangular matrix $\mathbf{L}$ as output. The outer loop$-j$ is used to compute ℓ_{ij}'s along a column using Eq. (2.47). After setting $sum1 = a_{jj}$ (initializing the accumulator variable), the accumulator loop over the k index variable, $-\ell_{ik}^2$

```
Pseudocode 2.12

Module CHOLESKY_DECOMP (n,A, L)
\ DESCRIPTION: A pseudomodule to decompose a symmetric matrix A as
\    LLᵀ by Cholesky Decomposition.
\ USES:
\    SQRT:: A built-in function computing the square-root of a value.
Declare: aₙₙ, ℓₙₙ                              \ Declare matrices as arrays
L ← 0                                          \ Initialize L with zero matrix
For [j = 1, n]                                 \ Loop j: Perform column sweep
    sum1 ← a_{jj}                                      \ Initialize sum1
    For [k = 1, (j − 1)]                        \ Loop k: Accumulating loop
        sum1 ← sum1 − ℓ²_{jk}                     \ Add −ℓ²_{jk} to accumulator
    End For
    ℓ_{jj} ← √sum1                                   \ Set diagonal element
    For [i = (j + 1), n]                \ Loop i: Find elements below diagonal
        sum2 ← a_{ij}              \ Initialize sum of off-diagonal elements
        For [k = 1, (j − 1)]                    \ Loop k: Accumulating loop
            sum2 ← sum2 − ℓ_{ik} * ℓ_{jk}        \ Add −ℓ_{ik}ℓ_{jk} to accumulator
        End For
        ℓ_{ij} ← sum2/ℓ_{jj}                            \ Save result on L
    End For
End For
End Module CHOLESKY_DECOMP
```

terms up to $k = (j-1)$ are accumulated. The diagonal element for the jth row then becomes $\ell_{jj} = \sqrt{sum1}$. The elements below the diagonals $(j < i)$ also require an accumulator, $sum2$, using the inner loop-i. This accumulator is first initialized as $sum2 = a_{ij}$. With the accumulator loop over index variable k, the $-\ell_{ik}\ell_{jk}$ terms are added to $sum2$. Lastly, an ℓ_{ij} is found as $sum2/\ell_{jj}$.

EXAMPLE 2.9: Cholesky Decomposition

In electronic structure calculations, Cholesky's algorithm can be efficiently utilized to remove linear dependencies in a given matrix. Decompose matrix $\mathbf{A}$ into $\mathbf{A} = \mathbf{LL}^T$.

$$\mathbf{A} = \begin{bmatrix} 4 & -2 & 4 & 2 \\ -2 & 10 & -5 & -4 \\ 4 & -5 & 9 & 1 \\ 2 & -4 & 1 & 7 \end{bmatrix}$$

SOLUTION:

Recall that $\mathbf{A}$ must be symmetric and positive-definite in order to apply the Cholesky decomposition. The given matrix is symmetric, and since all its eigenvalues are positive (≈ 17.5673, 7.05325, 4.29779, and 1.08164), the matrix meets the positive definiteness criteria. Therefore, $\mathbf{A}$ is suitable for the Cholesky decomposition.

The first row and first column elements are obtained as

$$\ell_{11} = \sqrt{a_{11}} = \sqrt{4} = 2$$

For $i = 2$, the second-row elements are computed from Eq. (2.47) as follows:

$$\ell_{21} = \frac{a_{21}}{\ell_{11}} = \frac{(-2)}{2} = -1, \qquad \ell_{22} = \sqrt{a_{22} - \ell_{21}^2} = \sqrt{10 - (-1)^2} = 3$$

For $i = 3$, similarly, we obtain

$$\ell_{31} = \frac{a_{31}}{\ell_{11}} = \frac{4}{2} = 2,$$

$$\ell_{32} = \frac{1}{\ell_{22}} (a_{32} - \ell_{21}\ell_{31}) = \frac{1}{(3)} (-5 - (-1)(2)) = -1$$

$$\ell_{33} = \sqrt{a_{33} - \ell_{31}^2 - \ell_{32}^2} = \sqrt{9 - (2)^2 - (-1)^2} = 2$$

The elements for the last row ($i = 4$) lead to

$$\ell_{4,1} = \frac{a_{4,1}}{\ell_{11}} = \frac{2}{2} = 1,$$

$$\ell_{4,2} = \frac{1}{\ell_{22}} (a_{4,2} - \ell_{4,1}\ell_{21}) = \frac{1}{(3)} (-4 - (1)(-1)) = -1$$

$$\ell_{4,3} = \frac{1}{\ell_{33}} (a_{4,3} - \ell_{31}\ell_{4,1} - \ell_{32}\ell_{4,2}) = \frac{1}{(2)} (1 - (2)(1) - (-1)(-1)) = -1$$

$$\ell_{44} = \sqrt{a_{44} - \ell_{4,1}^2 - \ell_{4,2}^2 - \ell_{4,3}^2} = \sqrt{7 - (1)^2 - (-1)^2 - (-1)^2} = 2$$

Finally, collecting all elements in **L**, we obtain

$$\mathbf{L} = \begin{bmatrix} 2 & 0 & 0 & 0 \\ -1 & 3 & 0 & 0 \\ 2 & -1 & 2 & 0 \\ 1 & -1 & -1 & 2 \end{bmatrix}$$

Discussion: Finding the **L** matrix, if it exists, requires two recursive relationships: one for establishing the diagonal elements ($i = j$), and the other for obtaining the elements below diagonals ($i > j$). The procedure is easy and straightforward; however, one must make sure that **A** is positive symmetric before attempting to solve the problem.

2.9 TRIDIAGONAL SYSTEMS

2.9.1 THOMAS ALGORITHM

Tridiagonal systems of linear equations are the most common type of linear system in practice. They are encountered in many science and engineering problems, such as interpolation problems, the numerical solution of ordinary and partial differential equations, and so on.

The tridiagonal matrix algorithm, also frequently referred to as *Thomas algorithm*, is a special and simplified case of the *Gauss elimination* algorithm. A general form of the tridiagonal system of linear equations is given in Eq. (2.11), where d_i, b_i, a_i are one-dimensional arrays of length n containing diagonal, lower- and upper-diagonal elements of

Thomas Algorithm

- The algorithm is simple and easy to implement;
- It is very fast owing to requiring minimum arithmetic operations;
- It requires minimum memory allocation (max. $5n - 2$ for five arrays of length n) in comparison to full matrix storage ($n^2 + 2n$).

- It can only be applied to diagonally dominant tridiagonal systems; that is, $|d_i| > |b_i| + |a_i|$ for all i;
- For large or ill-conditioned systems, round-offs can be a problem.

the matrix, respectively. Since a tridiagonal system has only one lower diagonal of non-zero elements, it is sufficient to employ the *forward elimination procedure* once for each row of the augmented matrix as follows:

$$\left[\begin{array}{ccccc|c} d_1 & a_1 & & & & c_1 \\ 0 & d_2' & a_2 & & & c_2' \\ & 0 & d_3' & a_3 & & c_3' \\ & & \ddots & \ddots & \ddots & \vdots \\ & & & 0 & d_{n-1}' & a_{n-1} & c_{n-1}' \\ & & & & 0 & d_n' & c_n' \end{array}\right] \begin{array}{l} \\ r_2' \leftarrow r_2 - r_1 b_2/d_1 \\ r_3' \leftarrow r_3 - r_2' b_3/d_2' \\ \vdots \\ r_{n-1}' \leftarrow r_{n-1} - r_{n-2}' b_{n-1}/d_{n-2}' \\ r_n' \leftarrow r_n - r_{n-1}' b_n/d_{n-1}' \end{array}$$

where the primes denote elements that have been modified from their original values. Note that only the diagonal and the rhs elements are modified in the forward elimination step. Hence, the forward elimination procedure leads to the following equations:

$$c_i' = c_i - \frac{b_i}{d_{i-1}} c_{i-1}, \qquad d_i' = d_i - \frac{b_i}{d_{i-1}} a_{i-1}, \qquad i = 2, 3, \ldots, n \tag{2.48}$$

The arithmetic operations on b_i's are not explicitly shown since the lower diagonal elements are zeroed out. In other words, the actual arithmetic operations on the lower diagonal elements that should result in $b_i^{(1)} = 0$ are not performed, as this would be a waste of time.

The resulting system is an upper-bidiagonal matrix, which is solved using the back substitution algorithm. Noting that $d_i' x_i + a_i x_{i+1} = c_i'$ for $i = (n-1), (n-2), \ldots, 2, 1$ we obtain:

$$x_n = \frac{c_n'}{d_n'}, \qquad x_i = \frac{c_i' - a_i x_{i+1}}{d_i'}, \qquad i = (n-1), (n-2), \ldots, 2, 1 \tag{2.49}$$

A pseudomodule, **TRIDIAGONAL**, solving a tridiagonal system of equations using the Thomas algorithm is presented in Pseudocode 2.13. The module inputs are $s1$ and sn (subscripts of the first and last equations), d_i, b_i, a_i, and c_i (diagonal, lower-, upper-diagonal, and rhs) elements. On exit, the solution is saved on **x**. The forward elimination loop modifies d_i and c_i, according to Eq. (2.48). By the end of the forward elimination, the last unknown becomes available, $x_{sn} = c_{sn}/d_{sn}$, which is used as a starting value for the back substitution step. The rest of the solution is obtained backward from Eq. (2.49) using the (back substitution) loop. Note that the memory requirement for the arrays used in this module is $5n - 2$. However, we can reduce the memory requirement to $4n - 2$ by eliminating the array **x** from the module and overwriting the solution on **c** as follows:

$$c_n' \leftarrow \frac{c_n'}{d_n'}, \qquad c_i' \leftarrow \frac{c_i' - a_i c_{i+1}'}{d_i'}, \qquad i = (n-1), (n-2), \ldots, 2, 1 \tag{2.50}$$

Pseudocode 2.13

Module TRIDIAGONAL $(s1, sn, \mathbf{b}, \mathbf{d}, \mathbf{a}, \mathbf{c}, \mathbf{x})$
\ DESCRIPTION: A pseudomodule employing the Thomas algorithm to a tri-
\ diagonal system. Caution! Arrays $\mathbf{c}$ and $\mathbf{d}$ are destroyed.
Declare: $b_{s1:sn}, d_{s1,sn}, a_{s1:sn}, c_{s1:sn}, x_{s1:sn}$ \ Declare arrays from $s1$ to sn
For $\left[i = s1 + 1, sn \right]$ \ Loop i: Forward elimination step
    $ratio \leftarrow b_i / d_{i-1}$ \ Set normalization factor
    $d_i \leftarrow d_i - ratio * a_{i-1}$ \ Modify diagonal element
    $c_i \leftarrow c_i - ratio * c_{i-1}$ \ Modify rhs element
End For
$x_{sn} \leftarrow c_{sn}/d_{sn}$ \ Get x_{sn} from last equation
For $\left[i = (sn - 1), s1, (-1) \right]$ \ Loop i: Back substitution step
    $x_i \leftarrow (c_i - a_i * x_{i+1})/d_i$ \ Find unknowns from $(sn - 1)$ to $s1$
End For
End Module TRIDIAGONAL

2.9.2 LU DECOMPOSITION

Crout's LU-decomposition (i.e., placing 1's to the diagonal of $\mathbf{U}$ instead of $\mathbf{L}$) of a tridiagonal matrix can be useful in solving $\mathbf{Tx} = \mathbf{c}_m$, where $\mathbf{T}$ is a tridiagonal matrix for $m = 1, 2, \ldots$, and so on. Consider a tridiagonal system in the form given by Eq. (2.11). We assume LU-decomposition in the following form:

$$\mathbf{T} = \mathbf{LU} = \begin{bmatrix} \ell_1 & & & & \\ b_2 & \ell_2 & & & \\ & b_3 & \ell_3 & & \\ & & \ddots & \ddots & \\ & & & b_{n-1} & \ell_{n-1} \\ & & & & b_n & \ell_n \end{bmatrix} \begin{bmatrix} 1 & u_1 & & & \\ & 1 & u_2 & & \\ & & 1 & u_3 & \\ & & & \ddots & \ddots \\ & & & & 1 & u_{n-1} \\ & & & & & 1 \end{bmatrix} \quad (2.51)$$

where the matrices $\mathbf{T}$ (with b_i, d_i, a_i, and c_i's), $\mathbf{L}$ (with ℓ_i and b_i's), and $\mathbf{U}$ (with u_i's) are stored as one-dimensional arrays to minimize the memory requirement.

Comparing and matching the elements of the $\mathbf{T}$ and $\mathbf{LU}$ matrices one-to-one using the same procedure applied to full matrices, the elements of the bidiagonal matrix are found as follows:

$$\begin{aligned} \ell_i &= d_i - b_i u_{i-1}, \quad \text{for } i = 2, \ldots, n \quad \text{with } \ell_1 = d_1, \\ u_i &= a_i / \ell_i, \quad \quad \text{for } i = 1, 2, \ldots, (n - 1). \end{aligned} \quad (2.52)$$

To further minimize memory allocation, if destroying $\mathbf{T}$ is unimportant, matrices $\mathbf{L}$ and $\mathbf{U}$ can be placed in $\mathbf{T}$ by overwriting $\ell_i \rightarrow d_i$ and $u_i \rightarrow a_i$. Then the symmetric tridiagonal system can be solved by employing a forward substitution procedure to solve $\mathbf{Ly} = \mathbf{b}$) and a back substitution procedure to solve $\mathbf{Ux} = \mathbf{y}$) yielding

Forward substitution: $x_1 = c_1/d_1$, $x_i = (c_i - b_i x_{i-1})/d_i$,   for $i = 2, \ldots, n$,
Back substitution: $x_j = x_j - a_j x_{j+1}$, $j = (n - 1), \ldots, 2, 1$

where the array $\mathbf{y}$ is (eliminated) replaced by $\mathbf{x}$ to reduce memory space.

The LU-decomposition requires about $3n$ arithmetic operations, while the forward- and back-substitutions altogether demand about $5n$ operations. Thus, the solution of a large tridiagonal system of equations by LU decomposition requires about $8n$ operations in total.

The Doolittle LU decomposition for the tridiagonal systems is not given here; however, an algorithm can easily be developed in a similar way. The advantages and disadvantages of Doolittle or Crout's decomposition methods are the same as those of the Thomas algorithm.

2.9.3 CHOLESKY DECOMPOSITION

Consider a symmetric tridiagonal matrix in the following:

$$
\mathbf{T} = \begin{bmatrix}
d_1 & b_2 \\
b_2 & d_2 & b_3 \\
& b_3 & d_3 & b_4 \\
& & \ddots & \ddots & \ddots \\
& & & b_{n-1} & d_{n-1} & b_n \\
& & & & b_n & d_n
\end{bmatrix}, \quad
\mathbf{L} = \begin{bmatrix}
\ell_1 \\
e_2 & \ell_2 \\
& e_3 & \ell_3 \\
& & \ddots & \ddots \\
& & & e_{n-1} & \ell_{n-1} \\
& & & & e_n & \ell_n
\end{bmatrix} \tag{2.53}
$$

where $\mathbf{T}$ has a Cholesky decomposition as $\mathbf{LL}^T$ with the given lower bidiagonal matrix $\mathbf{L}$. Then, the matrix product $\mathbf{LL}^T$ is obtained as

$$
\mathbf{LL}^T = \begin{bmatrix}
\ell_1^2 & e_2\ell_1 \\
e_2\ell_1 & e_2^2 + \ell_2^2 & e_3\ell_2 \\
& e_3\ell_2 & e_3^2 + \ell_3^2 & e_4\ell_3 \\
& & \ddots & \ddots & \ddots \\
& & & e_{n-1}\ell_{n-2} & e_{n-1}^2 + \ell_{n-1}^2 & e_n\ell_{n-1} \\
& & & & e_n\ell_{n-1} & e_n^2 + \ell_n^2
\end{bmatrix} = \mathbf{T} \tag{2.54}
$$

Inspecting and matching the non-zero elements of Eqs. (2.53) and (2.54), we arrive at the following relationships:

$$
\ell_1 = \sqrt{d_1}, \qquad e_k = \frac{b_k}{\ell_{k-1}}, \qquad \ell_k = \sqrt{d_k - e_k^2}, \qquad k = 2, 3, \ldots, n \tag{2.55}
$$

Having decomposed the matrix $\mathbf{T}$ as $\mathbf{LL}^T$, the linear system becomes $\mathbf{LL}^T\mathbf{x} = \mathbf{b}$. Setting $\mathbf{L}^T\mathbf{x} = \mathbf{y}$, the system can be written as $\mathbf{Ly} = \mathbf{b}$. Employing the forward substitution to $\mathbf{Ly} = \mathbf{b}$ and back-substitution to $\mathbf{L}^T\mathbf{x} = \mathbf{y}$ leads to

Forward substitution: $y_1 = c_1/d_1$, $y_i = (c_i - b_i y_{i-1})/d_i$, for $i = 2, 3, .., n$
Back-substitution: $x_n = y_n/d_n$, $x_i = (y_i - b_{i+1}x_{i+1})/d_i$, for $i = (n-1), \ldots, 2, 1$

Original principal diagonal and lower diagonal elements can be used to minimize the memory requirement by overwriting arrays as follows:

$$
d_1 \leftarrow \sqrt{d_1}, \quad b_k \leftarrow \frac{b_k}{d_{k-1}}, \quad d_k \leftarrow \sqrt{d_k - b_k^2}, \qquad k = 2, 3, \ldots, n
$$

Writing $\mathbf{y}$ on $\mathbf{c}$, then $\mathbf{x}$ on $\mathbf{c}$ (i.e., the solution is saved on the right-hand side vector), two arrays ($\mathbf{x}$ and $\mathbf{y}$) are eliminated, which makes the algorithm more memory efficient:

Forward substitution: $c_1 \leftarrow c_1/d_1$, $c_i \leftarrow (c_i - b_i c_{i-1})/d_i$, for $i = 2, 3, .., n$
Back substitution: $c_n \leftarrow c_n/d_n$, $c_i \leftarrow (c_i - b_{i+1}c_{i+1})/d_i$ for $i = (n-1), \ldots, 2, 1$

In Pseudocode 2.13, arrays **d** and **b** are destroyed (i.e., overwritten with the new values). When solving a tridiagonal system of the form $\mathbf{Ax}_m = \mathbf{b}_m$, where $\mathbf{b}_m$ changes, the original arrays **d** and **b** must be stored with different names (say $\mathbf{d}_0$ and $\mathbf{b}_0$) and retrieved ($\mathbf{d} \leftarrow \mathbf{d}_0$ and $\mathbf{b} \leftarrow \mathbf{b}_0$) each time before calling the module **TRIDIAGONAL**.

EXAMPLE 2.10: Application of Tridiagonal System of Equations

Five objects (W_1, W_2, W_3, W_4, and W_5) are connected in series with five springs (with spring constants of k_1, k_2, k_3, k_4, and k_5) that are suspended, as illustrated in the figure. We wish to determine the displacement of each object from its unstretched position. The equations for the steady-state system are expressed as follows:

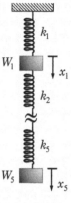

$$W_1 + k_2(x_2 - x_1) - k_1 x_1 = 0$$
$$W_2 + k_3(x_3 - x_2) - k_2(x_2 - x_1) = 0$$
$$W_3 + k_4(x_4 - x_3) - k_3(x_3 - x_2) = 0$$
$$W_4 + k_5(x_5 - x_4) - k_4(x_4 - x_3) = 0$$
$$W_5 - k_5(x_5 - x_4) = 0$$

Apply (a) Thomas algorithm to solve the resulting tridiagonal system of equations obtained for the case data given as $W_1 = W_3 = W_4 = W_5 = 10$ N, $W_2 = 15$ N, $k_1 = 100$, $k_2 = 300$, $k_3 = 180$, $k_4 = 200$, and $k_5 = 300$ N/m, and (b) Crout's algorithm to decompose the coefficient matrix into lower and upper tridiagonal matrices.

SOLUTION:

Substituting the numerical values of W's and k's into the steady-state equations, the augmented matrix for the linear system becomes

$$\left[\begin{array}{ccccc|c} 40 & -30 & & & & 1 \\ -60 & 96 & -36 & & & 3 \\ & -18 & 38 & -20 & & 1 \\ & & -20 & 50 & -30 & 1 \\ & & & -30 & 30 & 1 \end{array} \right]$$

The first step is to employ the forward substitution procedure. Performing the elementary row operations on the augmented matrix as indicated for each row, we find

$$\sim \left[\begin{array}{ccccc|c} 40 & -30 & & & & 1 \\ 0 & 51 & -36 & & & \dfrac{9}{2} \\ 0 & & \dfrac{430}{17} & -20 & & \dfrac{44}{17} \\ & 0 & & \dfrac{1470}{43} & -30 & \dfrac{131}{43} \\ & & 0 & & \dfrac{180}{49} & \dfrac{180}{49} \end{array} \right] \begin{array}{l} \\ r_2' \leftarrow r_2 + \dfrac{3}{2}r_1 \\ r_3' \leftarrow r_3 + \dfrac{6}{17}r_2' \\ r_4' \leftarrow r_4 + \dfrac{34}{43}r_3' \\ r_5' \leftarrow r_5 + \dfrac{43}{49}r_4' \end{array}$$

where r_i''s denote the altered rows due to prior row operations.

Now going back to the equation form, we have

$$40x_1 - 30x_2 = 1, \qquad 51x_2 - 36x_3 = \frac{9}{2}, \qquad \frac{430}{17}x_3 - 20x_4 = \frac{44}{17}$$

$$\frac{1470}{43}x_4 - 30x_5 = \frac{131}{43}, \qquad \frac{180}{49}x_5 = \frac{180}{49}$$

Next, the back substitution step is carried out in order to find the unknowns one by one from the last equation to the first one as follows:

$$\frac{180}{49}x_5 = \frac{180}{49} \;\to\; x_5 = 1$$

$$\frac{1470}{43}x_4 - 30x_5 = \frac{131}{43} \;\to\; x_4 = \frac{29}{30}$$

$$\frac{430}{17}x_3 - 20x_4 = \frac{44}{17} \;\to\; x_3 = \frac{13}{15}$$

$$51x_2 - 36x_3 = \frac{9}{2} \;\to\; x_2 = \frac{7}{10}$$

$$40x_1 - 30x_2 = 1 \;\to\; x_1 = \frac{11}{20}$$

(b) Notice that the tridiagonal matrix is not *symmetrical*, so Crout's algorithm is suitable for this example. To start decomposition, the first row leads to $\ell_1 = d_1 = 40$ and $u_1 = a_1/\ell_1 = -3/4$. Employing Eqs. (2.52) to the remaining rows and columns, we obtain

$$\ell_2 = d_2 - b_2 u_1 = 96 - (-60)\left(-\frac{3}{4}\right) = 51 \quad \text{and} \quad u_2 = \frac{a_2}{\ell_2} = \frac{(-36)}{51} = -\frac{12}{17}$$

$$\ell_3 = d_3 - b_3 u_2 = 38 - (-18)\left(-\frac{12}{17}\right) = \frac{430}{17} \quad \text{and} \quad u_3 = \frac{a_3}{\ell_3} = \frac{-20}{430/17} = -\frac{34}{43}$$

$$\ell_4 = d_4 - b_4 u_3 = 50 - (-20)\left(-\frac{34}{43}\right) = \frac{1470}{43} \quad \text{and} \quad u_4 = \frac{a_4}{\ell_4} = \frac{-30}{1470/43} = -\frac{43}{49}$$

$$\ell_5 = d_5 - b_5 u_4 = 30 - (-30)\left(-\frac{43}{49}\right) = \frac{180}{49}$$

Finally, by substituting the computed values into the **L** and **U** definitions, Eq. (2.51), the matrices are obtained in the desired forms.

$$\mathbf{L} = \begin{bmatrix} 40 & & & & \\ -60 & 51 & & & \\ & -18 & \dfrac{430}{17} & & \\ & & -20 & \dfrac{1470}{43} & \\ & & & -30 & \dfrac{180}{49} \end{bmatrix}, \quad \mathbf{U} = \begin{bmatrix} 1 & -\dfrac{3}{4} & & & \\ & 1 & -\dfrac{12}{17} & & \\ & & 1 & -\dfrac{34}{43} & \\ & & & 1 & -\dfrac{43}{49} \\ & & & & 1 \end{bmatrix}$$

Discussion: The number of arithmetic operations in employing the Thomas algorithm given in Pseudocode 2.13 are $(3n - 3)$ additions/subtractions and $(5n - 4)$ multiplications/divisions, yielding a total operation count of $8n - 7$. This is a significant reduction from the roughly $2n^3/3$ operations needed to solve a linear system with full matrix storage. The memory requirement can be reduced to $4n - 2$ from $5n - 2$ by eliminating intermediate arrays and writing the solution on the rhs vector. The LU-decomposition is preferred when multiple linear systems of the form $\mathbf{Tx} = \mathbf{b}_m$ are to be solved.

2.10 BANDED LINEAR SYSTEMS

The numerical solution of high-order ODEs or partial differential equations in two or three variables leads to *band diagonal* (or *banded*) linear systems of equations. A matrix with zero elements everywhere except on a narrow band about the main diagonal is referred to as a *banded matrix*. Clearly, if a large banded system of equations is solved with full $n \times n$ matrix storage, not only will the memory requirement increase but also the cpu-time due to excessive arithmetic operations with zeros. For this reason, solving a banded system or decomposing a banded matrix based on full matrix storage is inefficient and impractical.

An efficient way of solving banded linear systems involves storing and manipulating only the non-zero elements of the matrix. The Gauss elimination and/or LU-decomposition algorithms are specifically tailored for banded systems to significantly reduce memory allocation and cpu-time.

The definition of a banded matrix with elements a_{ij} is mathematically expressed as

$$a_{ij} = 0, \qquad j < i - m_L \quad \text{or} \quad j > i + m_U$$

where m_L and m_U ($m_L, m_U > 0$) are the lower and upper bandwidths, respectively.

The *bandwidth* of a banded matrix is defined as $n_b = m_L + m_U + 1$. In Fig. 2.8a, an 8×8 banded matrix is illustrated. The matrix has a lower and an upper bandwidth of 3 and 2, respectively, leading to $n_b = 6$. The compact form of the banded (rectangular) matrix, shown in Fig. 2.8b, excludes off-the-band zeros. It is evident that, if the total bandwidth is sufficiently small, considerable savings in memory requirements will be achieved by implementing the compact matrix storage scheme.

Several banded matrix indexing schemes are encountered in the literature. The compact matrix storage scheme adopted here is based on nonzero diagonals. Rotating the band 45° clockwise results in a compact matrix of size $n \times n_b$. Extra spaces created are padded with zeros, i.e., the zeros in bold font in the upper left and lower right corners of Fig. 2.8b. This allows the nonzero diagonals to be aligned as columns. In the compact matrix form denoted by $\mathbf{B} = [a_{ik}]$ ($n \times n_b$), the new indices represent row (i) and column (k, i.e., diagonal) numbers, respectively. In matrix $\mathbf{B}$, the indices of the main and uppermost diagonals become $d = m_L + 1$ and $n_b = m_L + m_U + 1$, respectively, as moving from left to right along columns. The relationship between the elements of the full banded and compact matrices can be expressed as $b_{ik} = a_{ij}$ where $k = d + j - i$ for $i, j = 1, 2, 3, \ldots, n$.

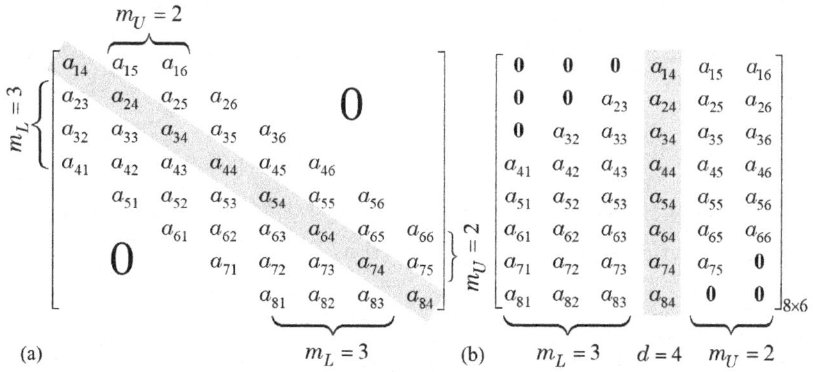

FIGURE 2.8: Matrix storage for (a) a banded, (b) a compact matrix.

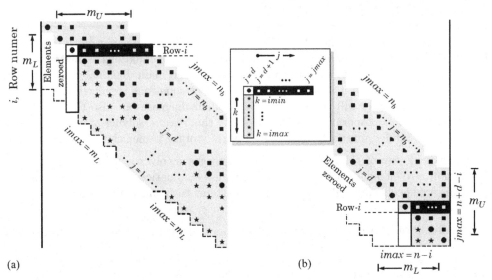

FIGURE 2.9: Depiction of naive Gauss Elimination procedure for a banded matrix with pivot at (a) $i < n - m_U$, (b) $i > n - m_U$.

2.10.1 NAIVE GAUSS ELIMINATION

We will employ the naive Gauss elimination procedure presented in Section 2.5 to a compact banded linear system. Consider a banded coefficient matrix with lower and upper bandwidths m_L and m_U, shown in Fig. 2.9. We will denote the maximum number of elements under a pivot by $imax$, which corresponds to m_L for most rows ($i \leqslant n - m_L$ in Fig. 2.9a), while it is $n - i$ for rows with $i > n - m_L$ (Fig. 2.9b). Thus, we may set $imax = \text{MIN}(m_L, n - i)$. Similarly, the column number of the rightmost non-zero element in row-i is $jmax = n_b$ for rows with $i \leqslant n - m_U$ (Fig. 2.9a) or $jmax = n + d - i$ otherwise (Fig. 2.9b), which can be generalized as $jmax = \text{MAX}(n_b, n + d - i)$.

In this procedure, the column elements below the diagonal (lower band) are to be zeroed out. Note that the indices k and j refer to the row and column indices of the pivot in the original matrix, respectively (*see* Fig. 2.9). In the *forward elimination* step, the elements in the pivot row (a_{ij}, $jmin \leqslant j \leqslant jmax$) are used in row reductions while clearing the elements below the pivot $a_{k+i,j-k}$ (in compact matrix notation) for $1 \leqslant k \leqslant imax$. The matrix coefficients and the right-hand side are modified as follows:

$$a_{i+k,j-k} = a_{i+k,j-k} - a_{ij} \left(\frac{a_{i+k, d-k}}{a_{id}} \right), \quad j = jmin, \dots, jmax \tag{2.56}$$

$$b_{i+k} = b_{i+k} - b_i \left(\frac{a_{i+k, d-k}}{a_{id}} \right), \quad k = imin, \dots, imax \tag{2.57}$$

where $imin = 1$, $imax = \text{MIN}(m_L, n - k)$, $jmin = d + 1$ and $jmax = \text{MIN}(n_b, n + d - i)$.

The *back substitution* procedure is employed as follows:

$$x_n = \frac{b_n}{a_{nd}}, \quad x_i = \frac{1}{a_{id}} \left(b_i - \sum_{j=jmin}^{jmax} a_{ij} \, x_{i+j-d} \right), \quad i = (n-1), (n-2), \dots, 2, 1 \tag{2.58}$$

where $jmin$ and $jmax$ have the same definitions as before.

Pseudocode 2.14

Module BAND_SOLVE $(n, m_L, m_U, \mathbf{A}, \mathbf{b})$
\ DESCRIPTION: A pseudomodule to solve a banded linear system by naive
\   Gauss elimination *without pivoting*.
\ USES:
\   BAND_BACK_SUBSTITUTE:: Back-substitution module (Pseudocodes 2.15);
\   MIN :: Built-in function to find the minimum of a set of integers.
Declare: $a_{n, m_L + m_U + 1}$, b_n \ Declare array variables
$d \leftarrow m_L + 1$ \ Find position of diagonal in compact matrix
$n_b \leftarrow d + m_U$ \ Find total bandwidth
For $\left[i = 1, n \right]$ \ Loop i: Forward elimination loop
    $imin \leftarrow 1$ \ First row below the diagonal
    $imax \leftarrow$ **MIN**$(m_L, n - i)$ \ Find number of elements below diagonal
    For $\left[k = imin, imax \right]$ \ Loop k: Sweep all elements below diagonal
        $ratio \leftarrow a_{i+k, d-k} / a_{i,d}$ \ Set normalization factor p
        $jmin \leftarrow d + 1$ \ Set index of pivot element in row
        $jmax \leftarrow$ **MIN**$(n_b, n + d - i)$ \ Set index of last non-zero element
        For $\left[j = jmin, jmax \right]$ \ Loop j: Perform elimination on row
            $a_{i+k, j-k} \leftarrow a_{i+k, j-k} - a_{i,j} * ratio$ \ Original matrix $\mathbf{A}$ is modified
        End For
        $b_{i+k} \leftarrow b_{i+k} - b_i * ratio$ \ Carry out elimination steps on rhs
    End For
End For
BAND_BACK_SUBSTITUTE$(n, m_L, m_U, \mathbf{A}, \mathbf{b})$ \ Solution is overwritten on $\mathbf{b}$
End Module BAND_SOLVE

A pair of pseudocodes that implement forward elimination and back substitution procedures of the Gauss elimination method on a band diagonal matrix are given in Pseudocodes 2.14 and 2.15. The module BAND_SOLVE requires a number of equations (n), the lower and upper bandwidths (m_L, m_U), the coefficient matrix in compact form ($\mathbf{A}$), and the right-hand side vector ($\mathbf{b}$) as input. On exit, the matrix $\mathbf{A}$ is an upper-banded matrix, and $\mathbf{b}$ is the altered rhs vector.

The forward elimination procedure for all rows is carried out in the outer loop over the i variable. Using the ith row as the "pivot row," the elimination of the elements below the pivot is accomplished in the inner loop-j that runs from $j = jmin$ to the last non-zero element to the right of the pivot at $jmax$. However, the arithmetic operations for the elements that give zero are not actually performed to save cpu-time. While the non-zero elements below the pivot are zeroed out row-by-row using the inner loop-k for $1 \leqslant i \leqslant imax$, the rest of the elements in row-k are also modified according to Eqs. (2.56) and (2.57). When the row reductions are completed for all i, j, and k's, the coefficient matrix becomes an upper banded matrix, which is now suitable for obtaining the solution vector with BAND_BACK_SUBSTITUTE. The module calls BAND_BACK_SUBSTITUTE to obtain the solution vector and returns the solution saved in $\mathbf{b}$.

The module BAND_BACK_SUBSTITUTE has the same argument list as BAND_SOLVE. However, the matrix $\mathbf{A}$ is an upper band matrix in compact form, which permits the back substitution algorithm to be employed. This module is not suitable to be used stand-alone unless matrix $\mathbf{A}$ is an upper diagonal band matrix. The unknown x_n is available from the last (nth) equation: $x_n = b_n / a_{n,d}$. Using the outer loop over i that signifies the rows and

Pseudocode 2.15

Module BAND_BACK_SUBSTITUTE $(n, m_L, m_U, \mathbf{A}, \mathbf{b})$
\ DESCRIPTION: A pseudomodule to solve an upper banded linear system.
\ USES:
\   MIN :: Built-in function to find the minimum of a set of integers.
Declare: $a_{n, m_L + m_U + 1}$, b_n \ Declare matrices as arrays
$d \leftarrow m_L + 1$ \ Find position of diagonal in compact matrix
$n_b \leftarrow d + m_U$ \ Find total bandwidth
$b_n = b_n / a_{nd}$ \ Find solution from last equation, $b_n \leftarrow x_n$
For $\left[i = (n-1), 1, (-1) \right]$ \ Loop i: Back substitution steps
$\quad jmin \leftarrow d + 1$ \ Set index of element next to diagonal
$\quad jmax \leftarrow$ **MIN**$(n_b, n + d - i)$ \ Set rightmost column index
$\quad$ **For** $\left[j = jmin, jmax \right]$ \ Loop j: Accumulating loop
$\qquad b_i \leftarrow b_i - a_{ij} * b_{i+j-d}$ \ Add $-a_{ij} * b_{i+j-d}$ to accumulator
$\quad$ **End For**
$\quad b_i \leftarrow b_i / a_{id}$ \ Solution is written on **b**
End For
End Module BAND_BACK_SUBSTITUTE

Eq. (2.58), the rest of the unknowns are obtained in the order from the last row ($i = n$) to the first row ($i = 1$). Note that at any ith row, the unknowns to the right of the pivot are known except for x_i, so the inner loop j runs from $jmin$ to $jmax$ to sum up the known products of '$a_{ij} * x_{i+j-d}$' to find the solution to an equation with one unknown. To reduce memory requirements, the solution vector **x** is written onto the rhs vector **b** without the need to use an extra array.

2.10.2   LU-DECOMPOSTION

In this section, it is assumed that a banded linear system results from the discretization of a partial differential equation and that it is diagonally dominant, which does not require pivoting. The naive Gauss elimination algorithm presented with Pseudocodes 2.14 and 2.15 does not implement any pivoting strategy. Hence, the LU-decomposition algorithm presented here also does not use pivoting.

Let **A** be an $n \times n_b$ real banded matrix defined as

$$
\mathbf{A} = \begin{bmatrix}
0 & \cdots & 0 & a_{1,d} & a_{1,d+1} & \cdots & a_{1,n_b} \\
\vdots & & a_{2,d-1} & a_{2,d} & a_{2,d+1} & \cdots & a_{2,n_b} \\
0 & & \vdots & \vdots & \vdots & & \vdots \\
a_{d,1} & \cdots & a_{d,d-1} & a_{d,d} & a_{d,d+1} & \cdots & a_{d,n_b} \\
\vdots & & \vdots & \vdots & \vdots & & \vdots \\
a_{k,1} & \cdots & a_{k,d-1} & a_{k,d} & a_{k,d+1} & \cdots & a_{k,n_b} \\
a_{k+1,1} & \cdots & a_{k+1,d-1} & a_{k+1,d} & a_{k,d+1} & & 0 \\
\vdots & & \vdots & \vdots & \vdots & & \vdots \\
a_{n,1} & \cdots & a_{n,d-1} & a_{n,d} & 0 & \cdots & 0
\end{bmatrix}, \tag{2.59}
$$

where $d = m_L + 1$ and $k = n_b - m_U$.

The lower and upper band matrices have a banded form similar to that of $\mathbf{A}$, and the decomposition is unique if $\mathbf{L}$ has unit diagonal elements. The compact forms of $\mathbf{L}$ and $\mathbf{U}$ can be written as

$$
\mathbf{L} = \begin{bmatrix}
0 & 0 & 0 & \ell_{1,d} \\
0 & 0 & \ell_{2,d-1} & \ell_{2,d} \\
0 & & \vdots & \vdots \\
\ell_{d,1} & \cdots & \ell_{2,d-1} & \ell_{d,d} \\
\vdots & & \vdots & \vdots \\
\ell_{i,1} & \cdots & \ell_{i,d-1} & \ell_{i,d} \\
\vdots & & \vdots & \vdots \\
\ell_{n,1} & \cdots & \ell_{n,d-1} & \ell_{n,d}
\end{bmatrix}, \quad
\mathbf{U} = \begin{bmatrix}
u_{1,d} & u_{1,d+1} & \cdots & u_{1,n_b} \\
\vdots & \vdots & & \vdots \\
u_{i,d} & u_{i,d+1} & \cdots & u_{i,n_b} \\
\vdots & \vdots & & \vdots \\
u_{k,d} & u_{k,d+1} & \cdots & u_{k,n_b} \\
\vdots & \vdots & & 0 \\
u_{n-1,d} & u_{n-1,d+1} & & \vdots \\
u_{n,d} & 0 & \cdots & 0
\end{bmatrix} \tag{2.60}
$$

where $\ell_{i,d} = 1$ for $i = 1, 2, \ldots, n$.

The elements of $\mathbf{A} = \mathbf{LU}$ can be generalized as

$$
a_{i,d+j-i} = \sum_{k=kmin}^{kmax} \ell_{i,d+k-i}\, u_{k,d+j-k}, \quad \left\{ \begin{matrix} 1 \leqslant i \leqslant n \\ \max(1, i-m_L) \leqslant j \leqslant \min(i+m_U, n) \end{matrix} \right\} \tag{2.61}
$$

where $kmin = \max(1, i-m_L, j-m_U)$, $kmax = \min(i,j) - 1$, and (i,j)'s denote the original matrix notations.

For $i \leqslant j \leqslant \min(i+m_U, n)$, the row sweep leads to

$$
u_{i,d+j-i} = a_{i,d+j-i} - \sum_{k=kmin}^{kmax} \ell_{i,d+k-i}\, u_{k,d+j-k}, \tag{2.62}
$$

On the other hand, for $j+1 \leqslant i \leqslant \min(j+m_L, n)$ column sweep yields

$$
\ell_{i,d+j-i} = \frac{1}{u_{j,d}} \left(a_{i,d+j-i} - \sum_{k=kmin}^{kmax} \ell_{i,d+k-i}\, u_{k,d+j-k} \right), \tag{2.63}
$$

Since the diagonals of $\mathbf{L}$ are $\ell_{i,d} = 1$, this property allows $\mathbf{L}$ and $\mathbf{U}$ to be stored exactly in place of $\mathbf{A}$, as shown below:

$$
\mathbf{A} = \begin{bmatrix}
0 & \cdots & 0 & u_{1,d} & u_{1,d+1} & \cdots & u_{1,n_b} \\
\vdots & & \ell_{2,d-1} & u_{2,d} & u_{2,d+1} & \cdots & u_{2,n_b} \\
0 & & \vdots & \vdots & \vdots & \vdots & \vdots \\
\ell_{d,1} & \cdots & \ell_{d,d-1} & u_{d,d} & u_{d,d+1} & \cdots & u_{d,n_b} \\
\vdots & & \vdots & \vdots & \vdots & \vdots & \vdots \\
\ell_{k1} & \cdots & \ell_{k,d-1} & u_{k,d} & u_{k,d+1} & \cdots & u_{k,n_b} \\
\ell_{k+1,1} & \cdots & \ell_{k+1,d-1} & u_{k+1,d} & u_{k,d+1} & & 0 \\
\vdots & & \vdots & \vdots & \vdots & & \vdots \\
\ell_{n1} & \cdots & \ell_{n,d-1} & u_{n,d} & 0 & \cdots & 0
\end{bmatrix}, \tag{2.64}
$$

The strategy of storing the computed ℓ_{ij} and u_{ij}'s on a_{ij} reduces the memory requirement.

A pair of pseudomodules, one for LU-decomposition (**BANDED_LU_ DECOMPOSE**) and another for the solution (**BANDED_LU_SOLVE**) of a real banded linear system, are presented in Pseudocodes 2.16 and 2.17. The module BANDED_LU_DECOMPOSE requires the number of equations (n), a banded matrix in compact form ($\mathbf{A}$), and the lower and upper bandwidths (m_L, m_U) as input. On exit, the triangular matrices $\mathbf{L}$ and $\mathbf{U}$ are written on the matrix $\mathbf{A}$.

For the first row, we obtain $u_{1,d+j} = a_{1,d+j}$ for $j = 1, \ldots, m_U$, while the second column below $a_{1,d+1}$ leads to $\ell_{i,d+1-i} = a_{i,d+1-i}/u_{1,d}$ for $i = 2$ to $i = m_L + 1$. A row sweep from the diagonal ($jmax = i$) to the rightmost non-zero element ($jmax$) gives a row of $\mathbf{U}$, while the inner loop over the k variable in Eq. (2.62) gives the summation terms. To obtain a column of $\mathbf{L}$, the columns are swept starting from the element below diagonal ($jmin = i+1$) and running down to the last non-zero element ($jmax$) with the summation in Eq.(2.63) obtained using the inner loop over k. This procedure is successively repeated until the last row.

Pseudocode 2.16

Module BANDED_LU_DECOMPOSE ($n, m_L, m_U, \mathbf{A}$)
\ DESCRIPTION: A pseudomodule to find $\mathbf{A} = \mathbf{LU}$ decomposition of
\ a banded matrix.
\ USES:
\ MAX/MIN:: Built-in functions to find the max/min of a set of integers.
Declare: a_{n,m_L+m_U+1}, b_n
$d \leftarrow m_L + 1$ \ Find position of diagonal in compact matrix
$n_b \leftarrow d + m_U$ \ Find total bandwidth
For $[i = 1, n]$ \ Loop i: Sweep from to to bottom
  $jmin \leftarrow i$ \ Find minimum column index in ith
  $jmax \leftarrow$ **MIN**($i + m_U, n$) \ Find maximum column index in ith
  For $[j = jmin, jmax]$ \ Loop j: Sweep through row-i left to right
    $p \leftarrow d + j - i$
    $kmin \leftarrow$ **MAX**($1, i - m_L, j - m_U$) \ Find $kmin$
    $kmax \leftarrow$ **MIN**(i, j) $- 1$ \ Find $kmax$
    For $[k = kmin, kmax]$ \ Loop k: Accumulating loop
      $s \leftarrow d + k - i; r \leftarrow d + j - k$
      $a_{ip} \leftarrow a_{ip} - a_{is} * a_{kr}$ \ Save u_{ip} on $\mathbf{A}$, $u_{ip} \rightarrow a_{ip}$
    End For
  End For
  $jmin \leftarrow i + 1$ \ First row below diagonal
  $jmax \leftarrow$ **MIN**($i + m_L, n$) \ Last row below diagonal
  For $[j = jmin, jmax]$ \ Loop j: Sweep through column-i below diagonal
    $p \leftarrow d + i - j$
    $kmin \leftarrow$ **MAX**($1, j - m_L, i - m_U$) \ Find $kmin$
    $kmax \leftarrow$ **MIN**(i, j) $- 1$ \ Find $kmax$
    For $[k = kmin, kmax]$ \ Loop k: Accumulating loop
      $s \leftarrow d + k - j$
      $r \leftarrow d + i - k$
      $a_{jp} \leftarrow a_{jp} - a_{js} * a_{kr}$ \ Add $-a_{js} * a_{kr}$ to a_{jp}
    End For
    $a_{jp} \leftarrow a_{jp}/a_{id}$ \ Save ℓ_{jp} on $\mathbf{A}$, $\ell_{jp} \rightarrow a_{jp}$
  End For
End For
End Module BANDED_LU_DECOMPOSE

Pseudocode 2.17

Module BANDED_LU_SOLVE $(n, m_L, m_U, \mathbf{A}, \mathbf{b})$
\ DESCRIPTION: A pseudomodule to solve $\mathbf{Ax} = \mathbf{b}$ system by LU-decomposition.
\ USES:
\   MAX/MIN:: Built-in functions to find the max/min of a set of integers.
Declare: a_{n,m_L+m_U+1}, b_n
$d \leftarrow m_L + 1$ \ Find position of diagonal in compact matrix
$n_b \leftarrow d + m_U$ \ Find total bandwidth length
For $[i = 1, n]$ \ Forward elimination step to solve $\mathbf{Ly} = \mathbf{b}$
   $kmin \leftarrow$ **MAX**$(1, \; i - m_L)$
   $kmax \leftarrow i - 1$
   For $[k = kmin, kmax]$ \ Applying row reduction to below diagonals
      $s \leftarrow d + k - i$
      $b_i \leftarrow b_i - a_{is} * b_k$ \ Apply Eq. (2.65)
   End For
End For
$b_n \leftarrow b_n/a_{nd}$ \ Save solution from the last equation on b_n
For $[i = (n-1), 1, (-1)]$ \ Backward elimination, $\mathbf{Ux} = \mathbf{y}$
   $kmin \leftarrow 1$
   $kmax \leftarrow$ **MIN**$(m_U, n - i)$
   For $[k = kmin, kmax]$
      $s \leftarrow k + d; \; r = k + i$
      $b_i \leftarrow b_i - a_{is} * b_r$ \ Apply Eq. (2.65)
   End For
   $b_i \leftarrow b_i/a_{id}$ \ Save the solution on b_i
End For
End Module BANDED_LU_SOLVE

The input argument list of BAND_LU _SOLVE includes the right-hand side vector $\mathbf{b}$ in addition to those of BANDED_LU_ DECOMPOSE. The task of finding the solution is broken into two parts: first, $\mathbf{y}$ is found such that $\mathbf{Ux} = \mathbf{y}$; next, $\mathbf{x}$ is found from $\mathbf{Ly} = \mathbf{b}$.

The forward substitution can be applied to $\mathbf{Ly} = \mathbf{b}$ to obtain $\mathbf{y}$ as follows:

$$y_i = b_i - \sum_{k=kmin}^{kmax} \ell_{i,d+k-i} \, y_k, \quad i = 1, 2, \ldots, n \tag{2.65}$$

where $kmin =$ MAX$(1, i - m_L)$ and $kmax = i - 1$. Note that the fact that $\ell_{i,d} = 1$ for all i's has been taken into account in Eq. (2.65), i.e., $\ell_{i,d}$'s have already been used in the formulation.

The back substitution procedure applied to $\mathbf{Ux} = \mathbf{y}$ yields

$$x_n = \frac{y_n}{u_{n,d}}, \qquad x_i = \frac{1}{u_{i,d}}\left(y_i - \sum_{k=kmin}^{kmax} u_{i,k+d} \, y_{k+i}\right), \qquad i = (n-1), \ldots, 3, 2, 1 \tag{2.66}$$

where $kmin = 1$ and $kmax = \min(m_U, n - i)$.

As a measure to reduce the memory requirement, $\mathbf{L}$ and $\mathbf{U}$ can be saved on matrix $\mathbf{A}$ by setting $\ell_{ij} = a_{ij}$ and $u_{ij} = a_{ij}$ in Eqs. (2.65) and (2.66). Similarly, setting $b_i = y_i$ in the forward elimination and $b_i = x_i$ in the back substitution algorithm eliminates the need for $\mathbf{y}$ and $\mathbf{x}$ vectors.

EXAMPLE 2.11: Gauss elimination of Banded Systems

Consider a cantilever beam of constant cross-section of length L that is fixed at one end and free at the other end, as shown in the figure below. A distributed load of $q(x) = q_0 \, (x/L)$ is applied to the beam.

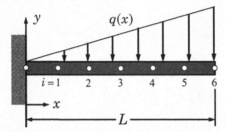

The model for the downward deflection, y, of the beam is given as

$$EI\frac{d^4y}{dx^4} = q(x), \quad y(0) = y'(0) = 0, \quad y''(L) = y''(L) = 0$$

where E is the modulus of elasticity and I is the cross-sectional area moment of inertia. For $EI = 500$ MPa, $L = 6$ m, $q_0 = 1$ MN/m and 6 intervals ($h = L/6$), solving the model differential equation using a second-order finite-difference scheme (*see* Section 10.9) yields the following linear system:

$$\begin{bmatrix} 7 & -4 & 1 & & & \\ -4 & 6 & -4 & 1 & & \\ 1 & -4 & 6 & -4 & 1 & \\ & 1 & -4 & 6 & -4 & 1 \\ & & 1 & -4 & 5 & -2 \\ & & & 2 & -4 & 2 \end{bmatrix} \begin{bmatrix} y_1 \\ y_2 \\ y_3 \\ y_4 \\ y_5 \\ y_6 \end{bmatrix} = \frac{h^4}{EI} \cdot \frac{q_0}{L} \begin{bmatrix} x_1 \\ x_2 \\ x_3 \\ x_4 \\ x_5 \\ x_6 \end{bmatrix}$$

Here $x_i = ih$ denotes the position of the nodal points in $i = 1, 2, \ldots, 6$ for which the deflection is computed, i.e., $y_i = y(x_i)$. (a) First apply the banded LU-decomposition algorithm to find $\mathbf{L}$ and $\mathbf{U}$ matrices, and (b) use $\mathbf{L}$ and $\mathbf{U}$ with the forward and backward substitution algorithms to solve the banded system of equations.

SOLUTION:

The $\mathbf{L}$ and $\mathbf{U}$ matrices are assumed to have the following compact forms:

$$\mathbf{L} = \begin{bmatrix} 0 & 0 & 1 \\ 0 & \ell_{22} & 1 \\ \ell_{31} & \ell_{32} & 1 \\ \ell_{4,1} & \ell_{4,2} & 1 \\ \ell_{5,1} & \ell_{5,2} & 1 \\ \ell_{6,1} & \ell_{6,2} & 1 \end{bmatrix}, \quad \mathbf{U} = \begin{bmatrix} u_{13} & u_{14} & u_{15} \\ u_{23} & u_{24} & u_{25} \\ u_{33} & u_{34} & u_{35} \\ u_{4,3} & u_{44} & u_{45} \\ u_{53} & u_{54} & 0 \\ u_{6,3} & 0 & 0 \end{bmatrix},$$

Employing Eq. (2.62) to the first row, we find

$$u_{13} = 7, \qquad u_{14} = -4, \qquad u_{15} = 1$$

Next, we apply Eq. (2.63) to the first column to get

$$\ell_{22}u_{13} = -4 \Rightarrow \ell_{22} = -\frac{4}{7}, \qquad \ell_{31}u_{13} = 1 \Rightarrow \ell_{31} = \frac{1}{7}$$

Revisiting Eq. (2.62), we find

$$\ell_{22}u_{14} + u_{23} = 6 \Rightarrow u_{23} = \frac{26}{7}, \qquad \ell_{22}u_{15} + u_{24} = -4 \Rightarrow u_{24} = -\frac{24}{7}$$

Then, by applying Eq. (2.63), we obtain

$$u_{14}\ell_{31} + u_{23}\ell_{32} = -4 \Rightarrow \ell_{32} = -\frac{12}{13} \qquad u_{23}\ell_{4,1} = 1 \Rightarrow \ell_{4,1} = \frac{1}{7}$$

In this manner, going back and forth between rows and columns (i.e., between Eqs. (2.62) and (2.63)), the following equations are obtained.

$$u_{25} = 1, \qquad\qquad u_{35} = 1,$$
$$u_{45} = 1, \qquad\qquad \ell_{6,1}u_{4,3} = 2,$$
$$\ell_{4,1}u_{23} = 1, \qquad\qquad \ell_{5,1}u_{33} = 1,$$
$$\ell_{32}u_{25} + u_{34} = -4, \qquad\qquad \ell_{5,2}u_{45} + u_{54} = -2,$$
$$\ell_{4,2}u_{35} + u_{44} = -4, \qquad\qquad \ell_{31}u_{14} + \ell_{32}u_{23} = -4,$$
$$\ell_{4,1}u_{24} + \ell_{4,2}u_{33} = -4, \qquad\qquad \ell_{5,1}u_{34} + \ell_{5,2}u_{4,3} = -4,$$
$$\ell_{6,1}u_{44} + \ell_{6,2}u_{53} = -4, \qquad\qquad \ell_{31}u_{15} + \ell_{32}u_{24} + u_{33} = 6,$$
$$\ell_{4,1}u_{25} + \ell_{4,2}u_{34} + u_{4,3} = 6, \qquad\qquad \ell_{5,1}u_{35} + \ell_{5,2}u_{44} + u_{53} = 5,$$
$$\ell_{6,1}u_{45} + \ell_{6,2}u_{54} + u_{6,3} = 2$$

Each time, solving one equation for one unknown leads to the following solution:

$$\mathbf{L} = \begin{bmatrix} 0 & 0 & 1 \\ 0 & -4/7 & 1 \\ 1/7 & -12/13 & 1 \\ 7/26 & -8/7 & 1 \\ 13/35 & -40/31 & 1 \\ 28/31 & -110/73 & 1 \end{bmatrix}, \qquad \mathbf{U} = \begin{bmatrix} 7 & -4 & 1 \\ 26/7 & -24/7 & 1 \\ 35/13 & -40/13 & 1 \\ 31/14 & -20/7 & 1 \\ 146/155 & -22/31 & 0 \\ 2/73 & 0 & 0 \end{bmatrix}$$

(b) Since $\mathbf{Ax} = \mathbf{b}$ can now be expressed as $\mathbf{Lz} = \mathbf{b}$ where $\mathbf{Uy} = \mathbf{z}$. Substituting the numerical values given in the problem, the matrix equation $\mathbf{Lz} = \mathbf{b}$ becomes

$$\begin{bmatrix} 0 & 0 & 1 \\ 0 & -4/7 & 1 \\ 1/7 & -12/13 & 1 \\ 7/26 & -8/7 & 1 \\ 13/35 & -40/31 & 1 \\ 28/31 & -110/73 & 1 \end{bmatrix} \begin{bmatrix} z_1 \\ z_2 \\ z_3 \\ z_4 \\ z_5 \\ z_6 \end{bmatrix} = \frac{1}{3000} \begin{bmatrix} 1 \\ 2 \\ 3 \\ 4 \\ 5 \\ 6 \end{bmatrix},$$

This system is solved using forward substitution, where the unknowns (the elements of vector $\mathbf{z}$) are calculated one by one, starting from the first to the last equation, which leads to

$$\mathbf{z} = \frac{1}{3000}\left[1, \; \frac{18}{7}, \; \frac{68}{13}, \; \frac{65}{7}, \; \frac{2331}{155}, \; \frac{1470}{73}\right]^T$$

Substituting $\mathbf{z}$ into $\mathbf{Uy} = \mathbf{z}$, we find

$$\begin{bmatrix} 7 & -4 & 1 \\ 26/7 & -24/7 & 1 \\ 35/13 & -40/13 & 1 \\ 31/14 & -20/7 & 1 \\ 146/155 & -22/31 & 0 \\ 2/73 & 0 & 0 \end{bmatrix}\begin{bmatrix} y_1 \\ y_2 \\ y_3 \\ y_4 \\ y_5 \\ y_6 \end{bmatrix} = \frac{1}{3000}\begin{bmatrix} 1 \\ 18/7 \\ 68/13 \\ 65/7 \\ 2331/155 \\ 1470/73 \end{bmatrix}$$

This linear system is solved using the back substitution algorithm, which results in

$$\mathbf{y} = \frac{1}{6000}\begin{bmatrix} 73 \\ 256 \\ 515 \\ 820 \\ 1147 \\ 1480 \end{bmatrix} = \begin{bmatrix} 0.012167 \\ 0.042667 \\ 0.085833 \\ 0.136667 \\ 0.191167 \\ 0.246667 \end{bmatrix}$$

Discussion: The solution of a single $\mathbf{Ax} = \mathbf{b}$ using LU-decomposition, whether $\mathbf{A}$ is a banded matrix or not, is not economical. The banded linear system avoids dealing with zeros, which can lead to significant savings in computation time. The elements of $\mathbf{L}$ and $\mathbf{U}$ matrices are easily obtained by successive row-column sweeps.

2.11 CLOSURE

Computer applications of linear algebra and matrices are encountered in many fields of science and engineering. Most applied problems are reduced to a set of simultaneous linear equations or a linear system. That is why manipulation of matrices and solving systems of linear algebraic equations have been the core of many methods. In this respect, the most commonly used direct methods have been covered in this chapter.

As the most popular methods, Gauss elimination and its variants *with* and *without* pivoting and Gauss-Jordan elimination algorithms are very effective and suitable for moderately sized simultaneous systems of linear equations. The direct elimination methods reduce equations to an upper triangular or diagonal matrix through a series of row reductions. Then, the unknowns are easily obtained by employing the back-substitution procedure. Gauss elimination is the method of choice for small systems (about a few hundred unknowns) with few zeros in the coefficient matrices. Specially tailored algorithms, such as symmetric or tridiagonal systems, and so on, are not only fast but also memory efficient.

The Gauss-Jordan can be economically used to obtain the inverse of a relatively small matrix as well. Though similar to Gaussian elimination, it zeros out the elements both above and below the principle diagonal of the augmented matrix. The Gauss-Jordan method can be successfully employed for overdetermined systems as well.

Decomposition methods factor a coefficient matrix into products of two (or three) matrices. It is more convenient to implement the LU decomposition methods when the right-hand side of a linear system changes while the coefficient matrix remains unchanged. Once the matrix is decomposed, the solution is obtained by first solving $\mathbf{Ly} = \mathbf{b}$ (forward substitution) for $\mathbf{y}$ and then solving $\mathbf{Ux} = \mathbf{y}$ (back substitution) for $\mathbf{x}$.

For large matrices or systems that are *not* diagonally dominant, round-off errors can be a serious problem. Errors in a computed solution are either due to existing errors in $\mathbf{A}$ and $\mathbf{b}$ or rounding errors made during the execution of the algorithm or both. The quality of matrix $\mathbf{A}$ strongly affects the error behavior. The condition of a matrix can be quantitatively determined by using the concept of norms. A large condition number implies ill-conditioning, which requires further treatment.

The numerical solution of high-order ODEs and elliptic PDEs in regular or irregular domains results in a banded linear system of equations. Finding numerical solutions to these kinds of problems is fast and very efficient for moderate-sized problems.

2.12 EXERCISES

Section 2.1 Fundamentals of Linear Algebra

E2.1 For given $\mathbf{A}$ and $\mathbf{B}$, find $\mathbf{A} + \mathbf{B}$ and $\mathbf{A} - \mathbf{B}$.

$$\mathbf{A} = \begin{bmatrix} 3 & 1 & -2 \\ 3 & -4 & 5 \end{bmatrix}_{2\times3} \qquad \mathbf{B} = \begin{bmatrix} 2 & -2 & -4 \\ -1 & 3 & -1 \end{bmatrix}_{2\times3}$$

E2.2 For given $\mathbf{A}$, $\mathbf{B}$, $\mathbf{C}$, and $\mathbf{D}$, find (a) $2\mathbf{A} - 3\mathbf{B} + \mathbf{C}$, and (b) determine x and y values that satisfy $\mathbf{D} = x\mathbf{B} + y\mathbf{C}$.

$$\mathbf{A} = \begin{bmatrix} 3 & 1 & -2 \\ 1 & -2 & -1 \end{bmatrix}, \quad \mathbf{B} = \begin{bmatrix} 1 & 2 & -1 \\ 4 & 3 & 1 \end{bmatrix}, \quad \mathbf{C} = \begin{bmatrix} 1 & 2 & 3 \\ 2 & -3 & 5 \end{bmatrix}, \quad \mathbf{D} = \begin{bmatrix} 6 & 12 & 10 \\ 16 & -6 & 22 \end{bmatrix}$$

E2.3 For given $\mathbf{A}$ and $\mathbf{B}$, find $\mathbf{AB}$ and $\mathbf{BA}$.

(a) $\mathbf{A} = \begin{bmatrix} 2 & 3 \\ 1 & -1 \\ 0 & 4 \end{bmatrix}, \quad \mathbf{B} = \begin{bmatrix} 5 & -2 & 4 & 7 \\ -6 & 1 & -3 & 0 \end{bmatrix},$ \qquad (b) $\mathbf{A} = \begin{bmatrix} 1 & 2 \\ 3 & 4 \end{bmatrix}, \quad \mathbf{B} = \begin{bmatrix} 1 & 1 \\ 4 & 1 \end{bmatrix}$

E2.4 For given $\mathbf{A}$ and $\mathbf{B}$, find $\mathbf{AB}$.

(a) $\mathbf{A} = \begin{bmatrix} 1 & 1 \\ 2 & 2 \end{bmatrix}, \quad \mathbf{B} = \begin{bmatrix} -1 & 1 \\ 1 & -1 \end{bmatrix},$ \qquad (b) $\mathbf{A} = \begin{bmatrix} 1 & 0 & 4 \\ 2 & 0 & 0 \\ 0 & 0 & 0 \end{bmatrix}, \quad \mathbf{B} = \begin{bmatrix} 0 & 0 & 0 \\ 0 & 3 & 5 \\ 0 & 0 & 0 \end{bmatrix}$

E2.5 For given $\mathbf{A}$, $\mathbf{B}$, and $\mathbf{C}$, find (a) $(\mathbf{A}^T\mathbf{C})\mathbf{B}$, (b) $(\mathbf{A}^T + \mathbf{B}^T)\mathbf{C}$, and (c) $(\mathbf{B}^T\mathbf{C})\mathbf{A}^T$.

$$\mathbf{A} = \begin{bmatrix} 2 & 1 & -1 & 1 \\ 3 & 0 & 2 & -3 \\ 1 & -2 & 3 & 0 \end{bmatrix}, \quad \mathbf{B} = \begin{bmatrix} 1 & 1 & 0 & 3 \\ 0 & 1 & -2 & 1 \\ 3 & 1 & 2 & 1 \end{bmatrix}, \quad \mathbf{C} = \begin{bmatrix} -2 & 0 & 1 \\ 1 & 1 & -3 \\ -2 & 1 & 1 \end{bmatrix}$$

E2.6 For given $\mathbf{A}$, $\mathbf{B}$, and $\mathbf{C}$, verify that (a) $\mathbf{AB} \neq \mathbf{BA}$; (b) $\mathbf{A}(\mathbf{BC}) = (\mathbf{AB})\mathbf{C}$; and (c) $\mathbf{A}^T\mathbf{B}^T = (\mathbf{BA})^T$.

$$\mathbf{A} = \begin{bmatrix} 2 & 1 & -2 \\ 3 & 1 & 2 \\ 1 & 4 & -3 \end{bmatrix}, \quad \mathbf{B} = \begin{bmatrix} 1 & 3 & -3 \\ 1 & 4 & -2 \\ 2 & -2 & 5 \end{bmatrix}, \quad \mathbf{C} = \begin{bmatrix} 2 & 1 & 0 \\ 2 & 1 & -1 \\ 1 & 0 & 2 \end{bmatrix}$$

E2.7 For given $\mathbf{A}$ and $\mathbf{B}$, determine a and b that satisfy the following matrix equation:

$$\mathbf{A} = \begin{bmatrix} a & b \\ c & d \end{bmatrix}, \quad \mathbf{B} = \begin{bmatrix} 4 & 2 \\ -1 & -4 \end{bmatrix}, \quad \left(3\mathbf{A} + \mathbf{B}^T\right)^T - 2\mathbf{A}^T = \begin{bmatrix} 7 & 1 \\ -3 & 0 \end{bmatrix}$$

E2.8 Find $\mathbf{A}$ such that the following matrix operations are satisfied:

$$\left(\mathbf{A}^T - 2\mathbf{I}\right)^{-1} = \begin{bmatrix} -3 & 8 \\ -1 & 3 \end{bmatrix} \quad \text{and} \quad \det(\mathbf{A}^T - 2\mathbf{I}) = -1$$

E2.9 Determine all the values a, b, c, and d that satisfy the following matrix equation:

$$\begin{bmatrix} a & b \\ c & d \end{bmatrix} \begin{bmatrix} 1 & 2 \\ 3 & 1 \end{bmatrix} - \begin{bmatrix} 1 & 2 \\ 3 & 1 \end{bmatrix} \begin{bmatrix} a & b \\ c & d \end{bmatrix} = \begin{bmatrix} 1 & 0 \\ -1 & -1 \end{bmatrix}$$

E2.10 For given $\mathbf{A}$ and $\mathbf{B}$, is there a pair of x and y that satisfy $\mathbf{A}^2 - \mathbf{B}^2 = (\mathbf{A} - \mathbf{B})(\mathbf{A} + \mathbf{B})$?

$$\mathbf{A} = \begin{bmatrix} 1 & 2 & -1 \\ x & 1 & -2 \\ -2 & -3 & 2 \end{bmatrix}, \quad \mathbf{B} = \begin{bmatrix} 4 & 1 & 3 \\ 2 & 0 & 1 \\ 7 & y & 5 \end{bmatrix}$$

E2.11 Use Sarrus's rule to calculate the determinants:

(a) $\begin{vmatrix} 1 & 2 & -1 \\ 1 & 1 & -2 \\ -1 & -1 & 1 \end{vmatrix}$, (b) $\begin{vmatrix} 1 & 1 & -1 \\ -1 & 1 & -1 \\ -1 & -1 & 1 \end{vmatrix}$, (c) $\begin{vmatrix} 3 & 2 & 1 \\ 4 & 1 & -3 \\ 1 & -5 & -2 \end{vmatrix}$, (d) $\begin{vmatrix} 1 & -5 & -1 \\ 6 & 3 & -1 \\ -4 & -1 & 3 \end{vmatrix}$

E2.12 Compute the determinants using the cofactor expansion across the first row:

(a) $\begin{vmatrix} 1 & -1 & -1 & -1 \\ 1 & 1 & -1 & 1 \\ 1 & 1 & 1 & -1 \\ 1 & -1 & 1 & 1 \end{vmatrix}$, (b) $\begin{vmatrix} 2 & -1 & -3 & 1 \\ 4 & 1 & -3 & 1 \\ 3 & 2 & 4 & -2 \\ 2 & -1 & 3 & 1 \end{vmatrix}$, (c) $\begin{vmatrix} 1 & 3 & -2 & 2 \\ -2 & -3 & -1 & 4 \\ 1 & 1 & 2 & 1 \\ 1 & -3 & 4 & 5 \end{vmatrix}$

E2.13 Find the unknowns that satisfy the following determinants:

(a) $\begin{vmatrix} 1 & 2 & x \\ -3 & 2 & 3 \\ 4 & 1 & 2 \end{vmatrix} = 4$, (b) $\begin{vmatrix} 5 & -3 & 1 \\ u & -2 & 1 \\ -2 & 3 & 2 \end{vmatrix} = 3$, (c) $\begin{vmatrix} 3 & -2 & 3 & 1 \\ 5 & 1 & -3 & 2 \\ 1 & 2 & 3 & 1 \\ 2 & 1 & z & -2 \end{vmatrix} = 0$

E2.14 Express the given system of linear equations in matrix equation form, $\mathbf{Ax} = \mathbf{b}$.

(a) $\begin{cases} 3x + 2y = 1 \\ x - 5y = 7 \end{cases}$, (b) $\begin{cases} 4y - 3x = 9 \\ 5x + 6y = 3 \end{cases}$ (c) $\begin{cases} a + 2b - c = 1 \\ a - 3c - 2b = 2 \\ c - 3a + 4b = 1 \end{cases}$

(d) $\begin{cases} 3u + 2v - w = 0 \\ w + 2u - 2v = 0 \\ 5v - w + 3u = 1 \end{cases}$ (e) $\begin{cases} 3x + y + 2z = 6 \\ -2y + 4z = 8 \\ 7z = 7 \end{cases}$ (f) $\begin{cases} 5a = 15 \\ 2a - 5b = -14 \\ 4a - 3b - 2c = 4 \end{cases}$

E2.15 Express the given system of linear equations in matrix equation form, $\mathbf{Ax} = \mathbf{b}$.

(a) $\begin{cases} 5x_1 - 4x_2 = 9 \\ -2x_1 + 4x_2 - x_3 = -8 \\ -2x_2 + 6x_3 - 3x_4 = 20 \\ -2x_3 + 4x_4 - x_5 = -15 \\ -2x_4 + 6x_5 - 2x_6 = 28 \\ -2x_5 + 4x_6 = -18 \end{cases}$ (b) $\begin{cases} 2x_1 + 3x_2 = 7, \\ x_1 + 2x_2 + x_3 = 7, \\ -x_1 - x_2 + 2x_3 + 4x_4 = 19, \\ x_2 + x_3 + 3x_4 - x_5 = 17, \\ -x_3 - x_4 + 3x_5 + 2x_6 = -14, \\ x_4 + x_5 + 4x_6 = -5 \end{cases}$

E2.16 Apply Cramer's rule to find solutions for the linear systems given below.

(a) $\begin{cases} 2x - 3y = -15 \\ 3x - 5y = -26 \end{cases}$, (b) $\begin{cases} 4x + 3y + 2z = 4 \\ 3x - 2y + 5z = 27 \\ -2x + 3y + z = -16 \end{cases}$, (c) $\begin{cases} x + y - 2z = -5 \\ 2x + y + 3z = 3 \\ x + 2y - 3z = -6 \end{cases}$

E2.17 Use Cramer's rule to find solutions for the linear systems given below.

(a) $\begin{cases} 3\sqrt{x} + 2/y = 11 \\ \sqrt{x} + 5/y = 8 \end{cases}$, (b) $\begin{cases} \sqrt{a} + 2\sqrt{b} - c^2 = 4 \\ \sqrt{a} - 3c^2 + 2\sqrt{b} = 2 \\ c^2 - 3\sqrt{a} + 4\sqrt{b} = 6 \end{cases}$, (c) $\begin{cases} 3xy + y^2 + 2e^z = 29 \\ 2y^2 - 4e^z = 14 \\ 3xy - e^z = 17 \end{cases}$

E2.18 Find all the eigenvalues and eigenvectors of the following matrices.

(a) $\begin{bmatrix} 2 & -3 \\ 3 & -8 \end{bmatrix}$, (b) $\begin{bmatrix} 3 & 5 \\ 16 & 1 \end{bmatrix}$, (c) $\begin{bmatrix} 12 & 4 & -18 \\ 13 & 6 & -21 \\ 11 & 4 & -17 \end{bmatrix}$, (d) $\begin{bmatrix} 3 & 4 & -8 \\ 2 & 7 & -10 \\ 2 & 4 & -7 \end{bmatrix}$

E2.19 Find all the eigenvalues and eigenvectors of the tridiagonal Toeplitz matrices given below.

(a) $\begin{bmatrix} -4 & 1 & & \\ 1 & -4 & 1 & \\ & 1 & -4 & 1 \\ & & 1 & -4 \end{bmatrix}$, (b) $\begin{bmatrix} 10 & -9 & \\ -1 & 10 & -9 \\ & -1 & 10 & -9 \\ & & -1 & 10 \end{bmatrix}$, (c) $\begin{bmatrix} 3 & -2 & & \\ -2 & 3 & -2 & \\ & -2 & 3 & -2 \\ & & -2 & 3 \end{bmatrix}$

E2.20 Calculate the 1-, 2-, and ∞-norms of the following vectors:

(a) $\mathbf{x} = \begin{bmatrix} 3 & 1 & -2 & 1 & 0 & 1 \end{bmatrix}^T$ (b) $\mathbf{y} = \begin{bmatrix} 1 & 1 & -1 & 1 & 1 \end{bmatrix}^T$, (c) $\mathbf{u} = \begin{bmatrix} -1 & 1 & 0 & 2 & 1 & -2 \end{bmatrix}^T$

E2.21 Calculate the 1-, 2-, and F-norms of the following matrices:

(a) $\mathbf{A} = \begin{bmatrix} 1 & 2 \\ 3 & 4 \end{bmatrix}$, (b) $\mathbf{A} = \begin{bmatrix} 1 & 0 & -2 \\ 2 & -3 & 1 \\ 4 & 1 & -1 \end{bmatrix}$, (c) $\mathbf{A} = \begin{bmatrix} 2 & 1 & -1 & 0 \\ 4 & -3 & 2 & 2 \\ 1 & 0 & 2 & -3 \\ 1 & 1 & -2 & 1 \end{bmatrix}$

E2.22 Compare 1-, 2-, and F-norms of $\|\mathbf{A}^2\|$ and $\|\mathbf{A}\|^2$ for the given matrices in **E2.21**.

E2.23 Use the adjoint method to find the inverse of the matrices given below:

(a) $\begin{bmatrix} 2 & -3 \\ 3 & -5 \end{bmatrix}$, (b) $\begin{bmatrix} 3 & -6 & 2 \\ 3 & 1 & -1 \\ 4 & 3 & -2 \end{bmatrix}$, (c) $\begin{bmatrix} 1 & -1 & 2 \\ 3 & -2 & 1 \\ -2 & 3 & 1 \end{bmatrix}$, (d) $\begin{bmatrix} 1 & 1 & -2 \\ 2 & -4 & 3 \\ 1 & 2 & -3 \end{bmatrix}$

E2.24 For given $\mathbf{A}$, $\mathbf{B}$, and $\mathbf{C}$, what should x and y be for these matrices to be singular?

(a) $\mathbf{A} = \begin{bmatrix} x & -1 & 1 \\ y & 1 & -1 \\ 4 & 2 & -1 \end{bmatrix}$, (b) $\mathbf{B} = \begin{bmatrix} 2 & x & -9 \\ 0 & y & x \\ 1 & 3 & -4 \end{bmatrix}$, (c) $\mathbf{C} = \begin{bmatrix} 2 & 2 & 3 \\ -2 & -1 & x \\ 4y & y & -y \end{bmatrix}$

E2.25 For given $\mathbf{A}$, verify $\det(\mathbf{A}^{-1}) = 1/\det(\mathbf{A})$.

$$\mathbf{A} = \begin{bmatrix} 2 & -2 & -3 \\ -2 & 2 & 2 \\ 4 & -5 & 2 \end{bmatrix}$$

E2.26 Find the condition(s) under which the following matrices are *not* invertible.

$\mathbf{A} = \begin{bmatrix} 1 & 2 & -1 \\ 2 & 4 & 2 \\ 3 & x & 2 \end{bmatrix}$, $\mathbf{B} = \begin{bmatrix} 3 & x+y & -2 \\ 2 & x-y & -1 \\ -3 & 1 & 5 \end{bmatrix}$, $\mathbf{C} = \begin{bmatrix} 3 & -8 & 2 \\ -2 & -y & 1 \\ 1 & 1 & y \end{bmatrix}$, $\mathbf{D} = \begin{bmatrix} x & -2 & 2 \\ -2 & x & 2 \\ 2 & 2 & x \end{bmatrix}$

E2.27 For given $\mathbf{B}$ and $\mathbf{P}$, find $\mathbf{PBP}^{-1}$ and $\mathbf{P}^{-1}\mathbf{BP}$.

$$\mathbf{P} = \begin{bmatrix} 1 & 2 & -1 \\ 1 & 1 & -2 \\ -1 & -1 & 1 \end{bmatrix}, \quad \mathbf{B} = \begin{bmatrix} 1 & 1 & -1 \\ -1 & 1 & -1 \\ -1 & -1 & 1 \end{bmatrix}$$

Section 2.2 Elementary Matrix Operations

E2.28 Propose a sequence of row operations on the given matrices to obtain the desired form.

(a) $\begin{bmatrix} 3 & 5 & 7 \\ x+2 & y+3 & z+4 \\ x+1 & y+2 & z+3 \end{bmatrix} \rightarrow \begin{bmatrix} x & y & z \\ 1 & 2 & 3 \\ 2 & 3 & 4 \end{bmatrix}$, (b) $\begin{bmatrix} x & -y & -x & y \\ y & x & -y & -x \\ x & -y & x & -y \\ y & x & y & x \end{bmatrix} \rightarrow \begin{bmatrix} x & -y & 0 & 0 \\ y & x & 0 & 0 \\ 0 & 0 & x & -y \\ 0 & 0 & y & x \end{bmatrix}$

E2.29 Apply elementary row operations to the homogeneous linear systems given below to find their reduced-row echelon forms.

(a) $\begin{cases} x_1 - 3x_2 - 11x_3 = 0 \\ 4x_1 + 5x_2 + 7x_3 = 0 \end{cases}$

(b) $\begin{cases} x_1 + 3x_2 - 2x_3 + 2x_4 = 0 \\ 2x_1 - x_2 + x_3 + 3x_4 = 0 \\ 7x_1 + 7x_2 - 4x_3 + 12x_4 = 0 \end{cases}$

(c) $\begin{cases} 2x_1 + x_2 - 3x_3 + 2x_4 - 7x_5 = 0 \\ 3x_1 - x_2 + 3x_3 + 2x_4 + 9x_5 = 0 \\ 2x_1 + 2x_2 - x_3 + 3x_4 - 6x_5 = 0 \\ -x_1 + 4x_2 + 5x_3 + x_4 + 2x_5 = 0 \end{cases}$

(d) $\begin{cases} x_1 + 2x_2 - x_3 + x_4 + x_5 = 0 \\ 4x_1 - 2x_2 + 3x_3 + x_4 + 5x_5 = 0 \\ 2x_1 + 2x_2 - 4x_3 + x_4 + 3x_5 = 0 \\ -7x_1 + 8x_2 - 5x_3 - 10x_5 = 0 \end{cases}$

E2.30 Find the augmented matrix and then reduce it to reduced-row echelon form to balance the following chemical equations:

(a) $x_1(C_2H_5OH) + x_2(O_2) \rightarrow x_3(CO_2) + x_4(H_2O)$

(b) $x_1(H_2(SO_4)) + x_2(Fe(OH)_3) \rightarrow x_3(Fe_2(SO_4)_3 + x_4(H_2O)$

(c) $x_1(Fe_2(SO_4)_3) + x_2(KOH) \rightarrow x_3(K_2SO_4) + x_4(Fe(OH)_3)$

(d) $x_1(HIO_3) + x_2(FeI_2) + x_3(HCl) \rightarrow x_4(FeCl_3) + x_5(ICl) + x_6(H_2O)$

(e) $x_1((KMnO_4) + x_2(HCl) \rightarrow x_3(KCl) + x_4(MnCl_2) + x_5(H_2O) + x_6(Cl)_2)$

(f) $x_1(C_2H_2Cl_4) + x_2(Ca(OH)_2) \rightarrow x_3(C_2HCl_3) + x_4(CaCl_2) + x_5(H_2O)$

E2.31 The traffic flow in a district with several one-way streets is depicted in **Fig. E2.31**. Obtain the general flow pattern for the network, assuming that the total flow into the network (or junction) is equal to the total flow out of the network (or junction). Apply *row reductions* to find the minimum number of vehicles for each road.

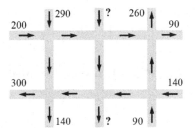

Fig. E2.31

E2.32 The traffic pattern for a roundabout is shown in **Fig. E2.32**. What is the smallest possible value for x_1? Note that the net flow rate in and out of the roundabout is 700 vph (vehicles per hour).

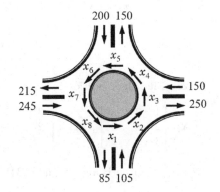

Fig. E2.32

E2.33 Find all solutions to the simultaneous equations by reducing the augmented matrix into reduced-row echelon form.

$$
\begin{aligned}
2x_1 &- 2x_2 + 4x_3 + 3x_4 + x_5 &= 5 \\
5x_1 &+ x_2 - x_4 - 3x_5 &= -7 \\
-x_1 &+ 4x_2 - 2x_3 + 3x_4 + x_5 &= 8 \\
-3x_1 &- 4x_2 + 5x_3 + 7x_4 + 10x_5 &= 35
\end{aligned}
$$

E2.34 Raw materials with different compositions from three sources are processed in an enrichment plant to obtain two highly concentrated products containing either B or C. The process flowchart with compositions in w% is given in **Fig. E2.34**.

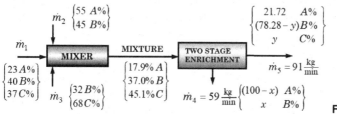

Fig. E2.34

The mass and species balance are given as

$$\text{Overall balance: } \dot{m}_1 + \dot{m}_2 + \dot{m}_3 = \dot{m}_4 + \dot{m}_5$$

$$\text{For A: } 23\dot{m}_1 + 55\dot{m}_2 + (0\%)\dot{m}_3 = (100 - x)59 + 21.72(91)$$

$$\text{For B: } 40\dot{m}_1 + 45\dot{m}_2 + 32\dot{m}_3 = 59x + 91(78.28 - y)$$

$$\text{For C: } 37\dot{m}_1 + (0)\dot{m}_2 + 68\dot{m}_3 = 91y$$

and at the enrichment stage:

$$\text{For A: } 17.9(\dot{m}_1 + \dot{m}_2 + \dot{m}_3) = \dot{m}_4(100 - x) + 21.72\dot{m}_5$$

$$\text{For B: } 37(\dot{m}_1 + \dot{m}_2 + \dot{m}_3) = x\dot{m}_4 + (78.28 - y)\dot{m}_5$$

$$\text{For C: } 45.1(\dot{m}_1 + \dot{m}_2 + \dot{m}_3) = x\dot{m}_4 + (78.28 - y)\dot{m}_5$$

(a) Write the augmented matrix for the linear (over-determined) system; (b) use row reduction to (if it is possible) obtain the steady flow rates of each feed and the composition of A and B at $\dot{m}_4$ and B and C at $\dot{m}_5$.

Section 2.3 Matrix Inversion

E2.35 Use the adjoint method to invert the following matrices:

(a) $\begin{bmatrix} 5 & -3 \\ 3 & -2 \end{bmatrix}$, (b) $\begin{bmatrix} 4 & 1 \\ 3 & 2 \end{bmatrix}$, (c) $\begin{bmatrix} 1 & 2 & 1 \\ 1 & 1 & -2 \\ 2 & 2 & -3 \end{bmatrix}$, (d) $\begin{bmatrix} 3 & 2 & 2 \\ 1 & 2 & -3 \\ 2 & 2 & -1 \end{bmatrix}$, (e) $\begin{bmatrix} 3 & 1 & 2 \\ 1 & 4 & 1 \\ 2 & 3 & 2 \end{bmatrix}$

E2.36 Use the Gauss-Jordan method to invert the following matrices:

(a) $\begin{bmatrix} 2 & -5 & -1 \\ -1 & -4 & 2 \\ -2 & 1 & 2 \end{bmatrix}$ (b) $\begin{bmatrix} 1 & 0 & -3 \\ 2 & 2 & -3 \\ 4 & 9 & 1 \end{bmatrix}$ (c) $\begin{bmatrix} 1 & 4 & 4 \\ 2 & 2 & 3 \\ -3 & 5 & 2 \end{bmatrix}$, (d) $\begin{bmatrix} 1 & 2 & 1 \\ 1 & 1 & -2 \\ 2 & 2 & -3 \end{bmatrix}$

(e) $\begin{bmatrix} 3 & 2 & 2 \\ 1 & -2 & -3 \\ 10 & 1 & -1 \end{bmatrix}$, (f) $\begin{bmatrix} 3 & 1 & 2 \\ 2 & 4 & 1 \\ 2 & 3 & 1 \end{bmatrix}$, (g) $\begin{bmatrix} 1 & 1 & -2 \\ 2 & 2 & -5 \\ 1 & 2 & -3 \end{bmatrix}$, (h) $\begin{bmatrix} 1 & 2 & 2 \\ 1 & 1 & 1 \\ 3 & 2 & 3 \end{bmatrix}$

E2.37 Use the Gauss-Jordan method to invert the following matrices:

(a) $\begin{bmatrix} 4 & 8 & 1 & 4 \\ 1 & 4 & -2 & 2 \\ 3 & 6 & 4 & -3 \\ 1 & 3 & 1 & -2 \end{bmatrix}$, (b) $\begin{bmatrix} 1 & 3 & -1 & 2 \\ 2 & 2 & -2 & -3 \\ -2 & -5 & 3 & -1 \\ 1 & 1 & 2 & 2 \end{bmatrix}$, (c) $\begin{bmatrix} 2 & -2 & -1 & -2 \\ -3 & 2 & 1 & 2 \\ 1 & 0 & 2 & 3 \\ 2 & 3 & 1 & 2 \end{bmatrix}$

E2.38 Invert the triangle matrix **A** using the Gauss-Jordan method (a) without rounding fractional numbers, (b) using the approximate matrix found by rounding fractional numbers, (c) Verify $\mathbf{AA}^{-1} = \mathbf{I}$ for both cases.

$$
\mathbf{A} = \begin{bmatrix} 1 & 0 & 0 & 0 \\ 1/2 & 1 & 0 & 0 \\ 1/3 & 1/4 & 1 & 0 \\ 1/4 & 1/5 & 1/6 & 1 \end{bmatrix}, \qquad
\mathbf{A} \cong \begin{bmatrix} 1 & 0 & 0 & 0 \\ 0.5 & 1 & 0 & 0 \\ 0.333 & 0.25 & 1 & 0 \\ 0.250 & 0.20 & 0.167 & 1 \end{bmatrix}
$$

Section 2.4 Triangular Systems of Linear Equations

E2.39 Apply the forward or backward substitution algorithms to find the solution to the following linear systems:

(a) $\begin{bmatrix} 4 & 0 & 0 \\ 1 & 4 & 0 \\ 3 & 6 & 4 \end{bmatrix} \begin{bmatrix} x_1 \\ x_2 \\ x_3 \end{bmatrix} = \begin{bmatrix} 12 \\ -5 \\ 13 \end{bmatrix}$,
(b) $\begin{bmatrix} 2 & 1 & 7 \\ 0 & 4 & 3 \\ 0 & 0 & 3 \end{bmatrix} \begin{bmatrix} x_1 \\ x_2 \\ x_3 \end{bmatrix} = \begin{bmatrix} 15 \\ -6 \\ 6 \end{bmatrix}$

(c) $\begin{bmatrix} 2 & -2 & 2 & 4 \\ 0 & 5 & 4 & -1 \\ 0 & 0 & 3 & 3 \\ 0 & 0 & 0 & -3 \end{bmatrix} \begin{bmatrix} x_1 \\ x_2 \\ x_3 \\ x_4 \end{bmatrix} = \begin{bmatrix} 16 \\ 2 \\ 9 \\ -3 \end{bmatrix}$,
(d) $\begin{bmatrix} 3 & 0 & 0 & 0 \\ 2 & -3 & 0 & 0 \\ 3 & 7 & 4 & 0 \\ 6 & -5 & 3 & 5 \end{bmatrix} \begin{bmatrix} x_1 \\ x_2 \\ x_3 \\ x_4 \end{bmatrix} = \begin{bmatrix} 6 \\ 10 \\ 8 \\ 19 \end{bmatrix}$

E2.40 Consider the truss system shown in **Fig. E2.40**. The equilibrium of the horizontal and vertical forces yields the following (9×9) linear system:

$$
\begin{bmatrix}
1 & c & 0 & 0 & 0 & 0 & 0 & 0 & 0 \\
0 & s & 0 & 0 & 0 & 0 & 0 & 0 & 0 \\
-1 & 0 & 0 & 1 & c & 0 & 0 & 0 & 0 \\
0 & 0 & 1 & 0 & s & 0 & 0 & 0 & 0 \\
0 & c & 0 & 0 & 0 & -1 & 0 & 0 & 0 \\
0 & s & 1 & 0 & 0 & 0 & 0 & 0 & 0 \\
0 & 0 & 0 & 1 & 0 & 0 & -1 & 0 & 0 \\
0 & 0 & 0 & 0 & 0 & 0 & 0 & 1 & 0 \\
0 & 0 & 0 & 0 & s & 0 & 0 & 1 & s
\end{bmatrix}
\begin{bmatrix} F_1 \\ F_2 \\ F_3 \\ F_4 \\ F_5 \\ F_6 \\ F_7 \\ F_8 \\ F_9 \end{bmatrix}
=
\begin{bmatrix} 0 \\ r \\ 0 \\ 0 \\ 0 \\ 0 \\ 0 \\ 0 \\ 0 \end{bmatrix}
$$

where $c = 0.6$, $s = 0.8$, and $r = -750$. Apply elementary row operations to convert the augmented matrix into a lower triangular matrix, and then solve the system of linear equations by employing the forward substitution algorithm.

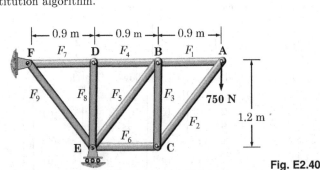

Fig. E2.40

Section 2.5 Gauss Elimination Methods

E2.41 The following linear system is from the discretization of an ordinary differential equation. Find the solution of the linear system using the Gauss elimination method.

$$y_1 + 3y_2 = 10, \quad y_1 + 4y_2 + 2y_3 = 23, \quad y_2 + 3y_3 + 5y_4 = 53, \quad y_3 + 7y_4 = 54$$

E2.42 Obtain the solution of the following systems of linear equations using the Gauss elimination method without pivoting.

(a) $\begin{cases} x + 3y - z = 8 \\ 2x - 2y + z = -11 \\ 3x + 4y - 5z = 11 \end{cases}$, (b) $\begin{cases} x + 2y - z = -5 \\ 4x + y + z = 6 \\ 2x + 2y - 3z = -5 \end{cases}$, (c) $\begin{cases} x + 3y - 2z = -6 \\ 2x - z + 3y = -3 \\ x + 2z + y = 2 \end{cases}$

(d) $\begin{cases} 3x + y + 2z = 7 \\ x + 3y + z = -6 \\ 4x + 3y - 2z = 9 \end{cases}$, (e) $\begin{cases} 4x + 3y + 2z = 4 \\ 3x - 2y + 5z = 27 \\ 2x - 3y - z = 16 \end{cases}$, (f) $\begin{cases} x + y - 2z = -5 \\ 2x + y + 3z = 3 \\ x + 2y - 3z = -6 \end{cases}$

E2.43 Consider the following partial fraction decomposition:

$$f(x) = \frac{5x^3 - x^2 + 14x + 4}{(x^2 + 2)(x^2 + 4)} = \frac{Ax + B}{x^2 + 2} + \frac{Cx + D}{x^2 + 4}$$

Clear the denominator and determine the set of four equations with four undetermined coefficients. Then, apply the Gauss elimination method to find the unknowns.

E2.44 Applying Kirchoff's current law to the currents flowing in each branch of the circuit given in **Fig. E2.44** yields

$$i_1 R_1 + R_3 (i_1 - i_2) + R_4 (i_1 - i_3) = 15\,V$$
$$R_3 (i_2 - i_1) + R_2 i_2 + R_5 (i_2 - i_3) = 0$$
$$R_4 (i_3 - i_1) + R_5 (i_3 - i_2) + R_6 i_3 = 0$$

Use the Gauss elimination method to solve the resulting system of equations for the currents. *Given:* $R_1 = R_2 = 1\,\Omega$, $R_3 = 6\,\Omega$, $R_4 = 3\,\Omega$, $R_5 = 4\,\Omega$, $R_6 = 10\,\Omega$.

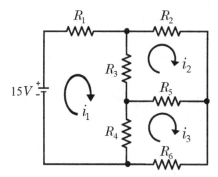

Fig. E2.44

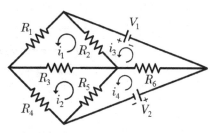

Fig. E2.45

E2.45 Consider the circuit shown in **Fig. E2.45**. Kirchoff's voltage law for the loop currents yields the following system of linear equations:

Loop 1: $R_1 i_1 + R_3 (i_1 - i_2) + R_2 (i_1 - i_3) = 0$
Loop 2: $R_3 (i_2 - i_1) + R_4 i_2 + R_5 (i_2 - i_4) = 0$
Loop 3: $R_2 (i_3 - i_1) + R_6 (i_3 - i_4) = V_1$
Loop 4: $R_6 (i_4 - i_3) + R_5 (i_4 - i_2) = V_2$

Find the solution to the linear system using the Gauss elimination method. *Given:* $R_1 = 3\,\Omega$, $R_2 = 5\,\Omega$, $R_3 = R_4 = 8\,\Omega$, $R_5 = R_6 = 2\,\Omega$, $V_1 = 14\,V$, $V_2 = 22\,V$.

E2.46 Apply the Gauss-Jordan method to obtain the solution to the following matrix equations:

(a) $\begin{bmatrix} 1 & -2 & -2 & -5 \\ 3 & 1 & -1 & 4 \\ -6 & 2 & 7 & 1 \\ 2.5 & 1 & -1.5 & 3 \end{bmatrix} \begin{bmatrix} x_1 \\ x_2 \\ x_3 \\ x_4 \end{bmatrix} = \begin{bmatrix} -20 \\ 18 \\ -6 \\ 15 \end{bmatrix}$,

(b) $\begin{bmatrix} 1 & 1 & 2 & -1 & 3 \\ -1 & 4 & -3 & 2 & 2 \\ 11 & -1 & 5 & -2 & 3 \\ 8 & 3 & -9 & 7 & -3 \\ 4 & 1 & -2 & 6 & 1 \end{bmatrix} \begin{bmatrix} x_1 \\ x_2 \\ x_3 \\ x_4 \\ x_5 \end{bmatrix} = \begin{bmatrix} 12 \\ 22 \\ 14 \\ 26 \\ 30 \end{bmatrix}$

E2.47 A third-degree polynomial passes through the points $(-2, -35)$, $(-1, -8)$, $(1, 10)$, and $(3, 20)$. Forcing the polynomial to satisfy all four points leads to a linear system of four equations and four unknowns. Establish the linear system and solve it using the Gauss-Jordan elimination method.

E2.48 Two weights are suspended with ropes, as shown in **Fig. E2.48**. The static equilibrium equations can be expressed as

$$-\frac{\sqrt{3}}{2}T_1 + T_2 = 0, \quad \frac{1}{2}T_1 - 25 = 0, \quad -T_2 + \frac{1}{2}T_3 = 0, \quad \frac{\sqrt{3}}{2}T_3 - 75 = 0$$

Apply elementary row operations to obtain the reduced-row echelon form of the augmented matrix to find the solution to the linear system. Does this system have a solution?

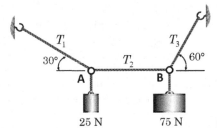

25 N   75 N   **Fig. E2.48**

E2.49 The three trolleys interconnected by springs are subjected to P_1, P_2, and P_3 loads, as shown in **Fig. E2.49**.

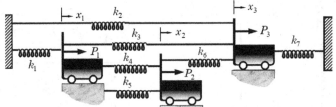

Fig. E2.49

The displacements of the trolleys from their equilibrium positions are governed by

$$P_1 + k_1(0 - x_1) + k_4(x_2 - x_1) + k_3(x_3 - x_1) = 0$$
$$P_2 + k_5(0 - x_2) + k_4(x_1 - x_2) + k_6(x_3 - x_2) = 0$$
$$P_3 + k_2(0 - x_3) + k_3(x_1 - x_3) + k_6(x_2 - x_3) + k_7(0 - x_3) = 0$$

where x's are displacements and k's are the spring constants. (a) Express the equilibrium equations in $\mathbf{Ax} = \mathbf{b}$ form; (b) find the displacements using the Gauss-Jordan method using the furnished data. *Given:* $k_1 = 750$ N/m, $k_2 = 5250$ N/m, $k_3 = 3000$ N/m, $k_4 = 1000$ N/m, $k_5 = 1670$ N/m, $k_6 = 3500$ N/m, $k_7 = 1500$ N/m, $P_1 = 825$ N, $P_2 = 3895$ N, and $P_3 = 1025$ N.

E2.50 Apply the Gauss elimination method with partial pivoting to solve the following linear systems:

(a) $\begin{bmatrix} 1 & 3 & 4 \\ 5 & 1 & -2 \\ -2 & 8 & 2 \end{bmatrix} \begin{bmatrix} x_1 \\ x_2 \\ x_3 \end{bmatrix} = \begin{bmatrix} 9 \\ -1 \\ -20 \end{bmatrix}$,

(b) $\begin{bmatrix} 3 & -3 & 1 \\ 3 & 2 & -2 \\ 4 & 1 & 2 \end{bmatrix} \begin{bmatrix} x_1 \\ x_2 \\ x_3 \end{bmatrix} = \begin{bmatrix} -7 \\ -8 \\ -3 \end{bmatrix}$

E2.51 Consider the separation process shown in **Fig. E2.51**. The inlet mass flow rate is $\dot{m}_1 = 18$ kg/h, and the mass fractions of each species at the inlet and outlet streams are given in the figure. Find the mass flow rate of each stream using the Gauss elimination method with partial pivoting. The mass conservation for the stream and the species is given below:

$$\dot{m}_2 + \dot{m}_3 + \dot{m}_4 = \dot{m}_1 = 18$$

$$0.09\dot{m}_2 + 0.48\dot{m}_3 + 0.18\dot{m}_4 = 0.18(18)$$

$$0.82\dot{m}_2 + 0.16\dot{m}_3 + 0.10\dot{m}_4 = 0.51(18)$$

Fig. E2.51

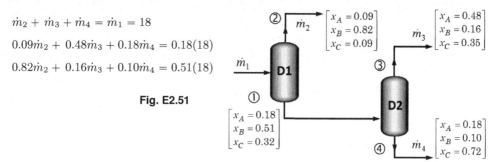

E2.52 (a) Use Kirchhoff's voltage law to obtain a linear system for the loop currents of the circuit in **Fig. E2.52**, (b) use the given data to solve the system using the Gauss elimination with partial pivoting. *Given:* $R_1 = 1\,\Omega$, $R_2 = R_3 = R_{10} = 2\,\Omega$, $R_6 = 5\,\Omega$, $R_4 = R_7 = R_8 = 3\,\Omega$, $R_5 = R_9 = 4\,\Omega$, $V_1 = 23\,V$, $V_2 = 18\,V$, $V_3 = 24\,V$, $V_4 = 48\,V$.

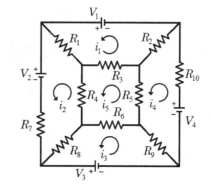

Fig. E2.52

E2.53 Apply Kirchhoff's current law to the electric circuit is given in **Fig. E2.53** to obtain a system of linear equations for the currents in each branch. Then solve the system of equations using a suitable direct method. *Given:* $R_1 = 5\,\Omega$, $R_2 = 2\,\Omega$, $R_3 = 4\,\Omega$, $R_4 = 3\,\Omega$, $R_5 = 2\,\Omega$, $R_6 = 5\,\Omega$.

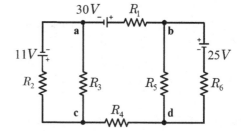

Fig. E2.53

Section 2.6 Computing Determinants

E2.54 Find the determinants of the matrices in **E2.11** by constructing the upper triangular matrix through elementary row operations.

E2.55 Find the determinants of the matrices in **E2.12** by constructing the upper triangular matrix through elementary row operations.

Section 2.7 Ill-conditioned Matrices and Linear Systems

E2.56 Find the ℓ_1, ℓ_2, and Frobenius norm-based condition numbers of the following matrices:

(a) $\begin{bmatrix} 1 & 0.3 \\ 3 & 1 \end{bmatrix}$, (b) $\begin{bmatrix} 99 & 98 \\ 98 & 97 \end{bmatrix}$, (c) $\begin{bmatrix} 2 & -0.019 & 4 \\ 2 & 4.99 & 3 \\ 1 & 5 & 1 \end{bmatrix}$, (d) $\begin{bmatrix} 2.9 & 1.9 & 3.89 \\ 1.07 & 4.05 & 1.9 \\ -2.1 & 4.9 & -1.95 \end{bmatrix}$

E2.57   Use the definition, Eq. (2.31), to calculate the spectral (ℓ_2−norm-based) condition number of the matrices in **E2.56**.

E2.58   Use the expression proposed by Guggenheimmer *et al.* [10], Eq. (2.32), to estimate the spectral condition numbers of the matrices in **E2.56**. Compare your estimates with those obtained in **E2.57**.

E2.59   Estimate the spectral (ℓ_2−norm-based) condition number of the matrices in **E2.56** using the two-step method, Eqs. (2.35) and (2.36), and compare your estimates with those obtained in **E2.58**. *Hint*: Choose vector elements of **A** as $u_i = (-1)^{i+1}$, $i = 1, 2, \ldots, n$.

E2.60   Consider the following tridiagonal Toeplitz matrix:

$$\mathbf{A} = \begin{bmatrix} 3.1 & 1.9 & & \\ 1.9 & 3.1 & 1.9 & \\ & 1.9 & 3.1 & 2 \\ & & 2 & 3.24 \end{bmatrix}$$

Calculate the ℓ_1−norm-based condition number by using (a) the definition—Eq. (2.31), (b) the two-step method, i.e., Eqs. (2.35) and (2.36), and (c) Eq. (2.33).

E2.61   Consider the matrix **A** in **E2.60**. (a) Apply row equilibration such that each row is a unit vector; (b) calculate the ℓ_1−norm-based condition number of the equilibrated matrix **M** using the definition; (c) estimate the same condition number using the two-step method; and (d) estimate the spectral condition number using Eq. (2.34).

E2.62   For given **A**, (a) calculate the Frobenius− and spectral (ℓ_2− norm-based) condition numbers; (b) estimate the spectral condition number using Eq. (2.33); (c) is the matrix ill-conditioned?

$$\mathbf{A} = \begin{bmatrix} 1 & 0 & 0 & 0 \\ 1/22 & 1 & 0 & 0 \\ 1/4 & 1/4 & 1 & 0 \\ 1/8 & 1/8 & 1/8 & 1 \end{bmatrix}$$

E2.63   Investigate the condition of **Ax** = **b** and the corresponding equilibrated system **Mx** = **d** for the following coefficient matrices:

(a) $\mathbf{A} = \begin{bmatrix} 11 & 22 & -6 \\ -125 & 350 & 55 \\ 0.03 & -0.1 & 0.05 \end{bmatrix}$,   (b) $\mathbf{A} = \begin{bmatrix} -187 & 66 & 99 \\ -0.05 & 0.1 & 0.05 \\ 6 & 19 & -3 \end{bmatrix}$,   (c) $\mathbf{A} = \begin{bmatrix} 0.1 & 66 & 99 \\ 0.02 & 0.3 & 1.2 \\ 0.1 & 190 & -2 \end{bmatrix}$

E2.64   To find the solution for the given linear system of equations, (a) use the simplified estimate for the spectral condition number of the coefficient matrix (**A**) to determine whether the linear system is ill-conditioned or not; (b) repeat Part (a) for the matrix generated by equilibrating the rows to a unit vector; (c) solve the system with the Gauss-Jordan elimination method, assuming exact arithmetic (*whole fractions without rounding*); (d) solve the system by rounding the fractions numbers and carrying out the arithmetic operations to four decimal places; (e) compare the solutions obtained in Parts (c) and (d) and interpret the results.

$$\begin{bmatrix} 1 & 1/2 & 1/3 & 1/4 & 1/5 \\ 1/2 & 1/3 & 1/4 & 1/5 & 1/6 \\ 1/3 & 1/4 & 1/5 & 1/6 & 1/7 \\ 1/4 & 1/5 & 1/6 & 1/7 & 1/8 \\ 1/5 & 1/6 & 1/7 & 1/8 & 1/9 \end{bmatrix} \begin{bmatrix} x_1 \\ x_2 \\ x_3 \\ x_4 \\ x_5 \end{bmatrix} = \begin{bmatrix} 59/60 \\ 11/20 \\ 169/420 \\ 34/105 \\ 689/2520 \end{bmatrix}$$

E2.65   (a) Solve the given linear system by the Gauss-Jordan elimination using the exact arithmetic (whole fractions without rounding); (b) repeat Part (a) after rounding the fractions to three-decimal places; (c) compare the solutions obtained in Parts (a) and (b) and interpret the results; (d) estimate the spectral condition number of the rounded matrix and the matrix whose rows were equilibrated by unit vector.

$$\begin{bmatrix} 1/3 & 4/9 \\ 1/7 & 3/16 \end{bmatrix} \begin{bmatrix} x_1 \\ x_2 \end{bmatrix} = \begin{bmatrix} 1/9 \\ 3/56 \end{bmatrix}$$

E2.66 Consider the linear system $\mathbf{Ax} = \mathbf{b}$, where

$$\mathbf{A} = \begin{bmatrix} 1 & 1/3 & 1/5 \\ 1/3 & 1/5 & 1/7 \\ 1/5 & 1/7 & 1/9 \end{bmatrix}, \qquad \mathbf{b} = \begin{bmatrix} 23/15 \\ 71/105 \\ 143/315 \end{bmatrix}$$

(a) Investigate if the coefficient matrix is ill-conditioned or not using the simple criterion for the spectral condition number; (b) solve the system of equations for three- and five-significant digits (chopped) numbers of $\mathbf{A}$ and $\mathbf{b}$; and (c) compare and explain the deviation of the results from the true solution of $\mathbf{x} = [\,1\ 1\ 1\,]^T$.

E2.67 Consider the linear system $\mathbf{Ax} = \mathbf{b}$, where

$$\mathbf{A} = \begin{bmatrix} 1 & 1/3 & 1/5 & 1/7 \\ 1/3 & 1/5 & 1/7 & 1/9 \\ 1/5 & 1/7 & 1/9 & 1/11 \\ 1/7 & 1/9 & 1/11 & 1/13 \end{bmatrix}, \qquad \mathbf{b} = \begin{bmatrix} 176/105 \\ 248/315 \\ 1888/3465 \\ 3800/9009 \end{bmatrix}$$

(a) Investigate if matrix $\mathbf{A}$ is ill-conditioned or not using the simple criterion for the spectral condition number; (b) solve the system of equations for three- and five-significant digits (chopped) numbers of $\mathbf{A}$ and $\mathbf{b}$; and (c) compare and explain the deviation of the results from the true solution of $\mathbf{x} = [\,1\ 1\ 1\ 1\,]^T$.

Section 2.8 Decomposition Methods

E2.68 Use the Doolittle method to find the LU-decomposition of the following matrices:

(a) $\begin{bmatrix} 2 & -2 & 4 \\ 1 & 8 & 8 \\ -3 & 9 & -1 \end{bmatrix}$, (b) $\begin{bmatrix} -3 & 2 & -1 \\ 6 & -6 & -2 \\ -9 & 9 & 7 \end{bmatrix}$, (c) $\begin{bmatrix} -2 & -2 & -3 \\ 8 & 6 & 14 \\ 4 & 10 & -6 \end{bmatrix}$,

(d) $\begin{bmatrix} 8 & 4 & -8 & 12 \\ 2 & 13 & -10 & 11 \\ 2 & 16 & 0 & 17 \\ 2 & 10 & -8 & 13 \end{bmatrix}$, (e) $\begin{bmatrix} -4 & 4 & 8 & 8 \\ -2 & 14 & -2 & 10 \\ 4 & 8 & -18 & -4 \\ -2 & 10 & 6 & 14 \end{bmatrix}$

E2.69 Use the Doolittle method to find the LU-decomposition of matrix $\mathbf{A}$ and then find the matrix product $\mathbf{UL}$.

$$\mathbf{A} = \begin{bmatrix} 2 & -1 & -2 & 2 \\ 6 & -2 & -9 & 7 \\ -2 & 3 & -2 & -2 \\ 2 & 3 & -16 & 12 \end{bmatrix}$$

E2.70 Use the Doolittle method to find the LU-decomposition of the following matrix, and then find the matrix $\mathbf{B}$ that satisfies $\mathbf{BA} = \mathbf{U}$.

$$\begin{bmatrix} a & 0 & s & 0 \\ 0 & 1 & 0 & s \\ ax & 1 & x(1+s) & s \\ as & b & s^2 & b(1+s) \end{bmatrix}$$

E2.71 The inverse of a matrix can also be obtained by $\mathbf{A}^{-1} = \mathbf{U}^{-1}\mathbf{L}^{-1}$, where $\mathbf{L}$ and $\mathbf{U}$ are factored matrices from LU-decomposition. Having decomposed $\mathbf{A}$, apply the Gauss-Jordan method to find the inverses of $\mathbf{U}$ and $\mathbf{L}$, and consequently calculate $\mathbf{U}^{-1}\mathbf{L}^{-1}$.

$$\mathbf{A} = \begin{bmatrix} 1 & 3 & -2 & 2 \\ -2 & -5 & -1 & -1 \\ 1 & 1 & 2 & 1 \\ 1 & -3 & 5 & 3 \end{bmatrix}$$

E2.72 The following systems of linear equations, in matrix form, yield the coefficient matrices in **E2.68**. To solve these systems, apply forward and backward substitution operations using $\mathbf{L}$ and $\mathbf{U}$ matrices found in **E2.68**.

(a) $\begin{cases} 2x_1 - 2x_2 + 4x_3 = 28 \\ x_1 + 8x_2 + 8x_3 = 26 \\ -3x_1 + 9x_2 - x_3 = -29 \end{cases}$,

(b) $\begin{cases} -2x_1 - 2x_2 - 3x_3 = -15 \\ 8x_1 + 6x_2 + 14x_3 = 58 \\ 4x_1 + 10x_2 - 6x_3 = 18 \end{cases}$,

(c) $\begin{cases} 8x_1 + 4x_2 - 8x_3 + 12x_4 = -72 \\ 2x_1 + 13x_2 - 10x_3 + 11x_4 = -70 \\ 2x_1 + 16x_2 + 17x_4 = 1 \\ 2x_1 + 10x_2 - 8x_3 + 13x_4 = -69 \end{cases}$,

(d) $\begin{cases} 8x_1 + 4x_2 - 8x_3 + 12x_4 = 0 \\ 2x_1 + 13x_2 - 10x_3 + 11x_4 = -78 \\ 2x_1 + 16x_2 + 0 + 17x_4 = -94 \\ 2x_1 + 10x_2 - 8x_3 + 13x_4 = -34 \end{cases}$

E2.73 Decompose the following symmetric matrices using Cholesky's method.

(a) $\begin{bmatrix} 4 & 6 & 4 \\ 6 & 10 & 4 \\ 4 & 4 & 12 \end{bmatrix}$,

(b) $\begin{bmatrix} 1 & 2 & -3 \\ 2 & 13 & 6 \\ -3 & 6 & 29 \end{bmatrix}$,

(c) $\begin{bmatrix} 4 & 2 & -4 \\ 2 & 10 & -11 \\ -4 & -11 & 14 \end{bmatrix}$,

(d) $\begin{bmatrix} 4 & 2 & -6 & 4 \\ 2 & 10 & -9 & -1 \\ -6 & -9 & 14 & -6 \\ 4 & -1 & -6 & 25 \end{bmatrix}$,

(e) $\begin{bmatrix} 1 & 2 & 0 & 1 \\ 2 & 8 & 6 & 6 \\ 0 & 6 & 13 & 4 \\ 1 & 6 & 4 & 7 \end{bmatrix}$,

(f) $\begin{bmatrix} 1 & 1 & -1 & -3 \\ 1 & 5 & -7 & 1 \\ -1 & -7 & 14 & -5 \\ -3 & 1 & -5 & 18 \end{bmatrix}$

E2.74 Use $\mathbf{A} = \mathbf{L}\mathbf{L}^T$ decomposition to solve the given system of linear equations. Note that the coefficient matrices are given in **E2.73**(a) and (f).

(a) $\begin{cases} 4x_1 + 6x_2 + 4x_3 = -9 \\ 6x_1 + 10x_2 + 4x_3 = -\frac{73}{5} \\ 4x_1 + 4x_2 + 12x_3 = -\frac{38}{5} \end{cases}$,

(b) $\begin{cases} x_1 + x_2 - x_3 - 3x_4 = -49 \\ x_1 + 5x_2 - 7x_3 + x_4 = -69 \\ -x_1 - 7x_2 + 14x_3 - 5x_4 = 81 \\ -3x_1 + x_2 - 5x_3 + 18x_4 = 170 \end{cases}$

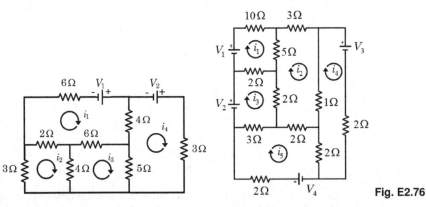

Fig. E2.75

Fig. E2.76

E2.75 Applying Kirchhoff's law of voltage (KLV) to the circuit in **Fig. E2.75** leads to the following system of equations for the loop currents:

$$\begin{bmatrix} 18 & -2 & -6 & -4 \\ -2 & 9 & -4 & 0 \\ -6 & -4 & 15 & -5 \\ -4 & 0 & -5 & 12 \end{bmatrix} \begin{bmatrix} i_1 \\ i_2 \\ i_3 \\ i_4 \end{bmatrix} = \begin{bmatrix} V_1 \\ 0 \\ 0 \\ V_2 \end{bmatrix}$$

where V_1 and V_2 voltages can be changed. Use Cholesky's method to decompose the coefficient matrix and solve the linear system using a forward and a backward elimination procedure for $V_1 = 16.8$ and $V_2 = 18.6$ V.

E2.76 Applying Kirchhoff's law of voltage (KLV) to the circuit shown in **Fig. E2.76** leads to the following system of equations for the loop currents:

$$\begin{bmatrix} 17 & -5 & -2 & 0 & 0 \\ -5 & 13 & -2 & -1 & -2 \\ -2 & -2 & 7 & 0 & -3 \\ 0 & -1 & 0 & 5 & -2 \\ 0 & -2 & -3 & -2 & 9 \end{bmatrix} \begin{bmatrix} i_1 \\ i_2 \\ i_3 \\ i_4 \\ i_5 \end{bmatrix} = \begin{bmatrix} V_1 \\ 0 \\ V_2 \\ -V_3 \\ V_4 \end{bmatrix}$$

where the voltages may be subject to change. Apply Cholesky's method to decompose the coefficient matrix, and then solve the linear system with forward and back substitution procedures for the case with $V_1 = 6$, $V_2 = 11$, $V_3 = 15$, and $V_4 = 11$ V.

E2.77 The truss shown in **Fig. E2.77** is applied vertically and horizontally loads for X and Y at node (2). The system members have the same cross section and elastic modulus, yielding the value of $EA = 2.5 \times 10^6$ kN. The nodal displacements of the system can be obtained by solving the equilibrium equations are given in matrix form as

$$\begin{bmatrix} 2.14443 & -0.44721 & -0.89443 & 0.44721 \\ -0.44721 & 0.22361 & 0.44721 & -0.22361 \\ -0.89443 & 0.44721 & 2.68330 & -0.44721 \\ 0.44721 & -0.22361 & -0.44721 & 3.17082 \end{bmatrix} \begin{bmatrix} d_{1x} \\ d_{1y} \\ d_{2x} \\ d_{2y} \end{bmatrix} = \begin{bmatrix} 0 \\ 0 \\ -X \\ -Y \end{bmatrix} \times 10^{-6}$$

where d_{mx} and d_{my} denote the x and y displacements of node m. Noting that the distances in the figure are given in meters, apply the Doolittle-LU decomposition once and determine the nodal displacements for the following loads: (a) $(X,Y) = (50,50)$ kN ; (b) $(X,Y) = (75,25)$ kN ;

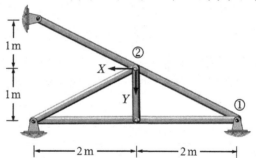

(c) $(X,Y) = (25,75)$ kN.

Fig. E2.77

E2.78 Find the inverse of the given matrices by $\mathbf{A}^{-1} = \mathbf{U}^{-1}\mathbf{L}^{-1}$.

(a) $\begin{bmatrix} 10 & -9 & 14 \\ 2 & -2 & 3 \\ 9 & -8 & 13 \end{bmatrix}$, (b) $\begin{bmatrix} -1 & -8 & 5 \\ 1 & 5 & -3 \\ 0 & -2 & 1 \end{bmatrix}$, (c) $\begin{bmatrix} -1 & -1 & 0 & 0 \\ 5 & 4 & 0 & 1 \\ 41 & 32 & 2 & 6 \\ -27 & -21 & -1 & -4 \end{bmatrix}$

Section 2.9 Tridiagonal Systems

E2.79 Apply the Thomas algorithm to solve the following tridiagonal system of equations:

(a) $\begin{cases} x_1 + 3x_2 = 10 \\ x_1 + 4x_2 + 2x_3 = 23 \\ x_2 + 3x_3 + 5x_4 = 53 \\ x_3 + 7x_4 = 54 \end{cases}$, (b) $\begin{cases} 2x_1 + x_2 = 1 \\ 2x_1 + 3x_2 - x_3 = -1 \\ x_2 + 4x_3 + 3x_4 = 5 \\ -2x_3 + 2x_4 + x_5 = 2 \\ -x_4 + 2x_5 = -6 \end{cases}$

E2.80 Apply the Thomas algorithm to solve the following tridiagonal system of equations:

$$\begin{bmatrix} 1 & 2 & & & & \\ 2 & 3 & -2 & & & \\ & 2 & 3 & -2 & & \\ & & 2 & 3 & -2 & \\ & & & 2 & 3 & -2 \\ & & & & 2 & 3 \end{bmatrix} \begin{bmatrix} x_1 \\ x_2 \\ x_3 \\ x_4 \\ x_5 \\ x_6 \end{bmatrix} = \begin{bmatrix} -1 \\ -5 \\ 8 \\ -8 \\ 11 \\ -3 \end{bmatrix}$$

E2.81 Find the Crout's LU-decomposition of the coefficient matrix of the linear system (by $\mathbf{Ax} = \mathbf{b}$) presented in **E2.80** and then solve the tridiagonal systems ($\mathbf{Ly} = \mathbf{b}$ and $\mathbf{Ux} = \mathbf{y}$) to obtain the solution.

E2.82 Use Crout's algorithm to decompose the following tridiagonal matrices as **LU**.

$$(a) \begin{bmatrix} 2 & 6 & 0 & 0 \\ 1 & 6 & -6 & 0 \\ 0 & -2 & 5 & -1 \\ 0 & 0 & 3 & -1 \end{bmatrix}, \quad (b) \begin{bmatrix} 1 & -2 & 0 & 0 \\ 2 & -1 & -9 & 0 \\ 0 & 2 & -8 & 8 \\ 0 & 0 & 1 & 0 \end{bmatrix}, \quad (c) \begin{bmatrix} 3 & 12 & 0 & 0 \\ 1 & 7 & 9 & 0 \\ 0 & 2 & 8 & 8 \\ 0 & 0 & 4 & 21 \end{bmatrix}$$

E2.83 Use Crout's algorithm to decompose the following tridiagonal matrices as **LU**.

$$(a) \begin{bmatrix} 3 & 6 & 0 & 0 & 0 \\ -1 & 0 & -2 & 0 & 0 \\ 0 & 2 & 2 & 4 & 0 \\ 0 & 0 & 1 & 5 & 8 \\ 0 & 0 & 0 & -2 & 1 \end{bmatrix}, \quad (b) \begin{bmatrix} 1 & 2 & 0 & 0 & 0 \\ 1 & 6 & 12 & 0 & 0 \\ 0 & 1 & 6 & 9 & 0 \\ 0 & 0 & 1 & 7 & 8 \\ 0 & 0 & 0 & 2 & 9 \end{bmatrix}, \quad (c) \begin{bmatrix} 2 & 8 & 0 & 0 & 0 \\ 2 & 11 & 6 & 0 & 0 \\ 0 & 1 & 3 & 2 & 0 \\ 0 & 0 & 1 & 4 & 8 \\ 0 & 0 & 0 & 2 & 11 \end{bmatrix}$$

E2.84 Use the following matrices (and notations) to develop the Doolittle algorithm to decompose a tridiagonal matrix as $\mathbf{A} = \mathbf{LU}$. Make the algorithm as cpu-time and memory-efficient as possible.

$$\begin{bmatrix} d_1 & a_1 & & & \\ b_2 & d_2 & a_2 & & \\ & \ddots & \ddots & \ddots & \\ & & b_{n-1} & d_{n-1} & a_{n-1} \\ & & & b_n & d_n \end{bmatrix} = \begin{bmatrix} 1 & & & & \\ \ell_2 & 1 & & & \\ & \ddots & \ddots & & \\ & & \ell_{n-1} & 1 & \\ & & & \ell_n & 1 \end{bmatrix} \begin{bmatrix} u_1 & a_1 & & & \\ & u_2 & a_2 & & \\ & & \ddots & \ddots & \\ & & & u_{n-1} & a_{n-1} \\ & & & & u_n \end{bmatrix}$$

E2.85 Apply the Doolittle algorithm to decompose the following tridiagonal matrices as $\mathbf{A} = \mathbf{LU}$.

$$(a) \begin{bmatrix} 4 & 1 & 0 & 0 \\ 4 & 3 & 1 & 0 \\ 0 & 8 & 11 & 1 \\ 0 & 0 & 21 & 5 \end{bmatrix}, \quad (b) \begin{bmatrix} 3 & 2 & 0 & 0 \\ 6 & 8 & 2 & 0 \\ 0 & 2 & 6 & 2 \\ 0 & 0 & 5 & 8 \end{bmatrix}, \quad (c) \begin{bmatrix} 3 & -2 & 0 & 0 \\ 6 & 5 & -1 & 0 \\ 0 & 18 & 1 & -2 \\ 0 & 0 & 15 & 2 \end{bmatrix}$$

E2.86 Apply the Doolittle algorithm to decompose the following tridiagonal matrices as $\mathbf{A} = \mathbf{LU}$.

$$(a) \begin{bmatrix} 3 & 3 & 0 & 0 & 0 \\ 6 & 8 & 2 & 0 & 0 \\ 0 & 6 & 10 & 1 & 0 \\ 0 & 0 & 16 & 8 & 2 \\ 0 & 0 & 0 & 12 & 11 \end{bmatrix}, \quad (b) \begin{bmatrix} 4 & 2 & 0 & 0 & 0 \\ 4 & 4 & 1 & 0 & 0 \\ 0 & 4 & 5 & 2 & 0 \\ 0 & 0 & 3 & 7 & 4 \\ 0 & 0 & 0 & 10 & 9 \end{bmatrix}, \quad (c) \begin{bmatrix} 4 & 1 & 0 & 0 & 0 \\ 8 & 3 & 1 & 0 & 0 \\ 0 & 4 & 8 & 2 & 0 \\ 0 & 0 & 8 & 6 & 4 \\ 0 & 0 & 0 & 4 & 10 \end{bmatrix}$$

E2.87 Consider the linear system $\mathbf{Ax} = \mathbf{b}$, where $\mathbf{b} = [1, 1, 1, 1, 1]^T$ and $\mathbf{A}$ is given in **E2.86**. Apply the **LU** decomposition to solve the linear system of equations.

E2.88 Use the Doolittle algorithm to decompose the following tridiagonal matrices as $\mathbf{A} = \mathbf{LU}$.

(a) $\begin{bmatrix} 9 & 6 & 0 & 0 \\ 6 & 5 & 3 & 0 \\ 0 & 3 & 13 & 8 \\ 0 & 0 & 8 & 41 \end{bmatrix}$, (b) $\begin{bmatrix} 4 & 6 & 0 & 0 \\ 6 & 18 & 3 & 0 \\ 0 & 3 & 2 & 3 \\ 0 & 0 & 3 & 25 \end{bmatrix}$, (c) $\begin{bmatrix} 9 & 6 & 0 & 0 \\ 6 & 13 & 9 & 0 \\ 0 & 9 & 25 & 12 \\ 0 & 0 & 12 & 25 \end{bmatrix}$

(d) $\begin{bmatrix} 4 & 2 & 0 & 0 & 0 \\ 2 & 10 & 12 & 0 & 0 \\ 0 & 12 & 17 & 1 & 0 \\ 0 & 0 & 1 & 5 & 4 \\ 0 & 0 & 0 & 4 & 13 \end{bmatrix}$, (e) $\begin{bmatrix} 1 & 2 & 0 & 0 & 0 \\ 2 & 20 & 4 & 0 & 0 \\ 0 & 4 & 10 & 6 & 0 \\ 0 & 0 & 6 & 5 & 1 \\ 0 & 0 & 0 & 1 & 26 \end{bmatrix}$, (f) $\begin{bmatrix} 9 & 3 & 0 & 0 & 0 \\ 3 & 2 & 4 & 0 & 0 \\ 0 & 4 & 17 & 4 & 0 \\ 0 & 0 & 4 & 25 & 3 \\ 0 & 0 & 0 & 3 & 26 \end{bmatrix}$,

E2.89 Consider a symmetric, positive-definite linear system $\mathbf{Ax} = \mathbf{b}$, where $\mathbf{b} = [1, 1, 1, 1, 1]^T$ and $\mathbf{A}$ is given in **E2.88**(d)-(f). Use the lower triangular matrix $\mathbf{L}$ obtained by applying the Cholesky decomposition to solve the linear system using forward and backward substitutions.

E2.90 Consider the five-stage separation process depicted in **Fig. E2.90**. Stream-1 V (kmole/s) is separated at a mole fraction of y_6. Stream-2 is L (kmole/s) at a concentration of x_0. The state variables are identified in the figure. The mass balance on each stage (i) leads to $V(y_{i+1} - y_i) = K(x_i - x_{i-1})$.

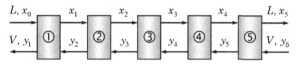

Fig. E2.90

Assuming the streams leaving the stage are in thermodynamic equilibrium, we may write $y_i = Kx_i$, where K is called the equilibrium ratio. The mass balances for each stage can be expressed in matrix form as

$$\begin{bmatrix} 1+\beta & -1 & & & \\ -1 & 1+\beta & -1 & & \\ & -1 & 1+\beta & -1 & \\ & & -1 & 1+\beta & -1 \\ & & & -1 & 1+\beta \end{bmatrix} \begin{bmatrix} x_1 \\ x_2 \\ x_3 \\ x_4 \\ x_5 \end{bmatrix} = \begin{bmatrix} \beta x_0 \\ 0 \\ 0 \\ 0 \\ y_6/K \end{bmatrix}$$

where $\beta = L/KV$. To solve the system for an arbitrary set of x_0, y_6, K, and β values, first find $\mathbf{L}$ by Cholesky decomposition of the coefficient matrix and then obtain the numerical solution for the case with $x_0 = 0.8$, $y_6 = 0.3$, $\beta = 1.25$, and $K = 1.2$.

Section 2.10 Banded Linear Systems

E2.91 A weight of 10 kN is suspended with ropes, as shown in **Fig. E2.91**. The system is in a state of static equilibrium. (a) derive the equilibrium equations and express them in matrix form; (b) solve the resulting system of banded linear equations by Gauss elimination.

E2.92 A weight of 50 N is suspended with ropes, as illustrated in **Fig. E2.92**. Assuming the system is in a state of static equilibrium, (a) derive all possible equilibrium equations, (b) express them in matrix equation form, and (c) solve the linear system with naïve Gauss elimination.

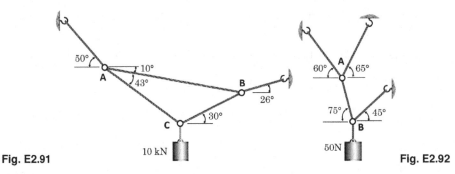

Fig. E2.91 10 kN 50N **Fig. E2.92**

E2.93 Consider the three-bar truss system shown in **Fig. E2.93**, subjected to internal and external forces. The truss is subjected to a vertical load of 5.78 kN at joint (1). In static equilibrium, the sum of the F_x- and F_y-components of the forces at each joint should be zero, so the equilibrium equations can be written as

At joint (1): $-\cos\alpha\, F_{12} + \cos\beta\, F_{13} = 0$ At joint (2): $\cos\alpha\, F_{12} + F_{23} + R_{2x} = 0$
$-\sin\alpha\, F_{12} - \sin\beta\, F_{13} - F = 0$ $\sin\alpha\, F_{12} + R_{2y} = 0$

At joint (3): $-F_{23} - \cos\beta\, F_{13} = 0$
$\sin\beta\, F_{13} + R_{3y} = 0.$

(a) Express the given six equations and six unknowns (the internal bar forces, F_{12}, F_{13}, and F_{23}, and the reaction forces, R_{2x}, R_{2y}, and R_{3y}) in matrix form, and (b) use the Gauss elimination method to find the direct solution of the resulting banded linear system. *Given:* $\ell_1 = 8$-m, $\ell_2 = 15$-m, and $\ell_3 = 17$-m.

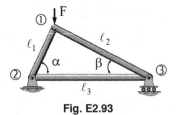

Fig. E2.93

E2.94 The matrix on the rhs is the compact form of the banded matrix **A**. Find a lower **L** and an upper **U** matrices such that **A** = **LU**:

$$\begin{bmatrix} 0 & 0 & 0 & 4 & -1 \\ 0 & 0 & -1 & 4 & -1 \\ 0 & -1 & -1 & 4 & -1 \\ -1 & -1 & -1 & 4 & -1 \\ -1 & -1 & -1 & 4 & -1 \\ -1 & -1 & -1 & 4 & 0 \end{bmatrix}$$

E2.95 Find the solution of a tridiagonal system using the compact banded system.

$$\begin{bmatrix} 4 & -1 & 0 & 0 & 0 & 0 \\ -1 & 4 & -1 & 0 & 0 & 0 \\ 0 & -1 & 4 & -1 & 0 & 0 \\ 0 & 0 & -1 & 4 & -1 & 0 \\ 0 & 0 & 0 & -1 & 4 & -1 \\ 0 & 0 & 0 & 0 & -2 & 4 \end{bmatrix} \begin{bmatrix} x_1 \\ x_2 \\ x_3 \\ x_4 \\ x_5 \\ x_6 \end{bmatrix} = \begin{bmatrix} -7 \\ 9 \\ 5 \\ 18 \\ 20 \\ 28 \end{bmatrix}$$

E2.96 For the given linear system, (a) use the Doolittle method to decompose the coefficient matrix as **A** = **LU**, and (b) apply forward (**Ly** = **b**) and backward (**Ux** = **y**) substitutions to find the solution of the linear system.

$$x_1 + 2x_2 + x_3 = 0$$
$$-2x_1 - 2x_2 - x_3 - 2x_4 = 0$$
$$-x_1 + 2x_2 + 3x_3 - 2x_4 + x_5 = 0$$
$$2x_2 - 3x_3 - 7x_4 - 4x_5 - 2x_6 = -\frac{1}{2}$$
$$2x_3 + 4x_4 + 3x_5 + 2x_6 = 0$$
$$-x_4 - 4x_5 - 3x_6 = 4$$

E2.97 Consider the compact banded systems below: (a) Use the Doolittle method to decompose the coefficient matrix as $\mathbf{A} = \mathbf{LU}$, and (b) apply forward $(\mathbf{Ly} = \mathbf{b})$ and backward $(\mathbf{Ux} = \mathbf{y})$ substitutions to find the solution of the linear system.

(a) $\begin{bmatrix} 0 & 0 & 1 & 2 \\ 0 & 1 & 1 & -2 \\ -1 & -4 & -2 & 2 \\ -1 & 0 & 3 & 1 \\ 4 & 2 & -3 & -2 \\ 1 & 3 & 6 & 2 \\ -2 & -8 & -3 & 0 \end{bmatrix} \mathbf{x} = \begin{bmatrix} 0 \\ -5 \\ -2 \\ 0 \\ 10 \\ 13 \\ -20 \end{bmatrix}$, (b) $\begin{bmatrix} 0 & 0 & 0 & 1 & 2 & 1 & 2 \\ 0 & 0 & 1 & 1 & 2 & 0 & 1 \\ 0 & -1 & -4 & 3 & -4 & 3 & 1 \\ 2 & 3 & 1 & 1 & -1 & -3 & 1 \\ 0 & 4 & 2 & 3 & 4 & -2 & 0 \\ 2 & 3 & -1 & -1 & 0 & 0 & 0 \\ 3 & -5 & -8 & 3 & 0 & 0 & 0 \end{bmatrix} \mathbf{x} = \begin{bmatrix} 0 \\ 2 \\ 1 \\ 0 \\ 4 \\ -4 \\ -13 \end{bmatrix}$

2.13 COMPUTER ASSIGNMENTS

CA2.1 Write a pseudocode, Module Function VECNORM$(n, p, \mathbf{x})$, to find $\|\mathbf{x}\|_p$; i.e., pth norm of a vector of length n. The integer input variable p should return ℓ_1, ℓ_2, or ℓ_∞-norm for the values 1, 2, or > 2, respectively.

CA2.2 Develop an algorithm to find the inverses of lower and upper triangular matrices using the Gauss-Jordan method, and write a computer program in a language of your choice to test your algorithm.

$$\begin{bmatrix} \ell_{11} & 0 & \cdots & 0 \\ \ell_{21} & \ell_{22} & & \vdots \\ \vdots & & \ddots & 0 \\ \ell_{n1} & \ell_{n2} & \cdots & \ell_{nn} \end{bmatrix}^{-1} , \quad \begin{bmatrix} u_{11} & u_{12} & \cdots & u_{1n} \\ 0 & u_{22} & \cdots & u_{2n} \\ \vdots & & \ddots & \vdots \\ 0 & 0 & \cdots & u_{nn} \end{bmatrix}^{-1}$$

CA2.3 A general matrix form of a backward tridiagonal system is given below. (a) Develop a simple memory-efficient algorithm to solve this system using the Gauss elimination method with no pivoting. (b) Write a computer program in a language of your choice to test your code for $n = 10$ with $a_i = b_i = -1$ and $d_i = 2$ with $c_1 = c_{10} = 1$, $c_i = 0$ for $i = 2, 3,\ldots, 9$. *Hint:* What would the coefficient matrix and the vector of unknowns (rhs) look like if you rearranged the equations backward from the bottom row to the top?

$$\begin{bmatrix} & & & & & a_1 & d_1 \\ & & & & a_2 & d_2 & b_2 \\ & & & a_3 & d_3 & b_3 \\ & & \cdot^{\displaystyle\cdot} & \cdot^{\displaystyle\cdot} & \cdot^{\displaystyle\cdot} \\ & a_{n-2} & d_{n-2} & b_{n-2} \\ a_{n-1} & d_{n-1} & b_{n-1} \\ d_n & b_n \end{bmatrix} \begin{bmatrix} x_1 \\ x_2 \\ x_3 \\ \vdots \\ x_{n-2} \\ x_{n-1} \\ x_n \end{bmatrix} = \begin{bmatrix} c_1 \\ c_2 \\ c_3 \\ \vdots \\ c_{n-2} \\ c_{n-1} \\ c_n \end{bmatrix}$$

CA2.4 The general matrix form of a penta-diagonal system of linear equations is given below.

$$
\begin{bmatrix}
d_1 & c_1 & f_1 \\
a_1 & d_2 & c_2 & f_2 \\
e_1 & a_2 & d_3 & c_3 & f_3 \\
 & e_2 & a_3 & d_4 & c_4 & f_4 \\
 & & \ddots & \ddots & \ddots & \ddots & \ddots \\
 & & & e_{m-2} & a_{m-1} & d_m & c_m & f_m \\
 & & & & \ddots & \ddots & \ddots & \ddots & \ddots \\
 & & & & & e_{n-4} & a_{n-3} & d_{n-2} & c_{n-2} & f_{n-2} \\
 & & & & & & e_{n-3} & a_{n-2} & d_{n-1} & c_{n-1} \\
 & & & & & & & e_{n-2} & a_{n-1} & d_n
\end{bmatrix}
\begin{bmatrix}
x_1 \\ x_2 \\ x_3 \\ x_4 \\ \vdots \\ x_m \\ \vdots \\ x_{n-2} \\ x_{n-1} \\ x_n
\end{bmatrix}
=
\begin{bmatrix}
b_1 \\ b_2 \\ b_3 \\ b_4 \\ \vdots \\ b_m \\ \vdots \\ b_{n-2} \\ b_{n-1} \\ b_n
\end{bmatrix}
$$

(a) Develop a simple and memory-efficient algorithm to solve this system using the Gauss elimination method with no pivoting, and (b) write a computer program in a language of your choice to test your algorithm for $n = 10$ with $a_i = e_i = c_i = f_i = -1$, $d_i = 4$ with the rhs vector set as $b_1 = b_{10} = 2$, $b_2 = b_9 = 1$, $b_i = 0$ for $i = 3, 4, \ldots, 8$. The true solution is $\mathbf{x} = \mathbf{1}$.

CA2.5 A general matrix form of a bordered tridiagonal system of linear equations is given below.

$$
\begin{bmatrix}
d_1 & a_1 & & & & & & f_1 \\
b_2 & d_2 & a_2 & & & & & f_2 \\
 & b_3 & d_3 & a_3 & & & & f_3 \\
 & & \ddots & \ddots & \ddots & & & \vdots \\
 & & & b_{n-2} & d_{n-2} & a_{n-2} & f_{n-2} \\
 & & & & b_{n-1} & d_{n-1} & a_{n-1} \\
e_1 & e_2 & \cdots & e_{n-1} & e_{n-2} & b_n & d_n
\end{bmatrix}
\begin{bmatrix}
x_1 \\ x_2 \\ x_3 \\ \vdots \\ x_{n-2} \\ x_{n-1} \\ x_n
\end{bmatrix}
=
\begin{bmatrix}
c_1 \\ c_2 \\ c_3 \\ \vdots \\ c_{n-2} \\ c_{n-1} \\ c_n
\end{bmatrix}
$$

Develop an algorithm to (a) perform LU decomposition to find matrices $\mathbf{L}$ and $\mathbf{U}$ that have the general forms given below; (b) solve the linear system using forward and backward substitutions ($\mathbf{Ly} = \mathbf{c}$ and $\mathbf{Ux} = \mathbf{y}$). (c) Write a computer program in a language of your choice to test the algorithms you developed in Parts (a) and (b). Make your programs as simple and memory-efficient as possible. Use the following as the test problem: $k = 5$, $\mathbf{d} = \mathbf{4}$, $\mathbf{a} = \mathbf{b} = \mathbf{f} = \mathbf{e} = \mathbf{1}$, and $\mathbf{c} = \{6,7,7,7,6,10\}$, and the true solution is $\mathbf{x} = \mathbf{1}$.

$$
\mathbf{L} =
\begin{bmatrix}
\ell_1 \\
b_2 & \ell_2 \\
 & b_3 & \ell_3 \\
 & & \ddots & \ddots \\
 & & & b_{n-2} & \ell_{n-2} \\
 & & & & b_{n-1} & \ell_{n-1} \\
t_1 & t_2 & \cdots & t_{n-1} & t_{n-2} & t_{n-1} & \ell_n
\end{bmatrix}
, \quad
\mathbf{U} =
\begin{bmatrix}
1 & u_1 & & & & & w_1 \\
 & 1 & u_2 & & & & w_2 \\
 & & 1 & u_3 & & & w_3 \\
 & & & \ddots & \ddots & & \vdots \\
 & & & & 1 & u_{n-2} & w_{n-2} \\
 & & & & & 1 & w_{n-1} \\
 & & & & & & 1
\end{bmatrix}
$$

CA2.6 Use Gauss-Jordan elimination with no pivoting to calculate the current in each branch of the circuit given in **Fig. CA2.6**. Note: The resistances are given in Ω.

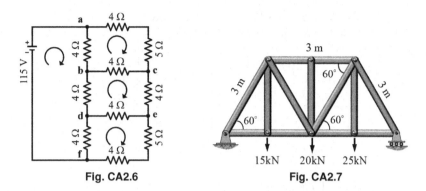

Fig. CA2.6   Fig. CA2.7

CA2.7 In order to determine the force in each member of the truss in **Fig. CA2.7**, (a) derive the equilibrium equations and express them in $\mathbf{Ax} = \mathbf{b}$ form; (b) apply elementary row operations to convert matrix $\mathbf{A}$ into lower triangular $\mathbf{L}$ to solve the system of linear equations using forward substitution. (c) State if the members are in tension (T) or compression (c).

CA2.8 Olive oil (O) is produced by an industrial extraction process that involves the preparation of a homogeneous olive paste (skin, pulp, and stone) by crushing and malaxing processes. Olive paste is made up primarily of olive oil, water, and carbohydrates, and olive oil is extracted from this paste by a series of centrifugation processes (primary CF and second-stage CF-1 and CF-2). The contents (in kg per kg slurry) of W (water), O (olive oil), and C (carbohydrates) at each stage of a production facility are depicted in the simplified flowchart in **Fig. CA2.8**. The facility will produce 304 kg of olive paste per hour. About half of the processed olive paste is converted into olive cake. Determine the steady-state flow rates at each branch of the flowchart in order to attain the specified contents (mass ratios). Use the furnished data, as well as the conservation of overall mass, CF-1 and CF-2, and the conservation of mass for olive oil at each stage, to construct an 8×8 system of linear equations for the flow rates. Then solve the system of linear equations using the Gauss-Jordan elimination algorithm.

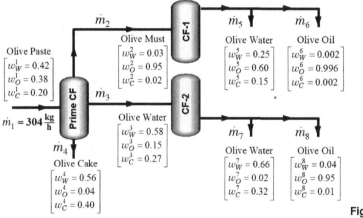

Fig. CA2.8

CA2.9 Consider balancing the following chemical reactions: Express all equations for the unknowns (chemical elements present in the equations) in matrix form. Then use the Gauss-Jordan elimination algorithm to solve the resulting system.

(a) $x_1 K_4 Fe(SCN)_6 + x_2 K_2 Cr_2 O_7 + x_3 H_2 SO_4 \rightarrow x_4 Fe_2(SO_4)_3 + x_5 Cr_2(SO_4)_3 + x_6 CO_2$
$$+ x_7 H_2 O + x_8 K_2 SO_4 + x_9 KNO_3$$

(b) $x_1 K_4 Fe(CN)_6 + x_2 KMnO_4 + x_3 H_2 SO_4 \rightarrow x_4 KHSO_4 + x_5 Fe_2(SO_4)_3 +$
$$x_6 MnSO_4 + x_7 HNO_3 + x_8 CO_2 + x_9 H_2 O$$

CA2.10 Modify the pseudomodule BAND_SOLVE (Pseudocodes 2.14) to carry out partial pivoting. Program this module in a language of your choice to solve the following system of equations given in compact form:

$$
\begin{bmatrix}
0 & 4 & -1 & -1 \\
-2 & 4 & -1 & -1 \\
-2 & 4 & -1 & -1 \\
-2 & 4 & -1 & -1 \\
-2 & 4 & -1 & -1 \\
-2 & 4 & -1 & 0 \\
-2 & 4 & 0 & 0
\end{bmatrix}
\begin{bmatrix}
x_1 \\ x_2 \\ x_3 \\ x_4 \\ x_5 \\ x_6 \\ x_7
\end{bmatrix}
=
\begin{bmatrix}
1 \\ -1 \\ -1 \\ 8 \\ 19 \\ -17 \\ -8
\end{bmatrix}
$$

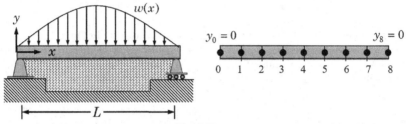

Fig. CA2.11

CA2.11 A simply supported beam of length L is resting on an elastic foundation of stiffness K (N/m^2), as shown in **Fig. CA2.11**. The displacement of a beam y (m) caused by a nonuniform (parabolic) load, $w(x) = w_0 x(L - x)$ (N/m), is described by the following fourth-order ODE:

$$
\frac{d^4 y}{dx^4} + \beta y = w(x) = \alpha \frac{x}{L}\left(1 - \frac{x}{L}\right), \qquad y(0) = y''(0) = y(L) = y''(L) = 0
$$

where $\alpha = w_0 L^2 / EI$ and $\beta = K/EI$.

The finite-difference method solution (*see* Chapter 10) corresponding to $\alpha = 50$, $\beta = 10$, and $h = \Delta x = L/8$ leads to the following penta-diagonal system of linear equations, given below.

$$
\begin{bmatrix}
5 + h^4\beta & -4 & 1 & & & & \\
-4 & 6 + h^4\beta & -4 & 1 & & & \\
1 & -4 & 6 + h^4\beta & -4 & 1 & & \\
& 1 & -4 & 6 + h^4\beta & -4 & 1 & \\
& & 1 & -4 & 6 + h^4\beta & -4 & 1 \\
& & & 1 & -4 & 6 + h^4\beta & -4 \\
& & & & 1 & -4 & 5 + h^4\beta
\end{bmatrix}
\begin{bmatrix}
y_1 \\ y_2 \\ y_3 \\ y_4 \\ y_5 \\ y_6 \\ y_7
\end{bmatrix}
=
\frac{\alpha h^4}{64}
\begin{bmatrix}
7 \\ 12 \\ 15 \\ 16 \\ 15 \\ 12 \\ 7
\end{bmatrix}
$$

(a) Use the program written in **Fig. CA2.7** to solve the linear system; (b) treat it as a banded system with bandwidths of 2 and solve it using Gauss elimination with no pivoting (Pseudocodes 2.14) and LU-decomposition for banded systems (Pseudocodes 2.16 and 2.17).

CA2.12 The steady-state heat conduction in the square geometry shown in **Fig. CA2.12** is solved using the finite-difference method. Three sides are kept at the same temperature, while one side is insulated, i.e., $(\partial T/\partial x)_{x=0} = 0$. The geometry is uniformly meshed, and the difference equations corresponding to the nodes are given as

$$
\begin{aligned}
\text{For } p = 1 \quad &: \quad -4T_1 + 2T_2 + T_5 + 400 = 0, \\
\text{For } p = 5 \quad &: \quad T_1 - 4T_5 + 2T_6 + T_9 = 0, \\
\text{For } p = 9 \quad &: \quad T_5 - 4T_9 + 2T_{10} + 500 = 0, \\
\text{For } p \ne 1, 5, 9 \quad &: \quad T_s + T_w - 4T_p + T_e + T_n = 0
\end{aligned}
$$

where s, w, e, and n denote the compass directions. Constructing the system of linear equations yields a banded system. Use compact banded forms to solve the system with LU-decomposition, etc.

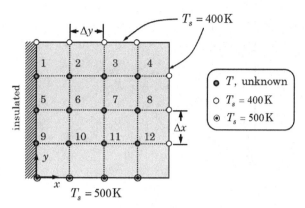

Fig. CA2.12

CA2.13 A special case of sparse matrices, as illustrated in **Fig. CA2.13** where the dots denote non-zero elements, are called the *skyline matrices*. The sparsity pattern of **L** or **U** is inherited from **A**, so it is easier to develop an algorithm.

A=   **L=**   **U=**

Fig. CA2.13

Consider the system of linear equations $\mathbf{Ax} = \mathbf{b}$ with a *symmetric skyline matrix* whose general form is given as below. Develop an algorithm to: (a) perform LU decomposition to give the **L** and **U** matrices that have the general forms given below; (b) solve the system using forward and backward substitutions ($\mathbf{Ly} = \mathbf{b}$ and $\mathbf{Ux} = \mathbf{y}$); and (c) solve the linear system using the Gauss elimination method with no pivoting. (d) Write a computer program in a language of your choice to test the algorithms you developed in Parts (a), (b), and (c). Use the following test problem: $i = 6$, $\mathbf{d} = \{1,2,3,2,4,1,1,2,5\}$, $\mathbf{p} = \{3,1,3,2,2\}$, $\mathbf{q} = \{2,1,1,3,1,2,1,1\}$, and $\mathbf{b} = \{2, 0, 2, 5, 11, 17, 9, -3, 19\}$. The true solution is $\mathbf{x} = \{1, -1, 2, -2, 3, -3, 4, -4, 5\}$. Make your algorithms as simple and memory-efficient as possible.

$$
\mathbf{A} = \begin{bmatrix}
d_1 & 0 & \cdots & 0 & p_1 & 0 & \cdots & 0 & q_1 \\
0 & d_2 & 0 & \vdots & p_2 & 0 & \cdots & 0 & q_2 \\
\vdots & \ddots & \ddots & 0 & \vdots & \vdots & \ddots & \vdots & \vdots \\
0 & \cdots & 0 & d_{i-1} & p_{i-1} & 0 & \cdots & 0 & q_{i-1} \\
p_1 & p_2 & \cdots & p_{i-1} & d_i & 0 & \cdots & 0 & q_i \\
0 & \cdots & 0 & 0 & 0 & d_{i+1} & 0 & \vdots & q_{i+1} \\
\vdots & 0 & \vdots & \cdots & \ddots & 0 & \ddots & 0 & \vdots \\
0 & 0 & 0 & \cdots & 0 & \cdots & 0 & d_{n-1} & q_{n-1} \\
q_1 & q_2 & \cdots & q_{i-1} & q_i & q_{i+1} & \cdots & q_{n-1} & d_n
\end{bmatrix} = \mathbf{LU}, \quad \mathbf{b} = \begin{bmatrix} b_1 \\ b_2 \\ \vdots \\ b_{i-1} \\ b_i \\ b_{i+1} \\ \vdots \\ b_{n-1} \\ b_n \end{bmatrix}
$$

$$
\mathbf{L} = \begin{bmatrix}
1 & 0 & 0 & \cdots & 0 & 0 & 0 \\
0 & 1 & 0 & \cdots & 0 & 0 & 0 \\
\vdots & 0 & \ddots & \ddots & \vdots & \vdots & \vdots \\
\ell_{p1} & \cdots & \ell_{p,i-1} & 1 & 0 & \cdots & 0 \\
\vdots & \cdots & 0 & 0 & \ddots & \ddots & \vdots \\
0 & 0 & \cdots & 0 & 0 & 1 & 0 \\
\ell_{q1} & \ell_{q1} & \cdots & \ell_{qi} & \cdots & \ell_{q,n-1} & d_n
\end{bmatrix},
$$

$$
\mathbf{U} = \begin{bmatrix}
u_1 & 0 & \cdots & u_{p1} & 0 & \cdots & u_{q1} \\
0 & u_2 & 0 & \vdots & 0 & 0 & u_{q2} \\
\vdots & 0 & \ddots & u_{p,i-1} & \vdots & \vdots & \vdots \\
0 & \cdots & 0 & u_i & 0 & \cdots & u_{qi} \\
\vdots & \cdots & 0 & \ddots & \ddots & 0 & \vdots \\
0 & 0 & \cdots & \vdots & 0 & u_{n-1} & u_{q,n-1} \\
0 & 0 & \cdots & 0 & \cdots & 0 & u_n
\end{bmatrix}
$$

Linear Systems: Iterative Methods

THE direct methods for solving systems of linear equations in various forms were discussed in Chapter 2. The direct methods require a finite number of arithmetic operations, regardless of the number of "*zero*" elements the matrix has. The direct methods also give the true solution in the absence of round-offs or other errors. Nonetheless, solving large linear systems by direct methods requires very large memory storage space and leads to excessive computation time. For this reason, direct methods are generally preferred in cases where a matrix is a *dense (matrix)* of a moderate size, i.e., a matrix with few zero elements.

DOI: 10.1201/9781003474944-3

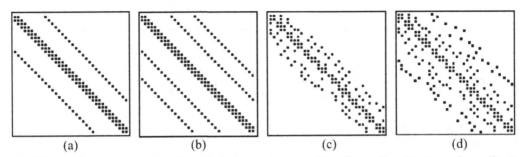

FIGURE 3.1: Sparcity patterns of coefficient matrices resulting from discretization of an elliptic PDE in (a) 2D and (b) 3D-rectangular, elliptic PDE in (c) 2D- and (d) 3D-nonrectangular domain. The non-zero elements are marked in black.

This chapter is mainly devoted to basic stationary iterative methods (Jacobi, Gauss-Seidel, and SOR) for solving large simultaneous systems of linear equations that cannot be effectively handled by direct methods. A substantial discussion on Conjugate Gradient Methods (CGMs) for symmetric and non-symmetric systems is also introduced. The implementations of stationary and CGM methods are presented with algorithms (also pseudocodes) and solved examples. The performance and convergence rates of the iterative methods covered in the text are also discussed.

3.1　OVERVIEW

The majority of large systems of linear equations are encountered in practice in the numerical solution of ordinary differential equations (ODEs) or partial differential equations (PDEs). The size of the matrices can be in the order of hundreds, thousands, or even millions. Furthermore, most of the non-zero elements are concentrated on or near the principal diagonal, while the remaining elements consist of zeros.

Numerical solutions of partial differential equations (PDEs) in two- and three-dimensional rectangular domains result in coefficient matrices with 5- and 7-diagonals, respectively, consisting of mostly non-zero elements (*see* Fig. 3.1a and b). Such linear systems with narrow bandwidths may be treated as *banded systems*, which do reduce the memory requirement and the computational efforts to some degree. Yet arithmetic operations have to be carried out with numerous zeros within the band. We also encounter resulting matrices with non-zero elements dispersed regularly or irregularly around the main diagonal, as depicted in Fig. 3.1c and d. These matrices generally appear in the numerical solution of linear or non-linear PDEs in non-rectangular domains. Such matrices with relatively few non-zero elements are called *sparse matrices*.

Digital computers and general multi-purpose algorithms are not designed to distinguish between zero and non-zero elements. In other words, even though humans can immediately spot zeros and skip arithmetic operations with them, digital computers do carry out the arithmetic operations regardless of the values of the numbers. When a linear system is treated as non-banded or non-sparse, the number of arithmetic operations required to obtain the solution is on the order of $\mathcal{O}(n^3)$, and the computational process can require excessive cpu-time, which can be prohibitive even for reasonably large matrices. Yet another vital constraint in dealing with large, dense matrices is the excessive memory requirement for storing the matrix elements, which can be extremely restrictive. For the above reasons, it should not be surprising that direct solution methods are *not* preferred when working with sparse systems.

Iterative methods are techniques that, unlike direct methods, begin with an *initial guess* for a system of linear (or nonlinear) equations and refine the solution by successively applying a suitably chosen estimation algorithm until a desired level of accuracy or convergence is achieved. Iterative methods require less memory allocation because only non-zero elements that are used in the computation are stored. This means that they are computationally cheaper as arithmetic calculations are carried out only with non-zero elements, reducing the cpu-time significantly.

To begin, consider a system of linear equations ($\mathbf{Ax} = \mathbf{b}$) given as

$$
\begin{aligned}
a_{11}x_1 &+ a_{12}x_2 &+ a_{13}x_3 &+ \cdots + a_{1n}x_n &= b_1 \\
a_{21}x_1 &+ a_{22}x_2 &+ a_{23}x_3 &+ \cdots + a_{2n}x_n &= b_2 \\
a_{31}x_1 &+ a_{32}x_2 &+ a_{33}x_3 &+ \cdots + a_{3n}x_n &= b_3 \\
\vdots & \vdots & \vdots & \ddots & \vdots \\
a_{n1}x_1 &+ a_{n2}x_2 &+ a_{n3}x_3 &+ \cdots + a_{nn}x_n &= b_n
\end{aligned}
\tag{3.1}
$$

where x_i's are the unknowns, a_{ij}'s and b_i's are the elements of the coefficient matrix ($\mathbf{A}$) and the right-hand side vector ($\mathbf{b}$), respectively.

This system can be solved iteratively after expressing Eq. (3.1) in the following general mathematical form:

$$
\mathbf{x}^{(p+1)} = \mathbf{F}(\mathbf{x}^{(p)}), \quad p \geqslant 0
\tag{3.2}
$$

where the superscript notation "(p)" denotes the iteration level, $\mathbf{x}$ is the vector of unknowns, $\mathbf{F} = [F_1, F_2, \ldots, F_n]^T$ is the *fixed-point form* of the linear equations that depends on the numerical algorithm, and $\mathbf{x}^{(p+1)}$ and $\mathbf{x}^{(p)}$ denote the vector of computed estimates at the $(p+1)$th and pth iteration steps, which will also be referred to throughout this text as *current* (new iterate) and *prior estimates* (previous iterate), respectively.

An *iterative method* is a procedure based on generating a sequence of improved approximations using an *initial guess*, $\mathbf{x}^{(0)}$, which should be supplied to start the iterative process. By setting $p = 0$ in Eq. (3.2) and using the initial guess on the rhs, a new estimate, $\mathbf{x}^{(1)}$, is obtained. Then, using $\mathbf{x}^{(1)}$, another "*hopefully better*" approximation, $\mathbf{x}^{(2)}$, is found, and subsequent approximations or estimates are found in this manner. In convergent systems, this process yields a sequence of estimates $\mathbf{x}^{(1)}, \mathbf{x}^{(2)}, \ldots, \mathbf{x}^{(p)}$ that improve with each iteration step. An iterative method is said to be *convergent* when the difference between the true solution $\mathbf{x}$ and the sequence $\mathbf{x}^{(p)}$ generated from an initial $\mathbf{x}^{(0)}$ tends to zero as the number of iterations increases, i.e., $\mathbf{x} - \mathbf{x}^{(p)} \to 0$ as $p \to \infty$.

3.1.1 STEPS OF AN ITERATIVE METHOD

All iterative algorithms consist of three basic steps: initialization, computation, and termination.

(i) Initialization: This is the first step in every iteration process where a suitable initial guess is made. In most problems, an arbitrary initial guess (i.e., $\mathbf{x}^{(0)}$=constant) is often set for the values of unknowns due to the simplicity and ease of implementation. Moreover, the linear systems of equations are not very sensitive to the initial guess, so there is no point in trying to choose a perfect initial guess for the unknowns. However, when dealing with the linearized form of a set of non-linear equations, the coefficients and the rhs of the linear systems consist of the prior iterates. In such cases, an initial guess in close proximity to the solution (also referred to as a "*smart guess*") speeds up the iteration process considerably.

(ii) **Computation:** This step involves the way current (new) estimates are computed, i.e., how $\mathbf{x}^{(p+1)}$ is computed. In this step, an *iterative scheme* which is expressed mathematically by the general equation Eq. (3.2) is implemented. This step is the core of an iterative algorithm and is different for each iterative method, as discussed in detail in the subsequent sections.

(iii) **Termination:** Any iterative solution procedure is terminated for basically two reasons:

(1) *When a current estimate converges to a solution within a predetermined tolerance.* This step is crucial to finding a quick solution while avoiding oversolving. The successive estimates of a converging iterative system will approach closer to the true solution with each subsequent iteration. Since the true solution $\mathbf{x}$ is unknown, it is impossible to verify how close $\mathbf{x}^{(p)}$ is to $\mathbf{x}$.

An iteration process is terminated when the current and prior estimates are sufficiently close to meet a preset *convergence tolerance* (or *stopping criterion*). The stopping criterion, ε, determines the level of accuracy sought in the solution and also affects the computation time. For general programming practices, analysts often use both *absolute* and *relative error* criteria for terminating an iteration process. The number of arithmetic operations depends not only on the accuracy required in the solution but also on the iteration scheme adopted. A modest number of iterations will be sufficient to obtain an approximate solution with acceptable accuracy when the accuracy requirement is not too strict. However, if the convergence tolerance is too small, then the total number of iterations and consequently the cpu-time will be substantially large.

Commonly used stopping criteria are based on the absolute error of the displacement vector at the pth iteration step, i.e., ℓ_1, ℓ_2, or ℓ_∞-norms of $\mathbf{d}^{(p)}$.

$$\|\mathbf{d}^{(p)}\|_1 = \sum_{i=1}^{n} |d_i^{(p)}| < \varepsilon_1, \quad \|\mathbf{d}^{(p)}\|_2 = \sqrt{\sum_{i=1}^{n} \left(d_i^{(p)}\right)^2} < \varepsilon_2, \quad \|\mathbf{d}^{(p)}\|_\infty = \max_{1 \leqslant i \leqslant n} |d_i^{(p)}| < \varepsilon_3 \quad (3.3)$$

where $\mathbf{d}^{(p)} = \mathbf{x}^{(p+1)} - \mathbf{x}^{(p)}$ is so-called the *displacement vector*. Another set of stopping criteria, based on the norms of $\mathbf{d}^{(p)}/\mathbf{x}^{(p+1)}$, provide a measure for relative errors:

$$\left\|\frac{\mathbf{d}^{(p)}}{\mathbf{x}^{(p+1)}}\right\|_1 = \sum_{i=1}^{n} \left|\frac{d_i^{(p)}}{x_i^{(p+1)}}\right| < \varepsilon_1, \qquad \left\|\frac{\mathbf{d}^{(p)}}{\mathbf{x}^{(p+1)}}\right\|_2 = \sqrt{\sum_{i=1}^{n} \left(\frac{d_i^{(p)}}{x_i^{(p+1)}}\right)^2} < \varepsilon_2,$$

$$\left\|\frac{\mathbf{d}^{(p)}}{\mathbf{x}^{(p+1)}}\right\|_\infty = \max_{1 \leqslant i \leqslant n} \left|\frac{d_i^{(p)}}{x_i^{(p+1)}}\right| < \varepsilon_3 \tag{3.4}$$

where ε_1, ε_2, and ε_3 are convergence tolerances that may be chosen differently.

(2) *When the number of iterations reaches a predetermined upper bound (MAXIT).* This criterion is very important in that we do not know in advance whether convergence with the desired tolerance can be achieved within a predetermined number of iterations. In the event that a system converges slowly or does not converge at all, the iteration process will go into an infinite loop unless the loop is terminated. Therefore, it is a good programming practice to set an upper bound (denoted by *maxit* throughout the text) for the maximum number of iterations permitted.

3.1.2 SOURCES OF ERROR IN ITERATIVE METHODS

There are basically two sources of error in iterative methods as applied to systems of linear equations: *truncation errors* and *round-off errors*.

All iterative methods are subjected to "truncation" because the iteration process eventually needs to be terminated after a finite number of iterations. In reality, it is impractical to perform an infinite number of iterations in an attempt to find the true solution, which is unknown. For this reason, the iterative solution of a linear system is "*approximate*." Nevertheless, the truncation (or approximation) error can be reduced by applying a smaller convergence tolerance, i.e., increasing the number of iterations.

The second source of error is round-off error, which is introduced when computing iterates over many iteration steps. In direct methods, a round-off error in the value of the last unknown propagates at every step of the solution, corrupting the values of the remaining unknowns, while a round-off error in an iterative method is an error that occurs only in the last iteration step. The reason for this is that the coefficient matrix is always intact throughout the iteration process, and the current estimates are always computed from scratch at each iteration using prior estimates as the initial guess vector.

In an iterative process, the magnitude of round-off error is usually not considered serious (except for ill-conditioned systems) because the truncation error introduced by terminating the iteration is generally larger. In this context, it is only natural to expect round-off errors to be a much less serious problem in iterative methods than in direct methods. Furthermore, there are two more factors that make iterative methods less sensitive to round-off errors. (i) The iterative methods are employed for diagonally dominant systems, which reduces the influence of round-off errors. The more dominant the system, the less the round-off error. (ii) The systems of equations dealt with are generally sparse, and the presence of the zeroes may remove the influence of some of the previous components.

3.2 STATIONARY ITERATIVE METHODS

3.2.1 JACOBI METHOD

Consider the system of linear equations given by Eq. (3.1). This system can be expressed in fixed-point form, i.e., $\mathbf{x} = \mathbf{F}(\mathbf{x})$, and it has a unique solution provided that $a_{ii} \neq 0$ for all i. However, there are several ways to express the system in a fixed-point form. The most common approach is to solve each equation for its diagonal component; that is, solve x_1 from the first equation, x_2 from the second equation, and so on. Employing this idea in Eq. (3.1) yields

$$x_1 = F_1(x_1, x_2, \ldots, x_n) = \frac{1}{a_{11}}(b_1 - a_{12}x_2 - a_{13}x_3 - \cdots - a_{1n}x_n)$$

$$x_2 = F_2(x_1, x_2, \ldots, x_n) = \frac{1}{a_{22}}(b_2 - a_{21}x_1 - a_{23}x_3 - \cdots - a_{2n}x_n)$$

$$x_3 = F_3(x_1, x_2, \ldots, x_n) = \frac{1}{a_{33}}(b_3 - a_{31}x_1 - a_{32}x_3 - \cdots - a_{3n}x_n)$$

$$\vdots \qquad\qquad\qquad \vdots$$

$$x_n = F_n(x_1, x_2, \ldots, x_n) = \frac{1}{a_{nn}}(b_n - a_{n1}x_1 - a_{n2}x_2 - \cdots - a_{n,n-1}x_{n-1})$$

Next, we introduce the superscript notation "(p)" to denote the estimates (or approximations) at the pth iteration step. With this notation, *Jacobi iteration equations* that are used to generate successive approximations are set up as

$$x_1^{(p+1)} = \frac{1}{a_{11}} \left(b_1 - a_{12}x_2^{(p)} - a_{13}x_3^{(p)} - a_{14}x_4^{(p)} - \cdots - a_{1n}x_n^{(p)} \right)$$

$$x_2^{(p+1)} = \frac{1}{a_{22}} \left(b_2 - a_{21}x_1^{(p)} - a_{23}x_3^{(p)} - a_{24}x_4^{(p)} - \cdots - a_{2n}x_n^{(p)} \right)$$

$$x_3^{(p+1)} = \frac{1}{a_{33}} \left(b_3 - a_{31}x_1^{(p)} - a_{32}x_2^{(p)} - a_{34}x_4^{(p)} - \cdots - a_{3n}x_n^{(p)} \right) \qquad (3.5)$$

$$\vdots \qquad\qquad \vdots$$

$$x_n^{(p+1)} = \frac{1}{a_{nn}} \left(b_n - a_{n1}x_1^{(p)} - a_{n2}x_2^{(p)} - a_{n3}x_3^{(p)} - \cdots - a_{n,n-1}x_{n-1}^{(p)} \right)$$

where $x_i^{(p)}$ and $x_i^{(p+1)}$ denote the estimates of the ith unknown at the pth (prior) and $(p+1)$th (current) iteration steps, respectively.

Jacobi method is also called the "*method of simultaneous displacements*" because each equation is simultaneously changed with prior estimates; in other words, all current estimates are calculated using the prior estimates. The Jacobi iteration equations, Eq. (3.5), can be expressed in the most general and most suitable form for programming as

$$x_i^{(p+1)} = \frac{1}{a_{ii}} \left\{ b_i - \sum_{j=1}^{i-1} a_{ij}x_j^{(p)} - \sum_{j=i+1}^{n} a_{ij}x_j^{(p)} \right\} \qquad (3.6)$$

where $a_{ii} \neq 0$ for all i is a mandatory requirement. It should, however, be pointed out that the Jacobi method, whose convergence issues are covered in greater detail in Section 3.3 does not converge for every linear system.

A set of pseudomodules for solving a linear system of equations with the Jacobi method is presented in Pseudocode 3.1. As input, the main module JACOBI requires the number of equations (n), the coefficient matrix ($\mathbf{A}$), the rhs vector ($\mathbf{b}$), an initial guess vector ($\mathbf{xo}$), a convergence tolerance (ε), and an upper bound for the number of iterations ($maxit$). The outputs of the module are the solution vector ($\mathbf{x}$), the total number of iterations performed ($iter$), and the ℓ_2-norm of the displacement vector ($error$).

The JACOBI_DRV driver module performs a one-step Jacobi iteration and provides the current estimate ($\mathbf{x}^{(p+1)}$) as well as the ℓ_2-norm of the displacement vector ($\delta = \|\mathbf{d}^{(p)}\|_2$) utilized in JACOBI to terminate the iteration process. On the first visit to JACOBI_DRV, $\mathbf{xo}$ is set to the initial guess ($\mathbf{xo} = \mathbf{x}^{(0)}$). In the subsequent visits, $\mathbf{xo}$ contains the prior estimates, $\mathbf{xo} = \mathbf{x}^{(p)}$. An accumulator, sum, is used to sum up all $a_{ij}x_i^{(p)}$ products (except $i = j$) in the ith row, and the current estimate $\mathbf{x}$ is found from Eq. (3.6). The number of iterations in JACOBI is counted with the counter variable p. After setting the current estimates as prior ($\mathbf{xo} \leftarrow \mathbf{x}$), the iteration process is continued until the convergence criterion ($\delta < \varepsilon$) is satisfied or the maximum number of iterations ($maxit$) is reached. On exit from JACOBI, the most recent displacement norm is set to $error$.

The ℓ_2-norm of the displacement vector, $\|\mathbf{d}^{(p)}\|_2$, is evaluated using Function Module ENORM, which has two arguments: a vector $\mathbf{x}$ and its length n. An accumulator loop and an accumulator variable δ are used to sum up all x_i^2. The square root of the accumulator is set to function ENORM. Note that, in the accumulator, x_i^2 is carried out as $x_i * x_i$, which is used to save cpu-time because the latter is about ten times faster.

Pseudocode 3.1

Module JACOBI $(n, \varepsilon, \mathbf{A}, \mathbf{b}, \mathbf{xo}, \mathbf{x}, maxit, iter, error)$
\ DESCRIPTION: A pseudomodule to solve $\mathbf{Ax} = \mathbf{b}$ using the Jacobi method.
\ USES:
\   JACOBI_DRV:: Driver module performing one step Jacobi iteration.
Declare: a_{nn}, b_n, x_n, xo_n \ Declare array variables
$p \leftarrow 0; \delta 0 \leftarrow 1$ \ Initialize iteration counter and ℓ_2-norm
Repeat   \ Iterate until convergence is achieved
   $p \leftarrow p + 1$ \ Count iterations
   JACOBI_DRV$(n, \mathbf{A}, \mathbf{b}, \mathbf{xo}, \mathbf{x}, \delta)$ \ Perform one-step Jacobi iteration
   Write: "For $p=$", p," Error=",δ," Rate=", δ/δ_0 \ Echo iteration progress
   $\mathbf{xo} \leftarrow \mathbf{x}$ \ Set current estimates as prior; $\mathbf{x}^{(p)} \leftarrow \mathbf{x}^{(p+1)}$
   $\delta_0 \leftarrow \delta$ \ Set current error as prior; $\delta^{(p)} \leftarrow \delta^{(p+1)}$
Until $\left[\delta < \varepsilon \text{ Or } p = maxit\right]$ \ Terminate iteration if $\delta < \varepsilon$ and $p \leqslant maxit$
$error \leftarrow \delta$ \ Set current ℓ_2-norm as error
$iter \leftarrow p$ \ Set current p as $iter$
\ If iteration limit $maxit$ is reached $with\ no$ convergence
If $\left[p = maxit\right]$ **Then**   \ Issue info and a warning
   Write: "Error=",$error$,"Jacobi failed to converge after",$maxit$,"iterations"
End If
End Module JACOBI

Module JACOBI_DRV $(n, \mathbf{A}, \mathbf{b}, \mathbf{xo}, \mathbf{x}, \delta)$
\ DESCRIPTION: A pseudomodule to perform one step Jacobi iteration.
\ USES:
\   ENORM:: Function Module calculating Euclidean (ℓ_2-)norm of a vector.
Declare: $a_{nn}, b_n, x_n, xo_n, d_n$ \ Declare array variables
For $\left[i = 1, n\right]$ \ Loop i: Sweep equations from top to bottom
   $sums \leftarrow 0$ \ Initialize accumulator, sums of off-diagonals
   For $\left[j = 1, n\right]$ \ Loop j: Sweep terms from left to right
     If $\left[i \neq j\right]$ **Then**   \ If $i = j$, skip diagonal term
       $sums \leftarrow sums + a_{ij} * xo_j$ \ Accumulate $a_{ij} * xo_j$ products
     End If
   End For
   $x_i \leftarrow (b_i - sums)/a_{ii}$ \ Estimate x_i from Eq. (3.6)
   $d_i \leftarrow x_i - xo_i$ \ Find displacement vector, $d_i \leftarrow x_i^{(p+1)} - x_i^{(p)}$
End For
$\delta \leftarrow$ **ENORM**$(n, \mathbf{d})$ \ Set ℓ_2-norm of $\mathbf{d}^{(p)}$ to δ
End Module JACOBI_DRV

Function Module ENORM $(n, \mathbf{x})$
\ DESCRIPTION: A function to calculate ℓ_2-norm, $\|\mathbf{x}\|_2$, of a vector of length n.
Declare: x_n
$delta \leftarrow 0$ \ Initialize ℓ_2-norm
For $\left[i = 1, n\right]$ \ Loop to find sums of squares of elements of $\mathbf{x}$
   $delta \leftarrow delta + x_i * x_i$ \ Accumulate x_i^2's
End For
ENORM$\leftarrow \sqrt{delta}$ \ Assign $\sqrt{\sum x_i^2}$ to ENORM
End Function Module ENORM

Jacobi Method

- The method is easy to understand and implement;
- Numerically it is robust;
- Iteration procedure can be carried out in a parallel environment.

- The method in practice is not preferred for large systems;
- It converges under certain conditions (it is generally used to solve linear systems whose coefficient matrix is "diagonally dominant");
- The rate of convergence, if it converges at all, is quite slow, making it unsuitable and unacceptable for large linear (or sparse) systems.

EXAMPLE 3.1: Application of the Jacobi method

Consider the circuit in the figure below.

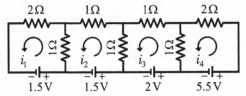

Kirchoff's voltage law for each loop yields a linear system for the loop currents. Apply the Jacobi method to estimate the loop currents. Use $\mathbf{i}^{(0)} = \mathbf{1}$ as the initial guess and $\|\mathbf{d}\|_2^{(p)} \leqslant 10^{-3}$ as the stopping criterion.

$$
\begin{bmatrix}
3 & -1 & 0 & 0 \\
-1 & 3 & -1 & 0 \\
0 & -1 & 3 & -1 \\
0 & 0 & -1 & 3
\end{bmatrix}
\begin{bmatrix}
i_1 \\ i_2 \\ i_3 \\ i_4
\end{bmatrix}
=
\begin{bmatrix}
1.5 \\ 1.5 \\ 2 \\ 5.5
\end{bmatrix}
$$

SOLUTION:

After expressing the system in explicit form and solving i_1 from the first equation, i_2 from the second equation, and so on, the Jacobi iteration equations are expressed as follows:

$$i_1^{(p+1)} = \frac{1}{3}(i_2^{(p)} + 1.5)$$

$$i_2^{(p+1)} = \frac{1}{3}(i_1^{(p)} + i_3^{(p)} + 1.5),$$

$$i_3^{(p+1)} = \frac{1}{3}(i_2^{(p)} + i_4^{(p)} + 2),$$

$$i_4^{(p+1)} = \frac{1}{3}(i_3^{(p)} + 5.5).$$

Setting $p = 0$ in the above equations and making use of the initial guess ($i_k^{(0)} = 1$, $k = 1, 2, 3, 4$), the first step estimates are obtained as

$$i_1^{(1)} = \frac{1}{3}\left(1 + 1.5\right) = 0.833333,$$

$$i_2^{(1)} = \frac{1}{3}\left(1 + 1 + 1.5\right) = 1.166666,$$

$$i_3^{(1)} = \frac{1}{3}\left(1 + 1 + 2\right) = 1.333333,$$

$$i_4^{(1)} = \frac{1}{3}(1 + 5.5) = 2.166666.$$

Now, setting $p = 1$ in the iteration equations and making use of $i_k^{(1)}$, the current step estimates are found as follows:

$$i_1^{(2)} = \frac{1}{3}(1.166666 + 1.5) = 0.888889,$$

$$i_2^{(2)} = \frac{1}{3}(0.833333 + 1.333333 + 1.5) = 1.222222,$$

$$i_3^{(2)} = \frac{1}{3}(1.166666 + 2.166666 + 2) = 1.777778,$$

$$i_4^{(2)} = \frac{1}{3}(1.333333 + 5.5) = 2.277778$$

The Jacobi iteration process is continued in this way to obtain new and better estimates. At the end of each iteration, the ratio of the ℓ_2−norms of the displacement vectors obtained in two consecutive iterations (i.e., $r = \|\mathbf{d}^{(p)}\|_2/\|\mathbf{d}^{(p-1)}\|_2$) is also calculated.

The estimates $\mathbf{i}^{(p)}$, displacements $\|\mathbf{d}^{(p)}\|_2$, and the ratio r are presented in Table 3.1. The method converged to an approximate set of solutions that met the prescribed error tolerance after 12 iterations. Notice that as the number of iterations increases, the estimated solution with each subsequent iteration approaches (i.e., converges) toward the true solution of the linear system: $\{i_1, i_2, i_3, i_4\} = \{1, 1.5, 2, 2.5\}$.

TABLE 3.1

p	$i_1^{(p)}$	$i_2^{(p)}$	$i_3^{(p)}$	$i_4^{(p)}$	$\|\mathbf{d}^{(p)}\|_2$	r
0	1	1	1	1		
1	0.833333	1.166667	1.333333	2.166667	1.236030	
2	0.888889	1.222222	1.777778	2.277778	0.464811	0.376051
3	0.907407	1.388889	1.833333	2.425926	0.230554	0.496016
4	0.962963	1.41358	1.938272	2.444444	0.122683	0.532122
5	0.971193	1.467078	1.952675	2.479424	0.066036	0.538268
6	0.989026	1.474623	1.982167	2.484225	0.035606	0.539187
7	0.991541	1.490398	1.986283	2.494056	0.019203	0.539322
8	0.996799	1.492608	1.994818	2.495428	0.010357	0.539341
9	0.997536	1.497206	1.996012	2.498273	0.005586	0.539344
10	0.999069	1.497849	1.998493	2.498671	0.003013	0.539345
11	0.999283	1.499187	1.99884	2.499498	0.001625	0.539345
12	0.999729	1.499374	1.999562	2.499613	0.000876	0.539345

Discussion: Note that the ratio of the norms also converges to the value of 0.539345, while $\|\mathbf{d}^{(p)}\|_2$ is steadily decreasing toward zero. This ratio represents the convergence rate, and its significance is covered in detail in Section 3.3.

3.2.2 GAUSS-SEIDEL (GS) METHOD

Gauss-Seidel method is probably one of the most common iterative methods used to solve a system of linear equations. Although it is similar to the Jacobi method, it utilizes the most current estimates in the subsequent equations as soon as they become available.

Gauss-Seidel Method

- The Gauss-Seidel method is easy to understand and program;
- It is a special case of the SOR method with $\omega = 1$;
- It is neither computationally complicated nor involved;
- It generally converges twice as fast as the Jacobi method;
- It converges to the correct solution regardless of the initial guess;
- It requires less memory compared to faster iterative methods.

- The method does not converge for every system of linear equations (applicable to strictly diagonally dominant, or symmetric positive definite matrices);
- It is, in general, not competitive with the non-stationary methods;
- Unlike the Jacobi method, computations generally cannot be carried out in parallel; parallelization depends on the properties of the coefficient matrix.

The *Gauss-Seidel iteration equations* are mathematically expressed as

$$x_1^{(p+1)} = \frac{1}{a_{11}} \left(b_1 - a_{12}\, x_2^{(p)} \quad - a_{13}\, x_3^{(p)} \quad - a_{14}\, x_4^{(p)} - \quad \cdots - a_{1n}\, x_n^{(p)} \right)$$

$$x_2^{(p+1)} = \frac{1}{a_{22}} \left(b_2 - a_{21} \boxed{x_1^{(p+1)}} - a_{23} x_3^{(p)} \quad - a_{24} x_4^{(p)} - \cdots - a_{2n} x_n^{(p)} \right)$$

$$x_3^{(p+1)} = \frac{1}{a_{33}} \left(b_3 - a_{31} \boxed{x_1^{(p+1)}} - a_{32} \boxed{x_2^{(p+1)}} - a_{34} x_4^{(p)} - \quad \cdots - a_{3n} x_n^{(p)} \right) \qquad (3.7)$$

$$\vdots \qquad \vdots \qquad \vdots$$

$$x_n^{(p+1)} = \frac{1}{a_{nn}} \left(b_n - a_{n1} \boxed{x_1^{(p+1)}} - a_{n2} \boxed{x_2^{(p+1)}} - \cdots - a_{n,n-1} \boxed{x_{n-1}^{(p+1)}} \right)$$

where the current estimates $x_i^{(p+1)}$ (for $i = 1, 2, \ldots, n-1$) on the rhs are boxed for visual clarity.

The generalized form of the iteration equations is expressed as

$$x_i^{(p+1)} = \frac{1}{a_{ii}} \left\{ b_i - \sum_{j=1}^{i-1} a_{ij} \boxed{x_j^{(p+1)}} - \sum_{j=i+1}^{n} a_{ij} x_j^{(p)} \right\} \quad \text{for } i = 1, 2, \ldots, n \qquad (3.8)$$

provided that $a_{ii} \neq 0$. Note that the presumably more accurate estimates (boxed components) from previous equations are available at the ith equation and can be readily incorporated in the first sum of Eq. (3.8). This procedure is convenient for computer calculations because a current estimate can be immediately stored in the location of the prior estimate, minimizing the number of storage locations.

A drawback of most iterative methods is that they may not converge for every linear system. In this regard, the applicability of the Gauss-Seidel method is limited. The terms and conditions required for the convergence of the iterative method are discussed in Section 3.3. For now, assuming that a linear system is convergent, the Gauss-Seidel method converges faster (i.e., reaches an approximate solution with fewer iterations) than the Jacobi method because the unknown components for $j < i$ are updated with the improved current estimates as soon as they become available.

EXAMPLE 3.2: Application of Gauss-Seidel method

Repeat Example 3.1 using the Gauss-Seidel method.

SOLUTION:

Using Eq. (3.8), the Gauss-Seidel iteration equations can be cast as follows:

$$i_1^{(p+1)} = \frac{1}{3}\left(i_2^{(p)} + 1.5\right)$$

$$i_2^{(p+1)} = \frac{1}{3}\left(\boxed{i_1^{(p+1)}} + i_3^{(p)} + 1.5\right)$$

$$i_3^{(p+1)} = \frac{1}{3}\left(\boxed{i_2^{(p+1)}} + i_4^{(p)} + 2\right)$$

$$i_4^{(p+1)} = \frac{1}{3}\left(\boxed{i_3^{(p+1)}} + 5.5\right)$$

Note that every current estimate from a prior unknown component (as shown in boxes) is used in the subsequent iteration equations.

Substituting the initial guess, $i_k^{(0)} = 1$, into the Gauss-Seidel iteration equations, the current step estimates are obtained as

$$i_1^{(1)} = \frac{1}{3}\left(1 + 1.5\right) = 0.833333$$

$$i_2^{(1)} = \frac{1}{3}\left(\boxed{0.833333} + 1 + 1.5\right) = 1.111111$$

$$i_3^{(1)} = \frac{1}{3}\left(\boxed{1.111111} + 1 + 2\right) = 1.370370$$

$$i_4^{(1)} = \frac{1}{3}\left(\boxed{1.370370} + 5.5\right) = 2.290123$$

where the boxed numbers are the most current estimates available.

TABLE 3.2

p	$i_1^{(p)}$	$i_2^{(p)}$	$i_3^{(p)}$	$i_4^{(p)}$	$\|\mathbf{d}^{(p)}\|_2$	r
0	1	1	1	1		
1	0.833333	1.111111	1.37037	2.290123	1.357098	
2	0.870370	1.246914	1.845679	2.448560	0.520418	0.383478
3	0.915638	1.420439	1.956333	2.485444	0.213927	0.411069
4	0.973480	1.476604	1.987349	2.495783	0.087001	0.406683
5	0.992201	1.493184	1.996322	2.498774	0.026736	0.307313
6	0.997728	1.498017	1.998930	2.499643	0.007840	0.293215
7	0.999339	1.499423	1.999689	2.499896	0.002283	0.291230
8	0.999729	1.499832	1.999909	2.499970	0.000664	0.290942

When this iteration procedure is continued in this manner, the Gauss-Seidel method converges to a set of estimates, meeting the prescribed error tolerance after 8 iterations. To depict the convergence history of the computed estimates, $\|\mathbf{d}^{(p)}\|_2$ and r ($\|\mathbf{d}^{(p)}\|_2/\|\mathbf{d}^{(p-1)}\|_2$) are presented in Table 3.2. Using the same set of uniform initial guesses and the same stopping criterion, the Gauss-Seidel iteration procedure converged to an approximate solution in 8 iterations, while the Jacobi method did so in 12 iterations (*see* Example 3.1).

Discussion: Savings of 4 iterations indicate 25% fewer calculations. Note that the $\|\mathbf{d}^{(p)}\|_2$ norm in Table 3.2 steadily decreases while the ratio of the norm in the last column converges to about 0.29, which is referred to as the *convergence rate* and is covered in detail in Section 3.3.

3.2.3   SOR METHOD

The SOR method, a faster variant of the Gauss-Seidel method, is widely employed to obtain the numerical solution of sparse linear systems. It is termed *Successive Over Relaxation* (*SOR*) because the current estimate is improved by a weighted average of the "*prior*" and "*Gauss-Seidel*" estimates as follows:

$$\mathbf{x}^{(p+1)} = (1 - \omega)\mathbf{x}^{(p)} + \omega\,\mathbf{x}_{\text{GS}}^{(p+1)},$$

where ω is referred to as the *relaxation* (or *acceleration*) *parameter*, and $\mathbf{x}_{\text{GS}}^{(p+1)}$ denotes the Gauss-Seidel estimate obtained by Eq. (3.8).

Substituting $\mathbf{x}_{\text{GS}}^{(p+1)}$ from Eq. (3.8) into the weighted-average, the *SOR iteration equations* yields

$$x_1^{(p+1)} = (1-\omega)x_1^{(p)} + \frac{\omega}{a_{11}}\left(b_1 - a_{12}x_2^{(p)} \quad - a_{13}x_3^{(p)} \quad - \cdots - a_{1n}x_n^{(p)}\right)$$

$$x_2^{(p+1)} = (1-\omega)x_2^{(p)} + \frac{\omega}{a_{22}}\left(b_2 - a_{21}x_1^{(p+1)} - a_{23}x_3^{(p)} \quad - \cdots - a_{2n}x_n^{(p)}\right)$$

$$x_3^{(p+1)} = (1-\omega)x_3^{(p)} + \frac{\omega}{a_{33}}\left(b_3 - a_{31}x_1^{(p+1)} - a_{32}x_2^{(p+1)} - \cdots - a_{3n}x_n^{(p)}\right) \tag{3.9}$$

$$\vdots \qquad \vdots$$

$$x_n^{(p+1)} = (1-\omega)x_n^{(p)} + \frac{\omega}{a_{nn}}\left(b_n - a_{n1}x_1^{(p+1)} - a_{n2}x_2^{(p+1)} - \cdots - a_{n,n-1}x_{n-1}^{(p+1)}\right)$$

A more generalized form of Eq. (3.9) may be written in a compact form as

$$x_i^{(p+1)} = (1-\omega)x_i^{(p)} + \frac{\omega}{a_{ii}}\left\{b_i - \sum_{j=1}^{i-1} a_{ij}x_j^{(p+1)} - \sum_{j=i+1}^{n} a_{ij}x_j^{(p)}\right\}, \quad i = 1, 2, \ldots, n \tag{3.10}$$

The relaxation parameter typically varies in the range $0 < \omega < 2$. For $\omega = 1$, Eq. (3.10) is reduced to the Gauss-Seidel method. The iteration process is termed *over-relaxation* for $1 < \omega < 2$ and *successive under-relaxation (SUR)* for $0 < \omega < 1$, which is usually applied when solving systems of non-linear equations. If the system is convergent, the method converges to an approximate solution regardless of the choice of ω; however, there is an *optimum value*, ω_{opt}, for which the system yields an approximate solution with the least number of iterations and faster convergence rate.

Factors like the number of equations, the strength of the diagonal dominance (covered in Section 3.3.1), the pattern of the coefficient (full, banded, sparse, etc.) matrix, and so on affect the optimum relaxation factor. Hence, the SOR method is generally applied to large systems with a "smart guess" for the ω_{opt}. If a system of linear equations is to be solved numerous times in an iterative algorithm, it is extremely important to estimate and use a ω_{opt} value close to the true optimum value (*see* Section 3.3.2 on estimating ω_{opt}). In near-optimal conditions, the problem-solving time can be significantly reduced.

Module SOR $(n, \varepsilon, \omega, \mathbf{A}, \mathbf{b}, \mathbf{xo}, \mathbf{x}, maxit, iter, error)$
\ DESCRIPTION: A pseudomodule to solve $\mathbf{Ax} = \mathbf{b}$ using the SOR method.
\ USES:
\   SOR_DRV:: Driver module performing one step SOR iteration.
Declare: a_{nn}, b_n, x_n, xo_n \ Declare array variables
$p \leftarrow 0$ \ Initialize iteration counter
$\delta0 \leftarrow 1$ \ Initialize ℓ_2-norm, $\delta^{(0)} = 1$
Repeat   \ Iterate until convergence is achieved
   $p \leftarrow p + 1$ \ Iteration counter
   SOR_DRV$(n, \omega, \mathbf{A}, \mathbf{b}, \mathbf{xo}, \mathbf{x}, \delta)$ \ Perform one-step SOR iteration
   Write: "For p=", p," Error=",δ," Rate=", δ/δ_0 \ Echo iteration progress
   $\delta_0 \leftarrow \delta$ \ Set current error as prior, $\delta^{(p)} \leftarrow \delta^{(p+1)}$
   $\mathbf{xo} \leftarrow \mathbf{x}$ \ Set current estimates as prior; $\mathbf{x}^{(p)} \leftarrow \mathbf{x}^{(p+1)}$
Until $\left[\delta < \varepsilon \text{ Or } p < maxit\right]$ \ Terminate iterations if $\delta < \varepsilon$ and $p < maxit$
$error \leftarrow \delta$ \ Set current ℓ_2-norm as error
$iter \leftarrow p$ \ Set current p as $iter$
\ If iteration limit $maxit$ is reached *with no* convergence
If $\left[p = maxit\right]$ **Then**   \ Issue info and a warning
   Write: "SOR method failed to converge after", $maxit$,"iterations"
   Write: "Error level reads is",$error$
End If
End Module SOR

Module SOR_DRV $(n, \omega, \mathbf{A}, \mathbf{b}, \mathbf{xo}, \mathbf{x}, \delta)$
\ DESCRIPTION: A pseudomodule to perform one step SOR iteration.
\ USES:
\   ENORM:: Function Module calculating Euclidean (ℓ_2-)norm of a vector (*see* Pseudocode 3.1).
Declare: $a_{nn}, b_n, x_n, xo_n, d_n$ \ Declare array variables
$\omega1 \leftarrow 1 - \omega$ \ Set $\omega1 = 1 - \omega$ to reduce arithmetic ops
For $\left[i = 1, n\right]$ \ Loop i: Sweep equations from top to bottom
   $sums \leftarrow 0$ \ Initialize accumulator
   For $\left[j = 1, n\right]$ \ Loop j: Accumulate off-diagonal terms of ith row
     If $\left[j > i\right]$ **Then**   \ For terms to the right of $a_{ii} * x_i$
      $sums \leftarrow sums + a_{ij} * xo_j$ \ Accumulate $a_{ij} * x_j^{(p)}$ products
     Else   \ Alternative condition is $i \geqslant j$
      If $\left[i > j\right]$ **Then**   \ For terms to the left of $a_{ii} * x_i$
       $sums \leftarrow sums + a_{ij} * x_j$ \ Accumulate $a_{ij} * x_j^{(p+1)}$ products
      End If
     End If   \ Note that case of $i = j$ is skipped
   End For
   $x_i \leftarrow \omega1 * xo_i + \omega * (b_i - sums)/a_{ii}$ \ Compute $x_i^{(p+1)}$ by Eq. (3.10)
   $d_i \leftarrow x_i - xo_i$ \ Find displacement vector, $d_i \leftarrow x_i^{(p+1)} - x_i^{(p)}$
End For
$\delta \leftarrow$ **ENORM**$(n, \mathbf{d})$ \ Set ℓ_2-norm of $\mathbf{d}^{(p)}$ to δ
End Module SOR_DRV

SOR Method

- The basic SOR algorithm is easy to understand and implement;
- It usually converges faster than the Jacobi or Gauss-Seidel methods;
- For linear systems, the iteration process is accelerated for $1 < \omega < 2$; however, the linear systems that diverge with SOR may converge for $0 < \omega < 1$;
- There is an optimum acceleration value (ω_{opt}) that results in the fewest iterations (or least cpu-time) and the highest convergence rate;
- Adaptive SOR algorithms that estimate ω_{opt} are also available;
- Parallelization properties depend on the structure of the coefficient matrix.

- The method, like Gauss-Seidel, may not converge for every linear system; the coefficient matrix must satisfy certain properties (covered in Section 3.3);
- Its performance can be sensitive to the choice of ω_{opt}, which is difficult and costly to estimate beforehand;
- Solving a linear system without a "smart estimate" for ω_{opt} can lead to a large number of iterations, which may be worse than that of the Gauss-Seidel method.

A pseudomodule, **SOR**, for solving a system of linear equations using the SOR method is presented in Pseudocode 3.2. In addition to **JACOBI**'s argument list, it requires a guess (preferably a *smart guess*) for the relaxation parameter ω. The structure of the **SOR** module is basically the same as that of **JACOBI**; however, the only difference is in the computation of the current estimates in **SOR_DRV**, carried out by Eq. (3.10). Note that a separate code for the Gauss-Seidel method is not given here as the SOR method becomes the GS method for $\omega = 1$.

The module **SOR_DRV** performs one-step SOR iteration and, as input, it requires the number of unknowns (n), the linear system (**A** and **b**), a near-optimal guess for ω, and an initial guess (**xo**). The outputs are the current estimates (**x**) and ℓ_2-norm of the displacement vector (δ). Setting $\omega 1 = 1 - \omega$ before moving on to calculate the current estimates helps save cpu-time, especially for large numbers of iterations and very large matrices. The outer loop i sweeps the equations from top to bottom, while the inner loop j sweeps the terms of the ith equation from left to right. If the current estimate of the jth unknown is available, $a_{ij} * x_j$ ($i > j$) or, if not, $a_{ij} * xo_j$ ($i < j$) terms are added to the accumulator (the sum of off-diagonal terms) using a couple of nested If-structures. The displacement vector **d** is calculated and used to estimate the error at the current iteration level by calculating the Euclidean norm, $\delta = \|\mathbf{d}\|_2$, obtained with the module **ENORM** and passed to **SOR** to aid in the convergence decision. The iteration procedure in **SOR** is terminated in the same manner as described for the Jacobi method.

From this point on, the **L** and **U** matrix notations will be used to denote the lower and upper triangular matrices, whose primary diagonal elements are zero. They should *not be confused* with those matrix definitions in the LU decomposition methods where $\ell_{ii} \neq 0$ and $u_{ii} \neq 0$ for all i.

3.3 CONVERGENCE OF STATIONARY ITERATIVE METHODS

For the formal theory of the stationary iterative methods, we will decompose the coefficient matrix $\mathbf{A}$ as $\mathbf{A} = \mathbf{D} - \mathbf{L} - \mathbf{U}$, where $\mathbf{D}$, $\mathbf{U}$, and $\mathbf{L}$ are diagonal, upper, and lower triangular matrices, respectively.

A system of linear equations may now be written in matrix form as

$$\mathbf{Dx} = (\mathbf{L} + \mathbf{U})\mathbf{x} + \mathbf{b} \tag{3.11}$$

Using superscript notation to incorporate the iteration steps, Eq. (3.11) can then be expressed as

$$\mathbf{Dx}^{(p+1)} = (\mathbf{L} + \mathbf{U})\mathbf{x}^{(p)} + \mathbf{b} \tag{3.12}$$

where all diagonal elements are kept on the left and off-diagonal elements on the rhs, which is equivalent to the matrix form of the Jacobi method. Note that if $a_{ii} \neq 0$ for all i, then the inverse of a diagonal matrix $\mathbf{D}$ exits, and it is simply a diagonal matrix consisting of the reciprocals of the elements on the diagonal, i.e., $d_{ii} = 1/a_{ii}$.

Solving Eq. (3.12) for $\mathbf{x}^{(p+1)}$ yields

$$\mathbf{x}^{(p+1)} = \mathbf{M}_J \mathbf{x}^{(p)} + \mathbf{D}^{-1}\mathbf{b} \tag{3.13}$$

where $\mathbf{M}_J = \mathbf{D}^{-1}(\mathbf{L} + \mathbf{U})$ is called the *Jacobi iteration matrix.*

The Gauss-Seidel iteration equations can be similarly expressed in matrix form as

$$\mathbf{Dx}^{(p+1)} = \mathbf{Lx}^{(p+1)} + \mathbf{Ux}^{(p)} + \mathbf{b} \tag{3.14}$$

where $\mathbf{Lx}^{(p+1)}$ denotes the terms in which are updated with the current estimates. Now, solving Eq. (3.14) for $\mathbf{x}^{(p+1)}$ gives

$$\mathbf{x}^{(p+1)} = \mathbf{M}_{\mathrm{GS}}\mathbf{x}^{(p)} + (\mathbf{D} - \mathbf{L})^{-1}\mathbf{b} \tag{3.15}$$

where $\mathbf{D} - \mathbf{L}$ is a lower triangular matrix, and $\mathbf{M}_{\mathrm{GS}} = (\mathbf{D} - \mathbf{L})^{-1}\mathbf{U}$ is referred to as the *Gauss-Seidel iteration matrix.*

Applying the same rationale to the SOR iteration equations, we obtain

$$\mathbf{x}^{(p+1)} = (\mathbf{D} - \omega\mathbf{L})^{-1}\big((1 - \omega)\mathbf{D} + \omega\mathbf{U}\big)\mathbf{x}^{(p)} + \omega(\mathbf{D} - \omega\mathbf{L})^{-1}\,\mathbf{b} \tag{3.16}$$

where the *SOR iteration matrix*, $\mathbf{M}_\omega$, takes the following form:

$$\mathbf{M}_\omega = (\mathbf{D} - \omega\mathbf{L})^{-1}\big((1 - \omega)\mathbf{D} + \omega\mathbf{U}\big) \tag{3.17}$$

In order to establish a general framework for the convergence properties of iterative methods, consider the following iteration equations in fixed-point matrix form:

$$\mathbf{x} = \mathbf{Mx} + \mathbf{c} \tag{3.18}$$

where $\mathbf{M}$ is the iteration matrix and $\mathbf{c}$ is a vector. Then any iteration scheme may be expressed as follows:

$$\mathbf{x}^{(p+1)} = \mathbf{Mx}^{(p)} + \mathbf{c} \tag{3.19}$$

where $\mathbf{x}^{(p+1)}$ and $\mathbf{x}^{(p)}$ have the usual meanings.

Subtracting Eq. (3.19) from Eq. (3.18) gives

$$\mathbf{e}^{(p+1)} = \mathbf{Me}^{(p)}, \quad p = 0, 1, 2, \ldots \tag{3.20}$$

where $\mathbf{e}^{(p)} = \mathbf{x} - \mathbf{x}^{(p)}$ is the true error at the pth iteration step.

Successively employing Eq. (3.20), the error at the pth iteration step becomes

$$\mathbf{e}^{(p)} = \mathbf{M}^p \mathbf{e}^{(0)}, \quad p = 0, 1, 2, \ldots \tag{3.21}$$

For an arbitrary initial guess $\mathbf{x}^{(0)}$ (or $\mathbf{e}^{(0)}$), the sequence of estimates should converge to $\mathbf{x}$ as $p \to \infty$ if $\lim_{p \to \infty} \mathbf{e}^{(p)} = \mathbf{0}$, where $\mathbf{0}$ is the zero vector. Hence, in order for an iterative algorithm to converge (i.e., for errors to decay) for any arbitrary initial guess vector, the *necessary* and *sufficient* condition is

$$\lim_{p \to \infty} \mathbf{M}^p \to \mathbf{O}$$

where $\mathbf{O}$ is the zero matrix.

Now suppose $\mathbf{M}$ has n real eigenvalues λ_k ($k = 1, 2, \ldots, n$) with $|\lambda_1| > |\lambda_2| \geqslant |\lambda_3| \geqslant \ldots \geqslant |\lambda_n|$ and corresponding linearly independent eigenvectors, $\mathbf{v}_k$. (Recall that the eigenvalues and eigenvectors of a square matrix are presented in Section 2.1.6, and they are covered in greater detail in Chapter 11.) The eigenvectors can then be used as the basis of n-dimensional vector space, and the arbitrary error vector $\mathbf{e}^{(0)}$ with its n components can also be uniquely expressed as a linear combination of them; specifically,

$$\mathbf{e}^{(0)} = \sum_{i=1}^{n} c_i \mathbf{v}_i \tag{3.22}$$

where c_i's are scalar constants. Hence, setting $p = 1$ in Eqs. (3.21) and (3.22) and using $\mathbf{M} \mathbf{v}_i = \lambda_i \mathbf{v}_i$, we may write

$$\mathbf{e}^{(1)} = \mathbf{M} \mathbf{e}^{(0)} = \sum_{i=1}^{n} c_i \mathbf{M} \mathbf{v}_i = \sum_{i=1}^{n} c_i \lambda_i \mathbf{v}_i \tag{3.23}$$

Multiplying Eq. (3.23) by $\mathbf{M}$ successively and similarly using the generalized identity $\mathbf{M}^p \mathbf{v}_i = \lambda_i^p \mathbf{v}_i$, we obtain

$$\mathbf{e}^{(p)} = \mathbf{M}^p \mathbf{e}^{(0)} = \sum_{i=1}^{n} c_i \mathbf{M}^p \mathbf{v}_i = \sum_{i=1}^{n} c_i \lambda_i^p \mathbf{v}_i = \lambda_1^p \left\{ c_1 \mathbf{v}_1 + \left(\frac{\lambda_2}{\lambda_1} \right)^p c_2 \mathbf{v}_2 + \ldots + \left(\frac{\lambda_n}{\lambda_1} \right)^p c_n \mathbf{v}_n \right\} \tag{3.24}$$

This expression indicates that $\mathbf{e}^{(p)}$ will tend to zero vector as p tends to infinity for any arbitrary $\mathbf{e}^{(0)}$ *if and only if* $|\lambda_i| < 1$ for all i. Also note that, for sufficiently large p, $(\lambda_k/\lambda_1)^p$ terms for $k = 2, 3, \ldots, n$ go to zero, yielding $\mathbf{e}^{(p)} \cong \lambda_1^p c_1 \mathbf{v}_1$ or likewise $\mathbf{e}^{(p+1)} \cong \lambda_1^{p+1} c_1 \mathbf{v}_1$. Thus, we can deduce $\mathbf{e}^{(p+1)} \cong \lambda_1 \mathbf{e}^{(p)}$, which by taking the ℓ_2−norm of both sides and rearranging, we get

$$\frac{\left\| \mathbf{e}^{(p+1)} \right\|_2}{\left\| \mathbf{e}^{(p)} \right\|_2} \cong |\lambda_1| = \rho(\mathbf{M}) \tag{3.25}$$

where $\rho(\mathbf{M})$ denotes the *spectral radius* of the iteration matrix $\mathbf{M}$, corresponding to the dominant (or largest magnitude) eigenvalue. In cases where a matrix has complex eigenvalues, the spectral radius can be interpreted as the radius of the smallest circle centered at the origin in the complex plane that contains all the eigenvalues of the matrix, either on the circle or inside.

Up until now, not much has been said about "under what conditions" iterative methods converge. However, it is also clear from Eq. (3.25) that the spectral radius of iteration matrix $\mathbf{M}$ must be less than "1" ($\rho(\mathbf{M}) < 1$) so that the errors decay as p increases; otherwise, the iterative method will diverge. In this respect, the necessary and sufficient condition for the convergence of the Jacobi, Gauss-Seidel, or SOR methods can be stated as $\rho(\mathbf{M}_J) < 1$, $\rho(\mathbf{M}_{GS}) < 1$, or $\rho(\mathbf{M}_\omega) < 1$. Having said that, it should be pointed out that the computation of the spectral radius can be very expensive and time-consuming. Instead, an estimate for the spectral radius of a matrix $\mathbf{A}$ can be obtained with the aid of the *Gerschgorin Theorem*, which follows:

$$\rho(\mathbf{A}) \leqslant \min\left(\max_i \sum_j |a_{ij}|, \ \max_j \sum_i |a_{ij}|\right) \tag{3.26}$$

This expression indicates that the dominant eigenvalue is less than or equal to the minimum of the maximum ℓ_1-norms of the row or column vectors of matrix $\mathbf{A}$.

In general, the diagonal dominance of a coefficient matrix $\mathbf{A}$ is a *sufficient condition* for convergence; that is, a matrix $\mathbf{A}$ is said to be *strictly diagonally dominant* if

$$|a_{ii}| > \sum_{j=1, j \neq i}^{n} |a_{ij}| \quad \text{for all } i \tag{3.27}$$

for every row with strict inequality for all i. On the other hand, the matrix $\mathbf{A}$ is said to be *weakly diagonally dominant*, with strict inequality at least for one i.

$$|a_{ii}| \geqslant \sum_{j=1, j \neq i}^{n} |a_{ij}| \quad \text{for all } i$$

In the literature, the term "diagonally dominant" is often used rather than "weakly diagonally dominant." Note that the diagonal dominance is a *sufficient condition*, not a *necessary condition* for the convergence of the stationary iterative methods. Usually, the more dominant the diagonal is, the more rapidly the system will converge. It should also be pointed out that even though this may seem like a very restrictive condition, the system of linear equations resulting from the numerical treatment of applied boundary value problems does satisfy the diagonal dominance condition.

 If $\mathbf{A}$ is diagonally dominant coefficient matrix, then the Jacobi or Gauss-Seidel methods do converge for any initial guess vector, $\mathbf{x}^{(0)}$. However, there are cases where Jacobi, Gauss-Seidel, or both methods may converge to an approximate solution even if $\mathbf{A}$ does not meet the *sufficient condition*.

3.3.1   RATE OF CONVERGENCE

When solving large linear systems with iterative methods, it is important that the methods be used under "conditions" that result in rapid convergence. The speed at which a sequence of estimates approaches its limit is called the *rate of convergence*. A faster rate indicates that fewer iterations are required to obtain a reasonably accurate estimate.

The dominant eigenvalue of an iteration matrix provides critical information on the rate of convergence. To quantify how fast an iterative algorithm will converge in terms of $\rho(\mathbf{M})$, we ask, "how many iterations does it take to reduce the error by a factor of 10?" For simplicity, suppose one-decimal-place accuracy is gained by a single iteration; for instance, accordingly and arbitrarily, we set $\|\mathbf{e}^{(p+1)}\|_2 = 10^{-3}$ and $\|\mathbf{e}^{(p)}\|_2 = 10^{-2}$. Using Eq. (3.25) and taking the base-10 logarithm of $\|\mathbf{e}^{(p)}\|_2 / \|\mathbf{e}^{(p+1)}\|_2$ gives

$$\log_{10} \frac{\|\mathbf{e}^{(p)}\|_2}{\|\mathbf{e}^{(p+1)}\|_2} = \log_{10}(10) = \log_{10}\left(\frac{1}{|\lambda_1|}\right) = \log_{10}\left(\frac{1}{\rho(\mathbf{M})}\right) = -\log_{10}(\rho(\mathbf{M})) = 1$$

EXAMPLE 3.3: Predicting if a linear system converges or not

Determine whether or not the following linear system converges with the Jacobi method.

$$\begin{bmatrix} 4 & -1 & 2 \\ -1 & 5 & 7 \\ 2 & -4 & 2 \end{bmatrix} \begin{bmatrix} x_1 \\ x_2 \\ x_3 \end{bmatrix} = \begin{bmatrix} 8 \\ 30 \\ -2 \end{bmatrix}$$

SOLUTION:

Checking for diagonal dominance for each row, we find:

$$r_1: \quad |4| > |-1| + |2|, \quad r_2: \quad |5| < |-1| + |7|, \quad r_3: \quad |2| < |-4| + |2|$$

Note that "as is" the matrix $\mathbf{A}$ is *not* diagonally dominant; however, by interchanging the second and third equations (rows), we get

$$\begin{bmatrix} 4 & -1 & 2 \\ 2 & -4 & 2 \\ -1 & 5 & 7 \end{bmatrix} \begin{bmatrix} x_1 \\ x_2 \\ x_3 \end{bmatrix} = \begin{bmatrix} 8 \\ -2 \\ 30 \end{bmatrix} \quad r_2 \leftrightarrow r_3$$

Now checking for diagonal dominance of the rows, we find

$$r_1: \quad |4| > |-1| + |2|, \quad r_2: \quad |-4| = |2| + |2|, \quad r_3: \quad |7| > |-1| + |5|,$$

The first and third rows are strictly diagonally dominant. However, the sum of the absolute values of the off-diagonal elements of the second row is equal to the absolute value of the diagonal element. This makes the system (*weakly*) *diagonally dominant*.

Discussion: A system of linear equations can be reordered (rows are interchanged) such that each diagonal element of the coefficient matrix is larger in magnitude than the sum of the magnitudes of the other coefficients in that row. In this example, we were able to interchange two rows, which illustrates that the system can be made diagonally dominant. However, the determination of diagonal dominance by visual inspection may not be so obvious for even fairly small matrices. Hence, it is better to resort to the Gerschgorin Theorem when in doubt.

---

which is the number of decimal digits by which the error is reduced after one iteration. In other words, mathematically speaking, the *rate of convergence* of the iterative scheme given by Eq. (3.19) is defined as

$$R = -\log_{10}\left(\rho(\mathbf{M})\right) \tag{3.28}$$

Since, for convergence, $0 < \rho(\mathbf{M}) < 1$, the number of decimal digits of accuracy gained per iteration increases as $\rho(\mathbf{M})$ decreases. Clearly, the smaller the optical radius, the higher the rate of convergence will be. Alternatively, for large p, using Eq. (3.25), we obtain

$$\|\mathbf{e}^{(p)}\|_2 \approx \lambda_1 \|\mathbf{e}^{(p-1)}\|_2 \approx \cdots \approx \lambda_1^p \|\mathbf{e}^{(0)}\|_2 = [\rho(\mathbf{M})]^p \|\mathbf{e}^{(0)}\|_2$$

The $[\rho(\mathbf{M})]^p$ term in this expression provides a measure of how the error norm decreases after p iterations. Now, we find the minimum value of p from $\rho^p(\mathbf{M}) \leqslant \varepsilon$ as

$$[\rho(\mathbf{M})]^p \leqslant 10^{-m} \quad \text{or} \quad p_{\min} = \left\lceil -\frac{m}{\log_{10}\rho(\mathbf{M})} \right\rceil = \left\lceil \frac{m}{R} \right\rceil \tag{3.29}$$

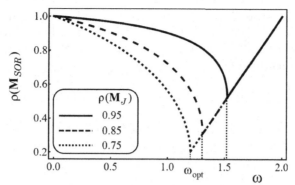

FIGURE 3.2: Variation of $\rho(\mathbf{M}_\omega)$ with ω and $\rho(\mathbf{M}_J)$.

where $\lceil\cdot\rceil$ denotes the ceiling function. Thus, it can be seen that the number of iteration steps required by a convergent iterative method to reduce the initial error vector by the factor $\varepsilon = 10^{-m}$ ($m > 0$) is inversely proportional to the rate of convergence, R. Consequently, the number of iterations to reduce the error by more than an order of magnitude is found by the inverse of the rate, i.e., $1/R$.

 For a sufficiently large number of iterations and asymptotic convergence rate R, the $1/R$ ratio gives approximately the number of iterations needed to reduce the error by a factor of 10. The smaller the spectral radius is, the larger the convergence rate becomes.

We now return to the SOR method, whose convergence rate is affected by the choice of ω. Recall that there exists an optimum ω for which the convergence of the SOR method is achieved with the minimum number of iterations. However, determining the best value for ω is the major difficulty of the SOR method. It turns out that there is a direct relationship between the spectral radii of the Jacobi and SOR iteration matrices resulting from the discretization of elliptic PDEs. The detailed analysis of this topic is beyond this book's scope, and at this stage, we will emphasize this here without going into much detail. However, the reader should consult Refs. [3, 12, 17, 29] for additional information on this topic.

The relationship between the spectral radii of the Jacobi and the SOR iteration matrices is

$$(\rho(\mathbf{M}_\omega) + \omega - 1)^2 = \omega^2\rho^2(\mathbf{M}_J)\rho(\mathbf{M}_\omega) \tag{3.30}$$

where $\rho(\mathbf{M}_J)$ and $\rho(\mathbf{M}_\omega)$ are the spectral radii of the Jacobi and the SOR iteration matrices, respectively.

Solving Eq. (3.30), a quadratic equation, for $\rho(\mathbf{M}_\omega)$ yields

$$\rho(\mathbf{M}_\omega) = \begin{cases} 1 - \omega + \frac{1}{2}\omega^2\rho^2(\mathbf{M}_J) + \omega\rho(\mathbf{M}_J)\sqrt{1 - \omega + \frac{1}{4}\omega^2\rho^2(\mathbf{M}_J)}, & 0 < \omega \leqslant \omega_{\text{opt}} \\ \omega - 1, & \omega_{\text{opt}} \leqslant \omega < 2 \end{cases} \tag{3.31}$$

where ω_{opt} is the optimum relaxation parameter estimated by

$$\omega_{\text{opt}} = \frac{2}{1 + \sqrt{1 - \rho^2(\mathbf{M}_J)}} \tag{3.32}$$

The variation of the spectral radius of the SOR iteration matrix, $\rho(\mathbf{M}_\omega)$, as a function of ω and $\rho(\mathbf{M}_J)$ is illustrated in Fig. 3.2. Notice that the slope of the curve approaches infinity as we approach ω_{opt} from the left side, whereas it is linear on the right side. This means that underestimating ω_{opt} by a small margin can result in a larger increase in $\rho(\mathbf{M}_\omega)$ than a small overestimation of equal magnitude. The convergence rate for the optimum is

$$\rho(\mathbf{M}_\omega) = \omega_{\text{opt}} - 1 \quad \text{for} \quad \omega_{\text{opt}} \leqslant \omega < 2$$

The ω_{opt} value depends on the size of the coefficient matrix, the strength of the diagonal dominance, and the structure of the linear system. In general, a larger linear system resulting from grid refining of elliptic PDEs yields a larger value of ω_{opt}. A best estimate for the optimum relaxation parameter is given by Eq. (3.32).

3.3.2 ESTIMATING OPTIMUM RELAXATION PARAMETER

There is not an easy and cheap way of determining ω_{opt} in advance; however, a *smart estimate* can be computed simultaneously while performing the SOR iterations. A relatively simple algorithm for estimating the optimum relaxation parameter given in Ref. [19] is described below.

When $\mathbf{A}$ is symmetric and positive definite, $\rho(\mathbf{M}_\omega)$ is minimum for $\omega = \omega_{\text{opt}}$. It can be shown that $\rho(\mathbf{M}_{\text{GS}}) = \rho^2(\mathbf{M}_J)$ from examining the eigenvalues of $\mathbf{M}_\omega$ at $\omega = 1$. Note that we can replace $\rho^2(\mathbf{M}_J)$ with $\rho(\mathbf{M}_{\text{GS}})$ in Eq. (3.32) as the maximum eigenvalue of $\mathbf{M}_{\text{GS}}$ can be easily computed from Eq. (3.25) during the SOR iterations simply by setting $\omega = 1$. But since the true error is not known, an error estimate at the mth iteration step is approximated by $\delta^{(m+1)} \cong \left|\lambda^{(m)}\right| \delta^{(m)}$, where $\delta^{(m)} = \|\mathbf{d}^{(m)}\|_2$, $\mathbf{d}^{(m)} = \mathbf{x}^{(m)} - \mathbf{x}^{(m-1)}$, and $\lambda^{(m)} \to \lambda_1$ for $m \to \infty$ (or sufficiently large m). Based on these assumptions, the algorithm is given as:

1. Perform m iterations ($m \approx$ 10-15 is sufficient in most cases) with $\omega =1$, or iterate until λ convergences to λ_1 within a preset tolerance value, i.e., $|1 - \lambda^{(m-1)}/\lambda^{(m)}| < \varepsilon$;

2. Compute ℓ_2-norm of the displacement vector at the end of the mth iteration step, i.e., $\delta^{(m)} = \|\mathbf{d}^{(m)}\|_2$;

3. Perform additional k iterations with $\omega =1$ ($k = 1$ is sufficient in most cases) and compute the ℓ_2-norm of the displacement vector, $\mathbf{d}^{(m+k)}$, i.e., $\delta^{(m+k)} = \|\mathbf{d}^{(m+k)}\|_2$;

4. Compute an estimate for the maximum eigenvalue of the Gauss-Seidel iteration matrix from $\lambda_1 \cong \left(\delta^{(m+k)}/\delta^{(m)}\right)^{1/k}$. Note that the ℓ_2-norms of $\mathbf{d}$ at $(m+k)$th and mth steps are now related to each other by $\delta^{(m+k)} \cong \lambda_1^k \delta^{(m)}$;

5. Estimate ω_{opt} after $(m + k)$th iterations by

$$\omega_{\text{opt}} \approx \frac{2}{1 + \sqrt{1 - \lambda_1}}$$

6. Perform the subsequent iterations with $\omega = \omega_{\text{opt}}$.

EXAMPLE 3.4: Application of the SOR method

Repeat Example 3.1 using the SOR method with $\omega = 1.1$ and estimate the optimum relaxation parameter.

SOLUTION:

The SOR iteration equations with $\omega = 1.1$ take the following form:

$$i_1^{(p+1)} = -0.10i_1^{(p)} + \frac{1.1}{3}\left(i_2^{(p)} + 1.5\right)$$

$$i_2^{(p+1)} = -0.10i_2^{(p)} + \frac{1.1}{3}\left(i_1^{(p+1)} + i_3^{(p)} + 1.5\right)$$

$$i_3^{(p+1)} = -0.10i_3^{(p)} + \frac{1.1}{3}\left(i_2^{(p+1)} + i_4^{(p)} + 2\right)$$

$$i_4^{(p+1)} = -0.10i_4^{(p)} + \frac{1.1}{3}\left(i_3^{(p+1)} + 5.5\right)$$

With $\mathbf{i}^{(0)} = 1$, the current estimates are found as

$$i_1^{(1)} = -0.1(1) + \frac{1.1}{3}\left(1 + 1.5\right) = 0.816667$$

$$i_2^{(1)} = -0.1(1) + \frac{1.1}{3}\left(\boxed{0.816667} + 1 + 1.5\right) = 1.116111$$

$$i_3^{(1)} = -0.1(1) + \frac{1.1}{3}\left(\boxed{1.116112} + 1 + 2\right) = 1.409241$$

$$i_4^{(1)} = -0.1(1) + \frac{1.1}{3}\left(\boxed{1.409241} + 5.5\right) = 2.433388$$

where the boxed values are current estimates that became available from prior equations.

The iteration procedure is continued in the same manner until the stopping criterion based on $|\mathbf{d}^{(p)}|_2$ is met in five iterations. The SOR iterations clearly converged to an approximate solution faster than the Jacobi and the Gauss-Seidel methods. The history of computed estimates and the ℓ_2−norm of the displacement vector are presented in Table 3.3. The fact that the SOR method converges to a solution in 5 iterations in comparison to 8 iterations in the Gauss-Seidel method (*see* Example 3.2) and 12 iterations in the Jacobi method (*see* Example 3.1) indicates that the $\omega = 1.1$ value is close to the optimum relaxation parameter.

TABLE 3.3

p	$i_1^{(p)}$	$i_2^{(p)}$	$i_3^{(p)}$	$i_4^{(p)}$	$\|\mathbf{d}^{(p)}\|_2$
0	1	1	1	1	
1	0.816667	1.116111	1.409241	2.433388	1.506377
2	0.877574	1.276888	1.952844	2.489371	0.572885
3	0.930435	1.479513	1.993306	2.498609	0.213481
4	0.999445	1.499391	1.999936	2.500116	0.072137
5	0.999832	1.499976	2.00004	2.500003	0.000718

To calculate a theoretical estimate for $\omega_{\mathbf{opt}}$, the spectral radius of the Jacobi iteration matrix, $\mathbf{M}_J$, is required:

$$\mathbf{M}_J = \mathbf{D}^{-1}(\mathbf{L+U}) = \begin{bmatrix} 3 & & & \\ & 3 & & \\ & & 3 & \\ & & & 3 \end{bmatrix}^{-1} \begin{bmatrix} 0 & 1 & 0 & 0 \\ 1 & 0 & 1 & 0 \\ 0 & 1 & 0 & 1 \\ 0 & 0 & 1 & 0 \end{bmatrix} = \begin{bmatrix} 0 & \frac{1}{3} & 0 & 0 \\ \frac{1}{3} & 0 & \frac{1}{3} & 0 \\ 0 & \frac{1}{3} & 0 & \frac{1}{3} \\ 0 & 0 & \frac{1}{3} & 0 \end{bmatrix}$$

The Jacobi iteration matrix is a tridiagonal Toeplitz matrix whose eigenvalues can easily be obtained as $\lambda = \pm(1\pm\sqrt{5})/6$ using Eq. (2.15). The maximum magnitude eigenvalue turns out to be $|\lambda|_{\max} = (1+\sqrt{5})/6 \approx 0.539$; thus, using Eq. (3.25), the spectral radius is found as $\rho(\mathbf{M}_J) = |\lambda_{\max}| = 0.539$.

Discussion: Recall that the spectral radius of an iteration matrix is estimated by Eq. (3.25) with $\|\mathbf{e}^{(p+1)}\|_2/\|\mathbf{e}^{(p)}\|_2$ where $\mathbf{e}^{(p)} = \mathbf{x}-\mathbf{x}^{(p)}$ is the true error, which cannot be determined exactly unless the true solution is known. Since $\lim_{p\to\infty} \mathbf{d}^{(p)} = \mathbf{e}^{(p)}$, the next best alternative for the error estimation is the displacement vector $\mathbf{d}^{(p)} = \mathbf{x}^{(p)} - \mathbf{x}^{(p-1)}$, which can be easily calculated at the end of every iteration.

Also recall that the $\|\mathbf{d}^{(p+1)}\|_2/\|\mathbf{d}^{(p)}\|_2$ ratios in Examples 3.1 and 3.2 had been computed and presented in Tables 3.1 and 3.2. As it should be clear from the above discussions, this ratio should converge to the spectral radius (or dominant eigenvalue) as p increases. The spectral radii for the Jacobi and Gauss-Seidel iteration matrices for Examples 3.1 and 3.2 have been found to be $\lambda_{J,\max} \cong 0.539$ and $\lambda_{GS,max} \cong 0.291$, respectively. These findings also verify the $\rho(\mathbf{M}_{GS}) = \rho^2(\mathbf{M}_J)$ relation, i.e., $\lambda_{GS,\max} = \lambda_{J,\max}^2$.

Now that $\rho(\mathbf{M}_J)$ is available, the optimum relaxation parameter can be estimated from Eq. (3.32) to find ω_{opt}, which yields 1.086 or rounded to 1.1. Hence, it should not be surprising that the SOR method converged to a solution with relatively fewer iterations than Gauss-Seidel since $\omega = 1.1 \approx \omega_{\mathrm{opt}}$. Nonetheless, it is clear that extending this analysis (of determining $\rho(\mathbf{M}_J)$ and ω_{opt}) to very large matrices is a very laborious and time-consuming task, even for computers. In practice, relatively rough estimates of the spectral radius can be given with minimal computation, using algorithms similar to those given in Section 3.3.2.

EXAMPLE 3.5: Estimating optimum relaxation parameter

Eleven springs with the same spring constants (k) and same unstretched lengths $(\ell = 15 \text{ cm})$ are attached to each other in series, as shown in the figure below.

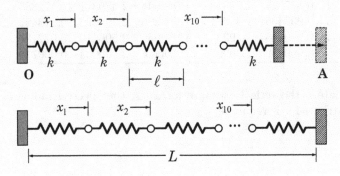

While fixed on the left-hand side, this system is stretched from the right side out to Point A, yielding a total length of $L = 2.2$ m. The positions $x_1, x_2, \ldots, x_{10}$ denote the distances with respect to Point O. The equilibrium equations can be written as

$$k(x_1 - \ell) = k\left((x_2 - x_1) - \ell\right),$$

$$k\left((x_i - x_{i-1}) - \ell\right) = k\left((x_{i+1} - x_i) - \ell\right), \quad i = 2, 3, \ldots, 9$$

$$k\left((x_{10} - x_9) - \ell\right) = k\left((L - x_{10}) - \ell\right)$$

(a) Apply the Jacobi and Gauss-Seidel methods to solve the resulting system of equations for the positions and compute the convergence rate for each case; (b) estimate the theoretical value of ω_{opt}; and (c) solve the linear system using the SOR method, starting with $\omega = 1$ and increasing in increments $\Delta\omega = 0.05$ up to $\omega = 1.95$ and then plot the relaxation parameter versus the total number of iterations, i.e., $\omega - N$. Use $\mathbf{x}^{(0)} = \mathbf{0}$ and $|\mathbf{d}^{(p)}|_2 < 10^{-5}$. The true solution of the linear system is given as $x_i = i/5$ for $i = 1, 2, \ldots, 10$.

SOLUTION:

(a) Simplifying and rearranging the equilibrium equations, following tridiagonal (Toeplitz) system is obtained:

$$
\begin{bmatrix}
2 & -1 \\
-1 & 2 & -1 \\
& -1 & 2 & -1 \\
& & \ddots & \ddots & \ddots \\
& & & -1 & 2 & -1 \\
& & & & -1 & 2
\end{bmatrix}
\begin{bmatrix}
x_1 \\ x_2 \\ x_3 \\ \vdots \\ x_9 \\ x_{10}
\end{bmatrix}
=
\begin{bmatrix}
0 \\ 0 \\ 0 \\ \vdots \\ 0 \\ L
\end{bmatrix}
$$

To estimate the convergence rate of the method, the maximum magnitude eigenvalue of the Jacobi iteration matrix is constructed as

$$
\mathbf{M}_J = \mathbf{D}^{-1}(\mathbf{L}+\mathbf{U}) = -\frac{1}{2}
\begin{bmatrix}
0 & 1 \\
1 & 0 & 1 \\
& 1 & 0 & 1 \\
& & \ddots & \ddots & \ddots \\
& & & 1 & 0 & 1 \\
& & & & 1 & 0
\end{bmatrix}
$$

The eigenvalues of $\mathbf{M}_J$ are found from Eq. (2.15) as $\lambda_i = \cos(i\pi/11)$ for $i = 1, 2, \ldots, 10$, and the spectral radius is found as $\rho(\mathbf{M}_J) = \lambda_{\max} = 0.95949$. Using Eq. (3.28), the convergence rate of the Jacobi method is found from $R = -\log (0.95949) = 0.017958$. The number of iteration steps required by the Jacobi method to reduce the maximum error by a factor of 10 is $p > 1/R \approx 56$. A total of 227 iterations are required for the linear system to converge to an approximate solution within the given tolerance.

The Gauss-Seidel method, on the other hand, converges to an approximate solution with 117 iterations. The Gauss-Seidel iteration matrix (not in tridiagonal form) yields

$$\mathbf{M}_{\text{GS}} = (\mathbf{D} - \mathbf{L})^{-1}\mathbf{U} = \begin{bmatrix} 0 & \frac{1}{2} & 0 & \cdots & 0 & 0 & 0 \\ 0 & \frac{1}{2^2} & \frac{1}{2} & \cdots & 0 & 0 & 0 \\ \vdots & \vdots & \vdots & \ddots & \vdots & \frac{1}{2} & 0 \\ 0 & \frac{1}{2^9} & \frac{1}{2^8} & \cdots & \frac{1}{2^3} & \frac{1}{2^2} & \frac{1}{2} \\ 0 & \frac{1}{2^{10}} & \frac{1}{2^9} & \cdots & \frac{1}{2^4} & \frac{1}{2^3} & \frac{1}{2^2} \end{bmatrix}$$

Finding the eigenvalues of this matrix, covered in Chapter 11, is not an easy task. At this point, the computation of the eigenvalues is left to the reader, and only the results are presented here. The maximum magnitude eigenvalue is found as $\lambda_{\max} = 0.062063$, i.e., $\rho(\mathbf{M}_{\text{GS}}) = 0.92063$. The convergence rate of the Gauss-Seidel method is similarly derived from Eq. (3.28) as $R = -\log(0.92063) = 0.035925$, and we find $p > 1/R \approx 28$, which is the number of iteration steps required by the Gauss-Seidel method to reduce the maximum error by a factor 10.

(b) Since $\rho(\mathbf{M}_J) = 0.95949$, the optimum relation parameter can be estimated by Eq. (3.32) as

$$\omega_{\text{opt}} = \frac{2}{1 + \sqrt{1 - \rho^2(\mathbf{M}_J)}} = \frac{2}{1 + \sqrt{1 - 0.95949^2}} = 1.56$$

It should also be pointed out that $\rho(\mathbf{M}_{\text{GS}}) = \rho^2(\mathbf{M}_J)$ relationship is also verified.

(c) The SOR solutions for ω's from 1 to 1.9 with 0.1 intervals were obtained, and the total number of iterations was plotted against the relaxation parameter ω and presented in Fig. 3.3. It is observed that the optimum value of the relaxation parameter occurred at about $\omega \approx 1.6$ (with 26 iterations). When we repeat the SOR solutions for $\omega = 1.56$, 1.57, 1.58, and 1.59 to narrow down the optimum value, we achieve convergence with 28, 24, 27, and 27 iterations, respectively. This indicates that the optimum relaxation parameter is realized at $\omega = 1.57$ with 24 iterations, which is consistent with the theoretical prediction.

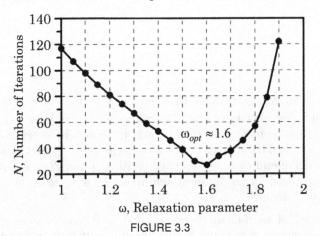

FIGURE 3.3

The SOR iteration matrix, $\mathbf{M}_\omega$, is also *not* in tridiagonal form, and it is too large and complicated; thus, it will not be reported here to save space. However, we will leave this task as an exercise for the reader to perform using the symbolic processor.

The dominant (maximum magnitude) eigenvalue can be determined by a suitable method presented in Chapter 11 as well. Nonetheless, we already know by Eq. (3.31) that the spectral radius of the SOR method is given as $\rho(\mathbf{M}_\omega) = \omega - 1$ for $\omega_{\text{opt}} \leqslant \omega \leqslant 2$. So we may use this expression to find $\rho(\mathbf{M}_\omega) = \omega - 1 = 0.56$ for $\omega_{\text{opt}} = 1.56$. The convergence rate of the SOR method with the optimum relaxation parameter is then found as $R = -\log_{10}\rho(\mathbf{M}_\omega) = -\log_{10}(0.56) = 0.57982$. In other words, the SOR with $\omega_{\text{opt}} = 1.56$ requires $1/R \approx 2$ steps to reduce the maximum error by a factor 10.

Discussion: Note that the Gauss-Seidel method in this example also converges twice as fast as the Jacobi method, and the number of iterations is half that of the Jacobi method, as the convergence rates of the Jacobi and Gauss-Seidel methods are found to be 0.017958 and 0.035925, respectively. The Gauss-Seidel and SOR iteration matrices are very complicated, even for simple tridiagonal (coefficient) matrices, and in practice, theoretical estimation of the ω_{opt} is computationally not feasible. It is clear that estimating ω_{opt} without the need for too many calculations and using the SOR method with ω_{opt} could provide a computational advantage. However, there are algorithms to estimate ω_{opt}, and an effective algorithm has been provided in this section.

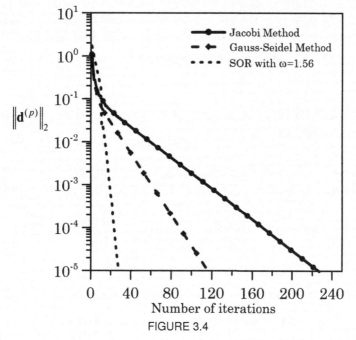

FIGURE 3.4

The Euclidean (ℓ_2) norm of the displacement vector is computed and comparatively depicted for the Jacobi, Gauss-Seidel, and SOR (with ω_{opt}) methods in Fig. 3.4. It is observed that the steady convergence rates are established very quickly (in ≈ 15 to 30 iterations). Fig. 3.4 clearly shows that the steady convergence rates for all three methods are linear, and the slopes of the straight lines also correspond to the theoretical convergence rates.

3.4 KRYLOV SPACE METHODS

3.4.1 CONJUGATE GRADIENT METHOD (CGM)

The conjugate gradient method (CGM) developed by Hestenes and Stiefel [14] is an effective iterative method used to solve *symmetric positive-definite* systems of linear equations. The method is usually applied to sparse systems that are impractical to handle by direct solution methods, such as Gauss elimination or Cholesky decomposition.

The method is based on minimizing the quadratic scalar function, defined as

$$f(\mathbf{x}) = \frac{1}{2}\mathbf{x}^T\mathbf{A}\mathbf{x} - \mathbf{b}^T\mathbf{x} \tag{3.33}$$

where $\mathbf{A}$ is a square symmetric positive definite matrix, and the gradient of the scalar function $f(\mathbf{x})$ is $\mathrm{grad} f = \mathbf{A}\mathbf{x} - \mathbf{b}$. Minimizing the gradient of $f(\mathbf{x})$ requires setting $\mathbf{A}\mathbf{x} - \mathbf{b} = \mathbf{0}$. In other words, minimizing $f(\mathbf{x})$ is equivalent to obtaining the solution of a system of linear equations given by $\mathbf{A}\mathbf{x} = \mathbf{b}$.

The gradient minimization is an iterative procedure that requires an initial guess, $\mathbf{x}^{(0)}$. The current step estimate is improved as follows:

$$\mathbf{x}^{(p+1)} = \mathbf{x}^{(p)} + \alpha^{(p)}\mathbf{d}^{(p)} \tag{3.34}$$

where $\mathbf{x}^{(p)}$ is the estimate at the pth iteration level, $\alpha^{(p)}$ is a scalar called the *line search parameter* which is chosen at the pth iteration step so that $f(\mathbf{x}^{(p)})$ is minimized in the vector of conjugate search direction, $\mathbf{d}^{(p)}$. This step is called the *line search*.

Substituting the current (improved) estimate, Eq. (3.34), into $\mathbf{A}\mathbf{x}^{(p+1)} = \mathbf{b}$ yields

$$\mathbf{A}\mathbf{x}^{(p)} + \alpha^{(p)}\mathbf{A}\mathbf{d}^{(p)} = \mathbf{b} \tag{3.35}$$

Noting that $\mathbf{b} - \mathbf{A}\mathbf{x}^{(p)}$ is the *residual vector* at the pth iteration step (denoted by $\mathbf{r}^{(p)}$), Eq. (3.35) can be rearranged as

$$\alpha^{(p)}\mathbf{A}\mathbf{d}^{(p)} = \mathbf{r}^{(p)} \tag{3.36}$$

Once the residual vector is known, the orthogonal (conjugate) search directions can be obtained by multiplying both sides of Eq. (3.36) with $\left(\mathbf{d}^{(p)}\right)^T$ and solving for $\alpha^{(p)}$:

$$\alpha^{(p)} = \frac{\left(\mathbf{d}^{(p)}\right)^T\mathbf{r}^{(p)}}{\left(\mathbf{d}^{(p)}\right)^T\mathbf{A}\mathbf{d}^{(p)}} = \frac{(\mathbf{d}^{(p)}, \mathbf{r}^{(p)})}{(\mathbf{d}^{(p)}, \mathbf{A}\mathbf{d}^{(p)})} \tag{3.37}$$

where we have now used the *inner product* notation, i.e., $(\mathbf{x}, \mathbf{y}) = \mathbf{x}^T\mathbf{y}$.

By left-multiplying Eq. (3.34) with $\mathbf{A}$ and subtracting $\mathbf{b}$ from both sides, we obtain

$$\mathbf{b} - \mathbf{A}\mathbf{x}^{(p+1)} = \mathbf{b} - \mathbf{A}\mathbf{x}^{(p)} - \alpha^{(p)}\mathbf{A}\mathbf{d}^{(p)}$$

After simplifications, this expression gives a recursive relationship involving the residual vectors of two successive iterations:

$$\mathbf{r}^{(p+1)} = \mathbf{r}^{(p)} - \alpha^{(p)}\mathbf{A}\mathbf{d}^{(p)} \tag{3.38}$$

It should be noted that the residuals are also orthogonal, i.e., $(\mathbf{r}^{(p)}, \mathbf{r}^{(p+1)}) = (\mathbf{r}^{(p+1)}, \mathbf{r}^{(p)}) = 0$. The next search direction is found from

$$\mathbf{d}^{(p+1)} = \mathbf{r}^{(p+1)} + \beta^{(p)}\mathbf{d}^{(p)} \tag{3.39}$$

where the *conjugation parameter* $\beta^{(p)}$ is also a scalar quantity determined such that $(\mathbf{d}^{(p+1)}, \mathbf{A}\mathbf{d}^{(p)}) = 0$, i.e., the two successive search directions are conjugate.

Substituting Eq. (3.39) into this inner product yields

$$(\mathbf{d}^{(p+1)}, \mathbf{A}\mathbf{d}^{(p)}) = 0 = (\mathbf{r}^{(p+1)}, \mathbf{A}\mathbf{d}^{(p)}) + \beta^{(p)}(\mathbf{d}^{(p)}, \mathbf{A}\mathbf{d}^{(p)}) \tag{3.40}$$

and solving Eq. (3.40) for $\beta^{(p)}$ leads to

$$\beta^{(p)} = -\frac{(\mathbf{r}^{(p+1)}, \mathbf{A}\mathbf{d}^{(p)})}{(\mathbf{d}^{(p)}, \mathbf{A}\mathbf{d}^{(p)})} \tag{3.41}$$

It turns out that the previous search direction and the current residual vector are also orthogonal, i.e., $(\mathbf{r}^{(p+1)}, \mathbf{d}^{(p)}) = 0$. Equation (3.41) can be further simplified by also recalling that $(\mathbf{r}^{(p+1)}, \mathbf{r}^{(p)}) = 0$. For the numerator of Eq. (3.41), the inner product of Eq. (3.38) with $\mathbf{r}^{(p+1)}$, leads to

$$(\mathbf{r}^{(p+1)}, \mathbf{A}\mathbf{d}^{(p)}) = -\frac{1}{\alpha^{(p)}}(\mathbf{r}^{(p+1)}, \mathbf{r}^{(p+1)}) \tag{3.42}$$

From Eq. (3.38), we may write $\mathbf{A}\mathbf{d}^{(p)} = (\mathbf{r}^{(p+1)} - \mathbf{r}^{(p)})/\alpha^{(p)}$. Also recalling $(\mathbf{d}^{(p)}, \mathbf{r}^{(p+1)}) = 0$, the denominator of Eq. (3.41) results in

$$(\mathbf{d}^{(p)}, \mathbf{A}\mathbf{d}^{(p)}) = \frac{1}{\alpha^{(p)}}\left((\mathbf{d}^{(p)}, \mathbf{r}^{(p+1)}) - (\mathbf{d}^{(p)}, \mathbf{r}^{(p)})\right) = -\frac{1}{\alpha^{(p)}}(\mathbf{d}^{(p)}, \mathbf{r}^{(p)}) \tag{3.43}$$

Finally, using Eq. (3.39) to get $\mathbf{d}^{(p)}$ and substituting it into Eq. (3.43) gives

$$(\mathbf{d}^{(p)}, \mathbf{A}\mathbf{d}^{(p)}) = -\frac{1}{\alpha^{(p)}}\left((\mathbf{r}^{(p)}, \mathbf{r}^{(p)}) + \beta^{(p-1)}(\mathbf{d}^{(p-1)}, \mathbf{r}^{(p)})\right) = -\frac{1}{\alpha^{(p)}}(\mathbf{r}^{(p)}, \mathbf{r}^{(p)}) \tag{3.44}$$

Now combining Eqs. (3.42) and (3.44) in Eq. (3.41), we arrive at

$$\beta^{(p)} = \frac{(\mathbf{r}^{(p+1)}, \mathbf{r}^{(p+1)})}{(\mathbf{r}^{(p)}, \mathbf{r}^{(p)})} \tag{3.45}$$

Initial Guess. If a rough estimate for the solution vector $\mathbf{x}$ is available, it can be used as the initial value. Otherwise, setting $\mathbf{x}^{(0)} = 0$ is sufficient for most cases since the CGM eventually converges in at most n iterations.

Stopping Criterion. In the CGM algorithm, residuals at two iteration levels ($\mathbf{r}^{(p)}$ and $\mathbf{r}^{(p+1)}$) are available, so there is no need to specify absolute or relative errors in terms of the displacement vector. The simplest and most common absolute and relative stopping criteria are based on the ℓ_2 norms of the available residuals at the end of each iteration step:

$$\|\mathbf{r}^{(p)}\|_2 = \left(\mathbf{r}^{(p)}, \mathbf{r}^{(p)}\right)^{1/2} < \varepsilon_1 \quad \text{or} \quad \frac{\|\mathbf{r}^{(p)}\|_2}{\|\mathbf{b}\|_2} < \varepsilon_2 \tag{3.46}$$

where ε_1 and ε_2 are the preset tolerances for the absolute and relative errors, respectively. One or both criteria can be employed at no additional cost, with the first criterion being cast as $\|\mathbf{r}^{(p)}\|_2 < \varepsilon_1 + \varepsilon_2\|\mathbf{b}\|_2$. Note that setting $\varepsilon_1 = 0$ and $\varepsilon_2 > 0$ and $\varepsilon_2 = 0$ and $\varepsilon_1 > 0$ give relative and absolute errors, respectively.

 Providing an initial guess is critical for nonlinear optimization problems, which may possess several local extrema. In such cases, the choice of the initial guess affects whether the method will converge or not, or to which solution set it will converge.

Conjugate Gradient Method

- The CGM is a simple method to program and easy to parallelize;
- Theoretically, in the absence of round-off errors, it yields the *true solution* after a finite number of iterations (as many or less than the number of equations);
- It converges, if it does so, with a linear convergence rate dictated by its *condition number*, i.e., the smaller the condition number is, the faster the convergence will be;
- It is particularly suitable and useful for large sparse systems with regular patterns.

- The method is applicable to linear systems with positive definite and symmetric coefficient matrices;
- It may exhibit instabilities even for small perturbations;
- It is computationally more expensive in comparison to stationary iterative methods per iteration step and requires more memory allocation;
- In large sparse systems, the rate of convergence may become quite slow unless an effective preconditioning is employed;
- In practice, the *true solution* may never be obtained as most directions are not conjugate, or round-off errors due to finite-precision arithmetic may result in a loss of orthogonality in the residuals.

A pseudomodule, **CGM**, employing the CGM on a linear system is presented in Pseudocode 3.3. As input, the module requires the number of unknowns (n), the linear system ($\mathbf{A}$ and $\mathbf{b}$), a vector of initial guess ($\mathbf{x}^{(0)}$), an upper bound for the number of iterations ($maxit$), and a convergence tolerance (ε). The approximate solution ($\mathbf{x}$) and the number of iterations performed ($iter$) are the outputs of the module. The module uses Pseudocodes 2.3 and 2.5 to compute the $\mathbf{x} \cdot \mathbf{y}$ and $\mathbf{Ax}$ products, respectively. The residual vector computed by $\mathbf{r}^{(0)} = \mathbf{b} - \mathbf{Ax}^{(0)}$ is set as the starting direction, $\mathbf{d}^{(0)} = \mathbf{r}^{(0)}$. The For-loop over p is the iteration loop. The ℓ_2-norm of the residual vector is found from $\rho^{(p)} = \sqrt{(\mathbf{r}^{(p)}, \mathbf{r}^{(p)})}$ and is used to terminate the iteration procedure when the convergence is achieved. For $\rho^{(p)} > \varepsilon$, $\alpha^{(p)}$ is computed by Eq. (3.37) to obtain $\mathbf{x}^{(p+1)}$ and corresponding residual $\mathbf{r}^{(p+1)}$ using Eqs. (3.34) and (3.38), respectively. The iteration process is repeated with a new direction determined by Eq. (3.39) along with Eq. (3.45) until the stopping criterion is satisfied.

In applying iterative methods to *sparse* linear systems, it is easier to perform computations with the non-zero matrix elements only. A programming strategy avoiding arithmetic operations with zeros results in a significant reduction in cpu-time.

As far as the memory requirement is concerned, the module needs additional memory space to be allocated for $\mathbf{c}$, $\mathbf{d}$, and $\mathbf{r}$, and they can be used for prior and current step quantities and may be overwritten to save the current step values. If $\mathbf{A}$ is a sparse or banded matrix, the memory requirement can be further reduced by employing banded or sparse matrix storage solutions.

Pseudocode 3.3

Module CGM $(n, \varepsilon, \mathbf{A}, \mathbf{b}, \mathbf{x}, maxit, iter, error)$
\ DESCRIPTION: A pseudomodule to solve $\mathbf{Ax} = \mathbf{b}$ using the CGM method.
\ USES:
\   Ax :: Module to perform matrix-vector multiplication (Pseudocode 2.5);
\   XdotY:: Function giving the dot product of two vectors (Pseudocode 2.3).
Declare: $a_{nn}, b_n, x_n, r_n, c_n, d_n$ \ Declare array variables
Ax$(n, \mathbf{A}, \mathbf{x}, \mathbf{r})$ \ Visit Ax to compute $\mathbf{Ax}^{(0)} = \mathbf{r}^{(0)}$
$\mathbf{r} \leftarrow \mathbf{b} - \mathbf{r}$ \ Find residual $\mathbf{r}^{(0)} = \mathbf{b} - \mathbf{Ax}^{(0)}$
$\mathbf{d} \leftarrow \mathbf{r}$ \ Initialize $\mathbf{d}^{(0)}$ with $\mathbf{r}^{(0)}$
$\rho 0 \leftarrow$ **XdotY**$(n, \mathbf{r}, \mathbf{r})$ \ Compute $\rho 0 = (\mathbf{r}^{(0)}, \mathbf{r}^{(0)})$
For $[p = 0, maxit]$ \ Begin iteration loop
    $R \leftarrow \sqrt{\rho 0}$ \ Find $\|\mathbf{r}^{(p)}\|_2$
    Write: "Iteration=", p,"$\|\mathbf{r}^{(p)}\|_2$=",R \ Printout iteration progress
    If $[R < \varepsilon]$ **Then**   \ Check for convergence, $\|\mathbf{r}^{(p)}\|_2 < \varepsilon$?
       Exit   \ Converged, EXIT loop
    Else   \ Not converged, continue iterations
       Ax$(n, \mathbf{A}, \mathbf{d}, \mathbf{c})$ \ Find Ax to compute $\mathbf{Ad}^{(p)} = \mathbf{c}^{(p)}$
       $\rho \leftarrow$**XdotY**$(n, \mathbf{d}, \mathbf{c})$ \ Compute $\rho = (\mathbf{d}^{(p)}, \mathbf{c}^{(p)})$
       $\alpha \leftarrow \rho 0/\rho$ \ Find $\alpha^{(p)}$
       $\mathbf{x} \leftarrow \mathbf{x} + \alpha * \mathbf{d}$ \ Find current estimate, $\mathbf{x}^{(p+1)}$
       $\mathbf{r} \leftarrow \mathbf{r} - \alpha * \mathbf{c}$ \ Find current residual, $\mathbf{r}^{(p+1)}$
       $\rho \leftarrow$**XdotY**$(n, \mathbf{r}, \mathbf{r})$ \ Find $\rho^{(p+1)} = (\mathbf{r}^{(p+1)}, \mathbf{r}^{(p+1)})$
       $\beta \leftarrow \rho/\rho 0$ \ Find $\beta^{(p)} = \rho^{(p+1)}/\rho^{(p)}$
       $\mathbf{d} \leftarrow \mathbf{r} + \beta * \mathbf{d}$ \ Find new direction $\mathbf{d}^{(p+1)}$
       $\rho 0 \leftarrow \rho$ \ Set $\rho^{(p-1)} \leftarrow \rho^{(p)}$
    End If
End For   \ End of iteration loop
$error \leftarrow R$ \ Set current ℓ_2−norm as error
$iter \leftarrow p$ \ Set current $p - 1$ as $iter$ on exit
\ If iteration limit $maxit$ is reached $with$ no convergence
If $[iter = maxit]$ **Then**   \ Issue info and a warning
    Write: "CGM failed to converge after",$maxit$,"iterations"
    Write: "The error at the last step =",$error$
End If
End Module CGM

EXAMPLE 3.6: Application of Conjugate Gradient Method

Apply the conjugate gradient method to estimate the solution of the following linear system. Use $\mathbf{x}^{(0)} = \mathbf{1}$ and $\|\mathbf{r}\|_2 < 10^{-5}$.

$$\begin{bmatrix} 8 & 1 & 2 & 1 \\ 1 & 6 & 1 & 2 \\ 2 & 1 & 7 & 1 \\ 1 & 2 & 1 & 8 \end{bmatrix} \begin{bmatrix} x_1 \\ x_2 \\ x_3 \\ x_4 \end{bmatrix} = \begin{bmatrix} -5 \\ 3 \\ 11 \\ -13 \end{bmatrix}$$

SOLUTION:

Notice that the coefficient matrix is symmetric and positive-definite. Starting with $\mathbf{x}^{(0)} = [\,1\ 1\ 1\ 1\,]^T$, the initial residual is found as

$$\mathbf{r}^{(0)} = \mathbf{b} - \mathbf{A}\mathbf{x}^{(0)} = [\,-17\quad -7\quad 0\quad -25\,]^T$$

Setting $\mathbf{d}^{(0)} = \mathbf{r}^{(0)}$, we find

$$\rho^{(0)} = (\mathbf{r}^{(0)}, \mathbf{r}^{(0)}) = (-17)^2 + (-7)^2 + 0^2 + (-25)^2 = 963$$

$$\mathbf{c}^{(0)} = \mathbf{A}\mathbf{d}^{(0)} = [\,-168 - 109 - 66 - 231\,]^T$$

$$(\mathbf{d}^{(0)}, \mathbf{c}^{(0)}) = (-17)(-168) + (-7)(-109) + 0(-66) + (-25)(-231) = 9394$$

Next, we calculate the step length and the current estimate as

$$\alpha^{(0)} = \frac{\rho^{(0)}}{(\mathbf{d}^{(0)}, \mathbf{c}^{(0)})} = \frac{963}{9394} = 0.102512$$

$$\mathbf{x}^{(1)} = \mathbf{x}^{(0)} + \alpha^{(0)}\mathbf{d}^{(0)} = [\,-0.742708\ 0.282414\ 1\ -1.562806\,]^T$$

The residual associated with the current estimate is obtained as follows:

$$\mathbf{r}^{(1)} = \mathbf{r}^{(0)} - \alpha^{(0)}\mathbf{c}^{(0)} = [\,0.222057\ 4.173834\ 6.765808\ -1.319672\,]^T$$

with $\rho^{(1)} = (\mathbf{r}^{(1)}, \mathbf{r}^{(1)}) = 64.9879$ and $\|\mathbf{r}^{(1)}\|_2 = \sqrt{\rho^{(1)}} = 8.06151 > \varepsilon$. Then the conjugation parameter is obtained as

$$\beta^{(0)} = \frac{\rho^{(1)}}{\rho^{(0)}} = \frac{64.9879}{963} = 0.0674848$$

To carry out the second iteration, we need to determine, $\mathbf{d}^{(1)}$ and $\mathbf{c}^{(1)}$:

$$\mathbf{d}^{(1)} = \mathbf{r}^{(1)} + \beta^{(0)}\mathbf{d}^{(0)} = [\,-0.925185\ 3.701440\ 6.765808\ -3.006792\,]^T$$

$$\mathbf{c}^{(1)} = \mathbf{A}\mathbf{d}^{(1)} = [\,6.824784\ 22.035679\ 46.204934\ -10.810833\,]^T$$

which yields $(\mathbf{d}^{(1)}, \mathbf{c}^{(1)}) = 420.369$. The step length is then calculated as

$$\alpha^{(1)} = \frac{\rho^{(1)}}{(\mathbf{d}^{(1)}, \mathbf{c}^{(1)})} = \frac{64.9879}{420.369} = 0.154597$$

The second estimate and its associated residuals are found as

$$\mathbf{x}^{(2)} = \mathbf{x}^{(1)} + \alpha^{(1)}\mathbf{d}^{(1)} = [\,0.885739\ 0.854647\ 2.045975\ -2.02765\,]^T$$

$$\mathbf{r}^{(2)} = \mathbf{r}^{(1)} - \alpha^{(1)}\mathbf{c}^{(1)} = [\,-0.833035\ 0.767180\ -0.377345\ 0.351652\,]^T$$

$$\rho^{(2)} = (\mathbf{r}^{(2)}, \mathbf{r}^{(2)}) = 1.54856 \quad \text{and} \quad \|\mathbf{r}^{(2)}\|_2 = \sqrt{\rho^{(2)}} = 1.24442 > \varepsilon$$

Next, the selection of the new, improved direction is determined as follows:

$$\beta^{(1)} = \frac{\rho^{(2)}}{\rho^{(1)}} = \frac{1.54856}{64.9879} = 0.023828$$

$$\mathbf{d}^{(2)} = \mathbf{r}^{(2)} + \beta^{(1)}\mathbf{d}^{(1)} = [\,-0.855081\ 0.855379\ -0.216127\ 0.280005\,]^T$$

The third estimate, in similar fashion, yields

$$\mathbf{c}^{(2)} = \mathbf{A}\mathbf{d}^{(2)} = [-6.137518, \ 4.621078, \ -2.087665, \ 2.879593]^T$$

$$(\mathbf{d}^{(2)}, \mathbf{c}^{(2)}) = 10.4584, \quad \alpha^{(2)} = \frac{\rho^{(2)}}{(\mathbf{d}^{(2)}, \mathbf{c}^{(2)})} = \frac{1.54856}{10.4584} = 0.1480685$$

$$\mathbf{x}^{(3)} = \mathbf{x}^{(2)} + \alpha^{(2)}\mathbf{d}^{(2)} = [-1.01235, \ 0.981302, \ 2.013973, \ -1.98619]^T$$

$$\mathbf{r}^{(3)} = \mathbf{r}^{(2)} - \alpha^{(2)}\mathbf{c}^{(2)} = [0.075738, \ 0.082943, \ -0.068228, \ -0.074725]^T$$

$$\rho^{(3)} = (\mathbf{r}^{(3)}, \mathbf{r}^{(3)}) = 0.0228547 \quad \text{and} \quad \|\mathbf{r}^{(3)}\|_2 = \sqrt{\rho^{(3)}} = 0.15118 > \varepsilon$$

$$\beta^{(2)} = \frac{\rho^{(3)}}{\rho^{(2)}} = \frac{1.54856}{64.9879} = 0.014759$$

$$\mathbf{d}^{(3)} = \mathbf{r}^{(3)} + \beta^{(2)}\mathbf{d}^{(2)} = [\ 0.063118 \ \ 0.095568 \ -0.071418 \ -0.070592\ ]^T$$

Finally, at the fourth iteration, we obtain

$$\mathbf{c}^{(3)} = [\ 0.387085 \ \ 0.423922 \ -0.348712 \ -0.381902\ ]^T,$$

$$(\mathbf{d}^{(3)}, \mathbf{c}^{(3)}) = 0.11681, \quad \alpha^{(3)} = 0.195659$$

$$\mathbf{x}^{(4)} = \mathbf{x}^{(3)} + \alpha^{(3)}\mathbf{d}^{(3)} = [-1 \ 1 \ 2 \ -2]^T \quad \text{and} \quad \mathbf{r}^{(4)} = [\ 0 \ 0 \ 0 \ 0\ ]^T$$

Discussion: The CGM is guaranteed to converge in at most n steps, where n is the number of equations. Using double precision computation, the CGM yielded the true solution accurate to 16 decimal places in four iterations (i.e., $n = 4$ steps). (The final estimates were rounded to the nearest value.) When the same system is solved with the Jacobi (Example 3.1) and Gauss-Seidel methods (Example 3.2), the results were obtained with 22 and 8 iterations, respectively. On the other hand, the SOR method with $\omega_{\mathrm{opt}} = 1.092$ converged to the solution in 4 iterations.

3.4.2 CONVERGENCE ISSUES OF THE CGM

For a symmetric positive definite $\mathbf{A}\mathbf{x} = \mathbf{b}$, an upper error bound for the pth estimate is given as

$$\frac{\|\mathbf{e}_A^{(p)}\|}{\|\mathbf{e}^{(0)}\|_A} \leqslant 2 \left(\frac{\sqrt{\kappa(\mathbf{A})} - 1}{\sqrt{\kappa(\mathbf{A})} + 1} \right)^p, \qquad p \geqslant 0, \quad \kappa(\mathbf{A}) > 1 \tag{3.47}$$

where $\|\mathbf{e}\|_A = \sqrt{(\mathbf{e}, \mathbf{A}\mathbf{e})}$, $\mathbf{e}^{(p)} = \mathbf{x} - \mathbf{x}^{(p)}$ denotes the true error vector, and $\kappa(\mathbf{A}) = \|\mathbf{A}\|_2 \|\mathbf{A}^{-1}\|_2$ denotes the spectral condition number of $\mathbf{A}$. In order to simplify this expression, we make use of $(x - 1)/(x + 1) < \exp(-2/x)$ in Eq. (3.47) as follows:

$$\frac{\|\mathbf{e}^{(p)}\|_A}{\|\mathbf{e}^{(0)}\|_A} < 2 \exp \left(-\frac{2p}{\sqrt{\kappa(\mathbf{A})}} \right), \qquad p \geqslant 0, \tag{3.48}$$

For a positive definite and symmetric $\mathbf{A}$, the condition number can be expressed as $\kappa(\mathbf{A}) = \lambda_{\max}/\lambda_{\min}$, where λ denotes the eigenvalues of $\mathbf{A}$. It should be noted that for $\kappa(\mathbf{A}) \approx 1$, the convergence is faster than the theoretical estimate as the eigenvalues are clustered closely together. The number of iterations required to reduce the error norm by a fixed amount is in the order of $\mathcal{O}(\sqrt{\kappa(\mathbf{A})})$. The systems with $\kappa(\mathbf{A}) \gg 1$ are the cases where the eigenvalues of $\mathbf{A}$ spread over a wide range, which leads to extremely slow convergence rates. The convergence of such systems can be accelerated by applying the procedure to an equivalent *preconditioned system*.

3.4.3 PRECONDITIONING

The performance of the CGM is very good when applied to well-conditioned systems, i.e., linear systems having $\kappa(\mathbf{A}) \approx 1$. If a linear system is ill-conditioned ($\kappa(\mathbf{A}) \gg 1$), then the numerical solution is plagued with round-off errors, affecting the convergence rate adversely. In fact, the error bound, Eq. (3.47), indicates that the CGM may converge very slowly even for a system that has a moderate condition number. In cases where the CGM and its variants are expected to converge slowly, the linear system is transformed into a problem with a smaller condition number so that rapid convergence is assured.

The technique of accelerating the convergence of a linear system by reducing its condition number is called *preconditioning*. A *preconditioner*, a matrix, is used to modify the original linear system so that it is easier and faster to solve iteratively.

The preconditioned form of $\mathbf{A}\mathbf{x} = \mathbf{b}$ is written as

$$\mathbf{P}^{-1}\mathbf{A}\mathbf{x} = \mathbf{P}^{-1}\mathbf{b} \tag{3.49}$$

where $\mathbf{A}$ is a symmetric positive-definite matrix and $\mathbf{P}$ is an invertible preconditioner matrix. The condition number of the modified matrix is reduced by selecting a reasonable $\mathbf{P}$, i.e., $\kappa(\mathbf{P}^{-1}\mathbf{A}) < \kappa(\mathbf{A})$. The effectiveness of a preconditioner depends on $\kappa(\mathbf{P}^{-1}\mathbf{A})$.

The residual form of the preconditioned system is written as $\mathbf{P}\mathbf{z} = \mathbf{r}$. This residual vector $\mathbf{r}$ is used to compute the scalar α and β parameters. The preconditioned linear system is solved at each iteration step; thus, the preconditioner should be selected such that the cost of solving $\mathbf{P}\mathbf{z}^{(p)} = \mathbf{r}^{(p)}$ is minimum. Preconditioning increases the memory requirement and computational effort to some degree. However, the additional cost incurred is compensated by the reduction in the number of iterations required to reach an acceptable solution.

A basic preconditioned conjugate gradient (PCGM) method algorithm is illustrated in Algorithm 3.1.

3.4.4 CGM FOR NONSYMMETRIC SYSTEMS

The CGM presented in the previous section cannot be applied to non-symmetric linear systems because the residual vectors will not be orthogonal with short recurrences. If $\mathbf{A}$ is a non-symmetric matrix, then an equivalent symmetric system is solved. For example, left multiplying the linear system with $\mathbf{A}^T$ results in a symmetric positive definite system:

$$\mathbf{A}^T\mathbf{A}\mathbf{x} = \mathbf{A}^T\mathbf{b} \tag{3.50}$$

and ℓ_2-norm of the residual is minimized, i.e., $\|\mathbf{b} - \mathbf{A}\mathbf{x}\|_2$. The algorithm derived from this formulation is often called the *conjugate gradient normal residual (CGNR)*. An alternative approach also makes use of $\mathbf{A}^T$ and requires setting $\mathbf{x} = \mathbf{A}^T\mathbf{y}$. The resulting linear system is now symmetric:

$$\mathbf{A}\mathbf{A}^T\mathbf{y} = \mathbf{b} \tag{3.51}$$

The method based on this formulation is referred to as the *conjugate gradient normal error (CGNE)*. Once Eq. (3.51) is solved, the unknown vector $\mathbf{x}$ is simply obtained from $\mathbf{x} = \mathbf{A}^T\mathbf{y}$.

\ ALGORITHM 3.1: Preconditioned Conjugate Gradient Method (PCGM)
$\mathbf{r}^{(0)} \leftarrow \mathbf{b} - \mathbf{A}\mathbf{x}^{(0)}$ \ Compute the initial residual
$\mathbf{z}^{(0)} \leftarrow \mathbf{P}^{-1}\mathbf{r}^{(0)}$ \ Solve linear system: $\mathbf{P}\mathbf{z}^{(0)} = \mathbf{r}^{(0)}$
$\mathbf{d}^{(0)} \leftarrow \mathbf{z}^{(0)}$ \ Initialize $\mathbf{d}^{(0)} = \mathbf{z}^{(0)}$
For $\left[p = 0, maxit\right]$ \ Iteration loop p
\ Find current estimate and its residual
   $\alpha^{(p+1)} \leftarrow (\mathbf{r}^{(p)}, \mathbf{z}^{(p)})/(\mathbf{d}^{(p)}, \mathbf{A}\mathbf{d}^{(p)})$ \ Find current $\alpha^{(p)}$
   $\mathbf{x}^{(p+1)} \leftarrow \mathbf{x}^{(p)} + \alpha^{(p)}\mathbf{d}^{(p)}$ \ Find current solution, $\mathbf{x}^{(p+1)}$
   $\mathbf{r}^{(p+1)} \leftarrow \mathbf{r}^{(p)} - \alpha^{(p)}\mathbf{A}\mathbf{d}^{(p)}$ \ Find current residuals, $\mathbf{r}^{(p+1)}$
   If $\left[\|\mathbf{r}^{(p+1)}\| < \varepsilon\right]$ **Then**   \ Check for convergence
      Exit   \ Exit the iteration loop
   Else
\ Not converged, compute new direction at pth step
      $\mathbf{z}^{(p+1)} \leftarrow \mathbf{P}^{-1}\mathbf{r}^{(p+1)}$ \ Solve $\mathbf{P}\mathbf{z}^{(p+1)} = \mathbf{r}^{(p+1)}$
      $\beta^{(p)} \leftarrow (\mathbf{r}^{(p+1)}, \mathbf{z}^{(p+1)})/(\mathbf{r}^{(p)}, \mathbf{z}^{(p)})$ \ Find new β
      $\mathbf{d}^{(p+1)} \leftarrow \mathbf{z}^{(p+1)} + \beta^{(p)}\mathbf{d}^{(p)}$ \ Find new direction, $\mathbf{d}$
   End If
End For   \ End iteration loop p

The CGNR and CGNE algorithms are presented in *Algorithms 3.2* and *3.3*, respectively. Both of these methods can be employed (or coded) easily by modifying the CGM algorithms. The scope of this topic is wide; Generalized Minimal Residual (GMRES), Biconjugate Gradient (BiCGM), Quasi-Minimal Residual (QMR), Conjugate Gradient Squared (CGS), and Biconjugate Gradient Stabilized (BiCGstab) are commonly encountered methods in the literature applicable to non-symmetric systems.

\ ALGORITHM 3.2: Conjugate Gradient Normal Residual method (CGNR)
$\mathbf{r}^{(0)} \leftarrow \mathbf{b} - \mathbf{A}\mathbf{x}^{(0)}$ \ Find initial residual
$\mathbf{z}^{(0)} \leftarrow \mathbf{A}^T\mathbf{r}^{(0)}$ \ Find $\mathbf{z}^{(0)} = \mathbf{A}^T\mathbf{r}^{(0)}$
$\mathbf{d}^{(0)} \leftarrow \mathbf{z}^{(0)}$ \ Set $\mathbf{d}^{(0)} = \mathbf{z}^{(0)}$
For $\left[p = 0, maxit\right]$ \ Iteration loop
ICommentCompute current estimate and its residual at $(p+1)$th step
   $\alpha^{(p+1)} \leftarrow (\mathbf{z}^{(p)}, \mathbf{z}^{(p)})/(\mathbf{A}\mathbf{d}^{(p)}, \mathbf{A}\mathbf{d}^{(p)})$
   $\mathbf{x}^{(p+1)} \leftarrow \mathbf{x}^{(p)} + \alpha^{(p)}\mathbf{d}^{(p)}$
   $\mathbf{r}^{(p+1)} \leftarrow \mathbf{r}^{(p)} - \alpha^{(p)}\mathbf{A}\mathbf{d}^{(p)}$
   If $\left[\|\mathbf{r}^{(p+1)}\| < \varepsilon\right]$ **Then**   \ Convergence achieved
      Exit   \ Exit the iteration loop
   Else
ICommentNot convergenced, compute new direction at pth step
      $\mathbf{z}^{(p+1)} \leftarrow \mathbf{A}^T\mathbf{r}^{(p+1)}$ \ Solve $\mathbf{P}\mathbf{z}^{(p+1)} = \mathbf{r}^{(p+1)}$
      $\beta^{(p)} \leftarrow (\mathbf{z}^{(p+1)}, \mathbf{z}^{(p+1)})/(\mathbf{z}^{(p)}, \mathbf{z}^{(p)})$ \ Find new β
      $\mathbf{d}^{(p+1)} \leftarrow \mathbf{z}^{(p+1)} + \beta^{(p)}\mathbf{d}^{(p)}$ \ Find new direction, $\mathbf{d}$
   End If
End For   \ End the iteration loop-p

GMRES yields the smallest residual for a fixed number of iteration steps; however, the memory requirements and computational costs are large. The BiCGM method requires $\mathbf{A}^T\mathbf{x}$ product besides $\mathbf{A}\mathbf{x}$, which increases the computational cost. Furthermore, minimization of the residuals is not assured, and extremely erratic convergence behavior may be observed.

Parallelization properties are similar to those for CGM. QMR was designed to overcome the irregular convergence behavior of BiCGM. The method requires $\mathbf{A}^\mathbf{T}\mathbf{x}$ product as well; hence, the computational cost is slightly higher than BiCGM. The CGS converges about twice as fast as the BiCGM, but its convergence behavior in some cases may be quite irregular, and it might diverge if the initial guess is close to the solution. The computational cost is similar to that of BiCGM. The BiCGstab method is designed as an alternative to that of CGS because it avoids the erratic convergence patterns of CGS. The speed of convergence is similar to that of CGS, and the computational cost is also similar to that of CGS or BiCGM. More information on the use of iterative methods for solving linear systems, particularly CGM and its derivatives, can be found in Refs. [4, 5, 11, 8, 18, 25, 29, 35, 37, 38].

 Note that the condition number of $\mathbf{A}\mathbf{A}^T$ is the square the condition number of $\mathbf{A}$, i.e., $\kappa(\mathbf{A}^T\mathbf{A}) = \kappa^2(\mathbf{A})$. Slow convergence rates can be observed, especially in the case of ill-conditioned systems.

\ ALGORITHM 3.3: Conjugate Gradient Normal Error method (CGNE)
$\mathbf{r}^{(0)} \leftarrow \mathbf{b} - \mathbf{A}\mathbf{x}^{(0)}$ \hspace{2cm} \ Find initial residual
$\mathbf{z}^{(0)} \leftarrow \mathbf{A}^T\mathbf{r}^{(0)}$ \hspace{2cm} \ Find $\mathbf{z}^{(0)} = \mathbf{A}^T\mathbf{r}^{(0)}$
For $[p = 0, maxit]$ \hspace{2cm} \ Iteration loop
\ Find current estimate and its residual at $(p+1)$th step
$\quad \alpha^{(p)} \leftarrow (\mathbf{r}^{(p)}, \mathbf{r}^{(p)})/(\mathbf{d}^{(p)}, \mathbf{d}^{(p)})$
$\quad \mathbf{x}^{(p+1)} \leftarrow \mathbf{x}^{(p)} + \alpha^{(p)}\mathbf{d}^{(p)}$
$\quad \mathbf{r}^{(p+1)} \leftarrow \mathbf{r}^{(p)} - \alpha^{(p)}\mathbf{A}\mathbf{d}^{(p)}$
$\quad$ **If** $\left[\|\mathbf{r}^{(p+1)}\| < \varepsilon\right]$ **Then** \hspace{1.5cm} \ Convergence achieved
$\quad\quad$ **Exit** \hspace{3cm} \ Exit the iteration loop
$\quad$ **Else**
\ Not convergenced, compute new direction at pth step
$\quad\quad \beta^{(p)} \leftarrow (\mathbf{r}^{(p+1)}, \mathbf{r}^{(p+1)})/(\mathbf{r}^{(p)}, \mathbf{r}^{(p)})$ \hspace{1cm} \ Find new β
$\quad\quad \mathbf{d}^{(p+1)} \leftarrow \mathbf{A}^T\mathbf{r}^{(p+1)} + \beta^{(p)}\mathbf{d}^{(p)}$ \hspace{1cm} \ Find new direction, d
$\quad$ **End If**
End For \hspace{3cm} \ End the iteration loop-p

3.4.5 CHOICE OF PRECONDITIONERS

In many applications, such as the numerical solution of Poisson's equation, the condition number of the coefficient matrix could become quite high. Hence, it is important to modify the system by using a suitable preconditioner to reduce the condition number. The preconditioner $\mathbf{P}$ should be chosen such that the cost of solving the preconditioned (modified) system is not too expensive in comparison to the original system.

The ideal preconditioner is the matrix itself ($\mathbf{P} = \mathbf{A}$) which would reduce the condition number to 1. However, inverting the matrix, $\mathbf{P}^{-1}$, would be cpu- and memory-intensive. On the other hand, the choice of the simplest preconditioner is a diagonal matrix ($\mathbf{P} = \mathbf{D}$) because it is easily constructed by the reciprocals of the diagonal elements. The *Jacobi* (or *diagonal*) *preconditioner* is a diagonal matrix having the diagonal entries of matrix $\mathbf{A}$; that is,

$$\mathbf{P} = \mathbf{D} = \begin{cases} a_{ii}, & i = j \\ 0, & i \neq j \end{cases}$$

where $a_{ii} \neq 0$ for all i. The Jacobi preconditioner can be very efficient if $\mathbf{A}$ is symmetric (positive definite) and strictly diagonally dominant. Additionally, a diagonal preconditioner may be damped as $\mathbf{P} = \mathbf{D}/\omega$, where ω is a damping parameter $(\omega \neq 0)$. For non-symmetric matrices, the common choice of $\mathbf{D}$ is

$$d_{ii} = \sqrt{\sum_{j=1}^{n} a_{ij}^2} \quad \text{for} \quad i = 1, 2, \dots, n$$

The Gauss-Seidel and the SOR preconditioners are, respectively, defined as $\mathbf{P} = \mathbf{D} + \mathbf{L}$ and $\mathbf{P} = \mathbf{D} + \omega\mathbf{L})$ with $\omega \neq 0$. For a symmetric matrix $\mathbf{A}$, it can be partitioned as $\mathbf{A} = \mathbf{L} + \mathbf{D} + \mathbf{L}^{\mathbf{T}}$, where $\mathbf{D}$ and $\mathbf{L}$ are the diagonal and lower-triangular matrices of $\mathbf{A}$, respectively. The Symmetric Successive Over Relaxation (SSOR) preconditioner is defined as

$$\mathbf{P} = (\mathbf{D} + \omega\mathbf{L}) \, \mathbf{D}^{-1} \left(\mathbf{D} + \omega\mathbf{L}^{T} \right), \quad 0 < \omega < 2$$

where ω is a relaxation parameter. An effective preconditioned system can be expressed as

$$\widetilde{\mathbf{A}}\widetilde{\mathbf{x}} = \widetilde{\mathbf{b}}$$

where $\widetilde{\mathbf{A}} = \mathbf{D}^{-1/2}\mathbf{A}\,\mathbf{D}^{-T/2}$, $\widetilde{\mathbf{x}} = \mathbf{D}^{T/2}\mathbf{x}$, $\widetilde{\mathbf{b}} = \mathbf{D}^{-1/2}\mathbf{b}$, and

$$\mathbf{D}^{1/2} = \begin{bmatrix} \sqrt{a_{11}} & & & \\ & \sqrt{a_{22}} & & \\ & & \ddots & \\ & & & \sqrt{a_{nn}} \end{bmatrix}, \quad \mathbf{D}^{-1/2} = \begin{bmatrix} 1/\sqrt{a_{11}} & & & \\ & 1/\sqrt{a_{22}} & & \\ & & \ddots & \\ & & & 1/\sqrt{a_{nn}} \end{bmatrix}$$

Since $\mathbf{D}$ is a diagonal matrix, its transpose is also equal to itself. The resulting preconditioned system is also symmetric. Note that matrix $\widetilde{\mathbf{A}}$ is constructed once at the beginning, and the true solution is obtained as $\mathbf{x} = \mathbf{D}^{-T/2}\widetilde{\mathbf{x}} = \mathbf{D}^{-1/2}\widetilde{\mathbf{x}}$.

The scope of this topic is restricted by the preceding discussions; however, there are a number of more sophisticated ways of preconditioning in the literature. For more information, the readers are referred to advanced texts on the topic for detailed discussions [29]. An effective preconditioner needs to be cheap to construct and implement (not too hard on the cpu or memory requirements), and the preconditioned system should be easy to solve, i.e., converge rapidly.

3.5 IMPROVING ACCURACY OF ILL-CONDITIONED SYSTEMS

In general, increasing computation precision significantly improves most ill-conditioned linear systems. In the event that this strategy fails, a corrective iterative scheme may be devised for an ill-conditioned system of linear equations to achieve considerable improvements.

Consider an ill-conditioned linear system: $\mathbf{Ax} = \mathbf{b}$. Suppose its first solution $(\mathbf{x}')$ is obtained with a direct or iterative method. The numerical solution is approximate (not exact), at least due to round-off errors, even if a direct method is applied. However, depending on the severity of ill-conditioning, we can expect the final estimate to deviate significantly from the true solution. To determine the effect on the solution of the system, we multiply $\mathbf{A}$ with $\mathbf{x}'$ to get

$$\mathbf{Ax}' = \mathbf{b}' \tag{3.52}$$

where $\mathbf{b}'$ will naturally be different than $\mathbf{b}$. Subtracting Eq. (3.52) from $\mathbf{Ax} = \mathbf{b}$ side-by-side gives

$$\mathbf{A}(\mathbf{x} - \mathbf{x}') = \mathbf{b} - \mathbf{b}' \tag{3.53}$$

The difference (or deviation) between the "estimated" and "correct" solutions is $\Delta x = x - x'$. Also, defining $\Delta b = b - b'$, Eq. (3.53) takes the form

$$A\,\Delta x = \Delta b \tag{3.54}$$

If we can find the exact values of Δb and Δx of Eq. (3.54), the system accuracy will surely improve in one step. Nevertheless, the round-off errors may inevitably prevent rapid convergence. Thus, upon numerically solving Eq. (3.54) with a suitable method, a correction to the approximate solution can be found from

$$x = x' + \Delta x \tag{3.55}$$

If the condition number is not too large, the improved solution can be obtained in a few steps; otherwise, the procedure is repeated a few more times until a convergence criterion, such as $\|\Delta x\|_\infty < \varepsilon$, is satisfied.

Another technique to improve the solution of a linear system is preconditioning, which was covered in Section 3.4.3. In which case, $PAx = Pb$ is solved instead of $Ax = b$, provided that the precondition matrix P is chosen such that the condition number of PA is smaller than that of A; that is, $\kappa(PA) < \kappa(A)$.

EXAMPLE 3.7: Improving solution of ill-Conditioned linear systems

Consider the following system of linear equations:

$$6x_1 + \quad x_2 = 10$$
$$2x_1 + 0.333x_2 = 3.332$$

(a) Calculate the spectral condition number to determine the condition of the given system; (b) Solve the system for $b_2 = 3.33$ and 3.3333 to determine the effect of rounding on the solution; (c) apply the corrective iterative technique to improve the solution accuracy for the $b_2 = 3.33$ case.

SOLUTION:

(a) Since the coefficient matrix is 2×2, the true spectral condition number can be calculated from $\kappa(A) = \|A\|_2 \|A^{-1}\|_2$ without much difficulty. The inverse matrix and pertinent spectral norms are found as

$$\|A\|_F = \|A\|_2 = 6.41178, \quad A^{-1} = \begin{bmatrix} -166.5 & 500 \\ 1000 & -3000 \end{bmatrix}, \quad \|A^{-1}\|_2 = 3205.89$$

which leads to $\kappa(A) = 20555 \gg 1$. This magnitude of the condition number indicates that the system is ill-conditioned; in other words, we can expect the solution to be highly susceptible to rounding.

Next, we find $|\det(A)| = 0.0020$ and estimate the condition number from Eq. (2.33):

$$\kappa(A) \approx \frac{2\,\|A\|_F^n}{\sqrt{n^n}\,|\det(A)|} = \frac{2\,(6.41178)^2}{\sqrt{2^2}\,(0.002)} = 20555$$

which also gives the true value of the condition number spot on.

(b) The solution of the original system (without any rounding) is found as $(x_1, x_2) = (1, 4)$. For $b_2 = 3.33$ and $b_2 = 3.3333$, the numerical solutions are obtained as $(x_1, x_2) = (0, 10)$ and $(1.65, 0.1)$, respectively. We would normally expect a small change in the final solution as a result of a small change in one of the coefficients of either **A** or **b**. However, this example reveals that the solution of a linear system with a single perturbed coefficient may be substantially different from the true solution. It is clear that the given linear system is very sensitive to fluctuations in coefficient values, and the system is ill-conditioned.

(c) To correct the ill-effects of rounding in the case $b_2 = 3.33$ leading to $\mathbf{x}' = (x_1, x_2) = (0, 10)$, we find the new rhs vector:

$$\mathbf{b}' = \mathbf{A}\mathbf{x}' = \begin{bmatrix} 6 & 1 \\ 2 & 0.333 \end{bmatrix} \begin{bmatrix} 0 \\ 10 \end{bmatrix} = \begin{bmatrix} 10 \\ 3.33 \end{bmatrix}$$

The difference (or deviation) in the rhs is obtained as

$$\Delta\mathbf{b} = \mathbf{b} - \mathbf{b}' = \begin{bmatrix} 10 \\ 3.332 \end{bmatrix} - \begin{bmatrix} 10 \\ 3.33 \end{bmatrix} = \begin{bmatrix} 0 \\ 0.002 \end{bmatrix}$$

Next, the deviations in the solution, $\Delta\mathbf{x}$, are found by solving the following system of equations:

$$\begin{bmatrix} 6 & 1 \\ 2 & 0.333 \end{bmatrix} \begin{bmatrix} \Delta x_1 \\ \Delta x_2 \end{bmatrix} = \begin{bmatrix} 0 \\ 0.002 \end{bmatrix}$$

The exact solution of this system is obtained likewise by the Cramer's rule with no round-off errors: $(\Delta x_1, \Delta x_2) = (1, -6)$. The improved solution is then obtained as

$$\mathbf{x} = \mathbf{x}' + \Delta\mathbf{x} = \begin{bmatrix} 0 \\ 10 \end{bmatrix} + \begin{bmatrix} 1 \\ -6 \end{bmatrix} = \begin{bmatrix} 1 \\ 4 \end{bmatrix}$$

Discussion: With a single corrective iteration, we achieved a significantly improved solution. However, it should be noted that no rounding errors were made when solving $\mathbf{A}\,\Delta\mathbf{x} = \Delta\mathbf{b}$. In the case of rounding on one or more of the elements of **A** and **b** (or in extremely ill-conditioned cases), this procedure may require several corrective steps. This example illustrates the ill effects of small errors or round-offs that easily arise from the way digital computers store and process real numbers. If not careful, a digital error or a user-introduced round-off error has the potential to lead to a completely different solution.

3.6 CLOSURE

Direct solution methods impose serious restrictions on memory requirements and cpu-time. However, as an alternative, iterative methods only require storing non-zero matrix elements and performing computations with only non-zero elements, which not only greatly resolves the memory requirement issue to a large extent but also significantly reduces the cpu-time.

The basic stationary iterative methods are discussed, and the pseudocodes are presented for each method. The performance of SOR depends on how close the relaxation parameter is to the optimum value. In this regard, an algorithm to estimate the optimum relaxation parameter for the SOR method is given. The SOR method with a smart estimate for the ω_{opt} parameter is the method of choice. The conditions for convergence and convergence rates are discussed.

The classic CGM and its convergence properties for positive symmetric systems are covered in-depth due to being a fundamental Krylov space method. The fact that acceleration parameters are calculated as part of the CGM and CGM-like algorithms eliminates the need to estimate the optimum relaxation parameter as in the SOR. The method, in the absence of round-offs, results in the true solution of a system of $n \times n$ linear equations in at most n iterations. The CGM performs very well for well-conditioned linear systems ($\kappa(\mathbf{A}) \approx 1$). Nonetheless, as a result of using finite precision arithmetic, the method may depict slow convergence, depending on the accumulation of the round-offs that plague orthogonality relations.

The advantages of the iterative methods can be summarized as follows:

- Iterative methods are suitable for solving large linear systems requiring extremely large memory allocation and cpu-time;

- Iterative methods are very efficient when applied to diagonally dominant sparse matrices, so they are relatively less sensitive to the growth of round-off errors;

- An iterative procedure can be terminated once a reasonably approximate solution is reached;

- Some algorithms only prefer approximate solutions with a low degree of accuracy. For instance, it is often sufficient to solve a linear system with inner iterations within a nonlinear outer iteration with a low degree of accuracy;

- In some cases, a "rough" or "sufficiently close" estimate for the solution may be available, which, when used, can dramatically accelerate the iteration process. For example, in transient problems, almost all algorithms require the solution of a linear system at every time step. In general, the solutions at the current time level $(t + dt)$ and the prior time level (t) are very close to each other, so a solution from the prior time level could be used as the initial guess for the current time level. This technique of initializing an algorithm with a solution from a prior iteration rather than starting from scratch is called a "warm start" and serves to obtain the current solution much more quickly.

The disadvantages of the iterative methods are summarized as follows:

- A major handicap of iterative methods is that not all linear systems can be solved with iterative methods;

- The error in the final solution depends on the method, the linear system, and the number of iterations performed;

- The effectiveness and speed of an iterative method depend on the numerical algorithm and its convergence rate;

- An iterative process, if it converges, gives an approximate solution, not the true solution. In general, some iterative algorithms may diverge, while others might converge so slowly that they are practically useless.

A relatively small change in the elements of the coefficient matrix of an ill-conditioned system results in unacceptably large deviations in the solution. Hence, the computations need to be carried out using high-precision computing. Central to the error analysis of a linear system, the condition number is a tool for measuring the relative magnitude of the ill-conditioning, and it provides a quantitative measure for the number of significant decimal digits that are likely to be lost when solving a system of equations. Additionally, a corrective iterative technique is presented to improve the accuracy of the solution for ill-conditioned systems.

3.7 EXERCISES

Section 3.1 Overview

E3.1   An iterative algorithm leads to the following prior and current estimates at the pth iteration level: Compute the $\|\mathbf{d}^{(p)}\|_1$, $\|\mathbf{d}^{(p)}\|_2$, and $\|\mathbf{d}^{(p)}\|_\infty$ norms of the displacement vector, which is defined as $\mathbf{d}^{(p)} = \mathbf{x}^{(p+1)} - \mathbf{x}^{(p)}$.

(a) $\mathbf{x}^{(p)} = \{0.0015,\ 0.0072,\ 0.0044,\ 0.0092\}$,   (b) $\mathbf{x}^{(p)} = \{0.351,\ 0.217,\ 0.347,\ 0.697\}$,

$\mathbf{x}^{(p+1)} = \{0.0019,\ 0.0063,\ 0.0048,\ 0.0081\}$ $\mathbf{x}^{(p+1)} = \{0.359,\ 0.233,\ 0.336,\ 0.722\}$

(c) $\mathbf{x}^{(p)} = \{35.12,\ 43.77,\ 62.73,\ 55.79\}$,

$\mathbf{x}^{(p+1)} = \{34.62,\ 44.13,\ 63.35,\ 53.98\}$

E3.2   Use the given two iteration steps given in **E3.1** to compute $\|\mathbf{d}^{(p)}/\mathbf{x}^{(p+1)}\|_1$, $\|\mathbf{d}^{(p)}/\mathbf{x}^{(p+1)}\|_2$, and $\|\mathbf{d}^{(p)}/\mathbf{x}^{(p+1)}\|_\infty$ norms.

E3.3   Consider the following iteration sequences: How many iterations will be required to get an estimate that satisfies the $\|\mathbf{d}^{(p)}\|_\infty < 0.01$ criterion?

(a) $\mathbf{x}^{(p)} = \left\{ \dfrac{1}{40p + 19e^{-p}},\ \dfrac{7p^2 - 1}{28p^2 + 99},\ \dfrac{\ln p}{155p},\ 3e^{-p} \right\}$,   (b) $\mathbf{x}^{(p)} = \left\{ \dfrac{3}{p^2 + 1},\ \dfrac{5 - p^2}{p^2 + 1},\ 3 - \dfrac{p + 1}{p^3} \right\}$,

(c) $\mathbf{x}^{(p)} = \left\{ \dfrac{p \sin p}{p + 9},\ \dfrac{\cos p}{p^2 + 9},\ \sqrt{p + 1} - \sqrt{p},\ \dfrac{1}{\sqrt{4p^2 + 1}} \right\}$

E3.4   Repeat **E3.3** if the convergence criterion is replaced by $\|\mathbf{d}^{(p)}/\mathbf{x}^{(p+1)}\|_\infty < 0.01$.

E3.5   Put the following system of equations into the fixed-point iteration form.

(a) $\left\{ \begin{array}{l} 2x - y - z = 8 \\ x + 4y - 2z = 5 \\ 3x - 2y + 5z = 2 \end{array} \right\}$,   (b) $\left\{ \begin{array}{l} 4x - 3y + 2z = 1 \\ -x + 4y - 3z = 5 \\ 2x - 2y - 7z = 4 \end{array} \right\}$,   (c) $\left\{ \begin{array}{l} 5x - 3y - 2z = 9 \\ 2x + 5y - 3z = 12 \\ 3x - 3y + 10z = 30 \end{array} \right\}$

Section 3.2 Stationary Iterative Methods

E3.6   For the given linear systems: (i) perform the first two Jacobi iterations by hand using $\mathbf{x}^{(0)} = \mathbf{0}$, (ii) use a spreadsheet program to find an approximate solution estimated with a maximum absolute error less than $\varepsilon = 10^{-2}$.

(a) $\left\{ \begin{array}{l} 2x + y + z = -0.08 \\ x + 3y + 2z = 0.11 \\ x + y + 3z = -0.23 \end{array} \right\}$,   (b) $\left\{ \begin{array}{l} 5x + y - 2z = 1.1 \\ x + 3y - z = 1.8 \\ x - y + 3z = 3 \end{array} \right\}$

E3.7   Find the approximate solution of the systems of equations in **E3.6** using the Jacobi method. How many iterations were required for convergence? Use $\mathbf{x}^{(0)} = \mathbf{0}$ and $\|\mathbf{d}^{(p)}\|_2 < 0.5 \times 10^{-5}$ as the stopping criterion.

E3.8   Consider the given linear system of equations. (a) Find the true solution of the system using a direct method; (b) Perform the first three Jacobi iterations by hand with $\mathbf{x}^{(0)} = \mathbf{1}$ as the initial guess. Does it look like it is converging? (c) Reorder the equations as $\{eq2, eq3, eq1\}$ and use a spreadsheet program to find a set of estimates with $\|\mathbf{d}^{(p)}\|_\infty < 10^{-2}$ as the convergence criterion.

$$x - 3y - 4z = 4, \qquad 4x + 2y - z = 3, \qquad x + 4y - 2z = -8.$$

E3.9   For the given linear system, (a) find the solution of the system using a direct method; (b) carry out the first four iterations of the Jacobi method by hand using $\mathbf{x}^{(0)} = \mathbf{0}$; (c) interchange the first and third equations ($r_1 \leftrightarrow r_3$) and repeat Part (b). Comment on the iteration progress.

$$2x - y + 6z = 1.6, \qquad x + 6y + z = -0.8, \qquad 5x - 2y + 2z = 1.95$$

E3.10 For the given linear system, (a) carry out the first four iterations of the given linear system with the Jacobi method using $\mathbf{x}^{(0)} = \mathbf{0}$ and a spreadsheet program; (b) how many iterations are required to obtain the solution with $\|\mathbf{d}^{(p)}\|_\infty < 5 \times 10^{-6}$ as the convergence criterion?

$$\begin{bmatrix} 4 & 0 & 0 & 1 & 2 & 0 \\ 1 & 5 & 1 & 0 & 0 & 3 \\ -2 & 0 & 4 & 0 & 0 & -3 \\ 1 & 0 & 2 & 8 & 0 & -1 \\ 0 & 1 & 0 & 3 & 10 & 4 \\ 2 & 0 & 3 & 0 & -1 & 6 \end{bmatrix} \begin{bmatrix} x_1 \\ x_2 \\ x_3 \\ x_4 \\ x_5 \\ x_6 \end{bmatrix} = \begin{bmatrix} 12.5 \\ 21 \\ -4.5 \\ 21.5 \\ 53 \\ 26 \end{bmatrix}$$

E3.11 For the given linear system: (a) find an approximate solution of the system using the Jacobi method with $\mathbf{x}^{(0)} = \mathbf{0}$ and $\|\mathbf{d}^{(p)}\|_\infty < 5 \times 10^{-6}$ as the convergence criterion; (b) compute $\|\mathbf{d}^{(p)}\|_\infty / \|\mathbf{d}^{(p-1)}\|_\infty$ after the first iteration and comment on the convergence of this ratio; (c) how many iterations were required to obtain the converged estimate?

$$\begin{bmatrix} 6 & 2 & & 1 & & \\ -2 & 6 & 2 & & 1 & \\ & -2 & 6 & 2 & & 1 \\ -1 & & -2 & 6 & 2 & \\ & -1 & & -2 & 6 & 2 \\ & & -1 & & -2 & 6 \end{bmatrix} \begin{bmatrix} x_1 \\ x_2 \\ x_3 \\ x_4 \\ x_5 \\ x_6 \end{bmatrix} = \begin{bmatrix} 9 \\ 7 \\ 7 \\ 5 \\ 5 \\ 3 \end{bmatrix}$$

E3.12 Repeat **E 3.6** using the Gauss-Seidel method.
E3.13 Repeat **E 3.7** using the Gauss-Seidel method.
E3.14 Repeat **E 3.8** using the Gauss-Seidel method.

E3.15 Repeat **E 3.9** using the Gauss-Seidel method.

E3.16 Repeat **E 3.10** using the Gauss-Seidel method.

E3.17 Repeat **E 3.11** using the Gauss-Seidel method.

E3.18 For the given linear system: (a) find the true solution using a direct method; (b) carry out (by hand) the *first three iterations* of the Jacobi method using $\mathbf{x}^{(0)} = \mathbf{0}$; (c) carry out (by hand) the first three iterations of the Gauss-Seidel method using the same initial guess and compare the $\|\mathbf{d}^{(p)}\|_2$ norms for both methods.

$$\begin{bmatrix} 4 & -2 & 0 & -1 \\ -2 & 4 & -2 & 0 \\ 0 & -2 & 4 & -2 \\ -1 & 0 & -2 & 4 \end{bmatrix} \begin{bmatrix} x_1 \\ x_2 \\ x_3 \\ x_4 \end{bmatrix} = \begin{bmatrix} -1.44 \\ 0.58 \\ 0.06 \\ 0.90 \end{bmatrix}$$

E3.19 For the systems of linear equations given in **E 3.6** (a) write out the iteration equations for the SOR method; (b) obtain approximate solutions with $\omega = 1, 1.1, 1.2, 1.3, 1.4$, and 1.5 to determine the best choice for the acceleration parameter. Start with $\mathbf{x}^{(0)} = \mathbf{0}$ and use $\|\mathbf{d}^{(p)}\|_\infty < 5 \times 10^{-4}$ as the convergence criterion.

E3.20 For the systems of linear equations given in Part (c) of **E 3.8** (a) write out the iteration equations for the SOR method; (b) obtain approximate solutions with $\omega = 1, 1.1, 1.2, 1.3$, and 1.4 to determine the best choice for the acceleration parameter. Start with $\mathbf{x}^{(0)} = \mathbf{0}$ and use $\|\mathbf{d}^{(p)}\|_\infty < 5 \times 10^{-4}$ as the convergence criterion.

E3.21 For the system of linear equations given **E 3.9** (a) write out the iteration equations for the SOR method; (b) obtain approximate solutions with $\omega = 1, 1.1, 1.2, 1.3$, and 1.4 to determine the best choice for the acceleration parameter. Start with $\mathbf{x}^{(0)} = \mathbf{0}$ and use $\|\mathbf{d}^{(p)}\|_\infty < 5 \times 10^{-4}$ as the convergence criterion.

E3.22 For the systems of linear equations given in **E 3.10** (a) write out the iteration equations for the SOR method; (b) obtain approximate solutions with $\omega = 1, 1.1, 1.2$, and 1.3 to determine the best choice for the acceleration parameter. Start with $\mathbf{x}^{(0)} = \mathbf{0}$ and use $\|\mathbf{d}^{(p)}\|_\infty < 5 \times 10^{-4}$ as the convergence criterion.

E3.23 For the systems of linear equations given in **E 3.11** (a) write out the iteration equations for the SOR method; (b) obtain approximate solutions with $\omega = 0.7$, 0.8, 0.9, 1, 1.1, 1.2, and 1.3 to determine the best choice for the acceleration parameter. Start with $\mathbf{x}^{(0)} = \mathbf{0}$ and use $\|\mathbf{d}^{(p)}\|_\infty < 5 \times 10^{-4}$ as the convergence criterion.

Section 3.3 Convergence of Stationary Iterative Methods

E3.24 Determine if the following linear systems meet the "diagonal dominance" condition for the Jacobi and Gauss-Seidel methods.

(a) $\begin{bmatrix} 4 & 2 & 3 \\ -3 & 4 & 4 \\ 2 & 4 & 5 \end{bmatrix} \begin{bmatrix} x_1 \\ x_2 \\ x_3 \end{bmatrix} = \begin{bmatrix} 9 \\ 5 \\ 11 \end{bmatrix}$, (b) $\begin{bmatrix} 3 & 2 & 1 \\ -3 & 2 & 1 \\ 2 & 7 & 3 \end{bmatrix} \begin{bmatrix} x_1 \\ x_2 \\ x_3 \end{bmatrix} = \begin{bmatrix} 6 \\ 0 \\ 12 \end{bmatrix}$,

(c) $\begin{bmatrix} 3 & 2 & 1 & -1 \\ 0 & -3 & 2 & 1 \\ 1 & 2 & 2 & -1 \\ 1 & -1 & -1 & 3 \end{bmatrix} \begin{bmatrix} x_1 \\ x_2 \\ x_3 \\ x_4 \end{bmatrix} = \begin{bmatrix} 5 \\ 0 \\ 4 \\ 2 \end{bmatrix}$, (d) $\begin{bmatrix} 1 & 1 & -3 & 1 \\ 1 & 3 & 1 & -2 \\ 2 & -1 & 4 & 2 \\ 2 & -2 & 3 & 5 \end{bmatrix} \begin{bmatrix} x_1 \\ x_2 \\ x_3 \\ x_4 \end{bmatrix} = \begin{bmatrix} 0 \\ 3 \\ 7 \\ 8 \end{bmatrix}$

E3.25 Find the Jacobi iteration matrices of the linear systems in **E 3.24** and verify whether or not the Jacobi method will converge with an arbitrary initial guess.

E3.26 For the *converging* linear systems in **E 3.25**, find approximate solutions using the Jacobi method. Start with $\mathbf{x}^{(0)} = \mathbf{0}$ and use $\|\mathbf{d}^{(p)}\|_\infty < 5 \times 10^{-3}$ as the convergence criterion.

E3.27 Find the Gauss-Seidel iteration matrices for the linear systems in **E 3.24** and verify whether the Gauss-Seidel method will converge for arbitrary initial guesses.

E3.28 For the converging linear systems in **E 3.27**, find approximate solutions using the Gauss-Seidel method. Start with $\mathbf{x}^{(0)} = \mathbf{0}$ and use $\|\mathbf{d}^{(p)}\|_\infty < 5 \times 10^{-3}$.

E3.29 For given matrix $\mathbf{A}$, find the spectral radius of $\mathbf{A}$ and $\mathbf{A}^{-1}$:

$$\mathbf{A} = \begin{bmatrix} 0 & 1/2 & 0 & 0 \\ 1/2 & 0 & 1/2 & 0 \\ 0 & 1/2 & 0 & 1/2 \\ 0 & 0 & 1/2 & 0 \end{bmatrix}$$

E3.30 The following linear systems are *not* diagonally dominant. For given linear systems, find the spectral radii of the Jacobi and Gauss-Seidel iteration matrices and determine whether or not the methods will converge with an arbitrary initial guess.

(a) $\begin{bmatrix} 1 & 2 & -2 \\ 1 & 1 & 1 \\ -2 & 2 & 1 \end{bmatrix} \begin{bmatrix} x \\ y \\ z \end{bmatrix} = \begin{bmatrix} 3 \\ 3 \\ 3 \end{bmatrix}$, (b) $\begin{bmatrix} 5 & 3 & 4 \\ 3 & 6 & 4 \\ 4 & 3 & 5 \end{bmatrix} \begin{bmatrix} x \\ y \\ z \end{bmatrix} = \begin{bmatrix} 12 \\ 13 \\ 12 \end{bmatrix}$, (c) $\begin{bmatrix} 2 & 2 & 4 \\ 1 & 3 & 3 \\ -3 & 2 & 5 \end{bmatrix} \begin{bmatrix} x \\ y \\ z \end{bmatrix} = \begin{bmatrix} 2 \\ -1 \\ 3 \end{bmatrix}$

E3.31 For the given matrix, find the values of β for which the Jacobi method converges.

$$\begin{bmatrix} 4 & \beta & \beta & 0 \\ \beta & 4 & \beta & \beta \\ \beta & \beta & 4 & \beta \\ 0 & \beta & \beta & 4 \end{bmatrix}$$

E3.32 For a given linear system, (a) find the solution of the system using the Gauss-Seidel and SOR (for $\omega = 1.3$) methods with $\|\mathbf{d}^{(p)}\|_\infty < 5 \times 10^{-3}$ as the convergence criterion, and (b) estimate the optimum relaxation parameter for the SOR method. Use $\mathbf{x}^{(0)} = \mathbf{1}$ for the initial guess.

$$\begin{bmatrix} 4 & 1 & 6 \\ 1 & 6 & 2 \\ -2 & 2 & -4 \end{bmatrix} \begin{bmatrix} x \\ y \\ z \end{bmatrix} = \begin{bmatrix} 22 \\ -16 \\ -24 \end{bmatrix}$$

E3.33 Consider the linear system in **E 3.18**; (a) estimate the theoretical value of ω_{opt}; (b) having rounded ω_{opt} to two-decimal places, carry out the first three SOR iterations (by hand) starting with $\mathbf{x}^{(0)} = \mathbf{0}$.

E3.34 Find the solution of the system of linear equations resulting from **E2.45** using the Jacobi and Gauss-Seidel methods with $\|\mathbf{d}^{(p)}\|_\infty < 5 \times 10^{-4}$ as the convergence criterion, and estimate the optimum relaxation parameter for the SOR method. Start with $\mathbf{x}^{(0)} = \mathbf{0}$.

E3.35 (a) Find the solution of the system of linear equations resulting from **E2.76** using the Jacobi and Gauss-Seidel methods with $\|\mathbf{d}^{(p)}\|_\infty < 10^{-5}$ as the convergence criterion and estimate the rate of convergence and the minimum number of iterations to reduce the error by a factor of 10. (b) Estimate the optimum relaxation parameter and the rate of convergence for the SOR method in the range $1 \leqslant \omega \leqslant 1.5$. Start with $\mathbf{x}^{(0)} = \mathbf{0}$.

E3.36 (a) Find the solution of the given system of linear equations using the Jacobi and Gauss-Seidel methods with $\|\mathbf{d}^{(p)}\|_\infty < 10^{-5}$ as the convergence criterion; (b) estimate the rate of convergence and the minimum number of iterations to reduce the error by a factor of 10; and (c) to estimate the optimum relaxation parameter, obtain the SOR solutions for ω values with 0.1 increments in the range $0.9 \leqslant \omega \leqslant 1.5$ and estimate the convergence rate. Start iterations with $\mathbf{x}^{(0)} = 0$.

$$\begin{bmatrix} 6 & -1 & -3 & 0 & 0 \\ -1 & 4 & -1 & -2 & 0 \\ -3 & -1 & 6 & 2 & 0 \\ 0 & -2 & -2 & 5 & -1 \\ 0 & 0 & 0 & -1 & 1 \end{bmatrix} \begin{bmatrix} x_1 \\ x_2 \\ x_3 \\ x_4 \\ x_5 \end{bmatrix} = \frac{1}{15} \begin{bmatrix} 1 \\ 1 \\ 1 \\ 0 \\ 1 \end{bmatrix}$$

E3.37 The linear system given below is *not* diagonally dominant. (a) Investigate whether the Jacobi and Gauss-Seidel methods will converge for this linear system; (b) for the converging method(s), determine the number of iterations required to achieve convergence with $\|\mathbf{d}^{(p)}\|_\infty < 10^{-4}$; (c) Verify whether the $\rho^2(\mathbf{M}_J) = \rho(\mathbf{M}_{GS})$ relation holds; (d) Is the SOR method applicable for solving the system? If yes, estimate the optimum relaxation parameter and the rate of convergence by varying ω value from 1 to 1.5 in increments of 0.1. Use $\mathbf{x}^{(0)} = 0$ as the initial guess.

$$\begin{bmatrix} 20 & 8 & 0 & 3 & 5 \\ 8 & 16 & 2 & 5 & 6 \\ 2 & 3 & 18 & 7 & 8 \\ 4 & 5 & 6 & 20 & 2 \\ 6 & 5 & 4 & 2 & 19 \end{bmatrix} \begin{bmatrix} x_1 \\ x_2 \\ x_3 \\ x_4 \\ x_5 \end{bmatrix} = \begin{bmatrix} 36 \\ 37 \\ 38 \\ 37 \\ 36 \end{bmatrix}$$

E3.38 Ten objects of equal weight (W) are connected in series with ten springs (k) and suspended as shown in **Fig. E3.38**. The steady-state equilibrium equations yield

$$2x_1 - x_2 = W/k,$$

$$-x_{i-1} + 2x_i - x_{i+1} = W/k, \quad i = 2, 3, \ldots, 9$$

$$-x_9 + x_{10} = W/k$$

where x_i denotes the displacement of the object i from its unstretched position. (a) For $W/k = 1$ cm, apply the Jacobi and Gauss-Seidel methods to solve the tridiagonal system of equations for the displacements, (b) compute the convergence rate for both methods, (c) estimate the theoretical optimum relaxation parameter, and (d) apply the SOR method to solve the linear system for values starting at $\omega = 1$ in $\Delta\omega = 0.05$ increments to $\omega = 1.95$ and estimate the optimum relaxation parameter. Start with $\mathbf{x}^{(0)} = \mathbf{0}$ and use $\|\mathbf{d}^{(p)}\|_\infty < 10^{-5}$ for convergence tolerance.

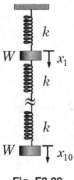

Fig. E3.38

Section 3.4 Krylov Space Methods

E3.39 Find approximate solutions to the given linear systems using the conjugate gradient method. Use $\mathbf{x}^{(0)} = \mathbf{0}$ as the initial guess and $\|\mathbf{d}^{(p)}\|_\infty < 5 \times 10^{-4}$ as the convergence criterion.

(a) $\begin{aligned} 4x_1 + x_2 + 2x_3 &= 17 \\ x_1 + 3x_2 + x_3 &= -2 \\ 2x_1 + x_2 + 4x_3 &= 19 \end{aligned}$ (b) $\begin{aligned} 9x_1 + x_2 + x_3 &= 21 \\ x_1 + 8x_2 + 2x_3 &= -7 \\ x_1 + 2x_2 + 12x_3 &= 63 \end{aligned}$ (c) $\begin{aligned} 9x_1 + 4x_2 + 3x_3 &= 32.75 \\ 4x_1 + 5x_2 + x_3 &= 22.25 \\ 3x_1 + x_2 + 4x_3 &= 20.5 \end{aligned}$

E3.40 Find approximate solutions to the given linear systems using the conjugate gradient method. Use $\mathbf{x}^{(0)} = \mathbf{0}$ as the initial guess and $\|\mathbf{d}^{(p)}\|_\infty < 5 \times 10^{-6}$ as the convergence criterion.

(a)
$\begin{aligned} 6x_1 + 2x_2 + x_3 + 3x_4 &= 15.60 \\ 2x_1 + 5x_2 + 2x_3 + 4x_4 &= 14.85 \\ x_1 + 2x_2 + 8x_3 + 5x_4 &= 19 \\ 3x_1 + 4x_2 + 5x_3 + x_4 &= 13.15 \end{aligned}$

(b)
$\begin{aligned} 4x_1 + 3x_2 + 4x_3 + 3x_4 &= 10.95 \\ 3x_1 + 9x_2 + 5x_3 + 2x_4 &= 13.90 \\ 4x_1 + 5x_2 + 11x_3 + 4x_4 &= 22.25 \\ 3x_1 + 2x_2 + 4x_3 + x_4 &= 8.75 \end{aligned}$

(c)
$\begin{aligned} 10x_1 + 2x_2 + x_3 + 6x_4 &= 15.15 \\ 2x_1 + 8x_2 + 4x_3 + 5x_4 &= 27.35 \\ x_1 + 4x_2 + 6x_3 + 4x_4 &= 19.25 \\ 6x_1 + 5x_2 + 4x_3 + 3x_4 &= 18.9 \end{aligned}$

E3.41 For the given linear systems: (i) Obtain the preconditioned system of equations, $\tilde{\mathbf{A}}\tilde{\mathbf{x}} = \tilde{\mathbf{b}}$, where $\tilde{\mathbf{A}} = \mathbf{D}^{-1/2}\mathbf{A}\mathbf{D}^{-T/2}$, $\tilde{\mathbf{x}} = \mathbf{D}^{T/2}\mathbf{x}$, and $\tilde{\mathbf{b}} = \mathbf{D}^{-1/2}\mathbf{b}$; (ii) Compute the infinity-based condition numbers of $\mathbf{A}$ and $\tilde{\mathbf{A}}$; (iii) solve the preconditioned system using the CGM. Start with $\mathbf{x}^{(0)} = \mathbf{0}$ and use $\|\mathbf{d}^{(p)}\|_\infty < 5 \times 10^{-6}$.

(a)
$\begin{aligned} x_1 + 3x_2 + 2x_3 &= 7.2 \\ 3x_1 + 4x_2 + 10x_3 &= 17 \\ 2x_1 + 10x_2 + 196x_3 &= 37.6 \end{aligned}$

(b)
$\begin{aligned} 0.25x_1 + x_2 + 3x_3 &= 2 \\ x_1 + 25x_2 + 16x_3 &= 46 \\ 3x_1 + 16x_2 + 81x_3 &= -13 \end{aligned}$

(c)
$\begin{aligned} 4x_1 + x_2 + 6x_3 &= 87 \\ x_1 + 0.04x_2 - 0.2x_3 &= 8 \\ 6x_1 - 0.2x_3 + 400x_3 &= 2043 \end{aligned}$

E3.42 Assume the linear systems in **E3.41** are conditioned by employing the Jacobi preconditioner $(\mathbf{P} = \mathbf{D})$. (i) What would be the infinity-based condition numbers of the preconditioned systems? (ii) Employ the preconditioned conjugate gradient method (PCGM) algorithm presented in Section 3.4 to estimate the solution of the system.

E3.43 Consider solving the following linear system using a preconditioner: (a) compute the infinity-based condition number to determine whether the system is ill-conditioned or not; (b) apply the Jacobi preconditioner $(\mathbf{P} = \mathbf{D})$ to obtain an approximate solution; (c) apply the SSOR preconditioner with $\omega = 1$ $(\mathbf{P} = (\mathbf{D} + \mathbf{L})\mathbf{D}^{-1}(\mathbf{D} + \mathbf{L}^T))$; (d) compute the infinity-based condition numbers of the preconditioned matrices in Parts (b) and (c); (e) vary ω between $0.4 \leqslant \omega \leqslant 1.9$ to determine ω_{opt}.

$$\begin{bmatrix} 1 & 1/3 & 1/5 & 1/7 \\ 1/3 & 1/5 & 1/7 & 1/9 \\ 1/5 & 1/7 & 1/9 & 1/11 \\ 1/7 & 1/9 & 1/11 & 1/13 \end{bmatrix} \begin{bmatrix} x_1 \\ x_2 \\ x_3 \\ x_4 \end{bmatrix} = \begin{bmatrix} 2/5 \\ -2/45 \\ -302/3465 \\ -278/3003 \end{bmatrix}$$

E3.44 Estimate the solution of the following linear systems using the CGNR (Conjugate Gradient Normal Residual) method. Use $\|\mathbf{d}^{(p)}\|_\infty < 5 \times 10^{-6}$ as the convergence criterion and $\mathbf{x}^{(0)} = \mathbf{0}$ as the initial guess vector.

(a) $\begin{aligned} 2x_1 + x_2 - 4x_3 &= 6 \\ x_1 + 3x_2 + x_3 &= 7 \\ 3x_1 + x_2 + 2x_3 &= -9 \end{aligned}$ (b) $\begin{aligned} 3x_1 + 4x_2 + 2x_3 &= 10 \\ 4x_1 + 3x_2 + 3x_3 &= 7 \\ 5x_1 + 4x_2 + x_3 &= 4 \end{aligned}$ (c) $\begin{aligned} 5x_1 - 2x_2 + 4x_3 &= 13 \\ 2x_1 - 3x_2 - 2x_3 &= 24 \\ 4x_1 + 3x_2 - 2x_3 &= 22 \end{aligned}$

E3.45 Estimate the solution of the following linear systems using the CGNR (Conjugate Gradient Normal Residual) method. Use $\|\mathbf{d}^{(p)}\|_\infty < 5 \times 10^{-6}$ as the convergence criterion and $\mathbf{x}^{(0)} = \mathbf{0}$ as the initial guess vector.

$$\text{(a)} \quad \begin{aligned} x_1 + 3x_2 + 2x_3 + x_4 &= 3 \\ 3x_1 + 2x_2 - 2x_3 - x_4 &= -1 \\ 2x_1 + x_2 - x_3 + 4x_4 &= 19 \\ -x_1 + 3x_2 - 3x_3 + x_4 &= -8 \end{aligned} \qquad \text{(b)} \quad \begin{aligned} 4x_1 + x_2 - x_3 + 3x_4 &= 12 \\ 2x_1 - 4x_2 - 3x_3 + x_4 &= 5 \\ x_1 - 3x_2 + x_3 + 3x_4 &= -12 \\ 2x_1 + 3x_2 + 2x_3 - x_4 &= 11 \end{aligned}$$

E3.46 Repeat **E3.44** with CGNE (Conjugate Gradient Normal Error) method.

E3.47 Repeat **E3.45** with CGNE (Conjugate Gradient Normal Error) method.

Section 3.5: Improving Accuracy of Ill-Conditioned Systems

E3.48 Find the approximate solution of the given linear system using the Gauss elimination method (a) *without* rounding the numbers; (b) with *rounding* the numbers and using three decimal places in the arithmetic. Compare your estimates with those obtained in Part (a). Is the system ill-conditioned? (Estimate the spectral condition number); (c) Apply the corrective algorithm to find an improved solution for the system in Part (b). Use high-precision arithmetic operations.

$$\frac{x_1}{3} + \frac{4x_2}{9} = \frac{7}{9}, \qquad \frac{x_1}{7} + \frac{3x_2}{17} = \frac{38}{119}$$

E3.49 For the given linear system below, (a) determine whether the coefficient matrix (**A**) is ill-conditioned or not by estimating $\kappa(\mathbf{A})$, (b) estimate the solution of the system using Gauss elimination by rounding the numbers to three decimal places; (c) repeat Part (b) using four decimal place arithmetic; (d) explain your findings in comparison to the true solution of $\mathbf{x} = \mathbf{1}$, (e) apply the one-step corrective algorithm to improve the solution in Part (b). Use high-precision arithmetic operations.

$$x_1 + \frac{x_2}{3} + \frac{x_3}{5} = \frac{23}{15}, \qquad \frac{x_1}{3} + \frac{x_2}{5} + \frac{x_3}{7} = \frac{71}{105}, \qquad \frac{x_1}{5} + \frac{x_2}{7} + \frac{x_3}{9} = \frac{143}{315}$$

3.8 COMPUTER ASSIGNMENTS

CA3.1 The following matrices are *not* diagonally dominant. For linear systems with given coefficient matrices, (a) find the spectral radius of the Jacobi and Gauss-Seidel iteration matrices (i.e., $\rho(\mathbf{M}_J)$ and $\rho(\mathbf{M}_{GS})$) and determine whether each method will converge with an arbitrary initial guess; (b) calculate $\rho(\mathbf{M}_\omega)$ for various acceleration parameters between 0.1 and 1.9 to determine for which values of ω the SOR or SUR scheme will converge. Is there an optimum value for ω?

$$\text{(a)} \begin{bmatrix} -4 & 4 & -7 \\ -3 & 6 & -9 \\ 5 & 1 & -9 \end{bmatrix}, \qquad \text{(b)} \begin{bmatrix} 3 & 4 & 5 \\ 6 & 3 & 1 \\ -1 & 1 & 2 \end{bmatrix}, \qquad \text{(c)} \begin{bmatrix} 2 & 4 & -1 \\ 1 & 3 & -2 \\ 3 & 2 & 4 \end{bmatrix},$$

$$\text{(d)} \begin{bmatrix} 4 & 1 & 3 & 1 \\ 1 & 1 & 1 & -1 \\ 2 & 2 & 5 & -4 \\ 1 & 1 & -4 & -3 \end{bmatrix}, \qquad \text{(e)} \begin{bmatrix} 8 & -3 & -9 & -2 \\ -4 & 5 & 2 & -2 \\ -2 & 4 & 10 & 6 \\ -5 & -4 & 9 & -9 \end{bmatrix}, \qquad \text{(f)} \begin{bmatrix} 3 & 2 & 1 & 3 \\ -1 & 2 & 1 & -2 \\ 2 & 1 & 3 & 2 \\ 4 & 1 & 2 & 4 \end{bmatrix}$$

CA3.2 Consider the following linear system:

$$\begin{bmatrix} 3 & 1 & & & 2 \\ 1 & 3 & 1 & & \\ & \ddots & \ddots & \ddots & \\ & & 1 & 3 & 1 \\ 2 & & & 1 & 3 \end{bmatrix} \begin{bmatrix} x_1 \\ x_2 \\ \vdots \\ x_{19} \\ x_{20} \end{bmatrix} = \begin{bmatrix} 9/2 \\ c_2 \\ \vdots \\ c_{19} \\ 81/10 \end{bmatrix}$$

where $c_i = (-1)^i + (i/2)$ for $i = 2, \ldots, 19$. (a) Estimate theoretical ω_{opt}, (b) find approximate solutions of the system using the SOR method in increments of $\Delta\omega = 0.1$ in the range of $0.8 \leqslant \omega \leqslant 1.7$. For each ω value, record the number of iterations required for convergence. Is there an ω_{opt} value? If yes, does it agree with the theoretical value? (d) Calculate the convergence rates for the numerical estimates obtained in Part (b). The true solution is given as $x_i = (-1)^i + (i/10)$, $i = 1, 2, \ldots, 20$. Use $\mathbf{x}^{(0)} = \mathbf{0}$ as the initial guess and $\|\mathbf{d}^{(p)}\|_\infty < 5 \times 10^{-4}$ for the stopping criterion.

CA3.3   Consider the following set of nonlinear equations:

$$2\sqrt{x} + \sqrt{y+1} + \sqrt{z+4} = 8, \qquad \sqrt{x+z} + 2y = 3, \qquad \sqrt{x+y} + 2z = 12$$

(a) Obtain the SOR iteration equations by isolating x, y, and z from the first, second, and third equations and applying the relaxation parameter, just as in the case of linear systems. (b) Solve the iteration equations for $\omega = 0.8, 0.9, 1$, and 1.1. How many iterations are required to satisfy the convergence criterion $\|\mathbf{d}^{(p)}\|_\infty < 5 \times 10^{-4}$, (c) compare and interpret your numerical estimates. Use $\mathbf{x}^{(0)} = \mathbf{0}$ for the starting guess.

CA3.4   Develop an algorithm (a pseudomodule) specifically to estimate the solution of the following penta-diagonal linear system, using the SOR method.

where $k = n - m + 1$ and the module should have the following argument list.

Module SOR5D$(n, m, \omega, \varepsilon, maxit, \mathbf{b}, \mathbf{d}, \mathbf{e}, \mathbf{f}, \mathbf{h}, \mathbf{q}, \mathbf{xo}, \mathbf{x}, iter, error)$

INPUT VARIABLES		
n	:	Number of (unknowns) equations;
m	:	Position of the 4th (below) and 5th (above) diagonals;
ω	:	A relaxation parameter $(1 \leqslant \omega < 2)$;
ε	:	Convergence tolerance;
$maxit$	:	Upper bound for the number of iterations;
$\mathbf{b}$	:	Array of below below diagonal elements, b_i, $i = m, (m+1), \ldots, n$;
$\mathbf{d}$	:	Array of below diagonal elements, d_i, $i = 2, 3, \ldots, n$;
$\mathbf{e}$	:	Array of diagonal elements, e_i, $i = 1, 2, \ldots, n$;
$\mathbf{f}$	:	Array of above diagonal elements, f_i, $i = 1, 2, \ldots, (n-1)$;
$\mathbf{h}$	:	Array of above above diagonal elements, h_i, $i = 1, 2, \ldots, k$;
$\mathbf{q}$	:	Array of right-hand-side, q_i, $i = 1, 2, \ldots, n$;
$\mathbf{xo}$	:	Array of initial guess, xo_i, $i = 1, 2, \ldots, n$;
OUTPUT VARIABLES		
$\mathbf{x}$	:	Array containing the solution, x_i, $i = 1, 2, \ldots, n$;
$iter$	:	Number of iterations performed;
$error$	:	Maximum error $(\|\mathbf{x} - \mathbf{xo}\|_\infty$ or $\|\mathbf{x} - \mathbf{xo}\|_2)$

CA3.5 The following linear system has a true solution, given by $x_i = (1+i/4)/5$ for $i = 1,\ldots,20$. Find the solution of the system by converting the module developed in **CA 3.4** into a real program of your choice. Run the program and record the number of iterations for the relaxation parameters in the range $1 \leqslant \omega < 1.9$ in increments of $\Delta\omega = 0.05$. Use $\mathbf{x}^{(0)} = \mathbf{0}$ as the initial guess and $\|\mathbf{d}^{(p)}\|_\infty < 10^{-6}$ for the stopping criterion.

$$
\begin{array}{c}
\quad\quad j=1 \quad j=2 \quad \cdots \quad j=7 \quad \cdots \quad j=14 \quad \cdots \quad j=20
\end{array}
$$

$$
\begin{array}{c}
i=1\\i=2\\ \vdots \\ i=7 \\ \vdots \\ i=14 \\ \vdots \\ \vdots \\ i=19 \\ i=20
\end{array}
\begin{bmatrix}
-6 & 2 & & & 1 & & & & & \\
2 & -6 & 2 & & & 1 & & & & \\
& \ddots & \ddots & \ddots & & & \ddots & & & \\
1 & & 2 & -6 & 2 & & & 1 & & \\
& \ddots & & \ddots & \ddots & \ddots & & & \ddots & \\
& & 1 & & 2 & -6 & 2 & & & 1 \\
& & & \ddots & & \ddots & \ddots & \ddots & & \\
& & & & \ddots & & \ddots & \ddots & \ddots & \\
& & & & & 1 & & 2 & -6 & 2 \\
& & & & & & 1 & & 2 & -6
\end{bmatrix}
\begin{bmatrix}x_1\\x_2\\\vdots\\x_7\\\vdots\\x_{14}\\\vdots\\\vdots\\x_{19}\\x_{20}\end{bmatrix}
=
\begin{bmatrix}q_1\\q_2\\\vdots\\q_7\\\vdots\\q_{14}\\\vdots\\\vdots\\q_{19}\\q_{20}\end{bmatrix}
$$

where $\mathbf{q} = (1/20)\,[-7,\,0,\,-1,\,-2,\,-3,\,-4,\,0,\,0,\,0,\,0,\,0,\,0,\,0,\,0,\,-25,\,-26,\,-27,\,-28,\,-29,\,-80]^T$.

CA3.6 Consider the following iteration scheme for the 20×20 penta-diagonal linear system given in **CA 3.4**:
$$\mathbf{T}\mathbf{x}^{(p+1)} = \mathbf{q} - \mathbf{M}\mathbf{x}^{(p)}$$
where $\mathbf{q} = [q_1, q_2, \cdots, q_m, \cdots, q_{n-1}, q_n]^T$, $\mathbf{x} = [x_1, x_2, \cdots, x_m, \cdots, x_{n-1}, x_n]^T$, $\mathbf{A} = \mathbf{T} + \mathbf{M}$ and

$$
\mathbf{T} = \begin{bmatrix}
e_1 & f_1 & & & & \\
d_2 & e_2 & f_2 & & & \\
& \ddots & \ddots & \ddots & & \\
& & d_m & e_m & f_m & \\
& & & \ddots & \ddots & \\
& & & d_{n-1} & e_{n-1} & f_{n-1} \\
& & & & d_n & e_n
\end{bmatrix},
\quad
\mathbf{M} = \begin{bmatrix}
& & h_1 & & & \\
& & & \ddots & & \\
b_m & & & & h_m & \\
& \ddots & & & & \ddots \\
& & b_k & & & & h_k \\
& & & \ddots & & \\
& & & & b_n &
\end{bmatrix}
$$

This scheme requires solving a tridiagonal system of equations at every iteration step. (a) Develop a pseudocode and then convert it into a running computer program; (b) solve the system given in **CA 3.4** and compare the number of iterations with those obtained with the SOR method. Use the initial guess $\mathbf{x}^{(0)} = \mathbf{0}$ and $\|\mathbf{d}^{(p)}\|_\infty < 10^{-6}$ for the stopping criterion.

CA3.7 Consider the following linear system:

$$
\begin{bmatrix}
-4 & 1 & 1 & & & & & \\
1 & -4 & 1 & 1 & & & & \\
1 & 1 & -4 & 1 & 1 & & & \\
& \ddots & \ddots & \ddots & \ddots & \ddots & & \\
& & 1 & 1 & -4 & 1 & 1 & \\
& & & 1 & 1 & -4 & 1 & \\
& & & & 1 & 1 & -4
\end{bmatrix}
\begin{bmatrix}x_1\\x_2\\x_3\\\vdots\\x_9\\x_{10}\\x_{11}\end{bmatrix}
= \frac{1}{10}
\begin{bmatrix}-17\\-18\\b_3\\\vdots\\b_9\\-62\\25\end{bmatrix}
$$

where $b_i = 4\,i\,(-1)^{i+1}$ for $i = 3,\ldots,9$, and the true solution is given by $x_i = 1 + (-1)^i(i/10)$ for $i = 1,\ldots,11$. (a) Develop a pseudocode and convert it into a running computer program to find an approximate solution of the system with the SOR method; (b) determine the optimum relaxation parameter by varying ω in $\Delta\omega = 0.05$ increments between 1 and 2 and making the plot of ω versus the number of iterations. Use $\mathbf{x}^{(0)} = \mathbf{0}$ as the initial guess and $\|\mathbf{d}^{(p)}\|_\infty < 10^{-6}$.

CA3.8 The Jacobi or Gauss-Seidel method can also be employed for some nonlinear systems of equations. Consider the following set of nonlinear equations:

$$15x + y^2 + z^2 = 11, \qquad x^2 + 28y - 3z = 24, \qquad 3x + y^2 + 20z = 11$$

(a) To find a set of (possible) suitable iteration equations, isolate x, y, and z from the first, second, and third equations, respectively; (b) apply the Jacobi and Gauss-Seidel methods to obtain approximate solutions with $\mathbf{x}^{(0)} = \mathbf{0}$ as the initial guess. The system has another solution near $(x,y,z) = (-21,-16,-8.75)$. Can you find the other solution set by changing the initial guess? Use $\|\mathbf{d}^{(p)}\|_\infty < 5 \times 10^{-5}$ for the stopping criterion.

CA3.9 Consider the following system of linear equations obtained for the loop currents of a circuit:

$$\begin{bmatrix} 20 & 0 & -12 & 0 & 0 \\ 0 & 14 & 0 & -4 & 0 \\ -12 & 0 & 42 & 0 & -29 \\ 0 & -4 & 0 & 8 & 0 \\ 0 & 0 & -29 & 0 & 31 \end{bmatrix} \begin{bmatrix} i_1 \\ i_2 \\ i_3 \\ i_4 \\ i_5 \end{bmatrix} = \begin{bmatrix} 0 \\ 48 \\ 0 \\ 0 \\ 82 \end{bmatrix}$$

(a) Solve the system iteratively using the Jacobi and Gauss-Seidel methods, (b) calculate the optimum relaxation parameter theoretically and also estimate it numerically by applying the SOR method for ω values from 1 in increments of 0.1 up to 1.9, (c) estimate the condition number of the coefficient matrix, and (e) solve the linear system using the Conjugate Gradient Method (CGM) with $\|\mathbf{r}^{(0)}\| < 10^{-4}$. Use $\mathbf{x}^{(0)} = \mathbf{0}$ as the initial guess and $\|\mathbf{d}^{(p)}\|_\infty < 10^{-6}$ for the convergence tolerance of the stationary methods.

CA3.10 Consider **E2.52**, for which the system of linear equations for the loop currents are

$$\begin{bmatrix} 5 & -1 & 0 & -2 & -2 \\ -1 & 10 & -3 & 0 & -3 \\ 0 & -3 & 12 & -4 & -5 \\ -2 & 0 & -4 & 12 & -4 \\ -2 & -3 & -5 & -4 & 14 \end{bmatrix} \begin{bmatrix} i_1 \\ i_2 \\ i_3 \\ i_4 \\ i_5 \end{bmatrix} = \begin{bmatrix} 23 \\ -18 \\ -24 \\ 48 \\ 0 \end{bmatrix}$$

(a) Solve the system iteratively using the Jacobi and Gauss-Seidel methods, (b) calculate the optimum relaxation parameter theoretically and also estimate it numerically by applying the SOR method for ω values from 1 in increments of 0.1 up to 1.9, (c) estimate the condition number of the coefficient matrix, and (e) solve the linear system using the Conjugate Gradient Method (CGM) with $\|\mathbf{r}^{(0)}\| < 10^{-4}$. Use $\mathbf{x}^{(0)} = \mathbf{0}$ as the initial guess and $\|\mathbf{d}^{(p)}\|_\infty < 10^{-6}$ for the convergence tolerance of the stationary methods.

CA3.11 A system of linear equations, $\mathbf{Ax} = \mathbf{b}$, can be converted to an equivalent system, $\hat{\mathbf{A}}\hat{\mathbf{x}} = \hat{\mathbf{b}}$, where $\hat{\mathbf{A}} = \mathbf{P}^{-1}\mathbf{A}$, $\mathbf{P}$ is a preconditioner, and $\hat{\mathbf{A}}$ is a symmetric positive-definite matrix. To ensure that $\hat{\mathbf{A}}$ is positive definite and symmetric, we take $\mathbf{P}^{-1} = \mathbf{LL}^T$ where $\mathbf{L}$ is a nonsingular lower triangular matrix. Then $\mathbf{P}^{-1}\mathbf{Ax} = \mathbf{P}^{-1}\mathbf{b}$ can be modified as follows:

$$\mathbf{L}^T\mathbf{Ax} = \mathbf{L}^T\mathbf{b} \quad \text{or} \quad \mathbf{L}^T\mathbf{ALL}^{-1}\mathbf{x} = \mathbf{L}^T\mathbf{b} \quad \text{or} \quad \hat{\mathbf{A}}\hat{\mathbf{x}} = \hat{\mathbf{b}}$$

where $\hat{\mathbf{A}} = \mathbf{L}^T\mathbf{AL}$, $\hat{\mathbf{x}} = \mathbf{L}^{-1}\mathbf{x}$ and $\hat{\mathbf{b}} = \mathbf{L}^T\mathbf{b}$. The CGM algorithm, of course, can be given in terms of "*hatted*" quantities; however, preconditioning can be directly incorporated into the iterations. The reader may find the details of this procedure in Ref. [29]. An algorithm to solve $\mathbf{Ax} = \mathbf{b}$ linear system using the Preconditioned Conjugate Gradient Method (PCGM) with the Jacobi preconditioner is given below. Write a computer program that incorporates the given algorithm.

\ **Algorithm PCGM: Preconditioned Conjugate Gradient method**

$\mathbf{r}^{(0)} \leftarrow \mathbf{b} - \mathbf{A}\mathbf{x}^{(0)}$	\ Compute the initial residual
$\mathbf{s}^{(0)} \leftarrow \mathbf{P}^{-1}\mathbf{r}^{(0)}$	\ Solve $\mathbf{P}\mathbf{s}^{(0)} = \mathbf{r}^{(0)}$ system
$\mathbf{d}^{(0)} \leftarrow \mathbf{s}^{(0)}$	\ Initialize search direction with $\mathbf{s}$
$\rho^{(0)} \leftarrow (\mathbf{r}^{(0)}, \mathbf{s}^{(0)})$	\ Compute $\rho^{(0)} = (\mathbf{r}^{(0)}, \mathbf{s}^{(0)})$
For $\left[p = 1, maxit\right]$	\ Begin the iteration loop p

\ *Compute current estimate and its residual at pth step*

$\alpha^{(p)} \leftarrow \rho^{(p-1)}/(\mathbf{d}^{(p-1)}, \mathbf{A}\mathbf{d}^{(p-1)})$ Compute $\alpha^{(p)}$	
$\mathbf{x}^{(p)} \leftarrow \mathbf{x}^{(p-1)} + \alpha^{(p)}\mathbf{d}^{(p-1)}$	\ Compute current solution, $\mathbf{x}^{(p)}$
$\mathbf{r}^{(p)} \leftarrow \mathbf{r}^{(p-1)} - \alpha^{(p)}\mathbf{A}\mathbf{d}^{(p-1)}$	\ Compute current residuals, $\mathbf{r}^{(p)}$
$\mathbf{s}^{(p)} \leftarrow \mathbf{P}^{-1}\mathbf{r}^{(p)}$	\ Solve $\mathbf{P}\mathbf{s}^{(p)} = \mathbf{r}^{(p)}$ system
If $\left[\|\mathbf{r}^{(p)}\| < \varepsilon\right]$ **Then**	\ Convergence achieved
Exit	\ Exit the iteration loop
Else	\ Linear system NOT converged yet!

\ *Not converged, compute new direction at pth step*

$\rho^{(p)} \leftarrow (\mathbf{r}^{(p)}, \mathbf{s}^{(p)})$	
$\beta^{(p)} \leftarrow \rho^{(p)}/\rho^{(p-1)}$	
$\mathbf{d}^{(p)} \leftarrow \mathbf{s}^{(p)} + \beta^{(p)}\mathbf{d}^{(p)}$	\ Compute current direction
End If	
End For	\ End the iteration loop p

CA3.12 Bi-Conjugate Gradient Method (BiCGM) is applied to any $\mathbf{A}\mathbf{x} = \mathbf{b}$ system. The method uses two residual vectors ($\hat{\mathbf{r}}$ and $\mathbf{r}$) and directions ($\hat{\mathbf{d}}$ and $\mathbf{d}$). The scalar α is chosen to force the bi-orthogonality condition, $(\mathbf{r}^{(p)}, \hat{\mathbf{r}}^{(p)}) = (\hat{\mathbf{r}}^{(p)}, \mathbf{r}^{(p)}) = 0$, and β is chosen to force the biconjugacy condition. There is also mutual orthogonality, i.e., $(\mathbf{d}^{(p)}, \hat{\mathbf{r}}^{(p)}) = (\hat{\mathbf{r}}^{(p)}, \mathbf{d}^{(p)}) = 0$. The reader may find the details of this procedure at Ref. [29]. The algorithm for the Bi-Conjugate Gradient Method (BiCGM) is given below. Write a computer program that incorporates the given algorithm.

\ **Algorithm BiCGM: Bi-Conjugate Gradient method**

$\mathbf{r}^{(0)} \leftarrow \mathbf{b} - \mathbf{A}\mathbf{x}^{(0)}$	\ Compute the initial residual
$\hat{\mathbf{r}}^{(0)} \leftarrow \mathbf{r}^{(0)}$	\ or any $\hat{r}^{(0)}, (\hat{r}^{(0)}, \hat{r}^{(0)}) \neq 0$,
$\mathbf{d}^{(0)} \leftarrow \mathbf{r}^{(0)}$	\ Initialize search direction with $\mathbf{r}$
$\hat{\mathbf{d}}^{(0)} \leftarrow \hat{\mathbf{r}}^{(0)}$	\ Initialize search direction with $\hat{\mathbf{r}}$
$\rho^{(0)} \leftarrow (\mathbf{r}^{(0)}, \hat{\mathbf{r}}^{(0)})$	\ Compute $\rho^{(0)} = (\mathbf{r}^{(0)}, \hat{\mathbf{r}}^{(0)})$
For $\left[p = 1, maxit\right]$	\ Begin the iteration loop p

\ *Compute current estimate and its residual at pth step*

$\alpha^{(p)} \leftarrow \rho^{(p-1)}/(\hat{\mathbf{d}}^{(p-1)}, \mathbf{A}\hat{\mathbf{d}}^{(p-1)})$ Compute $\alpha^{(p)}$	
$\mathbf{x}^{(p)} \leftarrow \mathbf{x}^{(p-1)} + \alpha^{(p)}\mathbf{d}^{(p-1)}$	\ Compute current estimate, $\mathbf{x}^{(p)}$
$\mathbf{r}^{(p)} \leftarrow \mathbf{r}^{(p-1)} - \alpha^{(p)}\mathbf{A}\mathbf{d}^{(p-1)}$	\ Compute current residual, $\mathbf{r}^{(p)}$
$\hat{\mathbf{r}}^{(p)} \leftarrow \hat{\mathbf{r}}^{(p-1)} - \alpha^{(p)}\mathbf{A}^T\hat{\mathbf{d}}^{(p-1)}$	
If $\left[\|\mathbf{r}^{(p)}\| < \varepsilon\right]$ **Then**	\ Convergence achieved
Exit	\ Exit the iteration loop
Else	\ Linear system NOT converged yet!

\ *Not converged, compute new direction at pth step*

$\rho^{(p)} \leftarrow (\mathbf{r}^{(p)}, \hat{\mathbf{r}}^{(p)})$	\ Compute current ρ
$\beta^{(p)} \leftarrow \rho^{(p)}/\rho^{(p-1)}$	
$\mathbf{d}^{(p)} \leftarrow \mathbf{r}^{(p)} + \beta^{(p)}\mathbf{d}^{(p)}$	
$\hat{\mathbf{d}}^{(p)} \leftarrow \hat{\mathbf{r}}^{(p)} + \beta^{(p)}\hat{\mathbf{d}}^{(p)}$	
End If	
End For	\ End the iteration loop p

CHAPTER 4

Nonlinear Equations

After completing this chapter, you should be able to
- explain and apply the bisection method;
- describe and implement the false-position method;
- explain and implement the fixed-point iteration method;
- apply the Newton-Raphson and modified Newton-Raphson methods;
- apply the Secant and modified Secant methods,
- understand and assess the error estimation of the iterative procedures;
- employ Aitken and Steffensen's acceleration techniques;
- list the pros and cons of the root-finding methods;
- describe and apply Newton's method to the system of nonlinear equations;
- explain Bairstow's algorithm and apply it to find the roots of a polynomial;
- understand and implement "Synthetic Division" and "Polynomial Reduction" as means for finding the real roots of polynomials.

MANY problems in science and engineering involve the solution of nonlinear algebraic equations, that is, finding the *roots* or *zeros* of a function. Functions whose roots or zeros are sought may be polynomials or transcendental functions containing trigonometric, logarithmic, exponential, or other functions or their combinations, and so on. Equations of this kind are referred to as *nonlinear algebraic equations*.

Finding the zeros of a univariant function, $y = f(x)$, is equivalent to determining its x-intercepts, i.e., solving $f(x) = 0$. The task of finding the intersections of two curves, $g(x) = h(x)$, also results in nonlinear algebraic equations. The latter can be expressed as a problem of finding the zeros of the function $f(x) = g(x) - h(x)$. Following are some examples of nonlinear algebraic equations with a single unknown:

$$\tan cx = x \qquad \text{or} \qquad x^2 \sinh(x) = 5$$

Collecting the expressions on one side of the equality sign gives

$$f(x) = \tan cx - x = 0 \quad \text{or} \quad g(x) = x^2 \sinh(x) - 5 = 0$$

A function may have a single, several, or no zeros at all. For example, in the examples above, $f(x)$ has an infinite number of zeros, whereas $g(x)$ has only one.

DOI: 10.1201/9781003474944-4

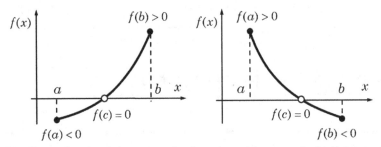

FIGURE 4.1: Graphical depiction of a function with a root in (a, b) interval.

Methods for explicitly determining the true roots of nonlinear equations are not generally available, except in a few special cases. For this reason, analysts have to resort to numerical methods that are usually iterative in nature.

In general, an *iteration equation* for a nonlinear equation is constructed as

$$x^{(p+1)} = F(x^{(p)}), \quad p = 0, 1, 2, \ldots$$

where $F(x)$ is referred to as the *iteration function*, p and $x^{(p)}$ denote the iteration step and the *estimate* (approximate value) for the root at the pth step, respectively.

Every iterative algorithm, including the ones for finding the roots of a nonlinear equation, requires an initial guess. Using the iteration function and the initial guess denoted by $x^{(0)}$, the first estimate is calculated as $x^{(1)} = F(x^{(0)})$. Next, using the latter estimate, a second improved second is found as $x^{(2)} = F(x^{(1)})$. A sequence of successive estimates, $x^{(3)}$, $x^{(4)}$, and so on, can be generated in the same manner by repeating this recursive procedure. After a sufficient number of iterations, we expect to obtain a "converged solution." In other words, if the iteration function is convergent, the subsequent estimates will approach the true root α; that is, $\lim_{p \to \infty} x^{(p)} = \alpha$.

 There may be several ways to define an *iteration function*. In such cases, each iteration equation affects the convergence of the iteration process and the convergence rate differently.

This chapter is mainly devoted to methods for finding the roots of nonlinear equations with a single variable. A brief introduction to the numerical solution of a system of nonlinear equations is also presented. Additionally, methods for finding the roots of polynomials are also covered. The most common numerical methods, along with their advantages and disadvantages, are presented.

4.1 BISECTION METHOD

The *bisection method*, also known as *interval halving*, is a numerical method for finding a root of a nonlinear equation within a specified interval. The nonlinear equation is first cast as $f(x) = 0$. Then, a suitable search interval (a, b) containing a unique root is determined. If $f(x)$ is continuous on (a, b) and $f(a)f(b) < 0$, then $f(a)$ and $f(b)$ have opposite signs (*see* Fig. 4.1). So $f(a)f(b) < 0$, in a way, provides a criterion for isolating or *bracketing* a root in (a, b). Henceforward, the search interval is halved successively until the interval within which the estimated root lies becomes sufficiently narrow.

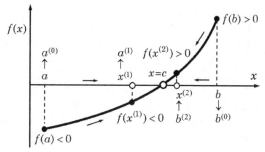

FIGURE 4.2: Depicting the convergence of the root with the Bisection method.

The interval-halving procedure is applied as follows: The function is evaluated at the midpoint of the interval $f(x^{(1)})$, where $x^{(1)} = (a+b)/2$ is also an intermediate estimate for the root. The midpoint divides the interval into two subintervals of equal width: $(a, x^{(1)})$ and $(x^{(1)}, b)$. If $f(x^{(1)}) = 0$, the procedure is terminated because $x^{(1)}$ is the root. Otherwise, we need to determine and discard the interval not containing the root. To accomplish this, the sign of $f(x^{(1)})$ is checked in relation to $f(a)$ or $f(b)$. For example, if $f(a)f(x^{(1)}) < 0$, the root is in $(a, x^{(1)})$, and the upper end of the search interval is updated to $b^{(1)} \leftarrow x^{(1)}$. If $f(a)f(x^{(1)}) > 0$, the root is bracketed as $(x^{(1)}, b)$, and the lower end of the search interval is updated to $a^{(1)} \leftarrow x^{(1)}$ (see Fig. 4.2). From this point onward, we will use $a^{(p)}$ and $b^{(p)}$ notations for the updated lower and upper ends of the search interval at the pth bisection step. Setting $(a^{(0)}, b^{(0)}) = (a, b)$, the midpoint becomes $x^{(p)} = (a^{(p-1)} + b^{(p-1)})/2$. Each time, the search interval is bracketed further down, and the endpoints are updated as $b^{(p)} \leftarrow x^{(p)}$ and $a^{(p)} \leftarrow a^{(p-1)}$ if $f(a^{(p-1)})f(x^{(p)}) < 0$ or $a^{(p)} \leftarrow x^{(p)}$ and $b^{(p)} \leftarrow b^{(p-1)}$ if $f(a^{(p-1)})f(x^{(p)}) > 0$. After a sufficient number of successive estimates, the search interval becomes small enough ($|b^{(p)} - a^{(p)}| < \varepsilon$) that the best estimate for the root becomes $x_{root} = (a^{(p)} + b^{(p)})/2$.

Alternatively, at the end of the foregoing algorithm, the estimate for the root can be further improved by employing a linear interpolation as follows:

$$x_{root} \cong \frac{a^{(p)} f(b^{(p)}) - b^{(p)} f(a^{(p)})}{f(b^{(p)}) - f(a^{(p)})} \cdot \tag{4.1}$$

Successive approximations, in the absence of truncation errors, will eventually converge to the true value after a sufficient number of iterations, i.e., $\lim_{p \to \infty} x^{(p)} = \alpha$. However, it is impossible to find the true root due to round-off errors that are unavoidable in digital computations. Hence, this iterative procedure is terminated when, at least, one of the following stopping criteria is met:

$$\left| x^{(p)} - x^{(p-1)} \right| < \varepsilon_1 \quad \text{or} \quad \left| \frac{x^{(p)} - x^{(p-1)}}{x^{(p)} + \delta} \right| < \varepsilon_2 \quad \text{or} \quad \left| f(x^{(p)}) \right| < \varepsilon_3$$

where ε_1, ε_2, and ε_3 are the prescribed tolerances, $x^{(p)}$ denotes the root estimate at the pth bisection step, and δ is a real small number ($\delta < \varepsilon_2$) added to the denominator to prevent "*division by zero*" in case $x^{(p)}$ becomes zero at some pth step.

The interval size can be determined in advance from $b^{(p)} - a^{(p)} = (b-a)/2^{p+1}$, which is a useful relationship that may also be used to determine the accuracy to be achieved after p bisections.

$$\left| x^{(p)} - \alpha \right| \approx b^{(p)} - a^{(p)} = \frac{b-a}{2^{p+1}} < \varepsilon \tag{4.2}$$

where ε is the tolerance and α is the best estimate for the root. This expression gives us a tool for predicting the number of bisections required to obtain the root within ε. If the interval estimate is quite good (sufficiently small), then the root is found within the prescribed tolerance very quickly.

 A solution is said to be correct within n decimal places if the error is less than $\varepsilon = 0.5 \times 10^{-n}$. Substituting this value into Eq. (4.2) and solving for p, we obtain an expression for the minimum number of iterations as $p > \log_2[10^n(b-a)]$.

Pseudocode 4.1

Module BISECTION (a, b, $maxit$, ε, $root$, $halves$)
\ DESCRIPTION: A pseudo module to estimate a real root of a nonlinear
\   equation in (a, b) using the Bisection method.
\ USES:
\   FUNC:: A user defined function providing the nonlinear equation, $f(x)$;
\   ABS:: A built-in function computing the absolute value.
$p \leftarrow 0$ \hfill \ Initialize counter
$interval \leftarrow b - a$ \hfill \ Find initial interval size
$fa \leftarrow$ **FUNC** (a) \hfill \ Evaluate $f(x)$ at endpoints
$fb \leftarrow$ **FUNC** (b)
Repeat \hfill \ Bisection loop
  $p \leftarrow p + 1$ \hfill \ Count bisections
  $xm \leftarrow (a + b)/2$ \hfill \ Find current midpoint
  $fxm \leftarrow$ **FUNC**(xm) \hfill \ Find $f(x)$ at midpoint
  Write: p, a, b, xm, fa, fb, fxm \hfill \ Print computation progress
  \ Is the root on either left or right interval?
  If $\left[fa * fxm > 0\right]$ **Then** \hfill \ Root is in the right interval
    $a \leftarrow xm$; $fa \leftarrow fxm$ \hfill \ Update a and fa
  Else \hfill \ Root is in the left interval
    $b \leftarrow xm$; $fb \leftarrow fxm$ \hfill \ Update b and fb
  End If
  $interval \leftarrow interval/2$ \hfill \ Halve current interval
Until $\left[(|fxm| < \varepsilon \ \textbf{And} \ \ interval < \varepsilon) \ \textbf{Or} \, p = maxit\right]$ \hfill \ End of loop
$root \leftarrow xm$ \hfill \ Current midpoint is the root estimate
$halves \leftarrow p$ \hfill \ Number of halves performed
\ If the iteration limit $maxit$ is reached *with no* convergence
If $\left[p = maxit\right]$ **Then** \hfill \ Issue info and a warning
  Write: "Failed to converge after",$maxit$,"iterations"
  Write: "Current estimate is accurate within", $interval$
End If
End Module BISECTION

Function Module FUNC (x)
\ DESCRIPTION: Non-linear equation suppled in $f(x) = 0$ form.
FUNC $\leftarrow x + 0.165 - 0.7\,e^{-0.1x}$ \hfill \ Set function to FUNC
End Function Module FUNC

Bisection Method

- The Bisection method is simple and reliable;
- It converges to a solution under any circumstances, as long as there is a root within the starting search interval;
- The estimates for the root improve as the number of bisections increases (the error bound decreases by half with each bisection).

- Determining the starting search interval (a, b) requires preliminary work to ensure that $f(a)$ and $f(b)$ have opposite signs;
- A repeated (double, triple, etc.) root cannot be determined because the search interval will yield $f(a)f(b) > 0$;
- It converges linearly with an average convergence rate of 0.5, which is considerably slow;
- It requires a large number of bisections to obtain an estimate with a reasonable degree of accuracy if the starting search interval is too large;
- Starting with an endpoint very close to the true root ($a \approx root$ or $b \approx root$) does not ensure fast convergence;
- Imaginary roots cannot be determined;
- It always converges to a root if $f(a)f(b) < 0$; however, this criterion implies the *existence* of a root rather than its *uniqueness*. If there is more than one root in the search interval, it will converge to any one of the roots.

A pseudomodule, **BISECTION**, implementing the bisection algorithm is presented in Pseudocode 4.1. The code requires a starting search interval (a, b), an upper bound on the number of iterations ($maxit$), and a convergence tolerance (ε) as input. The nonlinear equation whose root is sought is supplied as a function module, **FUNC**(x). The outputs of the module are the estimate for the root ($root$) and the number of bisections performed ($halves$). The bisection algorithm is implemented in a **For**-loop with index variable p, which terminates when either $p = maxit$ or both $|f(x^{(p)})| < \varepsilon$ and $interval < \varepsilon$ stopping criteria are met. During the bisections, the midpoint may happen to land exactly on the root or at a point very close to the root, $|f(x^{(p)})| < \varepsilon$; in this case, the loop must be terminated. This likelihood, though, has not been addressed in the present pseudocode and is left as an exercise for the reader. At every step, the midpoint of the current interval, $x^{(p)} = (a^{(p-1)} + b^{(p-1)})/2$, is evaluated as the current root estimate. If $f(a^{(p-1)})f(x^{(p)}) < 0$, then the root lies in the left subinterval: $(a^{(p-1)}, x^{(p)})$. The search interval is bracketed by updating the right endpoint as $b^{(p)} \leftarrow x^{(p)}$ and $f(b^{(p)}) \leftarrow f(x^{(p)})$. Otherwise, $f(a^{(p-1)})f(x^{(p)}) > 0$, the root lies in the right subinterval, $(x^{(p)}, b^{(p-1)})$, in which case the left endpoint values are updated as $a^{(p-1)} \leftarrow x^{(p)}$ and $f(a^{(p-1)}) \leftarrow f(x^{(p)})$. If the convergence criteria are not met, the interval size is halved before moving on to the next bisection procedure. So the final estimate is an approximate value within the tolerance prescribed.

Recall that checking for equality of two floating-point numbers as $a = b$ or $a - b = 0$ in programs is not a good practice! Hence, the root, in Pseudocode 4.1, is determined by satisfying the $|f(x^{(p)})| < \varepsilon$ condition instead of $|f(x^{(p)})| = 0$ or $f(x^{(p)}) = 0$.

EXAMPLE 4.1: Application of Bisection method

Signal delays are inherent in numerous engineering systems (control, electric transmission, remote control, etc.), for which delay differential equations (DDEs) constitute the basic mathematical models for real phenomena. A homogeneous first-order DDE is represented by

$$\frac{dy}{dt} + \alpha y(t - T) + \beta y(t) = 0, \quad 0 \leqslant t \leqslant T$$

where α and β are constants, and T is the delay time.

Assuming a homogeneous solution in the form of $y(t) = ce^{rt}$ and substituting it into the preceding DDE, the following nonlinear transcendental characteristic equation is obtained:

$$f(r) = r + \beta + \alpha e^{-rT} = 0$$

This characteristic equation has a root in (0.2, 0.6) for $\alpha = -0.7$, $\beta = 0.165$, and $T = 0.1$. Apply the bisection method to estimate the root accurately to within four decimal places.

SOLUTION:

The search interval is already given in the question as (0.2, 0.6). In practice, one should have a rough idea of where the root may be so that the search interval can be estimated as accurately as possible. The fact that $f(0.2) = -0.3211391$ and $f(0.6) = 0.1057648$ have opposite signs confirms that there is a root in (0.2, 0.6). Since a four-decimal place accurate solution is desired, we set the convergence tolerance to $\varepsilon = 0.5 \times 10^{-4}$.

TABLE 4.1

p	$a^{(p)}$	$b^{(p)}$	$f(a^{(p)})$	$f(b^{(p)})$	$r^{(p+1)}$	$f(r^{(p+1)})$
0	0.2	0.6	−0.3211391	0.1057648	0.4	−0.1075526
1	0.4	0.6	−0.1075526	0.1057648	0.5	−0.0008606
2	0.5	0.6	−0.0008606	0.1057648	0.55	0.0524604
3	0.5	0.55	−0.0008606	0.0524604	0.525	0.0258020
4	0.5	0.525	−0.0008606	0.0258020	0.5125	0.0124712
5	0.5	0.5125	−0.0008606	0.0124712	0.50625	0.0058054
6	0.5	0.50625	−0.0008606	0.0058054	0.503125	0.0024725
7	0.5	0.503125	−0.0008606	0.0024725	0.5015625	0.0008059
8	0.5	0.5015625	−0.0008606	0.0008059	0.5007813	−0.0000273
9	0.5007813	0.5015625	−0.0000273	0.0008059	0.5011719	0.0003893
10	0.5007813	0.5011719	−0.0000273	0.0003893	0.5009766	0.0001810
11	0.5007813	0.5009766	−0.0000273	0.0001810	0.5008789	0.0000768
12	0.5007813	0.5008789	−0.0000273	0.0000768	0.5008301	0.0000248

We begin by computing the midpoint $r^{(1)} = (0.2 + 0.6)/2 = 0.4$ and splitting the initial search interval into two subintervals: (0.2, 0.4) and (0.4, 0.6). The function at the midpoint yields $f(r^{(1)}) = -0.1075526$. Since $f(0.2)f(r^{(1)}) > 0$, the root lies in the right subinterval; thus, we can discard the left subinterval. This procedure is repeated for the new search interval, giving $r^{(2)} = (0.4 + 0.6)/2 = 0.5$ and $f(r^{(2)}) = -0.0008606$. From $f(0.5)f(r^{(2)}) > 0$, the root is determined to be in the

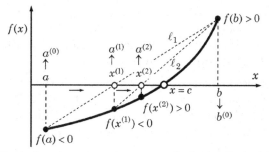

FIGURE 4.3: Depicting the root estimation with the method of false position.

right subinterval: (0.5, 0.6). Repeating the bisection procedure in this manner narrows down the search interval containing the root. Eventually, the search interval, as well as $f(r^{(n)})$, will approach zero with each bisection.

The results of the computations are presented in Table 4.1. It is worth noting that the values in the $f(r^{(p)})$ column approach zero with each iteration, and the stopping criteria are met at the 12th step, i.e., $|r^{(12)} - r^{(11)}| = 4.88 \times 10^{-5} < \varepsilon$ and $|f(r^{(12)})| = 2.48 \times 10^{-5} < \varepsilon$.

Discussion: The minimum number of bisections required to compute the root within four decimal places can be estimated by Eq. (4.2), which gives $n > \log_2(0.4 \times 10^4) = 11.966 \approx 12$. The computational effort can be reduced if the starting interval $(b - a)$ is selected as narrowly as possible.

4.2   METHOD OF FALSE POSITION

Method of False Position, or *Regula Falsi Method*, is also based on bracketing a root in a tight interval. It is similar to the bisection method; however, it differs only in computing the new estimate. The root is estimated by the x-intercept of a straight line passing through the endpoints of the current interval, i.e., $(a^{(p)}, f(a^{(p)}))$ and $(b^{(p)}, f(b^{(p)}))$.

Assuming that $f(x)$ is a continuous function with a root in (a, b), the equation of a straight line (ℓ_1) passing through the points $(a, f(a))$ and $(b, f(b))$ is

$$y = \frac{bf(a) - af(b)}{b - a} + \frac{f(b) - f(a)}{b - a}x \tag{4.3}$$

An approximation to the root can then be obtained from the x-intercept of ℓ_1, i.e., $x^{(1)}$ as illustrated in Fig. 4.3.

Bracketing the root is done in the same way as described for the bisection method. A second line (ℓ_2) for the new interval leads to a second x-intercept $(x^{(2)})$, which is closer to the root, and so on. This procedure is repeated until the interval size is sufficiently small, i.e., $|b^{(p)} - a^{(p)}| < \varepsilon$ or $|f(x^{(p)})| < \varepsilon$.

An expression for estimating the root (i.e., the intercept) in the $(a^{(p)}, b^{(p)})$ interval can be obtained by setting $y = 0$ in Eq. (4.3) and solving for $x^{(p+1)}$:

$$x^{(p+1)} = \frac{a^{(p)} f(b^{(p)}) - b^{(p)} f(a^{(p)})}{f(b^{(p)}) - f(a^{(p)})} \tag{4.4}$$

Figure 4.3 depicts how successive estimates approach the root. Note that in the course of bracketing, if $f(x)$ is concave up, the right end of the search interval is anchored while the left end moves toward the root. If $f(x)$ is concave down, this time the left end remains anchored while the right end moves toward the root.

The algorithm for the method is basically the same as the bisection method. The only difference is that the new estimate is replaced by Eq. (4.4), whereby the algorithm and the pseudocode presented for the bisection method can be used with a minor modification. For this reason, no pseudocode for this method has been given.

The advantages and disadvantages of the method of false position are pretty much the same as those of the bisection method. Nevertheless, since the method is based on finding the point that intersects the x-axis by connecting the end points of the interval containing the root with a straight line, it can converge faster than the bisection method. However, if the function is not suitable to be approximated with a line, the method may fail or stagnate.

EXAMPLE 4.2: Method of False Position

Repeat Example 4.1 using the method of False Position.

SOLUTION:

The first estimate for the root is calculated from Eq. (4.4)

$$r^{(1)} = \frac{(0.2)f(0.6) - (0.6)f(0.2)}{f(0.6) - f(0.2)} = \frac{0.213836}{0.426904} = 0.500901$$

which leads to $f(r^{(1)}) = 9.826 \times 10^{-5} > \varepsilon$.

We observe that the root is on the left subinterval, $(0.2, 0.500901)$, since $f(a)f(r^{(1)}) < 0$. The second estimate is similarly obtained as

$$r^{(2)} = \frac{(0.2)f(0.500901) - (0.500901)f(0.2)}{f(0.500901) - f(0.2)} = \frac{0.160878}{0.321237} = 0.500807$$

with $f(r^{(2)}) = 1.356 \times 10^{-7} < \varepsilon$.

The quantitative results of the first three iterations are summarized in Table 4.2. The method converged much faster (with only 2 iterations) than the bisection method (12 iterations in Example 4.1.) Due to introducing linear interpolation into this scheme, it should, and usually does, give better estimates of the root, especially when the approximation of the nonlinear function by a linear function is valid.

TABLE 4.2

p	$a^{(p)}$	$b^{(p)}$	$f(a^{(p)})$	$f(b^{(p)})$	$r^{(p+1)}$	$f(r^{(p+1)})$
0	0.2	0.6	-0.3211391	0.1057648	0.500901	9.83×10^{-5}
1	0.2	0.500901	-0.3211391	0.0000999	0.500807	1.36×10^{-7}
2	0.2	0.500807	-0.3211391	9.47×10^{-8}	0.500807	8.97×10^{-11}

Discussion: In this example, not only was the root estimate correct to within six decimal places due to $|f(r^{(2)})| < 0.5 \times 10^{-6}$, but it was also achieved in two iterations. Also note that the root, in this case, was bracketed from the left side, whereas bracketing in the bisection method alternates between both sides as it closes in on the root.

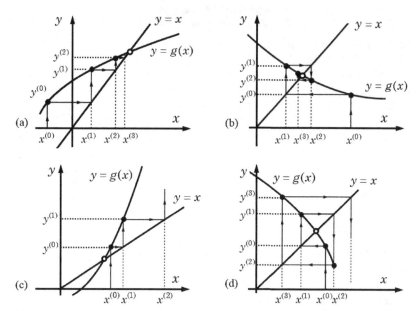

FIGURE 4.4: Depiction of the fixed-point iteration sequence of possible cases: (a) monotone convergence $(0 \leqslant g'(x^{(p)}) < 1)$; (b) oscillating convergence $(-1 < g'(x^{(p)}) \leqslant 0)$; (c) monotonic divergence $(g'(x^{(p)}) > 1)$; (d) oscillating divergence $(g'(x^{(p)}) < -1)$.

4.3 FIXED-POINT ITERATION

A nonlinear equation can be rearranged into an equivalent $x = g(x)$ form, which may be accomplished in a variety of ways. Here, x is referred to as the "fixed-point" for the iteration function $g(x)$. Then the iteration equation can be expressed as

$$x^{(p+1)} = g(x^{(p)}), \qquad p \geqslant 0 \tag{4.5}$$

The fixed-point iteration, as a technique, is equivalent to finding the geometric intersection of the line $y = x$ and the curve $y = g(x)$. The numerical procedure is illustrated in Fig. 4.4a. This iterative procedure begins with an initial guess, $x^{(0)}$. Then, the point on the curve's vertical projection is found: $y^{(0)} = g(x^{(0)})$. This value is then projected horizontally on the $y = x$ line to find an improved estimate, $x^{(1)} = y^{(0)}$. Note that the arrows in the figure indicate the direction of the projections. The procedure, which involves a vertical projection onto the curve followed by a horizontal projection onto $y = x$, is repeated until the points on the curve approach a fixed-point.

Several alternative forms of the iteration function $g(x)$, generally non-unique, can be obtained for a given nonlinear equation. For example, $x^2 + x = e^{-x}$ can be rewritten as

$$x = e^{-x} - x^2 \quad \text{or} \quad x = \sqrt{e^{-x} - x} \quad \text{or} \quad x = -\ln(x^2 + x) \quad \text{or} \quad x = e^{-x}(x+1)$$

Some of these may converge rapidly or slowly, or some may not converge at all. In general, the convergence behavior depends on the iteration function and the initial guess.

To assess the convergence behavior of an iteration equation, we assume that $g(x)$ and $g'(x)$ are continuous functions on (a, b) in which a root lies. The *sufficient* (but not *necessary*) *condition* is $|g'(x^{(p)})| \leqslant \lambda < 1$ for all x on (a, b), where λ is an upper bound provided that the initial estimate is close enough to the true root. After a sufficient number of iterations, λ gives the convergence rate.

```
                          Pseudocode 4.2

   Module FIXED_POINT (maxit, ε, root, iter)
   \ DESCRIPTION: A pseudomodule to find a real root of nonlinear equation
   \   using the Fixed-Point Iteration method.
   \ USES:
   \   FUNC:: A user-defined function supplying nonlinear equation f(x);
   \   ABS:: A built-in function computing the absolute value.
   x0 ← root                                      \ Initialize initial root
   e0 ← 1                              \ Initialize prior and current errors
   p ← 0                                     \ Initialize iteration counter
   Repeat                       \ Iterate until convergence conditions are met
       p ← p + 1                                      \ Count iterations
       x1 ← FUNC(x0)                       \ Find current estimate x^(p)
       e1 ← |x1 − x0|            \ Find current error, |x^(p) − x^(p−1)|
       λ ← e1/e0                           \ Estimate convergence rate
       Write: p, x0, x1, e1, λ               \ Print iteration progress
       x0 ← x1                          \ Set current estimate as prior
       e0 ← e1                             \ Set current error as prior
       e2 ← |FUNC(x1) − x1|      \ Find error-2, |g(x^(p)) − x^(p)|
   Until [(e1 < ε And e2 < ε) Or p = maxit]          \ Test for convergence
   root ← x1                          \ Root is found if p < maxit
   iter ← p                       \ Number of iterations performed
   \ If the iteration limit maxit is reached with no convergence
   If [p = maxit] Then                     \ Issue info and a warning
       Write: "Maximum number of iterations reached, root=", root
       Write: "Estimated root has NOT converged! errors=", e1, e2
   End If
   End Module FIXED_POINT
```

The iteration sequence for four possible cases is graphically depicted in Fig. 4.4. Figure 4.4a and b show that convergence is guaranteed for $|g'(x^{(p)})| < 1$ as the iteration sequence approaches the fixed-point of $y = x$ and $y = g(x)$. In other words, if $|g'(x^{(p)})| < 1$, the $x^{(p+1)} = g(x^{(p)})$ is said to be *locally convergent*, meaning that there is an interval containing $x^{(0)}$ such that the method converges for any starting value $x^{(0)}$ within the interval. On the other hand, in Fig. 4.4c and d for $|g'(x^{(p)})| > 1$, we observe that the iteration sequence leads away from the fixed (intersection) point. We can then say that the method diverges for any starting $x^{(0)}$ value other than the true root itself. Defining the error as $e^{(p)} = x^{(p)} - x^{(p-1)}$, it can be shown that the error is linearly proportional to the previous error by $e^{(p)}/e^{(p-1)} \rightarrow |g'(x^{(p)})| < \lambda$ as $p \rightarrow \infty$. Hence, the method of the fixed-point iterations is said to be *linearly convergent*. Even when $g(x)$ is convergent ($\lambda < 1$), its rate of convergence may turn out to be very slow in some cases. On the other hand, $g'(x^{(p)}) = 1$ could yield the root estimate of either a slow convergence rate or divergences.

A pseudomodule, FIXED_POINT, implementing the fixed-point method is given in Pseudocode 4.2. The code requires an initial estimate ($root$), an upper bound for the number of iterations ($maxit$), and a convergence tolerance (ε) as input. The iteration function $g(x)$ is provided separately as an external function module, FUNC(x). On exit, the variables $root$ and $iter$ are set to the estimated root and the total number of iterations performed, respectively. The current root estimate is computed from $x1 = x^{(p)} = g(x^{(p-1)})$ in a **Repeat-Until**

Fixed-Point Iteration Method

- The method is simple and easy to apply;
- Its cost per iteration is low;
- If it does converge, it may converge very slowly;
- Its convergence rate can be estimated simultaneously with the root.

- The method requires an initial guess that is sufficiently close to the root; even then, the convergence may not always be guaranteed;
- The convergence property depends on the choice of iteration function $g(x)$, i.e., finding a suitable iteration function can be a major problem.

---

loop as long as the number of iterations does not exceed $maxit$ ($p = maxit$) or stopping criteria, $e^{(p)} = e1 = |x^{(p)} - x^{(p-1)}| < \varepsilon$ and $e2 = |g(x^{(p)}) - x^{(p)}| < \varepsilon$, are not met concurrently. Absolute errors from prior ($e0 = |x^{(p)} - x^{(p-1)}|$) and current iterations ($e1 = |x^{(p+1)} - x^{(p)}|$) are tracked and used to estimate the rate of convergence, $\lambda^{(p)} = e1/e0$, at every iteration. Note that the error from the prior iteration level is initialized to $e0 = 1$ before the loop, and the current estimate along with the current error ($x1$ and $e1$) is assigned as the prior estimate and error ($x0 \leftarrow x1$ and $e0 \leftarrow e1$) once they have been used. Additionally, the agreement of the fixed-point form ($e2 = |g(x^{(p+1)}) - x^{(p+1)}|$) is verified in the convergence decision. When the number of iterations reaches $maxit$, it means that the estimated root has yet to converge. In this case, a warning message is printed out along with the current estimate and current error.

4.4 NEWTON-RAPHSON METHOD

The Newton-Raphson method is one of the most widely used methods for finding the roots of a nonlinear equation. To illustrate how the method works, we consider a nonlinear equation set up as $f(x) = 0$. The Taylor series of a real function $f(x)$ about a point x_0 is given by

$$f(x) = f(x_0) + (x - x_0)f'(x_0) + \frac{(x - x_0)^2}{2!}f''(x_0) + \ldots \tag{4.6}$$

Equation (4.6) will yield zero for the true root of $f(x)$. Thus, setting $f(x) = 0$, the rhs of Eq. (4.6) becomes a polynomial of infinitieth degree, one of whose roots is also the root of $f(x)$. Finding the root of this equation is very difficult. However, to find an approximation for this root, for convenience, we will replace this equation with a finite polynomial. The simplest approximation can be found by truncating the Taylor series into a two-term sum by neglecting the higher-order terms:

$$0 \cong f(x_0) + (x - x_0)f'(x_0) + \ldots \tag{4.7}$$

Solving Eq. (4.7) for x yields

$$x = x_0 - \frac{f(x_0)}{f'(x_0)} \tag{4.8}$$

which provides an expression for estimating the root. Computing a new estimate x_1 from Eq. (4.8) will not satisfy $f(x_1) = 0$ unless it happens to be the root. This estimate is used on the rhs of Eq. (4.8) to obtain a new and hopefully improved estimate. If the new estimate is not sufficiently close to the root, this procedure is repeated until successive estimates are close enough to the root with reasonable accuracy.

EXAMPLE 4.3: Application of Fixed-Point Iterations

Repeat Example 4.1 using the fixed-point iteration with $r^{(0)} = 1$ as the initial guess.

SOLUTION:

We define the iteration function as $g(r) = r = 0.7e^{-0.1r} - 0.165$ since it is easy to solve for r. Starting with the given initial guess, $r^{(0)} = 1$, and setting $p = 0, 1$ and 2 in $r^{(p+1)} = g(r^{(p)})$, the first three iterations are obtained as follows:

$$r^{(1)} = g(r^{(0)}) = 0.7e^{-0.1(1)} - 0.165 = 0.4683862,$$

$$r^{(2)} = g(r^{(1)}) = 0.7e^{-0.1(0.4683862)} - 0.165 = 0.5029690$$

$$r^{(3)} = g(r^{(2)}) = 0.7e^{-0.1(0.5029690)} - 0.165 = 0.5006629$$

The iterative estimation is repeated until the stopping criterion, $|e^{(p)}| < 0.5 \times 10^{-4}$, is met. Also, by computing $r^{(p+1)} = g(r^{(p)})$, $g'(r^{(p)})$, and $e^{(p)} = r^{(p+1)} - r^{(p)}$ in the same manner, the resulting estimated values are presented in Table 4.3. Notice that the absolute error is obtained as $|e^{(4)}| = 1.022 \times 10^{-5} < 0.5 \times 10^{-4}$ after 4 iterations. Also note that for $p > 2$, $|g'(r^{(p)})|$ approaches -0.0666, which satisfies the *sufficient condition* for convergence.

Similarly, the ratio of the absolute errors of two successive iterations yields

$$\frac{|e^{(4)}|}{|e^{(3)}|} = |g'(r^{(p)})| = 0.06658 = \lambda < 1$$

This ratio, which is directly proportional to $|g'(r^{(p)})|$, depicts a linear convergence rate.

TABLE 4.3

p	$r^{(p)}$	$g(r^{(p)})$	$|e^{(p)}|$	$r^{(p+1)}$	$g'(r^{(p)})$
0	1	0.4683862	0.5316138	0.4683862	−0.063339
1	0.4683862	0.5029690	0.0345828	0.5029690	−0.066797
2	0.5029690	0.5006629	0.0023060	0.5006629	−0.066566
3	0.5006629	0.5008165	0.0001535	0.5008165	−0.066582
4	0.5008165	0.5008062	1.02×10^{-5}	0.5008062	−0.066581

Discussion: In this example, there are two other candidates for the iteration function: (1) $g(r) = r = 0.7e^{-0.1r} - 0.165$ and (2) $g(r) = -10\ln((r + 0.165)/0.7)$. Investigating the convergence condition for the first iteration function, we obtain $g'(r) = -0.07e^{-0.1r}$. It is worth noting that $g'(r)$ is defined for all positive r and is valid within the range $0 < |g'(r)| < 1$. This means that the first function satisfies the convergence criterion for all positive r. On the other hand, the second iteration function with $g'(r) = -10/(r + 0.165)$ yields $|g'(r)| > 1$ for any initial guess value falling in the range $0 < r < 9.835$. It is clear that the second function will diverge with the initial value of $r^{(0)} = 1$.

It may be difficult to find the convergence rate or convergence interval when more than one fixed-point iteration equation exists. However, evaluating the convergence rate during the iteration process does not impose an extra computational cost, as done in Pseudocode 4.2.

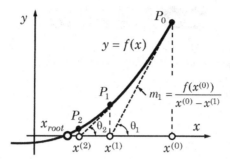

FIGURE 4.5: Graphical depiction of the Newton-Raphson method.

Introducing the superscript (iteration) notation, the *Newton-Raphson iteration equation* can be expressed as

$$x^{(p+1)} = x^{(p)} - \frac{f(x^{(p)})}{f'(x^{(p)})}, \qquad p \geqslant 0 \tag{4.9}$$

where $x^{(p)}$ is the estimate for the root at the pth iteration step.

Geometrical Interpretation: A root of $y = f(x)$ is also its x-intercept, as illustrated in Fig. 4.5. The slope of the tangent line to $f(x)$ at point P_0 is $f'(x^{(0)})$. The x-intercept of the tangent line, $x^{(1)}$, can be calculated from the slope, i.e., $m_1 = \tan\theta_1 = f'(x^{(0)}) = f(x^{(0)})/(x^{(0)} - x^{(1)})$. Similarly, a tangent line drawn to $f(x)$ at point P_1 intersects the x-axis at $x^{(2)}$. Likewise, the intercept of the tangent is determined from the slope of $f(x)$ at $x^{(1)}$; $m_2 = \tan\theta_2 = f'(x^{(1)}) = f(x^{(1)})/(x^{(1)} - x^{(2)})$. Repeating this procedure successively yields a sequence of estimates $(x^{(3)}, x^{(4)}, \dots)$ that approach the true root as $p \to \infty$. The generalized expression for the intercept at the pth iteration step can be written as

$$x^{(p+1)} = x^{(p)} - \frac{f(x^{(p)})}{f'(x^{(p)})}, \qquad p \geqslant 0 \tag{4.10}$$

It becomes clear that both Eq. (4.9) and Eq. (4.10) are identical. In practice, the Newton-Raphson method is based on finding the tangent lines, followed by calculating the intercepts that should eventually approach the true root.

Convergence issues: The method needs an initial guess, $x^{(0)}$, to begin the iteration process. However, it requires two function evaluations per iteration step, namely $f(x^{(p)})$ and $f'(x^{(p)})$, in comparison to bisection or fixed-point iteration methods. Noting that $f(x^{(p+1)})$ will tend to zero, Eq. (4.6) for any pth iteration step can be written as

$$f(x^{(p)}) + (\alpha - x^{(p)})f'(x^{(p)}) + \frac{(\alpha - x^{(p)})^2}{2!}f''(\xi) = 0 \tag{4.11}$$

where α is the true root, x_0 is replaced by $x^{(p)}$, and the third term is the Lagrange form of the remainder, where $\alpha < \xi < x^{(p)}$. Making use of Eq. (4.10) in the first term of Eq. (4.11) and simplifying, we obtain

$$x^{(p+1)} - \alpha = \frac{f''(\xi)}{2f'(x^{(p)})}(x^{(p)} - \alpha)^2 \quad \text{or} \quad \left| e^{(p+1)} \right| \leqslant K \left| e^{(p)} \right|^2 \tag{4.12}$$

where $e^{(p)} = x^{(p)} - \alpha$ is the true error at the pth iteration step, and K is the asymptotic error constant defined as $K = |f''(\alpha)|/2|f'(\alpha)|$. This result implies that if $f(x)$ is continuously

Newton-Raphson Method

- The method is simple, reliable, and easy to implement;
- Its convergence is quadratic when the initial guess is in close proximity to the root;
- Most divergence problems can be overcome by simply changing the initial guess.

- The method requires an analytical evaluation of the first derivative;
- It may not be suitable for functions whose derivatives are difficult to obtain and/or require too many arithmetic operations;
- Convergence cannot be guaranteed for any given function or any initial guess;
- A *Division By Zero* problem may arise in case the of $f'(x^{(p)}) \to 0$;
- If a nonlinear equation has a root of multiplicity m $(m > 1)$, it may diverge or converge very slowly with a linear convergence rate;
- Some functions, like periodic functions, may exhibit unexpected behavior for a range of estimates.

differentiable with $f'(\alpha) \neq 0$ and $f''(\alpha)$ exists, then convergence of the Newton-Raphson method is *quadratic* or faster if $f''(\alpha) = 0$. For $p \gg 1$, we can roughly estimate K from $\delta^{(p+1)} \cong K\left(\delta^{(p)}\right)^2$, where $\delta^{(n)} = x^{(n+1)} - x^{(n)}$.

There are cases when Newton-Raphson fails to converge. The convergence property of the method depends on the nature of the given function and the choice of the initial guess. Most of the divergence problems can be attributed to poor choice of an initial guess, which could simply be resolved by changing the initial guess. For instance, an initial guess or a new estimate may correspond to a local extremum point where the derivative is very small or zero, then the $\delta^{(p)}$ term will tend to be very large, which will result in throwing the current estimate away from the true root. Or, if a root is near an inflection point, the current estimate also tends to diverge from the true root. Also, if a function is not continuously differentiable in the vicinity of a root or has a vertical tangent line, the method will fail or diverge.

EXAMPLE 4.4: Application of Newton-Raphson method

Repeat Example 4.1 using the Newton-Raphson method with $r^{(0)} = 1$ as an initial guess.

SOLUTION:

In this method, we cast the nonlinear equation as $f(r) = 0$. The method, besides $f(r)$, requires the first derivative, which is given below:

$$f(r) = r + 0.165 - 0.7e^{-0.1r} \quad \text{and} \quad f'(r) = 1 + 0.07e^{-0.1r}$$

For $p = 0$, using the iteration equation, Eq. (4.10), the first estimate leads to

$$r^{(1)} = r^{(0)} - \frac{f(r^{(0)})}{f'(r^{(0)})} = 1 - \frac{0.5316138}{1.063339} = 0.5000522$$

Likewise, for $p = 1$, we obtain

$$r^{(2)} = r^{(1)} - \frac{f(r^{(1)})}{f'(r^{(1)})} = 0.5000522 - \frac{-0.000805}{1.066586} = 0.50080687$$

A sequence of estimates ($r^{(2)}$, $r^{(3)}$, and so on) is found by carrying out successive iterations in the same way. The results are summarized in Table 4.4. It is observed that with each subsequent iteration, the root estimates approach the value of $r = 0.5008069$. As we have seen, the Newton-Raphson method not only estimated the root with fewer iterations but also yielded an eight-decimal place-accurate answer. Also, note that the last column converges to 0.00312, which corresponds to the asymptotic error constant whose true value is found as $K = |f''(r)|/2|f'(r)| = -0.003121$.

TABLE 4.4

p	$r^{(p)}$	$f(r^{(p)})$	$f'(r^{(p)})$	$\delta^{(p)}$	$\left|\delta^{(p)}\right|/\left|\delta^{(p-1)}\right|^2$
0	1	0.53161381	1.0633390	-0.4999478	
1	0.50005220	-0.00080493	1.0665857	0.0007547	0.003019
2	0.50080687	-1.9×10^{-9}	1.0665807	1.78×10^{-9}	0.003121
3	0.50080687	$\mathcal{O}(10^{-16})$	1.0665810	$\mathcal{O}(10^{-16})$	

Discussion: Note that, after two iterations, we achieve $|\delta^{(2)}| = 0.178 \times 10^{-8}$ and $|f(r^{(2)})| = 0.19 \times 10^{-8} < 0.5 \times 10^{-8}$, i.e., the root estimate is accurate to at least eight digits. In the third iteration, the order of magnitude of the error becomes $\mathcal{O}(10^{-16})$. This is not surprising because the convergence rate of the method is quadratic, i.e., much more rapid than previously covered methods. Actually, the error after the first iteration is about three-thousandths of (corresponding to the asymptotic error constant) the square of the previous error.

A pseudomodule, **NEWTON_RAPHSON**, implementing the Newton-Raphson algorithm is presented in Pseudocode 4.3. The module requires an initial estimate ($root$), an upper bound for iterations ($maxit$), and a convergence tolerance (ε) as input. Two user-defined external functions, **FUNC**(x) and **FUNCP**(x), are used to supply $f(x)$ and $f'(x)$, respectively. The estimated root ($root$) and the total number of iterations performed ($iter$) are the module outputs. After initializing the root ($x0 \leftarrow root$), the iteration procedure is carried out in a Repeat-Until loop until the convergence criteria ($|\delta| < \varepsilon$ and $|fn| < \varepsilon$) are met.

The iteration procedure begins by evaluating $f(x0)$ and $f'(x0)$, and then the current estimate x is computed from Eq. (4.9). At every iteration step, the asymptotic convergence constant is also computed by $K \cong \delta/\delta0^2$, where $\delta0 = x0 - x^{(p-1)}$ and $\delta = x - x0$. The stopping criteria, $|\delta| < \varepsilon$ and $|f(x)| < \varepsilon$, must be satisfied concurrently. If the convergence criteria are not met while within the preset maximum number of iterations (i.e., for $p \leqslant maxit$), then, once computed and used, the current estimate and the associated error are set as prior before the next iteration, i.e., $x0 \leftarrow x$ and $\delta0 \leftarrow \delta$. On exit from the module, the current root estimate and the total iteration number are set to $root$ and $iter$ variables, respectively. If the upper bound $maxit$ is not sufficiently large enough, due to slow convergence or for some other reason, the iteration loop terminates before the convergence criteria are met. In which case, the user is issued a warning message along with the latest estimate and error.

Pseudocode 4.3

Module NEWTON_RAPHSON $(root, maxit, \varepsilon, iter)$
\ DESCRIPTION: A pseudomodule to find a root of nonlinear equation,
\ $f(x) = 0$, using Newton-Raphson method.
\ USES:
\   FUNC and FUNCP :: User defined functions for $f(x)$ and $f'(x)$.
$x0 \leftarrow root$ \ Initialize root with input $root$
$\delta 0 \leftarrow 1$ \ Initialize (prior) error
$p \leftarrow 0$ \ Initialize iteration counter
Repeat   \ Iterate until *convergence conditions* are met
  $fn \leftarrow$ FUNC$(x0)$ \ Compute $f(x^{(p)})$
  $fpn \leftarrow$ FUNCP$(x0)$ \ Compute $f'(x^{(p)})$
  $\delta \leftarrow -fn/fpn$ \ Finc current displacement, $\delta^{(p)}$
  $K \leftarrow |\delta|/\delta 0^2$ \ Estimate convergence constant
  Write: $p, x0, fn, fpn, |\delta|, K$ \ Print iteration progress
  $x \leftarrow x0 + \delta$ \ Find current estimate, Eq. (4.10)
  $x0 \leftarrow x$ \ Set current estimate as prior
  $\delta 0 \leftarrow |\delta|$ \ Set current displacement as prior
  $p \leftarrow p + 1$ \ Count iterations
Until $\left[(|\delta| < \varepsilon \textbf{ And } |fn| < \varepsilon) \textbf{ Or } p = maxit \right]$ \ Converged?
$root \leftarrow x$ \ Root is found, if $p < maxit$
$iter \leftarrow p$ \ Number of iterations = counter
\ If the iteration limit $maxit$ is reached *with no* convergence
If $\left[p = maxit \right]$ **Then**   \ Issue info and a warning
  Write: "Maximum number of iterations reached"
  Write: "Estimated root has NOT converged! $|\delta|, |f(x)|=$", $\delta 0, fn$
End If
End Module NEWTON_RAPHSON

A root of multiplicity m is defined as the number of repeating (multiple) roots of a nonlinear equation. When $f(x)$ has a root of multiplicity $m > 1$ at $x = \alpha$, $f(\alpha) = f'(\alpha) = \ldots = f^{(m-1)}(\alpha) = 0$, but $f^{(m)}(\alpha) \neq 0$. Hence, the Newton-Raphson method leads to a *linear convergence rate* for a root of multiplicity.

4.5 MODIFIED NEWTON-RAPHSON METHOD

The Newton-Raphson method fails when $f'(x^{(p)}) = 0$. Likewise, when $f(x)$ has a multiple root, say in $x = \alpha$, the function as well as its first derivative becomes zero, $f(\alpha) = f'(\alpha) = 0$. In the latter case, as the successive estimates approach the true root, the first derivative in Eq. (4.10) will tend to zero, i.e., $f'(x^{(p)}) \to 0$. Depending on $f(x)$, this may result in divergence or oscillations around the true value. Consequently, such cases require a different approach.

Let $f(x)$ be a continuous and differentiable function (up to at least second order) with at least one repeating root. We define a new function $u(x)$ as follows:

$$u(x) = \frac{f(x)}{f'(x)} \tag{4.13}$$

where $u(x)$ and $f(x)$ share the same root. Furthermore, suppose $f(x)$ has a root of multiplicity m. This can occur if $f(x)$ contains a factor $(x - \alpha)$ such as

$$f(x) = (x - \alpha)^m g(x), \quad m > 1 \tag{4.14}$$

where now $g''(\alpha) \neq 0$.

Upon substituting Eq. (4.14) into Eq. (4.13) and arranging, we obtain

$$u(x) = \frac{(x - \alpha)g(x)}{m\,g(x) + (x - \alpha)g'(x)} \tag{4.15}$$

which is a function with a simple root at $x = \alpha$. It can be shown that $u'(\alpha) = 1/m \neq 0$, where m indicates the multiplicity of the root. Since it is effective for simple roots, the Newton-Raphson method can also be employed for $u(x)$ instead of $f(x)$ by simply replacing $f(x)$ in Eq. (4.10) with $u(x)$:

$$x^{(p+1)} = x^{(p)} - \frac{u(x^{(p)})}{u'(x^{(p)})} \tag{4.16}$$

where $u'(x)$ can now be expressed in terms of $f(x)$ and $f'(x)$ as follows:

$$u'(x) = 1 - \frac{f(x)\,f''(x)}{(f'(x))^2}$$

Substituting $u'(x)$ and Eq. (4.13) into Eq. (4.16) and simplifying results in

$$x^{(p+1)} = x^{(p)} - \frac{f(x^{(p)})f'(x^{(p)})}{f'^2(x^{(p)}) - f(x^{(p)})f''(x^{(p)})} \tag{4.17}$$

Normally, we do not know in advance that a nonlinear equation has a root of multiplicity. However, if we knew that $f(x)$ had a multiplicity of m, we could speed up the Newton-Raphson method by replacing Eq. (4.10) with the following one:

$$x^{(p+1)} = x^{(p)} - m\frac{f(x^{(p)})}{f'(x^{(p)})} \tag{4.18}$$

which gives the conventional Newton-Raphson for $m = 1$.

The algorithm, as well as terminating the iteration process, is similar to that of the Newton-Raphson method. For this reason, a pseudocode for the modified Newton-Raphson method has not been given, as Pseudocode 4.3 could be used by simply replacing functions $f(x)$ and $f'(x)$ in FUNC(x) and FUNCP(x) with $u(x)$ and $u'(x)$ in the user-defined functions.

EXAMPLE 4.5: Finding a repeating root of a function

The function $f(x) = e^x - 2x - 2 + \ln 4$ has a root in $(-2, 2)$. Estimate the root correct to four decimal places, using $x^{(0)} = -1$ as the initial guess.

Modified Newton-Raphson Method

- The method is simple and easy to implement;
- The modified scheme given by Eq. (4.18) is computationally effective and fast if the multiplicity is known in advance;
- Both schemes of the modified method, like Newton-Raphson, converge quadratically.

- The variant of the modified Newton-Raphson method given by Eq. (4.17) requires deriving and calculating explicit expressions for $f'(x)$ and $f''(x)$, besides $f(x)$, which can be computationally expensive for complicated functions.

SOLUTION:

The graph of $f(x)$ in $-2 < x < 2$ is presented in Fig. 4.6, which reveals that $f(x)$ has a repeating (multiple) root about $x \approx 0.7$. Therefore, it is appropriate and necessary to apply modified Newton-Raphson to achieve stable and fast convergence.

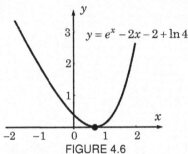

FIGURE 4.6

To apply Eq. (4.17), the first and second derivatives are obtained:

$$f(x) = e^x - 2x - 2 + \ln 4 = 0, \quad f'(x) = e^x - 2, \quad f''(x) = e^x$$

Starting with the initial guess of $x^{(0)} = -1$, we find

$$f(x^{(0)}) = 1.754174, \quad f'(x^{(0)}) = -1.632120, \quad f''(x^{(0)}) = 0.367879$$

which lead to

$$\delta^{(1)} = \frac{f(x^{(0)})f'(x^{(0)})}{f'^{\,2}(x^{(0)}) - f(x^{(0)})f''(x^{(0)})} = -1.418396$$

Upon substituting this result into Eq. (4.16), we find

$$x^{(1)} = -1 - (-1.418396) = 0.418396$$

When the iteration process is carried out in the same manner, it is observed that the estimates converge to the answer only after three iterations. The iteration progress is presented in Table 4.5. The rate of convergence is quadratic, i.e., $K \to \delta^{(p)}/(\delta^{(p-1)})^2 \to 1/6$. Since $|\delta^{(3)}| = 0.219 \times 10^{-4} < \varepsilon = 0.5 \times 10^{-4}$, we may predict at least four-decimal place accuracy in $x^{(3)}$ even though $|f(x^{(3)})| = 0.48 \times 10^{-9}$.

TABLE 4.5

p	$x^{(p)}$	$f(x^{(p)})$	$f'(x^{(p)})$	$f''(x^{(p)})$	$\delta^{(p)}$	$\delta^{(p)}/(\delta^{(p-1)})^2$
0	-1	1.754174	-1.632120	0.36788	1.418396	0.130858
1	0.418396	0.069025	-0.480478	1.51952	0.263266	0.130858
2	0.681662	0.000131	-0.022839	1.97716	0.011465	0.165388
3	0.693125	0.5×10^{-9}	-0.4×10^{-4}	1.99996	0.22×10^{-4}	0.166664

The given nonlinear equation has a root of multiplicity 2. Using this information, we can also use Eq. (4.18) with $m = 2$ to solve the problem. The iteration history is summarized in Table 4.6. This scheme also converges to an estimate accurate to at least seven decimal places: $|\delta^{(4)}| < 0.5 \times 10^{-7}$ and $|f(x^{(4)})| < 0.5 \times 10^{-7}$. The rate of convergence is also quadratic with a bounded K.

TABLE 4.6

p	$x^{(p)}$	$f(x^{(p)})$	$f'(x^{(p)})$	$\delta^{(p)}$	$\delta^{(p)}/(\delta^{(p-1)})^2$
0	-1	1.7541740	-1.632120	2.1495640	
1	1.149564	0.2439826	1.156816	0.4218174	0.0912901
2	0.7277466	0.0012110	0.070410	0.0344000	0.1933341
3	0.6933467	0.398×10^{-7}	0.399×10^{-3}	0.199×10^{-3}	0.1685966
4	0.6931472	-0.22×10^{-15}	0.133×10^{-7}	0.335×10^{-7}	0.8408510

When the problem is also solved using the conventional Newton-Raphson method, it yields the approximate solution after 14 iterations: $|\delta^{(14)}| = 0.309 \times 10^{-4} < \varepsilon$ and $|f(x^{(14)})| = 0.39 \times 10^{-8} < \varepsilon$. The iteration progress, along with the $\delta^{(p)}/\delta^{(p-1)}$ ratio, is tabulated in Table 4.7. Note that $|f(x^{(p)})|$ approaches zero faster than $|f'(x^{(p)})|$, which is why the Newton-Raphson method remains stable and converges to an approximate value. The convergence of the ratio $\delta^{(p)}/\delta^{(p-1)}$ to 0.5 verifies that the convergence rate is linear.

TABLE 4.7

p	$x^{(p)}$	$f(x^{(p)})$	$f'(x^{(p)})$	$\delta^{(p)}$	$\delta^{(p)}/\delta^{(p-1)}$
0	-1	1.754174	-1.6321210	1.074782	
1	0.0747820	0.3143796	-0.9223510	0.3408460	0.3171303
2	0.4156280	0.0703604	-0.4846780	0.1451695	0.4259093
3	0.5607974	0.0167686	-0.2479309	0.0676341	0.4658980
4	0.6284316	0.0040992	-0.1253320	0.0327067	0.4835838
5	0.6611384	0.0010137	-0.0630039	0.0160897	0.4919403
$\vdots$	$\vdots$	$\vdots$	$\vdots$	$\vdots$	$\vdots$
12	0.6928997	0.61×10^{-7}	-0.495×10^{-3}	0.124×10^{-3}	0.4999381
13	0.6930235	-0.15×10^{-7}	-0.247×10^{-3}	0.618×10^{-4}	0.4999691
14	0.6930853	0.39×10^{-8}	-0.123×10^{-3}	0.309×10^{-4}	0.4999845

Discussion: After three iterations, the estimated root with the modified Newton-Raphson method, Eq. (4.17), is accurate to at least four decimal places. Since the true value of the root is $\alpha = \ln 2$, we can determine the true error by $\alpha - x^{(3)} = 0.2218 \times 10^{-4} < 0.5 \times 10^{-4}$, which also verifies that this estimate is accurate to four decimal places. Note that the magnitudes of $|f(x^{(p)})|$ and $|\delta^{(p)}|$ can be significantly different; that is why it is important to ensure that both criteria are met.

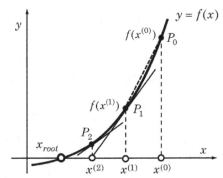

FIGURE 4.7: Depiction of the Secant method.

Since the multiplicity of the root is known, we are able to use Eq. (4.18) to find an estimate for the root. The method converged to an approximate value with a quadratic convergence rate after 4 iterations, with fewer arithmetic operations and function evaluations compared to using Eq. (4.17).

The conventional Newton-Raphson method with the same initial guess converged after 14 iterations with a linear convergence rate rather than quadratic, near a root of multiplicity. Although the modified Newton-Raphson method is given by Eq. (4.17) is preferred for cases with the roots of multiplicity, it is less efficient and requires more computational effort, largely due to $f''(x)$ evaluations, than the conventional Newton-Raphson method for simple roots. If the multiplicity of the root is known, Eq. (4.18) should be utilized instead of Eq. (4.17) because of its computational efficiency.

4.6 SECANT AND MODIFIED SECANT METHODS

A major handicap of the Newton-Raphson or modified Newton-Raphson method is that both methods require derivatives of $f(x)$. On the other hand, deriving analytical expressions for the derivatives of some functions can be difficult or may result in complicated expressions. In such cases, numerical methods that do not require analytical derivatives are more attractive.

One of the methods that suits this purpose is the *Secant method*. The algorithm is similar to that of the Newton-Raphson method. Basically, the same iteration equation, Eq. (4.10), can be used with the exception that $f'(x^{(p)})$ is replaced with a suitable approximation.

In Fig. 4.7, the two initial guess values and a line that passes through P_0 and P_1 are depicted. The slope of the secant line at P_1 is set equal to that of the tangent (dashed) line; that is,

$$f'(x^{(1)}) \approx \frac{f(x^{(0)}) - f(x^{(1)})}{x^{(0)} - x^{(1)}} \tag{4.19}$$

Substituting Eq. (4.19) into Eq. (4.10), a new estimate $x^{(2)}$ is found as

$$x^{(2)} = x^{(1)} - \frac{f(x^{(1)})}{f'(x^{(1)})} \cong x^{(1)} - \frac{f(x^{(1)})(x^{(0)} - x^{(1)})}{f(x^{(0)}) - f(x^{(1)})} \tag{4.20}$$

Equation (4.20) involves two abscissas ($x^{(0)}$, $x^{(1)}$) and two ordinates ($f(x^{(0)})$ and

Secant Method

- The Secant method is simple and easy to apply;
- It requires only one function evaluation as opposed to two evaluations in the Newton-Raphson method;
- It does not require derivatives;
- Starting with a set of initial guesses in close proximity to the root, the method converges very rapidly (faster than that of bisection and Regula-falsi methods) with an order of convergence of 1.62 (i.e., super-linear convergence).

- It requires two starting initial guesses;
- It suffers from many of the disadvantages of the Newton-Raphson method;
- Its convergence behavior depends on the function and the proximity of the initial guesses to the true value;
- The current estimates have no assured error boundaries;
- It may not always converge to a solution for the same reasons also pointed out for the Newton-Raphson method.

$f(x^{(1)}))$ to compute $x^{(2)}$. That is why a couple of initial guesses must be supplied to start the iteration process. However, unlike the bisection method, the signs of $f(x^{(0)})$ and $f(x^{(1)})$ need not be opposite.

To generalize the iteration equation, $f'(x^{(p)})$ in Eq. (4.10) is replaced by

$$f'(x^{(p)}) \cong \frac{f(x^{(p-1)}) - f(x^{(p)})}{x^{(p-1)} - x^{(p)}} \tag{4.21}$$

After simplifications, the iteration equation for the Secant method becomes

$$x^{(p+1)} = \frac{x^{(p)} f(x^{(p-1)}) - x^{(p-1)} f(x^{(p)})}{f(x^{(p-1)}) - f(x^{(p)})}, \qquad p \geqslant 1 \tag{4.22}$$

The $x^{(p-1)}$ and $f(x^{(p-1)})$ values must always be saved for the next iteration.

The *modified Secant method* is "modified" in the sense that it requires only one initial guess instead of two. To implement this method, $f'(x^{(p)})$ in Eq. (4.10) is replaced by the following approximation obtained from the definition of limit:

$$f'(x^{(p)}) \cong \frac{f(x^{(p)} + h) - f(x^{(p)})}{h} \tag{4.23}$$

where h should be chosen "small enough" so that the proposed approximation yields accurate estimates and "large enough" so that the current estimates are not contaminated by round-off errors.

The iteration equation for the modified Secant method can be expressed as

$$x^{(p+1)} = x^{(p)} - \frac{h f(x^{(p)})}{f(x^{(p)} + h) - f(x^{(p)})}, \qquad p \geqslant 0 \tag{4.24}$$

In Pseudocode 4.4, a pseudomodule implementing the secant method is presented. The SECANT_METHOD requires two initial guesses ($x0$ and $x1$), an upper bound for the number of iterations ($maxit$), and a convergence tolerance (ε) as input. The nonlinear equation $f(x)$ is provided to the module by a user-defined external function, FUNC(x). The outputs of the module are the $root$ and the total number of iterations performed, $iter$. As an iteration loop, a Repeat-Until-construct is used. First, the $f(x)$ is calculated for the initial guesses, $f(x^{(0)})$ and $f(x^{(1)})$. At the top of every iteration step, the estimate closest to the root is determined. Of these two, the $x^{(p)}$ assumed to be is closer to the root, which means $|f(x^{(p)})| < |f(x^{(p-1)})|$. The estimates are swapped ($x^{(p)} \rightleftarrows x^{(p-1)}$, $f(x^{(p)}) \rightleftarrows f(x^{(p-1)})$) if $|f(x^{(p)})| > |f(x^{(p-1)})|$ occurs at any time before a new estimate, $x^{(p+1)}$, is computed with Eq. (4.22). The convergence test is based on both $\delta^{(p)} = |x^{(p+1)} - x^{(p)}| < \varepsilon$ and $|f(x^{(p+1)})| < \varepsilon$. If no convergence is achieved, then the two most recent estimates are kept for the next iteration step by setting $x^{(p-1)} \leftarrow x^{(p)}$, $x^{(p)} \leftarrow x^{(p+1)}$ and $f(x^{(p-1)}) \leftarrow f(x^{(p)})$, $f(x^{(p)}) \leftarrow f(x^{(p+1)})$. To prevent runaway looping, the iteration count is monitored to ensure it does not exceed $maxit$. And if $maxit$ is reached with no convergence, the most current values are printed out with a warning message to the user.

EXAMPLE 4.6: Application of Secant method

Repeat Example 4.1 using (a) the Secant method with $r^{(0)} = 0$ and $r^{(1)} = 0.1$ and (b) the modified Secant method with $h = 0.05$ and $r^{(0)} = 0$.

SOLUTION:

(a) For the two initial guesses, we obtain

$$f(r^{(0)}) = -0.535, \qquad f(r^{(1)}) = 0.1 + 0.165 - 0.7e^{-0.1^2} = -0.428035$$

Using Eq. (4.22), the first estimate and associated ordinates are found as

$$r^{(2)} = \frac{r^{(1)}f(r^{(0)}) - r^{(0)}f(r^{(1)})}{f(r^{(0)}) - f(r^{(1)})} = \frac{0.1(-0.535) - (0)(-0.428035)}{(-0.535) - (-0.428035)} = 0.5001636$$

$$f(r^{(2)}) = -0.000686$$

The next estimate obtained as

$$r^{(3)} = \frac{r^{(2)}f(r^{(1)}) - r^{(1)}f(r^{(2)})}{f(r^{(1)}) - f(r^{(2)})} = \frac{0.5001636(-0.428035) - 0.1(-0.000686)}{(-0.428035) - (-0.000686)}$$

$$= 0.500806$$

which yields $|f(r^{(3)})| = 0.869 \times 10^{-6} < \varepsilon$ but $\delta^{(3)} = r^{(3)} - r^{(2)} > \varepsilon$. The iterative procedure is continued in this manner until the convergence criteria are met, i.e., the procedure is terminated after 4 iterations when $|f(r^{(4)})| < \varepsilon$ and $|\delta^{(4)}| < \varepsilon$. The iteration history is summarized in Table 4.8.

TABLE 4.8

p	$r^{(p)}$	$f(r^{(p)})$	$|\delta^{(p)}|$
0	0	−0.535000	
1	0.1	−0.428035	0.1
2	0.5001636	−6.867 × 10⁻⁴	0.4001631
3	0.5008061	−8.695 × 10⁻⁷	6.430 × 10⁻⁴
4	0.5008069	−1.747 × 10⁻¹²	8.153 × 10⁻⁷

Pseudocode 4.4

Module SECANT_METHOD ($x0$, $x1$, *maxit*, ε, *root*, *iter*)
\ DESCRIPTION: A pseudomodule to compute a root of nonlinear equation
\   using the Secant method.
\ USES:
\   ABS:: A built-in function computing the absolute value;
\   FUNC:: A user defined function giving nonlinear equation $f(x)$.
$fx0 \leftarrow$ **FUNC**($x0$)   \ Find $f(x^{(0)})$
$fx1 \leftarrow$ **FUNC**($x1$)   \ Find $f(x^{(1)})$
$p \leftarrow 1$ \ Initialize iteration counter
Repeat   \ Iterate until *convergence conditions* are met
  $p \leftarrow p+1$ \ Count iterations
  If $\left[\,|fx1| > |fx0|\,\right]$ **Then**   \ Swap to order as $|f(x^{(p-1)})| > |f(x^{(p)})|$
    $xt \leftarrow x1; fxt \leftarrow fx1$ \ Store $x1$ and $fx1$ on temporary variables
    $x1 \leftarrow x0$ \ Swap x's, $x^{(p)} \rightleftarrows x^{(p-1)}$
    $fx1 \leftarrow fx0$ \ Swap f's, $f(x^{(p)}) \rightleftarrows f(x^{(p-1)})$
    $x0 \leftarrow xt; fx0 \leftarrow fxt$ \ Set xt to $x^{(p-1)}$ and fxt to $f(x^{(p-1)})$
  End If
  $x2 \leftarrow (x1 * fx0 - x0 * fx1)/(fx0 - fx1)$ \ Find current estimate
  $fx2 \leftarrow$ **FUNC**($x2$)   \ Find $f(x^{(p+1)})$
  $\delta \leftarrow |x2 - x1|$ \ Find displacement $\delta^{(p)}$
  Write: p, $x2$, $fx2$, δ \ Print iteration progress
  \ Keep the last two estimates and corresponding ordinates
  $x0 \leftarrow x1; fx0 \leftarrow fx1$ \ Set $x^{(p-1)} = x^{(p)}$ and $f(x^{(p-1)}) = f(x^{(p)})$
  $x1 \leftarrow x2; fx1 \leftarrow fx2$ \ Set $x^{(p)} = x^{(p+1)}$ and $f(x^{(p)}) = f(x^{(p+1)})$
Until $\left[(|fx2| < \varepsilon$ **And** $\delta < \varepsilon)$ **Or** $p = maxit\right]$ \ Check for convergence
$root \leftarrow x2$ \ Root is found, if $p < maxit$
$iter \leftarrow p$ \ Number of iterations = Counter
If $\left[p = maxit\right]$ **Then**   \ Issue info and a warning
  Write: "Maximum number of iterations reached"
  Write: "Estimated root has NOT converged,! $|\delta|, |f(x)|=$", $\delta, fx2$
End If
End Module SECANT_METHOD

(b) For $r^{(0)} = 0$ and $h = 0.05$, we obtain $f(r^{(0)}) = -0.535$ and $f(r^{(0)} + h) = -0.481509$. Using Eq. (4.24) gives

$$r^{(1)} = r^{(0)} - \frac{h\,f(r^{(0)})}{f(r^{(0)} + h) - f(r^{(0)})} = 0 - \frac{0.05(-0.535)}{-0.481509 - (-0.535)} = 0.5000817$$

The iteration process converges to an approximate solution after three iterations when both convergence criteria are met, that is, $|f(r^{(4)})| = 1.851 \times 10^{-11} < \varepsilon$ and $|r^{(4)} - r^{(3)}| = 1.114 \times 10^{-7} < \varepsilon$. The iteration history is provided in Table 4.9.

TABLE 4.9

p	$r^{(p)}$	$f(r^{(p)})$	$\dfrac{f(r^{(p)} + h) - f(r^{(p)})}{h}$	$|\delta^{(p)}|$
0	0	-0.535	1.0698250	
1	0.5000817	-7.735×10^{-4}	1.0664193	0.5000817
2	0.5008070	1.188×10^{-7}	1.0664145	7.253×10^{-4}
3	0.5008069	-1.851×10^{-11}	1.0664145	1.114×10^{-7}

Discussion: Comparing to the iteration history of the Newton-Raphson method in Table 4.4, the Secant method (in Table 4.9) converges at $r^{(4)}$, whereas the Newton-Raphson method reached this accuracy by $r^{(3)}$. This is because the Secant method converges at a rate of 1.62, which is considerably faster than the linear convergence rate of the fixed-point iteration method but slightly slower than the quadratic convergence rate of the Newton-Raphson method. As a rule of thumb, if the computational effort required to evaluate $f'(x)$ is less than roughly half of the effort required to evaluate $f(x)$, the Newton-Raphson method is more efficient; otherwise, the secant method is.

The convergence of the modified secant method depends on the selection of the initial guess and the selection of h. The smaller the value of h, the more accurate $f'(x)$. In this example, $h = 0.05$ produced very good estimates for the derivatives.

4.7 ACCELERATING CONVERGENCE

From a given convergent sequence, one can build several iterative schemes that converge faster toward the same point. This process, known as *acceleration* has found an application in the root-finding algorithms. In this context, an iteration method with a linear convergence rate can be accelerated. *Aitken's acceleration* technique is one of several that suit this purpose. This technique transforms an original sequence $\{x^{(p)}\}$ into another sequence that converges with fewer iterations (faster) to the answer than the original sequence.

To illustrate the method, consider the sequence $\{x^{(p)}\}$ that converges to α, i.e., $\lim_{p \to \infty} x^{(p)} \to \alpha$. Recalling the proportionality of the error for sufficiently large p, we may write

$$\frac{e^{(p+1)}}{e^{(p)}} = \frac{e^{(p)}}{e^{(p-1)}} = \lambda \tag{4.25}$$

where $e^{(p)} = x^{(p)} - \alpha$ is the true error at the pth iteration step, α is the true root, and λ is the convergence rate ($|\lambda| < 1$ for converging series).

Using Eq. (4.25), for large p, we may write

$$\frac{x^{(p+1)} - \alpha}{x^{(p)} - \alpha} = \frac{x^{(p)} - \alpha}{x^{(p-1)} - \alpha} \tag{4.26}$$

Solving Eq. (4.26) for α yields

$$\alpha = \frac{x^{(p+1)} x^{(p-1)} - \left(x^{(p)}\right)^2}{x^{(p+1)} - 2x^{(p)} + x^{(p-1)}} \tag{4.27}$$

or alternatively

$$\alpha^{(p)} = x^{(p+1)} - \frac{\left(x^{(p+1)} - x^{(p)}\right)^2}{x^{(p+1)} - 2x^{(p)} + x^{(p-1)}} \tag{4.28}$$

Aitken's Acceleration Technique

- The method does not depend on the kind of root-finding method used to generate a sequence;
- It is not confined to sequences obtained by root-finding methods but can also be applied to convergent series.

- Its convergence is rather slow;
- Computation of the differences ($\Delta x^{(p)}$ and $\Delta^2 x^{(p)}$) can lead to "loss of significance";
- An original sequence should have a sufficient number of terms; otherwise, the acceleration procedure may lead to divergence or, at worst, give a solution that is grossly in error.

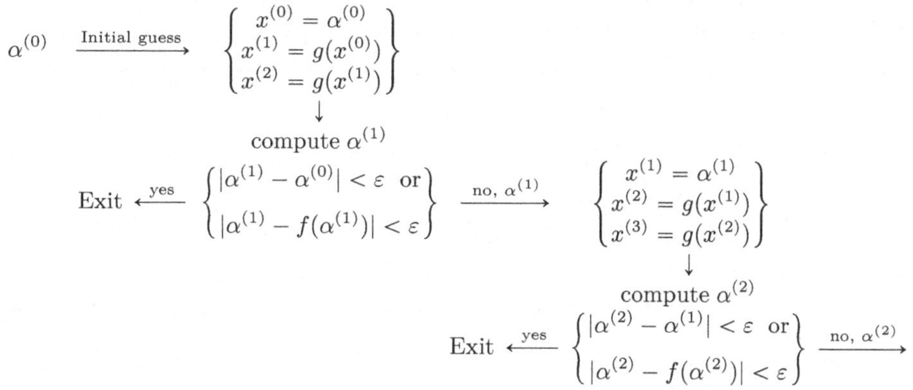

$$\alpha^{(0)} \xrightarrow{\text{Initial guess}} \left\{ \begin{array}{l} x^{(0)} = \alpha^{(0)} \\ x^{(1)} = g(x^{(0)}) \\ x^{(2)} = g(x^{(1)}) \end{array} \right\}$$

$$\downarrow$$

$$\text{compute } \alpha^{(1)}$$

$$\text{Exit} \xleftarrow{\text{yes}} \left\{ \begin{array}{l} |\alpha^{(1)} - \alpha^{(0)}| < \varepsilon \text{ or} \\ |\alpha^{(1)} - f(\alpha^{(1)})| < \varepsilon \end{array} \right\} \xrightarrow{\text{no, } \alpha^{(1)}} \left\{ \begin{array}{l} x^{(1)} = \alpha^{(1)} \\ x^{(2)} = g(x^{(1)}) \\ x^{(3)} = g(x^{(2)}) \end{array} \right\}$$

$$\downarrow$$

$$\text{compute } \alpha^{(2)}$$

$$\text{Exit} \xleftarrow{\text{yes}} \left\{ \begin{array}{l} |\alpha^{(2)} - \alpha^{(1)}| < \varepsilon \text{ or} \\ |\alpha^{(2)} - f(\alpha^{(2)})| < \varepsilon \end{array} \right\} \xrightarrow{\text{no, } \alpha^{(2)}}$$

FIGURE 4.8: Illustration of iteration sequence of implementing the Steffensen's method.

Defining $\Delta x^{(p)} = x^{(p)} - x^{(p-1)}$ and $\Delta^2 x^{(p)} = \Delta x^{(p)} - \Delta x^{(p-1)}$, Eq. (4.28) can then be recast as

$$\alpha^{(p)} = x^{(p+1)} - \frac{\left(\Delta x^{(p+1)}\right)^2}{\Delta^2 x^{(p+1)}}, \qquad p \geqslant 1 \tag{4.29}$$

which is used to generate a new and improved sequence of estimates, $\alpha^{(p)}$, with the aid of $x^{(p-1)}$, $x^{(p)}$ and $x^{(p+1)}$.

Steffensen's Acceleration is another acceleration method that is frequently used. It is an acceleration scheme that offers accelerated convergence for sequences that converge to a solution linearly. It is a combination of fixed-point iteration and Aitken's acceleration scheme and has a quadratic convergence property. The idea behind this technique is that $\alpha^{(p)}$ is thought to be a better approximation to the fixed-point estimate $x^{(p+1)}$, so it should be used as the next iterate for the fixed-point iteration process. In brief, the improved estimates are used in the fixed-point iteration equations as soon as they become available.

A pseudomodule, **STEFFENSEN**, for accelerating the convergence of a sequence with Steffensen's acceleration scheme is given in Pseudocode 4.5. An initial estimate ($x0$), an upper bound for iterations ($maxit$), and a convergence tolerance (ε) are supplied as input. As usual, the iteration function $g(x)$ is supplied to the module by **FUNC**(x), a user-defined external function. The iteration sequence is illustrated in the flowchart presented in Fig. 4.8.

Steffensen's Method

- The method has a quadratic convergence rate;
- It does convert diverging iterations into a converging sequence;
- Unlike Newton's methods, it does not require the first or second derivatives, which provide a considerable advantage in cases where a derivative is not easy to evaluate.

- The cost of faster convergence comes as a result of two function evaluations, namely $g(x^{(p)})$ and $g(x^{(p+1)})$ per iteration step, which increases the cpu time;
- The convergence behavior depends on the iteration function and the proximity of the initial guess to the true value;
- The method may fail to converge, or the sequence of estimates may either oscillate about the true value or diverge if the initial guess is not sufficiently proximate to the true root;
- It is difficult to extend and apply problems to multivariable systems.

The improved sequence of estimates is denoted by $\alpha^{(p)}$. At the start, the initial guess is set to $\alpha^{(0)} = x^{(0)}$. The intermediate values, $x^{(1)} = g(x^{(0)})$ and $x^{(2)} = g(x^{(1)})$, are successively computed within the **Repeat-Until** loop. An improved estimate, $\alpha^{(1)}$, is found from Eq. (4.29), and the convergence check is performed. The root estimation procedure is repeated until both $|\alpha^{(p)} - \alpha^{(p-1)}| < \varepsilon$ and $|\alpha^{(p)} - f(\alpha^{(p)})| < \varepsilon$ are simultaneously satisfied. So long as $p \leqslant maxit$, the improved estimate is set as the initial value $(x^{(p)} \leftarrow \alpha^{(p)})$ to compute the next set of subsequent intermediate values: $x^{(p+1)} = g(x^{(p)})$ and $x^{(p+2)} = g(x^{(p+1)})$. When the number of iterations reaches $p = maxit$ with no convergence, the most recent values are printed out along with a warning message to the user.

Pseudocode 4.5

Module STEFFENSEN ($x0$, $maxit$, ε, $root$, $iter$)
\ DESCRIPTION: A pseudomodule to compute a root of nonlinear equation
\   using the Steffensen's acceleration with fixed-point iteration method.
\ USES:
\   FUNC:: A user-defined function supplying $g(x)$ as FUNC(x)=$g(x)$;
\   ABS :: A built-in function computing the absolute value.
$\alpha0 \leftarrow x0$ \ Initialize root with initial guess
$p \leftarrow 0$ \ Initialize iteration counter
Repeat   \ Iterate until *convergence conditions* are met
   $p \leftarrow p + 1$ \ Count iterations
   $x1 \leftarrow$ **FUNC**($x0$)   \ Find estimate $x^{(p)}$
   $\Delta x1 \leftarrow x1 - x0$ \ Find displacement $\Delta x1$
   $x2 \leftarrow$ **FUNC**($x1$)   \ Find another estimate $x^{(p+1)}$
   $\Delta x2 \leftarrow x2 - x1$ \ Find displacement $\Delta x2$
   $\alpha \leftarrow x2 - (dx2)^2/(dx2 - dx1)$ \ Find improved estimate, $\alpha^{(n)}$
   $d\alpha \leftarrow \alpha - \alpha0$ \ Find $d\alpha$
   $dg \leftarrow \alpha -$ **FUNC**(α)   \ Find dg
   Write: p, $x0$, $x1$, $x2$, α, $d\alpha$, dg \ Print iteration progress

$\alpha 0 \leftarrow \alpha; \quad x0 \leftarrow \alpha 0$ \qquad \textbackslash Set current values as prior

Until $\left[(|d\alpha| < \varepsilon \textbf{ And } |dg| < \varepsilon) \textbf{ Or } p = maxit\right]$ \qquad \textbackslash Check for convergence

$root \leftarrow \alpha$ \qquad \textbackslash Root is found, if $p < maxit$

$iter \leftarrow p$ \qquad \textbackslash Counter gives total number of iterations

If $\left[p = maxit\right]$ **Then** \qquad \textbackslash Issue info and a warning

  Write: "Maximum number of iterations reached"

  Write: "Estimated root has NOT converged,! $\alpha, g(\alpha)=$", α, dg

End If

End Module STEFFENSEN

EXAMPLE 4.7: Acceleration techniques

Consider the iteration sequence $\{r^{(p)}\}$ generated in Example 4.3. (a) Apply Aitken's acceleration to the sequence to improve its convergence and calculate the convergence rates for both $\{r^{(p)}\}$ and $\{\alpha^{(p)}\}$ sequences; (b) Apply Steffensen's acceleration to estimate the root to four decimal places.

SOLUTION:

(a) The fixed-point iterates are generated using $g(r) = 0.7e^{-0.1r} - 0.165$ as the iteration function. The first five iterates, $\{r^{(p)}\}$, are read from Table 4.3. The approximate convergence rate of the sequence is computed from $\lambda^{(p)} \approx \Delta r^{(p+1)}/\Delta r^{(p)}$.

Starting with $p = 1$, we find

$$\Delta r^{(1)} = r^{(1)} - r^{(0)} = 0.44683862 - 1 = -0.5316138$$

$$\Delta r^{(2)} = r^{(2)} - r^{(1)} = 0.5029690 - 0.4683862 = 0.0345828$$

$$\Delta^2 r^{(2)} = \Delta r^{(2)} - \Delta r^{(1)} = 0.0345828 - (-0.5316138) = 0.5661966$$

Then, for the current step, we find

$$\alpha^{(1)} = r^{(2)} - \frac{\left(\Delta r^{(2)}\right)^2}{\Delta^2 r^{(2)}} = 0.5029690 - \frac{(0.0345828)^2}{0.566197} = 0.5008567$$

$$\lambda^{(1)} \approx \left|\frac{\Delta r^{(2)}}{\Delta r^{(1)}}\right| = \left|\frac{0.0345828}{0.5316138}\right| = 0.0650524$$

For $p = 2$, we obtain

$$\Delta r^{(3)} = r^{(3)} - r^{(2)} = 0.5006629 - 0.5029690 = -0.0023060$$

$$\Delta^2 r^{(3)} = \Delta r^{(3)} - \Delta r^{(2)} = -0.0023061 - (0.0345828) = -0.0368889$$

$$\alpha^{(2)} = r^{(3)} - \frac{\left(\Delta r^{(3)}\right)^2}{\Delta^2 r^{(3)}} = 0.5006629 - \frac{(-0.0023060)^2}{(-0.0368889)} = 0.5008071$$

$$\lambda^{(2)} \approx \left|\frac{\Delta r^{(3)}}{\Delta r^{(2)}}\right| = \left|\frac{-0.0023060}{0.0345828}\right| = 0.0666815$$

TABLE 4.10

p	$r^{(p)}$	$\Delta r^{(p)})$	$\lvert \lambda_r^{(p)} \rvert$	$\alpha^{(p)}$	$\Delta \alpha^{(p)}$	$\lvert \lambda_\alpha^{(p)} \rvert$
0	1					
1	0.4683862	-0.5316138	0.0650524	0.5008567		
2	0.5029690	0.0345828	0.0666815	0.5008071	-4.97×10^{-5}	0.0043963
3	0.5006629	-0.002306	0.0665740	0.5008069	-2.18×10^{-7}	0.0044340
4	0.5008165	1.54×10^{-4}	0.0665811	0.5008069	-9.7×10^{-10}	
5	0.5008062	-1.02×10^{-5}				

Repeating this procedure leads to a new sequence $\{\alpha^{(p)}\}$ with four values. Five iteration steps of Aitken's acceleration parameters are presented in Table 4.10. The actual computations were performed to 15 decimal places but rounded to 7 decimal places to fit in the tables. Note that the fixed-point iteration sequence $\{r^{(p)}\}$ depicts a convergence rate that is slightly oscillating as $\lambda_r \to 0.06658$. However, as the new sequence $\{\alpha^{(p)}\}$ decreases toward the true root with the convergence rate $\lambda_\alpha \to 0.04434$, we observe $\lambda_\alpha \approx (2/3)\lambda_r$, which is evidence of faster convergence than fixed-point iteration.

(b) To start Steffensen's acceleration procedure, we begin with the first three iterates of fixed-point iteration $(r^{(0)}, r^{(1)}, r^{(2)})$. Using Eq. (4.29), the first approximation with Aitken's acceleration, $\alpha^{(1)}$, is found as

$$\alpha^{(1)} = r^{(2)} - \frac{\left(\Delta r^{(2)}\right)^2}{\Delta^2 r^{(2)}} = 0.502969 - \frac{(0.0345828)^2}{0.566197} = 0.5008567$$

with $\Delta \alpha^{(1)} = \alpha^{(1)} - r^{(0)} = -0.4991433$.

We use this value to restart fixed-point iterations, $r^{(1)} = \alpha^{(1)}$, and calculate the next two iterates as

$$r^{(2)} = g(r^{(1)}) = g(0.5008567) = 0.5008036,$$

$$r^{(3)} = g(r^{(2)}) = g(0.5008036) = 0.5008071$$

Next, we compute

$$\Delta r^{(2)} = r^{(2)} - r^{(1)} = 0.5008036 - 0.5008567 = -5.31 \times 10^{-5}$$

$$\Delta r^{(3)} = r^{(3)} - r^{(2)} = 0.5008067 - 0.5008036 = 3.1 \times 10^{-6}$$

$$\Delta^2 r^{(3)} = \Delta r^{(3)} - \Delta r^{(2)} = 3.1 \times 10^{-6} - (-5.31 \times 10^{-5}) = 5.62 \times 10^{-5}$$

Applying Aitken's acceleration formula, Eq. (4.29), to $r^{(1)}$, $r^{(2)}$, and $r^{(3)}$ we find

$$\alpha^{(2)} = r^{(3)} - \frac{\left(\Delta r^{(3)}\right)^2}{\Delta^2 r^{(3)}} = 0.5008071 - \frac{(3.1 \times 10^{-6})^2}{5.62 \times 10^{-5}} = 0.5008069$$

The iteration history is summarized in Table 4.11. In this exercise, as shown, the Steffensen's acceleration required only two iterations and converged with $\lvert \alpha^{(2)} - \alpha^{(1)} \rvert = 0.498 \times 10^{-4} < \varepsilon$ and $\lvert \alpha^{(2)} - g(\alpha^{(1)}) \rvert = 0.549 \times 10^{-12} < \varepsilon$. The ratio $\Delta\alpha^{(2)}/\Delta\alpha^{(1)}$ at the end of the second iteration is about 10^{-4}, which indicates a convergence faster than the linear convergence rate.

TABLE 4.11

p	$r^{(p-1)}$	$r^{(p)})$	$r^{(p+1)}$	$|\alpha^{(p)}|$	$\Delta\alpha^{(p)}$	$|\alpha^{(p)} - g(\alpha^{(p)})|$
1	1	0.4683862	0.5029690	0.5008567	0.4991433	5.31×10^{-5}
2	0.5008567	0.5008036	0.5008071	0.5008069	5×10^{-5}	5.49×10^{-13}

Discussion: The fixed iterative point method was modified such that Steffensen's method converges quadratically while still keeping the cost low and the method simple.

In this example, we clearly observe that Steffensen's method converges faster than Aitken's method. Though a word of caution is in order, due to the subtractions of very close iterates, the "loss of significance" can occur, and it may ultimately corrupt the results. Hence, the Aitken acceleration process should be terminated immediately after the iterates become apparently stationary. But Steffensen's method combines the fixed-point iteration with Aitken's method by using the new iterates, $\alpha^{(p)}$, immediately since they are better estimates of the root than $r^{(p)}$. The method converges with a quadratic convergence rate but does not require the derivative of the nonlinear equation.

4.8 SYSTEM OF NONLINEAR EQUATIONS

Systems of nonlinear equations are frequently encountered in many branches of science and engineering since most physical systems occurring in nature are inherently nonlinear. Hence, the mathematical modeling of such systems also leads to nonlinear simultaneous equations.

Nonlinear systems of equations could be very complicated and difficult to solve if the system variables are highly interdependent, in which case the solution of such a system may depict abrupt changes with slight perturbations in the input variables. For this reason, the numerical solution of nonlinear equations is one of the major challenges facing a numerical analyst today.

Hence, the search for the numerical solution of nonlinear systems of equations has been an old and difficult problem. The bracketing methods are practically very difficult to implement in systems of nonlinear equations. The best alternative, for being very simple and effective, is Newton's method. That is why the numerical algorithm that we will be discussing in this section is an extension of the *Newton-Raphson Method* to multivariable functions.

For simplicity, we consider the following system of two nonlinear equations with two unknowns:

$$f_1(x, y) = 0, \qquad f_2(x, y) = 0 \tag{4.30}$$

where x and y are independent variables (unknowns), and f_1 and f_2 are two coupled nonlinear functions. In some cases, a 2×2 non-linear system can be reduced to a single nonlinear equation without much difficulty. Nevertheless, in this section, we will assume that the coupled nonlinear systems are irreducible to a single nonlinear equation.

Consider the Taylor series expansion of $f(x, y)$ about (x_0, y_0):

$$f(x, y) = f(x_0, y_0) + \frac{1}{1!}\left((x-x_0)\frac{\partial}{\partial x} + (y-y_0)\frac{\partial}{\partial y}\right)f(x_0, y_0) + \mathcal{O}\left((\delta x)^2 + (\delta y)^2\right) + \ldots \quad (4.31)$$

Now, consider the truncated Taylor series approximation of $f_1(x, y)$ and $f_2(x, y)$ with the first two terms in the vicinity of (x_0, y_0):

$$f_1(x, y) \cong f_1(x_0, y_0) + (x - x_0)\frac{\partial f_1}{\partial x}(x_0, y_0) + (y - y_0)\frac{\partial f_1}{\partial y}(x_0, y_0)$$

$$f_2(x, y) \cong f_2(x_0, y_0) + (x - x_0)\frac{\partial f_2}{\partial x}(x_0, y_0) + (y - y_0)\frac{\partial f_2}{\partial y}(x_0, y_0) \quad (4.32)$$

When (x, y) is the true solution, both equations will be satisfied. Thus, setting $f_1(x, y) = 0$ and $f_2(x, y) = 0$ in Eq. (4.32) and rearranging yields the following matrix equation:

$$\begin{bmatrix} \dfrac{\partial f_1}{\partial x}(x_0, y_0) & \dfrac{\partial f_1}{\partial y}(x_0, y_0) \\[2ex] \dfrac{\partial f_2}{\partial x}(x_0, y_0) & \dfrac{\partial f_2}{\partial y}(x_0, y_0) \end{bmatrix} \begin{bmatrix} \delta x \\[1ex] \delta y \end{bmatrix} = -\begin{bmatrix} f_1(x_0, y_0) \\[1ex] f_2(x_0, y_0) \end{bmatrix} \quad (4.33)$$

where $\delta x = x - x_0$ and $\delta y = y - y_0$, and the matrix evaluated at (x_0, y_0) is called the *Jacobian matrix*, which is usually denoted by $\mathbf{J}$. In order for Eq. (4.33) to have a unique set of solutions, the Jacobian matrix needs to be non-singular, i.e., $\det(\mathbf{J}) = J \neq 0$.

The solution of Eq. (4.33) provides a set of improved estimates. Adopting the super-script notation, the sequence of estimates and the absolute errors at any pth iteration step are denoted by $(x^{(p)}, y^{(p)})$ and $(\delta x^{(p)} = x^{(p+1)} - x^{(p)}, \delta y^{(p)} = y^{(p+1)} - y^{(p)})$, respectively.

The iteration equations from the solution of Eq. (4.33) can be written explicitly as

$$x^{(p+1)} = x^{(p)} - \frac{1}{J}\left\{ f_1(x^{(p)}, y^{(p)})\frac{\partial f_2}{\partial y}(x^{(p)}, y^{(p)}) - f_2(x^{(p)}, y^{(p)})\frac{\partial f_1}{\partial y}(x^{(p)}, y^{(p)}) \right\}$$

$$y^{(p+1)} = y^{(p)} - \frac{1}{J}\left\{ -f_1(x^{(p)}, y^{(p)})\frac{\partial f_2}{\partial x}(x^{(p)}, y^{(p)}) + f_2(x^{(p)}, y^{(p)})\frac{\partial f_1}{\partial x}(x^{(p)}, y^{(p)}) \right\} \quad (4.34)$$

where

$$J = J(x^{(p)}, y^{(p)}) = \left(\frac{\partial f_1}{\partial x}\frac{\partial f_2}{\partial y} - \frac{\partial f_1}{\partial y}\frac{\partial f_2}{\partial x}\right)_{(x^{(p)}, y^{(p)})} \quad (4.35)$$

Note that the algorithm requires the partial derivatives of $f_1(x, y)$ and $f_2(x, y)$ with respect to x and y variables. Hence, it is necessary to evaluate f_1 and f_2 as well as their partial derivatives, preferably analytically. The convergence criterion can be based on ℓ_∞−norm $\max\left(\delta x^{(p)}, \delta y^{(p)}\right) < \varepsilon$ or ℓ_2−norm $((\delta x^{(p)})^2 + (\delta y^{(p)})^2)^{1/2} < \varepsilon$.

Partial derivatives of a Jacobian matrix may be computed using finite difference formulas if nonlinear equations contain very long and/or complex mathematical expressions. However, this alternative may increase cpu-time and can have an adverse effect on convergence if the system is sensitive to small perturbations or the linear system is ill-conditioned.

Now, we extend the Newton's method to n simultaneous nonlinear equations with n unknowns. Let the system of nonlinear equations be given as

$$\mathbf{f}(\mathbf{x}) = 0 \qquad (4.36)$$

where

$$\mathbf{x} = \begin{bmatrix} x_1 \\ x_2 \\ \vdots \\ x_n \end{bmatrix}, \qquad \mathbf{f}(\mathbf{x}) = \begin{bmatrix} f_1(\mathbf{x}) \\ f_2(\mathbf{x}) \\ \vdots \\ f_n(\mathbf{x}) \end{bmatrix} = \begin{bmatrix} f_1(x_1, x_2, \ldots, x_n) \\ f_2(x_1, x_2, \ldots, x_n) \\ \vdots \\ f_n(x_1, x_2, \ldots, x_n) \end{bmatrix}$$

In matrix-vector notation, the generalized Taylor series expansion of $\mathbf{f}(\mathbf{x})$ about $\mathbf{x}^{(p)}$ in general space can be written as

$$\mathbf{f}(\mathbf{x}) \approx \mathbf{f}(\mathbf{x}^{(p)}) + \frac{d}{d\mathbf{x}}\mathbf{f}(\mathbf{x}^{(p)})\left(\mathbf{x} - \mathbf{x}^{(p)}\right) + \cdots$$

Higher-order terms of the Taylor series can be neglected if $\mathbf{x} = \mathbf{x}^{(p+1)}$ is sufficiently close to $\mathbf{x}^{(p)}$. Then, a two-term Taylor approximation becomes

$$\mathbf{f}(\mathbf{x}^{(p+1)}) \approx \mathbf{f}(\mathbf{x}^{(p)}) + \mathbf{J}\left(\mathbf{x}^{(p)}\right)\boldsymbol{\delta}^{(p)} + \mathcal{O}(\boldsymbol{\delta}^{(p)})^2 \qquad (4.37)$$

where $\boldsymbol{\delta}^{(p)} = \mathbf{x}^{(p+1)} - \mathbf{x}^{(p)}$ is the displacement (or error) vector, $\mathbf{x}^{(p+1)}$ and $\mathbf{x}^{(p)}$ are vectors containing the current and prior estimates at the pth iteration step, and $\mathbf{J}$ is an $n \times n$ Jacobian matrix expressed as

$$\mathbf{J}(\mathbf{x}) = \frac{d}{d\mathbf{x}}\mathbf{f}(\mathbf{x}) = \begin{bmatrix} \dfrac{\partial f_1}{\partial x_1}(\mathbf{x}) & \dfrac{\partial f_1}{\partial x_2}(\mathbf{x}) & \cdots & \dfrac{\partial f_1}{\partial x_n}(\mathbf{x}) \\ \dfrac{\partial f_2}{\partial x_1}(\mathbf{x}) & \dfrac{\partial f_2}{\partial x_2}(\mathbf{x}) & \cdots & \dfrac{\partial f_2}{\partial x_n}(\mathbf{x}) \\ \vdots & \vdots & \ddots & \vdots \\ \dfrac{\partial f_n}{\partial x_1}(\mathbf{x}) & \dfrac{\partial f_n}{\partial x_2}(\mathbf{x}) & \cdots & \dfrac{\partial f_n}{\partial x_n}(\mathbf{x}) \end{bmatrix} \qquad (4.38)$$

with the matrix elements $J_{ij} = \partial f_i / \partial x_j$ evaluated at $\mathbf{x}^{(p)}$.

Assuming $\mathbf{x}^{(p+1)}$ is sufficiently close enough to the true solution, thus setting $\mathbf{f}(\mathbf{x}^{(p+1)}) = 0$ in Eq. (4.37) and rearranging, the following system of linear equations for $\boldsymbol{\delta}^{(p)}$ is obtained:

$$\mathbf{J}(\mathbf{x}^{(p)})\,\boldsymbol{\delta}^{(p)} = -\mathbf{f}(\mathbf{x}^{(p)}) \qquad (4.39)$$

which should be solved with a direct method at every iteration step. Nevertheless, the characteristic of the Jacobian matrix is important in determining the numerical method, as it may be ill-conditioned. The Jacobian matrix is a dense matrix, and for a small number of nonlinear equations, it can be inverted using a symbolic processor to obtain the solution as $\boldsymbol{\delta}^{(p)} = -\mathbf{J}^{-1}(\mathbf{x}^{(p)})\mathbf{f}(\mathbf{x}^{(p)})$. The current estimates, $\mathbf{x}^{(p+1)}$, are then updated as

$$\mathbf{x}^{(p+1)} = \mathbf{x}^{(p)} + \boldsymbol{\delta}^{(p)} \qquad (4.40)$$

The Jacobian is a dense matrix and is usually not diagonally dominant. Hence, the iterative methods for solving Eq. (4.39) in nonlinear problems cannot be considered an alternative.

Newton's Method

- Newton's method is easy to understand and implement;
- It can be applied to solve a variety of problems;
- If a solution exists, it converges at a quadratic convergence rate with a good set of initial guesses, i.e., the error is squared (or the number of accurate-digit doubles) at each iteration step.

- It may fail to converge, or convergence difficulties may be observed if the system is very sensitive to initial values;
- The Jacobian should be constructed, preferably *analytically*, which can be tedious for large or complex systems;
- The Jacobian and its inverse have to be calculated at each iteration step, which can be quite time-consuming depending on how large the system is.
- The resulting linear systems are generally not suitable to be solved by iterative methods; thus, applying a direct method at each iteration step can also be cpu-time intensive;
- For ill-conditioned systems, $\kappa(\mathbf{J}) \gg 1$, a proper correction algorithm should also be employed.

A good set of initial guesses, $\mathbf{x}^{(0)}$, is required to start the iteration process. Nonetheless, the choice of a set of initial values is a determining factor for the convergence of the method. If the initial values are not in close proximity to the true solution, then the procedure may diverge or converge. In this context, finding a set of good initial guesses is important and a bit tricky as well, because it often requires some knowledge of the variables of the actual physical problem being undertaken. The current estimates are computed from Eqs. (4.39) and (4.40), preferably employing the Gauss elimination method with partial pivoting, which is the most common and generally the most robust method for this purpose. Note that the Jacobian matrix requires prior estimates ($\mathbf{J}^{(p)} = \mathbf{J}(\mathbf{x}^{(p)})$), and thus it needs to be computed at every iteration step. The process of solving the system of linear equations at every iteration step is repeated until convergence is achieved.

Pseudocode 4.6

Module NEWTON_SYSTEM ($n, \mathbf{x}0, \mathbf{x}, maxit, \varepsilon, Norm, iter$)
\ DESCRIPTION: A pseudomodule to find common solution of a system of
\   a coupled nonlinear equations using Newton's method.
\ USES:
\   FCN:: User-defined module supplying the nonlinear eqs and Jacobian;
\   GAUSS_ELIMINATION_P:: Gauss-Elimination module (Pseudocode 2.10;
\   ENORM:: Function Module calculating Euclidean (ℓ_2-)norm of a vector
   (Pseudocode 3.1)
Declare: $x_n, x0_n, \delta_n, f_n, j_{n,n}$ \ Declare array variables
$p \leftarrow 0$ \ Initialize iteration counter
Repeat   \ Iterate until convergence conditions are met
   $p \leftarrow p + 1$ \ Count iterations
   FCN($n, \mathbf{x}0, \mathbf{f}, \mathbf{J}$)   \ Compute $\mathbf{f}(x^{(p)})$ and $\mathbf{J}(x^{(p)})$
   GAUSS_ELIMINATION_P($n, \mathbf{J}, \mathbf{f}, \delta$)   \ Solve $\mathbf{J}(\mathbf{x}^{(p)})\delta^{(p)} = \mathbf{f}(\mathbf{x}^{(p)})$

```
        Norm ← ENORM(n, δ)                       \ Set ℓ₂−norm of δ⁽ᵖ⁾ to Norm
        Write: "Iteration=", p, "ℓ₂ norm=",Norm       \ Print iteration progress
        For [k = 1, n]                            \ Loop k: Find current estimates
          xₖ ← x0ₖ − δₖ                            \ Compute x⁽ᵖ⁺¹⁾
          x0ₖ ← xₖ                                 \ Set current estimates as prior
        End For
        Write: x, f
      Until [Norm < ε Or p = maxit]                \ Check for convergence
      iter ← p                              \ Total number of iterations performed
      \ If the iteration limit maxit is reached with no convergence
      If [p = maxit] Then      \ Print most current estimates and issue a warning
          Write: "Maximum number of iterations reached, xᵢ's=", x
          Write: "Estimated root has NOT converged! Errors δᵢ's=", δ
      End If
   End Module NEWTON_SYSTEM


   Module FCN (n, x, f, J)
   \ DESCRIPTION: A user-defined module to evaluate nonlinear Eqs & Jacobian.
   Declare: xₙ, fₙ, jₙ,ₙ                              \ Declare array variables
   \ Supply n-nonlinear eqs as f₁(x₁,...,xₙ) = 0, ..., fₙ(x₁,...,xₙ) = 0
   f₁ ← x₁² + 2x₂² − 10                \ Equation 1: f₁(x₁,x₂) = x₁² + 2x₂² − 10 = 0
   f₂ ← x₁²x₂ + x₂³ + 3                \ Equation 2: f₂(x₁,x₂) = x₁²x₂ + x₂³ + 3 = 0
   \ Construct the Jacobian, Jᵢⱼ = ∂fᵢ/∂xⱼ
   j₁,₁ ← 2x₁;   j₁,₂ ← 4x₂;               \ Row-1: ∂f₁/∂x₁ = 2x₁, ∂f₁/∂x₂ = 4x₂
   j₂,₁ ← 2x₁x₂;   j₂,₂ ← x₁² + 3x₂²       \ Row-2: ∂f₂/∂x₁ = 2x₁x₂, ∂f₂/∂x₂ = x₁² + 3x₂²
   End Module FCN
```

A *sufficient condition* for the convergence at every iteration step is $\det(\mathbf{J}^{-1}) < 1$. A convergence criterion is generally based on either ℓ_∞ or ℓ_2-norms or both. The most common criterion is based on ℓ_2-norm; that is,

$$\left\| \boldsymbol{\delta}^{(p)} \right\|_2 = \left\| \mathbf{x}^{(p+1)} - \mathbf{x}^{(p)} \right\|_2 < \varepsilon$$

A pseudo-module, **NEWTON_SYSTEM**, implementing Newton's algorithm for a system of nonlinear equations is presented in Pseudocode 4.6. The code requires the number of equations (n), an initial guess vector $(\mathbf{x}0)$, an upper bound for iterations $(maxit)$, and a convergence tolerance (ε) as input. The system of nonlinear equations $\mathbf{f}(\mathbf{x})$ and the Jacobian $\mathbf{J}(\mathbf{x})$ are supplied by **FCN**, a user-defined module. The estimated solution $(\mathbf{x})$, Euclidean norm of the displacement vector $(Norm)$, and the total number of iterations performed $(iter)$ are the outputs. As a standard procedure, a **Repeat-Until** (iteration) loop is checked against $maxit$ to avoid computations going into an infinite loop. By visiting the module **FCN** at the top of every iteration step, the nonlinear equations $\mathbf{f}(\mathbf{x}^{(p)})$ and the Jacobian $\mathbf{J}(\mathbf{x}^{(p)})$ are updated. Then, $\mathbf{J}(\mathbf{x}^{(p)})\boldsymbol{\delta}^{(p)} = -\mathbf{F}(\mathbf{x}^{(p)})$ is solved using the Gauss elimination procedure that implements a pivoting technique. The current estimates are found by Eqs. (4.40). The iteration procedure is terminated when the $\|\boldsymbol{\delta}^{(p)}\|_2 < \varepsilon$ condition is met; otherwise, the iteration procedure is repeated by setting the current estimates as prior $\mathbf{x}^{(p)} \leftarrow \mathbf{x}^{(p+1)}$.

EXAMPLE 4.8: Implementing Newton's method to two equations with two unknowns

A pair of reversible chemical reactions that occur simultaneously are given as follows:

$$A + 2B \rightleftarrows D, \qquad A + C \rightleftarrows D$$

where A, B, C, and D denote chemical compounds. A chemical reactor is fed with 15 moles of A, 9 moles of B, and 5 moles of C. At equilibrium, the first and second reactions produce x and y moles of D, respectively. The system leads to the following equilibrium equations:

$$K_1 = \frac{x+y}{(15-x-y)(9-2x)^2}, \qquad K_2 = \frac{x+y}{(15-x-y)(5-y)}$$

where $K_1 = 1$ and $K_2 = 50$ are the equilibrium constants. Determine x and y accurately to four decimal places using Newton's method. Use $x^{(0)} = y^{(0)} = 2$.

SOLUTION:

Rearranging the equilibrium equations yields the following nonlinear system of equations:

$$f_1(x,y) = x + y - (15-x-y)(9-2x)^2 = 0,$$

$$f_2(x,y) = x + y - 50(15-x-y)(5-y) = 0$$

To construct the Jacobian matrix, the Newton's method requires partial derivatives of f_1 and f_2 wrt x and y, which are obtained as

$$\frac{\partial f_1}{\partial x} = 1 + (9-2x)(69-6x-4y), \qquad \frac{\partial f_1}{\partial y} = 1 + (9-2x)^2,$$

$$\frac{\partial f_2}{\partial x} = 1 + 50(50-y), \qquad \frac{\partial f_2}{\partial y} = 1 + 50(20-x-2y).$$

For given initial guess set, we obtain $\mathbf{f}(x,y)$, $\mathbf{J}(x,y)$, and $J(x,y)$ as

$$\mathbf{f}^{(0)}(2,2) = \begin{bmatrix} -271 \\ -1646 \end{bmatrix}, \qquad \mathbf{J}^{(0)}(2,2) = \begin{bmatrix} 246 & 26 \\ 151 & 701 \end{bmatrix}, \qquad J^{(0)}(2,2) = 168520$$

where $J(x,y)$ is the determinant of the Jacobian matrix. Note that, with $\det(\mathbf{J}^{(0)}) > 1$, the system is not ill-conditioned.

Now we proceed to compute the current displacement vector as

$$\boldsymbol{\delta}^{(0)} = -\left[\mathbf{J}^{(0)}(2,2)\right]^{-1}\mathbf{f}^{(0)}(2,2) = \begin{bmatrix} 0.873338 \\ 2.159951 \end{bmatrix}$$

where $\boldsymbol{\delta}^{(p)} = [\delta_x^{(p)} \ \delta_y^{(p)}]^T$ is the displacement vector with $\delta_x^{(p)} = x^{(p+1)} - x^{(p)}$ and $\delta_y^{(p)} = y^{(p+1)} - y^{(p)}$.

Next, the current estimates are updated by $\mathbf{x}^{(1)} = \mathbf{x}^{(0)} + \boldsymbol{\delta}^{(0)}$ as follows:

$$\mathbf{x}^{(1)} = \begin{bmatrix} 2 \\ 2 \end{bmatrix} + \begin{bmatrix} 0.873338 \\ 2.159951 \end{bmatrix} = \begin{bmatrix} 2.873338 \\ 4.159951 \end{bmatrix}$$

Note that $\|\boldsymbol{\delta}^{(p)}\|_2$ is an indicator of how close the overall estimate is to the true solution. Thus, to obtain estimates accurate to four decimal places, we will seek $\|\boldsymbol{\delta}^{(p)}\|_2 < \varepsilon$, where the tolerance is set to $\varepsilon = 0.5 \times 10^{-4}$.

TABLE 4.12

p	$x^{(p)}$	$y^{(p)})$	$\delta_x^{(p)}$	$\delta_y^{(p)}$	$\|\boldsymbol{\delta}^{(p)}\|_2$
0	2	2	0.873338	2.159950	2.329829
1	2.873338	4.159951	0.601855	0.683618	0.910804
2	3.475193	4.843570	0.318869	0.119941	0.340681
3	3.794063	4.963511	0.095786	0.007497	0.096079
4	3.889849	4.971007	0.008659	0.000057	8.66×10^{-3}
5	3.898508	4.971064	6.91×10^{-5}	4.68×10^{-7}	6.91×10^{-5}
6	3.898577	4.971063			

The iterative procedure was continued in the same manner until $\|\boldsymbol{\delta}^{(0)}\|_2 < \varepsilon$ was achieved in six iterations. The summary of the iteration history is presented in Table 4.12. It should be pointed out that the converged solution is in fact one of the two solutions of the nonlinear system. The other set is found as (5.227185, 4.957696) by changing the initial guess.

Discussion: Nonlinear systems of equations usually do not have closed-form solutions. Consequently, various numerical methods have been developed for solving such systems of equations. In this section, we introduced Newton's method, which is a generalization of the Newton-Raphson method that has a quadratic convergence rate. In this respect, it is superior to other methods such as bisection, fixed-point, or secant methods, and so on.

With a "smart initial guess," we hope to have the system converge quite rapidly. In this problem, initially, the error norm is reduced by approximately one-third in the first few iterations, but then the convergence accelerates as the estimates approach the true solution. The advantage of Newton's method is that it may require fewer iterations to converge an approximate solution, compared to other methods with a lower rate of convergence. A nonlinear system that does not converge may be indicative of a non-existing solution or ill-condition.

4.9 BAIRSTOW'S METHOD

Bairstow's method is one of the earliest methods for finding all (real and imaginary) roots of a polynomial with real coefficients. It is an iterative method for determining a quadratic factor of a polynomial of nth degree $P_n(x)$, based on the idea of synthetic division of a polynomial by a quadratic polynomial.

Consider the following nth-degree $(n > 2)$ polynomial with real coefficients:

$$P_n(x) = x^n + a_1 x^{n-1} + a_2 x^{n-2} + \ldots + a_{n-1}x + a_n = 0 \qquad (4.41)$$

where $a_1, a_2, \ldots, a_n$ are the coefficients. A polynomial may have n simple, multiple (repeated), or complex roots coexisting in conjugate pairs.

This method, in essence, is based on the synthetic division of a polynomial by a quadratic factor $(x^2 + px + q)$, which is then used to compute its real or imaginary roots. Any polynomial can be expressed in terms of one of its quadratic factors as

$$P_n(x) = (x^2 + px + q)Q_{n-2}(x) + Rx + S \qquad (4.42)$$

where $Q_{n-2}(x) = x^{n-2} + b_1 x^{n-3} + \ldots + b_{n-3}x + b_{n-2}$ is the reduced polynomial, p and q (initially unknown) are the coefficients of the quadratic polynomial, and $Rx + S$ is the remainder term. Once a quadratic factor is obtained (i.e., p and q are found), the remainder should yield zero $(R = S = 0)$.

We assume that the coefficients of the remainder depend on p and q, i.e., $R = R(p, q)$ and $S = S(p, q)$, which will vanish when p and q satisfy the system of nonlinear equations defined as $R(p, q) = 0$ and $S(p, q) = 0$. Thus, in a way, Bairstow's method is reduced to simply finding the solution of a 2×2 nonlinear system, whose numerical solution is obtained iteratively applying Newton's method (see Section 4.8). The remainder for a set of initial guesses will most likely not result in non-zero R and S; thus, we require the solution of the following system of nonlinear equations:

$$R(p^{(0)} + \delta p, \ q^{(0)} + \delta q) = 0, \qquad S(p^{(0)} + \delta p, \ q^{(0)} + \delta q) = 0 \qquad (4.43)$$

where δp and δq are the improvements of p and q, respectively.

The next step is to determine a pair of δp and δq that ensures the elimination of the remainder. Using the Newton's method presented in Section 4.8 and replacing $(f_1, f_2) \rightarrow (R, S)$, $(x, y) \rightarrow (p, q)$, $(x_0, y_0) \rightarrow (p^{(0)}, q^{(0)})$ and $(\delta x, \delta y) \rightarrow (\delta p, \delta q)$ in Eqs. (4.32) and (4.33), the solution of the system of linear equations for $\delta p = p - p^{(0)}$ and $\delta q = q - q^{(0)}$ can be expressed as

$$\begin{bmatrix} \dfrac{\partial R}{\partial p}(p^{(0)}, q^{(0)}) & \dfrac{\partial R}{\partial q}(p^{(0)}, q^{(0)}) \\[2ex] \dfrac{\partial S}{\partial p}(p^{(0)}, q^{(0)}) & \dfrac{\partial S}{\partial q}(p^{(0)}, q^{(0)}) \end{bmatrix} \begin{bmatrix} \delta p \\[2ex] \delta q \end{bmatrix} = - \begin{bmatrix} R(p^{(0)}, q^{(0)}) \\[2ex] S(p^{(0)}, q^{(0)}) \end{bmatrix} \qquad (4.44)$$

Upon solving Eq. (4.44), the improvements become available, and then the next values (current estimates) of p and q can be computed. From this point on, we will use $p^{(k)}$ and $q^{(k)}$ notations to denote p and q values at the kth iteration step. Also, we define the improvements at the kth iteration step as $\delta p^{(k)} = p^{(k+1)} - p^{(k)}$ and $\delta q^{(k)} = q^{(k+1)} - q^{(k)}$. The current solution is then obtained as $p^{(k+1)} = p^{(k)} + \delta p^{(k)}$ and $q^{(k+1)} = q^{(k)} + \delta q^{(k)}$. In subsequent iterations, the improvements should yield zero if the initial guess is good.

In order to solve Eq. (4.44), we note that R, S, and the pertinent partial derivatives are unknown at this stage. The coefficients of the polynomials $P_n(x)$ and $Q_{n-2}(x)$ must also satisfy Eq. (4.42). So multiplying things out, collecting similar x-terms and constants, and equating the coefficients yields

$$a_1 = b_1 + p,$$
$$a_2 = b_2 + pb_1 + q,$$
$$a_3 = b_3 + pb_2 + qb_1,$$

$$\vdots$$

$$a_k = b_k + pb_{k-1} + qb_{k-2}, \tag{4.45}$$

$$\vdots$$

$$a_{n-2} = b_{n-2} + pb_{n-3} + qb_{n-4}$$
$$a_{n-1} = R + pb_{n-2} + qb_{n-3}$$
$$a_n = S + qb_{n-2}$$

where $b_1, b_2, \ldots, b_{n-2}, R, S$ can be determined step by step.

Equation (4.45) can be expressed as a single recurrence relationship by setting $b_{-1} = 0$ and $b_0 = 1$. This leads to the following single-recurrence relationship:

$$b_i = a_i - pb_{i-1} - qb_{i-2}, \quad i = 1, 2, \ldots, n \tag{4.46}$$

which should also yield $b_{n-1} = a_{n-1} - pb_{n-2} - qb_{n-3} = R$ and $b_n = a_n - pb_{n-1} - qb_{n-2} = S - pb_{n-1}$. Thus, we may write

$$R = b_{n-1} \quad \text{and} \quad S = b_n + pb_{n-1}, \tag{4.47}$$

Having derived the above expressions, the partial derivatives can now be found from

$$\frac{\partial R}{\partial p} = \frac{\partial b_{n-1}}{\partial p}, \qquad \frac{\partial R}{\partial q} = \frac{\partial b_{n-1}}{\partial q}$$
$$\frac{\partial S}{\partial p} = \frac{\partial}{\partial p}(b_n + pb_{n-1}) = \frac{\partial b_n}{\partial p} + p\frac{\partial b_{n-1}}{\partial p} + b_{n-1} \tag{4.48}$$
$$\frac{\partial S}{\partial q} = \frac{\partial}{\partial q}(b_n + pb_{n-1}) = \frac{\partial b_n}{\partial q} + p\frac{\partial b_{n-1}}{\partial q}$$

Substituting Eqs. (4.47) and (4.48) into Eq. (4.43) and simplifying, we get

$$\left(\frac{\partial b_{n-1}}{\partial p}\right)\delta p + \left(\frac{\partial b_{n-1}}{\partial q}\right)\delta q = -b_{n-1} \tag{4.49}$$

$$\left(\frac{\partial b_n}{\partial p} + p\frac{\partial b_{n-1}}{\partial p} + b_{n-1}\right)\delta p + \left(\frac{\partial b_n}{\partial q} + p\frac{\partial b_{n-1}}{\partial q}\right)\delta q = -b_n - pb_{n-1} \tag{4.50}$$

Multiplying Eq. (4.49) with p and subtracting it from Eq. (4.50) yields

$$b_n + \left(\frac{\partial b_n}{\partial p} + b_{n-1}\right)\delta p + \left(\frac{\partial b_n}{\partial q}\right)\delta q = 0 \tag{4.51}$$

This last expression may be used in place of Eq. (4.50). Furthermore, Eq. (4.46) is used to evaluate the partial derivatives of b_n and b_{n-1} in Eqs. (4.49) and (4.50) and to develop recurrence relationships for $\partial b_i/\partial p$ and $\partial b_i/\partial q$. Since the coefficients (a_i's) of the polynomial are constants, we may set $\partial a_i/\partial p = \partial a_i/\partial q = 0$. The partial derivatives in Eq. (4.46) become

$$\frac{\partial b_i}{\partial p} = -p\frac{\partial b_{i-1}}{\partial p} - b_{i-1} - q\frac{\partial b_{i-2}}{\partial p} \tag{4.52}$$

Bairstow's Method

- The method gives all real and imaginary roots of a polynomial;
- It does not require arithmetic operations with imaginary numbers;
- It does not require partial derivatives;
- It converges, when it does, with a quadratic rate of convergence.

- The method is hard to understand and implement;
- It provides the roots of only "polynomials;"
- It requires a good set of initial guesses; the closer the starting quadratic equation is to an actual one ($p^{(0)} \approx p$, $q^{(0)} \approx q$), the faster it converges;
- It is not suitable for every polynomial and has difficulty handling large polynomials;
- It can be sensitive to the initial guess and/or the coefficients of the polynomial.

$$\frac{\partial b_i}{\partial q} = -p\frac{\partial b_{i-1}}{\partial q} - q\frac{\partial b_{i-2}}{\partial q} \qquad (4.53)$$

Recalling that $b_{-1} = 0$ and $b_0 = 1$, we conclude the following:

$$\frac{\partial b_{-1}}{\partial p} = \frac{\partial b_0}{\partial p} = 0, \qquad \frac{\partial b_{-1}}{\partial q} = \frac{\partial b_0}{\partial q} = 0 \qquad (4.54)$$

Finally, the partial derivatives are computed from Eqs. (4.52), (4.53), and (4.54). However, to obtain simpler expressions, we adapt $\partial b_i/\partial p = -c_{i-1}$ $(i = 1, 2, \ldots, n)$ definition, in which case we may write

$$\frac{\partial b_{i-1}}{\partial p} = \frac{\partial b_i}{\partial q} = -c_{i-2} \qquad (4.55)$$

Using these, the recurrence relationship can be further simplified as follows:

$$c_{-1} = 0, \quad c_0 = 1, \quad c_i = b_i - pc_{i-1} - qc_{i-2}, \qquad i = 1, 2, \ldots, (n-1) \qquad (4.56)$$

where b_i's are calculated from a_i's, and c_i's are calculated from b_i's.

Using Eq. (4.49), (4.51), (4.55), and (4.56), we obtain

$$c_{n-2}\delta p + c_{n-3}\delta q = b_{n-1}, \qquad (c_{n-1} - b_{n-1})\delta p + c_{n-2}\delta q = b_n \qquad (4.57)$$

Defining $c_{n-1} - b_{n-1} = \bar{c}_{n-1}$, the system of equations for the correction terms becomes

$$\begin{bmatrix} \delta p \\ \delta q \end{bmatrix} = \begin{bmatrix} c_{n-2} & c_{n-3} \\ \bar{c}_{n-1} & c_{n-2} \end{bmatrix}^{-1} \begin{bmatrix} b_{n-1} \\ b_n \end{bmatrix} \qquad (4.58)$$

Pseudocode 4.7

Module BAIRSTOW (n, $p0$, $q0$, $\mathbf{a}$, ε, $maxit$, **xre**, **xim**)
\ DESCRIPTION: A pseudomodule to compute all (real and imaginary) roots
\   of an nth degree polynomial given by Eq.(4.41) with Bairstow's method.

```
\ USES:
\   QUADRATIC:: Module for finding all roots of a quadratic equation;
\   ABS:: A built-in function computing the absolute value.
Declare: a_{0:n}, b_{0:n}, c_{0:n}, xre_n, xim_n, xr_2, xi_2        \ Declare array variables
For [k = n, 0, (−1)]                                   \ The method works if a_0 = 1
    a_k ← a_k/a_0                  \ Normalize coefficients of the polynomial with a_0
End For
m ← n                                     \ Save the degree of polynomial for later use
kount ← 0                                \ Counter for the number of roots "found"
While [n > 1]                       \ Loop to find all quadratic factors until n = 1
    p ← p0;  q ← q0                                    \ Set initial guesses to p, q
    k ← 0                                            \ Initialize iteration counter
    Δ ← 1                                        \ Initialize overall error estimate
    \ b is an array of length n containing coefficients of quotient polynomial
    \ c is an array of length n containing coefficients of partial derivatives
    While [Δ > ε And k ⩽ maxit]               \ Loop to find a quadratic factor
        k ← k + 1                                              \ Count iterations
        b_0 ← 1;  c_0 ← 1                                \ Initialize b_0 and c_0
        b_1 ← a_1 − p;  c_1 ← b_1 − p                        \ Find b_1 and c_1
        For [i = 2, n]                        \ Loop i: compute b_i's and c_i's
            b_i ← a_i − p * b_{i−1} − q * b_{i−2}               \ Use Eq. (4.46)
            c_i ← b_i − p * c_{i−1} − q * c_{i−2}               \ Use Eq. (4.56)
        End For
        \ Construct & solve 2 × 2 system to find improved p & q
        c̄ ← c_{n−1} − b_{n−1};  δ ← c_{n−2} * c_{n−2} − c̄ * c_{n−3}
        δp ← (b_{n−1} * c_{n−2} − b_n * c_{n−3})/δ                    \ Find δp
        δq ← (b_n * c_{n−2} − b_{n−1} * c̄)/δ                         \ Find δq
        p ← p + δp;  q ← q + δq                     \ Find current estimates p, q
        Δ ← |δp| + |δq|                     \ Find current error by ‖δ^{(p)}‖_1
    End While
    If [k = maxit And Δ > ε] Then                      \ Check for convergence
        Write: "Quadratic factor did not converge after", maxit,"iterations"
        Write: "Recent values of p, q and Δ are", p, q, Δ
        Exit
    Else                                     \ Find all roots of a quadratic factor
        QUADRATIC(p, q, xr, xi)               \ Find roots of x^2 + px + q = 0
        \ Save real and imaginary parts on arrays xr, xi
        kount ← kount + 1                         \ Root 1 is xr_1 + i * xim_1
        xre_{kount} ← xr_1
        xim_{kount} ← xi_1
        kount ← kount + 1                         \ Root 2 is xr_2 + i * xim_2
        xre_{kount} ← xr_2
        xim_{kount} ← xi_2
        n ← n − 2                                \ Find the degree of quotient
    End If
    p0 ← p;  q0 ← q            \ Set current (p, q) as "prior" for next iterations
    For [i = 0, n]              \ Loop to reset coefficients of the new polynomial
        a_i ← b_i                                         \ Set b_i's as a_i's
    End For
```

If $\begin{bmatrix} n = 1 \end{bmatrix}$ **Then**   \ Case of a simple root, $x + a_1 = 0$
  $kount \leftarrow kount + 1$ \ Count last root
  $xre_{kount} \leftarrow -a_1$ \ Set $-a_1$ as the final root
  $xim_{kount} \leftarrow 0$
End If
End While
$n \leftarrow m$ \ Restore degree of polynomial for output
End Module BAIRSTOW

Module QUADRATIC $(p, q, \mathbf{xr}, \mathbf{xi})$
\ DESCRIPTION: A pseudomodule to find the real and imaginary roots of
\ a quadratic equation of the form $x^2 + px + q = 0$.
\ USES:
\ SQRT:: A built-in function computing the square-root of a value.
Declare: xr_2, xi_2 \ Declare array variables
$\Delta \leftarrow p^2 - 4*q$ \ Compute discriminant
If $\begin{bmatrix} \Delta < 0 \end{bmatrix}$ **Then**   \ Case of Imaginary roots
  $\Delta \leftarrow \sqrt{-\Delta}$ \ Set $\Delta \leftarrow \sqrt{4q - p^2}$
  $xr_1 \leftarrow -p/2;\ xi_1 \leftarrow \Delta/2$ \ Root 1: $(-p + i\sqrt{4q - p^2})/2$
  $xr_2 \leftarrow xr_1;\ xi_2 \leftarrow -xi_1$ \ Root 2: $(-p - i\sqrt{4q - p^2})/2$
Else   \ Case of Real roots $(\Delta \geqslant 0)$
  $\Delta \leftarrow \sqrt{\Delta}$ \ Set $\Delta \leftarrow \sqrt{p^2 - 4q}$
  $xr_1 \leftarrow (-p + \Delta)/2;\ xi_1 \leftarrow 0$ \ Root 1: $(-p + \sqrt{p^2 - 4q})/2$
  $xr_2 \leftarrow (-p - \Delta)/2;\ xi_2 \leftarrow 0$ \ Root 2: $(-p - \sqrt{p^2 - 4q})/2$
End If
End Module QUADRATIC

which gives

$$\delta p = \frac{b_{n-1}c_{n-2} - b_n c_{n-3}}{c_{n-2}^2 - \bar{c}_{n-1}c_{n-3}}, \qquad \delta q = \frac{b_n c_{n-2} - b_{n-1}\bar{c}_{n-1}}{c_{n-2}^2 - \bar{c}_{n-1}c_{n-3}} \tag{4.59}$$

Since this solution is obtained from a truncated two-term Taylor series approximation, the leading error is $\mathcal{O}[(\delta p)^2] + \mathcal{O}[(\delta q)^2]$. If the initial guess is in close proximity to the true solution, the method converges quadratically.

The pseudomodule, **BAIRSTOW**, designed to compute all roots of a polynomial, is presented in Pseudocode 4.7. The degree of the polynomial (n), a set of initial guesses $(p^{(0)}, q^{(0)})$, an array containing the coefficients of the polynomial $(a_i, i = 1, 2, \ldots, n)$, an upper bound for the maximum iterations $(maxit)$, and a convergence tolerance (ε) are supplied as input. Two internal arrays ($\mathbf{b}$ and $\mathbf{c}$) are used to keep the coefficients of the quotient polynomial (b_k''s) and the partial derivatives (c_k''s).

Since this algorithm is based on a polynomial in the form of Eq. (4.41), the coefficients of the input polynomial are normalized by a_0 so that $a_0 = 1$, i.e., $a_i \leftarrow a_i/a_0$ for $i = n, (n - 1), \ldots, 1, 0$. The iteration loop (innermost **While**-loop) is used to determine the coefficients of a quadratic factor (p and q). First, the coefficients b_i's and c_i's are calculated from Eqs. (4.46) and (4.56). Then, the improvements δp and δq are found from Eq. (4.59), and the current step estimates (p and q) are obtained by $p^{(k+1)} = p^{(k)} + \delta p$ and $q^{(k+1)} = q^{(k)} + \delta q$. The convergence criterion based on the ℓ_1-norm of $\boldsymbol{\delta}^{(p)}$ yields $\Delta = |\delta p| + |\delta q| < \varepsilon$. When p and q converge to a set of solutions, a quadratic factor (i.e., $x^2 + px + q$) is found. The roots of the quadratic equation are obtained by the module **QUADRATIC** giving the real and imaginary roots on two separate arrays **xre** and **xim**. The outermost While-loop is for repeating the procedure for another quadratic factor from the deflated polynomial, $Q_{n-2}(x)$, after setting $a_i \leftarrow b_i$, $i = 0, 1, 2, \ldots$, i.e., $P_{n-2}(x) \leftarrow Q_{n-2}(x)$. Note that if n is odd, then the last factor will be a linear factor, which yields $x \leftarrow -a_1$.

EXAMPLE 4.9: Finding all roots of a polynomial

The poles of a transfer function are the roots of the following 7th-degree polynomial:

$$s^7 + 1.5s^6 - 11s^5 + 15s^4 - 76s^3 + 43.5s^2 - 64s + 30 = 0$$

Find all real and imaginary poles of this transfer function using the Bairstow's method, starting with the initial values $p^{(0)} = q^{(0)} = 0$. For the convergence criterion, use $|\delta p| + |\delta q| < 10^{-4}$.

SOLUTION:

The coefficients of the 7th-degree polynomial are set as $a_0 = 1$, $a_1 = 1.5$, $a_2 = -11$, $a_3 = 15$, $a_4 = -76$, $a_5 = 43.5$, $a_6 = -64$, and $a_7 = 30$.

1st iteration: We start with $p^{(0)} = q^{(0)} = 0$. The b_k's and c_k's are computed from Eq. (4.46) and Eq. (4.56), respectively.

k	a_k	b_k	c_k	
0	1	1	1	$\bar{c} = c_6 - b_6 = 0$
1	1.5	1.5	1.5	
2	−11	−11	−11	$\delta p = -0.26635$
3	15	15	15	$\delta q = 0.689655$
4	−76	−76	−76	
5	43.5	43.5	43.5	$p^{(1)} = p^{(0)} + \delta p = -0.26635$
6	−64	−64	−64	$q^{(1)} = q^{(0)} + \delta q = -0.26635$
7	30	30		

2nd iteration: Continuing with $p^{(1)} = -0.26635$, $q^{(1)} = 0.689655$ we find

k	a_k	b_k	c_k	
0	1	1	1	$\bar{c} = c_6 - b_6 = 38.353$
1	1.5	1.76635	2.0327	
2	−11	−11.2192	−11.367	$\delta p = 0.35422$
3	15	10.7936	6.364	$\delta q = 0.24562$
4	−76	−65.3878	−55.853	
5	43.5	18.6401	−0.6253	$p^{(2)} = p^{(1)} + \delta p = 0.08787$
6	−64	−13.9402	24.4126	$q^{(2)} = q^{(1)} + \delta q = 0.93528$
7	30	13.4318		

3rd iteration: Restart with $p^{(2)} = 0.08787$, $q^{(2)} = 0.93528$:

k	a_k	b_k	c_k	
0	1	1	1	$\bar{c} = c_6 - b_6 = 49.1373$
1	1.5	1.41213	1.3242	
2	−11	−12.0594	−13.111	$\delta p = -0.0862$
3	15	14.7389	14.652	$\delta q = 0.05596$
4	−76	−66.016	−55.041	
5	43.5	35.5159	26.648	$p^{(3)} = p^{(2)} + \delta p = 0.00167$
6	−64	−5.3773	43.76	$q^{(3)} = q^{(2)} + \delta q = 0.99124$
7	30	−2.7447		

4th iteration: Continue with $p^{(3)} = 0.0016$, $q^{(3)} = 0.99124$.

k	a_k	b_k	c_k	
0	1	1	1	$\bar{c} = c_6 - b_6 = 50.8005$
1	1.5	1.4983	1.4967	
2	−11	−11.994	−12.987	$\delta p = -0.00168$
3	15	13.5348	12.073	$\delta q = 0.00873$
4	−76	−64.134	−51.28	
5	43.5	30.1906	18.309	$p^{(4)} = p^{(3)} + \delta p = -10^{-5}$
6	−64	−0.4784	50.322	$q^{(4)} = q^{(3)} + \delta q = 0.99997$
7	30	0.07476		

5th iteration: Continue with $p^{(4)} = 0$, $q^{(4)} = 0.99997$.

k	a_k	b_k	c_k	
0	1	1	1	$\bar{c} = c_6 - b_6 = 50.8005$
1	1.5	1.5	1.5	
2	−11	−12	−13	$\delta p = -1.04 \times 10^{-5}$
3	15	13.5	12	$\delta q = 3.27 \times 10^{-5}$
4	−76	−64	−51	
5	43.5	30	18	$p^{(5)} = p^{(4)} + \delta p = 0$
6	−64	0	51	$q^{(5)} = q^{(4)} + \delta q = 1$
7	30	0		

After five iterations, the convergence criterion yields $|\delta p| + |\delta q| = 4.31 \times 10^{-5} < 10^{-4}$; hence, the iterations are terminated. This yields the coefficients of the quadratic equation with $p = 0$ and $q = 1$, i.e., the quadratic factor is $(s^2 + 1)$. The coefficients of the deflated polynomial (b_k's) are

$$b_0 = 1, \quad b_1 = 1.5, \quad b_2 = -12, \quad b_3 = 13.5, \quad b_4 = -64, \quad b_5 = 30$$

Note that the degree of this polynomial is 5, and b_6 and b_7 representing the remainder terms (R and S) are zero. Then, the given polynomial can be factored as

$$(s^2 + 1)(s^5 + 1.5s^4 - 12s^3 + 13.5s^2 - 64s + 30)$$

Next, setting $a_i \leftarrow b_i$'s for $i = 1,2,...,5$ and applying Bairstow's algorithm once more with $p^{(0)} = q^{(0)} = 0$ yields

$$s^5 + 1.5s^4 - 12s^3 + 13.5s^2 - 64s + 30 = (s^2 + 4.5s - 2.5)(s^3 - 3s^2 + 4s - 12)$$

The method is applied for the last time by setting $a_i \leftarrow b_i$'s for $i = 1, 2, 3$ and using $p^{(0)} = q^{(0)} = 0$ as initial values. The procedure yields

$$s^3 - 3s^2 + 4s - 12 = (s^2 + 4)(s - 3)$$

Finally, by putting all the quadratic factors together, we obtain

$$s^7 + 1.5s^6 - 11s^5 + 15s^4 - 76s^3 + 43.5s^2 - 64s + 30 = (s^2 + 1)(s^2 + 4)(s + 5)(s - 3)(s - 0.5)$$

The final step is to find the roots of all quadratic or linear factors, which leads to

$$s_{1,2} = \pm i, \quad s_{3,4} = \pm 2i, \quad s_5 = 3, \quad s_6 = 0.5, \quad s_7 = -5$$

The closer the initial values are to the actual p and q of a quadratic factor, the faster convergence is achieved.

Discussion: The method is straightforward; however, calculating the roots of even low-degree polynomials by hand is quite complicated. The method works very well, as in this example, especially when the real and imaginary roots are of the same order. But polynomials tend to be ill-conditioned when they have a few roots that are too large or too small. In most real-world applications where polynomials are also used to model data, the coefficients of the polynomial are not exact because of imperfect primary data. As a result, a small perturbation in the coefficients of the polynomial results in large changes in the roots. It is also worth noting that this problem is not limited to the Bairstow method alone. The ill-condition problem can be remedied by increasing the computation precision, providing a degree of improvement, but it becomes impossible to avoid round-off errors in high-order polynomials.

4.10 POLYNOMIAL REDUCTION AND SYNTHETIC DIVISION

So far in this chapter, we have covered common numerical methods for finding the root of a nonlinear equation, which can be used to find the real root of a polynomial as well. However, while finding a single root of a polynomial with a suitable method is simple, determining the second, third, and remaining roots with different starting guesses can be somewhat problematic as the method may yield a previously determined root. Hence, once a root is found, we need to eliminate the possibility of ending up with the same root again.

Polynomial reduction or *polynomial deflation* is a technique to remove a root (say, $x = \alpha$) from the original polynomial. To achieve this, the polynomial $P_n(x)$ is divided by $(x - \alpha)$ which leads to a new polynomial $Q_{n-1}(x)$ of degree $n - 1$: $P_n(x) = (x - \alpha)Q_{n-1}(x)$. The new polynomial no longer has $x = \alpha$ as its root unless $P_n(x)$ has multiple roots at $x = \alpha$.

Horner's rule, also called *synthetic division*, has a variety of uses and saves work and time when evaluating a polynomial at a certain value ($x = \alpha$).

Consider the following nth-degree polynomial:

$$P_n(x) = a_n x^n + a_{n-1} x^{n-1} + a_{n-2} x^{n-2} + \ldots + a_1 x + a_0 \tag{4.60}$$

$P_n(x)$ is divided by a simple linear factor $(x - \alpha)$ to give the polynomial $Q_{n-1}(x)$:

$$\frac{P_n(x)}{x - \alpha} = Q_{n-1}(x) + \frac{R_0}{x - \alpha} \tag{4.61}$$

where $Q_{n-1}(x) = b_{n-1}x^{n-1} + b_{n-2}x^{n-2} + \ldots + b_1 x + b_0$ and R_0 is the remainder. If $x = \alpha$ is a root of $P_n(x)$, then $R_0 = 0$.

Multiplying both sides of Eq. (4.61) with $(x - \alpha)$ gets rid of the denominators:

$$P_n(x) = (x - \alpha) Q_{n-1}(x) + R_0 \tag{4.62}$$

Expanding the rhs of Eq. (4.62) and grouping the x-terms as well as the constants gives

$$\begin{aligned}
a_n x^n + a_{n-1}x^{n-1} + a_{n-2}x^{n-2} + \ldots + a_1 x + a_0 = b_{n-1}x^n + (b_{n-2} - cb_{n-1})x^{n-1} \\
+ (b_{n-3} - cb_{n-2})x^{n-2} + \ldots + (b_0 - cb_1)x + R_0 - cb_0
\end{aligned} \tag{4.63}$$

For the two sides to be equal, the coefficients of the two polynomials should be equal. Hence, by matching both sides of Eq. (4.63), the following sequence of constants is obtained:

$$\begin{aligned}
a_n &= b_{n-1} \\
a_{n-1} &= b_{n-2} - \alpha\, b_{n-1} &\rightarrow&\quad b_{n-2} = a_{n-1} + \alpha\, b_{n-1} \\
a_{n-2} &= b_{n-3} - \alpha\, b_{n-2} &\rightarrow&\quad b_{n-3} = a_{n-2} + \alpha\, b_{n-2} \\
&\ \ \vdots & & \qquad\vdots \\
a_1 &= b_0 - \alpha\, b_1 &\rightarrow&\quad b_0 = a_1 + \alpha\, b_1 \\
R_0(\alpha) &= a_0 + \alpha\, b_0
\end{aligned} \tag{4.64}$$

The coefficients of $Q_{n-1}(x)$ can be obtained from Eq. (4.64) and the following recurrence relationship:

$$b_n = 0, \quad b_k = a_{k+1} + \alpha\, b_{k+1} \quad \text{for} \quad k = (n-1), (n-2), \ldots, 1, 0 \tag{4.65}$$

Reducing $Q_{n-1}(x)$ by $x - \alpha$, following $(n-2)$th-degree polynomial is obtained:

$$\frac{Q_{n-1}(x)}{x - \alpha} = S_{n-2}(x) + \frac{R_1}{x - \alpha} \tag{4.66}$$

where $S_{n-2}(x) = c_{n-2}x^{n-2} + c_{n-3}x^{n-3} + \ldots + c_1 x + c_0$ and R_1 is also the remainder.

Combining Eqs. (4.62) and (4.66) yields

$$P_n(x) = (x - \alpha)^2 S_{n-2}(x) + (x - \alpha)R_1 + R_0 \tag{4.67}$$

The derivative of Eq. (4.67) with respect to x becomes

$$\frac{dP_n}{dx} = (x - \alpha)^2 \frac{dS_{n-2}}{dx} + 2(x - \alpha)S_{n-2}(x) + R_1 \tag{4.68}$$

Adopting the Newton-Raphson method to find a root of $P_n(x)$, $P_n(x^{(p)})$ and $P_n'(x^{(p)})$ are required to compute the successive estimates. Note that Eqs. (4.67) and (4.68) are reduced to $R_0 = P_n(\alpha)$ and $R_1 = P'_n(\alpha)$ for the true root, i.e., $x^{(p)} \to \alpha$. Then the Newton-Raphson iteration equation can be expressed as

$$x^{(p+1)} = x^{(p)} - \frac{R_0(x^{(p)})}{R_1(x^{(p)})}, \quad p = 0, 1, 2, \ldots \tag{4.69}$$

where as usual, subscript p denotes the iteration step, $R_0(x^{(p)})$ and $R_1(x^{(p)})$ are the remainders for an estimate $x^{(p)}$. Note that $x^{(p)} \to \alpha$ and $R_0(x^{(p)}) \to 0$ for $p \to \infty$.

Polynomial Reduction/Synthetic Division

- Evaluating polynomials using synthetic division is computationally cheaper and easier since it does not require powers of x;
- There is no risk of getting the same root (except for a root of multiplicity $m \geqslant 2$) since the root is factored out once it is found;
- The algorithm produces the coefficients of the reduced polynomial, $Q_{n-1}(x)$, as well as the remainder, R_0, which provides information on whether the root is found or not, i.e., $R_0 = 0$?

- A random, unintelligent first guess could result in a large number of iterations;
- Every new root is obtained with finite accuracy, and the numerical errors in the coefficients of the reduced polynomials could build up so that the computed roots thereafter could become more and more inaccurate;
- The order in which the roots are found can also affect the coefficients of reduced polynomials;
- The method may suffer from instabilities when the roots are sensitive to small changes in the coefficients of the polynomials;
- Polynomials with roots of multiplicity are also sensitive to small changes in their coefficients.

When a polynomial is ill-conditioned, polynomial reduction can lead to a degradation of accuracy in the coefficients of the reduced polynomial and, subsequently, in estimated roots. To reduce the effects of such errors, consider using the "approximate root" as the initial guess to find another root of the original polynomial.

As we have learned, an initial guess is critical to accelerating convergence. As a strategy to find a good estimate, it is desirable to isolate each root or determine the bounds of all real roots, i.e.,

$$L \leqslant \text{any real root of } P_n(x) \leqslant U$$

The subject of root isolation is rich in studies and diversity. In this context, as a simple approach, determining the upper or lower bounds of the roots of a polynomial will be briefly presented here. This alternative is helpful in narrowing down the search intervals. Estimating the bounds of a root (L, U) is followed by employing a rootfinding technique to determine the root. Once a root is found with acceptable accuracy, it is factored out, and the procedure is repeated for the reduced polynomial. With this approach, searching for the roots of a polynomial is reduced in a way to finding the bounds of the roots.

Lagrange's and Cauchy's estimates for the upper bounds of all roots are given by

$$U_\ell = \max\left\{1, \sum_{k=1}^{n-1}\left|\frac{a_k}{a_n}\right|\right\} \quad \text{and} \quad U_c = 1 + \max_{0 \leqslant k \leqslant n-1}\left\{\left|\frac{a_k}{a_n}\right|\right\}$$

The upper bound may be taken as the smaller of the two: $U = \min\{U_\ell, \ U_c\}$. The commonly used upper and lower bounds $(L \leqslant |x| \leqslant U)$ are given as follows:

Pseudocode 4.8

Module POLY_ROOT $(n, \mathbf{a}, maxit, \varepsilon, nroot, \mathbf{roots})$
\ DESCRIPTION: A pseudomodule to compute all *real roots* of a polynomial
\   given by Eq. (4.60) using the Horner's rule and Newton-Raphson method.
\ USES:
\   ONE_ROOT :: Module for determining a real root;
\   BOUNDS :: Module for determining the bounds of roots of a polynomial.
Declare: $a_{0:n}$, $b_{0:n}$, $roots_n$ \ Declare array variables
Declare: Logical: $found$ \ Declare logical variable
For $\left[k = 0, n \right]$ \ Loop to normalize the coefficients of P_n with a_n
    $a_k \leftarrow a_k/a_n$ \ Algorithm works if $a_n = 1$
End For
$m \leftarrow n$ \ Save the degree of polynomial for output
$found \leftarrow$ True   \ Locate a root
$r \leftarrow 0$ \ Initialize number of roots (found)
While $\left[found \textbf{ And } r < n \right]$ \ Main loop to find one root
    BOUNDS$(m, \mathbf{a}, Lbound, Ubound)$ \ Find (L, U) bounds of $P_m(x)$
    $x0 \leftarrow -Lbound$ \ Initialize lower bound
    ONE_ROOT$(m, x0, \mathbf{a}, \mathbf{b}, maxit, \varepsilon, found)$ \ A root exists if $found$=True
    If $\left[found \right]$ **Then**   \ If a root is found
        $r \leftarrow r + 1$ \ Count roots found
        $roots_r \leftarrow x0$ \ Save root on **roots**
    End If
    $m \leftarrow m - 1$ \ The degree of quotient
    For $\left[k = 0, m \right]$ \ Find coefficients of the reduced polynomial
        $a_k \leftarrow b_k$ \ Assing coefficients of Q_{m-1} to **a**
    End For
End While
$nroot \leftarrow r$ \ Total number of roots $=$ found
End Module POLY_ROOT

Module ONE_ROOT $(n, x0, \mathbf{a}, \mathbf{b}, maxit, \varepsilon, found)$
\ DESCRIPTION: A pseudomodule to compute a real roots of an nth degree
\   polynomial applying synthetic division to $P_n(x)$ twice in a row and finds
\   the coefficients of the reduced polynomials and remainders.
\ USES:
\   ABS:: A built-in function computing the absolute value.
Declare: $a_{0:n}$, $b_{0:n}$ \ Declare array variables
Declare: Logical: $found$ \ Declare logical variable
$p \leftarrow 0; \delta0 \leftarrow 1$ \ Initialize iteration counter and absolute error
$found \leftarrow$ True   \ Initialize logical variable
$kount \leftarrow 0$ \ Initialize counter for divergence decision
While $\left[\delta0 > \varepsilon \textbf{ And } p \leqslant maxit \right]$
    $p \leftarrow p + 1$ \ Count iterations
    $b_n \leftarrow 0$ \ Synthetic division of $P_n(x)/(x - x^{(p)})$
    For $\left[k = (n-1), 0, (-1) \right]$ \ Loop k to sweep index k backwards
        $b_k \leftarrow a_{k+1} + x0 * b_{k+1}$ \ Find coefficients of $Q_{n-1}(x)$
    End For
    $R0 \leftarrow a_0 + x0 * b_0$ \ Find remainder

$$m \leftarrow n - 1 \qquad \text{\\ Degree of reduced polynomial, } Q_{n-1}(x)$$

```
    m ← n − 1                          \ Degree of reduced polynomial, Q_{n-1}(x)
    c_m ← 0                            \ Apply synthetic division to Q_m(x)
    For [k = (m − 1), 0, (−1)]         \ Loop k to sweep index k backwards
        c_k ← b_{k+1} + x0 * c_{k+1}   \ Find coefficients of S_{n−2}(x)
    End For
    R1 ← b_0 + x0 * c_0                \ Remainder Q_m(x) = (x − x^{(p)})S_{m−1}(x) + R1
    δ ← R0/R1                          \ Compute δ^{(p)} ← P_n(x^{(p)})/P'_n(x^{(p)})
    x0 ← x0 − δ                        \ Use Newton-Raphson, x^{(p+1)} ← x^{(p)} − δ^{(p)}
    \ Check to see if subsequent displacements are increasing (diverging)
    If [ |δ| > δ0 ] Then               \ Current error is larger than prior
        kount ← kount + 1; δ0 ← |δ|    \ Count occurences
    End If
    If [kount = 10] Then               \ if |δ| > δ0 occurence is repeated 10 times
        Write: "Diverging sequence is obtained"   \ Issue a warning
        found ← False                  \ Root not found, perhaps imaginary pairs
        Exit                           \ Exit the module
    End If
    δ0 ← |δ|                           \ Set current displacement as prior, δ^{(p)} ← |δ^{(p+1)}|
End While
End Module ONE_ROOT


Module BOUNDS (n, a, Lbound, Ubound)
\ DESCRIPTION: A pseudomodule to estimate lower and upper bounds of the
\     roots of an nth degree polynomial: Lbound < |roots of P_n(x)| < Ubound.
\ USES:
\     MAX:: A built-in function returning the maximum of the agrument list.
Declare: a_{0:n}
Lbound ← 0;  Ubound ← 0               \ Initialize Lbound and Ubound
For [k = 1, n]
    Lbound ← MAX(Lbound, |a_k|^{1/k})      \ L ← max(L, |a_k|^{1/k})
    Ubound ← MAX(Ubound, |a_{n−k}|^{1/k})  \ U ← max(U, |a_{n−k}|^{1/k})
End For
Lbound ← 0.5/Lbound; Ubound ← 2 * Ubound   \ Apply Eq. (4.70)
End Module BOUNDS
```

$$L = \left(2 \max_{1 \leqslant k \leqslant n} \left\{ \left| \frac{a_k}{a_0} \right|^{1/k} \right\} \right)^{-1}, \qquad U = 2 \max_{1 \leqslant k \leqslant n} \left\{ \left| \frac{a_{n-k}}{a_n} \right|^{1/k} \right\} \tag{4.70}$$

More elaborate approximations for the upper/lower bounds exist and can be found in the literature, some of which have been summarized in Refs. [110, 111].

Pseudocode 4.8 is designed to find all real roots of an nth-degree polynomial using the synthetic division coupled with the Newton-Raphson method. The main module, POLY_ROOT, requires the coefficients of the nth-degree polynomial $P_n(x)$ (a_i, $i = 0, 1, 2, \ldots, n$), an upper bound for the maximum number of iterations ($maxit$), and a convergence tolerance (ε) as input. The outputs are the total number of roots found ($nroot$) and an array containing the roots ($roots_i$, $i = 0, 1, 2, \ldots, nroot$). The coefficients of the reduced polynomial, $Q_{n-1}(x)$, are stored on an internal array, b_i for $i = 0, 1, 2, \ldots, n - 1$. The algorithm is based on the normalized polynomial, i.e., $a_i \leftarrow a_i/a_n$ for $i = 0, 1, \ldots, n$. The module used the module **BOUNDS** to determine the global bounds that contain all roots of $P_n(x)$. The initial estimate is set to $x^{(0)} = L$ as priority is given to finding the smallest roots as they are most affected by rounding. The module calls **ONE_ROOT** to obtain a root and the coefficients of the reduced polynomial. If a (real) root exists, then module **ONE_ROOT** will return an estimate for the root $(x0)$, and the logical variable $found$ is set to True. Once a root is found, the counter (r) is incremented by one, and the root found $(x0)$ is written to the solution array, **roots**. The reduced polynomial is then set as $P_{n-1}(x)$, i.e., $a_i \leftarrow b_i$, $i = 0, 1, 2, \ldots, (n-1)$, and the above procedure is repeated for the reduced polynomial. If no root is found, then $found$ is set to $false$, and the While-loop is terminated.

The pseudomodule, **ONE_ROOT**, is designed to find a real root of an nth-degree polynomial using the Newton-Raphson method. The degree (n) and the coefficients $(a_i, i = 0, 1, 2, \ldots, n)$ of the polynomial, the initial guess for the root $(x0)$, an upper bound for the maximum number of iterations $(maxit)$, and a convergence tolerance (ε) are the inputs. If a root exists, $found$ is set to $true$ and the estimated root $(x0)$ is returned as output along with the array **b** containing the coefficients of the reduced polynomial. With $x0$ and a_k's, the first synthetic division yields b_k's and $P_n(x0) = R_0$. Applied to $Q_{n-1}(x)$, the second synthetic division with $x0$ gives $S_{n-2}(x)$ (i.e., c_k's) and R_1 corresponding to $P_n'(x0) = R_1$. Using $\delta = R_0(x0)/R_1(x0)$, the current (improved) estimate is computed from Eq. (4.69). The While-loop, in which improved estimates are computed, is terminated when $|\delta| < \varepsilon$ is met. If $P_n(x)$ or any one of its subsequent reduced polynomials does not have a real root, then $|\delta^{(p)}|$ will either diverge or oscillate; hence, $|\delta^{(p+1)}| > |\delta^{(p)}|$ occurrences are counted by $kount$ to determine if the iterations stabilize or diverge. If $kount = 10$, then $found$ is set to $false$ and the procedure is terminated.

The pseudomodule, **BOUNDS**, is used to estimate the lower and upper bounds of an nth-degree polynomial, $P_n(x)$. The degree (n) and the coefficients $(a_i, i = 0, 1, 2, \ldots, n)$ of the polynomial are the inputs of the module. On exit, the module provides the lower bound $(Lbound)$ and upper bound $(Ubound)$ as output. The module determines the bounds using the criteria provided in Eq. (4.70).

EXAMPLE 4.10: Application of polynomial reduction technique

Find all the real zeros of $P_4(x) = x^4 + 1.5x^3 + 3x^2 + 6x - 4$ using the Newton-Raphson method coupled with polynomial reduction. Use $\varepsilon = 10^{-6}$ as convergence tolerance and $x^{(0)} = 0$ as the initial guess.

SOLUTION:

The coefficients of the polynomial are: $a_4 = 1$, $a_3 = 1.5$, $a_2 = 3$, $a_1 = 6$, and $a_0 = -4$. Noting that $x^{(0)} = 0$ and applying the sequence given by Eq. (4.65) to $P_4(x)$ and $Q_3(x)$ we get:

$$
\begin{array}{ll}
b_4 = 0, & c_3 = 0, \\
b_3 = a_4 = 1 & c_2 = b_3 = 1, \\
b_2 = a_3 + (0)b_3 = 1.5 + 0(1) = 1.5, & c_1 = b_2 + (0)c_2 = 1.5 + 0(1) = 1.5, \\
b_1 = a_2 + (0)\,b_2 = 3.0 + 0(1.5) = 3, & c_0 = b_1 + (0)\,c_1 = 3.0 + 0(1.5) = 1, \\
b_0 = a_1 + (0)\,b_1 = 6.0 + 0(1) = 6, & c_0 = b_1 + (0)\,c_1 = 3.0 + 0(1.5) = 1,
\end{array}
$$

We can determine the remainders from $P_4(x)$ and $Q_3(x)$ as follows:

$$R_0 = a_0 + (0)\, b_0 = -4 + 0(6) = -4, \quad \text{and}$$
$$R_1 = b_0 + (0)\, c_0 = 6 + 0(1) = 6.$$

At the end of the first iteration, we obtain

$$x^{(1)} = x^{(0)} - \frac{R_0(x^{(0)})}{R_1(x^{(0)})} = 0 - \frac{(-4)}{6} = \frac{2}{3}$$

When this procedure is repeated for the new estimate $x^{(1)}$ and check for convergence (i.e., $|x^{(p)} - x^{(p-1)}| < \varepsilon$) each time a new estimate is found. The first root ($x = 0.5$) converges after 4 iterations. The iteration details are:

Iteration # 1, $x^{(1)} = 2/3$

k	a_k	b_k	c_k
4	1	0	0
3	1.5	1	0
2	3	1.5	1
1	6	3	1.5
0	-4	6	3
$R_0 = -4$	$R_1 = 6$		

Iteration # 2, $x^{(2)} = 0.516854$

k	a_k	b_k	c_k
4	1	0	0
3	1.5	1	0
2	3	2.1667	1
1	6	4.4444	2.83333
0	-4	8.963	6.33333
$R_0 = 1.9753$	$R_1 = 13.1852$		

Iteration # 3, $x^{(1)} = 0.5001797$

k	a_k	b_k	c_k
4	1	0	0
3	1.5	1	0
2	3	2.0169	1
1	6	4.0424	2.5337
0	-4	8.0893	5.352
$R_0 = 0.181$	$R_1 = 10.8555$		

Iteration # 4, $x^{(2)} = 0.5$

k	a_k	b_k	c_k
4	1	0	0
3	1.5	1	0
2	3	2.0002	1
1	6	4.0004	2.5004
0	-4	8.0009	5.2511
$R_0 = 0.0019$	$R_1 = 10.6274$		

The first step yields a third-degree polynomial: $Q_3(x) = x^3 + 1.5x^2 + 3x + 6$. In the second step we proceed to find the root of $Q_3(x)$ with the initial guess, $x^{(0)} = 0$. The second root converges to the zero of $x = 2$ after six iterations.

Iteration # 1, $x^{(1)} = 2/3$

k	a_k	b_k	c_k
3	1	0	0
2	2	1	0
1	4	2.5	1
0	8	5.25	3
$R_0 = 10.625$	$R_1 = 6.75$		

Iteration # 2, $x^{(2)} = 0.516854$

k	a_k	b_k	c_k
3	1	0	0
2	2	1	0
1	4	0.9259	1
0	8	3.0055	-0.1481
$R_0 = 4.7719$	$R_1 = 3.1646$		

Iteration # 3, $x^{(3)} = -2.127923$

k	a_k	b_k	c_k
3	1	0	0
2	2	1	0
1	4	-0.582	1
0	8	5.5026	-3.1639
$R_0 = -6.2076$	$R_1 = 13.6718$		

Iteration # 4, $x^{(4)} = -2.007676$

k	a_k	b_k	c_k
3	1	0	0
2	2	1	0
1	4	-0.1279	1
0	8	4.2722	-2.2558
$R_0 = -1.0909$	$R_1 = 9.0725$		

Iteration # 5, $x^{(3)} = -2.000029$			
k	a_k	b_k	c_k
3	1	0	0
2	2	1	0
1	4	-0.0077	1
0	8	4.0154	-2.0154
$R_0 = -0.0616$		$R_1 = 8.0616$	

Iteration # 6, $x^{(4)} = -2$			
k	a_k	b_k	c_k
3	1	0	0
2	2	1	0
1	4	-0.1279	1
0	8	4.2722	-2.2558
$R_0 = 0$		$R_1 = 8$	

Finally the polynomial is factored as

$$x^4 + 1.5x^3 + 3x^2 + 6x - 4 = (x - 0.5)(x + 2)(x^2 + 4) = 0$$

Note that the quadratic multiplier has imaginary roots.

Discussion: The synthetic division (or deflation) algorithm is a simple process applied to an array of polynomial coefficients. With this method, there is no need to determine the upper and lower bounds of the roots. The computational effort that goes into searching for all or several roots of a polynomial can be significantly reduced. More importantly, we can avoid encountering the problem of the method converging to a non-repeating root multiple times.

4.11   CLOSURE

A root of a nonlinear equation can be obtained using the methods or techniques presented in this chapter. All root-finding algorithms require iterative procedures. Fewer iterations imply fewer floating-point operations and faster convergence. The choice of a method depends on the nonlinear equation whose root is to be determined.

Bracketing methods such as bisection or false-position methods are suitable for searching the roots of a nonlinear equation within a specified interval. These methods converge rather slowly and require a clever estimation of the lower and upper bounds of the initial interval in which a root lies. These methods have the lowest convergence rate, and it may be difficult to find the desired root if the interval contains several roots. With the bracketing methods, convergence is guaranteed since a root is confined within a closed interval that is constantly narrowed down until its width is less than a predefined tolerance value. The false position method usually converges more rapidly than the bisection method.

Open-domain methods such as Newton-Raphson, Secant, and so on are normally preferred because they converge to a solution much more rapidly. Newton-Raphson or modified Newton methods with a quadratic convergence rate require an initial estimate (instead of two) and converge to a root faster than other bracketing methods. The initial guess can be chosen randomly; generally, the closer the initial guess is to the root, the fewer iterations are required for the accuracy desired in the estimates. If a non-linear equation has more than one root, all roots can be found with the Newton-Raphson method by varying the initial guess over a wide range. If the derivatives of $f(x)$ are long, complex, or require too many arithmetic operations, the Newton-Raphson methods may not be computationally favorable. In such cases, the Secant or modified Secant methods (having a convergence rate of 1.62), which do not require the derivative of the nonlinear equation, could be a suitable alternative that leads to an estimate as fast and effective as next to the Newton-Raphson methods. But the open-domain methods may diverge since the new estimates are not bracketed.

Aitken's and Steffensen's methods serve to accelerate the speed of the convergence of slowly converging sequences. Steffensen's method provides quadratic convergence, but the function $f(x)$ must be three times continuously differentiable in order to ensure quadratic convergence.

Many failures that may be encountered when using the root-finding algorithms can be traced back to a poor initial guess. Instead of making random estimates, it is possible in engineering applications to make a *smart estimate* by simply interjecting the physics of the problem. This reduces not only the computational effort but also the potential for round-off accumulation due to possibly numerous iterations.

The real roots of a polynomial can be determined with either bisection, Newton-Raphson, secant, or other methods by specifying the initial guess or interval. But these methods are clearly not equipped to handle finding the imaginary roots. The Bairstow method with a quadratic convergence rate is fast and very effective for finding all *real and imaginary* roots of a polynomial, and it is suitable for most applications. Polynomial reduction, coupled with bisection, Newton-Raphson, etc. methods, can also be effectively used to obtain all the *real* roots of a polynomial. However, polynomials that are ill-conditioned require high-precision computation, so avoid the accumulation of round-off errors that may lead to many inaccurate estimates.

Newton's method is extended to solve a system of nonlinear equations. The method converges quickly with a good set of initial guesses. The computation of the Jacobian that is done at each iteration can be done numerically or analytically; however, it may be difficult to find analytical expressions of the Jacobian for some problems.

4.12 EXERCISES

Section 4.1 Bisection Method

E4.1 Apply the bisection method for the given search intervals to estimate the following roots accurate to two decimal places. *Hint*: Assume $f(x) = x^n - a = 0$.

$$(a)\, \sqrt{15.1492} \quad [3, 4], \qquad (b)\, \sqrt[3]{15.1492}, \quad [2, 3], \qquad (c)\, \sqrt[4]{15.1492} \quad [1.5, 2.5]$$

E4.2 For each of the following nonlinear equations, (a) verify that the nonlinear equation has a root in the specified interval, and (b) estimate the number of bisections that would be required to locate the root to an accuracy of $\varepsilon = 10^{-3}$.

(a) $(3x - 2)\ln x = x + 1$, (i) $(0.1, 1)$, (ii) $(1, 3)$

(b) $x^2 - 3x \ln x = \sin \pi x$, (i) $(1, 3)$, (ii) $(3, 6)$

(c) $x \cos x + x^2 \sin x + 3 = \sqrt{x + 1}$, (i) $(2.6, 3.6)$, (ii) $(11, 14)$

(d) $\cos 2x \cosh x = 2$, (i) $(2, 3)$, (ii) $(3, 4)$

E4.3 Estimate the roots of the nonlinear equations in **E4.2** using the bisection method to an accuracy of $\varepsilon = 10^{-3}$.

E4.4 Given $f(x) = x^4 - 3.7x^3 - 1.5x^2 + 14.4x - 10.8 = 0$ has four real roots in $(-3,4)$. Propose search intervals enclosing each root.

E4.5 Apply five bisections to **E4.4** to estimate a root in $(-1, 2)$. How accurate is your estimate?

E4.6 Do the curves $y_1 = \cos(x/3)$ and $y_2 = \sin(4x/5)$ intersect in $(0, \pi/2)$? If yes, find the intersection point accurate to two decimal places $(\varepsilon = 0.5 \times 10^{-2})$.

E4.7 Given $f(x) = x \sin(\pi x) - 1$ has two roots near $x \approx 2.5$, use the bisection method with initial intervals of $(1.5, 2.5)$ and $(2.5, 3.5)$ to estimate the roots, accurate to three decimal places.

E4.8 The nonlinear equation $e^{-ax} = \sin(\omega x)$ is encountered in the design of a planetary gear system in an automatic transmission. Consider the case of $a = 1$ and $\omega = \pi/4$, which has a root in $(0,1)$. Apply the bisection method to estimate the root accurately to at least four decimal places.

E4.9 Given $f(x) = 6x - 2x^3 - 3\sin(2x) - 1$ has a local maximum point in $(1, 2)$. Use the bisection method to estimate the maximum point, accurate to two decimal places.

Section 4.2 Method of False Position

E4.10 Repeat **E4.1** with the method of false position.

E4.11 Repeat **E4.2** with the method of false position.

E4.12 Repeat **E4.5** with the method of false position.

E4.13 Repeat **E4.6** with the method of false position.

E4.14 Repeat **E4.7** with the method of false position.

E4.15 Repeat **E4.8** with the method of false position.

E4.16 Repeat **E4.9** with the method of false position.

E4.17 The following polynomial has a root in $(0,1)$: $f(x) = x^3 - x^2/2 + x/12 - 1/2216 = 0$. (a) use the method of false position to estimate the root, accurate to three decimal places; (b) compare your estimate with the true root, given that the polynomial is $f(x) = (x - 1/6)^3$; (c) How would you explain the difference between the estimated and the true root?

E4.18 Use the method of false position to estimate the root of the following equation in $(0, \pi/3)$ to an accuracy of $\varepsilon = 5 \times 10^{-4}$.

$$\tan\left(\frac{\pi}{5} - x\right) + \tan\left(\frac{x}{5}\right)\tan\left(x + \frac{\pi}{10}\right) - x = 0$$

E4.19 Use the method of False Position to estimate the root of the following equation in $(0, \pi/3)$ to an accuracy of $\varepsilon = 10^{-4}$.

$$2x + \sin(2x) - \cos(3x) = 3$$

E4.20 The maximum solubility of graphite in liquid Fe (in wt% C) as a function of temperature (in °C) is correlated with $T(x) = 2420\sqrt{x} - 80.1x^{4/3} - 3258.35$. Using this correlation, estimate the maximum graphite solubility at $1600°C$ and $2500°C$ accurately to *one decimal place*, using the method of false position. *Note*: The solubility is expected to be within $(4, 10)$ wt%.

E4.21 The following nonlinear equation is encountered in the design of helical gears:

$$\tan\theta - \theta = K$$

where θ denotes pressure angle and K is a constant. For $K = 0.005$, use the method of false position to estimate θ within $(0.2, 0.5)$ accurately to *four decimal places*.

Section 4.3 Fixed-Point Iteration

E4.22 Repeat **E4.8** using the fixed-point iteration with $x^{(0)} = 1$ as the initial guess.

E4.23 Repeat **E4.18** with the method of fixed-point iteration method using $x^{(0)} = 0.5$.

E4.24 Estimate the root of $e^{-3x} = 2x - 1$, using the fixed-point iteration method with $x^{(0)} = 0$ and $\varepsilon = 5 \times 10^{-4}$.

E4.25 The point of intersection of $y = \sqrt{6x + 5}$ and $y = x^2 - 4x$ is to be determined. The two possible fixed-point iteration functions are:

(i) $g(x) = \frac{1}{4}(x^2 - \sqrt{5 + 6x})$,   (ii) $g(x) = \sqrt{4x + \sqrt{5 + 6x}}$

(a) Find the intersection point using the fixed-point iteration method with $x^{(0)} = 0$ and $\varepsilon = 5 \times 10^{-4}$ (also compute $g'(x)$ at each iteration step); (b) which iteration function converged faster? why?

E4.26 There is a root of the equation $x^7 - 2x - 1 = 0$ in the interval $(1,2)$. Can we apply the fixed-point iteration method to estimate its root using the following two iteration functions? Explain why.

$$\text{(i) } g(x) = \frac{1}{2}(x^7 - 1), \qquad \text{(ii) } g(x) = \sqrt[7]{1 + 2x}$$

E4.27 For the iteration functions given below, (a) determine which ones will converge to a fixed-point, and (b) find the convergence rate for the convergent cases.

$$\text{(a) } g(x) = \frac{3x}{4} + \frac{1}{8x^3}, \quad \text{(b) } g(x) = \frac{1}{3x + 4}, \quad \text{(c) } g(x) = \frac{1}{3}\sin x + \frac{1}{4x}$$

E4.28 Repeat **E4.19** with the method of fixed-point iteration. An iteration function is given as

$$x^{(n+1)} = \frac{1}{2}\left(3 - \sin(2x^{(n)}) + \cos(3x^{(n)})\right)$$

(a) Compute the first five iterations starting with $x^{(0)} = 0.7$ and explain the convergence trend; (b) Repeat Part (a) by modifying the iteration equation by adding x to both sides of the iteration function to obtain:

$$x^{(n+1)} = \frac{1}{2}x^{(n)} + \frac{1}{4}\left(3 - \sin(2x^{(n)}) + \cos(3x^{(n)})\right)$$

E4.29 Given that $f(x) = 12x^3 - 23x^2 - 37x + 70$ has a root in the neighborhood of 1.5, (a) which one of the following iteration functions will converge to the root? (b) Using the convergent iteration functions, estimate the root to an accuracy of $\varepsilon = 5 \times 10^{-4}$ and calculate the linear convergence rate.

$$\text{(a) } g(x) = \frac{70}{37} - \frac{23x^2}{37} + \frac{12x^3}{37}, \quad \text{(b) } g(x) = \frac{\sqrt{70 - 37x + 12x^3}}{\sqrt{23}}, \quad \text{(c) } g(x) = \frac{23}{12} - \frac{35}{6x^2} + \frac{37}{12x}$$

E4.30 Repeat **E4.21** using the fixed-point iteration method with $\theta^{(0)} = 0.3$.

E4.31 A relationship between voltage and current flowing through a solar cell under the maximum power condition is given by the following equation:

$$\frac{V_m}{V_t} = \ln\left[\left(1 + \frac{I_{ph}}{I_s}\right) \Big/ \left(1 + \frac{V_m}{V_t}\right)\right]$$

where I_s is the saturation current of the diode ($=2.5 \times 10^{-12}$ A), I_{ph} is the photo current ($=25$ mA), V_t is the thermal voltage ($=28$ mV), and V_m is the voltage corresponding to the maximum power. Apply the fixed-point iteration method to find the maximum voltage, accurate to two decimal places. Use $(V_m/V_t)^{(0)} = 10$ V for the initial guess.

E4.32 The stiffness (k) of two plates fastened by a bolted joint is defined by

$$k = (\pi/2)Ed\tan\alpha/\ln\left(\frac{(\ell\tan\alpha + d_w - d)(d_w + d)}{(\ell\tan\alpha + d_w + d)(d_w - d)}\right)$$

where E is the Young's modulus, d is the diameter of the bolt, d_w is the washer face diameter, and ℓ is the plate thickness. Assuming $d_w = 1.5d$, $E = 180$ GPa, $\ell = 0.15$ m, and $\tan\alpha = 0.6$, apply the fixed-point iteration method to estimate to three decimal place accuracy the bolt diameter (d) corresponding to $k = 8.623 \times 10^9$ N/m. Use $d^{(0)} = 0.02$ m.

Section 4.4 Newton-Raphson Method

E4.33 Repeat **E4.1** using the Newton-Raphson method with $x^{(0)} = 1$.

E4.34 Repeat **E4.3** using the Newton-Raphson method. Use the midpoint as an initial guess.

E4.35 Repeat **E4.8** using the Newton-Raphson method. Use $x^{(0)} = 0$ as the initial guess.

E4.36 For each of the following nonlinear equations with specified initial values, apply the Newton-Raphson method to estimate the root, accurate to three decimal places.

(a) $x + 1 = 2e^{-x}$, $x^{(0)} = 0$

(b) $(3 - x)e^{-x} = 1$, $x^{(0)} = 0$

(c) $2x - \ln(x + 2) = e^{-x}$, $x^{(0)} = 0.2$

(d) $x \ln(1 + x) = \sin 2x$, $x^{(0)} = 0.8$

(e) $x^2 - 3x + 1 = x \sin x$, $x^{(0)} = 1$

(f) $x \ln x (2 + \ln x) - 3 + x = 0$, $x^{(0)} = 1$

(g) $x \cdot 4^x - 80 = 0$, $x^{(0)} = 2$

(h) $x \cdot 4^x - 4x^2 = 3$, $x^{(0)} = 1.3$

(i) $2x (1 - 3x^2 - 2x) \ln x + 3x = x^2 - 5$, $x^{(0)} = 1.2$

E4.37 Repeat **E4.20** with the Newton-Raphson method for $x^{(0)} = 4$ and $\varepsilon = 10^{-2}$.

E4.38 Repeat **E4.21** with the Newton-Raphson method for $x^{(0)} = 0.2$ and $\varepsilon = 5 \times 10^{-4}$.

E4.39 The nonlinear equation $\sin \theta = \alpha \theta$ is encountered in physics when studying Fraunhofer diffraction. Estimate the root of the nonlinear equation for $\alpha = \sqrt{2/3}$, accurate to four decimal places, using the Newton-Raphson method with $\theta^{(0)} = 1$.

E4.40 The following equation describes the transient current of a circuit:

$$i(t) = 2e^{-2t} \sin\left(\frac{\pi t}{3}\right) + 3\sin\left(2t - \frac{\pi}{3}\right)$$

Use the Newton-Raphson method to estimate (a) the moment when the current becomes zero and (b) the moment in the neighborhood of $t^{(0)} = 1.6$, when the current is at its maximum or minimum. The convergence tolerance is given as $\varepsilon = 5 \times 10^{-4}$.

E4.41 The spectral emissive power of a black body is given by

$$E(\lambda, T) = \frac{C_1}{\lambda^5 (\exp(C_2/\lambda T) - 1)}$$

where λ is the wavelength (μm), T is the temperature of the blackbody (K), C_1 and C_2 are constants defined respectively as 3.742×10^8 W.μ m^4/m^2 and 1.439×10^4 μm.K. Use the Newton-Raphson method to estimate the wavelength corresponding to the maximum emissive power within an absolute error tolerance of $\varepsilon = 10^{-3}$. *Hint:* Defining $x = C_2/\lambda T$, the spectral emissive power can be written as $E(x) = C x^5/(e^x - 1)$, where $C = C_1 T^5/C_2^5$, which is also constant. Use $x^{(0)} = 5$ for the initial guess.

E4.42 Water flows in an open, rectangular channel of constant width (w), as shown in **Fig. E4.42**. There is a small ramp (height δ) downstream. The Bernoulli equation, which relates the pressure, velocity, and height of fluid at two points (before and after the ramp), and conservation of mass ($Q = V_0 h_0 w = V_1 h_1 w$) yield the following equation:

$$\frac{Q^2}{2gw}\left(\frac{1}{h_0^2} - \frac{1}{h^2}\right) - (h - h_0) - \delta = 0$$

where w is the width of the channel and δ is the height of the ramp, g is the acceleration of gravity, h_0 and h are the water levels, and V_0 and V_1 are the velocities at the elevations z_0 and z_1. For an open channel with $Q = 0.4$ m^3/s, $w = 0.8$ m, $h_0 = 0.5$ m, $\delta = 0.07$ m, estimate the elevation of the water surface downstream of the ramp, i.e., $h + \delta = ?$ Use $\varepsilon = 10^{-3}$ and $h^{(0)} = h_0$ for the initial guess.

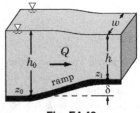

Fig. E4.42

E4.43 The pressure (P), volume (V), and temperature (T) of a real gas are related to each other through the van der Waals equation:

$$\left(P + \frac{n^2 a}{V^2}\right)(V - nb) = nRT$$

where $R = 0.08206$ atm-L/mol-K, a and b are constants, and n is the molarity of the gas. The pressure of 3 mol of BCl_3 (boron trichloride) gas ($a = 15.39$ atm.L^2/mol^2, $b = 0.1222$ L/mol) is 1.15 atm at 310 K. Using the absolute error tolerance of $\varepsilon = 10^{-3}$ and the initial guess of $V^{(0)} = 20$ L, estimate the volume of BCl_3 by applying the Newton-Raphson method. Compare your estimate with that obtained from the ideal gas law.

E4.44 The kinematic equations of a crank mechanism illustrated in **Fig. E4.44** are given as

$$\ell_1 \cos\theta + \ell_2 \cos\alpha - L = 0, \quad \ell_1 \sin\theta + \ell_2 \sin\alpha = 0$$

For $L = 0.625$ m, $\ell_1 = 18$ cm, $\ell_2 = 54$ cm, estimate α and θ accurately to two decimal places by applying the Newton-Raphson method. Use $\theta^{(0)} = 60°$. *Hint*: Reduce the equations into a nonlinear equation.

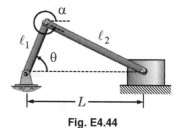

Fig. E4.44

E4.45 The current-voltage relationship of a semiconductor diode is given by Shockley's diode equation: $I = I_s(\exp(V/nV_T) - 1)$, where n is called the *ideality factor*, which can be assumed to be unity, I_s is the saturation current (A), V is the applied voltage (V), V_T is the thermal voltage. Consider a diode connected in series with a resistor ($R = 1.5$ kΩ) and a battery with $E = 9$ V as shown in **Fig. E4.45**. The applied voltage for the closed circuit can be written as

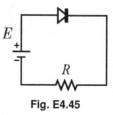

Fig. E4.45

$$E = IR + V = I_s R\left(e^{V/V_T} - 1\right) + V$$

Using $I_s = 10^{-8}$ A and $V_T = 0.026$ V, estimate the applied voltage across the diode to three decimal places by applying the Newton-Raphson method. Use $V^{(0)} = 0.4$ as the initial guess.

E4.46 The Gibbs free energy of one mole of a gas mixture in the temperature range above 200 K is given as

$$G(T) = -RT\left\{\ln\left(\frac{AT^{7/2}}{1 - e^{-B/T}}\right) + \frac{C}{T}\right\}$$

where $R = 8.314$ J/K, $A = 4.3 \times 10^{-4} K^{-7/2}$, $B = 6000$ K and $C = 52000$ K are the gas constants for H_2. Apply the Newton-Raphson method to estimate the temperature (accurate to one decimal place) that yields $G(T) = -5.19 \times 10^5$ J. Use $T^{(0)} = 260$ for the initial guess.

E4.47 The internal effectiveness factor for a first-order reaction (A→B) in a spherical catalyst pellet is given by

$$\eta = \frac{3}{\phi^2}\left(\phi\coth\phi - 1\right)$$

where ϕ is the *Thiele modulus* of the first-order reaction defined as $\phi^2 = k_1 R^2/D_e$, where R is the pellet radius, k_1 is the rate constant, and D_e is the diffusion coefficient. The magnitude of the effectiveness factor ($0 \leqslant \eta \leqslant 1$) indicates the relative importance of diffusion and reaction. Apply the Newton-Raphson method to estimate the Thiele modulus of a first-order reaction, accurate to four decimal places, that will achieve the effectiveness factor $\eta = 0.875$. Use $\phi^{(0)} = 2$ for the initial guess.

E4.48 The head loss in pipes may be described by $H = KQ^{1.75}$, where Q is the flow rate (m^3/s) and K is constant depending on the pipe properties. A water distribution system with two inlets and two outlets, as well as the flow rates and pipe constants, are given in **Fig. E4.48**. We obtain four equations from the continuity of flow at junctions; ❶, ❷, ❸, and ❹; that is,

At junction ❶, $0.20 = Q_1 + Q_2$ At junction ❷, $Q_2 = 0.10 + Q_4$

At junction ❸, $Q_1 + 0.05 = Q_3$ At junction ❹, $Q_3 + Q_4 = 0.15$

We have an additional equation from the head losses, which must be zero for the loop:

$$K_2 Q_2^{1.75} + K_4 Q_4^{1.75} - K_3 Q_3^{1.75} - K_1 Q_1^{1.75} = 0$$

The above equations constitute a system of nonlinear equations with five equations and four unknowns. Using the available equations, reduce this system to a single nonlinear equation with one variable. Then solve the resulting equation with the Newton-Raphson method to estimate the flow rates (accurate to four decimal places) at each pipe section. Use $Q_k^{(0)} = 0$ for the initial guess.

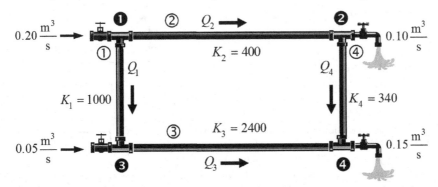

Fig. E4.48

E4.49 A typical normalized voltage waveform is depicted in **Fig. E4.49**. Such pulses are used in electrical engineering to characterize voltage surges on power lines and on circuits. For pulses, a series of definitions (t_{10}, t_{90}, t_m) are specified by IEEE Standards C62.41-2002. t_{50} is a parameter for measuring the width of the pulse. The rise time of a waveform is specified as the time taken for the waveform to go from 10% to 90% of its maximum value. The time for the peak is not specified, but it is clearly the point where the derivative of $V(t)$ is zero. A waveform is given as

$$V(t) = 1.44576 \, e^{-0.01t^2} (1 - e^{-0.415t})$$

Use the Newton-Raphson method to determine t_{10}, t_{90}, t_m and t_{50}, accurate to four decimal places.

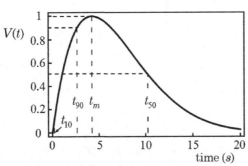

Fig. E4.49

Section 4.5 Modified Newton-Raphson Method

E4.50 Use the Newton-Raphson and modified Newton-Raphson methods with $x^{(0)} = 1$ to estimate the root of the following function with an absolute error less than $\varepsilon = 10^{-5}$. Comment on the convergence of both methods.

$$f(x) = x^2 + e^{-2x} - 2xe^{-x}$$

E4.51 Use the Newton-Raphson and modified Newton-Raphson methods with $x^{(0)} = 3$ to estimate the root of the following function with an absolute error less than $\varepsilon = 10^{-5}$. Comment on the convergence of both methods.

$$f(x) = 6 + x^2 - 4x \sin\left(\frac{\pi x}{4}\right) - 2\cos(\pi x)$$

E4.52 Use the modified Newton-Raphson method to estimate the roots of the following polynomials with the given initial guesses within the absolute error tolerance of $\varepsilon = 10^{-5}$.

(a) $25x^5 + 5x^4 - 111x^3 + 157x^2 - 322x + 294 = 0$, $x^{(0)} = 1$

(b) $x^5 - 8.1x^4 + 22.87x^3 - 27.783x^2 + 21.87x - 19.683 = 0$, $x^{(0)} = 1.5$

(c) $x^5 - 1.6x^4 - 1.88x^3 + 3.312x^2 + 0.864x - 1.728 = 0$, $x^{(0)} = 0.25$ and $x^{(0)} = -0.5$

E4.53 Apply the modified Newton-Raphson method with $x^{(0)} = 1$ to estimate the root of the following equation to an accuracy of $\varepsilon = 10^{-5}$. Is it a root with multiplicity? If yes, determine the degree of multiplicity.

$$f(x) = -1 + 4x - 4x^2 - 2\ln x + 8x^2 \ln x + (2x - 1)\ln 16$$

E4.54 Apply the modified Newton-Raphson method with $x^{(0)} = 1.5$ to estimate the root of $4x - 4x^{3/2} + x^2 = 0$ to an accuracy of $\varepsilon = 10^{-5}$. Is it a root with multiplicity? If yes, determine the degree of multiplicity.

E4.55 Apply the modified Newton-Raphson method with $x^{(0)} = 3$ to estimate the root of $2 + \cos 2x - \sin x = 0$ to an accuracy of $\varepsilon = 10^{-5}$. Is it a root with multiplicity? If yes, determine the degree of multiplicity.

E4.56 The dynamic properties of a machine may be obtained from the natural frequency ω_0 (in an undamped state) and damping ratio D. A factor α, which constitutes the cutting parameters, is proportional to cutting depth. When modeling the radial motion between a workpiece and a tool, the following nonlinear equation is encountered:

$$\lambda^2 + 2D\lambda + 1 + \alpha\left(1 - e^{-\omega_0 T_0 \lambda}\right) = 0$$

Estimate the root of this equation for the cases given below by applying the modified Newton-Raphson method with $\lambda^{(0)} = 1$ and $\varepsilon = 10^{-4}$.

(a) $D = 0.08$, $\alpha = 0.26$, $\omega_0 T_0 = -1.90$,   (b) $D = 0.10$, $\alpha = 1.52$, $\omega_0 T_0 = -1.44$

Section 4.6 Secant and Modified Secant Methods

E4.57 Repeat **E4.8** using (a) the Secant method with $x^{(0)} = 0$ and $x^{(1)} = 0.1$ as the starting values, and (b) the modified Secant method with $h = 0.05$ and $x^{(0)} = 0$.

E4.58 The quality of the initial guess (or starting values) could result in divergence, The quality of the initial guess (or starting values) could result in divergence, slow convergence, or fast convergence. To investigate the effect of the starting values on the Secant method, estimate the root of $f(x) = 6x - 2x^3 - 3\sin(2x) - 1$ within $\varepsilon = 10^{-6}$ using the following starting values:

(a) $x^{(0)} = -1$, $x^{(2)} = 2$, (b) $x^{(0)} = 0$, $x^{(2)} = 2$, (c) $x^{(0)} = 0$, $x^{(2)} = 1$,
(d) $x^{(0)} = 1$, $x^{(2)} = 1.1$, (e) $x^{(0)} = 2$, $x^{(2)} = 3$, (f) $x^{(0)} = 1$, $x^{(2)} = 2$

E4.59 Use the Secant method with $t^{(0)} = 0.5$ and $t^{(1)} = 1$ to estimate the maximum of the given function with an absolute error tolerance less than $\varepsilon = 10^{-5}$.

$$f(t) = 3e^{-1.5t} \sin 3t + 0.3\sin 4t$$

E4.60 Repeat **E4.59** using the Secant method with $t^{(0)} = 0.1$, $t^{(1)} = 0.2$ and $\varepsilon = 10^{-5}$.

E4.61 Repeat **E4.21** using the Secant method with $\theta^{(0)} = 0.3$, $\theta^{(1)} = 0.4$ and $\varepsilon = 10^{-5}$.

E4.62 The amount of money required to pay off a mortgage over a fixed period of time is calculated from the *ordinary annuity equation*:

$$P = A \frac{i(1+i)^n}{(1+i)^n - 1}$$

where A is the amount of the mortgage, P is the amount of each payment, and i is the interest rate per period for the n payment periods. Suppose that a 5-year home mortgage in the amount of \$100,000 is needed and the borrower can afford house payments of at most \$1750 per month. Use the secant method with $i^{(0)} = 0.005$, $i^{(0)} = 0.01$ and $\varepsilon = 10^{-4}$ to estimate the maximum interest rate that the borrower can afford to pay.

E4.63 The revised van der Waals equation of state, also known as the Redlich–Kwong equation of state, is an empirical algebraic equation that relates the volume (V), pressure (P), and temperature (T) of gases. Generally, it is more accurate than the van der Waals equation and the ideal gas equation at temperatures above the critical temperature. Redlich–Kwong proposed the following:

$$\left(P + \frac{a}{\sqrt{T}V(V+b)} \right) (V - b) = RT$$

where $R = 0.08206$ atm-L/mol-K is the universal gas constant, a and b are related to the critical properties (T_c, P_c) of the substance, and they are defined as

$$a = 0.42747 R^2 T_c^{2.5}/P_c \quad \text{and} \quad b = 0.08664 RT_c/P_c$$

Estimate the volume of 1 mole of boron trichloride gas at 310 K and 1.15 atm with the Secant method. Use $T_c = 455$ K, $P_c = 39.46$ atm, $V^{(0)} = 10$ L, $V^{(1)} = 15$ L and $\varepsilon = 10^{-4}$.

E4.64 Apply the Secant method with $x^{(0)} = 2$ and $x^{(1)} = 3$ to estimate the intersection point of $y = \sqrt{3x + 2} + 3\sqrt{x + 1}$ and $y = 2\sqrt{1 + x^2}$ with an absolute error tolerance of $\varepsilon = 10^{-3}$.

E4.65 Repeat **E4.64** using (a) the modified secant method with $h = 0.01$, (b) the Newton-Raphson method with $x^{(0)} = 2$, (c) compare both methods in terms of the total iteration numbers required for convergence.

E4.66 Estimate the roots of the following nonlinear equations using the Secant method with an absolute error tolerance of $\varepsilon = 10^{-4}$.

(a) $2\sqrt{x} + x - 3\ln x = 3$, $x^{(0)} = 1.8$, $x^{(1)} = 2$ (b) $xe^{-2x} + x = 2$, $x^{(0)} = 0.7$, $x^{(1)} = 1.1$
(c) $\tan^{-1}\left(\frac{2x}{1+x} \right) \ln(1 + x) = 2$, $x^{(0)} = 1$, $x^{(1)} = 1.5$

E4.67 Repeat **E4.66** with the modified secant method using $h = 0.1, 0.05$, and 0.01 and $x^{(0)} = 1.8$. How does the value of h affect convergence?

E4.68 Water is pumped from a reservoir at elevation $h_1 = 400$ m to another reservoir at elevation $h_2 = 430$ m through a pipeline with a diameter of $D = 27.5$ cm and a length of $L = 100$ m, as illustrated in **Fig. E4.68**. Neglecting local losses, the energy equation between the two reservoir surfaces can be written as

$$h_p = f \frac{L}{D} \frac{V^2}{2g} + h_2 - h_1$$

where h_p is the pump head (m), h_1 and h_2 are the elevations of the reservoirs, f is the friction factor, V is the mean velocity of water (m/s), and g is the acceleration of gravity (m/s^2). The velocity and the friction factor can be expressed in terms of the discharge Q as

$$V = Q/(\pi D^2/4), \quad f = 0.01(D/Q)^{1/4}$$

If the pump head-Q characteristic curve is given by $h_p = 33 - 78.6\,Q - 9060\,Q^2$ where Q and h_p are in units m^3/s and m, respectively, (a) derive an expression for the discharge Q in the pipeline; (b) find the discharge, accurate to four decimal places, using the modified secant method with $Q^{(0)} = 0.01$ and $h = 0.01$.

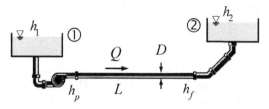

Fig. E4.68

E4.69 A thermistor is a resistor whose resistance is dependent on temperature. Thermistors are used as temperature sensors due to their high sensitivity. A change in the temperature of the thermistor results in a change in electrical resistance. By measuring the resistance of the thermistor, one can determine the temperature. The *Steinhart–Hart equation*, a third-order approximation, relates the temperature and the resistance as

$$T^{-1} = a + b \ln R + c(\ln R)^2$$

where R is the resistance (Ω), T is the absolute temperature (K), and the coefficients of the Steinhart–Hart equation of a thermistor are $a = 1.32 \times 10^{-3}$, $b = 2.34 \times 10^{-4}$, and $c = 8.8 \times 10^{-8}$. Apply the modified secant method to estimate the resistance of the thermistor at $40°$C. Use $h = 0.01$, $\varepsilon = 10^{-5}$ and $R^{(0)} = 100\ \Omega$.

E4.70 In modeling an AC-DC converter, the following nonlinear equation is encountered:

$$\sin(\beta - \phi) - e^{-\beta/\tan\phi} \sin\phi = 0$$

where β is the dead-lock angle ($\pi \leqslant \beta \leqslant 2\pi$) and ϕ is the load angle ($0 \leqslant \phi \leqslant \pi/2$). Estimate two roots, ϕ_1 and ϕ_2, for $\beta = 11\pi/9$ using the modified secant method with $h = 0.01$, $\phi^{(0)} = 0.5$ and $\phi^{(0)} = 1.3$. Use $\varepsilon = 10^{-3}$.

E4.71 In the study of like-charge conducting same-size spheres, identifying the regions of attraction or repulsion leads to the following nonlinear equation:

$$\left(\ln \frac{4a}{s} + 2\gamma \right) \sqrt{\frac{s}{a}} = \delta_0$$

where a is the capacitance coefficient, s is the surface-to-surface distance of the spheres, δ_0 is a parameter varying as a function of charge ratio, and $\gamma = 0.5772$ is Euler's constant. This nonlinear equation yields two s/a values (roots) for every δ_0. Estimate the two roots for $\delta_0 = 0.5$, 1.5, and 2.5 using the modified secant method with $h = 0.001$ and $\varepsilon = 10^{-3}$.

E4.72 In an isentropic turbine, a flammable gas expands at atmospheric pressure. If the exhaust temperature satisfies the following equation, find the flame temperature using the modified secant method with $h = 0.01$, $T^{(0)} = 400$ K, and $\varepsilon = 10^{-3}$.

$$3.45 \times 10^{-2}\,(T - 597) - 3.98 \times 10^{-6}\,(T^2 - 597^2) + 0.195 \ln\left(\frac{T}{597} \right) + \ln\left[\frac{93.3T - 157.2}{486} \right] = 0$$

Section 4.7 Accelerating Convergence

E4.73 Consider the fixed-point iteration with $g(x) = 1 + x/\sqrt{x+2}$ and $x^{(0)} = 1$. (a) Carry out the first 10 steps to compute the linear convergence rate from $\lambda^{(n)} \approx |g'(x^{(n)})|$ and the approximate convergence rate from $\lambda^{(n)} \approx |\Delta x^{(n+1)}/\Delta x^{(n)}|$, (b) apply the Aitken acceleration to the sequence of approximations in Part (a) and evaluate the approximate convergence rate from $\lambda^{(n)} \approx |\Delta \alpha^{(n+1)}/\Delta \alpha^{(n)}|$. Use $\varepsilon = 0.5 \times 10^{-4}$.

E4.74 Repeat **E4.73** for $g(x) = \ln(2 + e^{-x})$ with $x^{(0)} = 1/3$.

E4.75 Consider the following finite series:

$$S^{(n)} = \frac{1}{1 \cdot 3} + \frac{1}{3 \cdot 5} + \cdots + \frac{1}{(2n-1)(2n+1)} \quad \text{or} \quad S^{(n)} = S^{(n-1)} + \frac{1}{(4n^2-1)}, \quad S^{(0)} = 0, \ n \geqslant 1$$

which converges to $1/2$ for large n. (a) Sum up the sequence up to $S^{(10)}$, the true absolute error by $|1/2 - S^{(n)}|$, and the convergence rate from $\lambda^{(n)} \approx |S^{(n+1)} - S^{(n)}|/|S^{(n)} - S^{(n-1)}|$, and (b) apply Aitken's acceleration to the sequence generated in Part (a) and compare your results.

E4.76 Successive approximations produced by a linear iterative process yield the results presented below. Use the Aitken acceleration to compute $\{\alpha^{(n)}\}$ and estimate the convergence rate.

n	$x^{(n)}$	n	$x^{(n)}$
0	0.84089 64153	5	0.74517 97287
1	0.78323 89193	6	0.74324 85821
2	0.75966 08667	7	0.74239 71531
3	0.74957 52538	8	0.74202 11823
4	0.84089 64153	9	0.74185 50490

E4.77 Estimate the root of $x = \sqrt{x - x^5}$ with $x^{(0)} = 1/4$, (a) using the fixed-point iteration method, accurate to six decimal places. How many iterations are required to reach the answer? (b) How many iterations would be required if Aitken acceleration were used?

E4.78 For the following sequences, generate the first five terms of the sequence $\{\alpha^{(n)}\}$ using the Aitken acceleration and determine the decimal-place accuracy of $\alpha^{(4)}$.

(i) $x^{(n+1)} = 3 + 2e^{-x^{(n)}}$, $\quad x^{(0)} = 0$, $\quad$ (ii) $x^{(n+1)} = \sqrt{1 - \sin x^{(n)}}$, $\quad x^{(0)} = 0$,

(iii) $x^{(n+1)} = \sqrt{2 + \sqrt[3]{x^{(n)}}}$, $\quad x^{(0)} = 1$, $\quad$ (iv) $x^{(n+1)} = \exp(x^{(n)}/3)$, $\quad x^{(0)} = 2$

E4.79 The isentropic condensation temperature is the temperature at which saturation is reached when a parcel of moist air is lifted adiabatically with its vapor content (w) held constant. It can be estimated from the solution of the following nonlinear equation:

$$T_{ic} = \frac{C_1}{\ln(C_2 \varepsilon / w p_i) + (1/\kappa) \ln(T_i/T_{ic})}$$

where $C_1 = 5400$ K, $C_2 = 2.5 \times 10^9$ kPa, $\kappa = 0.286$, $\varepsilon = 0.622$, w is the water vapor mixing ratio, and T_i and p_i are the initial temperature (K) and pressure (kPa), respectively. Using the fixed-point iteration with Aitken's acceleration, estimate the isentropic condensation temperature for the case of $w = 0.04$ kg vapor/kg dry air, $T_i = 265$ K and $p_i = 100$ kPa. Use $T^{(0)} = T_i$ and $\varepsilon = 10^{-3}$.

E4.80 For the fixed-point iteration equation given by $x^{(n+1)} = \sin(x^{(n)} + \sin x^{(n)})$ with $x^{(0)} = 1$, apply Steffensen's acceleration to estimate $\alpha^{(1)}$ and $|\alpha^{(1)} - g(\alpha^{(1)})|$.

E4.81 For the fixed-point iteration equation given by $x^{(n+1)} = \cos(x^{(n)}/2)\cosh(x^{(n)}/3)$ with $x^{(0)} = 1/2$, apply Steffensen's acceleration to estimate $\alpha^{(1)}$ and $\left|\alpha^{(1)} - g(\alpha^{(1)})\right|$.

E4.82 For the fixed-point iteration equation given by $x^{(n+1)} = x^{(n)}/(5 + (x^{(n)})^2) + 3x^{(n)} - 1$ and $x^{(0)} = 0$, apply Steffensen's acceleration to estimate $\alpha^{(2)}$ and $\left|\alpha^{(2)} - g(\alpha^{(2)})\right|$.

E4.83 Apply Steffensen's acceleration to the iteration equation: $g(x) = 5 - 2x \ln(1 + 0.5x^2)$ using $x^{(0)} = 1$. (a) Perform 5 iterations to estimate $\alpha^{(5)}$ and $\left|\alpha^{(5)} - g(\alpha^{(5)})\right|$; and (b) what can you say about the accuracy of $\alpha^{(5)}$? (c) Show that Steffensen's method converges quadratically.

E4.84 Apply Steffensen's acceleration to estimate the solution of $2x = 5^{-x}$ to within $\varepsilon = 10^{-6}$. Use $x^{(0)} = 1$ as the starting value.

E4.85 Repeat **E4.78** with Steffensen's acceleration for four iterations.

E4.86 Repeat **E4.31** with Steffensen's acceleration for four iterations.

E4.87 A vertical metal panel ($A = 0.05$ m^2) suspended in a room at $T_\infty = 15°C$ is subjected to electric current, which causes the panel to produce $Q = 24$ W heat energy. The rate of heat transfer from the plate to the room is calculated by

$$Q = 0.0475A(T_s - T_\infty)\left(0.825 + 14.15\left(\frac{T_s - T_\infty}{T_s + T_\infty}\right)^{1/6}\right)^2$$

where T_s is the mean surface temperature (K) and A is the surface area of the plate ($2A$, losses from both sides). Apply Steffensen's acceleration to estimate the mean surface temperature of the panel, accurate to two decimal places. Use $T_s^{(0)} = 300$K for the initial value.

E4.88 Consider the iteration sequence $\{x^{(p)}\}$ generated in **Example 4.3**. (a) Apply Aitken's acceleration to the sequence to improve convergence and calculate the convergence rates for both $\{x^{(p)}\}$ and $\{\alpha^{(p)}\}$ sequences; (b) Starting with $x^{(0)} = 1$, apply Steffensen's acceleration to estimate the root to four decimal places.

Section 4.8 System of Nonlinear Equations

E4.89 Using the given initial guesses, estimate the common solution of the following systems of nonlinear equations accurate within $\varepsilon = 10^{-5}$.

(a) $x^2 + y^2 = 10,$ $x^2y - xy^3 - xy + 3 = 0,$ $(x^{(0)}, y^{(0)}) = (1, 1)$

(b) $x^2 + 2xy = 1.35,$ $x + 3y + y^2 - x^2 = 4.76,$ $(x^{(0)}, y^{(0)}) = (1, 1)$

(c) $2x^2 + x + 3y = 7,$ $3x + 4y + xy = 3,$ $(x^{(0)}, y^{(0)}) = (0, 0)$

(d) $e^{x-y} = 7,$ $xe^y + ye^x = 9,$ $(x^{(0)}, y^{(0)}) = (2, 1)$

(e) $\cos(x - y) + x = \pi,$ $\cos(x + y) + 2y = \pi,$ $(x^{(0)}, y^{(0)}) = (3, 1.5)$

(f) $2y/x + 2x^2/y = 17,$ $4y^2/x + 2x/y^3 = 5,$ $(x^{(0)}, y^{(0)}) = (1, 1)$

(g) $y\sqrt{x} + \sqrt{y} = 21,$ $xy + x + y = 49,$ $(x^{(0)}, y^{(0)}) = (5, 5)$

E4.90 The steady-state temperature (in °C) distribution of a large metal plate ($0 \leqslant x \leqslant 6$ m, $0 \leqslant y \leqslant 4$ m) is given by the following equation:

$$T(x, y) = x^3 + 4y^3 - 9x^2 - 24y^2 + 2xy + 15x + 36y + 65$$

Use Newton's method to estimate the critical points of the temperature distribution, accurate to two decimal places, and identify them as the local maximum, local minimum, or saddle point. Use (1,1), (1,3), (4,1), and (4,3) for the initial guesses.

E4.91 Find the intersection points of the ellipse $2(x - 1)^2 + (y + 1)^2 = 2$ and the circle $x^2 + y^2 = 1$. Use (0,1) and (1,1) as the initial guesses and $\varepsilon = 0.5 \times 10^{-2}$ as convergence tolerance.

E4.92 Apply Newton's method to find all intersection points of the parabola $y = 6 - 2x - x^2$ and the ellipse $(x + 1)^2/9 + (y - 3)^2/4 = 1$. Use (−2,5), (0,5), (−3,2), and (1,2) as the initial guesses and $\varepsilon = 0.5 \times 10^{-3}$ as convergence tolerance.

Section 4.9 Bairstow's Method

E4.93 Apply Bairstow's method to estimate all real and imaginary roots of the polynomials given below with a tolerance of $|\delta p| + |\delta q| < 0.001$. Use $p^{(0)} = q^{(0)} = 0$ as the starting values for all solutions.

(a) $f(x) = x^3 + 0.92x^2 - 0.674x - 3.45$

(b) $f(x) = x^4 + 2x^3 + 3.31x^2 + 2.32x + 1.375$

(c) $f(x) = x^5 - 0.58x^4 - 6.0845\,x^3 - 11.0546\,x^2 - 3.3959x + 1.63625$

(d) $f(x) = x^6 + 6.95x^5 + 20.42x^4 + 42.11x^3 + 67.48x^2 + 57.24x + 7.2$

(e) $f(x) = x^7 + 2.85\,x^6 + 3.4x^5 - 6.2\,x^4 - 31.45\,x^3 - 43.8\,x^2 + 7.4x + 92$

E4.94 A mixture of 1.5 mol of N_2 and 3 mol of H_2 is introduced into a 10 L reaction vessel at 500 K to make NH_3. At this temperature, the reaction equilibrium constant for $N_2(g)+3H_2(g) \rightleftarrows 2NH_3(g)$ is given as $K = 306$. Using the equilibrium equation below, estimate the equilibrium concentrations by applying Bairstow's method with $p^{(0)} = q^{(0)} = -1$ as the initial guess. Use $|\delta p| + |\delta q| < 0.001$ for the stopping criterion.

$$K = \frac{[NH_3]^2}{[N_2][H_2]^3} = \frac{(2x)^2}{(0.15 - x)(0.3 - x)^2}$$

E4.95 The chemical equilibrium of the ammonia synthesis reaction is given by

$$K = \frac{8x^2(4 - x)^2}{(6 - 3x)^2(2 - x)} = 0.19$$

where x is the ammonia concentration. Estimate the equilibrium concentration of ammonia using Bairstow's method with $p^{(0)} = q^{(0)} = 0$ as the initial guess. Use $|\delta p| + |\delta q| < 0.001$ for the stopping criterion.

E4.96 The azeotropic point of a binary solution is described by

$$\frac{0.6\,(1 - x)^2 - 0.4\,x^3}{(0.6 - 0.2\,x)^2} + 0.625 = 0$$

where x denotes the composition of the dominant component. Find the azeotropic point composition using Bairstow's method, starting with $p^{(0)} = -1$, $q^{(0)} = 2$. Use $|\delta p| + |\delta q| < 0.001$ for the stopping criterion.

E4.97 The combustion reaction of methane is given by $CH_4+2O_2 \rightarrow CO_2+2H_2O(g)$. Both methane and air enter the burner at $25°C$. The fundamental equation relating reaction heat to temperature is given by

$$\Delta H° = \Delta H°_{298} + \int_{298}^{T} \overline{C}_p^0 \, dT$$

where $\Delta H°$ is the standard heat of reaction at temperature T, and the standard heat of reaction at 298 K is given as $\Delta H°_{298} = -802\,\text{kJ}$. The mean heat capacity of the products $\overline{C}_p^o$ is given as

$$\overline{C}_p^o(T) = 361.4 + 0.079T - \frac{536000}{T^2} \quad \left(\frac{J}{kg \cdot K}\right)$$

Assume the combustion reaction goes to completion adiabatically and the overall energy balance for the process reduces to $\Delta H = 0$; that is,

$$\Delta H°_{298} + \int_{298}^{T} \overline{C}_p^0 \, dT = 0$$

Under these conditions, estimate the maximum temperature (*flame temperature*) that can be reached due to the combustion of methane by using Bairstow's method (use $p^{(0)} = -2000$, $q^{(0)} = 1000$). For the stopping criterion, use $|\delta p| + |\delta q| < 0.001$.

E4.98 0.55 mol N_2, 0.8 mol H_2, and 0.35 mol NH_3 gas are mixed in a container at 298 K. The pressure-scale equilibrium constant is given by $K_p = 6.0406 \times 10^5$ atm^{-2}. The pressure of the mixture is maintained at $P = 0.0025$ atm throughout the course of the chemical reaction stated by $N_2(g)+3H_2(g) \rightleftarrows 2NH_3(g)$. The equilibrium equation leads to

$$-85.935125x^4 + 116.01241875x^3 - 64.067615x^2 + 17.10638x - 0.70912 = 0$$

Estimate the amount of N_2, H_2, and NH_3 at equilibrium using Bairstow's method with $p^{(0)} = -1$, $q^{(0)} = 0$ as the initial guess. Use $|\delta p| + |\delta q| < 0.1$ for the stopping criterion.

E4.99 The characteristic equation of matrix $\mathbf{A}$ is given as

$$\lambda^8 - 64\lambda^7 + 1351\lambda^6 - 10420\lambda^5 + 21310\lambda^4 + 56\lambda^3$$
$$+ 1621\lambda^2 + 44000\lambda + 10^6 = 0$$

Find all eigenvalues (roots) using Bairstow's method. Start with $p^{(0)} = -2000$, $q^{(0)} = 1000$, and use $|\delta p| + |\delta q| < 0.1$ for the stopping criterion.

$$\mathbf{A} = \begin{bmatrix} 8 & 7 & 6 & 5 & 4 & 3 & 2 & 1 \\ 9 & 8 & 7 & 6 & 5 & 4 & 3 & 2 \\ & 9 & 8 & 7 & 6 & 5 & 4 & 3 \\ & & 9 & 8 & 7 & 6 & 5 & 4 \\ & & & 9 & 8 & 7 & 6 & 5 \\ & & & & 9 & 8 & 7 & 6 \\ & & & & & 9 & 8 & 7 \\ & & & & & & 9 & 8 \end{bmatrix}$$

Section 4.10 Polynomial Reduction and Synthetic Division

E4.100 Use Synthetic division to reduce the given polynomial by the specified root.

(a) $P_3(x) = x^3 - 2.9x^2 - 11.32x + 26.24$ $(x = 2)$

(b) $P_3(x) = x^3 - 2x^2 - 9.39x + 16.83$ $(x = -3)$

(c) $P_4(x) = 2x^4 + 5x^3 - 15x^2 - 10x + 8$ $(x = 1/2)$

(d) $P_4(x) = x^4 - 4x^3 - 7x^2 + 22x + 24$ $(x = -1)$

(e) $P_5(x) = x^5 - x^3 - 2x^2 + 32$ $(x = -2)$

E4.101 Estimate all real roots of the following polynomials, accurate to six decimal places. Reduce the polynomials down to quadratic equations and solve the quadratic equation to find the last two roots. Start with $x^{(0)} = 0$ to estimate a root, and then use this estimate as the initial guess to find the next root, and so on.

(a) $x^4 - 4.8x^3 - 9.49x^2 + 35.892x + 46.368 = 0$

(b) $x^4 - 3x^3 - 9.75x^2 + 15.5x + 30 = 0$

(c) $50x^4 + 180.5x^3 + 136.02x^2 - 40.788x + 2.6136 = 0$

(d) $x^5 + 3.32x^4 + 0.72x^3 - 0.75x^2 - 0.025x + 0.01 = 0$

E4.102 The composition-temperature $(x_B\text{–}T)$ phase diagram of a binary system of A and B is illustrated in **Fig. E4.102**. The temperatures, describing the solidus and liquidus curves, are given with the following third-degree polynomials:

$$T_s(x_B) = 33 + 8.39x_B + 34x_B^2 - 5.39x_B^3$$
$$T_l(x_B) = 33 + 80.14x_B - 53x_B^2 + 9.86x_B^3$$

Estimate the liquidus and solidus compositions (x_a and x_b) corresponding to 50°C with two decimal places of accuracy. Use $x^{(0)} = 0.5$ and polynomial reduction for both cases.

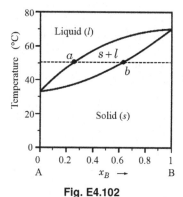

Fig. E4.102

E4.103 A uniform beam subjected to a varying distributed load is depicted in **Fig. E4.103**. The equation for the deflection is given as

$$y = K\xi(0.2 - 0.4\xi + 2.8\xi^2 - 4.42\xi^3 + 1.83\xi^4)$$

where $\xi = x/L$, L is the length of the beam, and K is a constant. Use the method of polynomial reduction to estimate the point of maximum deflection as well as the deflection. Start with $\xi^{(0)} = 0.55$.

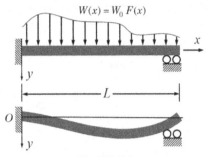

Fig. E4.103

E4.104 The system characteristic curve is the response of the pump head of the installation to a given liquid flow rate. The head h_p (m) is given below as a fourth-degree polynomial of flow rate Q (m³/h). Use the method of polynomial reduction to estimate the flow rate corresponding to $h_p(Q) = 9$ m, accurate to two decimal places. Use $Q^{(0)} = 50$ m³/h for the initial guess.

$$h_p(Q) = 10.89 + 0.00475\,Q - 0.000762\,Q^2 + 7.36 \times 10^{-6}Q^3 - 2.96 \times 10^{-8}Q^4$$

E4.105 A seven-component flash process is modeled by the following equation:

$$\sum_{i=1}^{7} \frac{(1 - k_i)\,C_i}{1 + \psi\,(k_i - 1)} = 0$$

where C_i and k_i are the feed composition and the equilibrium ratio provided in the table below, and ψ is the fraction of the feed that goes into the vapor phase. This equation leads to a polynomial of 6th degree for ψ. Apply Bairstow's method to estimate ψ in the range (0,1). Use $\varepsilon = 10^{-5}$ and $(p, q)=(0,0)$ for the starting guess.

i	1	2	3	4	5	6	7
k_i	1.71	3.19	0.74	0.41	0.22	0.18	0.067
C_i	0.005	0.835	0.04	0.016	0.006	0.008	0.09

4.13 COMPUTER ASSIGNMENTS

CA4.1 In fluid mechanics, the two correlations that are commonly used for estimating the friction factor in smooth pipe flows are given as

$$\text{Haaland Formula:} \quad \frac{1}{\sqrt{f}} = -1.8 \log_{10}\left(\frac{6.9}{\text{Re}}\right), \quad 4 \times 10^3 < \text{Re} < 10^8$$

$$\text{Colebrook Formula:} \quad \frac{1}{\sqrt{f}} = -2 \log_{10}\left(\frac{2.51}{\text{Re}\sqrt{f}}\right), \quad 4 \times 10^3 < \text{Re} < 10^8$$

where f and Re denote the friction factor and the Reynolds number, respectively. The Haaland formula is an explicit equation that allows the friction factor to be computed for a specified Reynolds number. On the other hand, the Colebrook formula is an implicit equation that requires solving a nonlinear equation. Apply a suitable numerical method to estimate the friction factor, accurate to at least six decimal places for Re=10^4, 10^5, 10^6, 10^7, 10^8, and 10^9 using Colebrook's equation. Compare your estimates to those computed using Haaland's formula.

CA4.2 In quantum physics, the solution of the Schrödinger equation for a finite square well with potential V_0 leads to the following nonlinear equation:

$$\tan z = \sqrt{\left(\frac{z_0}{z}\right)^2 - 1}, \quad z = \frac{a}{\hbar}\sqrt{2m(E + V_0)} \quad \text{and} \quad z_0 = \frac{a}{\hbar}\sqrt{\frac{2m}{V_0}}$$

where E is the energy, m is the mass of the particle, and $\hbar$ is the Plank's constant. Using a suitable numerical method, estimate the quantity z, accurate to five decimal places for $z_0 = 0.05, 0.1, 0.5, 1, 5, 10$, and 20 and obtain a plot of z_0 versus z.

CA4.3 The transient one-dimensional heat conduction from a plane wall, subjected to convection on both sides, requires the solution of the following nonlinear equation:

$$\beta_n \sin \beta_n - \text{Bi} \cos \beta_n = 0, \qquad n = 0, 1, 2, 3, ..$$

where $\text{Bi} = hL/k$ is referred to as the Biot number, h is the convection transfer coefficient, k is the conductivity of the plate, L is the thickness of the wall. This nonlinear equation has an infinite number of roots called eigenvalues. Use an appropriate numerical root finding method to estimate the first five eigenvalues $(\beta_n, n = 0,1,2,3,4)$ correct to six-decimal places for $\text{Bi}=0, 0.1, 0.5, 1, 2$ and 10.

CA4.4 When a mass m_2 is attached to the tip of a vertical rod, having a length L and a mass m_1, longitudinal vibrations in the rod are observed. The analytical solution for the vibration frequency leads to the following nonlinear equation:

$$\frac{m_1}{m_2} = \omega_n \sqrt{\frac{m_1}{k}} \tan\left(\omega_n \sqrt{\frac{m_1}{k}}\right)$$

where k is a constant and ω_n is the vibration frequency. Letting $\alpha_n = \omega_n \sqrt{m_1/k}$ and $R = m_1/m_2$ (the mass ratio), the above equation is expressed as $R = \alpha_n \tan \alpha_n$. Use a suitable numerical root-finding method to estimate the first five eigenvalues $(\alpha_n, n = 0,1,2,3,$ and $4)$ accurate to six decimal places for $R = 0.01, 0.1, 0.25, 0.5, 1, 2, 5$, and 10.

CA4.5 The *criticality equation* for a spherical nuclear reactor, surrounded by an infinite (very thick) reflector, is given as

$$BR \cot BR - 1 = -\frac{D_r}{D_c}\left(\frac{BR}{BL_r} + 1\right)$$

where R is the reactor radius, D_c and D_r are the diffusion coefficients of the core and reflector, L_r is the reflector diffusion length, and B is a parameter, referred to as *buckling*. In order for a reactor to be able to achieve self-sustained nuclear reactions, the foregoing nonlinear equation needs to be satisfied. This equation, which also has an infinite number of solutions (*eigenvalues*) can be rewritten in simplified dimensionless form as

$$\psi \cot \psi = 1 - \eta\left(\frac{\psi}{BL_r} + 1\right)$$

where $\psi = BR$ and $\eta = D_r/D_c$ is the diffusion coefficient ratio. Using a suitable numerical root-finding method, estimate the smallest positive root (ψ_{min}), accurate to four decimal places for $BL_r = 0.6$ and $\eta = 0.6, 0.7, 0.8, 0.9, 1, 1.2, 1.4$, and 1.6.

CA4.6 The following nonlinear equation has two positive roots for any $c > 0$. Apply a suitable numerical root-finding method to estimate both of the roots, accurate to five decimal places for $c = 0.3, 0.5, 1, 2, 5$, and 25.

$$e^{-3x} + 1.26\, e^{x/3} + 1.2 e^{-3x/2} = c\, x^2$$

CA4.7 The following nonlinear equation is encountered in the study of two-phase diffusion problems. Apply a suitable root-finding method to estimate the smallest four positive roots $(\beta_n, n = 1, 2, 3,$ and $4)$, accurate to five decimal places for $\lambda = 0.1, 0.5, \pi/4, 1$ and $\pi/2$.

$$\sin(\lambda\beta) \cos(\beta) + \cos(\lambda\beta) \sin \beta - \beta \sin \beta \sin(\lambda\beta) = 0$$

CA4.8 In structural engineering, slender members experience a mode of failure called *buckling*. When designing a slender member, the critical buckling load is calculated for safety purposes. If a load P acts through the centroid of the cross section, the critical buckling load is calculated from $P_{crit} = \pi^2 EI/L^2$, where E is the modulus of elasticity (Pa), L is the length (m), and I is the moment of inertia (m^4). However, in reality, the load P might be applied at an offset or the slender member may not be completely straight (*see* **Fig. CA4.8**). In such cases, the maximum stress is calculated with the *secant formula*, which is given as

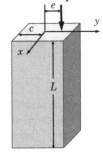

Fig. CA4.8

$$\sigma_{\max} = \frac{P}{A}\left(1 + \frac{ec}{R^2}\sec\left(\frac{L}{2}\sqrt{\frac{P}{EI}}\right)\right)$$

where e is the eccentricity (m), A is the cross-section area (m^2), c is the half depth of the column (m), r is the radius of gyration of the cross section (m), and σ is the stress (Pa). The secant formula can be rearranged to give

$$S_{\max} = p_c\left(1 + \phi\sec(\frac{\pi}{2}\sqrt{p_c})\right)$$

where three dimensionless quantities appear: $S_{\max} = \sigma_{\max}/(P_{crit}/A)$, $\phi = ec/R^2$, and $p_c = P/P_{crit}$. Use a suitable numerical method to estimate the buckling load ratio for $S_{\max} = 1, 3, 5, 7, 9$, and $\phi = 0.5, 1$, and 2, accurate to three decimal places.

CA4.9 The Peng-Robinson equation of state is used to estimate the volume of pure propane gas as a function of pressure and temperature.

$$P = \frac{RT}{V-b} - \frac{a\alpha}{V^2 + 2bV - b^2}$$

where T is the temperature in Kelvin, V is the volume in cm^3/gmol, P is the pressure MPa, $R = 8.3145$ J/mol-K, $a = 1.14 \times 10^6$ MPa.cm^6/gmol^2, $b = 56.29$ cm^3/gmol, and α is given by

$$\alpha = \left(1 + 0.6027\left(1 - \sqrt{\frac{T}{369.8}}\right)\right)^2$$

Write a computer program to estimate the volume (cm^3/gmol) of propane at a specified temperature (K) and pressure (MPa). Using this program, estimate the volume for $T = 300, 320, 340$, and 360 K and $P = 0.1, 0.5$, and 1 MPa.

CA4.10 Gibbs energy of mixing the two liquid phases A and B is given as

$$\Delta G_{mix} = nRT\left\{x_A \ln x_A + x_B \ln x_B + \beta x_A x_B\right\}$$

where $R = 8.314$ J/mol-K is the universal gas constant, T is the mixing temperature (K), x denotes the mole fraction of a phase, and β is a dimensionless parameter that is a measure of chemical interactions (if $\beta < 0$, mixing is exothermic or endothermic otherwise). For $\beta > 2$, the Gibbs energy of mixing has a double minimum, and we can expect phase separation to take place. Noting that $x_A + x_B = 1$, use the fixed-point iteration to estimate the compositions corresponding to the minimum points with three decimal place accuracy for $\beta = 2, 2.25, 2.5$, and 3.

CA4.11 The natural frequencies (ω_n) of radial vibrations of an elastic sphere are given as

$$\tan(\xi_n) = \frac{B\gamma^2\xi_n^2 - 4(A\xi_n + B)}{A\gamma^2\xi_n^2 + 4(B\xi_n - A)}$$

where $\gamma^2 = G/(\lambda + 2G)$, A and B are material dependent constants, R is the radius of the sphere, $\xi_n = k_n R$, and k_n is defined as $k_n^2 = \omega_n^2 \rho/(\lambda + 2G)$, ρ is the density, $G = E/2(1+\nu)$ is the shear modulus, E is the Young's modulus, ν is the Poisson ratio, and $\lambda = E\nu/(1+\nu)(1-2\nu)$. For a spheroidal rubber gum for which $E = 9$ MPa, $\nu = 0.45$, $R = 0.10$ m, $\rho = 1100$ kg/m^3, and $A/B = 10$, the frequency equation is reduced to

$$\tan(\xi_n) = \frac{\xi_n^2 - 440\xi_n - 44}{2(5\xi_n^2 + 22\xi_n - 220)}$$

Using a suitable root-finding method of your choice, estimate the natural frequency of the vibration (ω_1), accurate to four decimal places.

CA4.12 Bridgman analysis is used to estimate the stresses that would develop in the necking region of a cylindrical tension specimen. The analysis makes use of the following Bridgman correction factor (B):

$$B = \frac{\sigma_{TB}}{\sigma_T} = \left(\left(1 + \frac{2R}{a}\right) \ln\left(1 + \frac{a}{2R}\right) \right)^{-1}$$

where σ_T is the tensile strength, a is the radius of the tensile specimen at the thinnest section of the neck, and R is the radius of curvature of the neck profile as illustrated in **Fig. CA4.12**. Estimate a/R ratios that correspond to correction factors of $B = 0.776, 0.822, 0.9$, and 0.95.

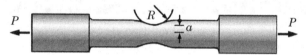

Fig. CA4.12

CA4.13 A reduction reaction of $M = 1$ mol of an oxide at $T = 300$ K is investigated. The electrode potential difference, which is described by the Nernst equation, takes the following form:

$$\Delta E = \frac{RT}{nF} \ln\left(\frac{nMF - Q}{Q}\right)$$

where $R = 8.3145$ J/K.mol is the universal gas constant, T is the temperature (K), $F = 96480$ C/mol is the Faraday's constant, Q is the discharge (C), and n is the number of electrons transferred. Using the Newton-Raphson method, estimate the number of electrons transferred when the discharge is at its maximum, which is accurate to two decimal places. The electrode potential difference is given as 1.2 mV.

CA4.14 When studying heat transfer in a plane parallel medium with anisotropic scattering, a transcendental equation similar to the one given below is encountered:

$$x \operatorname{arccoth} x = \frac{c x^3 - 2x - 1}{0.1 x^3 + c x + 1}$$

Use a numerical root-finding method to solve the given nonlinear equation, accurate to four decimal places for $c = 0.2, 0.4, 0.6, 0.8$, and 1.

CA4.15 The zeroth- and first-degree Legendre polynomials are defined as $P_0(x) = 1$ and $P_1(x) = x$. The nth-degree Legendre polynomial can be generated using the following recurrence relationship:

$$n P_n(x) = (2n - 1)xP_{n-1}(x) - (n - 1)P_{n-2}(x), \qquad \text{for } n \geqslant 2$$

On the other hand, a recursive expression for the derivative of the nth-degree Legendre polynomial is given by

$$\frac{dP_n(x)}{dx} = \frac{nP_{n-1}(x) - nxP_n(x)}{1 - x^2}, \quad n \geqslant 1$$

Using the Newton-Raphson method as well as the above relationships, write a computer program that estimates the positive roots of the Nth-degree Legendre polynomial up to $N = 10$, accurate to within specified ε. The number of positive roots is $N/2$ for which you can use the following expression as your initial guess:

$$x_k^{(0)} = \cos\left(\frac{(4k - 1)\pi}{4N + 2}\right), \quad k = 1, 2, .., \frac{N}{2}$$

Numerical Differentiation

A $UNIVARIANT$ *function*, $y = f(x)$, defines a relationship between two (x-independent and y-dependent) variables. In the physical world, each variable represents a physical quantity, such as density, distance, mass, time, temperature, etc. Functional relationships between any two quantities are defined as either *continuous functions* or *discrete functions*.

An explicitly defined (continuous) function $y = f(x)$ contains an infinite amount of information through all x-values in its domain. The numerical value of a function for an input value, say $x = c$ within its domain, can be obtained simply by direct substitution: $y = f(c)$.

Differentiation operation is encountered in all fields of science and engineering. Recall that freshman calculus largely deals with the differentiation and integration of continuous functions and their applications. However, in practice, most (even univariant) functions that scientists and engineers deal with are not always known explicitly. Functional relationships between some physical variables are obtained as a result of either sampling or experimentation. Hence, a set of data establishes a relationship, which may have a form or a shape, and the data points are not continuous or connected. A data set that describes a functional

DOI: 10.1201/9781003474944-5

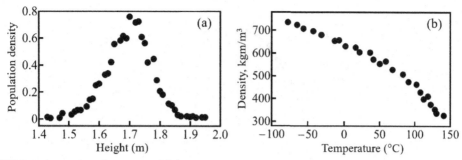

FIGURE 5.1: A graphic depiction of (a) the height distribution of individuals in a population and (b) the density-temperature relationship of a chemical solution.

relationship is referred to as a *discrete function*. In other words, the values of a discrete function are represented solely by a number of distinct and separate data points that are unconnected to each other and whose values are expressed by a table or list of ordered pairs (points). For instance, the height distribution of individuals in a large population or the variation of the experimentally measured density of a chemical solution with temperature, depicted in Fig. 5.1a and b, are two examples of discrete functions. Most physical properties (density, viscosity, heat capacity, enthalpy, saturation pressure, etc.) are assumed to be smooth functions of space and time whose explicit mathematical relationships are not known. The physical properties of solids, liquids, or gases are generally available as discrete functions of temperature, pressure, etc.

Discrete functions are also encountered in the numerical solution of the differential equation. An ordinary differential equation contains an infinite amount of information ($y(x)$ for any x), while a computer can handle only a finite amount of data (y_1, y_2, ..., y_n). For this reason, the numerical solution of a differential equation is carried out following a discretization procedure in which a finite set of points is created for which the solution is obtained.

A discrete function may be either *uniformly* or *non-uniformly spaced*. Data sampling with digital instruments or data collected periodically is generally uniformly spaced, which is computationally favorable. The mathematical operations with uniformly spaced discrete data sets are simpler and require less computation and cpu-time. The memory requirement is also halved since the abscissas do not need to be stored. Therefore, numerical analysts prefer dealing with uniformly spaced data whenever possible due to the computational advantages mentioned above.

In general, an explicitly (or implicitly) defined function can be differentiated exactly. Nevertheless, the differentiation of discrete functions requires a suitable approximation. This chapter is devoted to the numerical methods and techniques for approximately differentiating discrete and/or continuous functions.

5.1 BASIC CONCEPTS

Numerical differentiation is the process of evaluating the derivative of a function. It provides ways and means of estimating the derivative of any continuous or discrete function. The derivative of a smooth differentiable function $y = f(x)$ is the instantaneous rate of change of $f(x)$, or, geometrically speaking, it is the slope of the tangent line drawn to $y = f(x)$ at any point x. In order to approximate the slope of a tangent line at $x = a$, we attempt to

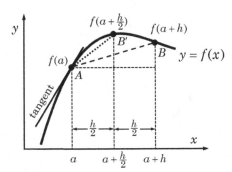

FIGURE 5.2: Determining the slope of a function at point A.

evaluate the slope of the secant line (*see* Fig. 5.2). To do just that, consider $A(a, f(a))$ a fixed point and $B(a + h, f(a + h))$ a nearby mobile point on $y = f(x)$. Next, we consider the secant line through points A and B. The slope of the secant line passing through points A and B can be written as

$$m_{|AB|} = \frac{f(a + h) - f(a)}{h}$$

This is clearly not exactly equal to the slope of the tangent line to $f(x)$ at the point $A(a, f(a))$. When we move point B toward A along $y = f(x)$, we observe that the slope of the secant line approaches the slope of the new tangent line. Mathematically, we achieve this by taking the limit of the slope as $h \to 0$. In Fig. 5.2, the point B moves along the curve toward B' as h decreases, ending up close enough to A. In other words, the slope of the secant line approaches the slope of the tangent line as h is reduced to below a sufficiently small value.

Derivative (Definition). Mathematically speaking, the first derivative (or true slope) of $y = f(x)$ at $x = a$ is the limit of the value of the difference quotient as the secant lines get closer and closer to being a tangent line. The derivative of a smooth continuous univariant function, $y = f(x)$, at $x = a$ is expressed as

$$f'(a) = \lim_{h \to 0} \frac{f(a + h) - f(a)}{h} \tag{5.1}$$

The limit operation in Eq. (5.1) can be performed if $f(x)$ is explicitly known. However, in practice, some functions that we deal with do not have explicit forms. For instance, data collected periodically are usually uniformly spaced with a fixed sampling interval size, h. Every data collection instrument has a sampling rate that is limited by default manufacturing settings, which cannot be changed for the most part. In other words, since the sampling interval is fixed, it is practically impossible to truly perform the limit operation as $h \to 0$. Therefore, the analyst has no choice but to use available data with the present interval size to approximately estimate the derivative of a given discrete function. In other words, since the numerical differentiation is approximate, the result naturally contains some error. Therefore, before using any approximation, it is important to be aware of the magnitude of the error in the differentiation process.

Truncation and Leading Error. If $f(x)$ is a continuous and infinitely differentiable function, it can be expanded into the Taylor series in the neighborhood of point a:

$$f(a + h) = f(a) + h f'(a) + \frac{h^2}{2!} f''(a) + \frac{h^3}{3!} f'''(a) + \cdots + \frac{h^n}{n!} f^{(n)}(a) + R_n(h) \tag{5.2}$$

where $R_n(h)$ is the remainder defined as

$$R_n(h) = \frac{h^{n+1}}{(n+1)!} f^{(n+1)}(\xi), \qquad a \leqslant \xi \leqslant a+h$$

Solving Eq. (5.2) for $f'(a)$ yields

$$f'(a) = \frac{f(a+h) - f(a)}{h} - \frac{h}{2!} f''(a) - \frac{h^2}{3!} f'''(a) - \cdots - \frac{h^n}{(n+1)!} f^{(n+1)}(\xi) \qquad (5.3)$$

All the terms on the rhs, except the first one, have a factor that includes the power of h. Hence, as $h \to 0$, the limit of Eq. (5.3) does give Eq. (5.1) due to the vanishing terms with the factor of h^k for $k \geqslant 1$. But, since the interval size of a discrete function is fixed, the limit operation cannot, in reality, be achieved for $h \to 0$. For this reason, the derivatives of discrete functions can only be determined approximately.

Neglecting the remaining terms that contain multiples of h^k, Eq. (5.3) yields

$$f'_{\text{approx}}(a) \approx \frac{f(a+h) - f(a)}{h} \qquad (5.4)$$

Besides approximating the derivatives of functions, it is equally important to predict the magnitude of the resulting approximation errors. Equation (5.4) contains a certain amount of error whose magnitude we cannot quantify at this stage. To estimate the approximation error, we first define the differentiation error as the difference between the true and the approximate values:

$$f'_{\text{true}}(a) - f'_{\text{approx}}(a) = -\frac{h}{2!} f''(a) - \frac{h^2}{3!} f''(a) - \frac{h^3}{4!} f^{(4)}(a) - \cdots \qquad (5.5)$$

This expression, which includes higher derivatives of $f(x)$, accounts for the *total error* (or *truncation error*) made when $f'(a)$ is approximated by Eq. (5.4). Since higher derivatives are also unknown, it is impossible to estimate the true truncation error. However, generally, we are interested in knowing the relative magnitude of the error rather than its true value. Note that the terms containing h^k ($k \geqslant 2$) vanish quickly with increasing k, and an approximate $f'(a)$ can be expressed as

$$\left| f'_{\text{true}}(a) - \frac{f(a+h) - f(a)}{h} \right| \approx \frac{h}{2} |f''(a)| \qquad (5.6)$$

This term, referred to as the *leading error*, is the first term on the rhs of Eq. (5.5). In other words, the rest of the terms in Eq. (5.5) are much smaller than the leading term, which makes up the bulk of the true error. Hence, whenever we speak of the "truncation error" to assess the magnitude of the error of a finite difference approximation, we will be referring to the "leading error."

In practice, Equation (5.5) is sufficient to predict the magnitude of the error, but $f''(a)$ is unknown when $f(x)$ is a discrete function. In this regard, an *upper bound* on truncation error can be obtained by replacing $f''(a)$ with $f''(\xi)$:

$$\left| f'_{\text{true}}(a) - \frac{f(a+h) - f(a)}{h} \right| = \frac{h}{2} |f''(\xi)| \qquad (5.7)$$

where $a \leqslant \xi \leqslant a+h$ and $|f''(\xi)|$ denotes its maximum value on $[a, a+h]$. Then, the *truncation error* (leading error) of the approximation for the first derivative, Eq. (5.4), becomes $-hf''(\xi)/2$.

Order of Accuracy. At this point, we make use of the big-O notation, $\mathcal{O}(\cdot)$, to symbolize the order of error. Recall that $\mathcal{O}(h)$ implies that the error is directly proportional to h; that is, $\mathcal{O}(h) \equiv C \cdot h$, where C is a constant. The exponent of h is referred to as the *order of accuracy*, which is also indicative of how rapidly the truncation error vanishes as the interval is refined. $\mathcal{O}(h^n)$ denotes *nth-order* approximation. The higher the order, the faster the truncation error vanishes. Note that the $f''(\xi)/2$ as a multiplier in Eq. (5.7) is independent of h regardless of ξ, and it can be assumed to be a constant. The truncation error can then be expressed as $hf''(\xi)/2 \equiv C \cdot h \equiv \mathcal{O}(h)$. Making use of the big-O notation, we express Eq. (5.3) as

$$f'(a) = \frac{f(a+h) - f(a)}{h} + \mathcal{O}(h) \tag{5.8}$$

which is referred to as the first-order accurate *finite-difference formula*.

Round-off Errors. Apart from truncation error, another type of error that plays an unfavorable role in numerical differentiation is the *round-off error*. The problem arises because the derivative of a function cannot be computed with the same precision as the function itself. To illustrate this point, consider an explicitly defined continuous function, $f(x)$. Theoretically, the true value of $f'(x)$ is obtained by making h in Eq. (5.4) as small as possible ($h \to 0$). The numerators of the finite difference formulas typically consist of $f(x)$, $f(x \pm h)$, $f(x \pm 2h)$, and so on. If we make the value of h too small, the numerator will be approximately equal to zero due to the restrictive nature of machine epsilon. This would be, of course, inaccurate! The sum of the coefficients of the terms in the numerator of any difference formula, without exception, is always zero, so when evaluating the numerator, several significant figures may be lost while performing multiplication, subtraction, and addition operations with coefficients. Nevertheless, we cannot make h too large either because the truncation error would dominate the results. Unfortunately, this problem cannot be fully rectified, but its adverse effects may be mitigated to some extent by either using high-precision arithmetic or adopting finite difference formulas of accuracy greater than or equal to 2.

In order to analyze the loss of accuracy during numerical differentiation, we use $\hat{f}(x)$ to denote the floating point representation of the function. The difference between the true and machine-computed value, $\hat{f}'_{\text{comp}}(x)$, from the approximate formula is

$$f'_{\text{true}}(x) - \hat{f}'_{\text{comp}}(x) = f'_{\text{true}}(x) - \frac{\hat{f}(x+h) - \hat{f}(x)}{h}$$

Next, we assume $\hat{f}(x)$ differs from the true value, $f(x)$, in relative error sense, by a small amount on the order of machine epsilon, i.e., $\hat{f}(x) = f(x)(1 + \epsilon)$ and $\epsilon \approx \mathcal{O}(\epsilon_{\text{mach}})$. Substituting the floating point values of the function into the approximation, we find

$$\begin{aligned} f'_{\text{true}}(x) &- \frac{f(x+h) + \epsilon_1 f(x+h) - f(x) - \epsilon_2 f(x)}{h} \\ &= \left(f'_{\text{true}}(x) - f'_{\text{approx}}(x) \right) + \frac{\epsilon_1 f(x+h) - \epsilon_2 f(x)}{h} \end{aligned} \tag{5.9}$$

where $f'_{\text{approx}}(x)$ is evaluated by Eq. (5.4). Note that the first and second terms in Eq. (5.9) account for the truncation and round-off errors, respectively. Then, the total error can be expressed as

$$E_{\text{total}}(h) = E_{\text{trunc}}(h) + E_{\text{rndoff}}(h) \tag{5.10}$$

where $E_{\text{trunc}}(h) = f'_{\text{true}}(x) - f'_{\text{approx}}(x)$, and $E_{\text{rndoff}}(h) = (\epsilon_1 f(x+h) - \epsilon_2 f(x))/h$.

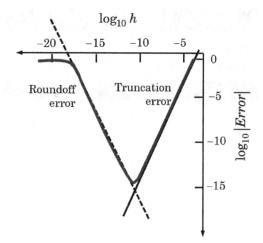

FIGURE 5.3: Depicting numerical differentiation errors as a function of h.

An upper bound for the truncation error, using Eq. (5.7), is found as

$$E_{\text{trunc}}(h) = \frac{h}{2}M \tag{5.11}$$

where $M = \max|f''(\xi)|$ on $x \leqslant \xi \leqslant x + h$. On the other hand, for very small h, the upper bound for the round-off error is reduced to

$$E_{\text{rndoff}}(h) = \left| \frac{\epsilon_1 f(x+h) - \epsilon_2 f(x)}{h} \right| \leqslant \frac{|\epsilon_1 - \epsilon_2|}{h}N \leqslant \frac{2\varepsilon^*}{h}N$$

where ε^* is the upper bound for the relative error in which no underflow or overflow occurs, N denotes $N = |f(x)|$, assuming $|f(x+h)| \approx |f(x)|$. In the above expression, however, we do not know the signs of ϵ_1 and ϵ_2; thus, we cannot accurately estimate $|\epsilon_1 - \epsilon_2|$. But if ϵ_1 and ϵ_2 have opposite signs, the magnitude of the difference could be doubled, so we replaced this difference with $|\epsilon_1 - \epsilon_2| \cong 2\varepsilon^*$.

The round-off and truncation errors are graphically depicted in Fig. 5.3, where we immediately observe two distinct regions that are dominated by either truncation or round-off errors. The truncation error, Eq. (5.11), is proportional to h, and it decreases linearly with decreasing h. On the other hand, the region dominated by round-off error is inversely proportional to h, as given by $2\varepsilon^*N/h$. For large h, it is initially very small (on the order of machine epsilon) but increases with decreasing h. The total error decreases as h is reduced, but the approximations worsen with further reduction in h since the round-off errors become more dominant than the truncation errors. This means that there is an optimum value where the total error is at a minimum, and we cannot reduce h below that to improve the difference approximation.

In order to find the minimum of the total error, we set $dE_{\text{total}}/dh = 0$ giving

$$\frac{dE_{\text{total}}}{dh} = \frac{1}{2}M - \frac{2\varepsilon^*}{h^2}N = 0 \rightarrow \quad h^* = 2\sqrt{\frac{\varepsilon^*N}{M}} \tag{5.12}$$

where h^* is the optimum value of h.

EXAMPLE 5.1: Effect of reducing h on approximate derivative.

Given $f(x) = x^5$, we wish to estimate $f'(0.4)$ using Eq. (5.4). (a) Investigate the effect of reducing h on the accuracy of $f'(0.4)$, and (b) find the optimum h^*.

SOLUTION:

(a) Since $f(x)$ is explicitly given, the true $f'(0.4)$ and total error can be calculated. For a specified h, the true error due to the use of the difference approximation is determined by

$$E_{\text{total}}(h) = \left| f'_{\text{true}}(0.4) - f'_{\text{approx}}(0.4) \right|$$

where $f'_{\text{comp}}(0.4) = (\hat{f}(0.4 + h) - \hat{f}(0.4))/h$ and since the machine-computed values are inexact (i.e., floating point values), f's in Eq. (5.4) are replaced with $\hat{f}$'s to emphasize this distinction.

TABLE 5.1

h	$f'_{\text{comp}}(0.4)$	$E_{\text{true}}(h)$
10^{-1}	0.2101	8.21000×10^{-2}
10^{-2}	0.134562010	6.56201×10^{-3}
10^{-3}	0.128641602	6.41602×10^{-4}
10^{-4}	0.128064016	6.40160×10^{-5}
10^{-5}	0.128006400	6.40016×10^{-6}
10^{-6}	0.128000640	6.39997×10^{-7}
10^{-7}	0.128000064	6.39856×10^{-8}
10^{-8}	0.128000006	6.08049×10^{-9}
10^{-9}	0.128000003	2.95798×10^{-9}
10^{-10}	0.128000006	6.42743×10^{-9}
10^{-11}	0.127999868	1.32350×10^{-7}
10^{-12}	0.127996572	3.42832×10^{-6}
10^{-13}	0.127935856	6.41436×10^{-5}
10^{-14}	0.128022593	2.25925×10^{-5}
10^{-15}	0.126634814	1.36519×10^{-3}
10^{-16}	0.138777878	1.07779×10^{-2}
$10^{-16.55}$	0.123100623	4.89938×10^{-3}
10^{-18}	0	0.128

In Table 5.1, the computed derivative and true error are presented for a wide range of increment size h. As h decreases down to $h \approx 10^{-9}$, the true error also decreases in parallel with the improvement in the computed approximation. Recall that real numbers in digital computers are stored as rounded or truncated to the nearest floating-point numbers, and they cannot be precisely represented on computers. Hence, it is observed that the computed approximations lose accuracy for very small h as the true differentiation error for $h < 10^{-9}$ begins to increase. The reason for this is primarily the loss of significance. As a result of rounding due to subtracting two nearly equal quantities, namely $f(0.4 + h) - f(0.4)$, the computed value loses significant digits. And then, to make matters worse, the inaccuracies are magnified when this difference is divided by a very small number such as h. The computation breaks down slightly below $h \approx 10^{-16.55} = 2.818 \times 10^{-17}$, so we may take this value as the upper bound ($\varepsilon^* \approx 2.818 \times 10^{-17}$) for which no underflow occurs.

In Fig. 5.4, a log-log plot of the total approximation error based on 200 values is plotted as a function of h. The solid line represents the error estimate $E_{\text{total}}(h)$ given by Eq. (5.10). On the right side of the figure, where the truncation error dominates the error, the theoretical straight line ($=Mh/2$) and the computed errors are in complete agreement. However, on the left side of the figure, where the round-off errors dominate, the computed values depict a jagged distribution just below the solid line, and the data points are clearly not in line with the theoretical expectations. The floating point arithmetic for the differences df (i.e., $f(0.4+h) - f(0.4)$) breaks down below ε^*. Because the small values df are set to "zero," the error of the computed data is horizontally aligned at $\log_{10}(0.128) \approx -0.893$ which is the logarithm of the true value of the derivative.

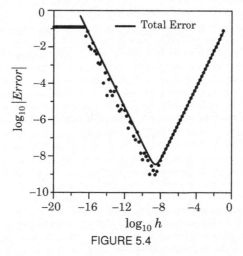

FIGURE 5.4

(b) Finally, in order to compute the optimum value of the increment, we note that $M = 20(0.4)^3 = 1.28$ and $N = (0.4)^5 = 0.01024$. Then, using Eq. (5.12), we find

$$h^* = 2\sqrt{\frac{\varepsilon^* N}{M}} = 2\sqrt{\frac{2.818 \times 10^{-17}(0.01024)}{1.28}} = 9.496 \times 10^{-10}$$

which is consistent with the $h^* \approx 10^{-9}$ value in Table 5.1.

Discussion: The numerical differentiation process is highly sensitive to round-off errors. Decreasing the interval size reduces the impact of the truncation error while limiting the magnitude of the total error by the round-off error. That is why, as h is reduced down to h^*, the finite difference approximation improves in accuracy. But as h is reduced further below h^*, the finite difference approximations lose significant digits due to the "loss of significance" phenomenon observed in such numerical subtraction operations. Then the round-off errors begin to dominate the arithmetic calculations, resulting in an increasing trend in the total error.

This analysis applies only to Eq. (5.4). You can investigate the effects of the round-off errors by performing similar analyses for other difference formulas that you would like to make use of in your work.

5.2 FIRST-ORDER FINITE DIFFERENCE FORMULAS

Consider a uniformly-spaced discrete function $f(x)$ on $[x_0, x_n]$: $f(x_0), f(x_1), \ldots, f(x_n)$ with $x_i = ih$ for $i = 0, 1, \ldots, n$. Before going any further, we will adopt the following short-hand notations:

$$f(x_i) = f_i, \quad f(x_i \pm h) = f(x_{i\pm 1}) = f_{i\pm 1}, \quad f(x_i \pm 2h) = f(x_{i\pm 2}) = f_{i\pm 2},$$

$$f'(x_i) = f_i', \quad f''(x_i) = f_i'', \quad f^{(4)}(x_i) = f_i^{(4)}, \quad \text{and so on}$$

This notation not only helps abbreviate the finite difference expressions but also has a convenient form that can be used to easily code in any computer language.

The finite difference formula for $f'(x_i)$, obtained by substituting x_i for a in Eq. (5.8), is referred to as the *Forward Difference Formula (FDF)* and is expressed in short-hand notation as:

$$f_i' = \frac{f_{i+1} - f_i}{h} + \mathcal{O}(h) \tag{5.13}$$

Finite-difference expressions can be further abbreviated and simplified with the use of *difference operators*, e.g., employing the *forward difference operator*, Eq. (5.13) becomes

$$f_i' = \frac{\Delta f_i}{h} + \mathcal{O}(h) \tag{5.14}$$

where $\Delta f_i = f_{i+1} - f_i$ involves values of the function at x_i and x_{i+1}.

Now, consider the Taylor series of $f(x_i - h)$ about x_i:

$$f(x_i - h) = f(x_i) - h\, f'(x_i) + \frac{h^2}{2!} f''(x_i) - \frac{h^3}{3!} f'''(x_i) + \ldots + (-1)^n \frac{h^n}{n!} f^{(n)}(x_i) + \cdots \tag{5.15}$$

Solving Eq. (5.15) for $f'(x_i)$ yields

$$f'(x_i) = \frac{f(x_i) - f(x_i - h)}{h} + \frac{h}{2!} f''(x_i) - \frac{h^2}{3!} f'''(x_i) + \frac{h^3}{4!} f^{(4)}(x_i) - \cdots \tag{5.16}$$

Equation (5.16) can be expressed as follows:

$$f'(x_i) = \frac{f(x_i) - f(x_i - h)}{h} + \mathcal{O}(h) \tag{5.17}$$

Noting that $x_{i-1} = x_i - h$ and using the subscript notation leads to

$$f_i' = \frac{f_i - f_{i-1}}{h} + \mathcal{O}(h) \tag{5.18}$$

which is called the *Backward Difference Formula (BDF)* due to the use of the discrete point f_{i-1} behind f_i. The truncation error of this alternative formula becomes $E(h) = h f''(\xi)/2$ for $x_{i-1} \leqslant \xi \leqslant x_i$.

The backward difference operator ∇ operating as $\nabla f_i = f_i - f_{i-1}$ abbreviates Eq. (5.18) as

$$f'_i = \frac{\nabla f_i}{h} + \mathcal{O}(h) \tag{5.19}$$

So far, only the finite difference formulas for the first derivative have been derived. Similarly, the difference formulas for $f''(x_i)$ or higher derivatives could be developed through the use of Taylor series.

To derive a difference formula for $f''(x)$, consider the Taylor series of $f(x_i + 2h)$ about x_i:

$$f(x_i+2h)=f(x_i)+2hf'(x_i)+\frac{(2h)^2}{2!}f''(x_i)+\frac{(2h)^3}{3!}f'''(x_i)+\cdots+\frac{(2h)^n}{n!}f^{(n)}(x_i)+\cdots \quad (5.20)$$

Setting $a = x_i$ in Eq. (5.2) and using Eq. (5.20) to eliminate $f'(x_i)$ leads to

$$f''(x_i) = \frac{f(x_i + 2h) - 2f(x_i + h) + f(x_i)}{h^2} - hf'''(x_i) + \dots \quad (5.21)$$

or, in shorthand notation,

$$f_i'' = \frac{f_{i+2} - 2f_{i+1} + f_i}{h^2} + \mathcal{O}(h) \quad (5.22)$$

which is first-order accurate with the truncation error $E(h) = -hf'''(\xi)$ for $x_i \leqslant \xi \leqslant x_i + 2h$. The numerator may be abbreviated as follows:

$$f_{i+2} - 2f_{i+1} + f_i = (f_{i+2} - f_{i+1}) - (f_{i+1} - f_i) = \Delta f_{i+1} - \Delta f_i = \Delta(\Delta f_i) = \Delta^2 f_i$$

Implementing the short-hand notation in Eq. (5.22) results in

$$f_i'' = \frac{\Delta^2 f_i}{h^2} + \mathcal{O}(h) \quad (5.23)$$

or f_i'' leads to $f_i'' = \nabla^2 f_i / h^2 + \mathcal{O}(h)$ for the backward difference formula.

Finite difference expressions for higher-order derivatives can be similarly developed. For instance, the forward and backward difference formulas of the nth derivative of $f(x)$ at $x = x_i$ with the order of accuracy of $\mathcal{O}(h)$ are expressed as

$$\frac{d^n f_i}{dx^n} = \frac{\Delta^n f_i}{h^n} + \mathcal{O}(h) \quad (5.24)$$

$$\frac{d^n f_i}{dx^n} = \frac{\nabla^n f_i}{h^n} + \mathcal{O}(h) \quad (5.25)$$

5.3   SECOND-ORDER FINITE-DIFFERENCE FORMULAS

So far, we have discussed the derivation of the first-order accurate (backward and forward) finite-difference formulas. It is possible to develop higher-order ($\mathcal{O}(h^2)$, $\mathcal{O}(h^3)$, etc.), more accurate finite difference formulas by simply including more terms (leading term and so on) from the Taylor series expansion.

Equation (5.3) is written for any discrete point, $x = x_i$, as

$$f'(x_i) = \frac{f(x_i + h) - f(x_i)}{h} - \frac{h}{2!}f''(x_i) - \frac{h^2}{3!}f'''(x_i) - \frac{h^3}{4!}f^{(4)}(x_i) - \cdots \quad (5.26)$$

Replacing $f''(x_i)$ by Eq. (5.21) leads to

$$f'(x_i) = \frac{f(x_i+h)-f(x_i)}{h} - \frac{h}{2}\left(\frac{f(x_i+2h)-2f(x_i+h)+f(x_i)}{h^2} - hf'''(x_i)+\dots\right)$$
$$-\frac{h^2}{6}f'''(x_i) + \dots \quad (5.27)$$

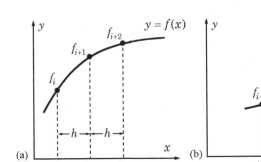

FIGURE 5.5: Graphical depiction of data points used in (a) forward and (b) backward differences of $f''(x_i)$.

Collecting the common terms and simplifying yields

$$f'(x_i) = \frac{-f(x_i + 2h) + 4f(x_i + h) - 3f(x_i)}{2h} - \frac{h^2}{3} f'''(x_i) + \dots \tag{5.28}$$

where the new leading error term is clearly $-h^2 f'''(x_i)/3$; in other words, this formulation is second-order accurate, i.e., $\mathcal{O}(h^2)$. But this formula requires two additional data points besides $f(x_i)$, namely $f(x_i + h)$ and $f(x_i + 2h)$ (see Fig. 5.5a). The truncation error is cast as $E(h) = -h^2 f'''(\xi)/3$, where $x_i \leqslant \xi \leqslant x_i + 2h$.

Finally, the second-order forward difference formula, in shorthand notation, for f'_i can be expressed as

$$f'_i = \frac{-f_{i+2} + 4f_{i+1} - 3f_i}{2h} + \mathcal{O}(h^2) \tag{5.29}$$

A second-order backward difference formula for f'_i requires $f(x_i - h)$ and $f(x_i - 2h)$ points on the left of x_i (see Fig. 5.5b). Similarly, we expand $f(x_i - h)$ and $f(x_i - 2h)$ into Taylor series about x_i to find a backward differentiation formula for $f''(x_i)$. Substituting the BDF for $f''(x_i)$ into Eq. (5.16) leads to

$$f'_i = \frac{3f_i - 4f_{i-1} + f_{i-2}}{2h} + \mathcal{O}(h^2) \tag{5.30}$$

where the truncation error is $E(h) = h^2 f'''(\xi)/3$ for $x_{i-2} \leqslant \xi \leqslant x_i$. Note that when the interval size is reduced by a factor of 2, the truncation error is reduced by a factor of 4.

First-order accurate difference formulas for $f'(x_i)$ involve a data point either on the right (x_{i+1}, f_{i+1}) or left side (x_{i-1}, f_{i-1}) of the point (x_i, f_i), where the derivative is evaluated. As the order of accuracy of a finite difference formula increases, the number of data points required in the formula also increases. For example, three data points about the point (x_i, f_i) are needed to develop a finite difference formula of order $\mathcal{O}(h^3)$. These points in forward or backward difference formulas are distributed on the right or left sides of x_i, respectively.

 When dealing with a set of discrete data, we do not know whether the data is generated from a continuous or piecewise continuous function. The function, in fact, may not be differentiable at some points in its domain. Hence, numerical differentiation must always be handled with caution!

5.4   CENTRAL DIFFERENCE FORMULAS

In central difference formulas, the points are distributed about x_i in symmetric pairs. Consider the Taylor series of $f(x_i + h)$ and $f(x_i - h)$ about x_i given by Eqs. (5.2) and (5.15), respectively. Subtracting Eq. (5.15) from Eq. (5.2) and solving for $f'(x_i)$ yields

$$f'(x_i) = \frac{f(x_i + h) - f(x_i - h)}{2h} - \frac{h^2}{6} f'''(x_i) - \frac{h^4}{120} f^{(5)}(x_i) \ldots \qquad (5.31)$$

or

$$f'_i = \frac{f_{i+1} - f_{i-1}}{2h} + \mathcal{O}(h^2) \qquad (5.32)$$

where the truncation error is $E(h) = -h^2 f'''(\xi)/6$ for $x_{i-1} \leqslant \xi \leqslant x_{i+1}$.

Now, adding Eqs. (5.2) and (5.15) side by side leads to

$$f(x_i + h) + f(x_i - h) = 2f(x_i) + h^2 f''(x_i) + \frac{h^4}{12} f^{(4)}(x_i) + \ldots \qquad (5.33)$$

Solving Eq. (5.33) for $f''(x_i)$ gives

$$f''(x_i) = \frac{f(x_i + h) - 2f(x_i) + f(x_i - h)}{h^2} - \frac{h^2}{12} f^{(4)}(x_i) - \ldots \qquad (5.34)$$

or

$$f''_i = \frac{f_{i+1} - 2f_i + f_{i-1}}{h^2} + \mathcal{O}(h^2) \qquad (5.35)$$

where the truncation error is $E(h) = -h^2 f^{(4)}(\xi)/12$ for $x_{i-1} \leqslant \xi \leqslant x_{i+1}$.

Equations (5.32) and (5.35), which are second-order accurate finite difference formulas, are referred to as the *Central Difference Formulas (CDFs)*. More compact expressions, similar to forward and backward differences, can be obtained by incorporating the *central difference operator* δ, which is defined as

$$\delta f_i = f_{i+1/2} - f_{i-1/2}$$

It is possible to show that $\delta^2 f_i = \delta f_{i+1/2} - \delta f_{i-1/2} = f_{i+1} - 2f_i + f_{i-1}$. Then, Eq. (5.35) can be expressed as

$$f''_i = \frac{\delta^2 f_i}{h^2} + \mathcal{O}(h^2) \qquad (5.36)$$

A graphical depiction of the data points used, along with the predicted slopes with the forward, backward, and central difference formulas for $f'(x_i)$, is presented in Fig. 5.6. The slopes of the secant lines, estimated by the forward and backward differences, provide a visual depiction of the significant deviations from the true slope. But the slope estimated with the central difference formula (i.e., the slope of the line passing through (x_{i-1}, f_{i-1}) and (x_{i+1}, f_{i+1}) points) clearly yields much better agreement with the slope of the true tangent line. Thus, so long as the number of data points is sufficient, the central difference formulas should be preferred due to their second-order accuracy. Note that the errors in the derivatives (i.e., slopes) can be spotted visually, but their magnitudes cannot.

For high-order derivatives, the central difference formulas can be generalized in terms of the forward and/or backward difference operators as

$$\frac{d^n f_i}{dx^n} = \begin{cases} \dfrac{1}{2h^n} \left(\nabla^n f_{i+n/2} + \Delta^n f_{i-n/2} \right), & \text{if } n \text{ is even} \\[2mm] \dfrac{1}{2h^n} \left(\nabla^n f_{i+(n-1)/2} + \Delta^n f_{i-(n-1)/2} \right), & \text{if } n \text{ is odd} \end{cases} + \mathcal{O}(h^2) \qquad (5.37)$$

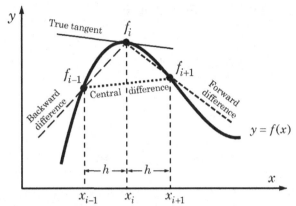

FIGURE 5.6: Graphical depiction of forward, backward, and central difference approximations for f_i'.

EXAMPLE 5.2: Order of accuracy and truncation error

Use Taylor series expansion to determine the derivative represented by the following approximation and find the truncation error and order of accuracy.

$$\frac{-3f_{i-2} - 25f_i + 30f_{i+1} - 2f_{i+3}}{30h} = ? \quad \mathcal{O}(h^?)$$

SOLUTION:

(a) The Taylor series of $f(x_{i+kh})$ about x_i can be generalized as

$$f_{i+k} = f_i + kh\, f_i' + \frac{(kh)^2}{2!}f_i'' + \frac{(kh)^3}{3!}f_i''' + \frac{(kh)^4}{4!}f_i^{(4)} + \cdots$$

where k is an integer denoting the number of increments $(k > 0)$ or decrements $(k < 0)$. Then the Taylor series of $f(x_{i-2})$, $f(x_{i+1})$, and $f(x_{i+3})$ about x_i are found by setting $k = 1$, 3, and -2 to give

$$f_{i+1} = f_i + h\, f_i' + \frac{h^2}{2}f_i'' + \frac{h^3}{6}f_i''' + \frac{h^4}{24}f_i^{(4)} + \frac{h^5}{120}f_i^{(5)} + \cdots$$

$$f_{i+3} = f_i + 3h\, f_i' + \frac{9h^2}{2!}f_i'' + \frac{9h^3}{2}f_i''' + \frac{27h^4}{8}f_i^{(4)} + \frac{81h^5}{40}f_i^{(5)} + \cdots$$

$$f_{i-2} = f_i - 2h\, f_i' + 2h^2 f_i'' - \frac{4h^3}{3}f_i''' + \frac{2h^4}{3}f_i^{(4)} - \frac{4h^5}{15}f_i^{(5)} + \cdots$$

Substituting the above expressions into the numerator and collecting the common terms leads to

$$-3f_{i-2} - 25f_i + 30f_{i+1} - 2f_{i+3} = 30h\, f_i' - \frac{15h^4}{2}f_i^{(4)} - 3h^5 f_i^{(5)} + \cdots$$

where the lowest order derivative on the rhs is f_i'. Dividing it by $30h$ gives

$$f_i' = \frac{-3f_{i-2} - 25f_i + 30f_{i+1} - 2f_{i+3}}{30h} + \boxed{\frac{h^4}{4}f_i^{(4)}} + \frac{h^5}{10}f_i^{(5)} + \cdots$$

The first term of this expression is identical to the finite difference formula given in the problem statement; thus, we conclude that the difference formula is that of the first derivative (f_i').

The remaining terms shed light on the truncation error. The leading (truncation) error is $E(h) = h^4 f^{(4)}(\xi)/4$ for $x_{i-2} \leqslant \xi \leqslant x_{i+3}$, and the difference formula is fourth-order accurate due to having h^4 factor, i.e., $\mathcal{O}(h^4)$.

Discussion: The Taylor series can be effectively used to derive approximations for any derivative and help determine the truncation error term and the order of accuracy. The finite difference formulas are generally referred to by their order of accuracy. A high value indicates faster convergence to the true value with decreasing h.

 Higher-order approximations for the higher derivatives could plague the solution with noise since higher-order derivatives tend to be larger, especially near sharp gradients.

5.5 FINITE DIFFERENCES AND DIRECT-FIT POLYNOMIALS

Finite difference formulas can alternatively be derived with the aid of polynomials. This approach requires fitting a set of function values to obtain an approximating polynomial and then differentiating the fitted polynomial.

A function to be fitted (or approximated) by a polynomial is assumed to be continuous, differentiable, and have exact values at the furnished discrete points. A polynomial fit can then be achieved in a variety of ways, such as by either directly fitting a set of discrete data to a polynomial, by using Lagrange polynomials (*see* Chapter 6), or by approximating with least squares polynomial fit (*see* Chapter 7). In this section, we will rather focus on applying the direct-fit polynomials to obtain finite difference formulas.

To generate numerical differentiation formulas by direct-fit polynomials, follow the three-step procedure given below:

1. Establish an nth-degree polynomial, $p_n(x) = a_n x^n + a_{n-1} x^{n-1} \ldots + a_1 x + a_0$;

2. Force $p_n(x)$ to be equal to $f(x)$ at $n + 1$ data points or nodes, i.e., setting $f(x_i) = p_n(x_i) = f_i$ for $i = 0, 1, 2, \ldots, n$ gives a linear system of $n + 1$ equations for the undetermined coefficients $(a_0, a_1, \ldots, a_n)$;

3. Find an approximation to $f^{(k)}(x_i)$ $(k \leqslant n)$ by setting it to the kth derivative of $p_n(x_i)$, i.e., $f^{(k)}(x_i) = p_n^{(k)}(x_i)$;

Here, the distribution of x_i's depends on the type of difference formulas desired to be derived, i.e., forward, backward, or central.

Once the type of difference formula is decided, the other parameter that we can control is the degree of the approximating polynomial. Deciding on the degree of an approximating polynomial is an important step in this process. In the preceding sections, it was shown that $f'(x_i)$ computed by the two-point difference formulas (FDF or BDF) led to first-order accurate approximations, i.e., $\mathcal{O}(h)$. On the other hand, difference formulas for $f''(x_i)$, even as $\mathcal{O}(h)$, required three points. This means that the second derivative cannot be

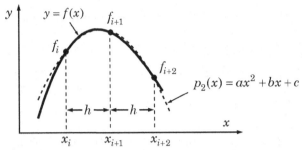

FIGURE 5.7: A polynomial approximation to a discrete data.

approximated with linear relationships like the first derivatives, and an approximating polynomial of the lowest degree representing $f(x)$ should be a parabola.

There is a relationship between the desired order of accuracy in the finite difference formulas and the degree of interpolating polynomials. For example, consider the nth-degree accurate forward difference approximation for $f'(a)$, Eq. (5.3), whose truncation error is expressed as

$$E(h) = \frac{h^n}{(n+1)!} f^{(n+1)}(\xi), \quad a \leqslant \xi \leqslant a + h$$

which implies that, if $f(x)$ is a polynomial of nth degree, the finite difference approximation to any kth derivative $f^{(k)}(x_i)$ (for $k \leqslant n$) will be exact since $d^{n+1}f/dx^{n+1} = 0$. This deduction can be used to determine finite difference formulas of the desired order of accuracy, i.e., the degree of approximating polynomial.

The order of truncation error of a function approximated by an nth-degree polynomial ($f(x) \cong p_n(x)$) is $\mathcal{O}(h^{n+1})$. The order of accuracy of $f'(x_i)$ becomes $\mathcal{O}(h^n)$, which also requires $(n+1)$ data points in the difference formula to attain this accuracy. Different relationships for higher-order derivatives apply; for instance, the degree of a direct fit approximating polynomial should be $n + k - 1$ in order to find a finite difference formula for the kth derivative with an accuracy order of $\mathcal{O}(h^n)$. Once the degree (n) of an approximating polynomial is decided, $p_n(x)$ should satisfy $(n+1)$ data points. This step leads to a system of simultaneous linear equations with $(n+1)$ unknowns for the coefficients of the polynomial. Finally, the $f^{(k)}(x)$ derivative of $f(x)$ is approximated by the $p_n^{(k)}(x)$ derivative.

Consider deriving a second-order accurate forward difference formula, $\mathcal{O}(h^2)$, for $f'(x_i)$ using a direct-fit polynomial. For this case, an approximating polynomial is a second-degree polynomial (i.e., a parabola).

$$f(x) \cong p_2(x) = ax^2 + bx + c, \quad x_i \leqslant x \leqslant x_{i+2} \tag{5.38}$$

In Figure 5.7, an arbitrary function $f(x)$ and an approximating polynomial $p_2(x)$ are depicted. The parabola is forced to pass through the uniformly spaced three data points: (x_i, f_i), (x_{i+1}, f_{i+1}), and (x_{i+2}, f_{i+2}). We apply a shifted coordinate system by setting $x_i = 0$, $x_{i+1} = h$, and $x_{i+2} = 2h$ to simplify the derivation process while maintaining generality. Matching the function and the parabola for all three data points leads to the following system of linear equations with three unknowns (a, b, c):

$$\begin{aligned}
f_i &= f(x_i) = p_2(0) = a(0)^2 + b(0) + c \\
f_{i+1} &= f(x_{i+1}) = p_2(h) = a(h)^2 + b(h) + c \\
f_{i+2} &= f(x_{i+2}) = p_2(2h) = a(2h)^2 + b(2h) + c
\end{aligned} \tag{5.39}$$

TABLE 5.2: The order of accuracy of derivatives with symmetrical (central) and non-symmetrical (forward/backward) finite difference formulas.

Non-symmetric Differences			Symmetric Differences		
$f_i =$	$p_n(x_i)$		$f_i =$	$p_n(x_i)$	
$f_i' \cong$	$p_n'(x_i)$	$\mathcal{O}(h^n)$	$f_i' \cong$	$p_n'(x_i)$	$\mathcal{O}(h^n)$
$f_i'' \cong$	$p_n''(x_i)$	$\mathcal{O}(h^{n-1})$	$f_i'' \cong$	$p_n''(x_i)$	$\mathcal{O}(h^n)$
$f_i''' \cong$	$p_n'''(x_i)$	$\mathcal{O}(h^{n-2})$	$f_i''' \cong$	$p_n'''(x_i)$	$\mathcal{O}(h^{n-2})$
$f_i^{(4)} \cong$	$p_n^{(4)}(x_i)$	$\mathcal{O}(h^{n-3})$	$f_i^{(4)} \cong$	$p_n^{(4)}(x_i)$	$\mathcal{O}(h^{n-2})$
$\vdots$	$\vdots$	$\vdots$	$\vdots$	$\vdots$	$\vdots$
$f_i^{(n-1)} \cong$	$p_n^{(n-1)}(x_i)$	$\mathcal{O}(h^2)$	$f_i^{(n-1)} \cong$	$p_n^{(n-1)}(x_i)$	$\mathcal{O}(h^2)$
$f_i^{(n)} \cong$	$p_n^{(n)}(x_i)$	$\mathcal{O}(h)$	$f_i^{(n)} \cong$	$p_n^{(n)}(x_i)$	$\mathcal{O}(h^2)$

Solving Eq. (5.39) for a, b, and c yields

$$a = \frac{f_i - 2f_{i+1} + f_{i+2}}{2h^2}, \quad b = \frac{-3f_i + 4f_{i+1} - f_{i+2}}{2h}, \quad c = f_i \quad (5.40)$$

Having found the direct-fit polynomial, we now turn our attention to finding the difference formulas for $f'(x_i)$ and $f''(x_i)$. We achieve this by matching the first and second derivatives of $f(x)$ and $p_2(x)$ at $x_i = 0$; that is, $f'(x_i) = f'(0) \cong p_2'(0) = b$ and $f''(x_i) = f''(0) \cong p_2''(0) = 2a$. Finally, using the coefficients from Eq. (5.40), we obtain

$$f_i' \cong b = \frac{-f_{i+2} + 4f_{i+1} - 3f_i}{2h} \quad (5.41)$$

$$f_i'' \cong 2a = \frac{f_i - 2f_{i+1} + f_{i+2}}{h^2} \quad (5.42)$$

Note that Eqs. (5.41) and (5.42) are the same finite difference formulas derived from the Taylor series—Eqs. (5.29) and (5.22). Also note that the order of accuracy of f_i' and f_i'' is $\mathcal{O}(h^2)$ and $\mathcal{O}(h)$, respectively. Each successive differentiation (of a direct-fit polynomial) reduces the order of the associated truncation error. The drop in the order of accuracy of difference formulas also depends on how the data points are distributed around the point where the numerical derivative is sought. Using the $p_n(x)$ to evaluate higher-order derivatives, one order of accuracy is lost for each increasing order for derivatives of non-symmetrical (forward or backward) difference formulas. In symmetrical (central) difference formulas, two orders of accuracy are lost for derivatives increasing by multiples of two due to points being added in symmetric pairs. The order of accuracy for the derivatives for symmetrical or non-symmetrical finite difference formulas is illustrated in Table 5.2.

Table 5.3 summarizes the central difference formulas of order of accuracy $\mathcal{O}(h^2)$ and $\mathcal{O}(h^4)$ for derivatives up to 4th order. The CDFs should be used whenever there is a sufficient number of (also symmetric) data points on the left and right sides of a point (x_i, f_i). The forward and backward difference formulas of order of accuracy $\mathcal{O}(h)$, $\mathcal{O}(h^2)$, and $\mathcal{O}(h^4)$ are presented for derivatives up to the 4th order in Tables 5.4 and 5.5. The FDFs are used when there are no data points on the left of (x_i, f_i), while the BDFs are used when data points are stacked on the left of (x_i, f_i).

TABLE 5.3: The central difference formulas of order of accuracy $\mathcal{O}(h^2)$ and $\mathcal{O}(h^4)$ for derivatives up to fourth order.

Difference Formula	$E(h)$
$f_i' \cong \dfrac{1}{2h}\left(f_{i+1}-f_{i-1}\right)$	$-\dfrac{1}{6}f'''(\xi)h^2$
$\cong \dfrac{1}{12h}\left(-f_{i+2}+8f_{i+1}-8f_{i-1}+f_{i-2}\right)$	$\dfrac{1}{30}f^{(5)}(\xi)h^4$
$f_i'' \cong \dfrac{1}{h^2}\left(f_{i+1}-2f_i+f_{i-1}\right)$	$-\dfrac{1}{12}f^{(4)}(\xi)h^2$
$\cong \dfrac{1}{12h^2}\left(-f_{i+2}+16f_{i+1}-30f_i+16f_{i-1}-f_{i-2}\right)$	$\dfrac{1}{90}f^{(6)}(\xi)h^4$
$f_i''' \cong \dfrac{1}{2h^3}\left(f_{i+2}-2f_{i+1}+2f_{i-1}-f_{i-2}\right)$	$-\dfrac{1}{4}f^{(5)}(\xi)h^2$
$\cong \dfrac{1}{8h^3}\left(-f_{i+3}+8f_{i+2}-13f_{i+1}+13f_{i-1}-8f_{i-2}+f_{i-3}\right)$	$\dfrac{7}{120}f^{(7)}(\xi)h^4$
$f_i^{(4)} \cong \dfrac{1}{h^4}\left(f_{i+2}-4f_{i+1}+6f_i-4f_{i-1}+f_{i-2}\right)$	$-\dfrac{1}{6}f^{(6)}(\xi)h^2$
$\cong \dfrac{1}{6h^4}\left(-f_{i+3}+12f_{i+2}-39f_{i+1}+56f_i-39f_{i-1}+12f_{i-2}-f_{i-3}\right)$	$\dfrac{7}{240}f^{(8)}(\xi)h^4$

EXAMPLE 5.3: Estimate first derivative with difference formulas

Given $f(x) = e^x$, for decreasing values of h, estimate $f'(1)$ and the associated truncation error with the first-order FDF, second-order CDF, and fourth-order CDF formulas. How do truncation errors compare to true errors?

SOLUTION:

First-, second-, and fourth-order difference formulas, as well as their truncation errors, for f_i' are given in Tables 5.3 and 5.4. To estimate the truncation error, $f''(\xi)$, $f'''(\xi)$, and $f^{(5)}(\xi)$ derivatives are required. Note that e^ξ is an increasing function for all ξ, and all high-order derivatives give $f(x) = e^x$. The absolute maximum of the derivatives occurs at the rightmost end of the pertinent intervals, i.e., $\xi = 1 + h$ for FDF and 2th-order CDF and $\xi = 1 + 2h$ for the 4th-order CDF. Then, the difference formulas and the corresponding truncation errors can be expressed as

$$f'_{\text{FDF}}(1) = \frac{e^{1+h}-e}{h}, \qquad\qquad E(h) = -\frac{h}{2}e^{1+h}$$

$$f'_{\text{CDF}}(1) = \frac{e^{1+h}-e^{1-h}}{2h}, \qquad\qquad E(h) = -\frac{h^2}{6}e^{1+h}$$

$$f'_{\text{CDF}}(1) = \frac{-e^{1+2h}+8e^{1+h}-8e^{1-h}+e^{1-2h}}{12h}, \quad E(h) = \frac{h^4}{30}e^{1+2h}$$

Since, in this example, $f(x)$ is explicitly given, the true value and the associated true error can be calculated as $f'_{\text{true}}(1) = e$ and $E_{\text{true}}(h) = |f'_{\text{true}}(h) - f'_{\text{approx}}(h)|$, respectively. In Table 5.6, the numerical estimates of $f'(1)$ obtained by the first-order FDF, second-order CDF, and fourth-order CDF, along with the associated

TABLE 5.4: The forward difference formulas of order of accuracy $\mathcal{O}(h)$, $\mathcal{O}(h^2)$, and $\mathcal{O}(h^4)$ for the derivatives up to fourth order.

Difference Formula	$E(h)$
$f_i' \cong \dfrac{1}{h}\left(f_{i+1}-f_i\right)$	$-\dfrac{1}{2}f''(\xi)h$
$\cong \dfrac{1}{2h}\left(-f_{i+2}+4f_{i+1}-3f_i\right)$	$\dfrac{1}{3}f'''(\xi)h^2$
$\cong \dfrac{1}{12h}\left(-3f_{i+4}+16f_{i+3}-36f_{i+2}+48f_{i+1}-25f_i\right)$	$\dfrac{1}{5}f^{(5)}(\xi)h^4$
$f_i'' \cong \dfrac{1}{h^2}\left(f_{i+2}-2f_{i+1}+f_i\right)$	$-f'''(\xi)\,h$
$\cong \dfrac{1}{h^2}\left(-f_{i+3}+4f_{i+2}-5f_{i+1}+2f_i\right)$	$\dfrac{11}{12}f^{(4)}(\xi)h^2$
$f_i''' \cong \dfrac{1}{h^3}\left(f_{i+3}-3f_{i+2}+3f_{i+1}-f_i\right)$	$-\dfrac{3}{2}f^{(4)}(\xi)h$
$\cong \dfrac{1}{2h^3}\left(-3f_{i+4}+14f_{i+3}-24f_{i+2}+18f_{i+1}-5f_i\right)$	$\dfrac{7}{4}f^{(5)}(\xi)h^2$
$f_i^{(4)} \cong \dfrac{1}{h^4}\left(f_{i+4}-4f_{i+3}+6f_{i+2}-4f_{i+1}+f_i\right)$	$-2f^{(5)}(\xi)\,h$
$\cong \dfrac{1}{h^4}\left(-2f_{i+5}+11f_{i+4}-24f_{i+3}+26f_{i+2}-14f_{i+1}+3f_i\right)$	$\dfrac{17}{6}f^{(6)}(\xi)\,h^2$

TABLE 5.5: The backward difference formulas of order of accuracy $\mathcal{O}(h)$, $\mathcal{O}(h^2)$, and $\mathcal{O}(h^4)$ for the derivatives up to fourth order.

Difference Formula	$E(h)$
$f_i' \cong \dfrac{1}{h}\left(f_i-f_{i-1}\right)$	$\dfrac{1}{2}f''(\xi)h$
$\dfrac{1}{2h}\left(3f_i-4f_{i-1}+f_{i-2}\right)$	$\dfrac{1}{3}f'''(\xi)h^2$
$\cong \dfrac{1}{12h}\left(25f_i-48f_{i-1}+36f_{i-2}-16f_{i-3}+3f_{i-4}\right)$	$\dfrac{1}{5}f^{(5)}(\xi)h^4$
$f_i'' \cong \dfrac{1}{h^2}\left(f_i-2f_{i-1}+f_{i-2}\right)$	$f'''(\xi)\,h$
$\cong \dfrac{1}{h^2}\left(2f_i-5f_{i-1}+4f_{i-2}-f_{i-3}\right)$	$\dfrac{11}{12}f^{(4)}(\xi)h^2$
$f_i''' \cong \dfrac{1}{h^3}\left(f_i-3f_{i-1}+3f_{i-2}-f_{i-3}\right)$	$\dfrac{3}{2}f^{(4)}(\xi)h$
$\cong \dfrac{1}{2h^3}\left(5f_i-18f_{i-1}+24f_{i-2}-14f_{i-3}+3f_{i-4}\right)$	$\dfrac{7}{4}f^{(5)}(\xi)h^2$
$f_i^{(4)} \cong \dfrac{1}{h^4}\left(f_i-4f_{i-1}+6f_{i-2}-4f_{i-3}+f_{i-4}\right)$	$2f^{(5)}(\xi)\,h$
$\cong \dfrac{1}{h^4}\left(3f_i-14f_{i-1}+26f_{i-2}-24f_{i-3}+11f_{i-4}-2f_{i-5}\right)$	$\dfrac{17}{6}f^{(6)}(\xi)\,h^2$

truncation errors and the true (absolute) errors, are tabulated for decreasing values of h. Notice that the truncation errors are always greater than the true errors, as the interval size h is kept in the region dominated by truncation errors $(h < h^*)$.

TABLE 5.6

h	$f'_{\text{approx}}(1)$	$|E(h)|$	True error
	First-order FDF		
10^{-1}	2.85884195	0.1502108	0.1405601
10^{-2}	2.73191866	0.0137280	0.0136368
10^{-3}	2.71964142	0.0013605	0.0013596
10^{-4}	2.71841775	0.0001359	0.0001359
	Second-order CDF		
10^{-1}	2.72281456	5.0069×10^{-3}	4.5327×10^{-3}
10^{-2}	2.71832713	4.5760×10^{-5}	4.5305×10^{-5}
10^{-3}	2.71828228	4.5350×10^{-7}	4.5304×10^{-7}
	Fourth-order CDF		
10^{-1}	2.71827276	1.1067×10^{-5}	9.0717×10^{-6}
10^{-2}	2.71828183	9.2440×10^{-10}	9.0609×10^{-10}

A log-log graph of the true absolute error as a function of h is presented in Fig. 5.8. Since the truncation error is proportional to h^m, a curve representing the true error should be a straight line with a slope giving the order of the difference formula, i.e., m. For instance, for $h = 0.1$, the 10-based logarithm of true absolute errors with the 1st, 2nd, and 4th-order difference formulas is -0.852, -2.344, and -5.042, respectively. These values, respectively, are realized as -1.855, -4.344, and -9.042 for $h = 0.01$. The slopes of the lines are found from $m = \log_{10} E_{\text{true}}(10^{-1}) - \log_{10} E_{\text{true}}(10^{-2})$ which verifies the order of each finite difference equation as 1, 2, and 4.

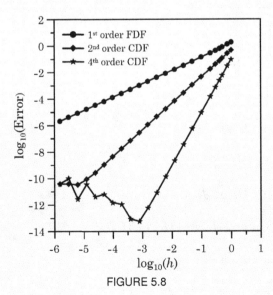

FIGURE 5.8

Highly accurate approximations for the derivatives can be achieved when using high-order difference formulas (order of accuracy > 2). As h is decreased (moving from right to left in Fig. 5.8) in the region dominated by truncation error, the log-errors decrease linearly, most rapidly for the fourth-order formula. But the effect of round-off errors becomes evident as h is further decreased. After passing a minimum point, the total error begins to rise in a jagged fashion and eventually dominates the total errors. As the order of accuracy of the approximation increases, the crossover point moves to the right (greater efficiency) and down (greater accuracy).

Discussion: The truncation error, if calculated fairly accurately, provides a conservative means of estimating the computational errors resulting from finite difference formulas. In practice, the true distribution of a discrete function is not known. Therefore, although it is not possible to calculate the true error, the magnitude of the truncation error can be roughly estimated.

In this example, the effects of rounding on the difference formulas are observed in Fig. 5.8. We note that the use of the second-order finite difference formula for $h \lesssim 10^{-5}$ and the fourth-order finite difference formula for $h \lesssim 10^{-3}$ is contaminated with round-off errors, which lead to an increase in the total error. Hence, overall, h should be chosen "small enough" so that the finite difference formulas produce accurate approximations and "large enough" so that the approximations are not too contaminated by round-off errors.

EXAMPLE 5.4: Calculate speed and acceleration from a dataset

The Vehicle Acceleration Test (VAT) result of a minivan is given below as distance-time data recorded at 5-second intervals. The van, initially at rest, reaches its maximum speed in 25 seconds. Estimate the speed (m/s) and acceleration (m/s^2) for each available data point using second-order accurate finite-difference formulas.

t (s)	0	5	10	15	20	25
s (m)	0.0	5.45	21.3	82.84	212.86	473.6

SOLUTION:

The distance, $s(t)$, is provided by a set of uniformly spaced ($\Delta t = 5$ s) discrete data points. The speed and acceleration can only be estimated for $t = 0$, 5, 10, 15, 20, and 25 seconds since no other data is available in the set. We first express the available in an array form (t_i, s_i). For $1 \leqslant i \leqslant 6$, we may write $(t_1, s_1) = (0, 0)$, $(t_2, s_2) = (5, 5.45)$, and so on. Note that since Δt is uniform, there is no need to establish the array t_i for the time variable.

By definition, the speed is the derivative of distance with respect to time, $v(t) = ds/dt$, and the acceleration is the second derivative of distance with respect to time, $a(t) = d^2s/dt^2$. To estimate the speed and acceleration with an accuracy of order $\mathcal{O}(h^2)$, the central difference formulas will be applied for the interior data points ($2 \leqslant i \leqslant 5$):

$$v_i = \frac{ds_i}{dt} = \frac{s_{i+1} - s_{i-1}}{2\,\Delta t}, \qquad a_i = \frac{d^2 s_i}{dt^2} = \frac{s_{i+1} - 2s_i + s_{i-1}}{(\Delta t)^2}$$

Ar $t = 0$ (or $i = 1$), when the minivan is at rest, we will apply the forward difference formula of order $\mathcal{O}(h^2)$ since no data prior to $t = 0$ ($i < 1$) is available. Thus, the speed and acceleration at $t = 0$ ($i = 1$) will be calculated as

$$v_1 = \frac{ds_i}{dt} = \frac{-s_3 + 4s_2 - 3s_1}{2\Delta t}, \qquad a_1 = \frac{d^2 s_i}{dt^2} = \frac{-s_4 + 4s_3 - 5s_2 + 2s_1}{(\Delta t)^2}$$

When the minivan reaches its maximum speed at $t = 25$ s (or $i = 6$), we will use the backward difference formula of order $\mathcal{O}(h^2)$ since no other data is available for $t > 25$ s. Hence, the speed and acceleration at $t = 25$ s will be calculated from

$$v_6 = \frac{ds_6}{dt} = \frac{s_4 - 4s_5 + 3s_6}{2\Delta t}, \qquad a_6 = \frac{d^2 s_6}{dt^2} = \frac{-s_3 + 4s_4 - 5s_5 + 2s_6}{(\Delta t)^2}$$

Substituting the numerical values into the difference formulas, the speeds at $t = 0$ and $t = 25$ s are found as

$$v_1 = \frac{-s_3 + 4s_2 - 3s_1}{2\Delta t} = \frac{-21.3 + 4(5.45) - 3(0)}{2(5)} = 0.05 \frac{m}{s}$$

$$v_6 = \frac{s_4 - 4s_5 + 3s_6}{2\Delta t} = \frac{82.84 - 4(212.86) + 3(473.6)}{2(5)} = 65.22 \frac{m}{s}$$

For the interior data points ($2 \leqslant i \leqslant 5$), the speed values are found as

$$v_2 = \frac{s_3 - s_1}{2\Delta t} = \frac{21.3 - 0}{2(5)} = 2.13 \frac{m}{s}$$

$$v_3 = \frac{s_4 - s_2}{2\Delta t} = \frac{82.84 - 5.45}{2(5)} = 7.739 \frac{m}{s}$$

$$v_4 = \frac{s_5 - s_3}{2\Delta t} = \frac{212.86 - 21.3}{2(5)} = 19.156 \frac{m}{s}$$

$$v_5 = \frac{s_6 - s_4}{2\Delta t} = \frac{473.6 - 82.84}{2(5)} = 39.076 \frac{m}{s}$$

The accelerations at $i = 1$ and $i = 6$ are computed with the 2nd-order FDF and BDF, respectively. Using the pertinent difference formulas from Tables 5.3 and 5.5 we find

$$a_1 = \frac{-s_4 + 4s_3 - 5s_2 + 2s_1}{(\Delta t)^2} = \frac{-82.84 + 4(21.3) - 5(5.45) + 2(0)}{5^2} = -0.9956 \frac{m}{s^2}$$

$$a_6 = \frac{-s_3 + 4s_4 - 5s_5 + 2s_6}{(\Delta t)^2} = \frac{-21.3 + 4(82.84) - 5(212.86) + 2(473.6)}{5^2} = 7.7184 \frac{m}{s^2}$$

For the interior data points ($2 \leqslant i \leqslant 5$), using the 2nd-order CDF, the acceleration values yield

$$a_2 = \frac{s_3 - 2s_2 + s_1}{(\Delta t)^2} = \frac{21.3 - 2(5.45) + 0}{5^2} = 0.416 \, \frac{\text{m}}{\text{s}^2}$$

$$a_3 = \frac{s_4 - 2s_3 + s_2}{(\Delta t)^2} = \frac{82.84 - 2(21.3) + 5.45}{5^2} = 1.8276 \, \frac{\text{m}}{\text{s}^2}$$

$$a_4 = \frac{s_5 - 2s_4 + s_3}{(\Delta t)^2} = \frac{212.86 - 2(82.84) + 21.3}{5^2} = 2.7392 \, \frac{\text{m}}{\text{s}^2}$$

$$a_5 = \frac{s_6 - 2s_5 + s_4}{(\Delta t)^2} = \frac{473.6 - 2(212.86) + 82.84}{5^2} = 5.2288 \, \frac{\text{m}}{\text{s}^2}$$

The estimated maximum speed and maximum acceleration are 65.22 m/s (or 234.8 km/h) and 7.6184 m/s, respectively. A sample pseudo-program to calculate the speed and acceleration for this example is provided in Pseudocode 5.1.

Discussion: When the vehicle is at rest, the speed and acceleration must be zero. However, at $t = 0$, we could not obtain the velocity and acceleration values, which should have been "zero," with the finite difference formulas. One of the reasons for this result is due to truncation errors stemming from a large data sampling interval ($\Delta t = 5$ s). A smaller data sampling interval ($\Delta t < 5$) would improve the accuracy of calculations.

Finite-Difference Formulas

- Finite-difference formulas are simple and easy to use;
- Higher-order finite difference formulas can be easily generated using difference operators;
- For a given h, 2nd- or 4th-order difference formulas offer a descent trade-off between accuracy and the number of function evaluations;
- Derivatives of explicit functions can be accurately computed by increasing the order of accuracy (i.e., data points).

- The truncation error is a function of interval size, $\mathcal{O}(h^n)$, and it has a strong influence on the accuracy of differentiation;
- Difference formulas are plagued with *round-off* and *loss of significance* errors when $h \approx 0$;
- When high-order difference formulas are used, there is no need to use a very small h to get a very good estimate; however, such formulas avoid floating-point problems at the expense of the increased number of function evaluations.

Higher derivatives can also be numerically calculated by successive differentiations: $f'' = (f')'$, $f''' = (f'')'$, etc. This approach causes the uncertainties already present in the original data, as well as the total errors in the computed quantities, to be propagated to the next level. In this regard, a prudent approach would be to use the original discrete data set with the suitable finite difference formulas of the desired order of accuracy.

Pseudocode 5.1

Program EXAMPLE_5.4
\ DESCRIPTION: A pseudocode to calculate the speed and acceleration using
\   a set of distance data given in Example 5.4.
\ VARIABLES:
\ n :: Number of data in the set;
\ Δt :: Time interval (s);
\ S :: Array of length n containing distance (m);
\ v :: Array of length n containing speed (m/s)
\ a :: Array of length n containing acceleration (m/s^2).
Declare: S_n, v_n, a_n \ Declare array variables
Read: $n, \Delta t, (S_i, i = 1, n)$ \ Read data into program
\ FD formulas of order $\mathcal{O}(h^2)$ are used for all data points.
$v_1 \leftarrow (-S_3 + 4S_2 - 3S_1)/(2\Delta t)$ \ Use FDF for v at $i = 1$
$a_1 \leftarrow (2S_1 - 5S_2 + 4S_3 - S_4)/(\Delta t)^2$ \ Use FDF for a at $i = 1$
For $\left[i = 2, (n-1) \right]$ \ Loop for finding v and a at interior points
$\quad v_i \leftarrow (S_{i+1} - S_{i-1})/(2\Delta t)$ \ Use CDF for v
$\quad a_i \leftarrow (S_{i+1} - 2S_i + S_{i-1})/(\Delta t)^2$ \ Use CDF for a
End For
$v_n \leftarrow (3S_n - 4S_{n-1} + S_{n-2})/(2\Delta t)$ \ Use BDF for v at $i = n$
$a_n \leftarrow (2S_n - 5S_{n-1} + 4S_{n-2} - S_{n-3})/(\Delta t)^2$ \ Use BDF for a at $i = n$
Write: $(t_i, S_i, v_i, a_i, i = 1, n)$ \ Print out the results
End Program EXAMPLE_5.4

5.6 DIFFERENTIATING NON-UNIFORMLY SPACED DISCRETE DATA

In practice, a set of discrete data collected or generated may not always be uniformly spaced, in which case the finite-difference formulas given in Tables 5.3–5.5 cannot be used.

In general, the approximations for differentiating non-uniformly spaced data sets are derived using polynomials: direct-fit, Lagrange, or divided differences. In this section, we will illustrate the use of direct-fit polynomials, which is just as in uniform cases, based on fitting the data directly with a polynomial of nth degree, $p_n(x)$, and differentiating the polynomial as many times as necessary.

There are situations, such as in the numerical solution of differential equations, where an analyst does not wish the discrete data to be uniform. It is efficient to work with uniformly

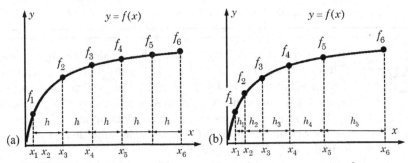

FIGURE 5.9: (a) Uniform, (b) non-uniform data set near steep changes.

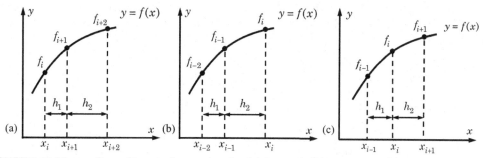

FIGURE 5.10: Non-uniform discrete data points for (a) forward, (b) backward, (c) central differences.

spaced discrete data when the rate of change in the solution of the differential equation is small over a wide range, in which case the change in a physical quantity can be well represented by low-order polynomials. A continuous function $f(x)$ containing a sharp rise near x_1 and discrete data sets generated from $f(x)$ with uniform and non-uniform intervals are shown in Fig. 5.9. Notice that $f(x)$ is almost linear for $x_3 < x < x_6$; hence, the rate of change can be estimated fairly accurately with low-order polynomials. However, the derivatives of functions that exhibit steep or sharp changes (e.g., for $x < x_3$), calculated based on uniform distributions, cannot be estimated very accurately by low-order finite difference formulas.

To adequately capture or represent important features of any discrete distribution (velocity, temperature, concentration gradients, etc.), analysts deliberately stack the data points in regions where steep or sharp changes are expected (*see* Fig. 5.9b). Hence, suitable difference formulas need to be generated to accommodate the use of non-uniformly spaced data sets.

Consider the non-uniformly spaced discrete data points shown in Fig. 5.10. To derive second-order finite difference approximations for the case depicted in Fig. 5.10a, we adopt a quadratic approximating polynomial: $p_2(x) = ax^2 + bx + c$. Using the shifted coordinate system and setting $x_i = 0$, $x_{i+1} = h_1$, and $x_{i+2} = h_1 + h_2$, the data points must satisfy the approximating polynomial:

$$f(x_i) = f_i \cong p_2(0) = c$$
$$f(x_{i+1}) = f_{i+1} \cong p_2(h_1) = ah_1^2 + bh_1 + c \tag{5.43}$$
$$f(x_{i+2}) = f_{i+2} \cong p_2(h_1 + h_2) = a(h_1 + h_2)^2 + b(h_1 + h_2) + c$$

which constitute a system of three equations with three unknowns: a, b, and c. Since $f_i' \cong p_2'(x_i) = p_2'(0) = b$, the forward-difference formula for the first derivative is obtained as

$$f_i' = -\frac{2h_1 + h_2}{h_1(h_1 + h_2)}f_i + \frac{h_1 + h_2}{h_1 h_2}f_{i+1} - \frac{h_1}{h_2(h_1 + h_2)}f_{i+2} \tag{5.44}$$

Similarly, the second derivative is found by matching $p_2''(0) = f_i'' = 2a$, leading to

$$f_i'' = \frac{2}{h_1(h_1 + h_2)}f_i - \frac{2}{h_1 h_2}f_{i+1} + \frac{2}{h_2(h_1 + h_2)}f_{i+2} \tag{5.45}$$

Note that setting $h_1 = h_2 = h$ yields the expressions for the uniformly spaced discrete data.

Equations (5.44) and (5.45) do not provide much information on the truncation error or the order of accuracy. This issue can be resolved by making use of the Taylor series, as in the case of uniform discrete functions. For this case, the truncation error and order of accuracy with the forward difference formulations of the first and second derivatives are given below:

$$\text{For } f_i': \quad E(h_1, h_2) = \frac{f'''(\xi)}{6} h_1(h_1 + h_2) \equiv \mathcal{O}(h_1^2 + h_1 h_2), \qquad x_i \leqslant \xi \leqslant x_{i+2}$$

$$\text{For } f_i'': \quad E(h_1, h_2) = -\frac{f'''(\xi)}{3}(2h_1 + h_2) \equiv \mathcal{O}(2h_1 + h_2)$$

The central and backward-difference formulas for f_i' and f_i'' are similarly obtained by using the shifted coordinates as $x_{i-1} = -h_1$, $x_i = 0$, $x_{i+1} = h_2$ (see Fig. 5.10b) and $x_i = 0$, $x_{i-1} = -h_1$, $x_{i-2} = -(h_1 + h_2)$ (see Fig. 5.10c), respectively. Matching the derivatives of the approximating polynomial, $f_i' \cong p_2'(0) = b$ and $f_i'' = p_2''(0) = 2a$, we find for the central difference formulas:

$$f_i' = -\frac{h_2}{h_1(h_1 + h_2)} f_{i-1} + \left(\frac{1}{h_1} - \frac{1}{h_2}\right) f_i + \frac{h_1}{h_2(h_1 + h_2)} f_{i+1} \tag{5.46}$$

$$f_i'' = \frac{2}{h_1(h_1 + h_2)} f_{i-1} - \frac{2}{h_1 h_2} f_i + \frac{2}{h_2(h_1 + h_2)} f_{i+1} \tag{5.47}$$

The truncation errors for the central difference formulas, Eqs. (5.46) and (5.47), which are derived using the Taylor series in a similar fashion, yield

$$\text{For } f_i': \quad E(h_1, h_2) = -\frac{f'''(\xi)}{6} h_1 h_2 \equiv \mathcal{O}(h_1 h_2) \qquad (x_{i-1} \leqslant \xi \leqslant x_{i+1})$$

$$\text{For } f_i'': \quad E(h_1, h_2) = \frac{f'''(\xi)}{3}(h_1 - h_2) - \frac{f^{(4)}(\xi)}{12}(h_1^2 - h_1 h_2 + h_2^2) + \cdots$$

$$\equiv \mathcal{O}(h_1 - h_2) \quad + \quad \mathcal{O}(h_1^2 - h_1 h_2 + h_2^2)$$

Note that the order of accuracy of f_i'' is $\mathcal{O}(h_1 - h_2)$ for $h_2 \neq h_1$ and $\mathcal{O}(h^2)$ for $h_2 = h_1 = h$.

An alternative central difference formula can be obtained by expanding f_{i+1} and f_{i-1} into Taylor series about x_i and eliminating f_i's. Again, using the shifted coordinate system and denoting $f(x_i) = f_i = f(0)$, $f_{i+1} = f(h_2)$, and $f_{i-1} = f(-h_1)$, we arrive at

$$f_i' = \frac{f_{i+1} - f_{i-1}}{h_1 + h_2} + \frac{1}{2}(h_1 - h_2) f_i'' + \frac{1}{6}\mathcal{O}\left(h_1^2 - h_1 h_2 + h_2^2\right)$$

which behaves nearly quadratic when $h_1 \approx h_2$.

The backward difference formulas are obtained as

$$f_i' = \frac{h_1 + 2h_2}{h_2(h_1 + h_2)} f_i - \frac{h_1 + h_2}{h_1 h_2} f_{i-1} + \frac{h_2}{h_1(h_1 + h_2)} f_{i-2} \tag{5.48}$$

$$f_i'' = \frac{2}{h_2(h_1 + h_2)} f_i - \frac{2}{h_1 h_2} f_{i-1} + \frac{2}{h_1(h_1 + h_2)} f_{i-2} \tag{5.49}$$

Employing the Taylor series analysis to obtain the truncation errors for the backward-difference formulas, Eqs. (5.48) and (5.49), we find

$$\text{For } f_i': \quad E(h_1, h_2) = \frac{f'''(\xi)}{6} h_2(h_1 + h_2) \equiv \mathcal{O}(h_1 h_2 + h_2^2) \qquad (x_{i-2} \leqslant \xi \leqslant x_i)$$

$$\text{For } f_i'': \quad E(h_1, h_2) = \frac{f'''(\xi)}{3}(h_1 + 2h_2) \equiv \mathcal{O}(h_1 + 2h_2)$$

Here, the order of accuracy of f_i'', is first order, whether the data set is uniform or not.

EXAMPLE 5.5: Deriving difference formulas with direct-fit polynomials

Consider the non-uniform discrete distribution depicted in Fig. 5.11. Using a third-order approximating polynomial, derive a forward difference formula for $f''(x_i)$ and $f'''(x_i)$.

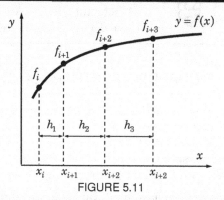

FIGURE 5.11

SOLUTION:

A 3rd-degree direct-fit polynomial is chosen as $p_3(x) = ax^3 + bx^2 + cx + d$. Using a shifted coordinate system, the direct-fit polynomial must satisfy every data point, i.e., $f(x_{i+k}) \cong p(x_{i+k})$ for $k = 0, 1, 2,$ and 3:

For $x_i = 0$, $f_i = d$

For $x_{i+1} = h_1$, $f_{i+1} = ah_1^3 + bh_1^2 + ch_1 + d$

For $x_{i+2} = h_1 + h_2$, $f_{i+2} = a(h_1 + h_2)^3 + b(h_1 + h_2)^2 + c(h_1 + h_2) + d$

For $x_{i+3} = h_1 + h_2 + h_3$, $f_{i+3} = a(h_1 + h_2 + h_3)^3 + b(h_1 + h_2 + h_3)^2$
$$+ c(h_1 + h_2 + h_3) + d$$

which is a system of four linear equations with four undetermined coefficients: a, b, c, and d.

The second derivative at x_i is set to the second and third derivatives of the fitted polynomial: $f_i'' \cong p_3''(0) = 2b$ and $f_i''' \cong p_3'''(0) = 6a$, which give

$$f_i'' = \frac{2(3h_1 + 2h_2 + h_3)f_i}{h_1(h_1 + h_2)(h_1 + h_2 + h_3)} - \frac{2(2h_1 + 2h_2 + h_3)}{h_1 h_2(h_2 + h_3)}f_{i+1} + \frac{2(2h_1 + h_2 + h_3)}{h_2 h_3(h_1 + h_2)}f_{i+2}$$
$$- \frac{2(2h_1 + h_2)f_{i+3}}{h_3(h_2 + h_3)(h_1 + h_2 + h_3)} + \mathcal{O}\left(3h_1^2 + 4h_2 h_1 + 2h_3 h_1 + h_2^2 + h_2 h_3\right)$$

$$f_i''' = -\frac{6f_i}{h_1(h_1 + h_2)(h_1 + h_2 + h_3)} + \frac{6f_{i+1}}{h_1 h_2(h_2 + h_3)} - \frac{6f_{i+2}}{h_2 h_3(h_1 + h_2)}$$
$$+ \frac{6f_{i+3}}{h_3(h_2 + h_3)(h_1 + h_2 + h_3)} + \mathcal{O}\left(3h_1 + 2h_2 + h_3\right)$$

The truncation errors in the above finite difference approximations were obtained from the Taylor series and presented with each formula.

Discussion: Note that when $h_1 \approx h_2 \approx h_3$, the finite difference formula for f_i'' is nearly second-order accurate, even though the derivation is carried out with a third-degree approximating polynomial. For notably different interval sizes, the second-order accuracy will be lost. On the other hand, the order of accuracy of the approximation for f_i''' is first order in that the error is proportional to the sum of the interval sizes from x_i to x_{i+1}, x_{i+2}, and x_{i+3}. The only approximation formula that has third-order accuracy is the formula for f_i'.

The finite difference formulas derived for non-uniformly spaced discrete functions have a lower order of accuracy compared to those derived for uniform discrete functions with the same stencil.

5.7   RICHARDSON EXTRAPOLATION

Frequently, numerical differentiation is applied to explicitly defined functions involving mathematical expressions that are either too long or too complicated. The first-, second-, or higher-order derivatives of such functions will likely be either too long or too complicated as well. In such cases, it may be preferable to compute derivatives numerically rather than dealing with long and complex expressions.

So far, we have used Taylor series or direct-fit polynomials to derive finite difference formulas to estimate any order derivative of a function ($f'(x_i)$, $f''(x_i)$, and so on) to any order of accuracy ($\mathcal{O}(h)$, $\mathcal{O}(h^2)$,...). Also, we have shown that, in most cases, the numerical differentiation is not "exact" and provides only an approximate answer. Nevertheless, the accuracy of a differentiation with finite differences can be improved by either decreasing the step size (i.e., $h \to 0$) or increasing the order of accuracy of the difference formula (i.e., $\mathcal{O}(h^m)$, $m > 2$, which requires more data points).

An alternative technique, referred to as *Richardson extrapolation*, can only be applied to explicitly defined, continuous, and sufficiently differentiable functions. It consists of using two derivative estimates obtained with different step sizes using a low-order finite difference formula to calculate a higher order, more accurate estimate.

To illustrate, consider a sufficiently differentiable continuous function $f(x)$. The central difference approximation of the first derivative, Eq. (5.31), can be expressed as

$$f'(x) = \frac{f(x+h) - f(x-h)}{2h} + K_1 h^2 + K_2 h^4 + K_3 h^6 + \ldots \tag{5.50}$$

where K_1, K_2, K_3, and so on are the coefficients containing higher-order derivatives of $f(x)$.

An approximation for the first derivative (as a function of h) is defined as

$$D_0(h) = \frac{f(x+h) - f(x-h)}{2h} \tag{5.51}$$

which is second-order accurate, $\mathcal{O}(h^2)$. Next, we rewrite Eq. (5.50) as follows:

$$D_0(h) = f'(x) - K_1 h^2 - K_2 h^4 - K_3 h^6 - \ldots \tag{5.52}$$

To increase the order of accuracy of Eq. (5.50), we shall eliminate the $K_1 h^2$ term. To do that, a second estimate for the first derivative will be needed. Equation (5.52) for $h/2$ (halving the interval size) leads to

$$D_0\left(\frac{h}{2}\right) = f'(x) - K_1 \frac{h^2}{4} - K_2 \frac{h^4}{16} - K_3 \frac{h^6}{64} - \ldots \tag{5.53}$$

Now, using Eq. (5.52) and Eq. (5.53), $\mathcal{O}(h^2)$ terms are eliminated to give

$$D_1(h) = \frac{4D_0(h/2) - D_0(h)}{3} = f'(x) + K_2 \frac{h^4}{4} + K_3 \frac{5h^6}{16} + \ldots \tag{5.54}$$

which is $\mathcal{O}(h^4)$ due to being the term with the lowest degree on the rhs.

We can similarly eliminate the term with h^4 in Eq. (5.54) to find another estimate with the order of accuracy $\mathcal{O}(h^6)$. The interval size of Eq. (5.54) is likewise halved to give

$$D_1\left(\frac{h}{2}\right) = f'(x) + K_2\frac{h^4}{64} + K_3\frac{5h^6}{1024} + \ldots \tag{5.55}$$

Next, eliminating K_2h^4 between Eq. (5.54) and (5.55) yields

$$D_2(h) = \frac{4^2 D_1(h/2) - D_1(h)}{4^2 - 1} = f'(x) + K_3\frac{h^6}{64} + \ldots \tag{5.56}$$

which is $\mathcal{O}(h^6)$.

Repeating the elimination procedure in the same manner, successive approximations to the first derivative can be generalized by the following expression:

$$f'(x) \cong D_m(h) = \frac{4^m D_{m-1}(h/2) - D_{m-1}(h)}{4^m - 1} + \mathcal{O}(h^{2m+2}), \quad m = 1, 2, \ldots \tag{5.57}$$

where $D_m(h)$ denotes the extrapolation value at the mth step.

Equation (5.57) can also be used to generate finite difference formulas of high accuracy. However, instead, we can apply a tabular approach using Richardson extrapolation. The extrapolation process can be repeated indefinitely, with each step of the extrapolation leading to a new estimate that is more accurate than the previous one. We start by using a difference formula, such as the centered difference formula in this section, where h is decreased by a factor of 2. So we employ Eq. (5.51) for $h = 1/2^k$ where $k = 0, 1, 2$, and so on. Then we employ the Richardson extrapolation formula, Eq. (5.57), to these values for $m \geqslant 1$.

To obtain an expression for the convergence criterion, we add and subtract $D_{m-1}(h/2)$ in the numerator of Eq. (5.57) to find

$$D_m(h) = D_{m-1}(h/2) + \frac{D_{m-1}(h/2) - D_{m-1}(h)}{4^m - 1} + \mathcal{O}(h^{2m+2}), \tag{5.58}$$

Note that the fractional term on the rhs of Eq. (5.58) is the error estimate for two successive approximations: $|D_m(h) - D_{m-1}(h/2)| < \varepsilon$. Yet this expression does not provide much information on the levels of h. To overcome this problem, Eq. (5.57) is expressed and coded as a double subscript array as follows:

$$D_{k,m} = D_{k,m-1} + \frac{D_{k,m-1} - D_{k-1,m-1}}{4^m - 1}, \quad m > 0 \tag{5.59}$$

where $D_{k,0}$ is the difference formula for the pertinent derivative (in this section evaluated solely by Eq. (5.51)), k denotes the row index (or exponent in $h = 1/2^k$), and m denotes the extrapolation step (column index).

Besides Eq. (5.58), a more conservative convergence criterion can be applied to diagonal elements at row k as follows:

$$|D_{k,k} - D_{k-1,k-1}| < \varepsilon \tag{5.60}$$

 The method can be applied not only to $f'(x)$ but also to forward, backward, and central differences of higher derivatives. In such cases, the suitably generated Richardson interpolation formula $D_0(h)$ must be used.

TABLE 5.7: Richardson extrapolation table for $f'(x)$ using the CDF.

		$m = 0$	$m = 1$	$m = 2$	$m = 3$
k	Step size	$\mathcal{O}(h^2)$	$\mathcal{O}(h^4)$	$\mathcal{O}(h^6)$	$\mathcal{O}(h^8)$
0	h	$D_{0,0} = D_0(h)$			
1	$\dfrac{h}{2}$	$D_{1,0} = D_0\left(\dfrac{h}{2}\right)$	$D_{1,1} = D_1(h)$		
2	$\dfrac{h}{2^2}$	$D_{2,0} = D_0\left(\dfrac{h}{2^2}\right)$	$D_{2,1} = D_1\left(\dfrac{h}{2}\right)$	$D_{2,2} = D_2(h)$	
3	$\dfrac{h}{2^3}$	$D_{3,0} = D_0\left(\dfrac{h}{2^2}\right)$	$D_{3,1} = D_1\left(\dfrac{h}{2^2}\right)$	$D_{3,2} = D_2\left(\dfrac{h}{2}\right)$	$D_{3,3} = D_3(h)$
$\vdots$	$\vdots$	$\vdots$	$\vdots$	$\vdots$	$\vdots$

A few steps of successive Richardson extrapolation of the numerical differentiation are illustrated in Table 5.7. The numerical estimates improve as we move down any column, mainly due to a reduction in the step size. The estimates also improve as we move to the right along the same row as a result of increasing order of accuracy. In the Richardson extrapolation, each improvement made for the forward (or backward) difference formula increases the order of solutions by one, whereas for the central difference formula, each improvement increases the order by two.

Pseudocode 5.2

Module RICHARDSON $(x0, h, \varepsilon, n, \mathbf{D}, deriv)$
\ DESCRIPTION: A pseudomodule to compute the first derivative of an
\   explicitly defined function using the Richardson's extrapolation technique.
\ USES:
\   FUNC:: A user-defined function supplying $f(x)$.
\   ABS:: A built-in function computing the absolute value.
Declare: $D_{0:10,0:10}$ \ Declare maximum table size
$m \leftarrow 0$ \ Initialize extrapolation step, m
$k \leftarrow 0$ \ Initialize interval size exponent, $h/2^k$
$err \leftarrow 1$ \ Initialize error
While $\left[err > \varepsilon\right]$ \ Start of the loop
$\quad D_{k,0} \leftarrow (\textbf{FUNC}(x0 + h) - \textbf{FUNC}(x0 - h))/(2h)$ \ Use CDF for $f'(x_0)$
$\quad$ **For** $\left[m = 1, k\right]$ \ Loop k: Compute estimates for kth row
$\quad\quad D_{k,m} \leftarrow D_{k,m-1} + (D_{k,m-1} - D_{k-1,m-1})/(4^m - 1)$ \ Apply Eq. (5.59)
$\quad$ **End For**
$\quad$ **If** $\left[k \geqslant 1\right]$ **Then**   \ Check for convergence
$\quad\quad err \leftarrow |D_{k,k} - D_{k-1,k-1}|$ \ Find abs. error for Eq. (5.60)
$\quad$ **End If**
$\quad h \leftarrow h/2$ \ Find next step size, $h/2$
$\quad k \leftarrow k + 1$ \ Find next row number, k
End While
$n \leftarrow k - 1$ \ Final size of the table
$deriv \leftarrow D_{n,n}$ \ Set final estimate to $f'(x_0)$
End Module RICHARDSON

Richardson Extrapolation Method

- Richardson's extrapolation is used to generate high-accuracy results while using low-order formulas;
- Successive computation of $D_{k,m}$ is faster and cheaper (due to fewer function evaluations) than using equivalent higher-order finite difference equations;
- It is suitable for adaptive computations, and any derivative (assuming it exists) can be computed within a specified tolerance;
- Round-off errors are less of a concern since relatively large h can be employed.

- The method can be applied only to explicit functions;
- It uses more points involving a wider range than perhaps necessary;
- It is based on the assumption that higher and bounded derivatives of $f(x)$ exist, which introduces vulnerability when this assumption is violated;
- It can be employed for uniform discrete functions only if the number of data points in a uniform set is multiples of 2.

In Pseudocode 5.2, the pseudomodule **RICHARDSON** is given for computing the first derivative of an explicitly defined continuous function using Richardson's extrapolation technique. The module requires the point at which the derivative will be calculated ($x0$), an initial interval size (h), and a tolerance (ε) as input. The output is the size of the Richardson extrapolation table (n), the table of the extrapolations **D**, and the first derivative ($deriv$) that meets the convergence criterion. The function $f(x)$ is supplied to the module by **FUNC**(x), a user-defined external function module. Because the row variable is incremented before the **End While** statement, its value becomes the table size plus one. Thus, the table size is set to $n = k - 1$.

Extrapolation step m, row index k, and the differentiation error (err) are initialized. Richardson's table, Table 5.7, is constructed using a **While**-loop. The first row consists of only $D_{0,0}$. Before going to the top of the loop, the interval size is halved ($h/2$) and the row index is increased by one, $k \leftarrow k + 1$. For the second row, after computing $D_{1,0}$, the extrapolation formula is applied using **For**-loop for as many columns as possible. A conservative criterion for the differentiation error, $|D_{k,k} - D_{k-1,k-1}| < \varepsilon$, is monitored at the end of each row.

5.8 ALTERNATIVE METHODS OF EVALUATING DERIVATIVES

A series of experimental measurements is one of the sources for generating discrete data. Recall that all experimental data also contains measurement errors or noise to a certain degree. The derivatives of such data are a bit tricky since a small error or noise in the measurements may lead to amplified errors or large non-physical fluctuations. These effects become increasingly significant as the order of the derivative increases.

Differentiation of discrete functions with finite differences that are based on polynomial approximations is unsatisfactory because the interval size cannot be reduced arbitrarily. Finite difference formulas of high order of accuracy tend to oscillate and/or amplify the errors (*see* Chapter 6). The greater the degree of the polynomial, the larger the amplifications.

EXAMPLE 5.6: Differentiation with Richardson extrapolation

Under specific conditions, the Van der Waals equation of state for a real gas takes the following form:

$$P(V) = \frac{25000}{V - 57} - \frac{5.2 \times 10^6}{V^2}$$

where V is the volume (cm³), and P is the pressure (bar). Starting with the CDF and $h = 5$ cm³, use Richardson's extrapolation to estimate dP/dV for $V = 150$ cm³ within $\varepsilon = 10^{-3}$ tolerance.

SOLUTION:

To be able to start the extrapolation procedure, we need two estimates of dP/dV with the CDF using $h = 5$ and $h = 2.5$ cm³:

$$D_{0,0} = \frac{P(150 + 5) - P(150 - 5)}{2(5)} = \frac{38.6608 - 36.7663}{2(5)} = 0.189454$$

$$D_{1,0} = \frac{P(150 + 2.5) - P(150 - 2.5)}{2(2.5)} = \frac{38.1843 - 37.2313}{5} = 0.190596$$

Using the estimates, we obtain an improved estimate by applying the (first-step) Richardson extrapolation as

$$D_{1,1} = \frac{4D_{1,0} - D_{0,0}}{3} = \frac{4(0.190596) - 0.189454}{3} = 0.190977$$

The estimated error at the end of this step is $|D_{1,1} - D_{0,0}| = 0.01523 > 10^{-3}$. To obtain an improved second step estimate, dP/dV is calculated with $h = 1.25$ cm³:

$$D_{2,0} = \frac{P(150 + 1.25) - P(150 - 1.25)}{2(1.25)} = \frac{37.9451 - 37.4679}{2.5} = 0.190880$$

Using $D_{0,1}$ and $D_{0,2}$, the first step extrapolation value $D_{1,2}$ is obtained as follows:

$$D_{2,1} = \frac{4D_{2,0} - D_{1,0}}{3} = \frac{4(0.190880) - 0.190596}{3} = 0.190975$$

Now that $D_{1,1}$ and $D_{1,2}$ are available, the second step Richardson extrapolation value can be calculated as

$$D_{2,2} = \frac{16D_{2,1} - D_{1,1}}{15} = \frac{16(0.190975) - 0.190977}{15} = 0.190974 \frac{\text{bar}}{\text{cm}^3}$$

The error estimate at the end of the second step is $|D_{2,2} - D_{1,1}| = 3.2 \times 10^{-6} < 10^{-3}$, which is sufficient to terminate the computation at this stage. The true derivative is found to be 0.19097391.

Discussion: The true error for the computed derivative is 9×10^{-8}, which is in fact much less than 10^{-3} or $3.2 \times 10^{-6} < 10^{-3}$.

The technique is a convenient and effective way to improve accuracy without much computational effort. Similar hierarchies can be constructed for forward, backward, and central differences for the first-, second-, and higher-order derivatives. It should also be pointed out that Richardson's extrapolation process can be used not only to improve the order of a given numerical differentiation formula but also to find the given basic numerical differentiation formulas.

An alternative to differentiation with finite differences is to fit the data to an appropriate mathematical model and then differentiate it analytically. Curve fitting and its applications, covered in Chapter 7, smooth out the noise or measurement errors to some extent. Curve fitting the data to a polynomial is equally *not* recommended unless the finite difference equations are equivalent to polynomial interpolation. The coefficients of finite difference formulas (e.g., Eqs. (5.44)-(5.49)) for non-uniform data are functions of interval size (h_1, h_2, ...), and hence the finite differences of such functions are computationally even more involved. That is why differentiation after curve fitting is very useful, especially for non-uniform data, as it provides a great deal of savings in cpu-time as well.

Cubic spline interpolation (covered in Chapter 6) makes use of nearby data points. One of the most important features of cubic spline interpolation is that it provides a smoothly varying interpolant as well as its first and second derivatives (*see* Section 6.4). That is why splines are better models, and the two-step spline approach requires (1) interpolation of the data by a spline and (2) differentiation of the splines. Unlike other methods, this approach allows computation of the first and second derivatives at locations other than the original data points.

5.9 CLOSURE

The goal of numerical differentiation is to develop formulas to estimate the derivatives of functions about a base point. The numerical differentiation formulas are based on approximating polynomials. Taylor polynomials, as well as direct-fit, Lagrange, and divided difference polynomials, work well for both uniformly and non-uniformly spaced data.

The order of accuracy of a particular derivative depends on the number of function points used. Using the polynomials, it is possible to derive difference formulas with varying orders of accuracy, i.e., $\mathcal{O}(h)$ with two points, $\mathcal{O}(h^2)$ with three points, and so on.

The accuracy of any numerical differentiation scheme can be increased by either (i) reducing the interval size (h or h_i's) as much as possible or (ii) using finite difference formulas with an order of accuracy of $\mathcal{O}(h^2)$, $\mathcal{O}(h^4)$, or higher if the number of data points in a distribution is sufficient.

(1) When dealing with uniformly or non-uniformly spaced discrete functions in $[a, b]$, the numerical differentiation can be carried out by forward, backward, or central difference formulas. Deriving difference formulas with direct-fit polynomials is the easiest avenue, given the functional points you want to use and the degree of accuracy you desire. In this case, we do not have the freedom to reduce the interval size because either experimentally collected or computed discrete data sets and their interval sizes are invariant. The best option we have is to use different formulas with a higher order of accuracy.

(2) An explicitly defined function can be differentiated exactly. However, when these functions consist of very long or complex expressions, higher derivatives also yield very long and complex expressions. In such cases, numerical differentiation may be preferred due to its computational speed and simplicity. The value of h can be varied by the analyst, and there is an optimum value for which one can reduce h without increasing the effect of round-off errors. In this respect, the Richardson extrapolation technique can also be applied to explicitly defined functions. It is a powerful tool for improving the accuracy of a finite-difference formula. This technique allows us to achieve greater precision with minimum effort (fewer function evaluations) in comparison to other methods. The underlying idea is of critical importance in numerical analysis, and it is often used in integrations, extrapolation, and so on.

5.10 EXERCISES

Section 5.1 Basic Concepts

E5.1 Use the definition of limit to find $f'(x)$ for the functions given below.

$$\text{(a) } f(x) = x^3, \qquad \text{(b) } f(x) = \sin x, \qquad \text{(c) } f(x) = e^{2x}$$

E5.2 For functions in **E5.1**, find approximations for $f'(x)$ for an arbitrary x using Eq. (5.4).

E5.3 For approximations in **E5.2**, identify the leading error term.

E5.4 For approximations in **E5.2**, identify the truncation error as well as its upper/lower bounds.

E5.5 Apply the first-order forward difference formula to the functions given below to (i) estimate $f'(1)$ and true errors by using $h = 10^{-m}$ where $m = 1, 2, \ldots, 20$, and (ii) interpret your results.

$$\text{(a) } f(x) = \frac{x^3}{1 + x^2}, \qquad \text{(b) } f(x) = \ln x, \qquad \text{(c) } f(x) = e^{x-1}$$

E5.6 Derive an expression for the optimum interval size for the following difference formulas:

$$\text{(a) } f'(x) \cong \frac{f(x + h) - f(x - h)}{2h}, \qquad \text{(b) } f''(x) \cong \frac{f(x + h) - 2f(x) + f(x - h)}{h^2}$$

Section 5.2 First-Order Finite Difference Formulas

E5.7 Derive a general difference formula for $\Delta^n f_i$.

E5.8 Derive a first-order forward difference formula for $d^5 f_i / dx^5$.

E5.9 Using the first-order forward and backward difference formulas, (a) estimate the first derivative for $f(x) = x^3$ at $x_0 = 2$ with $h = 0.25, 0.1, 0.025$, and 0.01, (b) compare the estimates and truncation errors with those of the true derivatives and errors, and discuss your results.

E5.10 Repeat **E5.9** for $f''(x_0)$.

E5.11 Using the first-order forward difference formulas, (a) estimate $f'(1)$ and $f''(1)$ for $f(x) = \ln^2(x + 1)$ with $h = 0.1, 0.05, 0.025$, and 0.01, (b) compare your estimates with those of the true values, and discuss your results.

E5.12 Repeat **E5.5** for $f''(1)$ with $h = 10^{-1}, 10^{-2}, \ldots, 10^{-8}$, and interpret your results.

Section 5.3 Second-Order Finite Difference Formulas

E5.13 (a) Use $h = 0.1$ and the second-order accurate forward and backward difference formulas to estimate $f'(1)$, where $f(x) = e^{2x}$; (b) compare your estimates and truncation errors with those of the true values and discuss your results.

E5.14 Using the second-order accurate finite difference formulas, investigate the existence of the first derivative of the given discrete function at $x = 1$. *Hint*: The first derivative of a continuous function at a point exists if its left- and right-hand derivatives are equal.

x_i	0	0.5	1	1.5	2
$f(x_i)$	-2	-1.25	3	13.75	34

E5.15 Using the second-order accurate finite difference formulas, investigate the existence of the first derivative of the given discrete function at $x = 2$.

x_i	1.5	1.75	2	2.25	2.5
$f(x_i)$	0.879	1.352	2	1.414	0

E5.16 Using the Taylor series expansion, determine (a) which derivatives the following difference formulas correspond to; (b) the truncation error term; and (c) the order of accuracy, i.e., $\mathcal{O}(h^?)$. *Hint*: Expand $f_{i\pm k}$ in the difference formulas into a Taylor series around x_i.

(i) $\dfrac{f_{i+3} - f_{i-2}}{5h}$,

(ii) $\dfrac{f_{i-2} - 4f_{i-1} + 6f_i - 4f_{i+1} + f_{i+2}}{h^4}$,

(iii) $\dfrac{f_{i-3} - 6f_{i-2} + 12f_{i-1} - 10f_i + 3f_{i+1}}{2h^3}$,

(iv) $\dfrac{-3f_{i-2} + 20f_i - 30f_{i+1} + 15f_{i+2} - 2f_{i+3}}{10h^3}$

(v) $\dfrac{f_{i-2} - 8f_i + 15f_{i+1} - 11f_{i+2} + 3f_{i+3}}{h^2}$

E5.17 Even-order derivatives of a function $f(x)$ are zero for all x, i.e., $f''(x) = f^{(4)}(x) = f^{(6)}(x) = \ldots = 0$. Using the Taylor series expansion, determine (a) which derivative the following difference formula corresponds to, (b) the truncation error term, and (c) the order of accuracy.

$$\frac{16f_{i+3} - 60f_{i+2} + 72f_{i+1} - 28f_i}{4h^3} = ? \quad \mathcal{O}(h^?)$$

E5.18 Use the Taylor series expansions to derive the truncation error for the following difference formulas and list them from the highest order of accuracy to the lowest.

(a) $\dfrac{5f_{i-2} - 3f_{i-1} - 10f_i + 3f_{i+1} + 5f_{i+2}}{6h}$,

(c) $\dfrac{-f_{i-3} + 3f_{i-2} - 5f_{i-1} + 5f_{i+1} - 3f_{i+2} + f_{i+3}}{4h}$,

(b) $\dfrac{f_{i-3} - 3f_{i-2} + 5f_{i-1} - 10f_i + 7f_{i+1}}{5h}$

(d) $\dfrac{-f_{i-2} + 2f_{i-1} - 2f_{i+1} + f_{i+2}}{2h^3}$

E5.19 Using the Taylor series, determine the truncation error and order of accuracy of the following finite difference formulas for f_i'.

(a) $f_i' = \dfrac{2f_{i+3} - 9f_{i+2} + 18f_{i+1} - 11f_i}{6h}$,

(b) $f_i' = \dfrac{-3f_{i+4} + 16f_{i+3} - 36f_{i+2} + 48f_{i+1} - 25f_i}{12h}$

E5.20 Using the Taylor series, determine the truncation error and order of accuracy of the following finite difference formulas:

(a) $f_i'' = \dfrac{1}{12h^2} \left\{ \begin{array}{l} 11f_{i+4} - 56f_{i+3} + 114f_{i+2} \\ -104f_{i+1} + 35f_i \end{array} \right\}$,

(b) $f_i''' = \dfrac{1}{4h^3} \left\{ \begin{array}{l} 7f_{i+5} - 41f_{i+4} + 98f_{i+3} \\ -118f_{i+2} + 71f_{i+1} - 17f_i \end{array} \right\}$

E5.21 The left- and right-biased difference formulas for the first derivative are defined as $(2(f_i')^- + (f_i')^+)/3$ and $((f_i')^- + 2(f_i')^+)/3$, respectively. For the given discrete data below, estimate $f'(1.5)$ using (a) the first- and second-order accurate forward and backward difference formulas, (b) the left- and right-biased formulas, and (c) compare the accuracy of the estimates in Parts (a) and (b) if the true derivative is given as 1.4375.

x_i	1.3	1.4	1.5	1.6	1.7	1.8
$f(x_i)$	2.53	2.35	2.41	2.61	2.9	3.24

Section 5.4 Central Difference Formulas

E5.22 The sum of the coefficients in any difference formula is zero. Explain why.

E5.23 Expand the following finite-difference operations to obtain explicit forms:

(a) $\Delta \nabla f_{i-1}$, (b) $\Delta \delta^2 f_{i+1}$, (c) $\Delta \delta \nabla f_{3/2}$

E5.24 Use Eq. (5.37) to find a second-order central difference formula for $d^5 f_i/dx^5$.

E5.25 For an arbitrary x and h and $f(x) = x^2$, obtain the central difference formulas for $f'(x)$ and $f''(x)$. Comment on your findings.

E5.26 Repeat **E5.25** for $f(x) = x^3$, and discuss the implications of your findings.

E5.27 Find the truncation error for **E5.25**.

E5.28 Find the truncation error for **E5.26**.

E5.29 Repeat **E5.5** using the CDFs with $h = 10^{-1}, 10^{-2}, \ldots, 10^{-12}$.

E5.30 Repeat **E5.12** using the CDFs with $h = 10^{-1}, 10^{-2}, \ldots, 10^{-8}$.

E5.31 Derive a third-order accurate finite difference formula for $f'(x)$ that uses the following points: $f(x - 2h)$, $f(x - h)$, $f(x)$, and $f(x + h)$.

E5.32 Given x_0 and h, estimate $f'(x_0)$ for the following functions using the central difference formulas of order of accuracy of $\mathcal{O}(h^2)$ and $\mathcal{O}(h^4)$.

(a) $f(x) = 2x^3 - x^2 + 4x - 3$, $x_0 = 1$, $h = 0.1$ (b) $f(x) = 3x^4 + x^3 + x$, $x_0 = 2$, $h = 0.05$
(c) $f(x) = x^2 e^{-2x}$, $x_0 = 0.1$, $h = 0.01$ (d) $f(x) = x \tan 2x$, $x_0 = \pi$, $h = 0.02$

E5.33 For given $f(x) = xe^{-x}$, (a) estimate $f'(0.7)$ using the second-order accurate forward, backward, and central difference formulas with $h = 0.2, 0.1, 0.05$, and 0.025; (b) compare your estimates with the true value.

E5.34 For given $f(x) = x^3$, (a) estimate $f'(2)$ using the central difference formula with $h = 0.25$, $0.1, 0.025$, and 0.010; (b) compare your estimates with the true value.

E5.35 Repeat **E5.34** for $f''(2)$.

E5.36 For given $f(x) = e^{2x}$ (a) estimate $f'(1)$ using three-point forward, backward, and central difference formulas, i.e., $\mathcal{O}(0.1^2)$; (b) compare your results with the true value.

E5.37 For the given discrete function, estimate $f'(0)$, $f'(2)$, $f'(4)$, $f''(0)$, and $f''(4)$ using suitable second-order accurate difference formulas.

x_i	0	1	2	3	4
y_i	30	36	30	12	−18

E5.38 A 20-second recording of a vehicle's motion is given below with 2.5-second intervals. Using the finite difference formulas of order of accuracy $\mathcal{O}(h^2)$, estimate the acceleration for each data point listed.

x_i	0	2.5	5	7.5	10	12.5	15	17.5	20
$f(x_i)$	0	11	19	28	37	40	42	35	16

E5.39 Shear stress, τ (N/m^2), due to the flow of a fluid over a solid surface, is defined by $\tau = \mu(du/dy)_{y=0}$, where μ is the dynamic viscosity (N-s/m^2), u is the velocity component (m/s) parallel to the surface, and y (mm) is the distance normal to the surface (*see* **Fig. E5.39**). Velocity has been measured with an LDV (Laser Doppler velocimetry), and the results are tabulated below. Use the finite difference formulas of order of accuracy $\mathcal{O}(h)$, $\mathcal{O}(h^2)$, $\mathcal{O}(h^3)$, and $\mathcal{O}(h^4)$ to estimate the shear stress for the given case and comment on the accuracy of your findings. The dynamic viscosity of the fluid is given as $\mu = 7.5 \times 10^{-4}$ N-s/m^2.

x_i (mm)	0	1.5	3	4.5	6	7.5
y_i (mm)	0	3.6	6.4	8.76	10.56	12.2

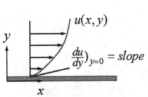

Fig. E5.39

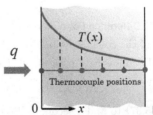

Fig. E5.40

E5.40 Heat applied to one side of a plate with a conductivity of $k = 50$ W/m-K results in a steady temperature distribution. The temperature along the wall is measured by a series of thermocouples, as shown in **Fig. E5.40**. The thermocouple position x (mm) and measured temperature T ($^\circ$C) are tabulated below. The heat flux in the plate, q/A (W/m^2), is related to the plate conductivity and the derivative of temperature as $q/A = -k\, dT/dx$. Estimate the heat flux at the thermocouple locations using the second-order accurate finite difference formulas.

x (mm)	0	5	10	15	20	25
T ($^\circ$C)	90	86	82	78	74	70

E5.41 Kinematic viscosity versus temperature data for a biodiesel is tabulated below. Compute $d\nu/dT$ at $T = 30°\mathrm{C}$ as accurately as possible using the second-order accurate (a) central difference; (b) forward difference; (c) backward difference formulas; (d) are all estimates obtained for the derivative consistent? Explain why.

T (°C)	10	20	30	40	50	60	70
ν (m^2/s)	5.4	4.2	3.3	2.7	2.3	1.9	1.6

E5.42 The data below depicts transient core temperature readings of a solidifying melt. (a) Is the number of data points sufficient to determine a steady cooling rate in °C per minute? (b) If yes, estimate the solidification rate.

t (min)	0	2	4	5	8	10	12
T (°C)	1575	1217	1019	903	880	860	840

E5.43 Newton's law of cooling is given as

$$\frac{dT}{dt} = -\frac{1}{\tau_t}(T(t) - T_\infty)$$

where $T(t)$ is the temperature of the object, T_∞ is the ambient temperature, and τ_t is the thermal time constant (in seconds). If the ambient temperature is $10°\mathrm{C}$, estimate the thermal time constant for the solidification of the melt given in **E5.42**.

E5.44 The monthly average of daylight hours (L) for a town is tabulated below. Use the second-order accurate finite difference formulas to determine the months at which the rate of change in daylight hours is at its maximum or minimum.

Month	Jan	Feb	Mar	Apr	May	Jun	Jul	Aug	Sep	Oct	Nov	Dec
L (hr)	9.35	10.42	12.07	13.19	14.32	15.05	14.47	13.46	12.24	11.02	9.50	9.17

E5.45 In an isobaric process, the heat capacity of gases is defined as the rate of change of enthalpy h (J/kg) with respect to temperature T (K), i.e., $c_p = (\partial h/\partial T)_p$. An experimental work yields the $h - T$ data furnished below. For the tabulated temperature range, estimate the heat capacity of the gas at every available temperature point with an order of accuracy of $\mathcal{O}\left((\Delta T)^2\right)$.

T (K)	250	300	350	400	450	500
h (J/kg)	19.75	28.61	39.11	51.25	65.04	80.45

E5.46 A uniform 1 m-long beam, simply supported at both ends, is subjected to a bending moment given by $M(x) = EI d^2y/dx^2$, where $y(x)$ is the beam deflection (m), $M(x)$ is the bending moment (N·m), E is the elasticity modulus (Pa), and I is the inertia moment (m^4). Use second-order-accurate finite difference formulas to estimate the bending moment at each available data point for a beam whose deflection is tabulated below. *Given:* $EI = 200$ N-m^2.

x_i (m)	0	0.125	0.250	0.375	0.500	0.625	0.750	0.875	1
y_i (m)	0	0.117	0.206	0.263	0.285	0.271	0.220	0.130	0

E5.47 A numerical simulation for a thick-walled cylindrical pressure vessel yields the tabulated data for the radial displacement. The relationships between the radial strain, radial and tangential stresses are given by:

$$\varepsilon_r = \frac{du}{dr}, \qquad \sigma_r = \frac{E}{1 - \nu^2}\left(\frac{du}{dr} + \nu\frac{u}{r}\right), \qquad \sigma_\theta = \frac{E}{1 - \nu^2}\left(\nu\frac{du}{dr} + \frac{u}{r}\right)$$

where $E = 200$ GPa is the modulus of elasticity and $\nu = 0.3$ is the Poisson ratio. Using second-order accurate finite difference formulas, estimate the radial strain, radial, and tangential stresses at each available radial location. Make sure you are using the correct units in your calculations.

x_i (m)	0.600	0.625	0.650	0.675	0.700	0.725	0.750	0.775	0.800
u (mm)	1.22	1.33	1.46	1.60	1.75	1.9	2.06	2.22	2.30

E5.48 In an isentropic process, where there is no entropy change, the following relationship between T (in K) and v (in m^3/kg) holds: $(\partial T/\partial P) = (T/C_p)(\partial v/\partial T)_P$. Using the given ideal gas data below and the second-order difference formulas, estimate $\partial T/\partial P$ at each available temperature point.

T (°C)	10	15	20	25	30	35	40	45	50
v (m^3/kg)	1	0.966	0.934	0.904	0.876	0.85	0.825	0.802	0.78
C_p(kJ/kg.K)	1.7	1.73	1.76	1.8	1.83	1.87	1.9	1.96	1.98

E5.49 The load deviation of a disc spring is estimated both theoretically and experimentally. The variation of the measured and computed load with the deviation is tabulated below. Use the second-order accurate finite difference equations, $\mathcal{O}(h^2)$, to estimate the rate of change in the load (dF/ds) for each theoretical and experimental data point listed.

s (mm)	0	0.2	0.4	0.6	0.8	1	1.2	1.4
F_{theo} (N)	50	1500	2400	3200	4000	4700	5350	6000
F_{theo} (N)	150	1000	2250	3175	3950	4850	6030	8000

E5.50 The volumetric expansion coefficient is defined by $\beta = -(1/\rho)(d\rho/dT)$. The density of a water-based solution as a function of temperature is available in 5 K increments from 290 to 335 K. Estimate the volumetric expansion coefficient for every available temperature data using finite difference equations of order of accuracy of $\mathcal{O}(h^2)$.

T (K)	290	295	300	305	310	315	320	325	330	335
ρ (kg/m^3)	999	998	997	995	993	991	989	987	984	981

E5.51 Experimental data to determine the variation of soil temperature (T) with depth (z) is collected at the same time of day from three locations considered dry, medium, and wet. The conductivities of the soil are $k = 0.4$, 1.1, and 1.6 W/m-K, respectively, at dry, medium, and wet soil locations. The heat loss from the soil to the surrounding air is governed by Fourier's law, defined as $q = -k\,dT/dz$, where q is the heat loss per unit area, k is the conductivity of soil, and dT/dz is the rate of change of soil temperature with depth. Use the furnished data below to estimate the rate of heat loss per square meter from each location, $-k(dT/dz)_{z=0}$, using the finite difference formulas of order of accuracy $\mathcal{O}(h)$, $\mathcal{O}(h^2)$, $\mathcal{O}(h^3)$, and $\mathcal{O}(h^4)$.

	T (°C)		
z (m)	Dry	Medium	Wet
0	5.8	5.5	4.9
1	8.7	7.6	7.1
2	10.3	9.2	9.5
3	11	10.4	12
4	11.3	10.9	14.5
5	11.4	11.3	14.6

E5.52 Spectral emissive power $(E_\lambda, \text{W/m}^2\mu\text{m})$ is defined as the amount of radiation energy emitted by a blackbody at an absolute temperature T per unit time, per unit surface area, and per unit wavelength about wavelength λ. The change in the spectral emissive power of a black body at 500 K with the wavelength $(\lambda, \mu\text{m})$ is tabulated below. Calculate the $dE_\lambda/d\lambda$ and $d^2E_\lambda/d\lambda^2$ derivatives for every available data points using finite difference equations of order of accuracy $\mathcal{O}(h^2)$.

λ (μm)	2	10	18	26	34	42	50
E_λ (W/m$^2\mu$m)	6.6	223	50	15.6	6.2	2.9	1.5

E5.53 The change in the spectral emissive power E_λ of a black body of 10 μm wavelength with the temperature (T, K) is tabulated below. Estimate dE_λ/dT and $d^2 E_\lambda/dT^2$ derivatives for each temperature value listed, using the finite differences of order of accuracy $\mathcal{O}(h^2)$.

T (K)	300	350	400	450	500	550	600
E_λ (W/m$^2\mu$m)	31	63	105	160	223	295	374

E5.54 The performance of an engine at 30% acceleration level is tested. The torque (T) is determined as a function of angular speed (N), and the results of the test runs are presented below. Estimate the dT/dN and d^2T/dN^2 derivatives for each angular speed value listed using the finite difference equations of order of accuracy $\mathcal{O}(h^2)$.

N (rpm)	1300	1800	2300	2800	3300	3800	4300
T (N·m)	133	141	147	153	152	150	155

E5.55 A circular disc is rotating about its axis. The angular position of a point on the outer ring of a disc, $\theta(t)$, is tabulated below as a function of time. For each data value listed, estimate the disk's angular speed, $\omega(t) = d\theta/dt$, and the angular acceleration, $\alpha(t) = d\omega/dt$, using the finite difference equations of order of accuracy $\mathcal{O}(h^2)$.

t (s)	0	0.5	1	1.5	2	2.5	3
θ (rad)	-1	-1.036	-0.572	0.392	1.855	3.819	6.283

E5.56 In a circuit, voltage $(V(t))$ is defined by Kirchoff's first law as $V(t) = L(di/dt) + R\,i(t)$, where L is the inductance, R is the resistance, and $i(t)$ is the current varying with time. The resistance and inductance are fixed at $R = 0.15$ Ω and $L = 0.95$ H. When the voltage is applied, the current is measured at 0.2-second intervals. Use the reported data below to estimate the voltage corresponding to every current listed, using the finite differences of order of accuracy $\mathcal{O}(h^2)$.

t (s)	0	0.2	0.4	0.6	0.8	1
i (A)	3	3.18	3.32	3.44	3.52	3.58

Section 5.5 Finite Differences and Direct-Fit Polynomials

E5.57 (a) Use the direct-fit polynomials technique to validate the following forward difference equation; (b) apply the Taylor series to show that the order of the difference equation is $\mathcal{O}(h^3)$.

$$f_i' \cong \frac{-11f_i + 18f_{i+1} - 9f_{i+2} + 2f_{i+3}}{6h}$$

E5.58 Find a finite difference formula for $f'(x)$ that has a truncation error of order $\mathcal{O}(h^3)$ and uses the points $f(x - 2h)$, $f(x - h)$, $f(x)$, and $f(x + h)$.

E5.59 The function $f(x)$ is defined at points f_i, f_{i+1}, and f_{i+2}, where $x_{i+1} - x_i = h$ and $x_{i+2} - x_{i+1} = 3h$. For $f'(x_i)$ and $f''(x_i)$, develop forward difference formulas of order of accuracy $\mathcal{O}(h^2)$.

E5.60 The function $f(x)$ is given as a non-uniform discrete function with a constant stretching ratio $r = h_{i+1}/h_i$ ($r > 1$), as shown in **Fig. E5.60**. For $f'(x_i)$ and $f''(x_i)$, develop forward difference formulas of order of accuracy $\mathcal{O}(h^2)$.

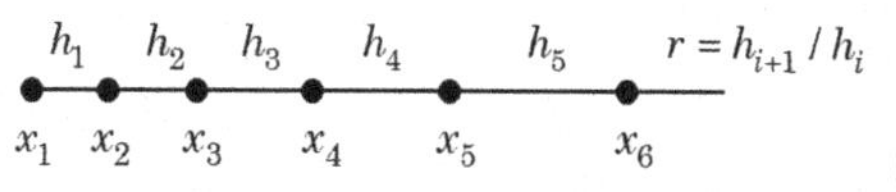

Fig. E5.60

Section 5.6 Differentiating Non-uniformly Spaced Discrete Functions

E5.61 Consider a non-uniformly spaced discrete function, $f(x)$. Employ the direct-fit polynomials to derive second-order accurate backward and central difference formulas for $f'(x_i)$ and $f''(x_i)$. Given: $x_i - x_{i-1} = h_1$, $x_{i+1} - x_i = h_2$, and $x_{i-1} - x_{i-2} = h_2$.

E5.62 A non-uniform discrete function is tabulated below. Estimate $f'(x_i)$ and $f''(x_i)$ for each data point listed using second-order accurate finite difference formulas.

x_i	0	0.5	0.9	1.1	1.5	1.8
$f(x_i)$	1.12	0.69	0.37	0.26	0.25	0.47

E5.63 A non-uniform discrete function is tabulated below. Estimate $f'(x_i)$ and $f''(x_i)$ for each data point listed using the second-order accurate finite difference formulas.

x_i	1.2	1.5	1.9	2.3	2.4	2.8
f_i	2.577	2.597	2.625	2.653	2.66	2.69

E5.64 The process of "icing water" (ice thickness) in a very large container is studied. The container is kept in a freezer maintained at a steady $-12°C$, and ice formation (thickness, h in mm) is recorded for over a period of 10 hours at different time intervals. Use the given data below to estimate the rate of icing (dh/dt, mm/min) by using the second-order accurate finite difference formulas.

t (min)	0	30	180	220	420	540	600
h (mm)	0	2	5	5.6	7.7	8.8	10.1

E5.65 The effect of plastic pellet additives on the strength of concrete is investigated. The change in the compression strength is measured as a function of the volumetric ratio of the pellets, x (%). For the given data below, estimate the rate of change in the compression strength with pellet ratio (dP/dx) for every data point listed. Use second-order-accurate finite difference formulas.

x (%)	0	3	7	10	15	25	35
P (kPa)	88	73	58	45	29	22	18

E5.66 The data on the annual steel production (in tons) of a plant since 2008 is presented below. Estimate the rate of steel production (dm/dt) using second-order-accurate finite difference formulas.

t	2008	2010	2011	2014	2016	2018
m	2280	2560	2660	2990	3110	3150

E5.67 For a reaction whose rate is rapid enough to achieve dynamic equilibrium, the *van't Hoff equation* is expressed as

$$\frac{d(\ln K_{eq})}{d(1/T)} = -\frac{\Delta H^0}{R}$$

where $R = 8.314$ J/K.mol is the ideal gas constant, ΔH^0 is the reaction standard enthalpy (J), K_{eq} is the reaction equilibrium constant, and T is the reaction temperature (in K). Experimental data on the temperature versus equilibrium constant of a chemical reaction is tabulated below. Using the finite differences, estimate an approximate value for ΔH^0. *Hint*: The van't Hoff relation gives the slope of a straight line when it is plotted as $1/T$ versus $\ln(K_{eq})$.

T (°C)	39.9	40.2	52.9	64	77.6	79.5
K_{eq} ($\times 10^5$)	0.2721	0.279	0.750	1.675	4.186	4.7265

Section 5.7 Richardson Extrapolation

E5.68 Apply the Richardson extrapolation to estimate $f'(x_0)$ accurately to five decimal places for the given functions below. Use the CDF with a starting value of $h = 0.1$. How accurate is the estimated value?

(a) $f(x) = \sqrt{x}$, $x_0 = 1$,   (b) $f(x) = (x^2 + 1)\sin x$, $x_0 = 0$
(c) $f(x) = 1/(5 - x^2)$, $x_0 = 1$,   (d) $f(x) = \ln(1 + \sqrt{x + 1})$, $x_0 = 0$

E5.69 Repeat **E5.68** to estimate $f''(x_0)$.

E5.70 Develop Richardson extrapolation steps, starting with the first-order-accurate forward difference formula, and write a program to adaptively evaluate the derivative of a function with an accuracy of $|D_{k,k} - D_{k-1,k-1}| < \varepsilon$.

E5.71 Consider the problems given in **E5.68**. Use the Richardson extrapolation with the forward difference formula as the starter and apply the algorithm developed in **E5.70**) to estimate the derivatives at the specified points. How accurate will the derivative be after three steps of Richardson extrapolation (i.e., how accurate is $D_{3,3}$?).

E5.72 Can you apply Richardson's extrapolation to the following discrete function to estimate $f'(3)$ using central difference formula? If yes, find the derivative as accurately as possible.

x_i	2.8	2.9	3	3.1	3.2
$f(x_i)$	4.883	4.941	5	5.061	5.124

5.11 COMPUTER ASSIGNMENTS

CA5.1 Consider the following central-difference formula:

$$f'_i \cong \frac{-f_{i+2} + 8f_{i+1} - 8f_{i-1} + f_{i-2}}{12h}$$

Using the Taylor series, (a) find the truncation error for the difference formula, (b) obtain an expression for the optimum spacing for which the total error is minimum, and (c) write a program to compute $f'(0)$, its truncation, and true errors for $f(x) = e^x$ with $h = 10^{-1}, 10^{-2}, \dots, 10^{-10}$.

CA5.2 Transient data for the concentration of the product of a chemical reaction is presented below. An nth-order law for a single reactant is described by the following differential equation: $dC/dt = -kC^n(t)$, where $k = 0.54 \, \mathrm{M}^{1-n}/\min$ is the reaction rate, and n is the order of reaction. Use second-order accurate differences to determine the order of the reaction.

t (min)	0	20	40	60	80	100	120
C (M)	0.55	0.067	0.041	0.029	0.022	0.018	0.015

CA5.3 A tube 0.2-m in diameter with a conductivity of $k = 10 \, \mathrm{W/m\text{-}^\circ C}$ is used to transport an exothermic solution (*see* **Fig. CA5.3**). A steady-state temperature reading along the radial position is in the table below. The heat conduction equation for the tube can be written as

$$\frac{1}{r}\frac{d}{dr}\left(r\frac{dT}{dr}\right) + \frac{\dot{q}_0}{k} = 0, \quad 0 \leqslant r \leqslant R$$

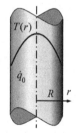

where $\dot{q}_0$ is the constant heat generation (W/m³), and R is the radius of the tube (m). Use second-order-accurate finite difference formulas to estimate the amount of heat generation.

Fig. CA5.3

r (m)	0	0.01	0.02	0.03	0.04	0.05	0.06	0.07	0.08	0.09	0.1
T (°C)	215	214	211	206	199	190	179	166	151	134	115

CA5.4 Consider $f(x) = \cos 20\pi x$. (a) Estimate $f'(0.025)$ using first- and second-order accurate, $\mathcal{O}(h)$ and $\mathcal{O}(h^2)$, forward-difference formulas with $h = 0.4, 0.2, 0.1,$ and 0.05; (b) Compare your estimates with the true answer and discuss the accuracy; (c) Plot the function for $0 \leqslant x \leqslant 0.5$, and using this plot, discuss the implications of the results obtained in Part (a); (d) Reduce the values of h to determine an optimum value h; (e) Using this optimum value and second-order-accurate difference formulas, generate a table of the first and second derivatives of the given function for a $0 \leqslant x \leqslant 0.5$ interval with increments of 0.025.

CA5.5 Consider a non-uniform discrete function containing a set of n data points. (a) Write a general computer program to compute $f'(x_i)$ and $f''(x_i)$ derivatives at every available data point; (b) using this program, evaluate the derivatives of the tabulated discrete function.

x_i	0	0.2	0.5	0.6	0.7	0.9	1	1.05	1.08	1.1
$f(x_i)$	1	1.004	1.25	1.7465	2.8824	9.503	17	22.44	26.39	29.345

CA5.6 Consider the following table of data: (a) Use the computer program prepared in **CA 5.5** to compute f'_i for every data point listed; (b) estimate df_i/dx using df_i/dy, which may be estimated using the difference formulas of uniform distributions. Note that the data is uniformly distributed with respect to the y-variable, where $x_i = y_i^2$ and $y = 1, 1.1, 1.2, \ldots, 2, 2.1$.

x_i	$f(x_i)$	x_i	$f(x_i)$	x_i	$f(x_i)$
1.0	1	1.96	14.75789	3.24	110.19961
1.21	2.143589	2.25	25.62891	3.61	169.8363
1.44	4.299817	2.56	42.949673	4.0	256
1.69	8.157307	2.89	69.757574	4.41	378.22859

CA5.7 A digitized outline of an axisymmetric jug in xy-plane as $y = f(x)$, is presented in the table below. Write a computer program (or use a spreadsheet) that uses any arbitrary data to calculate the curvature $\kappa(x_i) = |f''(x_i)| / (1 + [f'(x_i)]^2)^{3/2}$ of a discrete function at every listed data point x_i.

x_i	$f(x_i)$	x_i	$f(x_i)$	x_i	$f(x_i)$	x_i	$f(x_i)$
0	0	0.07	0.1225	0.14	0.0857	0.21	0.0450
0.01	0.043	0.08	0.1205	0.15	0.079	0.22	0.0406
0.02	0.0741	0.09	0.1167	0.16	0.0726	0.23	0.0365
0.03	0.0956	0.1	0.1116	0.17	0.0664	0.24	0.0328
0.04	0.1098	0.11	0.1056	0.18	0.0605	0.25	0.0294
0.05	0.1181	0.12	0.0992	0.19	0.0550		
0.06	0.1220	0.13	0.0925	0.2	0.0498		

CA5.8 The surface tension of a water-based solution, σ (mN/m), as a function of normalized temperature $\theta = (373 - T)/100$ (T in$^\circ$C) is tabulated below. Write a program to calculate the $d\sigma/dT$ and $d^2\sigma/dT^2$ at every available data point using the finite differences.

θ, $^\circ$C	$\sigma(\theta)$, mN/m	θ, $^\circ$C	$\sigma(\theta)$, mN/m	θ, $^\circ$C	$\sigma(\theta)$, mN/m
0	10.5	0.3	37.5	0.65	51.3
0.1	14.7	0.45	40.5	0.7	54.6
0.15	22.8	0.5	41.1	0.85	64.4
0.25	26.6	0.55	44.5	1	73.9

CA5.9 The voltage across a capacitor is given as a function of time as follows:

$$V(t) = \frac{1}{C} \int_0^t i(t)dt$$

For a capacitor of $C = 5 \times 10^{-5}$ F, use the following data to find the evolution of the transient current across the capacitor, i.e., current for every time value available.

t (s)	V (mV)	t (s)	V (mV)	t (s)	V (mV)
0	0	0.6	14.85	1.2	23.69
0.1	2.33	0.7	16.98	1.3	24.25
0.2	4.84	0.8	18.87	1.4	24.62
0.3	7.44	0.9	20.5	1.5	24.84
0.4	10	1	21.83		
0.5	12.51	1.1	22.89		

CA5.10 In order to determine the convection heat transfer coefficient (h) of an indoor ice-skating ring that is kept at $T_\infty = 280$ K at all times, a small spherical metal ball ($mc_p = 250$ J/kg) embedded with a thermocouple is heated to 500 K. It is then left to cool in a suspended position. The temperature of the ball is read at 2-minute intervals and reported in the table below. A simple model for cooling the ball can be given as

$$-mc_p\frac{dT}{dt} = h(T - T_\infty) + \varepsilon\sigma(T^4 - T_\infty^4)$$

where $\sigma = 5.67 \times 10^{-8}$ W/m²·K⁴ is the Stefan-Boltzmann constant, h is the convection heat transfer coefficient (W/m²K), and $\varepsilon = 1$ is the emissivity of the ball. Using the data given, estimate the indoor convection heat transfer coefficient.

t (min)	T (K)	t (min)	T (K)	t (min)	T (K)
0	500	8	352.2	16	306.5
2	442.7	10	335.9	18	300.7
4	402.6	12	323.5	20	296.2
6	373.8	14	313.9		

CA5.11 Consider a two-dimensional array ($f_{i,j}$) of a scalar physical quantity $f(x,y)$ (temperature, pressure, etc.) defined on a rectangular domain. The rate of change of $f(x,y)$ along the path s is referred to as the *directional derivative*, and it is determined as

$$\frac{df}{ds} = \mathbf{n} \cdot \text{grad } f = \cos\theta\frac{\partial f}{\partial x} + \sin\theta\frac{\partial f}{\partial y}$$

where $\mathbf{n} = \cos\theta\,\mathbf{i} + \sin\theta\,\mathbf{j}$ is the unit vector along the direction-s, $\partial f/\partial x$, and $\partial f/\partial y$ are obviously the partial derivatives (*see* **Fig. CA5.11**).

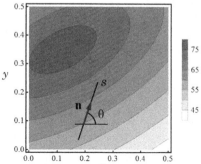

Fig. CA5.11

Write a computer program that reads the direction (θ) and a two-dimensional array $(f_{i,j}$ for $i = 0, 1, \ldots, n$, and $j = 0, 1, 2, \ldots, m)$ uniform in both directions-order, i.e., Δx and Δy are constant. The program should compute df/ds and d^2f/ds^2 derivatives for every available data point using second-order finite-difference formulas. Given data is the temperature distribution $T(x, y)$ on $[0,0.5] \times [0,0.5]$. Use the given true data below to validate your computer program. *Given*: Data for program verification: for $\theta = \pi/6$, $T(0.3, 0) = 3.2063$, $T(0.3, 0.1) = 0.367$, $T(0.3, 0.5) = -35.979$, $T(0, 0.1) = 23.634$, and $T(0.5, 0.1) = -7.1586$.

	y (m)					
x (m)	0	0.1	0.2	0.3	0.4	0.5
0	58.2	64.9	71.1	75	71.8	65.6
0.1	56.2	63.1	70.1	76.9	75.9	69.1
0.2	53.5	60.3	67.1	73.6	76.4	71.0
0.3	50.2	56.8	63.2	69	72.5	70.6
0.4	46.5	52.9	58.9	64.3	67.8	67.9
0.5	42.6	48.7	54.5	59.5	63.1	64.3

Interpolation and Extrapolation

DISCRETE functions, also discussed in Chapter 5, are the result of either sampling or experimentation. Most physical properties (density, viscosity, heat capacity, enthalpy, pressure, etc.) used in computations are assumed to be smooth functions whose explicit mathematical relationships are unknown. The physical properties of solids, liquids, or gases are usually available as discrete (tabular) data as a function of temperature, pressure, etc. In other words, the properties are only known at discrete points. On the other hand, even though an ordinary differential equation in $y(x)$ contains an infinite amount of information on $[a, b]$, its numerical solution is obtained only for a set of discrete data points because a computer can handle only a finite amount of data.

A set of discrete data may be either *uniformly* or *non-uniformly* spaced (*see* Fig. 6.1a or 6.1b). Periodically sampled or collected data with digital instruments is uniformly spaced. Differentiation, integration, or interpolation of a uniformly spaced set of discrete

DOI: 10.1201/9781003474944-6

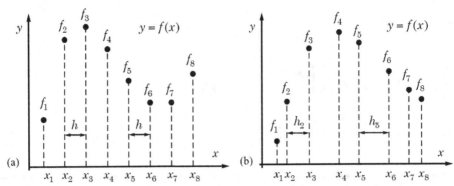

FIGURE 6.1: Discrete functions with (a) uniform or (b) nonuniform distribution.

data points is easier and requires less memory because abscissas need not be stored. That is why numerical analysts prefer dealing with uniformly spaced discrete functions due to their computational advantages.

Consider a smooth discrete function $f(x)$, which is defined by a set of discrete values $(x_i, f(x_i))$ for $i = 1, 2, \ldots, 8$ (*see* Fig. 6.1). We frequently need to estimate the value of the function for a point $x = c$ that lies in (x_1, x_8). The process of estimating $f(c)$, where c falls between $x_1 < c < x_8$ is called *interpolation*. The estimation of the function values lying outside the observed range, $c < x_1$ or $c > x_8$, is referred to as *extrapolation*.

When interpolation is mentioned, the first procedure that comes to mind is *linear interpolation*. Consider two points, $A(a, f(a))$ and $B(b, f(b))$, as illustrated in Fig. 6.2. We wish to estimate $f(c)$, where $a < c < b$. Linear interpolation is achieved by geometrically rendering a straight line between A and B. All estimates for $a < c < b$ lie on the straight line connecting the points A and B, i.e., the interpolates can be obtained from

$$\hat{f}(x) = f(a) + (x - a)\frac{f(b) - f(a)}{b - a} \tag{6.1}$$

where $\hat{f}(x)$ denotes all possible interpolates.

Linear interpolation yields satisfactory estimates in cases where the distribution of data points does not exhibit a strong curvature. As shown in Fig. 6.2, $y = f(x)$ has a strong curvature on $[a, b]$. The interpolation error, $f(c) - \hat{f}(c)$, is quite large for an interior c value, but for values of c near the end points, $x = a$ or $x = b$, the error decreases. In other words, linear interpolation is insufficient to estimate a wide range of values for a function with strong curvature and large interval sizes. A continuous function with strong curvature can

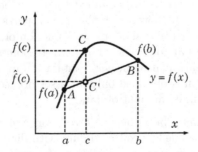

FIGURE 6.2: Graphical depiction of a linear interpolation.

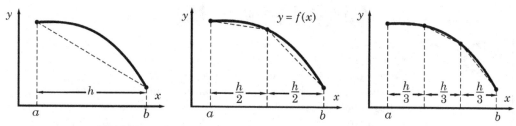

FIGURE 6.3: The effect of interval size on linear interpolation.

be better estimated by straight lines only if the interval size is small, as shown in Fig. 6.3. A remedy to improve the accuracy of interpolates is to decrease the interval size, i.e., increase the number of data points for the same range. However, this is not possible, especially when dealing with experimentally collected data. An alternative avenue is to represent the data with a polynomial, which has the potential to account for the actual curvature of the discrete function. Therefore, in most cases, it is necessary to use interpolating polynomials of $n \geqslant 2$ to obtain an improved estimate (interpolate) and to ensure better approximation.

In this chapter, numerical interpolation and extrapolation methods for uniformly and non-uniformly spaced discrete functions are discussed. As interpolation functions, in this text only polynomials and piecewise-defined polynomials (splines) are considered.

6.1 LAGRANGE INTERPOLATION

An alternative way of connecting a set of discrete data, aside from using straight lines, is to use a polynomial that passes through all data points. In this context, Lagrange polynomials play an important role in describing a set of discrete data points by a representative polynomial.

Consider a set of discrete data points (x_k, f_k) for $k = 0, 1, 2, \ldots, n$, called *interpolation points*. The abscissas x_k are assumed to be distinct and may not necessarily be uniformly spaced. A nth-degree polynomial can be used to represent the data set:

$$p_n(x) = c_0 + c_1 x + c_2 x^2 + \ldots + c_n x^n \tag{6.2}$$

which passes through all $(n+1)$ distinct data points should satisfy all interpolation points; that is,

$$p_n(x_k) = y_k, \qquad k = 0, 1, 2, \ldots, n \tag{6.3}$$

This condition is enforced to determine the unknown coefficients, c_0, c_1, c_2, ..., c_n. To do so, we seek a set of polynomials of degree n $(L_k(x)$ for $k = 0, 1, 2, \ldots, n)$ that satisfy the following property:

$$L_k(x_i) = \begin{cases} 0, & \text{if } k \neq i \\ 1, & \text{if } k = i \end{cases} \tag{6.4}$$

where x_k denotes $n + 1$ distinct abscissas for $k = 0, 1, 2, \ldots, n$, and $L_k(x)$ is so called the *Lagrange polynomials*. Then a *Lagrange interpolating polynomial* $p_n(x)$, passing through $n + 1$ points, can be constructed as follows:

$$f(x) \cong p_n(x) = L_0(x)f_0 + L_1(x)f_1 + \ldots + L_n(x)f_n = \sum_{k=0}^{n} L_k(x)f_k \tag{6.5}$$

Note that, owing to Eq. (6.4), Eq. (6.5) satisfies Eq. (6.3) for every x_k.

| Lagrange Interpolation Method |

- The method provides an easy and simple way of constructing an interpolating polynomial;
- It is very easy to adapt to computer programming;
- It is applicable to both uniform and non-uniform discrete functions;
- Once Lagrange polynomials are constructed, they can be used with different ordinate sets (f_k's) as long as the number of data points and the abscissas (x_k's) in the set are fixed.

- It is cpu-time intensive and unpractical for a large data set;
- Adding new data points to an existing data set increases the degree of the interpolating polynomial, which has to be reconstructed from scratch;
- Estimating the error bounds of an approximation is not easy;
- Errors outside the data range increase uncontrollably, which makes the method too risky to use for extrapolations.

Then, for every x_k, an nth-degree polynomial is defined as

$$L_k(x) = c_k(x-x_1)(x-x_2)\cdots(x-x_{k-1})(x-x_{k+1})\cdots(x-x_n) = c_k \prod_{\substack{i \neq k}}^{n}(x-x_i) \qquad (6.6)$$

where c_k is a constant, and the factor $(x - x_k)$ is skipped. Evaluating Eq. (6.6) at $x = x_k$ and applying the stipulated condition, Eq. (6.4), yields

$$1 = c_k(x_k-x_1)(x_k-x_2)\cdots(x_k-x_{k-1})(x_k-x_{k+1})\cdots(x_k-x_n) = c_k \prod_{\substack{i \neq k}}^{n}(x_k-x_i) \qquad (6.7)$$

Solving c_k from Eq. (6.7) and substituting it into Eq. (6.6), we find

$$L_k(x) = \prod_{\substack{i=0 \\ i \neq k}}^{n} \left(\frac{x - x_i}{x_k - x_i} \right), \quad k = 0, 1, 2, \ldots, n \qquad (6.8)$$

Note that $L_k(x) = 0$ for all $x = x_i$'s, except $x = x_k$, resulting in $L_k(x_k) = 1$.

ERROR ASSESSMENT. First, we assume the interpolation points (x_k, f_k) do in fact represent the true values of function $f(x)$ at the given abscissas, and $f(x)$ has $(n + 1)$ continuous derivatives for all x. Letting $p_n(x)$ to be a unique Lagrange interpolating polynomial of at most nth degree, passing the interpolating data points (x_k, f_k) for $k = 0, 1, 2, \ldots, n$, there exists a point $\xi(x)$ for any x in $[x_0, x_n]$ such that

$$e_n(x) = f(x) - p_n(x) = \frac{(x-x_0)(x-x_1)\cdots(x-x_n)}{(n+1)!} f^{(n+1)}(\xi(x)), \quad x_0 \leqslant \xi(x) \leqslant x_n \qquad (6.9)$$

where $e_n(x)$ is the error, defined as the difference between the true and interpolated values.

Equation (6.9) does not seem to be very useful since the true function that gives (x_k, f_k) is unknown. $f^{(n+1)}(\xi)$ and $\xi(x)$ are also the other two unknowns. However, Eq. (6.9) would provide bounds for the interpolation error if we can estimate the bounds of $f^{(n+1)}(\xi)$. For a smooth function $f(x)$, we can anticipate its higher derivatives to be smooth as well. But keep in mind that the higher derivatives of a function depicting sharp changes depict even sharper changes, which in turn lead to large interpolation errors.

One of the significant results of Eq. (6.9) is that if a discrete function $f(x)$ is truly a polynomial of nth degree or lower, i.e., $f(x) = p_m(x)$, $(m \leqslant n)$, then the interpolating polynomial, Eq. (6.5), reproduces the true polynomial, in which case the absolute error is zero since $f^{(n+1)}(\xi)$ will be zero for all x. Otherwise, the value estimated using a Lagrange interpolating polynomial will contain a certain amount of error. The magnitude of this error is influenced by a number of factors, such as the length of the interval, $[x_0, x_n]$; the interval size of the subintervals, Δx_k; the position of x in the interval; the degree of the interpolating polynomial; and the number of discrete data points in the set.

EXAMPLE 6.1: Lagrange Interpolation in Action

A set of data describing the efficiency (η) of the cross-flow turbine as a function of rotor inlet relative velocity (V) is tabulated below:

V	0.08	0.25	0.50	0.90
$\eta(V)$	0.25	0.625	0.81	0.43

(a) For efficiency, find a Lagrange interpolating polynomial that uses all data points; (b) Plot the interpolating polynomial and the true error for values in the range $0 \leqslant V \leqslant 1$ and discuss the accuracy of the polynomial approximation; (c) Use the interpolating polynomial in Part (a) to estimate $\eta(0.4)$, $\eta(0.7)$, and $\eta(0.85)$. The true theoretical efficiency is defined as

$$\eta(V) = 2V\left(\kappa b - V + \kappa b \sqrt{\kappa^2 + V^2 - 2\kappa b V}\right)$$

where $\kappa = 0.93$ and $b = \cos(\pi/10)$.

SOLUTION:

(a) Employing Eq. (6.8) to the relative velocity data leads to

$$L_0(V) = \frac{(V - 0.25)(V - 0.50)(V - 0.9)}{(0.08 - 0.25)(0.08 - 0.50)(0.08 - 0.9)},$$

$$L_1(V) = \frac{(V - 0.08)(V - 0.50)(V - 0.9)}{(0.25 - 0.08)(0.25 - 0.50)(0.25 - 0.9)},$$

$$L_2(V) = \frac{(V - 0.08)(V - 0.25)(V - 0.9)}{(0.50 - 0.08)(0.50 - 0.25)(0.50 - 0.9)},$$

$$L_3(V) = \frac{(V - 0.08)(V - 0.25)(V - 0.50)}{(0.90 - 0.08)(0.90 - 0.25)(0.90 - 0.50)}$$

Using Eq. (6.5), the interpolating polynomial is found as

$$\eta(V) \cong p_3(V) = (0.25)L_0(V) + (0.625)L_1(V) + (0.81)L_2(V) + (0.43)L_3(V)$$

which is a cubic polynomial that can be used to estimate the efficiency for any value of V in the range $0 \leqslant V \leqslant 1$.

(b) A comparative plot of the true efficiency, $\eta(V)$, with the third-degree interpolating polynomial and the distribution of the true error, $e_3(V)$, in the $0 \leqslant V \leqslant 1$

range are presented in Fig. 6.4a and b, respectively. For $V < 0.6$, the approximating polynomial agrees very well with the true efficiency. Only slight deviations are observed for $V > 0.6$. However, when the true error distribution is examined, some discrepancies between the true and interpolating polynomials do not go unnoticed. The true error $e_3(V)$ depends on the value of V, which depicts a maximum and two minimums on $[0,1]$. The errors are constrained near the interpolation points $[V_k, V_{k+1}]$ which anchor the approximation (the true error is zero at the interpolation points due to enforcing Eq. (6.3)).

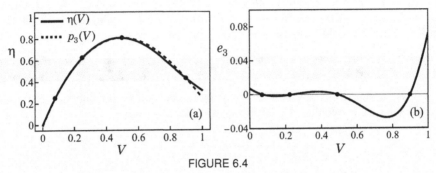

FIGURE 6.4

(c) The estimates with the interpolating polynomial are found as $\eta(0.4) \cong$ $p_3(0.4) = 0.78314$, $\eta(0.7) \cong p_3(0.7) = 0.70446$, and $\eta(0.85) \cong p_3(0.85) = 0.5116$. Note that in Fig. 6.4b, the deviation from horizontal "zero line" in the range $0 < V < 0.55$ is very small, resulting in a small interpolation error for $V = 0.4$ (=0.78617−0.78313=0.00304). However, the approximations for $\eta(0.7)$ and $\eta(0.85)$ yield relatively larger errors (0.02331 and 0.01776, respectively) since the interpolation points are close to $V \approx 0.77$, where the error is largest.

Discussion: An interpolating polynomial that passes all four data points is a cubic equation. Expanding and simplifying all-like expressions yield

$$\eta(V) \cong 1.0856\,V^3 - 4.39125\,V^2 + 3.55848\,V - 0.00713$$

which, using this form, will significantly reduce the computation time if it is to be used numerous times in a computation.

The true errors are small near interpolation (data) points but almost always depict an increase toward the midpoint of any two interpolation points. The absolute deviations in this example are ≈0.002 in [0.08,0.25], 0.0031 in [0.25,0.5], and −0.03 in [0.5,0.9], which indicates relatively good agreement. Note that the interpolating polynomial provides very good estimates when the distance between two data points is small; however, the errors tend to increase as the distance between data points increases.

If an interpolating polynomial is to be used numerous times, instead of applying Eq. (6.5), it is more convenient and practical to express it in simplified power form, $p_n(x) = a_n x^n + \cdots + a_1 x + a_0$, or preferably in the nested form as

$$p_n(x) = (((a_n x + a_{n-1})x + \cdots)x + a_1)x + a_0$$

```
                        Pseudocode 6.1

  Function Module LAGRANGE_EVAL (n, xval, x, f)
  \ DESCRIPTION: A pseudo-function module for evaluating the interpolant
  \    yval = y(xval) using Lagrange interpolation.
  \ USES:
  \    LAGRANGE_P:: A module to compute Lagrange polynomials (given below).
  Declare: x_{0:n}, f_{0:n}, L_{0:n}                    \ Declare array variables
  LAGRANGE_P(n, xval, x, L)                             \ Go to module to find L_k(xval)'s
  fval ← 0                                              \ Initialize accumulator
  For [k = 0, n]                                        \ Loop k: Accumulator for interpolant
      fval ← fval + L_k * f_k                           \ Accumulate L_k(xval) * f_k products
  End For
  LAGRANGE_EVAL← fval                                   \ Set fval to LAGRANGE_EVAL
  End Function Module LAGRANGE_EVAL


  Module LAGRANGE_P (n, xval, x, L)
  \ DESCRIPTION: A pseudomodule evaluating the Lagrange polynomials at xval.
  Declare: x_{0:n}, L_{0:n}                             \ Declare array variables
  For [k = 0, n]                                        \ Loop k: Accumulator to find L_k
      L_k ← 1                                           \ Initialize accumulating variable L_k
      For [i = 0, n]                                    \ Loop i: Loop to compute products
          If [i ≠ k] Then                               \ Skip block for k = i
              L_k ← L_k * (xval − x_i)/(x_k − x_i)      \ Product accumulation
          End If
      End For
  End For
  End Module LAGRANGE_P
```

A pseudomodule, **LAGRANGE_P**, evaluating the Lagrange polynomials, as well as a pseudomodule, **LAGRANGE_EVAL**, computing an approximation using Lagrange's method, are presented in Pseudocode 6.1. **LAGRANGE_P** evaluates all possible Lagrange polynomials for $x = xval$. (It is important to note that in this method the subscripts of the $n+1$ discrete data points range from 0 to n, which leads to Lagrange polynomials of nth degree.) As input, it requires the number of discrete data points n (i.e., the number of data points minus 1), the data points (x_k, f_k) for all k, and the abscissa ($xval$) at which interpolation is sought. The module **LAGRANGE_P** computes $L_k(xval)$ for all k. The **For**-loop over k computes L_k's using Eq. (6.8). A product accumulator L_k is initialized: $L_k \leftarrow 1$. The **For** loop over i is used to accumulate $(xval - x_i)/(x_k - x_i)$ products, except $i = k$, which is skipped with an If-construct. Once both loops are completed, the Lagrange polynomials calculated for $xval$ become available for interpolation and passed to the calling module.

The module **LAGRANGE_EVAL** evaluates an interpolant $yval = f(xval)$ using Lagrange polynomials. The module requires an input value of n, $xval$ for which interpolant is sought, and the discrete data points (x_k and y_k for $k = 0, 1, \ldots, n$). Although not included here to save space, it is always a good idea to do a data check at the beginning of every module to ensure the input (like $xval$) is within the valid data range. The actual interpolation procedure for $x = xval$ is carried out in this module. An accumulator $fval$ is used to accumulate $L_k(xval) * f_k$ products for all k, i.e., perform the summation in Eq. (6.5). Once the accumulation is completed, the result is set to **LAGRANGE_EVAL**.

EXAMPLE 6.2: Runge's phenomenon with a remedy

The energy dependence of neutron resonance crosssection, depicted in the figure below, is described by the Breit-Wigner formula,

$$\sigma(E) = \frac{\sigma_0}{1 + 4(E - E_0)^2/\Gamma^2}$$

where E_0 is the resonance energy (MeV), Γ is the width half maximum (MeV), σ_0 is the peak resonance cross section (barn). By defining $y = \sigma(E)/\sigma_0$ and $x = E - E_0$, the normalized cross section for $\Gamma = 2/3$ MeV can be expressed as $y = 1/(1 + 9x^2)$.

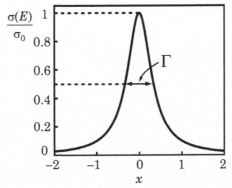

Establish a set of data with (i) $x = -2, -1.6, -1.2, 1.2, 1.6, 2$, (ii) $x = -2, -1.8, -1.6,$ $-1.4, -1.2, 0, 1.2, 1.4, 1.6, 1.8, 2$, and (ii) $x = -2, -1.9, -1.6, -1.2, -0.6, 0, 0.6, 1.2,$ $1.6, 1.9, 2$. For each case, (a) construct the Lagrange interpolating polynomial that passes through all data points; (b) compare the distributions of true and approximate values and discuss the results.

SOLUTION:

Using the abscissas provided in (i), (ii), and (iii), three sets of interpolation data (not listed here) are established by $y_k = y(x_k)$ for all k.

(i) The corresponding interpolating polynomials, respectively, are found as

$$p_4(x) = 0.06427x^4 - 0.367053x^2 + 0.46692$$
$$p_{10}(x) = -0.2883x^{10} + 2.569131x^8 - 7.83586x^6 + 9.898657x^4 - 5.082046x^2 + 1$$
$$q_{10}(x) = -0.032702x^{10} + 0.39718x^8 - 1.78882x^6 + 3.63707x^4 - 3.21813x^2 + 1$$

Note that the interpolating polynomials resulted in 4th and 10th-degree polynomials instead of the expected 5th and 11th degree polynomials. This is a natural consequence of working with a discrete function that is symmetric about $x = 0$.

The true function, interpolating polynomials, and interpolation points are comparatively presented for all three cases in Fig. 6.5. In Fig. 6.5a, the interpolating polynomial $p_4(x)$ yields a significant underestimation at $x = 0$ (the midpoint of the interval) and moderate deviations elsewhere. In Fig. 6.5b, on the other hand, the $p_{10}(x)$ (obtained with uniformly spaced 11 data points) clearly exhibits very poor performance in between each pair of interpolation points, especially near the endpoints of the data range. Also notice that increasing the number of interpolation points worsened the accuracy of the approximations rather than improving them.

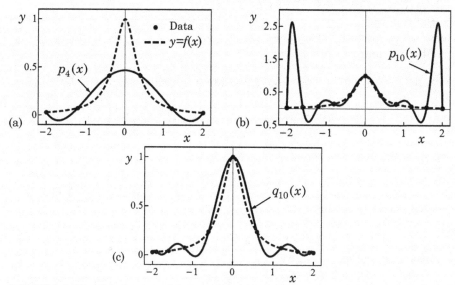

FIGURE 6.5: The distribution of the true function and the Lagrange interpolating polynomial for uniformly spaced (a) 6-interpolation points, (b) 11-interpolation points, and (c) non-uniformly spaced 11-interpolation points.

In Fig. 6.5c, the polynomial $q_{10}(x)$ obtained non-uniformly spaced interpolation points with finer spacings near the ends of the interval largely mitigates the deviations near the edges of the data interval.

Discussion: Generally, a sufficient number of data points is necessary in order for an interpolating polynomial to closely represent the true distribution. As a solution for improving the accuracy of an interpolating polynomial, beginners to numerical interpolation tend to increase the number of interpolation points, i.e., the degree of approximating polynomial. Nevertheless, high-degree polynomials are usually not suitable for interpolations because they lead to non-physical oscillations that cannot be eliminated by further increasing the degree of the polynomial. This observation is known as the *Runge's phenomenon* and it is observed when using a large number of "uniformly spaced data points," as in this example. It exhibits itself as oscillations at the endpoints of the uniformly spaced data set when a high-degree interpolating polynomial is employed.

One of the following remedies may be used to mitigate the adverse effects of Runge's phenomenon:

(i) use *piecewise* interpolating polynomials (i.e., instead of using a single interpolating polynomial for all, the data interval can be divided into several data sets and employ lower degree interpolation polynomials on each),

(ii) use non-uniformly spaced interpolation points that are more densely stacked by the endpoints of the interval (as in $q_{10}(x)$),

(iii) use spline curves that are piecewise polynomials (*see* Section 6.3), or

(iv) fit the data to a low-degree polynomial (or a suitable non-polynomial model) using the method of least squares (*see* Chapter 7).

6.2 NEWTON'S DIVIDED DIFFERENCES

Consider a set of discrete data points: (x_k, f_k) for $k = 0, 1, 2, \ldots, n$ and an interpolating polynomial of the form:

$$f(x) \cong p_n(x) = a_0 + a_1(x - x_0) + a_2(x - x_0)(x - x_1) + a_3(x - x_0)(x - x_1)(x - x_2)$$

$$+ \ldots + a_n(x - x_0)(x - x_1) \ldots (x - x_{n-1}) = \sum_{k=0}^{n} a_k \left(\prod_{i=0}^{k-1} (x - x_i) \right) \tag{6.10}$$

where a_n's are the unknown coefficients.

The polynomial satisfies the interpolation points, so imposing $f(x_k) = p_n(x_k) = f_k$ for $k = 0, 1, 2, \ldots, n$ leads to a system of linear equations with $(n + 1)$ unknowns. The solution to this system is obtained easily by applying the forward substitution procedure. For instance, when $x = x_0$, all but the first term of Eq. (6.10) vanishes, yielding $f_0 = a_0$. For $x = x_1$, all the terms in the polynomial except for the first two will vanish. Solving for a_1, after making use of $a_0 = f_0$, gives

$$a_1 = \frac{f_1 - f_0}{x_1 - x_0} \tag{6.11}$$

Setting $x = x_2$ in Eq. (6.10), utilizing a_0 and a_1, and solving for a_2 yields

$$a_2 = \frac{1}{x_2 - x_0} \left(\frac{f_2 - f_1}{x_2 - x_1} - \frac{f_1 - f_0}{x_1 - x_0} \right) \tag{6.12}$$

Continuing in this fashion, the rest of the coefficients are obtained as complicated expressions of x_k and f_k's. Nonetheless, there is an easy way to reproduce the coefficients by employing the concept of *divided differences*.

The divided differences are defined as differences starting from the 0th to nth divided difference. The *zeroth divided-difference* is the discrete function value at $x = x_k$; that is,

$$f[x_k] = f(x_k) = f_k$$

Substituting this value into Eq. (6.11) gives the *first divided-difference* of f:

$$f[x_k, x_{k+1}] = \frac{f[x_{k+1}] - f[x_k]}{x_{k+1} - x_k} \tag{6.13}$$

Equation (6.13), which is also symmetrical, can be written as

$$f[x_k, x_{k+1}] = f[x_{k+1}, x_k] = \frac{f[x_{k+1}]}{x_{k+1} - x_k} + \frac{f[x_k]}{x_k - x_{k+1}} \tag{6.14}$$

The second divided-difference is obtained from Eq. (6.12) as

$$f[x_k, x_{k+1}, x_{k+2}] = \frac{f[x_{k+1}, x_{k+2}] - f[x_k, x_{k+1}]}{x_{k+2} - x_k} \tag{6.15}$$

This process is continued in a recursive manner, yielding a general recurrence expression for the *nth divided difference* as follows:

$$f[x_0, x_1, \ldots, x_k] = \frac{f[x_1, x_2, \ldots, x_k] - f[x_0, x_1, \ldots, x_{k-1}]}{x_k - x_0} \tag{6.16}$$

TABLE 6.1: Construction of Newton's divided-difference table.

x_i	$f[x_i] = f_i$	$f[x_i, x_{i+1}]$	$f[x_i, x_{i+1}, x_{i+2}]$	$f[x_i, x_{i+1}, x_{i+2}, x_{i+3}]$
x_0	$f[x_0]$ ⟶	$f[x_0, x_1]$ ⟶	$f[x_0, x_1, x_2]$ ⟶	$f[x_0, x_1, x_2, x_3]$
x_1	$f[x_1]$ ⟶	$f[x_1, x_2]$ ⟶	$f[x_1, x_2, x_3]$	
x_2	$f[x_2]$ ⟶	$f[x_2, x_3]$		
x_3	$f[x_3]$			

Upon inspecting Eqs. (6.11) and (6.12), the first two coefficients of Newton's interpolating polynomial are found to be

$$a_1 = f[x_0, x_1], \qquad a_2 = f[x_0, x_1, x_2] = \frac{f[x_1, x_2] - f[x_0, x_1]}{x_2 - x_0} \tag{6.17}$$

Finally, the nth coefficient results in

$$a_n = f[x_0, x_1, \ldots, x_n] \tag{6.18}$$

A simple way to construct the divided differences for Newton's interpolation formula is to use the recursive relationships to generate the divided differences through the *Newton's Divided Difference Table*. This table is constructed by filling in $n+1$ entries of the first column with the abscissas, x_i's, and the second column with the zeroth divided difference (f_i's). Equation (6.16), is then used to systematically fill in the entries of the subsequent columns. That is, the divided-difference in the current column is calculated using the divided-differences from the previous column directed by arrows, as shown in Table 6.1. Once the table is filled out, the coefficients of the Newton interpolating polynomial (a_n for $i = 0, 1, \ldots, n$) are available in the first row. In practice, there is no need to store the entire table since only the first row entries are required in the interpolation formula.

The final form of the *Newton's interpolating polynomial* is written as

$$\begin{aligned}
f(x) &= f[x_0] \; + \; f[x_0, x_1](x - x_0) \; + \; f[x_0, x_1, x_2](x - x_0)(x - x_1) \\
&+ f[x_0, x_1, x_2, x_3](x - x_0)(x - x_1)(x - x_2) + \ldots \\
&+ f[x_0, x_1, \ldots, x_n](x - x_0)(x - x_1) \ldots (x - x_{n-1})
\end{aligned} \tag{6.19}$$

Equation (6.19) can also be expressed as a sequence of nested multiplications and additions, which is suitable for computer programming because it reduces computational efforts.

$$f(x) = a_0 + (x - x_0)\left(a_1 + (x - x_1)\left\{a_2 + (x - x_2)(\ldots (a_{n-1} + a_n(x - x_n))\right\}\right) \tag{6.20}$$

For a set of interpolation points, Newton's divided difference table depends on the ordering of the abscissas. The interpolating polynomial, unlike Lagrange interpolation, can be easily updated as interpolation points are removed or new points are added. The divided-difference table can be extended to include the new additional differences without requiring reconstructing the table from scratch. Strictly speaking, when the interpolation formula is expanded and simplified, all orderings yield the same unique interpolating polynomial.

TABLE 6.2: Data updating in the Newton's divided-difference table.

x_i	$f[x_i]=f_i$	$f[x_i,x_{i+1}]$	$f[x_i,x_{i+1},x_{i+2}]$	$f[x_i,x_{i+1},x_{i+2},x_{i+3}]$
x_0	$f[x_0]$ →	$f[x_0,x_1]$ →	$f[x_0,x_1,x_2]$ →	$f[x_0,x_1,x_2,x_3]$
x_1	$f[x_1]$ →	$f[x_1,x_2]$ →	$f[x_1,x_2,x_3]$ →	$f[x_1,x_2,x_3,x_4]$
x_2	$f[x_2]$ →	$f[x_2,x_3]$ →	$f[x_2,x_3,x_4]$	
x_3	$f[x_3]$ →	$f[x_3,x_4]$		
x_4	$f[x_4]$			

However, for a large data set, the best results are obtained with $x_n > x_{n-1} > \ldots > x_2 > x_1$. The reconstruction of the divided difference table upon adding new data, (x_4, f_4), to an existing set is depicted in Table 6.2. This process leads to computing the divided-differences and filling in the last spot of each column in the table.

Equation (6.19) may be used to cover all or partial interpolation points. This property of Newton's interpolating polynomials can be used to obtain all available interpolating polynomials. Starting with the first-degree (linear) interpolating polynomial for any x, the degree of the interpolating polynomial can be increased recursively by adding new terms. In this case, a series of low-degree interpolating polynomials can be expressed as

$$p_1(x) = f[x_0] + f[x_0, x_1](x - x_0),$$

$$p_2(x) = p_1(x) + f[x_0, x_1, x_2](x - x_0)(x - x_1),$$

$$p_3(x) = p_2(x) + f[x_0, \ldots, x_3](x - x_0)(x - x_1)(x - x_2),$$

$$\ldots \qquad \ldots$$

$$p_n(x) = p_{n-1}(x) + f[x_0, \ldots, x_n] \prod_{k=0}^{n-1} (x - x_k),$$

(6.21)

where $p_1(x)$, $p_2(x)$, and $p_3(x)$ are linear, quadratic, and cubic interpolating polynomials.

ERROR ASSESSMENT. A polynomial passing through $n+1$ interpolation points has a degree of n or less. As a matter of fact, when expanded into traditional polynomial form, Newton's interpolating polynomial exactly reproduces Lagrange's interpolating polynomial. Thus, the error term of Newton's interpolating polynomial is identical to that of the equivalent Lagrangian polynomial, and so the error term associated with the nth degree interpolating polynomial can also be explained by Eq. (6.9). This means that if a set of discrete data is truly represented by a polynomial of nth degree or less, the error will be zero regardless of x. This event can be visually identified from the divided difference table, where all entries in a column become the same. As with the Lagrangian interpolation polynomials, the interpolation errors are smaller near the endpoints of an interval, (x_k, x_{k+1}), between two consecutive interpolation points and larger roughly toward the middle.

Newton's Divided-Difference Method

- The method can be used to construct an interpolating polynomial of any degree, so long as there are sufficient data points;
- The interpolating polynomial resulting from the method provides greater advantage because it lends itself to nested multiplications;
- It is efficient and stable, and it allows adding more data points to obtain a higher-degree interpolating polynomial without discarding the existing data;
- A low-degree interpolating polynomial can yield very accurate results if the data is smooth.

- The method is a bit difficult to understand and implement;
- It is expensive and unpractical for interpolating large data sets;
- It is sensitive to the errors in the data set;
- It may lead to inaccurate results when the data points are closely spaced or when the discrete function is highly oscillatory;
- Estimating the error bounds is not easy;
- The errors outside the data interval increase uncontrollably, and it is too risky to use it for extrapolation purposes.

For smooth functions, the difference between two subsequent interpolating polynomials is expected to decrease, i.e., $|p_n(x) - p_{n-1}(x)| \to 0$. This difference, $R_n(x_0, \ldots, x_n)$, is the *error* of adding or "not" adding new terms to the interpolating polynomial.

$$p_n(x) - p_{n-1}(x) = R_n(x_0, \ldots, x_n) = f[x_0, \ldots, x_n] \prod_{k=0}^{n-1} (x - x_k) \qquad (6.22)$$

In general, any interpolation process should be carried out with an interpolating polynomial of the smallest degree possible. A linear $(p_1(x))$ or quadratic $(p_2(x))$ interpolation in the neighborhood of the interpolating points may be sufficient. In most cases, interpolating polynomials of at most 3rd or 4th degrees, constructed from the nearest data points, will yield very good results. Any high-degree interpolating polynomials derived from a large set of data should be treated with caution due to the implications that may also give rise to Runge's phenomenon.

Pseudocode 6.2

```
Module NEWTONS_TABLE (n, x, f, A)
\ DESCRIPTION: A pseudomodule to generate the Divided-Difference table.
Declare: x_{0:n}, f_{0:n}, a_{0:n,0:n}              \ Declare array variables
A ← 0                                              \ Initialize table entries
For [k = 0, n]                                     \ Place f_k's into the 1st column
    a_{k0} ← f_k                                   \ Set a_{k0} = f_k for all k
End For
For [j = 1, n]                          \ Loop j: Sweep column from top to bottom
    For [k = 0, n − j]                                       \ Loop k: Sweep
        a_{kj} ← (a_{k+1,j−1} − a_{k,j−1})/(x_{k+j} − x_k)   \ Compute coefficients
    End For
End For
End Module NEWTONS_TABLE
```

A pseudomodule, NEWTONS_TABLE, generating Newton's divided difference table is presented in Pseudocode 6.2. The module requires a set of $n+1$ discrete data points (x_i, f_i) as input. All divided differences, a_{ij}'s, are computed and stored on matrix $\mathbf{A}$, which is also the output of the module. The first column of the table is the zeroth divided difference (a_{k0}) and the ordinates are placed into the first column simply by substitution $(a_{k0} = f_k)$. Starting with the second column $(j = 1)$, the differences are computed from top to bottom using the entries of the prior column. The table is constructed column by column, and with each column, the number of successive differences decreases by one. Thus, the row sweep for the jth column covers from $k = 0$ to $(n-j)$. When the table is complete, the coefficients of the interpolating polynomial are available on the first row of the table. If the divided differences are to be computed only for constructing Newton's interpolating polynomials, then there is no need to store the entire table. This proposition has been incorporated in Pseudocode 6.3.

EXAMPLE 6.3: Interpolation with non-uniform data sets

Consider the following tabular data for the specific heat of a gas mixture.

T (K)	282	400	500	620	848
$c_p(T)$ (kJ/kg.K)	10.3	9.14	8.74	8.499	8.3

(a) Use Newton's divided differences to obtain an interpolating polynomial for $c_p(T)$ that passes through all data points; (b) given $c_p(T) = 8.14 - 100/T + 200000/T^2$ as the true specific heat, plot the interpolation error $e_4(T)$ for the interpolating polynomial in the range [270, 870] K and interpret your results.

SOLUTION:

(a) We note that the discrete data is non-uniformly spaced, and the interpolating polynomial passing all data points will be a 4th-degree polynomial $(n = 5 - 1 = 4)$. In other words, Newton's interpolating polynomial will have the following form:

$$p_4(T) = c_p[T_0] + c_p[T_0, T_1](T - T_0) + c_p[T_0, T_1, T_2](T - T_0)(T - T_1)$$
$$+ c_p[T_0, T_1, T_2, T_3](T - T_0)(T - T_1)(T - T_2)$$
$$+ c_p[T_0, T_1, T_2, T_3, T_4](T - T_0)(T - T_1)(T - T_2)(T - T_3)$$

The coefficients of the polynomial are easily determined once the divided difference table is constructed (*see* Table 6.3). The second and third columns of the table have been reserved for the original data set. The entries of subsequent columns are computed using the entries of the current column along with the recurrence formula—Eq. (6.16).

TABLE 6.3

i	T_i	$c_{p,i}$	$c_p[T_i, T_{i+1}]$	$c_p[T_i, T_{i+1}, T_{i+2}]$	$c_p[T_i, \ldots, T_{i+3}]$	$c_p[T_i, \ldots, T_{i+4}]$
0	282	10.3	-0.009831	2.6745×10^{-5}	-5.2344×10^{-8}	6.9647×10^{-11}
1	400	9.14	-0.004000	9.0530×10^{-6}	-1.2924×10^{-8}	
2	500	8.74	-0.002008	3.2630×10^{-6}		
3	620	8.499	-0.000873			
4	848	8.30				

Once the table is completed, the entries in the first row are the coefficients of the interpolating polynomial. Upon substituting the coefficients, we obtain a 4th-degree interpolating polynomial as

$$c_p(T) \cong p_4(T) = 10.3 - 0.00983051(T - 282)$$
$$+ 2.67455 \times 10^{-5}(T - 282)(T - 400)$$
$$- 5.23444 \times 10^{-8}(T - 282)(T - 400)(T - 500)$$
$$+ 6.96471 \times 10^{-11}(T - 282)(T - 400)(T - 500)(T - 620)$$

(b) The distribution of the true error, $e_4(T) = c_p(T) - p_4(T)$, is illustrated in Fig. 6.6. The magnitude of the interpolation error increases toward the middle of the subintervals, which are bounded on each side by an interpolation point that acts as an anchor. However, the error outside the original temperature range ($T < 282$ or $T > 848$) does grow uncontrollably.

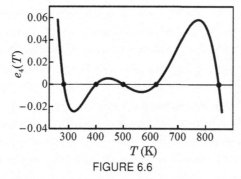

FIGURE 6.6

Discussion: The coefficients of Newton's divided difference polynomial are placed in the first row of the divided difference table. The interpolating polynomial produces exactly the data set for the interpolation points, which is why the errors at the original interpolation points are zero, just as in the Lagrange interpolation. Note that when dealing with experimental data, the true distribution (and the true error) is almost always unknown.

In Pseudocode 6.3, the pseudomodule, **NEWTONS_DD_COEFFS**, calculating the coefficients of Newton's divided difference interpolating polynomial and a function module, **NEWTONS_DD_EVAL**, evaluating the interpolant are presented. The divided differences in **NEWTONS_DD_COEFFS** are obtained in the same manner as in Pseudocode 6.2 after initializing the zeroth difference, $a_k = f_k$ for $k = 0, 1, 2, \ldots, n$. However, the coefficients are computed from the bottom to the top of each column using Eq. (6.16). In the mean time, the divided differences are overwritten each time from the last storage location backward so that, at the end, only the desired coefficients remain. This process is repeated column by column until all the data is used. The function module **NEWTONS_DD_EVAL** simply applies Eq. (6.10) to compute $fval$. Besides $xval$, this module requires the coefficients for the interpolating polynomial (a_{ij}) and the data set (x_i, f_i) as input. The program computes the difference from Eq. (6.22) to determine whether the interpolant has converged or not. If the discrete function is smooth and the number of data points is sufficient, the interpolant could converge within a preset tolerance level (i.e., $|R_n(x_0, \ldots, x_n)| < \varepsilon$) in which case the interpolating polynomial is truncated and the rest of the terms are discarded.

Pseudocode 6.3

Function Module NEWTONS_DD_EVAL $(n, xval, \mathbf{x}, \mathbf{f}, \mathbf{a})$
\ DESCRIPTION: A pseudofunction to find an interpolation $fval = f(xval)$, using
\   the Newton's Divided-Difference (NDD).
\ USES:
\   NEWTONS_DD_COEFFS:: Module computing NDD coefficients.
\   CAUTION: For ε, use MACH_EPS or set a value internally.
Declare: $x_{0:n}, f_{0:n}, a_{0:n}$ \ Declare array variables
NEWTONS_DD_COEFFS$(n, \mathbf{x}, \mathbf{f}, \mathbf{a})$ \ Get Newton's Divided Diff. coefficients
$fval \leftarrow a_0$ \ Initialize interpolant
$prod \leftarrow 1; \ k \leftarrow 0$ \ Initialize accumulator and counter
Repeat   \ Conditional accumulator
    $k \leftarrow k + 1$ \ Count terms
    $prod \leftarrow prod * (xval - x_{k-1})$ \ Accumulate, $prod = \prod_{i=0}^{k-1}(x - x_i)$
    $Rn \leftarrow prod * a_k$ \ Evaluating a term, Eq. (6.22)
    $fval \leftarrow fval + Rn$ \ Add term to interpolant
Until $\left[|Rn| < \varepsilon \textbf{ Or } k = n \right]$ \ Truncate the sum if $|Rn| < \varepsilon$ or $k = n$
NEWTONS_DD_EVAL$\leftarrow fval$ \ Set interpolant to function
End Function Module NEWTONS_DD_EVAL

Module NEWTONS_DD_COEFFS $(n, \mathbf{x}, \mathbf{f}, \mathbf{a})$
\ DESCRIPTION: A pseudomodule generating the NDD interpolation coefficients.
Declare: $x_{0:n}, f_{0:n}, a_{0:n}$ \ Declare array variables
$\mathbf{a} \leftarrow \mathbf{f}$ \ Initialize the coefficients, $a_i \leftarrow f_i$ for all i
For $\left[j = 1, n \right]$ \ Loop j: Sweep columns from left to right
    For $\left[k = n, j, (-1) \right]$ \ Loop k: Differences from last to 1st row
        $a_k \leftarrow (a_k - a_{k-1})/(x_k - x_{k-j})$ \ Divided differences
    End For
End For
End Module NEWTONS_DD_COEFFS

EXAMPLE 6.4: A unique feature of divided-difference table

The following data was obtained from a study investigating the effect of reaction time t (in hours) on the yield of a chemical product y (mol %). Use the given data to construct an interpolating polynomial for the yield that passes through all data points.

t (h)	1	2	4	8	11
$y(t)$ (mol%)	24.7	44	70	88.4	91.7

SOLUTION:

The number of data points in this data set is 5, so the interpolating polynomial passing through all points will be at most a 4th-degree polynomial ($n = 5 - 1 = 4$).

Using the data furnished, the Newton's divided differences are computed, and the resulting table is presented in Table 6.4. Notice that the third divided-differences (entries of the sixth column) are 0.1, and the fourth divided-difference (i.e., last column) is "zero." This indicates that the set of data can be exactly represented by a third-degree polynomial because the fourth divided difference is zero.

TABLE 6.4

i	t_i	y_i	$y[t_i, t_{i+1}]$	$y[t_i, t_{i+1}, t_{i+2}]$	$y[t_i, \ldots, t_{i+3}]$	$y[t_i, \ldots, t_{i+4}]$
0	1	24.7	19.3	−2.1	0.1	0
1	2	44.0	13.0	−1.4	0.1	
2	4	70.0	4.6	−0.5		
3	8	88.4	1.1			
4	11	91.7				

Finally, using the NDD coefficients in the first row of the table, the NDD interpolating polynomial yields

$$y(t) = 24.7 + (t-1)\,(19.3 + (t-2)\,(-2.1 + (t-4)\,(0.1)))$$
$$= 0.1\,t^3 - 2.8\,t^2 + 27\,t + 0.4$$

Discussion: This resulting interpolating polynomial is, in fact, the *true polynomial* from which the data was generated. Also notice that the interpolating polynomial turned out to be a 3rd-degree polynomial, even though we initially predicted that the interpolating polynomial would be at most 4th-degree.

If a discrete data set contains noise or experimental errors, the interpolating polynomial that passes through every available data point will be tainted due to the inclusion of the errors in the interpolating polynomial.

6.3 NEWTON'S FORMULAS FOR UNIFORMLY SPACED DATA

Newton's divided-difference or Lagrange methods can certainly be applied to uniformly spaced data. However, using interpolating polynomials or formulas especially suited for uniform data not only simplifies formulations but also reduces the computation time considerably. In this section, the derivation and use of interpolating polynomials are emphasized and customized for uniformly spaced data.

Consider a set of uniformly-spaced data: (x_k, f_k) for $k = 0, 1, \ldots, n$, where $x_k = x_0 + kh$ and $f_k = f(x_k)$. We make use of the (forward, backward, and central) difference operators (*see* Chapter 5) to obtain a more compact form for the interpolation formulas:

$$\Delta f_k = f_{k+1} - f_k, \qquad \nabla f_k = f_k - f_{k-1}, \qquad \delta f_k = f_{k+\frac{1}{2}} - f_{k-\frac{1}{2}} \qquad (6.23)$$

Also recall that the nth-order difference operators are related to lower orders with the following recurrence relationships:

$$\Delta^n f_k = \Delta(\Delta^{n-1} f_k), \qquad \nabla^n f_k = \nabla(\nabla^{n-1} f_k), \qquad \delta^n f_k = \delta(\delta^{n-1} f_k) \qquad (6.24)$$

Newton's interpolation formula can be used to obtain more simplified interpolating polynomials for uniformly spaced data points. Several versions of the interpolation formula are known as Gregory-Newton interpolation formulas. Now consider a set of uniform data

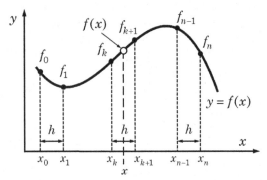

FIGURE 6.7: An example of uniform discrete data distribution.

on $[x_0, x_n]$ with n intervals. We seek a high-order interpolation formula for any x in $x_k \leqslant x \leqslant x_{k+1}$ as shown in Fig. 6.7. Noting that $a_0 = f[x_k] = f_k$ and making use of the forward difference operator, Eqs. (6.11) and (6.12) can be rewritten as

$$a_1 = \frac{f_{k+1} - f_k}{h} = \frac{\Delta f_k}{h} \tag{6.25}$$

$$a_2 = \frac{1}{2h}\left(\frac{\Delta f_{k+1}}{h} - \frac{\Delta f_k}{h}\right) = \frac{\Delta^2 f_k}{2h^2} \tag{6.26}$$

The remaining coefficients are obtained in the same way, leading to a generalized expression:

$$a_n = \frac{\Delta^n f_k}{n!\, h^n} \tag{6.27}$$

where $\Delta^n f_k$ is the nth forward-difference evaluated at the base point x_k, and $\Delta^0 f_k = f_k$.

Substituting Eqs. (6.25), (6.26), and (6.27) (i.e., the modified coefficients) into Eq. (6.10) yields

$$\begin{aligned} f(x) \cong p_n(x) = f_k &+ \frac{(x - x_k)}{h}\Delta f_k + \frac{(x - x_k)(x - x_k - h)}{2!h^2}\Delta^2 f_k \\ &+ \frac{(x - x_k)(x - x_k - h)(x - x_k - 2h)}{3!h^3}\Delta^3 f_k + \dots \\ &+ \frac{(x - x_k)(x - x_k - h)\dots(x - x_k - (n-1)h)}{n!h^n}\Delta^n f_k \end{aligned} \tag{6.28}$$

which can be expressed in a more compact form as

$$\begin{aligned} f(x) \cong p_n(s) = f_k &+ s\,\Delta f_k + \frac{s(s-1)}{2!}\Delta^2 f_k + \frac{s(s-1)(s-2)}{3!}\Delta^3 f_k + \dots \\ &+ \frac{s(s-1)\dots(s-n+1)}{n!}\Delta^n f_k = \sum_{m=0}^{n}\binom{s}{m}\Delta^m f_k \end{aligned} \tag{6.29}$$

where $s = (x - x_k)/h$ is referred to as the interpolating variable and

$$\binom{s}{m} = \frac{s(s-1)\cdots(s-m+1)}{m!}$$

Equation (6.29) is referred to as the *Gregory-Newton Forward Interpolation formula* (GN-FWD-IF). Note that the forward differences Δf_k, $\Delta^2 f_k$, $\Delta^3 f_k$, $\dots$, etc., correspond to the

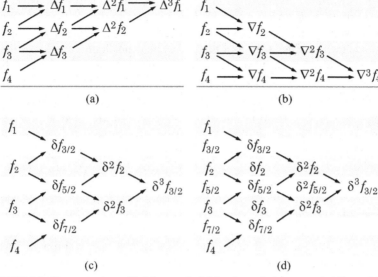

FIGURE 6.8: Construction of (a) forward, (b) backward, (c) central, (d) modified central difference tables.

values in row x_k. Likewise, an alternative interpolation formula, the *Gregory-Newton Backward Interpolation formula* (GN-BKWD-IF), can also be constructed by using backward differences as

$$
f(x) \cong p_n(s) = f_k + s\nabla f_k + \frac{s(s+1)}{2!}\nabla^2 f_k + \frac{s(s+1)(s+2)}{3!}\nabla^3 f_k + \dots
$$
$$
+ \frac{s(s+1)\dots(s+n-1)}{n!}\nabla^n f_k = \sum_{m=0}^{n} (-1)^m \binom{-s}{m} \nabla^m f_k
$$
(6.30)

where s has the usual definition. Both interpolation formulas, Eq. (6.29) and Eq. (6.30), satisfy $(n+1)$ data points.

It may be tempting to add as many terms as are available in the Gregory-Newton interpolation polynomials to obtain the most accurate approximation; however, if the discrete function is not smooth, adding up all or most terms does not guarantee improved accuracy. Adding each term in the interpolation polynomial increases the degree of the interpolating polynomial by one; that is, for the backward interpolation formula, we write

$$
\text{Linear} \quad : \quad p_1(s) = f_k + s\,\nabla f_k,
$$
$$
\text{Quadratic} \quad : \quad p_2(s) = p_1(s) + \frac{1}{2!}s(s+1)\,\nabla^2 f_k,
$$
$$
\text{Cubic} \quad : \quad p_3(s) = p_2(s) + \frac{1}{3!}s(s+1)(s+2)\,\nabla^3 f_k, \; \dots
$$

The forward and backward differences ($\Delta^n f_k$ and $\nabla^n f_k$) appearing in Eqs. (6.29) and (6.30) can be easily obtained by constructing difference tables similar to Newton's divided difference table. A forward difference table, in Fig. 6.8a, is constructed in ascending order with additional columns consisting of the differences between the values in the preceding column, i.e., $\Delta^n f_k = \Delta^{n-1} f_{k+1} - \Delta^{n-1} f_k$ (or $\nabla^n f_k = \nabla^{n-1} f_k - \nabla^{n-1} f_{k-1}$ for backward differences). It should be pointed out that the numerical values of the differences in the forward and backward difference tables are the same, but their notations (i.e., locations in the table) are different. In other words, $\nabla f_2 = f_2 - f_1$ and $\Delta f_1 = f_2 - f_1$ represent the same numerical value, but Δf_1 is placed into the first row while ∇f_2 is placed into the second row, as shown in Figs. 6.8a and 6.8b. On the other hand, the lower diagonal in the forward difference table or the upper diagonal in the backward difference table cannot be filled because an entry is lost for every new column for which a new set of differences is computed. The table size depends on the number of data points in the data set.

The GN-FWD or GN-BKWD interpolation formulas are used almost exclusively for interpolations near the beginning or end of a table, respectively. Thus, these formulas may not be appropriate when interpolating for x values near the middle of the tables. For such cases, interpolating formulas that make use of central differences are preferred. A *central difference table* is similarly constructed by using the central differences. In Fig. 6.8c, a new line (in plain background) between each line of data (with highlighted background) is inserted as shown. At this point, we define the lines containing the original data as *"full lines,"* and distinguish the new lines from the full lines as *"half lines."* The first central difference, $\delta f_{i+1/2} = f_{i+1} - f_i$, is computed using the values in the first column and written in the second column, exactly halfway between the two lines of values involved. This differencing procedure is repeated for subsequent columns by using the values of the previous column, i.e., $\delta^n f_{i+1/2} = \delta^{n-1} f_{i+1} - \delta^{n-1} f_i$. The final form of the table contains blank spaces on both half and full lines. These spaces are filled out by simply taking the arithmetic average of the numerical values above and below each blank space, as illustrated in Fig. 6.8d. The new table, the *modified central difference table*, which provides additional entries, is used to construct the central difference interpolating polynomials.

A number of interpolating formulas requiring central differences have been described in the literature with slightly different properties, two of the most important of which are presented here. The *Stirling formula* is used along with the full lines, while the *Bessel formula* uses the central differences along the half lines.

Stirling's Formula (full lines as the base line):

$$f(x) = f_k + s\,\delta f_k + \frac{s^2}{2!}\delta^2 f_k + \frac{s(s^2-1)}{3!}\delta^3 f_k + \frac{s^2(s^2-1)}{4!}\delta^4 f_k + \frac{s(s^2-1)(s^2-4)}{5!}\delta^5 f_k$$
$$+ \frac{s^2(s^2-1)(s^2-4)}{6!}\delta^6 f_k + \dots \tag{6.31}$$

Bessel's Formula (half lines as the base line):

$$f(x) = f_k + s(\delta f_k) + \frac{1}{2!}\left(s^2 - \frac{1}{4}\right)\delta^2 f_k + \frac{s}{3!}\left(s^2 - \frac{1}{4}\right)\delta^3 f_k + \frac{1}{4!}\left(s^2 - \frac{1}{4}\right)\left(s^2 - \frac{9}{4}\right)\delta^4 f_k$$
$$+ \frac{s}{5!}\left(s^2 - \frac{1}{4}\right)\left(s^2 - \frac{9}{4}\right)\delta^5 f_k + \dots \tag{6.32}$$

where $s = (x - x_k)/h$ and x_k is the corresponding abscissa of the base line (either a full or a half line).

Once forward, backward, or central difference tables are constructed, the GN-FWD or GN-BKWD and Bessel or Stirling interpolation formulas can be used to find interpolates for a discrete data set. The best estimates are obtained if the base line is chosen as the line in the table that has a large number of entries. This strategy allows for the construction of a high-degree interpolating polynomial. For a smooth function, successively higher differences along the chosen base line become smaller in magnitude. This implies that increasing the degree of the interpolating polynomial improves the accuracy of the interpolate. However, increasing the degree of the interpolation polynomial is not advised if higher successive differences depict an increase in magnitude or oscillate about the base line. A simple guideline for selecting the optimal base line is to choose the line in the table closest to the interpolation point so that the resulting value of s is the smallest, provided that the higher differences are decreasing in magnitude. In order to guarantee a fairly large number of entries in the base line, we generally prefer the GN-FWD-IF if the value to be interpolated is near the top of the table and the -BKWD-IF if it is near the bottom of the table. For the interpolations near the center of the data set, central difference interpolation formulas (Stirling or Bessel) are usually used.

Pseudocode 6.4

Module GN_FWD_EVAL $(n, xval, \mathbf{x}, \mathbf{f}, \mathbf{a}, fval)$
\ DESCRIPTION: A pseudomodule to interpolate uniformly spaced discrete data
\   using the Gregory-Newton Forward interpolation.
Declare: x_n, f_n, $a_{n,n+1}$ \ Declare array variables
BINARY_SEARCH$(n, xval, \mathbf{x}, k)$ \ Find first index k of $[x_k, < x_{k+1}]$
$h \leftarrow x_2 - x_1$ \ Find interval size
$s \leftarrow (xval - x_k)/h$ \ Find interpolation variable
$coef \leftarrow 1$ \ Initialize numerator
$factor \leftarrow 1$ \ Initialize denominator
$fval \leftarrow a_{k2}$ \ Initialize the interpolant
For $\left[i = 3, n + 1\right]$ \ Loop i: Evaluate interpolant
    $coef \leftarrow coef * (s + 3 - i)$ \ Find $s(s+1)\ldots(s+n)$
    $factor \leftarrow factor * (i - 2)$ \ Find $(i-2)!$
    $Rn \leftarrow coef * a_{ki}/factor$ \ Find a next term
    $fval \leftarrow fval + Rn$ \ Apply Eq. (6.29)
End For
End Module GN_FWD_EVAL

Module GN_FWD_TABLE $(n, \mathbf{x}, \mathbf{f}, \mathbf{a})$
\ DESCRIPTION: A pseudomodule generate GN Forward interpolation table.
Declare: x_n, f_n, $a_{n,n+1}$ \ Declare array variables
For $\left[i = 1, n\right]$ \ Loop i: Placa data into the table
    $a_{i1} \leftarrow x_i$ \ Place abscissas in 1st column
    $a_{i2} \leftarrow f_i$ \ Place ordinates in 2nd column
End For
For $\left[j = 3, n + 1\right]$ \ Generate jth column
    For $\left[i = 1, n - j + 2\right]$ \ Sweep column from top to bottom
        $a_{ij} \leftarrow a_{i+1,j-1} - a_{i,j-1}$ \ Find forward differences for jth column
    End For
End For
End Module GN_FWD_TABLE

Module BINARY_SEARCH $(n, xval, \mathbf{x}, k)$
\ DESCRIPTION: A pseudomodule to find index k the interval $x_k \leqslant xval < x_{k+1}$
\   using the Binary Search algorithm.
Declare: x_n \ Declare array variable
$First \leftarrow 1$; $Last \leftarrow n$ \ Set 1st & last index of array
While $\left[Last - first > 1\right]$ \ while two elements are apart from each other

$Middle \leftarrow (Last + First)/2$ \ Find index of mid point
If $\left[xval > x_{Middle}\right]$ **Then**   \ $xval$ is on the rhs $(Middle, Last)$
    $First \leftarrow Middle$ \ Narrow down from lhs, $First = Middle$
Else
    If $\left[xval < x_{Middle}\right]$ **Then**   \ $xval$ is on the lhs $(First, Middle)$
       $Last \leftarrow Middle$ \ Narrow down from rhs, $Last = Middle$
    Else   \ Case of $x_{Middle} = xval$
       $First \leftarrow Middle; Last \leftarrow Middle$
       Exit
    End If
  End If
End While
$k \leftarrow First$ \ Set first index k of interval to $First$
End Module BINARY_SEARCH

A pseudomodule, GN_FWD_TABLE, generating the forward difference table of a uniformly spaced discrete data set is presented in Pseudocode 6.4. As input, the module requires the number of data points (n) and the data set (x_i, f_i) for $i = 1, 2, \ldots, n$. First, the input data are placed on the 1st and 2nd columns of the generating table, i.e., $(x_i, f_i) \rightarrow (a_{i1}, a_{i2})$. Starting from the third column, the first forward difference and subsequent higher forward differences are computed using the prior column entries with $a_{ij} \leftarrow a_{i+1,j-1} - a_{i,j-1}$ and placed in the same column in the table. The module output is the generated forward difference table (**a**). It should be noted that this module should be called only once in the main program, and then numerous interpolations can simply be performed by using the evaluation module, GN_FWD_EVAL.

The pseudomodule, GN_FWD_EVAL, uses the forward difference table (**a**) in addition to n and **x**. The module for an input value of $xval$ computes the interpolant $fval$. The first subscript of the interval (k) containing $xval$ is determined (i.e., $x_k \leqslant xval < x_{k+1}$) using the module BINARY_SEARCH, implementing the binary search algorithm. Then, using the interpolation variable $s \leftarrow (xval - x_k)/h$, the forward interpolation formula, Eq. (6.29) is used to find the interpolate. This module computes the interpolant with a polynomial of the highest possible degree, i.e., all available differences along the nearest baseline are incorporated into the interpolation formula. Nevertheless, it is prudent to check for the smoothness of the discrete data before indulging in the computation of an estimate with an interpolation polynomial of the highest degree. The module can be easily modified by adapting the strategy implemented in the NEWTONS_DD_EVAL module presented in Pseudocode 6.3. The Pseudocode 6.4 can be used for Gregory-Newton backward interpolation as well as central difference interpolation formulas with minor modifications.

We can determine whether a set of discrete data is smooth or not by inspecting the entry values of a difference table. If the discrete data is smooth, the successively higher differences along the chosen base line will become smaller in magnitude. In such cases, a high-degree interpolating polynomial will produce very accurate interpolates.

ERROR ASSESSMENT. As we have seen, the GN-FWD or GN-BKWD interpolation formulas are essentially generated from Newton's divided differences; thus, the error bound for

TABLE 6.5: Error propagation in difference tables.

$$\delta^4 f_{k-2} + \varepsilon$$

$$\delta^3 f_{k-\frac{3}{2}} + \varepsilon$$

$$f_{k-1} \qquad \delta^2 f_{k-1} + \varepsilon \qquad \delta^4 f_{k-1} - 4\varepsilon$$

$$\delta f_{k-\frac{1}{2}} + \varepsilon \qquad \qquad \delta^3 f_{k-\frac{1}{2}} - 3\varepsilon$$

$$f_k + \varepsilon \qquad \delta^2 f_k - 2\varepsilon \qquad \delta^4 f_k + 6\varepsilon$$

$$\delta f_{k+\frac{1}{2}} - \varepsilon \qquad \qquad \delta^3 f_{k+\frac{1}{2}} + 3\varepsilon$$

$$f_{k+1} \qquad \delta^2 f_{k+1} + \varepsilon \qquad \delta^4 f_{k+1} - 4\varepsilon$$

$$\delta f_{k+\frac{3}{2}} - \varepsilon$$

$$\delta^4 f_{k+2} + \varepsilon$$

an nth degree interpolating polynomial, $p_n(x)$, can be obtained by

$$|e_n(x)| = |f(x) - p_n(x)| = \left| s(s \pm 1)(s \pm 2)\cdots(s \pm n)\frac{h^{n+1} f^{(n+1)}(\xi)}{(n+1)!} \right|, \quad x_0 \leqslant \xi \leqslant x_n$$

(6.33)

The order of magnitude of the *approximation error*, $|e_n(x)|$, is about the next difference not used in $|p_n(x)|$. The multiplying factor $|s(s \pm 1)(s \pm 2)\cdots(s \pm n)|$ is an nth-degree polynomial with decreasing magnitude near the data end points ($s = 0$ and $s = \pm n$). Since $e_n(x)$ is proportional to h^{n+1}, the magnitude of the error decreases for $h < 1$ with increasing n. On the other hand, if $f^{(n+1)}(\xi)$ derivative grows larger with n, the magnitude of the error decreases only if h is sufficiently small. Neglecting high-order differences could also give rise to *truncation errors*.

The accuracy of the data in the set *itself* is also crucial in terms of the overall accuracy of the interpolation. Sometimes the data contains errors that the interpolation formulas cannot be blamed for. In this context, the difference tables can serve as an important tool for checking the smoothness of the discrete data set or detecting isolated errors. To illustrate this, consider an error of "ε" in a certain tabular value; that is, we assume that all data in the set are correct except for one that is in $+\varepsilon$ error. As illustrated in Table 6.5, the isolated error is magnified and spread out in a triangular pattern as higher differences are generated. While the largest error occurs in the same row as the erroneous value, the algebraic sum of the errors in any column is zero. A single-entry error yields the binomial coefficient pattern for any column, and high-order differences greatly magnify the errors. For this reason, high-degree interpolating polynomials are very sensitive to errors in the data. As we have seen, a high-degree interpolating polynomial may yield estimates that significantly diverge from

Gregory-Newton Interpolation Formulas

- The method has the same advantages as Newton's divided differences;
- They require fewer arithmetic operations and less cpu-time since they are derived specifically for uniformly spaced data;
- For smooth distributions, the convergence (or desired accuracy) can be monitored by $|p_n(s) - p_{n-1}(s)| < \varepsilon$.

- As a method, it is no different from Newton's divided difference;
- They can only be applied to uniformly spaced discrete functions.

the true value, just as the Lagrange interpolating polynomial does. It is therefore advised to construct error propagation tables to carry out an exploratory data analysis. Also note that even if an original data (error-free) set is used, errors can still be introduced during the construction of the difference tables. For a large data set, the arithmetic operations may inevitably yield notable *round-off errors*.

EXAMPLE 6.5: Interpolation with uniform data sets

The following data are read from a bacterial growth curve:

t (h)	0.5	1.0	1.5	2.0	2.5
$f(t)$	0.75	2	3.375	4.8	6.25

where t is the time in hours, $f(t) = \log_{10} n(t)$, and $n(t)$ is the bacterial population. (a) Use the data to generate the difference tables and develop 2nd, 3rd, and 4th-degree GN-FWD, GN-BKWD, and central interpolating polynomials; (b) use the polynomials derived in Part (a) to compute $f(0.55)$, $f(0.7)$, $f(1.76)$, and $f(2.3)$. The true solution (growth curve) is given as $\log_{10} n(t) = 6t^2/(1 + 2t)$.

SOLUTION:

(a) In this exercise, a set of discrete data generated from a known function is used to illustrate the implementation and performance of the interpolating formulas. In reality, the true solution is unknown; thus, the true interpolation errors cannot be determined with absolute certainty. In general, the best we can do is estimate the trend of the true function based on our expertise in the physical or numerical problem.

Considering the given data set, at most a 4th-degree ($n = 5 - 1 = 4$) interpolating polynomial can be generated from the set. As a first step, pertinent difference tables should be constructed. The generated forward, backward, and modified central difference tables are presented, respectively, in Tables 6.6, 6.7, and 6.8.

TABLE 6.6

i	t_i	f_i	Δf_i	$\Delta^2 f_i$	$\Delta^3 f_i$	$\Delta^4 f_i$
1	0.5	0.75	1.25	0.125	−0.075	0.05
2	1	2	1.375	0.05	−0.025	
3	1.5	3.375	1.425	0.025		
4	2	4.8	1.45			
5	2.5	6.25				

TABLE 6.7

i	t_i	f_i	∇f_i	$\nabla^2 f_i$	$\nabla^3 f_i$	$\nabla^4 f_i$
1	0.5	0.75				
2	1	2	1.25			
3	1.5	3.375	1.375	0.125		
4	2	4.8	1.425	0.05	−0.075	
5	2.5	6.25	1.45	0.025	−0.025	0.05

TABLE 6.8

i	t_i	f_i	δf_i	$\delta^2 f_i$	$\delta^3 f_i$	$\delta^4 f_i$
1	0.5	0.75				
3/2	**0.75**	**1.3750**	1.25			
2	1	2	**1.3125**	0.125		
5/2	**1.25**	**2.6875**	1.375	**0.0875**	−0.075	
3	1.5	3.375	**1.4**	0.05	**−0.050**	0.05
7/2	**1.75**	**4.0875**	1.425	**0.0375**	−0.025	
4	2	4.8	**1.4375**	0.025		
9/2	**2.25**	**5.525**	1.45			
5	2.5	6.25				

The numerical values in the tables are the same, but their positions in the columns change. It should also be noted that the successively higher differences along any line (from left to right) are getting smaller in magnitude, which implies that the discrete data set is *smooth*. As a result, we can expect to make very accurate estimates with high-degree interpolating polynomials.

To develop interpolation polynomials, we first need to decide which interpolation formula to use. For $t = 0.55$ and 0.7, we will prefer the GN-FWD-IF since there are more entries in the baseline $x = 0.5$ in Table 6.6; these entries are $f_1 = 0.75$, $\Delta f_1 = 1.25$, $\Delta^2 f_1 = 0.125$, $\Delta^2 f_1 = -0.075$, and $\Delta^4 f_1 = 0.05$. For this case, Eq. (6.29) becomes

$$f(t) \cong p_4(t) = 0.75 + s(1.25) + \frac{s(s-1)}{2!}(0.125) + \frac{s(s-1)(s-2)}{3!}(-0.075)$$
$$+ \frac{s(s-1)(s-2)(s-3)}{4!}(0.05)$$

where $s = (t - t_1)/h = 2t - 1$.

A general interpolating polynomial in terms of t can be found by substituting $s = 2t - 1$ into the GN-FWD interpolation formula. Simplifying leads to a 4th-degree interpolating polynomial:

$$p_4(t) = \frac{1}{30}t^4 - \frac{4}{15}t^3 + \frac{101}{120}t^2 + \frac{197}{120}t - \frac{1}{4}, \qquad 0.5 \leqslant t \leqslant 2.5$$

Likewise, for $t_1 = 0.5$ as the baseline, linear, quadratic, and cubic interpolating polynomials are also obtained as follows:

$$p_1(t) = f_1 + s\,\Delta f_1 = \frac{1}{2}(5t - 1),$$

$$p_2(t) = p_1(t) + \frac{s(s-1)}{2!}\Delta^2 f_1 = \frac{1}{8}(2t^2 + 17t - 3),$$

$$p_3(t) = p_2(t) + \frac{s(s-1)(s-2)}{3!}\Delta^3 f_1 = \frac{1}{20}(-2t^3 + 11t^2 + 37t - 6)$$

It is also possible to generate GN-FWD (or GN-BKWD or central) interpolating polynomials using any one of the other rows (t_k, $k = 2$, 3, 4, or 5) as the baseline. For instance, the entries for baseline $t_2 = 1$ ($s = (t - t_2)/h = 2t - 2$) are read from

Table 6.6 as $f_2 = 2$, $\Delta f_2 = 1.375$, $\Delta^2 f_2 = 0.05$, and $\Delta^3 f_2 = -0.025$. Then, Eq. (6.29) for $n = 1$, 2, and 3 yields

$$p_1(t) = \frac{1}{4}(11t - 3),$$

$$p_2(t) = \frac{1}{10}(t^2 + 25t - 6),$$

$$p_3(t) = \frac{1}{60}(-2t^3 + 15t^2 + 137t - 30)$$

Note that a 4th-degree interpolating polynomial cannot be constructed with $t_2 = 1$ as a baseline due to having fewer entries.

In Table 6.6, contrary to Table 6.8, there are fewer entries corresponding to lines (rows) of $t = 1.7$ and 2.5. Thus, it is suitable to use GN-BKWD-IF since there are more entries in the $t = 2$ and 2.5 rows. Using $t_5 = 2.5$ as the baseline ($s = (t - t_5)/h = 2t - 5$), the baseline entries are read from Table 6.8 as $f_5 = 6.25$, $\nabla f_5 = 1.45$, $\nabla^2 f_5 = 0.0255$, $\nabla^3 f_5 = -0.025$, and $\nabla^4 f_5 = 0.05$. The GN-BKWD interpolating polynomial that passes from all five data points, Eq. (6.30), leads to

$$f(t) \cong p_4(t) = \frac{1}{30}t^4 - \frac{4}{15}t^3 + \frac{101}{120}t^3 + \frac{197}{120}t - \frac{1}{4}, \qquad 0.5 \leqslant t \leqslant 2.5$$

Notice that $p_4(t)$ is the same as the polynomial obtained from the GN-FWD-IF. The other lower-degree backward interpolating polynomials can also be constructed as follows:

$$p_1(t) = f_5 + s \nabla f_5 = \frac{1}{10}(29t - 10),$$

$$p_2(t) = p_1(t) + \frac{s(s+1)}{2!}\nabla^2 f_5 = \frac{1}{40}(2t^2 + 107t - 30),$$

$$p_3(t) = p_2(t) + \frac{s(s+1)(s+2)}{3!}\nabla^3 f_5 = \frac{1}{60}(-2t^3 + 15t^2 + 137t - 30)$$

For interpolations (t values) near the middle of the dataset, the number of row-wise data in the near middle of Tables 6.6 and 6.7 is fewer than that of Table 6.8. Since we prefer to include as many terms as possible for smooth distributions, the central difference (Stirling or Bessel) interpolation formulas can be used to interpolate points near the middle of a dataset. In this text, only Stirling and Bessel's interpolation formulas were introduced. Which one to use is best determined by how close the interpolation point is to a *full* or a *half line*, the number of entries in the baseline, whether a distribution is smoot or not, and so on.

The row with the most number of entries in Table 6.8 is observed to be for $t_3 = 1.5$ (a *full line*), whose entries from left to right are also decreasing in magnitude. The interpolation formula best suited for this case is Stirling's formula. Using $t_3 = 1.5$ as the base line ($s = (t - t_3)/h = 2t - 3$), the entries along the baseline are determined to be $f_3 = 3.375$, $\delta f_3 = 1.4$, $\delta^2 f_3 = 0.05$, $\delta^3 f_3 = -0.05$, and $\delta^4 f_3 = 0.05$. However, as expected, the interpolating polynomials (with degrees lower than four) presented below are different from those obtained with the forward and backward interpolating formulas.

$$p_1(t) = f_3 + s\,\delta f_3 = \frac{1}{40}(112t - 33),$$

$$p_2(t) = p_1(t) + \frac{s^2}{2!}\delta^2 f_3 = \frac{1}{10}(t^2 + 25t - 6),$$

$$p_3(t) = p_2(t) + \frac{s(s^2 - 1)}{3!}\delta^3 f_3 = \frac{1}{15}(-t^3 + 6t^2 + 31t - 6)$$

Stirling's formula that passes all data points yields the *same* 4th-degree polynomial obtained by the GN-FWD and -BKWD-IFs.

(b) The 1st to 4th-degree interpolating polynomials generated by taking $t_1 = 0.5$ as the baseline and the estimates computed for $t = 0.55$, 0.7, 1.76, and 2.30 are comparatively presented along with the true values in Table 6.9. It is observed that as the degree of the interpolating polynomial increases, the estimates approach the true values. Furthermore, the estimates calculated by $p_4(x)$ or $p_3(x)$ yield fairly good agreements due to the smoothness of the discrete distribution, i.e., $|\Delta f_0| > |\Delta^2 f_0| > |\Delta^3 f_0|$, and so on.

TABLE 6.9

t	$f(t)$	$p_1(t)$	$p_2(t)$	$p_3(t)$	$p_4(t)$
0.55	0.864286	0.875	0.869375	0.8672375	0.8642044
0.70	1.225000	1.250	1.2350	1.2302000	1.2281200
1.76	4.111858	3.900	4.1394	4.1145024	4.1125106
2.3	5.667857	5.250	5.8350	5.647800	5.666520

To determine how well the interpolating polynomials behave overall in the range $0.5 \leqslant t \leqslant 2.5$, the true error distribution, $e_n(t) = f(t) - p_n(t)$, is also presented graphically in Fig. 6.9a. The maximum error in $p_4(x)$ is about 0.003, so $e_4(x)$ is practically flat. Since it is constructed from the first three data points, the $p_3(x)$ does not pass through the rightmost data point, $t = 2.5$. Therefore, the $p_3(x)$ depicts relatively larger deviations for $2 < t \leqslant 2.5$. The $p_2(t)$ corresponds to the case where the last two data points are removed from the generated polynomial, which explains why the estimates for $t = 2$ and 2.5 deviate from the true values as much as they do.

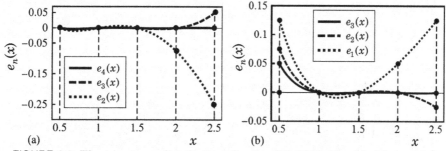

FIGURE 6.9: The true error distributions for the GN-FWD-IF with (a) $t_1 = 0.5$ and (b) $t_2 = 1$ as the baseline.

The GN-FWD interpolation polynomial that can be constructed with the number of data points available in row $t_2 = 1$ (as the baseline) can be at most a third degree. Hence, the true error distributions of the 1st, 2nd, and 3rd degree inter-polating polynomials are depicted in Fig. 6.9b. The agreement of the interpolating polynomials with the true distribution is observed to be better about the baseline, but this agreement deteriorates as t is moved further away from the baseline. On the other hand, the overall agreement improves with an increasing degree of interpolating polynomials.

The GN-BKWD interpolation polynomials are preferred when interpolations near the end of the data sets are desired. The 2nd, 3rd, and 4th-degree interpolating poly-nomials with $t_5 = 2.5$ as the baseline are different from those obtained by the GN-FWD-IFs. The true error distributions for $p_2(x)$, $p_3(x)$, and $p_4(x)$ are comparatively illustrated in Fig. 6.10a. In constructing $p_3(x)$ and $p_2(x)$, the data points $t_1 = 0.5$, and $t_1 = 0.5$ and $t_2 = 1$ have been removed, respectively. For this reason, the errors increase toward the lower end of the data $(t_1 = 0.5)$.

The agreement of the interpolating polynomials obtained by Stirling's formula with the true distribution is illustrated in Fig. 6.10b. The polynomials depict good agreement for estimates in the neighborhood of the baseline $(t_3 = 1.5)$ since the lower degree polynomials also use the interpolation point near the center, which serves as anchors.

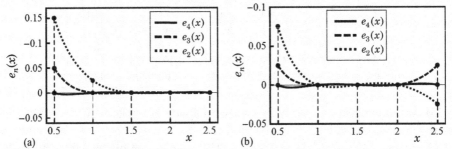

(a) (b)

FIGURE 6.10: Distribution of $e_n(t)$ with (a) the GN-BKWD-IF with $t_5 = 2.5$ as the baseline, and (b) the Stirling's formula with $t_3 = 1.5$ as the baseline.

The estimated interpolants computed by the 1st to 4th-degree interpolating poly-nomials generated with $t_5 = 2.5$ as the baseline, along with the true values, are comparatively presented for $t = 0.55, 0.7, 1.76$, and 2.30 in Table 6.10. For $p_3(t)$ and $p_4(t)$, the estimates depict agreements up to two-decimal places for values close to the baseline $(t > 2)$. Also note that the estimates with $p_4(t)$ are very good through-out the data interval, while $p_2(t)$ and $p_3(t)$ depict substantial deviations from the true values for $x < 1.5$ and $x < 1$, respectively.

TABLE 6.10

t	$f(t)$	$p_1(t)$	$p_2(t)$	$p_3(t)$	$p_4(t)$
0.55	0.864286	0.595	0.73638	0.825913	0.866204
0.70	1.225000	1.030	1.14700	1.209004	1.228120
1.76	4.111858	4.104	4.11288	4.111341	4.112511
2.3	5.667857	5.670	5.66700	5.668600	5.66652

Discussion: An interpolating polynomial generated by including all data points is the same regardless of the central, forward, or backward interpolation formula

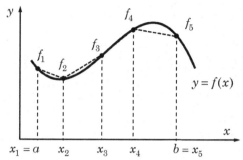

FIGURE 6.11: Illustration of linear splines employed to a data set of five.

used. That being the case, one might ask, "Should we construct an interpolation polynomial of the highest degree?" To answer this question, it is necessary to examine difference tables and determine whether the discrete distribution is smooth or not. For a smooth distribution, the magnitudes of the entries (higher differences) in any row should be decreasing. If the magnitudes of the higher differences oscillate or increase, use low-degree interpolating polynomials.

Generating an interpolation polynomial for a large dataset (even if it is smooth) can lead to Runge's phenomenon. Besides, using fewer data points in GN-FWD, GN-BKWD, or central interpolation formulas results in lower-order interpolation polynomials that are different (perhaps better) from each other depending on the baseline selected. In any case, a lower-degree interpolating polynomial should only be used in the vicinity of the preferred baseline. The best strategy for interpolation of large data sets is to divide the data set into smaller sets and generate a piecewise distribution by constructing low-order interpolation polynomials (at most 4th or 5th degrees) for each set.

6.4 CUBIC SPLINE INTERPOLATION

A polynomial interpolation interpolates between $n + 1$ points. As we have learned, a major problem with polynomial interpolation is that it often leads to unacceptably inaccurate estimates. The problem gets even worse when dealing with a large number of data points, which is, in practice, commonly encountered as Runge's phenomenon. Hence, interpolation with a single polynomial covering the whole data range is generally not advised. Alternatively, low-order piecewise polynomials can be employed throughout the data range to avoid Runge's phenomenon. Since only low-degree polynomials are used, the problem of over-fitting or excessive oscillations is eliminated.

A function composed of low-order piecewise polynomials is frequently referred to as a spline. A data point where two splines are joined is called a *knot*. Conventionally, a data interval is divided into more than one interval, where a linear, quadratic, or cubic polynomial is employed for each interval. In this respect, spline methods present a different approach than the polynomial interpolations covered so far.

The simplest spline method is a *linear spline*, where two data points are joined by a line. As shown in Fig. 6.11, a linear spline with a data set of five (n-points) consists of four lines ($y = m_k x + b_k$, $k = 1, \ldots, n - 1$). While $y = f(x)$ is continuous at the knots, its first

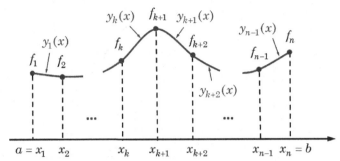

FIGURE 6.12: Cubic splines employed to a discrete data set.

derivative, $y'(x)$, is undefined. In other words, the linear splines lack smoothness, i.e., the left and right derivatives (slopes of the lines on each side of the knot) are not equal. In many cases, however, physical considerations require interpolating functions to be continuous and differentiable.

An alternative approach is to use quadratic (*quadratic spline*) or cubic polynomials (*cubic spline*) to achieve smoothness and greater accuracy. While only a couple of knots are sufficient to define a unique line, an infinite number of quadratic or cubic polynomials pass between any two knots. So to be able to define a unique high-degree spline, the continuity of the second derivative, $y''(x)$, in addition to $y(x)$ and $y'(x)$, must be satisfied at every knot.

The use of cubic polynomials is the most common choice in the literature. The reason for this is that cubic splines can be joined in different ways to produce an overall interpolating curve. At this stage, we shall consider non-uniformly spaced data and will employ a cubic polynomial $y_k(x)$, which has a different set of coefficients for each interval, $x_k \leqslant x \leqslant x_{k+1}$. Each cubic polynomial is then joined to its neighboring cubic polynomials $(y_{k-1}(x)$, $y_k(x)$, and $y_{k+1}(x))$ at the knots by matching the slopes and curvatures, i.e., $y'(x)$ and $y''(x)$. The treatments for the knots at both ends of the data interval will be covered in the subsequent discussions. Once the coefficients of the cubic polynomials are obtained, these piecewise polynomials are valid in the interval that they are specified.

Consider a nonuniform discrete data set consisting of (x_1, f_1), (x_2, f_2), ..., (x_n, f_n) data points (knots) as illustrated in Fig. 6.12. A cubic polynomial that will be imposed on $(n-1)$ intervals is expressed for the kth interval, $[x_k, x_{k+1}]$, as

$$y_k(x) = a_k(x - x_k)^3 + b_k(x - x_k)^2 + c_k(x - x_k) + d_k \quad [x_k, x_{k+1}] \quad (6.34)$$

In other words, the cubic polynomials valid in each interval are written as

$$y_1(x) = a_1(x - x_1)^3 + b_1(x - x_1)^2 + c_1(x - x_1) + d_1, \quad [x_1, x_2]$$
$$y_2(x) = a_2(x - x_2)^3 + b_2(x - x_2)^2 + c_2(x - x_2) + d_2, \quad [x_2, x_3]$$
$$\vdots \qquad \vdots$$
$$y_{n-1}(x) = a_{n-1}(x - x_{n-1})^3 + b_{n-1}(x - x_{n-1})^2 + c_{n-1}(x - x_{n-1}) + d_{n-1}, \quad [x_{n-1}, x_n]$$

This system consists of $(n-1)$ cubic polynomials (as the number of intervals) and $4 \times (n-1)$ unknowns (namely the coefficients of the cubic polynomials a, b, c, and d's). The following conditions will then be imposed to determine the unknowns:

(1) Continuity of $y(x)$: The cubic splines are continuous at *interior knots*; that is, interpolating polynomials must satisfy the data points:

$$y_k(x_k) = f_k, \qquad k = 1, 2, \ldots, (n-1), \qquad \text{and} \qquad y_{n-1}(x_n) = f_n; \tag{6.35}$$

$$y_k(x_{k+1}) = y_{k+1}(x_{k+1}) = f_{k+1}, \qquad k = 1, 2, \ldots, (n-2) \tag{6.36}$$

From the continuity condition, we find $(n-1)$ equations (or d_k's) as

$$y_k(x_k) = d_k = f_k, \qquad k = 1, 2, \ldots, (n-1) \tag{6.37}$$

Employing Eq. (6.37), the continuity of splines leads to

$$y_{k+1}(x_{k+1}) = y_k(x_{k+1}) = a_k \, \Delta x_k^3 + b_k \, \Delta x_k^2 + c_k \, \Delta x_k + f_k = f_{k+1} \tag{6.38}$$

where $\Delta x_k = x_{k+1} - x_k$.

(2) Continuity of $y'(x)$: The first derivatives of cubic splines are continuous at interior knots; that is,

$$y_k'(x_{k+1}) = y_{k+1}'(x_{k+1}), \qquad k = 1, \ldots, (n-2) \tag{6.39}$$

where

$$y_k'(x) = 3a_k(x - x_k)^2 + 2b_k(x - x_k) + c_k \tag{6.40}$$

By investigating the slopes at the knots from the right- and left-hand sides, we get

$$y_{k+1}'(x_{k+1}) = 3a_{k+1}(x_{k+1} - x_{k+1})^2 + 2b_{k+1}(x_{k+1} - x_{k+1}) + c_{k+1} = c_{k+1} \tag{6.41}$$

$$y_k'(x_{k+1}) = 3a_k \, (\Delta x_k)^2 + 2b_k \Delta x_k + c_k \tag{6.42}$$

The equality of Eq. (6.40) and (6.42) leads to

$$3a_{k-1} \, (\Delta x_{k-1})^2 + 2b_{k-1}\Delta x_{k-1} + c_{k-1} = c_k \tag{6.43}$$

Equation (6.43) is made up of $3 \times (n-1)$ equations, i.e., a_k, b_k, and c_k for $k = 1, 2, \ldots, n-2$.

(3) Continuity of $y''(x)$: The second derivatives (curvatures) of cubic splines are continuous at interior knots, that is,

$$y_k''(x_{k+1}) = y_{k+1}''(x_{k+1}), \qquad k = 1, 2, \ldots, (n-2) \tag{6.44}$$

The second derivative of Eq. (6.34) yields

$$y_k''(x) = 6a_k(x - x_k) + 2b_k, \tag{6.45}$$

Since $y''(x)$ of a cubic polynomial gives a linear relationship, its distribution on $[x_k, x_{k+1}]$ must also be linear. Defining $S_k = y''(x_k)$ for $k = 1, 2, \ldots, (n-1)$ and using this notation in subsequent derivations, we may write

$$y''(x_{k+1}) = S_{k+1} = 6a_k(x_{k+1} - x_k) + 2b_k \tag{6.46}$$

$$y''(x_k) = S_k = 6a_k(x_k - x_k) + 2b_k \tag{6.47}$$

Using Eqs. (6.46) and (6.47) to solve for a_k and b_k yields

$$b_k = \frac{S_k}{2}, \qquad a_k = \frac{S_{k+1} - S_k}{6 \, \Delta x_k} \tag{6.48}$$

Substituting a_k and b_k into Eq. (6.38) and solving for c_k results in

$$c_k = \frac{f_{k+1} - f_k}{\Delta x_k} - \Delta x_k \left(\frac{S_{k+1} + 2S_k}{6} \right) \tag{6.49}$$

Notice that we have introduced additional unknowns, namely the curvatures (i.e., S_k's). Next, to determine the curvatures at the knots, we substitute the coefficients of the polynomial, Eqs. (6.48) and (6.49), into Eq. (6.43)

$$\begin{aligned} y'_k = c_k &= \frac{f_{k+1} - f_k}{\Delta x_k} - \Delta x_k \left(\frac{S_{k+1} + 2S_k}{6} \right) \\ &= \frac{S_k - S_{k-1}}{2} \Delta x_{k-1} + S_{k-1} \Delta x_{k-1} + \left\{ \frac{f_k - f_{k-1}}{\Delta x_{k-1}} - \Delta x_{k-1} \left(\frac{S_k + 2S_{k-1}}{6} \right) \right\} \end{aligned} \tag{6.50}$$

Rearranging Eq. (6.50) gives

$$\Delta x_{k-1} S_{k-1} + 2(\Delta x_k + \Delta x_{k-1}) S_k + \Delta x_k S_{k+1} = 6 \left(\frac{\Delta f_k}{\Delta x_k} - \frac{\Delta f_{k-1}}{\Delta x_{k-1}} \right) \tag{6.51}$$

The above stipulations provide a set of $(3n - 5)$ equations: Eq. (6.34) gives $(n-1)$ equations, Eq. (6.35) and (6.36) yield $2 \times (n - 2)$ equations. However, the number of equations is still less than the number of unknowns. It is clear that two additional equations are required to complete a system of linear equations.

(4) End Conditions: Equation (6.51), which is valid for $k = 2, 3, \ldots, (n-1)$, provides $n - 2$ equations. Additionally, we need two equations to establish n equations with S_k as unknowns. These equations are supplied by estimating the curvature at the end points of the data range. Several spline formulations can be developed through the curvature estimations referred to as "end conditions." In this section, we consider only two cases: *natural splines* and *linear extrapolation* end conditions.

6.4.1   NATURAL SPLINE END CONDITION

The natural spline condition imposes the inflection point conditions at the end points. In other words, the discrete distribution is assumed to have *"inflection points"* at $x = x_1$ and $x = x_n$, where $y''(x_1) = 0$ and $y''(x_n) = 0$. Recalling the definition $S(x) = y''(x)$ in the preceding section, we may write

$$S_1 = 0 \quad \text{and} \quad S_n = 0 \tag{6.52}$$

Now that we have two more equations, we can construct the system of linear equations by using Eqs. (6.51) and (6.52) for $k = 2, 3, \ldots, (n-1)$ to give

$$\begin{aligned} \Delta x_1 (0) + 2(\Delta x_2 + \Delta x_1) S_2 + \Delta x_2 S_3 &= 6 \left(\frac{\Delta f_2}{\Delta x_2} - \frac{\Delta f_1}{\Delta x_1} \right) \\ \Delta x_2 S_2 + 2(\Delta x_3 + \Delta x_2) S_3 + \Delta x_3 S_4 &= 6 \left(\frac{\Delta f_3}{\Delta x_3} - \frac{\Delta f_2}{\Delta x_2} \right) \\ \vdots \qquad\qquad\qquad \vdots \\ \Delta x_{n-3} S_{n-3} + 2(\Delta x_{n-2} + \Delta x_{n-3}) S_{n-2} + \Delta x_{n-2} S_{n-1} &= 6 \left(\frac{\Delta f_{n-2}}{\Delta x_{n-2}} - \frac{\Delta f_{n-3}}{\Delta x_{n-3}} \right) \\ \Delta x_{n-2} S_{n-2} + 2(\Delta x_{n-1} + \Delta x_{n-2}) S_{n-1} + \Delta x_{n-1} (0) &= 6 \left(\frac{\Delta f_{n-1}}{\Delta x_{n-1}} - \frac{\Delta f_{n-2}}{\Delta x_{n-2}} \right) \end{aligned} \tag{6.53}$$

Solving the system of linear equations gives S_k's. When Eqs. (6.52) and (6.53) are combined, we obtain a tridiagonal system of linear equations given below:

$$
\begin{bmatrix}
\bar{d}_2 & \bar{a}_2 \\
\bar{b}_3 & \bar{d}_3 & \bar{a}_3 \\
& \bar{b}_4 & \bar{d}_4 & \bar{a}_4 \\
& & \ddots & \ddots & \ddots \\
& & & \bar{b}_{n-2} & \bar{d}_{n-2} & \bar{a}_{n-2} \\
& & & & \bar{b}_{n-1} & \bar{d}_{n-1}
\end{bmatrix}
\begin{bmatrix}
S_2 \\ S_3 \\ S_4 \\ \vdots \\ S_{n-2} \\ S_{n-1}
\end{bmatrix}
=
\begin{bmatrix}
c_2 \\ c_3 \\ c_4 \\ \vdots \\ c_{n-2} \\ c_{n-1}
\end{bmatrix}
\tag{6.54}
$$

where $\bar{a}_k = \Delta x_k$, $\bar{b}_k = \Delta x_{k-1}$, $\bar{d}_k = 2(\bar{a}_k + \bar{b}_k)$ and $c_k = 6(\Delta f_k/\Delta x_k - \Delta f_{k-1}/\Delta x_{k-1})$ which can also be expressed in terms of divided differences as $c_k = 6\left(f[x_k, x_{k+1}] - f[x_{k-1}, x_k]\right)$.

6.4.2 LINEAR EXTRAPOLATION END CONDITION

This is one of the most frequently used end-conditions. Instead of setting $S_1 = 0$ and $S_n = 0$ (natural splines), taking into account the fact that the second derivatives of cubic splines are linear, the end point values are estimated by linear extrapolation. Hence, linear extrapolation for the knot-2 yields

$$
\frac{S_2 - S_1}{\Delta x_1} = \frac{S_3 - S_2}{\Delta x_2} \quad \text{and} \quad S_1 = \left(1 + \frac{\Delta x_1}{\Delta x_2}\right)S_2 - \frac{\Delta x_1}{\Delta x_2}S_3
\tag{6.55}
$$

The linear extrapolation for knot number $(n-1)$ gives

$$
\frac{S_n - S_{n-1}}{\Delta x_{n-1}} = \frac{S_{n-1} - S_{n-2}}{\Delta x_{n-2}} \quad \text{and} \quad S_n = \left(1 + \frac{\Delta x_{n-1}}{\Delta x_{n-2}}\right)S_{n-1} - \frac{\Delta x_{n-1}}{\Delta x_{n-2}}S_{n-2}
\tag{6.56}
$$

Substituting $k = 2$ in Eq. (6.51), we find

$$
\Delta x_1 S_1 + 2(\Delta x_2 + \Delta x_1)S_2 + \Delta x_2 S_3 = 6\left(\frac{\Delta f_2}{\Delta x_2} - \frac{\Delta f_1}{\Delta x_1}\right)
\tag{6.57}
$$

The extrapolation value of S_1, Eq. (6.55), is substituted into Eq. (6.57) to get

$$
d_2 S_2 + a_2 S_3 = 6\left(\frac{\Delta f_2}{\Delta x_2} - \frac{\Delta f_1}{\Delta x_1}\right)
\tag{6.58}
$$

Similarly, substituting S_n from Eq. (6.56) into Eq. (6.51) for $k = n-1$, followed by simplifications, we arrive at

$$
b_{n-1} S_{n-2} + d_{n-1} S_{n-1} = 6\left(\frac{\Delta f_{n-1}}{\Delta x_{n-1}} - \frac{\Delta f_{n-2}}{\Delta x_{n-2}}\right)
\tag{6.59}
$$

The coefficients for $k = 3, 4, .., n-2$ equations and the system of linear equations are the same except for the first and last rows, which are modified with the following quantities:

$$
\begin{aligned}
a_2 &= \frac{(\Delta x_2)^2 - (\Delta x_1)^2}{\Delta x_2}, & d_2 &= \frac{(\Delta x_1 + \Delta x_2)(\Delta x_1 + 2\Delta x_2)}{\Delta x_2}, \\
b_{n-1} &= \frac{(\Delta x_{n-2})^2 - (\Delta x_{n-1})^2}{\Delta x_{n-2}}, & d_{n-1} &= \frac{(\Delta x_{n-1} + \Delta x_{n-2})(\Delta x_{n-1} + 2\Delta x_{n-2})}{\Delta x_{n-2}}
\end{aligned}
\tag{6.60}
$$

6.4.3 COMPUTING SPLINE COEFFICIENTS

Upon employing an end-condition, $S_2, S_3, \ldots, S_{n-1}$ are found from the solution of the system of linear equations. Before implementing the cubic spline interpolation, the coefficients (a, b, c, and d's) are computed for each $[x_k, x_{k+1}]$ interval (for $k = 1, 2, \ldots, (n-1)$), as follows:

$$a_k = \frac{S_{k+1} - S_k}{6\,\Delta x_k}, \quad b_k = \frac{S_k}{2}, \quad c_k = \frac{f_{k+1} - f_k}{\Delta x_k} - \Delta x_k \left(\frac{S_{k+1} + 2S_k}{6} \right), \quad d_k = f_k \quad (6.61)$$

The arithmetic operations are significantly reduced when the data set is uniform. For example, the tridiagonal system of linear equations for cubic spline interpolation with *linear extrapolation end conditions* takes the form:

$$\begin{bmatrix} 6 & 0 & 0 & 0 & \cdots & 0 \\ 1 & 4 & 1 & 0 & \cdots & 0 \\ 0 & 1 & 4 & 1 & & 0 \\ \vdots & & \ddots & \ddots & \ddots & \vdots \\ 0 & \cdot & 0 & 1 & 4 & 1 \\ 0 & \cdots & 0 & 0 & 0 & 6 \end{bmatrix} \begin{bmatrix} S_2 \\ S_3 \\ S_4 \\ \vdots \\ S_{n-2} \\ S_{n-1} \end{bmatrix} = 6 \begin{bmatrix} f_2'' \\ f_3'' \\ f_4'' \\ \vdots \\ f_{n-2}'' \\ f_{n-1}'' \end{bmatrix} \quad (6.62)$$

where $f_k'' = \delta^2 f_k / h^2 = (f_{k+1} - 2f_k + f_{k-1})/h^2$, and the linear extrapolation end conditions become $S_1 = 2S_2 - S_3$ and $S_n = 2S_{n-1} - S_{n-2}$.

EXAMPLE 6.6: Application of cubic splines with two end conditions

The plastic viscosity (μ) versus yield stress (σ) relationship for a concrete recipe with a special ingredient is investigated. The experimental results are presented below:

μ (Pa·s)	5	9	11	15	19	24
σ (Pa)	126	196	294	590	980	1550

Find interpolates for plastic viscosities of $\mu = 7, 10, 12, 18$, and 23 Pa· s using the cubic splines with (a) natural spline and (b) linear extrapolation end conditions.

SOLUTION:

(a) For the natural spline end condition (setting $S_1 = S_6 = 0$), Eq. (6.62) takes the form

$$\begin{bmatrix} 12 & 2 & & \\ 2 & 12 & 4 & \\ & 4 & 16 & 4 \\ & & 4 & 18 \end{bmatrix} \begin{bmatrix} S_2 \\ S_3 \\ S_4 \\ S_5 \end{bmatrix} = \begin{bmatrix} 189 \\ 150 \\ 141 \\ 99 \end{bmatrix}$$

The solution of the tridiagonal system of equations yields $S_2 = 14.3831$, $S_3 = 8.20148$, $S_4 = 5.70402$, and $S_5 = 4.23244$. The cubic splines (i.e., coefficients) corresponding each interval are calculated using Eq. (6.61) as follows:

$$\sigma_1(\mu) = 0.599295(\mu-5)^3 + 0 \times (\mu-5)^2 + 7.911275(\mu-5) + 126, \qquad [5, 9]$$

$$\sigma_2(\mu) = -0.515134(\mu-9)^3 + 7.191543(\mu-9)^2 + 36.677449(\mu-9) + 196, \qquad [9, 11]$$

$$\sigma_3(\mu) = -0.104061(\mu-11)^3 + 4.100739(\mu-11)^2 + 59.262015(\mu-11) + 294 \qquad [11, 15]$$

$$\sigma_4(\mu) = -0.061316(\mu-15)^3 + 2.852010(\mu-15)^2 + 87.073013(\mu-15) + 590, \qquad [15, 19]$$

$$\sigma_5(\mu) = -0.141081(\mu-19)^3 + 2.116220(\mu-19)^2 + 106.94593(\mu-19) + 980, \qquad [19, 24]$$

(b) For the linear extrapolation end-condition requirement Eq. (6.60), the system of linear equations yields

$$\begin{bmatrix} 24 & -6 & & \\ 2 & 12 & 4 & \\ & 4 & 16 & 4 \\ & & -2.25 & 29.25 \end{bmatrix} \begin{bmatrix} S_2 \\ S_3 \\ S_4 \\ S_5 \end{bmatrix} = \begin{bmatrix} 189 \\ 150 \\ 141 \\ 99 \end{bmatrix}$$

The solution of the system of equations gives $S_2 = 10.11016$, $S_3 = 8.940622$, $S_4 = 5.623055$, and $S_5 = 3.817158$.

Next, linear extrapolations are carried out as follows: $S_1 = 2S_2 - S_3 = 12.449222$ and $S_6 = 2S_5 - S_4 = 1.559787$. Having calculated the coefficients using Eq. (6.61), the cubic splines corresponding to each interval become

$$\sigma_1(\mu) = -0.097461(\mu-5)^3 + 6.224611(\mu-5)^2 - 5.839066(\mu-5) + 126, \qquad [5,9]$$

$$\sigma_2(\mu) = -0.097461(\mu-9)^3 + 5.055078(\mu-9)^2 + 39.27969(\mu-9) + 196, \qquad [9,11]$$

$$\sigma_3(\mu) = -0.138232(\mu-11)^3 + 4.470311(\mu-11)^2 + 58.33047(\mu-11) + 294, \qquad [11,15]$$

$$\sigma_4(\mu) = -0.075246(\mu-15)^3 + 2.811527(\mu-15)^2 + 87.45782(\mu-15) + 590, \qquad [15,19]$$

$$\sigma_5(\mu) = -0.075246(\mu-19)^3 + 1.908579(\mu-19)^2 + 106.33825(\mu-19) + 980, \qquad [19,24]$$

The cubic spline coefficients corresponding to the corresponding μ value are used as interpolating polynomials. For example, $\mu = 7$ falls into the first interval. Thus, we calculate the interpolation from $\sigma(7) = \sigma_1(7)$ and similarly the other interpolates as $\sigma(10) = \sigma_2(10)$, $\sigma(12) = \sigma_3(12)$, $\sigma(18) = \sigma_4(18)$, and $\sigma(23) = \sigma_5(23)$.

TABLE 6.11

μ (Pa·s)	Natural Spline	Linear Extrapolation	Difference
7	146.617	138.441	8.176
10	239.354	240.237	-0.883
12	357.259	356.662	0.597
18	875.232	875.646	-0.414
23	1432.614	1431.075	1.539

Interpolates computed in both cases, along with the difference between the two estimates, are comparatively presented in Table 6.11. For the μ values away from the end points, the cubic spline estimates with both end conditions are very good, yielding absolute errors less than 1%. It turns out that the difference between the estimates obtained with both end conditions for μ values near the lower end is the largest. When the data is visually inspected, the distribution has a large curvature at the left end ($\mu''(5) > 0$), and the inflection point condition imposed ($\mu''(5) = 0$) is naturally not suitable for the data.

Discussion: The overall performance of the spline interpolations with the linear extrapolation end condition is better than that obtained with the natural spline end conditions. This is because calculating the curvature at each end by linear extrapolation from the curvature of the neighboring interval leads to better curvature estimates for uniform distributions.

Pseudocode 6.5

Module SPLINE $(opt, n, \mathbf{x}, \mathbf{f}, \mathbf{a}, \mathbf{b}, \mathbf{c})$
\ DESCRIPTION: A pseudomodule to compute the cubic spline coefficients $(a_n,$
\ $b_n, c_n)$ for "natural spline" or "linear extrapolation" end conditions. The
\   cubic splines have the following form
\ $y_k(x) = a_x(x - x_k)^3 + b_k(x - x_k)^2 + c_k(x - x_k) + f_k, \qquad x_k \leqslant x \leqslant x_{k+1}$
\ USES:
\   TRIDIAGONAL:: Module to solve tridiagonal system (Pseudocode 2.13)
Declare: $x_n, f_n, a_n, b_n, c_n, dx_n, df_n, \bar{a}_n, \bar{b}_n, \bar{c}_n, \bar{d}_n, S_n$ \ Declare array variables
For $\left[k = 1, n - 1\right]$ \ Loop k: Calculate the differences
  $dx_k \leftarrow x_{k+1} - x_k$ \ Finde Δx_k's
  $df_k \leftarrow (f_{k+1} - f_k)/dx_k$ \ Find $f[x_k, x_{k+1}]$'s
End For
For $\left[k = 2, n - 1\right]$ \ Loop k: Construct tridiagonal matrix
  $\bar{b}_k \leftarrow dx_{k-1}$ \ Find lower diagonals
  $\bar{a}_k \leftarrow dx_k$ \ Find upper diagonals
  $\bar{d}_k \leftarrow 2 * (\bar{a}_k + \bar{b}_k)$ \ Find diagonals
  $\bar{c}_k \leftarrow 6 * (df_k - df_{k-1})$ \ Construct rhs
End For
If $\left[opt = 0\right]$ **Then** \ $opt = 0$: Case of natural spline end conditions
  $\bar{d}_1 \leftarrow 1; \ \bar{a}_1 \leftarrow 0; \ \bar{c}_1 \leftarrow 0$ \ Set coeffients of 1st row
  $\bar{d}_n \leftarrow 1; \ \bar{b}_n \leftarrow 0; \ \bar{c}_n \leftarrow 0$ \ Set coeffients of last row
  $s1 \leftarrow 1; \ sn \leftarrow n$ \ Indices of the 1st & last rows
Else \ $opt \neq 1$: Case of linear extrapolation end conditions
  $cf \leftarrow 1 + dx_1/dx_2$
  $\bar{a}_2 \leftarrow (dx_2 - dx_1) * cf$ \ Modify coeffients of 1st row
  $\bar{d}_2 \leftarrow (2 * dx_2 + dx_1) * cf$
  $cf \leftarrow 1 + dx_{n-1}/dx_{n-2}$
  $\bar{b}_{n-1} \leftarrow (dx_{n-2} - dx_{n-1}) * cf$ \ Modify coeffients of lastst row
  $\bar{d}_{n-1} \leftarrow (2 * dx_{n-2} + dx_{n-1}) * cf$
  $s1 \leftarrow 2; \ sn \leftarrow n - 1$ \ Indices of the 1st & last rows
End If
TRIDIAGONAL$(s1, s2, \bar{\mathbf{b}}, \bar{\mathbf{d}}, \bar{\mathbf{a}}, \bar{\mathbf{c}}, \mathbf{S})$ \ Use Pseudocode 2.13 to solve system
If $\left[opt \neq 0\right]$ **Then** \ Modify 1st and last rows by Eq. (6.60)
  $S_1 \leftarrow S_2 * (1 + dx_1/dx_2) - S_3 * dx_1/dx_2$
  $S_n \leftarrow S_{n-1} * (1 + dx_{n-1}/dx_{n-2}) - S_{n-2} * dx_{n-1}/dx_{n-2}$
End If
For $\left[k = 1, n - 1\right]$ \ Compute Spline coefficients from Eq. (6.61)
  $a_k \leftarrow (S_{k+1} - S_k)/(6 * dx_k)$
  $b_k \leftarrow S_k/2$
  $c_k \leftarrow df_k - dx_k * (S_{k+1} + 2 * S_k)/6$
End For
End Module SPLINE

Module SPLINE_EVAL $(n, xval, \mathbf{x}, \mathbf{f}, \mathbf{a}, \mathbf{b}, \mathbf{c}, \mathbf{fval})$
\ DESCRIPTION: A pseudomodule to evaluate $f(xval), f'(xval), f''(xval)$
\ USES:
\   BINARY_SEARCH:: Module performing binary search, Pseudocode 6.4.
Declare: $x_n, f_n, a_n, b_n, c_n, fval_{0:2}$ \ Declare array variables

Cubic Spline Interpolation

- The method provides stable and smooth interpolating polynomials;
- Once the cubic spline coefficients are determined, not only $y(x)$ but also $y'(x)$ and $y''(x)$ derivatives can be easily estimated at low cost;
- Cubic spline algorithm with linear extrapolation end-condition generally gives better results;
- Nonphysical oscillations associated with Runge's phenomenon are not observed.

- Since the curvature of the distribution at the endpoints is critical, the method may not produce a proper interpolation curve if the end conditions do not suit the true curvature;
- Splines are not suitable for extrapolation operations;
- The method is a bit more computationally involved in comparison to linear splines.

If $\left[xval < x_1 \text{ Or } xval > x_n \right]$ **Then**   \ Check if $xval$ is within data range
 Write: "$xval$ outside data range!"
 Exit   \ Issue a warning and exit the module
End If
BINARY_SEARCH$(n, xval, \mathbf{x}, k)$ \ Find first index k of $x_k \leqslant xval < x_{k+1}$ interval
$dx \leftarrow xval - x_k$
$fval_0 \leftarrow ((a_k * dx + b_k) * dx + c_k) * dx + f_k$ \ interpolant at $xval$
$fval_1 \leftarrow (3 * a_k * dx + 2 * b_k) * dx + c_k$ \ first derivative at $xval$
$fval_2 \leftarrow 6 * a_k * dx + 2 * b_k$ \ second derivative at $xval$
End Module SPLINE_EVAL

A pseudomodule, SPLINE, computing the cubic spline coefficients for a set of discrete data is presented in Pseudocode 6.5. The module requires the number of data points (n), the data set (x_i, f_i), and an option key for the cubic-spline end-condition type (opt) as input. The module computes and returns the cubic spline coefficients a_i, b_i, and c_i for $i = 1$ to n as output. Arrays dx_i and df_i are used to compute the interval spacing and first divided differences, and S_i denotes the second derivative. Since the procedure requires the solution of a tridiagonal system given by Eq. (6.54), the module **TRIDIAGONAL** (*see* Pseudocode 2.13) is utilized here. The first and last row of the tridiagonal system are specified according to the natural spline (Eq. (6.52) or linear extrapolation end conditions, Eq. (6.60), respectively. The intermediate arrays $\bar{a}_i$, $\bar{b}_i$, $\bar{c}_i$, and $\bar{d}_i$ are used for constructing the tridiagonal system. This module should be called once in the related module, so long as **f** is not altered.

The accompanying module, SPLINE_EVAL, is for evaluating the interpolant as well as its first and second derivatives. The module requires the number of discrete data points (n), an interpolation point ($xval$), a set of discrete data (x_i, f_i), and the cubic spline coefficients (a_i, b_i, and c_i generated by SPLINE) as input. An input validity check for $xval$ is carried out at the top of the module. Then the first subscript k of the interval containing $xval$ is determined *via* BINARY_SEARCH (*see* Pseudomodule 6.4). The module can be called as many times as necessary to calculate $f(xval)$, or $f'(xval)$ and $f''(xval)$ using Eqs. (6.34), (6.40), and (6.45), which are stored on variables $fval_0$, $fval_1$, and $fval_2$, respectively.

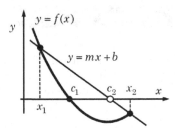

FIGURE 6.13: Graphical depiction of rootfinding by linear interpolation.

ERROR ASSESSMENT. A cubic-spline interpolant constructed by implementing any end-condition is unique. The uniqueness of the interpolants allows us to estimate the approximation errors (error bounds) associated with cubic splines by the polynomial approximations, i.e., the absolute error bounds for the cubic splines are given by

$$\max_{x\in[x_k,x_{k+1}]} |y(x) - y_k(x)| \leqslant \frac{h^4}{4!} \max_{x\in[x_k,x_{k+1}]} \left|y^{(4)}(x)\right|$$

where $h = \max|x_{k+1} - x_k|$. This expression indicates that if h is sufficiently small, the order of the interpolation error will be $\mathcal{O}(h^4)$. The order of error for the cubic splines method with linear extrapolation is $\mathcal{O}(h^4)$ as $h \to 0$, whereas the maximum error for the natural spline end condition is $\mathcal{O}(h^2)$.

6.5 ROOTFINDING BY INVERSE INTERPOLATION

The most frequently encountered problem in science and engineering is to find the value of x for a given y in a set of data—(x_i, y_i). This reverse process is known as *inverse interpolation*. In other words, inverse interpolation is the process of finding the value of an argument that corresponds to a given value of a function within a tabulated data set.

The simplest and easiest inverse interpolation method is linear interpolation, which requires a single interval and two data points. In this case, the interval of the root, $[x_1, x_2]$, is first bracketed, just as in the bisection method. As in Fig. 6.13, the endpoint values of the interval must have opposite signs ($y_1 > 0$, $y_2 < 0$ or $y_2 > 0$, $y_1 < 0$). The equation of a straight line connecting the end points is $y = y_1 + (y_2 - y_1)(x - x_1)/(x_2 - x_1)$. This line cuts the $x-$axis at $x = c_2$, which is obtained by simply setting $y = 0$ in the equation of straight line and solving for x:

$$c_2 \approx x = \frac{x_2 y_1 - x_1 y_2}{y_1 - y_2}$$

This approximation may be used as an estimate for the root ($c_2 \approx c_1$). If the curvature of the curve near the root is relatively small (i.e., the curve is near linear), then the estimate for the root will be close to the true value ($c_1 \approx c_2$). However, estimating a root with a linear interpolation could be off by a large margin for a discrete function with a large curvature, as seen in Fig. 6.13. In such cases, a higher-degree interpolating polynomial may be needed to obtain a better estimate. Unlike linear interpolation, a high-degree interpolating polynomial requires more than one interval and more data points to accurately mimic the curvature near the root.

Now consider $y = f(x)$ on $[x_0, x_n]$, which satisfies the inverse-function theorem; that is, $y = f(x)$ is differentiable and $f'(x) \neq 0$. Inverse interpolation, $x = f^{-1}(y)$, requires interchanging the data so that the discrete function is cast as (y_i, x_i) for $i = 0, 1, 2, \ldots, n$.

Inverse Polynomial Interpolation

- The method is simple and straightforward;
- It requires neither derivatives nor an iterative procedure;
- A quadratic or cubic polynomial is sufficient in most cases;
- It can also be used to find the roots of an explicit function in cases where the root-finding methods exhibit convergence difficulties or require a large number of iterations for convergence.

- The function, $y = f(x)$, must be an invertible function that allows inverse interpolation;
- The quality of the estimate depends on how well the inverse function is approximated by a polynomial.

Finding the value of $x = f^{-1}(0)$ is then equivalent to determining the root of $y = f(x)$. Even if the original data is uniformly spaced, the set obtained by interchanging x and y's will not be uniformly spaced. A second-, third-, or higher-order polynomial can be used for interpolation as long as a sufficient number of data points are available. Even if there is sufficient data, it is crucial to determine which points (among many) to use in the interpolation to estimate the root. Selecting four or five points from several intervals in the neighborhood of the root would be adequate in most cases. In this regard, the Lagrange interpolating polynomials can best serve this purpose. An approximating polynomial is obtained as $x(y) \cong \Sigma_{k=0}^{n} x_k L_k(y)$, where n is the degree of polynomial. The root is then estimated by setting $y = 0$ in the interpolating function: $x_{root} \approx x(0)$.

The inverse interpolation can also be used to estimate the value of x that corresponds to a particular value of $y(x) = y_x$. Instead of using all available data, the distribution of the data should be examined visually, and lower-order polynomials using localized data should be constructed for inverse interpolation.

ERROR ASSESSMENT. Inverse interpolation error depends basically on the numerical interpolation method used. The error of inverse Lagrangian interpolation is

$$e_n(x) = (x - x_0)(x - x_1) \cdots (x - x_n) \frac{f^{-1(n+1)}(\xi)}{(n + 1)!}, \qquad x_0 \leqslant \xi \leqslant x_n$$

The derivatives of $f^{-1}(x)$ can be calculated in terms of the derivatives of $f(x)$; for instance,

$$\frac{d}{dx}\left(f^{-1}(x)\right) = \frac{1}{f'(x)}, \qquad \frac{d^2}{dx^2}\left(f^{-1}(x)\right) = -\frac{f''(x)}{(f'(x))^3}, \cdots$$

which reveals the limitations of the inverse interpolation. That is, the error is clearly unbounded for $f'(x) \to 0$. Nonetheless, we can still estimate an interpolant through inverse interpolation, although the inverse function does not exist in $[x_0, x_n]$. But in this case, we should expect the accuracy of the estimate to be very poor.

When dealing with experimental data, knowing the accuracy of the data is important. If the accuracy of the data is unknown, then the true solution may be off by a large margin even if the root is estimated with high-degree interpolating polynomials.

EXAMPLE 6.7: Uses of inverse interpolation in practice

The density of plastic-based composite materials increases in proportion to the amount of fiber additive used. A correlation for the flexural modulus E (in GPa) of a composite material as a function of density ρ (in g/cm^3) is given as

$$E(\rho) = 3967\,\rho^4 - 15200\,\rho^3 + 16420\,\rho^2 - 5584, \qquad 1.3 \leqslant \rho \leqslant 1.8$$

Estimate what the density of the material should be so that the modulus of elasticity is at least 250 GPa.

SOLUTION:

The problem is finding the root of $E(\rho) - 250 = 0$, which can be obtained by employing a root-finding method (*see* Chapter 4). First, a set of four data points is generated, and the abscissas and ordinates are interchanged, as shown in Table 6.12. The problem is then reduced to finding $\rho(250)$.

TABLE 6.12

i	0	1	2	3
E_i	143.938	190.131	324.981	614.379
ρ_i	1.5	1.6	1.7	1.8

A Lagrange interpolating polynomial using all points is a cubic polynomial:

$$p_3(E) = (143.938)L_0(E) + (190.131)L_1(E) + (324.981)L_2(E) + (614.379)L_3(E)$$

Expanding and simplifying yields the approximation, which we find

$$\rho(E) = 1.47266 \times 10^{-8}E^3 - 1.75671 \times 10^{-5}E^2 + 0.00679294\,E + 0.842278$$

which for $E = 250$ GPa leads to $\rho = 1.67268$ g/cm^3.

Discussion: We can also construct a quadratic interpolation polynomial, for which we need three data points. If we choose the data points corresponding to $i = 0, 1$, and 2, the interpolating polynomial becomes

$$\rho(E) = -7.86148 \times 10^{-6}E^2 + 0.0047911\,E + 0.973253$$

which yields 1.67969 g/cm^3 for $E = 250$ GPa. This estimate also agrees with the cubic approximation. However, the linear interpolation between two bracketing points, (190.131,1.6) and (324.981,1.7), yields 1.6444 g/cm^3, which is considerably different from the estimates found by quadratic and cubic polynomials. When the data is plotted in the range $140 \leqslant E \leqslant 325$, the density depicts a strong curvature in this interval, which explains the large deviation from the other two estimates.

6.6 MULTIVARIATE INTERPOLATION

All interpolation methods discussed so far involved single-variable (univariate) polynomials, meaning that the dependent variable is a function of a single independent variable: $y = f(x)$. However, in many problems arising in science and engineering applications, the dependent variable is a function of two or more independent variables, e.g., $z = f(x, y)$. Such functions are usually referred to as *multivariate functions*. When multivariate discrete functions are

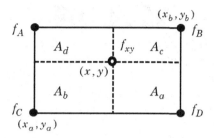

FIGURE 6.14: Graphical depiction of bilinear interpolation.

available as tabular data, multivariate interpolations are inevitable to perform calculus operations such as differentiation, integration, and so on.

Much of the theory for univariate interpolation can be generalized to multivariable interpolation problems. Polynomial interpolation of multivariate discrete functions is usually more difficult than that of univariate ones. Because there is a lack of uniqueness, particularly, it may not suffice to require that the interpolation points be distinct. In this section, for simplicity, we consider a discrete function of two independent variables, $z = f(x, y)$, containing known values on a regular grid (not necessarily uniformly spaced) to illustrate the implementation of multivariate interpolation.

6.6.1 BILINEAR INTERPOLATION

The method is generally applied to discrete functions sampled on two-dimensional rectangular grids. Consider a discrete function with two variables defined for a rectangular mesh, i.e., a set of nonuniform discrete data $z_{ij} = f(x_i, y_j) = f_{ij}$ for $i = 1, 2, \ldots, n$ and $j = 1, 2, \ldots, m$ representing the known discrete values corresponding to (x_i, y_j) in a rectangular ($a \leqslant x \leqslant b$ and $c \leqslant y \leqslant d$) domain.

Bilinear interpolation is based on the assumption that the variation between the neighboring four data points is linear, i.e., $z = f(x, y) = a + bx + cy$. The closer the points are to each other ($\Delta x = x_b - x_a \to 0$ and $\Delta y = y_b - y_a \to 0$), the better the accuracy of the interpolation is. In Fig. 6.14, the interpolant, $fval$, and its neighboring data points (f_A, f_B, f_C, and f_D) are illustrated. The linear interpolation in terms of neighboring nodes can be written as

$$fval = \frac{1}{A} \left(f_A A_a + f_B A_b + f_C A_c + f_D A_d \right) \qquad (6.63)$$

where $A_a = (x_b - x)(y - y_a)$, $A_b = (x - x_a)(y - y_a)$, $A_c = (x_b - x)(y_b - y)$, $A_d = (x - x_a)(y_b - y)$ and $A = (x_b - x_a)(y_b - y_a)$ are the areas of rectangles.

An alternative bi-linear interpolation equation has the form: $f(x, y) = a_0 + a_1 x + a_2 y + a_3 xy$, where the coefficients (a_0, a_1, a_2, and a_3) are obtained by solving a system of four equations from direct substitution, i.e., $f(x_a, y_b) = f_A$, $f(x_b, y_b) = f_B$, $f(x_a, y_a) = f_C$, and $f(x_b, y_a) = f_D$.

A pseudomodule, BILINEAR_INTERP, performing bilinear interpolation for the input of a non-uniformly spaced set of data, (x_i, y_j, f_{ij}) for $i = 1, 2, \ldots, m$ and $j = 1, 2, \ldots, n$, is presented in Pseudocode 6.6. The module requires a vector $\mathbf{x}$ of length m, containing the x-coordinates, and a vector $\mathbf{y}$ of length n, containing the y-coordinates of the data points, a two-dimensional array $\mathbf{f}$, containing the f_{ij} data at (x_i, y_j) for $i = 1, 2, \ldots, m$ and $j = 1, 2, \ldots, n$, and $(xval, yval)$ is the coordinate point for which the interpolation is to be performed. The output $fval$ is the interpolated value, $fval = f(xval, yval)$. Though not

---

Pseudocode 6.6

Module BILINEAR_INTERP $(m, n, xval, yval, \mathbf{x}, \mathbf{y}, \mathbf{f}, fval)$
\ DESCRIPTION: A pseudomodule to compute $f(xval, yval)$ of a bivariate
\   nonuniform discrete data set using bilinear interpolation.
\ USES:
\   BINARY_SEARCH:: A binary search module given in Pseudocode 6.4.
Declare: x_m, y_n, f_{mn} \ Declare array variables
\ Input validity check for $xval$ and $yval$ should be inserted here
BINARY_SEARCH$(m, xval, \mathbf{x}, i)$ \ Find first index i of $x_i \leqslant xval < x_{i+1}$
BINARY_SEARCH$(n, yval, \mathbf{y}, j)$ \ Find first index j of $y_j \leqslant yval < y_{j+1}$
$Area \leftarrow (x_{i+1} - x_i) * (y_{j+1} - y_j)$
$fa \leftarrow f_{i,j+1}; \; Aa \leftarrow (x_{i+1} - xval) * (yval - y_j)$
$fb \leftarrow f_{i+1,j+1}; \; Ab \leftarrow (xval - x_i) * (yval - y_j)$
$fc \leftarrow f_{ij}; \; Ac \leftarrow (x_{i+1} - xval) * (y_{j+1} - yval)$
$fd \leftarrow f_{i+1,j}; \; Ad \leftarrow (xval - x_i) * (y_{j+1} - yval)$
$fval \leftarrow (fa * Aa + fb * Ab + fc * Ac + fd * Ad) / Area$ \ Employ Eq. (6.63)
End Module BILINEAR_INTERP

---

given here to save space, the module should include an input validity check to determine whether $(xval, yval)$ is within the domain of the discrete function. Then, the first indices of the rectangular cell (i, j) where the point $(xval, yval)$ is located are determined through the binary search algorithm. Finally, the interpolate $fval$ is obtained upon employing Eq. (6.63) and returned as output.

6.6.2 BIVARIATE LAGRANGE INTERPOLATION

A second method involves the use of Lagrange polynomials, which also have the capability of representing nonuniform discrete data sets. The application of this technique requires defining the data on square or rectangular grids, typically on all four corners of an element.

Consider a rectangular, non-uniform grid in the $x-y$-plane, and data is defined at every intersecting grid line, as depicted in Fig. 6.14. Using the notations described in Section 6.1, the bivariate Lagrange interpolating polynomial can be written (in so-called tensor product) as

$$z = f(x, y) \cong p_{m,n}(x, y) = \sum_{i=1}^{m} L_i(y) \left(\sum_{j=1}^{n} L_j(x) f_{ij} \right) \tag{6.64}$$

where m and n are the number of data points in the x and y-directions, (x_m, y_n, f_{mn}) denotes a bivariate set of discrete data with $i = 1, 2, \ldots, m$ and $j = 1, 2, \ldots, n$ defined in a rectangular domain, and $L_m(x)$ and $L_n(y)$ are the Lagrange polynomials (basis functions) derived using Eq. (6.8) for x and y data, respectively.

In this approach, the multidimensional interpolation problem is decoupled into a product of one-dimensional ones. In fact, extending this bivariate approach to more dimensions is also, in principle, quite simple. The advantages and disadvantages of the Lagrange method pretty much apply here, but other interpolation methods should be considered if the discrete function is irregularly dispersed throughout the domain or defined for non-rectangular domains.

Pseudocode 6.7

Module BIVARIATE_LAGRANGE $(m, n, xval, yval, \mathbf{x}, \mathbf{y}, \mathbf{f}, fval)$
\ DESCRIPTION: A pseudomodule to compute $f(xval, yval)$ of a bivariate
\   nonuniform discrete data set using Lagrange interpolation.
\ USES:
\   LAGRANGE_P:: Module generating Lagrange polynomials, Pseudocode 6.1
Declare: $x_m, y_n, f_{mn}, Lx_m, Ly_n$ \ Declare arrays
\ Input validity checks for $xval$ and $yval$ should be inserted here
LAGRANGE_P$(m, xval, \mathbf{x}, \mathbf{Lx})$ \ Compute $Lx_i(x)$'s $i = 1, \ldots, m$
LAGRANGE_P$(n, yval, \mathbf{y}, \mathbf{Ly})$ \ Compute $Ly_i(x)$'s $i = 1, \ldots, n$
$fval \leftarrow 0$ \ Initialize Interpolate
For $\left[i = 1, m\right]$ \ Accumulator loops
    For $\left[j = 1, n\right]$
        $fval \leftarrow fval + Lx_i * Ly_j * f_{ij}$ \ Accumulate $Lx_i * Ly_j * f_{ij}$'s
    End For
End For
End Module BIVARIATE_LAGRANGE

A pseudomodule, BIVARIATE_LAGRANGE, performing bivariate Lagrange interpolation using a non-uniformly spaced data set, (x_i, y_j, f_{ij}) for $i = 1, 2, \ldots, m$ and $j = 1, 2, \ldots, n$ is presented in Pseudocode 6.7. The input and output variables are the same as those of the Pseudocode 6.6. The Lagrange polynomials ($Lx_m(xval)$ and $Ly_n(yval)$) are evaluated for the interpolation point $(xval, yval)$ using the set of $\mathbf{x}$ and $\mathbf{y}$ arrays. Then, the interpolant $fval$ is obtained from Eq. (6.64) and returned as output.

6.7 EXTRAPOLATION

So far, we have discussed the methods of estimating the values of a discrete function within its original observation range, $x_0 \leqslant x \leqslant x_n$, called "*interpolation.*" Estimating a discrete function outside of its original observation range ($x < x_0$ or $x > x_n$) is referred to as *extrapolation.*

An extrapolation procedure, in essence, is simply implementing a generated interpolating polynomial outside its range. In interpolation, the distribution is anchored at both ends of a subinterval where the interpolation point lies, whereas in extrapolation, the distribution is anchored only on one side and is loose on the other side. That is why the uncertainty in extrapolation is much greater, mainly due to the assumption that the existing trends will continue outside the data range. If a discrete function is uniformly spaced, then the Gregory-Newton forward or backward interpolation formulas can be employed for extrapolations. The choice of formula depends on the position of x in relation to the data range. To extrapolate for $x < x_0$, the first row corresponding to $x = x_0$ is used as the baseline together with the forward interpolation formula. To extrapolate for $x > x_n$, the backward interpolation formula with the last row ($x = x_n$) as the baseline is adopted.

ERROR ASSESSMENT. In Section 6.3, it was shown that the most accurate estimates were obtained when the interpolant was in the neighborhood of the original data points. In extrapolation, however, the first and last data points of the set are the outermost anchors. Since the extrapolating polynomial extends far out of its original interval, the extrapolation error can become very large. Hence, extreme care should be exercised when performing extrapolation.

EXAMPLE 6.8: Bivariate Lagrange Interpolation

Consider a ternary system (A, B, and C liquids) in equilibrium. Estimating the logarithm of the activity coefficient of C ($\log \gamma_C$) with the mole fractions of the liquids A and B (x_A and x_B) is needed. Experimentally generated data for $z = f(x_A, x_B) = \log\left(\gamma_C(x_A, x_B)\right)$ are presented in the table below. Use bi-Lagrange interpolation to estimate $f(0.22, 0.10)$ and $f(0.13, 0.27)$.

		x_B	
x_A	0.05	0.15	0.35
0	0.231	0.316	0.484
0.10	0.207	0.293	0.464
0.25	0.161	0.246	0.414
0.35	0.124	0.208	0.373

SOLUTION:

By inspection of tabular data, we determine that $m = 3$ and $n = 2$. The Lagrange polynomials for x_A's, $L_0(x_A)$, $L_1(x_A)$, $L_2(x_A)$ and $L_3(x_A)$, and x_B's $L_0(x_B)$, $L_1(x_B)$, $L_2(x_B)$ are constructed as follows:

$$L_0(x_A) = -\frac{800}{7}(x_A - 0.1)(x_A - 0.25)(x_A - 0.35),$$

$$L_1(x_A) = \frac{800}{3}(x_A - 0)(x_A - 0.25)(x_A - 0.35),$$

$$L_2(x_A) = -\frac{800}{3}(x_A - 0)(x_A - 0.10)(x_A - 0.35),$$

$$L_3(x_A) = \frac{800}{7}(x_A - 0)(x_A - 0.10)(x_A - 0.25),$$

$$L_0(x_B) = \frac{100}{3}(x_B - 0.15)(x_B - 0.35),$$

$$L_1(x_B) = -50(x_B - 0.05)(x_B - 0.35),$$

$$L_2(x_B) = \frac{100}{3}(x_B - 0.05)(x_B - 0.15)$$

Using $f(x_{A,i}, x_{B,j}) = f_{ij}$ notation, the data is denoted by an array $\mathbf{f}$ with its elements defined as $f_{00} = 0.231$, $f_{01} = 0.316$, $f_{10} = 0.207$, $f_{20} = 0.161$, and so on. Then the interpolating polynomial using Eq. (6.64) can be expressed as

$$f(x_A, x_B) \cong p_{m,n}(x_A, x_B) = \sum_{j=0}^{n} L_j(x_B)\left(\sum_{i=0}^{m} L_i(x_A)f_{ij}\right)$$

Using this interpolating polynomial we get $f(0.22, 0.10) = 0.21386$ and $f(0.13, 0.27) = 0.38742$. The true values are 0.213418 and 0.386813, respectively.

Discussion: Interpolation by univariate polynomials is a very classical and well-studied topic. However, interpolation by polynomials of several variables is much more intricate and is a subject that is currently an active area of research.

Interpolation to a point within the domain, where the data is defined at the nodes placed at each intersection of a rectangular grid, is the simplest case in this regard. The bilinear and bivariate Lagrange interpolation methods presented in this section

(a)

Δf_i	$\Delta^2 f_i$	$\Delta^3 f_i$	$\Delta^4 f_i$
0.55	−0.13	0.04	0.02
0.42	−0.09	0.06	

(b)

Δf_i	$\Delta^2 f_i$	$\Delta^3 f_i$	$\Delta^4 f_i$
0.55	0.87	1.65	3.98
1.42	2.52	5.63	

(c)

Δf_i	$\Delta^2 f_i$	$\Delta^3 f_i$	$\Delta^4 f_i$
0.55	−0.13	1.65	3.98
1.42	1.52	5.63	

FIGURE 6.15: Extrapolation for $x < x_0$ using Gregory-Newton forward difference table depicting three different cases, (a), (b), and (c).

can give satisfactory estimates for smooth functions. When deciding which method to use among the ones presented here, it is best to pick the method that matches the nature and behavior of the data while avoiding overfitting or underfitting.

In order to obtain a reasonable extrapolant, the discrete function should be smooth and suitable for a polynomial extrapolation. In this respect, the backward or forward interpolation tables offer a clue as to how good an extrapolation may turn out to be. In other words, for smooth functions, the magnitudes of successive higher-order differences along the chosen baseline should decrease, i.e.,

$$|\Delta f_k| > |\Delta^2 f_k| > |\Delta^3 f_k| > \cdots \quad \text{or} \quad |\nabla f_k| > |\nabla^2 f_k| > |\nabla^3 f_k| > \cdots$$

Otherwise, any extrapolation operation can be extremely risky in that the estimates may drastically deviate from the true trend.

For extrapolation of x values ($x < x_0$) with GN-FWD-IF, consider the snapshots from the baselines of three different tables, as given in Fig. 6.15. In Fig. 6.15a, the magnitude (in absolute value) of the successive differences decreases for smooth distributions as the order of differences increases. For this case, the highest degree extrapolating polynomial would be $f(x) \cong p_4(x)$ for extrapolations of $x < x_0$. However, as x shifts further away from x_0, predictions will tend to diverge from true expectations. On the other hand, in Fig. 6.15b, the successive differences depict a constantly increasing trend in magnitude, which indicates that the extrapolations will diverge when using high-order extrapolating polynomials. In such cases, the linear extrapolating polynomial, $f(x) \cong p_1(x)$, can be used as the best alternative. In Fig. 6.15c, the magnitude of the forward differences after the $|\Delta^2 f_k|$ term depicts a continuously increasing trend. It is tempting to use the second order $f(x) \cong p_2(x)$ polynomial for extrapolation, but it will be a prudent practice to check the $|p_n(x) - p_{n-1}(x)|$ difference before deciding whether to add an extra term to the interpolating polynomial.

The extrapolation process is much more delicate than the interpolation process. The function may have a discontinuity that is not obvious in the data but still has a fatal effect

Extrapolation

- A univariant extrapolation is the simplest way of data estimation;
- An extrapolation method allows estimating the quantities that are not directly measurable or observable, or that are too costly or cpu-time intensive;
- It helps filling in the gaps outside data range, making the data more complete and consistent.

- Quality of the extrapolation depends on the assumptions made (continuous, smooth, etc.) and/or approximations that may not always hold true or represent the reality of the data;
- It is significantly affected by the fluctuations in the existing data sets;
- It can introduce errors and uncertainties in the estimated values, especially when the data is noisy or extrapolated far beyond the data range.

on the extrapolation process. As a final word, in situations where extrapolation cannot be avoided, exercise the following precautionary actions: (1) use low-order (linear or quadratic) polynomials; (2) restrict the value of x as close as possible to the tabulated region, even if a low-order polynomial approximation is used; and (3) plot the data and verify whether the extrapolated value is meaningful or not.

EXAMPLE 6.9: Extrapolation of oscillating functions

Generate uniformly spaced discrete data points for $f(x) = \sin x$ with $x = 0.1$, 0.5, 0.9, 1.3, and 1.7. Using the data, estimate $f(2.1)$, $f(2.5)$, $f(2.9)$, $f(3.3)$, and $f(3.7)$ and discuss the accuracy of the extrapolants.

SOLUTION:

Since estimates for values greater than 1.7 are sought, the baseline for the extrapolations is chosen as $x = 1.7$. Also, note that the number of entries in the backward difference table is the largest in this row (baseline). Hence, for uniformly spaced data with $h = 0.4$, the backward difference table is generated and presented in Table 6.13.

TABLE 6.13

i	x_i	f_i	∇f_i	$\nabla^2 f_i$	$\nabla^3 f_i$	$\nabla^4 f_i$
1	0.1	0.099833				
2	0.5	0.479426	0.379592			
3	0.9	0.783327	0.303901	−0.075691		
4	1.3	0.963558	0.180231	−0.123670	−0.047979	
5	1.7	0.991665	0.028107	−0.152125	−0.028455	0.019525

Setting $x_5 = 1.7$ as the baseline, the entries in this row are $f_5 = 0.991665$, $\nabla f_5 = 0.028107$, $\nabla^2 f_5 = -0.152125$, $\nabla^3 f_5 = -0.028455$, and $\nabla^4 f_5 = 0.019525$. Notice that the magnitude of the second difference jumps to ≈ 0.152, but then the magnitudes of the higher (3rd and 4th) differences depict a steady decline. The extrapolation variable becomes $s = (x - x_5)/h = (x - 1.7)/0.4$. The abscissas of the points to be extrapolated are given in increments of h; hence, the extrapolation variable becomes $s = 1, 2, 3, 4,$ and 5 for $x = 2.1, 2.5, 2.9, 3.3,$ and 3.7, respectively.

The GN-BKWD interpolation formula will be used to generate extrapolation polynomials passing through all data points:

$$f(x) \cong p_4(s) = f_5 + s\,\nabla f_5 + \frac{s(s+1)}{2!}\nabla^2 f_5 + \frac{s(s+1)(s+2)}{3!}\nabla^3 f_5$$
$$+ \frac{s(s+1)(s+2)(s+3)}{4!}\nabla^4 f_5$$

The 4th-degree extrapolation polynomials for $x = 2.1, 2.5,$ and 2.9 are explicitly given as

$$f(2.1) \cong p_4(s = 1) = f_5 + \nabla f_5 + \nabla^2 f_5 + \nabla^3 f_5 + \nabla^4 f_5,$$

$$f(2.5) \cong p_4(s = 2) = f_5 + 2\,\nabla f_5 + 3\,\nabla^2 f_5 + 4\,\nabla^3 f_5 + 5\nabla^4 f_5,$$

$$f(2.9) \cong p_4(s = 3) = f_5 + 3\,\nabla f_5 + 6\,\nabla^2 f_5 + 10\,\nabla^3 f_5 + 15\nabla^4 f_5,\dots$$

Using the above extrapolation expressions, any low-degree extrapolation polynomial can also be deduced. The first two terms correspond to linear, the three terms to quadratic, and the four terms to cubic extrapolating polynomials.

TABLE 6.14

x	$\sin x$	$p_2(x)$	$p_3(x)$	$p_4(x)$
2.1	0.863209	0.86765 (0.44%)	0.83919 (2.40%)	0.858709 (0.45%)
2.5	0.598472	0.59151 (0.70%)	0.47769 (12.1%)	0.575284 (2.32%)
2.9	0.239249	0.16324 (7.60%)	−0.12131 (36.1%)	0.171499 (6.77%)
3.3	−0.157746	−0.41715 (25.9%)	−0.98625 (82.9%)	−0.303010 (14.5%)
3.7	−0.529836	−1.14966 (62.0%)	−2.14560 (162%)	−0.779090 (24.9%)

A summary of extrapolated estimates using 2nd, 3rd, and 4th-degree GN-BKWD interpolating polynomials along with the true errors is presented in Table 6.14. It is observed that the third-degree extrapolant, $p_3(x)$, produces worse estimates than that of the second-degree extrapolant $p_2(x)$.

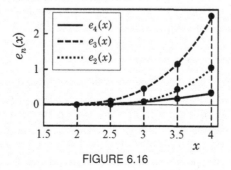

FIGURE 6.16

A graphical depiction of the absolute errors, $e_n(x) = |\sin x - p_n(x)|$, on the $1.5 \leqslant x \leqslant 4$ interval is presented in Fig. 6.16. The results of the linear extrapolations are not presented here because they gave the worst results. We note that the second and fourth-degree polynomials extrapolate fairly well up to $x \approx 3$, while the third degree polynomial deviates from the true distribution much earlier. The sine function has a maximum at $\pi/2(\approx 1.57)$, so a quadratic formula obtained from the last three interpolation points seems to better represent the curvature of the data beyond $x = 1.7$. While none of the extrapolated values could be called accurate by any means, the performance of the second-degree approximation is the "best."

Discussion: The further away the extrapolation point is from the baseline, the larger the extrapolation error will inevitably be. If a polynomial extrapolation must be done with non-smooth or poorly behaved functions, then a very low degree extrapolation is usually the safest but even this should be performed only for values of x very close to the baselines.

6.8   CLOSURE

Scientific experimentation, data collection, or numerical computation usually result in values for a function only at discrete points that are either *uniformly* or *non-uniformly* spaced. We often want to find the value of the function $f(x)$ at points that do not correspond to discrete data points. A number of numerical methods have been developed over the years for interpolation (estimating the unknown function within the data range) or extrapolation (estimating the unknown function outside the data range) purposes.

Lagrange interpolation makes use of Lagrangian polynomials, and it can be applied to uniformly or non-uniformly spaced discrete functions. A Lagrange interpolating polynomial always includes the original data points. The interpolant is anchored on both sides of an interval, but the error tends to be largest toward the middle of any subinterval. If the subinterval size is very large, the magnitude of the interpolation errors increases. Also, for a large data set, the number of floating-point arithmetic operations and rounding errors will increase considerably, yielding Runge's effect. Lagrange interpolating polynomials are not suitable for extrapolation as well.

Newton's divided difference can be applied to both uniformly and non-uniformly spaced discrete functions, while the Gregory-Newton interpolation polynomials are specifically suited for uniformly spaced discrete functions. Similar to the Lagrange interpolation, the interpolating polynomials are anchored at the interpolation points but yield the largest errors toward the middle of each interval. Unlike the Lagrange method, lower-order interpolating polynomials can be constructed by using fewer data points, which reduces some of the handicaps of high-order approximating polynomials. The tabulated differences (in absolute value) depicting a decreasing trend along a chosen baseline imply a "smooth function," which likely leads to convergence with fewer terms.

The cubic spline interpolation method is based on constructing a third-order polynomial for each interval instead of using a single interpolation polynomial for the whole range of data. The accuracy of cubic spline interpolation depends on the distribution of data and the suitability of the end condition. Calculating the coefficients of the cubic polynomials takes time and effort, but once they are evaluated, only a small number of arithmetic operations are needed to obtain the interpolate as well as its derivatives.

Multivariate interpolation is much more complex than univariant interpolations. Bilinear and bi-Lagrange interpolations for smooth functions and data defined on a regular rectangular grids are presented.

6.9 EXERCISES

Section 6.1 Lagrange Interpolation

E6.1 Find the Lagrange interpolating polynomial that passes through (x_0, y_0) and (x_1, y_1).

E6.2 Find the Lagrange interpolating polynomials of the highest degree for the given data sets.

(a)

x	1	7
y	7	1

(b)

x	0	3	11
y	6	0	12

(c)

x	1	2	4	5
y	8	19	71	124

E6.3 Use the following data to (a) construct a Lagrange interpolating polynomial and (b) show that it satisfies the interpolation points.

x	−1	0	2	3
$f(x)$	1	−1	19	77

E6.4 (a) Use Lagrange's method to construct a quadratic polynomial passing through $A(2, -4)$, $B(5, -1)$, and $C(8, 20)$; (b) find a quadratic polynomial by setting up a system of equations as $ax_i^2 + bx_i + c - y_i = 0$ for $i = 1$, 2, and 3; and (c) compare the polynomials obtained in Parts (a) and (b), and (d) plot the Lagrange polynomials on [2,8].

E6.5 (a) Construct a Lagrange interpolating polynomial for $\ln x$ on [1,4] using $\ln(1)=0$, $\ln(2)=0.693147$, and $\ln(4)=1.38629$; (b) estimate $\ln(3)$ as well as its interpolation error in percent.

E6.6 Find the Lagrange interpolating polynomial passing through $A(-1, -9)$, $B(1, -1)$, $C(2, 3)$, and $D(3, 23)$.

E6.7 Determine the interval width (h) of a set of uniformly spaced discrete data generated from $f(x) = 1/x^2$ in [2,3] so that a quadratic interpolating polynomial interpolates to five-decimal places accuracy.

E6.8 Use Lagrange's method to estimate $f(1.5)$ and the error bound using the following data generated from $f(x) = 1/\sqrt{x+1}$.

x	0	0.5	1	2
$f(x)$	1	0.8165	0.7071	0.57735

E6.9 Given (x_{i-1}, f_{i-1}), (x_i, f_i), and (x_{i-2}, f_{i-2}) are points uniformly spaced by h. (a) Find an approximation for $f(x)$ using the Lagrange interpolating polynomial passing through all three points; (b) use this approximation also to develop approximations for the first- and second-derivatives at $x = x_i$, i.e., $f_i' = f'(x_i)$, and $f_i'' = f''(x_i)$.

E6.10 A pump characteristic curve can be expressed for the pump head, h_p, with a second-degree polynomial: $h_p = aQ^2 + bQ + c$. The set of data for the expected range of operation of a certain brand of pump is given below. Use the given data and Lagrange's method to find the coefficients of the pump characteristic curve.

Q (m³/h)	5	65	140
h_p (m)	10.91	9.56	5.43

E6.11 The thermal conductivity of an alloy as a function of temperature is tabulated below. (a) Find the Lagrange interpolating polynomial for the thermal conductivity that passes through all data points; (b) plot the data and the interpolating polynomial together on the same graph and comment on how well the polynomial fits the data; (c) use the interpolating polynomial to estimate the thermal conductivity of the alloy for $T = 0$, 17, 100, and 450°C.

T (°C)	−23	77	327	727
k (W/m · °C)	406	396	379	352

E6.12 The given is a set of data from the lift coefficient measurements for a wing as a function of angle of attack (α). (a) Find the Lagrange interpolating polynomial for the lift coefficient that passes through all data points; (b) plot the data and the interpolating polynomial together on the same graph and discuss how well the polynomial fits the data; (c) use the interpolating polynomial to estimate the lift coefficient for $\alpha = 0$, 10, and 21.5°.

α (°)	−5	5	16	20	23
C_L	0.23	1.51	2.9	3	1

E6.13 The light transmittance of a polymer material was tested using polymer sheets in various thicknesses. Experimental results for the transmissivity (τ) as a function of sheet thickness (t) are presented below. (a) Find a Lagrange interpolating polynomial for the transmissivity that passes through all data points; (b) plot the data and the interpolating polynomial together on the same graph and comment on how well the polynomial fits the data; (c) use the interpolating polynomial to estimate the transmissivity for $t = 2$, 4, and 6 mm.

t (mm)	0	3	5	7
τ (%)	100	71	56	45

E6.14 The dissolution rate of a chemical substance in water was experimentally measured as a function of temperature. Using the tabulated results below, (a) find a Lagrange interpolating polynomial for the dissolution rate that passes through all data points; (b) plot the data and the interpolating polynomial together on the same graph and comment on how well the polynomial fits the data; (c) use the derived interpolating polynomial to estimate the dissolution rates for $T = 285$, 295, and 303 K.

T (K)	275	300	305
R (M/s)	0.35	2	2.75

E6.15 The Brinell hardness test is used to determine the hardness of metals and alloys. The test is conducted by applying a predetermined test load to a carbide ball of fixed diameter, which is held for a period of time and then removed. The resulting impression diameter, d, and hardness are related to each other. A specimen is subjected to a series of tests with various loads. The impression diameter and Brinell Hardness Number (BHN) of a series of tests are presented below. (a) Find the Lagrange interpolating polynomial for the Brinell hardness number that passes through all data points; (b) plot the data and the interpolating polynomial together on the same graph and comment on how well the polynomial fits the data; (c) use the interpolating polynomial to estimate the Brinell hardness numbers for the specimen with $d = 3.5$, 4.5, and 5.5 mm.

d (mm)	2.5	3	4	5	6
BHN	600	415	229	145	95

E6.16 The specific volume (m³/kg) of the vapor phase of a liquid at various temperatures is tabulated below. (a) Find the Lagrange interpolating polynomial for the specific volume that passes through all data points; (b) plot the data and the interpolating polynomial together on the same graph and comment on how well the polynomial fits the data; (c) estimate the specific volume for $T = 60$, 100, 135, and 175°C using the derived interpolating polynomial.

T (°C)	50	72	83	200
ν (m^3/kg)	12	5	3	0.125

Section 6.2 Newton's Divided Differences

E6.17 Complete the following divided difference tables, where f_{ij} is the shorthand notation for $f[x_i, .., x_j]$.

(a)

i	x_i	f_i	$f_{i,i+1}$	$f_{i,i+2}$	$f_{i,i+3}$	$f_{i,i+4}$
0	1	9	16	14	0	−0.7089
1	1.5	17	?	?	?	
2	2	?	?	?		
3	2.5	?	?			
4	3	?				

(b)

i	x_i	f_i	$f_{i,i+1}$	$f_{i,i+2}$	$f_{i,i+3}$
0	0.25	−5.25	7.8	4	0
1	1.2	?	?	?	
2	1	?	?		
3	2.2	?			

E6.18 Construct Newton's divided difference table for the following data set and find a polynomial approximation that passes through all data points.

x_i	1	2	4	5
$f(x_i)$	0	7	39	64

E6.19 Construct Newton's divided difference table for the following data set and find a polynomial approximation that passes through all data points.

x_i	−1	1	2	3
$f(x_i)$	−9	−1	3	23

E6.20 Construct Newton's divided difference table for the following data set and find a polynomial approximation that passes through all data points.

x_i	0.2	0.5	1	1.5	1.7
$f(x_i)$	0	0	0	0	0.14

E6.21 For the given data set below, (a) obtain the coefficients of Newton's divided difference interpolation formula, which passes through all data points; (b) obtain the coefficients for the data reordered as {(0,1), (3,52), (2,9), (4,145)} and {(4,145), (0,1), (2,9), (3,52)} and comment on your observations.

x_i	0	2	3	4
$f(x_i)$	1	9	52	145

E6.22 Using the following data, (a) construct Newton's divided difference table; (b) generate first- to fourth-degree interpolating polynomials; (c) plot the interpolating polynomials along with the data points; (d) estimate $f(0.5)$, $f(1.5)$, and $f(2.5)$ using the approximating polynomials obtained in Part (b). Given the true values of $f(0.5) = 1.040625$, $f(1.5) = 3.27812$, and $f(2.5) = 10.76562$, which interpolating polynomial gave better estimates? Explain why.

x_i	0	1	1.3	2	3
$f(x_i)$	1	1.55	2.4	6.6	13.15

E6.23 Using the lift coefficient-angle of attack data in **E6.12**, (a) generate first- to fourth-degree interpolating polynomials; (b) plot the interpolating polynomials along with the data points; (c) estimate C_L's for $\alpha = 0$, 10, and 21.5° using the interpolating polynomials found in Part (a). Which interpolating polynomial gave better estimates? Explain why.

E6.24 Using the dissolution rate-temperature data in **E6.14**, (a) derive first- and second-degree interpolating polynomials; (b) plot the interpolating polynomials along with the data points; (c) estimate the dissolution rates for $T = 285$, 295, and 303 K using the interpolating polynomials found in Part (a). Which interpolating polynomial gave better estimates? Explain why.

E6.25 Using the specific volume of the saturated vapor-temperature data in **E6.16**, (a) generate first- to third-degree interpolating polynomials; (b) plot the interpolating polynomials along with the data points; (c) estimate the specific volume for $T = 60$, 100, 135, and 175°C using the interpolating polynomials found in Part (a); (d) which interpolating polynomial results in better estimates? (e) Reorder the data backward to obtain first- and second-degree interpolating polynomials. Which interpolating polynomial gave better estimates? Explain why.

E6.26 The compressibility factor of a gas mixture at 50°C is given below as a function of pressure. Using the furnished data, (a) derive first- to third-degree Newton's interpolating polynomials; (b) plot the interpolating polynomials along with the data points; (c) estimate the compressibility factors at $p = 5$, 20, 30, 50, 75, and 90 bar using the interpolating polynomials derived in Part (a). Which interpolating polynomial gives better estimates? Explain why.

p (bar)	1	10	40	100
Z	0.996	0.96	0.83	0.6

E6.27 Repeat **E6.26** by reordering the data backward.

Section 6.3 Newton's Formulas for Uniformly Spaced Data

E6.28 For the given data below, (a) derive a Gregory-Newton forward interpolating polynomial that passes through all data points, and (b) estimate $f(0.23)$ and $f(1.37)$.

x	0.2	0.6	1	1.4	1.8	2.2
$f(x)$	0.008	1.512	13	52.136	145.8	330.088

E6.29 Repeat **E6.28** using the Gregory-Newton backward interpolation formula.

E6.30 Experimental data below depicts the solubility of a certain rock salt (x in g/100 g water) in water as a function of water temperature (T in °C). Using the data, (a) derive first- through fifth-degree Gregory-Newton forward interpolating polynomials; (b) plot the interpolating polynomials along with the data points and interpret the agreement; (c) using the interpolating polynomials in Part (a), estimate the solubility at $T = 15$, 30, 40, 60, and 75°C. Which interpolating polynomial gave better estimates for the given data range? Explain why. *Given*: The true values are $x(15°C)=217$, $x(30°C)=250.26$, $x(40°C)=264.07$, $x(60°C)=283.53$, and $x(75°C)=294.24$.

T (°C)	5	20	35	50	65	80
x (g)	164	231	258	275	287	297

E6.31 The dynamic viscosity of a water-based solution is given below as a function of temperature. Using this data, (a) derive first- through fourth-degree Gregory-Newton forward interpolating polynomials; (b) plot the interpolating polynomials along with the data points and interpret the results; (c) estimate the dynamic viscosity for $T = 15$, 45, 58, 85, and 95°C using the interpolating polynomials. Which interpolating polynomial gave estimates for the data range given? Explain why. *Given*: The true values are $\mu(15°C)=1.1404$, $\mu(45°C) =0.5955$, $\mu(58°C)=0.4788$, $\mu(85°C)=0.3307$, $\mu(95°C)= 0.2948$ cP.

T (°C)	10	30	50	70	90
μ (cP)	1.305	0.8	0.545	0.4013	0.312

E6.32 The experimental data used to determine the drag coefficient (C_D) of an airfoil as a function of angle of attack (α) are given below. Using the furnished data, (a) derive first- through fourth-degree Gregory-Newton forward and backward interpolating polynomials; (b) plot the interpolating polynomials along with the data points and interpret the results; (c) estimate the drag coefficient for $\alpha = -2$, 3, and 10° using the interpolating polynomials. Which interpolating polynomial gives better estimates for the data range given? Explain why.

α (deg)	-8°	-4°	0°	4°	8°	12°
C_D	0.0435	0.0302	0.0448	0.0707	0.1122	0.1508

E6.33 The Rockwell hardness (R) of a steel specimen (rods) heat-treated from one end was measured at a distance of x (in mm) from the treated end. Using the distance-hardness data presented below, (a) construct forward, backward, and central difference tables; (b) derive the first- through sixth-degree Gregory-Newton forward and backward interpolating polynomials; and (c) estimate the Rockwell hardness at $x = 3$, 8, 14, and 27 mm from the treated end using the interpolating polynomials. Which interpolating polynomials give better estimates for the data range given? Explain why.

x	0	5	10	15	20	25	30
R	60	35.3	26.1	23.8	23.3	22	20

E6.34 Saturation pressure (p) versus saturation temperature (T) data for a water-based solution is presented below. Using the data, (a) construct forward, backward, and central difference tables; (b) derive the first- through sixth-degree Gregory-Newton forward and backward interpolating polynomials; (c) estimate the saturation pressure for $T = 15$, 45, 90, 100, and 145°C using the interpolating polynomials. Which interpolating polynomials gave better estimates for the data range given? Explain why?

$T(°C)$	5	30	55	80	105	130	155
p (kPa)	0.87	4.25	15.76	47.42	120.9	270.3	543.5

E6.35 The compressive strength $(P$ in MPa) of concrete is improved by the process called "*aging*" (days). A new concrete recipe that includes an additive (labeled as XZ12) has been put through aging tests. Using the tabulated data, (a) construct forward, backward, and central difference tables; (b) derive the Gregory-Newton forward and backward interpolating polynomials from the first to fifth degree; and (c) estimate the compressive strength for $t = 7$, 14, and 45 days using the interpolating polynomials.

Aging	1	11	21	31	41	51
P	11.15	32.76	36.76	38.8	40.1	41

Section 6.4 Cubic-Spline Interpolation

E6.36 Using the data in **E6.12**, (a) employ the natural spline end condition to find cubic spline interpolating polynomials for the lift coefficient; (b) estimate the lift coefficients at $\alpha = 0$, 10, and 21.5°; (c) repeat Parts (a) and (b) with the linear extrapolation end condition.

E6.37 Using the data in **E6.15**, (a) employ the natural spline end condition to find cubic spline interpolating polynomials for the Brinell Hardness number (BHN); (b) estimate the BHN for $d = 3.5$, 4.5, and 5.5-mm; and (c) repeat Parts (a) and (b) with a linear extrapolation end condition.

E6.38 Using the data in **E6.16**, (a) employ the natural spline end condition to find cubic spline interpolating polynomials for the specific volume; (b) estimate the specific volume at $T = 60$, 100, 135, and 175°C; and (c) repeat Parts (a) and (b) with a linear extrapolation end condition.

E6.39 Using the data in **E6.26**, (a) employ the natural spline end condition to find cubic spline interpolating polynomials for the compressibility factor of CO; (b) estimate the compressibility factors at $p = 5$, 20, 30, 50, 75, and 90 bar; (c) repeat Parts (a) and (b) with a linear extrapolation end condition.

E6.40 The length parameter (ξ) of a triangular fin and its thermal efficiency (η) are presented below. Using cubic splines with a linear extrapolation end condition, estimate the fin thermal efficiency at $\xi = 0.25$, 0.5, 0.6, 0.7, 0.8, 1, 1.25, 1.5, 1.75, and 2.

ξ (-)	0	0.41	0.77	1.2	1.8	2.5
η (%)	100	90	78	62	49	38

E6.41 Using the data in **E6.40** and the cubic splines with a linear extrapolation end condition, estimate the length parameter at η = 55%, 65%, 75%, 85%, and 95%.

Section 6.5 Rootfinding by Inverse Interpolation

E6.42 For the given tabulated discrete function $f(x)$, estimate (a) the root in the given interval, and (b) find the x value corresponding to $f(x) = 0.075$.

x	3.6	3.7	3.8	3.9	4	4.1
$f(x)$	0.09546	0.05383	0.01281	−0.02724	−0.06604	−0.10327

E6.43 For the given tabulated discrete function $f(x)$, estimate (a) the root in the given interval, and (b) find the x value corresponding to $f(x) = 0.10$ and 0.35.

x	1.3	1.5	1.7	1.9	2.1
$f(x)$	−0.65893	−0.3918	−0.09793	0.21952	0.55807

E6.44 For the given tabulated discrete function $f(x)$, estimate its root in the given interval.

x	1.5	1.75	2	2.25	2.5
$f(x)$	−0.42324	−0.18777	−0.02456	0.05625	0.4962

E6.45 For the given tabulated discrete function $f(x)$, estimate its root in the given interval.

x	2	2.5	3	3.5
$f(x)$	8.1948	3.0164	−5.0755	−18.052

Section 6.6 Multivariate Interpolation

E6.46 The thermal conductivity, k (W/m·°C), of an alloy is tabulated below as a function of an impurity x (%) and temperature T (°C). Use Lagrange's interpolation to obtain an interpolating polynomial (passing through all data points) to estimate the thermal conductivity of the alloy with impurity contents of 3.75%, 7.5%, 18.5%, and 27.5% at 150, 425, and 875°C.

x (%)	T (°C)			
	0	200	600	1000
1	73	62	40	35
5	54	45	33	31
10	40	36	29	29
30	22	22	24	29

E6.47 The compressibility factor, $Z(T, p)$, of a water-based solution at various temperatures and pressures is tabulated below. To estimate the compressibility factor at 10 and 40 bar and 550, 875, 1250, and 1750 K, derive Lagrange's interpolation to obtain an interpolating polynomial that passes through all data points.

p (bar)	T (K)			
	400	700	1000	2000
5	0.003	0.994	0.999	1.002
15	0.009	0.984	0.995	1.002
50	0.029	0.941	0.987	1.002
250	0.143	0.618	0.935	1.008

E6.48 The density of an acidic solution ρ (g/m^3) as a function of acid content x (%) and temperature (°C) is tabulated below. Use Lagrange's interpolation to generate an interpolating polynomial that uses all data points to estimate the compressibility factor for 15%, 50%, and 85% acid solutions at 5, 25, and 55°C.

Acid	T (°C)				
x (%)	0	15	40	60	100
1	10.1	10.1	9.98	9.89	9.65
34	12.7	12.5	12.4	12.2	11.9
79	17.4	17.2	16.9	16.8	16.4
100	18.5	18.3	18.1	17.9	17.5

E6.49 In the grain drying process, the moisture loss (M) is determined by the initial (%w) and final moisture contents (%w). Use Lagrange's method to obtain an interpolating polynomial passing through all data points to estimate the moisture loss for initial moisture contents of 21, 25, and 28% that result in a final moisture content of 12, 14, 16, and 18%.

Initial	Final moisture content %w				
moisture	19	17	15	13	11
w%	Moisture loss (kg/ton)				
30	136	157	176	195	213
27	111	120	153	172	180
24	62	84	106	126	146
20	12	36	59	80	101

Section 6.7 Extrapolation

E6.50 Using **E6.14** data and Newton's divided differences, derive first- and second-degree extrapolation polynomials to estimate the dissolution rates at $T = 310$ and 315 K.

E6.51 Using **E6.16** data and Newton's divided differences, derive first-, second-, and third-degree extrapolation polynomials to estimate the specific volume at $T = 20, 40, 225$, and 250°C.

E6.52 Using **E6.26** data and Newton's divided differences, derive first-, second-, and third-degree extrapolation polynomials to estimate the compressibility factor at 50°C and 110, 130, and 200 bar.

E6.53 Using **E6.28** data (excluding $x = 2.2$) and Newton's divided differences, (a) derive first-, second-, third-, and fourth-degree extrapolation polynomials to estimate $f(0)$, $f(2.2)$, $f(2.6)$, and $f(3)$; and (b) discuss the accuracy of the extrapolated values if $f(x) = 15x^4 - 2x^3$ is the true function.

E6.54 Using **E6.30** data and Newton's divided differences, derive first- to fifth-degree extrapolation polynomials to estimate the dissolution rate of the salt at $T = 85$ and 90°C. What can you say about the accuracy of the estimates?

E6.55 The laptop sales of a computer company between 2018 and 2023 are presented in the table below. Assuming the trend will continue, forecast laptop sales in 2025, 2027, and 2030.

y (year)	2018	2019	2020	2021	2022	2023
S (thousand)	219.9	230.1	208	174.9	163.7	152

E6.56 Consider the saturation pressure versus saturation temperature data of a water-based solution given in **E6.34**. Estimate the saturation pressure of the solution at $T = 160$ and 175°C using linear, quadratic, and cubic approximations. Given that the true values are 609.2 and 849 kPa, respectively, which approximation produces better estimates? Explain.

E6.57 Consider the concrete aging data in **E6.35**. Estimate the compressive strength at $t = 90$ days. Given that the true value is 42.5, which approximation produces better estimates? Explain.

6.10 COMPUTER ASSIGNMENTS

CA6.1 Use the dynamic viscosity data of a water-based solution given in **E6.31** to obtain interpolating polynomials with the GN-FWD/-BKWD IFs. Calculate the absolute error using the following analytical expression for viscosity and discuss the results.

$$\mu(T) = 0.02414 \times 10^{247.8/(T+133)}$$

CA6.2 The Gamma function, $\Gamma(x)$, is defined by the integral given below.

$$\Gamma(x) = \int_0^\infty u^{x-1} e^{-u} du$$

The Gamma function for some x in $[0.1,1]$ is tabulated below. (a) Find the interpolating polynomials using Newton's divided difference method; (b) For each interpolating polynomial, plot the absolute error, $e_i = |\Gamma_i - P_n(x_i)|$.

x	0.1	0.2	0.3	0.45	0.75	1
$\Gamma(x)$	9.51351	4.59084	2.99157	1.96814	1.22542	1

CA6.3 The following discrete data on $[-3,3]$ is derived from $y = 3x/(1+x^2)$. Apply cubic spline interpolation to the data set for both the natural and linear extrapolation end conditions to calculate the values of the function as well as its first- and second-order derivatives for values in the given range with $\Delta x = 0.5$, 0.25, and 0.125 intervals. Discuss the accuracy of the results.

i	x_i	y_i	i	x_i	y_i	i	x_i	y_i
1	-3	-0.9	6	-0.5	-1.2	11	2	1.2
2	-2.5	-1.0345	7	0	0	12	2.5	1.0345
3	-2	-1.2	8	0.5	1.2	13	3	0.9
4	-1.5	-1.3846	9	1	1.5			
5	-1	-1.5	10	1.5	1.3846			

CA6.4 The enthalpy of a flue gas is calculated based on experiments at temperatures between 400 and 1600 K. According to thermodynamics, the heat capacity at constant pressure is related to $c_p = \partial h/\partial T$. Use the tabulated flue gas data to calculate the enthalpy, heat capacity, and its derivative for temperatures from 500 K to 1500 K in 50 K increments by cubic spline (linear extrapolation end condition) interpolation.

T (K)	400	600	800	1000	1200	1400	1600
h (kJ/kg)	413	627	851	1086	1331	1586	1853

CA6.5 Consider a uniformly spaced bivariate discrete function denoted by $f_{pq} = f(x_p, y_q)$. Generate a bi-quadratic interpolating polynomial using central differences $\mathcal{O}[(\Delta x)^3, (\Delta y)^3]$ and write a computer code to carry out bivariate interpolation. Use a nine-point grid, i.e., $x_{p+1} - x_p = \Delta x$, $y_{q+1} - y_q = \Delta y$, $x_{val} - x_p = h_x$ and $y_{val} - y_q = h_y$. Employ a binary search algorithm to find (p, q) and construct the computational grid.

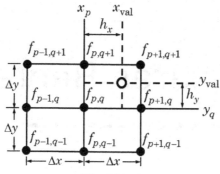

Fig. CA6.10

Hint: The Taylor series of $f(x, y)$ about (x_p, y_q) is given as

$$f(x_{\text{val}}, y_{\text{val}}) = \sum_{k=0}^{\infty} \frac{1}{k!} \left[(x_{\text{val}} - x_p)\frac{\partial}{\partial x} + (y_{\text{val}} - y_q)\frac{\partial}{\partial y} \right]^k f(x_p, y_q)$$

For partial derivatives, use the central difference formulas. Here, h_x and h_y denote spacings with respect to the $x-$ and $y-$ directions, respectively.

CA6.6 The partial pressure p (mmHg) of a water-sodium hydroxide solution is tabulated below as a function of temperature T ($^\circ$ C) and sodium hydroxide content x (g NaOH/100g water). Using the formulation developed in **CA 6.5**, write a program to compute the partial pressure and estimate $p(10^\circ$ C,5), $p(10^\circ$ C,25), $p(10^\circ$ C,35), $p(45^\circ$ C,5), $p(45^\circ$ C,25), $p(45^\circ$ C,35), $p(75^\circ$C,5), $p(75^\circ$C,25), $p(75^\circ$ C,35), $p(115^\circ$ C,5), $p(115^\circ$ C,25), $p(115^\circ$C,35).

g NaOH per	Temperature ($^\circ$ C)						
100 g water	0	20	40	60	80	100	120
0	4.6	17.5	55.3	149.5	355.5	760	1489
10	4.2	16	50.6	137	325.5	697	1365
20	3.6	13.9	44.2	120.5	288.5	621	1225
30	2.9	11.3	36.6	101	246	537	1070
40	2.2	8.7	28.7	81	202	450	920

CA6.7 Using the furnished data for $\Gamma(x)$ in **CA 6.2**, estimate the Gamma function $\Gamma(x)$ and its derivative $\Gamma'(x)$ with increments of 0.05 from $x = 0.1$ to 1 using cubic spline interpolation with (a) the natural cubic spline and (b) linear extrapolation end conditions.

CA6.8 Consider the population of a small town between 1960 and 2020.

Year	1960	1970	1980	1990	2000	2010	2020
Population ($\times 10^3$)	13.1	17.3	21.4	23.3	25.1	26.2	27.3

Use the population data to find a sixth-degree interpolating polynomial to extrapolate the population. Write a program to extrapolate the population for a 10-year increment and estimate the population for the years 2030, 2040, and 2050.

CHAPTER 7

Least Squares Regression

LEARNING OBJECTIVES

After completing this chapter, you should be able to

- understand and discuss the concept of least squares regression;
- apply least squares regression to polynomial models to approximate a set of discrete data;
- list the advantages and disadvantages of least squares regression;
- explain the goodness of fit analysis and apply it to fitted models;
- understand and apply the linearization procedure to nonlinear models;
- apply multivariate least squares regression to multivariate data;
- employ the least squares procedure under the integral sign to obtain simple approximate functions;
- utilize least squares regression to obtain an approximate solution to an over- or under-determined system.

REGRESSION is a method of finding the best-fit for a given set of discrete data. Least squares is one of the most common methods used for fitting data to straight lines or arbitrary curves. Recall that the data representing the discrete values (distribution) of a continuum may be collected digitally with digital instruments or obtained from experimental measurements or calculations based on experiments, as discussed in Chapters 5 and 6. The numerical values of a continuum between any two discrete values of a uniform or non-uniform discrete function are unknown. Moreover, experimentally derived data are not perfect, and even when perfectly calibrated instruments are used, the measured quantities may contain errors to some extent. Furthermore, keep in mind that instruments are not always in perfect condition because the accuracy of all measuring instruments degrades over time due to many factors such as wear and tear, electric or mechanical shock, or a non-ideal working environment (temperature, humidity, pressure, magnetism, fumes, etc.). All these factors contribute to the quality of the data.

A set of experimentally derived data for a continuum that is supposed to be linear is depicted in Fig. 7.1a. The data points, which should ideally correspond to points on the straight line, exhibit deviations from a linear relationship. These deviations, or scatter, are called *noise*. A prediction curve can be constructed with, for instance, a Lagrange interpolation polynomial to represent the data with an interpolating polynomial. Nevertheless, a high-order Lagrange interpolating polynomial will pass through all data points, including

DOI: 10.1201/9781003474944-7

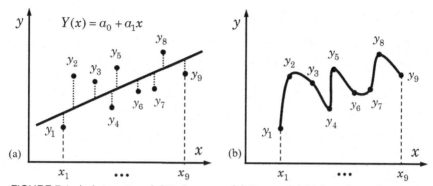

FIGURE 7.1: A data set and fitted curves: (a) linear, (b) high-order polynomial.

the data tainted by the noise, as shown in Fig. 7.1b. As a result, the prediction curve will also contain noise, and the reliability of the predictions will be questionable. For this reason, a method called the *least squares fit* is generally used to generate a function (a curve-fit) that is free from existing noise in the raw data. This method, in addition to devising an approximating function for interpolation purposes, is also used to find some experimentally determined physical quantities for materials or various physical processes.

In this chapter, least squares regression and applications in science and engineering fields are introduced. Overfitting issues, as well as the quality of fitting in terms of goodness of fit, are discussed. The linearization procedures of nonlinear models commonly used in practice are discussed. The method is extended to multivariate curve fits, over- or underdetermined linear systems, as well as minimization problems under the integral sign.

7.1 POLYNOMIAL REGRESSION

One of the most commonly used models in curve fitting is the *polynomial model*, where a set of discrete data is approximated by an nth-degree polynomial. In this class, the *linear regression* (i.e., fitting data to a *straight line*) is a commonly encountered procedure in almost every field of science and engineering.

Let (x_1, y_1), (x_2, y_2), ..., (x_n, y_n) be a set of n discrete data points, perhaps, contaminated with noise. The principle of least squares regression is to minimize the difference between the *predicted* and *measured* (or *observed*) values, and so we define the error as

$$e(x) = Y(x) - y(x) \tag{7.1}$$

where $Y(x)$ and $y(x)$, respectively, denote the predicted and observed values, and $e(x)$ is the error referred to as the *residual*.

In order to obtain a *model* (or a *prediction curve*) free of noise, minimizing the sum of the squares of residuals is considered an effective criterion; that is, the sum of the squares of residuals is the *objective function* defined as

$$E = \sum_{i=1}^{n} e^2(x_i) = \sum_{i=1}^{n} \left(Y(x_i) - y(x_i) \right)^2 \tag{7.2}$$

where n is the number of measurements, observations, or sample size. Notice that the objective function amplifies the effects of large residuals while ignoring the effects of small residuals.

Now consider a best-fit model to be an mth-degree polynomial:

$$Y(x) = a_0 + a_1 x + a_2 x^2 + \ldots + a_m x^m \tag{7.3}$$

Substituting Eq. (7.3) into Eq. (7.2), the objective function becomes

$$E(a_0, a_1, \ldots, a_m) = \sum_{i=1}^{n} \left(a_0 + a_1 x_i + a_2 x_i^2 + \ldots + a_m x_i^m - y_i \right)^2 \tag{7.4}$$

where (x_i, y_i) denotes the set of measured or observed data, $a_0, a_1, \ldots, a_m$ are the coefficients of the polynomial, also called *model parameters* and can be varied in order to minimize E.

The objective function is a single equation with $m+1$ unknowns: $a_0, a_1, \ldots, a_m$. We also require $m+1$ equations to determine the unknowns. We can supply the extra equations in a variety of ways, but keep in mind that we also wish to minimize E. Recall from the calculus lectures that the partial derivatives of a function with respect to its independent variables vanish at a local minimum point. Hence, we will derive $m+1$ equations by differentiating E with respect to every unknown and then setting each equation to zero as follows:

$$\frac{\partial E}{\partial a_0} = 0, \quad \frac{\partial E}{\partial a_1} = 0, \quad \frac{\partial E}{\partial a_2} = 0, \quad \ldots, \quad \frac{\partial E}{\partial a_m} = 0 \tag{7.5}$$

For instance, differentiating with respect to a_0 by applying the chain rule gives

$$\frac{\partial E}{\partial a_0} = \sum_{i=1}^{n} \frac{\partial}{\partial a_0} \left(a_0 + a_1 x_i + a_2 x_i^2 + \ldots + a_m x_i^m - y_i \right)^2$$
$$= \sum_{i=1}^{n} 2 (\cdots) \frac{\partial}{\partial a_0} (\cdots) = 2 \sum_{i=1}^{n} \left(a_0 + a_1 x_i + a_2 x_i^2 + \ldots + a_m x_i^m - y_i \right) = 0 \tag{7.6}$$

where $\partial(\ldots)/\partial a_0 = 1$. Since the rhs of Eq. (7.6) is zero, the multiplicative factor "2" in front of the summation sign may be dropped. So removing the parenthesis and rearranging yields

$$n\, a_0 + \left(\sum_{i=1}^{n} x_i \right) a_1 + \left(\sum_{i=1}^{n} x_i^2 \right) a_2 + \ldots + \left(\sum_{i=1}^{n} x_i^m \right) a_m = \sum_{i=1}^{n} y_i \tag{7.7}$$

Similarly, differentiation with respect to a_1 leads to

$$\frac{\partial E}{\partial a_1} = \sum_{i=1}^{n} 2 (\cdots) \frac{\partial}{\partial a_1} (\cdots) = 2 \sum_{i=1}^{n} \left(a_0 + a_1 x_i + a_2 x_i^2 + \ldots + a_m x_i^m - y_i \right) x_i = 0 \tag{7.8}$$

where $\partial(\ldots)/\partial a_1 = x_i$. After algebraic manipulations, Eq. (7.8) leads to

$$\left(\sum_{i=1}^{n} x_i \right) a_0 + \left(\sum_{i=1}^{n} x_i^2 \right) a_1 + \left(\sum_{i=1}^{n} x_i^3 \right) a_2 + \ldots + \left(\sum_{i=1}^{n} x_i^{m+1} \right) a_m = \sum_{i=1}^{n} x_i y_i \tag{7.9}$$

Likewise, evaluating the partial derivatives of E with respect to the other unknown coefficients and setting them to zero completes the $m+1$ equations. The collection of these equations altogether forms a system of linear equations.

The system of linear equations for the unknown coefficients is expressed as

$$\begin{bmatrix} n & \sum x_i & \sum x_i^2 & \cdots & \sum x_i^m \\ \sum x_i & \sum x_i^2 & \sum x_i^3 & \cdots & \sum x_i^{m+1} \\ \sum x_i^2 & \sum x_i^3 & \sum x_i^4 & \cdots & \sum x_i^{m+2} \\ \vdots & \vdots & \vdots & \ddots & \vdots \\ \sum x_i^m & \sum x_i^{m+1} & \sum x_i^{m+2} & \cdots & \sum x_i^{2m} \end{bmatrix} \begin{bmatrix} a_0 \\ a_1 \\ a_2 \\ \vdots \\ a_m \end{bmatrix} = \begin{bmatrix} \sum y_i \\ \sum x_i y_i \\ \sum x_i^2 y_i \\ \vdots \\ \sum x_i^m y_i \end{bmatrix} \tag{7.10}$$

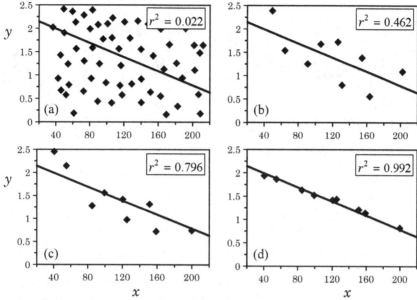

FIGURE 7.2: Correlatability of the observed to a linear model; (a) no correlation, (b) weak correlation, (c) fair correlation, (d) strong correlation.

where the summation signs encompass all n.

The resulting system of linear equations for $m = 1, 2,$ or 3 may be solved by the Cramer's rule. For high-degree polynomial models, the linear system tends to be ill-conditioned. In many cases, the degree of a polynomial is severely restricted due to rounding errors. For example, polynomial models of $m > 7$ generally produce absurd results using single-precision arithmetic. It is therefore recommended to use direct solution algorithms with pivoting and high-precision arithmetic to overcome the adverse effects of ill-conditioning (*see* Chapter 2).

7.1.1 DETERMINING REGRESSION MODEL

Consider a set of discrete data where x_i and y_i are the independent and dependent (observed) variables, respectively. In a set of data, a correlation between the independent variable and the observed variables may or may not exist. If a functional relationship can be established between two or more variables, it can be said that there is a correlation between the dependent and independent variables.

The correlatability of a linear model to four different sets of observed data is depicted in Fig. 7.2. In Fig. 7.2a, the observed data spreads out over a wide range, indicating "*no correlation*" between x and y. In Fig. 7.2b, the data is gathered around a straight line with large residuals, indicating "*weak correlation*." As depicted in Fig. 7.2c and d, the tighter the observed data points are around the model, the stronger the correlation.

A regression model (or best-fit curve) is best determined after inspecting the observed data distribution. If the observed data is suitable to be fitted into a polynomial, then we must address the following question: *what should the degree of the polynomial be?* To answer this question, a logical approach is to plot the observed data and determine the best-fit model (or the degree of the polynomial) by visual inspection. The two observed data sets are illustrated in Fig. 7.3. The visual inspection reveals that the data set in Fig.

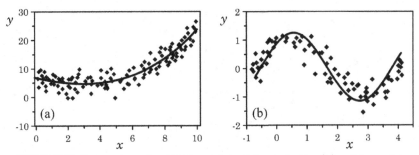

FIGURE 7.3: Data sets and suitable models; (a) quadratic, (b) cubic model.

7.3a is suitable for a quadratic model (or a higher-degree polynomial). Likewise, the set of data in Fig. 7.3b is a candidate for a cubic polynomial (or a sinusoidal) model. One should experiment with applicable models that give optimal goodness of fit.

The most important and critical factor in adopting a best-fit model should include consideration of the physical event and the relationship suggested by the theory for which the data was collected. For example, consider the experimental determination of the density of a substance using a batch of samples of different shapes and sizes. The density of a substance may be estimated from a single measurement using a single sample. While measuring the mass of a sample is fairly straightforward, measuring the volume in this example essentially conceals a major challenge. To determine the volume, we can make use of Archimedes' principle, which states that "*the volume of displaced fluid is equivalent to the volume of an object fully immersed in a fluid.*" Nonetheless, the volume of the substance cannot be precisely measured due to the presence of inhomogeneities, such as pores and cracks or trapped air bubbles within, etc., which introduce uncertainties. For this reason, multiple measurements (using samples of different sizes) should be performed to minimize measurement errors. Considering the physics of this phenomenon, the theory suggests a linear relationship between the mass and the volume of a substance, $m = \rho V$, where the density is the slope of the straight line. Hence, when the mass and volume data of the samples are plotted, the data should ideally be aligned exactly on a straight line. In reality, the measurements will most likely depict deviations that may give a discrete mass-volume relationship that does not look linear, as seen in Fig. 7.4. The deviations of the observed values from linearity are due to the so-called "noise" caused by uncertainties (pores, cracks, etc.) in the measurements. Hence, even though the data set appears to conform to a quadratic or cubic relationship (depicted with a dashed curve), this observation is inconsistent with the theory, and one should not be tempted to use a model other than a linear model.

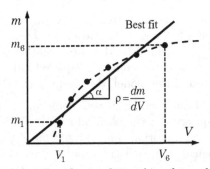

FIGURE 7.4: Measured mass-volume relationship of samples of a substance.

Zero-Intercept Model. In cases where the data is theoretically related by the $y = mx$ relationship, the regression line is forced to intercept the origin. Hence, the objective function is chosen as

$$E(m) = \sum_{i=1}^{n} (y_i - mx_i)^2$$

with m as the only (unknown) model parameter. Setting $\partial E/\partial m = 0$ yields

$$m = \left. \sum_{i=1}^{n} x_i y_i \middle/ \sum_{i=1}^{n} x_i^2 \right.$$

7.1.2 GOODNESS OF FIT

After deciding on a suitable model, the model constants are determined by applying the least squares method. Next, we seek answers to the following questions: *How well does this best model represent the data? How good will the estimates from this model be? Can the model be further improved?* and so on.

A plot of the model along with the data, as shown in Fig. 7.3, provides a visual aid regarding the suitability of the model to the data. However, a visual aid alone is not sufficient and provides a qualitative assessment only; a measurable quantitative criterion is also required to measure the *goodness of fit*. We know that there is a residual (i.e., $e_i = Y_i - y_i \neq 0$) for every point that does not correspond exactly to the best curve. The *sum of the squares of the residuals (SSR)* of the observed data will be minimal owing to the minimization requirement—Eq. (7.5). Under ideal conditions, Eq. (7.4) should be zero for any data that is completely free of noise.

The *sum of the squares of the mean deviation (SSMD)*, which is a measure of how much the data deviates from the mean value, is computed from

$$S = \sum_{i=1}^{n} (y_i - \bar{y})^2 \tag{7.11}$$

where $\bar{y}$ is the arithmetic mean of the data set. Unless all data are distributed along a horizontal line ($y_i =$ constant), S will not be zero even if an ideal curve fit is found.

Residual Plots: An assessment of the usefulness of a curve fit should start by examining the residuals *via* the residual plots that reveal the effectiveness and reliability of the model adopted. Residuals, $e_i = Y_i - y_i$, provide qualitative information about the goodness of the fit. A residual plot is obtained simply by placing the abscissas (x_i) on the horizontal axis and the residuals (e_i's) on the vertical axis. As an example, a data set along with its best fit and the residual plot corresponding to the best-fit line are respectively shown in Fig. 7.5a and b. Note that the observed data points are gathered along the best-fit line. If the residuals are small and unbiased, we can claim that the model is suitable for this data set. In this context, what we mean by "*unbiased*" is that the predicted values are not systematically too high or too low within the regression range.

Before examining quantitative metrics of the goodness-of-fit, one should qualitatively assess the residual plots because they can reveal a biased model far more effectively. A biased model cannot be trusted, and a new model should be proposed.

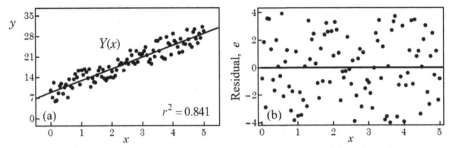

FIGURE 7.5: (a) Observed data along with the best fit, (b) residual plot.

A residual plot of a linear regression depicts certain features that hint at the suitability or appropriateness of the regression. Some examples of *residual plots* that might be encountered as a result of a linear regression procedure are depicted in Fig. 7.6. The residual data is termed *homoscedastic* if the variance is constant or depicts homogeneous scatter. For unbiased and homoscedastic data (*seen* in Fig. 7.6a), the residuals are randomly scattered around the $e = 0$ line, and the average of the residuals is zero in each thin vertical band, $x_i - \delta < x < x_i + \delta$ where δ is a small number. The residuals form the same size horizontal bands below and above the $e = 0$ line, indicating that the variances (or standard deviations) of the error are the same all across the x-axis. Also note that there are no *outliers* in the residual distribution; in other words, no one residual "stands out" from the random pattern of the residuals. The residuals in the biased and homoscedastic cases are illustrated in Fig. 7.6b yield a linear pattern, which is most likely due to an independent variable that was overlooked or not taken into account in the experiment. The residuals in Fig. 7.6c forming a quadratic pattern are most likely indicative of a nonlinear relationship; in other words, a linear relationship is forced onto nonlinear data. A linearization (i.e., transformation) of the data covered in Section 7.3 might help mitigate this type of problem. In *heteroscedastic* data, the variance of the residuals increases or decreases along the horizontal axis. As illustrated in Fig. 7.6d, in an unbiased heteroscedastic case, while the average of the residuals yields zero for every thin vertical band, the variance increases from the left of the plot to the right. Biased and heteroscedastic linear and quadratic patterns are shown in Fig. 7.6e and f are also signs of trouble owing to using an unsuitable regression model.

 In multi-regressions, plotting the residuals with respect to each independent variable may help determine the kind of relationship between the independent and dependent variables. However, in general, the knowledge, expertise, and experience of the analyst in the field are the most valuable assets that contribute to finding relationships between variables and developing a suitable model.

In multi-regressions, sometimes the bias and/or heteroscedasticity of the residuals cannot be mitigated. This is largely due to implementing an unsuitable regression model that does not take into account one or more of the "critical" parameters or variables. For instance, modeling the development of a storm as a function of pressure and wind speed may reveal that the temperature, humidity, or both (additional variables) need to be taken into account as well.

R-Squared: R-Squared is one of the quantitative metrics for assessing the goodness of a regression. The difference between E (the sum of the squares of the residuals of the observed data) and S (the sum of the squares of the mean deviation) is accepted as a

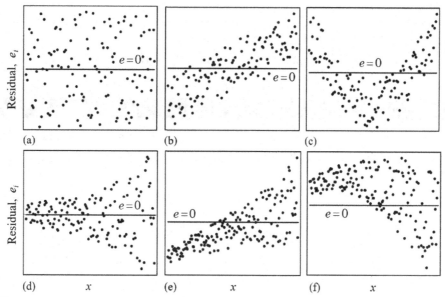

FIGURE 7.6: Possible residual patterns: (a) unbiased and homoscedastic, (b), (c) biased and homoscedastic, (d) unbiased and heteroscedastic, (e), (f) biased and heteroscedastic.

measure of the accuracy of the curve fit. Because this difference is scale-dependent, most commonly a normalized form of the difference, obtained by dividing by S, is used:

$$r^2 = (S - E)/S = 1 - E/S \qquad (7.12)$$

where r^2 is referred to as the *r-squared* or *coefficient of determination*, while r is termed the *correlation coefficient*. In practice, r^2 is more commonly reported instead of r.

Ideally, for any regression, r-squared (or r) should be calculated and interpreted because it provides a quantitative measure for the strength of the relationship between the dependent and independent variable(s). Notice that r-squared approaches "1" when E is small (or S is large), hinting that the model explains a large fraction of the variation in the data. In other words, when the observed and predicted values are in perfect agreement, the sum of the residuals becomes zero, yielding $r^2 = 1$, meaning that the regression model explains 100% of the variability of the data. When the data, as shown in Fig. 7.2a, is dispersed, the r-squared approaches zero because $E \to S$, in which case the variability of the data cannot be explained by the model. The variation of r-squared in connection with the distribution of data is illustrated in Fig. 7.2b, c, and d. Notice that as r-squared approaches one, the data cluster more tightly around the best-fit. Also, r-squared represents the percentage of the data close to the fit; for instance, $r^2 = 0.81$ means 81% of the data are represented by the best fit and 19% are not.

EXAMPLE 7.1: Application of Linear Regression

The effect of salinity of water x (parts per thousand) in an agricultural land on the electrical conductivity y (miliSiemens per cm) of the soil was investigated. The results of the salinity and conductivity measurements of samples taken from five locations in the field are tabulated below. Assuming that the data can be described by a linear regression model, apply least squares regression to determine the *best-fit parameters* and discuss the *goodness of the fit*.

x_i (ppt)	2	6	7	10	14
y_i (mS/cm)	3	4	6	7	8

SOLUTION:

The model is $Y(x) = a_0 + a_1 x$, and the model parameters are a_0 and a_1, respectively. The objective function is defined as the sum of the squares of the residuals as

$$E(a_0, a_1) = \sum_{i=1}^{5} (a_0 + a_1 x_i - y_i)^2$$

Differentiating the objective function with respect to the model parameters a_0 and a_1 gives a 2×2 system of linear equations. However, essentially having to perform these operations, we will make use of Eq. (7.10) by including the first two rows and two columns, leading to

$$\begin{bmatrix} n & \sum x_i \\ \sum x_i & \sum x_i^2 \end{bmatrix} \begin{bmatrix} a_0 \\ a_1 \end{bmatrix} = \begin{bmatrix} \sum y_i \\ \sum x_i y_i \end{bmatrix}$$

The coefficient matrix and the rhs of this system require $\sum x_i$, $\sum x_i^2$, $\sum y_i$, and $\sum x_i y_i$ quantities. The original data and pertinent calculational details are presented in Table 7.1.

TABLE 7.1

i	x_i	y_i	x_i^2	$x_i y_i$	$(y_i - \bar{y})^2$	$(y_i - Y_i)^2$
1	2	3	4	6	6.76	0.00198
2	6	4	36	24	2.56	0.65110
3	7	6	49	42	0.16	0.56626
4	10	7	100	70	1.96	0.18553
5	14	8	196	112	5.76	0.10998
Σ	**39**	**28**	**385**	**254**	**17.2**	**1.51485**

The values typed in bold numbers in the last row of the table are the sum of the values in that column. Substituting the computed quantities into the matrix equation, we obtain

$$\begin{bmatrix} 5 & 39 \\ 39 & 385 \end{bmatrix} \begin{bmatrix} a_0 \\ a_1 \end{bmatrix} = \begin{bmatrix} 28 \\ 254 \end{bmatrix}$$

which yields $(a_0, a_1) = (2.16337, 0.44059)$. The squares of both the deviations from the mean value and the residues are also presented on the 6th and 7th columns of Table 7.1, respectively. Using the tabulated data, we find

$$S = \sum_{i=1}^{5} (y_i - 5.6)^2 = 17.20, \qquad E = \sum_{i=1}^{5} (Y(x_i) - y_i)^2 = 1.51485$$

Finally, the r-squared value is obtained as $r^2 = (S - E)/S = 0.912$.

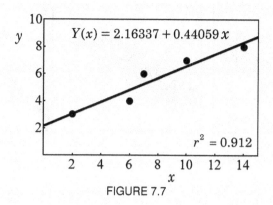

FIGURE 7.7

Discussion: It is important to know quantitative measurements of model coefficients and model accuracy, and we must also understand visual approaches to evaluate our model. In this regard, the distribution of the measured data as well as the best-fit model are comparatively depicted in Fig. 7.7. It is observed that the measured data spreads in close proximity to the (linear) model, and 91.2% of the data is well represented by the linear model. The model parameters, which are the slope and intercept, are found to be 0.44059 and 2.16337, respectively.

The correlation coefficient for a linear model, $Y(x) = a_0 + a_1 x$, is reduced to

$$r = \frac{n \sum x_i y_i - (\sum x_i)(\sum y_i)}{\sqrt{n \sum x_i^2 - (\sum x_i)^2} \sqrt{n \sum y_i^2 - (\sum y_i)^2}} \tag{7.13}$$

Adjusted R-Squared: A danger of relying solely on r-squared is that it can lead an analyst to misconclusions, especially when assessing a set with very few data points. Similarly, using a large data set or adding extra variables to the data set may also lead to illusive conclusions. In other words, when more data is added, r^2 does not decrease but rather increases, giving a false sense of confidence. Similar arguments can be made for multivariable regression models as well. For instance, when an independent variable is added to the model, r^2 increases, and an increase in r^2 approaching unity can also be misleading for the analyst. In this context, variables having very little or no effect on the main variable(s) should be excluded from the model to reduce model complexity and to build understandable and interpretable models. In this regard, to penalize adding extra variables, the adjusted r^2 is defined as

$$r_{adj}^2 = 1 - (1 - r^2)\left(\frac{n - 1}{n - k - 1}\right) \tag{7.14}$$

where n is the number of data points, k is the number of variables, including the model constants. When $(1 - r^2)$, i.e., E/S, is multiplied by $(n - 1)/(n - 1 - k)$, the new factor will always be smaller than r^2 since $1 + k/(n - 1 - k) > 1$. This way, adding extra data and variables is penalized, and a new index of compliance is established.

RMSE: Another quantitative metric for measuring the quality of the model fit is the *rmse* (abbreviation for *root mean square error*). A best-fit curve, $Y(x)$, estimates a prediction for any x, while the *rmse* measures the spread of y's. It is defined as the square

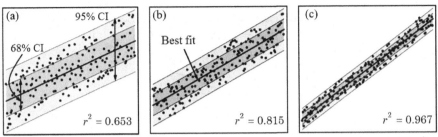

FIGURE 7.8: Variation of 68% and 95% $rmse$ bands (CI: confidence interval) with (a) $r^2 = 0.653$, (b) $r^2 = 0.815$, and (c) $r^2 = 0.967$.

root of the arithmetic mean of the sum of the squares of the residuals and is given as

$$\text{rmse} = \sqrt{\frac{1}{n} \sum_{i=1}^{n} (Y_i - y_i)^2} = \sqrt{\frac{E}{n}} \tag{7.15}$$

where n is the number of measurements (or observations).

An alternative definition of $rmse$ is the *standard deviation* of residuals. Using Eq. (7.11), the standard deviation of the measured (or observed) data set is computed as

$$\sigma = \sqrt{\frac{1}{n} \sum_{i=1}^{n} (y_i - \bar{y})^2} = \sqrt{\frac{S}{n}} \tag{7.16}$$

and the variance, denoted by s, is obtained from $s = \sigma^2$.

The standard deviation, as a measure of dispersion, is a quantitative metric indicating how spread out the measured (or observed) points are. Using Eq. (7.12), the $rmse$ can be written as

$$\frac{E}{n} = (1 - r^2)\frac{S}{n} \qquad \text{or} \qquad \text{rmse} = \sqrt{\frac{E}{n}} = \sqrt{1 - r^2}\,\sigma \tag{7.17}$$

which can be obtained in terms of r^2 and σ. For example, $rmse = 0$ (i.e., $r = 1$) indicates that the measured data are 100% represented by the fitted (predicted) model. Since $rmse$ is measured on the same scale and in the same units as y, 68% of y-values should be within $\pm rmse$. In Fig. 7.8, the data set is represented by confidence intervals of 68% and 95% ($\pm 2 \times rmse$) around the best-fit. Notice that the band shrinks as r-squared increases.

Predicted-Observed (PO) Data Plots. Predicted versus observed (PO) plots *do* provide qualitative and quantitative information about the regression models. When assessing the predictions of multi-regression models, the use of residual plots is not that practical. So instead, the predicted-observed values can be plotted by placing the predicted values (Y's) on the horizontal axis and the observed values (y's) on the vertical axis. The slope (m) and intercept (a) of the PO line, $Y = a + my$, describe the *consistency* and the *model bias*, respectively. In Fig. 7.9, the results of a linear and a nonlinear regression model, as well as their PO plots, are depicted. It is evident that for identical observations or predictions, the resulting PO plot is perfectly aligned on the diagonal (i.e., $y = Y$ line and $rmse = 0$). Likewise, having data points close to the diagonal line ($rmse \approx 0$) indicates a reasonably good fit for the overall data. The data points closely spread about the diagonal line, as seen in Fig. 7.9a, are an indication that the predicted and observed data are correlated. Any nonlinear trends in the plots indicate no correlation. The unit slope validates the linear relationship between y and Y, i.e., the linear regression model is suitable for the data. On the

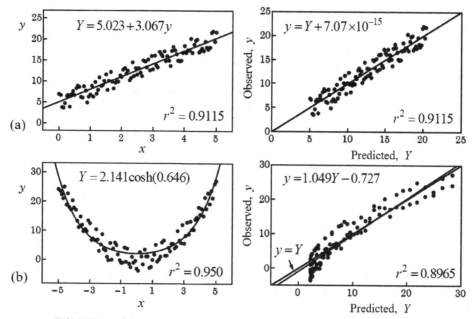

FIGURE 7.9: (a) Linear and (b) nonlinear fits with data and PO plots.

other hand, the deviation of the slope from unity reflects a relative scaling factor between the two sets. A non-zero intercept indicates that some data are shifted relative to others by a constant offset. The model bias (shift) is determined (or quantified) by regressing the PO values onto a straight line and comparing the slope and intercept of the best-fit line against the diagonal line. This way, any bias in the data can be identified. Notice that, in Fig. 7.9b, a nonlinear fit leads to a relatively small shift (≈ 0.727) to the right.

When linearly correlated data are fitted to a line, the r-squared value of the $x - Y$ and $Y - y$ regressions are the same ($r = 0.9115$) for Fig. 7.9a. However, this is not the case for nonlinear regression models, as noted in Fig. 7.9b ($r^2 = 0.95$ for $x - Y$ and 0.8965 for $Y - y$). The r-squared, the slope and intercept of the PO line, and so on provide parameters for judging and building confidence in the model's performance.

EXAMPLE 7.2: Quadratic model and goodness of fit

Experimental results for the efficiency η (%) of a turbine-type centrifugal pump versus the flowrate Q (m³/min) are presented below:

Q (m³/min)	1.1	2.3	4.3	6.6	9.1
η (%)	26	57	78	90	42

Assuming the data is suitable for a quadratic regression model, apply least squares regression to determine the *best-fit parameters* and discuss the *goodness of the fit*.

SOLUTION:

This regression model, $\eta(Q) = a_0 + a_1 Q + a_2 Q^2$, has three model parameters, which are obtained by solving the system of linear equations.

TABLE 7.2

i	Q_i	η_i	Q_i^2	Q_i^3	Q_i^4	$Q_i\eta_i$	$Q_i^2\eta_i$
1	1.1	26	1.21	1.331	1.4641	28.6	31.46
2	2.3	57	5.29	12.167	27.984	131.1	301.53
3	4.3	78	18.49	79.507	341.88	335.4	1442.22
4	6.6	90	43.56	287.496	1897.5	594	3920.4
5	9.1	42	82.81	753.571	6857.5	382.2	3478.02
Σ	**23.4**	**293**	**151.36**	**1134.07**	**9126.3**	**1471.3**	**9173.63**

The linear system is simply constructed by replacing (x,y) with (Q, η) in the first three rows and three columns of Eq. (7.10) yielding

$$\begin{bmatrix} n & \Sigma Q_i & \Sigma Q_i^2 \\ \Sigma Q_i & \Sigma Q_i^2 & \Sigma Q_i^3 \\ \Sigma Q_i^2 & \Sigma Q_i^3 & \Sigma Q_i^4 \end{bmatrix} \begin{bmatrix} a_0 \\ a_1 \\ a_2 \end{bmatrix} = \begin{bmatrix} \Sigma \eta_i \\ \Sigma Q_i \eta_i \\ \Sigma Q_i^2 \eta_i \end{bmatrix}$$

The computational details of the terms required to construct the coefficient matrix are summarized in Table 7.2. Substituting the pertinent numerical sums in the last row of the table, the linear system becomes

$$\begin{bmatrix} 5 & 23.4 & 151.36 \\ 23.4 & 151.36 & 1134.072 \\ 151.36 & 1134.072 & 9126.298 \end{bmatrix} \begin{bmatrix} a_0 \\ a_1 \\ a_2 \end{bmatrix} = \begin{bmatrix} 293 \\ 1471.3 \\ 9173.63 \end{bmatrix}$$

which, for the model parameters, yields $a_0 = -10.4986$, $a_1 = 36.3711$, and $a_2 = -3.34032$. And computing E and S results in

$$E = \sum_{i=1}^{5} (\eta(Q_i) - \eta_i)^2 = 79.1432, \qquad S = \sum_{i=1}^{5} (\eta_i - 58.6)^2 = 2703.2$$

A graphical depiction of the observed data as well as the best fit (prediction) curve is presented in Fig. 7.10. In Figure 7.11, the residual and rmse confidence interval plots are presented. Using E and S, we obtain $r^2 = 0.97072$, which indicates that 97.07% of the observed data are explained by the best-fit curve. This result supports the conclusion that the quadratic equation exhibits very good agreement between the observed and predicted data, as is also evident from Fig. 7.10. On the other hand, the residual plot (Fig. 7.11a) has two points with large residuals ($\approx \pm 6$) while the *rmse* plot ($rmse = \sqrt{(79.1432/5)} \approx 4$) reveals that all these points lie within 68% CI, i.e., $Y(x_i) \pm rmse$.

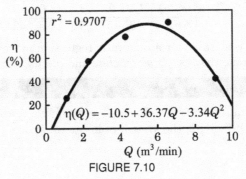

FIGURE 7.10

least squares Regression

- least squares regression is an effective tool for numerical analysts;
- Estimated model parameters are the optimal values;
- When a model compatible with the theory is used, a very good result can be obtained even with a relatively small amount of data;
- Regression allows important physical parameters (slope, intercept, etc.) to be determined;
- The theory of computing confidence intervals, predictions, goodness of fit, and so on is well established;
- Numerous software programs are available, allowing the use of a broad range of single- or multi-regression models.

- Regression models can justifiably be used within the range of data; hence, they have poor extrapolation properties;
- The procedure can be sensitive to outliers or some data anomalies;
- The data is assumed to be suitable for the adopted regression model;
- Overfitting can be a major issue when using polynomial models;
- Nonlinear models may require iterative techniques to solve the system of nonlinear equations.

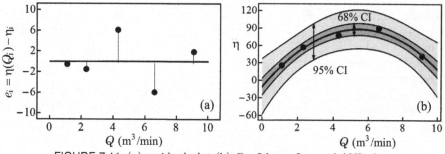

FIGURE 7.11: (a) residual plot (b) Confidence Interval (CI) plot.

Discussion: A regression model is trained to find model parameters that minimize the difference between predicted values and observed (actual) values in its data. The qualitative and quantitative tools should be exploited to the fullest to make sure the model can be safely used to make predictions on new, unseen data.

A pseudomodule, **POLYNOMIAL_FIT**, calculating the model parameters and r-squared of an mth-degree polynomial model by least squares regression is presented in Pseudocode 7.1. The module requires the number of data points (n), the degree of the polynomial model (m), and the set of data points (x_i, y_i, $i = 1, 2, \ldots, n$) as input. The module outputs are the model parameters **a**, the sum of the squares of the residuals (E), the sum of the squares of the mean deviation (S), and the r-squared. In the first block of loops ($i, j = 1, 2, \ldots, m + 1$), the elements of the coefficient matrix c_{ij} are constructed. In the second block of loops ($i = 1, 2, \ldots, m + 1$), the rhs elements, b_i, are obtained using Eq. (7.10). As a precaution against ill-conditioning, the resulting system of linear equations is solved using the Gauss elimination with pivoting algorithm, Pseudocode 2.10, to minimize the adverse effects of the roundoff errors.

Pseudocode 7.1

Module POLYNOMIAL_FIT $(n, m, \mathbf{x}, \mathbf{y}, \mathbf{a}, E, S, r2)$
\ DESCRIPTION: A pseudomodule fitting data to an mth degree polynomial.
\ USES:
\ GAUSS_ELIMINATION_P:: Gauss elimination module, Pseudocode 2.10
Declare: $x_n, y_n, c_{m+1,m+1}, b_{m+1}, a_{m+1}$ \ Declare array variables
For $\left[i = 1, m+1 \right]$ \ Loop i: Sweep rows from top to bottom
   For $\left[j = 1, m+1 \right]$ \ Loop j: Sweep a row from left to right
      $c_{ij} \leftarrow 0$ \ Initialize accumulator
      For $\left[k = 1, n \right]$ \ Loop k: Accumulator loop
         $p \leftarrow i + j - 2$ \ Find exponent p
         $c_{ij} \leftarrow c_{ij} + x_k^p$ \ Add x_k^p to accumulator
      End For
   End For
End For
For $\left[i = 1, m+1 \right]$ \ Loop i: Construct the *rhs*
   $b_i \leftarrow 0$ \ Initialize accumulator
   For $\left[k = 1, n \right]$ \ Loop k: Accumlator loop
      $p \leftarrow i - 1$ \ Find exponent p
      $b_i \leftarrow b_i + y_k * x_k^p$ \ Add $y_k * x_k^p$ to accumulator
   End For
End For
\ Solve $\mathbf{C}r = \mathbf{b}$ using Gauss elimination with pivoting
GAUSS_ELIMINATION_P$(m+1, \mathbf{C}, \mathbf{b}, \mathbf{a})$ \ System is solved by Pseudocode 2.10
$yavg \leftarrow 0$ \ Initialize accumulator, $yavg$
For $\left[k = 1, n \right]$
   $yavg \leftarrow yavg + y_k$ \ Accumulate y_k's
End For
$yavg \leftarrow yavg/n$ \ Find average y
$E \leftarrow 0; \ S \leftarrow 0$ \ Initialize E and S
For $\left[k = 1, n \right]$ \ Loop k: Loop to find E and S
   $yk \leftarrow a_1$ \ Initialize y_k as $y_k \leftarrow a_1$
   For $\left[j = 1, m \right]$ \ Loop j: Compute predictons, Y_k's
      $yk \leftarrow yk + a_{j+1} * x_k^j$ \ Add $a_{j+1} * x_k^j$'s to accumulator
   End For
   $S \leftarrow S + (y_k - yavg)^2$ \ Accumulate squares of deviations from $yavg$
   $E \leftarrow E + (yk - y_k)^2$ \ Accumulate squares of residues
End For
$r2 \leftarrow 1 - E/S$ \ Find r-squared
End Module POLYNOMIAL_FIT

7.1.3 OVERFITTING

When a polynomial is chosen as a regression model, it is tempting to increase the degree of the polynomial to improve the model and its prediction ability. As the degree polynomial (assuming the number of data points is sufficient) is gradually increased, the residuals decrease and r-squared improves until the observed and predicted data become identical, i.e., $r^2 \rightarrow 1$. One can increase the degree of a polynomial as long as there is a statistically significant reduction in the variance computed by $\sigma^2 = \Sigma e_i^2/(n - m - 1)$, where e_i is the

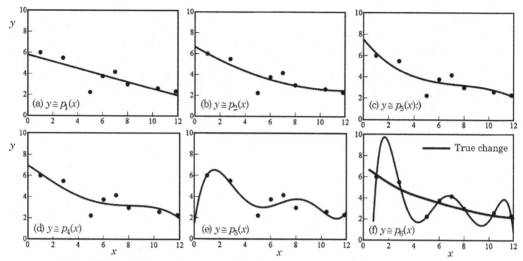

FIGURE 7.12: Effects of increasing the degree of polynomial model on the fit.

individual residual, n and m are the number of data points and the degree of the polynomial, respectively. Note that if the data is corrupted with significant noise, the best-fit polynomials tend to contain the noise as well.

The effect of increasing the degree of the polynomial on the least squares fit is illustrated in Fig. 7.12. For linear and quadratic models, as shown in Fig. 7.12a and b, the residuals are noticeably large. For $n = 3$, 4, and 5, the polynomials approach closer to the observed data points, except for a few, and the residuals decrease further, as seen in Fig. 7.12c, d, and e. However, the 6th-degree polynomial model, shown in Fig. 7.12f, perfectly satisfies the observed data points but depicts large oscillations leading to abnormal deviations. It should be recalled from Chapter 6 that oscillations cannot be mitigated simply by increasing the degree of the approximating polynomial. Doing so, on the contrary, increases and spreads the magnitude of the oscillations over the whole data range. Thus, one should not be fooled by the improvements attained with r^2 (i.e., $r^2 \to 1$). To avoid falling into this pitfall, the visual inspection tools should also be utilized to the fullest.

 The linear systems resulting from implementing polynomial regression models can be ill-conditioned, and the round-off errors also tend to distort the solution. Polynomials up to the third or fourth degree can be managed relatively without a problem; however, polynomial models of degrees higher than four are rarely used.

7.2 TRANSFORMATION OF VARIABLES

In data analysis, a *transformation* is frequently applied to a set of data to obtain a linear or polynomial model. Transforming variables in regression is a technique applied to improve the linear model (and the data) and to meet the underlying assumptions of regression analysis. A successful transformation model can better satisfy the assumptions of normality, linearity, and homoscedasticity.

TABLE 7.3: Typical transformation examples.

Model	Transformation
$\sqrt{y} = A + B\,x$	$Y = \sqrt{y}, \ X = x,$
$y = A + B\sqrt{x}$	$Y = y, \ X = \sqrt{x},$
$\log y = A + B\,x$	$Y = \log y, \ X = x$
$\log y = A + B \log x$	$Y = \log y, \ X = \log x$
$\log y = A + B/x$	$Y = \log y, \ X = 1/x,$

A transformation is applied to either dependent or independent variables, or both. This process involves changing the variable(s) of the data with a suitable mathematical function of an independent or dependent variable so that the final relationship between the transformed variables is linear. For instance, we may change the measurement scales of a set of data by replacing variable x with e^x or $\ln x$, or similarly replacing variable y with $\log y$ or $1/y$, and so on.

A transformation changes the shape of a distribution or relationship. In this process, the model parameters are exactly conserved; hence, the regression between the dependent and independent variables remains the same. For example, consider a model given by $\ln y = A + B\sqrt{x}$. Redefining a set of new variables as $Y = \ln y$ and $X = \sqrt{x}$, we transform the model to a linear relationship: $Y = A + BX$. Note that the model parameters (A and B) have been preserved in the new model. Some typical transformation examples are presented in Table 7.3.

In univariate distributions, transformations are basically applied for the following reasons: (i) obtain linear relationships, (ii) reduce skewness, and (iii) achieve convenience. We have already covered ways and means of achieving linear regression models. In general, we also seek to reduce the *skewness* of the distribution because it is easier to treat and interpret a non-skewed distribution. For this purpose, skewness reduction can be achieved by taking the square, cubic, etc., roots, or powers, or logarithms, or reciprocals, and so on. A transformation of data in percentages to ratios, or in radians to degree scales, or in non-dimensional quantities (T/T_r, U/U_r where quantities with the r subscript denote a reference value), and so on, may be more convenient. This type of transformation does not affect the shape of a distribution.

Confidence intervals computed on transformed variables should be recomputed by transforming back to the original units of interest. Models (goodness of fit, etc.) can only be compared to the original unit of the dependent variable.

7.3 LINEARIZATION OF NON-LINEAR MODELS

Nonlinear regression is a very powerful alternative to linear regression in that it provides greater flexibility in fitting a curve (i.e., a nonlinear function). Besides, most physical events in science and engineering are nonlinear in nature, and the observed nonlinear data should satisfy pertinent theoretical relationships. Nonetheless, applying the least squares regression procedure to a nonlinear model leads to a nonlinear system of equations (the *nonlinear least squares problem*) that can be solved by iterative methods (*see* Chapter 3 and Chapter 4).

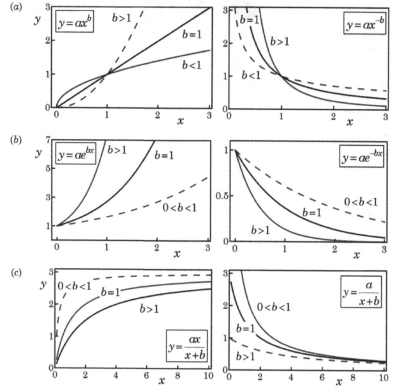

FIGURE 7.13: Influence of b on (a) power, (b) exponential, (c) saturation models.

However, the nature of equations and the number of variables can make the numerical solution very difficult and/or dependent on the initial guess.

A scatter plot of a set of data may reveal a nonlinear distribution, in which case transforming the variable(s) may become necessary to facilitate an easy solution while avoiding the solution of a nonlinear system of equations. Thankfully, some of the most common nonlinear models can be easily linearized or transformed; in other words, it is possible to "*transform*" the original data to make it preferably "*linear*," which then allows the linear regression model to be employed. These models will be referred to as *transformable* or *linearizable models*. The linearizable models are classified into three categories.

Power Model: Some physical quantities encountered in a wide range of science and engineering fields are related by $y = a\,x^b$. The magnitude of b determines the rate of change. The model results in growth of the dependent variable y ($b > 0$) or decay ($b < 0$) (*see* Fig. 7.13a). This model is also suitable for cases where the underlying model is unknown.

Exponential Growth/Exponential Decay Model: A significant number of physical laws are also represented by the exponential form $y = a\,e^{bx}$ ($b \neq 0$). The exponential model, like the power model, yields exponential growth ($b > 0$) or exponential decay ($b < 0$) in the independent variable (*see* Fig. 7.13b).

Saturated Growth/Saturated Decay Rate Model: This model can be expressed as either $y = ax/(x+b)$ or $y = a/(x+b)$. Both models generally observed in chemical reaction kinetics depict growth or decay rate under limiting conditions in which the dependent variable levels off (or saturates) as $x \to \infty$ ($ax/(x + b) \to a$, $a/(x+b) \to 0$) (*see* Fig. 7.13c).

Pseudocode 7.2

Module LINEARIZE_REGRESS $(n, \mathbf{x}, \mathbf{y}, model, a, b, E, S, r2)$
\ DESCRIPTION: A pseudomodule employing least squares regression to obtain
\ the nonlinear best fit parameters for the model $y = ax^b$.
Declare: $x_n, y_n, c_{22}, b_2, X_n, Y_n$
$\mathbf{X} \leftarrow \ln(\mathbf{x}); \mathbf{Y} \leftarrow \ln(\mathbf{y})$ \hfill \ Linearize data set for $model = 1$
For $[i = 1, 2]$ \hfill \ Loop i: Construct 2×2 linear system
   For $[j = 1, 2]$ \hfill \ Loop j : Sweep row from left to right
      $c_{i,j} \leftarrow 0$ \hfill \ Initialize accumulator c_{ij}
      For $[k = 1, n]$ \hfill \ Accumulato loop k
         $c_{ij} \leftarrow c_{ij} + X_k^{i+j-2}$ \hfill \ Accumulate X_k^p's
      End For
   End For
   $b_i \leftarrow 0$ \hfill \ Initialize accumulator for rhs
   For $[k = 1, n]$ \hfill \ Loop k: Construct the rhs
      $b_i \leftarrow b_i + X_k^{i-1} * Y_k$ \hfill \ Accumulate $X^p * Y$'s
   End For
End For
$dd \leftarrow c_{11} * c_{22} - c_{21} * c_{12}$ \hfill \ Use Cramer's rule to solve 2×2 system
$d1 \leftarrow b_1 * c_{22} - b_2 * c_{12}$
$d2 \leftarrow b_2 * c_{11} - b_1 * c_{21}$
$A \leftarrow d1/dd; \ m \leftarrow d2/dd$ \hfill \ Find A and m, linearized case.
$a \leftarrow e^A; \ b \leftarrow m$ \hfill \ Find a and b, nonlinear case.
$yavg \leftarrow 0$ \hfill \ Initialize average Y
For $[k = 1, n]$ \hfill \ Loop k: Calculate average Y
   $yavg \leftarrow yavg + y_k$ \hfill \ Accumulate y_k's
End For
$yavg \leftarrow yavg/n$ \hfill \ Find $yavg$
$E \leftarrow 0; \ S \leftarrow 0$ \hfill \ Initialize E and S
For $[k = 1, n]$ \hfill \ Loop k: Accumulator for E and S
   $S \leftarrow S + (y_k - yavg)^2$ \hfill \ Accumulate mean deviations
   $E \leftarrow E + (a * x_k^b - y_k)^2$ \hfill \ Accumulate residuals
End For
$r2 \leftarrow 1 - E/S$ \hfill \ Calculate r-squared
End Module LINEARIZE_REGRESS

Linearization of the "*power*" and "*exponential*" models is carried out by taking the natural logarithm of the model, resulting in linear relationships in terms of $\ln y$ versus $\ln x$, and $\ln y$ versus x, respectively. Defining new variables as $X = x$ or $X = \ln x$ and $Y = \ln y$ (and model parameters as $A = \ln a$ and $m = b$), the nonlinear model is transformed to $Y = A + mX$.

On the other hand, when saturated growth or saturated decay rate models are linearized by inverting the original data and defining $X = x$ or $1/x$ and $Y = 1/y$, these models are reduced to the same linear relationship. The model parameters are then found once the linear least squares procedure is employed on the transformed data set, (X_i, Y_i). Henceforth, the original model parameters can be computed from the transformed model parameters. The linearization procedure of common nonlinear models is presented in Table 7.4.

TABLE 7.4: Linearization of common nonlinear models.

Model (Curve)	Linearization, $Y = A + mX$
$y = a x^b$	$\ln y = \ln a + b \ln x$, $Y = \ln y$, $X = \ln x$, $A = \ln a$, $m = b$
$y = a e^{bx}$	$\ln y = \ln a + b x$, $Y = \ln y$, $X = x$, $A = \ln a$, $m = b$
$y = \dfrac{ax}{b+x}$	$\dfrac{1}{y} = \dfrac{b+x}{ax}$, $Y = \dfrac{1}{y}$, $X = \dfrac{1}{x}$, $A = \dfrac{1}{a}$, $m = \dfrac{b}{a}$
$y = \dfrac{a}{b+x}$	$\dfrac{1}{y} = \dfrac{b+x}{a}$, $Y = \dfrac{1}{y}$, $X = x$, $A = \dfrac{b}{a}$, $m = \dfrac{1}{a}$

In most cases, the original nonlinear model parameters (a and b) have physical significance. The objective of linear regression is sometimes to determine these physical parameters. For instance, $N(t) = N_0 e^{-\lambda t}$ represents the decay of a radioactive isotope, where $\lambda = \ln(2)/t_{1/2}$ is the decay constant and $t_{1/2}$ is the half-life of the isotope. The decay constant appears in the transformed linear model parameter as $m = -\lambda$. To determine the half-life of any radioactive isotope, the radioactivity measurements of the isotope (decay counts per unit time) are carried out, and time versus decay rate data is obtained (t_i, N_i). When the data is fitted to the linearized (exponential growth) model, the half-life of the isotope is obtained from $t_{1/2} = \ln(2)/\lambda = -\ln(2)/m$.

A pseudomodule, **LINEARIZE_REGRESS**, performing the nonlinear fit to the model $y = ax^b$ is given in Pseudocode 7.2. As input, the module requires the number of data points (n), the data (x_i, y_i, $i = 1, 2, \ldots, n$), and the model flag ($model$) for selecting the nonlinear model. The module outputs are the model parameters ($a0$ and $b0$), the sum of the squares of residuals (E), the sum of the squares of mean deviation (S), and r-squared ($r2$). The model flag is used for the model (indicating the type of linearization procedure) to be applied, as given in Table 7.4. To save space, however, the implementation of only $y = ax^b$ ($model = 1$ case) is shown here, but the module can be generalized to include the other linearizable models without much difficulty. First, the data is linearized as $X_k = \ln x_k$ and $Y_k = \ln y_k$ for $k = 1, 2, \ldots, n$. The model parameters (A and m for $model = 1$) are obtained by solving a 2×2 system of linear equations (c_{ij}, $i, j = 1, 2$ and b_i, $i = 1, 2$). Since they involve the sums of X_k^p or $Y_k X_k^p$, the elements of the coefficient matrix and the rhs are set as accumulating variables, and the resulting system is solved by Cramer's rule. The model parameters are determined: $a = e^A$ and $b = m$, and using the original data, the sums of the squares of residuals and the sum of the squares of mean deviation (S) are computed by Eqs. (7.4) and (7.11), respectively.

EXAMPLE 7.3: Linearization of power model

The following strain-stress data was obtained from the tensile testing of an alloy steel. The theoretical relationship between strain and stress is dictated by the Hollomon equation, stated as $\sigma = K \cdot \varepsilon^n$, where σ is the stress, ε is the strain, K is the strength coefficient, and n is a material-dependent constant. Apply the method of least squares to determine the model parameters (K and n) of the alloy steel.

ε	0.002	0.003	0.004	0.006	0.01	0.015	0.02	0.024
σ (MPa)	535	550	565	585	610	630	645	650

SOLUTION:

TABLE 7.5

i	X_i	Y_i	X_i^2	X_iY_i	$(Y_i - \bar{Y}_i)^2$	e_i^2
1	−6.21461	6.28227	38.6213	−39.0418	0.01124	1.45×10^{-7}
2	−5.80914	6.30992	33.7461	−36.6552	0.00614	1.53×10^{-5}
3	−5.52146	6.33683	30.4865	−34.9885	0.00265	6.76×10^{-6}
4	−5.11599	6.37161	26.1734	−32.5971	0.00028	3.05×10^{-7}
5	−4.60517	6.41346	21.2076	−29.5351	0.00063	5.31×10^{-6}
6	−4.19970	6.44572	17.6375	−27.0701	0.00330	3.20×10^{-6}
7	−3.91202	6.46925	15.3039	−25.3078	0.00655	4.27×10^{-6}
8	−3.72970	6.47697	13.9107	−24.1572	0.00786	2.45×10^{-5}
Σ	**−39.10781**	**51.106**	**197.0872**	**−249.353**	**0.03866**	**5.72×10^{-5}**

The Hollomon equation is a nonlinear equation suitable for the power model. Taking the natural logarithm of both sides, we find

$$\ln \sigma = \ln K + n \ln \varepsilon$$

Upon setting $Y = \ln \sigma$, $X = \ln \varepsilon$, and $A = \ln K$, it is transformed to $Y = A + nX$. Next, the strain-stress data are transformed according to the transformation variables: $(\varepsilon_i, \sigma_i) \to (X_i, Y_i)$. Using Eq. (7.10) and the transformed data, the linear least squares leads to 2×2 system of linear equations given as

$$\begin{bmatrix} n & \sum X_i \\ \sum X_i & \sum X_i^2 \end{bmatrix} \begin{bmatrix} A \\ n \end{bmatrix} = \begin{bmatrix} \sum Y_i \\ \sum X_iY_i \end{bmatrix}$$

The numerical sums for ΣX_i, ΣX_i^2, ΣY_i, and $\Sigma(Y_iX_i)$ are required to construct and solve the linear system. The details for the construction of the required sums are tabulated in Table 7.5. Substituting these numerical values into the matrix equation yields

$$\begin{bmatrix} 8 & -39.1078075 \\ -39.1078075 & 197.0871520 \end{bmatrix} \begin{bmatrix} A \\ n \end{bmatrix} = \begin{bmatrix} 51.106024 \\ -249.35294 \end{bmatrix}$$

The determinant of the coefficient matrix is 47.276, which implies that *ill-conditioning* is not going to be a problem (*see* Section 2.7). The solution of the linear system is found to be $A = 6.78335387$ and $n = 0.080822913$. Then, we evaluate $K = e^A \cong 883$ and take $n \simeq 0.081$, yielding the Hollomon equation:

$$\sigma = 883 \, \varepsilon^{0.081}, \qquad 0.002 \leqslant \varepsilon \leqslant 0.024$$

To assess the goodness of the fit, we calculate the mean value $\bar{\sigma} = 596.25$, the sum of the squares of the mean deviations, $S = \Sigma(\sigma_k - \bar{\sigma})^2 = 13587.5$, and the sum of the squares of the residuals $E = \Sigma(\sigma(\varepsilon_k) - \sigma_k)^2 = 23.8716$. The r-squared is then determined as $r^2 = 1 - E/S = 0.99825$, which implies that 99.824% of the measured or observed data are explained by the best-fit model.

Discussion: The least square regression of the data to the Hollomon equation is simple and straight forward. The conformity of the measured experimental and predicted data from the best-fit curve is depicted in Fig. 7.14a. The visual inspection of the figure reveals an "excellent agreement" between the data and the best-fit curve.

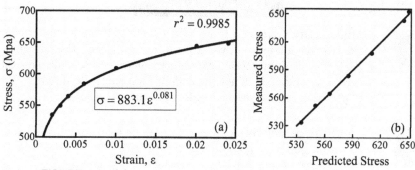

FIGURE 7.14: (a) Best-fit curve and original data, (b) PO curve.

The *predicted* and *measured* (observed) data are depicted in the PO plot shown in Fig. 7.14b. Applying the linear least squares fit to the PO data leads to

$$\sigma_{obs.} = 2.11231 + 0.99722\sigma_{pred.}, \qquad r^2 = 0.9984$$

Note that the slope of the line is 0.99722 (or 44.92° is the angle of slope, i.e., almost 45°), and the data has a bias of 2.11, which is small in comparison to the stress scale ranging from 500 to 700 MPa. Hence, we conclude that the curve fit can be confidently used within the $0.002 \leqslant \varepsilon \leqslant 0.024$ interval.

7.4 MULTIVARIATE REGRESSION

In most practical problems, the independent variables that determine or affect the dependent variable (outcome) are often not limited to a single variable. A multivariate regression model must be adopted in a process where the outcome is affected by more than one independent variable. The least squares regression can very well be applied to linear and nonlinear regression models, referred to as *multivariate regression* or *multivariate least squares*.

Linear model: It is the most common multivariate (with m independent variables) linear regression model and results in

$$Y = a_0 + a_1 X_1 + a_2 X_2 + \cdots + a_m X_m \tag{7.18}$$

The sum of the squares of the residuals, Eq. (7.2), for this model yields

$$E(a_0, a_1, \ldots, a_m) = \sum_{i=1}^{n} (a_0 + a_1 X_{1i} + a_2 X_{2i} + \cdots + a_m X_{mi} - Y_i)^2 \tag{7.19}$$

where $a_0, a_1, \ldots, a_m$ are the model parameters.

In order to minimize it, the partial derivatives of Eq. (7.19) with respect to model parameters are set to zero, i.e., $\partial E/\partial a_0 = \partial E/\partial a_1 = \ldots = \partial E/\partial a_m = 0$. After algebraic manipulations and simplification, a system of m equations with m unknowns is obtained as

$$\begin{bmatrix} n & \sum X_{1i} & \sum X_{2i} & \cdots & \sum X_{mi} \\ \sum X_{1i} & \sum X_{1i}^2 & \sum X_{2i} X_{1i} & \cdots & \sum X_{1i} X_{mi} \\ \sum X_{2i} & \sum X_{1i} X_{2i} & \sum X_{2i}^2 & \cdots & \sum X_{2i} X_{mi} \\ \vdots & \vdots & \vdots & \ddots & \vdots \\ \sum X_{mi} & \sum X_{1i} X_{mi} & \sum X_{2i} X_{mi} & \cdots & \sum X_{mi}^2 \end{bmatrix} \begin{bmatrix} a_0 \\ a_1 \\ a_2 \\ \vdots \\ a_m \end{bmatrix} = \begin{bmatrix} \sum Y_i \\ \sum X_{1i} Y_i \\ \sum X_{2i} Y_i \\ \vdots \\ \sum X_{mi} Y_i \end{bmatrix} \tag{7.20}$$

Multivariate Regression

- Multivariate regression provides a tool to determine relationships or the relative influence of the predictor variables on the outcome;
- It allows better prediction capability due to having multiple variables (predictors), i.e., it is not dependent on a single predictor;
- It provides the capability to identify outliers or anomalies.

- Multivariate (non-linear) regression can be a bit complicated and may require high-level mathematics and programming techniques;
- The output of multivariate regression models may be a bit more difficult to interpret;
- It does not have much scope for small data sets.

---

where the coefficient matrix and the rhs can be readily found using the observed data. However, it should be kept in mind that such linear systems tend to be ill-conditioned.

The non-linear model: Another frequently encountered nonlinear model involving multiple variables has the following power form:

$$y(x_1, x_2, \ldots, x_m) = A\, x_1^{a_1} x_2^{a_2} \cdots x_m^{a_m} \tag{7.21}$$

This model can be linearized similarly to the univariate power model. Taking the natural logarithm of Eq. (7.21) yields

$$\ln y = \ln A + a_1 \ln x_1 + a_2 \ln x_2 + \cdots + a_m \ln x_m \tag{7.22}$$

which gives Eq. (7.18) form after setting $a_0 = \ln A$, $(X_{ik}, Y_i) = (\ln x_{ik}, \ln y_i)$ for $k = 1, 2, \ldots, n$ and $i = 1, 2, \ldots, m$ in Eq. (7.22).

EXAMPLE 7.4: Applying Bivariate regression

The specific heat of a chemical compound has been experimentally determined as a function of temperature. The experimental results are presented in the table below:

T (°C)	5	25	100	185	260	320
c_p (kJ/kg°C)	5	22	25	30	34	41

An algebraic model for the specific heat is suggested as $c_p(T) = A + BT + C/T^2$, where T is in °C units. Apply a bivariate linear regression model to determine model parameters and discuss the goodness of the fit.

SOLUTION:

Instead of a single variable T, two independent variables may be introduced by defining $x_1 = T$ and $x_2 = 1/T^2$. The bivariate linear model now has the form:

$$c_p(x_1, x_2) = A + Bx_1 + Cx_2,$$

Employing the least squares procedure leads to a 3×3 system of linear equations (first three rows and three columns of Eq. (7.20)):

TABLE 7.6

i	x_{1i}	x_{2i}	c_{pi}	x_{1i}^2	$x_{1i}x_{2i}$	x_{2i}^2	$x_{1i}c_{pi}$	$x_{2i}c_{pi}$
1	5	0.04	5	25	0.2	1.60×10^{-3}	25	0.200
2	25	0.0016	22	625	0.04	2.56×10^{-6}	550	0.0352
3	100	1×10^{-4}	25	10000	0.01	1.00×10^{-8}	2500	0.0025
4	185	2.9×10^{-5}	30	34225	0.00541	9×10^{-10}	5550	9×10^{-4}
5	260	1.5×10^{-5}	34	67600	0.00385	2×10^{-10}	8840	5.0×10^{-4}
6	320	9.8×10^{-6}	41	102400	0.00313	1×10^{-10}	13120	4.0×10^{-4}
Σ	**895**	**0.041754**	**157**	**214875**	**0.262377**	**0.0016026**	**30585**	**0.23948**

$$
\begin{bmatrix}
n & \sum x_{1i} & \sum x_{2i} \\
\sum x_{1i} & \sum x_{1i}^2 & \sum x_{1i}x_{2i} \\
\sum x_{2i} & \sum x_{1i}x_{2i} & \sum x_{2i}^2
\end{bmatrix}
\begin{bmatrix} A \\ B \\ C \end{bmatrix}
=
\begin{bmatrix}
\sum c_{pi} \\
\sum x_{1i}c_{pi} \\
\sum x_{2i}c_{pi}
\end{bmatrix}
$$

Using the observed data, Table 7.6 is prepared to determine the coefficient matrix and the rhs. Substituting the pertinent numerical values into the matrix equation results in

$$
\begin{bmatrix}
6 & 895 & 0.041754 \\
895 & 214875 & 0.262377 \\
0.041754 & 0.262377 & 0.0016026
\end{bmatrix}
\begin{bmatrix} A \\ B \\ C \end{bmatrix}
=
\begin{bmatrix}
157 \\
30585 \\
0.23948
\end{bmatrix}
$$

The solution of the linear system yields the following best-fit model:

$$
c_p(T) = 19.7356 + 0.0606\,T - \frac{374.68}{T^2}
$$

with $E = 9$, $S = 762.834$, and $r^2 = 0.9882$, obtained from the original regression model.

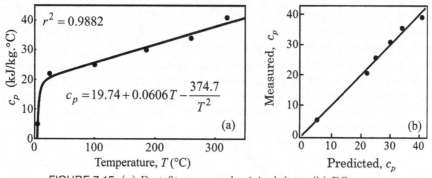

FIGURE 7.15: (a) Best-fit curve and original data, (b) PO curve.

Discussion: The presence of very large and very small values in the coefficient matrix should lead us to question the existence of ill-conditioning. In this context, the Frobenius-based condition number is found to be 2.458×10^8 implying the requirement of high precision arithmetic to avoid the adverse effects of ill-conditioning.

The measured data and the prediction model (best-fit curve) are comparatively depicted in Fig. 7.15a. The model and the measured data are observed to be consistent, with no significant residuals. Besides a visual inspection, a high coefficient of determination ($r^2 = 0.9882$) validates the very good agreement between the data and the model. In Fig. 7.15b, the predicted-measured c_p data are clearly tightly aligned along the diagonal. Fitting the predicted-measured data to a straight line gives a slope of 1 and negligible bias. Hence, we conclude that the model can be confidently used in $5 \leqslant T \leqslant 320 \ °C$.

7.5 CONTINUOUS LEAST SQUARES APPROXIMATION

In science and engineering, one often has to deal with complicated functions. Sometimes a simpler approximation is used to replace a function to simplify calculations or to more clearly understand or explain the relationships between dependent and independent variables. Of course, in the meantime, the approximation should yield results of reasonable and acceptable accuracy.

The "best" approximations for continuous functions should minimize the maximum-minimum error on the interval they are prescribed. But this is oftentimes very difficult to achieve, so we must settle for a "*better*" approximation. A better approximation of a continuous function oscillates about the function (under- and over-estimates) in the interval of interest.

In Fig. 7.16, linear and quadratic approximations replacing a continuous function are depicted. The areas between the true and approximate functions corresponding to the parts above and below the true function should be approximately equal. Ideally, it is desirable to have the deviations on both sides be in the same order of magnitude as well. Recall that we used the Taylor polynomials to approximate continuous functions. However, a Taylor polynomial is good only in the vicinity of $x = a$, and the degree of the polynomial needs to be increased to improve its accuracy. For instance, consider the linear and quadratic Taylor polynomial approximations of $y = e^x$. Note that in Fig. 7.17, both polynomial approximations are in very good agreement in the neighborhood of $x = 0$. Even though high-degree polynomials yield better approximations, eventually all approximations will deteriorate with increasing x. Hence, the Taylor polynomials are *not* suitable if we are looking for a simple approximation for relatively large intervals.

The least squares method, serving the same purpose, can be alternatively applied to continuous functions, $y = f(x)$, say on $[a, b]$, to obtain a simpler approximating function $Y(x)$ with a small *average error*. The average error of an approximation is defined by the

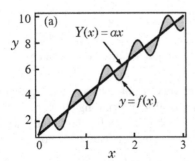
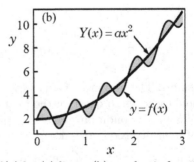

FIGURE 7.16: Approximation functions of $f(x)$ by (a) linear (b) quadratic functions.

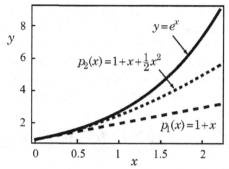

FIGURE 7.17: Linear and quadratic Taylor approximations for $y = e^x$.

root-mean-square error (rmse) expressed as

$$Rmse(Y; f) = \left[\frac{1}{b-a} \int_a^b (Y(x) - f(x))^2 dx \right]^{1/2} \qquad (7.23)$$

where $Rmse(Y; f)$ denotes the rmse error of approximating $f(x)$ by $Y(x)$. Note that minimizing Eq. (7.23) is equivalent to minimizing the sum of the squares of the continuous residual, $\Delta y = Y(x) - f(x)$; in other words, our objective function will be

$$E = \int_a^b \left(Y(x) - f(x) \right)^2 dx \qquad (7.24)$$

Equation (7.24) is referred to as the *least squares approximation problem*.

Now, consider an mth-degree approximating polynomial, $Y(x)$. The objective function, Eq. (7.24), is written as

$$E(a_0, a_1, \ldots, a_m) = \int_a^b \left(a_0 + a_1 x + a_2 x^2 + \ldots + a_m x^m - f(x) \right)^2 dx \qquad (7.25)$$

To obtain $m+1$ equations, the partial derivatives of the objective function are evaluated with respect to a_k for $k = 0, 1, 2, \ldots, m$ and set to zero to determine the solution set that makes Eq. (7.25) minimum.

$$\frac{\partial E}{\partial a_0} = 2 \int_a^b \left(a_0 + a_1 x + a_2 x^2 + \ldots + a_m x^m - f(x) \right) (1) dx = 0$$

$$\frac{\partial E}{\partial a_1} = 2 \int_a^b \left(a_0 + a_1 x + a_2 x^2 + \ldots + a_m x^m - f(x) \right) (x) dx = 0 \qquad (7.26)$$

$$\vdots \qquad \qquad \vdots$$

$$\frac{\partial E}{\partial a_m} = 2 \int_a^b \left(a_0 + a_1 x + a_2 x^2 + \ldots + a_m x^m - f(x) \right) (x^m) dx = 0$$

After simplifications, Eq. (7.26) becomes

$$a_0 \int_a^b dx + a_1 \int_a^b x dx + a_2 \int_a^b x^2 dx + \ldots + a_m \int_a^b x^m dx = \int_a^b f(x) dx$$

$$a_0 \int_a^b x dx + a_1 \int_a^b x^2 dx + a_2 \int_a^b x^3 dx + \ldots + a_m \int_a^b x^{m+1} dx = \int_a^b x f(x) dx \qquad (7.27)$$

$$\vdots$$

$$a_0 \int_a^b x^m dx + a_1 \int_a^b x^{m+1} dx + a_2 \int_a^b x^{m+2} dx + \ldots + a_m \int_a^b x^{2m} dx = \int_a^b x^m f(x) dx$$

The definite integrals in Eq. (7.27) can be evaluated analytically (or numerically if necessary), and then the resulting system of equations is solved for the coefficients.

 The resulting system of linear equations, Eq. (7.27), is ill-conditioned for $m > 5$. Therefore, approximating functions with high-degree polynomials is not recommended.

EXAMPLE 7.5: Regression under the integral sign

Find a continuous linear approximation to $y = 3x^2$ on $[0,2]$ using the least squares approximation.

SOLUTION:

We are seeking a linear approximation, $Y(x) = a + bx$, to $y = 3x^2$ whose residual can be written as $\Delta y = a + bx - 3x^2$. The objective function to be minimized becomes

$$E(a,b) = \int_0^2 \left(a + bx - 3x^2\right)^2 dx$$

where a and b are the unknown parameters.

Taking the partial derivatives of E with respect to a and b and using the chain rule, we obtain

$$\frac{\partial E}{\partial a} = 2\int_0^2 \left(a + bx - 3x^2\right)(1)dx = 0 \quad \Rightarrow \quad \int_0^2 (a + bx)dx = \int_0^2 3x^2 dx$$

$$\frac{\partial E}{\partial b} = 2\int_0^2 \left(a + bx - 3x^2\right)(x)dx = 0 \quad \Rightarrow \quad \int_0^2 (ax + bx^2)dx = \int_0^2 3x^3 dx$$

Definite integrals in the above expressions yield

$$\int_0^2 dx = 2, \qquad \int_0^2 x\,dx = 2, \qquad \int_0^2 x^2 dx = \frac{8}{3}, \qquad \int_0^2 x^3 dx = 4$$

Upon substituting the numerical values into the preceding equations, we obtain a 2×2 system of linear equations: $a + b = 4$, $a + 4b/3 = 6$. Having solved this system of equations, we find the linear approximation as $Y(x) = 6x - 2$.

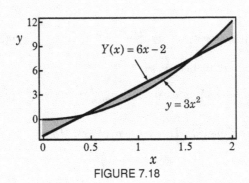

FIGURE 7.18

Discussion: The true function and its linear approximation are comparatively plotted in Fig. 7.18. We can see that the residual, Δ, is not uniform throughout [0,2]. The linear approximation at $x = 0$ and 2 underestimates e^x by $\Delta y = 2$, which is also the point where the absolute (maximum) errors occur. Notice that the integrals of the true and approximate functions on [0,2] are 8 (i.e., the same), while the error (i.e., the objective function) corresponds to $E = 1.6$.

7.6 OVER- OR UNDER-DETERMINED SYSTEMS OF EQUATIONS

In analyzing experimentally collected data, the number of equations is sometimes deliberately set higher than the number of unknowns in order to minimize the experimental errors. A linear system with more equations than unknowns is called an *overdetermined system*. In reality, such systems do not have exact solutions, but their solutions are obtained such that the *approximate solution error* is minimal.

Another type of system of linear equations encountered in the fields of electronics, statistics, geology, mechanics, and so on is called an *underdetermined system*. In these systems, the number of unknowns is greater than the number of equations. The solution of an underdetermined system can be obtained by employing the method of least squares if it has a common solution.

Consider a system of linear equations with m unknowns and n equations ($n > m$ or $n < m$):

$$\sum_{j=1}^{m} a_{ij}x_j = b_i, \quad i = 1, 2, \ldots, n \tag{7.28}$$

where a_{ij}'s and b_i's are known.

The residual for the ith equation can be expressed as

$$r_i = a_{i1}x_1 + a_{i2}x_2 + a_{i3}x_3 + \cdots + a_{im}x_m - b_i, \quad i = 1, 2, \ldots, n \tag{7.29}$$

Since the number of equations and the number of unknowns are *not* the same, an exact solution will not be achieved; in other words, the goal of 'zero residuals' will not be realized. Hence, we force the sum of the squares of the residuals (the objective function) to be a minimum as follows:

$$E(x_1, x_2, \ldots, x_m) = \sum_{i=1}^{n} (a_{i1}x_1 + a_{i2}x_2 + a_{i3}x_3 + \cdots + a_{im}x_m - b_i)^2 \tag{7.30}$$

Next, to find the minimum of the objective function, the partial derivatives of the sum of the squares of the residuals, Eq. (7.30), with respect to model parameters will be set to zero as follows:

$$\frac{\partial E}{\partial x_j} = 2 \sum_{i=1}^{n} (a_{i1}x_1 + a_{i2}x_2 + a_{i3}x_3 + \cdots + a_{im}x_m - b_i)a_{ij} = 0, \quad j = 1, 2, \ldots, m \tag{7.31}$$

This procedure yields a system of m linear equations with m unknowns, defined in this case as x_i's for $i = 1, 2, \ldots, m$.

EXAMPLE 7.6: Solving overdetermined system by least-square regression

The management of a spark-plug manufacturer believes that the sales figures (R_k piece) can be correlated by the state's population (A_{k1}) and gross domestic product per capita (GDP) (A_{k2}). The pertinent data for five states is summarized in the accompanying table. The manufacturer, based on past experience, believes that the sales income is related to the population and GDP per capita as

$$R_k = x_1 + A_{k1}x_2 + A_{k2}x_3$$

where x_1, x_2, and x_3 are correlation parameters. Use the method of least squares to determine the model parameters.

k	State	Sales, R_k (million \$)	Population, A_{k1} (million)	GDP per capita, A_{k2} (1000 \$)
1	California	33.3	39.5	67.9
2	Texas	18.2	28.3	58.4
3	Utah	2.1	3.1	49.7
4	Arizona	5.4	7.1	43.1
5	Idaho	1.3	1.7	39.8

SOLUTION:

The available data can be used to construct the following equations:

$$33.3 = x_1 + 39.5\,x_2 + 67.9\,x_3, \quad 5.4 = x_1 + 7.1\,x_2 + 43.1\,x_3$$
$$18.2 = x_1 + 28.3\,x_2 + 58.4\,x_3, \quad 1.3 = x_1 + 1.7\,x_2 + 39.8\,x_3$$
$$2.1 = x_1 + 3.1\,x_2 + 49.7\,x_3$$

These five equations with three unknowns constitute an overdetermined system, which can be rearranged in the form of residuals as follows:

$$r_1 = x_1 + 39.5\,x_2 + 67.9\,x_3 - 33.3 = 0,$$
$$r_2 = x_1 + 28.3\,x_2 + 58.4\,x_3 - 18.2 = 0,$$
$$r_3 = x_1 + 3.1\,x_2 + 49.7\,x_3 - 2.1 = 0,$$
$$r_4 = x_1 + 7.1\,x_2 + 43.1\,x_3 - 5.4 = 0,$$
$$r_5 = x_1 + 1.7\,x_2 + 39.8\,x_3 - 1.3 = 0$$

The objective is to minimize the sums of the squares of the residuals, so the objective function is constructed as follows:

$$E(x_1, x_2, x_3) = \sum_{i=1}^{5} r_i^2$$

Now, taking the partial derivatives of E with respect to x_1, x_2, and x_3 and setting the resulting equations to zero results in

$$\frac{\partial E}{\partial x_1} = 10\,x_1 + 159.4\,x_2 + 517.8\,x_3 - 120.6 = 0$$

$$\frac{\partial E}{\partial x_2} = 159.4\,x_1 + 4848.1\,x_2 + 9725.02\,x_3 - 3754.94 = 0$$

$$\frac{\partial E}{\partial x_3} = 517.8\,x_1 + 9725.02\,x_2 + 27865.4\,x_3 - 7425.6 = 0$$

Discussion: Solving the preceding 3×3 linear system yields

$$x_1 = -5.04608, \quad x_2 = 0.726145, \quad x_3 = 0.106824$$

and

$$E = \sum_{i=1}^{5} r_i^2 = 19.73854, \qquad S = \sum_{i=1}^{5} (r_i - \bar{r})^2 = 748.174$$

Finally, the coefficient of determination is found to be $r^2 = 0.97362$.

Discussion: Least squares regression is an alternative technique for solving an overdetermined system of equations based on the principle of least squares minimization of observation residuals. It is used extensively in the disciplines of engineering to minimize experimental measurement errors.

7.7 CLOSURE

In this chapter, the least squares technique for developing regression curves for discrete functions is presented. The method is one of the most common methods used to approximate a set of observations, i.e., discrete data sets. With this method, the best curve fit to a given discrete data set is achieved by minimizing the sum of the squares of the residuals (i.e., deviations) of the data points from the curve. The method is successfully applied to data sets that are not very accurate or precise. Often, the data collected from the instruments or devices through experiments may have a significant amount of noise, and this method helps smooth out the noise present in the data sets. As a result, unlike interpolating polynomials, the best-fit curve may not pass through every individual data point.

The method requires the selection of a regression model, $Y(x)$, and the process, termed "regression analysis," quantifies the relationship between independent and dependent variables. The linear problems, $Y(x) = a_0 + a_1 x$, are often encountered in regression analysis in statistics and engineering applications, while nonlinear problems are also commonly encountered in science and engineering problems. On the other hand, approximation functions in polynomial form, $Y(x) = p_m(x)$, can be generated for most experimental observations. By increasing the degree of the polynomial, the sum of the squares of the residuals decreases and r-squared improves; nevertheless, increasing the degree of an approximating polynomial yields Runge's phenomenon.

An analyst has several tools to assess the goodness of a regression. The so-called "best-fit" curve should not be assumed to be the "best-fit curve" without confirmation. Plotting and visually comparing the data and the selected model is one of the most important and obvious tools to verify the suitability of a model. On the other hand, quantitative tools such as r-squared, adjusted r-squared, $rmse$, PO-plots, and so on provide the analysts with ways and means of assessment. The goodness or fit of a model should be assessed using both numerical and visual tools.

When selecting a best-fit model, the underlying theory of the physical event should be taken into account. The physical properties (such as density, viscosity, heat capacity, conductivity, etc.) of a substance that vary with temperature can be described by polynomials. The discrete data for these properties (tabulated in reference books) are free of experimental uncertainties, making them less likely to contain interference. Some of the most common nonlinear relationships observed in science and engineering, such as power

(ax^b), exponential (ae^{bx}), saturation growth, and so on, can be easily linearized by applying a suitable transformation. The model parameters of the nonlinear regressions generally have physical meaning. The regression analysis, which can be very sensitive to outliers, might be unreliable if the discrete data is small in number or is not uniformly distributed. When the dependent variable is a function of two or more independent variables, multivariate linear or non-linear regression models can be employed in the same way. Multivariate power models are also easily linearized, just as univariate power models are.

The least squares method can be applied to replace complicated functions over a given range with continuous but simpler mathematical expressions by minimizing the squares of the difference between the original function and the approximation under the integral sign. Also, in applications of linear systems of equations, sometimes a linear system of equations has fewer equations (underdetermined systems) or more equations (overdetermined systems) than the unknowns. In such a case, the system may have no solution or an infinite number of solutions. The least squares method can often be applied to establish a problem that has a unique solution in the least squares sense.

7.8 EXERCISES

Section 7.1 Polynomial Regression

E7.1 The thermal conductivity k (in W/cm.K) of a substance is measured as a function of temperature T (in K), and the results are presented below. Assuming that the conductivity varies linearly with temperature ($k = a_0 + a_1 T$), apply least squares to determine the best-fit parameters and calculate r^2.

T	226	285	335	362	400	470
k	0.88	1	1.1	1.15	1.22	1.35

E7.2 A series of experiments are conducted to determine the torque required to turn a spring mechanism with an applied angle. The proposed relationship between the torque and applied angle is $T(\theta) = a_0 + a_1 \theta$. Considering the experimental results furnished below, (a) determine the least squares best-fit parameters; (b) calculate the coefficient of determination (r^2); and (c) comparatively plot the data and the best fit and interpret the results.

θ (rad)	0	0.39	0.78	1.18	1.57	1.96	2.35
T (N·m)	0.243	0.29	0.35	0.34	0.44	0.49	0.51

E7.3 The table below contains the results of five experiments conducted to determine the specific heat of an alloy, c_p (kJ/kg·K), as a function of temperature, T (K). The specific heat is assumed to be linear in the temperature range tested. Use the following experimental data to determine the least squares best-fit parameters and calculate r^2.

T	275	350	500	750	1250
c_p	0.38	0.39	0.41	0.43	0.48

E7.4 The wet-bulb T_w (°C) and cold-water temperature T_c (°C) data tabulated below will be used to develop a cooling-tower performance curve. Assuming that the relationship between the temperatures is linear, apply the linear regression to find the model parameters and comment on the goodness of the fit.

T_w (°C)	15.5	17.8	20.1	22.8	25.1	27.3	29.5
T_c (°C)	24.5	25.4	27.3	28.8	30.7	31.7	33.8

E7.5 The relative humidity ϕ (%) versus temperature T (°C) relationship of a dairy powder is tabulated below. Assuming that the relative humidity varies linearly with the temperature, (a) apply the linear least squares to find the best-fit parameters; (b) plot the residuals; (c) compute r^2 and comment on the goodness of the fit.

ϕ	15	22	29	33	36	45	52
T (°C)	94	80	73	68	60	55	45

E7.6   The following log-log data is to be used to establish the *power curve* (i.e., the relationship of the Reynolds number, Re, versus the power number, Po) of a rotationally reciprocating impeller under assumed operating conditions. It is known that the power curve on the log-log scale behaves linearly for $Re < 10^3$. In other words, the power curve can be expressed as $y = A + B\,x$, where $x = \log_{10}$ Re and $y = \log_{10}$ Po. Use the linear regression model to find the best-fit parameters and comment on the goodness of the fit.

$\log_{10}$Re	−0.39	0	0.7	1.4	2	2.5	3
$\log_{10}$Po	2.7	2.34	1.63	1.02	0.36	0.028	−0.35

E7.7   The braking distance of a vehicle (S) is investigated as a function of its pre-braking speed (v). The experimental data is known to be well described with a quadratic model, $S(v) = av^2 + bv + c$. Apply least squares regression to determine the model parameters and calculate r^2.

v (km/h)	10	40	85	115	145
S (m)	2.1	5.6	13.8	21	31

E7.8   A pump characteristic curve for the head (h_p in m) can be expressed by an nth-degree polynomial, i.e., $h_p = P_n(Q)$, where Q is the volumetric flowrate of the fluid in m^3/h. Consider the pump whose head-versus-flowrate test results are given below. Use the data and the method of least squares to obtain the pump characteristic curve with second-, third-, and fourth-degree polynomial approximations. Which approximation would you recommend? and why?

Q	5	25	65	90	120	140
h_p	10.91	10.60	9.56	8.47	7.13	5.43

E7.9   The lift coefficient (C_L) is a dimensionless parameter that relates the lift generated by a lifting body. A wind tunnel is used for determining a relationship between the lift coefficient and the angle of attack (the angle between the horizontal line and the line of motion). The wind tunnel test results for an airfoil are presented below. Use the method of least squares regression to develop linear and quadratic models. Which model is better? and Why?

C_L	−5	0	5	10	15	20
θ (deg)	0.02	0.56	1.06	1.50	1.71	1.65

E7.10   Consider two elements, A and B, which are completely soluble in each other in solid and liquid states. The following solidus (T_s in °C) and liquidus temperature (T_ℓ in °C) data are obtained from the binary A-B phase diagram as a function of B content (x_B, %). Use least squares regression to obtain the best-fit polynomials for the solidus and liquidus curves with $r^2 > 0.99$.

x_B	0	20	40	60	80	100
T_s	1110	1138	1186	1256	1353	1500
T_ℓ	1110	1291	1390	1452	1488	1500

E7.11   Hook's law states that the force applied (F) is directly proportional to the length (x) of a spring, $F = kx$, where k is the spring constant. Experimental data conducted to determine the spring constant of a new product is presented below. Apply the least squares to determine the spring constant.

x (cm)	1	3	4	5	8	10
F (N)	175	550	700	900	1450	1830

Section 7.2 Transformation of Variables

E7.12 The specific heat of a substance as a function of temperature is to be determined using the experimental results presented below. The temperature dependence of the specific heat is given as $c_p(T) = A + B/T$. Apply a linear regression model to determine A and B and calculate r^2.

T (K)	300	400	500	600
c_p (Kcal/kg°C)	1.267	1.351	1.402	1.434

E7.13 Experiments were carried out to determine the relationship between sugar solubility (x in g sugar/100g solution) in a water-based solution and solution temperature (T in °C). The data is assumed to be well-described by $x = a + b \ln T$. Using a linear regression model with the data furnished below, determine the best-fit parameters and the coefficient of determination.

T	5	10	30	50	75	90
x	160	200	260	275	295	298

E7.14 Grain-boundary strengthening is a method of strengthening materials that involves changing the average grain (crystallite) size. The relation between yield stress and grain size is described by the Hall-Petch equation given as $\sigma_y = \sigma_0 + k/\sqrt{d}$, where σ_y is the yield stress (MPa), σ_0 is a material constant (MPa), k is the strengthening coefficient ($\sqrt{\mu m}$), and d is the average grain diameter (μm). Experimental results for the yield stress and the grain size of a heat-treated alloy are tabulated below. (a) Apply least squares regression to the observed data to find the model parameters; (b) make a residual plot and calculate r^2. What can you say about the goodness of the fit?

d (μm)	2	5	10	16	20	30
σ_y (MPa)	156.6	145.2	135.4	134.5	129.1	126.9

E7.15 When a particular bacterial culture is introduced into milk, its population remains more or less the same for a short time (time lag), but then begins to increase exponentially. A suitable mathematical model is proposed as $\log_{10} n(t) = a + b e^t$. The observed bacterial population with time is tabulated below. (a) Use least squares regression to determine the model parameters; (b) calculate the coefficient of determination; and (c) plot residuals and discuss the goodness of the fit.

t (min)	0.2	0.5	0.9	1.2	1.8	2.5	3.3
$\log_{10} n$	4.7	4.8	5.1	5.4	6.1	10.8	13.9

E7.16 The variation of the aerodynamic drag coefficient (C_D) with speed (V) of a high-speed train model was determined as a result of a series of wind tunnel tests. The test results are tabulated below. The data is known to obey the $C_D(V) = a + b \ln(V)$ relationship. (a) Use least squares regression to find the model parameters; (b) calculate the coefficient of determination; and (c) plot residuals and discuss the goodness of the fit.

V (km/h)	50	90	110	140	190	270	440
C_D	0.37	0.36	0.35	0.344	0.335	0.32	0.31

E7.17 The data collected to determine the solubility of glycerin (y in g glycerin/100 g acid) in an acid solution (x in mole) is tabulated below. The relationship between the solubility and acid solution is given as $y = a + b\sqrt{x}$. Compute the best-fit parameters using the method of least squares and discuss the goodness of the fit.

x (mole)	1	4	9	16	25
y (g/100 g acid)	4.6	6.5	8.8	10.1	13.2

E7.18 The tabulated data is taken from an analog input signal. A suitable mathematical model for the data is given by $Y(t) = a + b \cos(2\pi t)$. (a) Use the data and least squares regression to estimate the model parameters, and (b) discuss the goodness of the fit.

t	0	0.1	0.2	0.3	0.4	0.5	0.6
Y	1.67	1.62	1.50	1.36	1.23	1.19	1.24

E7.19 A thermistor is an electrical resistor whose resistance changes in response to temperature, and it is used as temperature sensor due to its high sensitivity. By measuring the resistance of the thermistor, one can determine the temperature of the medium. A relation between the temperature and the resistance of a thermistor is given by the *Steinhart–Hart equation*, defined as

$$\frac{1}{T} = a + b \ln R + c(\ln R)^3$$

where R is the resistance (Ω) and T is the absolute temperature (K). Apply least squares regression to the furnished data to determine the coefficients of the Steinhart–Hart equation (a, b, and c) and the goodness of the fit.

T (K)	275	300	320	350	380
R (kΩ)	80.4	38.6	21.2	9.1	4.4

E7.20 The data for the solubility of graphite (x) in liquid iron as a function of temperature (T) is given below. A mathematical model relating the temperature of the iron to its graphite content (%) is given by $T(x) = a + b\sqrt{x} + cx$. (a) Apply least squares regression to determine the model parameters and r-squared; (b) plot the residuals and measured-predicted values; and (c) apply linear regression to the predicted-observed data to find the slope and bias of the PO-line. What can you say about the trustworthiness of the model?

x	4.3	5	6.5	8	10	11	12
T (°C)	1020	1710	1860	2590	2480	2690	3340

E7.21 The solubility of an unknown gas in an aqueous solution is studied experimentally. A correlation (model) for the Henry's law constant is given by

$$\ln K_H = \sum_{k=0}^{n} a_k T^k$$

(a) Use least squares regression with the tabulated data to determine linear, quadratic, and cubic regression model parameters and find r^2 for each case; (b) plot the residuals and PO values; and (c) apply linear regression to the PO data to find the slope and bias of the PO-line. What can you say about the reliability of the model?

T (°C)	80	100	150	195	260	350
K_H	5.94	6.17	6.22	6.18	5.99	4.82

E7.22 The effect of an additive on the drying time of an oil-based paint is investigated. The additive concentration (m in g/100 mL) and drying time (t in hours) are assumed to be described by the following model:
$$\sqrt{t} = a + bm + cm^2$$
(a) Use the data presented below and the least squares regression to find model parameters and calculate r^2; (b) plot the residuals and PO values; and (c) do you think that the model is suitable for the furnished data?

m (g/100 mL)	0	0.5	1	2	4	6
t (h)	4	3.5	3.1	2.6	2.4	3

Section 7.3 Linearization of Nonlinear Models

E7.23 Using the given data, apply least squares regression to the $Y(x) = ax^{bx}$ model. (a) Plot the data and the best-fit curve together for a visual inspection; (b) plot the residuals and calculate

r-squared; and (c) apply linear regression to the PO data to find the slope and bias of the PO line and the goodness of the fit.

x	1.2	2.8	4.3	5.4	6.8	7.9
$Y(x)$	2.1	11.5	28.1	41.9	72.3	91.4

E7.24 Repeat **E7.23** for $Y(x) = ax\,e^{bx}$.

E7.25 Electrical resistivity (ρ) and salt content (C) of soil at several locations in an agricultural region are measured. The relationship between the resistivity and soil content is assumed to obey the power law, i.e., $\rho = mC^n$. Using the tabulated data and least squares regression, determine the model parameters and comment on the goodness of the fit.

C (g/L)	0.3	1	2	4.7	7.2	10	12
ρ (Ω·m)	144.1	89.8	19.5	23.2	5.3	10.5	6.8

E7.26 The discharge-time relationship for a capacitor obeys the exponential decay model: $Q(t) = Q_0\,e^{-\lambda t}$, where R is the resistance, C is the capacitance, Q_0 is the initial charge, and $\lambda = 1/RC$. Use the tabulated experimental data to (a) estimate Q_0 and λ using least squares regression; (b) compute the coefficient of determination. (c) Discuss the goodness of the fit.

t (s)	10	20	30	40	50	75	130
Q (C)	14	13	11	10.3	7	4	1

E7.27 The variation of the viscosity of liquids with temperature is modeled by the so-called the *Andrade equation*: $\mu(T) = ae^{b/T}$, where T is the temperature (K). The results of the viscosity measurements of an acidic solution at different temperatures are presented below. Use the method of least squares to estimate model parameters and discuss the goodness of the fit.

T (K)	280	300	310	330	360
μ (N·s/m^2)	0.1	0.082	0.075	0.063	0.05

E7.28 The solubility of oxygen (m, mol/ton water) is a function of temperature (T, K) and is described by the *Setschenow equation*: $m = ae^{bT}$. The solubility of oxygen in a solution at different temperatures is tabulated below. Using least squares regression, (a) estimate the best-fit parameters; (b) compute r^2; (c) plot the residuals and PO plot; and (d) discuss the goodness of the fit.

T (K)	272	276	285	294	306	318	345
m (mol/ton)	2.17	1.97	1.65	1.37	1.14	0.99	0.83

E7.29 Steel-bearing balls removed from an annealing furnace are left to cool in the air. The temperature readings taken from the balls are tabulated below. Assuming that the conditions for the lumped capacitance apply, the cooling temperature obeys

$$T(t) - T_\infty = (T_i - T_\infty)\exp(-t/\tau)$$

where T_∞ is the ambient temperature (=10°C), T_i is the initial temperature of the bearings, and τ is a time constant. Use the method of least squares to estimate the time constant τ and initial temperature T_i. What can you say about the goodness of the fit?

t (min)	1	2	4	6	8	10
T (°C)	330	230	115	50	26	10

E7.30 A concrete recipe has been developed for use in nuclear reactor containment structures for protection from radioactivity. The relationship between the radiation intensity and the concrete thickness is given by the following relationship: $I(x) = I_0 e^{-\mu x}$, where I_0 is the incident intensity (cps, counts per second), x is the thickness (cm), and μ is the mass attenuation coefficient (cm^{-1}). Concrete slabs of varying thickness are irradiated from one end, and the intensity of the penetrating radiation is measured from the other end. Applying least squares regression to the tabulated data, (a) estimate the mass attenuation coefficient and incident intensity, and (b) determine the goodness of the fit.

x (cm)	1	16	25	30	50
I ($\times 10^{-6}$ cps)	1.86	0.33	0.023	0.01	0.0003

E7.31 In an isentropic process, the relationship between the pressure (P) and volume (V) of an ideal gas is known to obey $PV^\gamma = C=$constant, where γ is the heat capacity ratio. Apply least squares regression to the furnished experimental data to estimate γ and C and discuss the goodness of the fit.

V (m^3)	0.1	0.22	0.31	0.44	0.7
P (Pa)	75	25	16	10	5

E7.32 In an optimization project, the cost of a horizontal centrifugal pump as a function of power is required. The cost (C in \$) of the pump is related to its power (P in kW) by the $C = aP^n$ relationship. Market research yields the tabulated data presented below. Use least squares regression to estimate the best-fit parameters and discuss the goodness of the fit.

P (kW)	0.2	2.5	5.0	12	16
C (\$)	725	1400	1600	2000	2300

E7.33 Atmospheric pressure (P in mmHg) varies with altitude (z in km), which may be described by an exponential model $P = ae^{bz}$. Apply least squares regression to the tabulated data to estimate the best-fit parameters and discuss the goodness of the fit.

z (km)	0	0.15	0.46	1.07	2.45	6.1	10.7
P (mmHg)	760	746	720	668	565	350	180

E7.34 To determine the relationship between the volumetric flow rate (Q in m^3/h) and pressure drop (ΔP in kPa) in a hydraulic circuit, a series of measurements were performed. The pressure drop and flow rate are known to obey a power law model: $\Delta P = aQ^n$. Apply the method of least squares to the collected data tabulated below to estimate the best-fit parameters. What can you say about the goodness of the fit?

Q (m^3/h)	5	10	15	20	25
ΔP (kPa)	500	750	1100	1800	2750

E7.35 The tabulated data presented below was obtained from experimental work to determine the variation of the dynamic viscosity (μ) of an engine oil with temperature. The relationship of viscosity with temperature is known to obey $\mu = a/(b+T)$. Apply least squares regression to the data to estimate the model parameters and discuss the goodness of the fit.

t(K)	275	290	325	350	400
μ (N.s/m)	0.42	0.07	0.025	0.015	0.01

E7.36 The model for the dynamic viscosity of gases as a function of temperature is given by the *Sutherland equation*: $\mu = aT^{3/2}/(T + b)$ (in N·s/m). Apply the method of least squares fit to the furnished data to estimate the model parameters and discuss the goodness of the fit.

T (in K)	270	300	320	350	375
$\mu \times 10^6$	16.7	18.6	19.2	20.7	21.8

E7.37 A type of bacteria called *Acidithio-bacillus* is used for the processing of low-grade copper ores. A series of experiments with copper solutions containing the bacteria resulted in the tabulated results for the copper precipitation with residence time. The amount of precipitate as a function of time is thought to be governed by the relationship: $[c]\% = at/(b+t)$. Use the method of least squares to estimate the best fit parameters and discuss the goodness of the fit.

t (hrs)	18	66	120	190	250	320	390	475	570
[c] (Cu%)	4	14	25	36	45	55	65	72	80

E7.38 Impurities in a pure metal lower its compressive strength. A relationship between the impurity content (x in %) and the compressive strength (σ in MPa) is suggested as $\sigma = a/(b + x)$. Apply least squares regression to the furnished experimental data to estimate the best-fit parameters and discuss the goodness of the fit.

x (%)	0.01	0.02	0.04	0.045	0.055
σ (MPa)	150	130	100	95	88

E7.39 The following data are obtained from the molecular weight of a gas mixture (M) versus the *entrainment ratio* (weight of gas to weight of air, ω) curve. Assuming that the data is well represented by $\omega(M) = M/(aM + b)$, apply the least squares method to determine the best-fit parameters and the coefficient of correlation.

M	1.5	5	15	25	35	60	95	120	140
ω	0.12	0.42	0.75	0.92	1.1	1.33	1.47	1.53	1.6

E7.40 The half-life of an irradiated radioactive specimen is to be determined. The number of radioactive nuclides (cph, counts per hour) obeys the exponential decay law: $N(t) = N_0 e^{-\lambda t}$, where N_0 is the initial radioactivity (cph) and $\lambda = \ln(2)/t_{1/2}$, where $t_{1/2}$ is the half-life of the specimen. Apply least squares regression to estimate the half-life of this radioactive specimen and discuss the goodness of the fit.

t (h)	1	4	7	11	15	20	30
N (cph)	960	430	190	55	20	5	0.25

E7.41 The vapor pressure of an aqueous solution at a given temperature is determined experimentally. The vapor pressure (p in mmHg) is related to the temperature (T in °C) with the following relationship: $\log_{10} p = aT/(b + T)$, where a and b are solution-dependent constants. Use least squares regression to estimate model parameters and discuss the goodness of the fit.

T (°C)	1	3	4	6	8	9.9
p (mmHg)	1.3	2.1	2.6	4.1	6.1	9.0

Section 7.4 Multivariate Regression

E7.42 The solubility of N_2 in an aqueous solution (x in g gas/kg aq) is determined as a function of temperature (T in K) by a series of experiments. The $x - T$ relationship obeys the $\ln x = a + bT + c\ln T$ model. Apply least squares to the reported data to estimate the best-fit parameters and discuss the suitability of the model.

T	275	290	310	330	345
x	0.029	0.021	0.016	0.014	0.0135

E7.43 An interpolating function that relates the vapor pressure (p in Pa) of a liquid melt with temperature (T in K) is to be determined using the tabulated data. Use the following regression model: $\ln p(T) = A + B \ln T + C/T$, apply the method of least squares to determine the best-fit parameters, and discuss the goodness of the fit.

T (K)	350	400	450	500	550
p (Pa)	9	110	760	3900	14300

E7.44 As liquids A and B dissolve in water, the thermal conductivity of the resulting aqueous solution increases. The experimental results for the thermal conductivity measurements with differing amounts of A and B (vol% x_A and x_B) are tabulated below. Using the given data, apply least squares regression to the proposed model, $k(x_A, x_B) = a + bx_A + cx_B$ (in W/m·K), to determine the best-fit parameters and discuss the goodness of the fit.

x_A	3	7	15	4	3	23	9
x_B	0	1	4	8	16	12	27
k	11.8	12	12.6	14.5	17.2	14.9	20.6

E7.45 CO_2 levels (C in ppm) at a rural location are collected periodically. The set of data since the year 2000 ($t = year - 2000$) is tabulated below. The CO_2 concentration data with time is suitable for the following mathematical model:

$$C(t) = A + Bt + C\sin(2\pi t)$$

Use the least squares method to estimate the model parameters and comment on the goodness of the fit.

t	C	t	C	t	C	t	C
0.25	372.935	1.75	368.434	3.25	380.769	4.75	374.970
0.5	368.519	2	373.378	3.5	378.607	5	381.065
0.75	363.326	2.25	380.523	3.75	372.36	5.25	387.712
1	369.530	2.5	375.467	4	380.817	5.5	385.426
1.25	376.010	2.75	368.530	4.25	383.846	5.75	377.855
1.5	370.302	3	375.424	4.5	379.400	6	383.766

E7.46 In an open channel flow, the volumetric flow rate (Q in m³/h), channel inclination (m in radian), and hydraulic radius (R in m) are assumed to be related by $Q = a\, m^b R^c$. Use the furnished experimental data and the method of least squares to estimate the best-fit parameters and discuss the goodness of the fit.

m	R (m)			
(radian)	0.25	0.5	1.2	2
0.5	4.9	7.6	12.8	18.3
1.0	6.6	10.5	17.3	24.0
1.5	7.1	12.2	21.4	30.9

E7.47 The chemical affinity $f(x, y)$ of a mixture of Salt-1 and Salt-2 varies with the mole ratio of the salts in the mixture, $[x]$ and $[y]$. The chemical affinity of the salt solutions is anticipated to be well represented by $f([x], [y]) = a[x]^b[y]^c$. Apply linear regression to the experimental data below to estimate the best-fit parameters and discuss the goodness of the fit.

$[x]$ (mol/L)	0.1	0.5	0.9	0.5	1.1	1.1
$[y]$ (mol/L)	0.3	0.4	0.6	0.7	0.5	1.4
$f(x, y)$	0.009	0.021	0.032	0.027	0.031	0.047

E7.48 The light reflection coefficient (ρ in m^{-1}) of an optically anisotropic composite glassy material is to be determined as a function of the angle of incidence (θ in rad). The model proposed for the reflection coefficient is $\rho(\theta) = a + b\cos\theta + c\sin\theta$. Use the results of the light reflection experiments given below and the method of least squares to estimate the best-fit parameters and discuss the goodness of the fit.

θ (rad)	10	20	30	50	60	80
ρ (m^{-1})	4.05	3.80	3.40	2.09	1.42	0.12

E7.49 The rate equation for a chemical reaction taking place while being continuously stirred is assumed to have the form

$$r = \exp\left(A - \frac{B}{t}\right)C^n(t)$$

where A, B, and n are the model parameters. Use the multivariate linear regression model to determine the model parameters using the experimental data given below.

t	0.5	1	1.5	2	2.5	3	3.5	4	4.5	5
C	2.9	4.6	5.7	6.5	7.1	7.6	8	8.3	8.6	8.8
r	0.1	9	68	191	362	560	770	980	1180	1380

Section 7.5 Continious Least Squares Approximation

In Exercises **E7.50**–**E7.57**, (a) Use the method of least squares to approximate the given functions in the specified intervals with a continuous polynomial of a specified degree. (b) Calculate the residual error for the approximations.

E7.50 $y = \cos x$, quadratic on $[-\pi/2, \pi/2]$.

E7.51 $y = 1/(2 + x^2)$, quadratic on $[-2, 2]$.

E7.52 $y = |\sin \pi x|$, quadratic on $[-1, 1]$.

E7.53 $f(x) = \sqrt{2x}$, linear on $[1/2, 2]$.

E7.54 $f(x) = \sqrt{x}$, piecewise linear on intervals $[0, 1/4]$ and $[1/4, 1]$.

E7.55 $f(x) = \ln(1 + x)$, quadratic on $[0, 2]$.

E7.56 $y = e^x$, quadratic on $[-1, 1]$.

E7.57 $y = x^4$, quadratic approximation of the form $Y(x) = a + bx^2$ on $[-1, 1]$.

E7.58 Find an approximation in the form of $y = ax + b/(x + 1)$ to a cubic polynomial defined as $f(x) = 3 - x + 2x^2 - 3x^3/4$ on the interval $[0, 1]$ using the least squares method. Calculate the residual error for the approximation.

E7.59 Using the least squares method under the integral sign, fit the given data to a continuous $f(x) = ax^2 + b\sin x$ function. *Note:* Apply the trapezoidal rule for numerical integrations.

x_i	0.5	0.65	0.8	0.95	1.1	1.25	1.4
$f(x_i)$	1.6487	1.9155	2.2255	2.5857	3.0041	3.4903	4.0552

E7.60 For a solution, the relationship of vapor pressure with temperature, in the $250 \leqslant T \leqslant 900$ K range, is given below:

$$\log_{10} P_v = 32 - \frac{1600}{T} - 6\log_{10}T + 15 \times 10^{-6}T^2$$

Find the linear approximation $\log_{10} P_v = A + BT$ for the vapor pressure in the interval $250 \leqslant T \leqslant 300$ K. Estimate the relative magnitude of the error for the interval.

E7.61 The Nusselt number, which is a dimensionless heat transfer parameter in pipes, is given by the Dittus-Boelter correlation as

$$\text{Nu}(\text{Re}, \text{Pr}) = 0.023 Re^{0.8} \text{Pr}^n, \quad Re \geqslant 10000$$

where Re and Pr are also dimensionless parameters called the Reynolds and Prandtl numbers, respectively. For a flow optimization problem in pipes, a linear relationship (Nu= $A + B$ Re) in the range $10^4 \leqslant \text{Re} \leqslant 2 \times 10^4$ is required. Find a linear approximation for any Pr number in the indicated Re number range, i.e., find A and B as functions of Pr.

Section 7.6 Over and Underdetermined Systems

E7.62 Use the method of least squares to find the common solution of the following over- or under-determined system of linear equations.

(a) $x_1 - x_2 = 5$, $\quad 2x_1 - 3x_2 = 2$, $\quad x_1 + x_2 = 8$, $\quad 2x_1 + x_2 = 12$
(b) $2x_1 + x_2 + 3x_3 = 24$, $\quad 3x_1 + 4x_2 - 2x_3 = 14$
(c) $3x_1 + 2x_2 + x_3 + 3x_4 = 10$, $\quad x_1 + 4x_2 + 2x_3 - x_4 = 6$
(d) $x_1 + 2x_2 + x_3 = 1$, $\quad 3x_1 - x_2 + 4x_3 = -1$,
$\quad -3x_1 + 5x_2 - 3x_3 = 2$, $\quad 2x_1 + 3x_2 + 5x_3 = 4$
(e) $2x_1 + x_2 - x_3 + 3x_4 = 2.5$, $\quad 5x_1 + 3x_2 + 4x_3 - 2x_4 = 7$,
$\quad 5x_1 + 3x_2 + 4x_3 - 2x_4 = 1.5$

E7.63 A ball is projected into the air with an angle of θ_0 and an initial velocity of V_0, leading to a parabolic trajectory: $y = ax - bx^2$. During the ball's flight, at different instances, we obtain

$$2.7 - 4a + 16b = 0, \quad 4.8 - 8a + 64b = 0, \quad 6.9 - 14a + 196b = 0,$$
$$6 - 27a + 729b = 0, \quad 3.9 - 33a + 1089b = 0, \quad 1.4 - 37a + 1369b = 0$$

Recalling that $a = \tan \theta_0$ and $b = (g/2)/(V_0 \cos \theta_0)^2$, estimate the initial velocity and the projection angle by solving the overdetermined system.

7.9 COMPUTER ASSIGNMENTS

CA7.1 The Nusselt number for fully developed laminar flow in a circular tube annulus with the outer surface insulated and the inner surface at constant temperature is given below (*see* Ref. [27]). A model for the Nusselt number as a function of diameter ratio has the following form: Nu $= a + b (D_o/D_i) + c (D_o/D_i)^2$, where D_i and D_o are the inner and outer diameters, respectively. Apply the method of least squares to the available data to determine the proposed model parameters.

D_i/D_o	0.02	0.05	0.1	0.25	0.5	1
Nu_i	32.337	17.46	11.56	7.3708	5.7382	4.8608

CA7.2 Experimental results for the reduced pressure with the compressibility factor of a real gas at $T_R = 1.30$ are tabulated below. To determine a suitable relationship between Z and P_R, use polynomial models ($n = 2, 3, \ldots, 7$) to relate Z to P_R as accurately as possible. For polynomial models, plot the data and the polynomial approximations on the same graph to make both a visual and a numerical determination by calculating the r^2 value. Do you observe oscillations associated with overfitting?

P_R	0.05	0.5	1	1.5	2	2.5	3	3.5	4	4.5	5	5.5	6	6.5	7
Z	0.98	0.92	0.84	0.76	0.7	0.65	0.64	0.65	0.68	0.7	0.73	0.78	0.82	0.85	0.89

CA7.3 Friction factor (f) in pipes is known to vary with the relative roughness of the pipe wall (ε/D). The flow measurements in pipes for different diameter ratios yield the tabulated friction factors given below. The relationship of the friction factor with the relative roughness is known to obey $1/\sqrt{f} = -1.80\log_{10}(A + B(\varepsilon/D)^{0.36})$. Use the method of least squares to determine the best-fit parameters.

ε/D	10^{-2}	10^{-3}	5×10^{-3}	10^{-4}	10^{-6}
f	0.0072	0.0037	0.0057	0.0022	0.0012

CA7.4 A representative plot of the frequency (f) and amplitude (I) of a signal is presented below. The amplitude versus frequency model is known to fit the Gauss (bell-shaped) distribution:

$$I(f) = I_0 \exp\left[-\frac{(f - f_0)^2}{2\sigma^2}\right]$$

where I_0 and f_0 are, respectively, the maximum amplitude and corresponding frequency, and σ represents the width of the signal. Apply the method of least squares to estimate I_0, f_0, and σ.

f	0.1	0.45	0.8	1.15	1.5	1.85	2.2	2.55	2.9
$I(f)$	18	40	65	86	98	95	80	55	30

Hint: Linearize the model similarly to the exponential models; however, the number of equations is not sufficient to determine the parameters. Defining a new quantity Q as the ratio of the amplitudes between two data points results in the elimination of I_0 as follows:

$$Q(f_i) = \frac{I(f_{i-1})}{I(f_{i+1})} = \exp\left[\frac{2(f_i - f_0)}{\sigma^2}\right]$$

Next, the above expression is linearized to estimate f_0 and σ. The maximum amplitude is computed from

$$\ln I_0 = \frac{1}{n}\sum_{k=1}^{n}\left\{\ln I_k + \frac{(f_k - f_0)^2}{2\sigma^2}\right\}$$

CA7.5 The emissivity of a metal oxide (ε_λ) at 1400 K is experimentally determined as a function of wavelength. The wavelength vs. emissivity data is predicted to well fit the Gaussian distribution, i.e.,

$$\varepsilon_\lambda = \varepsilon_0 \exp\left[-\frac{(\lambda - \lambda_0)^2}{2\sigma^2}\right]$$

where ε_0 is the maximum emissivity, λ_0 is the frequency corresponding to the maximum emissivity, and σ is the deviation width of the frequency. Use the method of least squares to determine the best-fit parameters.

λ (μm)	1	1.5	3.25	4.9	6.5	11	13.5	14.9	18	23
ε_λ	0.17	0.175	0.3	0.63	0.9	0.94	0.7	0.56	0.43	0.34

CA7.6 Apply the method of least squares to the tabulated data below to obtain the best-fit approximating functions ($N = 2, 3$, and 4), which are defined in terms of Legendre polynomials.

x	0	0.8	1.6	2.4	3.2	4
y	0.1	1.9	2.1	3.4	4.3	4.9

The general form of an Nth-degree approximating polynomial is given as

$$Y(x) = \sum_{n=1}^{N} c_n P_n(x)$$

where $P_n(x)$ denotes the nth-degree Legendre polynomial, and the first four polynomials are given as

$$P_1(x) = x, \quad P_2(x) = \frac{1}{2}(3x^2 - 1), \quad P_3(x) = \frac{x}{2}(5x^2 - 3), \quad P_4(x) = \frac{1}{8}(35x^4 - 30x^2 + 3)$$

CA7.7 The Nusselt number (a dimensionless heat transfer rate) is considered to be the function of the Reynolds (a dimensionless number that describes the flow characteristics) and Prandtl (describing the thermal behavior of a fluid) numbers. For a pipe with specially designed turbulator inserts, the experimental measurements lead to the following tabulated results: Employ the method of least squares to the data to determine the model parameters for $Nu = a\mathrm{Re}^n\mathrm{Pr}^m$ and discuss the goodness of the fit.

Re	Pr	Nu	Re	Pr	Nu	Re	Pr	Nu
100	0.14	10	100	0.59	13	100	1.12	13.8
500	0.13	31	500	0.62	39	500	1.24	42.1
1220	0.14	57	1220	0.66	72	1220	1.16	77
2400	0.26	98	2400	0.71	115	2400	1.48	127
3500	0.28	128	3500	0.68	148	3500	1.33	162
7000	0.32	207	7000	0.69	234	7000	1.19	256
11000	0.27	284	11000	0.73	326	11000	1.42	263

CA7.8 Monthly average temperature T (°C) data collected from a local weather station is tabulated below.

Month	Temperature	Month	Temperature
Jan	11.1	Jul	29.4
Feb	12.8	Aug	30
Mar	16.4	Sep	26.7
Apr	20.8	Oct	21.9
May	25	Nov	16.1
Jun	27.8	Dec	11.4

Three mathematical models (A, B, and C) are considered for modeling the monthly average temperatures.

$$\begin{aligned} \text{Model A} \quad & T(m) = A + B\sin^2\left[(m-1)\pi/12\right] \\ \text{Model B} \quad & T(m) = A + B\exp\left[-C(m-6)^2\right] \\ \text{Model C} \quad & T(m) = A\exp\left[-B(m-C)^2\right] \end{aligned}$$

Use least squares regression to determine the model parameters and the most suitable model. *Note*: Models B and C are nonlinear equations that cannot be easily linearized. Use a trial-and-error approach where you make a guess for C to later obtain the estimates for A and B. Calculate r-squared each time until you find the optimum value.

CA7.9 The reaction of methanation on a catalyst is given as

$$CO + 3H_2 \rightarrow CH_4 + H_2O$$

Initial rates obtained at constant temperature at a variety of partial pressures (in atm) of reactants and products are tabulated below:

P_{CO}	P_{H2}	r_A	P_{CO}	P_{H2}	r_A	P_{CO}	P_{H2}	r_A
1	1	0.1219	1	2	0.1056	2	2	0.1056
1	1	0.0944	1	4	0.1203	2	4	0.1552
1	1	0.0943	1	8	0.1189	4	1	0.0533
1	1	0.0753	2	1	0.0782	4	2	0.0911
1	1	0.0753	2	2	0.1204	8	1	0.0317
1	1	0.0512	2	2	0.1057	8	8	0.1476
1	2	0.1274	2	2	0.1056			

Linearize the following models and apply multivariate least squares regression to find the model parameters and comment on the suitability of the models. Can you perform nonlinear regression? Does it affect model parameters?

$$(a) \quad r_A = kP_{CO}^n P_{H2}^m, \qquad (b) \ r_A = \frac{kP_{CO}P_{H2}}{(1 + K_{CO}P_{CO} + K_{H2}P_{H2})^2}$$

Numerical Integration

INTEGRATION is one of the most important mathematical operations encountered in many fields of science and engineering. Definite integration is used to calculate arc length, work, area, volume, centroid, and numerous other physical quantities.

Some functions can be integrated using the exact methods and techniques covered in calculus lectures. Furthermore, indefinite integrals of common functions and/or combinations of functions are available in the form of tables in mathematical handbooks as a reference. On the other hand, today there are computer software systems that can perform a wide range of symbolic computations on functions. When the definite integral of a function is evaluated numerous times in a computation, the true solution (if it exists) can be coded in the most general form (i.e., $\int_0^x e^{au} du = (e^{ax} - 1)/a$) to save time and avoid the accumulation of round-off errors.

Definite integrals of some functions are either impossible or very difficult to obtain by analytical means. This is where numerical integration enters the picture as an "*alternative*" for estimating a definite integral. A numerical integration does not give the "*true value*" of the integral; it is essentially an approximation to the integral itself. In other words, an analyst can only estimate an integral within a desired tolerance.

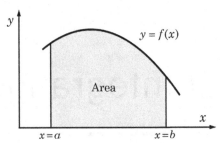

FIGURE 8.1: The true area under the curve (definite integral).

The nature of an integrand (the function you wish to integrate) may necessitate alternative numerical treatments. For instance, integrands may have singularities at their endpoints or within the integration domain, or integrals may cover a semi-infinite domain, $[a, \infty)$ or $(-\infty, b]$, or an infinite domain, $(-\infty, \infty)$, etc. Numerical integrations in such cases require suitable numerical techniques or treatments.

In practice, not all integrands are defined as explicit continuous functions. Some integrands are uniformly or nonuniformly spaced discrete functions, such as data digitally sampled or collected from experiments or obtained numerically from solutions of differential equations (*see* Chapters 9 and 10), and so on. In cases where the integrand is a discrete function, it is not possible to calculate the true integral using the tools and techniques available in calculus. Such integrals can only be calculated approximately using a suitable numerical method.

This chapter presents the most common numerical integration methods applied to both explicitly given continuous and discrete functions. The methods and techniques for dealing with singular and improper integrals are also covered. Extending and applying the numerical methods to estimate double integrals in rectangular and non-rectangular domains have also been presented.

8.1 TRAPEZOIDAL RULE

Let $f(x)$ be a continuous and integrable function defined on the closed interval $[a, b]$. The definite integral of $f(x)$ from a to b is expressed as

$$I = \int_a^b f(x)\,dx \tag{8.1}$$

which corresponds basically to the area of the region bounded by the function (curve) $y = f(x)$, x–axis, and $x = a$ and $x = b$ lines (*see* Fig. 8.1). When the true value of a definite integral cannot be determined, it is instead approximated by the area of the bounded region under the curve, which is the basis of numerical integration.

The trapezoidal rule, as a numerical method, is based on estimating the area under a curve by approximating the bounded domain with the areas of trapezoids. When implementing this idea, the curve defined by $y = f(x)$ on $[a, b]$ is approximated by a straight line connecting its end points, as shown in Fig. 8.2a. Then, the area under the curve is estimated by the area of the resulting trapezoid (i.e., the average of the long and short bases times the altitude), that is,

$$\int_{x_0}^{x_1} f(x)\,dx \approx (b - a)\left(\frac{f_0 + f_1}{2}\right) \tag{8.2}$$

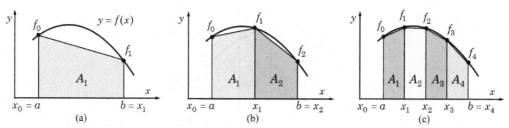

FIGURE 8.2: Graphical depiction of the effect of approximations by (a) one, (b) two, (c) four panels.

where the ordinates are denoted in shorthand notation as $f(x_i) = f_i$. This expression, which is an approximation to Eq. (8.1), is referred to as the *Trapezoidal Rule*.

When a single trapezoid is used for the entire integration interval of a function with strong curvature, the resulting error from this approximation can be very large since, geometrically speaking, the unshaded area between the curve and the trapezoid will be large, as depicted in Fig. 8.2a. It is evident that an approximation by a single trapezoid is not going to be a good representation of the area of the bounded region. To overcome this problem, the integration interval, $[a, b]$, is divided into N-subregions (also referred to as *panels*) of uniform size, $h = \Delta x = (b-a)/N$. Then, the endpoints of the panels are connected by straight-line segments, yielding trapezoids. The area of a trapezoid is an approximation to the area of a panel or subregion. A definite integral is then approximated by summing up the areas of the trapezoids corresponding to the panels. For example, when the interval $[a, b]$ is divided into two panels, the sum of the unshaded areas (i.e., the approximation error) is substantially reduced (Fig. 8.2b). The approximation error with four panels is further reduced (Fig. 8.2c). It is clear that as the number of panels increases, the arc segments bounded by the corresponding panels will approach a straight line, leading to an improved numerical estimate for the area. Hence, a numerical integration will theoretically tend to approach the true value (area) for $N \to \infty$.

To formulate this idea, we first consider the two-panel numerical integral:

$$\int_{x_0}^{x_2} f(x)dx = \int_{x_0}^{x_1} f(x)dx + \int_{x_1}^{x_2} f(x)dx \tag{8.3}$$

The integrals on the rhs of Eq. (8.3) are approximated by the areas of the trapezoids (shown in Fig. 8.2b as

$$\int_{x_0}^{x_1} f(x)dx \approx A_1 = h\left(\frac{f_0 + f_1}{2}\right) \tag{8.4}$$

and

$$\int_{x_1}^{x_2} f(x)dx \approx A_2 = h\left(\frac{f_1 + f_2}{2}\right) \tag{8.5}$$

Substituting these approximations, Eqs. (8.4) and (8.5), into Eq. (8.3) yields

$$\int_{x_0}^{x_2} f(x)dx \approx h\left(\frac{1}{2}f_0 + f_1 + \frac{1}{2}f_2\right) \tag{8.6}$$

In order to find a more general result, consider the interval $[a, b]$ divided into uniform N panels. The abscissas and ordinates are obtained by $x_i = a + ih$ and $f_i = f(x_i)$ for $i = 0, 1, \ldots, N$, respectively. The segment of the curve in each panel is approximated with a straight line that connects the endpoints, resulting in a trapezoid (Fig. 8.3). The lines are

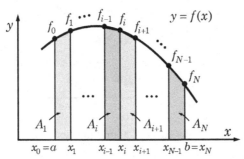

FIGURE 8.3: Graphical depiction of integration with multiple panels.

essentially simple linear interpolations for the curve on $[x_{i-1}, x_i]$. The area under the curve can then be estimated by finding the areas of the trapezoids. Finally, summing up the areas of the trapezoids in all panels yields

$$\int_a^b f(x)dx \approx \sum_{i=1}^N A_i = h\left(\frac{1}{2}f_0 + f_1 + f_2 + \ldots + f_{N-1} + \frac{1}{2}f_N\right) \tag{8.7}$$

Equation (8.7) is referred to as the *composite* (or $N-$*panel*) *Trapezoidal Rule*, and it can be generalized as

$$\int_a^b f(x)dx \approx T_N = h\left(\frac{f_0 + f_N}{2} + \sum_{i=1}^{N-1} f_i\right) \tag{8.8}$$

where T_N denotes the *approximation* or *estimate* obtained by the composite (or $N-$panel) Trapezoidal rule. Note that Eq. (8.8) can also be expressed as a weighted sum as

$$T_N \approx \sum_{i=0}^N w_i f(x_i) \tag{8.9}$$

where w_i's are the *weights*, defined as $w_0 = w_N = h/2$ and $w_i = h$ for $i = 1, 2, \ldots, N-1$.

It is evident that as the number of panels is increased, under the limiting condition ($N \to \infty$ or $h \to 0$), the numerical integral approaches the true value:

$$\int_a^b f(x)dx = \lim_{N \to \infty} \left(\sum_{i=1}^N A_i\right)$$

However, it is neither possible nor practical to use an infinitely large number of panels to estimate the definite integral of a function due to requiring excessive cpu-time (not to mention the round-off errors acquired). For this reason, the composite rule is limited to a finite number of panels that provide an approximate value. The approximation error can be reduced by increasing the number of panels, as illustrated in Fig. 8.2. Contrary to the integration of explicit functions, the panel width of a discrete function is fixed and cannot be arbitrarily adjusted. A larger data set with a smaller panel size means reproducing a discrete data set, which may not always be technically possible. Thus, the numerical analysts have to deal with the consequences of using the available discrete data set and then figure out the associated numerical errors.

Equation (8.8) does not provide any quantitative information about the accuracy of the values estimated by the composite trapezoidal rule or the magnitude of the errors made.

However, it is important that the resulting error be associated with a measurable and comparable numerical value. To be able to do that, we resort to the Taylor series, where we define the integral of a continuous and integrable function $f(x)$ as

$$I(x) = \int_a^x f(u)du \tag{8.10}$$

Assuming it is analytic, $I(x)$ is expanded to the Taylor series about x_i as follows:

$$I(x_i + h) = I(x_i) + h\,I'(x_i) + \frac{h^2}{2!}I''(x_i) + \frac{h^3}{3!}I'''(x_i) + \mathcal{O}(h^4) \tag{8.11}$$

Employing the Leibniz rule of differentiation to Eq. (8.10) and using a shorthand notation, we find

$$I(x_i + h) = I_{i+1}, \quad I'(x_i) = f(x_i) = f_i, \quad I''(x_i) = f'(x_i) = f'_i, \quad I'''(x_i) = f''(x_i) = f''_i,$$

Equation (8.11) can now be expressed in terms of $f(x)$ as

$$I_{i+1} = I_i + h\,f_i + \frac{h^2}{2!}f'_i + \frac{h^3}{3!}f''_i + \mathcal{O}(h^2) \tag{8.12}$$

Then, recalling the following forward difference formula for f'_i (*see* Chapter 5)

$$f'_i \cong \frac{f_{i+1} - f_i}{h} - \frac{h}{2}f''_i + \mathcal{O}(h^2)$$

Substituting it into Eq. (8.12) and simplifying yields

$$I_{i+1} = I_i + \frac{h}{2}(f_{i+1} + f_i) - \frac{h^3}{12}f''_i + \mathcal{O}(h^4) \tag{8.13}$$

Using Eq. (8.10), the definite integral of $f(x)$ on $[x_i, x_{i+1}]$ can now be expressed as

$$\int_{x_i}^{x_{i+1}} f(x)dx = \int_a^{x_{i+1}} f(x)dx - \int_a^{x_i} f(x)dx = I_{i+1} - I_i$$

Letting $A_{i+1} = I_{i+1} - I_i$ and substituting it into Eq. (8.13) results in

$$A_{i+1} = \frac{h}{2}(f_{i+1} + f_i) - \frac{h^3}{12}f''_i + \mathcal{O}(h^4) \tag{8.14}$$

Notice that $h(f_{i+1} + f_i)/2$ is the area of $f(x)$ estimated by the trapezoidal rule in the interval $[x_i, x_{i+1}]$. The other term in Eq. (8.14) provides a concrete measure for the magnitude of the incurred error. Applying this result to all panels on $[a, b]$, we get

$$T_N = \sum_{i=1}^{N} A_i = h\left(\frac{f_0 + f_N}{2} + \sum_{i=1}^{N-1} f_i\right) - \frac{h^3}{12}\sum_{i=0}^{N-1} f''_i + \mathcal{O}(h^4) \tag{8.15}$$

where the first term of Eq. (8.15) corresponds to the composite trapezoidal rule, and the second term represents the leading error.

To develop a simpler and more manageable expression, the mean value theorem is applied to the error term as follows:

$$\sum_{i=0}^{N-1} f''(x_i) = N f''(\xi) \equiv \frac{b-a}{h}f''(\xi), \quad a \leqslant \xi \leqslant b \tag{8.16}$$

where $N = (b-a)/h$.

Making use of Eq. (8.16) in Eq. (8.15) gives rise to

$$R_N \equiv -\frac{h^3}{12} \sum_{i=0}^{N-1} f_i'' = -\frac{h^2}{12}(b-a)f''(\xi), \quad a \leqslant \xi \leqslant b \tag{8.17}$$

where R_N will be referred to as the *dominant* or *global error* of the N-panel Trapezoidal rule. Notice that the magnitude of the error is second order, i.e., $R_N \equiv \mathcal{O}(h^2)$.

Finally, the $N-$panel trapezoidal rule with the order of global error is expressed as

$$T_N = h\left(\frac{f_0 + f_N}{2} + \sum_{i=1}^{N-1} f_i\right) + \mathcal{O}(h^2) \tag{8.18}$$

It is clear that the order of global error will be $\mathcal{O}(h^4)$ if the dominant error, Eq. (8.17), is somehow incorporated into the composite rule:

$$T_N = h\left(\frac{f(a) + f(b)}{2} + \sum_{i=1}^{N-1} f(x_i)\right) + R_N + \mathcal{O}(h^4) \tag{8.19}$$

Trapezoidal Rule with End Correction: To include the dominant error in Eq. (8.18), we note that the second derivative in Eq. (8.17) is unknown; however, applying the mean value theorem to $f'(\xi)$ gives

$$f''(\xi) = \frac{f'(b) - f'(a)}{b - a} \tag{8.20}$$

Substituting Eq. (8.20) first into Eq. (8.17) and then into Eq. (8.19) leads to

$$CT_N = \frac{h}{2}\left(f(a) + f(b) + 2\sum_{i=1}^{N-1} f(x_i)\right) - \frac{h^2}{12}(f'(b) - f'(a)) + \mathcal{O}(h^4) \tag{8.21}$$

which is a fourth-order accurate approximation referred to as the *N-panel Trapezoidal rule with end correction*. This name derives from the inclusion of $f'(a)$ and $f'(b)$ in the approximation. Equation (8.21) can be applied without much additional computation, provided that the true derivatives of the integrand at the end points are known. Clearly, CT_N can be used for estimating the integral of functions explicitly defined so that the true values of the first derivatives at the end points can be evaluated exactly. The first derivatives can be approximated with the finite differences, provided that at least fourth-order accurate finite difference formulas are used. However, fourth order accuracy cannot be assured.

In numerical integration of smooth functions, the composite trapezoidal rule may be sufficient to achieve the desired level of accuracy in most engineering applications. If highly accurate results are sought, the composite trapezoidal rule T_N with a large N can demand a great deal of computational effort. If $f'(a)$ and $f'(b)$ can be easily found, the composite rule with end correction CT_N may offer a remedy. To estimate an integral only once, the computational cost of using a very large N may be bearable and justified. However, when integrating many integrals or a single integral numerous times or integrands that are time-consuming to evaluate, we need to implement efficient numerical methods of high-order accuracy, i.e., $\mathcal{O}(h^4)$, $\mathcal{O}(h^6)$, and so on.

Trapezoidal Rule

- The trapezoidal rule is simple and easy to apply;
- It is a weighted sum of the integrand for a specified number of integration points;
- Cpu-time is the minimum for uniformly spaced discrete functions;
- There is no restriction on the number of panels to be used;
- As the number of panels increases, the numerical estimates of a continuous function always approach the true value;
- Since they are 2nd and 4th order accurate methods, T_N and CT_N give the true integral value for polynomials of degree $m \leqslant 1$ and $m \leqslant 3$, respectively.
- Error analysis is straightforward.

- The integrands must be continuous;
- The rate of convergence of T_N is rather slow;
- Integrands depicting oscillations and/or sharp changes require a large number of panels to yield sufficiently accurate estimates.

EXAMPLE 8.1: Application of Trapezoidal Rule in Thermodynamics

The tendency of a gas to escape or expand is explained by the *fugacity* property of the gas. For ideal gases, fugacity f is equal to its pressure, but in real gases, it is computed by the following integral:

$$\ln\left(\frac{f}{P}\right) = \int_0^P \frac{Z(x) - 1}{x}\,dx$$

where P is the pressure, Z is the *compressibility factor*, and f/P is referred to as the *fugacity coefficient*. The data on the compressibility factor of a real gas at a constant temperature are fitted to the curve given below:

$$Z(p) = 1 - 5 \times 10^{-4}\, p\, e^{-p/50}, \quad 0 < p < 400\,\text{atm}$$

Estimate $\ln(f/P)$ for $P = 400$ atm using the 8-panel Trapezoidal rule *with* and *without* end correction. Calculate the true error and the global error bounds for both cases.

SOLUTION:

Substituting $Z(p)$ into the integral and simplifying gives

$$\int_0^P \frac{Z(x) - 1}{x}\,dx = -5 \times 10^{-4} \int_0^P x\, e^{-x/50}\,dx$$

The true value of the above integral is $0.025(P + 50)\, e^{-P/50} - 1.25$ for $P = 400$ atm, which results in 2492.4521. Dividing the interval (0,400) into 8 panels yields $h = (P-0)/N = 400/8 = 50$ atm.

To estimate the integral with the N-panel trapezoidal rules, a set of discrete data is generated from $F(x) = x e^{-x/50}$. To use Eq. (8.9), the weights are defined as $w_0 = w_8 = 25$ and $w_i = 50$ for $i = 1$ to 7.

TABLE 8.1

i	x_i	F_i	w_i	$w_i F_i$
0	0	0	25	0
1	50	18.394	50	919.6986
2	100	13.5335	50	676.6764
3	150	7.46806	50	373.4030
4	200	3.66313	50	183.1564
5	250	1.68449	50	84.2243
6	300	0.74363	50	37.1813
7	350	0.31916	50	15.9579
8	400	0.13418	25	3.3546
			$\Sigma w_i F_i =$	**2293.6526**

The abscissas ($x_i = 50\,i$), the integrand for the integration points ($F_i = F(x_i)$), and the $w_i F_i$ products are calculated and tabulated in Table 8.1. The sum of the last column is the approximation obtained with the 8-panel trapezoidal rule:

$$T_8 = \sum_{i=0}^{8} w_i F_i = 2293.6526$$

To estimate the global error with Eq. (8.17), we require $F''(\xi)$. However, the $F''(\xi)$ cannot be truly determined because the specific value of ξ is unknown. Nonetheless, we can anticipate that it will be somewhere between its maximum and minimum values within the interval of integration.

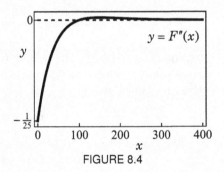

FIGURE 8.4

The distribution of $F''(\xi)$ on the interval $0 \leqslant x \leqslant 400$ is plotted in Fig. 8.4. It is observed that $F''(\xi)$ has an absolute minimum value of $-1/25$ at $x = 0$ and a local maximum value of $1/50e^3$ at $x = 150$. The upper and lower error bounds for $h = 50$ and $N = 8$ are found from Eq. (8.17) as follows:

$$R_{5,\text{max}} = -\frac{(50)^2}{12}(400 - 0)\left(-\frac{1}{25}\right) = 3333.33$$

$$R_{5,\text{min}} \equiv -\frac{(50)^2}{12}(400 - 0)\left(\frac{1}{50e^3}\right) = -82.98$$

which indicates that the true error ($2492.4521 - 2293.6526 = 198.7996$) is within the bounds.

Now we apply the end-correction formula. Substituting $F'(0) = 1$ and $F'(400) = -7/e^3$ into the correction term and then into Eq. (8.21), we find

$$\text{correction} = -\frac{h^2}{12}\left(F'(400) - F'(0)\right) = -\frac{(50)^2}{12}\left(-\frac{7}{e^8} - 1\right) = 208.823$$

and finally

$$CT_8 = 2293.6526 + 208.823 = 2502.475$$

which yields the true error of -10.02291.

Substituting these estimates into the integral in question, we obtain

$$\int_0^P \frac{Z(x) - 1}{x}\,dx = -5 \times 10^{-4} \int_0^P x\,e^{-x/50}\,dx = \begin{cases} -1.2462260, & \text{True integral} \\ -1.1468125, & \leftarrow T_8 \\ -1.2512375, & \leftarrow CT_8 \end{cases}$$

Discussion: We find that the true absolute errors for the trapezoidal rule *without* and *with* the end correction are 0.0994135 and 0.005011, respectively. The true error has been reduced about 20-fold with the use of the end correction formula. It is also seen that the true error of the estimate with the trapezoidal rule is within the theoretical error bounds given by Eq. (8.17). However, determining the error bound is often not an easy task because higher-order derivatives of many functions can be difficult to derive or impossible to analyze, even when derived to determine the local extremums.

Pseudocode 8.1

```
Module TRAPEZOIDAL_RULE_RF (a, b, n, intg, intgc)
 \ DESCRIPTION: A pseudo-module to estimate the definite integral of y = f(x)
 \   on [a, b] using composite Trapezoidal rule with/without end correction.
 \ USES:
 \   FX and FU :: User-defined functions providing f(x) and f'(x), respectively.
 h ← (b − a)/n                              \ Find panel width
 intg ← 0.5 * (FX(a) + FX(b))               \ Initialize intg with avg of endpoints
 xi ← a                                     \ Initialize abscissa, x_0 = a
 For [i = 1, n − 1]                         \ Accumulator loop
    xi ← xi + h                             \ Accumulate h to get x_i
    intg ← intg + FX(xi)                    \ Accumulate F(x_i)'s to get intg
 End For
 intg ← h * intg                            \ Estimate integral by Eq. (8.18)
 corr ← −h * h * (FU(b) − FU(a))/12         \ Compute the correction term
 intgc ← intg + corr                        \ Estimate integral by Eq. (8.21)
 End Module TRAPEZOIDAL_RULE_RF


 Function Module FX (x)                      \ User-defined Function module giving f(x)
 FX← x * EXP(−x/50)                          \ Built-in function EXP(x) is used
 End Function Module FX


 Function Module FU (x)                      \ User-defined Function module giving f'(x)
 FU← (50 − x) * EXP(−x/50)/50
 End Function Module FU
```

In Pseudocode 8.1, a pseudomodule, TRAPEZOIDAL_RULE_RF, is provided for estimating the definite integral of an explicitly given continuous function using the composite Trapezoidal rule *with* and *without* end correction. The module requires the number of panels (n) and the integration interval (a, b) as input. The integrand, as well as its derivative, which is required in the end correction formula, are supplied by a couple of user-defined functions, FX and FU, which are exemplified here for Example 8.1. The module generates uniformly spaced n panels and calculates the abscissas and the ordinates for $n + 1$ integration points. The algorithm is straightforward in that the integral of FX(x) on (a, b) is estimated with the trapezoidal rule without and with the end correction, Eqs. (8.19) and Eq. (8.21). Note that no array variables are used for abscissas and the integrand in order to reduce memory allocation.

A separate module is not given here for uniformly spaced discrete functions as integrands. A typical argument list requires the number of integration points n (note that n is the number of integration points, not the number of panels), the panel width h, and the discrete function $\mathbf{f}$ (f_i for $i = 1, 2, \ldots, n$) as input. In the module, the abscissas are no longer required because the discrete function is already provided as an array (i.e., $f(x_i) = f_i$). Thus, it is sufficient to specify the panel width to carry out the numerical integration. Likewise, for the same reasons, the function modules FX and FU are not required. The integral (*intg*) of f on $[x_1, x_n]$ is the only output of the module, and it is computed by Eq. (8.18). The trapezoidal rule with end correction requires accurate computation of the first derivatives at end points, which is *not* recommended for the integrands described by discrete functions unless the derivatives are evaluated with a high order (higher than 4th order) finite-difference approximation.

8.2 SIMPSON'S RULE

An analyst can arbitrarily determine the number of panels (or panel width) when numerically integrating a continuous explicit function. In other words, it is up to the analyst to use 100, 1000, or millions of panels so that an integral is approximated to a specified accuracy level. However, the number of panels (or the interval size) in a discrete function is fixed; that is, the number of data points can neither be arbitrarily increased nor the panel width reduced. Hence, to obtain estimates with improved accuracy, alternative integration formulas using the available data sets are required.

From this point on, we will seek higher-order integration schemes yielding smaller global errors for a fixed number of panels. We have so far learned that connecting the endpoints of a panel with a straight line and calculating the area of the resulting trapezoid led to the trapezoidal rule. By any means, a line is not well suited to approximate a curve in a large interval; therefore, the approximation error (the gap between the curve and the line) can be large. As an interpolating function, a parabola may be used to reduce this gap, but fitting a curve with a unique parabola requires three data points (or *two panels*, *see* Fig. 8.5a). As depicted qualitatively in Fig. 8.2b, the integration error is markedly reduced in comparison to the trapezoidal rule.

The fitting of three points to a parabola has been covered in Section 5.5. In Fig. 8.5a, the equation of a parabola centered about x_1 can be expressed as

$$f(x) \cong \frac{f_1''}{2} x^2 + f_1' x + f_1, \tag{8.22}$$

where f_1' and f_1'' denote central finite difference approximations.

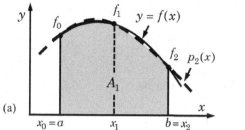 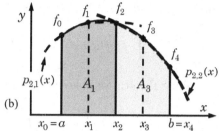

FIGURE 8.5: (a) Fitting a parabola to a two-panel interval, (b) fitting of two parabolas in a four-panel case.

Now, integrating the parabola on $[x_0, x_2]$ yields the following approximation for the area under the curve shown in Fig. 8.5a:

$$\int_{x_0}^{x_2} f(x)dx \cong \frac{h^3}{3}f_1'' + 2hf_1 = \frac{h}{3}(f_0 + 4f_1 + f_2) \qquad (8.23)$$

where $f_1'' \approx (f_2 - 2f_1 + f_0)/h^2$ is incorporated in Eq. (8.23). This approximation is referred to as the *Simpson's Rule* or *Simpson's 1/3 rule* due to the "1/3" factor.

To improve the accuracy of the numerical integration, we double the number of panels, i.e., halve the panel width. This time, two interpolating parabolas are required to integrate over $[x_0, x_4]$. As shown in Fig. 8.5b, the polynomials $p_{2,1}(x)$ and $p_{2,2}(x)$ are defined on the intervals $[x_0, x_2]$ and $[x_2, x_4]$, respectively. The area on $[x_0, x_4]$ is then obtained by employing Simpson's rule on the areas A_1 and A_3 as follows:

$$A_1 + A_3 = \int_{x_0}^{x_2} p_{2,1}(x)dx + \int_{x_2}^{x_4} p_{2,2}(x)dx \cong \frac{h}{3}(f_0 + 4f_1 + 2f_2 + 4f_3 + f_4) \qquad (8.24)$$

where $A_1 = (h/3)(f_0 + 4f_1 + f_2)$ and $A_3 = (h/3)(f_2 + 4f_3 + f_4)$.

In order to extend Simpson's rule to many panels, the number of panels obviously needs to be "*even*." Also, note that the integrands with odd and even subscripts are multiplied by 4 and 2, respectively. With this observation, a generalized approximation is obtained as

$$S_N = \frac{h}{3}\sum_{i=1}^{N/2}\left(f_{2i-2} + 4f_{2i-1} + f_{2i}\right) = \frac{h}{3}\left(f_0 + f_N + 4\sum_{\substack{i=1 \\ \text{odd}}}^{N-1} f_i + 2\sum_{\substack{i=2 \\ \text{even}}}^{N-2} f_i\right) \qquad (8.25)$$

where $f_i = f(x_i)$ and $x_i = a + ih$ for $i = 0, 1, 2, \ldots, N$. This equation, Eq. (8.25), is called *N-panel* (or *Composite*) *Simpson's (1/3-)Rule*.

We anticipate the numerical integration with Simpson's rule to provide a marked improvement over the trapezoidal rule. Nonetheless, at this stage, we are not yet able to quantify either the approximation error or its magnitude. In order to assess the numerical error incurred, we resort to the Taylor series again. We expand $I(x)$, defined by Eq. (8.10), to Taylor series at x_i about x_{i-1} and x_{i+1}:

$$I(x_{i+1}) = I_{i+1} = I_i + hf_i + \frac{h^2}{2!}f_i' + \frac{h^3}{3!}f_i'' + \frac{h^4}{4!}f_i''' + \frac{h^5}{5!}f_i^{(4)} + \frac{h^6}{6!}f_i^{(5)} + \mathcal{O}(h^7) \qquad (8.26)$$

$$I(x_{i-1}) = I_{i-1} = I_i - hf_i + \frac{h^2}{2!}f_i' - \frac{h^3}{3!}f_i'' + \frac{h^4}{4!}f_i''' - \frac{h^5}{5!}f_i^{(4)} + \frac{h^6}{6!}f_i^{(5)} + \mathcal{O}(h^7) \qquad (8.27)$$

where we made use of $I'(x) = f(x)$, $I''(x) = f'(x)$, and so on.

Subtracting Eq. (8.27) from Eq. (8.26), we get

$$A_i = \int_{x_{i-1}}^{x_{i+1}} f(x)dx = I_{i+1} - I_{i-1} = 2hf_i + \frac{h^3}{3}f_i'' + \frac{h^5}{60}f_i^{(4)} + \mathcal{O}(h^7) \tag{8.28}$$

Next, recalling the 4th-order-accurate CDF for f_i'' is given by

$$f_i'' = \frac{f_{i+1} - 2f_i + f_{i-1}}{h^2} - \frac{h^2}{12}f^{(4)}(\eta_i) + \mathcal{O}(h^4), \qquad x_{i-1} \leqslant \eta_i \leqslant x_{i+1} \tag{8.29}$$

We substitute Eq. (8.29) into Eq. (8.28) and simplify to obtain

$$A_i = \frac{h}{3}\left(f_{i+1} + 4f_i + f_{i-1}\right) - \frac{h^5}{90}f^{(4)}(\eta_i) + \mathcal{O}(h^7) \tag{8.30}$$

Note that the first term of Eq. (8.30) corresponds to the numerical integration of $f(x)$ on $[x_{i-1}, x_{i+1}]$ with the Simpson's rule; that is, A_i accounts for the area of the subinterval with two panels. Thus, to find the approximation of the integral in $[x_0, x_N]$, all we need to do is sum up the odd A_i's over $N/2$ subintervals.

A general expression for N-panels (provided that N is even) on $[a, b]$ yields

$$S_N = \sum_{i=1}^{N/2} A_{2i-1} = \frac{h}{3}\left(f_0 + f_N + 4\sum_{i=1}^{N/2} f_{2i-1} + 2\sum_{i=1}^{N/2-1} f_{2i}\right) - \frac{h^5}{90}\sum_{i=1}^{N/2} f^{(4)}(\eta_{2i-1}) + \frac{N}{2}\mathcal{O}(h^7) \tag{8.31}$$

The dominant error term in Eq. (8.31) can be treated in the same way as the dominant error term in the trapezoidal rule. Thus, the dominant error term yields

$$R_N = -\frac{h^5}{90}\sum_{i=1}^{N/2} f^{(4)}(\eta_{2i-1}) = -\frac{N}{2}\left(\frac{h^5}{90}f^{(4)}(\xi)\right) = -\frac{(b-a)}{180}h^4 f^{(4)}(\xi) \tag{8.32}$$

This last expression is the global error, which covers the entire range $[a, b]$. Also note that η_{2i-1} $(x_{2i-2} \leqslant \eta_{2i-1} \leqslant x_{2i})$ refers to an abscissa shared by two panels; therefore, the number of abscissas (terms) in the sum is $N/2$. As a result, the sum has been replaced by $(N/2)f^{(4)}(\xi)$ where $a \leqslant \xi \leqslant b$. By inspection of Eq. (8.32), the order of global error is said to be fourth order, $\mathcal{O}(h^4)$.

Finally, incorporating Eq. (8.32) in Eq. (8.31), an approximation for the definite integral of $f(x)$ on $[a, b]$ with the N-panel Simpson's rule can be written as

$$S_N = \frac{h}{3}\left(f_0 + f_N\right) + \frac{4h}{3}\sum_{\substack{i=1 \\ \text{odd}}}^{N-1} f_i + \frac{2h}{3}\sum_{\substack{i=2 \\ \text{even}}}^{N-2} f_i + \mathcal{O}(h^4) \tag{8.33}$$

which has the appropriate form to be applied to discrete functions. Equation (8.33) can also be cast for explicit continuous functions as

$$S_N = \frac{h}{3}\left(f(a) + f(b)\right) + \frac{4h}{3}\sum_{i=1}^{N/2} f(x_{2i-1}) + \frac{2h}{3}\sum_{i=1}^{N/2-1} f(x_{2i}) + \mathcal{O}(h^4) \tag{8.34}$$

Simpson's 1/3-rule can be expressed as a weighted-sum by definings $w_0 = w_N = h/3$, $w_i = 4h/3$ for odd i's, and $w_i! = 2h/3$ for even i's.

Simpson's Rule with End Correction: Since the dominant error term in Eq. (8.31) involves the fourth derivative, attempting to find an end correction by approximating this error term is impractical for several reasons. Nevertheless, an improved Simpson's formula can be found by constructing a fourth-degree polynomial approximation $p_4(x)$, which passes through the points (x_0, f_0), (x_1, f_1), and (x_2, f_2) (*see* Fig. 8.5a). To be able to define a unique polynomial, two additional constraints not requiring extra integration points are needed. To derive new constraints, we note that the trapezoidal rule with the end correction, Eq. (8.21), involves the first derivatives at the end points. Thus, we may also require that the first derivatives of the integrand at the end points of each double panel (f_0' and f_2') be known. Finally, the constraints sufficient to obtain a fourth-degree polynomial approximation can be summarized as follows:

$$f_0 = p_4(x_0), \quad f_1 = p_4(x_1), \quad f_2 = p_4(x_2), \quad f_0' = p_4'(x_0), \quad f_2' = p_4'(x_2),$$

The system of linear equations for five unknowns (i.e., the coefficients of $p_4(x)$) is solved by satisfying the five constraint equations above. Integrating the approximating polynomial, $p_4(x)$, over the two panels, $[x_0, x_2]$, yields

$$\int_{x_0}^{x_2} f(x)dx \cong \int_{x_0}^{x_2} p_4(x)dx = \frac{h}{15}\left(7f_0 + 16f_1 + 7f_2 + h\left(f_0' - f_2'\right)\right) + \frac{h^7 f^{(6)}(\eta)}{4725} \quad (8.35)$$

where $x_0 \leqslant \eta \leqslant x_2$. However, the error term in this equation has been derived from the Taylor series analysis. Note that this formula is analogous to the trapezoidal rule with end correction, which is derived by including the error term of the trapezoidal rule, whereas Eq. (8.35) is *not* based on an estimation of the dominant error term [21].

To obtain the composite rule, we construct $N/2$ continuous subintervals, each containing two panels. When the integrals over $[x_{2i-2}, x_{2i}]$ subintervals are summed up, the first derivatives pairwise cancel out at all interior points with the exception of the first and the last one. Using the shorthand notation, we finally obtain

$$CS_N = \frac{h}{15}\left(7\left(f_0 + f_N\right) + 16\sum_{\substack{i=1 \\ \text{odd}}}^{N-1} f_i + 14\sum_{\substack{i=2 \\ \text{even}}}^{N-2} f_i - h\left(f'(b) - f'(a)\right)\right) + \mathcal{O}(h^6) \quad (8.36)$$

which is referred to as *N-Panel* (or *Composite*) *Simpson's rule with end correction*. Note that N-panel Simpson's rule with end correction, which also requires an even number of panels, is *sixth-order* accurate, and it can be readily applied without much additional work.

A note on Simpson's Second Rule (3/8 rule): Another version of Simpson's rule is developed with three-panels (or equally spaced four integration points), as shown in Fig. 8.6. A cubic interpolating polynomial passing through all four points is constructed and then integrated over $[x_0, x_3]$ to yield

$$\widehat{S}_3 = \frac{3h}{8}\left(f_0 + 3f_1 + 3f_2 + f_3\right) - \frac{3h^5}{80}f^{(4)}(\xi), \quad x_0 < \xi < x_3 \quad (8.37)$$

where $h = (b-a)/3$ and $\widehat{S}_3$ is an approximation referred to as *Simpson's 3/8 rule*, named after the "3/8" factor.

This approximation can be extended to $N/3$ intervals (each containing 3 panels), which leads to *N-panel* (or *composite*) *Simpson's 3/8 rule*:

$$\widehat{S}_N = \frac{3h}{8}\left(f_0 + f_N + 3\sum_{i=1,4,7,.}^{N-2} f_i + 3\sum_{i=2,5,8,.}^{N-1} f_i + 2\sum_{i=3,6,9,.}^{N-3} f_i\right) + \mathcal{O}(h^4) \quad (8.38)$$

Simpson's 1/3 Rule

- Simpson's rule is relatively simple and easy to employ;
- Due to being 4th and 6th-order accurate methods, the S_N and CS_N yield the true integral values for polynomials of degree $m \leqslant 3$ and $m \leqslant 5$, respectively;
- Error analysis is straightforward.

- Integrands must be continuous;
- The number of panels is restricted to even numbers for the 1/3 rule and multiples of 3 for the 3/8 rule;
- The formula with the end correction cannot be applied to discrete functions unless the first derivatives are computed using a high-order finite difference formula ($\geqslant 6$).

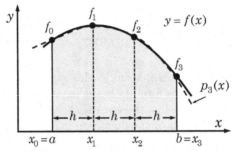

FIGURE 8.6: Simpson's three-panel integration.

which is a fourth-order formula with a global error of $R_N = -(b-a)h^4 f^{(4)}(\xi)/80$ that is larger than that of N-panel Simpson's 1/3 rule, $R_N = -(b-a)h^4 f^{(4)}(\xi)/180$. For this reason, whenever possible, Simpson's 1/3 rule should be preferred, instead of the 3/8 rule using four integral points, as it provides better accuracy with three integral points.

The total number of panels in the 3/8 rule should be in multiples of "3." In practice, when N is even, the 1/3 rule is preferred due to its smaller global error. When N is odd, the Simpson's 3/8 rule is only applied to the last three panels, while the 1/3 rule is employed for the rest.

A pseudomodule, **SIMPSONS_RULE_DF**, to estimate the integral of a uniformly spaced discrete function using composite Simpson's rule is given in Pseudocode 8.2. The module requires the number of panels (n), the panel width (h), and the discrete function f_i) for $i = 0, 1, 2, \ldots, n$ as input. The Simpson's 1/3 rule requires the number of panels to be "*even*." As an input validity check, it is prudent to test whether n is even or not right at the top of the module. (In cases where the number of panels is odd, the 1/3 rule is applied to all panels except the last three panels, and the 3/8 rule is adapted to the last three panels. Implementing this strategy in the module is left as an exercise for the reader.) The integral of the discrete function over (x_0, x_n) with the 1/3 rule (*intg*) is calculated by Eq. (8.33) using two accumulator loops (*odd* and *even* as the accumulator variables) for the odd- and even-subscripted integrands.

Pseudocode 8.2

Module SIMPSONS_RULE_DF $(n, h, \mathbf{f}, intg)$
\ DESCRIPTION: A pseudocode to estimate the integral of a uniformly spaced
\ discrete function $f(x)$ on $[x_1, x_n]$ using the composite Simpson's Rule.
Declare: f_n \ Declare array variables
$odd \leftarrow 0$ \ Initialize accumulator variable, odd
For $\left[i = 1, n-1, 2\right]$ \ Accumulator loop to accumulate f_i's for $i = 1, 3, 5, \ldots$
    $odd \leftarrow odd + f_i$ \ Accumulate f_i's with odd
End For
$even \leftarrow 0$ \ Initialize accumulator variable, $even$
For $\left[i = 2, n-2, 2\right]$ \ Accumulator loop to accumulate f_i's for $i = 2, 4, 6, \ldots$
    $even \leftarrow even + f_i$ \ Accumulate f_i's with, $even$
End For
$intg \leftarrow f_0 + f_n + 4 * odd + 2 * even$ \ Apply Eq.(8.33)
$intg \leftarrow intg * h/3$ \ Estimated integral value
End Module SIMPSONS_RULE_DF

A separete module is not presented here for the numerical integration of an explicitly defined continuous function. Such a module will require the number of panels (n) and the integration interval $[a, b]$ as input. Similar to Pseudocode 8.1, the integrand and its first derivative should be supplied to the module by a couple of user-defined functions. The abscissas and a set of discrete data points (i.e., integrand) should be prepared before the *even* and *odd*-subscripted f_i's are accumulated in accordance with Eq. (8.34), using the *odd* and *even* accumulator variables. If desired, the end correction formualtion, Eq. (8.36), can be implemented here.

EXAMPLE 8.2: Trapezoidal/Simpson's 1/3 rules with/without end correction

The surface area of the three-dimensional body generated by rotating the $y = 2\sqrt{\sin x}$ curve about the x-axis over $[0, \pi]$ is to be estimated. The problem setup leads to the following integral:

$$I = 4\pi \int_0^\pi \sqrt{\sin x + \cos^2 x}\, dx$$

Excluding the 4π factor, estimate the definite integral using uniform 4-, 6-, and 10-panel Trapezoidal and Simpson's rules *with* and *without* end correction. Discuss the accuracy of T_{10} and S_{10}? The true integral is 3.346868487781.

SOLUTION:

For $N = 10$, we have 11 integration points, whose abscissas are $x_i = ih$, where $h = (b-a)/N = \pi/10$. Since the number of panels is *even*, the Simpson's 1/3 rule is suitable for this exercise.

The T_{10} and S_{10} can be explicitly written as

$$T_{10} = h\left(\frac{f_0 + f_{10}}{2} + f_1 + f_2 + \ldots + f_8 + f_9\right)$$

$$S_{10} = \frac{h}{3}\left(f_0 + f_{10}\right) + \frac{4h}{3}\left(f_1 + f_3 + f_5 + f_7 + f_9\right) + \frac{2h}{3}\left(f_2 + f_4 + f_6 + f_8\right)$$

Both formulas can be expressed as weighted-sums, i.e., $I \approx \Sigma w_i f_i$.

TABLE 8.2

			Trapezoidal Rule		Simpson's Rule	
i	x_i	f_i	w_i	$w_i f_i$	w_i	$w_i f_i$
0	0	1	$h/2$	0.157079	$h/3$	0.104720
1	$\pi/10$	1.101601	h	0.346078	$4h/3$	0.461438
2	$2\pi/10$	1.114582	h	0.350156	$2h/3$	0.233438
3	$3\pi/10$	1.074481	h	0.337558	$4h/3$	0.450077
4	$4\pi/10$	1.023009	h	0.321388	$2h/3$	0.214259
5	$5\pi/10$	1	h	0.314159	$4h/3$	0.418879
6	$6\pi/10$	1.023009	h	0.321388	$2h/3$	0.214259
7	$7\pi/10$	1.074481	h	0.337558	$4h/3$	0.450077
8	$8\pi/10$	1.114582	h	0.350156	$2h/3$	0.233438
9	$9\pi/10$	1.101601	h	0.346078	$4h/3$	0.461438
10	π	1	$h/2$	0.157079	$h/3$	0.104720
			$\Sigma w_i f_i =$	**3.33867954**	$\Sigma w_i f_i =$	**3.34674092**

For $N = 10$, the abscissas x_i, corresponding integrands $f_i = f(x_i)$, and $w_i f_i$ products are calculated and tabulated in Table 8.2. The approximations for each method, expressed as weighted sums $\Sigma_i w_i f_i$, are available in the last row as 3.33867954 and 3.34674092 for the trapezoidal and Simpson's 1/3 rules, respectively. The corresponding true errors are 0.0081889 and 0.0001276 for T_{10} and S_{10}, respectively.

TABLE 8.3

Method	N	Estimate	Absolute Error
Trapezoidal Rule, T_N	4	3.29660530	0.0502632
	6	3.32428153	0.0225870
	10	3.33867954	0.0081889
Trapezoidal Rule, CT_N	4	3.34800949	0.0011410
(with end correction)	6	3.34712783	0.0002594
	10	3.34690421	0.0000357
Simpson's 1/3 Rule, S_N	4	3.34827618	0.0014077
	6	3.34578851	0.0010800
	10	3.34674092	0.0001276
Simpson's 1/3 rule, CS_N	4	3.34806282	0.0011943
(with end correction)	6	3.34685997	8.52×10^{-6}
	10	3.34687155	3.06×10^{-6}

The first derivative of the integrand at the end points, required for Eqs. (8.21) and (8.36), are respectively obtained as $f'(0) = 1/2$, $f'(\pi) = -1/2$. Using Eqs. (8.18), (8.21), (8.33), and (8.36), the T_N, CT_N, S_N, and CS_N are computed and presented in Table 8.3 for $N = 4$, 6, and 10. It is observed that as the number of panels is increased (or panel width is decreased), the estimates approach the true value regardless of the method used.

Discussion: As long as the number of panels is even (as in this case), the performance of Simpson's 1/3 rule for the same number of panels is better than that of the trapezoidal rule due to its fourth-order accuracy. On the other hand, the CT_N or CS_N approximations (with the end corrections) depict a significant improvement over the ordinary T_N or S_N approximations. The Simpson's 1/3 rule with end correction, having an order of error of $\mathcal{O}(h^6)$, yields superior estimates.

In Example 8.1, it was relatively easy to obtain $F''(x)$ to calculate the dominant error term. In this example, we have not attempted to determine the error bounds because the high-order derivatives of the integrand here are difficult to obtain and analyze. In reality, the typical error estimation of an approximation is determined by using two or more estimates with different numbers of panels, i.e., $|T_6 - T_4|$, $|T_{10} - T_6|$, or $|T_{10} - T_4|$, and so on. We could also use, say, T_{10}, S_{10}, or CT_{10} to compare with a better solution such as CS_{10} and use the difference to estimate the error, i.e., $|CT_{10} - T_{10}|$ or $|CT_{10} - S_{10}|$, or $|CT_{10} - CS_{10}|$, where theoretically CT_{10} is the most accurate one. If the error predicted in this manner satisfies a predetermined convergence criterion, then the integration procedure is considered complete.

 Even though Simpson's 1/3 rule is $\mathcal{O}(h^4)$, it may not always be superior to the trapezoidal rule. Recall that the global error of the Trapezoidal and Simpson's rules depends on the 2nd and 4th derivatives, respectively. In cases where the integrand is *not* smooth, higher derivatives and thus the truncation error may become very large. That is why, before applying *end correction* formulas, make sure that the integrand is sufficiently smooth; otherwise, there may be little or no advantage at all in using a so-called the "*better rule.*"

8.3 ROMBERG'S RULE

Romberg's rule is a powerful numerical technique to approximate a definite integral. It is a recursive procedure based on the trapezoidal rule that improves the numerical approximations by employing the Richardson extrapolation technique.

Before introducing the method, it is important to analyze the truncation error of the trapezoidal rule. For this purpose, consider a sufficiently differentiable continuous function, $f(x)$. The N-panel trapezoidal rule, Eq. (8.19), can be rewritten as

$$I = \frac{h}{2}\left(f(a) + f(b) + 2\sum_{i=1}^{N-1} f(a+ih)\right) + K_1 h^2 + K_2 h^4 + K_3 h^6 + \dots \tag{8.39}$$

where I is the true integral value, K_1, K_2, $K_3,\dots$ are the coefficients containing higher order derivatives of $f(x)$, and the first term on the rhs is the N-panel Trapezoidal rule, which is a function of panel width, i.e., $T_N(h)$.

Equation (8.39) can then be rewritten as

$$T_N(h) = I - K_1 h^2 - K_2 h^4 - K_3 h^6 + \dots \tag{8.40}$$

Now consider two numerical estimates, $T_N(h_1)$ and $T_M(h_2)$, with panel widths of h_1 and

h_2:

$$T_N(h_1) = I - K_1 h_1^2 - K_2 h_1^4 - K_3 h_1^6 + \ldots \tag{8.41}$$

$$T_M(h_2) = I - K_1 h_2^2 - K_2 h_2^4 - K_3 h_2^6 + \ldots \tag{8.42}$$

Next, using Eq. (8.41), we estimate the integral with twice the panel width (i.e., $h_1 = 2h_2$):

$$T_N(2h_2) = I - 4K_1 h_2^2 - 16K_2 h_2^4 - 64K_3 h_2^6 + \ldots \tag{8.43}$$

For h_2 case, setting $M = 2N$, we can then use a linear combination of Eq. (8.42) and Eq. (8.43) to eliminate $\mathcal{O}(h^2)$ terms as follows:

$$\frac{4T_{2N}(h_2) - T_N(2h_2)}{3} = I + 4K_2 h_2^4 + 20K_3 h_2^6 + \ldots = S_{2N}(h_2) \tag{8.44}$$

which becomes $\mathcal{O}(h^4)$. It should be noted that this approximation corresponds to the composite Simpson's 1/3 rule, which is $S_{2N}(h_2)$. The technique of improving the order of accuracy using the two estimates of I, is known as the *Richardson Extrapolation* (*see* Section 5.7).

This integral can also be extrapolated by setting $h_3 = h_2/2$:

$$\frac{4T_{4N}(h_3) - T_{2N}(2h_3)}{3} = I + 4K_2 h_3^4 + 20K_3 h_3^6 + \ldots = S_{4N}(h_3) \tag{8.45}$$

which is also fourth-order accurate.

The Richardson extrapolation technique can be applied to Simpson's rule at any stage to obtain an approximation of higher accuracy. For instance, in Eq. (8.45), setting $h_3 = h_2/2$ and using Eq. (8.44), the elimination of $\mathcal{O}(h^4)$ terms yields

$$\frac{16S_{4N}(h_2/2) - S_{2N}(h_2)}{15} = I - K_3 h_2^6 + \ldots$$

which is sixth-order accurate, i.e., $\mathcal{O}(h^6)$.

The Richardson extrapolation technique can be indefinitely repeated, with each step of the extrapolation yielding a new numerical estimate that is more accurate than the previous one, i.e., $\mathcal{O}(h^8)$, $\mathcal{O}(h^{10})$, etc. The successive evaluation of a numerical integral in this manner is known as *Romberg's Rule*. Now, we seek to express the preceding formulation in a suitable form that can be incorporated into a numerical algorithm. Since the first step of Romberg's rule is the trapezoidal rule, the first step will be expressed as

$$R_{k,1} = \frac{h}{2}\left(f(a) + f(b) + 2\sum_{i=1}^{M} f(a + ih)\right) \tag{8.46}$$

where $h = (b-a)/M$ is the panel width, M denotes the number of panels, which is defined as $M = 2^{k-1}$, and the subscript "k" controls the panel width. Note that for increasing k, the number of panels increases by two folds (or the panel width is cut by half), which is necessary to employ the Richardson extrapolation. For $k = 1$, 2 and 3 (or $M = 1$, 2 and 4), Eq. (8.46) can be expressed as

$$R_{1,1} = T_1 = \frac{b-a}{2}\Big(f(a) + f(b)\Big)$$

$$R_{2,1} = T_2 = \frac{b-a}{4}\left(f(a) + f(b) + 2f\left(a + \frac{b-a}{2}\right)\right) \tag{8.47}$$

$$R_{3,1} = T_4 = \frac{b-a}{8}\left(f(a) + f(b) + 2f\left(a + \frac{b-a}{4}\right) + 2f\left(a + \frac{b-a}{2}\right) + 2f\left(a + \frac{3(b-a)}{4}\right)\right)$$

It is clear that some of the function evaluations in $R_{k,1}$ overlap with those of $R_{k-1,1}$. To reduce the computational effort, for instance, Equation (8.47) can be rearranged to allow recursive representation as

$$R_{2,1} = \frac{R_{1,1}}{2} + \frac{b-a}{2} f\left(a + \frac{b-a}{2}\right) \tag{8.48}$$

$$R_{3,1} = \frac{R_{2,1}}{2} + \frac{b-a}{4}\left(f\left(a + \frac{b-a}{4}\right) + f\left(a + \frac{3(b-a)}{4}\right)\right) \tag{8.49}$$

Comparing Eqs. (8.48) and Eq. (8.49), a general recurrence relation for the trapezoidal rule is found as

$$R_{k,1} = \frac{R_{k-1,1}}{2} + h \sum_{i=1}^{M/2} f\left(a + (2i-1)h\right), \quad k = 2, 3, 4 \ldots \tag{8.50}$$

which provides an efficient way of incrementally computing the trapezoidal rule.

The Richardson extrapolation formula can now be generalized as follows:

$$R_{k,n} = \frac{4^{n-1} R_{k,n-1} - R_{k-1,n-1}}{4^{n-1} - 1} = R_{k,n-1} + \frac{R_{k,n-1} - R_{k-1,n-1}}{4^{n-1} - 1}, \quad n \geqslant 2 \tag{8.51}$$

where index n signifies the level of extrapolation, i.e., $n=1$ is linear (Trapezoidal), $n=2$ is quadratic (Simpson), and so on.

For $n = 2$ and 3, the Richardson extrapolations yield

$$R_{2,2} = \frac{4R_{2,1} - R_{1,1}}{3} + \mathcal{O}(h^4), \quad R_{3,2} = \frac{4R_{3,1} - R_{2,1}}{3} + \mathcal{O}(h^4), \ldots$$

$$R_{3,3} = \frac{16R_{3,2} - R_{2,2}}{15} + \mathcal{O}(h^6), \quad R_{4,3} = \frac{16R_{4,2} - R_{3,2}}{15} + \mathcal{O}(h^6), \cdots$$

The Romberg estimates, as depicted in Table 8.4, can be arranged in a tabular form, referred to as the *Romberg Table*. A Romberg table is a lower triangular form that extends down and across. It can be built either column-by-column or row-by-row. First, the estimates from the trapezoidal rule ($R_{k,1}$) are placed in the first column in the order of increasing k. The number of rows placed in the first column determines how large the table will be. The subsequent columns are obtained from Eq. (8.51) with only a few arithmetic operations. In fact, the majority of the computations (i.e., cpu-time) in Romberg's rule are devoted to construct the first column. The panel width for the first row is $h = b - a$, and in the subsequent rows, the panel width is halved. Because the first column corresponds to the composite trapezoidal rule, the order of accuracy of the first column is $\mathcal{O}(h^2)$. Then, $R_{k,2}$'s corresponding to the composite Simpson's rule with $\mathcal{O}(h^4)$ are placed into the second column. With each subsequent column added, the order of error increases by 2. Notice that the order of accuracy of each column has been identified at the top of the Romberg table in Table 8.4. Arrows indicate the direction of improvements in the estimates. The accuracy of the estimates increases as we go down each column due to decreasing h. As we move to the right along each row (or diagonally), the accuracy increases due to the increasing order of accuracy. Hence, a criterion can be established by taking advantage of the fact that the order of accuracy increases in a specific row to estimate an integral within a certain tolerance (ε) according to Romberg's rule. Using Eq. (8.51), the absolute difference between two subsequent estimates in a row can be written as

$$|R_{n,n} - R_{n,n-1}| = \frac{|R_{n,n-1} - R_{n-1,n-1}|}{4^{n-1} - 1} < \varepsilon$$

Romberg's Rule

- Romberg's rule is simple and easy to use;
- It is applied to explicitly given continuous functions;
- It is ideal for adaptive integration;
- It converges quickly for smooth functions;
- The number of function evaluations is very small for the level of accuracy achieved;
- The table can be constructed one row at a time, so the analyst does not have to determine the order of accuracy in advance.

- It is customarily not employed for uniformly spaced discrete functions unless the number of panels is multiples of two ($M = 2^n$) in which case the largest size of a Romberg table is limited by the available data;
- It cannot handle non-uniformly spaced discrete functions;
- For a large integration interval ($h = b - a$), the table size can become very large, making it susceptible to the ill-effects of round-off errors;
- It is not suitable for improper integrals (integrands with discontinuities or integrals with infinite or semi-infinite domains).

TABLE 8.4: Construction of Romberg table.

h	$\mathcal{O}(h^2)$	$\mathcal{O}(h^4)$	$\mathcal{O}(h^6)$	$\mathcal{O}(h^8)$	$\cdots$	$\mathcal{O}(h^{2n-2})$	$\mathcal{O}(h^{2n})$
$(b-a)/2^0$	$R_{1,1}$						
$(b-a)/2^1$	$R_{2,1}$	$R_{2,2}$					
$(b-a)/2^2$	$R_{3,1}$	$R_{3,2}$	$R_{3,3}$	$\searrow$			
$(b-a)/2^3$	$R_{4,1}$	$R_{4,2}$	$R_{4,3}$	$R_{4,4}$			
$\vdots$	$\downarrow$	$\vdots$	$\vdots$	$\vdots$	$\ddots$		
$\vdots$	$\vdots$	$\vdots$	$\rightarrow$	$\vdots$	$\cdots$	$R_{n-1,n-1}$	
$(b-a)/2^{n-1}$	$R_{n,1}$	$R_{n,1}$	$R_{n,3}$	$R_{n,4}$	$\cdots$	$R_{n,n-1}$	$R_{n,n}$

The extrapolated estimates along the diagonal approach the true answer much more rapidly. As a more conservative criterion, the difference of the diagonal values is computed; that is,

$$\frac{|R_{n,n} - R_{n-1,n-1}|}{2^{n-1}} < \varepsilon$$

The size of the resulting table for a particular integral depends on the size of the integration interval $[a, b]$ and the tolerance desired (ε).

A pseudomodule, ROMBERGS_RULE, estimating the definite integral of $f(x)$ on $[a, b]$ using Romberg's rule is presented in Pseudocode 8.3. As input, this module requires the integration interval $[a, b]$ and a tolerance (ε). The integrand, $f(x)$, is supplied as a user-defined function, FX(x). The module output, *intg*, is the final estimate for the definite integral. First, a one-panel Trapezoidal estimate, $R_{1,1}$, is found, and the rest of the Romberg table is constructed on a row-by-row basis using a **While**-loop (k-loop, $k = 2, 3, \ldots$). The $R_{k,1}$ is evaluated by Eq. (8.50), and the next level estimates corresponding to the kth row are computed by the Richardson extrapolation formula, Eq. (8.51). The **While**-loop

Pseudocode 8.3

Module ROMBERGS_RULE $(a, b, \varepsilon, intg)$
\ DESCRIPTION: A pseudocode to estimate the integral of $f(x)$ on $[a, b]$ within
\ ε tolerance using Romberg's rule.
\ USES:
\   ABS :: A built-in function computing the absolute value;
\   FX:: A user-defined function module supplying $f(x)$.
Declare: $R_{10,10}$ \ Declare max. table size as 10×10
\ **R** is an internal double-subscripted array (table) containing the Romberg table
$k \leftarrow 1$ \ Initialize row number
$h \leftarrow b - a$ \ Initialize panel width
$R_{1,1} \leftarrow (\mathbf{FX}(a) + \mathbf{FX}(b)) * h/2$ \ Apply one-panel Trapezoidal rule
$err \leftarrow 1$ \ Initialize error
While $\left[err > \varepsilon \right]$ \ Execute loop until $err < \varepsilon$
    $k \leftarrow k + 1$ \ Find current row number
    $m \leftarrow 2^{k-2}$ \ Number of panels for kth row
    $h \leftarrow h/2$ \ Halve the panel width
    $sums \leftarrow 0$ \ Initialize accumulator, $sums$
    For $\left[i = 1, m \right]$ \ Loop i: Apply Trapezoidal rule to 1st column
        $xi \leftarrow a + (2i - 1) * h$ \ Evaluate abscissas
        $sums \leftarrow sums + \mathbf{FX}(xi)$ \ Accumulate non-overlapping terms
    End For
    $R_{k,1} \leftarrow R_{k-1,1}/2 + h * sums$ \ Apply Eq. (8.50)
    For $\left[n = 2, k \right]$ \ Apply Richardson extrapolation
        $R_{k,n} \leftarrow R_{k,n-1} + (R_{k,n-1} - R_{k-1,n-1})/(4^{n-1} - 1)$
    End For
    $err \leftarrow |R_{k,k} - R_{k-1,k-1}|/2^{k-1}$ \ Find estimated error
End While
$intg \leftarrow R_{k,k}$ \ Set $R_{k,k}$ as the final estimate
End Module ROMBERGS_RULE

operates until the error term is reduced below the user-specified tolerance value, $err = |R_{k,k} - R_{k-1,k-1}|/2^{k-1} < \varepsilon$.

EXAMPLE 8.3: Employing the Romgerg's rule to estimate area

Consider a lamina that is bounded by Ar-
cihmedes' spiral $(r = \theta, y \geqslant 0)$ and $y = 0$
line corresponding to Region (D) in the xy-
coordinate system, as depicted in the figure.
The polar inertia moment for the lamina leads
to the following integral:

$$\int_0^\pi \theta^2 \sqrt{1 + \theta^2} d\theta$$

For the given integral, construct Romberg's table up to $\mathcal{O}(h^{10})$.

SOLUTION:

Recall that, in the Romberg table, the order of global error increases by 2 as we move from one column to another.

We roughly estimate that the table should be 5×5 by observing $\mathcal{O}(h^{2n}) \equiv \mathcal{O}(h^{10})$. Setting $f(\theta) = \theta^2\sqrt{1+\theta^2}$, one-panel trapezoidal rule gives

$$R_{1,1} = \frac{\pi-0}{2}\left(f(\pi)+f(0)\right) = \frac{\pi}{2}(32.53918+0) = 51.112425$$

Then, using Eq. (8.50), the first column of the Romberg table is constructed as follows:

$$R_{2,1} = \frac{R_{1,1}}{2} + \frac{b-a}{2}f\left(a+\frac{b-a}{2}\right)$$
$$= \frac{51.112425}{2} + \frac{\pi-0}{2}f\left(\frac{\pi}{2}\right) = 25.556212 + 7.217083 = 32.773295$$
$$R_{3,1} = \frac{R_{2,1}}{2} + \frac{b-a}{4}\left[f\left(a+\frac{b-a}{4}\right)+f\left(a+\frac{3(b-a)}{4}\right)\right]$$
$$= \frac{32.773295}{2} + \frac{\pi-0}{4}\left[f\left(\frac{\pi}{4}\right)+f\left(\frac{3\pi}{4}\right)\right] = 28.163282$$

Likewise, $R_{4,1}$ and $R_{5,1}$ are also evaluated from Eq. (8.50).

The second column ($n=2$) corresponding to Simpson's rule is obtained using Eq. (8.51) as follows:

$$R_{2,2} = \frac{4R_{2,1}-R_{1,1}}{3} = \frac{4(32.773295)-51.112426}{3} = 26.660252$$
$$R_{3,2} = \frac{4R_{3,1}-R_{2,1}}{3} = \frac{4(28.163282)-32.773295}{3} = 26.626611$$

and so on.

The third column entries are similarly calculated using Eq. (8.51) with $n=3$ as

$$R_{3,3} = \frac{4^2 R_{3,2}-R_{2,2}}{4^2-1} = \frac{16(26.626611)-26.660252}{15} = 26.624369$$
$$R_{4,3} = \frac{4^2 R_{4,2}-R_{3,2}}{4^2-1} = \frac{16(26.618893)-26.626611}{15} = 26.618379$$

TABLE 8.5

h	$\mathcal{O}(h^2)$	$\mathcal{O}(h^4)$	$\mathcal{O}(h^6)$	$\mathcal{O}(h^8)$	$\mathcal{O}(h^{10})$
π	51.112426				
$\pi/2$	32.773295	26.660252			
$\pi/2^2$	28.163282	26.626611	26.624369		
$\pi/2^3$	27.004991	26.618893	26.618379	26.618284	
$\pi/2^4$	26.714874	26.618168	26.618120	26.618116	26.618115

Finally, the fourth and fifth column entries are filled out using Eq. (8.51), with $n=4$ and $n=5$:

$$R_{4,4} = \frac{4^3 R_{4,2}-R_{3,3}}{4^3-1} = \frac{64(26.618379)-26.624369}{63} = 26.618284$$

All entries in the Romberg table are calculated and presented in Table 8.5. The numerical values, computed with high precision, are reported for convenience only in six-decimal places. The best estimate is 26.618115.

Discussion: The order of error for $R_{5,5}$ is $\mathcal{O}(h^{10}) \equiv \mathcal{O}(8.517 \times 10^{-8}) = K \times 10^{-8}$, where K is a constant involving $f^{(20)}(\xi)$. If $K < 1/2$, then $R_{5,5}$ will be correct to 8-decimal places but the magnitude of error for $R_{5,5}$ increases for $K > 1/2$. We also note that $f(\theta)$ is an increasing function on $[0,\pi]$, $0 \leqslant f(\theta) < 32.54$. High-order derivatives of $f(\theta)$ will tend to be large as well, making it more likely that K will be much larger than $1/2$.

To assess the accuracy of the best estimate, we compute the absolute differences between estimates along the last row and diagonal, which give $|R_{5,5} - R_{4,4}| = 0.1685 \times 10^{-3} < 0.5 \times 10^{-3}$ and $|R_{5,5} - R_{5,4}| = 0.658 \times 10^{-6} < 0.5 \times 10^{-5}$, respectively. These error estimates suggest that $R_{5,5}$ is accurate by at least three decimal places in the worst scenario. The true solution is

$$I_{\text{true}} = \frac{1}{8}\left(\pi(1 + 2\pi^2)\sqrt{1 + \pi^2} - \sinh^{-1}\pi\right) = 26.6181187059562$$

Using the true value, we can now find the true error of the best estimate as $E = |I_{true} - R_{5,5}| = 0.352 \times 10^{-5} < 0.5 \times 10^{-5}$, which indicates that the best estimate is in fact five-decimal place accurate. This example illustrates that it is not possible to predict in advance the size of the Romberg table corresponding to a certain level of accuracy.

8.4 ADAPTIVE INTEGRATION

The majority of the computation time of any numerical integration is devoted to function evaluations, which are associated with the number of panels or integration points used. By now, we have learned that the numerical estimate of an integral improves as the number of panels is increased, i.e., panel width is reduced. Moreover, when an integrand is smooth (varies slowly) over its integration range, highly accurate estimates can be achieved with a relatively small number of panels. But an integrand with sharp changes in short subintervals requires more panels to accurately approximate rapid or sharp variations. High-order derivatives of such integrands depict rapid changes as well. For this reason, it is difficult to determine the adequate number of panels (or optimum panel width) needed to ensure the desired accuracy. Consequently, using a small and uniform panel width everywhere may result in a very large number of panels over the entire integration domain, including the parts of the domain where the integrand varies slowly or mildly.

Adaptive integration technique is a powerful tool to estimate a definite integral within a desired tolerance with minimum computation time. The numerical technique discussed in this section is based on Simpson's 1/3 rule. However, this procedure can be easily extended to the trapezoidal rule or other suitable numerical integration methods.

Consider the numerical integration of $f(x)$ over the entire interval $[a, b]$ using the two-panel Simpson's 1/3 rule (Fig. 8.5a):

$$I(a,b) = \int_a^b f(x)dx = S(a,b) - \frac{h^5}{90}f^{(4)}(\xi_0) + \dots \quad (a < \xi_0 < b) \tag{8.52}$$

and

$$S(a,b) = \frac{b-a}{6}\left(f(a) + 4f(c) + f(b)\right) \tag{8.53}$$

where $h = (b-a)/2$ and $c = (a+b)/2$ is the midpoint. Notice that the truncation error in Eq. (8.52) involves the fourth derivative of the integrand. Evaluating the pertinent derivatives is not an efficient way of estimating the integration error. That is why a simple criterion is required to determine if the panel width is suitable for a particular subpanel by estimating the integration error without requiring $f^{(4)}(x)$.

By halving the integral and approximating each half with Simpson's rule (*see* Fig. 8.5b), we find

$$\int_a^c f(x)dx + \int_c^b f(x)dx = S(a,c) - \frac{(h/2)^5}{90}f^{(4)}(\xi_1) + S(c,b) - \frac{(h/2)^5}{90}f^{(4)}(\xi_2) + \dots$$

$$= S(a,c) + S(c,b) - \frac{h^5}{16}\frac{f^{(4)}(\xi_3)}{90} + \dots, \qquad a < \xi_3 < b \tag{8.54}$$

where the generalized intermediate value theorem has been applied in the second line to merge the two error terms.

The difference between the errors of the single-interval (two-panel) estimate $S(a,b)$ and the two-subinterval (four-panel) estimate $(S(a,c) + S(c,b))$ is found by subtracting Eq. (8.54) from Eq. (8.52):

$$S(a,b) - S(a,c) - S(c,b) = \frac{h^5}{90}f^{(4)}(\xi_0) - \frac{h^5}{16}\frac{f^{(4)}(\xi_3)}{90} + \dots \approx 15\left(\frac{h^5}{16}\frac{f^{(4)}(\xi_0)}{90}\right) + \dots \tag{8.55}$$

where we assumed $f^{(4)}(\xi_0) \approx f^{(4)}(\xi_3)$. The error term in Eq. (8.55) is 15 times that of Eq. (8.54). This observation provides a suitable *stopping criterion* that does not require $f^{(4)}(x)$. By incorporating a user-supplied tolerance (ε), a criterion for terminating the numerical integration for $[a,b]$ can be expressed as

$$\frac{1}{15}\left|S(a,c) + S(c,b) - S(a,b)\right| < \varepsilon \tag{8.56}$$

 In general, the 1/15-factor in Eq. (8.56) is replaced with a more conservative 1/10-factor to compensate for the error in the $f^{(4)}(\xi_0) \approx f^{(4)}(\xi_3)$ assumption.

A pseudo-recursive function module, **ADAPTIVE_SIMPSON**, for recursively evaluating the definite integral of an explicitly defined function within a tolerance value is presented in Pseudocode 8.4. The module arguments are the integration interval (a,b) and the accuracy tolerance desired (ε). The module uses the accompanying function module **SIMPSON** for estimating the definite integral of $f(x)$ on (a,b) using a two-panel Simpson's 1/3 rule, i.e., Eq. (8.53). The integration procedure starts with an estimate for the entire (initial) interval: $S(a,b)$. Then the integral is split into two subintegrals of equal width $(S(a,c)$ and $S(c,b))$, each computed within $\varepsilon/2$ tolerance. When the stopping criterion, Eq. (8.56), is satisfied for $[a,b]$, the integral has been estimated within the prescribed tolerance. We can then pass on to the next subinterval. If the stopping criterion is not met, then the integral, as before, is split into two subintegrals of equal width until convergence is achieved for each subintegral. This procedure is repeated for the remaining subintegrals.

Adaptive Integration Methods

- Adaptive integration is suitable for recursive programming;
- Trapezoidal or Simpson's rules can be efficiently implemented into adaptive integration algorithms;
- Functions depicting sharp changes can be effectively integrated;
- The accuracy of the final estimate can be predetermined by imposing a tolerance value.

- It is applied to explicitly defined continuous functions;
- It is not suitable for highly oscillating integrands, improper integrals over infinite-, or semi-infinite intervals, or discrete functions.

Pseudocode 8.4

Recursive Function Module ADAPTIVE_SIMPSON (a, b, ε)
\ DESCRIPTION: A recursive pseudo-function module to estimate the integral
\   of $f(x)$ on $[a, b]$ using adaptive Simpson algorithm.
\ USES:
\   SIMPSON:: A function module applying a two-panel Simpson's rule.
$c \leftarrow (a + b)/2$ \ Find midpoint of $[a, b]$
$Sab \leftarrow$ **SIMPSON**(a, b) \ Two-panel estimate on $[a, b]$
$Sac \leftarrow$ **SIMPSON**(a, c) \ Two-panel estimate on $[a, c]$
$Scb \leftarrow$ **SIMPSON**(c, b) \ Two-panel estimate on $[c, b]$
$S2ab \leftarrow Sac + Scb$ \ Four-panel estimate on $[a, b]$
$error \leftarrow |S2ab - Sab|/15$ \ Find error estimate
If $\left[error < \varepsilon\right]$ **Then**   \ Converged; Terminate integration
     ADAPTIVE_SIMPSON$\leftarrow S2ab$
Else   \ Not converged, halve interval & continue with $\varepsilon/2$ in each half
     ADAPTIVE_SIMPSON$\leftarrow$ ADAPTIVE_SIMPSON$(a, c, \varepsilon/2)$
               $+$ADAPTIVE_SIMPSON$(c, b, \varepsilon/2)$
End If
End Recursive Function Module ADAPTIVE_SIMPSON

Function Module SIMPSON (a, b)
\ DESCRIPTION: A pseudo-function to estimate the integral of $f(x)$ on $[a, b]$
\   using a two-panel Simpson's 1/3 rule.
\ USES:
\   FX:: A user-defined function supplying $f(x)$.
$c \leftarrow (a + b)/2$ \ Find midpoint of [a,b]
SIMPSON$\leftarrow$ (**FX**$(a) + 4 *$ **FX**$(c) +$ **FX**$(b)) * (b - a)/6$
End Function Module SIMPSON

EXAMPLE 8.4: Adaptive integration of Maxwell-Boltzmann distribution

The normalized Maxwell-Boltzmann distribution, for energy, is given by

$$f(E) = \frac{2\pi\sqrt{E}}{(\pi kT)^{3/2}} e^{-E/kT}, \qquad 0 \leqslant E < \infty$$

where k is Boltzmann's constant, T is the temperature, and E is the energy of a molecule. Setting $x = E/kT$, the fraction of the molecules having energies between 0 and $4kT$ can be expressed by the following integral:

$$\frac{2}{\sqrt{\pi}} \int_{x=0}^{4} \sqrt{x}\, e^{-x} dx$$

Use adaptive Simpson's rule to estimate the integral within $\varepsilon = 5 \times 10^{-3}$.

SOLUTION:

The starting interval is $[0,4]$. Defining $f(x) = \sqrt{x}\, e^{-x}$, we first estimate the integral with two-panel and four-panel Simpson's rule (Eqs. (8.53) and (8.54)):

$$S(0,4) = \frac{4-0}{6}(f(0) + 4f(2) + f(4)) = 0.5348022$$

$$S(0,2) + S(2,4) = \frac{2-0}{6}(f(0) + 4f(1) + f(2)) + \frac{4-2}{6}(f(2) + 4f(3) + f(4))$$
$$= 0.5543036 + 0.1909864 = 0.745290$$

Then, the first approximation error is found as

$$\frac{1}{15}|S(0,4)-S(0,2)-S(2,4)| = \frac{1}{15}|0.5348022-0.745290| = 0.0140325 > \varepsilon$$

The first approximation fails to be within the prescribed tolerance, so we divide $[0,2]$ into two subintervals, $[0,1]$ and $[1,2]$. Since $S(0,2)$ was previously computed, we need to evaluate only $S(0,1) + S(1,2)$, which gives

$$S(0,1) + S(1,2) = \frac{1-0}{6}(f(0) + 4f(0.5) + f(1)) + \frac{2-1}{6}(f(1) + 4f(1.5) + f(2))$$
$$= 0.3472345 + 0.2753971 = 0.6226316$$

The computed approximation error for $S(0,2)$ is

$$\frac{1}{15}|S(0,2)-S(0,1)-S(1,2)| = \frac{1}{15}|0.5543036-0.6226316| = 0.0045552 > \frac{\varepsilon}{2}$$

which also fails to be within the desired tolerance.

Next, dividing the interval $[0,1]$ into two subintervals, $[0,0.5]$ and $[0.5,1]$, and estimating the integral from $S(0,0.5) + S(0.5,1)$ yields

$$S(0,0.5)+S(0.5,1)=0.1655403+0.2027573 = 0.3682976$$

$$\frac{1}{15}|S(0,1)-S(0,0.5)-S(0.5,1)| = \frac{1}{15}|0.3472345-0.3682976| = 1.404\times10^{-3} > \frac{\varepsilon}{4}$$

Now, dividing the $[0, 0.5]$ interval into $[0, 0.25]$ and $[0.25, 0.5]$ intervals, and estimating the integral from the results in

$$S(0,0.25)+S(0.25,0.50)=0.0682266+0.1042413=0.1724679$$

$$\frac{1}{15}|S(0,0.5)-S(0,0.25)-S(0.25,0.5)| = \frac{1}{15}|0.1655403-0.1724679| = 4.62\times10^{-4} < \frac{\varepsilon}{8}$$

which satisfies the stopping criterion. This means that we have achieved the integration on $[0,0.5]$ with sufficient accuracy, and there is no need to further subdivide this interval.

Now, we proceed to compute the integral on $[0.5,1]$, which is obtained by $S(0.5, 0.75) + S(0.75, 1)$ as

$$\frac{1}{15}\left|S(0.5,1)-S(0.5,0.75)-S(0.75,1)\right| = \frac{1}{15}\left|0.2027573-0.2028052\right| = 3.19 \times 10^{-6} < \frac{\varepsilon}{8}$$

Since it meets the stopping criterion, we proceed to compute the integral on $[1,2]$ by $S(1, 1.5) + S(1.5, 2)$ as

$$S(1,1.5)+S(1.5,2) = 0.1602038+0.1153496 = 0.2755534$$

$$\frac{1}{15}\left|S(1,2)-S(1,1.5)-S(1.5,2)\right| = \frac{1}{15}\left|0.2753971-0.2755534\right| = 1.04 \times 10^{-5} < \frac{\varepsilon}{2}$$

The numerical integration on $[1,2]$ also satisfies the tolerance, so we can advance to the integration on $[2,4]$. The two-panel estimate for $[2,4]$ was computed before; thus, the four-panel estimate and the error of approximation are obtained by

$$S(2,3)+S(3,4) = 0.0794647+0.0533392 = 0.1328039$$

$$\frac{1}{15}\left|S(2,4)-S(2,3)-S(3,4)\right| = \frac{1}{15}\left|0.1327963-0.1328039\right| = 3.32 \times 10^{-6} < \frac{\varepsilon}{2}$$

Because the approximation error is well below the prescribed tolerance, the numerical integration procedure is terminated.

Finally, the estimate for the integral is obtained by summing up the more accurate estimates obtained for sub-integrals, leading to

$$S(0,0.5) + S(0.5,1) + S(1,2) + S(2,4) = 0.8417632$$

The true integral is $\sqrt{\pi}/2\,\mathrm{erf}(2) - 2e^{-4} = 0.845450112984953$, where erf denotes the error function. The true error is then obtained as

$$\left|I - S(0,0.5) - S(0.5,1) - S(1,2) - S(2,4)\right| = 3.687 \times 10^{-3} < \varepsilon$$

which is more accurate than the tolerance desired.

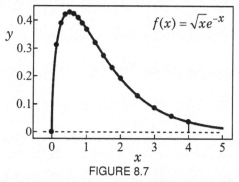

FIGURE 8.7

Discussion: In this example, the two-panel Simpson's rule, $S(a,b)$, was applied 15 times, which resulted in 45 function evaluations. As marked on the curve in Fig. 8.7, this procedure places extra integration points as needed in the intervals of the domain where the integrand exhibits sharp variation, as in $0.5 < x < 2$.

Note that the order of error for $Sab = S(a,b)$ is $\mathcal{O}(h^4)$, while for $S2ab = S(a,c) + S(c,b)$ it is $\mathcal{O}(h^4/16)$. Accordingly, using Sab and $S2ab$, the Richardson extrapolation may also be employed to further improve the accuracy of the integration, as follows:

$$I(a,b) \cong \frac{16 \times S2ab - Sab}{15} + \mathcal{O}(h^6)$$

In Pseudocode 8.4, the term $S2ab$ in the ADAPTIVE_SIMPSON← $S2ab$ statement should be replaced with the above expression.

8.5 NEWTON-COTES RULES

The Newton-Cotes rules or formulas have an important place among numerical integration methods. There are basically two kinds of rules (*open* or *closed*) that make up a family of formulas for the numerical integration of uniformly spaced integration points. This distinction is based on the choice of abscissas used in constructing the approximation polynomial for the interval $[x_0, x_n]$.

8.5.1 CLOSED NEWTON-COTES RULES

Any *closed* formula can be derived by constructing a Lagrange interpolating polynomial of nth degree, from a set of $n+1$ uniformly spaced discrete integration points, i.e., (x_0, f_0), (x_1, f_1), ..., (x_n, f_n), where $x_k = x_0 + kh$ for $k = 0, 1, \ldots, n$ with $h = (x_n - x_0)/n$. An approximation formula is then obtained by integrating the interpolating polynomial over $[x_0, x_n]$. For example, the procedure for a set of two points, (x_0, f_0) and (x_1, f_1), leads to

$$\int_{x_0}^{x_1} f(x)dx = \int_{x_0}^{x_1} p_1(x)dx = \int_{x_0}^{x_1} \left(\frac{x - x_1}{x_0 - x_1} f_0 + \frac{x - x_0}{x_1 - x_0} f_1 \right) dx = \frac{h}{2}(f_0 + f_1)$$

which is the trapezoidal rule with $h = x_1 - x_0$.

A general family of resulting numerical approximations is called *closed Newton-Cotes rules* because they include the integrands at both ends, i.e., f_0 and f_n. A few of the closed Newton-Cotes rules, along with the local error terms, are as follows:

$$\int_{x_0}^{x_2} f(x)dx = \frac{h}{3}\left(f_0 + 4f_1 + f_2\right) - \frac{h^5}{90}f^{(4)}(\xi) \qquad \text{Simpson's 1/3 } (n = 2)$$

$$\int_{x_0}^{x_3} f(x)dx = \frac{3h}{8}\left(f_0 + 3f_1 + 3f_2 + f_3\right) - \frac{3h^5}{80}f^{(4)}(\xi) \qquad \text{Simpson's 3/8 } (n = 3)$$

$$\int_{x_0}^{x_4} f(x)dx = \frac{2h}{45}\left(7f_0 + 32f_1 + 12f_2 + 32f_3 + 7f_4\right) - \frac{8h^7}{945}f^{(6)}(\xi) \qquad \text{Boole's rule } (n = 4)$$

where $x_0 < \xi < x_n$. Higher-order formulas can be found in Ref. [1].

We notice that the trapezoidal, Simpson's-1/3, and -3/8 rules are indeed closed Newton-Cotes formulas, derived by interpolating polynomials of degree $n = 1, 2,$ and 3, respectively. To obtain Boole's rule, a 4th-degree interpolating polynomial is used. The order of accuracy for even n is $n+3$, whereas for odd n is $n+2$. Upon employing the composite rule for integrals of uniformly spaced discrete functions over a general interval $[a, b]$, the order of accuracy decreases by one.

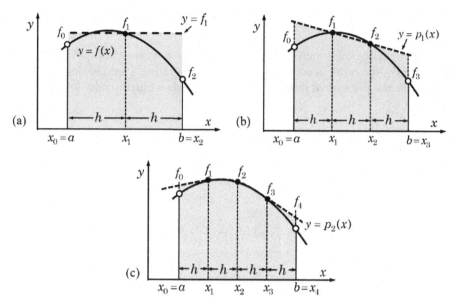

FIGURE 8.8: Graphical depiction of (a) two- (b) three- and (c) four-panel open Newton-Cotes abscissas and approximating polynomials (in black).

8.5.2 OPEN NEWTON-COTES RULES

Some integrands may be discontinuous at the lower, upper, or both ends of the integration interval. These integrals cannot be estimated with the "*closed*" Newton-Cotes formulas since they require the integrand at both ends, i.e., f_0 and f_n. Alternative formulas that avoid the integrands at end points are referred to as the *open Newton-Cotes rules* [1].

An *open* rule is derived by constructing a Lagrange interpolating polynomial that passes through a set of $n-1$ uniformly spaced discrete integration points, avoiding the end points, i.e., (x_1, f_1), (x_2, f_2), $\ldots$, (x_{n-1}, f_{n-1}). As illustrated earlier, the interpolating polynomial is integrated over $[x_0, x_n]$ to find an approximation formula.

The simplest case in this family, also known as the *midpoint rule*, requires two rectangular panels of width h. The approximate integral of $f(x)$ over $x_0 \leqslant x \leqslant x_2$ is set to the mid-point value $p_0(x) = f_1$ (i.e., height) times the interval width, $2h$ (*see* Fig. 8.8a):

$$\int_{x_0}^{x_2} f(x)dx = 2hf_1 + \frac{h^3}{3}f''(\xi), \qquad x_0 < \xi < x_2 \tag{8.57}$$

where $h^3 f''(\xi)/3$ is the leading truncation error. Notice that the integration error would be zero if $f''(\xi) = 0$ implying that the numerical estimate is exact if the integrand is constant or a straight line.

Another rule can also be obtained by dividing the integration interval into three panels of uniform width (*see* Fig. 8.8b). Excluding the end points, the interpolating polynomial passing through (x_1, f_1) and (x_2, f_2) is a straight line: $p_1(x) = (f_2(x - x_1) + f_1(x_2 - x))/h$. Integrating $p_1(x)$ over $[x_0, x_3]$ gives a second-order formula (i.e., *two-point rule*):

$$\int_{x_0}^{x_3} f(x)dx = \frac{3h}{2}(f_1 + f_2) + \frac{3h^3}{4}f''(\xi), \qquad x_0 < \xi < x_3 \tag{8.58}$$

where $3h^3 f''(\xi)/4$ is the leading error term. Note that even though the three-panel approximates the integrand with a straight line, the order of error is the same as the two-panel rule (i.e., Eq. (8.58)).

As the number of panels increases, the degree of the Lagrange interpolating polynomials and, consequently, the order of accuracy increases. Integrating a second-degree interpolating polynomial over the prescribed interval (Fig. 8.8c) yields a fourth-order formula (i.e., *three-point rule*):

$$\int_{x_0}^{x_4} f(x)dx = \frac{4h}{3}(2f_1 - f_2 + 2f_3) + \frac{14h^5}{45}f^{(4)}(\xi), \qquad x_0 < \xi < x_4 \qquad (8.59)$$

Higher-order formulas can be similarly developed; *four-point* (4th order) and *five-point* (6th order) *rules* are given as

$$\int_{x_0}^{x_5} f(x)dx = \frac{5h}{24}(11f_1 + f_2 + f_3 + 11f_4) + \frac{95h^5}{144}f^{(4)}(\xi), \qquad x_0 < \xi < x_5 \qquad (8.60)$$

$$\int_{x_0}^{x_6} f(x)dx = \frac{3h}{10}(11f_1 - 14f_2 + 26f_3 - 14f_4 + 11f_5) + \frac{41h^7}{140}f^{(6)}(\xi), \qquad x_0 < \xi < x_6 \qquad (8.61)$$

The dominant error in "open" or "closed" rules can be expressed as $K h^{n+1} f^{(n)}(\xi)$, where K is a constant and $h = (x_n - x_0)/n$. For formulas having the same h^n as a factor, h, the error term clearly becomes smaller for $h < 1$. But it is the coefficient K that provides a guide to the accuracy of the formulas, since the accuracy with a smaller K is better. Although Eqs. (8.59) and (8.60) have the same order of accuracy, Eq. (8.59) is more accurate because of a smaller coefficient K, i.e., $K = 14/45 < 95/144$. On the other hand, to prevent the error term from growing, attention must also be paid to $f(x)$ to ensure that its higher derivatives, $f^{(n)}(\xi)$, do not become excessively large within the integration interval.

Many versions of the Newton-Cotes formulas that use a large number of panels have been developed. For larger n, one might expect the numerical estimates with the Newton-Cotes formulas to improve, however, but that is not the case. Employing an open or closed Newton-Cotes formula derived from a single high-degree interpolating polynomial over the entire $[a, b]$ range does not assure good results unless the integration domain is quite small, i.e., small $b - a$. On the other hand, to estimate the definite integral of a function over a large interval with reasonable accuracy, a very large number of integration points (or a high-order interpolation polynomial) spanning the entire range is required.

For any n, only the weights of the Newton-Cotes formula are determined because the abscissas are uniformly spaced. The weights turn out to be symmetrical with respect to the midpoint of the integration interval, and the sum of the weights is equal to the integration interval. In closed formulas (for $n \geqslant 8$) and in open formulas (for $n \geqslant 2$), the weights turn out to be of mixed sign and, moreover, grow without bound. Besides being laborious and having poor round-off error properties, these undesirable features lead to Runge's phenomenon, which causes the error to grow. (Recall that fitting uniformly spaced discrete points to a high-degree interpolating polynomial tends to exhibit Runge's phenomenon; *see* Section 6.1). Hence, the Newton-Cotes formulas for large n are rarely used in numerical computations.

An alternative approach to using high-order formulas in order to increase the accuracy of a numerical integration process is to take advantage of the *composite rule*, as in the cases of the N-panel composite closed Newton rules. In practice, it is preferable to use composite (open or closed) Newton-Cotes formulas rather than applying high-order formulas that cover larger intervals. To accomplish this, the original interval $[a, b]$ is divided into M subregions,

Open Newton-Cotes Formulas

- Open Newton-Cotes formulas are simple and easy to program;
- Formulae of high order of accuracy are available;
- They are effective in treating the integrands that have singularities at the end point(s);
- They can be easily adapted to composite rules;
- A Newton-Cotes rule based on n integration points yields the exact value of the integral for polynomials of degree at most $n - 1$;
- They are used in multistep integration procedures in the numerical treatment of ordinary differential equations (*see* Section 9.4).

- An integrand must be smooth to ensure the specified order of accuracy;
- Very high-order-accurate integration formulas are rarely used in practice because they also suffer from Runge's phenomenon;
- Errors can be very large for large integration intervals, which necessitates the use of a composite rule in most cases;
- The convergence of any open formula can be very slow for integrands with singular points;
- Weights can take large values and alternate in sign for $n \geqslant 2$, which can be a source of instability, in particular, due to the propagation of rounding errors.

and then the integral is piecewise estimated by applying a low-order Newton-Cotes formula to each subregion. Numerical integration carried out in this manner is called *composite Newton-Cotes rules*. The global error terms for the open Newton-Cotes composite rules are given as $(b-a)h^2 f''(\xi)/6$, $(b-a)h^2 f''(\xi)/4$, $7(b-a)h^4 f^{(4)}(\xi)/90$, $19(b-a)h^5 f^{(4)}(\xi)/144$, and $41(b-a)h^6 f^{(6)}(\xi)/840$ for Eqs. (8.57)-(8.61), respectively.

Open Newton-Cotes rules are generally *not* applied to regular smooth functions. They are useful only when estimating certain improper integrals where the integrand is discontinuous at the lower, upper, or both ends of the integration interval.

8.6 INTEGRATION OF NON-UNIFORM DISCRETE FUNCTIONS

The numerical methods, such as trapezoidal, Simpson's, and Romberg's rules, are generally formulated for uniformly spaced discrete functions. In cases where the integrand is explicitly known, the analyst has the freedom to choose the numerical method or number of panels, and so on, in which case uniformly spaced discrete data points are generated to simplify the computation procedure and reduce the cpu-time.

When dealing with non-uniformly spaced discrete functions that are encountered as a result of random data sampling or experimental observations, the analyst has no choice but to use the available data "as is." In this section, numerical integration formulations for the composite trapezoidal or Simpson's rules that allow non-uniform panel widths are presented.

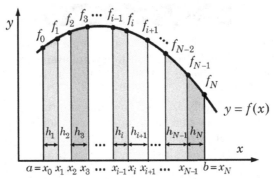

FIGURE 8.9: A non-uniform distribution

Consider a non-uniformly spaced discrete function on $[a, b]$ (*see* Fig. 8.9) whose abscissas are given as $a = x_0 < x_1 < \ldots < x_{N-1} < x_N = b$. Let $f(x_i) = f_i$ and $h_i = x_i - x_{i-1}$ denote the integrand at x_i and preceding interval size, respectively. The area of the trapezoid bounded by panel $[x_{i-1}, x_i]$ is $A_i = (f_{i-1} + f_i)h_i/2$. Then the *N-panel trapezoidal rule* can be cast as the sum of the areas of all trapezoids:

$$\int_a^b f(x)dx \cong T_N = \frac{1}{2} \sum_{n=1}^{N} h_i(f_{i-1} + f_i)$$

For Simpson's rule, a second-degree Lagrange interpolating polynomial passing through three data points, (x_{i-1}, f_{i-1}), (x_i, f_i), (x_{i+1}, f_{i+1}), is constructed (*see* Section 6.1) as follows:

$$f(x) \cong \frac{x(x - h_{i+1})}{h_i(h_i + h_{i+1})} f_{i-1} - \frac{(x + h_i)(x - h_{i+1})}{h_i h_{i+1}} f_i + \frac{x(x + h_i)}{h_{i+1}(h_i + h_{i+1})} f_{i+1} \quad (8.62)$$

Integrating the approximation for $f(x)$, Eq. (8.62), on $[x_{i-1}, x_{i+1}]$ yields

$$A_i = \int_{x_{i-1}}^{x_{i+1}} f(x)dx \cong \frac{(h_{i+1} + h_i)}{6} \left\{ \frac{(2h_i - h_{i+1})}{h_i} f_{i-1} + \frac{(h_{i+1} + h_i)^2}{h_{i+1}h_i} f_i + \frac{(2h_{i+1} - h_i)}{h_{i+1}} f_{i+1} \right\} \quad (8.63)$$

Employing Eq. (8.63) to an even number of panels results in

$$\int_a^b f(x)dx \cong A_1 + A_3 + \ldots + A_{N-1} = \sum_{i=1}^{[N/2]} A_{2i-1}$$

$$= \sum_{i=1}^{[N/2]} \frac{(h_{2i} + h_{2i-1})}{6} \left\{ \frac{(2h_{2i-1} - h_{2i})}{h_{2i-1}} f_{2i-2} + \frac{(h_{2i} + h_{2i-1})^2}{h_{2i-1}h_{2i}} f_{2i-1} + \frac{(2h_{2i} - h_{2i-1})}{h_{2i}} f_{2i} \right\}$$

which is the composite rule for non-uniform discrete functions.

For an "odd" number of panels, the integral of the interpolating polynomial, Eq. (8.62), over the last panel $[x_{n-1}, x_n]$ is obtained as follows and added to Eq. (8.63).

$$\int_{x_{n-1}}^{x_n} f(x)dx \cong \frac{h_{n-1}(2h_{n-1} + 3h_{n-2})}{6(h_{n-2} + h_{n-1})} f_n + \frac{h_{n-1}(h_{n-1} + 3h_{n-2})}{6h_{n-2}} f_{n-1}$$
$$- \frac{h_{n-1}^3}{6h_{n-2}(h_{n-2} + h_{n-1})} f_{n-2} \quad (8.64)$$

```
Module NONUNIFORM_SIMPSON (panel, x, f, intg)
 \ DESCRIPTION: A pseudocode to estimate the integral of a non-uniformly
 \    spaced discrete function using Simpson's rule.
 \ USES:
 \    MOD:: A built-in function returning the remainder of x divided by y.
 Declare: x_{0:panel}, f_{0:panel}, h_{0:panel}          \ Declare array variables
 For [i = 0, panel]                                       \ Loop: Find panel widths
     h_i ← x_{i+1} − x_i
 End For
 m ← MOD(n − 1, 2)                        \ Find remainder of n − 1 divided by 2
 If [m = 0] Then                          \ Determine if number of panels is even or not
     np ← panel + 1                        \ if EVEN, apply Eq. (8.63) to all panels
 Else
     np ← panel                           \ if ODD, apply Eq. (8.63) to all but last panel
 End If
 intg ← 0                                  \ Initialize integration accumulator variable
 For [i = 0, (np − 2), 2]                  \ Loop: Compute and accumulate Eq. (8.63)
     dx ← (h_{i+1} + h_i)/6
     aw ← (2h_i − h_{i+1})/h_i
     ap ← (h_{i+1} + h_i)^2/(h_{i+1} * h_i)
     ae ← (2h_{i+1} − h_i)/h_{i+1}
     intg ← intg + dx * (aw * f_i + ap * f_{i+1} + ae * f_{i+2})
 End For
     n = panel
 If [m = 1] Then                           \ For odd number panels, apply Eq. (8.64)
     intg ← intg + f_n h_{n−1}(2h_{n−1} + 3h_{n−2})/(6(h_{n−2} + h_{n−1}))
     intg ← intg + f_{n−1} h_{n−1}(h_{n−1} + 3h_{n−2})/(6h_{n−2})
     intg ← intg − f_{n−2} h_{n−1}^3/(6h_{n−2}(h_{n−2} + h_{n−1}))
 End If
End Module NONUNIFORM_SIMPSON
```

A pseudomodule, NONUNIFORM_SIMPSON, estimating the integral of a non-uniformly spaced discrete function is given in Pseudocode 8.5. The module requires the number of data points (n) and the non-uniform discrete data set, (x_i, f_i) for $i = 1, 2, \ldots, n$, as input. Using the array of abscissas, the panel widths are computed within a For-loop as $h_i = x_{i+1} − x_i$ for $i = 1, 2, \ldots, n−1$. The module first checks whether the number of panels is even using the built-in MOD (modulo) function. In the case of $m = 0$, the $n−1$ is evenly divided by 2 with no remainder; that is, the number of panels is even, so the number of panels is set to $np = n$. For $m = 1$, the number of panels is odd, and $np = n−1$ becomes even, for which Eq. (8.63) is applied. On the other hand, Eq. (8.64), which is the integration over the last panel, is added to the previous sum. The result is saved to the output (accumulator) variable $integ$.

8.7 GAUSS-LEGENDRE METHOD

The Gauss-Legendre method is a powerful numerical tool to compute the definite integral of an explicitly defined function. The method requires the evaluation of the integrand at non-uniformly spaced integration points to achieve the highest accuracy, which is much higher than Newton-Cotes formulas.

Integration of Non-Uniformly Spaced Discrete Functions

- It is essential for estimating the numerical integral of non-uniform functions;
- Numerical integration procedures based on open- or closed- Newton-Cotes formulas can be easily employed for non-uniformly spaced discrete functions;

- Estimating the error and its bounds is not an easy task;
- The order of accuracy is dominated by the largest h values;
- Cpu-time increases due to computing the coefficients.

The objective of the method is to determine a set of abscissas (integration points) and corresponding weights such that the numerical approximation of an integral on $[-1, 1]$ can be expressed as a weighted sum:

$$\int_{-1}^{+1} f(x)\,dx \cong G_N = \sum_{i=1}^{N} w_i f(x_i) \tag{8.65}$$

where N is the number of quadrature points, x_i and w_i's are the abscissas and corresponding weights, respectively, together referred to as *Gauss-Legendre quadrature*.

Since the numerical estimate is a weighted sum, it is easy to program it once the quadrature for a specified N is known. The main difficulty in this method, however, is obtaining highly accurate quadrature. To illustrate the determination of the quadrature, we set $f(x) = x^k$ in Eq. (8.65) and integrate the left-hand side exactly. This process yields the following set of $2N$ nonlinear equations with $2N$ unknowns:

$$\sum_{i=1}^{N} w_i x_i^k = \int_{-1}^{+1} x^k\,dx = \frac{1 + (-1)^k}{k+1}, \quad k = 0, 1, 2, \ldots, (2N-1) \tag{8.66}$$

or explicitly

$$
\begin{aligned}
w_1 + w_2 + \cdots + w_N &= 2 \\
x_1 w_1 + x_2 w_2 + \cdots + x_N w_N &= 0 \\
x_1^2 w_1 + x_2^2 w_2 + \cdots + x_N^2 w_N &= \frac{2}{3} \\
\vdots \qquad \vdots \\
x_1^{2N-1} w_1 + x_2^{2N-1} w_2 + \cdots + x_N^{2N-1} w_N &= 0
\end{aligned}
\tag{8.67}
$$

The solution of Eq. (8.67) for specified N gives the Gauss-Legendre quadrature [1]. Although it would be theoretically possible to solve Eq. (8.67), in practice, it is extremely difficult to find a numerical solution to this set of nonlinear equations, especially when N is fairly large. To get around this problem, alternatively, a set of abscissas is proposed. The problem is then reduced to determining only the weights from Eq. (8.67), which is now a set of linear equations with weights as unknowns. It turns out that the abscissas are the roots of the Nth-degree Legendre polynomial: $P_N(x) = 0$. Evaluating the corresponding weights is a bit involved, so before we do that, we will briefly visit the so-called *Legendre polynomials* and some of their properties [1].

The nth-degree Legendre polynomial is defined as

$$P_n(x) = \frac{1}{2^n n!} \frac{d^n}{dx^n}(x^2 - 1)^n \tag{8.68}$$

For $n = 0$ and $n = 1$, Equation (8.68) yields $P_0(x) = 1$ and $P_1(x) = x$. An easy way to generate higher-degree Legendre polynomials is to make use of the recurrence relationship that relates Legendre polynomials of three subsequent degrees:

$$n P_n(x) = (2n - 1)x P_{n-1}(x) - (n - 1)P_{n-2}(x), \qquad \text{for } n \geqslant 2 \tag{8.69}$$

which leads up to 4th degree are as follows:

$$P_2(x) = \frac{1}{2}(3x^2 - 1), \quad P_3(x) = \frac{x}{2}(5x^2 - 3), \quad P_4(x) = \frac{1}{8}(35x^4 - 30x^2 + 3)$$

The roots of the Legendre polynomials for a small N can be easily calculated by hand. However, for a large N, the roots in general are calculated with the Newton-Raphson method, leading to the following iteration equation:

$$x^{(p+1)} \cong x^{(p)} - \frac{P_N(x^{(p)})}{P'_N(x^{(p)})} \tag{8.70}$$

Here, $P_N(x)$ is evaluated by Eq. (8.69), and $P'_N(x)$, which is the first derivative of $P_N(x)$, is evaluated by applying the following recursive relationship:

$$(1 - x^2)\frac{dP_n(x)}{dx} = nP_{n-1}(x) - nxP_n(x), \tag{8.71}$$

It turns out that the weights can be computed from

$$w_i = \frac{2}{(1 - x_i^2)[P'_N(x_i)]^2} \tag{8.72}$$

The proof of Eq. (8.72) will be skipped here as it is out of the scope of this text.

To compute the abscissas, it is critical to start the iteration process with an initial guess sufficiently close to a zero of $P_N(x)$. An initial guess for an abscissa is obtained by $x_i^{(0)} = \cos((4i - 1)\pi/(4N + 2))$ [22]. Once an abscissa (i.e., a zero) is iteratively found using Eq. (8.70), the corresponding weight is calculated by Eq. (8.72). It should be noted that the distribution of abscissas is symmetrical with respect to $x = 0$, and the weights are all positive. The abscissas are non-uniformly spaced and stacked closer toward the end points as N increases. This feature, unlike Newton-Cotes formulas, allows the use of more points in a single interval by avoiding Runge's phenomenon. The complete quadrature sets for various N are tabulated in Appendix B.1.

The global error for the Gauss-Legendre quadrature is given by

$$R_N = \frac{2^{2N+1}(N!)^4}{(2N + 1)[(2N)!]^3}f^{(2N)}(\xi), \qquad -1 < \xi < 1 \tag{8.73}$$

Note that Eq. (8.73), which involves $f^{(2N)}(\xi)$ derivative, will be zero for polynomials of up to the $(2N-1)$th degree since $f^{(2N)}(\xi) = 0$. This result indicates that the Gauss-Legendre method yields the true values of the integrands of polynomials of degree up to $2N-1$ if no round-off errors were made. This property makes the method very powerful since, even for small N, numerical approximations with higher accuracy can be achieved. Nevertheless, an integrand sometimes may not be smooth and might behave differently at different intervals. In such cases, the most rational approach is to divide the integral into several subintervals, then integrate each region separately with a low-order quadrature rule, and finally add all estimates together.

GENERALIZATION: The Gauss-Legendre quadrature method can be extended to an arbitrary definite integral. To do this, a definite integral is transformed into an integral on $[-1, 1]$. Upon invoking $x = (b - a)u/2 + (b + a)/2$ substitution with $u \in [-1, 1]$, we get

$$\int_{x=a}^{b} f(x)dx = \frac{b-a}{2} \int_{u=-1}^{+1} f\left(\frac{b-a}{2}u + \frac{b+a}{2}\right) du \tag{8.74}$$

Employing the Gauss-Legendre sum to the right-hand side of Eq. (8.74) yields

$$\int_{a}^{b} f(x)dx = \sum_{i=1}^{N} w_i f(x_i) + R_N \tag{8.75}$$

where

$$x_i = \frac{b-a}{2}u_i + \frac{b+a}{2}, \qquad w_i = \frac{b-a}{2}w_{ui} \tag{8.76}$$

and (u_i, w_{ui}) is the quadrature set on $[-1, 1]$ (Appendix B.1), and R_N is the global error given by

$$R_N = \frac{(b-a)^{2N+1}(N!)^4}{(2N+1)[(2N)!]^3} f^{(2N)}(\xi), \qquad a < \xi < b \tag{8.77}$$

whose upper and lower bounds can be estimated using the maximum and minimums of $f^{(2N)}(\xi)$.

The global error clearly involves the derivative $f^{(2N)}(x)$. Equations (8.73) and Eq. (8.77) are generally not very useful unless the integrand is a polynomial, a simple exponential or trigonometric function, and so on. Otherwise, they are impractical for assessing the accuracy of an estimated value. The best way to assess the accuracy is to compare the estimates for several different values of N, such as G_{N_1} and G_{N_2}. In certain cases, these comparisons may yield substantially different results, which may be indicative of an integrand with singular point(s) or a highly oscillatory character.

The foregoing comments regarding accuracy apply equally to any numerical integration method. The Gauss-Legendre method is unique in that it yields very accurate estimates for a variety of integrands with very small N. The accuracy of the method may inspire confidence in an analyst, who might attempt to do one evaluation with a single (perhaps large) value of N and accept the estimate blindly.

The *true error* of an integral can be determined only when its *true solution* is known. The reader may wonder why the global error bounds have been computed in some examples presented here even when the true error can be readily calculated. It should be kept in mind that we resort to numerical integration in cases where the true solution cannot be found. In this text, examples with known true solutions are used to illustrate the importance and effectiveness of the error estimation process.

> **Pseudocode 8.6**

Module GAUSS_LEGENDRE_QUAD $(N, \varepsilon, \mathbf{x}, \mathbf{w})$
\ DESCRIPTION: A pseudomodule generating N-point Gauss-Legendre quad-
\   rature for $N \geqslant 2$; abscissas and weights are on $[-1, 1]$.
\ USES:
\   ABS :: A built-in function computing the absolute value.
\   COS :: Trigonometric cosine function.
Declare: x_N, w_N \ Declare array variables
$m \leftarrow (N+1)/2$ \ Find number of abscissas in symmetric half
For $\left[i = 1, m \right]$ \ Loop i: find x_i and w_i
$\quad u \leftarrow \cos\left(\pi\left(4i - 1 \right)/\left(4N + 2 \right)\right)$ \ Apply initial guess estimator x_i
$\quad \delta \leftarrow 1$ \ Initialize absolute error
$\quad$ **While** $\left[|\delta| > \varepsilon \right]$ \ Newton-Raphson iteration loop to find x_i
$\quad\quad P0 \leftarrow 1$ \ Initialize P_0
$\quad\quad P1 \leftarrow u$ \ Initialize P_1
$\quad\quad$ **For** $\left[k = 2, N \right]$ \ Loop to find P_N
$\quad\quad\quad P2 \leftarrow ((2k - 1)uP1 - (k - 1)P0)/k$ \ Apply Eq. (8.69)
$\quad\quad\quad P0 \leftarrow P1$ \ Assign $P_{k-2} \leftarrow P_{k-1}$
$\quad\quad\quad P1 \leftarrow P2$ \ Assign $P_{k-1} \leftarrow P_k$
$\quad\quad$ **End For**
$\quad\quad dP \leftarrow n(uP2 - P0)/(u^2 - 1)$ \ Compute $P'_N(u)$ by Eq. (8.71)
$\quad\quad \delta \leftarrow P2/dP$ \ Difference between two successive iterates
$\quad\quad u \leftarrow u - \delta$ \ Apply Newton-Raphson, Eq. (8.70)
$\quad$ **End While**   \ Exits loop with converged x_i
$\quad x_i \leftarrow -u$ \ Save root as the negative abscissa too
$\quad w_i \leftarrow 2/(1 - u^2)/dP^2$ \ Find corresponding weight by Eq. (8.72)
$\quad x_{N+1-i} \leftarrow u$ \ Save the symmetric abscissa
$\quad w_{N+1-i} \leftarrow w_i$ \ Save the symmetric weight
End For
End Module GAUSS_LEGENDRE_QUAD

A pseudomodule, **GAUSS_LEGENDRE_QUAD**, generating the N-point Gauss-Legendre quadrature on $[-1, 1]$ is presented in Pseudocode 8.6. As input, the module requires the number of integration points (N) and a tolerance value (ε) for the desired accuracy in the quadrature. The output is the quadrature set: (x_i, w_i) for $i = 1, 2, .., N$.

A For-loop over i-variable covers the computation of *positive* abscissas only, i.e., x_i for $i = 1, 2, .., (N+1)/2$. In other words, the zeros with the opposite sign (as well as their weights) are not computed due to symmetry. In order to speed up and to ensure convergence, the initial guess for the ith zero is obtained by $(x_i)^{(0)} = \cos\left(\pi\left(4i-1\right)/\left(4N+2\right)\right)$. Then, using a While-loop, the Newton-Raphson iteration procedure is employed until the zero (i.e., x_i) is found within the preset tolerance ($< \varepsilon$). To find an improved estimate of x_i using Eq. (8.70), we need to evaluate the Nth-degree Legendre polynomial and its derivative ($P_N(x)$ and $P'_N(x)$). This is achieved with a For-loop over k-variables from $k = 2$ to N, which applies the recurrence relations to find $P_N(x)$ using Eq. (8.69). Once $P_N(x)$ is obtained, $P'_N(x)$ is found by using (8.71). The iteration procedure is terminated when the absolute error between two consecutive iterates is less than the prescribed tolerance ($\delta = |u^{(p+1)} - u^{(p)}| < \varepsilon$). Having found x_i, the corresponding weight is determined from Eq. (8.72). The negative abscissas of the Gauss-Legendre quadrature are produced from the known positive zeros, and they are assigned the same corresponding weights.

EXAMPLE 8.5: Deriving and employing 3-point Gauss-Legendre quadrature

The solar heat flux (kW/m^2) incident on an absorber tube wall is approximated by

$$q(x) = (22x^3 - 5.5x^2 - 2x + 30)e^{x^3/40} \quad \text{for} \quad -1 \leqslant x \leqslant +1$$

where $x = \cos\theta$, and θ corresponds to the radial angle along the symmetric half-perimeter of the tube, i.e., $0 \leqslant \theta \leqslant \pi$. The mean solar heat flux for the tube may be calculated using the following integral:

$$q_m = \frac{1}{2}\int_{-1}^{1} q(x)dx$$

(a) Generate the 3-point Gauss-Legendre quadrature (abscissas and weights), and
(b) find G_3 as well as the corresponding error bounds and the true error.

SOLUTION:

(a) For $N=3$, the abscissas (x_i's) are determined by setting $P_3(x)=0$; that is,

$$P_3(x) = \frac{1}{2}x(5x^2 - 3) = 0$$

which yields $x_1 = 0$ and $x_{2,3} = \pm\sqrt{3/5}$.

Noting $P_3'(x) = (15x^2 - 3)/2$, the quadrature weights are obtained from Eq. (8.72) as follows:

$$w_1 = \frac{2}{(1-0^2)(P_3'(0))^2} = \frac{8}{9}$$

$$w_{2,3} = \frac{2}{(1-(\pm\sqrt{3/5})^2)(P_3'(\pm\sqrt{3/5}))^2} = \frac{5}{9}$$

(b) Using Eq. (8.65), the numerical estimate is obtained as

$$\int_{-1}^{+1} q(x)dx \approx G_3 = \sum_{i=1}^{3} w_i q(x_i) = \frac{8}{9}q(0) + \frac{5}{9}q\left(\sqrt{\frac{3}{5}}\right) + \frac{5}{9}q\left(-\sqrt{\frac{3}{5}}\right)$$

$$= \frac{8}{9}(30) + \frac{5}{9}(35.788906) + \frac{5}{9}(17.816303)$$

$$= 56.44733833$$

Finally, the mean solar heat flux is found as $q_m = 56.44733/2 = 28.2364 \ kW/m^2$.

Discussion: To assess the global error, we make use of Eq. (8.73):

$$R_3 = \frac{2^7(3!)^4}{7(6!)^3}q^{(6)}(\xi) = 6.3492 \times 10^{-5}q^{(6)}(\xi), \quad -1 < \xi < 1$$

For determining the upper and lower bounds of the error, we require the maximum and minimum values of $q^{(6)}(\xi)$ on $[-1, 1]$.

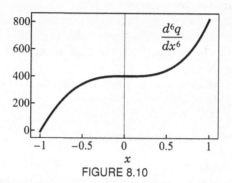

FIGURE 8.10

The graph of $q^{(6)}(x)$ illustrated in Fig. 8.10 depicts an absolute minimum and an absolute maximum value of $q^{(6)}(-1) \approx -8.3$ and $q^{(6)}(1) \approx 819.5)$, respectively, at the endpoints of the integration interval. Substituting these values into the global error term yields

$$6.3492 \times 10^{-5}(-8.3) < R_3 < 6.3492 \times 10^{-5}(819.5)$$

or

$$-0.000527 < \text{True Error} < 0.052032$$

Discussion: It is possible to obtain a highly accurate value for the integral by increasing the number of integration points, i.e., G_5, G_9, and so on. The true integral of this example exits and is found as $I_{\text{true}} = 56.47277954$, which is reported correct to eight decimal places. The absolute error can now be found to be $56.472779538 - 56.447338333 = 0.025441205$. Notice that the true error is within the lower and upper bounds of the global error.

EXAMPLE 8.6: Applying Gauss-Legendre method to an arbitrary definite integral

Repeat Example 8.2 using 4-point Gauss-Legendre quadrature and determine the true error.

SOLUTION:

For now, if we exclude the factor 4π, we will transform the integral into an integral over $[-1, 1]$ so that the Gauss-Legendre quadrature can be applied.

With $x = \pi(u+1)/2$ substitution and $dx = (\pi/2)du$, the integral becomes

$$\int_0^\pi \sqrt{\sin x + \cos^2 x}\, dx = \frac{\pi}{2} \int_{u=-1}^1 \sqrt{\sin x + \cos^2 x}\, du \cong G_4 = \frac{\pi}{2} \sum_{i=1}^N w_i f(x_i)$$

where $x_i = \pi(u_i + 1)/2$ and (u_i, w_i) are the 4-point Gauss-Legendre quadrature on $[-1, 1]$ (*see* Appendix B.1).

Gauss-Legendre Quadrature Method

- The method is easy to implement as it yields a simple weighted sum;
- The degree of precision of G_N is $2N-1$, and it gives the exact value for the integral of a polynomial of degree $2N-1$ or less;
- It is superior to Newton-Cotes rules because very high levels of accuracy can be achieved with a small number of function evaluations (even for $N < 10$);
- It offers, in some cases, a remedy for computing improper integrals since abscissas avoid the end points.

- It is limited to explicit functions, i.e., unsuitable for discrete functions;
- Abscissas and weights are not static; They change as N is increased, which makes the method unsuitable for recursive algorithms;
- To obtain highly accurate estimates, high-precision quadrature sets must be generated or should be available in tabular form;
- Numerical approximations with $N > 20$ may be justified if the integrand (having singularities, rapid oscillations, etc.) behaves badly;
- Theoretical error estimation is not an easy task.

TABLE 8.6

i	u_i	w_i	x_i	$f(x_i)$	$w_i f(x_i)$
1	-0.861136	0.347855	0.218127	1.081467	0.376194
2	-0.339981	0.652145	1.036755	1.058232	0.690121
3	0.339981	0.652145	2.104837	1.058232	0.690121
4	0.861136	0.347855	2.923466	1.081467	0.376194
				$\sum w_i f_i =$	**2.132630**

The Gauss quadrature (x_i, w_i), integrands $f(x_i)$, and weighted products $w_i f(x_i)$ are calculated and presented in Table 8.6. The sum of the last column gives the weighted sum, Eq. (8.65). Substituting this result into the above expression yields

$$G_4 = \frac{\pi}{2}(2.132630) = 3.349925$$

Discussion: The true integral, correct to eight decimal places, is 3.34686849. The true (absolute) error is determined to be 0.0030567, which is accurate to two decimal places. For G_6, G_8, and G_{10}, the true errors are calculated to be 1.1555×10^{-4}, 5.6852×10^{-6}, and 3.1621×10^{-7}, respectively. This example illustrates how fast the Gauss-Legendre method converges. Especially note that the G_{10} yielded a six-decimal place accurate estimate with only ten function evaluations.

8.8 COMPUTING IMPROPER INTEGRALS

So far, we have dealt with the numerical computation of definite integrals with a finite domain of integration and a finite range on this domain. However, in practice, we encounter two types of integrals that do not meet the foregoing conditions. In this regard, we refer to an integral as "*improper integral*" when the interval of integration or the integrand itself is unbounded. Improper integrals can present challenges to numerical computation.

Type I (improper) integrals are definite integrals that are infinite at one or both end-points, that is, the integration interval is infinite or semi-infinite (*see* Fig. 8.11). Another class of improper integrals (*Type II*) have integrands with a singularity within or on one or both end points of the integration interval, i.e., the integration interval is finite, but the integrand is unbounded (*see* Fig. 8.12).

Type I improper integrals have the following general forms:

$$\int_a^\infty f(x)dx \quad \text{or} \quad \int_{-\infty}^{-b} f(x)dx \quad \text{or} \quad \int_{-\infty}^\infty f(x)dx$$

where $a \geqslant 0$ and $b \geqslant 0$.

An improper integral can be estimated numerically, provided that it is convergent. If we recall the physical interpretation of a definite integral, Type I integrals can similarly be interpreted as the area bounded by the x-axis, the integral interval, and the curve, provided that $f(x) > 0$. It is important to determine whether a definite integral is convergent or not before attempting to estimate it. If an improper integral is divergent, the true integral (area under the curve) will be infinite; in other words, the numerical estimate will differ as the number of integration points (or panels) continuously increases.

Coordinate transformation: Before moving on to discuss the numerical integration of Type I and Type II integrals, we should point out that, in some cases, it is possible to avoid integrals with discontinuities by applying a suitable coordinate transformation.

In order to avoid Type I integrals over the semi-infinite interval, the two integrals presented above can be transformed into the following integrals upon $x = 1/u$ (or $x = -1/u$) substitution:

$$\int_0^{1/a} g\left(\frac{1}{u}\right) \frac{du}{u^2} \quad \text{or} \quad \int_0^{1/b} g\left(-\frac{1}{u}\right) \frac{du}{u^2}$$

If $g(\pm 1/u)/u^2$ is not singular (or has a removable singularity) at $u = 0$, the numerical integration can be safely and efficiently carried out with one of the previously covered numerical integration methods.

One may devise various u-substitutions to transform such an improper integral into one with a finite interval. For instance, $[0, \infty)$ can be transformed to $[0, L]$ by $x = Lu/(u+c)$ (where c is a real constant), or to $[0,1]$ by $x = \tanh u$, and so on. Alternatively, when the

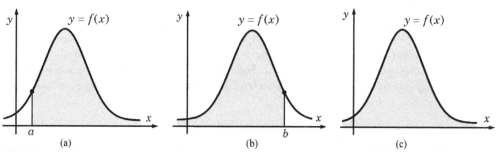

FIGURE 8.11: Graphical depiction of integration domains for Type I integrals with infinity at (a) the upper limit, (b) the lower limit, or (c) both limits.

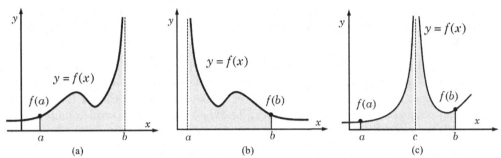

FIGURE 8.12: Graphical depiction of integrands for Type II integrals with discontinuities at the (a) upper end, (b) lower end, or (c) inside the integration interval.

transformations do not work, an integral over $(0, \infty)$ can be split into two integrals as

$$\int_0^\infty f(x)dx = \int_0^a f(x)dx + \int_a^\infty f(x)dx$$

Typically, $f(x)$ is finite in a bounded interval, $[0, a]$, while $f(x) \approx 0$ everywhere else. Taking this into account, the first integral can be estimated by any of the methods discussed earlier. Then, a coordinate transformation such as $x = 1/u$ or $x = a/(1 - u)$ can be employed for the second integral so that the integration interval is finite.

The integration over $(-\infty, -b)$ $(b \geqslant 0)$ can be handled in the same manner as the ones over (a, ∞) $(a \geqslant 0)$. Depending on the numerical method to use, the integration of functions over $(-\infty, \infty)$ is carried out by first splitting the integral into either two or three convenient parts, i.e., $(-\infty, 0)$ and $(0, \infty)$ or $(-\infty, -b)$, $(-b, a)$, and (a, ∞). Any suitable method can then be applied to each integral separately.

 Avoid calculating Type I integrals directly if possible. First, investigate whether the integral can be transformed into a definite integral (having a continuous integrand) with a suitable u-substitution. Whenever possible, prefer approximating definite integrals to improper integrals.

Improper integrals of Type II have a discontinuity at the lower, upper, or both end points, or at an interior point, as illustrated in Fig. 8.12. Recall that the definite integral of a function with closed Newton-Cotes formulas is based on the assumption that the integrand is continuous and defined on $[a, b]$. There is, however, an exception where the integrand has a multiplying factor, the so-called weight function. The integral can be handled numerically if the weight function is singular.

Some improper integrals of Type II can be transformed to standard definite integrals in a similar manner as for Type I, by using a suitable change of variable that completely eliminates the discontinuity of the integral. For example, the following integrands having a discontinuous point at either lower or upper end points are transformed into standard form with the indicated u-substitutions:

$$\int_0^1 \frac{f(x)}{\sqrt[4]{x}}dx \quad (x = u^4) \quad \rightarrow \quad 4\int_0^1 u^2 f(u^4)du$$

$$\int_0^1 \frac{f(x)}{\sqrt{1-x}}dx \quad (1 - x = u^2) \quad \rightarrow \quad 2\int_0^1 f(1 - u^2)du$$

$$\int_{-1}^{1} \frac{f(x)}{\sqrt{1-x^2}} dx \quad (x = \sin u) \quad \rightarrow \quad \int_{-\pi/2}^{\pi/2} f(\sin u) du$$

where $f(x)$ is a bounded function over the entire integration range.

For a bounded $f(x)$ on $[0,1]$, a discontinuity at $x = 0$ of the following integrals is eliminated with the indicated change of variable, while the transformed integrals are now the ones over a semi-infinite interval, i.e., $(0, \infty)$ or $(-\infty, 0)$. These integrals can then be calculated using a suitable quadrature method.

$$\int_{0}^{1} f(x) \ln x \, dx \quad (\ln x = -u) \quad \rightarrow \quad -\int_{0}^{\infty} u e^{-u} f(e^{-u}) du$$

$$\int_{0}^{1} f(\ln x) dx \quad (\ln x = u) \quad \rightarrow \quad \int_{-\infty}^{0} e^{u} f(u) du$$

Subtraction of singularity: In cases where integral transformations cannot offer a remedy, a singularity can be eliminated by a technique referred to as *subtraction of singularity*. The discontinuity is removed by adding and subtracting an integrable function from the integrand, thereby forming two integrals, one with a *removable singularity* and the other that can be evaluated analytically or numerically. For example, in the definite integral of $e^{cx}/\sqrt{x}$ on $[0, 1]$, the integrand has a non-removable singular point at $x = 0$ that behaves as $1/\sqrt{x}$. By subtracting and adding $1/\sqrt{x}$ to the integrand, the result can be expressed in two parts as follows:

$$\int_{0}^{1} \frac{e^{cx}}{\sqrt{x}} dx = \int_{0}^{1} \frac{e^{cx} - 1}{\sqrt{x}} dx + \int_{0}^{1} \frac{dx}{\sqrt{x}}$$

The second integral is easily integrated analytically, which gives "2." On the other hand, the first integral can be evaluated numerically using N-panel composite trapezoidal or Simpson's rules, expressed as

$$\int_{0}^{1} \frac{e^{cx} - 1}{\sqrt{x}} dx \cong w_0 \left(\frac{e^{cx_0} - 1}{\sqrt{x_0}} \right) + \sum_{i=1}^{N} w_i \left(\frac{e^{cx_i} - 1}{\sqrt{x_i}} \right)$$

where $x_i = ih$ for $i = 0, 1, 2, \ldots, N$, and the w_i's are the weights whose values depend on the method selected. For $x_0 = 0$, the first term of this approximation results in $0/0$-indeterminate form. Applying L'Hopital's rule to remove the indeterminate form yields $(e^{cx_0} - 1)/\sqrt{x_0} \rightarrow 0$ as $x_0 \rightarrow 0$. Using this result, the numerical estimate (i.e., weighted sum) is reduced to

$$\int_{0}^{1} \frac{e^{cx} - 1}{\sqrt{x}} dx \cong \sum_{i=1}^{N} w_i \left(\frac{e^{cx_i} - 1}{\sqrt{x_i}} \right)$$

which is now finite.

Next, consider the following example with a singularity at $x = a$:

$$\int_{0}^{a} \frac{f(x) dx}{\sqrt{a^2 - x^2}}$$

where $f(x)$ is bounded on $[0, a]$, and the singularity arises only from the $1/\sqrt{a^2 - x^2}$ term. After adding and subtracting the $f(a)/\sqrt{a^2 - x^2}$ term from the integrand, the integral can be rearranged as a sum of two integrals as follows:

$$\int_{0}^{a} \frac{f(x)}{\sqrt{a^2 - x^2}} dx = \int_{0}^{a} \frac{f(a)}{\sqrt{a^2 - x^2}} dx + \int_{0}^{a} \frac{f(x) - f(a)}{\sqrt{a^2 - x^2}} dx$$

The first integral is convergent despite its singularity at $x = a$, and it can be easily integrated analytically to give $\pi f(a)/2$. The second integral has the $0/0$-indeterminate form at $x = a$. Employing then N-panel composite trapezoidal or Simpson's rules to the integral results in

$$\int_0^a \frac{f(x) - f(a)}{\sqrt{a^2 - x^2}} dx \cong w_0 F(x_0) + w_1 F(x_1) + \ldots + w_{N-1} F(x_{N-1}) + w_N F(x_N)$$

where $F(x) = \big(f(x) - f(a)\big)/\sqrt{a^2 - x^2}$. Note that $F(x_N)$ leads to $0/0$-indeterminate form for $x = a$, which can be resolved by applying L'Hopital's rule: $F(x_N) = \lim_{x \to a} F(x)$.

> Here we use the term *"removable singularity"* to refer to an integrand $f(x)$ that has a discontinuity at $x = a$ but $\lim_{x \to a} f(x)$ exists. Whenever an integrand has a removable singularity, the best way to approach it is to remove it analytically and specify the limit value in the function module accordingly.

Ignoring singularity: This approach is often applied to Type II integrals with a singularity at the lower, upper, or both endpoints. It is a relatively primitive way of treating singular integrals, as we circumvent the singular point by applying a numerical integration rule while disregarding or ignoring the singularity. Inasmuch as the convergence of the numerical integration rule is ordinarily established only for bounded integrable functions, the analyst, in a way, blindly applies a numerical procedure and hopes for the best. The best results are obtained when the *Gauss-Legendre quadrature* or *open Newton-Cotes rules* are applied because the methods inherently avoid endpoints. However, this approach does not work well for the integrands depicting oscillatory behaviors.

Improper integrals of Type II having a singular point at an interior point $x = c$ ($a < c < b$) are handled by simply splitting the integral into two integrals (each integral having a discontinuity at either endpoint) as shown below, and then a suitable method is applied to each integral separately.

$$\int_a^b f(x) dx = \int_a^{c^-} f(x) dx + \int_{c^+}^b f(x) dx \tag{8.78}$$

For Type II integrals, the foregoing techniques discussed are applicable, provided that the integrals are convergent.

Truncating interval: This technique is usually employed for estimating Type I integrals. Let $f(x)$ be an integrable function on any finite interval $[a, b]$. An improper integral of $f(x)$ on $[a, \infty)$ can be broken up into two parts:

$$\int_a^\infty f(x) dx = \int_a^b f(x) dx + \int_b^\infty f(x) dx$$

where b is a sufficiently large number. Then the first integral is approximated using a suitable numerical method. The second integral corresponds to the approximation error resulting from truncating the integral to $[a, b]$; consequently, an upper bound for this error, $R(b)$, can be estimated by evaluating the integral of $f(x)$ over $[b, \infty)$. To illustrate this approach, consider the following Type I integral:

$$\int_1^\infty \frac{dx}{(x^4 + 1)\sqrt{x}} = \int_1^b \frac{dx}{(x^4 + 1)\sqrt{x}} + \int_b^\infty \frac{dx}{(x^4 + 1)\sqrt{x}}$$

Here, however, the second integral is not integrable analytically. Hence, to estimate the error bound, the integrand is approximated to be simple and easily integrable. Note that for all large real x, we may reduce the integrand as follows:

$$\frac{1}{(x^4+1)\sqrt{x}} < \frac{1}{x^{9/2}}$$

Then an expression for the upper bound of the approximation error is found as

$$\int_b^\infty \frac{dx}{(x^4+1)\sqrt{x}} < \int_b^\infty \frac{dx}{x^{9/2}} = R(b) = \frac{2}{7b^{7/2}}$$

Now it is possible to determine a suitable value of b that minimizes $R(b)$, while maintaining the desired accuracy. The error estimates for $b = 5$, 10, and 20 are obtained as 1.0222×10^{-3}, 9.04×10^{-5}, and 7.98×10^{-6}, respectively. The error bound of $R(20) = 7.98 \times 10^{-6} < 0.5 \times 10^{-5}$ implies that five decimal place accuracy can be achieved if the first integral is also calculated better than five decimal places.

8.9 GAUSS-LAGUERRE METHOD

The Gauss-Laguerre method is another powerful quadrature method that serves specifically to handle Type I improper integrals of the following form:

$$\int_0^\infty e^{-x} f(x) dx \tag{8.79}$$

Just as in the Gauss-Legendre method, Eq. (8.79) is also expressed as a weighted sum:

$$\int_0^\infty e^{-x} f(x) dx \cong GL_N = \sum_{i=1}^N w_i f(x_i) \tag{8.80}$$

where N is the number of quadrature points, and (x_i, w_i) are the so-called *Gauss-Laguerre quadrature* [1]. Here, $f(x)$ is assumed to be bounded in the prescribed semi-infinite domain, in which case the Gauss-Laguerre sum will converge to an approximate value for increasing N.

The Laguerre polynomials are defined as

$$L_n(x) = e^x \frac{d^n}{dx^n}(x^n e^{-x}), \qquad n \geqslant 0 \tag{8.81}$$

With $L_0(x) = 1$ and $L_1(x) = 1 - x$ as starters, any high-degree Laguerre polynomial can be found by the following recurrence relationship [1]:

$$L_n(x) = (2n - 1 - x)L_{n-1}(x) - (n-1)^2 L_{n-2}(x), \qquad n \geqslant 2 \tag{8.82}$$

The N-point *Gauss-Laguerre quadrature points* (or *abscissas*) are chosen as the zeros of the Nth-degree Laguerre polynomial; that is, $L_N(x) = 0$. Then, the Gauss-Laguerre quadrature weights (w_i's) are computed from

$$w_i = \frac{(N!)^2}{x_i [L_N'(x_i)]^2} \tag{8.83}$$

where $L_N'(x_i)$ is the first derivative of the Nth-degree Laguerre polynomial, and it can be calculated from the following relationship:

$$\frac{dL_n(x)}{dx} = \frac{n}{x}\big(L_n(x) - nL_{n-1}(x)\big), \qquad n \geqslant 1 \tag{8.84}$$

The Gauss-Laguerre quadrature sets for selected N values are tabulated in Appendix B.2.

The global error is given by

$$R_N = \frac{(N!)^2}{(2N)!}f^{(2N)}(\xi), \qquad 0 < \xi < \infty \tag{8.85}$$

where the upper and lower bounds for this can be determined from the maximum and minimum values of $f^{(2N)}(\xi)$ on $(0, \infty)$.

The improper integrals that fall into this category do not always come in standard form (Eq. (8.79)), but they can be expressed in the standard form with a suitable change of variable followed by a bit of algebra. Some common cases are presented below:

(1) $\int_0^\infty f(x)dx$ integrals are transformed to the standard form as follows:

$$\int_0^\infty f(x)dx \equiv \int_0^\infty e^{-x}F(x)dx \tag{8.86}$$

where $F(x) = e^x f(x)$. Equation (8.86) can then be expressed as a weighted sum:

$$\int_0^\infty e^{-x}F(x)dx \cong \sum_{i=1}^N w_i F(x_i) = \sum_{i=1}^N W_i f(x_i) \tag{8.87}$$

where W_i denotes the modified weights defined as $W_i = w_i e^{x_i}$ (*see* Appendix B.2).

(2) $\int_a^\infty f(x)dx$ integral is transformed to the standard form by introducing the $u = x - a$ substitution, leading to

$$\int_0^\infty f(u+a)du \quad \text{or} \quad \int_0^\infty e^{-x}F(x)du$$

where $F(x) = e^x f(x+a)$. Note that $\int_{-\infty}^{-b} f(x)dx$ can be treated in the same manner.

The *advantages* and *disadvantages* of the Gauss-Laguerre method are pretty much the same as those of the Gauss-Legendre method. The fact that the integration interval of the Gauss-Laguerre quadrature is semi-infinite makes it a very powerful method that the Newton-Cotes methods cannot match.

A pseudomodule, **GAUSS_LAGUERRE_QUAD**, for generating N-point Gauss-Laguerre quadrature is provided in Pseudocode 8.7. The module requires the number of quadrature points (N) and a convergence tolerance (ε) as input, and the quadrature set (x_i, w_i) is the output. This module computes the abscissas by finding the zeros of the Nth-degree Laguerre polynomial ($L_n(x) = 0$) that are non-uniformly distributed on $(0, \infty)$. All roots (abscissas) are covered sequentially from the smallest to the largest within the outermost **For**-loop over index variable k. Each root (denoted by the variable u) is found by the Newton-Raphson iterations, performed within the **While**-loop as

$$x^{(p+1)} = x^{(p)} - \frac{L_N(x^{(p)})}{L_N'(x^{(p)})}, \quad \text{for } p = 0, 1, 2, \ldots$$

```
                        Pseudocode 8.7

Module GAUSS_LAGUERRE_QUAD (N, ε, x, w)
  \ DESCRIPTION: A pseudomodule to generate N-point Gauss-Laguerre
  \    quadrature abscissas and weights for any N ⩾ 2.
  \ USES:
  \    ABS :: A built-in function computing the absolute value;
  \    FACTORIAL :: A function computing N!, Pseudocode 1.4.
  Declare: x_N, w_N                            \ Declare array variables
  factor ← N!                                  \ Calculate N! only once
  For [k = 1, N]                               \ Loop k: Find kth zero, x_k
     If [k ⩽ 3] Then          \ Apply initial guess estimator for first three zeros
        u ← π²(k − 0.25)²/(4N + 1)
     Else                     \ Case of k > 3, extrapolates using most recent three zeros
        u ← 3x_{k−1} − 3x_{k−2} + x_{k−3}
     End If                                    \ u denotes x_k^{(p)}
     δ ← 1                                     \ Initialize error for x_k
     While [|δ| > ε]                           \ Newton-Raphson iteration loop
        L0 ← 1; L1 ← 1 − u       \ Initialize Lagrange polynomials, L0 and L1
        For [m = 2, N]                         \ Apply Eq. (8.82) to find L_N
           L2 ← (2m − 1 − u) * L1 − (m − 1)² * L0
           L0 ← L1; L1 ← L2                    \ Keep two most recent polynomials
        End For
        dL ← n * (L1 − n * L0)/u               \ Compute L'_N(u) by Eq. (8.84)
        δ ← L2/dL                 \ Find difference between two-successive iterates
        u ← u − δ                              \ Finc current estimate u^{(p+1)}
     End While
     x_k ← u                                   \ Save converged root as x_k
     w_k ← (factor/dL)²/u                       \ Compute weight from Eq. (8.83)
  End For
End Module GAUSS_LAGUERRE_QUAD
```

where $L_N(x)$ and $L'_N(x)$ are computed recursively using Eq. (8.82) and (8.84). In this algorithm, however, it is crucial to start the iteration process with an initial guess close enough to each zero of $L_N(x)$ to avoid finding the same zero over and over again. To serve this purpose, an initial guess value in close proximity to the true abscissa is calculated with the following asymptotic approximation:

$$x_i^{(0)} = \frac{(i − 0.25)^2 \pi^2}{4N + 1}$$

which works well for $i < 4$. Nonetheless, the initial guesses become worse for large x. Hence, the initial guess for $i \geqslant 4$ is estimated by using the following extrapolation formula by Takemasa [30]:

$$x_i^{(0)} = 3x_{i-1}^{(0)} − 3x_{i-2}^{(0)} + x_{i-3}^{(0)}$$

The iteration process is terminated when the stopping criterion ($|\delta^{(p)}| < \varepsilon$) is met. Following the initial guess estimation, the Newton-Raphson algorithm converges to a root very quickly. However, the number of maximum iterations ($maxit$) may be set to prevent an infinite loop for large N. The quadrature weights are then obtained from Eq. (8.83).

EXAMPLE 8.7: Computing improper integrals of Type I

The Debye's model for the heat capacity, C_V, for the low-temperature limit ($T \ll T_D$) leads to the following expression:

$$\frac{C_V}{kN_a} \cong 9\left(\frac{T}{T_D}\right)^3 \int_0^\infty \frac{x^4 e^x \, dx}{(e^x - 1)^2}$$

where k is the Boltzmann's constant, N_a is the number of atoms in the solid, and T and T_D are the absolute and Debye temperatures, respectively. Apply $N = 2$, 4, 6, and 8-point Gauss-Laguerre quadrature to estimate the above integral over $(0, \infty)$ and compare the estimates with the true value of $4\pi^4/15$.

SOLUTION:

The integral in question is an improper integral of Type I that is integrated over a semi-infinite interval, so the Gauss-Laguerre method can be applied to calculate its approximate value. However, the integral is not in the standard Gauss-Laguerre form but can be easily converted to standard form as follows:

$$\int_0^\infty \frac{x^4 e^x \, dx}{(e^x - 1)^2} = \int_0^\infty e^{-x} f(x) \, dx \cong GL_N = \sum_{i=1}^N w_i f(x_i)$$

where (x_i, w_i) are the Gauss-Laguerre quadrature abscissas and weights, and the integrand takes the form $f(x) = x^4 e^{2x}/(e^x - 1)^2$.

TABLE 8.7

N	2	4	6	8
GL_N	21.7878911	25.9906374	25.9746594	25.9761192
True Error, e_N	4.1878665	0.01487975	0.00109822	0.00036155

The numerical estimates of the transformed integral with $N = 2$, 4, 6, and 8-point Gauss-Laguerre quadrature sets are computed, and the estimated results along with the true (approximation) errors are presented in Table 8.7. It is observed that the sequence of the numerical estimates is approaching the true value of 25.9757576 very fast. The ratio of the subsequent true absolute errors (indicative of convergence speed) for the estimated sequence, e_{N_1}/e_{N_2}, is $e_2/e_4 \approx 281.4$, $e_4/e_6 \approx 13.55$, and $e_6/e_8 \approx 3.04$. While the convergence speed is very high at small N, it decreases with increasing N.

Discussion: Note that the original integrand $x^4 e^x/(e^x - 1)^2$ is indeterminate at $x = 0$, and the integrand is bounded on $(0, \infty)$. However, the singularity at $x = 0$ does not pose a problem in this example because none of the Laguerre abscissas coincide with $x = 0$. In other words, no special treatment is required for $x = 0$. Nonetheless, the integrand for the transformed case, $f(x)$, is unbounded on $(0, \infty)$ as it is likewise indeterminate at $x = 0$. While the numerical integration using the Gauss-Laguerre quadrature works for the unbounded $f(x)$ from an analytical perspective, the integration in the transformed form is not always numerically stable. In this example, even though $f(x)$ is unbounded, we observe that the convergence rate slows down with increasing N as the largest abscissa, x_N, grows larger.

8.10 GAUSS-HERMITE METHOD

The Gauss-Hermite method is another powerful quadrature technique used in the numerical approximation of the following types of improper integrals:

$$\int_{-\infty}^{\infty} e^{-x^2} f(x) dx \tag{8.88}$$

Similar to the other quadrature methods, the integral is replaced with a numerical approximation expressed as a weighted sum:

$$\int_{-\infty}^{\infty} e^{-x^2} f(x) dx \cong GH_N = \sum_{i=1}^{N} w_i f(x_i) \tag{8.89}$$

where N is the number of quadrature points and (x_i, w_i) is a set of parameters referred to as *Gauss-Hermite quadrature* [1]. The quadrature abscissas are the zeros of the nth-degree Hermite polynomial, which is defined as

$$H_n(x) = (-1)^n e^{x^2} \frac{d^n}{dx^n} (e^{-x^2}), \qquad n \geqslant 0 \tag{8.90}$$

where $H_0(x) = 1$ and $H_1(x) = 2x$. Higher-degree Hermite polynomials can also be constructed with the aid of the following recurrence relationship:

$$H_n(x) = 2x H_{n-1}(x) - 2(n-1) H_{n-2}(x), \qquad n \geqslant 2 \tag{8.91}$$

The first derivative of the nth-degree Hermite polynomial satisfies the following relationship:

$$\frac{dH_n(x)}{dx} = 2n H_{n-1}(x), \qquad n \geqslant 1 \tag{8.92}$$

The computation of the Gauss-Hermite quadrature set requires the solution of a system of $2N$ nonlinear equations, as in the case of the Gauss-Legendre quadrature (*see* Section 8.7). The zeros of an Nth-degree Hermite polynomial are set as the abscissas, i.e., the solution of $H_N(x) = 0$ gives x_i's. Accordingly, the non-linear system with $2N$ unknowns is reduced to a linear system with the weights (w_i's) as unknowns. It turns out that the weights also satisfy the following expression:

$$w_i = \frac{\sqrt{\pi} N! 2^{N+1}}{\left(H_N'(x_i) \right)^2} \tag{8.93}$$

Gauss-Hermite quadrature sets are tabulated for selected N values in Appendix B.3. The global error for the Gauss-Hermite method is given as

$$R_N = \frac{\sqrt{\pi}(N!)}{2^N (2N)!} f^{(2N)}(\xi), \qquad -\infty < \xi < \infty \tag{8.94}$$

A pseudomodule, **GAUSS_HERMITE_QUAD**, generating N-point Gauss-Hermite quadrature is presented in Pseudocode 8.8. The module requires the number of points (N) and convergence tolerance (ε) as input. The module output is the quadrature set (x_i, w_i) for $i = 1, 2, \ldots, N$. The zeros of $H_N(x)$ are symmetrically distributed on $(-\infty, \infty)$. Hence, the duplicate zeros with the opposite sign are not computed. However, the Newton-Raphson procedure may end up with a previously detected zero if the initial guess is not sufficiently close to the root.

Pseudocode 8.8

Module GAUSS_HERMITE_QUAD $(N, \varepsilon, \mathbf{x}, \mathbf{w})$
\ DESCRIPTION: A pseudomodule to generate N-point Gauss-Hermite quadrature
\   abscissas and weights for $N \geqslant 2$.
\ USES:
\   MOD :: A built-in function returning the remainder of x divided by y;
\   ABS :: A built-in function computing the absolute value;
\   SQRT:: A built-in function computing the square-root of a value;
\   FACTORIAL:: Function Module computing factorial, Pseudocode 1.4.
Declare: x_N, w_N \ Declare array variables
$io \leftarrow$ **MOD**$(N, 2)$ \ Find remainder of $N/2$.
If $\left[io = 0 \right]$ **Then**   \ Number of zeros is *even*
    $m = N/2$ \ Find half of the zeros
Else   \ $io = 1$, Number of zeros is *odd*
    $m = (N + 1)/2$ \ Find half of the zeros (includes $x = 0$)
End If
$factor \leftarrow 2^{N+1} N! \sqrt{\pi}$ \ Precalculate the factor for weights
For $\left[k = 1, m \right]$ \ Loop k: Find the abscissa x_k
    If $\left[k \leqslant 4 \right]$ **Then**   \ Apply initial guess estimator for first four zeros
       If $\left[io = 0 \right]$ **Then**   \ Case of even N
          $u \leftarrow (k - 0.5)\pi/\sqrt{2N + 1}$
       Else   \ Case of odd N
          $u \leftarrow (k - 1)\pi/\sqrt{2N + 1}$
       End If
    Else   \ Apply extrapolation formula for $k > 4$
       $j \leftarrow m + k - io$ \ Find location of new zero
       $u \leftarrow 3x_{j-1} - 3x_{j-2} + x_{j-3}$ \ Extrapolate using most recent 3-zeros
    End If
    $\delta \leftarrow 1$ \ Initialize successive difference
    While $\left[|\delta| > \varepsilon \right]$ \ Newton-Raphson iteration loop for x_k
       $H0 \leftarrow 1$ \ Initialize $H0$
       $H1 \leftarrow 2u$ \ Initialize $H1$
       For $\left[i = 2, N \right]$ \ Apply Eq. (8.91) to find H_N
          $H2 \leftarrow 2 * (u * H1 - (i - 1) * H0)$
          $H0 \leftarrow H1$ \ Assign $H_{i-2} \leftarrow H_{i-1}$
          $H1 \leftarrow H2$ \ Assign $H_{i-1} \leftarrow H_i$
       End For
       $dH \leftarrow 2 * N * H0$ \ Compute $H'_N(u)$ by Eq. (8.92)
       $\delta \leftarrow H2/dH$ \ Difference between two successive iterates
       $u \leftarrow u - \delta$ \ Find current estimate $u^{(p+1)}$
    End While
    $j \leftarrow m + k - io$ \ Find location of positive root
    $x_j \leftarrow u$ \ Save converged root as x_k
    $w_j \leftarrow factor/dH^2$ \ Compute weight by Eq. (8.93)
    $x_{N+1-j} \leftarrow -u$ \ Find (negative) abscissas
    $w_{N+1-j} \leftarrow w_j$ \ Find corresponding weights
End For
End Module GAUSS_HERMITE_QUAD

Thus, to get around this problem, the initial guess that is sufficiently close to the true root is estimated for $i \leqslant 4$ as:

$$x_i^{(0)} = \begin{cases} \dfrac{(2i-1)\pi}{2\sqrt{2N+1}}, & N = \text{even}, \\[3mm] \dfrac{(i-1)\pi}{\sqrt{2N+1}}, & N = \text{odd} \end{cases}$$

(8.95)

which are predictions sufficiently close to true ones. For $i \geqslant 5$, an initial guess is computed from the extrapolation formula $x_i^{(0)} = 3x_{i-1}^{(0)} - 3x_{i-2}^{(0)} + x_{i-3}^{(0)}$, introduced by Takemasa [31]. For odd N, one abscissa (zero= of $H_n(x)$ is $x = 0$. For even N, the positive abscissas ($m = (n+1)/2$) are computed in the outermost For-loop (x_i, $k = 1, 2, \ldots, m$). The Newton-Raphson iteration loop to find x_k is carried out by the While-loop):

$$u^{(p+1)} = u^{(p)} - \frac{H_N(u^{(p)})}{H_N'(u^{(p)})}$$

where $H_N(x)$ and $H_N'(x)$ are evaluated recursively using Eqs. (8.91) and (8.92). The iteration process is terminated when the stopping criterion, $|u^{(p+1)} - u^{(p)}| < \varepsilon$, is met, i.e., While-loop is terminated. Once an abscissa (zero) is found, its corresponding weight is computed from Eq. (8.95) [28]. Negative abscissas are determined from known positive abscissas, while the corresponding weights are the same as the weights of positive abscissas.

 The reader is warned that computing the weights for cases with a large N ($N > 20$) will result in "*arithmetic overflow.*" As a remedy to avoid arithmetic overflow, an orthonormal Hermite polynomial set is used (*see* Ref. [31]).

Any improper integral of $f(x)$ bounded on $(-\infty, \infty)$ can be evaluated once it is transformed to the standard form, Eq. (8.88), as follows:

$$\int_{-\infty}^{\infty} f(x)dx = \int_{-\infty}^{\infty} e^{-x^2} F(x)dx$$

(8.96)

where $F(x) = e^{x^2} f(x)$. Then Eq. (8.96) can be expressed as a weighted sum as follows:

$$\int_{-\infty}^{\infty} e^{-x^2} F(x)dx \cong \sum_{i=1}^{N} w_i e^{x_i^2} f(x_i) = \sum_{i=1}^{N} W_i f(x_i)$$

(8.97)

where $W_i = w_i e^{x_i^2}$ are the modified weights, also presented in Appendix B.3.

EXAMPLE 8.8: Computing improper integrals of Type I

The standard normal distribution has the following probability density:

$$F(x) = \frac{1}{\sqrt{2\pi}} e^{-x^2/2}, \qquad -\infty < x < \infty$$

(a) Use 2, 4, 6, 8, and 10-point Gauss-Hermite quadrature to estimate the integral of $F(x)$ over the entire domain, and (b) can you estimate the upper bound of the global error for any N? Recall that the true value of the integral of $F(x)$ over the entire domain is "1."

SOLUTION:

(a) This integral is not in standard Gauss-Hermite form, but it can be transformed into standard form as follows:

$$\int_{-\infty}^{\infty} F(x)\,dx = \int_{-\infty}^{\infty} e^{-x^2} f(x)\,dx \cong GH_N = \sum_{i=1}^{N} w_i f(x_i)$$

where $f(x) = e^{x^2/2}/\sqrt{2\pi}$. Note that $f(x)$, an even function, has a local minimum at $x = 0$, and it is unbounded as $x \to \pm\infty$. Nevertheless, $F(x)$ itself is a bounded function on $(-\infty, \infty)$, so the desired integral is convergent. The unboundedness of the integrand of the transformed integral, $f(x)$, however, can lead to slow convergence.

TABLE 8.8

N	2	4	6	8	10
GH_N	0.90794308	0.97038621	0.99896336	0.99988715	0.99998764
e_N	0.09205692	0.02961379	0.00103664	0.00011285	1.236×10^{-5}

Using the Gauss-Hermite quadrature sets given in Appendix B.3, the numerical estimates GH_N, along with the true approximation errors $e_N = 1 - GH_N$, are presented in Table 8.8 for $N = 2$, 4, 6, 8, and 10. It is observed that the absolute errors are about 9.2% and 2.96%, for GH_2 and GH_4, respectively. But the numerical estimates improve as N is increased, leading to absolute errors less than 0.1% for $N \geqslant 6$.

(b) In order to estimate the global error bounds, e.g., for $N = 2$ and 4, the fourth and eighth derivatives of the integrand are required:

$$f^{(4)}(x) = (3 - 24x^2 + 8x^4)e^{x^2/2},$$
$$f^{(8)}(x) = (105 + 420x^2 + 210x^4 + 28x^6 + x^8)e^{x^2/2}$$

We observe that both $f^{(4)}(x)$ and $f^{(8)}(x)$ are unbounded as $x \to \pm\infty$ and have a local minimum at $x = 0$. Hence, it is not possible to find an upper bound for the numerical estimates since the higher derivatives are unbounded. However, we can find the lower bounds for the estimates since $f^{(4)}(0) = 3$, and $f^{(8)}(0) = 105$; in other words, $f^{(4)}(\xi) \geqslant 3$ and $f^{(8)}(\xi) \geqslant 105$. Then the lower bounds of the global error are found by $R_2 > 2!\sqrt{\pi}/2^2 4! = 0.1108$ and $R_4 > 4!\sqrt{\pi}/2^4 8! = 0.00692$.

Discussion: The original integrand $F(x)$ is bounded, and its integral is convergent over the entire integration domain because $F(x)$ decays very fast as $x \to \pm\infty$. However, when the integral is transformed to suit the general Gauss-Hermit form, the transformed form of the integral can become unstable or cause slow convergence problems due to the unboundedness of the new integrand, $f(x)$. Yet the integrals are likely to converge for sufficiently large N. Nonetheless, given that GH_{10} is computed with only 10 function evaluations with an absolute error of 1.236×10^{-5}, the method is superior to other methods such as trapezoidal or Simpson's rules, which can only be applied after truncating the integral over a bounded range.

8.11 GAUSS-CHEBYSHEV METHOD

The Gauss-Chebyshev method is also a powerful quadrature method that can handle the numerical integration of integrands that have discontinuities in both the lower and upper integral limits. The standard form of the integral is

$$\int_{-1}^{1} \frac{f(x)dx}{\sqrt{1-x^2}} \tag{8.98}$$

where $f(x)$ is bounded and $(1-x^2)^{-1/2}$ term is singular at $x = \pm 1$.

The integral is approximated by a weighted sum as follows:

$$\int_{-1}^{1} \frac{f(x)dx}{\sqrt{1-x^2}} \cong C_N = \sum_{i=1}^{N} w_i f(x_i) \tag{8.99}$$

where C_N denotes the N-point Gauss-Chebyshev approximation and (x_i, w_i) are so-called the *Gauss-Chebyshev quadrature*.

The *abscissas* for an N-point quadrature are given by the zeros of the Chebyshev polynomial of the first kind, $T_n(x)$, which are symmetrical about $x = 0$. Fortunately, in this method, the abscissas (x_i) and the *weights* (w_i) can be calculated explicitly as follows:

$$x_i = \cos\frac{(2i-1)\pi}{2N} \quad \text{and} \quad w_i = \frac{\pi}{N}, \qquad i=1,2,\ldots,N \tag{8.100}$$

The global error term for the method is given as

$$R_N = \frac{\pi}{(2N)!2^{2N-1}} f^{(2N)}(\xi), \qquad -1 < \xi < 1 \tag{8.101}$$

A pseudocode for generating the Gauss-Chebyshev quadrature has not been given here since the quadrature abscissas and weights are explicitly available, Eq. (8.100), which can be easily calculated practically at no cost.

GENERALIZATION: This method, similar to Gauss-Legendre, can also be extended to estimate a definite integral over an arbitrary interval after transforming it to the standard Gauss-Chebyshev form. Consider the following general integral with discontinuities at both ends:

$$\int_{a}^{b} \frac{f(x)dx}{\sqrt{(x-a)(b-x)}} \tag{8.102}$$

Upon applying the $x = (b-a)u/2 + (b+a)/2$ substitution, the integral is transformed to standard Gauss-Chebyshev form, which can then be approximated by the weighted sum as follows:

$$\int_{u=-1}^{1} f\left(\frac{b-a}{2}u + \frac{b+a}{2}\right)\frac{du}{\sqrt{1-u^2}} \cong \sum_{i=1}^{N} w_i f(x_i) \tag{8.103}$$

where (u_i, w_i) are the Gauss-Chebyshev quadrature defined by Eq. (8.100) and $x_i = ((b-a)u_i + (a+b))/2$ are the abscissas. The global error term, R_N, takes the following form:

$$R_N = \frac{\pi(b-a)^{2N+1}}{(2N)!2^{4N}} f^{(2N)}(\xi), \quad a < \xi < b \tag{8.104}$$

EXAMPLE 8.9: Application of Gauss-Chebyshev method to Type II integrals

Estimate the following integral using the three-point Gauss-Chebyshev quadrature and compare the result with the true value of 3π.

$$\int_{-1}^{3} \frac{x^2 dx}{\sqrt{(x+1)(3-x)}}$$

SOLUTION:

The integral is already in the standard form, Eq. (8.102), with $f(x) = x^2$. For $N = 3$, the abscissas and weights on $(-1, 1)$ are found from Eq. (8.100) as

$$u_1 = \cos\frac{\pi}{6} = \frac{\sqrt{3}}{2}, \quad u_2 = \cos\frac{\pi}{2} = 0, \quad u_3 = \cos\frac{5\pi}{6} = -\frac{\sqrt{3}}{2}, \quad w_i = \frac{\pi}{3} \text{ (for all } i)$$

The abscissas in the $[-1, 3]$ interval are calculated by $x_i = 2u_i + 1$ as

$$x_1 = \sqrt{3} + 1, \quad x_2 = 1, \quad x_3 = -\sqrt{3} + 1$$

The estimate for the given integral is obtained by Eq. (8.103) as

$$C_3 = \frac{\pi}{3}\left(x_1^2 + x_2^2 + x_3^2\right) = \frac{\pi}{3}\left[\left(\sqrt{3}+1\right)^2 + (1)^2 + \left(-\sqrt{3}+1\right)^2\right] = 3\pi$$

which happens to give the true integral value. Note that the global error term R_3, Eq. (8.104), is

$$R_3 = \frac{\pi}{180} f^{(6)}(\xi), \quad -1 < \xi < 3$$

For $f(x) = x^2$, the Gauss-Chebyshev sum, C_3, should theoretically yield the true value because $f^{(6)}(x) = 0$ and $R_3 = 0$. Moreover, it is also clear that the global error R_3 will be zero for $f(x)$ being a polynomial of degree $m \leqslant 5$, i.e., $f(x) = p_m(x)$, provided that the numbers are not rounded or truncated.

Discussion: As in all Gaussian quadrature methods, the Gauss-Chebyshev approximation obtained by a weighted sum involves a fixed number of integration points and provides the true value for polynomials of degree $2N - 1$, where N is the number of integration points. In this example, the quadrature abscissas and weights were determined in their exact forms and used in the arithmetic computations. This process completely eliminated the possibility of any truncation and/or round-off errors.

Some integrals may partially have Gauss-Chebyshev form; that is, the integrand can have a discontinuity only at $x = a$ or $x = b$. For example, consider the integration of $g(x) = f(x)/\sqrt{b-x}$ and $g(x) = f(x)/\sqrt{x-a}$ on $[a, b]$. These can be expressed in the standard Gauss-Chebyshev form as follows:

$$\int_a^b g(x)dx = \int_a^b \frac{f(x)dx}{\sqrt{b-x}} = \int_a^b \frac{f(x)\sqrt{x-a}dx}{\sqrt{(x-a)(b-x)}} = \int_a^b \frac{F(x)dx}{\sqrt{(x-a)(b-x)}} \tag{8.105}$$

where $F(x) = f(x)\sqrt{x-a}$.

Gauss-Laguerre and Gauss-Hermite Quadrature Methods

- High-order accuracy can be achieved using GL_N and GH_N;
- They converge very quickly for relatively small N if the integrals are in standard improper forms.

- The theoretical error estimation procedure is not an easy task;
- They are not suitable for discrete functions;
- They are very poor at integrating non-exponentially decreasing integrands;
- Highly accurate quadrature sets are required to compute integrals with high accuracy.

EXAMPLE 8.10: Alternative ways to calculate Type II integrals

Estimate the following integral using (a) 2-, 3-, 4-, and 6-panel open Newton-Cotes formulas; (b) 5-, 8-, 16-, 32-, and 64-point Gauss-Chebyshev quadrature; (c) 4-, 8-, 16-, 32-, and 64-point Gauss-Legendre quadrature; (d) 4-, 8-, 16-, 32-, and 64-point Gauss-Laguerre quadrature after transforming the integral into the one on $(0, \infty)$; and (e) after truncating it to an improper integral of Type I.

$$\int_0^1 \frac{dx}{\sqrt{x}}$$

SOLUTION:

This integral is one of the easiest integrals in calculus, whose true value is "2." However, we encounter a problem in its numerical integration because the integrand is discontinuous at $x = 0$, which makes it a Type II improper integral.

(a) The *closed* Newton-Cotes formulas are not suitable for estimating this integral because the integrand is not defined at $x = 0$, i.e., $f(x) \to \infty$ for $x \to 0^+$, which is why we need to resort to *open*-type formulas. Applying the 2-, 3-, 4-, and 6-panel composite open Newton-Cotes rules, Eqs. (8.57), (8.58), (8.59), (8.60), and (8.61) avoid the singularity at $x = 0$. The orders of error for the composite rules are $\mathcal{O}(h^2)$, $\mathcal{O}(h^2)$, $\mathcal{O}(h^3)$, and $\mathcal{O}(h^4)$ for the 2-, 3-, 4-, and 6-panel schemes, respectively. The numerical results, keeping the total number of panels (N) the same in each method, are presented in Table 8.9.

TABLE 8.9

N	Eq. (8.57)	Eq. (8.58)	Eq. (8.59)	Eq. (8.61)
24	1.825525	1.811233	1.849491	1.859210
48	1.876562	1.866423	1.924745	1.900447
96	1.912699	1.905522	1.924745	1.929605
192	1.938265	1.933187	1.946787	1.950223
6912	1.989710	1.988864	1.991131	1.991704

We observe that the three-panel formula Eq. (8.58) with $K = 3f''(\xi)/4$ exhibits worse performance than that of the two-panel formula Eq. (8.57) with $K = f''(\xi)/3$, even though they both are second-order accurate. The explanation for this situation lies in the magnitude of the truncation error constants.

(b) The Gauss-Chebyshev method, which inherently avoids the end points, is a potential remedy for such an integral with an endpoint singularity. However, the integral is not in the standard Gauss-Chebyshev form. Nevertheless, it could be put in the standard form by multiplying both the numerator and denominator with $\sqrt{1-x}$, giving:

$$\int_0^1 \frac{dx}{\sqrt{x}} = \int_0^1 \frac{\sqrt{1-x}}{\sqrt{x(1-x)}} dx$$

Now, defining $f(x) = \sqrt{1-x}$ and using the N-point Gauss-Chebyshev quadrature, the integral can be approximated as

$$C_N = \sum_{i=1}^N w_i \sqrt{1-x_i}$$

where $x_i = (u_i + 1)/2$ and (u_i, w_i) are the Gauss-Chebyshev quadratures on $(-1, 1)$.

The numerical estimates for $N = 5, 8, 16, 32$, and 64 are presented in Table 8.10. It is observed that the C_N estimates converge very quickly toward the true value of "2" for increasing N since the discontinuity at the end points is avoided by $f(x)$.

TABLE 8.10

N	5	8	16	32	64
C_N	2.008248	2.003216	2.000803	2.0002	2.00005

(c) The Gauss-Legendre method can be employed for Type II integrals since the abscissas never coincide with the upper or lower endpoints. The numerical approximation using the Gauss-Legendre quadrature can be expressed as

$$\int_0^1 \frac{dx}{\sqrt{x}} \approx G_N = \sum_{i=1}^N \frac{w_i}{\sqrt{x_i}}$$

where (u_i, w_i) is the quadrature set on $[-1, 1]$ and $x_i = (u_i + 1)/2$. The numerical estimates with $N = 4, 8, 16, 32$, and 64 are computed and tabulated in Table 8.11. It is observed that the Gauss-Legendre estimates slowly converge toward the true value of 2 for increasing N with a small convergence rate.

TABLE 8.11

N	4	8	16	32	64
G_N	1.806342	1.89754	1.947225	1.973201	1.986421

(d) To avoid the singular point at $x = 0$, we may employ the $x = 1/y^2$ substitution, which yields the following Type I improper integral:

$$\int_0^1 \frac{dx}{\sqrt{x}} = 2 \int_1^\infty \frac{dy}{y^2}$$

The new integral has a discontinuity at $y = 0$, outside the integration domain $[1, \infty)$, and moreover, it is not in the standard Gauss-Laguerre form either. Next, the integral is transformed to the standard form by multiplying the integrand with $e^{-u}e^{u}$ under the integral sign and applying the $u = y - 1$ substitution as follows:

$$2\int_1^\infty \frac{dy}{y^2} = 2\int_0^\infty e^{-u}\frac{e^u\,du}{(u+1)^2}$$

Having defined $f(x) = e^x/(x+1)^2$, the 4-, 8-, 16-, 32-, and 64-point Gauss-Laguerre quadratures are used to estimate the transformed integral. The tabulated results in Table 8.12 indicate that this alternative integration technique also converges slowly.

TABLE 8.12

N	4	8	16	32	64
GL_N	1.852473	1.930778	1.96633	1.983509	1.991889

(e) The integrand of the improper integral over the $1 \leqslant y < \infty$ interval obtained in Part (d) is continuous in this interval since $y = 0$ lies outside the integration range. Hence, this integral can be truncated into a proper definite integral, which can then be approximated by either the trapezoidal or, preferably, Simpson's rules. To determine an interval that minimizes the truncation error, the integral is split into two integrals:

$$\int_1^\infty \frac{dy}{y^2} = \int_1^b \frac{dy}{y^2} + \int_b^\infty \frac{dy}{y^2}$$

where the first integral is the truncated integral and the second one, denoted by $R(b)$, is the approximation error due to truncation. In this example, the analytical integration of $R(b)$ is quite simple and yields $R(b) = 1/b$. It is clear that to get an estimate that is accurate to four significant digits (i.e., $R(b) = 1/b < 10^{-4}$), the value of b should be greater than 10^4. Hence, we can conclude that this alternative technique is computationally impractical.

Discussion: Even though the numerical approximations with the open Newton-Cotes rules and Gauss-Legendre method improve as the number of integration points increases, the convergence rate is slow. However, the integrand in the Chebyshev sum avoids the endpoint discontinuities. The approximation C_{64} (estimate with $N = 64$) yields an answer correct to four decimal places.

Another technique that can be used to avoid singularity in the integrals is to transform an integral into a new one whose integrand is not singular in the new range. In this regard, even though the integrand $1/x^2$ is bounded on $[1, \infty)$, the integrand in the transformed space $f(x) = e^x/(x+1)^2$ is not bounded. This feature accounts for the slow convergence with increasing N.

Finally, resorting to blindly truncating improper integrals is not recommended unless the truncation error is estimated, as demonstrated in this example. In cases where $R(b)$ is difficult to obtain analytically, find an upper bound by approximating the integrand to a form that can be easily integrated (*see* Section 8.8). Another alternative is to obtain the numerical estimates for several b values ($b_1 < b_2 < b_3$). Then, for instance, using Simpson's rule for a fixed h, calculate the difference ($|S_N(b_1) - S_N(b_2)|$, $|S_N(b_2) - S_N(b_3)|$, and so on) and determine the value of b, which will be acceptable.

8.12 COMPUTING INTEGRALS WITH VARIABLE LIMITS

Definite integrals with variable upper or lower integration limits are frequently encountered in science and engineering. Such integrals may be generalized as follows:

$$I(x) = \int_{y=u(x)}^{v(x)} f(x,y)\, dy \tag{8.106}$$

where $u(x)$ and $v(x)$ are functions of x. Hence, in such cases, it may be difficult to determine in advance the number of integration points required to ensure that the computed integral is estimated with the desired accuracy. To overcome this problem, the definite integral is preferably transformed into an integral with fixed (constant) upper and/or lower limits. For instance, upon applying the following substitution to Eq. (8.106)

$$y(x,t) = u(x) + (v(x) - u(x))\frac{t-c}{d-c} \tag{8.107}$$

the integral is transformed into an integral over an arbitrary interval $[c,d]$:

$$I(x) = \int_{t=c}^{d} F(x,t)\, dt \quad \text{with} \quad F(x,t) = \frac{v(x)-u(x)}{d-c} f\big(x, y(x,t)\big) \tag{8.108}$$

Once the transformation is completed, a suitable numerical method is applied to estimate Eq. (8.108) for an arbitrary x. The Gauss-Legendre or Gauss-Chebyshev quadrature methods, for instance, can be applied due to the considerable computational savings they offer. It is often more convenient to transform Eq. (8.106) to an integral on $[-1,1]$ to take advantage of tabulated quadrature in this range so that no further processing is necessary.

EXAMPLE 8.11: Computing integrals with variable limits

Propose a suitable numerical method to estimate the following integral, and transform the integral to one with constant upper and lower endpoints.

$$g(x) = \int_{-x}^{x} F(x,t) \quad \text{with} \quad F(x,t) = \frac{\sin^2(t)}{\sqrt{t^2 - x^2}}\, dt$$

SOLUTION:

(a) The integrand $F(x,t)$ is discontinuous at the endpoints, $x = \pm t$. Having $\sqrt{t^2 - x^2}$ in the denominator makes it a suitable candidate for the Gauss-Chebyshev method. Therefore, it is more convenient to transform the integral to the one on $[-1,1]$.

(b) To obtain the standard Gauss-Chebyshev form, we apply Eq. (8.107) to obtain the following transformation:

$$t = -x + x(y+1) = x\,y \quad \rightarrow \quad dt = x\, dy$$

Substituting these into the integral and simplifying leads to

$$g(x) = \int_{-x}^{x} \frac{\sin^2(t)}{\sqrt{x^2 - t^2}}\, dt = \frac{x}{|x|} \int_{y=-1}^{+1} \frac{\sin^2(xy)}{\sqrt{1-y^2}}\, dy$$

Defining $f(y) = x \sin^2(xy)/|x|$, the integral is in standard Gauss-Chebyshev form, which may be replaced by an N-point Gauss-Chebyshev sum:

$$\int_{y=-1}^{+1} \frac{f(y)}{\sqrt{1-y^2}} dy \cong \frac{\pi}{N} \sum_{n=1}^{N} f(y_n)$$

Finally, substitution of the quadrature sum into the transformed integral yields

$$g(x) = \int_{-x}^{x} \frac{\sin^2(t)}{\sqrt{x^2-t^2}} dt \cong C_N = \frac{\pi}{N} \frac{x}{|x|} \sum_{i=1}^{N} \sin^2(x \, y_i)$$

where y_i's are the abscissas of the Gauss-Chebyshev quadrature on $[-1, 1]$.

Discussion: It can be difficult to determine and/or apply a suitable method or treatment (number of panels or quadrature points, error estimation, etc.) for the integrals with variable limits. However, it becomes easier once such an integral for an arbitrary x is transformed into one with constant integration limits.

8.13   DOUBLE INTEGRATION

Numerical approximation of double integrals is relatively easy when the integration domain is rectangular (Domain D: $a \leqslant x \leqslant b$, $c \leqslant y \leqslant d$, see Fig. 8.13a). To illustrate the integration of $f(x, y)$ over a rectangular domain, consider the following double integration:

$$I = \int_{x=a}^{b} \int_{y=c}^{d} f(x,y) dy dx \qquad (8.109)$$

where $f(x, y)$ is assumed to be an *integrable function* that cannot be expressed as a product of two single-variable functions, i.e., $f(x, y) \neq X(x)Y(y)$. In the case of $f(x, y) = X(x)Y(y)$, the double integral is reduced to the product of two single integrals, $(\int_a^b X(x) dx)(\int_c^d Y(y) dy)$, for which any suitable method or technique applicable to single integrals can be used.

Fubini's theorem states that the double integral of any continuous function $f(x, y)$ over a rectangular domain (D as shown in Fig. 8.13a) can be calculated as an iterated integral

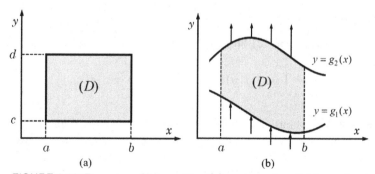

FIGURE 8.13: Domains of integration (a) rectangular, (b) irregular.

in either order of integration:

$$\iint_D f(x,y)dA = \int_a^b \int_c^d f(x,y)dydx = \int_c^d \int_a^b f(x,y)dxdy$$

In this regard, we choose to compute the integral with respect to y first:

$$F(x) = \int_{y=c}^d f(x,y)dy \tag{8.110}$$

Equation (8.109) can then be cast as a single definite integral as

$$I = \int_{x=a}^b F(x)dx \tag{8.111}$$

Any suitable numerical method can be employed for Eq. (8.111), assuming it is not singular in or on the boundaries of domain D. However, recall that numerical integration of some integrands with removable singularities can be obtained using a suitable technique.

In generalizing the numerical integration procedure, we exploit the idea that all numerical estimates can be expressed as a weighted sum. So we approximate Eq. (8.111) as

$$\int_a^b F(x)\,dx \cong \sum_{i=1}^N w_{xi}F(x_i) \tag{8.112}$$

where N is the number of integration points and (x_i, w_{xi}) for $i = 1,2,\ldots,N$ is a set of abscissas and weights depending on the chosen method.

To complete the numerical calculation of Eq. (8.109), Equation (8.110) is also expressed as a weighted sum for any x in $[a,b]$ as follows:

$$F(x) \approx \sum_{j=1}^M w_{yj}f(x,y_j) \tag{8.113}$$

where M is the number of integration points, and (y_i, w_{yi}) for $i = 1,2,\ldots,M$ is a set of abscissas and weights depending on the method selected. It should be pointed out that the number of integration points (or panels) or the numerical method for the integration over each independent variable need not be the same. In other words, the analyst is free to select the most suitable numerical method or the number of integration points for each integration variable. It should be kept in mind, however, that the accuracy of the numerical estimates will be affected by the number of integration points and the applied methods, which will contribute to the global error.

Combining with Eqs. (8.112) and (8.113) for x_i's, we arrive at

$$\int_{x=a}^b \int_{y=c}^d f(x,y)dydx \approx \sum_{i=1}^N w_{xi} \sum_{j=1}^M w_{yj}f(x_i,y_j) \tag{8.114}$$

In the case of double integration in an irregular domain, as depicted in Fig. 8.13b, the range of integration over the y variable is $u(x) \leqslant y \leqslant v(x)$, which is transformed to an integral with a fixed integration interval, as discussed in Section 8.13. Then, the double integral can be expressed by the following approximation:

$$\int_{x=a}^b \int_{y=u(x)}^{v(x)} f(x,y)\,dydx \cong \frac{1}{d-c}\sum_{i=1}^N w_{xi}(v_i-u_i)\sum_{j=1}^M w_{tj}\,f\left(x_i,u_i+(t_j-c)\frac{v_i-u_i}{d-c}\right) \tag{8.115}$$

where the integral over the variable y is transformed to an integral over the variable t in an arbitrary interval $[c, d]$ using Eq. (8.107). The lower and upper endpoints are abbreviated as $u_i = u(x_i)$, $v_i = v(x_i)$, and (t_i, w_{ti}) is the set of abscissas and weights on $[c, d]$ depending on the chosen numerical method.

EXAMPLE 8.12: Computing hydrostatic pressure force by double integration

A shipwreck is lying upside down on a flat seafloor at a depth of H m, as illustrated in the figure below.

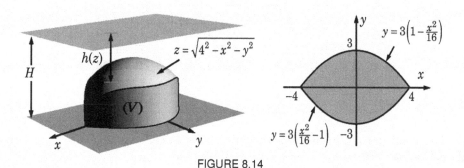

FIGURE 8.14

The ship, a three-dimensional space region (V), is bounded by the $z = 0$ plane (the seafloor), a sphere (the top: $z = \sqrt{4^2 - x^2 - y^2}$) and on the sides by parabolic planes (the hulls: $y = \pm 3(x^2/16 - 1)$). The ship is subjected to hydrostatic pressure force, $dF = \rho g h(z)$ per differential area dA. The total force is found by integrating over the bottom surface area of the ship, leading to the following double integral:

$$F = \iint \rho g h(z) dA = \rho g \int_{x=-4}^{4} \int_{y=3(x^2/16-1)}^{3(1-x^2/16)} \left(H - \sqrt{16 - x^2 - y^2} \right) dy \, dx$$

First transform both integrals to the ones on $[-1, 1]$, and then, for $H = 100$ m, estimate the double integral using 2-, 3-, and 4-point Gauss-Legendre quadrature on both the x- and y-integrals and assess the accuracy of the integrals. The lengths in the figure are given in meters.

SOLUTION:

By making the $x = 4x'$ substitution ($dx = 4dx'$), the integral over x is easily transformed to an integral on $-1 \leqslant x' \leqslant 1$. On the other hand, to transform the integral over the y-variable to the one on $-1 \leqslant y' \leqslant 1$, we employ Eq. (8.108) to convert $y = 3(x^2/16 - 1)$ to $y' = -1$ and $y = 3(1 - x^2/16)$ to $y' = 1$ as

$$y = 3\left(\frac{x^2}{16} - 1\right) + \frac{3}{2}\left(1 - \frac{x^2}{16}\right)(y' + 1)$$

Also, substituting $x' = x/4$ leads to $y = 3(1 - x'^2)y'$ and $dy = 3(1 - x'^2)dy'$ for $-1 \leqslant y' \leqslant 1$. We can now express the integration over the y-variable as

$$I_1(x') = 3(1 - x'^2) \int_{y'=-1}^{+1} \left(H - \sqrt{16 - 16x'^2 - 9(1 - x'^2)^2 y'^2} \right) dy'$$

Next, this integral is replaced with the N-point Gauss-Legendre sum:

$$I_1(x') \cong 3(1 - x'^2) \sum_{i=1}^{N} w_i \left(H - \sqrt{16 - 16x'^2 - 9(1 - x'^2)^2 y_i^2} \right)$$

where (y_i, w_i) are the N-point Gauss-Legendre quadrature set on $[-1, 1]$ (*see* Appendix B.1).

TABLE 8.13

i	j	x_i'	w_i'	y_j'	w_j'	$w_i' w_j' F_{i,j}$
1	1	-0.57735	1	-0.57735	1	775.5596
1	2	-0.57735	1	0.57735	1	775.5596
2	1	0.57735	1	-0.57735	1	775.5596
2	2	0.57735	1	0.57735	1	775.5596
						3102.2384

The x'-integration is also replaced by the N-point Gauss-Legendre sum. Using the same quadrature set, $(x_i, w_i) \equiv (y_i, w_i)$, we write

$$I_2 = \int_{x=-4}^{4} I_1(x)dx = 4 \int_{x'=-1}^{1} I_1(x')dx' \cong 4 \sum_{j=1}^{N} w_j I_1(x_j')$$

Finally, substituting $I_1(x')$ into the above expression yields

$$I_2 = \sum_{j=1}^{N} w_j \sum_{i=1}^{N} w_i F(x_j', y_i')$$

where

$$F(x', y') = 12(1 - x'^2)\left(H - \sqrt{16 - 16x'^2 - 9(1 - x'^2)^2 y'^2} \right)$$

The numerical details for only the $N = M = 2$ case are tabulated in Table 8.13. For $N = M = 2$, 3, and 4, the numerical estimates are found to be 3102.2384, 3097.5511, and 3096.7293, respectively.

Discussion: The error estimation for the Gauss-Legendre method in practice can be very difficult, even for the integration of single-variable functions, since it requires the $2N$th derivative. Yet, the true error may be much less than a bound established by the R_N. Another approach to estimating the error in numerical approximation is to use two Gaussian-Legendre quadrature sets of different orders and to obtain the error as the difference between the two estimates. In this example, the absolute error between the estimates with $N = 2$ and $N = 3$ is 4.6873 (or 0.15%), while the estimates with $N = 3$ and $N = 4$ are 0.8218 (or 0.0265%), which indicates that the integral computed even with $N = M = 2$ is a good approximation for the double integral.

8.14   CLOSURE

In science and engineering, we basically encounter the definite integration of two kinds of functions: (a) explicit or (b) discrete.

(a) Numerical estimation of a definite integral of an explicitly stated continuous function $f(x)$ should be carried out if the analytical integration is either impossible or too difficult to obtain. In this case, the user has the freedom to choose (i) a numerical method and (ii) a suitable h.

Low-order closed Newton-Cotes formulas such as Trapezoid and Simpson rules, usually applied in composite formulations, are preferred to higher-order methods due to their ability to provide satisfactory solutions to many problems. In this regard, the computation time and accuracy of the Trapezoidal and Simpson's rules can be further improved with a little computational effort by employing the *end corrected* formulas, leading to the orders of error of $\mathcal{O}(h^4)$ and $\mathcal{O}(h^6)$, respectively. The choice of the interval size h in the compound rules is another challenge of the numerical integration process. In an effort to reduce the truncation error that is a function of h, the interval size should be chosen as "small," but not so small that it requires unacceptably excessive cpu-time. The explicit expressions derived for the truncation errors can guide the analyst in selecting a suitable method and an optimal h.

Romberg's rule allows the estimation of an integral to a desired degree of accuracy, which is why it is used for the adaptive integration of well-behaved functions. Most of the cpu-time is devoted to generating the first column (the trapezoidal rule with minimum functional evaluations) of Romberg's table, and the rest of the table entries are obtained by Richardson's extrapolation.

Quadrature methods are quite powerful and effective, as they provide highly accurate estimates of the integrals of smooth functions with much fewer integration points. Gauss-Legendre quadrature can be applied to the integration of any explicit function (except for those depicting highly oscillatory behavior), including those with discontinuities at endpoints. There are several efficient quadrature methods, each of which corresponds to a special case: Type II integrals with singular points on both ends (Gauss-Chebyshev), Type I improper integrals over a semi-infinite (Gauss-Laguerre) or infinite domain (Gauss-Hermite), and so on. However, very accurate computations of abscissas and weights are required.

The open Newton-Cotes rules avoid the integrand at end points and thus should be preferred only for Type II integrals with discontinuities at one or both endpoints. Even then, low-order open Newton-Cotes formulas implemented with the composite rule should be preferred to a high-order formula that leads to Runge's phenomenon. In many cases, improper integrals can be transformed into ones in more convenient finite ranges where the integrals can be computed with a standard numerical method of choice. If an improper integral can be transformed into a proper integral, then this avenue should be taken first.

(b) Discrete functions arise from experiments and data collection instruments, where the data (integration) points and the interval size are fixed. Estimating the definite integral of a discrete function is carried out using the composite trapezoidal rule or Simpson's rules. In special cases, Romberg's rule may be used only if the number of panels is a multiple of two (2^m) and the number of data points is sufficient. Since the analyst does not have the freedom to modify the interval size h, a high-order numerical method should be exploited.

Finally, the numerical methods or techniques applied to the integration of univariant, explicit continuous, or discrete functions can easily be extended to multiple integrals.

8.15 EXERCISES

Section 8.1 Trapezoidal Rule

E8.1 Find the theoretical global error bounds if the N-panel trapezoidal rule is to be applied to the following integrals:

$$\text{(a)} \int_0^1 x^4 dx, \quad \text{(b)} \int_{-1}^1 e^{2x} dx, \quad \text{(c)} \int_0^{1/2} \frac{dx}{x+1}, \quad \text{(d)} \int_0^{\pi/2} \sin(e^x) dx$$

E8.2 Find the minimum number of panels required to evaluate the following integrals using the N-panel trapezoidal rule so that the approximation error is less than 10^{-4}.

$$\text{(a)} \int_{-1}^1 x^5 dx, \quad \text{(b)} \int_1^2 \ln(x) dx, \quad \text{(c)} \int_0^3 \frac{dx}{x^2+9}, \quad \text{(d)} \int_0^{\pi/2} \cos \frac{x}{2} dx$$

E8.3 Estimate the following integral: (i) analytically; (ii) numerically using the trapezoidal rule with 1-, 2-, 4-, and 8-panels; (iii) determine the theoretical global error bounds; (iv) compare the true error with the approximation errors; and (v) discuss your findings.

$$\text{(a)} \int_0^3 (9x - 3x^2) dx, \quad\quad\quad \text{(b)} \int_0^1 xe^{-x^2} dx$$

E8.4 Estimate the following integral: (a) using the 6- and 10-panel trapezoidal rule; (b) determining the global error bounds; and (c) comparing the approximation errors with the maximum global error bound. The integral value, correct to nine decimal places, is given as 5.8091946665521.

$$\int_0^\pi \sqrt{1 + 4\sin x} \, dx$$

E8.5 Consider the uniformly spaced discrete function $f(x)$ tabulated below. Use the trapezoidal rule with (a) $h = 0.4$; (b) $h = 0.2$; and (c) $h = 0.1$ to estimate the integral of $f(x)$ between $x = 0.4$ and 1.2.

x	0.4	0.5	0.6	0.7	0.8	0.9	1	1.1	1.2
$f(x)$	15	12	10	8.6	7.5	6.7	6	5.5	5

E8.6 Consider the uniformly spaced discrete function $f(x)$ tabulated below. Use the trapezoidal rule to (a) estimate the integral below and (b) compare your numerical estimate with the true value of 9.542592.

$$\int_{0.5}^{1.4} (1 + x^2) f(x) dx$$

x	0.5	0.65	0.8	0.95	1.1	1.25	1.4
$f(x)$	2	2.69	3.56	4.61	5.84	7.25	8.84

E8.7 The heat capacity of gases in an isobaric process is defined as the rate of enthalpy change with temperature as $c_p = (\partial h / \partial T)_p$. The experimental data for the heat capacity (in kJ/kg·K) of a gas mixture as a function of temperature (in K) is tabulated below.

T	250	300	350	400	450	500
c_p	0.161	0.194	0.226	0.259	0.292	0.324

The enthalpy of the gas mixture at a given temperature T is determined as follows:

$$h(T) - h(250\,\text{K}) = \int_{T=250}^{T} c_p(y) dy$$

Given $h(250\text{K})$=19.8 kJ/kg, apply the trapezoidal rule to estimate the enthalpy of the gas for every temperature value available in the table.

E8.8 Estimate the following integrals (i) analytically and (ii) numerically with the specified number of panels using the trapezoidal rule with and without end-correction. (iii) Determine the errors for both numerical estimates and discuss your findings.

(a) $\int_0^1 x^3\, dx$ $(N = 5)$, (b) $\int_0^2 x^7\, dx$ $(N = 5, 10, 20)$, (c) $\int_0^1 x(x^2 + 3)^3 dx$ $(N = 10)$

E8.9 Estimate the given integrals (a) analytically and (b) numerically by using the 10-panel trapezoidal rule. (c) Determine the true error and the theoretical global error bounds to discuss the accuracy of your estimate. (d) Can you apply the trapezoidal rule with end-correction to this integral?

$$\int_{-2}^0 \sqrt{4 - x^2}\, dx$$

E8.10 A discrete function, $f(x)$, is tabulated below. (a) Estimate the definite integral of $f(x)$ between $x = 1$ and 2.6 using the trapezoidal rule; (b) compare your estimate with the true value of 45.943466667. (c) Can you apply the trapezoidal rule with end-correction to this integral? Explain.

x	1	1.2	1.4	1.6	1.8	2	2.2	2.4	2.6
$f(x)$	0	2.016	4.288	8.352	15.744	28	46.656	73.248	109.312

Section 8.2 Simpson's Rule

E8.11 Repeat **E8.1** using Simpson's 1/3 rule.

E8.12 Repeat **E8.2** using Simpson's 1/3 rule.

E8.13 Repeat **E8.3** using Simpson's 1/3 rule.

E8.14 Repeat **E8.4** using Simpson's 1/3 rule.

E8.15 Repeat **E8.5** using Simpson's 1/3 rule.

E8.16 Repeat **E8.6** using Simpson's 1/3 rule.

E8.17 Repeat **E8.8** using Simpson's 1/3 rule.

E8.18 Repeat **E8.9** using Simpson's 1/3 rule.

E8.19 Repeat **E8.10** using Simpson's 1/3 rule.

E8.20 Estimate the following integrals: (a) using the 8-panel Simpson's 1/3 rule with and without the end-correction; (b) What can you say about the accuracy of your estimates? Use a symbolic processor for plotting and evaluating the required derivatives.

(a) $\int_0^{\pi/2} \dfrac{u\cos u\, du}{u^4 + u^3 + 1}$, (b) $\int_0^\pi \sqrt{\cos\dfrac{x}{4}}\, dx$, (c) $\int_0^1 e^{\tan^{-1}(2x)}\, dx$, (d) $\int_0^1 e^{(x/2)^x}\, dx$.

E8.21 Estimate the following integral using $N=4$, 6, and 10-panel Trapezoidal and Simpson's 1/3 rules with and without end-correction, and compare your estimates with the true value of 0.5048545941137.

$$\int_0^1 \sin(\pi x^2)\, dx$$

E8.22 Estimate the following definite integrals as well as the global error for the specified N using (i) the trapezoidal rule, (ii) Simpson's 1/3 rule, and (iii) Simpson's 3/8 rule. (iv) Determine which method yields a more accurate estimate. Why?

(a) $\int_0^{\pi/2} \ln(3 + 2\sin x)dx$ $(N = 11)$ $I_{\text{true}} = 2.263925463697169$

(b) $\int_0^1 e^{x^3} dx$ $(N = 10)$, $I_{\text{true}} = 1.3419044179774196$

E8.23 Estimate the following definite integral using the 4- and 10-panel trapezoidal and Simpson's 1/3 rules. If the true value of the integral is 1.0000211, which method is a more accurate estimate?

$$\int_0^2 \frac{\ln(2.41\,x + 1)}{x^2 + 1}\,dx$$

E8.24 The current passing through a circuit is described by the following equation:

$$i(t) = \frac{1}{L}e^{-tR/L} \int_{u=0}^{t} V(u)e^{uR/L}\,du$$

where $i(t)$ is the current in amperes, $V(t)$ is the applied voltage in volts, L is the inductance ($L = 5$ Henry), and R is the resistance ($R = 10\ \Omega$). The transient voltage applied is given in the table below. Use the trapezoidal rule to estimate the current for each value of time provided in the data set.

t (s)	0	0.2	0.4	0.6	0.8	1	1.2	1.4
$V(t)$	0	89	105	34	−65	−110	−65	33
t (s)	1.6	1.8	2	2.2	2.4	2.6	2.8	3
$V(t)$	104	88	0	−89	−105	−34	66	110

E8.25 The tidal height recorded at a tidehouse for a 24-hour period is tabulated below. We wish to determine the root-mean-square (rms) value of the tidal height, which denotes the potential of tidal power. The rms value is computed as

$$H_{rms} = \sqrt{\frac{1}{T} \int_{t=0}^{T} H^2(t)\,dt}$$

where T is the recording duration (in hours) and H is the instantaneous tidal height in meters. Use the given data to estimate the rms value with the trapezoidal and Simpson's 1/3 rules.

t	H	t	H	t	H	t	H	t	H
0	0.30	5	−1.24	10	2.94	15	−2.70	20	2.81
1	−1.20	6	0.25	11	1.88	16	−2.36	21	3.30
2	−2.29	7	1.75	12	0.40	17	−1.32	22	2.98
3	−2.70	8	2.86	13	−1.11	18	0.16	23	1.96
4	−2.31	9	3.30	14	−2.24	19	1.67	24	0.49

E8.26 A digital recording of the speed (V, km/min) of a rocket is presented below. (a) Determine a numerical method to estimate the distance that the rocket traveled with the least possible error, and (b) estimate the distance the rocket traveled during its 22-minute flight.

t	V	t	V	t	V	t	V	t	V
0	0	5	6.09	10	2.46	15	0.62	20	0.07
1	4.09	6	5.36	11	1.93	16	0.44	21	0.03
2	6.07	7	4.56	12	1.49	17	0.3	22	0
3	6.73	8	3.79	13	1.13	18	0.2		
4	6.62	9	3.08	14	0.85	19	0.12		

E8.27 (a) Estimate the following integral involving the tabulated discrete function $f(x)$ using the trapezoidal and Simpson's 1/3 rules. (b) Which method gave a better estimate? why? (c) Can you estimate the theoretical global error for this integral?

$$\int_{x=1}^{2} \frac{f(x)}{x}\,dx$$

x	1	1.125	1.25	1.375	1.5	1.625	1.75	1.875	2
$f(x)$	5	5.75	6.65	7.84	9.25	10.95	12.97	15.4	18

E8.28 (a) Estimate the following integral involving the tabulated discrete function $f(x)$ using the trapezoidal and Simpson's 1/3 rules. (b) Which method gave a better estimate? why? (c) Can you apply the end-point corrected formulas for this integral?

$$\int_{x=0}^{1.2} \frac{f(x)dx}{1+0.3\,x^2}$$

x	0	0.15	0.3	0.45	0.6	0.75	1.05	1.2
$f(x)$	3	3.1	3.2	3.4	3.7	4.2	5.2	5.8

E8.29 The velocity $v(x)$ and temperature $T(x)$ distributions of water fluid flowing between two large heated plates, depicted in **Fig. E8.29**, are obtained from the numerical solution of pertinent governing equations. The numerical solution for selected points is presented in the table below. Use a suitable numerical method to estimate the mean fluid velocity and mean temperature.

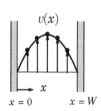

x (m)	0	0.2	0.4	0.6	0.8	1
v (m/s)	0	0.236	0.351	0.358	0.241	0
T (°C)	60	57	46	51	55	62

Compute the mean velocity and mean temperature from

$$v_m = \frac{1}{W}\int_{x=0}^{W} v(x)dx, \quad T_m = \frac{1}{v_m}\int_{x=0}^{W} v(x)T(x)dx$$

Fig. E8.29

where W is the distance between the plates.

E8.30 A particle is moving from point A(1,0) to point B(6.485,0.866) under the influence of a force field given is by $\mathbf{F}(x,y,z) = (x^2 + y^2)\mathbf{i} + (x^2 - y^2)\mathbf{j}$ along the path defined by the uniformly spaced discrete data set presented in the table below.

t	$x(t)$	$y(t)$	t	$x(t)$	$y(t)$	t	$x(t)$	$y(t)$
0	0	0	2.8	0.794	0.208	5.6	3.177	−0.407
0.4	0.016	0.407	3.2	1.038	−0.208	6	3.648	0
0.8	0.065	0.743	3.6	1.313	−0.588	6.4	4.150	0.407
1.2	0.146	0.951	4	1.621	−0.866	6.8	4.685	0.743
1.6	0.259	0.995	4.4	1.962	−0.995	7.2	5.252	0.951
2	0.405	0.866	4.8	2.334	−0.951	7.6	5.852	0.995
2.4	0.584	0.588	5.2	2.74	−0.743	8	6.485	0.866

Estimate the work done by the particle using the trapezoidal and Simpson's 1/3 rules. *Clue*: Apply second-order difference formulas for the derivatives and compute the work by

$$W = \int_{A}^{B} \mathbf{F} \cdot d\mathbf{r} = \int_{t=0}^{8} \left\{ (x^2 + y^2)\frac{dx}{dt} + (x^2 - y^2)\frac{dy}{dt}\right\}dt$$

E8.31 Estimate the following integral using (a) a 6-panel Simpson's 1/3 and 3/8 rules; (b) compare the theoretical maximum and the true errors for both methods; and (c) explain the difference in the numerical estimates. The true integral value is 1.054593421075752.

$$\int_{0}^{1} (5x + 1)e^{-4x^2} dx$$

E8.32 Consider **E8.31**: what is the minimum number of panels that should be used with the Simpson's 1/3 rule so that the order of approximation error is less than 10^{-4}?

Section 8.3 Romberg's Rule

E8.33 Consider the first column of Romberg's table for the definite integral, presented below. Complete the table and compute $R_{4,4} - R_{4,3}$, $R_{4,4} - R_{3,3}$, and the absolute error in the final estimate.

$$\int_0^2 xe^{-x}\,dx = 1 - \frac{3}{e^2}$$

0.2706705665
0.5032147244
0.5705876472
0.5880964505

E8.34 Consider the first column of Romberg's table for the definite integral presented below. Complete the table and compute $R_{4,4} - R_{4,3}$, $R_{4,4} - R_{3,3}$, and the absolute error in the final estimate.

$$\int_0^{\sqrt{3}} x(1+x^2)^{3/2}\,dx = \frac{31}{5}$$

12.0
7.7362742979
6.5890218645
6.2975565330

E8.35 Consider the first column of Romberg's table for the definite integral presented below. Complete the table and compute $R_{5,5} - R_{5,4}$, $R_{5,5} - R_{4,4}$, and the absolute error in the final estimate.

$$\int_0^\pi (2-x)\cos x\,dx = 2$$

4.9348022005
2.4674011003
2.1060585751
2.0259014934
2.0064379289

E8.36 Consider the first column of Romberg's table for the definite integral presented below. Complete the table and compute $R_{5,5} - R_{5,4}$, $R_{5,5} - R_{4,4}$, and the absolute error in the final estimate.

$$\int_0^\pi \frac{x\,dx}{\sin x} = 2G$$

2.0190987135
1.8819073817
1.8446859979
1.8351378037
1.8327339925

where $G = 0.915965594177219$ is the Catalan's constant.

E8.37 Use Romberg's rule to evaluate the following integrals within $\varepsilon = 10^{-6}$.

(a) $\int_0^{0.8} e^{-1.5625\,x^2}\,dx$,   (b) $\int_0^{\pi/4} \ln(\sec x)\,dx$,   (c) $\int_0^1 \sqrt[3]{x^3+1}\,dx$

E8.38 Repeat **E8.22** with Romberg's rule to find $R_{2,2}$ and $R_{3,3}$. Comment on the accuracy of the integrations.

E8.39 Repeat **E8.10**, by constructing the largest Romberg's table as possible. How many decimal places of accuracy are attained in the worst scenario?

E8.40 Repeat **E8.27**, by constructing the largest Romberg's table as possible. How many decimal places of accuracy are attained in the worst scenario?

E8.41 Repeat **E8.28**, by constructing the largest Romberg's table as possible. How many decimal places of accuracy are attained in the worst scenario?

E8.42 Estimate the following integrals using the Romberg's rule by constructing tables of a specified size. How many decimal places of accuracy are attained in the worst scenario?

(a) $\int_1^2 \frac{e^{-x}}{x}\,dx$ (4×4),   (b) $\int_0^\pi \frac{\sin x}{x}\,dx$ (3×3),   (c) $\int_0^{2\pi} \frac{e^{-x}dx}{1+\sin^2 x}$ (5×5)

E8.43 Consider the shaded region (D) that lies between $y = x^2$ and $y = 2\cos x$ curves on $0 \leqslant x \leqslant \alpha$ as shown in **Fig. E8.43**. The area of a region R can be determined by the following definite integral:

$$\text{Area} = \int_0^\alpha \left(2\cos x - x^2\right) dx$$

where α is the solution of the $2\cos x = x^2$ non-linear equation on $x > 0$. Estimate the area with at least three decimal places of accuracy.

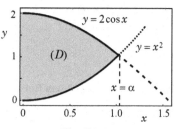

Fig. E8.43

E8.44 A bar of variable circular cross section, as illustrated in **Fig. E8.44**, is given. The axial displacement (u) of the bar under a load P can be computed from

$$u = \int_{x=0}^{L} \frac{P}{EA} dx$$

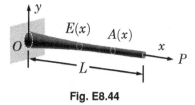

Fig. E8.44

where L is the length (m), E is the Young's modulus (Pa), and A is the cross-sectional area (m^2). Assuming $P = 2000$ kg and using the tabulated radius and Young's modulus data below, use Romberg's method to determine the axial displacement of the bar with the least error.

x (cm)	0	4	8	12	16	20	24	28	32
r (cm)	11	7	4.5	3	2.2	2	2	2	2
E (MPa)	35	33	31	30	29	27	24	21	18

E8.45 A weather station records precipitation (in cm/m^2) in 15-minute intervals. A data set collected over a 4-hour period is tabulated below. Using the furnished data, (a) estimate the total precipitation, $H = \int h\,dt$, during a 4-hour rainfall using the Simpson's 1/3 rule. (b) Can you estimate the total precipitation using Romberg's rule? (c) How many decimal places of accuracy will be attained in the worst scenario?

t (h)	h (cm/m^2)	t (h)	h (cm/m^2)	t (h)	h (cm/m^2)
0	0	1.50	0.25	3.00	0.44
0.25	0.65	1.75	0.12	3.25	0.52
0.50	0.51	2.00	0.35	3.50	0.28
0.45	0.38	2.25	0.32	3.75	0.18
1.00	0.55	2.50	0.35	4.00	0
1.25	0.12	2.75	0.36		

Section 8.4 Adaptive Integration

E8.46 Develop an algorithm for the adaptive trapezoidal rule.

E8.47 Estimate the integral below (a) analytically and (b) numerically using adaptive Simpson's 1/3 rule to within 10^{-3}.

$$\int_0^3 \frac{dx}{\sqrt{x+1}}$$

E8.48 Repeat **E8.47** with the improved extrapolation formula.

E8.49 Repeat **E8.37**, using adaptive Simpson's 1/3 rule to within 10^{-3}.

E8.50 Repeat **E8.49** with the improved extrapolation formula.

E8.51   Estimate the integral below (a) numerically using adaptive Simpson's 1/3 rule to within 10^{-3}, and (b) determine the accuracy if the true value of the integral is 0.82811632884.

$$\int_0^{\pi/2} \sin(x^2)$$

Section 8.5 Newton-Cotes Rules

E8.52   Estimate the following integral: (a) analytically; (b) numerically, by applying the composite trapezoidal rule with 4- and 8-panels over the entire range; (c) repeat Part (b) with the composite midpoint rule; and (d) compare the error estimates for Parts (b) and (c) and discuss the accuracy of both methods.

$$\int_{-\pi/2}^{\pi/2} |\sin x| dx$$

E8.53   For the following piecewise functions, estimate the integral of $f(x)$ on $[0,1]$, (i) analytically; (ii) numerically, by applying the composite trapezoidal and midpoint rules using 4-panels over the entire range; and (iii) is $f(x)$ and $f'(x)$ continuous at $x = 1/2$? Comment on the accuracy of both methods.

(a) $f(x) = \begin{cases} x^2, & x \leq \frac{1}{2} \\ x - x^2, & x > \frac{1}{2} \end{cases}$ (b) $f(x) = \begin{cases} x^2, & x \leq \frac{1}{2} \\ x - x^2, & x > \frac{1}{2} \end{cases}$

E8.54   Estimate the following integral: (a) analytically; (b) numerically by employing the composite midpoint rule with 4-, 8-, 16-, 32-, 64-, and 128-panels over the entire range. (c) Discuss the convergence of the midpoint rule. (d) Can you apply the trapezoidal or Simpson's 1/3 rules to this integral?

$$\int_{-1}^{1} \frac{dx}{\sqrt{1-x^2}}$$

E8.55   Estimate the following integral: (a) analytically and (b) numerically by applying the composite midpoint, trapezoidal, and Simpson's 1/3 rules with 4-, 8-, 16-, 32-, and 64-panels over the entire range; (c) compare and discuss your estimates.

$$\int_0^{1/2} \ln(1 + 4x^2) dx$$

E8.56   To estimate the following integral, apply the composite midpoint, 2-point, 3-point, and 5-point open Newton-Cotes formulas with 24-, 48-, and 96-panels over the entire range and discuss your estimates.

$$\int_0^1 \frac{e^x}{\sqrt{1-x^2}} dx = 3.10437901785555$$

E8.57   Consider the following integrals, whose true solutions are provided: (a) Estimate the integrals by applying the composite midpoint, 2-point, 3-point, and 5-point open Newton-Cotes formulas with 24-, 48-, and 96-panels over the entire range. (b) Compare and discuss the numerical estimates.

(a) $\int_0^{\pi/4} \frac{x}{\sin 2x} dx = 0.4579827971,$ (b) $\int_0^2 \frac{\tan^{-1}(x/2)}{x} dx = 0.9159655942,$

(c) $\int_0^{\pi/2} e^{-2\sec x} \sec x \, dx = 0.1138938727,$ (d) $\int_0^1 \frac{dx}{2e^{\sqrt{x}} + e^{-\sqrt{x}} - 3} = 1.18340407703954$

Section 8.6 Integration of Non-uniform Discrete Functions

E8.58   The boron concentration in a water stream is measured over a 10-hour period. Use the trapezoidal rule and the tabulated data below to calculate the average Boron concentration (ppm) and the average time, which are determined from:

$$\overline{C} = \frac{1}{24}\int_{t=0}^{24} C(t)dt, \quad \bar{t} = \frac{1}{\overline{C}}\int_{t=0}^{24} t\,C(t)dt$$

t (h)	0	1	3	4	6	7	10	11	14	16	19	20	22	24
C (ppm)	15	21	29	32	39	41	45	46	41	35	24	20	15	13

E8.59 The current applied across an electric circuit is tabulated below. Using the trapezoidal rule, estimate the average current and the rms current, which are calculated as follows:

$$\bar{I} = \frac{1}{T}\int_{t=0}^{T} I(t)dt, \quad I_{\text{rms}} = \sqrt{\frac{1}{T}\int_{t=0}^{T} I^2(t)dt}$$

where T is the time interval.

t (msec)	I	t (msec)	I	t (msec)	I
0	0	3.5	26.7	7.0	24.3
0.5	4.7	4.0	28.5	7.5	21.2
1.0	9.3	5.0	30.0	8.0	17.6
2.0	17.6	5.5	29.6	9.0	9.3
2.5	21.2	6.5	26.7	10	0

E8.60 The stress-strain data obtained from a tensile test of a specimen is tabulated below. The total energy dissipation is calculated from

$$U(\varepsilon) = \int_0^\varepsilon \sigma(u)du$$

(a) Estimate the energy dissipation for every available ε-value using the trapezoidal rule. (b) Which method (Trapezoidal or Simpson's 1/3 rule) is more suitable for this problem? (c) How can we improve the estimate over the trapezoidal rule?

ε (mm)	σ (MPa)	ε (mm)	σ (MPa)	ε (mm)	σ (MPa)
0	0	0.1869	32.976	0.3907	43.098
0.0237	3.256	0.2072	36.573	0.4553	43.621
0.0543	8.982	0.2243	39.183	0.5130	43.991
0.0883	15.342	0.2615	40.488	0.6116	44.085
0.1223	21.055	0.3058	41.628		
0.1495	26.933	0.3432	42.616		

E8.61 The thermal expansion coefficient of a steel specimen as a function of temperature is tabulated below. The contraction of a specimen is evaluated from

$$\Delta D(T) = D_0 \int_{T_a}^{T} \alpha(\bar{T})\,d\bar{T}$$

where $\Delta D(T)$ is the contraction, D_0 is the initial diameter, α is the thermal expansion coefficient, and T_a is the ambient temperature. If a specimen 10 cm in diameter is immersed in dry ice at temperature T (in °C), estimate the contraction in the specimen for every available temperature listed in the table.

T(°C)	α (μm/m/°C)	T(°C)	α (μm/m/°C)	T(°C)	α (μm/m/°C)
-240	2.80	-160	6.50	-50	10.15
-220	3.81	-130	7.66	-20	10.85
-200	4.76	-100	8.70	-15	10.96
-180	5.66	-80	9.32	40	11.90

E8.62 Enthalpy and entropy are two important thermodynamic properties of a substance that are calculated from

$$H(T) = H^0 + \int_{275}^{T} C_p(x)dx, \qquad S(T) = S^0 + \int_{275}^{T} \frac{C_p(x)}{x}dx$$

where C_p is the specific heat, $H^0 = 47.37$ kJ/kg, and $S^0 = 0$ J/kg.K are the enthalpy and entropy of the substance at 275 K. The specific heat vs. temperature data is presented below. Use the data to estimate the enthalpy and entropy of the substance for every available temperature listed in the table.

T (K)	275	285	300	325	330	350	370
C_p (J/kg $\cdot$ K)	208.66	211.38	215.45	222.24	223.60	229.03	234.46

E8.63 Fugacity is a thermodynamic property of a real gas, and it is defined as

$$\ln \frac{f}{P} = \int_0^P \frac{Z(p) - 1}{p}dp$$

where f is the fugacity, P is the pressure, and Z is the compressibility. The compressibility factor of an ideal gas is tabulated below for various pressures. Estimate the fugacity of the gas for each available pressure value in the data.

P (atm)	Z	P (atm)	Z	P (atm)	Z
0	1	20	0.906	250	0.327
0.1	0.993	50	0.675	300	0.382
0.5	0.985	100	0.15	400	0.491
1	0.982	150	0.21	500	0.595
10	0.952	200	0.265		

E8.64 The cylinder (in diameter $D_p = 20$ cm) pressure (P) of an internal combustion engine is tabulated below in relation to the piston position (x) from the top dead center. The work done can be calculated from

$$W = \int \mathbf{F} \cdot d\mathbf{r} = A_p \int P dx$$

where A_p is the piston area (m^2), P is the pressure (Pa), and W is the work. Estimate the work done by the piston using the trapezoidal and Simpson's 1/3 rules.

x (cm)	P (MPa)	x (cm)	P (MPa)	x (cm)	P (MPa)
0	130	2.5	163	6	127
0.5	150	3	156	7	120
1	160	3.5	150	8	113
1.5	163	4	140	9	105
2	166	4.5	135		

E8.65 The diameter (d) of a circular shaft varies with axial position (x). The digitally sampled diameter data is tabulated below. The axial load of $P = 50$, kN is applied to one end of the shaft, whose modulus of elasticity is $E = 200$ GPa. The axial elongation of the shaft is computed by

$$\Delta z = \frac{P}{E} \int_0^L \frac{dz}{A(z)}$$

where $A(z) = \pi d^2(z)/4$ is the cross-section area of the shaft. Estimate the axial elongation of the piston using the trapezoidal and Simpson's 1/3 rules.

z (cm)	d (cm)	z (cm)	d (cm)	z (cm)	d (cm)	z (cm)	d (cm)
0	2	60	2.30	110	2.90	150	2.10
10	2.10	65	2.44	120	2.66	180	2.10
20	2.10	80	2.66	135	2.44	190	2.10
50	2.10	90	2.90	140	2.30	200	2
55	2.12	100	3	145	2.12		

Section 8.7 Gauss-Legendre Method

E8.66 Estimate the following integral: (a) analytically and (b) numerically, using the 2-point Gauss-Legendre quadrature ($x_{1,2} = \pm 1/\sqrt{3}$, $w_{1,2} = 1$, do not approximate the quadrature points and weights). (c) Calculate the global error and discuss the accuracy of the numerical estimates.

$$\int_{-1}^1 (x^3 + 3x^2 + x + 1)dx$$

E8.67 Estimate the following integral: (a) analytically and (b) numerically, using the 2-, 3-, and 4-point Gauss-Legendre quadrature. (c) Calculate the global error and discuss the accuracy of the numerical estimates.

$$\int_{-1}^1 (7x^2 + 5)^3 dx$$

E8.68 Estimate the following integral: (a) Using the 3-, 4-, and 5-point Gauss-Legendre quadrature. (b) Calculate and discuss the accuracy of the numerical estimates. The integral, correct to ten decimal places, is given as 1.3658660636140654.

$$\int_{-1}^1 \frac{\cos x}{1 + x^2} dx$$

E8.69 Estimate the following integral using the 2-, 3-, 4-, and 5-point Gauss-Legendre quadrature and discuss the accuracy of the estimates.

$$\int_0^\pi (\sin x)^{1/3} dx$$

E8.70 To estimate the following integrals: (a) use 5-point Gauss-Legendre quadrature; and (b) calculate the global error and discuss the accuracy of your estimates.

(a) $\int_0^{\pi/4} \cos(e^{x/2}) dx$, (b) $\int_0^1 \sinh(x^3)dx$, (c) $\int_0^{\pi/2} \cos(\sqrt{1 + x^2})dx$,

(d) $\int_1^3 \frac{e^x dx}{1 + x^5}$, (e) $\int_1^2 e^{-x^2} dx$, (f) $\int_0^1 x^{x^2} dx$

E8.71 Estimate the following integral: (a) analytically, and (b) numerically, using the 5-point Gauss-Legendre quadrature over the whole range. (c) Transform the integral by $u = \cos x$ substitution and then use the 5-point Gaussian quadrature to estimate the resulting integral. (d) Which integral yields a better estimate? why?

$$\int_0^\pi \frac{\sin 2x}{4 - 3\cos x} dx$$

E8.72 To estimate the following integral with a removable singularity at $x = 0$ (a) use 3-, 4-, and 5-point Gauss-Legendre quadrature. (b) Calculate the global errors and discuss the accuracy of your estimates.

$$\int_0^1 \frac{e^{x^2} - 1}{x} \, dx$$

E8.73 Consider the following integral, which has non-removable singularities at both endpoints. (a) Estimate the integral using the 2-, 3-, 4-, and 5-point Gauss-Legendre quadrature, and (b) discuss the accuracy of your estimates.

$$\int_{-1}^1 \frac{e^x}{\sqrt{1 - x^2}} \, dx$$

E8.74 Consider the integral in **E8.57c**, which has a non-removable singularity at $x = \pi/2$. (a) Estimate the integral using the 2-, 3-, 4-, and 5-point Gauss-Legendre quadrature, and (b) discuss the accuracy of your numerical estimates.

E8.75 Consider the integral in **E8.57d**, which has a non-removable singularity at $x = 0$. (a) Estimate the integral using the 2-, 3-, 4-, and 5-point Gauss-Legendre quadrature, and (b) discuss the accuracy of your numerical estimates.

E8.76 Estimate the following integral: (a) using a 2-point Gauss-Legendre quadrature over the whole range; (b) applying a two-point Gaussian quadrature to each integral (i.e., *composite rule*) after splitting it into three integrals on [0,1], [1,2], and [2,3]. (c) Calculate the true errors for Parts (a) and (b), discuss the accuracy of your estimates. The true integral is given as 5.251805081.

$$\int_0^3 \sqrt{e^{x/3} + x} \, dx$$

E8.77 Estimate the following integral: (a) analytically; (b) numerically, using 2-point Gauss-Legendre quadrature over the whole range; (c) by applying the 2-point Gaussian quadrature to each interval and having the integral split into four parts, $[-2, -1]$, $[-1.0]$, $[0,1]$, and $[1,2]$. (d) Calculate the true errors for Parts (a) and (b) and discuss the accuracy of your estimates.

$$\int_{-2}^2 \frac{dx}{x^2 + 4}$$

E8.78 Estimate the following integral: (a) analytically; (b) numerically, using 2-, 3-, 4-, and 5-point Gauss-Legendre quadrature. (c) Discuss the accuracy of your estimates.

$$\int_0^\pi \sin(10x) \cos x \, dx$$

Section 8.8 Computing Improper Integrals

E8.79 Estimate the following integrals using the *subtraction of singularity* technique with the 10-panel Simpson's 1/3 rule and discuss the accuracy of your estimate.

(a) $\displaystyle\int_0^{\pi/2} \frac{\sin x dx}{\sqrt{\pi - 2x}} = 1.38232508,$ (b) $\displaystyle\int_0^1 \frac{e^{-x} dx}{\sqrt{1 - x}} = 1.076159014,$

(c) $\displaystyle\int_0^1 \frac{\cosh x dx}{\sqrt{x + x^2}} = 1.92353087,$ (d) $\displaystyle\int_0^1 \frac{x \sin(\pi x/2)}{\sqrt{1 - x^2}} dx = 0.89036520,$

(e) $\displaystyle\int_0^1 \frac{\ln(1/x)}{1 + x} dx = \frac{\pi^2}{12},$ (f) $\displaystyle\int_0^1 \ln(\csc x) dx = 1.05672021.$

E8.80 Using 5-, 10-, 15-, and 20-point Gauss-Legendre quadrature, estimate **E8.79** integrals by *ignoring the singularity* and discuss the accuracy of your estimates.

E8.81 Apply the *truncating the intervals* technique to estimate the following integrals using the Simpson's 1/3 rule with $h = 0.1$ and $b = 5$, 10, and 15 as infinity, and discuss the accuracy of your estimates.

(a) $\int_1^\infty \dfrac{e^{-x}}{x^2} dx = 0.14849551,$

(b) $\int_1^\infty \dfrac{\sin x}{\sqrt{1 + x^2}} dx = 0.49037744,$

(c) $\int_0^\infty \dfrac{dx}{(1 + e^{x/5})^2} = 0.96573590,$

(d) $\int_0^\infty e^{-2x} \ln(1 + x) = 0.18066431.$

E8.82 In each of the following integrals, apply a suitable substitution to first transform the improper integral into a proper integral. Then, apply the 100-panel Simpson's 1/3 rule to estimate the resulting integrals and their true errors.

(a) $\int_{2/\pi}^\infty \dfrac{1}{x^4} \sin\left(\dfrac{1}{x}\right) dx = \pi - 2,$

(b) $\int_0^1 \dfrac{\sin^{-1}\sqrt{x}}{\sqrt{1 - x^2}} dx = 1.51219929379,$

(c) $\int_{-1}^1 \dfrac{\cos x}{\sqrt{1 - x^2}} dx = 2.403939430634,$

(d) $\int_0^{\pi/2} \dfrac{\cos 2x}{\sqrt{x}} dx = 0.9374359444665,$

(e) $\int_0^1 \dfrac{\sqrt{1 - x^2}}{\sqrt{x}} dx = 1.748038369528,$

(f) $\int_0^1 \sqrt{\dfrac{1 + \cos^{-1} x}{1 - x^2}} dx = 2.08129232270.$

E8.83 Estimate the following improper integrals using *open Newton-Cotes formulas* (midpoint rule, two-point rule, and three-point rule). Apply 10-, 50-, and 100-panels for each formula and discuss the accuracy of your estimates.

(a) $\int_0^1 \dfrac{dx}{(1 - x)^{1/4}} = \dfrac{4}{3},$

(b) $\int_0^1 \ln\left(\dfrac{1}{x}\right) e^{-2x} dx = 0.65963168,$

(c) $\int_0^{\ln 2} \sqrt{\dfrac{e^x + 1}{e^x - 1}} dx = \dfrac{\pi}{3} + \ln(2 + \sqrt{3}),$

(d) $\int_0^{\pi/2} \sqrt{\dfrac{\sec x}{x}} dx = 3.42312699$

(e) $\int_{-1}^1 \dfrac{dx}{\sqrt{1 - x^2}} = \pi,$

(f) $\int_0^{\pi/2} \ln(1 + \tan x) dx = 1.46036212,$

(g) $\int_0^1 \dfrac{\ln(x) dx}{x^2 - 1} dx = \dfrac{\pi^2}{8},$

(h) $\int_0^{\pi/3} \dfrac{\cos x dx}{\sqrt{4\cos^2 x - 1}} = \dfrac{\pi}{4}.$

E8.84 Estimate the following improper integrals using open Newton-Cotes formulas. (a) Divide the integration interval into 200-, 400-, and 800 subregions and apply the *three-point rule* for each subregion, and (b) discuss the accuracy of your estimates.

(a) $\int_0^2 \dfrac{dx}{(x - 1)^{2/3}} = 6,$

(b) $\int_{-1}^1 \ln|x|\, dx = -2,$

(c) $\int_{-1}^1 \dfrac{1 - x^{2/3} + x^2}{x^{2/3}(1 + x^2)} dx = 6 - \dfrac{\pi}{2},$

(d) $\int_0^8 \dfrac{dx}{\sqrt[3]{(\sqrt[3]{x} - 1)^2}} = \dfrac{144}{7},$

Section 8.9 Gauss-Laguerre Method

E8.85 Estimate the following integral using the 2-point Gauss-Laguerre quadrature.

$$\int_0^\infty e^{-x} dx$$

E8.86 Estimate the following integrals: (i) analytically and (ii) numerically using the Gauss-Laguerre quadrature with the specified number of points. (iii) Also calculate the upper- and lower-error bounds and discuss the accuracy of your estimates.

$$\text{(a)} \int_0^\infty e^{-x}\cos x\, dx, \ (N=4), \quad \text{(b)} \int_0^\infty e^{-3x} dx, \ (N=2),$$

$$\text{(c)} \int_0^\infty e^{-x}\sinh\frac{x}{3} dx, \ (N=3), \quad \text{(d)} \int_0^\infty e^{-x}\ln(x+1)dx. \ (N=4),$$

E8.87 Estimate the following integrals using the 5-point Gauss-Laguerre quadrature.

$$\text{(a)} \int_0^\infty \frac{e^{-2x}dx}{x^2+9}, \quad \text{(b)} \int_0^\infty e^{-x-2x^2} dx, \quad \text{(c)} \int_0^\infty e^{-3x}\sqrt{2+\sin(\tfrac{\pi x}{2})}dx$$

$$\text{(d)} \int_0^\infty \frac{e^{-x}\sqrt{x}}{x+3}dx, \quad \text{(e)} \int_0^\infty e^{-(x^2+3x+4)}dx.$$

E8.88 Consider the following improper integrals: (a) Apply a suitable substitution to transform the integrals into Gauss-Laguerre type integrals, and (b) estimate the integrals using the 5-point Gauss-Laguerre quadrature.

$$\text{(a)} \int_1^\infty \frac{e^{-x/2}}{x+3}dx, \quad \text{(b)} \int_\pi^\infty e^{-x}\sin^2 x dx, \quad \text{(c)} \int_{-1}^\infty e^{-x}\ln(1+x^2)dx$$

E8.89 Consider the following improper integrals: (a) Apply a suitable substitution to transform the integrals into Gauss-Laguerre type integrals, and (b) estimate the integrals using the 5-point Gauss-Laguerre quadrature.

$$\text{(a)} \int_0^\infty \frac{dx}{\cosh x}, \quad \text{(b)} \int_1^\infty \frac{\sin x}{x}dx, \quad \text{(c)} \int_2^\infty \frac{xdx}{x^3+3x^2-1}, \quad \text{(d)} \int_0^\infty \frac{e^{-x^2/5}}{1+x^3} dx$$

E8.90 Consider the following improper integral: Apply the 2-, 4-, 6-, and 8-point Gauss-Laguerre quadrature to estimate the integral and compare your estimates with the true value of 1.198705652879.

$$\int_0^\infty \frac{dx}{1+e^{x^2/5}}$$

E8.91 Consider the following improper integral: (a) Apply the 2-, 4-, 6-, and 8-point Gauss-Laguerre quadrature to estimate the integral, and compare your estimates with the true value of 1/2. (b) Explain how the estimates converge on the solution.

$$\int_0^\infty e^{-x}x^2\sin x\, dx$$

E8.92 Consider the following improper integral with a removable discontinuity at $x=0$: (a) Estimate the integral using the 2-, 3-, 4-, and 5-point Gauss-Laguerre quadrature, and (b) comment on the accuracy of the estimates.

$$\int_0^\infty \frac{e^{-x}-e^{-2x}}{x}dx = \ln 3$$

E8.93 Consider the following improper integrals with a removable discontinuity at $x=0$: (i) Estimate the integrals using the 2- through 8-point Gauss-Laguerre quadrature, and (ii) comment on the accuracy of the estimates.

$$\text{(a)} \int_0^\infty \frac{1-\cos x}{x^2}dx = \frac{\pi}{2}, \quad \text{(b)} \int_0^\infty \frac{\sin^3(x)}{x^3}dx = \frac{3\pi}{8}, \quad \text{(c)} \int_0^\infty \frac{x^2dx}{(x^2+1)^3(x^2/9+1)} = \frac{27\pi}{512}$$

E8.94 A power surge in an electric circuit across a resistor results in a change in the current, according to

$$I(t) = I_0 \exp\left(-\frac{(t - t_0)^2}{2\sigma^2}\right)$$

where $I_0 = 50$ A is the peak current, $t_0 = 0.05$ s is the instance of surge, and $\sigma = 0.008$ s is a parameter giving the peak width. The dissipated energy by the resistor is computed as

$$E = \int_{t=0}^{\infty} RI^2(t)dt$$

Estimate the energy dissipation using the 3-, 4-, and 5-point Gauss-Laguerre quadrature if $R = 20$ Ω. Are you able to obtain a converged solution?

E8.95 The spectral blackbody emissive power $(E_{b,\lambda})$ is defined by Planck's law.

$$E_{b\lambda}(\lambda, T) = \frac{C_1}{\lambda^5 \left[\exp(C_2/\lambda T) - 1\right]} \quad \left(\frac{W}{m^2 \cdot \mu m}\right)$$

where $C_1 = 3.742 \times 10^8$, W·μm^4/m^2, $C_2 = 1.439 \times 10^4$, μm^4·K, T is the absolute temperature of the solid surface, and λ is the wavelength of the radiation emitted by the solid (μm). The radiation emitted, including all wavelengths, is obtained from

$$\int_{\lambda=0}^{\infty} E_{b\lambda}(\lambda, T)d\lambda = \sigma T^4$$

Using 3-, 4-, and 5-point Gauss-Laguerre quadrature, estimate the Stefan-Boltzmann constant, which is 5.670374×10^{-8} W/m^{2}K^4. *Hint*: Apply the $x = 1/\lambda T$ substitution to the integral.

Section 8.10 Gauss-Hermite Method

E8.96 Estimate the following integral using the two-point Gauss-Hermite quadrature.

$$\int_{-\infty}^{\infty} e^{-x^2} dx$$

E8.97 (a) Derive the 3-point Gauss-Hermite quadratures; (b) Estimate the following integral with the Gauss-Hermite quadrature (without rounding) from Part (a); and (c) Discuss the accuracy of your estimates.

$$\int_{-\infty}^{+\infty} x^4 e^{-x^2} dx = \frac{3\sqrt{\pi}}{4}$$

E8.98 (i) Use the Gauss-Hermite quadrature with a specified N to estimate the following integrals; and (ii) calculate the global error bounds and discuss the accuracy of your estimates. *Note*: $K_0(x)$ is the modified Bessel function of the second kind, and $\Gamma(x)$ is the gamma function.

(a) $\int_{-\infty}^{\infty} e^{-x^2 - 2x} dx = e\sqrt{\pi}, \quad (N = 4)$, (b) $\int_{-\infty}^{\infty} e^{-x^2} \cos x \, dx = \frac{\sqrt{\pi}}{e^{1/4}}, \quad (N = 3)$,

(c) $\int_{-\infty}^{\infty} e^{-x^2} \sqrt{|x|} dx = \Gamma\left(\frac{3}{4}\right), \quad (N = 3)$, (d) $\int_{-\infty}^{\infty} e^{-x^2} \sin x \, dx = 0, \quad (N = 4)$,

E8.99 (a) Use the 2-, 3-, 4-, and 5-point Gauss-Hermite quadrature to estimate the following integral, and (b) comment on the accuracy of the estimates. *Note*: erfc(x) is the complementary error function.

(a) $\int_{-\infty}^{\infty} e^{-x^2} \cos 2x \, dx = \frac{\sqrt{\pi}}{e}$, (b) $\int_{-\infty}^{\infty} \frac{e^{-x^2} dx}{x^2 + 4} = \frac{\pi}{2} e^4 \text{erfc}(2)$,

(c) $\int_{-\infty}^{\infty} e^{-x^2} \sin^2 x dx = \frac{\sqrt{\pi}}{2}\left(1 - \frac{1}{e}\right)$, (d) $\int_{-\infty}^{\infty} e^{2x - x^2} dx = e\sqrt{\pi}$

E8.100 Consider the following improper integral: (i) Estimate the integral using the 2-, 4-, 6-, 8-, and 10-point Gauss-Hermite quadrature and compare your estimates with the true value; (ii) Find the upper and lower bounds of the global error for $N = 2$ and 4.

$$\text{(a)} \int_{-\infty}^{\infty} \frac{e^{-x^2} dx}{\sqrt{1+x^2}} = \sqrt{e}\, K_0\left(\frac{1}{2}\right), \qquad \text{(b)} \int_{-\infty}^{\infty} \frac{dx}{4x^4 + 1} = \frac{\pi}{2}$$

where $K_0(x)$ is the modified Bessel function of the second kind.

E8.101 (a) Evaluate the following integral using the 2-, 3-, 4-, and 5-point Gauss-Hermite quadrature; (b) Using a suitable substitution, transform the integral to the one on $[-\pi/2, \pi/2]$ and apply the 2-, 3-, 4-, and 5-point Gauss-Legendre quadrature; and (c) Discuss the accuracy of your estimates in Parts (a) and (b).

$$\int_{-\infty}^{+\infty} \frac{dx}{(x^2+1)\sqrt{3+4x^2}} = \ln 3$$

E8.102 Consider estimating the following improper integral: (a) Use the 2-, 3-, 4-, and 5-point Gauss-Legendre quadrature; (b) Apply the 2-, 3-, 4-, and 5-point Gauss-Hermite quadrature after transforming the integral to the one on $(-\infty, \infty)$; and (c) Discuss the accuracy of your estimates in Parts (a) and (b).

$$\int_{-\pi/2}^{\pi/2} \sin(2\tan\theta)\tan\theta d\theta = \frac{\pi}{e^2}$$

E8.103 Consider estimating the following improper integral: (a) use the 2-, 3-, 4-, and 5-point Gauss-Hermite quadrature; (b) apply the 2-, 3-, 4-, and 5-point Gauss-Laguerre quadrature after transforming the integral to the one on $(0, \infty)$; and (c) discuss the accuracy of your estimates.

$$\int_{-\infty}^{+\infty} \frac{x dx}{(e^{-x}+1)(e^x+2)} = \frac{(\ln 2)^2}{2}$$

Section 8.11 Gauss-Chebyshev Method

E8.104 Estimate the following improper integral: (a) analytically and (b) numerically using the 2-point Gauss-Chebyshev quadrature. (c) Also calculate the global error bounds and discuss the accuracy of your estimate.

$$\int_{-1}^{1} \frac{dx}{\sqrt{1-x^2}}$$

E8.105 Estimate the following integrals: (a) Use the 5-point Gauss-Chebyshev quadrature; and (b) Calculate the global error bounds and discuss the accuracy of your estimates.

$$\text{(a)} \int_{-1}^{1} \frac{x^6 dx}{\sqrt{1-x^2}} = \frac{5\pi}{16}, \qquad \text{(b)} \int_{-1}^{1} \frac{e^{-x} dx}{\sqrt{1-x^2}} = \pi I_0(1), \qquad \text{(c)} \int_{-1}^{1} \frac{\cos 2x dx}{\sqrt{1-x^2}} = \pi J_0(2).$$

where $J_0(x)$ and $K_0(x)$ are the Bessel functions.

E8.106 Consider estimating the following integrals: (a) use the 5-point Gauss-Chebyshev quadrature; and (b) calculate the global error bounds and discuss the accuracy of your estimates. *Note:* $I_0(x)$ is the modified Bessel function of the first kind.

$$\text{(a)} \int_0^2 \frac{\sinh(x/2)dx}{\sqrt{x(2-x)}} = \pi \sinh\left(\frac{1}{2}\right) I_0\left(\frac{1}{2}\right), \qquad \text{(b)} \int_0^1 \frac{e^{1-2x}dx}{\sqrt{x(1-x)}} = \pi I_0(1),$$

$$\text{(c)} \int_1^2 \frac{\ln x\, dx}{\sqrt{(x-1)(2-x)}} = 2\pi \ln\left(\frac{1+\sqrt{2}}{2}\right)$$

E8.107 Consider estimating the following integrals: (a) Use the 5-point Gauss-Chebyshev quadrature; and (b) Calculate the global error bounds and discuss the accuracy of your estimates.

(a) $\displaystyle\int_0^\pi \frac{\sin 2x}{\sqrt{x}}\,dx = 0.608688442,$ (b) $\displaystyle\int_0^1 \frac{e^x}{\sqrt{x}}\,dx = 2.925303492,$

(c) $\displaystyle\int_0^2 \frac{e^x}{\sqrt{2-x}}\,dx = 12.500854858,$ (d) $\displaystyle\int_{-1}^3 \frac{\ln(1+x^2)dx}{\sqrt{3-x}} = 5.472304123,$

(e) $\displaystyle\int_0^1 \frac{\cos(e^{-x})}{\sqrt{1-x}}\,dx = 1.692305349,$ (f) $\displaystyle\int_0^1 \frac{\sqrt[3]{x}\,dx}{(1+x^2)\sqrt{1-x}} = 1.104688454$

E8.108 Estimate the following integral using the 2-, 3-, 4-, 5-, 8-, and 10-point Gauss-Chebyshev quadrature and discuss the accuracy of your estimates.

(a) $\displaystyle\int_{-1}^1 \frac{dx}{\sqrt{1-x^4}} = 2\sqrt{\pi}\frac{\Gamma(5/4)}{\Gamma(3/4)} = 2.6220575543,$ (b) $\displaystyle\int_0^1 \frac{\tan^{-1}x}{x\sqrt{1-x^2}}\,dx = \frac{\pi}{2}\ln(1+\sqrt{2})$

where $\Gamma(x)$ is the Gamma function.

Section 8.12 Computing Integrals with Variable Limits

E8.109 Transform the following to integrals on the [0,1] interval.

(a) $\displaystyle\int_0^x \frac{\sin t}{t}\,dt,$ (b) $\displaystyle\int_x^\infty \frac{e^{-t}}{t^2}\,dt,$ (c) $\displaystyle\int_{\sqrt{x}}^\infty \frac{\cos t}{t}\,dt,$ (d) $\displaystyle\int_x^\infty (t-x)^{3/2}e^{-t}\,dt,$

(e) $\displaystyle\int_0^{\sin x} e^{-t^2}\,dt,$ (f) $\displaystyle\int_0^x \frac{\sin t\,dt}{\sqrt{x^2-t^2}},$ (g) $\displaystyle\int_x^\infty \frac{dt}{\sqrt{1+t^4}},$ (h) $\displaystyle\int_{-\infty}^{-x} \frac{dt}{\sqrt{1-t^3}}\ (x>0)$

(i) $\displaystyle\int_x^1 \frac{e^t\,dt}{\sqrt{t-x}},$ (j) $\displaystyle\int_{x-1}^{x+1} \frac{e^t\,dt}{t+1},$ (k) $\displaystyle\int_{x^2}^x \sin(t-x^2)dt,$ (l) $\displaystyle\int_x^{1/x} \tan^{-1}\sqrt{t-x}\,dt$

Section 8.13 Double Integration

E8.110 Consider estimating the following double integrals: Use (i) the trapezoidal rule, (ii) Simpson's 1/3 rule with 10-, 20-, 40-, and 80-panels for x- and y-intervals, and (c) discuss the accuracy achieved in both methods.

(a) $\displaystyle\int_{x=-1}^1 \int_{y=-2}^2 \frac{dydx}{x^2+y^2+1} = 3.72286450603,$ (b) $\displaystyle\int_{x=0}^1 \int_{y=0}^1 \frac{xy}{1+xy}\,dydx = 1 - \frac{\pi^2}{12}$

(c) $\displaystyle\int_{x=0}^2 \int_{y=0}^2 \frac{ydydx}{\sqrt{x^2+y^2+1}} = 2.05391167545,$ (d) $\displaystyle\int_{x=3}^6 \int_{y=2}^5 (x^2+y^2)dydx = 306$

(e) $\displaystyle\int_{x=0}^1 \int_{y=0}^1 \sqrt{x^2+y^2}dydx = \frac{1}{3}(\sqrt{2}+\ln(1+\sqrt{2}))$

E8.111 Repeat **E8.110** with 3-, 4-, and 5-point Gauss-Legendre quadratures for both x- and y-variables.

E8.112 Consider the following double integrals: (a) Transform the integrals with variable intervals to the specified (fixed) intervals, and estimate the integrals (b) using (i) the trapezoidal and Simpson's 1/3 rule with 10- and 20-panels for x- and y-intervals and (ii) 3- and 5-point Gauss-Legendre quadrature for x- and y-intervals, and (c) compare your solutions with those given true values.

(a) $\displaystyle\int_0^2 \int_0^{\sqrt{x}} (x^2 + xy + y^2)dydx = 5.32006868,$ $\{0 \leqslant y \leqslant \sqrt{x} \to 0 \leqslant u \leqslant 1\}$

(b) $\displaystyle\int_0^{\pi/2} \int_{\sin^2 x}^{\cos^2 x} (1 - 6x^2 y^2)dydx = 1.48352986,$ $\left\{\sin^2 x \leqslant y \leqslant \cos^2 x \to 0 \leqslant u \leqslant \dfrac{\pi}{2}\right\}$

(c) $\displaystyle\int_0^1 \int_0^x xye^{-xy}dydx = 0.08223952,$ $\{0 \leqslant y \leqslant x \to 0 \leqslant u \leqslant 1\}$

(d) $\displaystyle\int_{-1}^1 \int_x^{x^2} \sqrt{x^2 + y^2}e^{x^2 + x - 2y}dydx = 0.86697785,$ $\{x \leqslant y \leqslant x^2 \to -1 \leqslant u \leqslant 1\}$

E8.113   Estimate the following integrals using a suitable 5-point Gauss quadrature on both intervals and discuss the accuracy of your estimates.

(a) $\displaystyle\int_{x=0}^\infty \int_{y=0}^1 \frac{dydx}{(1 + x + y)^3} = \frac{1}{4},$

(b) $\displaystyle\int_{x=0}^\infty \int_{y=0}^\pi \sin(x+y)e^{-(x+y)}dydx = \frac{1 + e^{-\pi}}{2},$

(c) $\displaystyle\int_{x=0}^\infty \int_{y=0}^1 \frac{\exp\left(-x - x^2 y/8\right)dydx}{\sqrt{y - y^2}} = e\sqrt{2\pi}\,K_0(1),$

(d) $\displaystyle\int_{x=0}^\infty \int_{y=0}^1 \tan^{-1}\left(\frac{y}{x+1}\right)\frac{ydydx}{(x+1)^2} = \frac{(1 - \ln 2)}{2},$

(e) $\displaystyle\int_{x=-\infty}^\infty \int_{y=0}^\infty \frac{\exp\left\{-(x^2 + y^2)\right\}dydx}{\sqrt{x^2 + y^2}} = \frac{\pi\sqrt{\pi}}{2},$

(f) $\displaystyle\int_{x=0}^1 \int_{y=0}^1 \frac{e^{-xy}dydx}{\sqrt{y}} = 2\left(\sqrt{\pi}\,\mathrm{erf}(1) + \frac{1-e}{e}\right),$

(g) $\displaystyle\int_{x=-\infty}^\infty \int_{y=0}^1 (x+y)e^{y^2 - x^2}dydx = \frac{\sqrt{\pi}(e - 1)}{2},$

(h) $\displaystyle\int_{x=0}^{\pi/2} \int_{y=-x}^x \frac{y\cos(x - y)}{\sqrt{x^2 - y^2}}dydx = \frac{\pi^3}{12}J_1\left(\frac{\pi}{2}\right),$

where $\mathrm{erf}(x)$ is the error function, $J_1(x)$ and $K_0(x)$ are the Bessel and Modified Bessel functions, respectively.

E8.114   The polar inertia moment of a homogeneous rectangular plate is computed by the following double integral:

$$\iint_R (x^2 + y^2)dydx$$

where $R : 3 \leqslant x \leqslant 6,\ 2 \leqslant y \leqslant 5$. Estimate the polar inertia moment of the plate using (a) the trapezoidal rule and (b) the Simpson's 1/3 rule with 10-panels in x- and y-intervals for both cases.

E8.115   (i) Transform the following double integrals into a polar coordinate system; (ii) Estimate the integrals with the 5-point Gauss quadrature method and discuss the accuracy of your estimates.

(a) $\displaystyle\int_0^2 \int_0^{\sqrt{4 - x^2}} x^2 ye^{-x\sqrt{x^2 + y^2}}dydx = \frac{3\sqrt{\pi}}{4}\mathrm{erf}(2) + \frac{2}{e^4} - 1$

(b) $\displaystyle\int_{-1}^{1}\int_{0}^{\sqrt{1-x^2}}\frac{x^2e^{-x}dydx}{(x^2+y^2)^{3/2}}=\pi I_1(1)$

(c) $\displaystyle\int_{0}^{1}\int_{0}^{\sqrt{1-x^2}}\frac{ye^{-(x+y^2/x)}dydx}{\sqrt{x^2+y^2}}=\frac{1}{3}\left(1-\frac{1}{e}-E_1(1)\right),$

(d) $\displaystyle\int_{0}^{4}\int_{0}^{\sqrt{4-x^2}}\frac{\tan^{-1}(y/x)dydx}{\sqrt{x^2+y^2}}=\frac{\pi^2}{2}$

where $\mathrm{erf}(x)$ is the error function, $I_1(x)$ is the modified Bessel function of the first kind, and $E_1(x)$ is the exponential integral function.

E8.116 Consider a unit circular disk, whose density is given as $\rho(x,y)=\exp\{x-y-x^2-y^2\}$. Recall that the mass of a planar object is calculated with the following double integral: (a) Transform the double integral into the polar coordinate system; (b) estimate the mass of the disk using Trapezoidal and Simpson's 1/3 rules with 10-panels for radial integration and 20-panels for angular integration.

$$m=\iint_{x^2+y^2\leqslant 1}\rho(x,y)dxdy$$

E8.117 A three-dimensional solid object is bound by the $z=0$ plane, the $z=\exp(-x^2-y^2)$ surface, and the $x^2+y^2=1$ cylinder. The polar inertia moment of the object is reduced to the following double integral in the polar coordinate system:

$$I_z=\iiint_{V}(x^2+y^2)dV=\int_{\theta=0}^{2\pi}\int_{r=0}^{1}r^3e^{-r^2}drd\theta$$

Use the trapezoidal and Simpson's -1/3 rules with $N=M=10$, 20, 40, and 100 to estimate I_z.

E8.118 (a) Transform the following into double integrals with constant limits, and (b) then estimate the double integrals numerically using the trapezoidal and Simpson's 1/3 rules with $N=M=10$, 20, 40, and 100 panels.

(a) $\displaystyle\int_{0}^{1}\int_{0}^{x}xye^{x^2-y^2}dydx=\frac{e-2}{4},$ (b) $\displaystyle\int_{0}^{\pi}\int_{x}^{2x}(x+y)^2\sin(x+y)dydx=8\pi,$

(c) $\displaystyle\int_{0}^{2}\int_{-x}^{\sqrt{8x}}\frac{xydydx}{\sqrt{x^2+y^2}}=8\left(1-\frac{\sqrt{2}}{3}\right),$ (d) $\displaystyle\int_{0}^{1}\int_{0}^{e^x}\ln(e^x+y)dydx=2-e+(e-1)\ln 4$

8.16 COMPUTER ASSIGNMENTS

CA8.1 The digitized two-dimensional outline of a jug $(x,f(x))$ (measurements in meters), shown in **CA8.1**, is given in the table below:

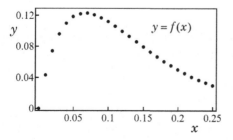
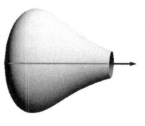

Fig. CA8.1

The surface area of the surface generated by revolving the curve $y = f(x)$ around the x–axis is calculated with

$$S = \int_a^b 2\pi f(x)\sqrt{1 + (f'(x))^2}\,dx$$

Write a computer program (or use a spreadsheet) to estimate the surface area of the jug with the trapezoidal and Simpson's 1/3 rules. Compute the surface area with both rules and discuss the accuracy of your estimates. *Hint*: Estimate $f'(x)$ using second-order finite difference formulas.

x	$f(x)$	x	$f(x)$	x	$f(x)$	x	$f(x)$
0	0	0.07	0.1225	0.14	0.0857	0.21	0.0450
0.01	0.0430	0.08	0.1205	0.15	0.0790	0.22	0.0406
0.02	0.0741	0.09	0.1167	0.16	0.0726	0.23	0.0365
0.03	0.0956	0.10	0.1116	0.17	0.0664	0.24	0.0328
0.04	0.1098	0.11	0.1056	0.18	0.0605	0.25	0.0294
0.05	0.1181	0.12	0.0992	0.19	0.0550		
0.06	0.1220	0.13	0.0925	0.20	0.0498		

CA8.2 The volume of a three-dimensional body, obtained by revolving $y = f(x)$ about the x–axis, is found by

$$V = \pi \int_a^b \left(f(x) \right)^2 dx$$

Write a computer program (or use a spreadsheet) to estimate the volume of the jug (presented in **CA 8.1**) by revolving the discrete data about the x-axis. Apply the trapezoidal and Simpson's 1/3 rules to estimate the volume of the jug and discuss the accuracy of your solutions.

CA8.3 The position of a flying object is available as a set of discrete data with uniform time steps, $\{x(t), y(t), z(t)\}$. (a) Write a computer program that reads the set of data and estimates the cumulative distance traveled (by the trapezoidal rule) corresponding to every available time value by

$$s(t) = \int_{t=0}^t \sqrt{\left(\frac{dx}{dt}\right)^2 + \left(\frac{dy}{dt}\right)^2 + \left(\frac{dz}{dt}\right)^2}\,dt$$

(b) Use $x(t) = 200t^2$, $y(t) = 350t^2$, $z(t) = 55te^{t/4}$ as the equations of motion to test your program. Generate time history data with $\Delta t = 0.1$ s up to 10 seconds and compare your answer with the true answer of 40.942 km. *Hint*: Evaluate the time derivatives using second-order finite-difference formulas.

CA8.4 Write a computer program for estimating the integral of a continuous function $f(x)$ on $[0, a]$ using the trapezoidal and Simpson's 1/3 rules and Gauss-Legendre quadrature (quadrature read from an input file). Use your program to estimate the following integral:

$$\int_0^1 \frac{dx}{1 + e^x}$$

(a) with the 6-, 12-, 24-, and 48-panel trapezoidal rule; (b) with the 5-, 10-, 20-, and 50-panel Simpson's 1/3 rule; (c) with the 3-, 5-, 7-, and 9-point Gauss-Legendre quadrature; and (d) discuss the accuracy of your estimates. The true integral value is $1 + \ln(2/(1 + e))$

CA8.5 The Arias intensity, used to measure the magnitude of earthquake shaking in earthquake engineering, is defined as the total energy stored per unit weight by a series of undamped simple oscillators at the end of the earthquake and is calculated with

$$I_h = \frac{\pi}{2g} \int_{t=0}^T \left(a_x^2(t) + a_y^2(t) \right) dt$$

where g is the acceleration of gravity (in m/s^2), T is the duration of ground shaking (in seconds), and $a_x(t)$ and $a_y(t)$ are the acceleration time histories in EW and NS directions (in m/s^2), respectively. (a) Write a computer program that reads strong-motion data (i.e., displacements $x(t)$

and $y(t)$ in EW and NS directions) to calculate the velocity and acceleration, as well as the Arias intensity, with the trapezoidal rule; (b) Test your program using the data below. *Hint*: Note that the strong motion data is not smooth; thus, you may use first-order finite-difference formulas to compute the velocity and acceleration components ($v_x(t) = dx/dt$, $a_x(t) = dv_x/dt$, and so on).

t (s)	$x(t)$ (cm)	$y(t)$ (cm)	t (s)	$x(t)$ (cm)	$y(t)$ (cm)
0	0	0	0.06	0.32	-1.3
0.01	-0.24	-1.45	0.07	-0.26	-0.63
0.02	0.45	1.26	0.08	-1.38	1.1
0.03	-1.09	0.53	0.09	0.12	-1.06
0.04	1.14	1.12	0.1	0.75	-0.66

CA8.6 A wire extends from $x = -a$ to $x = b$. The charge contained within a wire of length dx is $dq = \lambda(x)dx$, where $\lambda(x)$ denotes the charge density at position x. The electric potential at distance r between points $(x, 0)$ on the wire and (x_0, y_0) is given by

$$V = \int \frac{dq}{r} = \int_{x=-a}^{b} \frac{\lambda(x)dx}{\sqrt{(x - x_0)^2 + y_0^2}}$$

where $\lambda(x) = \lambda_0(1 + c\sin^2 x)$ with $\lambda_0 = 1$ and $a = b = 1$, $(x_0, y_0) = (0, d)$. Estimate the electric potential, correct to four decimal places, for $c = 0.1k$ ($k = 0, 1, 2, \ldots, 10$) and $d = 0.25, 0.5, 1$, and 2 using Simpson's 1/3 rule.

CA8.7 A speed boat leaves the harbor and follows the route (shown in **CA8.7**), which is well represented by $x(t) = 3\alpha t/(1 + t^3)$, $y(t) = 3\beta t^2/(1 + t^3)$ where t denotes time (in minutes), $\alpha = 26000$ m·min² and $\beta = 26000$ m·min. The force realized as a result of the propulsion of the engine, seawater, wind drags, etc. is assumed to be represented by

$$\mathbf{F}(x, y) = m\left(\frac{x^3 + y^3}{\alpha^3}\mathbf{i} + \frac{2x^2 y^2}{\alpha^2 \beta^2}\mathbf{j}\right)$$

Fig. CA8.7

where m is the mass of the boat. Estimate the work done (per kg) for the first 5 minutes of travel; that is, compute the given integral *correct to four-significant-place* using Romberg's rule.

$$\frac{W}{m} = \int_C \frac{\mathbf{F}(x, y)}{m} \cdot d\mathbf{r}$$

CA8.8 Wind turbine towers are subjected to strong winds. The wind pressures are measured from various heights on the tower, as shown in **CA8.8**. The data is reported in the table below.

Location	h (m)	p (kPa)
0	0	288
1	5	460
2	15	530
3	20	544
4	27	556
5	40	567
6	45	570
7	50	572
8	55	574

Fig. CA8.8

Using the trapezoidal rule and the following expression to estimate the center of pressure (cop) of the tower:

$$H_{\text{cop}} = \int_{x=0}^{H} hp(h)dh / \int_{x=0}^{H} p(h)dh$$

CA8.9 The error function, encountered in many fields of science and engineering, is defined as

$$\text{erf}(x) = \frac{2}{\sqrt{\pi}} \int_{0}^{x} e^{-t^2} dt$$

(a) Transform the integral to one on $[-1, 1]$; (b) write a computer program (or use a spreadsheet) to estimate the error function for any x using 2-, 3-, 4-, and 5-point Gauss-Legendre quadrature; (c) compare your estimates for $x = 0.1, 0.5, 1$, and 2 with the true values of 0.1124629160182849, 0.5204998778130466, 0.8427007929497149, and 0.9953222650189527, respectively.

CA8.10 The distribution of molecular speed is described by the well-known Maxwell-Boltzmann distribution, which is a function of the mass and temperature of molecules.

$$f(v) = 4\pi v^2 \left(\frac{m}{2\pi kT}\right)^{3/2} \exp\left[-\frac{mv^2}{2kT}\right]$$

where m is the mass of the molecule, k is the Boltzmann constant, v is the speed of the molecule, and T is the absolute temperature. The average and root-mean-square (rms) speeds and the average kinetic energy of the molecules are computed from

$$v_{avg} = \int_{v=0}^{\infty} v f(v) dv \qquad v_{rms} = \sqrt{\int_{v=0}^{\infty} v^2 f(v) dv}, \qquad E_{avg} = \int_{v=0}^{\infty} \frac{mv^2}{2} f(v) dv$$

Write a computer program (or use a spreadsheet) to estimate the average and rms speeds of gas molecules using a suitable quadrature method with $N = 2, 3, 4, 5, 10, 20, 30$, and 40 points.

CA8.11 In radiation heat transfer, the fraction of blackbody emissive power contained between 0 and λT (wavelength × temperature) leads to the following integral:

$$f(\lambda T) = \frac{15}{\pi^4} \int_{14390/\lambda T}^{\infty} \frac{x^3 dx}{e^x - 1}$$

Noting that $0 \leqslant \lambda T \leqslant 10^5$, how would you estimate the improper integral using available quadrature methods? Compare your estimates with the true value of 6.493 for $\lambda T = 10^5$.

CA8.12 The exponential integral function of first order is defined as

$$E_1(x) = \int_{x}^{\infty} \frac{e^{-t}}{t} dt$$

(a) Transform the integral to an improper integral on $(0, \infty)$ and estimate $E_1(x)$ for $x = 0.5, 1$, and 2 with $N = 2$-, 4-, 6-, and 10-point Gauss-Laguerre quadrature; (b) Transform the integral to a definite integral on $[0,1]$ and estimate $E_1(x)$ for $x = 0.5, 1$, and 2 with $N = 3$-, 5-, 7-, and 9-point Gauss-Legendre quadrature.

CA8.13 (a) Write a computer program to estimate the following double integral with the trapezoidal and Simpson's 1/3 rules by applying N and M panels for x- and y-intervals; (b) Estimate the double integral for $N = M = 8, 16$, and 32 and compare your numerical estimates with those of the true value, which is $(1 - 4\sqrt{2} + 3\sqrt{3})/3$.

$$\int_{x=0}^{1} \int_{y=0}^{1} \frac{xy \, dydx}{\sqrt{x^2 + y^2 + 1}}$$

CA8.14 Repeat **CA 8.13** using the Gauss-Legendre quadrature with $N = 3, 5, 7$, and 10 points on both x- and y-intervals.

CA8.15 The temperature distribution of a 1m×1m plate exposed to a heat source is obtained by thermal imaging, and the collected data (in $C°$) is presented below.

x (m)	y (m)					
	0	0.2	0.4	0.6	0.8	1
0	120.0	120.20	120.80	121.80	123.20	125.0
0.2	120.4	120.76	121.52	122.68	124.24	126.2
0.4	121.6	122.12	123.04	124.36	126.08	128.2
0.6	123.6	124.28	125.36	126.84	128.72	131.0
0.8	126.4	127.24	128.48	130.12	132.16	134.6
1	130.0	131.00	132.40	134.20	136.40	139.0

An average temperature value for any plate is computed by

$$T_{avg} = \frac{1}{A_{plate}} \iint\limits_{A_{plate}} T(x,y)dA$$

Estimate the average temperature of the plate (a) using the trapezoidal rule; (b) repeat Part (a) by employing the midpoint rule; and (c) comment on how you can apply Simpson's 1/3 rule.

ODEs: Initial Value Problems

A MATHEMATICAL model gives a theoretical prediction of a real-world physical event in mathematical terms. A set of governing differential equations describing a physical phenomenon or a system in the fields of science and engineering is an inevitable part of a mathematical model. Besides differential equations, every mathematical model consists of additional constraints such as *initial* or *boundary conditions*. In this regard, scientists and engineers perform simulations of a physical problem or a design by solving the pertinent mathematical model to understand the problem or predict its behavior under certain circumstances. Some ODEs are simple enough to have true solutions, while the

DOI: 10.1201/9781003474944-9

majority of ODEs either do not have true solutions or are difficult or too complicated to obtain. In the latter case, numerical solutions to ODEs are sought almost exclusively as a tool to find approximate solutions.

Differential equations encountered in science and engineering can be classified as (1) *Initial Value Problems (IVPs)*, (2) *Boundary Value Problems (BVPs)*, and (3) *Characteristic or Eigenvalue Problems (EVPs)*. This chapter is dedicated solely to the numerical solution of initial value problems. Boundary and Eigenvalue value problems are covered in Chapters 10 and 11, respectively.

Initial Value Problems (IVPs) describe a *transient* or *unsteady* behavior of a physical event for $t > t_0$, provided that the solution is known at $t = t_0$ (i.e., the initial value). In other words, the solution interval of the IVPs is semi-infinite. But numerical solutions are typically sought for a finite interval $t_0 < t \leqslant T$ or until a physical event reaches the steady-state. Some of the most common numerical methods for approximating linear or nonlinear IVPs are presented and discussed in this chapter.

9.1 FUNDAMENTAL PROBLEM

Consider a first-order ODE (the fundamental IVP) expressed as

$$\frac{dy}{dx} = f(x, y), \qquad y(x_0) = \alpha_0, \quad x > x_0 \tag{9.1}$$

where x is the independent, y is the dependent variable, $f(x, y)$ is a function describing the differential equation, and $y(x_0) = \alpha_0$ is a starting value called the *initial condition (IC)*.

Initial value problems are not confined to first-order ODEs. A second-order IVP is a second-order ODE with two initial conditions at $x = x_0$:

$$a\,y'' + b\,y' + c\,y = g(x), \quad y(x_0) = \alpha_0, \quad y'(x_0) = \alpha_1, \quad x > x_0 \tag{9.2}$$

High-order IVPs require as many initial conditions as the order of the ODE. For instance, the following IVP requires three initial conditions to find its unique solution.

$$ay''' + by'' + cy' + d\,y(x) = h(x), \quad y(x_0) = \alpha_0, \quad y'(x_0) = \alpha_1, \quad y''(x_0) = \alpha_2 \tag{9.3}$$

where α_0, α_1, and α_2 are constants describing some physical quantities at the initial state.

The coefficients of an IVP may consist of constants (a, b, c, and d) as in the example above, or they may be functions of the independent variable ($a(x)$, $b(x)$, $c(x)$, and so on), as in the example below:

$$(x^2 + 3)y'' + 2xy' + 5y(x) = g(x), \quad y(0) = 0, \quad y'(0) = 3$$

The differential equations presented above are termed *linear ordinary differential equations*. If at least one of the coefficients of the ODE involves a dependent variable and/or its derivatives (i.e., y, y', y'', ...), the ODE is referred to as a *non-linear ordinary differential equation*. The following is an example of a non-linear ODE (or IVP):

$$2\frac{d^2y}{dx^2} + e^{x+y}\frac{dy}{dx} + \sqrt{y + y'^2} = 0, \quad y(0) = 1, \quad y'(0) = -1$$

where $a = 2$, $b = e^{x+y}$, and $c = \sqrt{y + y'^2}$. Note that in this IVP, b is a function of y in addition to x, and c is a function of both y and y'.

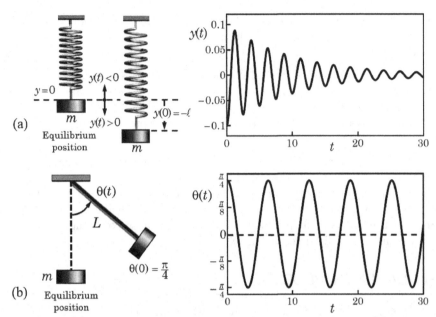

FIGURE 9.1: (a) Swinging pendulum system and a typical solution; (a) Damped mass-spring system and a typical solution.

In Fig. 9.1a, a damped motion of a mass m attached to a spring of length L is depicted. The mathematical model of this event is expressed with the following second-order ODE:

$$\frac{d^2y}{dt^2} + \frac{b}{m}\frac{dy}{dt} + \frac{k}{m}y(t) = 0$$

where k is the spring constant, b is the damping coefficient, and y is the vertical displacement relative to the equilibrium position. Before the motion is set ($t < 0$), the spring-mass system is in equilibrium, i.e., $y(t) = 0$, which is referred to as the trivial case. This IVP has a unique solution that depends on the imposed initial conditions. A motion is set when the attached mass is pulled down by an amount ℓ and then released at $t = 0$. Hence, the vertical displacement and velocity at rest, before the mass is released, are set as the initial condition values, i.e., $y(0) = -\ell$ and $v(0) = dy/dt)_{t=0} = 0$. The mass-spring system leads to a damped oscillatory motion about the equilibrium position. The mass eventually comes to rest at the equilibrium position after a sufficient amount of time (as $t \to \infty$, $y(t) \to 0$).

In Fig. 9.1b, a pendulum with mass m is displaced from its equilibrium position by $\theta_0 = \pi/4$. When the pendulum is released, a periodic (swinging) motion is set. Assuming the motion is undamped, a second-order ODE describing the angular position of the pendulum as a function of time, $\theta(t)$, is expressed as

$$\frac{d^2\theta}{dt^2} + \frac{g}{L}\sin\theta = 0$$

where t is time, g is the acceleration of gravity, L is the length of the pendulum, and θ is the angular displacement. This ODE, as is, has an infinite number of solutions. Since the IVP is a second-order ODE, the mathematical description of the model will be complete when two initial conditions (IC) specific to this motion are specified. Note that since the pendulum is initially at rest ($t = 0$), the angular displacement and angular speed at the onset are known, i.e., $\theta(0) = \pi/4$ and $\omega(0) = d\theta/dt)_{t=0} = 0$. By initializing the simulation time at $t = 0$, the rest of the motion will be relative to the initial time. The angular displacement varies in the range $-\pi/4 \leqslant \theta(t) \leqslant \pi/4$ during the swinging motion, as shown in Fig. 9.1b.

As we have seen in the preceding examples, the physical quantities attached to y and y' are defined at the initial time; that is, $y(t_0) = y_0$ and $y'(t_0) = y'_0$ must be given or known in order to find the solution of the IVPs for $t > 0$. The solution interval of an IVP is $[t_0, \infty)$; however, the numerical solution is frequently sought for a finite interval $(0 < t \leqslant T$ or until the *steady-state* is reached within a tolerance value).

9.2 ONE-STEP METHODS

One-step methods are numerical schemes in which numerical estimates or approximations of a first-order ODE are obtained by advancing one step at a time. In other words, $y(x_i + h)$ denoting the $(i+1)$th step estimate (also referred to as *current step estimate*) depends on $y(x_i)$ denoting the ith step estimate (referred to as *prior step estimate*).

9.2.1 EXPLICIT EULER METHOD

Consider the following first-order IVP:

$$\frac{dy}{dx} = f(x, y), \quad y(x_0) = \alpha_0, \quad x > x_0 \tag{9.4}$$

The objective is to find an approximate solution of this ODE on $[x_0, \infty)$. To obtain the approximate value, y' in Eq. (9.4) is replaced by the forward difference formula. Adapting a uniform step size throughout (i.e., $x_{i+1} - x_i = h$ for all i), this step, referred to as the *discretization* procedure, results in

$$\frac{y_{i+1} - y_i}{h} + \mathcal{O}(h) = f(x_i, y_i) \tag{9.5}$$

Solving Eq. (9.5) for the current step estimate, y_{i+1}, yields

$$y_{i+1} = y_i + h f(x_i, y_i) + \mathcal{O}(h^2), \quad i = 0, 1, 2, \ldots \tag{9.6}$$

where $x_i = x_0 + ih$ and $y_i = y(x_i)$ denote the abscissa and the prior step estimate, respectively. This numerical scheme is called the *Explicit Euler* or *Forward Euler Method* because the current estimate y_{i+1} is explicitly obtained using the known prior estimates. Inspecting Eqs. (9.5) and (9.6), it is clear that the method has a *local truncation error order* of $\mathcal{O}(h^2)$ and a *global error order* of $\mathcal{O}(h)$.

The solution strategy is graphically depicted in Fig. 9.2. Since y' is replaced by the first-order forward difference formula (FDF), the current projection is determined along the tangent line based on the derivative at the prior point. For any x on $[x_i, x_{i+1}]$, Eq. (9.6) can be rewritten as

$$y(x) = y_i + (x - x_i)f(x_i, y_i)$$

which corresponds to the equation of a straight line with slope $f(x_i, y_i) = y'_i$. The first estimate, y_1, is obtained after moving a small step, h. At x_1, a new tangent line is constructed, and a second estimate, y_2, is found along the new tangent line. This procedure, in which estimates are obtained step by step, is also referred to as *marching* or *forward advancing*.

By setting $i = 0$ in Eq. (9.6), the first step estimate is found as

$$y_1 = y(x_0 + h) = y_0 + h f(x_0, y_0) \tag{9.7}$$

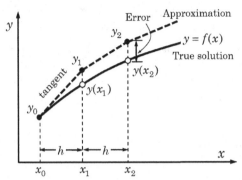

FIGURE 9.2: Graphical depiction of forward advancing with explicit Euler's method.

where x_0 and y_0 are prior values already known due to the initial condition, and y_1 is the current step estimate.

Now that y_1 is available, setting $i = 1$ in Eq. (9.6) yields

$$y_2 = y(x_1 + h) = y_1 + h\,f(x_1, y_1) \qquad (9.8)$$

where y_1 and y_2 denote the prior and current step estimates, respectively. As the solution is advanced one step at a time, the quantities on the rhs of Eq. (9.6), consisting of prior estimates, are always known. Hence, this procedure leads to a simple and straightforward explicit algorithm. The recursive procedure for finding the approximate solution at any x_i is obtained as follows:

$$y(x_2 + h) = y_3 = y_2 + h\,f(x_2, y_2)$$
$$y(x_3 + h) = y_4 = y_3 + h\,f(x_3, y_3)$$
$$\cdots \qquad \cdots$$
$$y(x_i + h) = y_{i+1} = y_i + h\,f(x_i, y_i)$$

Before proceeding any further, it should be noted that, from now on, the *numerical solution* of an IVP will be referred to as an *"estimate"* (or *"approximation"*) since the solution is *not* going to be exact.

Essentially, two types of errors are inevitably encountered in the numerical solution of an IVP: (*i*) *round-off error*, which is due to the finite precision of floating-point arithmetic, and (*ii*) *truncation error*, which is due to the discretization of the ODE.

Round-off errors can be remedied to some extent by increasing the arithmetic precision. Since the explicit Euler method is first-order accurate, $\mathcal{O}(h)$, a very small step size is required to achieve very high accuracy. Nonetheless, a digital computer can, of course, handle very small intervals in the order of billions, but then we would have to face the consequences of the round-off errors (*see* Section 1.2). Even if an extremely small step size is used, the round-off errors will accumulate over a large number of steps, and the computed estimates will tend to diverge from the true value. In other words, round-off errors *do* impose a limitation on the number of intervals that one can efficiently deal with.

Truncation error is inherited from the method, i.e., the way an ODE is discretized. Truncation error cannot be completely eliminated in the numerical solution of any ODE, but its magnitude can be reduced only by adopting a different numerical method. Truncation error can basically be viewed in two parts: (i) *global error*, which is the difference between the *true* and *estimated value*, $e_i = y_i - \hat{y}(x_i)$, and (ii) *local truncation error*, which is the

```
                      Pseudocode 9.1

Module EXPLICIT_EULER (h, x0, y0, xlast)
\  DESCRIPTION: A pseudomodule to solve y' = f(x, y) on [x0, xlast] using the
\    Explicit Euler method.
\  USES:
\    FCN:: User-defined function supplying f(x, y) = y'.
Write: "x, y =", x0, y0                          \ Print initial value
x ← x0                                           \ Initialize x
While [x < xlast]              \ Marching loop: Advance one step if x < xlast
    x ← x0 + h                          \ Find current step abscissa, x_{i+1}
    y ← y0 + h * FCN(x0, y0)      \ Compute current step estimate, y_{i+1}
    Write: "x, y =", x, y                         \ Print x_{i+1}, y_{i+1}
    x0 ← x;  y0 ← y                     \ Set current values as prior
End While
End Module EXPLICIT_EULER
```

error resulting from a one-step computation, $\varepsilon_i = y_i - y_{i-1}$, where y_{i-1} is the true solution of the prior point. The accumulation of all local errors with each step determines the global error observed after numerous steps. Note that global errors are not necessarily the sum of local errors. Global error generally becomes greater than the sum of the local errors if a numerical scheme is *unstable* (*see* Section 9.3). However, if the numerical scheme is *stable*, then the global error becomes less than the sum of the local errors. The goal of any numerical scheme is to minimize the global error; however, keep in mind that a numerical analyst can control only the local error.

A pseudomodule, EXPLICIT_EULER, implementing the explicit Euler scheme for the solution of the first-order IVP is presented in Pseudocode 9.1. The module requires a step size (h), a solution interval $[x0, xlast]$, and the initial value $y0$ as input. The module is general in that it can be applied to any first-order IVP. In other words, the solver module is separate, and the user only needs to supply the first derivative, $y' = f(x, y)$, as a user-defined function, FCN.

The numerical estimates are obtained step by step in the While-loop using Eq. (9.6),

```
                   Explicit Euler Method
```

- Explicit Euler method is the simplest and easiest algorithm;
- Memory requirement is minimal, as there is no need for arrays;
- It can be employed for linear and nonlinear IVPs;
- It requires one function evaluation per step.

- It is less accurate because it has a global error of $\mathcal{O}(h)$;
- The step size should be chosen sufficiently small in order to obtain accurate and meaningful solutions;
- It is *conditionally stable* (*see* Section 9.3), i.e., it is unstable under certain conditions, giving rise to a restriction on the size of h;
- Using a very small step size leads to excessive cpu-time as well as the accumulation of round-off errors.

until $xlast$ is reached. Note that the subscripted variables (x_i, y_i) in Eq. (9.6) are *not* translated to one-dimensional array variables. If array variables were used for x and y variables, the memory requirement would increase substantially in problems that require a large number of steps. Besides, the estimates calculated at the end of each step need not be kept in memory once they are printed out or saved to an output file. As a strategy to minimize memory usage, the current estimates (x, y) are printed out at the end of each step, and then the current estimates are assigned as prior estimates, $(x0, y0) \leftarrow (x, y)$, in preparation for the next step calculations. Thus, only the prior and current step estimates are known at any time step.

EXAMPLE 9.1: Implementing explicit Euler method

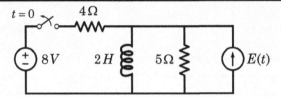

For $t < 0$, the RL circuit shown in the figure is in steady-state, and the current source E is off. When the switch is opened at $t = 0$, the source generating current at an exponential rate is turned on. The transient current response is subjected to the following initial value problem:

$$2\frac{dI}{dt} + 5I(t) = E(t), \quad i(0) = 2A, \quad E(t) = \begin{cases} 0, & t < 0 \\ 16e^{-t/2} & t \geqslant 0 \end{cases}$$

Solve the resulting IVP on [0,2] with the explicit Euler method using $\Delta t = 0.2$, 0.1, and 0.01 s. Compare your estimates with the true solution given as $I(t) = 4e^{-t/2} - 2e^{-5t/2}$.

SOLUTION:

For $t \geqslant 0$, rearranging the IVP as $dI/dt = f(t, I)$ leads to $f(t, I) = 8e^{-t/2} - 2.5I$. The initial condition at $t_0 = 0$ is given in the problem statement as $I(0) = I_0 = 2A$.

The computational details of the first three steps with $h = \Delta t = 0.2$ s are explicitly presented here. Starting with $i = 0$ ($t_1 = t_0 + h = 0.2$) in Eq. (9.6), the first step estimate, $I_1 = I(t_1) = I(0.2)$, is found as

$$I(0.2) = I_1 = I_0 + hf(0, I_0) = 2 + (0.2)f(0, 2) = 2 + (0.2)(3) = 2.6$$

Now, for $i = 1$ ($t_2 = t_1 + h = 0.4$), the second step estimate, $I_2 = I(2h) = I(0.4)$, is obtained as

$$I_2 = I_1 + hf(0.2, I_1) = 2.6 + (0.2)(7.38699) = 2.747740$$

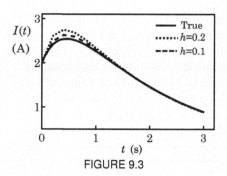

FIGURE 9.3

Next, setting $i = 2$ $(t_3 = 0.6)$, the third step estimate $I_3 = I(0.6)$ leads to

$$I_3 = I_2 + hf(t_2, I_2) = 2.747740 + (0.2)(-0.319504) = 2.683839$$

The estimates for the subsequent steps, like $I(0.8) = I_4$, $I(1) = I_5$, and so on, are obtained in a recursive manner. The true solution as well as the estimates $I(t_i)$ obtained with the explicit Euler for $h = 0.1$ and $h = 0.01$, are comparatively depicted in Fig. 9.3. Note that the deviations of the estimates from the true current can be clearly distinguished between $t = 0.2$ and $t = 1.5$, and the maximum deviations appear to occur around $t = 0.5$. For $t > 1.5$, as the estimates with both $h = 0.2$ and $h = 0.1$ recover and approach the true solution, the global errors (deviations) also approach zero. However, the magnitude of the deviations is smaller for the $h = 0.1$ case since a smaller value of h naturally leads to a smaller local truncation error.

TABLE 9.1

t	True Solution	$h = 0.2$		$h = 0.1$		$h = 0.01$	
		I_i	e_i	I_i	e_i	I_i	e_i
0	2	2		2		2	
0.2	2.406288	2.60000	1.94E-01	2.485984	7.97E-02	2.413221	6.93E-03
0.4	2.539164	2.74774	2.09E-01	2.629835	9.07E-02	2.547322	8.16E-03
0.6	2.517013	2.683839	1.67E-01	2.593561	7.66E-02	2.52414	7.13E-03
0.8	2.410610	2.527229	1.17E-01	2.46712	5.65E-02	2.416062	5.45E-03
1	2.261953	2.336126	7.42E-02	2.300049	3.81E-02	2.265774	3.82E-03
1.2	2.095672	2.138512	4.28E-02	2.119256	2.36E-02	2.098149	2.48E-03
1.4	1.925946	1.947355	2.14E-02	1.939005	1.31E-02	1.927408	1.46E-03
1.6	1.760685	1.768214	7.53E-03	1.766535	5.85E-03	1.761423	7.39E-04
1.8	1.604061	1.603033	1.03E-03	1.605205	1.14E-03	1.604308	2.47E-04
2	1.458042	1.452028	6.01E-03	1.456263	1.78E-03	1.45797	7.19E-05

The numerical estimates of the IVP for $h = 0.2$, 0.1, and 0.01, as well as the true errors $(e_i = |I_{\text{true}}(t_i) - I_i|)$ at each time step, are comparatively tabulated in Table 9.1. It is clear that the absolute errors are reduced by half on average as the time step size is reduced from $h = 0.2$ to $h = 0.1$. Similarly, when the step size is reduced from $h = 0.1$ to 0.01 (i.e., h is reduced by a factor of ten), the errors also decrease by a factor of ten on average, confirming that the global error is directly proportional to h. Note that the absolute error, while smaller in magnitude at first, steadily increases up to $t \approx 0.5$, where it goes through a local maximum and then decreases as t increases.

Discussion: In this example, we can rule out round-off errors as a source of error accumulation since moderate step sizes ($h \approx 0.01-0.2$) and high precision arithmetic were used to obtain the estimates in Table 9.1. The main source of error here is primarily due to discretization (global) error, which is directly proportional to h, i.e., $\mathcal{O}(h)$. Additionally, for any h, the global errors depict an increase at the beginning of the solution ($0.2 < t < 1.5$), but then the errors steadily decrease. This behavior indicates that the estimates with this method are stable for the h values at which the solution is obtained.

 Each time step contributes a new error to the current estimate, which accumulates over a large number of steps. The rate of accumulation depends on the ODE, i.e., $y' = f(x, y)$.

9.2.2 METHOD OF TAYLOR POLYNOMIAL

In calculus courses, we have learned that every univariate function can be expanded into a power series about $x = x_0$. In other words, a function $f(x)$ can be expressed in the form of a power series if $f(x_0)$ and $f'(x_0)$, $f''(x_0)$, and so on derivatives can be evaluated. In this respect, the Taylor series is a robust tool for obtaining a power series approximation of a continuous and differentiable function.

Consider the first-order model IVP, Eq. (9.1). The first derivative of the function, $y' = f(x, y)$, and its initial value, $y(x_0) = y_0$, are already given in the problem. This information is sufficient to construct an nth-degree Taylor polynomial, $T_n(x)$, of $y(x)$ in the vicinity of $x = x_0$; that is,

$$T_n(x) = y(x_0) + (x-x_0)y'(x_0) + \frac{(x-x_0)^2}{2!}y''(x_0) + \ldots + \frac{(x-x_0)^n}{n!}y^{(n)}(x_0) + R_n(x) \qquad (9.9)$$

where $R_n(x)$ denotes the remainder, which is given as

$$R_n(x) = \frac{(x-x_0)^{n+1}}{(n+1)!}y^{(n+1)}(\xi), \qquad x_0 < \xi < x$$

For any $x = x_i$, $T_n(x)$ can be generalized as

$$T_n(x_i) = y_{i+1} = y_i + hf(x_i, y_i) + \frac{h^2}{2!}f'(x_i, y_i) + \ldots + \frac{h^n}{n!}f^{(n-1)}(x_i, y_i) + R_n \qquad (9.10)$$

Note that for order $n = 1$, Taylor's method gives the explicit Euler method. An upper bound can be estimated from

$$|R_n| < \frac{h^{n+1}}{(n+1)!}M, \qquad x_i < \xi < x_{i+1}$$

with

$$M \geqslant \left|y^{(n+1)}(\xi)\right|_{\max} = \left|f^{(n)}(\xi, y(\xi))\right|_{\max}, \qquad x_i < \xi < x_{i+1}$$

Even though the algorithm is straightforward, it does require the user to derive and supply an appropriate number of derivatives of y or f, which can become cumbersome very quickly. The above error estimation strategy can be useful when high-order derivatives are simple and easy to derive.

Pseudocode 9.2

Module TAYLOR_POLY (*degree*, *h*, *x0*, *y0*, *xlast*)
\ DESCRIPTION: A pseudomodule to solve $y' = f(x, y)$ on $[x0, xlast]$ using the
\   Taylor polynomials of *degree* $\geqslant 2$.
\ USES:
\   FCNS:: A user-defined function supplying y', y'', y''' and so on.
$h2 \leftarrow h * h/2$; $h3 \leftarrow h2 * h/3$ \ Pre-set $h^2/2!$, $h^3/3!$
Write: "$x, y=$", $x0, y0$ \ Print initial value
$x \leftarrow x0$ \ Initialize x
While $\left[x < xlast\right]$ \ Marching Loop : Advance one step if $x < x_{last}$
     $x \leftarrow x0 + h$ \ Find current step abscissa, x_{i+1}
     $yp1 \leftarrow$ **FCNS**$(1, x0, y0)$ \ Compute $y'(x_0)$ using FCNS
     $yp2 \leftarrow$ **FCNS**$(2, x0, y0)$ \ Compute $y''(x_0)$ using FCNS
     $y \leftarrow y0 + h * yp1 + h2 * yp2$ \ 2nd degree approximation, $y_{i+1} = T_2(x_{i+1})$
     If $\left[degree = 3\right]$ **Then**
          $yp3 \leftarrow$ **FCNS**$(3, x0, y0)$ \ Compute $y'''(x_0)$ using FCNS
          $y \leftarrow y + h3 * yp3$ \ 3nd degree approximation, $y_{i+1} = T_3(x_{i+1})$
     Else
          \ You can nest higher-order terms of the approximation here
     End If
     Write: "$x, y=$", x, y \ Print current estimates
     $x0 \leftarrow x$; $y0 \leftarrow y$ \ Set current step values as prior
End While   \ End the integration loop
End Module TAYLOR_POLY

Function Module FCNS $(n, x0, y0)$
\ DESCRIPTION: A pseudo Function for computing the derivatives up to the 4th
\   degree. All pertinent derivatives must be supplied here.
If $\left[n = 1\right]$ **Then**
     FCNS$\leftarrow f(x0, y0)$ \ Define $y'(x0) = f(x0, y0)$ here
Else
     If $\left[n = 2\right]$ **Then**
          FCNS$\leftarrow y''$ \ Define $y''(x0)$ at $(x0, y0)$ here
     Else
          \ You can nest all $y^{(n)}$'s at $(x0, y0)$ here
     End If
End If
End Function Module FCNS

Subsequently, higher-order derivatives of $y(x)$ can be generated by simply differentiating $y' = f(x, y)$ as many times as necessary. Using the chain rule, the second and third derivatives are obtained as follows:

$$y_i'' = f'(x_i, y(x_i)) \; = \left(\frac{\partial f}{\partial x}\frac{dx}{dx} + \frac{\partial f}{\partial y}\frac{dy}{dx}\right)_{x=x_i} = \left(\frac{\partial f}{\partial x} + \frac{\partial f}{\partial y}f(x, y)\right)_{x=x_i}$$

$$y_i''' = f''(x_i, y(x_i)) = \left(\frac{\partial^2 f}{\partial x^2} + 2y'\frac{\partial^2 f}{\partial x\partial y} + (y')^2\frac{\partial^2 f}{\partial y^2} + y''\frac{\partial f}{\partial y}\right)_{x=x_i}$$

$$= \left(\frac{\partial^2 f}{\partial x^2} + 2f\frac{\partial^2 f}{\partial x\partial y} + f^2\frac{\partial^2 f}{\partial y^2} + \left(\frac{\partial f}{\partial x} + f\frac{\partial f}{\partial y}\right)\frac{\partial f}{\partial y}\right)_{x=x_i},$$

> ### Method of Taylor Polynomials
>
> - It is a semi-analytical method for solving IVPs;
> - It is an explicit method that only requires knowledge of the prior step to estimate the current step;
> - It is capable of providing highly accurate solutions if a high-degree polynomial approximation can be derived, i.e., the order of accuracy is $\mathcal{O}(h^n)$ where n is the degree of polynomial approximation;
> - An optimal degree of a polynomial can be determined by ensuring that the remainder satisfies the desired accuracy ($|R_n| < \varepsilon$) in the solution range.
>
> - It may diverge when the ODE is "stiff" or $f(x, y)$ is not analytic near x_0 or has discontinuous points, and so on;
> - It requires explicit expressions for high-order derivatives of $f(x, y)$, which may be cumbersome or very difficult to find since the coefficients of the polynomials depend on $f(x, y)$;
> - Computationally, it can be expensive per step due to the larger number of function and high-order derivative evaluations.

Once the required derivatives are substituted into Eq. (9.10), the construction of the polynomial approximation of the nth degree is complete.

The module, **TAYLOR_POLY**, introduced in Pseudocode 9.2 computes the approximate solutions of the first-order IVP using Taylor polynomials up to 3rd degree. The module requires the degree of polynomial approximation (*degree*), the initial value ($y0$), as well as the solution interval $[x0, xlast]$ as input. Before starting the marching procedure, the multipliers $h^2/2!$ and $h^3/3!$ are computed and saved on the variables $h2$ and $h3$ as a cpu-time-saving strategy. Since Eq. (9.10) involves high-order derivatives, the IVP-specific derivatives should be derived by the analyst and provided to the main module *via* the user-defined external function module, **FCNS**. If-Then-Else structures are used to add new terms to Taylor polynomials. Using Eq. (9.10), the forward marching procedure is executed step by step in the While-loop until $xlast$ is reached. As with Pseudocode 9.1, the use of array variables is avoided to keep the memory requirement at a minimum. Hence, the current estimates are assigned as prior estimates, $(x0, y0) \leftarrow (x, y)$, before returning to the top of the loop. If the user wishes, for instance, to obtain only, say, a third-degree polynomial approximation to solve the IVP, then If-Then-Else statements within the While-loop can be removed (in Pseudocode 9.2) to make the program more cpu-time efficient. The module does not have an output argument as **EXPLICIT_EULER** since the solutions at intermediate steps are printed out or saved on an external file.

The arguments of **FCNS** are the order of the derivative (n) and the point ($x0$, $y0$) where the nth derivative is evaluated. For $n = 1, 2, 3$, and so on, the derivatives $y'(x0)$, $y''(x0)$, $y'''(x0)$, and so on are respectively defined using conditional statements, as shown in the pseudocode presented in Pseudocode 9.2.

> The order of a numerical method is determined by its order of global error. An nth-order method has a global error of $\mathcal{O}(h^n)$.

EXAMPLE 9.2: Applying Taylor polynomial method

Using $h = 0.1$, obtain 2nd, 3rd, and 4th-degree Taylor polynomial approximations for the IVP given in Example 9.1.

SOLUTION:

The initial condition and the first derivative are $(t_0, I_0) = (0, 2)$ and $f(t, I) = 8e^{-t/2} - 2.5I$, respectively. Our next step is to find analytical expressions for higher derivatives. It turns out that the derivatives up to the nth order can be easily obtained by sequential differentiation of dI/dt, as shown below:

$$\frac{d^2I}{dt^2} = \frac{d}{dt}\left(\frac{dI}{dt}\right) = \frac{df}{dt} = -4e^{-t/2} - \frac{5}{2}\frac{dI}{dt} = -24e^{-t/2} + \frac{25}{4}I(t)$$

$$\frac{d^3I}{dt^3} = \frac{d}{dt}\left(\frac{d^2I}{dt^2}\right) = \frac{d}{dt}\left(-24e^{-t/2} + \frac{25}{4}I\right) = 62e^{-t/2} - \frac{125}{8}I(t)$$

$$\frac{d^4I}{dt^4} = \frac{d}{dt}\left(\frac{d^3I}{dt^3}\right) = \frac{d}{dt}\left(62e^{-t/2} - \frac{125}{8}I\right) = -156e^{-t/2} + \frac{625}{16}I(t)$$

Substituting the above expressions into Eq. (9.10) leads to the nth-degree Taylor polynomial:

$$I_{i+1} = T_n(t_i) = I_i + \sum_{k=1}^{n} \frac{h^k}{k!}\frac{d^k}{dt^k}I(t_i) + \mathcal{O}(h^{n+1})$$

The order of approximation depends on the level of accuracy desired, which can be determined by ensuring that the truncation error is less than a preset tolerance. For $n \geqslant 2$, noting that $M = \max|I^{(n+1)}(\xi)|$, we may write

$$|R_n| = \frac{h^{n+1}}{(n+1)!}M < 0.5 \times 10^{-d}, \qquad t_i < \xi < t_{i+1}$$

where d is the decimal-place accuracy desired.

Even though the high-degree derivatives are obtained easily and simplified to $I_i^{(n+1)} = C_1e^{-t_i/2} + C_2I(t_i)$, determining the value of M, for instance, for $n = 3$, requires the investigation of the d^4I_i/dt^4 distribution over the solution range, which is not an easy task.

TABLE 9.2

t_i	True Solution	T_2		T_3		T_4	
		I_i	e_i	I_i	e_i	I_i	e_i
0	2	2		2		2	
0.2	2.406288	2.37	3.63E-02	2.411000	4.71E-03	2.405808	4.80E-04
0.4	2.539164	2.494668	4.45E-02	2.544864	5.70E-03	2.538582	5.82E-04
0.6	2.517013	2.476146	4.09E-02	2.522182	5.17E-03	2.516483	5.30E-04
0.8	2.410610	2.377308	3.33E-02	2.414776	4.17E-03	2.410181	4.28E-04
1	2.261953	2.236576	2.54E-02	2.265099	3.15E-03	2.261628	3.25E-04
1.2	2.095672	2.077174	1.85E-02	2.097951	2.28E-03	2.095436	2.36E-04
1.4	1.925946	1.912903	1.30E-02	1.927550	1.60E-03	1.925779	1.67E-04
1.6	1.760685	1.751740	8.94E-03	1.761788	1.10E-03	1.760569	1.16E-04
1.8	1.604061	1.598086	5.97E-03	1.604807	7.46E-04	1.603982	7.89E-05
2	1.458042	1.454162	3.88E-03	1.458538	4.96E-04	1.457989	5.31E-05

For $h = 0.1$, the numerical estimates with the 2nd, 3rd, and 4th-degree Taylor polynomial approximations, along with the absolute errors ($e_i = |I_{i,\text{true}} - I_i|$), are comparatively tabulated in Table 9.2. It is clear that as the degree of the polynomial (or order of the approximation $\mathcal{O}(h^n)$) increases, the numerical estimates approach the true solution. In other words, the numerical (truncation) errors are reduced with increasing the degree of the Taylor polynomial.

Discussion: Table 9.2 shows that the errors in the Taylor approximations obtained with $T_2(t)$, $T_3(t)$, and $T_4(t)$ increase up to $x \approx 0.5$ (just as in Example 9.1), and then the errors tend to decrease with increasing x. It should also be noted that the order of global errors is consistent with $\mathcal{O}(h^n)$, i.e., for $h = 0.1$, the order of errors are $\mathcal{O}(10^{-2})$, $\mathcal{O}(10^{-3})$, and $\mathcal{O}(10^{-4})$ for $T_2(t)$, $T_3(t)$, and $T_4(t)$, respectively.

9.2.3 IMPLICIT EULER METHOD

As an alternative to the explicit Euler method, which uses the slope at point x_i, $y_i' = f(x_i, y_i)$, the slope in the implicit Euler method is estimated along the tangent line at point x_{i+1}: $y_{i+1}' = f(x_{i+1}, y_{i+1})$. The first derivative is discretized at $x = x_{i+1}$ using the first-order backward difference formula, while the right-hand side of Eq. (9.5) is evaluated at x_{i+1} as follows:

$$\frac{y_{i+1} - y_i}{h} + \mathcal{O}(h) = f(x_{i+1}, y_{i+1}) \tag{9.11}$$

This method is also referred to as *backward Euler method* in the literature.

By solving Eq. (9.11) for y_{i+1}, a general recursive relation marching in the x-direction is obtained, as in the explicit method. Having said that, note that when $f(x, y)$ is nonlinear, which is generally the case, this scheme may not permit us to explicitly isolate y_{i+1}. Thus, in cases where the IVP is nonlinear, the y_{i+1} has to be essentially solved iteratively using a root-finding method such as the fixed-point or Newton-Raphson method at every step. For instance, employing the fixed-point iteration method, the iteration equation can be expressed as

$$z^{(p+1)} = y_i + h f(x_{i+1}, z^{(p)}) \tag{9.12}$$

where $z = y_{i+1}$ and superscript p denotes the iteration step.

To employ the Newton-Raphson method, we express Eq. (9.11) as

$$G(z) = z - y_i - h f(x_{i+1}, z) = 0 \tag{9.13}$$

then the Newton-Raphson iteration equation becomes

$$z^{(p+1)} = z^{(p)} - \frac{G(z^{(p)})}{\frac{dG}{dz}(z^{(p)})} = z^{(p)} - \frac{G(z^{(p)})}{1 - h\frac{df}{dz}(z^{(p)})} \tag{9.14}$$

where $z^{(0)} = y_i$ may be taken as a suitable initial estimate.

Clearly, if f is nonlinear, then the implicit Euler method is not only computationally more involved but also time-consuming, in contrast to the explicit Euler method. Nevertheless, a *linearization* technique may be applied to avoid solving nonlinear equations. To demonstrate this technique, consider the Taylor series of $f(x_{i+1}, y_{i+1})$ about (x_{i+1}, y_i):

$$f(x_{i+1}, y_{i+1}) = f(x_{i+1}, y_i) + (y_{i+1} - y_i)\left(\frac{\partial f}{\partial y}\right)_{(x_{i+1}, y_i)} + \frac{(y_{i+1} - y_i)^2}{2!}\left(\frac{\partial^2 f}{\partial y^2}\right)_{(x_{i+1}, y_i)} + \dots \tag{9.15}$$

Pseudocode 9.3

Module IMPLICIT_EULER ($h,\ \varepsilon, maxit, x0,\ y0, xlast$)
\ DESCRIPTION: A pseudomodule to solve $y' = f(x, y)$ on $[x0,\ xlast]$ using the
\   Implicit Euler method. Fixed-Point iteration is used to solve the resulting
\   nonlinear equations.
\ USES:
\   ABS:: A built-in function computing the absolute value;
\   FCN:: A User-defined function supplying $f(x, y) = y'$.
Write: "$x, y=$", $x0, y0$ \ Print initial value
$x \leftarrow x0$ \ Initialize x
While $\left[x < xlast\right]$ \ Marching loop, march until $x <= xlast$
    $x \leftarrow x0 + h$ \ Advance one step, x_{i+1}
    $p \leftarrow 0$ \ Initialize counter for FP iterations
    $z \leftarrow y0$ \ Set initial estimate, $z^{(0)} = y_i$
    $y \leftarrow z$ \ Set current estimate to $z^{(0)}$
    Repeat   \ FP iteration loop to solve nonlinear eqs.
        $p \leftarrow p + 1$ \ Count iterations
        $z \leftarrow y0 + h * $**FCN**$(x, z)$ \ Find current estimate
        $aerr \leftarrow |z - y|$ \ Find absolute error
        $y \leftarrow z$ \ Set current step estimate as prior
        If $\left[p = maxit\right]$ **Then**   \ Check no. iterations to prevent infinite loop
            Write: "Max Iterations reached. Computation is terminated."
            Stop   \ FPI did not converge with current $maxit$ and h
        End If
    Until $\left[aerr < \varepsilon\right]$ \ When converged $y_{i+1} = z^{(p+1)}$
    Write: "$x, y=$", x, y \ Print current estimates
    $x0 \leftarrow x;\ y0 \leftarrow y$ \ Set current step values as prior
End While   \ End marching loop
End Module IMPLICIT_EULER

Replacing $f(x_{i+1}, y_{i+1})$ in Eq. (9.11) with the two-term approximation from Eq. (9.15) yields

$$\frac{y_{i+1} - y_i}{h} + \mathcal{O}(h) = f(x_{i+1}, y_i) + (y_{i+1} - y_i)\left(\frac{\partial f}{\partial y}\right)_{(x_{i+1}, y_i)} + \dots \qquad (9.16)$$

Note that since $y_{i+1} - y_i \equiv \mathcal{O}(h)$, the order of accuracy in this scheme remains the same as in the explicit method. Rearranging and solving Eq. (9.16) for y_{i+1} results in

$$y_{i+1} = y_i + \frac{hf(x_{i+1}, y_i)}{1 - h\dfrac{\partial f}{\partial y}(x_{i+1}, y_i)} + \mathcal{O}(h^2) \qquad (9.17)$$

Equation (9.17) now permits us to march forward step by step without the need for any iteration.

A pseudomodule IMPLICIT_EULER, implementing the implicit Euler scheme on a first-order IVP, is presented in Pseudocode 9.3. The module requires a step size (h), an upper bound for the maximum number of iterations ($maxit$), a convergence tolerance (ε) for the solution of the nonlinear equations, the initial value ($y0$), and the solution interval $[x0, xlast]$ as input. The module is basically the same as EXPLICIT_EULER, with the exception that the current estimate (z) is obtained by solving Eq. (9.12) iteratively using

Implicit Euler Method

- The implicit Euler method is a stable scheme that allows the use of large step sizes;
- It can be applied to both linear and nonlinear IVPs;
- The linearized scheme may offer some advantages if sufficiently large step sizes can be used.

- The method has a first-order global error, $\mathcal{O}(h)$;
- For nonlinear IVPs, the algorithm is a bit involved, and the cpu-time can be significantly higher due to the implementation of a root-finding scheme at each step.

the fixed-point iteration method within the **Repeat-Until** loop. In this loop, z and y denote the current $(z^{(p+1)})$ and prior $(z^{(p)})$ estimates, respectively. For sufficiently small h, the fixed-point iteration (FPI) generally converges very quickly; however, it should be noted that the Newton-Raphson method in general exhibits better convergence properties. An absolute error criterion, $|z - y| < \varepsilon$, is used to terminate the FPI procedure. Once the current step estimate y_{i+1} is obtained and printed out, it is assigned as the prior step estimate $(x0, y0) \leftarrow (x, y)$ before proceeding with the next step. The computations are repeated until $xlast$ is reached.

9.2.4 TRAPEZOIDAL RULE

The explicit Euler scheme linearizes IVP at x_i with the slope $f(x_i, y_i)$, while the implicit Euler method does that at x_{i+1} with the slope $f(x_{i+1}, y_{i+1})$. If the step size is not sufficiently small, the estimates with either method over many steps will lead to substantial deviation from the true solution, as illustrated in Fig. 9.4.

An alternative scheme can be established by replacing the slope with the arithmetic average of $f(x_i, y_i)$ and $f(x_{i+1}, y_{i+1})$, that is,

$$\frac{y_{i+1} - y_i}{h} + \mathcal{O}(h^2) = \frac{1}{2}\big(f(x_i, y_i) + f(x_{i+1}, y_{i+1})\big) \tag{9.18}$$

which is an *implicit scheme*, and it yields a nonlinear equation when the IVP is nonlinear. Note that y_i' in Eq. (9.18) is discretized by the central difference approximation at $x_{i+1/2}$, i.e., the mid-interval.

To find an expression for the truncation error of the trapezoidal rule, Eq. (9.4) is integrated from x_i to x_{i+1}:

$$y_{i+1} = y_i + \int_{x_i}^{x_{i+1}} f\big(u, y(u)\big) du \tag{9.19}$$

Replacing the integral with the trapezoidal rule yields

$$y_{i+1} = y_i + \frac{h}{2}\big(f(x_i, y_i) + f(x_{i+1}, y_{i+1})\big) - \frac{h^3}{12}y'''(\xi), \quad x_i < \xi < x_{i+1} \tag{9.20}$$

Comparing Eqs. (9.20) and (9.18), the two equations are clearly the same. However, Equation (9.20) reveals an expression for the local truncation error, $E = -h^3 y'''(\xi)/12$. Moreover, Eq. (9.20) indicates that the trapezoidal rule should give more accurate estimates than the

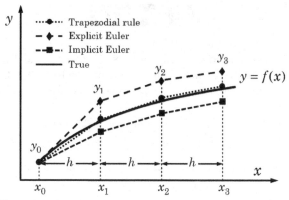

FIGURE 9.4: Graphical depiction of marching with the Explicit Euler, Implicit Euler methods and Trapezoidal rule.

Trapezoidal Rule

- The trapezoidal rule is a *stable* implicit method;
- It has a second-order global error, $\mathcal{O}(h^2)$;
- A linearized scheme with a suitable h can be successfully employed to solve nonlinear IVPs.

- The method suffers from the same drawbacks as the implicit Euler's method, i.e., since it is implicit, obtaining the current step estimate for nonlinear IVPs requires the solution of nonlinear algebraic equations, which increases the cpu-time.

explicit or implicit Euler methods because its global error is $\mathcal{O}(h^2)$. Also, note that the slope at each step is the average of those of explicit and implicit Euler schemes, which follows more closely to the true slope, as seen in Fig. 9.4. This method is referred to as the *Trapezoidal rule* since the trapezoidal rule is used in the numerical integration.

For nonlinear IVPs, the fixed-point iteration or Newton-Raphson methods can be applied (*see* Section 9.2.3). Equation (9.20) can likewise be linearized using the same procedure illustrated for the implicit Euler method. The linearization procedure for the trapezoidal scheme yields

$$y_{i+1} = y_i + \frac{h}{2} \frac{f(x_i, y_i) + f(x_{i+1}, y_i)}{1 - \dfrac{h}{2} \dfrac{\partial f}{\partial y}(x_{i+1}, y_i)} + \mathcal{O}(h^3) \tag{9.21}$$

Note that this expression retains the second-order global accuracy and also allows the numerical estimates to be obtained recursively and explicitly.

A pseudocode for the trapezoidal rule is *not* given here simply because replacing Eq. (9.12) in IMPLICIT_EULER with Eq. (9.18) is sufficient to modify the code for this purpose. The linearized implicit Euler and linearized trapezoidal methods are easier to implement since these schemes are explicit in nature, so the EXPLICIT_EULER module can be easily modified to accommodate the linearized methods. However, in this case, another user-defined function will be necessary to evaluate the partial derivative $\partial f / \partial y(x_{i+1}, y_i)$, which needs to be explicitly derived and supplied to the module.

Midpoint Method

- It is simple and easy to program;
- It can be employed for both linear and nonlinear IVPs;
- The global error for the midpoint method is second order, $\mathcal{O}(h^2)$;
- It is comparable to the trapezoidal rule in terms of cost.

- It requires two function evaluations per step, i.e., its computational cost is about twice as much as that of the explicit Euler method.

9.2.5 MIDPOINT METHOD

In this method, the first-order IVP, Equation (9.4), is evaluated at the midpoint of the next step, $y'(x_{i+1/2}) = f(x_{i+1/2}, y_{i+1/2})$. Then, using the central difference approximation for the first derivative results in

$$\frac{y_{i+1} - y_i}{h} + \mathcal{O}(h^2) = f(x_{i+1/2}, y_{i+1/2})$$

Solving for y_{i+1} gives

$$y_{i+1} = y_i + hf(x_{i+1/2}, y_{i+1/2}) + \mathcal{O}(h^3) \tag{9.22}$$

On the other hand, by replacing the integral in Eq. (9.19) with the midpoint rule, we also arrive at Eq. (9.22), which is why it is referred to as the *midpoint method*. In practice, Eq. (9.22) is applied as follows:

$$y_{i+1} = y_i + hf\left(x_i + \frac{h}{2}, y_i + \frac{h}{2}f(x_i, y_i)\right) + \mathcal{O}(h^3) \tag{9.23}$$

where $x_{i+1/2} = x_i + h/2$ and $y_{i+1/2} = y_i + (h/2)f(x_i, y_i)$.

9.2.6 MODIFIED EULER METHOD

The modified Euler's method is based on the trapezoidal rule. To make this an explicit method, the current estimate y_{i+1} on the rhs of the trapezoidal rule is replaced by the explicit Euler approximation $(y_{i+1} \leftarrow y_i + hf(x_i, y_i))$, as follows:

$$\frac{y_{i+1} - y_i}{h} + \mathcal{O}(h^2) = \frac{1}{2}\Big(f(x_i, y_i) + f(x_{i+1}, y_i + hf(x_i, y_i))\Big)$$

Hence, the modified Euler's scheme can be expressed as

$$y_{i+1} = y_i + \frac{h}{2}\big(f_{i,j} + f(x_{i+1}, y_i + hf_{i,j})\big) \quad \text{or} \quad y_{i+1} = y_i + \frac{1}{2}(k_1 + k_2) \tag{9.24}$$

where $f_{i,j} = f(x_i, y_i)$, $k_1 = hf(x_i, y_i)$, and $k_2 = hf(x_i + h, y_i + k_1)$.

The advantages and disadvantages of the method are practically the same as those of the midpoint method. In this section, pseudocodes are *not* given for the midpoint rule and modified Euler because they are explicit methods. The difference equation for the explicit Euler needs to be simply replaced by either Eq. (9.23) or (9.24).

EXAMPLE 9.3: Comparing implicit Euler, trapezoidal, midpoint, modified Euler methods

Solve Example 9.1 using the implicit Euler, trapezoidal, midpoint, and modified Euler methods with $h = 0.1$. Compare your numerical estimates with the true solution. Use $\varepsilon = 0.5 \times 10^{-6}$.

SOLUTION:

The difference equation for the implicit Euler method, Eq. (9.11), in this example, takes the following form:

$$I_{i+1} = I_i + hf(t_{i+1}, I_{i+1}) = I_i + h\left(8e^{-t_{i+1}/2} - 2.5I_{i+1}\right)$$

Since $f(t, I) = 8e^{-t/2} - 2.5I$, the IVP is a linear IVP. Hence, a root-finding method is *not* required to find the current estimate, I_{i+1}.

Next, solving for I_{i+1} yields the following explicit equation:

$$I_{i+1} = \frac{I_i + 8he^{-t_{i+1}/2}}{1 + 2.5h}, \quad i = 0, 1, 2, \ldots$$

Similarly, using Eqs. (9.18) and solving for I_{i+1}, the trapezoidal rule gives

$$I_{i+1} = (1 + 1.25h)^{-1}\left\{(1 - 1.25h)I_i + 4h\left(e^{-t_i/2} + e^{-t_{i+1}/2}\right)\right\}, \quad i = 0, 1, 2, \ldots$$

where $t_{i+1} = t_i + h$.

TABLE 9.3

t	True Solution	Implicit Euler	Trap. Rule	Midpoint Method	Modified Euler
0	2	0	0		
0.2	2.406288	6.02E-02	3.12E-03	7.07E-03	6.65E-03
0.4	2.539164	7.30E-02	3.76E-03	8.43E-03	7.79E-03
0.6	2.517013	6.58E-02	3.40E-03	7.49E-03	6.75E-03
0.8	2.410610	5.20E-02	2.72E-03	5.85E-03	5.09E-03
1	2.261953	3.78E-02	2.03E-03	4.22E-03	3.47E-03
1.2	2.095672	2.56E-02	1.45E-03	2.86E-03	2.15E-03
1.4	1.925946	1.60E-02	1.00E-03	1.81E-03	1.14E-03
1.6	1.760685	8.89E-03	6.73E-04	1.05E-03	4.35E-04
1.8	1.604061	3.87E-03	4.37E-04	5.25E-04	4.15E-05
2	1.458042	4.66E-04	2.74E-04	1.73E-04	3.45E-04

Discussion: Applying the midpoint and modified Euler schemes, Eqs. (9.23) and (9.24), is straightforward because they are explicit methods. Starting with the initial condition $(t_0, I_0) = (0, 2)$, the numerical estimates for $i = 1, 2, 3, \ldots$ (i.e., $t_i > 0$) are obtained recursively using the pertinent difference equations. In Table 9.3, the true errors by the method for selected points are presented comparatively. It is observed that the implicit Euler method yields the largest errors among all four methods. This result is not surprising since the global error of the implicit Euler is first order, i.e., $\mathcal{O}(10^{-2})$.

The trapezoidal, the midpoint, and the modified Euler methods, on the other hand, have second-order global errors, i.e., $\mathcal{O}(10^{-4})$. For this reason, the trapezoidal rule results in better estimates than those of the midpoint and modified Euler methods. The average error for the midpoint rule and the modified Euler method is about twice as large as that of the trapezoidal rule.

9.2.7 RUNGE-KUTTA (RK) METHODS

Using Taylor's polynomials to obtain the numerical solution $y' = f(x, y)$ in order to get a high-order approximation is not very practical. As we have seen earlier, obtaining explicit expressions for high-order derivatives of $y(x)$ depends on the function $f(x, y)$, and in most cases, the derivatives can become very complicated very quickly.

A class of *Runge-Kutta methods* are one-step *explicit schemes* that only involve $f(x, y)$ evaluations. Hence, it is easier to generate second-, third-, or fourth-order-accurate schemes.

All Runge-Kutta schemes have the following general form:

$$y_{i+1} = y_i + \phi(h, x_i, y_i) \tag{9.25}$$

where ϕ is referred to as the *increment function*, which is simply chosen to represent a suitable approximation for the slope on $[x_i, x_{i+1}]$.

The increment function may be expressed in terms of a weighted average as

$$\phi(h, x_i, y_i) = w_1 k_1 + w_2 k_2 + \cdots + w_n k_n \tag{9.26}$$

where n denotes the order of the Runge-Kutta method, $w_1, w_2, \ldots, w_n$ are the weights, and $k_1, k_2, \ldots, k_n$ are relationships defined as

$$\begin{aligned}
k_1 &= hf(x_i, y_i) \\
k_2 &= hf(x_i + \alpha_2 h, y_i + \beta_{21} k_1) \\
k_3 &= hf(x_i + \alpha_3 h, y_i + \beta_{31} k_1 + \beta_{32} k_2) \\
&\vdots \\
k_n &= hf(x_i + \alpha_n h, y_i + \sum_{m=1}^{n-1} \beta_{nm} k_m)
\end{aligned} \tag{9.27}$$

Equations (9.25) through (9.27) can be expressed in a compact form as

$$y_{i+1} = y_i + \sum_{m=1}^{n} w_m k_m \tag{9.28}$$

To determine the aforementioned weights and coefficients, consider the Taylor series expansion of $y(x_i + h)$ about x_i:

$$y_{i+1} = y_i + hy_i' + \frac{h^2}{2!} y_i'' + \frac{h^3}{3!} y_i''' + \cdots \tag{9.29}$$

The goal is to obtain the higher derivatives in Eq. (9.29) in terms of $f(x, y)$ by successive differentiations as follows:

$$y_i' = f(x_i, y_i)$$

$$y_i'' = \frac{df}{dx}\bigg|_i = (f_x + f_y y')_i = (f_x + f_y f)_i$$

$$y_i''' = \frac{d^2 f}{dx^2}\bigg|_i = \left(f_{xx} + 2f f_{xy} + f^2 f_{yy} + f_x f_y + f(f_y)^2\right)_i \quad (9.30)$$

$$\cdots$$

where subscripts x and y denote partial differentiations.

Substituting Eq. (9.30) into Eq. (9.29) yields

$$y_{i+1} = y_i + hf(x_i, y_i) + \frac{h^2}{2!}\left(f_x + f_y f\right)_i + \frac{h^3}{3!}\left(f_{xx} + 2f f_{xy} + f^2 f_{yy} + f_x f_y + f f_y^2\right)_i + \cdots \quad (9.31)$$

which also involves many partial derivatives. To avoid the partial derivatives and make it comparable to Eq. (9.28), the function evaluations in Eq. (9.27) are replaced with their Taylor series approximations.

Recall that the Taylor series of $f(x, y)$ about the point (a, b) is

$$f(a+h, b+k) = f(a, b) + \left(h\frac{\partial}{\partial x} + k\frac{\partial}{\partial y}\right)f(a, b) + \frac{1}{2!}\left(h\frac{\partial}{\partial x} + k\frac{\partial}{\partial y}\right)^2 f(a, b) + \cdots \quad (9.32)$$

Now the function evaluations in Eq. (9.27) are approximated by the Taylor series as follows:

$$f(x_i + \alpha_2 h, y_i + \beta_{21} k_1) \cong f(x_i, y_i) + \alpha_2 h f_x(x_i, y_i) + \beta_{21} k_1 f_y(x_i, y_i) + \dots$$

$$f(x_i + \alpha_3 h, y_i + \beta_{31} k_1 + \beta_{32} k_2) \cong f(x_i, y_i) + \alpha_3 h f_x(x_i, y_i) + (\beta_{31} k_1 + \beta_{32} k_2)$$
$$\times f_y(x_i, y_i) + \dots \quad (9.33)$$

$$\cdots \qquad\qquad \cdots$$

Next, substituting Eq. (9.33) into Eq. (9.27) and comparing with Eq. (9.31), a set of linear equations is obtained where w_n, α_n and β_{nm}'s are unknowns. In this section, only the derivation of the first- and second-order Runge-Kutta methods is presented to save space.

The first-order Runge-Kutta method yields

$$y_{i+1} = y_i + k_1 + \mathcal{O}(h^2) = y_i + hf(x_i, y_i) + \mathcal{O}(h^2) \quad (9.34)$$

where $k_1 = hf(x_i, y_i)$, and it corresponds to the *explicit Euler method*.

The increment function for quadratic Runge-Kutta is expressed as:

$$y_{i+1} = y_i + w_1 k_1 + w_2 k_2 + \mathcal{O}(h^3) \quad (9.35)$$

where

$$k_1 = hf(x_i, y_i) \quad \text{and} \quad k_2 = hf(x_i + \alpha h, y_i + \beta k_1) \quad (9.36)$$

The parameters α, β, w_1, and w_2 are determined in such a way that Eq. (9.35) satisfies the following 3rd-degree Taylor polynomial approximation of $y(x)$:

$$y_{i+1} = y_i + hf(x_i, y_i) + \frac{h^2}{2!}(f_x + f_y f)_{x_i} + \mathcal{O}(h^3) \quad (9.37)$$

Runge-Kutta Method

- Runge-Kutta schemes are explicit and relatively easy to apply;
- They are self-starting, i.e., only an initial condition is required to start the forward marching procedure;
- They have better stability characteristics (*see* Section 9.2);
- Unlike the method of Taylor polynomials, partial derivatives of $f(x, y)$ are not needed;
- Fourth- and fifth-order RK-schemes are very popular and widely used.

- RK methods do not provide any means of error estimation based on the interval size h;
- They require a relatively large amount of cpu-time due to multiple function evaluations;
- They are insufficient and ineffective in cases where the IVP is "*stiff.*"

where $k_1 = hf(x_i, y_i)$. On the other hand, k_2 is approximated by the following Taylor series approximation:

$$
\begin{aligned}
k_2 &= hf(x_i + \alpha h, y_i + \beta k_1) \\
&= hf(x_i, y_i) + h^2 \alpha \frac{\partial f}{\partial x}(x_i, y_i) + h^2 \beta \, f(x_i, y_i) \frac{\partial f}{\partial y}(x_i, y_i) + \mathcal{O}(h^3)
\end{aligned}
\tag{9.38}
$$

Then substituting Eq. (9.38) into Eq. (9.35) yields

$$
\begin{aligned}
y_{i+1} &= y_i + w_1 hf(x_i, y_i) + w_2 \left(hf + h^2 \alpha \, f_x + h^2 \beta \, f \, f_y \right)(x_i, y_i) + \mathcal{O}(h^3) \\
&= y_i + (w_1 + w_2)hf(x_i, y_i) + h^2 \left(w_2 \alpha \, f_x + w_2 \beta \, f \, f_y \right)(x_i, y_i) + \mathcal{O}(h^3)
\end{aligned}
\tag{9.39}
$$

Now equating the coefficients of like terms in Eq. (9.37) and Eq. (9.39) leads to:

$$
\begin{aligned}
hf(x_i, y_i) &: \quad w_1 + w_2 = 1, \\
h^2 f_x(x_i, y_i) &: \quad \alpha w_2 = \frac{1}{2}, \\
h^2 f_y f(x_i, y_i) &: \quad \beta w_2 = \frac{1}{2},
\end{aligned}
\tag{9.40}
$$

Equation (9.40) represents an underdetermined system–four unknowns and three equations. One of the parameters can be chosen arbitrarily; for instance, the system can be solved in terms of w_2. With the choice of $w_2 = 1/2$, Eq. (9.40) yields $w_1 = 1/2$ and $\alpha = \beta = 1$, the most common second-order RK-scheme. For $w_2 = 3/4$, the solution is $w_1 = 1/4$, $\alpha = \beta = 2/3$, which is known as the *method of Ralston*.

Higher-order Runge-Kutta schemes are derived in the same manner; however, the derivation can get extremely complicated. Among the Runge-Kutta methods, the fourth-order scheme (RK4) is the one that is most commonly used in the literature. The derivation of RK4 uses the fourth-order Taylor series expansion, which results in 11 equations and 13 unknowns. The system of linear equations is solved by arbitrarily choosing two unknowns. The most popular 2nd, 3rd, and 4th order Runge-Kutta schemes are presented below:

The second-order Runge-Kutta scheme

$$
\left.
\begin{aligned}
k_1 &= h \, f(x_i, \, y_i) \\
k_2 &= h \, f\left(x_{i+1}, \, y_i + k_1\right)
\end{aligned}
\right\}
\qquad y_{i+1} = y_i + \frac{1}{2}\left(k_1 + k_2\right) + \mathcal{O}(h^3)
\tag{9.41}
$$

which is also known as the *Improved Euler's Method*.

The third-order Runge-Kutta scheme

$$\left.\begin{array}{l} k_1 = h\,f(x_i, y_i) \\[2mm] k_2 = h\,f\left(x_{i+1/2}, y_i + \dfrac{k_1}{2}\right) \\[2mm] k_3 = h\,f(x_{i+1}, y_i - k_1 + 2k_2) \end{array}\right\} \quad y_{i+1} = y_i + \frac{1}{6}\left(k_1 + 4k_2 + k_3\right) + \mathcal{O}(h^4) \quad (9.42)$$

The fourth-order Runge-Kutta scheme

$$\left.\begin{array}{l} k_1 = h\,f(x_i, y_i) \\[2mm] k_2 = h\,f\left(x_{i+1/2}, y_i + \dfrac{k_1}{2}\right) \\[2mm] k_3 = h\,f\left(x_{i+1/2}, y_i + \dfrac{k_2}{2}\right) \\[2mm] k_4 = h\,f(x_{i+1}, y_i + k_3) \end{array}\right\} \quad y_{i+1} = y_i + \frac{1}{6}\left(k_1 + 2k_2 + 2k_3 + k_4\right) + \mathcal{O}(h^5) \quad (9.43)$$

where $x_{i+1/2} = x_i + h/2$, $x_{i+1} = x_i + h$, and y_{i+1} is the current step estimate. Note that the order of the local error is one degree higher than the global error.

Pseudocode 9.4

Module RUNGE_KUTTA $(n,\ h, x0,\ y0, xlast)$
\\ DESCRIPTION: A pseudomodule to solve a first-order IVP on $[x0,\ xlast]$
\\   with 2nd, 3rd, or 4th order Runge-Kutta method.
\\ USES:
\\   DRV_RK:: Module performing one-step Runge-Kutta scheme.
Write: "$x, y=$",$x0, y0$ \\ Print initial value
$x \leftarrow x0$ \\ Initialize x
While $\left[x < xlast\right]$ \\ Marching loop
    DRV_RK$(n, h,\ x0, y0,\ x, y)$ \\ Apply one-step RK scheme
    Write: "$x, y=$",x, y \\ Print current step
    $x0 \leftarrow x;\ \ y0 \leftarrow y$ \\ Set current values as prior
End While
End Module RUNGE_KUTTA

Module DRV_RK $(n,\ h, x0,\ y0,\ x,\ y)$
\\ DESCRIPTION: A driver Module employing one-step RK2, RK3, and RK4 for
\\ $n = 2$, 3, and 4, respectively. Implementation of RK4 only is presented.
\\ USES:
\\   FCN:: A user-defined function module supplying $f(x, y)$.
Declare: k_4
$xh \leftarrow x0 + h/2$ \\ Find $x_{i+1/2}$
$x1 \leftarrow x0 + h$ \\ Find x_{i+1}
If $\left[n = 4\right]$ **Then**   \\ Apply RK4 scheme
    $k_1 \leftarrow h * \textbf{FCN}(x0,\ y0)$
    $k_2 \leftarrow h * \textbf{FCN}(xh,\ y0 + k_1/2)$
    $k_3 \leftarrow h * \textbf{FCN}(xh,\ y0 + k_2/2)$
    $k_4 \leftarrow h * \textbf{FCN}(x1,\ y0 + k_3)$
    $ym \leftarrow (k_1 + 2k_2 + 2k_3 + k_4)/6$
    $y \leftarrow y0 + ym$ \\ Find current step estimate

```
    x ← x1
  Else
    If [n = 3] Then                                    \ Apply RK3 scheme
        \ You can nest other lower-order schemes
    End If
  End If
  End If
  End Module DRV_RK
```

A pseudomodule, RUNGE_KUTTA, solving a first-order IVP with a nth-order Runge-Kutta method is presented in Pseudocode 9.4. The module requires the order of the method (n), a step size (h), an initial value ($y0$), and a solution interval $[x0, xlast]$ as input. The ODE (y') is supplied to the module with the user-defined external function, FCN. The marching procedure is implemented within the While-construct until $xlast$ is reached. The driver module, DRV_RK, accompanying the RUNGE_KUTTA (main module), computes a single-step RK estimate based on the order of scheme selected (i.e., n). The DRV_RK input arguments are the same as those of RUNGE_KUTTA. The abscissa x ($= x_{i+1}$) and the estimate for the current step y ($= y_{i+1}$) computed by Eq. (9.43) are the outputs of the DRV_RK. In Pseudocode 9.4, however, the module is presented only for RK4 to save space since RK2 and RK3 (cases of $n = 2$ or 3) can be easily integrated into the module using nested If-constructs. As in previously given codes, no array variables for x and y have been used here, which is why, after each step, the current values are printed out and assigned as the prior, $(x0, y0) ← (x, y)$.

EXAMPLE 9.4: On the application of Runge-Kutta Methods

Solve Example 9.1 using 2nd, 3rd, and 4th order Runge-Kutta schemes with $h = 0.1$, and compare your estimates with the true solution.

SOLUTION:

For step size $h = 0.1$, the first three steps of the fourth-order Runge-Kutta scheme are given below.

First, using $(t_0, I_0) = (0, 2)$ and Eq. (9.43), the k_i's of the 4th order Runge-Kutta method is calculated as follows:

$$k_1 = hf(t_0, I_0) = (0.1)f(0, 2) = (0.1)(3) = 0.3,$$

$$k_2 = hf(t_{1/2}, I_0 + \frac{k_1}{2}) = hf(0.05, 2.15) = (0.1)(2.42748) = 0.242748,$$

$$k_3 = hf(t_{1/2}, I_0 + \frac{k_2}{2}) = hf(0.05, 2.12137) = (0.1)(2.499044) = 0.249904,$$

$$k_4 = hf(t_1, I_0 + k_3) = hf(0.10, 2.249904) = (0.1)(1.985074) = 0.198507$$

Substituting the k_i's into Eq. (9.13) yields

$$I(0.10) = I_1 = I_0 + \frac{1}{6}(k_1 + 2k_2 + 2k_3 + k_4)$$

$$= 2 + \frac{1}{6}(0.3 + 2(0.242748) + 2(0.249904) + 0.198507) = 2.247302$$

For the steps $i = 2, 3$, and 4, the corresponding estimates for $t_2 = 0.2$, $t_3 = 0.3$, and

$t_4 = 0.4$ are found as

$k_1 = 0.199158$, $k_2 = 0.155475$, $k_3 = 0.160935$, $k_4 = 0.121811$, and $I_2 = 2.406267$

$k_1 = 0.122303$, $k_2 = 0.089143$, $k_3 = 0.093288$, $k_4 = 0.063678$, and $I_3 = 2.498074$

$k_1 = 0.064048$, $k_2 = 0.039041$, $k_3 = 0.042167$, $k_4 = 0.019924$, and $I_4 = 2.539139$

The estimates for the subsequent steps are easily and recursively computed.

TABLE 9.4

t	True Solution	Runge-Kutta Method		
		RK2	RK3	RK4
0	2	0	0	
0.2	2.406288	6.65E-03	4.38E-04	2.17E-05
0.4	2.539164	7.79E-03	5.18E-04	2.56E-05
0.6	2.517013	6.75E-03	4.55E-04	2.23E-05
0.8	2.410610	5.09E-03	3.51E-04	1.70E-05
1	2.261953	3.47E-03	2.48E-04	1.19E-05
1.2	2.095672	2.15E-03	1.63E-04	7.59E-06
1.4	1.925946	1.14E-03	9.82E-05	4.38E-06
1.6	1.760685	4.35E-04	5.20E-05	2.09E-06
1.8	1.604061	4.15E-05	2.03E-05	5.32E-07
2	1.458042	3.45E-04	5.03E-07	4.75E-07

Discussion: In Table 9.4, the numerical estimates obtained by the 2nd, 3rd, and 4th-order Runge-Kutta methods, the true errors along with the true solution are comparatively presented for selected points. It is observed that as the order of the method increased, the numerical estimates in the solution interval improved. For example, the average error of the solutions obtained with the RK3 scheme is about 15 times less than that of the RK2. Similarly, comparing the true errors incurred with RK3 and RK4, the average error due to RK3 is about 20 times that of RK4. Also notice that the fourth-order Runge-Kutta method with a relatively large step size ($h = 0.1$) yielded numerical estimates correct to at least four significant places.

9.3 NUMERICAL STABILITY

Up until now, we focused on increasing the accuracy of the numerical estimates by either decreasing the step size or increasing the order of the numerical scheme. To solve an IVP, it may be tempting to apply any of the numerical methods discussed so far with an arbitrary step size without much thought. However, it should be noted that some numerical methods may not always behave as they should. This is where the concept of *numerical stability*, the most critical property of a numerical method, enters the picture.

In a numerical method, it is desirable to produce estimates within the range of the true solution when the initial value of an IVP is perturbed. Even in the absence of round-off or truncation errors, a numerical method rather than dampening the numerical errors could magnify them to a point where the errors continue to grow unboundedly. Of course, there are many cases where the true solution grows unboundedly, but for now, we shall restrict our discussion to the cases having bounded true solutions.

The numerical solution of an IVP with an *unstable numerical method* leads to unbounded growth with any choice of step size, whereas a *stable numerical method* does not cause numerical instabilities or unbounded growth. On the other hand, the stability of a method may depend on the choice of step size; such methods are referred to as *conditionally stable* methods. In some cases, the step size required for stability may be too small, and thus the cost of the calculations may become prohibitively large. It is therefore important to determine the stability properties of a numerical method using stability analysis.

9.3.1 STABILITY OF EULER METHODS

Consider $y' = \lambda y$ with $y(0) = 1$ (a real λ), for which the solution is $y(x) = e^{\lambda x}$. It is clear that the true solution is stable and bounded for $\lambda < 0$, i.e., $y \to 0$ as $x \to \infty$. Applying the explicit Euler scheme to the IVP leads to

$$y_{i+1} = y_i + h\left(\lambda y_i\right) = (1 + \lambda h)\, y_i, \qquad i = 0, 1, 2, \ldots \tag{9.44}$$

The numerical estimates for successive steps are found as follows:

$$y_i = (1 + \lambda h)\, y_{i-1} = (1 + \lambda h)^2\, y_{i-2} = \ldots = (1 + \lambda h)^i y_0 \tag{9.45}$$

Here, the term $1 + \lambda h$ is called the *amplification factor*, and it will be denoted by ξ.

Equation (9.45) implies that the amplification factor must be bounded ($|\xi| < 1$) in order to prevent errors from growing. In other words, for $|\xi| \geqslant 1$, numerical instabilities will be observed beyond a certain (critical) step size, that is, solving $|1 + h\lambda| \leqslant 1$ gives $h \leqslant -2/\lambda$, which is the upper bound for a stable solution. In the explicit Euler method, the step size must be less than a critical step size to ensure "*numerical stability*," and hence the method is said to be *conditionally stable*. Besides being a first-order accurate method, the conditional stability of the explicit method is its major handicap.

Employing the implicit Euler method to $y' = \lambda y$ yields

$$y_{i+1} = y_i + \lambda h\, y_{i+1} \quad \text{or} \quad y_{i+1} = \xi\, y_i \quad \text{or} \quad y_{i+1} = \xi^i y_0, \qquad i = 0, 1, 2, \ldots$$

where $\xi = 1/(1 - \lambda h)$. The condition of numerical stability ($|\xi| \leqslant 1$) results in $|1 - \lambda h| \geqslant 1$, which is satisfied for all $\lambda < 0$.

Similarly, applying the trapezoidal rule to $y' = \lambda y$, the stability condition for the amplification factor becomes $\xi = (2 + \lambda h)/(2 - \lambda h)$. Letting $\lambda = -a < 0$ with $a > 0$, the stability analysis for the implicit Euler and Trapezoidal rules yields $\xi = 1/(1 + ah) < 1$ and $\xi = (2 - ah)/(2 + ah) < 1$, respectively. Consequently, the implicit Euler and Trapezoidal rules are *unconditionally stable* for $\lambda < 0$. Moreover, linearized implicit Euler and trapezoidal methods are also *unconditionally stable*. In summary, all implicit methods are unconditionally stable, but the main drawback of these methods is their high computational cost per step.

The selection of the step size is important because the numerical method should be chosen such that it is not only *accurate* but also *stable*. Even though linearized schemes are unconditionally stable, in practice they may lead to some loss of total stability for nonlinear IVPs.

9.3.2 STABILITY OF RUNGE-KUTTA METHODS

Runge-Kutta methods are inherently *conditionally stable* because they are explicit schemes. First, we investigate the stability range of the second-order Runge-Kutta method (RK2) by revisiting the model IVP, $y' = \lambda y$ (for $\lambda < 0$).

The difference equation of the model with RK2 can be expressed as

$$k_1 = \lambda h y_i, \quad k_2 = \lambda h(y_i + k_1) = \lambda h(1 + \lambda h) y_i$$

substituting into Eq. (9.41) and rearranging yields

$$y_{i+1} = \left(1 + \lambda h + \tfrac{1}{2}\lambda^2 h^2\right) y_i + \mathcal{O}(h^3)$$

where $1 + \lambda h + \lambda^2 h^2/2 = \xi$ is the amplification factor.

The numerical stability requires $|\xi| \leqslant 1$, yielding the same (restriction) stability interval as the explicit Euler method, i.e., $\lambda h \leqslant -2$. For higher-order RK schemes, the stability interval improves slightly, yielding $\lambda h \leqslant -2.513$ and $\lambda h \leqslant -2.785$ for the RK3 and RK4, respectively. However, it should be noted that "numerical stability" does not necessarily translate to "numerical accuracy!" A numerical method can be *stable* yet *inaccurate*. We strive to obtain numerical estimates of an IVP as accurately as possible by reducing the step size and/or using a high-order numerical scheme. But it should be kept in mind that the computational cost of a numerical solution is dominated by the number of function evaluations, which is directly proportional to the cpu-time. An analyst's goal is to apply the maximum allowable step size h to reach the final destination ($x = xlast$), as a large step size leads to a fewer function evaluations and lower cpu-time. Thus, the optimum step size and cost-effectiveness of a numerical scheme depend not only on its accuracy but also on stability considerations.

Recall that the RK4 requires four function evaluations per step, while the explicit Euler method requires only one. Nevertheless, it provides better stability ($\lambda h \leqslant 2.785$) and accuracy characteristics over the explicit Euler, RK2, and RK3 methods. These stability and accuracy characteristics of the RK4 method make it one of the most popular schemes for the numerical solution of IVPs.

9.3.3 STIFFNESS

In Section 9.3.2, the explicit Euler method was demonstrated to be conditionally stable with a stability condition of $h \leqslant -2/\lambda$ for the model IVP. This requirement results in an extremely small step size for a large λ ($|\lambda| \gg 1$). This phenomenon observed in IVPs is referred to as *stiffness*.

Stiffness depends basically on the IVP itself, the accuracy criterion, the length of the solution interval, and the absolute stability interval of the applied method. There is no precise definition of *stiffness* in the literature. All we know is that differential equations containing widely varying time scales (i.e., eigenvalues) exhibit the stiffness phenomenon. Some components of the solution decay much more rapidly than others. The accuracy restriction depends mainly on the slowly varying component of the solution; however, the rapidly varying component dictates the use of very small step sizes over the entire solution interval to maintain stability. Even if we apply an adaptive step size procedure to overcome this problem (i.e., by using a small step size for the interval where the solution changes rapidly and larger steps elsewhere), the stability requirement may still force us to use a small step size across the entire solution domain.

EXAMPLE 9.5: Stability of the explicit Euler method

Under certain conditions, heat transfer analysis is idealized as a *lumped system*, in which case the temperature of an object can be assumed to vary with time only. The energy balance in a convection-dominated heat transfer process yields the following IVP:

$$\frac{d\Theta}{dt} = -a\,\Theta, \qquad \Theta(0) = 1$$

where $\Theta(t) = (T(t)-T_\infty)/(T_0-T_\infty)$ is the dimensionless temperature, T_∞ is the ambient temperature, T_0 and $T(t)$ are respectively the initial and instantaneous temperatures of the object, and a is a constant depending on the material and geometry of the object as well as environmental conditions. Apply the explicit Euler method with the step sizes $h = 0.40$, and 0.45 to obtain numerical estimates up to $t = 8$. Compare your results with the true solution, given as

$$\Theta(t) = e^{-at}$$

SOLUTION:

This is the model IVP with $\lambda = -a$, whose true solution decays fairly quickly. The numerical estimates on $[0,8]$ with $h = 0.40$, and $h = 0.45$, as well as the true solution, are comparatively depicted in Fig. 9.5. For $h > 0.5$, the numerical estimates (not shown here) oscillate widely at startup and quickly diverge with increasing t. Recall that the stability analysis for the explicit Euler method dictates the choice of step size to be $h \leqslant 0.5$. Nevertheless, the numerical estimates for values of $h = 0.40$, and $h = 0.45$ (i.e., for $h \to 0.5$) not only oscillated about the true solution at the start but also the amplitude of the oscillations increased as h approached 0.5.

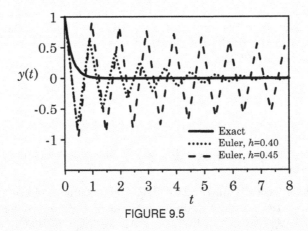

FIGURE 9.5

Discussions: Although a numerical solution, depending on the proximity of h to 0.5, initially oscillates more violently, it recovers and decays toward the true solution as t increases. Nevertheless, the recovery period can become very large as $h \to 0.5$. This example illustrates that to achieve the desired accuracy, the step size h may need to be further restricted by staying far enough away from the boundary of the stability region (i.e., in explicit Euler $h \cong 0.5$) so that the numerical solution does not oscillate in cases where the method is stable for a specified h.

EXAMPLE 9.6: Nonlinear problems with and without linearization

Consider the following third-order chemical reaction rate equation:

$$\frac{dC}{dt} = -5C^3(t), \quad C(0) = 1$$

where $C(t)$ is the concentration of the reactant as a function of time t. Solve the reaction rate equation *with* and *without* linearization for $t \leqslant 1$ with the explicit Euler, implicit Euler, and trapezoidal rule using $h = \Delta t = 0.1$. Compare your numerical estimates with the true solution given as $C(t) = 1/\sqrt{10t+1}$. Use $\varepsilon = 0.5 \times 10^{-6}$ for convergence tolerance.

SOLUTION:

The problem is a first-order nonlinear IVP with $C(0) = C_0 = 1$ as the initial condition and $f(t,C) = -5C^3$.

Using Eq. (9.6), the explicit Euler method leads to the following equation:

$$C_{i+1} = C_i - 5hC_i^3 = (1 - 5hC_i^2)C_i, \quad i = 0, 1, 2, \ldots$$

where $\xi = 1 - 5h\,C_i^2$ is the amplification factor.

Since the stability requirement for the explicit method dictates that the step size should satisfy $|\xi| \leqslant 1$, we deduce $h \leqslant 2/5C_i^2$ (or $C_i^2 \leqslant 2/5h$). For $h = 0.1$, the stability restriction leads to a limitation on the numerical estimates at any given step, i.e., the stability condition is $C_i^2 \leqslant 4$. Noting that $C(t_i) \leqslant 1$ for the true solution, it is clear that the step size of $h = 0.1$ will not violate the stability condition. Then the current step estimates can be computed recursively since the rhs of the difference equation contains the prior step estimates.

For the linearized difference equations, we need $\partial f/\partial C = -15C^2$. Using Eqs. (9.17) and (9.21), the linearized implicit Euler and trapezoidal schemes, respectively, yield the following difference equations:

$$C_{i+1} = C_i - \frac{5hC_i^3}{1 + 15hC_i^2} = \left(\frac{1 + 10hC_i^2}{1 + 15hC_i^2} \right) C_i$$

$$C_{i+1} = C_i - \frac{10hC_i^3}{2 + 15hC_i^2} = \left(\frac{2 + 5hC_i^2}{2 + 15hC_i^2} \right) C_i$$

where the expressions in brackets on the rhs are the amplification factors. It is clear that in both cases, the amplification factors are restrained against an increase in hC_i^2. Hence, both schemes are *unconditionally stable*.

The true solution, as well as the true (absolute) errors obtained from the solutions using explicit Euler, linearized implicit Euler, and linearized trapezoidal schemes for $h = 0.1$, are comparatively presented in Table 9.5. In all three methods, the errors are considerably large in the initial steps but tend to decrease steadily as t increases. The explicit Euler and linearized implicit Euler methods yield errors of the same magnitude since both methods are first-order accurate, i.e., $\mathcal{O}(0.1)$. Notice that the errors from the linearized trapezoidal method are smaller than those obtained with the linearized implicit Euler method, which is an expected outcome since the trapezoidal scheme is a second-order accurate method, i.e., $\mathcal{O}(0.1^2)$.

TABLE 9.5

t	True Solution	Absolute error, $|e_i|$		
		Explicit Euler	Linearized Implicit Euler	Linearized Trapezoidal
0	1	0	0	0
0.1	0.707107	0.20711	0.09289	0.00718
0.2	0.577350	0.13985	0.09204	0.00515
0.3	0.500000	0.10437	0.07970	0.00372
0.4	0.447214	0.08255	0.06773	0.00282
0.5	0.408248	0.06783	0.05785	0.00222
0.6	0.377964	0.05727	0.04995	0.00181
0.7	0.353553	0.04935	0.04362	0.00150
0.8	0.333333	0.04321	0.03851	0.00128
0.9	0.316228	0.03831	0.03432	0.00110
1	0.301511	0.03433	0.03085	0.00096

Employing the implicit Euler method (without linearization) leads to

$$C_{i+1} = C_i + hf(t_{i+1}, C_{i+1}) = C_i - 5hC_{i+1}^3$$

which is a cubic equation for which C_{i+1} is the unknown, and it needs to be solved at every time step. Defining $z = C_{i+1}$, the difference equation becomes $z = C_i - 5hz^3$. Adapting the fixed-point iteration method to solve this, the FPI equation becomes

$$z^{(p+1)} = C_i - 5h\left(z^{(p)}\right)^3, \qquad p = 0, 1, 2, \ldots$$

where, as usual, the superscript "(p)" denotes the iteration step. To speed up the convergence, the numerical estimate from the prior time step is simply chosen as the initial guess for the current state, i.e., $z^{(0)} = C_i$.

TABLE 9.6

t	True Solution	Absolute error $|e_i|$	
		Implicit Euler	Trap. Rule
0	1	0	0
0.1	0.707107	0.06381	0.03351
0.2	0.577350	0.06255	0.02280
0.3	0.500000	0.05461	0.01636
0.4	0.447214	0.04703	0.01241
0.5	0.408248	0.04074	0.00981
0.6	0.377964	0.03564	0.00800
0.7	0.353553	0.03151	0.00668
0.8	0.333333	0.02812	0.00569
0.9	0.316228	0.02530	0.00492
1	0.301511	0.02294	0.00431

The nonlinear algebraic equation resulting from applying the trapezoidal rule is similarly obtained, yielding

$$z^{(p+1)} - C_i + \frac{5h}{2}\left(C_i^3 + \left(z^{(p+1)}\right)^3\right) = 0, \qquad p = 0, 1, 2, ..$$

In Table 9.6, the true solution as well as the true (absolute) errors obtained with the implicit Euler method and the trapezoidal rule are comparatively tabulated for $h = 0.1$. The errors due to the trapezoidal rule are about half of the implicit Euler method in the initial steps of the marching. However, as t increases, the implicit method, due to being a first-order scheme, yields errors about 4-5 times larger than those obtained from the trapezoidal rule.

Discussion: The fact that the true solution satisfies the stability condition for all t made it possible to obtain approximate solutions with the explicit method. However, in most nonlinear problems, the approximate solution with explicit schemes may reach magnitudes that violate the stability condition for $t \geqslant t_c$. In cases where we do not have a general idea about the solution trend, the most reliable way is to apply an implicit scheme *via* a root-finding method. Nonetheless, solving a nonlinear equation in this manner aside from being restrictive in terms of cpu-time, the linearized schemes offer an alternative economical solution.

9.4 MULTISTEP METHODS

One-step methods covered in Section 9.2 are also referred to as *self-starting methods* in that the marching process can start with the initial value (Fig. 9.6a). In the steps that follow, only the prior step estimate y_i is used to compute the current step estimate y_{i+1}. However, after a few steps, several prior estimates, y_i, y_{i-1}, y_{i-2}, ..., and their derivatives, y_i', y_{i-1}', y_{i-2}', ... (i.e., f_i, f_{i-1}, f_{i-2}, ...), become available. As depicted in Fig. 9.6b, the prior estimates can provide considerable information regarding the curvature of the true solution, which can then be utilized to extrapolate the estimate for the current step.

Methods making use of *more than one prior step* estimates are referred to as *multistep methods*. The main idea behind the multistep methods is to make use of y and y' from the prior steps to construct a high-degree interpolating polynomial to estimate the current step. This means that multistep methods are not self-starting; they require one or more prior step estimates of y and y' besides the initial value. The higher the degree of the interpolating polynomial, the more points with known prior step estimates are required.

A linear n-step scheme is generalized as

$$y_{i+1} = \alpha_1 y_i + \alpha_2 y_{i-1} + \ldots + \alpha_n y_{i-n+1} + h\{\beta_0 f_{i+1} + \beta_1 f_i + \ldots + \beta_n f_{i-n+1}\} \tag{9.46}$$

where $f_i = f(x_i, y_i)$ denotes the first derivative at (x_i, y_i), and α_k and β_k are the coefficients of the interpolating polynomial. The scheme is explicit for $\beta_0 = 0$, implicit otherwise. The implicit formulas are generally more accurate and stable than the explicit ones. However, as we have seen in one-step implicit methods, they require a root-finding algorithm to obtain the current step estimate. Adams-Bashforth (explicit) and Adams-Moulton (implicit) formulas are two of the most popular families of multistep methods.

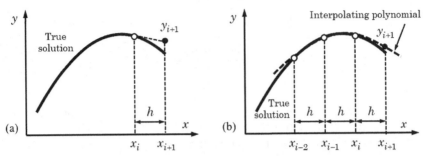

FIGURE 9.6: Marching strategy for (a) one-step, (b) multistep schemes.

9.4.1 ADAMS-BASHFORTH METHOD (AB)

Consider the first-order model IVP, $y' = f(x, y)$, with n known prior estimates, (x_i, y_i), $(x_{i-1}, y_{i-1}), \ldots, (x_{i-n+1}, y_{i-n+1})$, computed by a one-step method of the same order or better. Integrating Eq. (9.4) between x_i and x_{i+1} points gives

$$\int_{x_i}^{x_{i+1}} dy = y_{i+1} - y_i = \int_{x_i}^{x_{i+1}} f\big(u, y(u)\big)\, du \qquad (9.47)$$

To begin, the integrand is approximated by an $(n-1)$th-degree polynomial passing through n uniformly spaced points: (x_i, y_i), (x_{i-1}, y_{i-1}), $\ldots$, (x_{i-n+1}, y_{i-n+1}). A Lagrange interpolating polynomial for the integrand is constructed as follows:

$$f(x, y(x)) = \sum_{k=0}^{n-1} L_k(x) f\big(x_{i-k}, y(x_{i-k})\big) + e_n(x) \qquad (9.48)$$

where $L_k(x)$ denotes the Lagrange polynomials, defined as

$$L_k(x) = \prod_{\substack{j=0 \\ j \neq k}}^{n-1} \left(\frac{x - x_{i-j}}{x_{i-k} - x_{i-j}} \right), \qquad k = 0, 1, 2, \ldots, (n-1)$$

and $e_n(x)$ is the interpolation error given by

$$e_n(x) = \frac{f^{(n)}(x, \bar{\xi})}{n!} \prod_{j=0}^{n-1} (x - x_{i-j}), \qquad x_i < \bar{\xi} < x_{i+1}$$

Substituting Eq. (9.48) into Eq. (9.47) and integrating yields so-called *Adams-Bashforth Formulas*. Marching with uniformly spaced steps is preferred mainly due to the ease with which the interpolating polynomials can be constructed, resulting in less computational effort and cpu-time.

Adams-Bashforth formulas, using Eq. (9.46), are expressed as

$$y_{i+1} = y_i + h \sum_{k=1}^{n} \beta_k f_{i-k+1} + \mathcal{O}(h^{n+1}) \qquad (9.49)$$

where β_k's are the coefficients for Adams-Bashforth formulas, which are tabulated in Table 9.7 along with the local truncation error terms for orders $n = 1$ through 6.

TABLE 9.7: The coefficients (β_k) and the local truncation errors of n-step Adams-Bashforth Formulas $(n = 1, \ldots, 6)$.

n	β_1	β_2	β_3	β_4	β_5	β_6	$\mathcal{O}(h^{n+1})$
1	1						$\dfrac{1}{2}h^2 f'(\xi)$
2	$\dfrac{3}{2}$	$-\dfrac{1}{2}$					$\dfrac{5}{12}h^3 f''(\xi)$
3	$\dfrac{23}{12}$	$-\dfrac{16}{12}$	$\dfrac{5}{12}$				$\dfrac{9}{24}h^4 f'''(\xi)$
4	$\dfrac{55}{24}$	$-\dfrac{59}{24}$	$\dfrac{37}{24}$	$-\dfrac{9}{24}$			$\dfrac{251}{720}h^5 f^{(4)}(\xi)$
5	$\dfrac{1901}{720}$	$-\dfrac{2774}{720}$	$\dfrac{2616}{720}$	$-\dfrac{1274}{720}$	$\dfrac{251}{720}$		$\dfrac{475}{1440}h^6 f^{(5)}(\xi)$
6	$\dfrac{4277}{1440}$	$-\dfrac{7923}{1440}$	$\dfrac{9982}{1440}$	$-\dfrac{7298}{1440}$	$\dfrac{2877}{1440}$	$-\dfrac{475}{1440}$	$\dfrac{19087}{60480}h^7 f^{(6)}(\xi)$

Figure 9.7 depicts the use of prior points in $n = $ 2-, 3-, and 4-step schemes using the Adams-Bashforth Method (ABn). Each step requires the computation of the current estimate from the known prior estimates of y_i and f_i, f_{i-1}, $\ldots$, f_{i-n+1}.

The pseudomodule, **ADAMS_BASHFORTH**, solving a first-order IVP with the AB4 scheme is presented in Pseudocode 9.5. In the generalized form, the module requires the order of method (n), step size (h), initial value $(y0)$, and the solution interval $[x0, xlast]$ as input. Even though the input argument list includes n, only AB4 is incorporated into the module to save space, but it can be easily extended to include AB2 and AB3. As usual, the derivative should be supplied with a user-defined function module, **FCN**.

Using Table 9.8, the coefficients of the interpolation polynomial are assigned according to the selected n. (Should the module implement only AB4, the argument n can be omitted from the input list and, correspondingly, the If-construct.) The initial derivative is $y'_n = f(x_0, y_0)$, and the starting values y'_{n-1}, y'_{n-2}, $\ldots$, y'_1 are obtained using **DRV_RK** (i.e., the 4th-order Runge-Kutta scheme; *see* Pseudocode 9.4). Using Eq. (9.49) in a marching loop (**While**-loop), the current step estimates are obtained step by step until $xlast$ is reached. Once a current estimate is found, the starting values (i.e., prior estimates) are shifted one

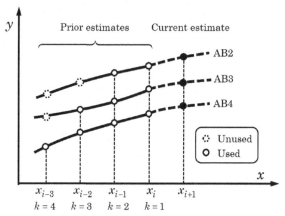

FIGURE 9.7: Estimates used in 2nd, 3rd, and 4th order in explicit multistep schemes.

Adams-Bashforth Method

- ABs require only one derivative evaluation per step, as opposed to four evaluations per step in RK methods, making AB schemes considerably faster;
- ABn gives the true solution for the IVPs, whose (true) solution is at most an $(n + 1)$th-degree polynomial.

- The AB schemes require multiple prior step estimates (starting values) in addition to the initial value to advance forward;
- Starting values are obtained with a one-step scheme, which should have the same or better (higher) order of error as the AB scheme used;
- With an increasing order of AB, accuracy increases while the stability region shrinks;
- AB formulas based on non-uniformly spaced interpolation points are much more complicated and cpu-time demanding.

---

step up to accommodate the marching step: $f_k \leftarrow f_{k-1}$ for $k = n, n-1, \ldots, 2$. This way, at any step, only the required number of prior estimates are preserved in memory. Before advancing to the next step, the current step abscissa and corresponding estimate are assigned as prior values ($y0 \leftarrow y$ and $x0 \leftarrow x$) before returning to the beginning of the **While**-loop.

Pseudocode 9.5

Module ADAMS_BASHFORTH (n, h, $x0$, $y0$, $xlast$)
\\ DESCRIPTION: A pseudomodule program to solve first-order IVP on $[x0, xlast]$
\\   using AB4 formulas. Starting values are computed using RK4.
\\ USES:
\\   FCN:: User-Defined Function Module supplying $f(x, y) = y'$;
\\   DRV_RK:: Module performing one-step RK scheme (Pseudocode 9.4).
Declare: f_4, β_4
\\ The coefficients of AB2, AB3, etc., can be nested below:
If $[n = 4]$ **Then**   \\ Set coefficients of AB4
    $\beta_1 \leftarrow 55/24$; $\beta_2 \leftarrow -59/24$; $\beta_3 \leftarrow 37/24$; $\beta_4 \leftarrow -9/24$
End If
$f_n \leftarrow$ **FCN**($x0, y0$)   \\ Set initial derivative, $f_n = f(x_0, y_0)$
Write: "$x, y=$",$x0, y0$ \\ Print initial values
For $[k = n - 1, 1, (-1)]$ \\ Find starting values
    DRV_RK($4, h, x0, y0, x, y$)   \\ Apply RK4 to find a starting value
    Write: "$x, y=$",x, y \\ Print starting values
    $f_k \leftarrow$ **FCN**(x, y)   \\ Find derivative from **FCN**(x, y)
    $x0 \leftarrow x$; $y0 \leftarrow y$ \\ Assign current step values as prior
End For
While $[x < xlast]$ \\ Marching loop
    $sums \leftarrow 0$ \\ Initialize accumulator $sums$
    For $[k = 1, n]$ \\ Apply nth-order AB polynomial
        $sums \leftarrow sums + \beta_k * f_k$

TABLE 9.8: The coefficients (β_k) and the local truncation errors of n-step Adams-Moulton formulas ($n = 1, \ldots, 6$).

n	β_0	β_1	β_2	β_3	β_4	β_5	$\mathcal{O}(h^{n+1})$
1	1						$-\dfrac{1}{2}h^2 f'(\xi)$
2	$\dfrac{1}{2}$	$\dfrac{1}{2}$					$-\dfrac{1}{12}h^3 f''(\xi)$
3	$\dfrac{5}{12}$	$\dfrac{8}{12}$	$-\dfrac{1}{12}$				$-\dfrac{1}{24}h^4 f'''(\xi)$
4	$\dfrac{9}{24}$	$\dfrac{19}{24}$	$-\dfrac{5}{24}$	$\dfrac{1}{24}$			$-\dfrac{19}{720}h^5 f^{(4)}(\xi)$
5	$\dfrac{251}{720}$	$\dfrac{646}{720}$	$-\dfrac{264}{720}$	$\dfrac{106}{720}$	$-\dfrac{19}{720}$		$-\dfrac{27}{1440}h^6 f^{(5)}(\xi)$
6	$\dfrac{475}{1440}$	$\dfrac{1427}{1440}$	$-\dfrac{798}{1440}$	$\dfrac{482}{1440}$	$-\dfrac{173}{1440}$	$\dfrac{27}{1440}$	$-\dfrac{863}{60480}h^7 f^{(6)}(\xi)$

```
End For
y ← y0 + h * sums                          \ Find current step estimate
x ← x0 + h                                  \ Find current step abscissa
For [k = n, 2, (−1)]                        \ Shift derivatives up one step
    f_k ← f_{k−1}
End For
f_1 ← FCN(x, y)                             \ Find current step derivative
Write: "x, y=",x, y                         \ Print current step estimate
x0 ← x;  y0 ← y                             \ Set current step values as prior
End While
End Module ADAMS_BASHFORTH
```

9.4.2 ADAMS-MOULTON METHOD

In this method, the Adams-Moulton Formulas (AMs) are derived in the same manner as the AB, with only one exception. The current step estimate y_{i+1} is also used in constructing the interpolating polynomial in addition to the prior step estimates generated from uniformly spaced points (x_i, y_i), (x_{i-1}, y_{i-1}), $\ldots$, (x_{i-n+2}, y_{i-n+2}). Hence, the AMn formulas are implicit by nature.

Lagrange interpolating polynomials in this case take the following form:

$$f(x, y(x)) = \sum_{k=0}^{n-1} L_k(x) f\big(x_{i-k+1}, y(x_{i-k+1})\big) + e_n(x) \tag{9.50}$$

where $L_k(x)$ and $e_n(x)$, in this case, are defined as

$$L_k(x) = \prod_{\substack{j=0 \\ j \neq k}}^{n-1} \left(\frac{x - x_{i-j+1}}{x_{i-k+1} - x_{i-j+1}} \right), \quad k = 0, 1, 2, \ldots, (n-1)$$

$$e_n(x) = \frac{f^{(n)}(x, \bar{\xi})}{n!} \prod_{j=0}^{n-1} (x - x_{i-j+1}), \quad x_i < \bar{\xi} < x_{i+1}$$

Substituting Eq. (9.50) into Eq. (9.47) and integrating yields *Adam-Moulton Formulas*, which are generalized as

$$y_{i+1} = y_i + h \sum_{k=0}^{n-1} \beta_k f_{i-k+1} + \mathcal{O}(h^{n+1}), \qquad (9.51)$$

where the coefficients (β_k's) and the local truncation errors are presented for $n = 1$ through 6 in Table 9.8. Note that for $n = 1$ the AM1 corresponds to the implicit Euler scheme. The AMs are also not self-starting; that is, a high-order one-step method is used to generate the required number of initial values to proceed forward marching. An AM can only be implemented once a sufficient number of starting values become available. Since AM is implicit, an iterative procedure similar to that described for the implicit Euler method can be applied (*see* Example 9.3).

Adams-Moulton Method

- AMs are implicit schemes;
- High-order AM formulas are available (*see* Table 9.8);
- AM schemes are more accurate than AB schemes of the same order due to smaller truncation error constants (*see* Tables 9.7 and 9.8);
- AMs have larger stability regions than ABs of the same order;
- To minimize the cpu-time, function evaluations can be reduced to a single function evaluation per step, as implemented in Pseudocode 9.6, by shifting (copying) the estimates from prior solutions.

- AMs require a high-order single-step scheme (such as RK4) to generate a sufficient number of starting values to start AM marching;
- AMs require larger memory allocation due to the need to save the required number of prior estimates;
- Cpu time increases due to iterative algorithms applied in cases where the use of a root-finding algorithm is required to find the root (current step estimate) of the nonlinear equation;
- The convergence of the root-finding methods depends on $f(x, y)$.

A pseudomodule, **ADAMS_MOULTON**, solving a first-order IVP with an AM2, AM3, or AM4 is presented in Pseudocode 9.6. The module requires the order of the method (n), a step size (h), a convergence tolerance (ε), an initial value ($y0$), and a solution interval [$x0, xlast$] as input. The derivative, $f(x, y)$, is supplied by the user-defined function, **FCN**(x, y). The required number of starting values is generated using **DRV_RK** (4th-order RK driver module, *see* Pseudocode 9.4). The accuracy of the starting values is critical to the accuracy of the estimates obtained in subsequent marching steps. Thus, it is recommended to find the starting values with RK4, even if the AM2 or AM3 schemes are used. Before applying the procedure, the coefficients of the interpolating polynomial given in Table 9.8 are set according to the order of the AM scheme. In the module, the scheme is fixed by AM4; however, it can be easily modified or extended to include AM2 and AM3. The marching procedure is carried out within the **While**-loop until *xlast* is reached. The current step estimates of AM4 are obtained step by step using Eq. (9.51) along with the fixed point iteration method performed in the **Repeat-Until**-loop. Once the current step estimate y (i.e., y_{i+1}) is found, the starting values are shifted one step up in preparation for the next step. To advance to the next step, the current step estimates are also assigned to the prior step estimates, $x0 \leftarrow x$; $y0 \leftarrow y$, before returning to the top of the **While**-loop.

Pseudocode 9.6

Module ADAMS_MOULTON $(n,\ h,\ \varepsilon,\ x0,\ y0, xlast)$
\ DESCRIPTION: A pseudomodule to solve a first-order IVP on $[x0, xlast]$ with
\   AM4. FPI method is used to solve the resulting nonlinear equations.
\ USES:
\   FCN:: A user-defined function module supplying $f(x, y) = y'$;
\   DRV_RK:: Module performing one-step RK (Pseudocode 9.4);
\   ABS :: A built-in function computing the absolute value.
Declare: f_4, β_4 \ $\beta_1 \leftarrow \beta_0$ beta's are shifted
\ The coefficients of AB2, AB3, etc., can be nested below:
If $\begin{bmatrix} n = 4 \end{bmatrix}$ **Then**   \ Set coefficients of AB4
    $\beta_1 \leftarrow 9/24;\ \ \beta_2 \leftarrow 19/24;\ \ \beta_3 \leftarrow -5/24;\ \ \beta_4 \leftarrow 1/24$
End If
$f_n \leftarrow$ **FCN**$(x0, y0)$ \ Find initial derivative
Write: "$x, y =$",$x0, y0$ \ Print initial values
For $\begin{bmatrix} k = n - 1, 2, (-1) \end{bmatrix}$ \ Find starting values
    DRV_RK$(4, h, x0, y0, x, y)$ \ Apply RK4 to find a starting value
    Write: "$x, y=$",x, y \ Print starting values
    $f_k \leftarrow$ **FCN**(x, y) \ Find derivative from FCN(x, y)
    $x0 \leftarrow x;\ y0 \leftarrow y$ \ Set current values as prior
End For
While $\begin{bmatrix} x < xlast \end{bmatrix}$ \ Marching Loop
    $x \leftarrow x0 + h$ \ Find current step abscissa
    $z \leftarrow y0$ \ Set initial guess for FPI
    $y \leftarrow z$ \ Assign current estimate as prior
    Repeat   \ Solve nonlinear equation for y_{i+1}
        $f_1 \leftarrow$ **FCN**(x, z) \ Find $y'(x)$
        $sums \leftarrow 0$ \ Initialize $sums$
        For $\begin{bmatrix} k = 1, n \end{bmatrix}$ \ Apply nth-order AB polynomial
            $sums \leftarrow sums + \beta_k * f_k$
        End For
        $y \leftarrow y0 + h * sums$ \ Find current step estimate
        $errz \leftarrow |y - z|$ \ Find absolute error
        $z \leftarrow y$ \ Assign current step estimate as prior
    Until $\begin{bmatrix} errz < eps \end{bmatrix}$ \ Converged if $|y - z^{(p)}| < \varepsilon$
    For $\begin{bmatrix} k = n, 2, -1 \end{bmatrix}$ \ Shift derivatives up one step
        $f_k \leftarrow f_{k-1}$
    End For
    $f_1 \leftarrow$ **FCN**(x, y) \ Find $y'(x)$ for current step
    Write: "$x, y=$",x, y \ Print current estimate
    $x0 \leftarrow x;\ y0 \leftarrow y$ \ Assign current values as prior
End While
End Module ADAMS_MOULTON

9.4.3 BACKWARD DIFFERENTIATION FORMULAS

Backward differentiation formulas (BDFs), introduced by Curtiss and Hirschfelder [6], are a class of linear multistep (implicit) formulas for numerically solving IVPs. The BDFs have become very popular in the last few decades, primarily due to their ability to handle the

TABLE 9.9: The coefficients and the local truncation errors for BDFs ($n = 1, \ldots, 6$).

n	β_0	α_1	α_2	α_3	α_4	α_5	α_6	$\mathcal{O}(h^{n+1})$
1	1	1						$-\dfrac{1}{2}h^2 f'(\xi)$
2	$\dfrac{2}{3}$	$\dfrac{4}{3}$	$-\dfrac{1}{3}$					$-\dfrac{1}{3}h^3 f''(\xi)$
3	$\dfrac{6}{11}$	$\dfrac{18}{11}$	$-\dfrac{9}{11}$	$\dfrac{2}{11}$				$-\dfrac{1}{4}h^4 f'''(\xi)$
4	$\dfrac{12}{25}$	$\dfrac{48}{25}$	$-\dfrac{36}{25}$	$\dfrac{16}{25}$	$-\dfrac{3}{25}$			$-\dfrac{1}{5}h^5 f^{(4)}(\xi)$
5	$\dfrac{60}{137}$	$\dfrac{300}{137}$	$-\dfrac{300}{137}$	$\dfrac{200}{137}$	$-\dfrac{75}{137}$	$\dfrac{12}{137}$		$-\dfrac{1}{6}h^6 f^{(5)}(\xi)$
6	$\dfrac{60}{147}$	$\dfrac{360}{147}$	$-\dfrac{450}{147}$	$\dfrac{400}{147}$	$-\dfrac{225}{147}$	$\dfrac{72}{147}$	$-\dfrac{10}{147}$	$-\dfrac{1}{7}h^7 f^{(6)}(\xi)$

numerical solution of stiff IVPs. The BDFs also have better stability properties than those of the ABs and AMs.

Constructing a BDF is quite simple and straightforward. The model first-order IVP, Eq. (9.1), is discretized at $x = x_{i+1}$ as follows:

$$y'(x_{i+1}) = f(x_{i+1}, y_{i+1}) \tag{9.52}$$

Afterward, $y'(x_{i+1})$ is replaced with the backward difference formula of the desired order. Recalling the differentiations with finite differences, the backward difference formulas, along with the truncation errors, from the first derivative up to the fourth-order (*see* Chapter 5) are presented below:

$$y'(x_{i+1}) = \begin{cases} \dfrac{y_{i+1} - y_i}{h} + \dfrac{h}{2}y''(\xi) \\[2mm] \dfrac{3y_{i+1} - 4y_i + y_{i-1}}{2h} + \dfrac{h^2}{3}y'''(\xi) \\[2mm] \dfrac{11y_{i+1} - 18y_i + 9y_{i-1} - 2y_{i-2}}{6h} + \dfrac{h^3}{4}y^{(4)}(\xi) \\[2mm] \dfrac{25y_{i+1} - 48y_i + 36y_{i-1} - 16y_{i-2} + 3y_{i-3}}{12h} + \dfrac{h^4}{5}y^{(5)}(\xi) \end{cases} \tag{9.53}$$

The BDF1 gives the *implicit Euler method*. Substituting Eq. (9.53) into Eq. (9.52) and solving for y_{i+1}, the second-, third-, and fourth-order (BDF2, BDF3, and BDF4) formulas are obtained as follows:

$$y_{i+1} = \dfrac{1}{3}(-y_{i-1} + 4y_i) + \dfrac{2h}{3}f(x_{i+1}, y_{i+1}) + \mathcal{O}(h^3)$$

$$y_{i+1} = \dfrac{1}{11}(2y_{i-2} - 9y_{i-1} + 18y_i) + \dfrac{6h}{11}f(x_{i+1}, y_{i+1}) + \mathcal{O}(h^4) \tag{9.54}$$

$$y_{i+1} = \dfrac{1}{25}(-3y_{i-3} + 16y_{i-2} - 36y_{i-1} + 48y_i) + \dfrac{12h}{25}f(x_{i+1}, y_{i+1}) + \mathcal{O}(h^5)$$

In general, the BDFn's can be expressed as follows:

$$y_{i+1} = \sum_{k=1}^{n} \alpha_k y_{i+k-1} + \beta_0 h f(x_{i+1}, y_{i+1}) + \mathcal{O}(h^{n+1}) \tag{9.55}$$

Backward Differentiation Formulas

- A BDF scheme of any order can be constructed very easily;
- BDFs of orders 2 to 6 are *conditionally stable*;
- BDFs are suitable for solving stiff IVPs;

- BDFs require the use of a root-finding algorithm to find the current estimate when the IVP is a nonlinear equation, which leads to increased cpu-time;
- BDFs need starting values (y_{i-1}, y_{i-2}, y_{i-3}, ...), which should be computed by another method with the same or better order of error before the marching procedure can be started;
- BDFs for $n \geqslant 7$ are *not convergent* due to their rigidity, resulting from overfitting of the data;
- Difficulties may arise if a fixed step size is adopted throughout the solution interval; thus, the step size may need to be refined.

---

where β_0 and α_k's are the corresponding coefficients presented in Table 9.9. Note that among the starting values under the summation sign in Eq. (9.55), only y_0 is known. As with the ABs and AMs, the rest of the starting values need to be generated with a one-step scheme before applying the BDFs. Since the BDFs are implicit, the current step estimate y_{i+1} is obtained by applying a suitable root-finding scheme when the IVP is nonlinear.

In Pseudocode 9.7, the pseudomodule, **BDFS**, is presented for solving a first-order IVP with an nth-order BDF. The module requires the order of the method (n), step size (h), initial value ($y0$), the solution interval $[x0, xlast]$, a convergence tolerance (ε) for the solution of non-linear equations, and the maximum number of FPIs allowed ($maxit$) as input. The IVP $y' = f(x, y)$ is supplied to the module with the user-defined external function **FCN**. Although the order of BDF (n) is present in the argument list, this module involves the solution only for BDF3 to save space, but it can be easily modified and extended to include other BDFn formulas.

The starting values for high-order BDFs are generated using the **DRV_RK** (RK4 of the module given in Pseudocode 9.4). The starting values as well as the numerical estimates are saved on a temporary array: $\hat{y}_n \leftarrow y0$, $\hat{y}_{n-1} \leftarrow y(x_0 + h)$, ..., $\hat{y}_1 \leftarrow y(x_0 + nh)$. The more accurate the starting values, the better the accuracy of the current step estimates. Thus, the RK4 should be employed here regardless of the order of the BDF. For $x < xlast$, a **While**-loop is used to perform forward marching with uniform steps. The current estimate at each step, Eq. (9.55), is obtained with a **Repeat-Until**-loop (a root-finding loop) using the fixed-point iteration method. Once the summation term of Eq. (9.55) is evaluated with a **While**-loop, the procedure is carried out within a **Repeat-Until** construct. If the iteration procedure reaches the maximum number of iterations allowed, the procedure is terminated with a warning. When the root of the nonlinear equation (current estimate) is successfully obtained, then, as in the cases of AM and AB, the starting values are shifted one step up in preparation for the next step. To advance up another step, the current step estimates are assigned as prior step estimates, $(x0, y0) \leftarrow (x, y)$, before returning to the top of the marching loop.

Module BDFS $(n,\ h,\ \varepsilon,\ x0,\ y0,\ xlast,\ maxit)$
\ DESCRIPTION: A pseudomodule to solve $y' = f(x, y)$ on $[x0, xlast]$ with
\   the BDF3. Fixed-Point iteration is used for solving nonlinear equation.
\ USES:
\   FCN:: A user-defined function module supplying $f(x, y) = y'$.
\   DRV_RK:: Module performing one-step Runge-Kutta scheme;
\   ABS :: A built-in function computing the absolute value.
Declare: $\hat{y}_{n+1}, \alpha_n$ \ Declare array variables
If $\left[n = 3 \right]$ **Then**   \ Set coefficients of BDF3
   $\alpha_1 \leftarrow 18/11;\ \alpha_2 \leftarrow -9/11;\ \alpha_3 \leftarrow 2/11;\ \beta \leftarrow 6/11$
End If
Write: "$x,\ y=$",$x0, y0$ \ Print a starting value
$\hat{y}_n \leftarrow y0$ \ Set initial value
For $\left[k = n - 1, 1, (-1) \right]$ \ Find starting values
   DRV_RK$(4, h, x0, y0, x, y)$ \ Apply RK4 to find a starting value
   Write: "$x, y=$",x, y \ Print starting value
   $\hat{y}_k \leftarrow y$ \ Add the estimate as starting value
   $x0 \leftarrow x;\ y0 \leftarrow y$ \ Assign current step values as prior
End For
While $\left[x < xlast \right]$ \ Marching loop
   $sums \leftarrow 0$ \ Initialize accumulator $sums$
   For $\left[k = 1, n \right]$ \ Apply BDFn
      $sums \leftarrow sums + \alpha_k * \hat{y}_k$
   End For
   $x \leftarrow x0 + h$ \ Find current step abscissa
   $z \leftarrow y0$ \ Set prior step estimate as initial guess for FPI
   $p \leftarrow 0$ \ Initialize iteration counter
   Repeat   \ Solve nonlinear eqs using FPIs
      $p \leftarrow p + 1$ \ Count iterations
      $y \leftarrow sums + \beta * h * \textbf{FCN}(x, z)$ \ Find current iteration step estimate
      $Err \leftarrow |y - z|$ \ Find absolute error
      $z \leftarrow y$ \ Assign current estimate to prior
   Until $\left[Err < \varepsilon\ \textbf{Or}\, p = maxit \right]$ \ Check for convergence
   If $\left[p = maxit \right]$ **Then**   \ Print a warning message and error
      Write: "Max. no of iteration reached for $x=$",x
      Write: "Absolute Error is $=$",Err
      Exit
   End If
   For $\left[k = n, 2, (-1) \right]$ \ Shift the estimates up one step
      $\hat{y}_k = \hat{y}_{k-1}$
   End For
   $\hat{y}_1 = y$
   Write: "$x, y=$",x, y \ Print current estimate
   $x0 \leftarrow x;\ y0 \leftarrow y$ \ Assign current step values as prior
End While
End Module BDFS

EXAMPLE 9.7: Comparison of AB, AM, and BDF methods

Repeat Example 9.1 using the 2nd, 3rd, and 4th-order Adams-Bashforth (AB), Adams-Moulton (AM), and backward differentiation formulas (BDF) with $h = 0.1$. Compare your estimates with the true solution.

SOLUTION:

In this example, only the numerical solutions for a few steps of AB2 and AM2 are illustrated explicitly, but the numerical results of higher-order schemes are presented as tabular data.

The initial condition and the corresponding derivative are $I(0) = I_0 = 2$ and $I_0' = f_0 = f(0, 2) = 3$. Using Eq. (9.49), together with the coefficients of the interpolating polynomial from Table 9.7, the difference equation for the AB2 is expressed as

$$I_{i+1} = I_i + \frac{h}{2}\left(3f_i - f_{i-1}\right), \quad \text{for } i \geqslant 1$$

Note that I_1 and $f_1 = I_1' = f(t_1, I_1)$ are unknown, which should be generated in order to be able to start the AB marching process. Normally, a suitable RK method (of the same or higher order of error) is used to obtain the starting value; however, since it is available, we will use the true solution here. This way, errors incurred during the computation of the starting values with the one-step method will not be carried to the AB, AM, or BDF marching steps, which will allow these methods to be compared based on the true errors.

For $t_1 = 0.1$, we find

$$I_1 = I(0.1) = 4e^{-0.05} - 2e^{-0.25} = 2.247316$$
$$f_1 = f(t_1, I_1) = f(0.1, 2.247316) = 1.991545$$

Now that the starting values (f_0 and f_1) are available, we can move on to implementing the AB2 scheme. The current estimate for the first step ($i = 1$) becomes:

$$I_2 = I_1 + \frac{h}{2}\left(3f_1 - f_0\right) = 2.247316 + \frac{0.1}{2}\left(3(1.991545) - (3)\right) = 2.396048$$

Before advancing to the next step, the starting value is computed as

$$f_2 = f(t_2, I_2) = f(0.2, 2.396048) = 1.248579$$

Then the current estimate ($i = 2$) is obtained from

$$I_3 = I_2 + \frac{h}{2}(3f_2 - f_1) = 2.396048 + \frac{0.1}{2}\left(3(1.248579) - 1.991545\right) = 2.483758$$

Similarly, for $i = 3$, we find $f_3 = f(t_3, I_3) = f(0.3, 2.483758) = 0.676269$ and

$$I_4 = I_3 + \frac{h}{2}(3f_3 - f_2) = 2.483758 + \frac{0.1}{2}\left(3(0.676269) - 1.248579\right) = 2.522769$$

This recursive procedure is repeated up to $t = 2$. The absolute errors of the estimates obtained with the 2nd, 3rd, and 4th-order ABs for the selected points are comparatively presented in Table 9.10 together with the true solutions. As the circuit current rises to a local maximum of 2.545 A at $t \approx 0.46$ s, all AM's exhibit an increasing trend up to $t \approx 0.5$ s, though the true errors are initially small. However, as t continues to increase after the circuit current reaches its maximum, the errors in the estimates decrease with the circuit current.

Using the set of coefficients from Table 9.8, the AM2 scheme can be written as

$$I_{i+1} = I_i + \frac{h}{2}\left(f_{i+1} + f_i\right), \quad i \geqslant 0$$

TABLE 9.10

t_i	True Solution	True absolute error, e_i		
		AM2	AM3	AM4
0	2			
0.2	2.406288	1.02E-02	0	0
0.4	2.539164	1.64E-02	2.67E-03	4.22E-04
0.6	2.517013	1.62E-02	3.37E-03	7.33E-04
0.8	2.410610	1.36E-02	3.10E-03	7.32E-04
1	2.261953	1.05E-02	2.51E-03	6.16E-04
1.2	2.095672	7.72E-03	1.90E-03	4.79E-04
1.4	1.925946	5.44E-03	1.38E-03	3.56E-04
1.6	1.760685	3.72E-03	9.70E-04	2.56E-04
1.8	1.604061	2.47E-03	6.67E-04	1.80E-04
2	1.458042	1.58E-03	4.50E-04	1.25E-04

For $i = 0$, we already have $I_0 = 2$ and $f_0 = 3$ from the initial value. The second starting value, $f_1 = f(t_1, I_1)$, comprises the unknown current step estimate, I_1. In this example, $y = f(t, I)$ is linear, which eliminates the need for a root-finding algorithm to find I_{i+1}. Hence, an explicit expression for I_{i+1} can easily be found as

$$I_{i+1} = \frac{4 - 5h}{4 + 5h}I_i + \frac{16h}{4 + 5h}\left(1 + e^{-h/2}\right)e^{-t_i/2}, \quad i \geqslant 0$$

TABLE 9.11

t_i	True Solution	True absolute error, e_i		
		AM2	AM3	AM4
0	2			
0.2	2.406288	3.12E-03	2.22E-04	0
0.4	2.539164	3.76E-03	3.95E-04	4.67E-05
0.6	2.517013	3.40E-03	3.98E-04	5.64E-05
0.8	2.410610	2.72E-03	3.37E-04	5.13E-05
1	2.261953	2.03E-03	2.62E-04	4.14E-05
1.2	2.095672	1.45E-03	1.93E-04	3.14E-05
1.4	1.925946	1.00E-03	1.38E-04	2.28E-05
1.6	1.760685	6.73E-04	9.60E-05	1.61E-05
1.8	1.604061	4.37E-04	6.54E-05	1.12E-05
2	1.458042	2.74E-04	4.39E-05	7.60E-06

The absolute true errors resulting from using the 2nd, 3rd, and 4th-order AMs and the true solutions for selected points are presented comparatively in Table 9.11. For increasing t, the numerical estimates with the AM show significant improvement as the errors decay rapidly. Similar to the AMs, the numerical estimates initially have small absolute errors, but as a result of the accumulation of global errors, the errors increase up to $t \approx 0.46$, where they reach their maximum and then decrease with increasing t.

TABLE 9.12

t_i	True Solution	True absolute error, e_i		
		BDF2	BDF3	BDF4
0	2			
0.2	2.406288	4.28E-03	0	0
0.4	2.539164	1.07E-02	1.33E-03	9.43E-05
0.6	2.517013	1.21E-02	2.08E-03	3.39E-04
0.8	2.410610	1.07E-02	1.99E-03	3.70E-04
1	2.261953	8.47E-03	1.62E-03	3.15E-04
1.2	2.095672	6.26E-03	1.24E-03	2.53E-04
1.4	1.925946	4.43E-03	9.05E-04	1.89E-04
1.6	1.760685	3.02E-03	6.41E-04	1.36E-04
1.8	1.604061	1.99E-03	4.43E-04	9.60E-05
2	1.458042	1.27E-03	3.00E-04	6.61E-05

In this example, the BDFs also yield explicit difference equations for I_{i+1}. Using the coefficients from Table 9.9, the difference equation for the BDF2 is expressed as

$$I_{i+1} = \frac{1}{3 + 5h} \left(4I_i - I_{i-1} + 16he^{-t_{i+1}/2} \right), \quad i \geqslant 1$$

This recursive procedure is repeated until $t = 2$ is reached. In Table 9.12, the true solution, along with the true errors, are comparatively depicted for 2nd, 3rd, and 4th-order BDFs for selected points. In this method, the distribution of the errors with each method follows a similar trend as in AB and AM. That is, the errors increase up until the circuit current reaches its maximum at $t = 0.46$ and then decrease rapidly with increasing t.

Discussion: The AMs, ABs, and BDFs are considerably faster than the RK methods of the same order due to requiring only one function evaluation once past the computation of the starting values. Notice that, for $n > 2$, the numerical solutions with AMs are significantly better than those of the same order as ABs and BDFs. The errors for the BDFs are smaller than those for the ABs and slightly worse than those for the AMs. This outcome is not surprising when the local truncation errors of the ABs, AMs, and BDFs are examined (*see* Tables 9.7, 9.8, and 9.9).

The main shortcoming of the AMs and BDFs is the necessity of solving the resulting nonlinear algebraic equations at each step when the IVP is nonlinear. The BDFs also suffer from the accumulation of global errors, which may grow with increasing t or die out with converging IVPs.

9.5 ADAPTIVE STEP-SIZE CONTROL

In all the methods discussed so far, the step size has been kept uniform throughout the solution range. Also, we have already learned that the choice of the step size strongly affects the accuracy of any numerical solution. For a large step size, the numerical solution tends to deviate from the true solution as x increases, while a step size that is too small leads not only to an increase in cpu-time but also to the accumulation of round-off errors. Furthermore, when the numerical simulation interval $[x_0, x_0 + T]$ is very large, it is not uncommon to have a true solution that varies slowly in some subintervals and rapidly in others. Large step sizes can be used where the solution varies slowly, primarily to reduce cpu-time without sacrificing accuracy.

A variable step size becomes necessary when the numerical simulation range is very large or when the true solution has very sharp and/or slow changes. But applying a variable step size brings about fluctuations in the magnitude of truncation and global errors, as they are proportional to the step size. As a result, the desired level of accuracy may not be achieved if one is not careful. Therefore, the truncation error is also estimated at each step to ensure that the numerical solution meets a predetermined accuracy criterion and to adjust the step size if necessary. If the truncation error is extremely small, the step size may be increased. Conversely, if the truncation error is large, the step size is reduced to prevent the errors from accumulating.

The fourth-order Runge-Kutta (RK4) is a very reliable and popular method in many fields because it is not only a higher-order method but also has a wider stability range. Most adaptive ODE solvers are based on the RK4 and fourth-order Adam's methods.

An adaptive numerical scheme presented here is based on the *Runge-Kutta-Fehlberg method*, which provides an efficient procedure for solving IVPs. In this method, a suitable step size is determined at each step. The current step estimates using the 4th and 5th order Runge-Kutta schemes (denoted by RKF45) are computed and compared with each other. The following six parameters are computed:

$$
\begin{aligned}
k_1 &= hf(x_i, y_i) \\
k_2 &= hf\left(x_i + \frac{1}{4}h, y_i + \frac{1}{4}k_1\right) \\
k_3 &= hf\left(x_i + \frac{3}{8}h, y_i + \frac{3}{32}k_1 + \frac{9}{32}k_2\right) \\
k_4 &= hf\left(x_i + \frac{12}{13}h, y_i + \frac{1932}{2197}k_1 - \frac{7200}{2197}k_2 + \frac{7296}{2197}k_3\right) \\
k_5 &= hf\left(x_i + h, y_i + \frac{439}{216}k_1 - 8k_2 + \frac{3680}{513}k_3 - \frac{845}{4104}k_4\right) \\
k_6 &= hf\left(x_i + \frac{1}{2}h, y_i - \frac{8}{27}k_1 + 2k_2 - \frac{3544}{2565}k_3 + \frac{1859}{4104}k_4 - \frac{11}{40}k_5\right)
\end{aligned}
\tag{9.56}
$$

Next, a current step estimate is computed using the following fourth-order Runge-Kutta scheme:

$$
\tilde{y}_{i+1} = y_i + \frac{25}{216}k_1 + \frac{1408}{2565}k_3 + \frac{2197}{4104}k_4 - \frac{1}{5}k_5 + \mathcal{O}(h^5)
\tag{9.57}
$$

Note that k_2 and k_6 have not been used here.

An improved estimate is obtained with the 5th-order Runge-Kutta scheme as follows:

$$
y_{i+1} = y_i + \frac{16}{135}k_1 + \frac{6656}{12825}k_3 + \frac{28561}{56430}k_4 - \frac{9}{50}k_5 + \frac{2}{55}k_6 + \mathcal{O}(h^6)
\tag{9.58}
$$

Pseudocode 9.8

Module ADAPTIVE_RKF45 (h, $x0$, $y0$, $xlast$, ε, $h_{\min}$, $h_{\max}$)
\ DESCRIPTION: A pseudomodule to solve $y' = f(x, y)$ on $[x0, xlast]$ with
\   adaptive Runge-Kutta-Fehlberg 45 scheme.
\ USES:
\   DRV_RK45:: Module performing one-step Runge-Kutta 45 scheme;
\   MIN :: Built-in function to find the minimum of a set of integers;
\   MAX:: A built-in function returning the maximum of the agrument list;
\   ABS :: A built-in function computing the absolute value.
$x \leftarrow x0$; $y \leftarrow y0$ \ Initialize solution
Write: "$x, y=$",$x0, y0$ \ Print initial solution
While $\big[x \leqslant xlast\big]$ \ Marching loop
   $h \leftarrow$ **MIN**($h, xlast - x$)   \ Close to the end of solution range?
   DRV_RKF45(h, $x0$, $y0$, x, y, R)   \ Apply one-step RKF45
   If $\big[R \leqslant \varepsilon\big]$ **Then**   \ Accept solution if $error \leqslant \varepsilon$
      $x \leftarrow x0 + h$ \ Find current step abscissa
      Write: "$x, y=$",x, y \ Print current step values
      $x0 \leftarrow x$; $y0 \leftarrow y$ \ Assign current step values as prior
   End If
   $\delta \leftarrow 0.84(\varepsilon/R)^{1/4}$ \ Find step size adjustment factor
   $ha \leftarrow \delta * h$ \ Adjust step size
   $h1 \leftarrow$ **MAX**(ha, $h_{\min}$)   \ If $ha < h_{min}$, set $h \leftarrow h_{min}$
   $h \leftarrow$ **MIN**($h1$, $h_{\max}$)   \ If $ha > h_{max}$, set $h \leftarrow h_{max}$
End While
End Module ADAPTIVE_RKF4

Module DRV_RKF45 (h, $x0$, $y0$, x, y, R)
\ DESCRIPTION: A pseudomodule applying a single step the Runge-Kutta-
\   Fehlberg 45 scheme with an error estimate R.
\ USES:
\   FCN:: A user-defined function module supplying $f(x, y) = y'$;
\   ABS :: A built-in function computing the absolute value.
Declare: k_6
$k_1 \leftarrow h *$ **FCN**($x0$, $y0$)
$k_2 \leftarrow h *$ **FCN** ($x0 + h/4$, $y0 + k_1/4$)
$k_3 \leftarrow h *$ **FCN** ($x0 + 3h/8$, $y0 + 3k_1/32 + 9k_2/32$)
$k_4 \leftarrow h *$ **FCN** ($x0 + 12h/13$, $y0 + 1932k_1/2197 - 7200k_2/2197 + 7296k_3/2197$)
$k_5 \leftarrow h *$ **FCN** ($x0 + h$, $y0 + 439k_1/216 - 8k_2 + 3680k_3/513 - 845k_4/4104$)
$yn \leftarrow y0 - 8k_1/27 + 2k_2 - 3544k_3/2565 + 1859k_4/4104 - 11k_5/40$
$k_6 \leftarrow h *$ **FCN**($x_0 + h/2$, yn)
\ Calculate y_{i+1} and the error estimate R
$R \leftarrow |k_1/360 - 128k_3/4275 - 2197k_4/75240 + k_5/50 + 2k_6/55|/h$
$y \leftarrow y0 + 25k_1/216 + 1408k_3/2565 + 2197k_4/4104 - k_5/5$
End Module DRV_RKF45

If the $\tilde{y}_{i+1}$ and y_{i+1} are in agreement within a preset accuracy criterion, then the computed estimate $\tilde{y}_{i+1}$ is accepted; otherwise, the step size is reduced. If the estimates agree to more significant digits than desired, then the step size is increased. An error estimate to adjust the step size is obtained by computing the difference between Eqs. (9.58) and (9.57) as

$$R = \frac{|y_{i+1} - \tilde{y}_{i+1}|}{h} = \frac{1}{h}\left|\frac{1}{360}k_1 - \frac{128}{4275}k_3 - \frac{2197}{75240}k_4 + \frac{1}{50}k_5 + \frac{2}{55}k_6\right| \tag{9.59}$$

The user only specifies the step size (h) and a tolerance (ε) for the error control. A strategy is then implemented to determine how to modify h so that the local truncation error is approximately equal to a preset tolerance. One of the following techniques is usually employed to ensure preset accuracy in an IVP solution while changing the step size:

(a) If R is within an acceptable range (say, $\varepsilon/4 \leqslant R \leqslant \varepsilon$), then the current estimate $\tilde{y}_{i+1}$ is accepted. However, if $R > \varepsilon$, then the step size is halved ($h \leftarrow h/2$) because it is large, and the current estimate $\tilde{y}_{i+1}$ is recomputed. If $R < \varepsilon/4$, then $\tilde{y}_{i+1}$ is accepted, and the step size is doubled ($h \leftarrow 2h$) because it is small.

(b) This second procedure is similar to the one presented above. The only difference is in the selection of the step size. When the current step size is to be modified, the new optimal step size is determined as $h \leftarrow \delta * h$, where $\delta = 0.84(\varepsilon/R)^{1/4}$ is the step size adjustment factor. Thus, in this procedure, the step size is automatically reduced (if $\varepsilon/R < 1$) or increased (if $\varepsilon/R > 1$) depending on the size of ε/R.

As a prudent practice, usually acceptable upper and lower bounds are assigned to prevent the computed step size from being too large or too small.

A pseudomodule, **ADAPTIVE_RKF45**, solving a first-order IVP with the Runge-Kutta-Fehlberg 4th/5th-order scheme is presented in Pseudocode 9.8. As input, the module requires a solution interval ($x0, xlast$), an initial step size (h), along with the upper and lower bounds for the step size ($h_{\min}, h_{\max}$), an initial value ($y0$), and a tolerance (ε). The external user-defined function module **FCN** supplies the first derivative. The module makes use of a driver module, **DRV_RK45**, to compute the current step estimate y, as well as an estimate for the error, R. If $R \leqslant \varepsilon$, then the current step estimate y is accepted as the solution. Otherwise, the step size is adjusted by $ha = \delta * h$ to prevent it from remaining small or becoming even smaller. Before proceeding to the next step, the current values are assigned as prior ($x0 \leftarrow x$ and $y0 \leftarrow y$). If $R > \varepsilon$, the step size is reduced by $\delta * h$ since $\varepsilon/R < 1$. The step size is assured to remain within the predefined lower and upper limits by the $h \leftarrow$ **MIN**($h, h_{\max}$) and $h \leftarrow$ **MAX**($\delta * h, h_{\min}$) statements. At the very last step, the step size is found from $h \leftarrow$ **MIN**($h, xlast - x$).

9.6 PREDICTOR-CORRECTOR METHODS

In previous sections, explicit and implicit one-step and multi-step methods have been discussed. These methods can also be used in combination as a pair of formulas known as *predictor-corrector methods*. Such methods involve two computational steps: (i) the *predictor* step employs an explicit method to obtain a reasonable estimate (*prediction*) for the current step y_{i+1} using an unknown function and its derivatives at some prior points; (ii) the *corrector* (next) step improves the estimate and its accuracy using an implicit formula. In this section, some of the most common predictor-corrector methods are discussed.

A graphical illustration of marching with a typical predictor-corrector method is presented in Fig. 9.8. The arrows depict the order of numerical computation for the predicted and corrected estimates, denoted by solid and hollow circles, respectively. The predictions $\hat{y}$'s (at points A, B, C) generally deviate from the true solution by a small margin. After the correction step, the estimates (at points a, b, c) approach the true distribution.

Heun's Method

- Heun's method is simple and easy to employ;
- It is computationally less involved;
- It is self-starting, i.e., the initial condition is sufficient to start the forward-marching procedure.

- Its global error is $\mathcal{O}(h^2)$, so the step size needs to be chosen sufficiently small to achieve estimates of reasonable accuracy;
- It has the same stability properties as the Euler method;
- The cpu-time increases compared to the explicit Euler method because it requires two function evaluations instead of one.

9.6.1 HEUN'S METHOD

Heun's method applies the *explicit Euler* method for the predictor and the *trapezoidal rule* for the corrector steps.

$$
\begin{aligned}
\text{Predictor:} \quad & \hat{y}_{i+1} = y_i + hf_i + \mathcal{O}(h^2), \\
\text{Corrector:} \quad & y_{i+1} = y_i + \frac{h}{2}(f_i + \hat{f}_{i+1}) + \mathcal{O}(h^3)
\end{aligned}
\tag{9.60}
$$

where $f_i = f(x_i, y_i)$, $\hat{f}_{i+1} = f(x_{i+1}, \hat{y}_{i+1})$, $\hat{y}_{i+1}$, and y_{i+1} denote, respectively, *predicted* and *corrected* estimates for the current step. Note that by replacing the implicit term, $f(x_{i+1}, y_{i+1})$, in the corrector step with $\hat{f}_{i+1} = f(x_{i+1}, \hat{y}_{i+1})$, the *corrected* step also becomes an explicit scheme. In fact, the method can be recast as a one-step scheme by substituting $\hat{y}_{i+1}$ in the corrector equation.

In Pseudocode 9.9, a pseudomodule, **PC_HEUN**, is presented for solving the first-order IVP with the Heun's (Predictor-Corrector) method. The module requires a step size (h), a solution interval $(x0, xlast)$, and an initial value $(y0)$ as input. The derivative $f(x, y)$ is supplied with the user-defined function module, **FCN**. The marching procedure for $x < xlast$ is carried out within a **While**-construct. In the predictor y_{i+1} and corrector $\hat{y}_{i+1}$ step estimates for the current step are computed by Eq. (9.60) using the prior estimate. Finally, the current values are set as prior $(x0 \leftarrow x; \; y0 \leftarrow y)$ before returning to the top of the loop.

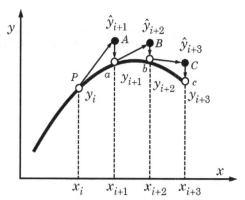

FIGURE 9.8: A graphical depiction of marching in predictor-corrector methods.

Pseudocode 9.9

Module PC_HEUN $(h,\ x0, y0,\ xlast)$
\ DESCRIPTION: A pseudomodule program to solve first-order IVP on $[x0, xlast]$
\   with the Heun's (Predictor-Corrector) method.
\ USES:
\   FCN:: A user-defined function module supplying $f(x, y) = y'$.
Write: "$x, y=$", $x0, y0$ \ Print initial values
$x \leftarrow x0$ \ Initialize abscissa
While $\big[x < xlast\big]$ \ Marching loop
     $k1 \leftarrow h * \textbf{FCN}(x0, y0)$ \ Find $k1 = h * f(x_i, y_i)$ for predictor
     $y \leftarrow y0 + k1$ \ Find "Predicted' estimate, $\hat{y}_{i+1}$)
     $x \leftarrow x0 + h$ \ Find next abscissa
     $k2 \leftarrow h * \textbf{FCN}(x, y)$ \ Find $k2 = h * f(x_{i+1}, \hat{y}_{i+1})$ for corrector
     $y \leftarrow y0 + (k1 + k2)/2$ \ Find current (corrected) step estimate
     Write: "$x, y=$", x, y \ Print current step values
     $x0 \leftarrow x;\ y0 \leftarrow y$ \ Assign current step values as prior
End While
End Module PC_HEUN

9.6.2   ADAMS-BASHFORTH-MOULTON METHOD (ABM4)

The *Adams-Bashforth-Moulton fourth-order method* employs the fourth-order Adams-Bashforth formula for the *predictor* step and the fourth-order Adams-Moulton formula for the *corrector* step:

$$\text{Predictor step:} \quad \hat{y}_{i+1} = y_i + \frac{h}{24}\left(55f_i - 59f_{i-1} + 37f_{i-2} - 9f_{i-3}\right) + \frac{251}{720}h^5 y^{(5)}(\xi) \quad (9.61)$$

$$\text{Corrector step:} \quad y_{i+1} = y_i + \frac{h}{24}\left(9\hat{f}_{i+1} + 19f_i - 5f_{i-1} + f_{i-2}\right) - \frac{19}{720}h^5 y^{(5)}(\bar{\xi}) \quad (9.62)$$

where $f_i = f(x_i, y_i)$, $\hat{f}_{i+1} = f(x_{i+1}, \hat{y}_{i+1})$, $x_{i-3} < \xi < x_{i+1}$ and $x_{i-2} < \bar{\xi} < x_{i+1}$.

Note that the corrector step is made explicit by replacing f_{i+1} with $\hat{f}_{i+1}$ in Eq. (9.62). For the predictor and corrector steps, we require f_0, f_1, f_2, and f_3 to initiate the forward-marching scheme. Since (x_0, y_0) is available as the initial condition, f_0 is computed with no trouble. However, the other *starting values*, y_1, y_2, and y_3, are generally obtained by a high-order one-step method such as RK4. Once f_1, f_2, and f_3 are computed, the forward-marching can begin.

Adams-Bashforth-Moulton Fourth-Order Method

- ABM4 requires fewer function evaluations, i.e., one function for the predictor f_i and one for the corrector $\hat{f}_{i+1}$ step;
- There is no need to evaluate derivatives for all prior points (quantities no longer needed can be shifted out);
- Its global error is $\mathcal{O}(h^4)$.

- It can lead to stability problems since it is an explicit method;
- An implicit scheme can be applied in the corrective step, but then cpu-time can increase significantly.

```
                        Pseudocode 9.10

  Module PC_ABM4 (h, x0, y0, xlast)
  \ DESCRIPTION: A pseudomodule program to solve a first-order IVP on
  \    [x0, xlast] with the ABM4 predictor-corrector method.
  \ USES:
  \    DRV_RK:: Module performing one-step RK4 scheme (see Pseudocode 9.4);
  \    FCN:: A user-defined function module supplying f(x, y) = y'.
  Declare: f4                                    \ Declare array variables
  Write: "x, y=", x0, y0                              \ Print initial values
  f4 ← FCN(x0, y0)                                \ Find derivative at x0
  For [k = 3, 1, (−1)]                     \ Compute other starting values
      DRV_RK(4, h, x0, y0, x, y)      \ Find estimate with one-step RK4 scheme
      Write: "x, y=",x, y                       \ Print starting estimates
      fk ← FCN(x, y)                      \ Compute derivative at (x, y)
      x0 ← x;  y0 ← y              \ Assign current step values as prior
  End For
  While [x < xlast]                                      \ Marching loop
      yh ← y0 + h ∗ (55f1 − 59f2 + 37f3 − 9f4)/24       \ Find prediction
      x ← x0 + h                                       \ Find abscissa
      fh ← FCN(x, yh)                               \ Find f̂ = f(x, yh)
      y ← y0 + h ∗ (9fh + 19f1 − 5f2 + f3)/24          \ Find correction
      Write: "x, y=",x, y                     \ Print current step estimate
      f4 ← f3;  f3 ← f2;  f2 ← f1;              \ Shift starting values up
      f1 ← FCN(x, y)                       \ Update nearest starting value
      x0 ← x;  y0 ← y              \ Assign current step values as prior
  End While
  End Module PC_ABM4
```

A pseudomodule, **PC_ABM4**, presented in Pseudocode 9.10 is designed to solve a first-order IVP using the ABM4. The module requires a step size (h), a solution interval $(x0, xlast)$, and the initial value $(y0)$ as input. The derivative $y' = f(x, y)$ is supplied with the user-defined function module, **FCN**. In preparation to apply the ABM4 method, the starting values for y_1, y_2, and y_3 (likewise, f_1, f_2, and f_3) are obtained with the **DRV_RK** module (in Pseudocode 9.4). For $x < xlast$, all predictions and corrections are respectively obtained by Eqs. (9.61) and (9.62) in the **While**-loop. Only four initial values $(f_0, f_1, f_2,$ and $f_3)$ are needed to start and maintain the forward march. As an efficient programming trick, we only keep the last four derivatives $(f_0, f_1, f_2,$ and $f_3)$ as well as prior and current step estimates $(y0$ and $y)$ to reduce memory allocation. Hence, we shift the derivatives one step back sequentially as follows: $f_0 \leftarrow f_1$, $f_1 \leftarrow f_2$, and $f_2 \leftarrow f_3$, and the current estimate is assigned to the prior estimate, $(x0, y0) \leftarrow (x, y)$, before returning to the top of the loop.

It is possible to use adaptive step size control in the ABM4 method. Consequently, we need an approximation for the local truncation error for the ABM4, which is given as follows:

$$E_t = \frac{19}{270} \frac{|y_{i+1} - \hat{y}_{i+1}|}{h}$$

Then, a procedure with a variable step size strategy may be applied, provided that the size

Milne's Method

- Milne's method is simple and easy to implement;
- It requires fewer function evaluations, i.e., only f_i and $\hat{f}_{i+1}$;
- It has a fourth-order global error, $\mathcal{O}(h^4)$, and is more accurate than the ABM4 scheme.

- Being an explicit method, it suffers from stability issues under certain conditions, stemming essentially from the corrector step. For this reason, the ABM4 method is often favored over Milne's method.

adjustment factor is calculated from

$$\delta = 0.84 \left(\frac{\varepsilon}{E_t} \right)^{1/4}$$

If the estimated truncation error is less than the desired tolerance ($E_t \leqslant \varepsilon$), the current estimate is accepted. The step size is also adjusted as $h \leftarrow \delta * h$ to prevent it from becoming either too small or too large. If $E_t > \varepsilon$, the step size is reduced according to $h \leftarrow \delta * h$. However, varying the step size in multistep methods is much more expensive than in one-step methods because it requires establishing starting values that are compatible with the adapted step sizes. In the latter case, the prior step estimates (or shifted starting values) cannot be used, and a new set of starting values must be computed at every step, making it a much more expensive method than one-step methods.

9.6.3   MILNE'S FOURTH-ORDER METHOD

Milne's predictor-corrector method employs three-point open Newton-Cotes formulas for the first derivative in the predictor and a three-point closed Newton-Cotes formula in the corrector step.

$$\text{Predictor step:} \quad \hat{y}_{i+1} = y_{i-3} + \frac{4h}{3} \left(2f_i - f_{i-1} + 2f_{i-2} \right) + \frac{28}{90} h^5 y^{(5)}(\xi) \quad x_{i-3} < \xi < x_{i+1} \quad (9.63)$$

$$\text{Corrector step:} \quad y_{i+1} = y_{i-1} + \frac{h}{3} \left(\hat{f}_{i+1} + 4f_i + f_{i-1} \right) - \frac{1}{90} h^5 y^{(5)}(\bar{\xi}) \quad x_{i-1} < \bar{\xi} < x_{i+1} \quad (9.64)$$

where $f_i = f(x_i, y_i)$ and $\hat{f}_{i+1} = f(x_{i+1}, \hat{y}_{i+1})$.

In Eq. (9.64), the implicit term f_{i+1} is replaced with $\hat{f}_{i+1}$, making the corrector step also explicit. Note that the truncation error constants for this method are smaller than those of ABM4 (Eqs. (9.61) and (9.62)), so theoretically, we can expect more accurate numerical estimates from this scheme. The numerical algorithm is also identical to that of the ABM4. The starting values are obtained with a high-order one-step method, and then f_1, f_2, and f_3 are evaluated to start the forward march. Like the ABM4, there is no need to store all derivatives during the forward-marching step; only the required derivatives are kept by shifting the current derivatives one step down.

In Pseudocode 9.11, a pseudomodule, **PC_MILNE**, is presented for solving a first-order model IVP using Milne's method. This module is basically identical to **PC_ABM4**. A major difference is in the predictor and corrector equations, which can be easily replaced with those of ABM4. Milne's predictor-corrector equations involve y_{i-3} and y_{i-1} rather than y_i as in the case of ABM4; hence, only the last four estimates (y_{i-3}, y_{i-2}, y_{i-1}, y_i) need to be stored at each step as well before returning to the top of the While-loop. The last four estimates of y are stored on a temporary storage array **g** of length 4.

Pseudocode 9.11

Module PC_MILNE $(h,\ x0, y0, xlast)$
\ DESCRIPTION: A pseudomodule to solve $y' = f(x, y)$ on $[x0, xlast]$ with the
\   ABM4 (predictor-corrector) method.
\ USES:
\   DRV_RK:: Module performing one-step RK4 (*see* Pseudocode 9.4);
\   FCN:: A user-defined function module supplying $f(x, y) = y'$.
Declare: $f_4,\ g_4$ \ Declare array variables
Write: "$x, y=$", $x0, y0$ \ Print initial values
$g_4 \leftarrow y0;\ f_4 \leftarrow$ **FCN**$(x0, y0)$ \ Set initial derivative as starter
For $\left[k = 3, 1, (-1)\right]$ \ Find starting values
   DRV_RK$(4, h,\ x0,\ y0,\ x, y)$ \ Apply one-step RK4
   Write: "$x, y=$",x, y \ Print starting value
   $f_k \leftarrow$ **FCN**(x, y) \ Find derivative at (x, y)
   $g_k \leftarrow y$ \ Assign solution to array **g**
   $x0 \leftarrow x;\ \ y0 \leftarrow y$ \ Set current step values as prior
End For
While $\left[x < xlast\right]$ \ Marching loop
   $yh \leftarrow g_4 + 4h * (2f_1 - f_2 + 2f_3)/3$ \ Find prediction
   $x \leftarrow x0 + h$ \ Find abscissa
   $fh \leftarrow$ **FCN**(x, yh) \ Corrector step
   $y \leftarrow g_2 + h * (fh + 4f_1 + f_2)/3$ \ Find correction (current estimate)
   Write: "$x, y=$",x, y \ Print current estimate
   $f_4 \leftarrow f_3;\ f_3 \leftarrow f_2;\ f_2 \leftarrow f_1;$ \ Shift starting values
   $f_1 \leftarrow$ **FCN**(x, y) \ Update nearest derivative
   $g_4 \leftarrow g_3;\ g_3 \leftarrow g_2;\ g_2 \leftarrow g_1;$
   $g_1 \leftarrow y;\ x0 \leftarrow x;\ y0 \leftarrow y$ \ Assign current values as prior
End While
End Module PC_MILNE

EXAMPLE 9.8: Comparison of Heun, ABM4, and Milne's methods

Repeat Example 9.1 using (a) Heun's, (b) the ABM4, and (c) Milne's (predictor-corrector) methods with $h = 0.1$. Compare the estimates with the true solution.

SOLUTION:

The initial condition and corresponding derivative are $I_0 = 2$ and $f_0 = 3$.

(a) We start with Heun's method by applying Eqs. (9.60) for $i = 0$, which yields the following predictor $(\hat{I})$ and corrector (I) values:

$$\hat{I}_1 = I_0 + hf_0 = 2 + 0.1(3) = 2.3$$

$$I_1 = I_0 + \frac{h}{2}\left(f_0 + f(t_1, \hat{I}_1)\right) = 2 + 0.05\left(3 + f(0.1, 2.3)\right) = 2.242992$$

Applying Eqs. (9.60) for $i = 1$, the second-step prediction and correction values are found as

$$\hat{I}_2 = I_1 + hf_1 = 2.242993 + 0.1f\left(0.1, 2.242993\right) = 2.443227$$

$$I_2 = I_1 + \frac{h}{2}\left(f_1 + f(x_2, \hat{I}_2)\right)$$

$$= 2.242993 + 0.05\left(2.002355 + f(0.2, 2.443227)\right) = 2.399641$$

In the third, fourth, and subsequent steps, the current prediction and correction values are obtained using the recursive difference equations, Eqs. (9.63) and (9.64). The true solution as well as the absolute (true) errors are presented in Table 9.13. The variation of errors across the solution range is similar to that obtained by the methods applied in the previous examples; that is, the error increases up to $t \approx 0.46$, where the true solution also has a local maximum and decays with increasing t.

(b) To apply the ABM4 method, starting values are also generated from the true solution. The fact that the starting values do not initially contain any errors allows for an ideal comparison with Milne's method. From the initial condition, we have $f_0 = 3$, and the rest of the starting values are obtained as follows:

$$I_1 = I_{\text{true}}(0.1) = 2.247316, \quad f_1 = f(t_1, I_1) = f(0.1, 2.247316) = 1.991545$$

$$I_2 = I_{\text{true}}(0.2) = 2.406288, \quad f_2 = f(t_2, I_2) = f(0.2, 2.406288) = 1.222979$$

$$I_3 = I_{\text{true}}(0.3) = 2.498099, \quad f_3 = f(t_3, I_3) = f(0.3, 2.498099) = 0.640417$$

For the predictor step, applying Eq. (9.61) yields

$$\hat{I}_4 = I_3 + \frac{h}{24}\left(55f_3 - 59f_2 + 37f_1 - 9f_0\right)$$

$$= 2.498099 + \frac{0.1}{24}\left(55(0.640416) - 59(1.222979) + 37(1.991545) - 9(3)\right)$$

$$= 2.538742$$

Then the intermediate value $f(t_4, \hat{I}_4)$ becomes

$$f(t_4, \hat{I}_4) = f(1.4, 2.538742) = 0.202991$$

Using Eq. (9.62), the corrector step yields

$$I_4 = I_3 + \frac{h}{24}\left(9f(t_4, \hat{I}_4) + 19f_3 - 5f_2 + f_1\right)$$

$$= 2.498099 + \frac{0.1}{24}\left(9(0.202991) + 19(0.640416) - 5(1.222979) + (1.991545)\right)$$

$$= 2.539230$$

Before moving on to the next step, calculating the derivative of the current step is necessary because it will be added to the list of starting values. So the required quantities are found as follows:

$$t_4 = 0.4, \quad I_4 = 2.539230, \quad f_4 = f(t_4, I_4) = 0.201771$$

The fifth-step predictor yields

$$\hat{I}_5 = I_4 + \frac{h}{24}\left(55f_4 - 59f_3 + 37f_2 - 9f_1\right)$$

$$= 2.539230 + \frac{0.1}{24}\left(55(0.201771) - 59(0.640416) + 37(1.222979) - 9(1.991545)\right)$$

$$= 2.541893$$

and $f(t_5, \hat{I}_5) = f(0.5, 2.541893) = -0.124326$.

Finally, employing the corrector step results in

$$I_5 = I_4 + \frac{h}{24}\left(9f(t_5, \hat{I}_5) + 19f_4 - 5f_3 + f_2\right)$$

$$= 2.539230 + \frac{0.1}{24}\left(9(-0.124326) + 19(0.201771) - 5(0.640416) + (1.222979)\right)$$

$$= 2.542295$$

By applying the foregoing recursive procedure in the same manner, the ABM4 estimates were obtained and presented comparatively in Table 9.13. Since the starting values (for the first four steps) are obtained from the true solution, the accumulation of truncation error is due only to the ABM4 method. For this reason, it is observed that the instant when the error reaches a maximum shifts to $t \approx 0.8$ due to the accumulation of errors.

TABLE 9.13

t_i	True Solution	True absolute error, e_i		
		Heun's Method	ABM4	Milne's Method
0	2			
0.2	2.406288	6.65E-03		
0.4	2.539164	7.79E-03	6.59E-05	4.12E-05
0.6	2.517013	6.75E-03	1.17E-04	5.59E-05
0.8	2.410610	5.09E-03	1.18E-04	5.57E-05
1	2.261953	3.47E-03	1.00E-04	5.00E-05
1.2	2.095672	2.15E-03	7.80E-05	4.32E-05
1.4	1.925946	1.14E-03	5.78E-05	3.72E-05
1.6	1.760685	4.35E-04	4.14E-05	3.25E-05
1.8	1.604061	4.15E-05	2.89E-05	2.91E-05
2	1.458042	3.45E-04	1.98E-05	2.69E-05

(c) Applying Milne's method (Eqs. (9.63) and (9.64)) for the first step yields

$$\hat{I}_4 = I_0 + \frac{4h}{3}\left(2f_3 - f_2 + 2f_1\right)$$

$$= 2 + \frac{4(0.1)}{3}\left(2(0.640416) - (1.222979) + 2(1.991545)\right) = 2.538793$$

$$I_4 = I_2 + \frac{h}{3}\left(f(t_4, \hat{I}_4) + 4f_3 + f_2\right)$$

$$= 2.406288 + \frac{0.1}{3}\left(0.202864 + 4(0.640416) + (1.222979)\right) = 2.539205$$

where the intermediate value is found as $f(t_4, \hat{I}_4) = 0.202864$.

Evaluating the derivative for the current step gives

$$f_4 = f(t_4, I_4) = f(0.4, 2.539205) = 0.201834$$

Proceeding with the second-step calculations results in

$$\hat{I}_5 = I_1 + \frac{4h}{3}\left(2f_4 - f_3 + 2f_2\right)$$

$$= 2.247316 + \frac{4(0.1)}{3}\left(2(0.201834) - 0.640416 + 2(1.222979)\right) = 2.541877$$

$$I_5 = I_3 + \frac{h}{3} \left(f(t_5, \hat{I}_5) + 4f_4 + f_3 \right)$$

$$= 2.498099 + \frac{0.1}{3} \left(-0.124286 + 4(0.201834) + 0.640416 \right) = 2.542214$$

where $f(t_5, \hat{I}_5) = -0.124286$.

The estimates computed with Milne's method have been repeated in the same manner for the subsequent steps and are presented in Table 9.13. In this example, the first four steps are derived from the true solution, just as in Part (b). For $t \approx 0.6$, a slight increase in errors is observed.

Discussion: Examining the true errors incurred by the three methods, the Heun's method has yielded the largest errors among all three methods. This result is understandable since this method is second-order accurate. The ABM4 and Milne's methods are both fourth-order methods, but Milne's method gives more accurate estimates. In this example, the errors are on average about 18 times smaller. The reason for this is that the truncation error constants in Milne's method, Eqs. (9.63) and (9.64), are smaller in comparison to those of the ABM4; Eqs. (9.61) and (9.62).

9.7 SYSTEM OF FIRST-ORDER IVPS (ODES)

Mathematical models of real-life problems generally involve two or more first- or higher-order coupled ODEs. So far, we have discussed the numerical methods for the first-order model IVP given in the form $y' = f(x, y)$. In this section, we discuss extending these numerical methods to first-order simultaneous linear and non-linear IVP-ODEs.

Consider a system of n first-order linear IVPs (ODEs) with n initial conditions:

$$\frac{dY_1}{dx} = f_1(x, Y_1, Y_2, \ldots, Y_n), \qquad Y_1(x_0) = Y_{1,0},$$

$$\frac{dY_2}{dx} = f_2(x, Y_1, Y_2, \ldots, Y_n), \qquad Y_2(x_0) = Y_{2,0},$$

$$\vdots \qquad\qquad\qquad \vdots \qquad\qquad\qquad \vdots \qquad\qquad (9.65)$$

$$\frac{dY_n}{dx} = f_n(x, Y_1, Y_2, \ldots, Y_n), \qquad Y_n(x_0) = Y_{n,0}$$

where x is the independent variable, and $Y_1(x)$, $Y_2(x)$, $\ldots$, $Y_n(x)$ are the dependent variables.

This system of ODEs can be expressed in vector notation as

$$\frac{d\mathbf{y}}{dx} = \mathbf{F}(x, \mathbf{y}), \qquad \mathbf{y}(x_0) = \mathbf{y}_0 \qquad\qquad (9.66)$$

where

$$\mathbf{F}(x, \mathbf{y}) = \begin{bmatrix} f_1(x, \mathbf{y}) \\ f_2(x, \mathbf{y}) \\ \vdots \\ f_n(x, \mathbf{y}) \end{bmatrix}, \qquad \mathbf{y} = \begin{bmatrix} Y_1 \\ Y_2 \\ \vdots \\ Y_n \end{bmatrix} \qquad \text{and} \qquad \mathbf{y}(x_0) = \mathbf{y}_0 = \begin{bmatrix} Y_{1,0} \\ Y_{2,0} \\ \vdots \\ Y_{n,0} \end{bmatrix}$$

An explicit or implicit numerical method similar to a single IVP can be readily generalized by replacing the scalar variable y and the scalar function f by vector functions $\mathbf{y}$ and $\mathbf{F}$, respectively. For instance, the explicit Euler scheme can be expressed in vector notation as:

$$\mathbf{y}_{i+1} = \mathbf{y}_i + h\,\mathbf{F}(x_i, \mathbf{y}_i), \qquad \mathbf{y}_0 = \mathbf{y}(x_0) \tag{9.67}$$

Similarly, the RK4 scheme for a system of first-order coupled ODEs is cast as

$$\mathbf{y}_{i+1} = \mathbf{y}_i + \frac{h}{6}\left(\mathbf{k}_1 + 2\mathbf{k}_2 + 2\mathbf{k}_3 + \mathbf{k}_4\right), \qquad \mathbf{y}_0 = \mathbf{y}(x_0) \tag{9.68}$$

and the RK parameters are also vectorized as follows:

$$
\begin{aligned}
\mathbf{k}_1 &= h\,\mathbf{F}\left(x_i,\, \mathbf{y}_i\right), \\
\mathbf{k}_2 &= h\,\mathbf{F}\left(x_{i+1/2}, \mathbf{y}_i + \mathbf{k}_1/2\right), \\
\mathbf{k}_3 &= h\,\mathbf{F}\left(x_{i+1/2}, \mathbf{y}_i + \mathbf{k}_2/2\right), \\
\mathbf{k}_4 &= h\,\mathbf{F}\left(x_{i+1}, \mathbf{y}_i + \mathbf{k}_3\right)
\end{aligned} \tag{9.69}
$$

where $k_{j,i}$'s for $i = 1, 2, \ldots, n$ are the elements of $\mathbf{k}_j$ for $j = 1, 2, 3$, and 4.

Starting with the initial conditions, all dependent variables for the current step, $\mathbf{y}_{i+1}$, can be explicitly computed using the prior step estimates—$\mathbf{y}_i$. However, the computation of $\mathbf{k}$'s is carried out sequentially because every $\mathbf{k}_m$ requires the information from the previous k (i.e., $\mathbf{k}_{m-1}$).

Special case: Consider a system of linear IVPs that can be expressed as follows:

$$\frac{d\mathbf{y}}{dx} = \mathbf{A}(x)\mathbf{y} + \mathbf{g}(x), \qquad \mathbf{y}(x_0) = \mathbf{y}_0 \tag{9.70}$$

where $\mathbf{A}$ and $\mathbf{g}$ can be functions of the independent variable x as

$$
\mathbf{A}(x) = \begin{bmatrix}
a_{11}(x) & a_{12}(x) & \cdots & a_{1n}(x) \\
a_{21}(x) & a_{22}(x) & \cdots & a_{2n}(x) \\
\vdots & \vdots & \ddots & \vdots \\
a_{21}(x) & a_{n2}(x) & \cdots & a_{nn}(x)
\end{bmatrix}
\quad \text{and} \quad
\mathbf{g}(x) = \begin{bmatrix}
g_1(x) \\
g_2(x) \\
\vdots \\
g_n(x)
\end{bmatrix}
$$

A linear IVP in the given form allows one to obtain the numerical solution with implicit schemes without having to resort to iterations. For instance, the implicit Euler method can be written as

$$\mathbf{y}_{i+1} = \mathbf{y}_i + h\left(\mathbf{A}(x_{i+1})\mathbf{y}_{i+1} + \mathbf{g}(x_{i+1})\right) \tag{9.71}$$

Solving Eq. (9.71) for $\mathbf{y}_{i+1}$ yields

$$\mathbf{y}_{i+1} = \left(\mathbf{I} - h\mathbf{A}_{i+1}\right)^{-1}\left(\mathbf{y}_i + h\,\mathbf{g}_{i+1}\right) \tag{9.72}$$

where $\mathbf{A}_{i+1} = \mathbf{A}(x_{i+1})$ and $\mathbf{g}_{i+1} = \mathbf{g}(x_{i+1})$.

When simulating unsteady physical phenomena (IVPs), it is not unusual to deal with systems of hundreds or possibly thousands of first-order ODE-IVPs. Hence, it is crucial to make sure that any numerical method employed is both accurate and efficient.

Pseudocode 9.12

Module BDF2_SE (n, h, $x0$, $\mathbf{y}0$, $xlast$)
\ DESCRIPTION: A pseudomodule to solve a first-order coupled linear ODEs
\   of the form; $d\mathbf{y}/dx = \mathbf{A}(x)\mathbf{y}(x) + \mathbf{g}(x)$ using the BDF2 method.
\ USES:
\   EQNS:: A user-defined module supplying coefficients of $\mathbf{A}$ and $\mathbf{g}$;
\   Ax:: Module to perform matrix-vector multiplication (Pseudocode 2.5);
\   INV_MAT:: Module to invert a matrix (Pseudocode 2.6);
\   DRV_RK:: Module performing one-step RK scheme; a vectorized version of
\   Pseudocode 9.9, which is not provided here.
Declare: $y0_n$, $y1_n$, y_n, $a_{n,n}$, g_n, $b_{n,n}$, rhs_n, c_3 \ Declare array variables
$c_1 \leftarrow -1/3$; $c_2 \leftarrow 4/3$; $c_3 \leftarrow 2h/3$ \ Set BDF2 coefficients
Write: "x,y=", $x0$,$\mathbf{y}0$ \ Print initial values
DRV_RK($4, h$, $x0$, $\mathbf{y}0$, x, $\mathbf{y}$)   \ Find a starting value using RK4
$\mathbf{y}1 \leftarrow \mathbf{y}$ \ Assign estimates to $\mathbf{y}1 \leftarrow \mathbf{y}(x_0 + h)$
Write: "x,y=", x,$\mathbf{y}1$ \ Print starting values
$x0 \leftarrow x$ \ Initialize abscissa
While $\left[x < xlast\right]$ \ Marching loop
    $x \leftarrow x0 + h$ \ Find current step abscissa
    EQNS($n, x, \mathbf{A}, \mathbf{g}$)   \ Construct $\mathbf{A}$ and $\mathbf{g}$ at x
    For $\left[i = 1, n\right]$ \ Construct $\mathbf{B} = \mathbf{I} - (2h/3)\mathbf{A}$
        For $\left[j = 1, n\right]$
            If $\left[i = j\right]$ **Then**
                $b_{ij} \leftarrow 1 - c3 * a_{ij}$ \ $b_{ii} = 1 - (2h/3) * a_{ii}$
            Else
                $b_{ij} \leftarrow -c3 * a_{ij}$ \ $b_{ij} = -(2h/3) * a_{ij}$
            End If
        End For
    End For
    $\mathbf{rhs} \leftarrow c_1 * \mathbf{y}0 + c_2 * \mathbf{y}1 + c_3 * \mathbf{g}$ \ Compute $\mathbf{rhs} \leftarrow c_1\mathbf{y}0 + c_2\mathbf{y}1 + c_3\mathbf{g}$
    INV_MAT($n, \mathbf{B}, \mathbf{B}$)   \ Invert matrix $\mathbf{B}$, $\mathbf{B} \leftarrow \mathbf{B}^{-1}$
    Ax($n, \mathbf{B}, \mathbf{rhs}, \mathbf{y}$)   \ Find current estimate, $\mathbf{y}_{i+1} \leftarrow \mathbf{B}^{-1} \cdot \mathbf{rhs}$
    Write: "x,y=", x,$\mathbf{y}$ \ Print current estimates
    $x0 \leftarrow x$; $\mathbf{y}0 \leftarrow \mathbf{y}1$; $\mathbf{y}1 \leftarrow \mathbf{y}$ \ Set current estimates as prior
End While
End Module BDF2_SE

Module EQNS (n, x $\mathbf{A}$, $\mathbf{g}$)   \ Declare array variables
\ DESCRIPTION: User-defined module supplying 1st-order linear ODEs.
Declare: a_{nn}, g_n
$a_{ij} \leftarrow \ldots$ $(i, j = 1, 2, \ldots, n)$ \ Define elements of $\mathbf{A}$
$g_i \leftarrow \ldots$ $(i = 1, 2, \ldots, n)$ \ Define elements of $\mathbf{g}$
End Module EQNS

In Pseudocode 9.12, the module BDF2_SE solving a first-order coupled linear IVPs with BDF2 is presented. The module requires the number of ODEs (n), a step size (h), a solution interval ($x0, xlast$), and the initial value ($\mathbf{y}0$) as input. The set of ODEs is supplied to the module with the user-defined function module EQNS that returns the coefficients of

A and **g** that may be x-dependent. The main module also requires matrix multiplication and inversion modules. To begin forward marching, the starting estimate $\mathbf{y}_1$ is obtained by the RK4 scheme using DRV_RK. For $x < xlast$, the current estimates are obtained by using Eq. (9.74) within a While-loop. The current step estimates are assigned as prior estimates $(x0 \leftarrow x; \mathbf{y}_0 \leftarrow \mathbf{y}_1; \mathbf{y}_1 \leftarrow \mathbf{y})$ before returning to the top of the loop.

In Pseudocode 9.12, the pseudomodule, BDF2_SE, is given for solving n simultaneous first-order linear IVP equations of the form of Eq. (9.70) using the BDF2 scheme.

$$\frac{d\mathbf{y}_i}{dx} = \mathbf{A}(x_{i+1})\mathbf{y}_{i+1} + \mathbf{g}(x_{i+1}), \qquad \mathbf{y}(x_0) = \mathbf{y}_0 \tag{9.73}$$

Replacing the derivative with the second-order BDF yields

$$\mathbf{y}_{i+1} = \frac{1}{3}\left(\mathbf{I} - \frac{2h}{3}\mathbf{A}_{i+1}\right)^{-1}(4\mathbf{y}_i - \mathbf{y}_{i-1} + 2h\,\mathbf{g}_{i+1}), \tag{9.74}$$

where $\mathbf{A}_{i+1} = \mathbf{A}(x_{i+1})$ and $\mathbf{g}_{i+1} = \mathbf{g}(x_{i+1})$.

9.8 HIGHER-ORDER ODES

Initial value problems are not solely confined to first-order ODEs. Second- or higher-order linear or nonlinear IVPs are commonly encountered. These problems have no true solutions except in very special cases and are predominantly solved numerically. A common approach to the numerical solution of higher-order IVPs is to reduce them into an equivalent system of first-order simultaneous ODEs.

Consider the following nth-order ODE-IVP:

$$y^{(n)}(x) = F\left(x, y', y'', \ldots, y^{(n-1)}\right) \tag{9.75}$$

subjected to the following n initial conditions:

$$y(x_0) = \alpha_0, \quad y'(x_0) = \alpha_1, \quad y''(x_0) = \alpha_2, \ldots, \quad y^{(n-1)}(x_0) = \alpha_{n-1}.$$

Equation (9.75) can be reduced into a system of first-order IVP by defining a set of new independent variables:

$$Y_1(x) = y(x), \quad Y_2(x) = y'(x), \quad Y_3(x) = y''(x), \quad \ldots \quad Y_n(x) = y^{(n-1)}(x)$$

As a result, the original IVP becomes equivalent to coupled n-first-order ODEs

$$
\begin{aligned}
\frac{dY_1}{dx} &= Y_2, & Y_1(x_0) &= \alpha_0, \\
\frac{dY_2}{dx} &= Y_3, & Y_2(x_0) &= \alpha_1, \\
&\;\;\vdots & &\;\;\vdots \\
\frac{dY_{n-1}}{dx} &= Y_n, & Y_{n-1}(x_0) &= \alpha_{n-2} \\
\frac{dY_n}{dx} &= F(x, Y_1, Y_2, \ldots, Y_n), & Y_n(x_0) &= \alpha_{n-1}
\end{aligned}
\tag{9.76}
$$

9.8.1 STIFF ODEs

In the numerical solution of coupled ODE-IVPs, just as in the case of a single ODE, a well-known problem called *stiffness* may be encountered. Also recall that this problem arises from the true solution of a system consisting of ODEs with components that grow and/or decay at different rates. Hence, the systems of ODEs that depict stiffness are often referred to as *stiff ODEs* or *stiff systems*.

These stiff ODEs are encountered in chemical reaction kinetics, control systems, electrical networks, and so on, and can be very difficult or impossible to solve numerically using the explicit methods discussed previously. This difficulty stems mainly from the small stability regions of explicit methods, which forces the analyst to use a very small step size to achieve a stable solution. On the other hand, using an extremely small step size (smaller than the desired accuracy, $\Delta x \leqslant \varepsilon$) can yield serious problems due to round-off errors.

Stiffness is a relative concept that depends on certain features of an IVP, such as the stability domain, integration interval, and desired accuracy. Then the question arises: How do we know that a system of coupled IVP is stiff? Mathematically speaking, if the Jacobian matrix has eigenvalues that differ in magnitude, then it can be said that the stiffness problem exists.

As a measure of the stiffness of a given problem, the stiffness ratio is generally defined in terms of the largest and smallest eigenvalues as

$$R = \frac{\max_i |\text{Re}(\lambda_i)|}{\min_i |\text{Re}(\lambda_i)|}$$

where λ_i are the eigenvalues of the Jacobian of the system, R is the stiffness ratio, and Re denotes the real part of an imaginary number. This definition of stiffness is invalid for nonlinear systems. Therefore, a verbal definition of stiffness that can be applied to both linear and nonlinear problems is given as follows:

Definition (Stiffness): *If a numerical method with a finite region of absolute stability, applied to a system with any initial conditions, is forced to use in a certain interval of integration a step length that is extremely small in comparison to the smoothness of the true solution in that interval, then the system is said to be stiff in that interval* [20].

If a numerical method or scheme with a larger stability region can be devised, then one can expect the numerical method to perform better in solving stiff ODEs.

 The stiffness ratio may not always be a good measure of stiffness, even in linear systems, because the problem has infinite stiffness if the minimum eigenvalue is zero. Thus, whenever a set of simultaneous linear first-order ODEs is solved, beware of the stiffness problem.

9.8.2 NUMERICAL STABILITY

Consider the following system of linear ODEs:

$$\frac{dY_1}{dx} = f_1(x, Y_1, Y_2) = a_{11}Y_1(x) + a_{12}Y_2(x)$$
$$\frac{dY_2}{dx} = f_2(x, Y_1, Y_2) = a_{21}Y_1(x) + a_{22}Y_2(x)$$

(9.77)

or, in matrix notation, the system can be expressed as

$$\frac{d\mathbf{y}}{dx} = \mathbf{Ay} \tag{9.78}$$

where

$$\mathbf{y} = \begin{bmatrix} Y_1 \\ Y_2 \end{bmatrix}, \quad \mathbf{A} = \begin{bmatrix} a_{11} & a_{12} \\ a_{21} & a_{22} \end{bmatrix}, \quad \text{and} \quad \mathbf{y}_0 = \begin{bmatrix} Y_{1,0} \\ Y_{2,0} \end{bmatrix}$$

Applying the explicit Euler scheme to Eq. (9.78) yields

$$\frac{\mathbf{y}_{i+1} - \mathbf{y}_i}{h} + \mathcal{O}(h) = (\mathbf{Ay})_i \tag{9.79}$$

which leads to

$$\begin{bmatrix} Y_1 \\ Y_2 \end{bmatrix}_{i+1} = \begin{bmatrix} 1 + ha_{11} & ha_{12} \\ ha_{21} & 1 + ha_{22} \end{bmatrix} \begin{bmatrix} Y_1 \\ Y_2 \end{bmatrix}_i + \mathcal{O}(h^2) \tag{9.80}$$

or

$$\mathbf{y}_{i+1} = \mathbf{G}_{\text{exp}}\mathbf{y}_i \tag{9.81}$$

where $\mathbf{G}_{\text{exp}} = \mathbf{I} + h\mathbf{A}$ matrix denotes the amplification matrix of the explicit method, and $\mathbf{I}$ is the identity matrix. Consequently, ignoring the global error term, the numerical estimate for any ith step can be written as

$$\mathbf{y}_i = \mathbf{G}_{\text{exp}}^i \mathbf{y}_0 \tag{9.82}$$

In order to avoid instabilities and have a bounded numerical solution (if it exits), the amplification matrix should approach the zero matrix for increasing powers of i. In other words, a suitable step size should be chosen to ensure a bounded numerical solution as $i \to \infty$, i.e., $\lim_{i\to\infty}\|\mathbf{y}_i\| \to \mathbf{0}$. It turns out that to obtain a bounded solution, the eigenvalues of the amplification matrix should satisfy the $|\lambda|_k < 1$ condition for all k.

Now, we apply the stability analysis to the implicit Euler scheme, which leads to

$$\mathbf{y}_{i+1} = \mathbf{G}_{\text{imp}}\mathbf{y}_i \tag{9.83}$$

where $\mathbf{G}_{\text{imp}} = (\mathbf{I} - h\mathbf{A})^{-1}$ is the amplification matrix for the implicit scheme. The eigenvalues of $\mathbf{G}_{\text{imp}}$ can be obtained as $\mu_k = 1/(1 - h\lambda_k)$, where λ_k's denote the eigenvalues of $\mathbf{A}$ (*see* Chapter 11). Thus, it becomes evident that all λ_k's must have negative real parts $(\text{Re}(\lambda_k) < 0$ for all $k)$ to ensure the stability of this model IVP problem (Eq. 9.78). This constraint also ensures $|\mu_k| < 1$ for all k and consequently satisfies $\lim_{i\to\infty}\|\mathbf{y}_i\| \to \mathbf{0}$.

Numerical methods devised for a single IVP, such as AB, AM, BDF, and predictor-corrector, can be generalized in the same way, and stability analysis can be performed, although it may be relatively difficult. The global error is calculated from the expression obtained by replacing the term $|y(x_i) - y_i|$ in univariate IVPs with $\|\mathbf{y}(x_i) - \mathbf{y}_i\|_\infty$, where $y(x_i)$ is the true solution at the ith step. Backward differentiation formulas are among the most widely used numerical schemes for stiff problems.

A system of coupled ODEs is said to be *stable* if its numerical solution is bounded for all x. Moreover, the stability of the numerical method is assured if the real parts of the Jacobian matrix are all negative. Note that even if the real part of a single eigenvalue were positive, the numerical solution would grow unboundedly.

EXAMPLE 9.9: Solving coupled system of first-order ODEs

Consider the following coupled-system of IVP:

$$\frac{dY_1}{dx} = 396Y_1(x) + 796\,Y_2(x), \qquad Y_1(0) = 1$$

$$\frac{dY_2}{dx} = -398Y_1(x) - 798\,Y_2(x), \quad Y_2(0) = 0$$

Obtain the numerical solution on [0,1] with (a) explicit, (b) implicit Euler, and (c) BDF2 schemes using $h = 0.1$. Compare your estimates with the true solutions given below:

$$Y_1(x) = 2\,e^{-2x} - e^{-400x} \quad \text{and} \quad Y_2(x) = e^{-400x} - e^{-2x}$$

SOLUTION:

(a) For the given IVP, we can write

$$\mathbf{y} = \begin{bmatrix} Y_1 \\ Y_2 \end{bmatrix}, \qquad \mathbf{A} = \begin{bmatrix} 396 & 796 \\ -398 & -798 \end{bmatrix} \quad \text{and} \quad \mathbf{y}_0 = \begin{bmatrix} 1 \\ 0 \end{bmatrix}$$

and employing the explicit Euler scheme, Eq. (9.80), leads to

$$\begin{bmatrix} Y_1 \\ Y_2 \end{bmatrix}_{i+1} = \begin{bmatrix} 1 + 396h & 796h \\ -398h & 1 - 798h \end{bmatrix} \begin{bmatrix} Y_1 \\ Y_2 \end{bmatrix}_i$$

For $h = 0.1$, the first two-step estimates yield

$$Y_{1,1} = Y_1(0.1) = 40.6, \qquad Y_{2,1} = Y_2(0.1) = -39.8,$$
$$Y_{1,2} = Y_1(0.2) = -1519.72, \qquad Y_{2,2} = Y_2(0.2) = 1520.36$$

In subsequent steps, the computed estimates quickly get out of control and diverge from the true solution.

(b) The implicit Euler scheme leads to

$$\begin{bmatrix} Y_1 \\ Y_2 \end{bmatrix}_{i+1} = \begin{bmatrix} 1 - 396h & -796h \\ 398h & 1 + 798h \end{bmatrix}^{-1} \begin{bmatrix} Y_1 \\ Y_2 \end{bmatrix}_i$$

Similarly, the first two steps with $h = 0.1$ yield

$$Y_{1,1} = Y_1(0.1) = 1.6422764, \qquad Y_{2,1} = Y_2(0.1) = -0.8089431$$
$$Y_{1,2} = Y_1(0.2) = 1.3882940, \qquad Y_{2,2} = Y_2(0.2) = -0.6938495$$

These estimates are close enough, given the magnitude of the truncation errors.

(c) In matrix form, the BDF2 can be written as

$$\mathbf{y}_{i+1} = \left(\mathbf{I} - \frac{2h}{3}\mathbf{A} \right)^{-1} \frac{1}{3}(4\mathbf{y}_i - \mathbf{y}_{i-1})$$

or

$$\begin{bmatrix} Y_1 \\ Y_2 \end{bmatrix}_{i+1} = \frac{1}{3} \begin{bmatrix} 1 - 264h & -\dfrac{1592}{3}h \\ \dfrac{796}{3}h & 1 + 532h \end{bmatrix}^{-1} \left(4\begin{bmatrix} Y_1 \\ Y_2 \end{bmatrix}_i - \begin{bmatrix} Y_1 \\ Y_2 \end{bmatrix}_{i-1} \right)$$

In this example, the starting value $\mathbf{y}_1$ is obtained from the true solution. Then, the first two steps with $h = 0.1$ result in

$$Y_{1,1} = Y_1(0.1) = 1.3502382, \qquad Y_{2,1} = Y_2(0.1) = -0.6811432$$
$$Y_{2,1} = Y_1(0.2) = 1.0933155, \qquad Y_{2,2} = Y_2(0.2) = -0.5469481$$

The true errors due to implementing the implicit Euler (IE) and 2nd-order Backward Difference Formula (BDF2) and the true solutions are comparatively presented in Table 9.14. Note that the true solutions have two exponential components that decay at different rates ($\lambda = 2$ and $\lambda = 400$); hence, the stiffness ratio is 200. This implies that the system is fairly stiff. When the tabulated errors are examined, it is observed that the implicit Euler method yields, as expected, larger errors than the second-order BDF2 since it is a first-order method. The errors for both methods initially depict an increase, but then they recover and continue to decrease as x increases.

TABLE 9.14

x	True, $Y_{1,i}$	Abs. error, $|e_i|$ IE	BDF2	True, $Y_{2,i}$	Abs. Error, $|e_i|$ IE	BDF2
0	1			0		
0.1	1.637462	0.00481		-0.818731	0.00979	
0.2	1.340640	0.04765	0.00960	-0.670320	0.02353	0.01082
0.3	1.097623	0.05977	0.00431	-0.548812	0.02988	0.00186
0.4	0.898658	0.06585	0.00679	-0.449329	0.03292	0.00345
0.5	0.735759	0.06799	0.00777	-0.367879	0.03399	0.00389
0.6	0.602388	0.06741	0.00826	-0.301194	0.03370	0.00413
0.7	0.493194	0.06497	0.00834	-0.246597	0.03248	0.00417
0.8	0.403793	0.06134	0.00812	-0.201897	0.03067	0.00406
0.9	0.330598	0.05702	0.00771	-0.165299	0.02851	0.00385
1	0.270671	0.05234	0.00717	-0.135335	0.02617	0.00359

Discussion: It is possible to uncouple some ODEs, in which case each ODE can be solved by a method suitable to its own characteristics. However, when a system of ODEs cannot be uncoupled, the stiffness of the system becomes a critical issue. The implicit methods, particularly BDFs, are useful. However, when the IVP is nonlinear, the system of nonlinear implicit BDFs can be solved by Newton's method.

9.9 SIMULTANEOUS NONLINEAR ODES

As we have learned so far, explicit methods can be employed on any IVP to obtain the numerical solution, whether the IVP is linear or nonlinear. However, the stability criterion of the explicit schemes may be more restrictive in nonlinear and stiff IVPs.

In nonlinear IVPs, the implicit methods require the implementation of a root-finding algorithm. At each step, a system of nonlinear algebraic equations must be solved iteratively using a suitable method discussed in Chapter 4. Among them, Newton's method is a popular and efficient way to solve most nonlinear algebraic equations.

Consider Eq. (9.66), employing the implicit Euler scheme gives

$$\mathbf{y}_{i+1} = \mathbf{y}_i + h\,\mathbf{F}(x_{i+1}, \mathbf{y}_{i+1}) \tag{9.84}$$

for which the residual vector is defined as

$$\mathbf{R}(x_{i+1}, \mathbf{y}_{i+1}) = \mathbf{y}_{i+1} - \mathbf{y}_i - h\,\mathbf{F}(x_{i+1}, \mathbf{y}_{i+1}) \tag{9.85}$$

which constitutes a system of nonlinear algebraic equations. Expanding the residual into a two-term Taylor series about $\mathbf{y}_{i+1}^{(p)}$ leads to

$$\mathbf{R}(x_{i+1}, \mathbf{y}_{i+1}^{(p+1)}) \cong \mathbf{R}(x_{i+1}, \mathbf{y}_{i+1}^{(p)}) + \left(\mathbf{I} - h\frac{\partial \mathbf{F}}{\partial \mathbf{y}_{i+1}}(x_{i+1}, \mathbf{y}_{i+1}^{(p)})\right)(\mathbf{y}_{i+1}^{(p+1)} - \mathbf{y}_{i+1}^{(p)}) \tag{9.86}$$

where $\partial \mathbf{F}/\partial \mathbf{y}_{i+1}$ is the Jacobian matrix and the superscript "(p)" denotes the iteration step.

We set $\mathbf{R}(x_{i+1}, \mathbf{y}_{i+1}^{(p+1)}) = 0$, which will eventually be satisfied within a preset tolerance value when the system of equations converges. Then, using Eq. (9.85) and solving Eq. (9.86) for $\mathbf{y}_{i+1}^{(p+1)}$, we get

$$\mathbf{y}_{i+1}^{(p+1)} = \mathbf{y}_{i+1}^{(p)} - \left(\mathbf{I} - h\,\mathbf{J}(x_{i+1}, \mathbf{y}_{i+1}^{(p)})\right)^{-1}\left(\mathbf{y}_{i+1}^{(p)} - \mathbf{y}_i - h\,\mathbf{F}(x_{i+1}, \mathbf{y}_{i+1}^{(p)})\right) \tag{9.87}$$

where $\mathbf{y}_{i+1}$ and $\mathbf{y}_i$, respectively, denote the current and prior step estimates.

Equation (9.87) is iterated to compute the current step estimate, $\mathbf{y}_{i+1}$. However, as a word of caution, it should be noted that the Jacobian matrix must be updated at each iteration if it is not constant (i.e., elements being functions of x and/or $\mathbf{y}_i$'s). Once the convergence criterion, $\|\mathbf{y}_{i+1}^{(p+1)} - \mathbf{y}_{i+1}^{(p)}\| < \varepsilon$, is met, the numerical solution is advanced to the next step.

9.10   CLOSURE

This chapter is largely devoted to numerical methods for solving the model IVP: $y' = f(x, y)$ with $y(x_0) = y_0$. The explicit and implicit Euler methods, requiring the solution along the tangent lines, have first-order global errors and are the simplest of all numerical methods in this class of IVPs. However, these most elementary numerical methods are seldom used in practice.

Methods of practical importance are the high-order Runge-Kutta methods, the multi-step Adams-Bashforth method, and the Adams-Moulton method. The popularity of the fourth-order Runge Kutta method is due to its accuracy, stability, and ease of programming, besides being an explicit method. The Adams-Bashforth methods, which are a bit more complicated to integrate into programming, are also explicit methods that provide an alternative to the Runge-Kutta methods. On the other hand, Adams-Moulton methods, which are implicit, have smaller truncation errors (in comparison to Adams-Bashford methods) and require integration of root-finding algorithms into programming, which in general yields larger cpu-time.

The two-step schemes, predictor-corrector methods, require a suitable combination of a predictor (explicit) and a corrector (implicit) scheme to establish a method with better convergence characteristics. They also provide appealing algorithms for the numerical solution of IVPs due to the fact that they require relatively few function evaluations. The fourth-order predictor-corrector methods require two function evaluations per step, while the RK4 requires four function evaluations per step. However, in addition to the initial value, the starting values need to be supplied by another method to apply these methods.

When selecting and using a numerical method, "stability" is a major issue that needs to be addressed. A numerical method is either *unstable*, *conditionally stable*, or *unconditionally stable*. An unstable method tends to diverge, while stable methods, involving implicit schemes, always converge to an approximate solution. On the other hand, explicit methods are conditionally stable and may require a very small step size to satisfy the "convergence condition." Another problem that arises in the numerical solution of IVPs is the stiffness phenomenon. Stiff differential equations also require small step sizes, often smaller than what is required for accuracy. In this respect, the backward difference formulas (BDFs) are very efficient at handling stiff IVPs.

Almost all methods presented in this chapter involve a uniform step size throughout the solution interval. Adaptive methods allow the step size to be varied (as large as possible) in order to accommodate sharp or slow changes in the solution while retaining errors below a specified tolerance. Among all adaptive methods, the Runge-Kutta-Fehlberg method, which uses a fourth- and fifth-order explicit RK method (RKF45), is the most popular due to the ease of its use.

Many science and engineering problems involve two or more variables, which lead to a set of coupled ODEs. Also, high-order ODEs need to be reduced to a set of first-order simultaneous ODEs. Any one of the one-step or multi-step methods presented in this chapter can be applied to numerically solve a system of simultaneous first- or higher-order ODEs. The scalar quantities are replaced with vector notations.

9.11 EXERCISES

Section 9.1 Fundamental Problem

E9.1 For Exercises (a)-(g), determine (i) the dependent and independent variables, (ii) the order of the ODE, and (iii) the problem type as first- or second-order linear or nonlinear IVP.

(a) $\dfrac{dy}{dx} = \sin(xy)$, $\quad y(0) = 1$ $\qquad$ (b) $\dfrac{d^2r}{dt^2} = t^2\dfrac{dr}{dt} - (3t+1)r + 2$, $\quad r(0) = 0$, $\dfrac{dr}{dt}(0) = 0$

(c) $\dfrac{dx}{dt} = tx - \cos t$, $\quad y(1) = 0$ $\qquad$ (d) $\dfrac{d^3u}{dx^3} = xu\dfrac{du}{dx} - 1$, $\quad u(0) = -1$, $u'(0) = u''(0) = 0$

(e) $y^2\dfrac{d^2x}{dy^2} + 2y\dfrac{dx}{dy} + x = 0$, $\quad x(0) = 0$, $\dfrac{dx}{dy}(0) = 1$

(f) $\dfrac{d^4y}{dx^4} = \dfrac{dy}{dx}e^{-xy}$, $\quad y(0) = \frac{1}{2}$, $y'(0) = y''(0) = y'''(0) = 0$

(g) $\dfrac{d^2p}{dz^2} + p\left(\dfrac{dp}{dz}\right)^3 - zp^2 = 0$, $\quad p(1) = 0$, $\dfrac{dp}{dz}(1) = 0$

Section 9.2 One-Step Methods

E9.2 Solve the following IVPs for the given interval and initial conditions using the explicit Euler method, and compare your results with the true solutions.

(a) $y' = x(1+y)$, $\quad x_0 = 0$, $\quad y_0 = 0$, $\quad h = 0.1$, $[0,1]$; $\quad y_{\text{true}}(x) = -1 + e^{x^2/2}$

(b) $y' = 3x^2y^2$, $\quad x_0 = 0$, $\quad y_0 = 1/2$, $\quad h = 0.1$, $[0,1]$; $\quad y_{\text{true}}(x) = 1/(2 - x^3)$

(c) $y' + y^2\cos x = y$, $\quad x_0 = 0$, $\quad y_0 = 2/3$, $\quad h = \pi/20$, $[0,\pi/4]$; $y_{\text{true}}(x) = 2/(2e^{-x} + \sin x + \cos x)$

(d) $y' = y^2e^{-x}$, $\quad x_0 = 0$, $\quad y_0 = 1/2$, $\quad h = 0.2$, $[0,2]$; $\quad y_{\text{true}}(x) = e^x/(1 + e^x)$

(e) $y' = x/(1+y)$, $\quad x_0 = 0$, $\quad y_0 = 2$, $\quad h = 0.05$, $[0,0.5]$; $\quad y_{\text{true}}(x) = -1 + \sqrt{9 + x^2}$

E9.3 Solve the following IVPs for $h = 0.1$ and 0.01 using the explicit Euler method on $[0,1]$ and compare your results with the true solution for both cases.

$$\text{(a) } y' + 2y = x^2 y^3, \quad y(0) = 1, \quad y_{\text{true}}(x) = 4/\sqrt{8x^2 + 4x + 1 + 15e^{4x}}$$

$$\text{(b) } y' + y = 2/y^3, \quad y(0) = 1, \quad y_{\text{true}}(x) = (2 - e^{-4x})^{1/4}$$

$$\text{(c) } y' + 2xy = x\sqrt{y}, \quad y(0) = 1, \quad y_{\text{true}}(x) = (1 + e^{-x^2} + 2e^{-x^2/2})/4$$

$$\text{(d) } y' - y/(x+1) = y^2, \quad y(0) = -1, \quad y_{\text{true}}(x) = -2(x+1)/(x^2 + 2x + 2)$$

$$\text{(e) } y' + 3\,e^{y-x} = 0, \quad y(0) = 0, \quad y_{\text{true}}(x) = -\ln(4 - 3e^{-x})$$

$$\text{(f) } y' \cos y = xe^{-\sin y}, \quad y(0) = 0, \quad y_{\text{true}}(x) = \sin^{-1}(\ln(1 + x^2/2))$$

$$\text{(g) } (1 + x^2)y' = 2x\cos^2 y, \quad y(0) = 0, \quad y_{\text{true}}(x) = \tan^{-1}\left(\ln(1 + x^2)\right)$$

E9.4 Apply Taylor's method to the following first-order ODEs to obtain a third-degree polynomial approximation.

$$\text{(a) } y' = x + y, \quad \text{(b) } y' = x/y, \quad \text{(c) } y' = -xy^2, \quad \text{(d) } y' = x^3 - y, \quad \text{(e) } y' = e^{-xy}, \quad \text{(f) } y' = e^{\sin y}$$

E9.5 Repeat **E9.2** using Taylor's method: (i) obtain 2nd- and 3rd-degree polynomial approximations; (ii) estimate the solution of the IVPs; and (iii) compare your estimates with those of the true solutions.

E9.6 Apply Taylor's method to obtain a fourth-degree polynomial approximation to the following first-order ODEs:

$$\text{(a) } y' = 5x^5 e^{-y}, \quad \text{(b) } y' = 2x\cos^2 y/(1 + x^2), \quad \text{(c) } y' = e^{-\cos y}$$

E9.7 Find expressions for the upper error bound of the second-degree Taylor polynomial approximations of **E9.5**.

E9.8 The motion of a damped spring-mass system is described with the following second-order ODE:

$$m\frac{d^2y}{dt^2} + c\frac{dy}{dt} + ky(t) = 0$$

where c is the damping coefficient, k is the spring constant, and m is the mass. The spring is released from a point d cm above its equilibrium position, i.e., $y(0) = d$ and $y'(0) = 0$. Apply Taylor's method to find a fourth-degree polynomial approximation.

E9.9 Apply the implicit Euler method to **E9.2**. (a) Find an explicit expression for y_{i+1}; (b) obtain numerical estimates with $h = 0.1$ and 0.01; and (c) compare these estimates with the true solution.

E9.10 Solve the IVPs in **E9.2** using the trapezoidal rule and compare your estimates with the true solution. Use $\varepsilon = 10^{-6}$ as a tolerance value for convergence.

E9.11 Solve the IVPs in **E9.2** using the linearized implicit Euler method and compare your estimates with the true solution.

E9.12 Solve the IVPs in **E9.2** using the linearized trapezoidal rule and compare your estimates with the true solution.

E9.13 Solve the IVPs in **E9.2** using the midpoint rule and compare your estimates with the true solution.

E9.14 Solve the IVPs in **E9.2** using the modified Euler method and compare your estimates with the true solution.

E9.15 Solve the IVPs in **E9.3** using the implicit Euler method and compare your estimates with the true solution. Use $\varepsilon = 10^{-6}$.

E9.16 Solve the IVPs in **E9.3** using the trapezoidal rule and compare your estimates with the true solution. Use $\varepsilon = 10^{-6}$.

E9.17 Solve the IVPs given in **E9.3** using the midpoint rule and compare your estimates with the true solution.

E9.18 Solve the IVPs given in **E9.3** using the midpoint rule and compare your estimates with the true solution.

E9.19 Solve the IVPs in **E9.2** using the 3rd- and 4th-order Runge-Kutta schemes and compare your estimates with the true solution.

E9.20 Solve the IVPs in **E9.3** using the 3rd-order Runge-Kutta scheme and compare your estimates with the true solution.

E9.21 Water flows out of a discharge tube ($d = 5$ cm) located at the bottom of a square-section tank with a side $a = 2$ m; *see* **Fig. E9.21**. The change of water depth in the tank with time, $h(t)$, is described with the following first-order IVP:

$$-\frac{dh}{dt} = C\frac{\pi}{4}\left(\frac{d}{a}\right)^2 \sqrt{2gh}, \quad H(0) = 3.5\,\text{m}$$

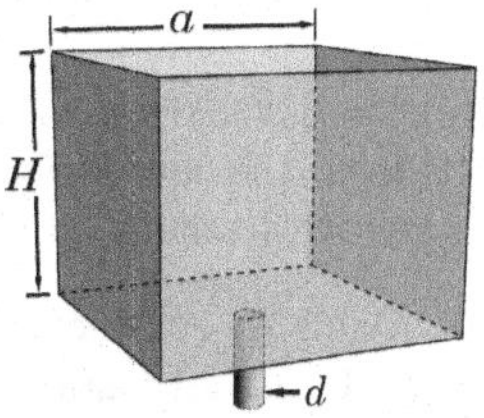

Fig. E9.21

where C is the discharge coefficient ($C = 0.58$), g is the acceleration of gravity (9.81 m/s^2), and d is the diameter of the discharge tube. Solve the given IVP with the 4th-order Runge-Kutta method using $\Delta t = 5$ seconds until the tank is completely empty. Plot the $h - t$ distribution. How long does it take to empty the tank?

E9.22 The following first-order ODE describes the current $I(t)$ passing an RL circuit:

$$\frac{dI}{dt} = \frac{E}{L}\sin\omega t - \frac{R}{L}I(t), \quad i(0) = 0$$

Note that, initially, no current flows through the circuit when it is off. For a circuit with $E = 110$ V, $L = 1$ H, $\omega = 250$, and $R = 50\ \Omega$, use the 4th-order Runge-Kutta method to estimate the circuit current at intervals of $\Delta t = 0.002$ up to $t = 0.2$ seconds.

E9.23 Water will be transported from a reservoir ($D \gg d$) to a destination much further away ($L \gg D$). When the valve at the end of the pipe is brought to a fully open position, the transient behavior of the water exit velocity $V(t)$ is described with the following IVP:

$$\frac{dV(t)}{dt} + \frac{1}{2L}V^2(t) = g\frac{h}{L}, \quad V(0) = 0$$

where g is the acceleration of gravity, h is the water elevation in the reservoir, and L is the length of the pipe. For $L = 10$ m and $h = 1.5$ m, use the 4th-order Runge-Kutta method and an increment of $\Delta t = 0.01$ second to estimate the time required for water exit velocity to reach its steady-state value of $V = \sqrt{2gh}$.

E9.24 A sack of a salt compound is poured into a large tank containing 200 liters of water. The salt concentration in the tank, $C(t)$ in percent as a function of time, is described by the following differential equation:

$$(120 - 5.55\,C)\frac{dC}{dt} = 0.235\,(48 - 4.95C)(23 - 1.2C), \quad C(0) = 0$$

Initially, no salt is present $C(0) = 0$; i.e., the tank is filled with water only. Using the RK4 method with $\Delta t = 1$ s, estimate the transient concentration of the solution until the solution is 99% saturated. Plot the $C - t$ distribution.

E9.25 The rate equation for a reaction involving two gases (A and B) is expressed by the following

ODE:

$$\frac{dx}{dt} = k(P_A - 2x)^2(P_B - x), \qquad x(0) = 0$$

where x is the partial pressure of the product, k is a constant given as 1.12×10^{-7} mm^{-2}s^{-2}, and P_A and P_B are the partial pressures of A and B, respectively. For $P_A = 359$ mmHg and $P_B = 400$ mmHg, estimate the partial pressure up to $t = 250$ s using the RK4 with $\Delta t = 1$ s.

E9.26 Corn from a storage tank is pneumatically fed into a cylindrical spray dryer with a radius of $R = 0.5$ m. A grain of corn enters the dryer through a tube at $r = 0$ (the center of the dryer) with a speed of 50 m/s, and it travels toward the outer wall of the dryer under the influence of the centrifugal forces. The governing differential equation for the speed V of a single grain of corn as a function of the radial position r is dictated by the following first-order non-linear differential equation:

$$V(r)\frac{dV}{dr} + A V^3(r) = B r, \qquad V(0) = 50$$

where $A = 0.000285$, $B = 5.96 \times 10^5$, and $0 \leqslant r \leqslant R$. Estimate the speed of the grain of corn as it travels from the center point $r = 0$ where it enters the dryer, until it hits the outer wall $(r = R)$, using the RK4 scheme with $\Delta r = 0.025$ m.

E9.27 A highly conducting plate $(A_s = 1$ m^2, $\rho = 800$ kg/m$^3)$ is initially at ambient temperature $T_\infty = 20°$C. At an instant $t = 0$, a constant heat flux $(q_0'' = 8000$ W/m$^2)$ is applied to one side of the plate, while heat loss takes place from the other side through convection. The first-order differential equation describing the mean transient temperature of the plate of thickness L is given by:

$$\frac{dT}{dt} + \frac{hA_s}{mc}(T(t) - T_\infty) = \frac{q_0'' A_s}{mc}$$

where $h = 40$ W/m^{2}K is the convection transfer coefficient, $c = 500$ J/kg·K is the specific heat, and $L = 1$ cm is the thickness of the plate. (a) Use the RK4 scheme to estimate the mean temperature of the plate in intervals of $\Delta t = 5$ s for up to 6 minutes. (b) How long does it take for the plate to reach $100°$C?

E9.28 An aluminum ball $(\rho = 2700$ kg/m^3, $c = 1035$ J/kg·K, $\varepsilon = 0.75)$ with a diameter of 5 cm is maintained in the annealing furnace at a uniform temperature of 800 K. At $t = 0$, it is removed from the furnace and placed in a room $(T_\infty = 300$ K, $h = 10$ W/m^{2}K$)$ to cool off. The heat balance, including the convection and radiation heat losses, yields the following nonlinear ODE:

$$mc\frac{dT}{dt} = -hA_s(T - T_\infty) - \varepsilon\sigma A_s(T^4 - T_\infty^4)$$

where $\sigma = 5.67 \times 10^{-8}$ W/m^{2}K^4 is the Stefan-Boltzmann constant, A_s is the surface area, m is the mass of the ball, c is the specific heat, and ε is the emissivity of aluminum. Use the RK4 with $\Delta t = 60$ s to estimate (a) the temperature variation of the ball for a period of 1 hour and (b) the time required for the sphere to reach 400 K.

E9.29 A parachutist $(m = 66$ kg$)$ deploys his parachute as soon as he jumps from a helicopter at 3500 meters altitude. The net equilibrium force acting on the parachutist gives the following differential equation:

$$F_{\text{net}} = m\frac{dv}{dt} = mg - C v$$

where v is the free fall speed, $g = 9.81$ m/s^2 is the acceleration of gravity, and $C = 13.2$ kg/s is the average drag coefficient experienced during the fall. Assuming that the speed of the parachutist just before he jumps is $v = 0$ (at $t = 0$), estimate the parachutist's terminal speed using the RK4 scheme with $\Delta t = 0.1$ s.

Section 9.3 Numerical Stability

E9.30 Consider the following difference equations for the model IVP: $y' = f(x, y) = \lambda y$. Find the order of the truncation error and determine the stability condition for each equation.

(a) $y_{i+2} = 4y_{i+1} - 3y_i - 2hf(x_i, y_i)$, (b) $y_{i+1} = \dfrac{1}{3}(4y_i - y_{i-1}) + \dfrac{2h}{3}f(x_{i+1}, y_{i+1})$

(c) $y_{i+2} = 2y_i - y_{i+1} + 3hf(x_i, y_i)$, (d) $y_{i+1} = y_i + hf(x_{i+1/2}, y_{i+1/2})$ (midpoint method)

(e) $y_{i+1} = y_i + \dfrac{h}{2}\{f(x_i, y_i) + f(x_i, y_i + hf(x_i, y_i))\}$ (Heun's method)

E9.31 Consider employing the RK4 method to get $y' = f(x, y) = \lambda y$. Find the amplification factor for the RK4 and investigate its stability.

E9.32 Find the analytical solution for each of the IVPs given below and discuss whether they are stiff or not.

(a) $y' = \lambda(y - \sin 2x) + 2\cos 2x$, $y(0) = 1$, (b) $y' = y - y^9$, $y(0) = 1/10$,

(c) $y' = \lambda(y - x^2) + 2x$, $y(0) = 1$, (d) $y' = 1 - 10^4 y^2$, $y(0) = 0$,

(e) $y' = -\lambda(y - e^{-x}) - e^{-x}$, $y(0) = 0$

Section 9.4 Multistep Methods

E9.33 Repeat **E9.2** with the Adams-Bashforth methods (AB2 and AB3) using $h = 0.1$. Calculate the starting values using the true solution, and compare your estimates with the true solution.

E9.34 Repeat **E9.3** with the Adams-Bashforth methods (AB2 and AB3) using $h = 0.1$. Calculate the starting values using the true solution, and compare your estimates with the true solution.

E9.35 A simplified heat transfer model for an isothermal cylinder leads to the following first-order nonlinear IVP:

$$\frac{d\Theta}{dt} + \lambda\Theta^{4/3} = 0, \quad \Theta(0) = \Theta_0 = T_0 - T_\infty$$

where $\Theta(t) = T(t) - T_\infty$ and $T(t)$, T_∞, and T_0 are respectively the cylinder, ambient, and initial temperatures, and λ is a constant. (a) Find the analytical solution of the IVP; (b) For $T_0 = 100°C$, $T_\infty = 15°C$, and $\lambda = 10^{-4}\,\mathrm{K}^{-1/3}/\mathrm{s}$, estimate the temperature over a 1-hour period with $\Delta t = 100$ s intervals using the AB2, AB3, and AB4 schemes; (c) Estimate the time required for the cylinder to cool to $30°C$. *Note:* Use the equivalent order Runge-Kutta method to obtain the starting values.

E9.36 Repeat **E9.35(b,c)** with the AM2, AM3, and AM4 schemes.

E9.37 When a spherically shaped glass particle is dropped into a non-Newtonian fluid, the equation of motion of the particle results in the following IVP:

$$\frac{du}{dt} + a(n)u^n - b = 0, \quad u(0) = u_0$$

where u is the vertical velocity of the particle (m/s), u_0 is the initial velocity, n is the flow behavior index, and a and b are fluid and particle-dependent constants. For a fluid with $n = 0.65$ ($a = 74.08$ and $b = 10.979$), the glass particle is released into the fluid with $u_0 = 10^{-3}$ m/s. Estimate its terminal velocity using the AB2, AB3, and AB4 schemes with $\Delta t = 2.5 \times 10^{-3}$s. *Note:* Use the equivalent order Runge-Kutta method to obtain the starting values.

E9.38 Repeat **E9.37** with the AM2, AM3, and AM4 schemes.

Section 9.5 Adaptive Step-Size Control

E9.39 Perform five steps of the adaptive RK45 method on the IVPs. Use $h_{min} = 0.01$, $h_{max} = 0.5$, $\varepsilon = 10^{-5}$, and $h = 0.2$ as the initial step size.

(a) $y' = y - xy^2$, $\quad y(0) = 1$, $\quad y_{true}(x) = (2e^{-x} + x - 1)^{-1}$
(b) $y' = (8 - 40x)y$, $\quad y(0) = 1$, $\quad y_{true}(x) = e^{8x-20x^2}$

E9.40 Given below is a simple neutron population model (*kinetic equation*) that takes into account delayed neutrons during a nuclear reactor start-up:

$$\frac{dn}{dt} = \frac{\rho(t) - \beta}{\Lambda} n(t) + \lambda C(t), \quad n(0) = 1 \quad \text{and} \quad C(t) = 4(e^{0.24t} - e^{-35t})$$

where Λ, $\rho(t)$, and β denote mean generation time, reactivity, and delayed neutron fraction, and $C(t)$ is delayed neutron population. For the case of $\beta = 0.007$, $\Lambda = 5 \times 10^{-5}$s, $\lambda = 0.08\,\text{s}^{-1}$, the kinetics equations are stiff. Use the adaptive RK45 scheme with $h = 0.1$ as the starting time step to obtain the first four steps of the numerical solution for $\rho = 0.75\beta$. The true solution is given as $n(t) = 0.992188\, e^{-35t} + 0.008513\, e^{0.24t}$. *Note:* Impose $h_{min} = 0.001$, $h_{max} = 0.1$, $\varepsilon = 10^{-5}$.

E9.41 An unsteady model for the mean temperature of a chemical reactor with reaction heat generation and convective cooling leads to

$$\frac{d\theta}{d\tau} = \alpha \exp\left(-\beta\tau - \frac{\gamma}{\theta(\tau)}\right) - \lambda\theta(\tau), \quad \theta(0) = 1$$

where $\alpha = 5$, $\gamma = 0.075$, and $\lambda = 2$ are the model constants, and $\theta(\tau)$ and τ denote dimensionless temperature and time, respectively. Use the adaptive RK45 scheme with $h = 0.05$ as the starting time step to obtain the first four steps of the numerical solution for (a) $\beta = 0.05$ and (b) $\beta = 0.5$. *Note:* Impose $h_{min} = 0.01$, $h_{max} = 0.25$, and $\varepsilon = 10^{-5}$.

E9.42 A mathematical model based on the net forces (gravity and drag) acting on a falling spherical raindrop can be expressed as

$$\frac{du}{dt} = g - \kappa u^2(t), \quad u(0) = 0, \quad \kappa = \frac{3}{4d}\frac{\rho_a}{\rho_w}C$$

where u is the velocity (m/s), $\rho_a = 1.2$ and $\rho_w = 1000$ kg/m^3 are respectively the densities of air and water, d is the raindrop diameter (m), $g = 9.81$ m/s^2 is the acceleration of gravity, and $C = 0.55$ is the drag coefficient. Use the adaptive RK45 scheme with the starting time step $h = 0.25$ to estimate the falling speed of a raindrop ($d = 5$ mm) for the first four time steps. The true solution is given as $u(t) = \sqrt{g/\kappa}\tanh(t\sqrt{\kappa g})$. *Note:* Impose $h_{min} = 0.01$, $h_{max} = 0.25$, and $\varepsilon = 10^{-5}$.

E9.43 The probability of failure of a machine part subjected to a cyclic load is given by

$$\frac{dp}{dt} = -ap(t)\ln\left(\frac{p(t)}{K}\right), \quad p(0) = p_0$$

where t is the operation time (hours), p is the probability (%), a and K are constants. For $p_0 = 10^{-5}$, $a = 1.25 \times 10^{-3}$, $K = 100$, apply the adaptive RK45 scheme with the starting time step $h = 200$ hours to estimate the first four steps. Impose $h_{min} = 1$, $h_{max} = 500$, and $\varepsilon = 10^{-5}$. The true solution is given as $p(t) = K(p_0/K)^{\exp(-at)}$

Section 9.6 Predictor-Corrector Methods

E9.44 Repeat **E9.2** using (i) Heun's, (ii) AMB4, and (iii) Milne's predictor-corrector methods and compare your estimates with the true solutions.

E9.45 Repeat **E9.3** using (i) Heun's, (ii) AMB4, and (iii) Milne's method predictor-corrector methods and compare your estimates with the true solutions.

E9.46 A simple RC circuit consists of a capacitor (C), a resistor (R), and a voltage source. $E(t)$ denotes the input voltage across the voltage source as a function of time. The voltage across the capacitor satisfies the following differential equation:

$$RC\frac{dV_c}{dt} + V_c(t) = E(t), \qquad V_c(0) = V_{c0}$$

The voltage is constant $E(t) = E_0$ for $0 \leqslant t \leqslant t_0$, but then the voltage source is turned off $E(t) = 0$ $(t \geqslant t_0)$. For $t_0 = 2$ s, $E_0 = 8$ V, $R = 2.5\,\Omega$, and $C = 0.125$ F, use (a) the Heun's and (b) the AMBF predictor-corrector methods with $\Delta t = 0.25$ s to estimate the transient capacitor voltage over a period of 5 s for $V_{c0} = 0$ and $V_{c0} = 10$ V cases. Compare your solutions with the true solutions given below and assess the accuracy of the methods.

For $V_{c0} = 0\,\text{V}$
$$V_c(t) = \begin{cases} 8(1 - e^{-3.2t}), & 0 \leqslant t \leqslant 2 \\ 4816.76e^{-3.2t}, & t \geqslant 2 \end{cases}$$

For $V_{c0} = 10\,\text{V}$
$$V_c(t) = \begin{cases} 8 + 2e^{-3.2t}, & 0 \leqslant t \leqslant 2 \\ 4806.76e^{-3.2t}, & t \geqslant 2 \end{cases}$$

E9.47 A simplified mathematical model of the spread of an epidemic, also known as the Kermack-McKendrick model, is given as follows:

$$\frac{dR}{dt} = \alpha\{P - R(t) - H(t)\}, \quad H(t) = H_0 \exp\left(-\frac{\beta}{\alpha}R(t)\right),$$

where α is the recovery rate of infected individuals, β is the rate of infection in healthy individuals, and H, R, I, and P denote the healthy, recovered, infected, and total number of individuals in the population, respectively. During the COVID-19 pandemic, in a country, no definite deaths are attributed to the pandemic at a time $t = 0$ and 143 per 10000 are infected, i.e., $R(0) = 0$ and $I(0) = 143$. For given $P = 10000$, $\alpha = 0.185/\text{day}$, $\beta = 3.25 \times 10^{-5}$ /(day.person), use (a) Heun's, (b) Adams' 4th, and (c) Milne's 4th-order methods to estimate the number of infected and recovered individuals at 1-week intervals ($\Delta t = 7$ days) for a 3-month period. Use the RK4 scheme to obtain the starting values. *Note*: At any time, the total number of healthy, recovered/deceased, and infected individuals is constant over the simulation period, i.e., $I(t) + H(t) + R(t) = P$.

E9.48 *Air quenching* is the process of attaining the desired properties in a product by cooling metal objects in an air or inert gas environment. After reaching a uniform temperature of 1000K in an annealing furnace, a steel ball $(D = 6$ cm, $m = 0.89$ kg) is removed from the furnace $(t = 0)$ and left to cool in the ambient air. Assuming that the heat loss is primarily due to the convection process, the conservation of energy yields

$$m\frac{d}{dt}(c_p(T)T(t)) = -UA(T(t) - T_\infty), \qquad T(0) = 1000\,K$$

where T, m, and A are respectively the temperature (K), mass (kg), and surface area (m^2) of the ball, $c_p(T)$ is the temperature-dependent specific heat (J/kg.K), $T_\infty = 290$ K is the ambient temperature, and $U = 90$ W/m^2·K is the overall heat transfer coefficient. Use Heun's, Adam's, and Milne's methods with $\Delta t = 1$ minute to estimate the transient temperature change of the steel ball during the cooling. How long does it take for the temperature of the ball to reach 310 K? Obtain the starting values using the RK4 method. *Given*: $c_p(T) = 80.7 + 0.945\,T(t)$ J/kg·K.

E9.49 Repeat **E9.41** up to $\tau = 2.5$ using Heun's, AB4, and Milne's 4th-order methods.

E9.50 A spherical tank of radius R has a drainage tap of radius r at its bottom (*see* **Fig. E9.50**). The tank has an inlet tap at the top, which maintains atmospheric pressure in the tank at all times. The mathematical model that gives the liquid level in a spherical tank that discharges its liquid is defined by the following first-order IVP:

$$\frac{dh}{dt} = -\frac{A\sqrt{2gh}}{\sqrt{\pi^2 h^2 (h - 2R)^2 - A^2}}, \qquad h(0) = h_0$$

Fig. E9.50

where A is the cross-sectional area of the drainage tap. The drainage velocity can be computed from

$$v = \sqrt{2gh + (dh/dt)^2}$$

The tank ($R = 3$ m) is filled up to 99% of its height (i.e., $h_0 = 2R(0.99)$), which will be fully drained by opening the discharge valve. Estimate (a) the liquid level and the exit (drainage) velocity until the tank is fully drained; (b) the elapsed time to drain the tank using the Heun's predictor-corrector method with $\Delta t = 1$, 5, and 10 s; (c) discuss the effects of changing the time steps.

Section 9.7 System of First-Order IVPs (ODEs)

E9.51 Consider the RLC circuit shown in **Fig. E9.51**. The current I through the inductor and the voltage V across the capacitor C satisfy the following coupled first-order ODEs:

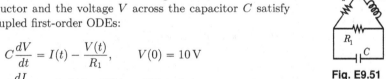

$$C\frac{dV}{dt} = I(t) - \frac{V(t)}{R_1}, \qquad V(0) = 10\,\text{V}$$

$$L\frac{dI}{dt} = -R_2 I(t) - V(t), \qquad I(0) = 2\,\text{A}$$

Fig. E9.51

(a) Express the IVP in matrix form; (b) use the explicit Euler method with $\Delta t = 1$ and 0.5 s to estimate the voltage and current for a period of 4 s; (c) compare and discuss the accuracy of the numerical estimates. The true solution of the system is given as $I(t) = (2 + \alpha_1)\,e^{-\lambda_1 t} - \alpha_1\,e^{-\lambda_2 t}$, $V(t) = -\alpha_2\,e^{-\lambda_1 t} + (10 + \alpha_2)\,e^{-\lambda_2 t}$. *Given*: $R_1 = 10\,\Omega$, $R_2 = 5\,\Omega$, $L = 2\,\text{H}$, $C = 50\,\text{F}$, $\lambda_1 = 2.49599$, $\lambda_2 = 0.00601$, $\alpha_1 = 2.0112683$, and $\alpha_2 = 0.032168$.

E9.52 Consider the following consecutive reaction, in which the first step is reversible.

$$X \underset{k_{-1}}{\overset{k_1}{\rightleftharpoons}} Y \overset{k_2}{\longrightarrow} Z$$

The reaction rates constitute a coupled first-order IVP, which can be expressed as follows:

$$\left\{ \begin{array}{l} \dfrac{dX}{dt} = -k_1 X + k_{-1} Y \\ \dfrac{dY}{dt} = k_1 X - (k_{-1} + k_2) Y \end{array} \right\}, \qquad \begin{array}{l} X(0) = 50\,\text{M}, \\ Y(0) = \ \ 0\ \text{M}. \end{array}$$

where $k_1 = 1$, $k_{-1} = 0.5$, and $k_2 = 0.1\,\text{s}^{-1}$. (a) Express the IVP in matrix form; (b) Use the explicit Euler method with $\Delta t = 1$ and 0.5 s to estimate X and Y concentrations for a period of 4 s; and (c) compare and discuss the accuracy of your numerical estimates. *Given*: The true solution is $X(t) = 31.805\,e^{-\alpha_1 t} + 18.195\,e^{-\alpha_2 t}$, $Y(t) = 34.02\left(-e^{-\alpha_1 t} + e^{-\alpha_2 t}\right)$, where $\alpha_1 = 1.53485$ and $\alpha_2 = 0.065153$.

E9.53 Two bodies (m_1 and m_2) of equal size, with heat capacities c_1 and c_2, are brought into thermal contact and suspended in a room with a convection transfer coefficient h and ambient temperature $T_\infty = 5°$ C. The area and conductivity of the contact surface are denoted by A_c and *k.* The conservation of energy leads to the following first-order coupled ODEs (IVP).

$$m_1 c_1 \frac{dT_1}{dt} = -\frac{kA_c}{L}(T_1 - T_2) - hA(T_1 - T_\infty), \qquad T_1(0) = 100°C$$

$$m_2 c_2 \frac{dT_2}{dt} = -\frac{kA_c}{L}(T_2 - T_1) - hA(T_2 - T_\infty), \qquad T_2(0) = -10°C$$

where L is the thickness of the contact material, and T_1 and T_2 are the bulk temperatures of the bodies. (a) Express the IVP in matrix form; (b) use the explicit Euler method $\Delta t = 1$ and 0.5 s to estimate the temperatures T_1 and T_2 for a period of 4 seconds if $m_1 c_1 = 100$ J/K, $m_2 c_2 = 120$ J/K, $kA_c/L = 5$ W/K, and $hA = 30$ W/K; (c) Compare and discuss the accuracy of the numerical estimates. The set of true solutions is given as

$$T_1(t) = 5 + 80\,e^{-\alpha_1 t} + 15\,e^{-\alpha_2 t} \quad \text{and} \quad T_2(t) = 5 - 40\,e^{-\alpha_1 t} + 25\,e^{-\alpha_2 t}$$

where $\alpha_1 = 3/8$, and $\alpha_2 = 4/15$.

E9.54 Repeat **E9.51**; (a) obtain 2nd and 3rd-degree Taylor polynomial approximations, and (b) estimate X and Y up to $t = 4$ s using $\Delta t = 1$ and 0.5 s.

E9.55 Repeat **E9.52**; (a) obtain 2nd and 3rd-degree Taylor polynomial approximations; (b) estimate V and I up to $t = 4$ s using $\Delta t = 1$ and 0.5 s.

E9.56 Repeat **E9.53**; (a) obtain 2nd and 3rd-degree Taylor polynomial approximations; (b) estimate T_1 and T_2 temperatures up to $t = 4$ s using $\Delta t = 1$ and 0.5 s.

E9.57 Use a Taylor polynomial approximation to derive a third-order explicit scheme to solve the given coupled first-order ODEs.

$$\frac{dY_1}{dt} = -2Y_1 - Y_2 + x + 1, \qquad \frac{dY_2}{dt} = Y_1 + 2Y_2 + 3 - x$$

E9.58 Below are coupled first-order linear IVPs with constant coefficients. (a) Express the system in matrix form; (b) use the RK4 and BDF4 schemes to estimate the solution with the specified step size and specified range; (c) compare your estimates with the true solutions and discuss the accuracy of the methods. *Note:* Use the true solution to find the required starting values.

(a) $Y_1' = -2Y_1 - 3Y_2 + 9e^x, \quad Y_1(0) = -10 \quad Y_{1,\text{true}}(x) = 2e^x - 12e^{-x}, \quad h = 0.2\ [0,2]$
$\quad\ Y_2' = -Y_1 + 3Y_2 - 28e^{-x}, \quad Y_2(0) = 5, \quad Y_{2,\text{true}}(x) = e^x + 4e^{-x}$

(b) $Y_1' = Y_1 - 2Y_2 + 31x^2 - 18x, \quad Y_1(0) = -2, \quad h = 0.1\ [0,1]$
$\quad\ Y_2' = -3Y_1 + 6Y_2 + 28x^3 + 62x - 10, \quad Y_2(0) = -1,$
$\quad\ Y_{1,\text{true}}(x) = 2x^4 + 10x^3 + x^2 - 2, \quad Y_{2,\text{true}}(x) = x^4 + x^3 + x^2 - 10x - 1$

(c) $Y_1' = -2Y_1 + 2Y_2 - 8x, \quad Y_1(0) = 2, \quad h = 0.1\ [0,1], \quad Y_{1,\text{true}}(x) = 3 - 2x - 2e^{-3x} + e^{2x},$
$\quad\ Y_2' = 2Y_1 + Y_2 + 2x - 6, \quad Y_2(0) = 5, \quad Y_{2,\text{true}}(x) = 2 + 2x + e^{-3x} + 2e^{2x}$

(d) $Y_1' = -Y_1 - Y_2 + (x^2 + 1)e^{-x}, \quad Y_1(0) = 1, \quad h = 0.1\ [0,1], \quad Y_{1,\text{true}}(x) = xe^{-x} + e^{-2x}$
$\quad\ Y_2' = -Y_1 - Y_2 + 3xe^{-x} \quad Y_2(0) = 1, \quad Y_{2,\text{true}}(x) = x^2 e^{-x} + e^{-2x}$

(e) $Y_1' = -Y_1 + Y_2 + \left(1 - \frac{1}{2}x^2\right)e^{-x}, \quad Y_1(0) = \frac{1}{4}, \quad h = 0.1\ [0,1], \quad Y_{1,\text{true}}(x) = e^{-x}x + \frac{1}{4}e^x$
$\quad\ Y_2' = Y_1 - Y_2 + \frac{3}{4}e^x, \quad Y_2(0) = \frac{1}{2}, \quad Y_{2,\text{true}}(x) = \frac{1}{2}(x^2 e^{-x} + e^x)$

(f) $Y_1' = -Y_1(x) + 8 - e^{-x}, \quad Y_1(0) = 2, \quad h = 0.1\ [0,1], \quad Y_{1,\text{true}}(x) = 8 - (x + 6)e^{-x}$
$\quad\ Y_2' = -Y_1(t) - Y_2(t) + e^{-x} - 2, \quad Y_2(0) = 0, \quad Y_{2,\text{true}}(x) = \frac{1}{2}(x^2 + 14x + 20)e^{-x} - 10$

E9.59 Below are coupled first-order linear IVPs with variable coefficients. (a) Express the system in matrix form; (b) use the RK4 and BDF4 schemes to estimate the solution with the specified step size and specified range; (c) compare your estimates with the true solutions and discuss the accuracy of the methods. *Note:* Use the true solution to find the required starting values.

(a) $Y_1' = xY_1 - Y_2 - x^3$, $\quad Y_1(0) = 3$, $\quad h = 0.1$ $[0,1]$, $\quad Y_{1,\text{true}}(x) = x^3 + x^2 + 3x + 3$

$\quad\ \ Y_2' = 4Y_1 - 4xY_2 + 4x^5 - 24x - 11$, $\quad Y_2(0) = -3$, $\quad Y_{2,\text{true}}(x) = x^4 + x - 3$

(b) $Y_1' = -xY_1 + Y_2 + 3$, $\quad Y_1(0) = 2$, $\quad h = 0.2$ $[0,2]$, $\quad Y_{1,\text{true}}(x) = 2 + 3x$

$\quad\ \ Y_2' = (x^2 - 2)Y_1 - xY_2 + 12x + 6$, $\quad Y_2(0) = 0$, $\quad Y_{2,\text{true}}(x) = 2x + 3x^2$

(c) $Y_1' = (1 + e^{-x})Y_1 - e^x Y_2 + e^x$, $\quad Y_1(0) = 0$, $\quad h = 0.1$ $[0,1]$, $\quad Y_{1,\text{true}}(x) = xe^x$

$\quad\ \ Y_2' = e^{-x}Y_1 - Y_2 + e^{-x} - x$, $\quad Y_2(0) = 0$, $\quad Y_{2,\text{true}}(x) = xe^{-x}$

(d) $Y_1' = -5Y_1 - 2xY_2 + 8e^{-x}$, $\quad Y_1(0) = 2$, $\quad h = 0.1$ $[0,1]$, $\quad Y_{1,\text{true}}(x) = (2 - x^2)e^{-x}$

$\quad\ \ Y_2' = -4Y_1 - 2xY_2 + 9e^{-x}$ $\quad Y_2(0) = 1$, $\quad Y_{2,\text{true}}(x) = (2x + 1)e^{-x}$

E9.60 One-delayed group kinetics equations for a nuclear reactor are given as

$$\frac{dn}{dt} = \frac{\rho(t) - \beta}{\Lambda}n(t) + \lambda C(t), \qquad n(0) = n_0$$
$$\frac{dC}{dt} = \frac{\beta}{\Lambda}n(t) - \lambda C(t). \qquad\qquad C(0) = C_0 = \frac{\beta}{\lambda\Lambda}n_0$$

where $n(t)$ is the neutron density, $C(t)$ is the delayed neutron concentration, Λ is the mean generation time, $\rho(t)$ is called the reactivity, and β is the delayed neutron fraction. The reactor initially operates with a steady neutron population of n_0, and a step reactivity change of $\rho(t) = \rho_0$ =constant is introduced at $t = 0$. Having normalized the neutron density ($n_0 = 1$), solve the kinetics equations to estimate n and C for a period of 1 second using the RK2 and RK4 with intervals $\Delta t = 0.1$, 0.05, and 0.025 s for the input of $\rho_0 = 0.75\beta$. The U-235 data are $\beta = 0.007$, $\Lambda = 5 \times 10^{-5}$ s, and $\lambda = 0.08$ s^{-1}. The true solutions are given as

$$n(t) = 3.946425\,e^{\alpha_1 t} - 2.946425\,e^{-\alpha_2 t} \quad C(t) = 1738.294\,e^{\alpha_1 t} + 11.70615\,e^{-\alpha_2 t}$$

where $\alpha_1 = 0.23784$ and $\alpha_2 = 35.31784$.

E9.61 Repeat **E9.60** using the same data using the AB4 with $\Delta t = 0.025$ s for (a) step insertions of $\rho(t) = 0.25\beta$, 0.50β, 0.75β, and (b) ramp insertions of $\rho(t) = 0.25\beta t$, $0.50\beta t$, and $0.75\beta t$. Make xy-plots for each case and determine the time for neutron density to reach $n(t)/n_0 = 10^3$.

E9.62 A simple mathematical (linear) model for simulating a car crush (*see* **Fig. E9.62**) is given below:

$$\frac{dv}{dt} = -\frac{k}{m}x(t) - \frac{b}{m}v(t) \qquad x(0) = 0$$
$$\frac{dx}{dt} = v(t) \qquad\qquad\quad v(0) = v_0$$

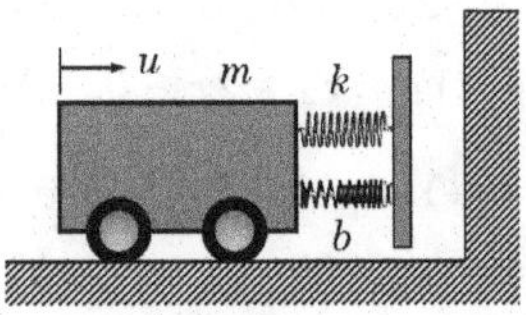

Fig. E9.62

where x denotes the displacement of the car and the deformation of the spring, v is the velocity of the car and the relative velocity across the damper, b is the viscous damping coefficient, and k is the stiffness given as $k = 1206$ N/m. Initially, the car ($m = 890$ kg) is moving at a constant speed of $v_0 = 50$ km/h. (a) Obtain the analytical solution of the model. (b) For over a period of 0.2 s, simulate the car crash for (i) undamped ($b = 0$) and (ii) damped ($b = 25000$ kg/s) cases using Heun's predictor-corrector method; plot the displacement ($x(t)$) and deceleration ($a =$

$(kx + bv)/m)$ of the car and the total force $(F = ma = kx + bv)$ acting on the barrier as a function of time; and (c) compare your estimates with the true solution you found in Part (a).

E9.63 Use the RK4 with $\Delta t = 0.2$ to estimate (a) the altitude of the parachutist in **E9.29** during the first 27 seconds of descent and (b) the time elapsed to land on the ground.

Section 9.8 High-Order ODEs

E9.64 The ODEs given below are second-order IVPs. (a) Convert the IVPs into a set of coupled first-order IVPs, and (b) use the RK4 method with $h = 0.1$ to estimate the solution.

(a) $y'' + 25y' + 25y = 0$, $y(0) = 0$, $y'(0) = 1$ $[0,3]$, $y_{true}(x) = \frac{1}{24}(e^{-x} - e^{-25x})$

(b) $y'' + xy' - 2y = 0$, $y(0) = 1$, $y'(0) = 0$, $[0,1]$, $y_{true}(x) = 1 + x^2$

(c) $yy'' + (y')^2 = 3x$, $y(1) = 1$, $y'(1) = 3/2$, $[1,2]$, $y_{true}(x) = x^{3/2}$

(d) $(y')^2 - y'' - 4y + 1 = 0$, $y(0) = 0$, $y'(0) = -1$, $[0,1]$, $y_{true}(x) = x^2 - x$

(e) $x\cos^2(1/x)y'' + 2\cos^2(y)y' = 0$, $y(1) = 1$, $y'(1) = -1$ $[1,2]$, $y_{true}(x) = 1/x$

(f) $y'' - 2xyy' - y^2 + 1 = 0$, $y(0) = 0$, $y'(0) = 0$, $[0,1]$, $y_{true}(x) = (1 - e^{x^2})(1 + e^{x^2})$

E9.65 Consider a mass m attached to the end of a pendulum (*see* **Fig. E9.65**). Neglecting the air resistance, the motion of the pendulum is governed by the following second-order nonlinear differential equation:

$$\frac{d^2\theta}{dt^2} = -\frac{g}{L}\sin\theta, \qquad \theta(0) = \theta_0, \qquad \frac{d\theta(0)}{dt} = \omega_0$$

where L, ω_0, and $\theta(t)$ denote the length, initial angular velocity, and angular displacement, respectively, and g is the acceleration of gravity. Derive an explicit third-order scheme using the Taylor polynomial approximation.

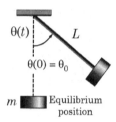

Fig. E9.65

E9.66 Consider the motion of the pendulum in **E9.65**, where $L = 0.25$ m, $\theta(0) = \pi/4$, and $\omega(0) = 0$. Using the RK4 method with $\Delta t = 0.1$ s increments, (a) estimate the angular velocity and displacement of the pendulum for a period of 2 seconds (tabulate the results as $\theta - t$ and $\theta' - t$), and (b) find the period of the pendulum using the results found in Part (a) and compare it with the theoretical value found from $T = 2\pi\sqrt{L/g}$.

E9.67 The charge, q, in a simple RLC circuit is governed by the following differential equation:

$$\frac{d^2q}{dt^2} + \frac{R}{L}\frac{dq}{dt} + \frac{1}{LC}q = \frac{1}{L}E(t), \qquad q(0) = 2, \quad i(0) = C\frac{dq}{dt} = \frac{1}{3}$$

where $L = 1$ is the inductor, C is the capacitor, R is the resistor, and E is the potential. For a circuit with $L = 1$ H, $C = 1/9$ F, $R = 10\,\Omega$, and $E(t) = 2\sin(t/2)$ V, estimate the charge and the current using the RK4 method with $\Delta t = 0.25$ s for a period of 9 s after the circuit is turned on. Tabulate the results as $q - t$ and $dq/dt - t$.

E9.68 The motion of a nonlinear spring is governed by the so-called Duffing equation.

$$\frac{d^2x}{dt^2} + \alpha\frac{dx}{dt} + \beta x(t) + \gamma x^3(t) = F_0\cos\omega t, \qquad x(0) = A, \quad x'(0) = B$$

where x is the displacement. For $\alpha = 1$, $\beta = \gamma = 20$, $F_0 = 0$, $A = 0.1$, and $B = 0$, use the RK4 method with $\Delta t = 0.1$ s to solve the Duffing equation up to $t = 9$ s and tabulate the results as $x - t$ and $V - t$.

E9.69 The motion of a nonlinear, damped Van der Pol oscillator is given by the following second-order nonlinear differential equation:

$$\frac{d^2x}{dt^2} - \mu(1-x^2)\frac{dx}{dt} + x = 0, \quad x(0) = 0, \quad x'(0) = V(0) = v$$

where x is the displacement, μ is a damping coefficient, and V is the velocity. For $\mu = 2$ and $v = 0.5$, solve the Van der Pol equation using the RK4 method with $\Delta t = 0.2$ s for a period of 30 s and tabulate the estimates as $x - t$ and $V - t$.

Section 9.9 Simultaneous Nonlinear ODEs

E9.70 Below are the coupled first-order nonlinear ODEs with variable coefficients. Use the RK4 scheme to estimate the solution for the specified interval and step size.

(a) $Y_1' = x + \sqrt{Y_1^2 + Y_2^2}, \quad Y_2' = Y_1(1 + Y_1), \quad Y_1(0) = 0, \quad Y_2(0) = 0, \quad h = 0.2 \ [0,2]$

(b) $Y_1' = Y_1 e^{-xY_2} + 1, \quad Y_2' = Y_1 \sin x - Y_2 \cos x, \quad Y_1(0) = 0, \quad Y_1(0) = 1, \quad h = 0.2 \ [0,2]$

(c) $Y_1' = Y_1 \sin(2x + Y_1 Y_2), \quad Y_2' = \pi - Y_2 \cos(x - Y_1 Y_2), \quad Y_1(1) = 1, \quad Y_2(1) = -1, \quad h = 0.1 \ [1,2]$

(d) $Y_1' = \sin(Y_1 + x) + \sin(Y_2 + x), \quad Y_2' = \cos(Y_1 + x) + \cos(Y_2 + x)$
$\quad\quad Y_1(0.5) = 0.25, \quad Y_2(0.5) = 0.5, \quad h = 0.2 \ [0.5, 2.5]$

E9.71 The following simple two-variable (Healthy and Infected) population model has been proposed to simulate the COVID-19 outbreak in a small country:

$$\left\{ \begin{array}{l} \dfrac{dH}{dt} = -0.009H(t)I(t), \\[2mm] \dfrac{dI}{dt} = 0.009H(t)I(t) - 0.22I(t) \end{array} \right\} \quad \begin{array}{l} H(0) = 99.9 \ (99.9\%) \\[2mm] I(0) = 0.1 \ (0.10\%) \end{array}$$

where H and I denote the percentage of the healthy and infected populations, and t denotes time (in days). Assuming that only one in a thousand is infected at the start of the simulation ($t = 0$), forecast the healthy and infected populations over a 30-day period. Use the Euler, RK2, and RK4 methods with time steps of $\Delta t = 1$ day.

E9.72 Solve the given second-order initial value problem numerically on $0 < x \leqslant 3$ using the fourth-order Runge-Kutta scheme with a step size of $h = 0.2$.

$$y'' - y\,y' = 0.1\,x^2, \quad y(0) = 1, \quad y'(0) = 0$$

E9.73 Solve the given second-order IVP on $0 < x \leqslant 2$ using the fourth-order Runge-Kutta scheme with a step size of $h = 0.2$.

$$Y_1' = 1 + Y_1(1 - Y_1 Y_2), \quad Y_2' = e^{-x} - Y_2(Y_1 + 2), \quad Y_1(0) = 1, \quad Y_2(0) = 1$$

9.12 COMPUTER ASSIGNMENTS

CA9.1 Write a computer program to solve the IVP given in **E9.57**, subject to $Y_1(0) = 0$ and $Y_2(0) = 0$ conditions using the 2nd and 3rd-degree Taylor polynomial approximations with $h = 0.1$

increments up to $x = 2$, and compare your estimates with the true solution given below.

$$Y_{1,\text{true}}(x) = \frac{1}{9}(2\sqrt{3}\sinh(\sqrt{3}x) - 12\cosh(\sqrt{3}x) + 3x + 12),$$

$$Y_{2,\text{true}}(x) = \frac{1}{9}(8\sqrt{3}\sinh(\sqrt{3}x) + 18\cosh(\sqrt{3}x) + 3x - 18)$$

CA9.2 Consider a cylindrical tank ($H = D = 4$ m) with a discharge pipe of diameter $d = 5$ cm, as shown in **Fig. CA9.2**. The tank is initially empty and is filled with water at a rate of 1.8 m^3/min. The water level in the tank, h, is governed by the following first-order nonlinear differential equation:

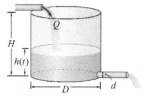

Fig. CA9.2

$$\frac{dh}{dt} = \frac{Q}{A_t} - \left(\frac{d}{D}\right)^2\sqrt{2g\,h(t)}, \qquad h(0) = 0$$

where $A_t = \pi D^2/4$ is the tank cross section, Q is the rate of inflow water (m^3/s), and g is the acceleration of gravity. Write a program using the RK4 method ($\Delta t = 0.5$ min) (a) to estimate the filling time, instantaneous water level, and discharge rate until the water level reaches 95% H. Once 95% H is reached, turn off the filling tab and let the water drain from the tank. How long does it take to discharge water completely from the tank?

CA9.3 The air-quenching problem presented in **E9.48** is extended to include radiation heat losses. At $t = 0$, a steel ball ($D = 6$ cm, $m = 0.89$ kg) with a uniform temperature of 1000 K is removed from the furnace and left to cool in the ambient air. The specific heat and convection heat transfer coefficients are given as quadratic functions of temperature. The conservation of energy, including radiation and convection heat losses, yields

$$m\frac{d}{dt}\left(c_p(T)T(t)\right) = -U(T_f)A\left(T(t) - T_\infty\right) - \sigma A\left(T^4(t) - T_\infty^4\right), \qquad T(0) = 1000\,K$$

where T, m, and A are respectively the temperature (K), mass (kg), and surface area (m^2) of the ball, $c_p(T)$ is the temperature-dependent specific heat (J/kg.K), T_f is the film temperature computed as $(T(t) + T_\infty)/2$, T_∞ is the ambient temperature (290K), and U is the overall heat transfer coefficient (W/m^2·K). Use the 4th-order Adam's method with $\Delta t = 60$ s (1 min) to simulate the cooling of the steel ball. Determine the following: (a) How long does it take for the ball to reach 310 K? Plot the cooling curve (temperature vs. time), (b) compute and plot the instantaneous convection and radiation heat transfer rates ($q_{conv}(t) = U(T_f)A(T(t) - T_\infty)$ and $q_{rad}(t) = \sigma A\left(T^4(t) - T_\infty^4\right)$), (c) obtain the cooling curve for the constant property assumption (with mean values of $\bar{c}_p = 644$ J/kg·K and $\bar{U} = 202$ W/m^2·K) and compare the results with those in Part (a).

Given: $c_p(T) = 769 - 1.589T + 0.00196T^2$ J/kg·K and $U(T) = 80 + 0.2226T - 4.9 \times 10^{-5}T^2$, W/m^2·K.

CA9.4 Consider the following coupled first-order system of nonlinear equations:

$$Y_1' = cY_2(x), \quad Y_2' = c\left(1 - Y_1(x)\right), \qquad Y_1(0) = 0, \quad Y_2(0) = 0$$

where c is a constant. Use the trapezoidal rule, RK4, BDF2, BDF4, and AM4 methods to estimate the solutions for $c = 10$, 10^2, 10^3, and 10^3 on $[0,0.1]$ and discuss the accuracy of the solutions. The true solutions are $Y_1(x) = 1 - \cos(cx)$ and $Y_2(x) = \sin(cx)$.

CA9.5 A simple three-variable (healthy, infected, and recovered) mathematical model for the COVID-19 epidemic can be expressed as

$$dH/dt = -0.0061H(t)I(t) + 0.0014R(t), \quad H(0) = 99.9,\ (99.9\%)$$

$$dI/dt = 0.0061H(t)I(t) - 0.44H(t) \qquad\qquad I(0) = 0.1\ (0.10\%)$$

$$dR/dt = 0.44I(t) - 0.0014R(t) \qquad\qquad\quad R(0) = 0.$$

where t is the time (days), H, I, and R denote the percentage of the Healthy, Infected & deceased, and infected & Recovered population, respectively. Assuming that only one in a thousand is infected, use the implicit Euler's, RK2, and RK4 methods with time steps of $\Delta t = 1$ day to forecast the healthy, infected-deceased, and infected-recovered populations over a period of 120 days.

CA9.6 Consider a bidirectional $2X + Y_2 \rightleftarrows 2Z$ reaction, which may be represented by a couple of uni-directional (forward and backward) reactions:

$$2X + Y_2 \xrightarrow{k_1} 2Z \qquad 2Z \xrightarrow{k_2} 2X + Y_2$$

where $k_1 = 0.5$ and $k_2 = 0.1$ s^{-1} are the reaction rate constants. This leads to the following set of non-linear ordinary differential equations:

$$dC_x/dt = 2k_2 C_z^2(t) - 2k_1 C_x^2(t)C_y(t), \quad C_x(0) = 1$$
$$dC_y/dt = k_2 C_z^2(t) - k_1 C_x^2(t)C_y(t), \qquad C_y(0) = 1$$
$$dC_z/dt = 2k_1 C_x^2(t)C_y(t) - 2k_2 C_z^2(t), \quad C_{xy}(0) = 0$$

where C_x, C_y, and C_z are the concentrations of X, Y, and Z, respectively. Perform a numerical simulation of the chemical process for a period of 15 seconds. (a) Use a suitable numerical method to solve the pertinent equations and obtain the C_x, C_y, and $C_z - t$ distributions. (b) Investigate the effect of k_1 using $k_1 = 0.001, 0.5$, and 250 1/M.s. (c) Investigate the effect of k_2 using ($k_2 = 0.001$, 0.1, and 1000).

CA9.7 The mass of an engine block (*see* **Fig. CA9.7**), whose stiffness and damping coefficients are given respectively as $k = 2.714 \times 10^6$ N/m and $b = 325$ N/m·s, is 110 kg. Upon ignition, the block is subjected to an input force of $F(t) = F_0 \sin \omega t$ with $F_0 = 3\pi$ N and $\omega = \omega_0 = 50\pi$ rad/s. The displacement $y(t)$ of the engine block relative to its equilibrium position is expressed by the following second-order differential equation:

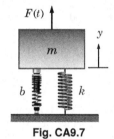

Fig. CA9.7

$$m\frac{d^2y}{dt^2} + b\frac{dy}{dt} + ky(t) = F(t)$$

Using the given data and the AB4 predictor-corrector method to estimate the displacement y, vertical velocity of the block $v = dy/dt$, power input $P_{\text{input}} = F(t)v(t)$, power dissipated $P_{\text{diss}} = bv^2(t)$, and power stored $P_{\text{stored}} = P_{\text{input}} - P_{\text{diss}}$ in the range $[0,4]$ seconds. (a) First, find a suitable Δt that gives a consistent solution for the $[0,4]$ interval. Plot $t-y$, $t-P_{\text{input}}$, $t-P_{\text{diss}}$, and $t-P_{\text{stored}}$. Repeat the solution for $\omega = \omega_0/3$ rad/s.

CA9.8 Consider the following chemical reactions and their kinetic behavior:

$$A \underset{k_{-1}}{\overset{k_1}{\rightleftarrows}} B, \qquad B \xrightarrow{k_2} C$$

The reaction rate equations and the initial conditions are given by

$$dC_A/dt = -k_1 C_A + k_{-1}C_B, \qquad C_A(0) = 0.10,$$
$$dC_B/dt = k_1 C_A - (k_{-1} + k_2)C_B, \quad C_B(0) = 0,$$
$$dC_C/dt = k_2 C_B, \qquad C_C(0) = 0.$$

where $k_1 = 0.02$, $k_{-1} = 0.01$, and $k_2 = 0.15$ s^{-1}. (a) Express the pertinent equations of motion as a first-order coupled ODE-IVP. (b) Use a suitable method to perform a numerical simulation to determine the effects of (i) k_1 (i.e., simulate $k_1 = 0.005, 0.02$, and 0.10); (ii) k_{-1} (i.e., simulate $k_{-1} = 0.01, 0.1$, and 0.5); (iii) k_2 (i.e., simulate $k_2 = 0.15, 0.60$, and 1.8); Plot $\theta - t$, $\omega - t$, $v - t$ and $\ell - t$ distributions for each case.

CA9.9 A suspension system of a wheel of an automobile, depicted in **Fig. CA9.9**, can be expressed by the following coupled second-order differential equations:

$$\frac{d^2x_1}{dt^2} + \frac{d_1}{m_1}\left(\frac{dx_1}{dt} - \frac{dx_2}{dt}\right) + \frac{k_1}{m_1}(x_1 - x_2) = 0$$

$$\frac{d^2x_2}{dt^2} + \frac{d_1}{m_2}\left(\frac{dx_2}{dt} - \frac{dx_1}{dt}\right) + \frac{k_1}{m_2}(x_2 - x_1) + \frac{k_2}{m_2}(x_2 - x_3) = 0$$

where m_1 is the quarter of the mass of the automobile ($m_1 = 245$ kg), m_2 is the mass of the wheel-axle ($m_2 = 25$ kg), $k_1 = 15000$ N/m is the main spring constant of the automobile, $k_2 = 55000$ N/m is the spring constant of the wheel, $d_1 = 500$ N.s/m is the shock absorber constant, x_1 and x_2 are respectively the vertical displacements of the automobile body and the wheel, and x_3 is the input road disturbance. Initial conditions are given as

$$x_1(0) = x_2(0) = \frac{dx_1}{dt}(0) = \frac{dx_2}{dt}(0) = 0, \quad x_3(t) = \begin{cases} \dfrac{h_1}{2}(1 - \cos 2\pi t), & 1 \leqslant t \leqslant 2 \\[2mm] \dfrac{h_2}{2}(1 - \cos \pi t), & 6 \leqslant t \leqslant 8 \\[2mm] 0, & \text{otherwise} \end{cases}$$

where $h_1 = 0.125$ m and $h_2 = 0.10$ m denote the amplitudes of the bump disturbances.

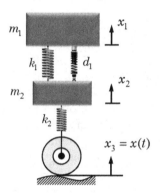

Fig. CA9.9

Perform a numerical simulation of this problem for a period of 20 seconds. (a) Use a suitable numerical method to solve the pertinent equations and obtain $x_1, x_2 - t$ and $\dot{x}_1, \dot{x}_2 - t$ distributions. How does the system react to the road disturbance? (b) Investigate the effect of the spring constant of the wheel by simulating with $k_2 = 40000$, 55000, and 70000 N/m; (c) Investigate the effect of the spring constant of the automobile by simulating with $k_1 = 7000$, 15000, and 22500 N/m; (d) Investigate the effect of the spring constant of the automobile by simulating with $d_1 = 250$, 500, and 1000 N.s/m.

ODEs: Boundary Value Problems

BOUNDARY *Value Problems (BVP)* arise generally when attempting to find *steady-state* solutions to physical events. Mathematical models of events in steady-state depend only on spatial variables (x, y, z). Since steady-state events do not change with time, the time derivatives of the dependent variables are zero. In broader terms, boundary value problems consist of partial differential equations (PDEs). However, in this chapter, the discussion on the BVPs will be restricted only to one dimension that leads to an ordinary differential equation (ODE). Problems, in which the solution, its derivatives, or a combination of both are specified at only two points (i.e., boundary points) are called *two-point boundary value problems*.

In this chapter, the finite difference method for the solution of two-point boundary value problems (BVPs) will be discussed. Using the finite difference and finite volume methods, the numerical solution of the second-order linear ordinary differential equations subjected to Dirichlet, Neumann, and/or Robin boundary conditions is discussed in detail. The implications of uniform and non-uniform gridding and the solution of the two-point BVP problem with non-uniform grids are examined. The solution of nonlinear two-point BVPs with a finite difference and shooting methods are discussed in detail. Finally, the solution of two-point fourth-order linear BVP is covered.

DOI: 10.1201/9781003474944-10

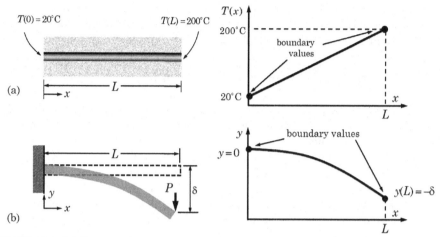

FIGURE 10.1: (a) Heat conduction in a rod and a predicted solution; (b) Lateral deflection of a beam and a predicted solution.

10.1 INTRODUCTION

A two-point boundary value problem in one dimension is an ODE with two boundary conditions. Consider the following second-order ordinary differential equation (ODE):

$$p(x)y''(x) + q(x)y'(x) + r(x)y(x) = g(x), \qquad a \leqslant x \leqslant b$$

where x and y denote the physical quantities (independent and dependent variables, respectively). It should be noted throughout this chapter that the term *boundary value problem* is used to refer to *two-point boundary value problem*.

Having an ODE for a physical event alone is not sufficient to obtain a unique solution to the problem it represents. Boundary value problems may have no solution, only one solution, or an infinite number of solutions. In the case of having many solutions, additional equations are required to be able to find its unique and complete solution. These equations, referred to as *boundary conditions* (BCs), arise as a result of the system requirements or constraints, which specify the behavior of the physical phenomenon at the boundary of a system. In fact, boundary conditions represent the influence of the environment on the isolated model.

In two-point BVPs, the independent variable is the space variable, commonly denoted by x. The number of BCs required to obtain a unique solution to a two-point BVP is equal to the order of the ODE. Boundary conditions for system boundaries are constructed by specifying the dependent variable and/or a linear combination with its derivative. In a two-point BVP, the BCs are imposed at the end points of the solution interval, and the solution is acquired only in the interval restricted with these two end points. In practice, most physical phenomena occurring on boundaries of the system can be expressed mathematically with the following generalized boundary conditions:

$$\alpha_1 y'(a) + \beta_1 y(a) = \gamma_1, \qquad \alpha_2 y'(b) + \beta_2 y(b) = \gamma_2,$$

where α_1, α_2, β_1, β_2, γ_1 and γ_2 are constants.

To illustrate how BCs are determined, consider the example of a heat-conducting rod extending along a wall ($0 \leqslant x \leqslant L$), as shown in Fig. 10.1a. It is assumed that the side surface of the rod is insulated, and the heat is transmitted longitudinally along the side surface of the rod without any loss to the surrounding wall. The temperature distribution $T(x)$ for this event is governed by the following ordinary differential equation:

$$\frac{d^2T}{dx^2} = 0, \qquad 0 \leqslant x \leqslant L$$

where the independent variable x and the dependent variable T denote the axial (longitudinal) coordinate and longitudinal temperature, respectively. The analytical solution for this ODE is $T(x) = c_1 + c_2 x$, where c_1 and c_2 are arbitrary constants. This solution implies that the temperature variation depicted in Fig. 10.1a should be linear, which is. Obviously, the temperature distribution represents a family of solutions for all possible c_1 and c_2 pairs. To obtain a unique solution for this specific case, we need to impose conditions that represent the physics of the event as realistically as possible. For this example, we can set the temperatures at the boundaries to $T(0) = 20°C$ (left BC) and $T(L) = 200°C$ (right BC), as shown in Fig. 10.1a. Then, with the aid of the BCs, we can proceed to obtain a unique solution (i.e., arbitrary constants, c_1 and c_2) for axial rod temperature.

Consider a beam with constant cross section and length L as an example of two-point BVP, shown in Fig. 10.1b. It is fixed at one end and subjected to a singular lateral load P at the other (free) end. The differential equation describing the lateral deflection of the beam is described with

$$EI\frac{d^2y}{dx^2} = M(x), \qquad 0 \leqslant x \leqslant L$$

where y is the lateral deflection, x is the axial coordinate, EI is referred to as flexural rigidity, and M is the bending moment.

The general solution of this ODE also results in an infinite number of solutions due to the arbitrary constants in its homogeneous solution. To determine the lateral deflection of this problem under consideration (i.e., a unique solution), the BCs describing the physical phenomenon need to be specified. Notice that there is no deflection at the left end where the beam is fixed to the wall, so we can set $y(0) = 0$. On the other hand, the amount of downward deflection on the free end of the beam is specified as δ, so we can set $y(L) = -\delta$. (Pay attention to how the coordinate system is defined.) Now that we have obtained the boundary conditions for both ends, after defining problem-specific EI and $M(x)$, it is possible to find the solution for the deflection, $y(x)$, throughout $0 \leqslant x \leqslant L$.

10.2 TWO-POINT BOUNDARY VALUE PROBLEMS

An ordinary differential equation with BCs defined at each endpoint of the solution interval is referred to as a *two-point BVP*. There are numerous methods for solving such problems; however, this section covers the most widely and commonly used finite-difference and finite-volume methods.

The *Finite Difference Method* (FDM) is one of the oldest and most widely used numerical methods for solving ODE-BVPs. In this section, the implementation of the method on two-point linear BVPs subjected to various BCs will be illustrated step by step.

Consider the following two-point BVP:

$$\begin{aligned}
p(x)y''(x) + q(x)y'(x) + r(x)y(x) &= g(x), \qquad a < x < b \\
\alpha_1 y'(a) + \beta_1 y(a) = \gamma_1, \qquad \alpha_2 y'(b) &+ \beta_2 y(b) = \gamma_2
\end{aligned} \qquad (10.1)$$

where α_1, α_2, β_1, β_2, γ_1, and γ_2 are real constants, and $p(x)$, $q(x)$, $r(x)$, and $g(x)$ may be continuous functions of x on $[a, b]$. Also, we assume that $r(x)/p(x) \leqslant 0$ on $[a, b]$, which is a necessary condition for the existence of a unique solution. Note that Eq. (10.1) is defined in a closed interval, $a < x < b$, where the BCs are defined only for both endpoints.

A *general solution* of an ODE has an infinite number of solutions. A general solution can be found if an ODE can be solved analytically, but numerical solutions can only be obtained when all the required BCs are stated. In any case, boundary conditions are necessary to find a unique solution to an ODE. In practice, one of the four types of boundary conditions is applied to a boundary: Dirichlet, Neumann, Robin, and periodic BCs. It is the physics of a physical phenomenon that dictates the type of boundary condition.

10.2.1   DIRICHLET BOUNDARY CONDITION

A boundary condition is referred to as *Dirichlet BC* when the solution of a differential equation, $y(x)$, is specified at the boundary. If the solution is known at one of the boundary points, say $x = a$, the BC is expressed as $y(a) = \delta_1$, where δ_1 is the known solution at the left boundary. Using the general form of the BCs given in Eq. (10.1), the Dirichlet BC in computer programs can be expressed by setting $\delta_1 = \gamma_1/\beta_1$ and $\alpha_1 = 0$. In the example depicted in Fig. 10.1a, the temperatures at the boundary points are known. When the coordinate system is specified as the direction from $x = 0$ to $x = L$, the temperatures on both sides of the wall (boundaries) can be expressed as $T(0) = 20°C$ and $T(L) = 200°C$. In the beam example in Fig. 10.1b, the beam does not deflect at the fixed end, i.e., left BC is $y(a) = 0$ at $x = a$. On the right end, the deflection is specified as $y(b) = -\delta$, which is the right BC.

10.2.2   NEUMANN BOUNDARY CONDITION

A boundary condition is referred to as a *Neumann BC* when only the derivative of $y(x)$ is specified at a boundary, that is, $y'(a) = \lambda_1$ or $y'(b) = \lambda_2$. Neumann BC can be expressed in terms of the general BC form by setting $\lambda_1 = \gamma_1/\alpha_1$ or $\lambda_2 = \gamma_2/\alpha_2$ with the corresponding $\beta_1 = 0$ or $\beta_2 = 0$.

Numerous BCs with physical events fall into this category. For instance, $y'(a) = 0$ is also called the *symmetry BC*, which implies the physical property $y(x)$ is symmetrical with respect to $x = a$. In other words, the physical property $y(x)$ has a local maximum or local minimum at $x = a$. If a physical model and its BCs are symmetric, the numerical solution is sought for only half of the system by applying the $y' = 0$ condition to the symmetry axis or symmetry plane. In many cases, the derivative of the physical quantity, also called the *gradient*, is known. For example, in heat transfer analysis, the flow of heat in the x-direction is governed by Fourier's law ($q = -k\, dT/dx$), where k is the conductivity of the medium, T is the temperature, and q is the heat flux. Consider the rod depicted in Fig. 10.1a. If a heat flux is specified (q_0) at the right end of the rod ($x = L$), then the constant temperature gradient condition is known at the right end, $dT(L)/dx = -q_0/k$, but the temperature $T(L)$ is unknown.

10.2.3   ROBIN (MIXED) BOUNDARY CONDITION

A boundary condition is termed *Mixed BC* or *Robin BC* when a linear combination of $y(x)$ and $y'(x)$ is specified at a boundary, e.g., $\alpha_1 y'(a) + \beta_1 y(a) = \gamma_1$ or $\alpha_2 y'(b) + \beta_2 y(b) = \gamma_2$. This type of BC is also referred to as *convection BC*, where neither $y(x)$ nor $y'(x)$ is known at the boundary. In heat transfer, a convection BC at $x = a$ is imposed as $-kT'(a) = h(T(a) - T_\infty)$, which can be expressed as $kT'(a) + hT(a) = hT_\infty$, i.e., $\alpha_2 = k$, $\beta_2 = h$, $\gamma_2 = hT_\infty$.

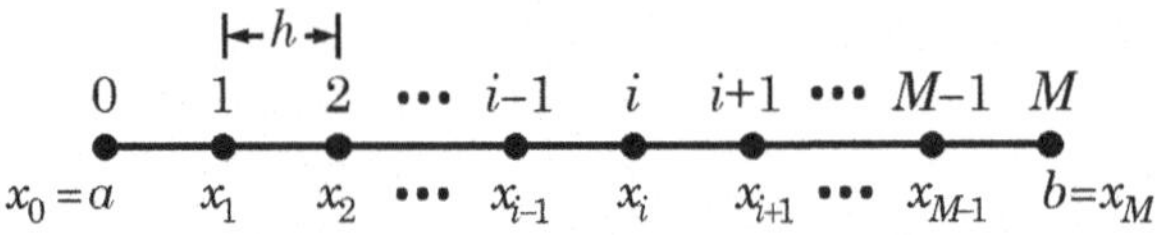

$$\mid\!\!\leftarrow h\rightarrow\!\!\mid$$

0   1   2 $\cdots$ $i{-}1$ i $i{+}1$ $\cdots$ $M{-}1$ M

$x_0 = a$ x_1 x_2 $\cdots$ x_{i-1} x_i x_{i+1} $\cdots$ x_{M-1} $b = x_M$

FIGURE 10.2: Typical grid construction on $[a, b]$.

10.2.4 PERIODIC BOUNDARY CONDITION

Up to this point, we have addressed the types of BCs that are expressed in terms of $y(x)$ and/or $y'(x)$ at a boundary. Another important type of BC encountered in practice is called a *periodic boundary condition*. This BC type does not require additional equations for the physical boundaries. A periodic boundary condition implies that the ODE's solution on $[a, b]$ is periodic with a period $T = b - a$. That is, the BCs can be simply stated as $y(b) = y(a)$, $y'(b) = y'(a)$, and so on.

> In the subsequent illustrations, we will denote the symbol ● for a node whose solution is known (i.e., case of Dirichlet BC) and the symbol $\otimes$ for a node whose solution is unknown (i.e., cases of Neumann or Robin BC).

10.3 FINITE DIFFERENCE SOLUTION OF LINEAR BVPs

The finite difference method (FDM) is one of the most common numerical methods used to solve two-point BVPs. The objective of the FDM is to find an approximate solution to $y(x)$ that satisfies the ODE and its BCs through a discrete approximation. The numerical solution of Eq. (10.1) using the finite difference method is explained step by step below:

Step 1. Gridding: *Gridding* (or *meshing*) is a process of constructing discrete points for which approximate solutions are sought. In this procedure, the solution interval ($a \leqslant x \leqslant b$) is first divided into M subintervals. In most cases, a uniform interval size is applied, i.e., $h = x_{i+1} - x_i = (b - a)/M$. Next, *grid points* (or *nodes*) are placed at the end points of every subinterval. The nodes are then enumerated, starting from either 0 or 1, as depicted in Fig. 10.2. The abscissas of the nodes are calculated from $x_i = a + ih$ ($i = 0, 1, 2, \ldots, M$), but the corresponding ordinates, $y_i = y(x_i)$'s, are the unknowns to be determined as the objective of FDM.

> If a physical property, $y(x)$, is expected to depict sharp changes at either one or both end points of the solution interval, then gridding should be carried out such that the nodal points are clustered where sharp changes are expected.

Step 2. Discretizing: Discretization is the process of transforming a continuous function satisfying an ODE, as well as its BCs, into a discrete function. An ODE is discretized at a node x_i, and the derivatives appearing in the ODE (y'_i, y''_i, and so on) are replaced by finite difference approximations.

Consider the two-point BVP given with Eq. (10.1), which is satisfied for every x on $[a, b]$. The ODE for the ith nodal point ($x = x_i$) is expressed as

$$p(x_i)y''(x_i) + q(x_i)y'(x_i) + r(x_i)y(x_i) = g(x_i)$$

or, in shorthand notation,

$$p_i y_i'' + q_i y_i' + r_i y_i = g_i, \quad i = 0, 1, 2, \ldots, M \tag{10.2}$$

where $y''(x_i) = y_i''$, $y'(x_i) = y_i'$, $p(x_i) = p_i$, etc.

To ensure that the order of global error is second order, we must use second-order finite difference formulas for the first-, second-, or other derivatives, that is,

$$y_i' = \frac{y_{i+1} - y_{i-1}}{2h} + \mathcal{O}(h^2), \qquad y_i'' = \frac{y_{i+1} - 2y_i + y_{i-1}}{h^2} + \mathcal{O}(h^2) \tag{10.3}$$

Substituting the CDFs in Eq. (10.3) into Eq. (10.2) yields

$$p_i \left(\frac{y_{i+1} - 2y_i + y_{i-1}}{h^2} \right) + q_i \left(\frac{y_{i+1} - y_{i-1}}{2h} \right) + r_i y_i = g_i, \qquad i = 0, 1, \ldots, M$$

and rearranging similar terms, we get

$$\left(\frac{p_i}{h^2} - \frac{q_i}{2h} \right) y_{i-1} + \left(r_i - 2\frac{p_i}{h^2} \right) y_i + \left(\frac{p_i}{h^2} + \frac{q_i}{2h} \right) y_{i+1} = g_i, \qquad i = 0, 1, \ldots, M \tag{10.4}$$

Equation (10.4), referred to as the *general difference equation*, gives a relationship between the discrete values of three consecutive nodes.

 In the discretization of an ODE, all derivatives approximated by the corresponding (forward, backward, and/or central) finite difference formulas should have the same order of truncation error. If approximations with different orders are used (i.e., $\mathcal{O}(h)$, $\mathcal{O}(h^2)$, $\mathcal{O}(h^3)$, and so on), the smallest order of truncation error dominates the order of global error of the ODE.

Step 3. Implementing BCs: A general difference equation is satisfied for every nodal point whose ordinate (solution) is unknown. Such a node will hereinafter be referred to as an *unknown node*. Difference equations for the unknown nodes are obtained by applying Eq. (10.4) node as follows:

$$
\begin{aligned}
x = a \quad &\text{(for } i = 0) & b_0\,\boxed{y_{-1}} + d_0 y_0 + a_0 y_1 &= g_0, \\
x = x_1 \quad &\text{(for } i = 1) & b_1 y_0 + d_1 y_1 + a_1 y_2 &= g_1, \\
x = x_2 \quad &\text{(for } i = 2) & b_2 y_1 + d_2 y_2 + a_2 y_3 &= g_2, \\
\cdots \quad & \cdots & \cdots & \\
x = x_{M-1} \quad &\text{(for } i = M-1) & b_{M-1} y_{M-2} + d_{M-1} y_{M-1} + a_{M-1} y_M &= g_{M-1}, \\
x = b \quad &\text{(for } i = M) & b_M y_{M-1} + d_M y_M + a_M\,\boxed{y_{M+1}} &= g_M.
\end{aligned}
\tag{10.5}
$$

where $b_i = (p_i/h - q_i/2)/h$, $a_i = (p_i/h + q_i/2)/h$, and $d_i = r_i - 2p_i/h^2$.

Equation (10.5) consists of $M + 1$ equations and $M + 3$ unknowns (y_{-1}, y_0, y_1, ..., y_{M-1}, y_M, y_{M+1}). An unknown node at a specified point (x_i) corresponds to a numerical approximation of the true solution $y(x_i)$. We note that the discretization procedure yields two additional nodes, (x_{-1}, y_{-1}) and (x_{M+1}, y_{M+1}), marked by $\otimes$ in Fig. 10.3, which fall outside the solution interval. These nodes are referred to as *fictitious nodes* and are eliminated using the BCs. The interior nodes (marked by "**o**") are the unknowns whose values are to be determined.

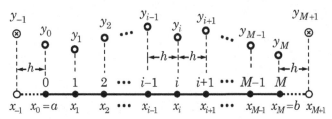

FIGURE 10.3: Discretization and fictitious nodes.

In this step, the number of equations and the number of unknowns are balanced, thus creating a set of algebraic equations with a unique solution. In this regard, two equations can be derived with the help of the BCs. For instance, if Dirichlet BCs are imposed on both boundary points ($y(a) = A$ and $y(b) = B$), then y_0 and y_M are known. In this case, the nodal values from $i = 1$ to $(M - 1)$ are the unknowns, and the values of the fictitious nodes (y_{-1} and y_{M+1}) do not enter this picture. However, if one or both BCs are specified as either Neumann or Robin BC, then the solutions at the boundary nodes (either y_0 or y_M or both) are unknowns, and the discretization of the BCs provides additional difference equations to eliminate the fictitious, y_{-1} and y_{M+1}, nodes.

Step 4. Constructing a linear system: After incorporating the boundary conditions, a unique system of linear equations can be constructed. The size of the linear system ($M - 1$, M, or $M + 1$) depends on the number of subintervals and the type of BCs applied to both boundaries. Assuming a Neumann or Robin BC is applied to either boundary, the difference equations corresponding to unknown nodes, including the boundary ones, can be expressed in matrix equation form as

$$\mathbf{Ax} = \mathbf{r} \tag{10.6}$$

where $\mathbf{x}$ and $\mathbf{r}$ are column vectors defined as

$$\mathbf{x} = \begin{bmatrix} y_0 & y_1 & y_2 & \cdots & y_{M-1} & y_M \end{bmatrix}^T, \quad \mathbf{r} = \begin{bmatrix} g_0 & g_1 & g_2 & \cdots & g_{M-1} & g_M \end{bmatrix}^T$$

and $\mathbf{A}$ is a matrix in tridiagonal form expressed as follows:

$$
\begin{array}{c}
\\
i = 0 \\
i = 1 \\
i = 2 \\
\vdots \\
i = M-1 \\
i = M
\end{array}
\quad
\mathbf{A} =
\begin{array}{c}
\begin{array}{ccccccc}
y_{-1} & y_0 & y_1 & y_2 & \cdots & y_{M-1} & y_M & y_{M+1}
\end{array} \\
\boxed{b_0}
\begin{bmatrix}
d_0 & a_0 & & & & \\
b_1 & d_1 & a_1 & & & \\
& b_2 & d_2 & a_2 & & \\
& & \ddots & \ddots & \ddots & \\
& & & b_{M-1} & d_{M-1} & a_{M-1} \\
& & & & b_M & d_M
\end{bmatrix} \boxed{a_M}
\end{array}
$$

Note that eliminating fictitious nodes modifies only the first and last rows of $\mathbf{A}$ and $\mathbf{r}$.

Step 5. Solving the linear system: Once a system of linear equations is constructed (the coefficient matrix and the rhs vector are properly defined), the numerical approximations corresponding to the nodal points established in **Step 1** are obtained by solving the system of linear equations with a suitable *direct* or *iterative* method (*see* Chapters 2 and 3). A second-order linear two-point BVP always results in a tridiagonal system, which is effectively solved by Thomas' algorithm (*see* Section 2.10); however, non-linear two-point BVPs also require iterative algorithms (*see* Section 10.7).

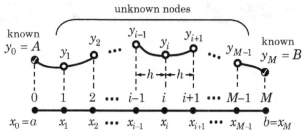

FIGURE 10.4: Illustration of grids with Dirichlet BCs.

10.3.1 IMPLEMENTING DIRICHLET BCs

Consider applying Dirichlet BCs on both boundaries of Eq. (10.1):

$$y(a) = A \quad \text{and} \quad y(b) = B \tag{10.7}$$

where A and B are real constants.

A discrete set (x_i and y_i, $i = 0, 1, 2, \ldots, M$) corresponding to a uniformly spaced grid structure containing M subintervals in the interval $[a, b]$ is illustrated in Fig. 10.4. In this conjectural distribution, the boundary nodes have been marked with the symbol ✅ because they are specified with the Dirichlet BCs, i.e., $y(a) = y_0 = A$ and $y(b) = y_M = B$. Thus, the difference equations corresponding to the boundary nodes ($i = 0$ and $i = M$) can be discarded, leading to $M - 1$ equations with $M - 1$ unknowns (y_i, $i = 1, 2, \ldots, M-1$).

Even though the difference equations for $i = 0$ and $i = M$ have been discarded, the boundary nodal values (y_0 and y_M) are present (boxed) in the $i = 1$ and $i = M - 1$th equations:

$$\text{For } i = 1, \qquad b_1 \boxed{y_0} + d_1 y_1 + a_1 y_2 = g_1$$
$$\text{For } i = M - 1, \quad b_{M-1} y_{M-2} + d_{M-1} y_{M-1} + a_{M-1} \boxed{y_M} = g_{M-1}$$

Substituting $y_0 = A$ and $y_M = B$ into the difference equations and carrying them to the rhs, only the unknowns remain on the left. Consequently, we obtain

$$d_1 y_1 + a_1 y_2 = g_1 - b_1 A$$
$$b_2 y_1 + d_2 y_2 + a_2 y_3 = g_2$$
$$\vdots \tag{10.8}$$
$$b_{M-2} y_{M-3} + d_{M-2} y_{M-2} + a_{M-3} y_{M-1} = g_{M-2}$$
$$b_{M-1} y_{M-2} + d_{M-1} y_{M-1} = g_{M-1} - a_{M-1} B$$

Finally, the linear system, Equation (10.8), can be expressed in matrix form as follows:

$$
\begin{array}{c}
\\
i = 1 \\
i = 2 \\
i = 3 \\
\vdots \\
i = M - 2 \\
i = M - 1
\end{array}
\begin{array}{c}
\begin{array}{cccccc}
y_1 & y_2 & y_3 & \cdots & y_{M-2} & y_{M-1}
\end{array} \\
\begin{bmatrix}
d_1 & a_1 & & & & \\
b_2 & d_2 & a_2 & & & \\
& b_3 & d_3 & a_3 & & \\
& & \ddots & \ddots & \ddots & \\
& & & b_{M-2} & d_{M-2} & a_{M-2} \\
& & & & b_{M-1} & d_{M-1}
\end{bmatrix}
\end{array}
\begin{bmatrix}
y_1 \\ y_2 \\ y_3 \\ \vdots \\ y_{M-2} \\ y_{M-1}
\end{bmatrix}
=
\begin{bmatrix}
g_1 - b_1 A \\ g_2 \\ g_3 \\ \vdots \\ g_{M-2} \\ g_{M-1} - a_{M-1} B
\end{bmatrix}
\tag{10.9}
$$

where $g(x_i) = g_i$. The size of the resulting system of linear equations is $(M - 1) \times (M - 1)$, and it is efficiently solved using Thomas' algorithm.

EXAMPLE 10.1: Boundary value problem with Dirichlet BCs

A beam of length $L = 2$ m with a non-uniform cross section is supported at both ends and carries a uniformly distributed load of $w_0 = 4000$ N/m, as shown in the figure below. The displacement, $y(x)$, in the beam is governed by the following BVP:

$$EI(x)\frac{d^2y}{dx^2} = w_0\frac{x}{2}(L - x), \qquad y(0) = y(L) = 0$$

where $E = 50$ GPa is the elasticity modulus, and $I(x)$ is the inertia moment given as $I(x) = I_0\, e^{-0.2x}$, where $I_0 = 10^{-6}$ m^4. Obtain the numerical approximations using the FDM with uniform spacings of $h = 0.4$, 0.2, and 0.1. Also, find the true solution to compare with your estimates.

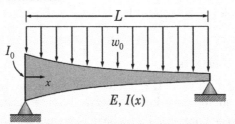

SOLUTION:

Step 1: First, the solution domain, $(0,2)$, is divided into uniformly spaced M subintervals $(h = (2 - 0)/M)$, as shown in Fig. 10.5. The nodes are placed on both sides of the subintervals and enumerated. The nodal solutions are denoted with $y(x_i) = y_i$ for $i = 0, 1, 2, \ldots, M$, where $x_i = ih$'s are the abscissas. Since the BCs are of Dirichlet type, the nodal solutions at the boundaries are already known, i.e., $y(0) = y_0 = 0$ and $y(1) = y_M = 0$. Thus, the total number of unknowns becomes $M - 1$.

unknown values

$y_0 = 0 \qquad y_1 \quad y_2 \ \cdots\ y_{i-1} \quad y_i \quad y_{i+1} \ \cdots\ y_{M-1} \quad y_M = 0$

$$\begin{array}{ccccccccc} 0 & 1 & 2 & \cdots & i-1 & i & i+1 & \cdots & M-1 & M \end{array}$$

$x_0 = 0 \quad x_1 \quad x_2 \ \cdots\ x_{i-1} \quad x_i \quad x_{i+1} \ \cdots\ x_{M-1} \quad 2 = x_M$

FIGURE 10.5

Step 2: The ODE can be written for an ith node within the solution range ($1 \leqslant i \leqslant M - 1$) as

$$EI_0 e^{-0.2x_i} y_i'' = w_0\frac{x_i}{2}(L - x_i), \qquad 0 < x_i < 2 \tag{10.10}$$

Substituting the CDF of y_i'' in Eq. (10.10) gives

$$EI_0 e^{-0.2x_i}\left(\frac{y_{i+1} - 2y_i + y_{i-1}}{h^2}\right) = w_0\frac{x_i}{2}(L - x_i), \tag{10.11}$$

Rearranging Eq. (10.11) and collecting similar terms, the *general difference equation* is obtained:

$$y_{i+1} - 2y_i + y_{i-1} = \frac{w_0 h^2}{2EI_0}x_i(L - x_i)e^{0.2x_i}, \qquad i = 1, 2, \ldots, (M - 1) \tag{10.12}$$

Step 3: Making use of Eq. (10.12) for $i = 1, 2, \ldots, (M-1)$, we get

For $i = 1$ $\boxed{y_0} - 2y_1 + y_2 = g_1$

For $i = 2$ $y_1 - 2y_2 + y_3 = g_2$

$$\vdots \qquad\qquad \vdots \tag{10.13}$$

For $i = M - 2$ $y_{M-3} - 2y_{M-2} + y_{M-1} = g_{M-2}$

For $i = M - 1$ $y_{M-2} - 2y_{M-1} + \boxed{y_M} = g_{M-1}$

where $b_i = a_i = 1$, $d_i = -2$, and $g_i = w_0 h^2 x_i (L - x_i) e^{0.2x_i}/2EI_0$. Notice that the boxed quantities are the nodal boundary values whose solutions are known, i.e., $y_0 = 0$ and $y_M = 0$.

Step 4: Upon substituting $y_0 = 0$ and $y_M = 0$ into Eq. (10.13) and collecting the known quantities on the rhs, we obtain an $(M-1) \times (M-1)$ linear system, which can be expressed in matrix form as follows:

$$\begin{bmatrix} -2 & 1 & & & & \\ 1 & -2 & 1 & & & \\ & 1 & -2 & 1 & & \\ & & \ddots & \ddots & \ddots & \\ & & & 1 & -2 & 1 \\ & & & & 1 & -2 \end{bmatrix} \begin{bmatrix} y_1 \\ y_2 \\ y_3 \\ \vdots \\ y_{M-2} \\ y_{M-1} \end{bmatrix} = \begin{bmatrix} g_1 \\ g_2 \\ g_3 \\ \vdots \\ g_{M-2} \\ g_{M-1} \end{bmatrix} \tag{10.14}$$

Step 5: The resulting system is tridiagonal, which can be easily solved using Thomas' algorithm. The solution of the linear system for $h = 0.4$, 0.2, and 0.1 (i.e., for $M = 5$, 10, and 20 intervals) yields 4, 9, and 19 unknowns. On the other hand, the BVP is a second-order linear ODE whose true solution is obtained as $y(x) = 170 - (x^2 - 22x + 170)e^{0.2x} + 11.96860535x$.

In Table 10.1, the true solution, as well as the true errors, are comparatively tabulated for $M = 5$, 10, and 20. Notice that the estimates turned out to be negative, indicating that displacement is directed downward. Since the beam has a non-uniform cross section, the solution is not exactly symmetrical with respect to $x = L/2$, which is not the result of round-off and/or truncation errors.

TABLE 10.1

	True	True Absolute Error		
x	Solution	$h = 0.4$	$h = 0.2$	$h = 0.1$
0	0			
0.2	−0.00617557		4.99E-05	1.25E-05
0.4	−0.01175910	3.69E-04	9.24E-05	2.31E-05
0.6	−0.01624199		1.26E-04	3.16E-05
0.8	−0.01921942	6.00E-04	1.50E-04	3.76E-05
1	−0.02040562		1.63E-04	4.07E-05
1.2	−0.01965035	6.50E-04	1.63E-04	4.07E-05
1.4	−0.01695682		1.48E-04	3.71E-05
1.6	−0.01250115	4.69E-04	1.17E-04	2.93E-05
1.8	−0.00665334		6.88E-05	1.72E-05
2	0			

Discussion: The numerical estimates improve as the number of subintervals is increased. For instance, as M is doubled, the global error decreased by about four folds, which should not come as a surprise since the order of error in the general difference equation is $\mathcal{O}(h^2)$. The average errors are 5.22×10^{-4}, 1.2×10^{-4}, and 3×10^{-5} for $M = 5$, 10, and 20, respectively. It is also worth noting that the numerical solution of an ODE is not known for any x other than those discrete values (y_i) corresponding to x_i's. To find solutions for any x other than x_i's, the procedure to be applied is interpolation.

10.3.2 IMPLEMENTING NEUMANN AND ROBIN BCs

The implementation of Neumann and Robin BCs is essentially the same, in that Neumann BC can be easily obtained by setting $\beta = 0$ in Robin BC. Therefore, the implementation of Robin BC should suffice for applying both BCs. Consider imposing Robin BC on both boundaries in Eq. (10.5). In this case, the nodal values at the boundaries (y_0 and y_M) are unknown. Thus, we end up with $M + 1$ equations and $M + 3$ unknowns, namely y_{-1}, y_0, ... y_M, y_{M+1}. Also, the difference equations for boundary nodes include fictitious nodes, i.e., y_{-1} and y_{M+1}. So to complete the linear system, we make use of the BCs for the two additional equations.

Our next step is to eliminate the fictitious nodes with the aid of the BCs. First, let us consider the left BC, $\alpha_1 y'(a) + \beta_1 y(a) = \gamma_1$, which is discretized at $x_0 = a$ (or $i = 0$). Next, expressing $y(a) = y_0$ and $y'(a) = y_0' \approx (y_1 - y_{-1})/(2h)$ and substituting in the left BC yields

$$\alpha_1 \left(\frac{y_1 - y_{-1}}{2h} \right) + \beta_1 y_0 = \gamma_1 \tag{10.15}$$

where $y_{-1} = y(x_{-1})$.

Solving Eq. (10.15) for the fictitious node gives $y_{-1} = y_1 - 2h(\gamma_1 - \beta_1 y_0)/\alpha_1$. Substituting this expression into the difference equation obtained by setting $i = 0$ in Eq. (10.5) and collecting the unknowns on the left and knowns on the right-hand side, the difference equation for the left boundary node takes the following form:

$$\left(d_0 + 2h \frac{\beta_1}{\alpha_1} b_0 \right) y_0 + (a_0 + b_0) y_1 = g_0 + 2h \frac{\gamma_1}{\alpha_1} b_0 \tag{10.16}$$

The right BC, $\alpha_2 y'(b) + \beta_2 y(b) = \gamma_2$, is likewise discretized at $x_M = b$ ($i = M$), replacing $y(b) = y_M$ and $y'(b) = y_M' \approx (y_{M+1} - y_{M-1})/(2h)$ leads to

$$\alpha_2 \left(\frac{y_{M+1} - y_{M-1}}{2h} \right) + \beta_2 y_M = \gamma_2$$

Solving for y_{M+1} yields

$$y_{M+1} = y_{M-1} + \frac{2h}{\alpha_2}(\gamma_2 - \beta_2 y_M) \tag{10.17}$$

where y_{M+1} denotes an approximation for the fictitious node in terms of y_{M-1} and y_M.

Substituting Eq. (10.17) into the difference equation obtained by setting $i = M$ in Eq. (10.5) and collecting the unknowns on the left and knowns on the right-hand side, we arrive at the following difference equation:

$$(a_M + b_M) y_{M-1} + \left(d_M - 2h \frac{\beta_2}{\alpha_2} a_M \right) y_M = g_M - 2h \frac{\gamma_2}{\alpha_2} a_M \tag{10.18}$$

The difference equations of the interior nodes are not affected by the implementation of the BCs.

Finally, combining Eqs. (10.5), (10.16), and (10.18), a system of $(M+1) \times (M+1)$ equations is obtained as

$$
\begin{array}{lll}
\text{For} & i=0 & \bar{d}_0 y_0 + \bar{a}_0 y_1 = \bar{g}_0 \\
\text{For} & i=1 & b_1 y_0 + d_1 y_1 + a_1 y_2 = g_1, \\
\text{For} & i=2 & b_2 y_1 + d_2 y_2 + a_2 y_3 = g_2, \\
& \vdots & \qquad \vdots \\
\text{For} & i=M-1 & b_{M-2} y_{M-2} + d_{M-1} y_{M-1} + a_{M-1} y_M = g_{M-1}, \\
\text{For} & i=M & \bar{b}_M y_{M-1} + \bar{d}_M y_M = \bar{g}_M
\end{array}
\tag{10.19}
$$

where

$$
\begin{aligned}
\bar{d}_0 &= d_0 + 2h\frac{\beta_1}{\alpha_1} b_0, & \bar{a}_0 &= a_0 + b_0, & \bar{g}_0 &= g_0 + 2h\frac{\gamma_1}{\alpha_1} b_0, \\
\bar{d}_M &= d_M - 2h\frac{\beta_2}{\alpha_2} a_M, & \bar{b}_M &= a_M + b_M, & \bar{g}_M &= g_M - 2h\frac{\gamma_2}{\alpha_2} a_M,
\end{aligned}
\tag{10.20}
$$

EXAMPLE 10.2: Applying FDM to heat transfer from trapezoidal fin

In the study of heat transfer, *"fins"* made of conducting metals or alloys are attached to a metallic object to increase the heat transfer rate by increasing its total surface area. Consider the trapezoidal fin shown in the figure.

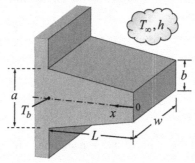

The governing ODE for the temperature excess, $\Theta(x) = T(x) - T_b$, is given as

$$
(x+\delta)\frac{d^2\Theta}{dx^2} + \frac{d\Theta}{dx} - m^2 L\,\Theta(x) = 0, \qquad \Theta'(0) = 0, \quad \Theta(L) = \Theta_b
$$

where $\Theta_b = T_b - T_\infty$, $\delta = aL/(b-a)$, and $m^2 = 2h_c/k(b-a)$ are constants, T_∞ and T_b denote the ambient and base temperatures, h_c is the convection heat transfer coefficient, k is the conductivity of the fin, L and w are respectively the fin length and width, and b and a denote the dimensions of the fin as shown in the figure.

A fin configuration is given as $a = 5$ mm, $b = 2$ mm, $L = 15$ cm, $\Theta_b = 100°\text{C}$ and $m^2 = 250$ m^{-2}. Use the FDM to obtain the numerical solution by dividing the fin length into $M = 5, 10, 20$, and 40 uniform intervals. Compare the estimates with the true solution is given with

$$
\Theta(x) = \Theta_b \frac{I_1(u)K_0(u\xi_x) + K_1(u)I_0(u\xi_x)}{I_1(u)K_0(u\xi_L) + K_1(u)I_0(u\xi_L)}
$$

where $u = 2m\sqrt{\delta L}$ and $\xi_x = \sqrt{1 + x/\delta}$, $I_0(x)$, $I_1(x)$, $K_0(x)$, and $K_1(x)$ are the zeroth and first-order modified Bessel functions of the first and second kinds.

SOLUTION:

Step 1: A one-dimensional uniform grid, depicted in Fig. 10.6, is obtained by dividing the solution interval $(0 \leqslant x < L)$ into uniformly spaced M subintervals $(h = L/M)$. Note that, for convenience, the coordinate system in the figure is chosen in the direction from the tip toward the base, thereby reversing the positions of the tip $(x = 0)$ and base $(x = L)$. Since the nodal value at the right boundary is known $(\Theta(L) = \Theta_b)$, for the right (base) node, we may set $\Theta_M = \Theta_b$. However, the nodal value of the left (fin tip) boundary is unknown due to the Neumann BC. The number of unknowns for this problem is determined to be M.

The numerical approximations for the excess temperature corresponding to the nodal points are labeled as $\Theta(x_i) = \Theta_i$ for $i = 0, 1, 2, \ldots, (M-1)$, where $x_i = ih$.

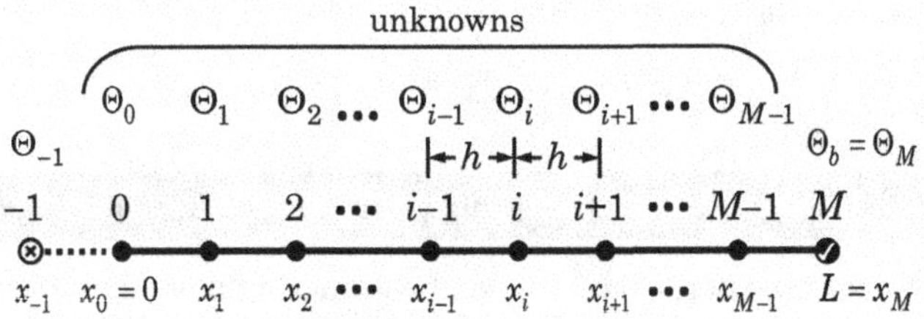

FIGURE 10.6

Step 2: The differential equation is satisfied for all nodes on $[0,L]$; hence, for any x_i, we may write

$$(x_i + \delta)\frac{d^2\Theta_i}{dx^2} + \frac{d\Theta_i}{dx} - m^2 L\, \Theta(x_i) = 0, \qquad x_i \in [0, L] \tag{10.21}$$

Substituting the CDFs for Θ_i' and Θ_i'' leads to

$$(x_i + \delta)\frac{\Theta_{i+1} - 2\Theta_i + \Theta_{i-1}}{h^2} + \frac{\Theta_{i+1} - \Theta_{i-1}}{2h} - m^2 L\, \Theta_i = 0, \quad i = 0, 1, \ldots, M-1 \tag{10.22}$$

Rearranging Eq. (10.22), the general difference equation is found as

$$b_i\Theta_{i-1} + d_i\Theta_i + a_i\Theta_{i+1} = c_i, \quad i = 0, 1, \ldots, M-1 \tag{10.23}$$

where

$$b_i = \frac{x_i + \delta}{h^2} - \frac{1}{2h}, \quad a_i = \frac{x_i + \delta}{h^2} + \frac{1}{2h}, \quad d_i = -\left(m^2 L + \frac{2(x_i + \delta)}{h^2}\right), \quad c_i = 0$$

Step 3: The difference equations for all unknown nodes are derived from Eq. (10.23) as follows:

For $i = 0$ $\qquad b_0\boxed{\Theta_{-1}} + d_0\Theta_0 + a_0\Theta_1 = 0$

For $i = 1$ $\qquad b_1\Theta_0 + d_1\Theta_1 + a_1\Theta_2 = 0$

$$\vdots \qquad \vdots$$

For $i = M - 2$ $\quad b_{M-2}\Theta_{M-3} + d_{M-2}\Theta_{M-2} + a_{M-2}\Theta_{M-1} = 0$

For $i = M - 1$ $\quad b_{M-1}\Theta_{M-2} + d_{M-1}\Theta_{M-1} + a_{M-1}\boxed{\Theta_M} = 0$

The fictitious node value, Θ_{-1}, at the left boundary is eliminated by discretizing the left BC with the CDF as $\Theta'_0 \cong (\Theta_1 - \Theta_{-1})/2h = 0$, which gives $\Theta_{-1} = \Theta_1$. Substituting this into the difference equation with $i = 0$ (after collecting the unknown nodes on the left) leads to

$$d_0\Theta_0 + (a_0 + b_0)\Theta_1 = 0$$

At the right boundary, the nodal value of $\Theta_M = \Theta_b$ is plugged into the difference equation obtained with $i = M - 1$. Likewise, rearranging this difference equation gives

$$b_{M-1}\Theta_{M-2} + d_{M-1}\Theta_{M-1} = -a_{M-1}\Theta_b$$

Step 4: The total of M equations and M unknowns can be expressed in matrix form as

$$
\begin{bmatrix}
d_0 & a_0 + b_0 & & & & \\
b_1 & d_1 & a_1 & & & \\
& b_2 & d_2 & a_2 & & \\
& & \ddots & \ddots & \ddots & \\
& & & b_{M-2} & d_{M-2} & a_{M-2} \\
& & & & b_{M-1} & d_{M-1}
\end{bmatrix}
\begin{bmatrix}
\Theta_0 \\
\Theta_1 \\
\Theta_2 \\
\vdots \\
\Theta_{M-2} \\
\Theta_{M-1}
\end{bmatrix}
=
\begin{bmatrix}
0 \\
0 \\
0 \\
\vdots \\
0 \\
-a_{M-1}\Theta_b
\end{bmatrix}
\tag{10.24}
$$

which can be easily programmed for any value of M.

Step 5: The resulting linear system of equations is tridiagonal in form and can be solved with Thomas' algorithm. The true errors of the numerical estimates obtained with $M = 5$, 10, 20, and 40 and the true solution are comparatively tabulated in Table 10.2. The average of the errors is about 0.2, 0.043, 0.011, and 0.03°C for $M = 5$, 10, 20, and 40, respectively.

TABLE 10.2

x	True Solution	True Absolute Error			
		$M = 5$	$M = 10$	$M = 20$	$M = 40$
0	23.5472	0.6588	0.1796	0.0460	0.0116
0.015	24.4572		0.1038	0.0266	0.0067
0.030	26.9522	0.2014	0.0563	0.0145	0.0037
0.045	30.8371		0.0256	0.0067	0.0017
0.060	36.0522	0.0116	0.0056	0.0016	0.0004
0.075	42.6217		0.0072	0.0017	0.0004
0.090	50.6264	0.0622	0.0144	0.0035	0.0009
0.105	60.1908		0.0170	0.0042	0.0010
0.120	71.4745	0.0636	0.0155	0.0038	0.0010
0.135	84.6699		0.0099	0.0025	0.0006
0.150	100				

Discussion: Results of very good quality were obtained even with a very small number of nodes. In this example, the maximum true error in the estimated solution was below 1°C even for $M = 5$. When M was doubled, the global error decreased by about fourfold since $\mathcal{O}((h/2)^2) \approx \mathcal{O}(h^2/4)$.

Also, since no approximation is introduced for the right BC, the least error is observed at the base ($x = L$), while the largest error is observed at the tip ($x = 0$)

due to approximating the right BC, $\Theta_0' = 0$, with CDF. However, in problems where more accurate solutions are sought, the density of nodes should be increased at the boundary where large gradients in the solution are expected.

Pseudocode 10.1

Module LINEAR_BVP $(M, \mathbf{x}, \boldsymbol{\alpha}, \boldsymbol{\beta}, \boldsymbol{\gamma}, \mathbf{y})$
\ DESCRIPTION: A pseudomodule to solve a two-point linear BVP given as:
\ $p(x)y'' + q(x)y' + r(x)y(x) = g(x)$, $a \leqslant x \leqslant b$ subject to following BCs:
\ $\alpha_1 y'(a) + \beta_1 y(a) = \gamma_1$, $\alpha_2 y'(b) + \beta_2 y(b) = \gamma_2$ with the FDM.
\ USES:
\   COEFFS:: Module supplying $p(x)$, $q(x)$, $r(x)$, and $g(x)$;
\   TRIDIAGONAL:: Tridiagonal system solver, Pseudocode 2.13)
Declare: $\alpha_2, \beta_2, \gamma_2, a_M, b_M, c_M, d_M, x_M, y_M$ \ Declare array variables
$h \leftarrow x_2 - x_1$ \ Find interval size
For $\left[k = 1, M\right]$ \ Setup the tridiagonal matrix row-wise
    COEFFS(x_k, px, qx, rx, gx) \ Get coefficients of ODE at $x = x_k$
    $d_k \leftarrow rx - 2 * px/h^2$ \ Setup diagonal element
    $a_k \leftarrow (px/h + qx/2)/h$ \ Setup above-diagonal element
    $b_k \leftarrow (px/h - qx/2)/h$ \ Setup below-diagonal element
    $c_k \leftarrow gx$ \ Setup rhs element
End For
\ Implement the left BC
If $\left[\alpha_1 = 0\right]$ **Then**   \ Case of Dirichlet BC on $x = a$
    $d_1 \leftarrow 1; c_1 \leftarrow \gamma_1/\beta_1;$
    $a_1 \leftarrow 0; b_1 \leftarrow 0$
Else   \ Case of Robin BC on $x = a$
    $a_1 \leftarrow a_1 + b_1$
    $d_1 \leftarrow d_1 + b_1 * 2h\beta_1/\alpha_1$
    $c_1 \leftarrow c_1 + b_1 * 2h\gamma_1/\alpha_1$
End If
\ Implement the right BC
If $\left[\alpha_2 = 0\right]$ **Then**   \ Case of Dirichlet BC on $x = b$
    $d_M \leftarrow 1; c_M \leftarrow \gamma_2/\beta_2;$
    $a_M \leftarrow 0; b_M \leftarrow 0$
Else   \ Case of Robin BC on $x = b$
    $b_M \leftarrow a_M + b_M$
    $d_M \leftarrow d_M - a_M * 2h\beta_2/\alpha_2$
    $c_M \leftarrow c_M - a_M * 2h\gamma_2/\alpha_2$
End If
TRIDIAGONAL $(1, M, \mathbf{b}, \mathbf{d}, \mathbf{a}, \mathbf{c}, \mathbf{y})$ \ Solve tridiagonal system to find $\mathbf{y}$
End Module LINEAR_BVP

Module COEFFS (x, p, q, r, g)
\ DESCRIPTION: A user-defined module supplying p, q, r, and g for any x.
$p \leftarrow p(x); \ q \leftarrow q(x); \ \ldots$ \ Define coefficients of linear ODE here
End Module COEFFS

A general pseudomodule, **LINEAR_BVP**, for numerically solving a two-point linear BVP

Finite Difference Method

- It is very easy to discretize ODEs using the finite difference method;
- For a linear second-order BVP, the finite difference procedure with the global error of order $\mathcal{O}(h^2)$ results in a tridiagonal system of equations whose direct solution with the Thomas algorithm is fast and accurate;
- Accuracy of the numerical estimates improves as the interval size is refined, i.e., $h \to 0$.

- Finite difference schemes of order of accuracy $\mathcal{O}(h^3)$, $\mathcal{O}(h^4)$, and so on are unreliable for the BVPs of conservative forms;
- The method is quite laborious to apply to non-uniform grids;
- When applied to 2D and 3D BVPs, it is generally limited to structured grids.

---

given with Eq. (10.1) is presented in Pseudocode 10.1). As input, the module requires the number of nodal points (M), the coefficients of the BCs as α_i, β_i, and γ_i for $i = 1,2$, and the abscissas $(x_i, \ i = 1, 2, \ldots, M)$ of the uniformly spaced nodal points established in the interval $[a, b]$. On output, the numerical solution is saved on the nodal values, y_i, $i = 1, 2, \ldots, M$. The **LINEAR_BVP** also makes use of two additional modules: (1) **COEFFS**, a user-defined module, supplying $p(x)$, $q(x)$, $r(x)$, and $g(x)$ for any x; (2) **TRIDIAGONAL** solving a tridiagonal system of equations, Pseudocode 2.13.

The interval size is set to $h = x_2 - x_1$ since all nodal points are uniformly spaced. The coefficients p_k, q_k, r_k, and g_k for a specified x_k (row) are computed in **COEFFS** at the top of the **For**-loop. Then the diagonal arrays of the tridiagonal matrix $(b_k, d_k, \text{and } a_k)$ and the rhs (c_k) are computed in accordance with Eq. (10.4). However, the first $(k = 1)$ and last $(k = M)$ rows of the tridiagonal system are modified to implement the Dirichlet or Robin the BCs using **If-Then-Else** constructs. If the Dirichlet BC is prescribed on $x = a$ $(\alpha_1 = 0)$, then the first row entries of the arrays are set to $d_1 = 1$, $a_1 = b_1 = 0$, and $c_1 = \gamma_1/\beta_1$. (Recall that it is more convenient to test the equality of floating point numbers as $|\alpha_1| < \epsilon$.) If the Neumann or Robin BC is prescribed on $x = a$ $(\alpha_1 \neq 0)$, then the first entries of the arrays are modified according to Eq. (10.16). Similarly for $x = b$, when $\alpha_2 = 0$, the Dirichlet condition to the last row is applied as $d_M = 1$, $a_M = b_M = 0$, and $c_M = \gamma_M/\beta_M$. The Robin BC on $x = b$ $(\alpha_2 \neq 0)$ modifies the last row entries according to Eq. (10.18). Once the setting up of the tridiagonal system is complete, it is solved using Thomas' algorithm with the module **TRIDIAGONAL** and saved on array **y**.

10.4 NUMERICAL SOLUTIONS OF HIGH-ORDER ACCURACY

In this section, we focus on reducing the discretization error in the numerical solution by increasing the order of accuracy. Since the numerical solution of a BVP requires solving a linear system of equations, a direct numerical solution (y_i) inevitably contains rounding errors due to numerous arithmetic operations or insufficient computational precision. Nevertheless, assuming that the computations are carried out using high-precision arithmetic with negligible round-off errors, we let $y_0, y_1, y_2, \ldots, y_M$ represent the true solutions of the system of linear equations.

Let $y(x_i)$ be the true solution of the BVP at node x_i. Then the *global discretization error* is defined as

$$\max_{\text{all } i} |y(x_i) - y_i|$$

The local discretization error is just the truncation error in approximating y_i' and y_i'' derivatives. Since the central differences of order $\mathcal{O}(h^2)$ were used in the preceding analysis, the local discretization error is also $\mathcal{O}(h^2)$.

As far as relating the local discretization errors to global errors, so long as second-order finite difference formulas are used for all derivatives, the global discretization error will also be second order. It can be shown that the resulting system of linear equations from the discretization of second-order BVPs is diagonally dominant, which is a *necessary condition* for the convergence of a tridiagonal system.

Higher-order ODEs result in a banded system of linear equations that can be solved using an iterative method or Gauss elimination (*see* Section 2.5). An advantage of using second-order central difference formulas is that discretizing a two-point BVP always results in a tridiagonal system from which a direct solution can be obtained. Nevertheless, dealing with difference equations with global error $\mathcal{O}(h^2)$ may result in a very large number of unknowns to attain the desired accuracy. Hence, it is sometimes tempting to use higher-order finite difference approximations for the derivatives to avoid dealing with large linear systems. In practice, a major difficulty in applying this strategy to a two-point BVP is encountered when implementing the BCs. For instance, the central difference formulas for y_i' and y_i'' of order $\mathcal{O}(h^4)$ are:

$$y_i' \cong \frac{-y_{i+2} + 8y_{i+1} - 8y_{i-1} + y_{i-2}}{12h},$$

$$y_i'' \cong \frac{-y_{i+2} + 16y_{i+1} - 30y_i + 16y_{i-1} - y_{i-2}}{12h^2}$$

Note that when implementing the approximations, for example, by setting $i = 0$ for the left boundary node, two fictitious nodes, namely y_{-1} and y_{-2}, will appear in the difference equation. To eliminate the fictitious nodes, two BCs need to be specified.

Another approach to obtaining high order, more accurate numerical solutions is to utilize the Richardson extrapolation technique, which was used to increase the order of accuracy of numerical differentiations (*see* Section 5.7) or integrations (*see* Section 8.3). Applying this technique to the numerical solution of the ODEs does not require discretization. For instance, to obtain fourth-order accurate solutions, two sets of numerical solutions with subinterval sizes h and $h/2$ are required. For a sixth-order accurate approximation, three sets of numerical approximations of order $\mathcal{O}(h^2)$ with interval sizes h, $h/2$, and $h/4$ are obtained. The extrapolation procedure is then employed, as follows:

$$\text{1st extrapolation} \quad \tilde{y}_{1,i} = \frac{1}{3}\left(4y_i\left(\frac{h}{2}\right) - y_i(h)\right) + \mathcal{O}(h^4)$$

$$\text{2nd extrapolation} \quad \tilde{y}_{2,i} = \frac{1}{3}\left(4y_i\left(\frac{h}{4}\right) - y_i\left(\frac{h}{2}\right)\right) + \mathcal{O}(h^4)$$

$$\text{3rd extrapolation} \quad \tilde{y}_i = \frac{1}{15}\left(16\tilde{y}_{2,i} - \tilde{y}_{1,i}\right) + \mathcal{O}(h^6)$$

where $y_i(h)$, $y_i(h/2)$, and $y_i(h/4)$ are the numerical approximations obtained with the step sizes of h, $h/2$, and $h/4$, respectively, and $\tilde{y}_{1,i}$, $\tilde{y}_{2,i}$, and $\tilde{y}_i$ are the numerical approximations obtained with the Richardson extrapolation.

EXAMPLE 10.3: Applying Richardson extrapolation to BVPs

A rocket body ($t = 1$ cm in thickness, $D_o = 20$ cm in outer diameter) is made of an aluminum alloy ($E = 64$ GPa, $\nu = 0.33$). For cylinders, the radial stress is given in terms of the displacement by $\sigma_r = E/(1 - \nu^2)(du/dr + \nu u/r)$, where E is the elasticity modulus, ν is the Poisson ratio, r is the radial coordinate, and $u = u(r)$ is the radial displacement satisfying the following second-order ODE:

$$r^2 \frac{d^2u}{dr^2} + r\frac{du}{dr} - u = 0, \qquad R_i \leqslant r \leqslant R_o$$

Assuming negligible external pressure on the body of the rocket ($P_o = 0$) in flight conditions, the internal pressure reaches $p_i = 25.6$ kPa (*see* figure below). The BCs for the event can be expressed as

$$r = R_i, \quad \sigma_r(R_i) = \frac{E}{1 - \nu^2}\left(\frac{du}{dr} + \nu\frac{u}{r}\right)_{r=R_i} = -P_i$$

$$r = R_o, \quad \sigma_r(R_o) = \frac{E}{1 - \nu^2}\left(\frac{du}{dr} + \nu\frac{u}{r}Big\right)_{r=R_o} = 0$$

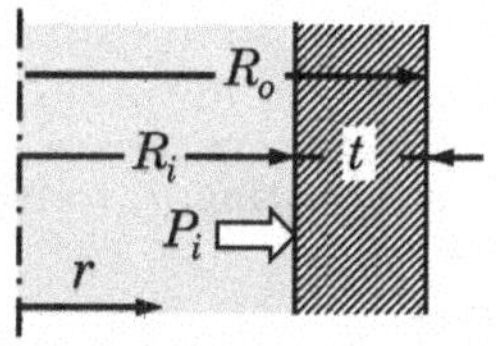

Apply FDM with uniform $M = 5$ and 10 intervals to (a) obtain 2nd-order numerical estimates; (b) apply Richardson's extrapolation to find numerical estimates of order $\mathcal{O}(h^4)$; (c) compare the results with the true solution given by $u(r) = 1.14253r + 0.02268/r$ (in mm).

SOLUTION:

(a) The underlying assumption in this analysis is that the rocket is long enough to permit one-dimensional analysis, $L/R_o \gg 1$. To be able to employ the Richardson extrapolation, we need two discrete sets of numerical estimates of order $\mathcal{O}(h^2)$. The FDM steps are presented below:

Step 1: Uniform grids are established by dividing the $R_i \leqslant r \leqslant R_o$ interval into M subintervals, $h = (R_o - R_i)/M = t/M$, as shown in Fig. 10.7. Nodal points are placed at the end points of each subinterval and are enumerated from 0 to M. The nodal values, unknowns, are labeled as $u(r_i) = u_i$ for $i = 0, 1, 2, \ldots, M$, where $r_i = R_i + ih$.

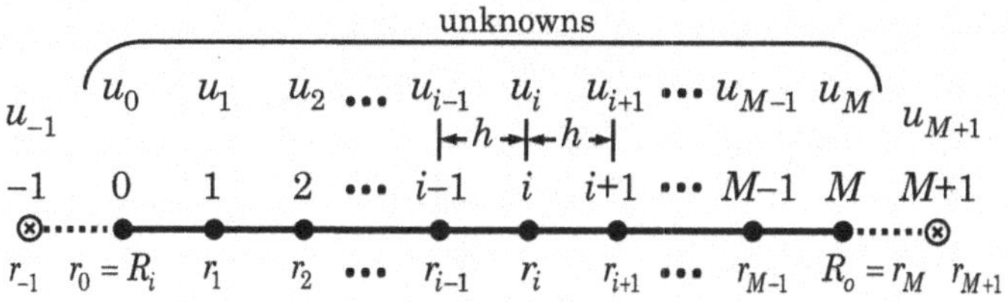

FIGURE 10.7

Step 2: The differential equation is satisfied for every nodal point $r_i \in [R_i, R_o]$. For an arbitrary r_i, the BVP is written as

$$r_i^2 \frac{d^2r_i}{dr^2} + r_i\frac{du_i}{dr} - u(r_i) = 0, \qquad R_i \leqslant r_i \leqslant R_o$$

or substituting the CDFs for du_i/dr and d^2u_i/dr^2 yields

$$r_i^2\left(\frac{u_{i+1} - 2u_i + u_{i-1}}{h^2}\right) - r_i\left(\frac{u_{i+1} - u_{i-1}}{2h}\right) - u_i = 0 \qquad (10.25)$$

Rearranging Eq. (10.25), the *general difference equation* is obtained as

$$b_i u_{i-1} + d_i u_i + a_i u_{i+1} = 0, \qquad i = 0, 1, 2, \ldots, M \tag{10.26}$$

where

$$b_i = \frac{r_i}{h}\left(\frac{r_i}{h} + \frac{1}{2}\right), \quad d_i = -\left(1 + 2\frac{r_i^2}{h^2}\right), \quad a_i = \frac{r_i}{h}\left(\frac{r_i}{h} - \frac{1}{2}\right), \quad c_i = 0$$

Step 3: To implement BCs, we start by setting $i = 0$ in Eq. (10.26), giving

$$b_0 u_{-1} + d_0 u_0 + a_0 u_1 = 0 \tag{10.27}$$

Discretizing the left BC at r_0 ($i = 0$) with the CDF yields

$$\frac{u_1 - u_{-1}}{2h} + \nu \frac{u_0}{r_0} = -(1 - \nu^2)\frac{P_i}{E}$$

and isolating u_{-1} gives

$$u_{-1} = u_1 + 2h\nu\frac{u_0}{r_0} + 2h(1 - \nu^2)\frac{P_i}{E}$$

Substituting this result into Eq. (10.27) and collecting similar terms results in

$$\left(d_0 + \frac{2h\nu}{r_0}b_0\right)u_0 + \left(a_0 + b_0\right)u_1 = -2hb_0(1 - \nu^2)\frac{P_i}{E},$$

For the right BC, setting $i = M$ in Eq. (10.26) gives

$$b_M u_{M-1} + d_M u_M + a_M u_{M+1} = 0 \tag{10.28}$$

where u_{M+1} corresponds to a fictitious nodal value. To eliminate this node, the right BC is discretized using the central difference formula as

$$\sigma_r(r_M) = \frac{E}{1 - \nu^2}\left(\frac{u_{M+1} - u_{M-1}}{2h} + \nu\frac{u_M}{r_M}\right) = 0$$

Solving for u_{M+1} results in

$$u_{M+1} = u_{M-1} - 2h\nu\frac{u_M}{r_M} \tag{10.29}$$

Substituting this into Eq. (10.28) and collecting similar terms yields

$$\left(a_M + b_M\right)u_{M-1} + \left(d_M - 2h\nu\frac{a_M}{r_M}\right)u_M = 0 \tag{10.30}$$

Step 4: Combining the difference equations in matrix form, a linear system of $(M + 1) \times (M + 1)$ equations is obtained as

$$\begin{bmatrix} \bar{d}_0 & \bar{a}_0 & & & & \\ b_1 & d_1 & a_1 & & & \\ & b_2 & d_2 & a_2 & & \\ & & \ddots & \ddots & \ddots & \\ & & & b_{M-1} & d_{M-1} & a_{M-1} \\ & & & & \bar{b}_M & \bar{d}_M \end{bmatrix} \begin{bmatrix} u_0 \\ u_1 \\ u_2 \\ \vdots \\ u_{M-1} \\ u_M \end{bmatrix} = \begin{bmatrix} c_1 \\ 0 \\ 0 \\ \vdots \\ 0 \\ 0 \end{bmatrix} \tag{10.31}$$

where $\bar{d}_0 = d_0 + 2h\nu b_0/r_0$, $\bar{a}_0 = a_0 + b_0$, $\bar{d}_M = d_M - 2h\nu a_M/r_M$, and $\bar{b}_M = a_M + b_M$.

Step 5: The tridiagonal system, Eq. (10.31), is solved for $M = 5$ ($h = 0.002$) and 10 ($h = 0.001$) as a first step for obtaining the numerical estimates of order $\mathcal{O}(h^4)$. Then, the Richardson extrapolation procedure is carried out as follows:

$$\tilde{u}_i = \frac{1}{3}\left(4\,u2_i - u1_i\right) + \mathcal{O}(h^4) \quad \text{for } i = 0, .., 5$$

where $u2_i$ and $u1_i$ are the numerical estimates corresponding to the ith node for $M = 10$ and 5, respectively.

TABLE 10.3

		True Solution	Estimates, $\mathcal{O}(h^2)$		Estimate, $\mathcal{O}(h^4)$
i	x_i	u_i	$h = 0.002$	$h = 0.001$	$\tilde{u}_i$
0	0.090	0.354828	0.354538	0.354755	0.354827
1	0.092	0.351634	0.351349	0.351563	0.351634
2	0.094	0.348674	0.348393	0.348604	0.348674
3	0.096	0.345933	0.345656	0.345863	0.345932
4	0.098	0.343397	0.343124	0.343328	0.343396
5	0.100	0.341053	0.340784	0.340985	0.341052

The true solution (in mm), second-order numerical estimates with $h = 0.002$ and 0.001, and the extrapolations are comparatively tabulated in Table 10.3. The average displacement in the rocket body is 0.35 mm, and the maximum displacement is on the inner surface. The average error for estimates with $h = 0.002$, 0.001, and Richardson extrapolation is about 2.8×10^{-4}, 7×10^{-5}, and 5×10^{-7}, respectively. Notice that the error of the estimates with the $h = 0.001$ case is reduced by $(1/2)^2$ compared to that of the $h = 0.002$ case. The fourth-order estimate obtained by the Richardson extrapolation yields five- to six-decimal-place accurate estimates with very few grid points.

Discussion: In this example, it is shown that Richardson extrapolation is a method that can be used to improve the order of accuracy of numerical solutions of ODEs uniformly discretized by the finite difference method. By combining the results of two sets of numerical solutions obtained by discretizing with different interval sizes (h and $h/2$), the leading truncation error terms can be eliminated, resulting in highly accurate results.

10.5 NON-UNIFORM GRIDS

As we have already seen, the first step in obtaining a numerical solution to an ODE is the gridding of the physical domain. A suitably designed grid results in good-quality numerical estimates; conversely, a poorly constructed grid yields poor numerical estimates. In many instances, the difficulties with numerical solutions may be attributed to poorly constructed grids. For this reason, gridding is a very important first step that should not be underestimated to obtain quality solutions.

In general, the finite difference method with uniform grids is simple and very accurate, and the numerical solution of most ODEs can be carried out using uniform grids. On the other hand, using a uniform grid does not always yield satisfactory estimates for cases with boundary layers. For example, if the solution to an ODE describing the behavior of a physical

FIGURE 10.8: Grid stretching from left to right, or right to left.

property (temperature, velocity, concentration, etc.) depicts a sharp gradient at one or both ends of the solution interval, then these changes cannot be adequately resolved with uniform grids. In order to accurately estimate the gradients within a boundary layer, a very small or fine grid should be adopted. If a uniform fine grid is applied throughout the domain, then the number of nodes (i.e., the size of the resulting matrix) increases undesirably. The purpose of using non-uniform grids is, in essence, to increase the grid density near the boundaries, where sharp changes are expected. With the computational power of present-day computers, a one-time solution of a very large system of linear equations may not be viewed as a big problem, but in cases where a system of linear equations is solved many times in an iterative procedure, the cpu time can be seriously reduced by adopting non-uniform grids, leading to smaller linear systems.

Grid generation for ODEs (one-dimensional cases) is one of the simplest tasks of gridding. Depending on the physics of the problem, grids can be concentrated at one or both ends of the interval or at an interior point. There are many techniques for grid generation, covering a considerable portion of computational work. This section provides a brief introduction to analytical means of grid generation.

Generating arbitrary grids can be done manually by the user or, generally, using a grid generation algorithm. For instance, stretched grids can be obtained by specifying the first interval size and a growth factor. Letting h be the first interval size, then the size of the subsequent intervals is determined as hr, hr^2, hr^3, and so on, where r is the growth factor. If $r > 1$, then grid points are clustered near $x = a$ or near $x = b$ for $r < 1$ (see Fig. 10.8). To generate a grid with M subintervals, only a first step size (h) and a growth factor (r) are sufficient. The first step size for an arbitrary interval $[a, b]$ can be determined from

$$h = \frac{1-r}{1-r^M}(b-a)$$

and the abscissas are determined from $x_i = a+h(1-r^i)/(1-r)$ for $i = 1, 2, \ldots, M$. The ODE is then discretized by replacing the derivatives with the finite difference formulas derived for *non-uniform grids*. This avenue increases the computational burden of constructing the coefficient matrix (i.e., due to the coefficients of a generalized difference equation containing complicated expressions).

An alternative technique requires the coordinate transformation of the physical domain (x-space) into a computational domain (ξ-space), to which uniform grids can be applied. First, the ODE is transformed from x-space to an equivalent ODE in ξ-space. Then, the transformed ODE is solved with uniform grids ($\Delta\xi$). In this technique, the grid stretching function $x = x(\xi)$ dictates the nodal distribution in x-space. For instance, a simple stretching function involving power functions such as $x = a + (b-a)\,\xi^n$, where $0 \leqslant \xi \leqslant 1$, may be used to construct non-uniform grids on $[a, b]$. The nodes are clustered near the left boundary for $n > 1$ and near the right boundary for $n < 1$ (see Fig. 10.9). More elaborate and specialized grid stretching functions involving logarithmic, exponential, hyperbolic, or trigonometric functions are available [7, 32, 36].

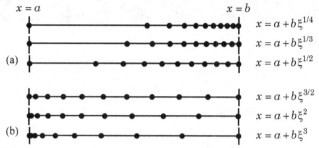

FIGURE 10.9: Grid stretching using $x = a + c\,\xi^n$ as stretching function examples of (a) $n < 1$, and (b) $n > 1$.

Use of Non-Uniform Grids

- Numerical treatment of a BVP using a suitable grid stretching function can lead to very satisfactory estimates;
- Properly designed non-uniform grids may result in a significant reduction in the number of nodes.

- Numerical treatment of the original BVP with non-uniform grids may not yield second-order global errors (*see* Section 5.6),
- A BVP in a transformed space can be very complicated.

Consider transforming the two-point BVP given by Eq. (10.1) into a BVP in ξ-space using the grid stretching function $x = a + c\,\xi^n$. The y' and y'' are transformed into derivatives with respect to the variable ξ. Applying the chain rule leads to

$$\frac{dy}{dx} = \frac{dy}{d\xi}\frac{d\xi}{dx} = \frac{dy}{d\xi}\left(\frac{1}{cn\,\xi^{n-1}}\right) \tag{10.32}$$

Next, applying the chain rule one more time and making use of Eq. (10.32) gives

$$\frac{d^2 y}{dx^2} = \frac{d}{d\xi}\left(\frac{dy}{dx}\right)\frac{d\xi}{dx} = \frac{d}{d\xi}\left(\frac{dy}{d\xi}\frac{1}{cn\xi^{n-1}}\right)\left(\frac{1}{cn\xi^{n-1}}\right) \tag{10.33}$$

Applying the product rule to differentiation and simplifying leads to

$$\frac{d^2 y}{dx^2} = \frac{1}{c^2 n^2 \xi^{2n-2}}\frac{d^2 y}{d\xi^2} + \frac{1-n}{c^2 n^2 \xi^{2n-1}}\frac{dy}{d\xi} \tag{10.34}$$

Substituting $x = a + c\,\xi^n$ and Eqs. (10.32) and (10.34) into Eq. (10.1) gives the BVP in the transformed coordinate:

$$\frac{p(\xi)}{c^2 n^2 \xi^{2n-2}}\frac{d^2 y}{d\xi^2} + \left(\frac{p(\xi)(1-n)}{c^2 n^2 \xi^{2n-1}} + \frac{q(\xi)}{cn\,\xi^{n-1}}\right)\frac{dy}{d\xi} + r(\xi)y(\xi) = f(\xi),$$

$$\bar{\alpha}_1\frac{dy}{d\xi}(0) + \bar{\beta}_1 y(0) = \bar{\gamma}_1, \quad \bar{\alpha}_2\frac{dy}{d\xi}(1) + \bar{\beta}_2 y(1) = \bar{\gamma}_2 \quad (0 \leqslant \xi \leqslant 1) \tag{10.35}$$

where $\bar{\alpha}_1$, $\bar{\beta}_1$, $\bar{\gamma}_1$, $\bar{\alpha}_2$, $\bar{\beta}_2$, and $\bar{\gamma}_2$ denote the transformed coefficients. However, a word of caution is in order: the stretching function should be chosen such that the ODE and its BCs are defined in the transformed space.

EXAMPLE 10.4: Coordinate transformation with a stretching function

Consider the following two-point BVP:

$$2xy'' + (20x + 1)y' + 10y = 0, \qquad y(0) = 1, \quad y'(1) + 10y(1) = 0$$

(a) Use $x = \xi^2$ as a stretching function to transform the BVP in x-space to a BVP in ξ-space; (b) solve the transformed BVP using the FDM with $M = 10$, 20, and 40 uniform intervals; (c) solve the original BVP using the FDM with the same number of uniform intervals. The true solution is given as $y(x) = e^{-10x}$.

SOLUTION:

(a) Employing the proposed stretching function, $x = \xi^2$, with uniform $\Delta\xi$ (*see* Fig. 10.10) leads to the concentration of nodes at the left boundary as illustrated in the figure below:

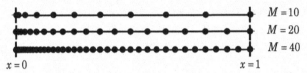

To recast the BVP in ξ-space, the derivatives are also transformed into ξ-space. Applying the chain rule and noting that $dx = 2\xi d\xi$, the first derivative is easily obtained as

$$\frac{dy}{dx} = \frac{dy}{d\xi}\frac{d\xi}{dx} = \frac{1}{2\xi}\frac{dy}{d\xi}$$

Next, applying the chain rule to dy/dx yields

$$\frac{d^2y}{dx^2} = \frac{d}{dx}\left(\frac{dy}{dx}\right) = \frac{d}{d\xi}\left(\frac{dy}{dx}\right)\frac{d\xi}{dx} = \frac{d}{d\xi}\left(\frac{1}{2\xi}\frac{dy}{d\xi}\right)\frac{1}{2\xi} = \frac{1}{4\xi^2}\left(\frac{d^2y}{d\xi^2} - \frac{1}{\xi}\frac{dy}{d\xi}\right)$$

Now, substituting y' and y'' into the ODE and BCs and simplifying the terms gives

$$\frac{d^2y}{d\xi^2} + 20\xi\frac{dy}{d\xi} + 20y(\xi) = 0$$

$$y(0) = 1, \quad \frac{dy}{d\xi}(1) + 20y(1) = 0$$

Note that the original domain is also transformed to $0 \leq \xi \leq 1$ in ξ-space.

(b) The step-by-step solution with the FDM is detailed below:

Step 1: First, the ξ-domain is divided into M uniform subintervals, and nodes are placed at the end of every subinterval. The nodes are enumerated, and the abscissas are calculated as $\xi_i = i\Delta\xi$ $(i = 0, 1, 2, \ldots, M)$, as depicted in Fig. 10.10.

FIGURE 10.10

Step 2: The transformed ODE is satisfied for every ξ_i:

$$\left(\frac{d^2y}{d\xi^2}\right)_{\xi=\xi_i} + 20\xi_i\left(\frac{dy}{d\xi}\right)_{\xi=\xi_i} + 20y_i = 0, \qquad \xi_i \in (0,1]$$

Replacing $dy/d\xi$ and $d^2y/d\xi^2$ with their CDFs approximations yields

$$\frac{y_{i+1} - 2y_i + y_{i-1}}{(\Delta\xi)^2} + 20\xi_i\frac{y_{i+1} - y_{i-1}}{2\,\Delta\xi} + 20y_i = 0, \qquad i = 1, 2, \ldots, M$$

Rearranging and collecting similar terms, the general difference equation becomes

$$\frac{1}{\Delta\xi}\left(\frac{1}{\Delta\xi} - 10\xi_i\right)y_{i-1} + 2\left(10 - \frac{1}{(\Delta\xi)^2}\right)y_i + \frac{1}{\Delta\xi}\left(\frac{1}{\Delta\xi} + 10\xi_i\right)y_{i+1} = 0$$

Step 3: Notice that the value of the left boundary node is known, i.e., $y_0 = 1$. Substituting $i = 1$ and y_0 into the general difference equation and carrying the known quantities to the rhs gives

$$2\left(10 - \frac{1}{(\Delta\xi)^2}\right)y_1 + \frac{1}{\Delta\xi}\left(\frac{1}{\Delta\xi} + 10\xi_1\right)y_2 = \frac{1}{\Delta\xi}\left(\frac{1}{\Delta\xi} - 10\xi_1\right)$$

Likewise, the difference equation for the right boundary node is obtained by setting $i = M$ in the general difference equation:

$$\frac{1}{\Delta\xi}\left(\frac{1}{\Delta\xi} - 10\xi_M\right)y_{M-1} + 2\left(10 - \frac{1}{(\Delta\xi)^2}\right)y_M + \frac{1}{\Delta\xi}\left(\frac{1}{\Delta\xi} + 10\xi_M\right)y_{M+1} = 0$$

which contains the fictitious node y_{M+1}.

Discretizing the right BC with the CDFs, we find

$$\frac{y_{M+1} - y_{M-1}}{2\,\Delta\xi} + 20y_M = 0 \quad \text{or} \quad y_{M+1} = y_{M-1} - 40\Delta\xi\,y_M$$

Substituting y_{M+1} into the difference equation for $i = M$ and simplifying the terms leads to

$$\frac{2}{(\Delta\xi)^2}y_{M-1} + \left(20 - \frac{2}{(\Delta\xi)^2} - \frac{40}{\Delta\xi} - 400\xi_M\right)y_M = 0,$$

The difference equations for $2 \leqslant i \leqslant (M-1)$ (interior nodes) are easily obtained from the general difference equation without the need for further action.

Step 4: By putting all the difference equations together, a matrix equation with a tridiagonal coefficient matrix is obtained.

Step 5: The resulting tridiagonal system is solved using Thomas' algorithm.

The true solution, as well as the true (absolute) errors arising from the FDM solution obtained with uniform $M = 10$, 20, and 40 grids, are tabulated in Table 10.4. This is the case when the nodes are concentrated at the left boundary. The solutions agree well with those of the case with $M = 10$ yielding an average error of 0.0107. As the number of grids increases by two folds, the errors also decrease by four folds. For instance, the estimates depict noteworthy improvements for $M = 20$ and 40, yielding average errors of 2.62×10^{-3} and 6.52×10^{-4}, respectively. The true errors are also observed to be small at the left boundary.

TABLE 10.4

x	ξ	True Solution	Absolute true error		
			$M = 10$	$M = 20$	$M = 40$
0	0	1	0	0	0
0.01	0.1	0.904837	4.83E-03	1.16E-03	2.88E-04
0.04	0.2	0.670320	1.58E-02	3.81E-03	9.44E-04
0.09	0.3	0.406570	2.47E-02	5.98E-03	1.48E-03
0.16	0.4	0.201897	2.57E-02	6.22E-03	1.54E-03
0.25	0.5	0.082085	1.91E-02	4.70E-03	1.17E-03
0.36	0.6	0.027324	1.05E-02	2.67E-03	6.71E-04
0.49	0.7	0.007447	4.29E-03	1.16E-03	2.95E-04
0.64	0.8	0.001662	1.29E-03	3.84E-04	1.00E-04
0.81	0.9	0.000304	2.77E-04	9.40E-05	2.50E-05
1	1	0.000045	3.90E-05	1.20E-05	3.00E-06

(c) The steps of the FDM solution of the original BVP with uniform Δx are left as an exercise for the reader. The true solution, as well as the true errors arising from the FDM estimates obtained with uniform $M = 10$, 20, and 40 subintervals, are tabulated in Table 10.5. The numerical estimates with $M = 10$ depict large deviations (average absolute error 2.11×10^{-1}) near the left boundary. Even though the numerical solution improves with increasing M, the resulting average errors are still larger than those of the non-uniform case.

TABLE 10.5

ξ	x	True Solution	Absolute true error		
			$M = 10$	$M = 20$	$M = 40$
0	0	1	0	0	0
0.316	0.1	0.367879	1.10E-01	2.65E-02	6.97E-03
0.447	0.2	0.135335	5.75E-02	1.47E-02	3.84E-03
0.548	0.3	0.049787	2.54E-02	6.86E-03	1.80E-03
0.632	0.4	0.018316	1.05E-02	2.99E-03	7.90E-04
0.707	0.5	0.006738	4.23E-03	1.26E-03	3.34E-04
0.775	0.6	0.002479	1.66E-03	5.15E-04	1.38E-04
0.837	0.7	0.000912	6.44E-04	2.07E-04	5.60E-05
0.894	0.8	0.000336	2.46E-04	8.10E-05	2.30E-05
0.949	0.9	0.000123	9.10E-05	3.00E-05	8.00E-06
1	1	0.000045	3.20E-05	1.00E-05	3.80E-04

Discussion: While the true distribution depicts a very rapid change up to $x \approx 0.5$, the rate of change slows down for $x > 0.5$. In Part (b), we obtain better numerical estimates by solving the BVP in ξ-space because more nodes are staggered at the left boundary, where the rapid change is observed. On the other hand, in Part (c), the uniform grids in the original BVP lead to fewer nodes near the left boundary, where a sharp change occurs. It is clear that wisely placed nodes make a big difference in obtaining very good estimates.

It should be kept in mind that the numerical solution of an ODE, regardless of the true distribution of the ODE, will approach the true solution provided that fine enough grids are used. In practice, when a numerical solution with a moderate number of subintervals is sought, non-uniform gridding could be most helpful.

10.6   FINITE VOLUME METHOD

Finite Volume Method (FVM) is also a widely used numerical method to discretize BVPs. The basis of this method is the integral conservation law.

Consider the following two-point BVP, subjected to the following mixed BCs:

$$\frac{d}{dx}\left(p(x)\frac{dy}{dx}\right) + q(x)\frac{dy}{dx} + r(x)y(x) = g(x), \qquad a < x < b$$

$$\alpha_1 y'(a) + \beta_1 y(a) = \gamma_1, \qquad \alpha_2 y'(b) + \beta_2 y(b) = \gamma_2 \tag{10.36}$$

where $p(x)$, $q(x)$, $r(x)$, and $g(x)$ may be piecewise continuous functions.

The BVPs in the form of Eq. (10.36) are also encountered in steady-state diffusion problems in layered media, where each layer consists of a different material of homogeneous composition. As a result, p, q, r, and g are typically material-dependent (step-wise varying) quantities, which may depict jump discontinuities at the interfaces. In such cases, the physical property $y(x)$ as well as the flux $p(x)dy/dx$ must satisfy what are called the *interface conditions*. Mathematically speaking, the continuity of the physical property and flux at an interface point $x = x_c$ is expressed as

$$\text{Continuity of physical property:} \quad y(x_c^-) = y(x_c^+)$$

$$\text{Continuity of physical flux :} \quad p(x_c^-)\frac{dy}{dx}\bigg|_{x_c^-} = p(x_c^+)\frac{dy}{dx}\bigg|_{x_c^+} \tag{10.37}$$

It is noted that the FVM is especially ideal for the numerical treatment of problems with jump discontinuities. The steps for the numerical solution of FVM are given as follows:

Step 1. Gridding: The gridding of the computational interval is similar to that of FDM. First, the solution interval is divided into M subintervals called *cells*, which are fine enough to ensure the resulting difference equations adequately approximate the ODE (*see* Fig. 10.11). Then, physical nodes are placed at the endpoints of each cell. A set of abscissas $\{x_i\}$ denoting these nodes is generated with the cell size $\Delta x_i = x_{i+1} - x_i$ that is allowed to be nonuniform. In the subsequent derivations, for generality, the cell size is assumed to be nonuniform.

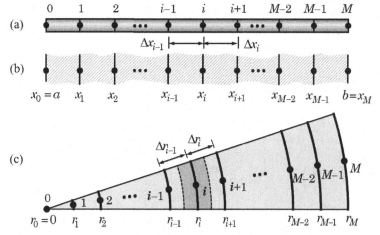

FIGURE 10.11: General grid configuration and finite volume cells for one-dimensional systems (a) rod, (b) plane parallel layers, (c) cylindrical or spherical shells.

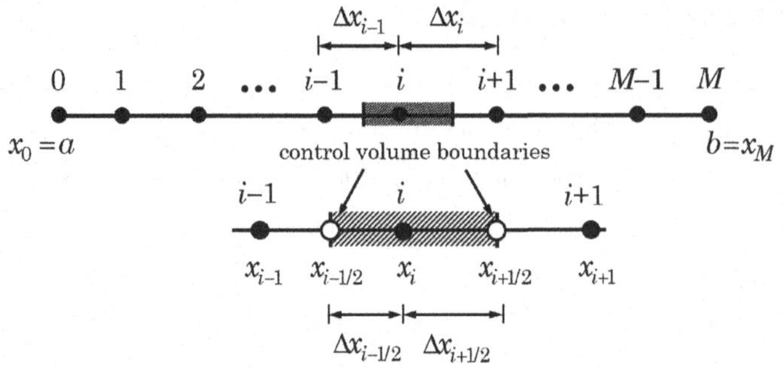

Δx_{i-1} Δx_i

0 1 2 ... i–1 i i+1 ... M–1 M

$x_0 = a$

control volume boundaries

$b = x_M$

i–1 i i+1

x_{i-1} $x_{i-1/2}$ x_i $x_{i+1/2}$ x_{i+1}

$\Delta x_{i-1/2}$ $\Delta x_{i+1/2}$

FIGURE 10.12: General grid configuration and a typical control volume surrounding a node at x_i.

A *control volume* (CV) is designated as $x_{i-1/2} \leqslant x \leqslant x_{i+1/2}$ by placing nodal (boundary) points into the middle of the cells (*see* Fig. 10.12). The abscissas of boundary nodes on either side of x_i are obtained with $x_{i-1/2} = x_i - \Delta x_{i-1}/2$ and $x_{i+1/2} = x_i + \Delta x_i/2$. Near the physical boundaries ($x = a$ and $x = b$), the CVs are created in such a way that the physical boundaries coincide with the CV boundaries, i.e., $x_0 \leqslant x \leqslant x_{1/2}$ and $x_{M-1/2} \leqslant x \leqslant x_M$.

> The volume element for the cylindrical and spherical geometries with radial symmetry is $dV = 2\pi r dr$ or $dV = 4\pi r^2 dr$, respectively. In the discretization of ODEs using the FVM, both sides of the ODE are multiplied by dV. Therefore, the constant multipliers in dV are usually dropped, leaving only $dV = r dr$ or $dV = r^2 dr$.

Step 2. Discretizing: The essence of the FVM is to integrate the governing ODE over each control volume, i.e., $x_{i-1/2} \leqslant x \leqslant x_{i+1/2}$. The quantities p, q, r, and g are assumed to be constant (uniform) in each CV, allowing stepwise variations. In subsequent discussions, we also assume Eqs. (10.36) and (10.37) make up a mathematical model (BVP) of a physical event in a rod or a plane parallel geometry, where a finite cell volume is proportional to the cross-section area, i.e., $dV = A\,dx$.

Equation (10.36) is integrated over ith cell, $x_{i-1/2} \leqslant x \leqslant x_{i+1/2}$, that is,

$$\int_{x_{i-1/2}}^{x_{i+1/2}} \frac{d}{dx}\left(p(x)\frac{dy}{dx}\right)dx + \int_{x_{i-1/2}}^{x_{i+1/2}} q(x)\frac{dy}{dx}dx + \int_{x_{i-1/2}}^{x_{i+1/2}} r(x)y(x)dx = \int_{x_{i-1/2}}^{x_{i+1/2}} g(x)dx \quad (10.38)$$

where $p(x)dy/dx$ is continuous. Each term in Eq. (10.38) is then replaced with a suitable approximation.

The first integral is split into two parts:

$$\int_{x_{i-1/2}}^{x_{i+1/2}} \frac{d}{dx}\left(p(x)\frac{dy}{dx}\right)dx = \int_{x_{i-1/2}}^{x_i^-} d\left(p(x)\frac{dy}{dx}\right) + \int_{x_i^+}^{x_{i+1/2}} d\left(p(x)\frac{dy}{dx}\right)$$

$$= p(x_i^-)\frac{dy}{dx}\bigg)_{x_i^-} - p(x_{i-1/2})\frac{dy}{dx}\bigg)_{x_{i-1/2}} + p(x_{i+1/2})\frac{dy}{dx}\bigg)_{x_{i+1/2}} - p(x_i^+)\frac{dy}{dx}\bigg)_{x_i^+}$$

Note that this equation holds whether x_i is an interface node or not.

Next, the continuity of flux at the ith node is invoked, and the derivatives at cell midpoints ($x_{i-1/2}$ and $x_{i+1/2}$) are replaced by their central difference approximations as follows:

$$\int_{x_{i-1/2}}^{x_{i+1/2}} \frac{d}{dx}\left(p(x)\frac{dy}{dx}\right)dx \cong p_{i+1/2}\left(\frac{y_{i+1}-y_i}{\Delta x_i}\right) - p_{i-1/2}\left(\frac{y_i-y_{i-1}}{\Delta x_{i-1}}\right) \qquad (10.39)$$

where $p(x_{i+1/2}) = p_{i+1/2}$ and $p(x_{i-1/2}) = p_{i-1/2}$.

The second term of Eq. (10.38) is also split into two parts as

$$\int_{x_{i-1/2}}^{x_{i+1/2}} q(x)\frac{dy}{dx}dx = \int_{x_{i-1/2}}^{x_i} q(x)\frac{dy}{dx}dx + \int_{x_i}^{x_{i+1/2}} q(x)\frac{dy}{dx}dx$$

For sufficiently small cells, the first derivatives are replaced by the CDFs, and $q(x)$ is calculated at the cell boundaries ($q_{i+1/2}$ and $q_{i-1/2}$) as

$$\int_{x_{i-1/2}}^{x_{i+1/2}} q(x)\frac{dy}{dx}dx \cong q_{i+1/2}\left(\frac{y_{i+1}-y_i}{\Delta x_i}\right)\frac{\Delta x_i}{2} + q_{i-1/2}\left(\frac{y_i-y_{i-1}}{\Delta x_{i-1}}\right)\frac{\Delta x_{i-1}}{2} \qquad (10.40)$$

This approximation also allows for a piecewise variation of $q(x)$ to be included.

The integrals of the third and fourth terms are simply expressed as follows:

$$\int_{x_{i-1/2}}^{x_{i+1/2}} r(x)y(x)dx = \int_{x_{i-1/2}}^{x_i} r(x)y(x)dx + \int_{x_i}^{x_{i+1/2}} r(x)y(x)dx \cong \frac{1}{2}\left(r_{i+1/2}\Delta x_i + r_{i-1/2}\Delta x_{i-1}\right)y_i$$
$$(10.41)$$

$$\int_{x_{i-1/2}}^{x_{i+1/2}} g(x)dx = \int_{x_{i-1/2}}^{x_i} g(x)dx + \int_{x_i}^{x_{i+1/2}} g(x)dx \cong \frac{1}{2}\left(g_{i+1/2}\Delta x_i + g_{i-1/2}\Delta x_{i-1}\right) \qquad (10.42)$$

Substituting the approximations Eqs. (10.39)-(10.42) into Eq. (10.38), for an interior ith cell, gives

$$p_{i+1/2}\left(\frac{y_{i+1}-y_i}{\Delta x_i}\right) - p_{i-1/2}\left(\frac{y_i-y_{i-1}}{\Delta x_{i-1}}\right) + q_{i+1/2}\left(\frac{y_{i+1}-y_i}{2}\right) + q_{i-1/2}\left(\frac{y_i-y_{i-1}}{2}\right)$$
$$+ \frac{1}{2}\left(r_{i+1/2}\Delta x_i + r_{i-1/2}\Delta x_{i-1}\right)y_i = \frac{1}{2}\left(g_{i+1/2}\Delta x_i + g_{i-1/2}\Delta x_{i-1}\right) \qquad (10.43)$$

or in a compact form as

$$b_i y_{i-1} + d_i y_i + a_i y_{i+1} = c_i, \qquad (10.44)$$

where

$$b_i = \frac{p_{i-1/2}}{\Delta x_{i-1}} - \frac{q_{i-1/2}}{2}, \qquad d_i = -a_i - b_i + \frac{1}{2}\left(r_{i+1/2}\Delta x_i + r_{i-1/2}\Delta x_{i-1}\right),$$

$$a_i = \frac{p_{i+1/2}}{\Delta x_i} + \frac{q_{i+1/2}}{2}, \qquad c_i = \frac{1}{2}\left(g_{i+1/2}\Delta x_i + g_{i-1/2}\Delta x_{i-1}\right)$$

Step 3. Implementing BCs: Implementing Dirichlet BC at node $i = 0$ or $i = M$ is straightforward. We consider the case of left BC ($\alpha_1 = 0$). This procedure can likewise be extended to the right BC ($\alpha_2 = 0$). The nodal value y_0, just as in FDM, is substituted into the difference equation obtained by setting $i = 1$ into Eq. (10.44) and carried to the rhs. Alternatively, since the solution at the boundary node is known, $y_0 = \gamma_1$ can be treated as a difference equation. Then the coefficients of the discretized equation are determined simply by setting $\beta_1 = 1$, $d_0 = 1$, $c_0 = \gamma_1$, and $a_0 = 0$, i.e., y_0 is incorporated into the linear system as if it were an unknown.

In the case of the Neumann or Robin BCs ($\alpha_1 y_0' + \beta_1 y_0 = \gamma_1$, $\alpha_1 \neq 0$), Eq. (10.36) is integrated over a control volume extending from x_0 to $x_{1/2}$, i.e., over a half-cell.

$$\int_{x_0}^{x_{1/2}} d\left(p(x)\frac{dy}{dx}\right) + \int_{x_0}^{x_{1/2}} q(x)\frac{dy}{dx}dx + \int_{x_0}^{x_{1/2}} r(x)y(x)dx = \int_{x_0}^{x_{1/2}} g(x)dx, \qquad (10.45)$$

The first term of Eq. (10.45) gives

$$\int_{x_0}^{x_{1/2}} d\left(p(x)\frac{dy}{dx}\right) = p(x_{1/2})\frac{dy}{dx}\Big)_{x_{1/2}} - p(x_0)\frac{dy}{dx}\Big)_{x_0}$$

Replacing $(dy/dx)_{x_{1/2}}$ with the central difference approximation and evaluating $(dy/dx)_{x_0}$ from the left BC as $y_0' = (\gamma_1 - \beta_1 y_0)/\alpha_1$ yields

$$\int_{x_0}^{x_{1/2}} d\left(p(x)\frac{dy}{dx}\right) \cong p_{1/2}\left(\frac{y_1 - y_0}{\Delta x_1}\right) - p_0\left(\frac{\gamma_1 - \beta_1 y_0}{\alpha_1}\right) \qquad (10.46)$$

Note that, unlike in the FDM, no fictitious nodes arise in the FVM discretization of Neumann or Robin BCs.

The second, third, and fourth terms of Eq. (10.45) are approximated by

$$\int_{x_0}^{x_{1/2}} q(x)\frac{dy}{dx}dx \cong q_{1/2}\left(\frac{y_1 - y_0}{\Delta x_0}\right)\frac{\Delta x_0}{2} \qquad (10.47)$$

$$\int_{x_0}^{x_{1/2}} r(x)y(x)dx \cong \frac{\Delta x_0}{2}r_0 y_0 \quad \text{and} \quad \int_{x_0}^{x_{1/2}} g(x)dx \cong \frac{\Delta x_0}{2}g_0 \qquad (10.48)$$

where $(dy/dx)_{x_{1/2}}$ is replaced by its central difference approximation.

Using approximations by Eq. (10.46), (10.47), and (10.48) yields

$$d_0 y_0 + a_0 y_1 = c_0, \qquad (10.49)$$

where

$$a_0 = \frac{p_{1/2}}{\Delta x_0} + \frac{q_{1/2}}{2}, \quad d_0 = -a_0 + \frac{\beta_1}{\alpha_1}p_0 + \frac{\Delta x_0}{2}r_0, \quad c_0 = \frac{\Delta x_0}{2}g_0 - \frac{\gamma_1}{\alpha_1}p_0$$

The foregoing derivation is extended to the right boundary node over the control volume $x_{M-1/2} \leqslant x \leqslant x_M$. For the right boundary node, the difference equation results in

$$b_M y_{M-1} + d_M y_M = c_M, \qquad (10.50)$$

where

$$b_M = \frac{p_{M-1/2}}{\Delta x_{M-1}} - \frac{q_{M-1/2}}{2}, \quad d_M = -b_M - \frac{\beta_2}{\alpha_2}p_M + \frac{\Delta x_{M-1}}{2}r_M, \quad c_M = \frac{\Delta x_{M-1}}{2}g_M - \frac{\gamma_2}{\alpha_2}p_M$$

Step 4. Constructing a linear system: The FVM equations for all interior nodes, Eq. (10.44), and those equations derived for the boundary nodes, Eqs. (10.49) and (10.50), constitute a tridiagonal system of linear equations that can be expressed as

$$\mathbf{Ay} = \mathbf{b}$$

Pseudocode 10.2

Module LBVP_FVM $(M, \mathbf{x}, \boldsymbol{\alpha}, \boldsymbol{\beta}, \boldsymbol{\gamma}, \mathbf{y})$
\ DESCRIPTION: A pseudomodule to solve a two-point linear BVP, defined in
\   conservation form: $(p(x)y')' + q(x)y' + r(x)y = g(x)$, $a \leqslant x \leqslant b$
\   subject to BCs: $\alpha_1 y'(a) + \beta_1 y(a) = \gamma_1$, $\alpha_2 y'(b) + \beta_2 y(b) = \gamma_2$ using the FVM.
\ USES:
\   COEFFS:: Module supplying $p(x)$, $q(x)$, $r(x)$, $g(x)$;
\   TRIDIAGONAL:: Tridiagonal system solver, Pseudocode 2.13
Declare: α_2, β_2, γ_2, a_M, b_M, c_M, d_M, x_M, y_M dx_M \ Declare array variables
For $[k = 1, M - 1]$
    $dx_k \leftarrow x_{k+1} - x_k$ \ Find cell sizes, $\Delta x_k = x_{k+1} - x_k$
End For
For $[k = 1, M]$ \ Setup of tridiagonal matrix row-wise
    $xk \leftarrow x_k$; $hp \leftarrow dx_k$; $xp \leftarrow xk + hp/2$ \ Scalarize x_k, dx_k, $x_{k+1/2}$
    COEFFS(xp, pp, qp, rp, gp) \ Find $p_{k+1/2}$, $q_{k+1/2}$, $r_{k+1/2}$, $g_{k+1/2}$
    If $[k = 1]$ **Then**   \ Modify coefficients for left boundary
       If $[\alpha_1 = 0]$ **Then**   \ Set up Dirichlet BC
          $d_k \leftarrow 1$; $c_k \leftarrow \gamma_1/\beta_k$; $a_k \leftarrow 0$; $b_k \leftarrow 0$
       Else   \ Set up Neumann or Robin BC
          $a_k \leftarrow pp/hp + qp/2$
          $d_k \leftarrow -a_k + rp * hp/2 + pp * \beta_1/\alpha_1$ \ Apply Eq. (10.49)
          $b_k \leftarrow 0$; $c_k \leftarrow gp * hp/2 - pp * \gamma_1/\alpha_1$
       End If
    Else
       If $[k = M]$ **Then**   \ Modify coefficients for right boundary
         If $[\alpha_2 = 0]$ **Then**   \ Set up Dirichlet BC
            $d_k \leftarrow 1$; $c_k \leftarrow \gamma_2/\beta_2$; $a_k \leftarrow 0$; $b_k \leftarrow 0$
         Else   \ Set up Neumann or Robin BC
            $b_k \leftarrow pm/hm - qm/2$;
            $d_k \leftarrow -b_k + rm * hm/2 - pm * \beta_2/\alpha_2$ \ Apply Eq. (10.50)
            $a_k \leftarrow 0$; $c_k \leftarrow gm * hm/2 - pm * \gamma_2/\alpha_2$
         End If
       Else
         \ Set up coefficients for interior nodes, $2 \leqslant k \leqslant m - 1$
         $b_k \leftarrow pm/hm - qm/2$ \ Apply Eq. (10.44)
         $a_k \leftarrow pp/hp + qp/2$
         $d_k \leftarrow -a_k - b_k + (rp * hp + rm * hm)/2$
         $c_k \leftarrow (gp * hp + gm * hm)/2$
       End If
    End If
    $hm \leftarrow hp$ \ Set $hm = hp$ for the next cell ($\Delta x_k = \Delta x_{k+1}$)
    $pm \leftarrow pp$; $qm \leftarrow qp$ \ Set coefficients, $p_{k-1/2} = p_{k+1/2}$, $q_{k-1/2} = q_{k+1/2}$)
    $rm \leftarrow rp$; $gm \leftarrow gp$ \ $r_{k-1/2} = r_{k+1/2}$, $g_{k-1/2} = g_{k+1/2}$ for the next cell
End For
TRIDIAGONAL $(1, M, \mathbf{b}, \mathbf{d}, \mathbf{a}, \mathbf{c}, \mathbf{y})$ \ Solve the tridiagonal system of equations
End Module LBVP_FVM

> **Finite Volume Method**
>
>
> - Discretization procedure requires integration of the ODE over a finite control volume, allowing the interface conditions of the discontinuous source terms as well as the Neumann or Robin BCs to be implemented naturally;
> - Unlike finite difference and finite element methods, the FVM does not require coordinate transformation to ODEs;
> - It is suitable for both uniform and non-uniform grids;
> - The discretized equations are conservative, i.e., mass, momentum, and energy are conserved for each finite (cell) volume.
>
>
> - Order of accuracy of FVM is $\mathcal{O}(h^2)$, and it is very difficult to achieve higher orders of accuracy;
> - Computational cost of BVPs with variable coefficients increases.

where $\mathbf{A}$ is the coefficient matrix, $\mathbf{r}$ is the rhs vector, and $\mathbf{y}$ is the numerical solution defined as

$$
\mathbf{A} = \begin{bmatrix} d_0 & a_0 & & & & \\ b_1 & d_1 & a_1 & & & \\ & b_2 & d_2 & a_2 & & \\ & & \ddots & \ddots & \ddots & \\ & & & b_{M-1} & d_{M-1} & a_{M-1} \\ & & & & b_M & d_M \end{bmatrix}, \quad \mathbf{y} = \begin{bmatrix} y_0 \\ y_1 \\ y_2 \\ \vdots \\ y_{M-1} \\ y_M \end{bmatrix}, \quad \mathbf{b} = \begin{bmatrix} c_0 \\ c_1 \\ c_2 \\ \vdots \\ c_{M-1} \\ c_M \end{bmatrix}
$$

Step 5. Solving the linear system: The resulting system of linear algebraic equations is solved by using Thomas' algorithm. Recall that the necessary condition for the convergence of a tridiagonal system (Section 2.10) is that the coefficient matrix must be diagonally dominant. The matrix equations resulting from FVM discretization are diagonally dominant; thus, convergence problems should not be expected.

A pseudomodule, **LBVP_FVM**, solving a linear BVP in conservation form using the FVM is presented in Pseudocode 10.2. The module requires the number of grid points (M), coefficients of the BCs $(\alpha_i, \beta_i, \gamma_i$ for $i = 1,2)$, and abscissas of the nodal points $(x_i, i = 1, 2, \ldots, M)$ as input. The numerical solution is saved in the output array $\mathbf{y}$. This module uses two additional modules: (1) **COEFFS**, a user-defined module supplying $p(x)$, $q(x)$, $r(x)$, and $g(x)$ for any x; and (2) **TRIDIAGONAL**, a module solving a tridiagonal system of equations in Pseudocode 2.13. This module allows the use of non-uniform cell sizes with ease. But, the array containing the abscissas $(\mathbf{x})$ needs to be prepared as input to the module before invoking it.

After computing Δx_k's, the tridiagonal system elements are set up row by row using the **For**-construct from $k = 1$ to M. Using the **COEFFS**, the coefficients of the BVP $(pp, qp, rp,$ and $gp)$ are computed at $x_{k+1/2}$. For $1 < k < M$, the coefficients of the difference equation (row entries) are calculated from Eq. (10.44). The computation of coefficients of difference equations (row elements) according to node number (left, right, or interior) and BC type (Dirichlet, Neumann, or Robin) is provided by a series of **If-Then-Else** constructs. The Dirichlet BCs are incorporated into the linear system by treating the boundary nodes as unknowns, as done in Pseudocode 10.2. The Neumann or Robin boundary conditions are implemented with Eqs. (10.49) and (10.50) for the left ($k = 1$) and right ($k = M$) boundary nodes, respectively. Once all elements of a row are calculated, the coefficients of the ODE and cell size at $x_{k+1/2}$ are no longer needed. So these are assigned as the quantities of the cell on the left, i.e., $pm \leftarrow pp$, $qm \leftarrow qp$, $rm \leftarrow rp$, $gm \leftarrow gp$, and $hm \leftarrow hp$ correspond to the values at $x_{k-1/2}$. The solution of the tridiagonal system is carried out in TRIDIAGONAL using the Thomas algorithm, and the solution is returned in $\mathbf{y}$.

EXAMPLE 10.5: Application of FVM to heat transfer from a fuel rod

Consider a cylindrical nuclear fuel pin ($k_f = 16$ W/m°C) with radius $R = 5$ cm, enclosed by a 2 cm thick steel cladding ($k_c = 50$ W/m°C, outer radius $R_o = 7$ cm), as shown in the figure. During steady operation, heat is generated in the nuclear fuel at a uniform rate of $q_0''' = 10^6$ W/m³. The fuel assembly in the pressurized reactor is cooled by water at $T_\infty = 132$°C, which adjoins the outer surface, and it is characterized by a convection coefficient of $h = 38000$ W/m²°C.

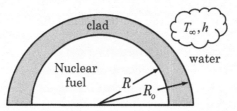

Assuming a one-dimensional conduction heat transfer condition exists, the temperature distribution for this system is governed by the following ODE:

$$\frac{1}{r}\frac{d}{dr}\left(rk(r)\frac{dT}{dr}\right) + q'''(r) = 0, \qquad 0 \leqslant r \leqslant R_o$$

where r is the radial coordinate (m), k is the thermal conductivity (W/m·°C), $q'''(r)$ is the internal heat generation (W/m³), and T is the temperature (°C). The conductivity and heat generation depict a jump discontinuity at the fuel-clad interface, i.e.,

$$k(r) = \begin{cases} k_f, & r < R \\ k_c, & r > R \end{cases}, \qquad q'''(r) = \begin{cases} q_0''', & r < R \\ 0, & r > R \end{cases}$$

The corresponding boundary and interface conditions are given as follows:

Radial symmetry at $r = 0$ $(dT/dr)_{r=0} = 0$

Continuity of $T(r)$ at $r = R$ $T(R^-) = T(R^+)$

Continuity of $q(r)$ at $r = R$ $-k_f(dT/dr)_{r=R^-} = -k_c(dT/dr)_{r=R^+}$

Convection transfer at $r = R_o$ $-k_c(dT/dr)_{r=R_o} = h(T(R_o) - T_\infty)$

Apply the FVM to obtain the temperature distribution within the fuel and clad using uniform $M = 7, 14, 28$, and 56 cells. Comment on the accuracy of the distribution.

SOLUTION:

An ODE of this type is very common in fluid flow and heat transfer problems. The steady-state diffusion equation (neglecting the transient and convective terms) yields

$$\text{div}(D\,\text{grad}\,\phi) + S_\phi = 0 \qquad (10.51)$$

where D is the diffusion coefficient, S_ϕ is called the source, and ϕ is a transport property.

Equation (10.51) for one-dimensional geometries can be expressed as

$$\frac{1}{r^g}\frac{d}{dr}\left(r^g D(r)\frac{d\phi}{dr}\right) + S_\phi(r) = 0 \tag{10.52}$$

where $g = 0$, 1, and 2 denote plane parallel, cylinder, and sphere geometry, respectively, and r denotes the axial coordinate for the plane parallel coordinate or the radial coordinate for the cylinder or sphere. In this example, the property ϕ represents the temperature.

Step 1. Gridding: Gridding is carried out by dividing the fuel with cladding regions into M uniform cells of size $\Delta r = R_o/M$. Nodal points are placed on cell boundaries at r_i, as shown in Fig. 10.13. The positions of the cell boundaries and midpoints are obtained by $r_i = i\Delta r$ and $r_{i+1/2} = r_i + \Delta r/2$, respectively, and M is chosen such that $r_N = N\Delta r$. Note that the cell boundary nodes coincide with the physical boundaries.

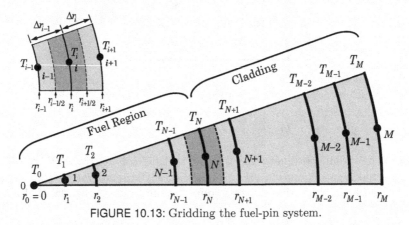

FIGURE 10.13: Gridding the fuel-pin system.

Step 2. Discretizing: To discretize, first the governing ODE is multiplied by the cylindrical volume element $dV = r\,dr$ (omitting 2π) and integrated over the finite cell volume, $r_{i-1/2} \leqslant r \leqslant r_{i+1/2}$:

$$\int_{r_{i-1/2}}^{r_{i+1/2}} \frac{d}{dr}\left(rk(r)\frac{dT}{dr}\right)dr = -\int_{r_{i-1/2}}^{r_{i+1/2}} rq'''(r)dr$$

The material conductivity varies step-wise at $r = r_N$; thus, the integral is split into two parts, where the heat flux $-k(r)dT/dr$ is continuous on both sides of the interface (i.e., interface condition). This equality holds for any node placed at r_i, whether it is an interface point or not. Keeping this in mind, the term on the left of the equality sign can be integrated as follows:

$$\int_{r_{i-1/2}}^{r_{i+1/2}} \frac{d}{dr}\left(rk(r)\frac{dT}{dr}\right)dr = \int_{r_{i-1/2}}^{r_i} \frac{d}{dr}\left(rk(r)\frac{dT}{dr}\right)dr + \int_{r_i}^{r_{i+1/2}} \frac{d}{dr}\left(rk(r)\frac{dT}{dr}\right)dr$$

$$= r_i\left(k\frac{dT}{dr}\right)_{r_i^-} - r_{i-1/2}k_{i-1/2}\frac{dT}{dr}\bigg)_{r_{i-1/2}} + r_{i+1/2}k_{i+1/2}\frac{dT}{dr}\bigg)_{r_{i+1/2}} - r_i\left(k\frac{dT}{dr}\right)_{r_i^+}$$

where $k_{i\pm1/2} = k(r_{i\pm1/2})$ denote the right- and left-hand cell conductivities, respectively.

Invoking the continuity of heat flux at $r = r_i$, $(kdT/dr)_{r_i^-} = (kdT/dr)_{r_i^+}$, and replacing the derivatives at $r_{i-1/2}$ and $r_{i+1/2}$ with the CDFs yields

$$\int_{r_{i-1/2}}^{r_{i+1/2}} \frac{d}{dr}\left(kr\frac{dT}{dr}\right)dr \cong r_{i+1/2}k_{i+1/2}\frac{T_{i+1} - T_i}{\Delta r} - r_{i-1/2}k_{i-1/2}\frac{T_i - T_{i-1}}{\Delta r}$$

Making use of Eq. (10.42), the term on the right-hand side becomes:

$$\int_{r_{i-1/2}}^{r_{i+1/2}} rq'''(r)dr = \int_{r_{i-1/2}}^{r_i} rq'''(r)dr + \int_{r_i}^{r_{i+1/2}} rq'''(r)dr \cong \frac{\Delta r}{2}\left(r_{i-1/2}q'''_{i-1/2} + r_{i+1/2}q'''_{i+1/2}\right)$$

Finally, the general form of the discretized equations for interior nodes can be written as

$$b_i T_{i-1} + d_i T_i + a_i T_{i+1} = c_i, \quad (i = 1, 2, \ldots, M - 1)$$

where

$$a_i = \left(\frac{r_i}{\Delta r} + \frac{1}{2}\right)k_{i+1/2}, \quad c_i = -\left(r_{i-1/2}q'''_{i-1/2} + r_{i+1/2}q'''_{i+1/2}\right)\frac{\Delta r}{2}$$

$$b_i = \left(\frac{r_i}{\Delta r} - \frac{1}{2}\right)k_{i-1/2}, \quad d_i = -a_i - b_i$$

Step 3. Implementing BCs: The left and right BCs are Neumann and Robin types, respectively. To obtain the finite volume equations for the boundary cells, we integrate the cells over the half-control volumes.

Noting that the conductivity and heat generation are $k = k_f$ and $q''' = q'''_0$ at the left boundary node, integration of Eq. (10.52) over $[r_0, r_{1/2}]$ yields

$$\int_{r_0}^{r_{1/2}} \frac{d}{dr}\left(rk(r)\frac{dT}{dr}\right)dr + \int_{r_0}^{r_{1/2}} rq'''(r)dr = r_{1/2}k_f\left(\frac{dT}{dr}\right)_{r_{1/2}} - r_0 k_f\left(\frac{dT}{dr}\right)_{r_0} + r_{1/2}q'''_0\frac{\Delta r}{2} = 0$$

In this expression, the second term on the rhs is zero since $r_0 = 0$ and also $(dT/dr)_{r=0} = 0$. The integral over the source term is approximated with the midpoint rule. The temperature gradient at $r_{1/2}$ is replaced with the central difference approximation as $(dT/dr)_{r_{1/2}} \cong (T_1 - T_0)/\Delta r$.

Substituting the foregoing approximations into the above expression and simplifying leads to

$$d_0 T_0 + a_0 T_1 = c_0$$

where $a_0 = k_f/2$, $d_0 = -a_0$, and $c_0 = -r_{1/2} q'''_0 \Delta r/2$.

For the right boundary cell (with $k = k_c$ and $q''' = 0$), integration of Eq. (10.52) over $[r_{M-1/2}, r_M]$ gives

$$\int_{r_{M-1/2}}^{r_M} \frac{d}{dr}\left(rk(r)\frac{dT}{dr}\right)dr = r_M k_c\left(\frac{dT}{dr}\right)_{r_M} - r_{M-1/2}k_c\left(\frac{dT}{dr}\right)_{r_{M-1/2}} = 0$$

Note that $k_c(dT/dr)_{r_M}$ can be obtained from the convection transfer BC, which can be expressed as $k_c dT_M/dr = h(T_\infty - T_M)$. The temperature gradient at $r_{M-1/2}$ is approximated with the CDF as $(dT/dr)_{r_{M-1/2}} \cong (T_M - T_{M-1})/\Delta r$. Substituting these approximations into the above equation yields

$$-hr_M(T_M - T_\infty) - r_{M-1/2}k_c\frac{T_M - T_{M-1}}{\Delta r} = 0$$

or rearranging and simplifying similar terms leads to

$$b_M T_{M-1} + d_M T_M = c_M$$

where $b_M = k_c r_{M-1/2}/\Delta r$, $d_M = -a_M - hr_M$, and $c_M = -hr_M T_\infty$.

Step 4. Constructing a linear system: We find $M - 1$ equations for the interior nodes $(0 < i < M)$, an equation for the left node $(i = 0)$, and one for the right node $(i = M)$, which together constitute a linear system of $M + 1$ equations. They can be expressed in matrix form as

$$\begin{bmatrix} d_0 & a_0 & & & & & \\ b_1 & d_1 & a_1 & & & & \\ & \ddots & \ddots & \ddots & & & \\ & & b_N & d_N & a_N & & \\ & & & \ddots & \ddots & \ddots & \\ & & & & b_{M-1} & d_{M-1} & a_{M-1} \\ & & & & & b_M & d_M \end{bmatrix} \begin{bmatrix} T_0 \\ T_1 \\ \vdots \\ T_N \\ \vdots \\ T_{M-1} \\ T_M \end{bmatrix} = \begin{bmatrix} c_0 \\ c_1 \\ \vdots \\ c_N \\ \vdots \\ c_{M-1} \\ c_M \end{bmatrix}$$

Step 5. Solving the linear system: The solution of this tridiagonal system of equations for $M = 7, 14, 28$, and 56 gives numerical estimates for the ODE. (Finding the true solution of this ODE is quite an easy task and is left to the reader as an exercise.)

TABLE 10.6

	True	True Error, absolute			
x	Solution	$M = 7$	$M = 14$	$M = 28$	$M = 56$
0	320.944	2.773	0.828	0.241	0.069
0.01	319.382	1.210	0.307	0.077	0.019
0.02	314.694	0.689	0.174	0.044	0.011
0.03	306.882	0.376	0.094	0.023	0.006
0.04	295.944	0.154	0.039	0.010	0.003
0.05	281.881	0.019	0.004	0.001	0
0.06	277.323	0.007	0.001	0	0.001
0.07	273.469	0.001	0.001	0.001	0.001

Discussion: The true solution, as well as the true errors, are comparatively presented in Table 10.6. The maximum absolute errors are obtained at the center of the fuel pin as 2.773, 0.828, 0.241, and 0.069°C for $M = 7, 14, 28$, and 56, respectively. The FVM is a second-order accurate method; thus, when doubling the grid size, the global errors are reduced by a factor of four on average. The numerical estimates improve with an increasing number of cells due to the reduction in the truncation error as the cell size decreases.

10.7   FINITE DIFFERENCE SOLUTION NONLINEAR BVPS

Consider the following nonlinear two-point BVP subject to Dirichlet BCs:

$$y'' = f(x, y, y'), \qquad a \leqslant x \leqslant b, \quad y(a) = A, \quad y(b) = B \qquad (10.53)$$

where A and B are known real constants. Furthermore, assume that f is continuous on the domain $D = \{(x, y, y') \in [a, b], y, y' \in \mathbb{R}\}$, $\partial f / \partial y$, and $\partial f / \partial y'$ exist and are continuous functions on D, $\partial f / \partial y > 0$, and $\partial f / \partial y'$ is bounded on D. The preceding assumptions guarantee that the BVP defined by Eq. (10.53) has a unique solution.

The finite difference method introduced in this section can be used to numerically solve nonlinear two-point BVPs. However, the resulting system of discretized equations yields a simultaneous system of nonlinear equations. In this section, we will present in detail the steps for solving nonlinear two-point BVPs subjected to Dirichlet BCs. The numerical procedure can be similarly and easily extended to nonlinear two-point BVPs with mixed BCs. The numerical solution procedure follows the same basic steps: gridding, discretizing, implementing BCs, constructing, and solving the system of equations.

Step 1. Gridding. The computational domain ($a \leqslant x \leqslant b$) is divided into M-subintervals to create uniformly or non-uniformly spaced grid nodes. Nevertheless, for the sake of simplicity, we adopt uniform subintervals: $h = (b - a)/M$. The nodes are enumerated starting from 0 to M. For given grids similar to those shown in Fig. 10.2, the abscissas are calculated: $x_i = a + ih$, where $i = 0, 1, 2, \ldots, M$.

Step 2. Discretizing. Equation (10.53) is satisfied for an arbitrary node x_i in (a, b): $y_i'' = f(x_i, y_i, y_i')$. The discretization is carried out by approximating the derivatives of y with the CDFs:

$$\frac{y_{i+1} - 2y_i + y_{i-1}}{h^2} = f\left(x_i, y_i, \frac{y_{i+1} - y_{i-1}}{2h}\right), \qquad (10.54)$$

which satisfies both true and approximate solutions of the nonlinear ODE. In this context, from now on, we will use y_i to denote a numerical (estimate) solution.

Step 3. Implementing the BCs. Implementing the Dirichlet BCs is as simple as expressing them as difference equations such that $y_0 - A = 0$ and $y_M - B = 0$.

Step 4. Constructing a nonlinear system. Rearranging Eq. (10.54) yields the following simultaneous system of non-linear equations:

$$\mathbf{g(y)} = \mathbf{0} \qquad (10.55)$$

where $\mathbf{y} = [y_0, y_1, \cdots, y_{M-1}, y_M]^T$ and $\mathbf{g} = [g_0, g_1, \cdots, g_{M-1}, g_M]^T$, and

$$g_0 = y_0 - A$$
$$g_i = y_{i+1} - 2y_i + y_{i-1} - h^2 f\left(x_i, y_i, \frac{y_{i+1} - y_{i-1}}{2h}\right), \quad i = 1, 2, \ldots, M-1$$
$$g_M = y_M - B$$

Step 5. Solving the nonlinear system. This nonlinear system is solved iteratively using a method such as fixed-point iteration, Newton's method, etc. Here, we will describe the implementation of two techniques: (1) Tridiagonal Iteration and (2) Newton's methods.

In nonlinear problems, regardless of the numerical method employed, the initial guess is crucial for the convergence of the BVPs. Fortunately, a nonlinear two-point BVP subjected to Dirichlet BCs is unique in that the distribution of the numerical solution can be roughly estimated with the aid of the boundary values. For instance, in Eq. (10.53), the numerical solution varies between $y(a) = A$ and $y(b) = B$. Hence, a *smart initial guess* for each nodal point can be obtained through linear interpolation between (a, A) and (b, B) as

$$y_i = A + \frac{B - A}{b - a}(x_i - a), \qquad i = 1, 2, \ldots, (M - 1) \tag{10.56}$$

10.7.1 TRIDIAGONAL ITERATION METHOD (TIM)

The left-hand side of Eq. (10.54) yields a tridiagonal matrix whose diagonal elements are constant. Hence, an iterative scheme is established by casting Eq. (10.55) as

$$\mathbf{T}\mathbf{y}^{(p+1)} = \mathbf{r}(\mathbf{y}^{(p)}) \tag{10.57}$$

where $\mathbf{y}$ denotes the estimates of the nodal points, $\mathbf{T}$ and $\mathbf{r}(\mathbf{y})$ are defined as

$$\mathbf{T} = \begin{bmatrix} 1 & 0 & & & \\ 1 & -2 & 1 & & \\ & \ddots & \ddots & \ddots & \\ & & 1 & -2 & 1 \\ & & & 0 & 1 \end{bmatrix}, \quad \mathbf{r}(\mathbf{y}) = \begin{bmatrix} A \\ h^2 f\left(x_1, y_1^{(p)}, \frac{y_2^{(p)} - y_0^{(p)}}{2h}\right) \\ \vdots \\ h^2 f\left(x_{M-1}, y_{M-1}^{(p)}, \frac{y_M^{(p)} - y_{M-2}^{(p)}}{2h}\right) \\ B \end{bmatrix}$$

Note that when the Robin BC is imposed to the left, right, or both boundaries, then only the first, last, or both rows will need to be modified.

Pseudocode 10.3

Module NONLINEAR_TIM (M, ε, $\mathbf{x}$, $\mathbf{yo}$, $\mathbf{y}$, $\boldsymbol{\alpha}$, $\boldsymbol{\beta}$, $\boldsymbol{\gamma}$, $maxit$)
\ DESCRIPTION: A pseudomodule to solve a nonlinear two-point BVP of the
\   form: $y'' = f(x, y, y')$, $\quad a \leqslant x \leqslant b$ subject to BCs:
\ $\alpha_1 y'(a) + \beta_1 y(a) = \gamma_1$, $\quad \alpha_2 y'(b) + \beta_2 y(b) = \gamma_2$, using TIM method.
\ USES:
\   FUNC:: A user supplied Function Module defining $f(x, y, y')$;
\   ENORM:: A function calculating ℓ_2-norm of a vector, Pseudocode 3.1;
\   TRIDIAGONAL:: Tridiagonal system solver, Pseudocode 2.13.
Declare: α_2, β_2, γ_2, a_M, b_M, c_M, d_M, x_M, y_M yo_M \ Declare array variables
$h \leftarrow x_2 - x_1$ \ Set interval size
$h2 \leftarrow 2 * h;$ $\quad hsqr \leftarrow h * h$ \ Save $2h$ & h^2 on scalar variables
$p \leftarrow 0$ \ Initialize iteration counter
Repeat   \ Iteration loop
$\quad$ **For** $\big[k = 1, M\big]$ \ Construct coefficient (tridiagonal) matrix
$\quad\quad$ **If** $\big[k = 1\big]$ **Then**   \ Set up coefficients for left boundary
$\quad\quad\quad$ **If** $\big[\alpha_1 = 0\big]$ **Then**   \ Set up Dirichlet BC
$\quad\quad\quad\quad d_k \leftarrow 1;$ $\quad c_k \leftarrow \gamma_1/\beta_1$
$\quad\quad\quad\quad a_k \leftarrow 0;$ $\quad b_k \leftarrow 0$
$\quad\quad\quad$ **Else**   \ Set up Neumann or Robin BC
$\quad\quad\quad\quad yp \leftarrow (\gamma_1 - \beta_1 * yo_k)/\alpha_1$ \ Compute y_1'

$$a_k \leftarrow 2; \quad b_k \leftarrow 0; \; d_k \leftarrow -2 + h2 * \beta_1/\alpha_1$$
$$c_k \leftarrow hsqr * \textbf{FUNC}(x_k, yo_k, yp) + h2 * \gamma_1/\alpha_1$$
    End If
   Else
    If $\left[k = M\right]$ **Then**   \ Set up coefficients for right boundary
     If $\left[\alpha_2 = 0\right]$ **Then**   \ Set up Dirichlet BC
$$d_k \leftarrow 1; \quad c_k \leftarrow \gamma_2/\beta_2; \quad a_k \leftarrow 0; \quad b_k \leftarrow 0$$
     Else   \ Set up Neumann or Robin BC
$$yp \leftarrow (\gamma_2 - \beta_2 * yo_k)/\alpha_2 \qquad\qquad \text{\textbackslash Evaluate } y_n'$$
$$b_k \leftarrow 2; \quad a_k \leftarrow 0; \; d_k \leftarrow -2 - h2 * \beta_2/\alpha_2$$
$$c_k \leftarrow hsqr * \textbf{FUNC}(x_k, yo_k, yp) - h2 * \gamma_2/\alpha_2$$
     End If
    Else   \ Set up coefficients for interior nodes by Eq. (10.57)
$$yp \leftarrow (yo_{k+1} - yo_{k-1})/h2 \qquad\qquad \text{\textbackslash Evaluate } y_k'$$
$$c_k \leftarrow hsqr * \textbf{FUNC}(x_k, yo_k, yp) \qquad\qquad\qquad \text{\textbackslash Find rhs}$$
$$a_k \leftarrow 1; \quad d_k \leftarrow -2; \; b_k \leftarrow 1 \qquad \text{\textbackslash Set up lhs–coefficients of TDM}$$
    End If
   End If
  End For
  TRIDIAGONAL$(1, M, \mathbf{b}, \mathbf{d}, \mathbf{a}, \mathbf{c}, \mathbf{y})$ \ Solve tridiagonal system
  $error \leftarrow \textbf{ENORM}(M, \mathbf{y} - \mathbf{yo})$ \ Find Euclidean norm, $|\mathbf{y}^{(p)} - \mathbf{y}^{(p-1)}|_2$
  $p \leftarrow p + 1$ \ Count iterations
  Write: "p, error=", p, $error$ \ Print out iteration progress
  yo $\leftarrow$ **y**   \ Set current estimates as prior
 Until $\left[error < \varepsilon \text{ Or } p = maxit\right]$
 If $\left[p = maxit\right]$ **Then**   \ Issue info and a warning
  Write: "Failed to converge after", $maxit$, "iterations"
  Write: "Current error is ", $error$
 End If
 End Module NONLINEAR_TIM

A pseudo-module, NONLINEAR_TIM, solving a non-linear BVP given with Eq. (10.53) using the TIM is presented in Pseudocode 10.3. As input, the module requires the number of grid points (M), an array of abscissas (x_i), an initial guess $(yo_i, i = 1, 2, \ldots, M)$, an upper bound for the maximum number of iterations allowed $(maxit)$, a convergence tolerance (ε), and the BCs $(\alpha_i, \beta_i, \gamma_i, i = 1,2)$. On exit, the numerical solution is saved in the array $\mathbf{y}$. The module also makes use of two external modules: (1) FUNC, supplying $y'' = f(x_i, y_i, y_i')$; and (2) TRIDIAGONAL, a module for solving a tridiagonal system of equations (Pseudocode 2.14). Only the first and last rows of the difference equations are modified to incorporate the BCs. A Dirichlet BC (e.g., $y_i = value$) can be easily incorporated into the system by simply setting $d_i = 1$, $a_i = b_i = 0$, and $c_i = value$, where $i = 1$ or M. In the case of Neumann or Robin BC $(\alpha_1 \neq 0$ or $\alpha_2 \neq 0)$, the fictitious node will appear in the difference equations. A fictitious node is eliminated by applying the same usual procedure as for the linear BVPs. The nonlinear terms are kept on the rhs of Eq. (10.57). The construction of $\mathbf{T}$ is easy since all rows consist of 1's and -2's. The tridiagonal system of equations is solved at each iteration using the Thomas algorithm. It is important to start with a "smart guess," $\mathbf{y}^{(0)}$ in order for the system to converge or make the system converge faster. After updating the rhs $\mathbf{r}(\mathbf{y})$ with the updated estimates, the procedure by which Eq. (10.57) is solved is repeated until the current and prior estimates converge within a predetermined tolerance value, i.e., $\|\mathbf{y}^{(p+1)} - \mathbf{y}^{(p)}\|_2 < \varepsilon$.

Finite Difference Methods for Nonlinear ODEs

- Finite difference methods are ideal for treating unstable nonlinear problems;
- The resulting tridiagonal system of equations can be efficiently solved using Thomas's algorithm;
- If a BVP converges, it converges quadratically.

- FDM may lead to a large number of simultaneous nonlinear equations, requiring iterative schemes to achieve an approximate solution with the desired accuracy;
- A smart initial guess (or a rough prediction) may be necessary to ensure convergence.

EXAMPLE 10.6: Solving a nonlinear BVP

A gas, labeled as X, in a sealed container diffuses into the liquid Y (shown in the figure) that undergoes the second-order reaction $X + Y \rightarrow XY$. Assuming the amount of the reaction product XY is negligible, a simplified mathematical model for the concentration of diffusing X into Y can be expressed by the following ODE:

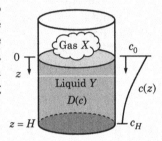

$$\frac{d}{dz}\left(D(c)\frac{dc}{dz}\right) - kc^2 = 0$$

where c is the concentration of X, k is the rate constant of the reaction, D is the diffusion coefficient, which is linearly proportional to the concentration ($D(c) = D_0\, c(z)$), and z is the coordinate axis. Introducing dimensionless quantities, the BVP and the BCs are rewritten as

$$\frac{d}{dZ}\left(C\frac{dC}{dZ}\right) - \phi^2 C^2 = 0, \qquad C(0) = 1, \qquad \frac{dC}{dZ}(1) = 0$$

where $C = c(z)/c_0$, $Z = z/H$, and $\phi^2 = h^2 k/D_0$, c_0 is the gas concentration at $z = 0$ (liquid surface), H is the depth of the liquid, and ϕ is a parameter called the Thiele modulus. For $\phi^2 = 1.25$, use the TIM iteration scheme to obtain the numerical solution for uniformly spaced $M = 10, 20, 40$, and 80 grids. The true solution is given as

$$C(Z) = \sqrt{\cosh(\phi\sqrt{2}(1 - Z))/\cosh(\phi\sqrt{2})}$$

SOLUTION:

To facilitate understanding of the application of the theoretical formulation presented, in this example we replace the dependent variable C with y (i.e., $C(Z) \rightarrow y(x)$) and the independent variable Z with x (i.e., $Z \rightarrow x$). Then the BVP can be expressed in the standard nonlinear form as

$$y'' = f(x, y, y') = \phi^2 y - (y')^2/y, \qquad y(0) = 1, \qquad y'(1) = 0$$

Step 1: Gridding. The interval $[0,1]$ is divided into M subintervals $(h = 1/M)$, with nodes placed on both sides of the subintervals. The nodes are enumerated from 0 to M, and the abscissas are determined by $x_i = ih$ for all i.

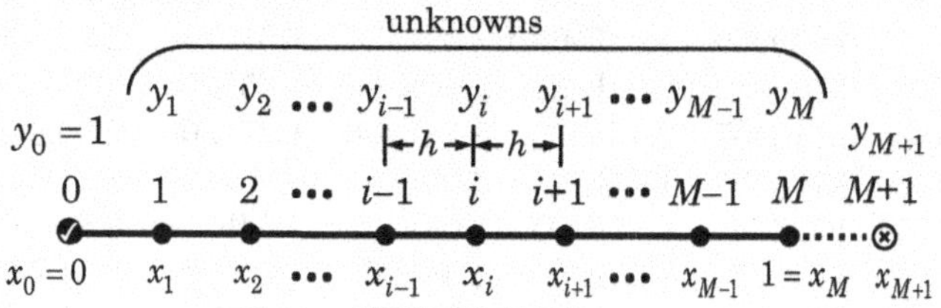

FIGURE 10.14

Step 2. Discretizing. The ODE, which is valid for any arbitrary $x = x_i$, is discretized using the CDFs as follows:

$$\frac{y_{i+1} - 2y_i + y_{i-1}}{h^2} = f\left(x_i, y_i, \frac{y_{i+1} - y_{i-1}}{2h}\right), \quad i = 1, 2, \ldots, M$$

Step 3. Implementing BCs. For the left and right boundary nodes, the BCs are expressed as $y_0 = 1$ and $y'_M = 0$, respectively. Applying the CDF to $y'_M = 0$, we obtain $y_{M+1} = y_{M-1}$.

Step 4. Constructing a nonlinear system. Substituting the known values into the first and last equations yields a system of $(M-1) \times (M-1)$ equations, expressed in matrix form as

$$\mathbf{T}\mathbf{y}^{(p+1)} = \mathbf{r}(\mathbf{y}^{(p)})$$

where $\mathbf{x} = [y_1, y_2, \cdots, y_{M-1}, y_M]^T$, and

$$\mathbf{T} = \begin{bmatrix} -2 & 1 & & & \\ 1 & -2 & 1 & & \\ & \ddots & \ddots & \ddots & \\ & & 1 & -2 & 1 \\ & & & 2 & -2 \end{bmatrix}, \quad \mathbf{r} = \begin{bmatrix} -\boxed{1} + h^2 f\left(x_1, y_1^{(p)}, \dfrac{y_2^{(p)} - \boxed{1}}{2h}\right) \\ h^2 f\left(x_2, y_2^{(p)}, \dfrac{y_3^{(p)} - y_1^{(p)}}{2h}\right) \\ \vdots \\ h^2 f\left(x_{M-1}, y_{M-1}^{(p)}, \dfrac{y_M^{(p)} - y_{M-2}^{(p)}}{2h}\right) \\ h^2 f\left(x_M, y_M^{(p)}, \boxed{0}\right) \end{bmatrix}$$

where the boxed values are the BCs, i.e., $y_0 = 1$ and $y'_M = 0$.

Step 5. Solving the system. An initial guess of a fixed value between 0 and 1 is assigned. Then the current estimate $\mathbf{y}^{(1)}$ is found from the solution of the tridiagonal system. In the first iteration, the difference between the prior and current estimates will likely be large. The current estimates are set as prior estimates before the next iteration: $\mathbf{y}^{(p)} \leftarrow \mathbf{y}^{(p+1)}$. The iteration procedure is repeated until the current and prior estimates converge within $\varepsilon < 10^{-6}$.

TABLE 10.7

x	True Solution	True (absolute) error			
		$M = 10$	$M = 20$	$M = 40$	$M = 80$
0	1				
0.1	0.930939	1.58E-05	3.92E-06	1.01E-06	2.32E-07
0.2	0.868912	2.24E-05	5.57E-06	1.45E-06	3.23E-07
0.3	0.813807	2.10E-05	5.18E-06	1.37E-06	2.90E-07
0.4	0.765603	1.26E-05	3.09E-06	8.65E-07	1.51E-07
0.5	0.724360	9.85E-07	3.08E-07	3.04E-08	6.67E-08
0.6	0.690212	1.77E-05	4.48E-06	1.00E-06	3.31E-07
0.7	0.663341	3.49E-05	8.78E-06	2.07E-06	6.03E-07
0.8	0.643952	5.00E-05	1.25E-05	3.00E-06	8.40E-07
0.9	0.632231	6.03E-05	1.51E-05	3.64E-06	1.00E-06
1	0.628308	6.39E-05	1.60E-05	3.87E-06	1.06E-06

The true solution, along with the true absolute errors of the numerical estimates obtained using the tridiagonal iteration technique with uniform intervals of $M = 10, 20, 40$, and 80, are tabulated in Table 10.7. The average errors are 3×10^{-5}, 7.5×10^{-6}, 1.83×10^{-6}, and 4.9×10^{-7}, respectively, for $M = 10, 20, 40$, and 80. When the intervals are doubled, the global errors decrease by about fourfold due to the difference scheme and second-order global error.

Discussion: The convergence of most nonlinear two-point BVPs depends on $\mathbf{r}(\mathbf{y})$. Convergence can be achieved fairly quickly as long as the initial guess is in the vicinity of the true solution. However, if there are rapidly growing terms on the rhs, such as e^y, y^m (large m), and so on, convergence issues may be experienced.

10.7.2 NEWTON'S METHOD

Consider the nonlinear system given by Eq. (10.55). This system can be solved with Newton's method, which was also applied to solve nonlinear systems of equations (*see* Section 4.8). The method leads to an improved solution of the following form:

$$\mathbf{y}^{(p+1)} = \mathbf{y}^{(p)} + \boldsymbol{\delta}^{(p)} \tag{10.58}$$

where $\boldsymbol{\delta}^{(p)}$ is the displacement vector, which is obtained from solving

$$\mathbf{J}(\mathbf{y}^{(p)})\,\boldsymbol{\delta}^{(p)} = -\mathbf{g}(\mathbf{y}^{(p)}) \tag{10.59}$$

where $\mathbf{J}(\mathbf{y}^{(p)})$ is the Jacobian matrix for this system, leading to a tridiagonal matrix given as

$$\mathbf{J}(\mathbf{y}^{(p)}) = \begin{bmatrix} 1 & 0 \\ b_1 & d_1 & a_1 \\ & \ddots & \ddots & \ddots \\ & & b_i & d_i & a_i \\ & & & \ddots & \ddots & \ddots \\ & & & & b_{M-1} & d_{M-1} & a_{M-1} \\ & & & & & 0 & 1 \end{bmatrix}$$

and for $i = 1, 2, \ldots, (M-1)$ the diagonal arrays are obtained from

$$b_i = \frac{\partial g_i}{\partial y_{i-1}} = 1 + \frac{h}{2}\frac{\partial f}{\partial y'}\left(x_i, y_i^{(p)}, \frac{y_{i+1}^{(p)} - y_{i-1}^{(p)}}{2h}\right)$$

$$d_i = \frac{\partial g_i}{\partial y_i} = -2 - h^2\frac{\partial f}{\partial y}\left(x_i, y_i^{(p)}, \frac{y_{i+1}^{(p)} - y_{i-1}^{(p)}}{2h}\right) \qquad (10.60)$$

$$a_i = \frac{\partial g_i}{\partial y_{i+1}} = 1 - \frac{h}{2}\frac{\partial f}{\partial y'}\left(x_i, y_i^{(p)}, \frac{y_{i+1}^{(p)} - y_{i-1}^{(p)}}{2h}\right)$$

For a simple function f, the Jacobian matrix can be easily constructed analytically and solved efficiently. For functions containing more complex expressions, numerical differentiation is applied to construct the Jacobian, which also contributes to numerical errors.

Pseudocode 10.4

Module NONLINEAR_NEWTON $(M, \varepsilon, \mathbf{x}, \mathbf{yo}, \mathbf{y}, \boldsymbol{\alpha}, \boldsymbol{\beta}, \boldsymbol{\gamma}, maxit)$
\ DESCRIPTION: A pseudomodule to solve a nonlinear two-point BVP of the
\   form: $y'' = f(x, y, y')$, $a \leqslant x \leqslant b$ subject to BCs:
\ $\alpha_1 y'(a) + \beta_1 y(a) = \gamma_1$, $\alpha_2 y'(b) + \beta_2 y(b) = \gamma_2$ with Newton's method.
\ USES:
\   FUNCS:: A user-defined module supplying $f(x, y, y')$, $\partial f/\partial y$, $\partial f/\partial y'$;
\   ENORM:: A function calculating ℓ_2–norm of a vector, Pseudocode 3.1;
\   TRIDIAGONAL:: Tridiagonal system solver, Pseudocode 2.13
Declare: $\alpha_2, \beta_2, \gamma_2, a_M, b_M, c_M, d_M, x_M, y_M, yo_M$ \ Declare array variables
$h \leftarrow x_2 - x_1$ \ Set interval size
$h2 \leftarrow 2*h;\ hsqr \leftarrow h*h$ \ Precompute multipliers to save time
$C1d \leftarrow h2*\beta_1/\alpha_1;\ C1r \leftarrow h2*\gamma_1/\alpha_1$
$C2d \leftarrow h2*\beta_2/\alpha_2;\ C2r \leftarrow h2*\gamma_2/\alpha_2$
$p \leftarrow 0$ \ Initialize iteration counter
Repeat   \ Iterate until convergence is achieved
   For $\begin{bmatrix} k = 1, M \end{bmatrix}$ \ Construct tridiagonal system
     If $\begin{bmatrix} k = 1 \end{bmatrix}$ **Then**   \ Set up coefficients for left boundary
       If $\begin{bmatrix} \alpha_1 = 0 \end{bmatrix}$ **Then**   \ Set up Dirichlet BC
         $a_k \leftarrow 0;\ b_k \leftarrow 0;\ c_k \leftarrow 0;\ d_k \leftarrow 1;\ yo_k \leftarrow \gamma_1/\beta_1$
       Else   \ Set up Neumann or Robin BC
         $yx \leftarrow (\gamma_1 - \beta_1*yo_k)/\alpha_1$ \ Evaluate y_1'
         FUNCS$(x_k, yo_k, yx, f, fy, fp)$ \ Find f, f_y, $f_{y'}$ at x_1
         $a_k \leftarrow 2;\ b_k \leftarrow 0$
         $d_k \leftarrow -2 + C1d - hsqr*(fy - fp*\beta_1*yo_k/\alpha_1)$
         $c_k \leftarrow -C1r + (C1d - 2)*yo_k + 2*yo_{k+1} - hsqr*f$
       End If
     Else
       If $\begin{bmatrix} k = M \end{bmatrix}$ **Then**   \ Set up coefficients for right boundary
         If $\begin{bmatrix} \alpha_2 = 0 \end{bmatrix}$ **Then**   \ Set up Dirichlet BC
           $a_k \leftarrow 0;\ b_k \leftarrow 0;\ c_k \leftarrow 0;\ d_k \leftarrow 1;\ yo_k \leftarrow \gamma_2/\beta_2$
         Else   \ Set up Robin BC
            $yx \leftarrow (\gamma_2 - \beta_2*yo_k)/\alpha_2$ \ Evaluate y_n'
            FUNCS$(x_k, yo_k, yx, f, fy, fp)$ \ Find f, f_y, $f_{y'}$ at x_M
            $a_k \leftarrow 0;\ b_k \leftarrow 2$ \ Modify coefficients of the last row
            $d_k \leftarrow -2 - C2d - hsqr*(fy - fp*\beta_2*yo_k/\alpha_2)$

$$c_k \leftarrow C2r - (C2d + 2) * yo_k + 2 * yo_{k-1} - hsqr * f$$
        End If
      Else   \ Set up coefficients for interior nodes
        $yx \leftarrow (yo_{k+1} - yo_{k-1})/h2$ \ Evaluate y'_k
        $yxx \leftarrow yo_{k+1} - 2 * yo_k + yo_{k-1}$ \ Find lhs
        FUNCS$(x_k, yo_k, yx, f, fy, fp)$ \ Find $f_k, (f_y)_k, (f_{y'})_k$
        $a_k \leftarrow 1 + fp/h2; \quad d_k \leftarrow -2 - hsqr * fy$
        $b_k \leftarrow 1 - fp/h2; \quad c_k \leftarrow yxx - hsqr * f$
      End If
    End If
  End For
  TRIDIAGONAL$(1, M, \mathbf{b}, \mathbf{d}, \mathbf{a}, \mathbf{c}, \mathbf{y})$ \ Solve tridiagonal system for correction
  $error \leftarrow$ **ENORM**$(M, \mathbf{c})$ \ Find Euclidean norm
  $\mathbf{y} \leftarrow \mathbf{yo} - \mathbf{c}$ \ Find current estimates
  $p \leftarrow p + 1$ \ Count iterations
  Write: "p, error=", p, $error$ \ Print out iteration progress
  $\mathbf{yo} \leftarrow \mathbf{y}$ \ Set current estimates as prior
Until $\left[error < \varepsilon \text{ Or } p = maxit\right]$
If $\left[p = maxit\right]$ **Then**   \ Issue info and a warning
  Write: "Failed to converge after",$maxit$,"iterations"
End If
End Module NONLINEAR_NEWTON

Module FUNCS $(x, y, yx, fun, dfdy, dfdp)$
\ DESCRIPTION: A user-defined module to evaluate f, $\partial f/\partial y$, $\partial f/\partial y'$ at (x, y, y').
$fun \leftarrow f(x, y, yx)$ \ Define ODE here
$dfdy \leftarrow \partial f/\partial y$ \ Define partial derivatives here
$dfdp \leftarrow \partial f/\partial y'$
End Module FUNCS

A pseudo-module, NONLINEAR_NEWTON, solving a nonlinear two-point BVP given with Newton's method is presented in Pseudocode 10.4. This module also allows Robin BCs to be imposed on both boundaries. The argument list is the same as that in Pseudocode 10.3. It requires (1) FUNCS, which provides $f(x_i, y_i, y'_i)$, as well as $\partial f/\partial y$ and $\partial f/\partial y'$ partial derivatives at (x_i, y_i, y'_i), and (2) TRIDIAGONAL, a tridiagonal linear system solver applying the Thomas' algorithm (*see* Pseudocode 2.13). The coefficients of the Jacobian are computed from Eq. (10.60). Only the first and last rows of $\mathbf{J}$ are modified to incorporate the left and right BCs. The Dirichlet BC ($y_i^{(p)} = \gamma_i/\beta_i$) is imposed by simply setting the coefficients as $d_i = 1$, $a_i = b_i = c_i = 0$, where $i = 1$ and 2 for the left and right boundaries, respectively. In Neumann or Robin BCs ($\alpha_1 \neq 0$ or $\alpha_2 \neq 0$), the fictitious nodes are eliminated by applying the same procedure employed as in the cases of two-point BVPs; hence, only the first or last row ($i = 1$ or M) coefficients (a_i, b_i, c_i, and d_i's) are modified accordingly. A smart initial estimate is helpful for speedy convergence. When the Dirichlet BC is applied on both boundaries, a smart initial guess can be found from Eq. (10.56). The initial guess is improved by solving Eqs. (10.59) using Thomas' algorithm, and the current estimates are updated by Eq. (10.58). This procedure is successively repeated to obtain better estimates. The convergence (or termination) criterion is based on the displacement; that is, when $\|\delta^{(p)}\|_2 = \|\mathbf{y}^{(p+1)} - \mathbf{y}^{(p)}\|_2 < \varepsilon$.

> **EXAMPLE 10.7: Solving a nonlinear BVP with Newton's method**

Repeat Example 10.6 for $\phi^2 = 16$ using the Newton's method.

> **SOLUTION:**

The gridding and discretization steps are the same as those presented in Example 10.6. The resulting nonlinear difference equations are expressed as $\mathbf{g}(\mathbf{y}) = 0$, which is explicitly given as

$$g_1 = -2y_1 + y_2 + \boxed{1} - h^2 f\left(x_1, y_1, \frac{y_2 - \boxed{1}}{2h}\right) = 0$$

$$g_k = y_{k-1} - 2y_k + y_{k+1} - h^2 f\left(x_k, y_k, \frac{y_{k+1} - y_{k-1}}{2h}\right) = 0, \quad k = 2, \dots, M-1$$

$$g_M = y_{M-1} - 2y_M + \boxed{y_{M-1}} - h^2 f\left(x_M, y_M^{(p)}, \boxed{0}\right) = 0$$

where $f(x, y, y') = \phi^2 y - (y')^2/y$. Note that the known BCs ($y_0 = 1$ and $y_M' = 0$) are boxed. Since $y_M' = 0$, we set $y_{M+1} = y_{M-1}$ in the final difference equation and highlighted it by boxing it too.

As in Example 10.6, an initial guess, $\mathbf{y}^{(0)}$, is obtained by a linear interpolation. Using Newton's method, the numerical approximations are estimated from

$$\mathbf{y}^{(p+1)} = \mathbf{y}^{(p)} + \boldsymbol{\delta}^{(p)}$$

where $\boldsymbol{\delta}^{(p)}$ is found from the solution of a tridiagonal system of linear equations. Noting that $\partial f/\partial y' = f_1(y, y') = -2y'/y$ and $\partial f/\partial y = f_2(y, y') = \phi^2 + (y'/y)^2$, the displacement vector, is obtained by solving the following system of linear equations:

$$\mathbf{J}(\mathbf{y}^{(p)})\,\boldsymbol{\delta}^{(p)} = -\mathbf{g}(\mathbf{y}^{(p)})$$

where $\mathbf{J}$ is the Jacobian, and using the tridiagonal matrix notations, the matrix elements are

$$b_i = \frac{\partial g_i}{\partial y_{i-1}} = 1 + \frac{h}{2} f_1\left(y_i^{(p)}, \frac{y_{i+1}^{(p)} - y_{i-1}^{(p)}}{2h}\right)$$

$$d_i = \frac{\partial g_i}{\partial y_i} = -2 - h^2 f_2\left(y_i^{(p)}, \frac{y_{i+1}^{(p)} - y_{i-1}^{(p)}}{2h}\right), \quad i = 1, 2, \dots, M$$

$$a_i = \frac{\partial g_i}{\partial y_{i+1}} = 1 - \frac{h}{2} f_1\left(y_i^{(p)}, \frac{y_{i+1}^{(p)} - y_{i-1}^{(p)}}{2h}\right)$$

For $i = 1$, we find

$$d_1 = -2 - h^2 f_2\left(y_1^{(p)}, \frac{y_2^{(p)} - 1}{2h}\right), \quad a_1 = 1 - \frac{h}{2} f_1\left(y_1^{(p)}, \frac{y_2^{(p)} - 1}{2h}\right)$$

For $i = M$, having eliminated the resulting fictitious node in the usual manner, we obtain

$$b_M = 2, \quad d_M = \frac{\partial g_M}{\partial y_M} = -2 - h^2 f_2\left(y_M^{(p)}, 0\right)$$

Starting with a uniform initial guess of $y_i^{(0)} = 0.3$, this system converges in five iterations. At the end of each iteration, the ℓ_2−norm is found as

$$\|\boldsymbol{\delta}^{(1)}\|_2 = 1.112720 \qquad \|\boldsymbol{\delta}^{(2)}\|_2 = 0.215433 \qquad \|\boldsymbol{\delta}^{(3)}\|_2 = 0.025728$$
$$\|\boldsymbol{\delta}^{(4)}\|_2 = 2.1727 \times 10^{-4} \qquad \|\boldsymbol{\delta}^{(5)}\|_2 = 1.3790 \times 10^{-8}$$

The true solution and the absolute errors resulting from numerical estimates for uniformly spaced $M = 10$, 20, 40, and 80 intervals are presented in Table 10.8. The average errors are 2.2×10^{-3}, 5.5×10^{-4}, 1.4×10^{-4}, and 3.5×10^{-5} for $M = 10$, 20, 40, and 80, respectively.

TABLE 10.8

x	True Solution	Absolute true error			
		$M = 10$	$M = 20$	$M = 40$	$M = 80$
0	1				
0.1	0.753648	1.74E-03	4.42E-04	1.11E-04	2.77E-05
0.2	0.568001	2.63E-03	6.65E-04	1.67E-04	4.18E-05
0.3	0.428120	2.97E-03	7.51E-04	1.88E-04	4.71E-05
0.4	0.322771	2.98E-03	7.53E-04	1.89E-04	4.72E-05
0.5	0.243539	2.79E-03	7.04E-04	1.76E-04	4.41E-05
0.6	0.184210	2.48E-03	6.26E-04	1.57E-04	3.93E-05
0.7	0.140381	2.11E-03	5.32E-04	1.33E-04	3.34E-05
0.8	0.109345	1.70E-03	4.29E-04	1.08E-04	2.69E-05
0.9	0.090194	1.32E-03	3.35E-04	8.42E-05	2.11E-05
1	0.083588	1.15E-03	2.93E-04	7.37E-05	1.85E-05

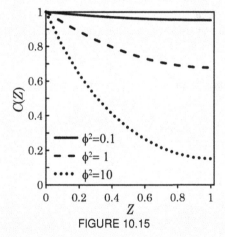

FIGURE 10.15

Discussion: Notice that the absolute errors are much larger in $\phi^2 = 16$ in comparison to the $\phi^2 = 1.25$ case presented in Example 10.6. To understand the reason for this, we refer to the concentration plot for $\phi^2 = 0.1$, 1, and 10 presented in Fig. 10.15. It is clear that $C(Z) \to 1$ as $\phi \to 0$. In other words, the gas penetration into the liquid by diffusion is faster and easier; that is why the $C(Z)$ variation tends to be linear or almost flat. As ϕ increases, gas diffusion becomes more difficult, leading to sharp concentration gradients and concave distribution. Hence, using second-order accurate, $\mathcal{O}(h^2)$, finite difference formulation, more accurate estimates for small ϕ can be obtained with a smaller or moderate number of nodes, while for a large ϕ, a large number of grid points should be utilized to accurately represent sharp changes in the concentration.

10.8   SHOOTING METHOD

There are many circumstances where a linear or nonlinear two-point BVP is too complex or impractical to be solved by conventional numerical methods. It can be extremely difficult to find numerical solutions to such problems with conventional FDM or FVM methods. Also, BVPs involving a semi-infinite (semi-bounded) interval pose another numerical difficulty since one boundary is at infinity. In this regard, shooting methods for complex or nonlinear two-point BVPs offer great convenience and advantages to the numerical analyst.

Consider the following two-point BVPs:

(1) $y'' - xyy' = e^{xy}$ $\qquad$ $y(a) = A, \quad y(b) = B$

(2) $y'' - 4xy = x^2 + 1$ $\qquad$ $y(a) = A, \quad y(\infty) = B$

The first BVP is a nonlinear ODE due to the yy' and e^{xy} terms. The FDM discretization leads to a set of nonlinear equations, which is not desirable, especially when the number of unknowns is large. The numerical solution of nonlinear two-point BVPs requires an iterative algorithm followed by a linearization procedure, which is generally very sensitive to an initial guess. This algorithm may diverge primarily because the iteration procedure is started with an unsuitable initial guess.

The second BVP is a linear ODE; however, the solution interval is semi infinite, and the Dirichlet BC is applied at $x = \infty$. In order to obtain the numerical solution, it is necessary to make a numerical prediction for "$x = \infty$." In some cases, the value of *infinity* may be in the order of $x \approx 1$, $x \approx 100$s, or larger. Hence, even though a BVP may be linear, it could be difficult to find a reliable solution or assess the accuracy of the solution because we cannot truly define ∞.

For the reasons mentioned, a BVP of this type is generally treated and solved as an IVP, which makes nonlinear ODEs easier to handle.

10.8.1   LINEAR ODES

Consider the following two-point BVP, subjected to Dirichlet BCs:

$$p(x)y'' + q(x)y' + r(x)y = f(x), \qquad y(a) = A, \quad y(b) = B \qquad (10.61)$$

where the coefficients $p(x)$, $q(x)$, $r(x)$, and $f(x)$ are continuous and defined on $[a, b]$ and $r(x)/p(x) \leqslant 0$.

Rearranging Eq. (10.61), the BVP may also be cast as

$$y'' = F(x, y, y'), \qquad y(a) = A, \quad y(b) = B \qquad (10.62)$$

Equations (10.61) or (10.62) can be reduced to a set of coupled first-order IVPs. Defining $Y_1(x) = y(x)$ and $Y_2(x) = y'(x)$, Eq. (10.62) leads to the following IVPs:

$$\frac{dY_1}{dx} = Y_2, \quad \frac{dY_2}{dx} = F(x, Y_1, Y_2),$$
$$Y_1(a) = A, \quad Y_1(b) = B, \qquad (10.63)$$

To solve Eq. (10.63) as an IVP, $Y_1(a)$ and $Y_2(a)$ ($y(a)$ and $y'(a)$) need to be known beforehand. However, the initial value $y'(a)$ in Eqs. (10.61) and (10.62) is undefined. Since initial conditions are required in a typical IVP problem, a suitable guess for $y'(a)$ (i.e.,

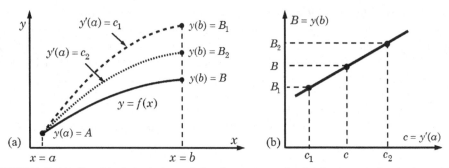

FIGURE 10.16: Graphical depiction of the shooting method (a) numerical solution with different B's, (b) guess versus target values.

$Y_2(a)$) should be made to start the forward marching process. A reasonable "*smart*" guess, which may be deduced from the physics of the problem or an arbitrary guess, is assigned to $Y_2(a)$ to start the process. For example, setting $Y_2(a) = c_1$ (a first guess), the IVP Eq. (10.63) is solved up to, say, $x = b$ using a numerical method to solve the IVPs. If the initial guess ($y'(a)$, which is also the initial slope) is much different than the true value, the target estimate, $y(b)$, will deviate from the true value by a large margin. After the first trial, the target value of $y(b) = Y_1(b) = B_1$ is recorded. Then the IVP is resolved with a second (smaller or larger) guess, e.g., $Y_2(a) = c_2$. We let the second target value be $Y_1(b) = B_2$, which is also recorded (*see* Fig. 10.16a). This technique is called the *shooting method* because of its resemblance to trying to hit a distant target. Normally, it is very unlikely that the target is hit with the first or second trials. Therefore, we continue shooting by adjusting the guess until the boundary value $Y_1(b) = B$ is satisfied.

In linear ODEs, the target values vary linearly with the guesses (*see* Fig. 10.16b). Using the results of the two trials, a straight line that passes through points (c_1, B_1) and (c_2, B_2) can be found to estimate the initial slope, $Y_2(a)$. A linear interpolation procedure leads to

$$c = c_2 + (c_1 - c_2)\frac{B - B_2}{b_1 - B_2} \tag{10.64}$$

where c is the true initial slope.

Of course, the true values of B and c depend on the accuracy (truncation and round-off errors) of the numerical procedure. In order to obtain highly accurate numerical estimates, the step size must be sufficiently small to ensure negligible truncation and round-off errors. Once B and c are obtained with sufficient accuracy, the numerical solution of the IVP is carried out one last time for $y'(a) = c$.

Boundary at infinity. A linear or non-linear two-point BVP may be given on a semi-infinite interval. Consider imposing the following BCs on the BVP given by Eq. (10.62):

$$y(x = a) = A \quad \text{and} \quad y(x \to \infty) = B, \qquad a < x < \infty \tag{10.65}$$

where the boundary at "∞" is, numerically speaking, only achievable asymptotically. In order to tackle it as a two-point BVP, it is essential to predict and assign a suitable value to "∞," which is not an easy task. But such problems are also more convenient to treat as IVPs using the shooting method.

The numerical solution of an ODE eventually yields $y \to B$ as $x \to \infty$. In Fig. 10.17, the numerical solutions of three BVPs in a semi-infinite interval are shown graphically. It is observed that the solutions approach their boundary (asymptotic) values at different x

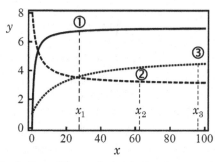

FIGURE 10.17: Evolution of numerical solution in a semi-infinite interval.

values ($x_1 < x_2 < x_3$). When the shooting method is used, an estimate for x_∞ value may be obtained based on the order of magnitude of the physical phenomenon under study. If an estimate is not possible, two trial solutions (with c_1 and c_2) must be obtained for a sufficiently large x value, denoted by x_∞. The solution should become asymptotic within a preset ε value, which depends on the physical problem and the choice of the analyst. The initial condition $y'(a) = Y_2(a) = c$ is found using Eq. (10.64) and the most recent two target values.

Shooting Method

- The shooting method can be easily implemented for both linear and nonlinear two-point BVPs;
- Numerical solutions of ODEs with an accuracy higher than $\mathcal{O}(h^2)$ can be obtained easily with either RK4, ABF4, or AMF4.

- A good estimate is required for the missing initial condition so that the iterative procedure converges quickly;
- High-order numerical schemes are susceptible to round-off errors;
- Nonlinear BVPs may lead to stability problems, especially over long intervals;
- In nonlinear two-point BVPs, the $c - B$ relationship is also nonlinear, and the secant method or similar root-finding methods may be used to estimate the initial conditions.

EXAMPLE 10.8: Solving BVP in a semi-infinite medium

A viscoelastic fluid is a non-Newtonian fluid that combines a viscous and an elastic component. The flow of linear viscoelastic fluid about a large horizontal plate leads to the following boundary value problem [26]:

$$\nu \frac{d^2 F_\beta}{d\eta^2} - (1 + \lambda\beta)\beta F_\beta(\eta) = 0, \qquad F_\beta(0) = U_0, \quad F_\beta(\infty) = 0$$

where $F_\beta = u/e^{\beta t}$ is the normalized velocity, u is the velocity, ν is the kinematic viscosity, λ is the relaxation time, and β is a real constant. Fluid attached to the plate moves quickly, whereas fluid particles sufficiently away from the plate remain stationary, i.e., $u \to 0$ as $\eta \to \infty$.

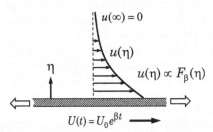

Apply the shooting method to the given two-point linear BVP to obtain a numerical solution and solve the resulting system of IVPs using the RK4 with $h = 0.1$.

SOLUTION:

Before attempting to apply the RK4 method, let us introduce the following transformation:

$$x = \eta\sqrt{(1 + \lambda\beta)\beta/\nu}, \qquad y(x) \equiv y(\eta) = F_\beta(\eta)/U_0$$

Then, the BVP can be expressed as

$$y'' - y = 0, \qquad y(0) = 1, \quad y(\infty) = 0$$

which has the true solution $y(x) = e^{-x}$.

Next, the problem is converted to a coupled first-order IVP by setting $Y_1(x) = y(x)$ and $Y_2(x) = y'(x)$, which results in

$$Y_1' = Y_2, \quad Y_1(0) = 1, \qquad Y_2' = Y_1, \quad Y_2(0) = c$$

As the unknown initial slope $y'(0)$ is unknown, we set $Y_2(0) = c$. We also assume a value for x_∞, which will be $x_\infty = 8$. Using the RK4 method for the step size of $h = 0.1$, the numerical solution is obtained sequentially from the initial value up to $x_\infty = 8$. The upper bound can be determined by achieving the asymptotic value within a tolerance. Noting that $y(0) = 1 > y(\infty) = 0$, it is clear that the solution overall is a decreasing function (i.e., $y'(x) < 0$), though it may depict some local extremums of the solution interval. Hence, the guess for the initial slope is chosen to satisfy $y'(0) < 0$.

TABLE 10.9

x	$Y_1(x) = y(x)$	$Y_2(x) = y'(x)$
0	1	-1
0.1	0.904838	-0.904838
0.2	0.818731	-0.818731
0.3	0.740818	-0.740818
0.4	0.670320	-0.670320
0.5	0.606531	-0.606531
1	0.406570	-0.406570
2	0.367880	-0.367880
3	0.135336	-0.135336
4	0.049787	-0.049787
5	0.018316	-0.018316
6	0.006738	-0.006738
7	0.002479	-0.002479
8	0.000912	-0.000912

Numerical solutions with $c = -0.1$ and $c = -1$ yield the following target values:

$$\text{For} \quad c_1 = -0.1, \quad B_1 = Y_1(8) = 1341.423$$
$$\text{For} \quad c_2 = -0.5, \quad B_2 = Y_1(8) = 745.2352$$

A new estimate for the initial slope is then calculated from Eq. (10.64) as

$$c = c_2 + (c_1 - c_2)\frac{B - B_2}{b_1 - B_2} = -0.5 + (-0.1 + 0.5)\frac{0 - 745.2352}{1341.423 - 745.2352} = -1$$

In other words, the true initial slope is $y'(0) = -1$. Since this problem is a linear BVP, it is solved as an IVP one last time for $y'(0) = -1$ to obtain its numerical solution, tabulated in Table 10.9. We observe that both y and y' decrease as x increases and asymptotically approach zero as x is further increased.

Discussion: We note that the numerical solutions for y and y' approach zero within $\epsilon < 10^{-3}$ as $x \to 8$, and our initial estimate of $x_\infty = 8$ is sufficiently large enough to yield fairly good estimates. If it turns out that the predicted x_∞ value is not large enough, the final solution is taken as the initial value and the IVP is solved up to a larger x_∞ value. Of course, we expect a numerical solution with a smaller h to yield better estimates. When tackling BVPs in a semi-infinite interval, the analyst should continue marching until a suitable x_∞ is found.

10.8.2   NON-LINEAR ODEs

Consider the second-order nonlinear two-point BVP expressed by Eq. (10.62). Nonlinear BVPs can also be converted into a set of coupled first-order IVPs, just like linear BVPs. Hence, finding the numerical solution for an initial slope $y'(B_i) = c_i$ is similar, but the $c - B$ relationship is no longer linear.

We start out with two independent guesses for the initial slope: $Y_2(a) = c_1$ and $Y_2(a) = c_2$. The IVP is then solved by a suitable numerical method, and the achieved target values are recorded, e.g., $Y_1(b) = B_1$ and $Y_1(b) = B_2$. Although there is no linear relationship between the target values $(B_1, B_2, \ldots, B_n)$ and the initial slopes of nonlinear two-point BVPs, the slope estimate can be improved with Eq. (10.64), which will unlikely satisfy $Y_1(b) = B$ in the first attempt. For this reason, to determine improved estimates for c, Equation (10.64) is recast as follows:

$$c_{p+1} \cong c_p + (c_{p-1} - c_p)\frac{B - B_p}{B_{p-1} - B_p}, \qquad p \geqslant 2 \tag{10.66}$$

where p denotes the iteration number, c_p and B_p are the initial slope and corresponding target estimate at the pth iteration, respectively.

EXAMPLE 10.9: Solving a BVP as an IVP

The steady-state energy balance for an absorbing planar medium is depicted below $(A_s \gg L)$ with no internal heat generation and specified surface temperatures is reduced to the following second-order nonlinear two-point BVP:

$$k\frac{d^2T}{dx^2} = \kappa\left(4\sigma T^4(x) - G(x)\right), \qquad T(0) = T_1, \quad T(L) = T_2$$

where κ is the absorption coefficient (m^{-1}), k is the thermal conductivity (W/m·K), σ is the Stefan-Boltzmann constant $(\text{W/m}^2\cdot\text{K}^4)$, G is the incident energy (W/m^2), L is the plate thickness (m), and T_1 and T_2 are the specified surface temperatures (K).

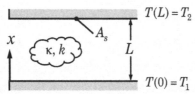

Introducing the following parameters and quantities

$$\Theta = T/T_1, \quad g = G/4\sigma T_1^4, \quad N = \kappa k/\sigma T_1^3, \quad \tau = \kappa x, \quad \tau_L = \kappa L$$

where Θ and g are dimensionless temperature and incident energy, N is called the radiation-conduction parameter, and τ and τ_L are the optical distance and optical thickness, respectively. Then the dimensionless form of the BVP becomes

$$N\frac{d^2\Theta}{d\tau^2} = \Theta^4(\tau) - g(\tau), \quad \Theta(0) = 1, \quad \Theta(L) = \Theta_2 = T_2/T_1$$

For $\Theta_2 = 0.5$, $N = 0.5$, and $g(\tau) = (\pi^2/18)\cos(\pi\tau/3) + \cos^4(\pi\tau/3)$, find the dimensionless temperature distribution in the medium. Apply the shooting method using the RK4 in increments of $\Delta\tau = 0.1$ to obtain a numerical solution. Compare your estimates with the true solution given with $\Theta(\tau) = \cos(\pi\tau/3)$.

Setting $Y_1(\tau) = \Theta(\tau)$ and $Y_2(\tau) = d\Theta/d\tau$, the two-point BVP is reduced to the following set of coupled first-order IVP:

$$\frac{dY_1}{d\tau} = Y_2, \quad \frac{dY_2}{d\tau} = \frac{1}{N}(Y_1^4 - g(\tau)), \quad Y_1(0) = 1, \quad Y_2(1) = 0.5$$

Since $\Theta'(0)$ is unknown, we set $Y_2(0) = c$ to find the numerical solution for an abitrary c. The coupled system of nonlinear IVP is solved using the RK4 method with $\Delta\tau = 0.1$. Numerical solutions for $c = 0.5$ and $c = -0.5$ (two independent initial guesses) yield the following target values:

$$\text{For} \quad c_1 = \quad 0.5, \quad B_1 = Y_1(1) = \quad 2.523334$$

$$\text{For} \quad c_2 = \quad -0.5, \quad B_2 = Y_1(1) = \quad -0.317103$$

Equation (10.66) is used to obtain a new prediction (c_p) and its corresponding numerical solution (B_p) is found by applying the RK4 method. The subsequent predictions and the target estimates are obtained iteratively until convergence is achieved, i.e., $|c_p - c_{p-1}| < \varepsilon$. The iteration history of the initial slope and the target estimates is summarized in Table 10.10. The converged value of $\Theta'(0)$ was found to be zero after six iterations.

Next, we find the numerical solution for the BVP on $[0,1]$ using the RK4 scheme with the initial slope $\Theta'(0) = 0$. The true solution and the true (absolute) errors are tabulated in Table 10.11. The numerical estimates Θ and Θ' are found to be accurate to at least four decimal places.

TABLE 10.10

p	$c_p = \Theta'(0)$	$B_p = \Theta(1)$
1	0.5	2.523334
2	−0.5	−0.317103
3	−0.212332	0.111556
4	0.048348	0.603751
5	−0.006601	0.486432
6	-2.46×10^{-5}	0.499986

TABLE 10.11

	True Solution		Absolute error	
τ	$Y_1(\tau)$	$Y_2(\tau)$	$\lvert Y_1 - \check{Y}_1 \rvert$	$\lvert Y_2 - \check{Y}_2 \rvert$
0	1	0		
0.1	0.994522	−0.109462	1.05E-07	8.95E-06
0.2	0.978148	−0.217725	1.40E-06	1.76E-05
0.3	0.951057	−0.323602	2.48E-06	2.58E-05
0.4	0.913545	−0.425934	5.54E-06	3.36E-05
0.5	0.866025	−0.523599	8.60E-06	4.18E-05
0.6	0.809017	−0.615527	1.30E-05	4.93E-05
0.7	0.743145	−0.700712	1.82E-05	5.59E-05
0.8	0.669131	−0.778219	2.34E-05	6.24E-05
0.9	0.587785	−0.847201	2.98E-05	6.76E-05
1	0.5	−0.906899	3.70E-05	7.17E-05

Discussion: In this example, the numerical solution of the given BVP was obtained with increments of $\Delta\tau = 0.1$, i.e., equivalent to $M = 10$ uniform subintervals. While non-linear FDM results in a quadratic global error, the global error of the solution with RK4 as IVP is $\mathcal{O}(0.1^4)$.

10.9   FOURTH-ORDER LINEAR DIFFERENTIAL EQUATIONS

Fourth-order ODEs are generally encountered in the areas of applied mathematics, beam theory, viscoelastic and inelastic flows, and electric circuits. The common fourth-order linear differential equation has the following general form:

$$p(x)y^{(iv)} + q(x)y'' + r(x)y(x) = g(x), \qquad a \leqslant x \leqslant b \qquad (10.67)$$

where p, q, r, and g are continuous functions of x. Four BCs (two for the left and two for the right boundary) are required to find its unique solution. The boundary conditions may be in the form of Dirichlet, Neumann, or Robin BC, or may include linear combinations including second or third derivatives as well.

In this section, we will discuss the numerical solution of such BVPs with the finite difference method. The solution steps of Eq. (10.67) are similar to the two-point BVPs, which are detailed below:

Step 1. Gridding: A one-dimensional grid structure is created by dividing the solution range into M uniform (or nonuniform) subintervals. The abscissas are found from $x_i = a + ih$ for $i = 0, 1, 2, \ldots, M$ if uniform grids ($h = x_{i+1} - x_i = (b-a)/M$) are adopted. As usual, nodal points are placed at both end points of each subinterval and are enumerated from 0 to M.

Step 2. Discretizing: For an arbitrary x_i on $[a, b]$, the ODE can be written as

$$p_i y_i^{(iv)} + q_i y_i'' + r_i y_i = g_i, \qquad i = 0, 1, \ldots, M \tag{10.68}$$

where short-hand notations ($y''(x_i) = y_i''$, $p(x_i) = p_i$, etc.) have been adopted.

All derivatives in Eq. (10.68) are approximated by CDFs, resulting in

$$p_i\left(\frac{y_{i+2} - 4y_{i+1} + 6y_i - 4y_{i-1} + y_{i-2}}{h^4}\right) + q_i\left(\frac{y_{i+1} - 2y_i + y_{i-1}}{h^2}\right) + r_i y_i = g_i \tag{10.69}$$

Rearranging and collecting similar terms of Eq. (10.69) gives

$$\frac{p_i}{h^4}y_{i-2} - \left(\frac{4p_i}{h^4} - \frac{q_i}{h^2}\right)y_{i-1} + \left(r_i + \frac{6p_i}{h^4} - \frac{2q_i}{h^2}\right)y_i - \left(\frac{4p_i}{h^4} - \frac{q_i}{h^2}\right)y_{i+1} + \frac{p_i}{h^4}y_{i+2} = g_i \tag{10.70}$$

which, for $i = 0, 1, \ldots, M$, leads to an $(M+1) \times (M+1)$ system.

Step 3. Implementing BCs: Implementing the Dirichlet BC is simple and straightforward. A Dirichlet BC ($y_0 - A = 0$ or $y_M - B = 0$) can be most easily implemented by setting the diagonals to "1," the off-diagonal coefficients to zero, and the rhs's to the BC values ($g_0 = A$, $g_M = B$). Using the compact banded matrix notation, the Dirichlet BCs are incorporated into the pentadiagonal linear system as $a_{03} = a_{M3} = 1$, $a_{04} = a_{05} = 0$, and $a_{M1} = a_{M2} = 0$. In the case of a Neumann or Robin BC (or other BCs involving either $y''(a)$ or $y''(b)$, or both), the discretization of both the ODE and the BCs at x_0 and x_M will result in fictitious node(s), i.e., y_{-1} and/or y_{M+1}. If a BC involves $y'''(a)$ or $y'''(b)$ or in a linear combination with y, y', or y'', then two fictitious nodes will appear on the opposite side (outside of the solution range) of the boundary node. These fictitious nodes are, in the same manner, eliminated with the aid of the BCs.

Step 4. Constructing a linear system: The difference equations can now be expressed in matrix form:

$$\mathbf{Ay} = \mathbf{g} \tag{10.71}$$

where $\mathbf{A}$ is a banded matrix (or a pentadiagonal system) with a band size of 5. The general difference equation is adapted to compact band matrix form as follows:

$$a_{i1}y_{i-2} + a_{i2}y_{i-1} + a_{i3}y_i + a_{i4}y_{i+1} + a_{i5}y_{i+2} = g_i, \quad i = 0, 1, \ldots, M$$

where a_{i1}, a_{i2}, a_{i3}, a_{i4}, and a_{i4} are sub-lower-diagonal, sub-diagonal, diagonal, upper-diagonal, and super-upper diagonal elements arranged as columns, respectively. Finally, the system can be expressed in compact banded matrix form as follows:

$$\begin{bmatrix} 0 & 0 & a_{03} & a_{04} & a_{05} \\ 0 & a_{12} & a_{13} & a_{14} & a_{15} \\ a_{21} & a_{22} & a_{23} & a_{24} & a_{25} \\ \vdots & \vdots & \vdots & \vdots & \vdots \\ a_{(M-2)1} & a_{(M-2)2} & a_{(M-2)3} & a_{(M-2)4} & a_{(M-2)5} \\ a_{(M-1)1} & a_{(M-1)2} & a_{(M-1)3} & a_{(M-1)4} & 0 \\ a_{M1} & a_{M2} & a_{M3} & 0 & 0 \end{bmatrix} \begin{bmatrix} y_0 \\ y_1 \\ y_2 \\ \vdots \\ y_{M-2} \\ y_{M-1} \\ y_M \end{bmatrix} = \begin{bmatrix} g_1 \\ g_2 \\ g_3 \\ \vdots \\ g_{M-2} \\ g_{M-1} \\ g_M \end{bmatrix} \tag{10.72}$$

Step 5. Solving the linear system: The system of banded (specifically penta-diagonal) equations can be solved with the Gauss elimination method (*see* Section 2.1.1, Pseudocodes 2.14 and 2.15). In this case, the memory storage requirement for the coefficient matrix in either banded or pentadiagonal form is the same.

Pseudocode 10.5

Module LBVP4 $(M, \mathbf{x}, AB, \boldsymbol{\alpha}, \boldsymbol{\beta}, \boldsymbol{\gamma}, \mathbf{y})$
\\ DESCRIPTION: A pseudomodule to solve a 4th-order linear BVP of the form:
\\ $\quad p(x)y^{(iv)} + q(x)y'' + r(x)y(x) = g(x), \quad a \leqslant x \leqslant b$, subjected to BCs:
\\ $\quad y(a) = A, \quad \alpha_1 y'(a) + \beta_1 y(a) = \gamma_1, \quad y(b) = B, \quad \alpha_2 y'(b) + \beta_2 y(b) = \gamma_2,$
\\ USES:
\\ $\quad$ COEFFS:: A user-defined module supplying $p(x), q(x), r(x), g(x)$
\\ $\quad$ BAND_SOLVE:: A banded linear system solver, Pseudocode 2.14
Declare: $x_M, AB_{M,5}, y_M$
$h \leftarrow x_2 - x_1; h2 \leftarrow h * h; h4 \leftarrow h2 * h2$ $\qquad$ \\ Set interval size & $h2, h4$
$m_L \leftarrow 2; m_U \leftarrow 2$ $\qquad$ \\ Set the upper & lower bandwidths
For $[k = 1, M]$ $\qquad$ \\ Construct banded linear system
$\quad$ **COEFFS**(x_k, px, qx, rx, gx) $\qquad$ \\ Find coefficients for $x = x_k$
$\quad AB_{k,1} \leftarrow px; AB_{k,5} \leftarrow AB_{k,1}$ $\qquad$ \\ Set up matrix elements and rhs
$\quad AB_{k,2} \leftarrow qx * h2 - 4 * px; AB_{k,4} \leftarrow AB_{k,2}$
$\quad AB_{k,3} \leftarrow 6 * px - 2 * qx * h2 + rx * h4; y_i \leftarrow gx * h4$
$\quad$ **If** $[k = 1]$ **Then** $\qquad$ \\ Set up Dirichlet BC
$\quad\quad AB_{k,1} \leftarrow 0; AB_{k,2} \leftarrow 0;$ $\qquad$ \\ Modified coefficients of row $k = 1$
$\quad\quad AB_{k,4} \leftarrow 0; AB_{k,5} \leftarrow 0; AB_{k,3} \leftarrow 1; y_k \leftarrow A$
$\quad$ **Else**
$\quad\quad$ **If** $[k = 2]$ **Then** $\qquad$ \\ Set up Robin BC
$\quad\quad\quad AB_{k,3} \leftarrow AB_{k,3} + AB_{k,1}$ $\qquad$ \\ Modify coefficients of row $k = 2$
$\quad\quad\quad AB_{k,2} \leftarrow AB_{k,2} + 2h * \beta_1 * AB_{k,1}/\alpha_1$
$\quad\quad\quad y_k \leftarrow y_k + 2h * \gamma_1 * AB_{k,1}/\alpha_1; AB_{k,1} \leftarrow 0$
$\quad\quad$ **Else**
$\quad\quad\quad$ **If** $[k = M - 1]$ **Then** $\qquad$ \\ Set up Robin BC
$\quad\quad\quad\quad AB_{k,3} \leftarrow AB_{k,3} + AB_{k,5}$ $\qquad$ \\ Modify coefficients of row $k = M-1$
$\quad\quad\quad\quad AB_{k,4} \leftarrow AB_{k,4} - 2h * \beta_2 * AB_{k,5}/\alpha_2$
$\quad\quad\quad\quad y_k \leftarrow y_k - 2h * \gamma_2 * AB_{k,5}/\alpha_2; AB_{k,5} \leftarrow 0$
$\quad\quad\quad$ **Else**
$\quad\quad\quad\quad$ **If** $[k = M]$ **Then** $\qquad$ \\ Set up Dirichlet BC
$\quad\quad\quad\quad\quad AB_{k,1} \leftarrow 0; AB_{k,2} \leftarrow 0;$ \\ Modified coefficients of row $k = M$
$\quad\quad\quad\quad\quad AB_{k,4} \leftarrow 0; AB_{k,5} \leftarrow 0; AB_{k,3} \leftarrow 1; y_k \leftarrow B$
$\quad\quad\quad\quad$ **End If**
$\quad\quad\quad$ **End If**
$\quad\quad$ **End If**
$\quad$ **End If**
End For
BAND_SOLVE$(M, m_L, m_U, \mathbf{AB}, \mathbf{y})$ $\qquad$ \\ Use Pseudocode 2.14
End Module LBVP4
Module COEFFS (x, p, q, r, g)
\\ DESCRIPTION: A user-defined module supplying $p(x), q(x), r(x),$ and $g(x)$.
$P \leftarrow p(x); Q \leftarrow q(x); \ldots$ $\qquad$ \\ Define coefficients here
End Function Module COEFFS

A pseudomodule, **LBVP4**, numerically solving a fourth-order linear BVP defined by Eq. (10.67) (with Dirichlet or Robin BCs) is presented in Pseudocode 10.5. The module requires the number of nodes (M), the Dirichlet BCs (A and B), the Robin BCs (α_i, β_i, γ_i, $i = 1$ and 2), and an array containing the abscissas of the nodes (x_i, $i = 1, 2, \ldots, M$) as input. Note that in the Robin BC, $\alpha_i \neq 0$. On exit, the numerical estimate $\mathbf{y}$ is stored on an array of length M. The module requires two additional modules: (1) **COEFFS**, a user-defined module supplying the coefficients of the 4th-order linear ODE for an arbitrary input x, and (2) **BAND_SOLVE**, a banded linear system solver, given in Pseudocode 2.14. The structure of the code is similar to Pseudocode 10.1, i.e., the coefficients of the difference equation and the rhs, Eq. (10.70) are computed. To complete the construction of the banded matrix, the first two and last two rows of the matrix and the corresponding rhs entries are modified. (For Example 10.10, the pertinent difference equations are Eqs. (10.74), (10.75), (10.77), and (10.78).) Finally, the banded system is solved, and the solution estimate is returned in the array variable $\mathbf{y}$.

EXAMPLE 10.10: Solution of fourth-order BVP

Consider a beam of length $L = 1$ m with the flexural rigidity $EI = 25$ kN·m^2, resting on an elastic foundation with stiffness $K = 150$ kN/m^2. As shown in the figure, both free ends are embedded. The beam is subjected to a uniform loading of $w_0 = 2500$ kN/m.

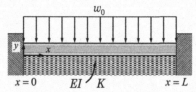

The deflection (y) of the beam is expressed with the following BVP:

$$EI\frac{d^4y}{dx^4} + Ky = w_0, \qquad y(0) = y'(0) = y(L) = y'(L) = 0$$

Use the finite difference method to obtain the numerical solution of the BVP with uniform grid intervals of $M = 5, 10, 20,$ and 40, and discuss the results.

SOLUTION:

The problem is a fourth-order linear BVP with four BCs: two Dirichlet and two Neumann BCs. Substituting the numerical values into the differential equation and simplifying yields

$$y^{(iv)} + 6y = 100, \qquad y(0) = y'(0) = y(1) = y'(1) = 0$$

We follow the numerical solution procedure for the fourth-order linear BVPs:

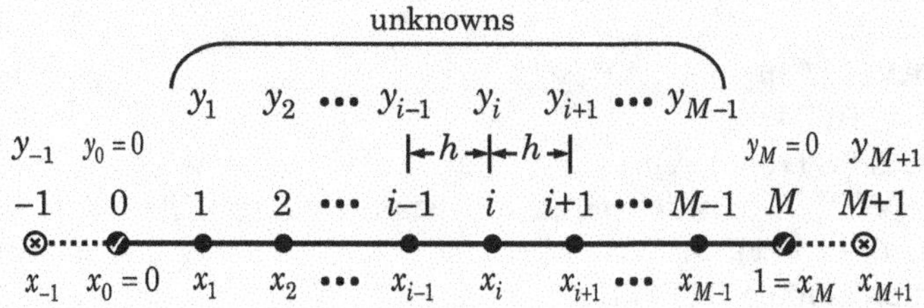

unknowns

$$y_1 \quad y_2 \quad \cdots \quad y_{i-1} \quad y_i \quad y_{i+1} \quad \cdots \quad y_{M-1}$$

$$y_{-1} \quad y_0 = 0 \qquad\qquad |\!\leftarrow\! h \!\rightarrow\!|\!\leftarrow\! h \!\rightarrow\! | \qquad\qquad y_M = 0 \quad y_{M+1}$$

$$-1 \quad 0 \quad 1 \quad 2 \quad \cdots \quad i-1 \quad i \quad i+1 \quad \cdots \quad M-1 \quad M \quad M+1$$

$$x_{-1} \quad x_0 = 0 \quad x_1 \quad x_2 \quad \cdots \quad x_{i-1} \quad x_i \quad x_{i+1} \quad \cdots \quad x_{M-1} \quad 1 = x_M \quad x_{M+1}$$

FIGURE 10.18

Step 1. Gridding: A set of uniformly spaced grid points is established by dividing the beam into M uniform subintervals, giving $h = 1/M$. The abscissas for the resulting nodal points are $x_i = ih$ for $i = 0, 1, 2, \ldots, M$ (*see* Fig. 10.18).

Step 2. Discretization: The fourth-order ODE is satisfied for all $x_i \in (0,1)$, thus we may write

$$y_i^{(iv)} + 6y_i = 100$$

Next, approximating $y^{(iv)}$ by the CDF results in

$$\frac{y_{i+2} - 4y_{i+1} + 6y_i - 4y_{i-1} + y_{i-2}}{h^4} + 6y_i = 100$$

Rearranging and simplifying gives

$$y_{i-2} - 4y_{i-1} + 6(1 + h^4)y_i - 4y_{i+1} + y_{i+2} = 100h^4, \quad i = 1, 2, \ldots, M-1 \qquad (10.73)$$

where the difference equations for $i = 0$ and $i = M$ are excluded since the Dirichlet BCs are imposed on the corresponding boundary nodes.

Step 3. Implementing BCs: Although the boundary nodes are known ($y_0 = y_M = 0$), the fictitious nodes (y_{-1} and y_{M+1}) also appear in the difference equations for $i = 1$ and $i = M-1$ due to the Neumann BCs (*see* Fig. 10.18). The fictitious nodes are eliminated by applying the same procedure employed as in the cases of two-point BVP. The left boundary BC is discretized using the CDF as $y'(0) \cong (y_1 - y_{-1})/2h = 0$, which leads to $y_{-1} = y_1$. Similarly, for the right boundary, we obtain $y_{M+1} = y_{M-1}$.

Setting $i = 1$ in Eq. (10.73), the difference equation for the left node becomes

$$y_{-1} - 4y_0 + 6(1 + h^4)y_1 - 4y_2 + y_3 = 100h^4$$

In light of our findings from the treatment of left BCs ($y_0 = 0$ and $y_{-1} = y_1$), the difference equation takes the form

$$(7 + 6h^4)y_1 - 4y_2 + y_3 = 100h^4 \qquad (10.74)$$

Now, substituting $y_0 = 0$ in the difference equation for node $i = 2$ gives

$$-4y_1 + 6(1 + h^4)y_2 - 4y_3 + y_4 = 100h^4 \qquad (10.75)$$

Thus, y_0 and y_{-1} are eliminated from the difference equations for $i = 1$ and $i = 2$.

Next, we move on to the right boundary. Setting $i = M-1$ in Eq. (10.73) yields

$$y_{M-3} - 4y_{M-2} + 6(1 + h^4)y_{M-1} - 4y_M + y_{M+1} = 100h^4 \qquad (10.76)$$

Substituting $y_M = 0$ and $y_{M+1} = y_{M-1}$ into Eq. (10.76) gives

$$y_{M-3} - 4y_{M-2} + (7 + 6h^4)y_{M-1} = 100h^4 \qquad (10.77)$$

Finally, imposing $y_M = 0$ in the difference equation for node $i = M-2$ yields

$$y_{M-4} - 4y_{M-3} + 6(1 + h^4)y_{M-2} - 4y_{M-1} = 100h^4 \qquad (10.78)$$

Step 4. Constructing a linear system: The general difference equation, Eq. (10.73), for interior nodes ($i = 3, 4, \ldots, M-4$, and $M-3$) is unaffected by the BCs. However, the finite difference equations corresponding to the first two and last two nodes ($i = 1, 2, M - 1$, and $M - 2$) have been modified as a result of implementing the BCs. Combining all difference equations for all unknown nodes, Eqs. (10.73)-(10.75), (10.77), and (10.78), the system of linear equations is expressed in matrix form as

$$
\begin{bmatrix}
E+1 & -4 & 1 \\
-4 & E & -4 & 1 \\
1 & -4 & E & -4 & 1 \\
& \ddots & \ddots & \ddots & \ddots & \ddots \\
& & 1 & -4 & E & -4 & 1 \\
& & & 1 & -4 & E & -4 \\
& & & & 1 & -4 & E+1
\end{bmatrix}
\begin{bmatrix}
y_1 \\ y_2 \\ y_2 \\ \vdots \\ y_{M-3} \\ y_{M-2} \\ y_{M-1}
\end{bmatrix}
= 100h^4
\begin{bmatrix}
1 \\ 1 \\ 1 \\ \vdots \\ 1 \\ 1 \\ 1
\end{bmatrix}
$$

where $E = 6(1 + h^4)$.

Step 5. Solving the linear system: This system is a banded matrix (penta-diagonal) with upper and lower bandwidths of 2, which is solved using the Gauss elimination method designed specifically for banded systems. The BVP has an analytical solution, but it is not given here explicitly because it contains long and complicated expressions. Finding the true solution is left to the readers as a practice problem.

TABLE 10.12

x	True Solution	Numerical estimates, M			
		5	10	20	40
0	0	0	0	0	0
0.1	0.033377		0.04077	0.03523	0.03384
0.2	0.105447	0.15758	0.11852	0.10872	0.10627
0.3	0.181594		0.19870	0.18587	0.18266
0.4	0.237139	0.31486	0.25664	0.24202	0.23836
0.5	0.257296		0.27759	0.26237	0.25857

The solution of the BVP is symmetrical with respect to $x = 0.5$; hence, the numerical solutions for $M = 5, 10, 20$, and 40 are comparatively tabulated in Table 10.12 for the first half only. The average error is in the order of 0.065, 0.016, 0.004, and 0.001 for $M = 5, 10, 20$, and 40, respectively.

Discussion: Notice that the numerical error for $M = 5$ is quite high, but the estimates depict improvement as M increases. From the numerical results, we observe that the error decreased by a factor of four as the number of intervals was halved.

10.10 CLOSURE

Many steady-state problems in science and engineering lead to linear or nonlinear two-point ODEs (*two-point BVPs*). The most common two-point BVPs consist of a second-order linear ODE and two BCs, which may be of the Dirichlet, Neumann, or Robin type.

The *Finite Difference Method* is the easiest and most common technique to find the approximate solution of a linear BVP. The method is based on converting the ODE-BVP into a system of algebraic equations by gridding and discretizing procedures. When the derivatives are approximated with finite difference formulas, the method leads to a tridiagonal system of linear equations, traditionally solved with the Thomas algorithm. The global error in this case is $\mathcal{O}(h^2)$. However, apart from reducing the interval size, high-accuracy numerical solutions can be achieved with Richardson's extrapolation technique, which allows numerical estimates of the $\mathcal{O}(h^4)$ or higher order to be obtained. For this, we need two sets of solutions with grid intervals of h and $h/2$.

Higher-order two-point ODEs lead to a banded system of linear equations, which may also be solved using a band matrix solver to obtain an approximate solution. However, the numerical solution of a higher-order ODE also requires as many BCs as the order of the ODE, i.e., a fourth-order ODE requires four BCs, and so on.

The *Finite Difference Method* using uniform grids is the simplest and cheapest, but it is not satisfactory for problems with boundary layers. If the number of nodal points is not sufficient to resolve the boundary layer (at least several nodes within it), then the numerical solution can result in gross errors even for the interior nodes. The use of large enough grid points to resolve the boundary layer may yield an unacceptably large cpu-time. The problem can be solved by employing non-uniform grids with smaller spacing near the boundary or by transforming the ODE with a suitable stretching function.

The *Finite Volume Method* is another discretization method suitable for those that especially arise from physical conservation laws. In other words, it is particularly useful for BVPs that are cast in conservation forms because boundary and interface conditions can naturally and easily be implemented without requiring fictitious nodes. The two-point BVP is then converted into a system of algebraic equations by integrating it over a finite control volume, centering a nodal point. The method is also second-order accurate.

An ODE or a system of coupled ODEs as BVPs can be treated as an IVP, which can then be solved by employing the shooting method, which uses the methods developed for IVPs, by guessing the initial condition(s) and iterating until the BCs are satisfied. Solving a linear or nonlinear two-point BVP with a BC at infinity is very difficult with the finite difference method since the magnitude of *infinity* for any particular problem is unknown. In this regard, the shooting method is suitable for these problems, where the BVP is treated as an initial value problem by guessing and interpolating the "initial slope." High-order accurate numerical solutions (3rd or 4th-order RK methods, etc.) of two-point BVPs can be achieved with the shooting method, while FDM leads to at most second-order accurate solutions.

The numerical solution of nonlinear two-point BVPs is often a very difficult task. The finite difference method is generally employed when the shooting method fails. The methods presented here (Newton's and tridiagonal iteration methods) require an iterative procedure to obtain an approximate solution. Nonetheless, these methods can be very sensitive to initial guesses and can lead to convergence problems.

10.11   EXERCISES

Section 10.1 Introduction

E10.1   Identify the following ODEs as either *linear* or *nonlinear*.

(a) $x^2\,u'' + x\,u' - x^2 u = e^x$,   (b) $y'' + x\,y' + (1 - \sin x)y = 0$,   (c) $z'' = aze^{z'}$,

(d) $Z'' = (1 - e^{x^2})ZZ' + 1$,   (e) $(1 + \cos y)\,y'' - (2 - \sin y)\,y = 0$,   (f) $u'' - (1 + x^2)u = 0$

Section 10.2 Two-Point Boundary Value Problems

E10.2 Define the following BCs subject to the ODE $y'' + y = g(x)$, as Dirichlet, Neumann, or Robin.

(a) $y(-1) = 3$, $y(1) = -2$,
(b) $y'(0) = 0$, $y(5) = 1$,
(c) $y'(-1) + 2y(-1) = 0$, $y'(1) - 2y(0) = 0$,
(d) $y(0) = y(1)$, $y'(0) = y'(1)$,
(e) $y(-2) = 1$, $2y'(1) + 3y(1) = 4$

E10.3 Verify that $y(x) = c_1 \sin x + c_2 \cos x$ satisfies $y'' + y = 0$ and determines the solution of the BVP (i.e., c_1 and c_2) that satisfy the following BCs:

(a) $y(0) = 0$, $y(\pi) = 0$,
(b) $y'(0) = 0$, $y(\pi) = 1$,
(c) $y'(0) = 1$, $y'(\pi/2) = 2$,
(d) $2y'(0) + y(0) = 1$, $2y'(\pi/2) - y(\pi/2) = 1$,
(e) $y(-\pi) = 1$, $y'(\pi/4) + y(\pi/4) = 1$

E10.4 Verify that $y(x) = c_1\sqrt{x} + c_2\sqrt{x^3}$ satisfies $4x^2 y'' - 4xy + 3y = 0$ and determines the solution of the BVP (i.e., c_1 and c_2) that satisfies the following BCs:

(a) $y(1) = 3$, $y(4) = 6$,
(b) $y'(0) = 0$, $y(4) = 1$,
(c) $y'(1) - y(1) = 1$, $y'(4) + y(4) = 1$,
(d) $y'(1) = 1$, $y'(4) = 2$

Section 10.3 Finite Difference Solution Linear BVPs

E10.5 Consider the following BVPs with Dirichlet BCs applied to both sides. (i) Divide the solution interval into M equally spaced subintervals and use the FDM to derive the general difference equation with a second-order global error; (ii) implement the BCs and obtain a linear system of equations in matrix form; (iii) solve the resulting linear system for $M = 5$.

(a) $x^2 y'' + xy' - x^2 y = 0$, $y(1) = 1$, $y(2) = 0$
(b) $(x^2 + 2)y'' + (3x - 1)y' - 2y = e^{3x}$, $y(-1) = 0$, $y(1) = 2$
(c) $(1 + x^2)y'' + xy' + 3y = (\ln x)^2$, $y(1) = 2$, $y(2) = -1$

E10.6 Consider the given BVPs with Neumann or Robin BC on the left and Dirichlet BC on the right boundary. (i) Divide the solution interval into M equally spaced subintervals and use the FDM to derive the general difference equation with a second-order global error; (ii) implement the BCs and obtain a linear system of equations in matrix form; (iii) solve the resulting linear system for $M = 5$.

(a) $y'' + (1 + x)y' + 5e^{-x}y = e^x$, $y'(0) = 1$, $y(5/2) = 0$
(b) $(1 + 2x)y'' + 2y' - 2y = 2x$, $y'(1) = 4$, $y(3) = 2$
(c) $(1 + \sin x)y'' + \cos x\, y' + y = x$, $y'(0) - 2y(0) = 0$, $y(1) = 3$
(d) $y'' + (2 - x^2)y = 2x^2 + 1$, $y'(0) - y(0) = 2$, $y(2) = 5$

E10.7 Consider the given BVPs with Dirichlet BC on the left and Neumann or Robin BC on the right boundary. (i) Divide the solution interval into M equally spaced subintervals and use the FDM to derive the general difference equation with a second-order global error; (ii) implement the BCs and obtain a linear system of equations in matrix form; (iii) solve the resulting linear system for $M = 5$.

(a) $y'' - 2xy' + y = \ln(1 + x)$, $y(0) = 1$, $y'(1) = 0$
(b) $(2x^2 + 5)y'' + 2xy' - 3y = 0$, $y(1) = 3$, $y'(6) = y(6)$
(c) $y'' + 8y' + xy = 14x + 2$, $y(2) = 1$, $y'(3) = 0$
(d) $-(x^2 + 4)y'' + xy' + (1 + 4x)y = \sqrt{x + 0.8}$, $y(0.2) = 0$, $y'(1.2) - y(1.2) = 0$,
(e) $e^x y'' + y' - e^{-x}y = \sin x$, $y(-\pi/4) = 1/5$, $2y'(\pi/4) + y(\pi/4) = 1$

E10.8 Consider the following BVPs with Robin BCs on both boundaries: (i) Divide the solution interval into M equally spaced subintervals and use the FDM to derive the general difference equation with a second-order global error; (ii) implement the BCs and obtain a linear system of equations in matrix form; (iii) solve the resulting linear system for $M = 5$.

(a) $(1 + 2x)y'' - 2y' + (2e^x - x)y = \sec x$, $\quad y'(0) + y(0) = 0$, $\quad y'(1) - y(1) = 0$,

(b) $(x^2 + 1)y'' + (x - 2)y' + (1 + \sin^2 x)y = 0$, $\quad y'(-1) + 3y(-1) = 2$, $\quad y'(1) - 3y(1) = -2$,

(c) $3y'' + (3x + 2)y' + 3y = e^{-x^2}$, $\quad 2y'(1) - y(1) = 4$, $\quad y'(2) + 3y(2) = 6$

E10.9 Use the FDM to solve the following two-point linear BVPs with $M = 5, 10, 20$, and 40 equispaced subintervals. Compare your numerical estimates with the given true solution.

(a) $x^2 y'' - 2y = 3x^2 - 1$, $\quad y(0) = 1/2$, $\quad y'(1) = 1$, $\quad y_{true}(x) = 1/2 + x^2 \ln x$

(b) $x^2 y'' - 5xy' + 8y = 0$, $\quad y'(1) + 2y(1) = 4$, $\quad y'(2) - 2y(2) = 2$, $\quad y_{true}(x) = x^4 - x^2/2$

(c) $x^2 y'' + 4xy' + 2y = 2 \ln x$, $\quad y(1/22) = -3/2 - \ln 2$, $\quad y(1) = 12$, $\quad y_{true}(x) = 4/x - 2/x^2 - 3/2 + \ln x$

(d) $y'' - (x^2 + 1)y = 0$, $\quad y'(0) = 0$, $\quad y(1) = \sqrt{e}$, $\quad y_{true}(x) = e^{x^2/2}$

E10.10 Use the FDM to solve the following two-point linear BVPs with $M = 5, 10, 20$, and 40 equispaced subintervals. Compare your numerical estimates with the given true solution.

(a) $(1 + x^2)y'' + xy' - y = 3x^2$, $\quad y(0) = 0$, $y(1) = 1$, $\quad y_{true}(x) = 2 + 2(\sqrt{2} - 1)x + x^2 - 2\sqrt{1 + x^2}$

(b) $2xy'' - (4x + 3)y' + (3 - 8x^3)y = 0$, $\quad y(0) = 1$, $\quad y(1) = 1$, $\quad y_{true}(x) = e^{x(1-x)}$

E10.11 A rectangular fin attached to a hot wall is shown in **Fig. E10.11**. The temperature distribution as a function of x (distance from the base) is governed by the ODE given as

$$\frac{d^2 T}{dx^2} - \frac{hP}{kA}(T(x) - T_\infty) = 0, \quad (0 \leqslant x \leqslant L)$$

where h is the convection heat transfer coefficient (W/m^{2}K), k is the conductivity (W/m·K), P is the perimeter (m), and A is the cross-section area of the fin (m^2). The base temperature is $T(0) = T_b = 420$ K, and the fin tip is insulated (i.e., $T'(L) = 0$). (a) Discretize the governing ODE with FDM by dividing the fin length into M equal subintervals; (b) for given $hP/kA = 22$ and 70, $L = 0.1$ m, and $T_\infty = 290$ K, obtain the numerical solution for $M = 5$, 10, and 20 intervals; (c) compare and comment on the convergence of the numerical solutions.

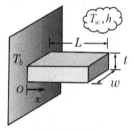

Fig. E10.11

E10.12 When two chemicals in liquid form are mixed ($k_{mix} = 1$ W/m·K) in a cylindrical reactor ($L/R \gg 1$), exothermic reactions take place, resulting in a heat generation of 1500 W/m^3 (*see* **Fig. E10.12**). The reactor diameter, ambient temperature, and convection heat transfer coefficient are given as 1.2 m, $T_\infty = 10°$ C, and $h = 20$ W/m^{2}K, respectively. Using the governing ODE and BCs given below, (a) discretize the ODE using the FDM by dividing the radius into M equal subintervals; (b) implement the BCs and express the general difference equations in matrix equation form; and (c) solve the resulting linear system for $M = 5$.

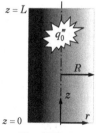

Fig. E10.12

$$\frac{1}{r}\frac{d}{dr}\left(rk\frac{dT}{dr}\right) + q'''(r) = 0, \quad \left(\frac{dT}{dr}\right)_{r=0} = 0, \quad \left(\frac{dT}{dr}\right)_{r=R} = \frac{h}{k}(T_\infty - T(R))$$

E10.13 Fluid flow in an annular cylindrical duct ($R_1 \leqslant r \leqslant R_2$), extending in the z-direction, is shown in **Fig. E10.13**. The governing ODE for the axial velocity is given by

$$\frac{1}{r}\frac{d}{dr}\left(r\frac{du}{dr}\right) = \frac{1}{\mu}\frac{dP}{dz} = C$$

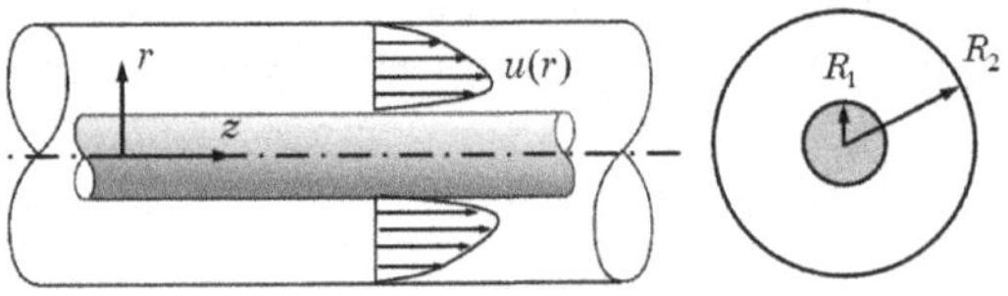

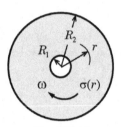

Fig. E10.13

where u is the axial velocity component, r is the radial coordinate, dP/dx is the pressure drop across the duct, and μ is the kinematic viscosity of the fluid. The cylinder walls are impermeable, i.e., $u(R_1) = 0$ and $u(R_2) = 0$. To solve the ODE using FDM with $C = -0.01$ and $C = -10$, (a) obtain the general difference equation by dividing the $R_1 \leqslant r \leqslant R_2$ interval into M equal subintervals; (b) implement the BCs and express the difference equations in matrix form; (c) solve the linear system of equations for $R_1 = 0.1$, $R_2 = 0.35$, and $M = 5$; and (d) find the true solution of the BVP and compare it with your estimates.

E10.14 The governing differential equation of radial stresses of a thick rotating annular disc made of aluminum (*see* **Fig. E10.14**) is given as

$$r\frac{d^2\sigma}{dr^2} + 3\frac{d\sigma}{dr} + (3 + \nu)\,\rho\omega^2 r = 0, \quad R_1 \leqslant r \leqslant R_2$$

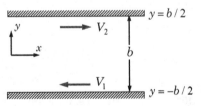

Fig. E10.14

where r is the radial coordinate, σ is the radial stress, ω is the angular speed, $\rho = 2700$ kg/m^3, and $\nu = 0.35$ are the density and Poisson ratio of aluminum, respectively. Assuming that there are no radial stresses at the inner and outer surfaces (i.e., $\sigma(R_1) = \sigma(R_2) = 0$), (a) obtain a general difference equation by dividing $R_1 \leqslant r \leqslant R_2$ into M equal subintervals; (b) implement the BCs and express the difference equations in matrix form; (c) solve the linear system of equations for $R_1 = 0.1$, $R_2 = 0.35$, $\omega = 2$ rad/s, $M = 5$ and 10, and (d) compare your estimates with the true solution given as

$$\sigma(r) = \omega^2\rho(3 + \nu)\frac{(R_1^2 - r^2)(r^2 - R_2^2)}{8r^2}$$

E10.15 Consider the fluid flow between two large parallel plates, as shown in **Fig. E10.15**. The gap between the plates is filled with grease ($\mu = 1.8$ N·/m^2). The plates are moving in opposite directions at velocities V_1 and V_2. The governing ODE and the BCs are given as

$$\frac{d^2u}{dy^2} = \frac{1}{\mu}\frac{dp}{dx} = C = \text{constant}, \quad (-\frac{b}{2} \leqslant y \leqslant \frac{b}{2})$$
$$u(b/2) = -V_1, \quad u(-b/2) = V_2$$

Fig. E10.15

where u is the y-velocity component, b is the gap thickness, dp/dx is the pressure drop, and μ is the kinematic viscosity of grease. To solve the governing ODE with the FDM, divide $-b/2 \leqslant y \leqslant b/2$ into M equispaced subintervals and (a) derive the general difference equation; (b) implement the BCs and express the difference equations in matrix equation form; (c) solve the linear system of equations for $dp/dx = -7.2$ Pa/m, $b = 5$ mm, $V_1 = 0.0075$ m/s, $V_2 = 0.005$ m/s, and $M = 5$; (d) obtain the true solution of the BVP and compare your estimates with that of the true solution.

E10.16 A mathematical model for steady-state axial diffusion in a tubular reactor is described with the following BVP:

$$D\frac{d^2c}{dz^2} - u\frac{dc}{dz} + r = 0$$

where c is the gas concentration, D is the diffusion coefficient, u is the axial velocity, z is the axial coordinate, and r is the reaction rate. The reaction rate equation is given with $r = -kc$, where k is the reaction rate coefficient. In order to obtain a more general solution to such problems, it is customary to express differential equations in terms of dimensionless quantities. Introducing the following dimensionless quantities: $C = c/c_0$, $Z = z/L$, Pe$= uL/D$, and Da$= kL/u$, where Pe and Da are referred to as Peclet and Damköhler numbers. The BVP can then be expressed with the dimensionless quantities as

$$\frac{1}{\text{Pe}}\frac{d^2C}{dZ^2} - \frac{dC}{dZ} - \text{Da}\,C = 0, \quad C = C(Z), \quad 0 \leqslant Z \leqslant 1$$

subjected to dispersion (at the entry) and continuity (at the exit) BCs, which, in dimensionless forms, are expressed as

$$\left(-\frac{1}{\text{Pe}}\frac{dC}{dZ} + C\right)_{Z=0} = 1 \quad \text{and} \quad \left(\frac{dC}{dZ}\right)_{Z=1} = 0$$

Solve the pertinent BVP using the FDM for Pe=10 and Da=1, dividing the tube length into $M = 10$, 20, and 40 intervals, and compare your estimates with the true solution given by

$$C(Z) = \frac{(b - \text{Pe})e^{m_1 Z} + (b + \text{Pe})e^{b+m_2 Z}}{b - a + (a + b)e^b}$$

where $a = \text{Pe} + 2\text{Da}$, $b = \sqrt{\text{Pe}^2 + 4\text{PeDa}}$, $m_1 = (\text{Pe} + b)/2$, and $m_2 = (\text{Pe} - b)/2$.

E10.17 Consider an infinitely long plane pad, having a slight inclination, and a plate moving with a velocity V, as depicted in **Fig. E10.17**.

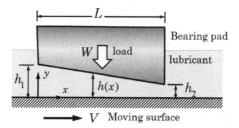

Fig. E10.17

The thickness of an oil film between the two surfaces varies linearly as $h(x) = h_1 - (h_1 - h_2)x/L$, where L is the length of the pad in the direction of the plane motion. The pressure in the lubricant, due to the pad, is governed by the following BVP:

$$\frac{d}{d\xi}\left(H^3(\xi)\frac{dP}{d\xi}\right) = 1 - m, \quad P(0) = P(1) = 0$$

where the following dimensionless quantities have been introduced: $\xi = x/L$, $P = ph_2^2/6\mu V L$, $m = h_1/h_2$, $H(\xi) = h/h_2 = m - (m - 1)\xi$. Use the FDM to solve the given BVP to obtain the pressure distribution: (a) for $m = 2$ by dividing $0 \leqslant \xi \leqslant 1$ into $M = 5$ equispaced subintervals; (b) repeat Part (a) for $m = 3$; and (c) repeat Parts (a) and (b) with $M = 10$. Compare your numerical estimates with the true solution given by

$$P(\xi) = \frac{(m - 1)(\xi - \xi^2)}{(m + 1)(m + (1 - m)\xi)^2}$$

E10.18 A mathematical model for steady-state radial diffusion or reaction in a spherical catalyst is described by the following BVP:

$$D\left(\frac{d^2C}{dr^2} + \frac{2}{r}\frac{dC}{dr}\right) - kC = 0$$

where k is the reaction rate constant, D is the diffusion coefficient, C is the concentration, and r is the radial coordinate. Introducing the dimensionless quantities $C = c/c_0$, $\eta = r/R$, $\phi^2 = R^2 k/D$, where c_0 is the maximum concentration, R is the radius of the catalyst, and ϕ is so-called the Thiele modulus. The BVP and the BCs in dimensionless quantities are expressed as

$$\frac{d^2 C}{d\eta^2} + \frac{2}{\eta}\frac{dC}{d\eta} - \phi^2 C = 0, \quad \left(\frac{dC}{d\eta}\right)_{\eta=0} = 0, \quad C(1) = 1$$

Use the FDM to solve the BVP for $\phi = 1$ and 3 by dividing $0 \leqslant \eta \leqslant 1$ into $M = 5$ and 10 intervals, and compare your numerical solutions with the true solution given by $C(\eta) = \sinh \phi\eta/(\eta \sinh \phi)$.

Section 10.4 Numerical Solutions of High-Order Accuracy

E10.19 Repeat **E10.9** to find numerical estimates of order $\mathcal{O}(h^6)$. Use FDM estimates for $M = 10$, 20, and 40 subintervals and compare your estimates with those of the true solution.

E10.20 Repeat **E10.10** to find numerical estimates of order $\mathcal{O}(h^6)$. Use FDM estimates for $M = 10$, 20, and 40 subintervals, and compare your estimates to the true solution.

Section 10.5 Non-Uniform Grids

E10.21 Consider the following coordinate transformation:

$$x = \alpha + (1 - \alpha) \ln\left(\frac{(\beta - 1)h + 2\xi}{(\beta + 1)h - 2\xi}\right) / \ln\left(\frac{\beta + 1}{\beta - 1}\right)$$

which maps $0 \leqslant \xi \leqslant h$ onto $0 \leqslant x \leqslant 1$. (a) Plot the node distribution for $\beta = 1$, $\alpha = 0.1$, and 0.5 using 20 equispaced subintervals in the ξ-domain; (b) obtain expressions for y' and y'' derivatives as a function of ξ.

E10.22 Consider the following BVP with Dirichlet BCs on both sides. (a) Convert the problem to the ξ-domain by clustering the grid points near $x = 0$ with the $x = (e^{a\xi}-1)/(e^a-1)$ transformation; (b) Plot the distribution of nodal points for $a = 0.01$, 0.5, 1, and 2 by using 20 equispaced subintervals in the ξ-domain.

$$y'' + y' + y = -x^2, \quad y(0) = 1, \quad y(1) = 0$$

E10.23 Consider the following BVP with Dirichlet BCs on both sides. Using $M = 4$ and $h_1 = 8\Delta x$, $h_2 = 4\Delta x$, $h_3 = 2\Delta x$, and $h_4 = \Delta x$ (nodes clustered by $x = 1$), (a) discretize the BVP with the FDM to obtain the difference equations; (b) express the resulting difference equations in matrix form; (c) solve and compare your estimates to the true solution given as $y(x) = 2x^4$.

$$xy'' + (x - 3) y' - 4y = 0, \quad y(0) = 0, \quad y(1) = 2$$

E10.24 Consider the following BVP with Dirichlet BCs on both sides. Using $M = 4$ and $h_1 = \Delta x$, $h_2 = 2\Delta x$, $h_3 = 4\Delta x$, and $h_4 = 8\Delta x$ (nodes clustered by $x = 0$), (a) discretize the BVP with the FDM to obtain the difference equations; (b) express the resulting difference equations in matrix form; (c) solve and compare your estimates to the true solution given as $y(x) = 2x/(x + 1)$.

$$2(x + 1)^2 y'' + 5(x + 1)y' + y = 2, \quad y(0) = 0, \quad y(9) = 9/5$$

E10.25 Consider the following BVP: Using $M = 8$ and $h_1 = \Delta x$, $h_2 = r\Delta x$, $h_3 = r^2\Delta x$, $h_4 = r^3\Delta x$, $h_5 = r^3\Delta x$, $h_6 = r^2\Delta x$, $h_7 = r\Delta x$, and $h_8 = \Delta x$ (nodes clustered on both sides), (a) discretize the BVP with the FDM and implement the BCs; (b) write out and solve the resulting finite-difference equations; (c) compare the estimates with the true solution given as $y(x) = 4x(1 - x)$.

$$y'' + 2y' + 4y = -16x^2, \quad y(0) = 0, \quad y(1) = 0$$

E10.26 Consider **E10.23**. (a) Transform the BVP to the ξ-computational domain using the transformation equation $x = \sqrt{\xi}$; (b) solve the transformed BVP using the FDM with $M = 4$ equispaced subintervals; (c) comment on the accuracy of your numerical estimates.

E10.27   Consider **E10.24** (a) Transform the BVP to the ξ-computational domain using the transformation equation $x = \xi^2$; (b) solve the transformed BVP using the FDM with $M = 4$ equispaced subintervals; (c) comment on the accuracy of your numerical estimates.

E10.28   Consider **E10.25** (a) Transform the BVP to the ξ-computational domain using the transformation equation given below; (b) solve the transformed BVP using the FDM with $M = 8$ equispaced subintervals; (c) comment on the accuracy of your numerical estimates.

$$x = \frac{1}{2}\left(1 + \beta \tanh\left[(\xi - \tfrac{1}{2})\ln[(\beta+1)/(\beta-1)]\right]\right)$$

Section 10.6 Finite Volume Method

E10.29   Solve the following BVPs using FVM with $M = 10$ uniformly sized cells and compare your estimates with the true solution.

(a) $-\left((x^2+1)y'\right)' + 12(x+1)y = 24x^4 + 12, \quad y'(0) = 0, \quad y(1) = 3, \quad y_{true}(x) = 2x^3 + 1$

(b) $-\left(4(x+1)y'\right)' + (x-1)y = 0, \quad y(0) = 1, \quad 2y'(2) + y(2) = 0, \quad y_{true}(x) = e^{-x/2}$

(c) $(xy')' + xy = 0, \quad y'(0) = 0, \quad y(4) = 2, \quad y_{true}(x) = 2J_0(x)/J_0(4)$

(d) $e^x\left(e^x y'\right)' + y' + 2y = 6, \quad y(0) = 3, \quad y'(1) + 2y(1) = 0, \quad y_{true}(x) = 3e^{-2x}$

(e) $(x^2 y')' + (x^2 - x)y' - y = 0, \quad y(0) = 0, \quad y'(1) + 2y(1) = 1, \quad y_{true}(x) = (x - 1 + e^{-x})/x$

where $J_0(x)$ is the Bessel function.

E10.30   Repeat **E10.17** using the FVM with uniformly sized cells.

E10.31   Repeat **E10.18** using the FVM with uniformly sized cells.

E10.32   A research nuclear reactor operates with plate-type UO_2 fuel elements are enveloped by a steel cladding. Making use of the physical symmetry, the half-fuel and clad thicknesses are given as $a = 0.6$ cm and $b = 0.3$ cm, respectively (*see* **Fig. E10.32**). Heat is generated in the fuel region ($q_0''' = 75$ W/cm³) as a result of nuclear reactions. The conductivities of fuel and cladding are $k_f = 0.6$ W/cm·K and $k_c = 0.15$ W/cm·K, respectively. At steady-state operating conditions, the temperature distribution in the fuel and cladding can be modeled with the following set of ODEs:

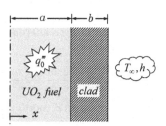

Fig. E10.32

$$k_f \frac{d^2 T_f}{dx^2} + q_0''' = 0 \quad (0 \leqslant x \leqslant a), \qquad k_c \frac{d^2 T_c}{dx^2} = 0, \quad (a \leqslant x \leqslant a + b)$$

subjected to the following boundary and interface conditions:

Symmetry at the center	$(dT_f/dx)_{x=0} = 0$
Convection from cladding	$k_c (dT_c/dx)_{x=a+b} = h\left(T_\infty - T_c(a+b)\right)$
Continuity of temperature at interface	$T_f(a) = T_c(a)$
Continuity of heat flux at interface	$k_f (dT_f/dx)_{x=a} = k_c (dT_c/dx)_{x=a}$

where $h = 1$ W/cm²·K is the convection transfer coefficient, and T_∞ is the fluid bulk temperature. Use the FVM to solve the given problem using $M = 9$ uniformly sized cells and compare your estimates with those of the true solution, which is given as

$$T(x) = \begin{cases} 277.5 - 62.5x^2, & 0 \leqslant x \leqslant 0.6\,\text{cm} \\ 435 - 300x, & 0.6 \leqslant x \leqslant 0.9\,\text{cm} \end{cases}$$

E10.33 Consider a thick-walled spherical pressure vessel, shown in **Fig. E10.33**. A strain gauge bonded tangentially at the inner wall to measure the normal tangential strain gives a reading of $\varepsilon_\theta(R_i) = 0.00065$ at maximum pressure. The strain at the outer wall is zero, i.e., $\varepsilon\theta(R_o) = 0$. Since the radial displacement and tangential strain are related with $\varepsilon_\theta = u/r$, these conditions lead to the BCs in terms of the displacement. The governing ODE for the radial displacement is given by

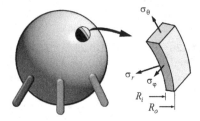

Fig. E10.33

$$\frac{d^2u}{dr^2} + \frac{2}{r}\frac{du}{dr} - 2\frac{u}{r^2} = 0, \quad R_i \leqslant r \leqslant R_o \quad u(R_i) = R_i\varepsilon_\theta(R_i) \quad \text{and} \quad u(R_o) = 0$$

where $R_i = 50$ cm and $R_o = 66$ cm are the inner and outer radii of the vessel under internal pressure. To solve the BVP for radial displacement using the FVM, (a) divide the wall thickness into 8 cells and obtain the displacement at the surfaces of each cell; (c) calculate the maximum normal stress at the inner pressure vessel wall, which is given by

$$\sigma_{\max} = \frac{E}{(1+\nu)(1-2\nu)}\left[2\nu\frac{u}{r} + (1-\nu)\frac{du}{dr}\right]_{r=R_0}$$

where $E = 200$ GPa is the Young's modulus of steel and $\nu = 0.3$ is the Poisson's ratio.

E10.34 Consider the two-layer composite spherical pressure vessel ($R_i = 0.4$m, $R_I = 0.6$, $R_o = 0.75$) shown in **Fig. E10.34**. The differential equation for the radial displacement in the vessel subjected only to internal pressure is governed by the following ODE:

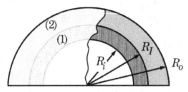

Fig. E10.34

$$\frac{d^2u}{dr^2} + \frac{2}{r}\frac{du}{dr} - 2\frac{u}{r^2} = 0, \quad R_i \leqslant r \leqslant R_o$$

where r is the radial coordinate and u is the radial displacement. The radial stress-displacement relationship is given by

$$\sigma_r = \frac{E}{(1+\nu)(1-2\nu)}\left(2\nu\frac{u}{r} + (1-\nu)\frac{du}{dr}\right)$$

where σ_r is the radial stress and E is the modulus of elasticity. The material properties of the inner and outer shells are $E_1 = 200$ GPa, $\nu_1 = 0.3$, $E_2 = 80$ GPa, and $\nu_2 = 0.36$, respectively. The corresponding boundary conditions are given as

$$\text{At } r = R_i, \quad \sigma_r(R_i) = -P \quad \text{(internal wall pressure)}$$

$$\text{At } r = R_I, \quad u(R_I^-) = u(R_I^+) \quad \text{and}$$

$$\sigma_r(R_I^-) = \sigma_r(R_I^+) \quad \text{(interface continuity)}$$

$$\text{At } r = R_o, \quad \sigma_r(R_i) = 0 \quad \text{(external wall pressure)}$$

where $P = 10$ MPa is the applied internal pressure. Apply the FVM to solve the BVP for the radial displacement subjected to the prescribed BCs. Obtain numerical estimates for $\Delta r = 0.05$, 0.025, 0.0125, and 0.00625 m. (b) Solve the ODE using second-order difference formulas. The difference equation for the continuity of radial stress will contain five nodal points; thus, the resulting system of equations will no longer be a tridiagonal system. Use the Gauss-Seidel method to solve the system of equations.

E10.35 A steel pipe shown in **Fig. E10.35**, carrying steam at $T_{\infty,1} = 260°\text{C}$, has an internal diameter of $D_i = 100$ mm and a thickness of 5 mm. The conductivity of the pipe is $k_{pipe} = 15$ W/m°C, and the convection heat transfer coefficient for the internal flow is $h_{in} = 32$ W/m² °C. The pipe is insulated with a 30-mm-thick insulation material having a conductivity of $k_{ins} = 0.5$ W/m°C. The pipe is exposed to air at $T_{\infty,2} = 20°\text{C}$ with $h_{out} = 80$ W/m²°C.

The governing ODE for this system is given as

$$\frac{1}{r}\frac{d}{dr}\left(k(r)\,r\,\frac{dT}{dr}\right) = 0, \qquad R_i \leqslant r \leqslant R_o$$

where r is the radial coordinate (m), k is the thermal conductivity (W/m°C), and T is the temperature (°C).

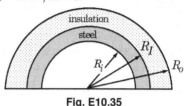

Fig. E10.35

Note that the conductivity has a jump discontinuity at the pipe-insulation interface, which is expressed as $k(r) = k_{pipe}$ for $r < R_I$ and $k(r) = k_{ins}$ for $r > R_I$. The corresponding boundary and interface conditions are given as follows:

$$\text{At } r = R_i, \quad k_{\text{steel}}\frac{dT}{dr}(R_i) = h_{in}\left(T(R_i) - T_{\infty 1}\right)$$

$$\text{At } r = R_I, \quad T(R_I^-) = T(R_I^+) \quad \text{and} \quad k_{\text{steel}}\frac{dT}{dr}(R_I^-) = -k_{ins}\frac{dT}{dr}(R_I^+)$$

$$\text{At } r = R_o, \quad -k_{ins}\frac{dT}{dr}(R_o) = h_{out}\left(T(R_o) - T_{\infty 2}\right)$$

Apply the FVM to numerically estimate the temperature under the prescribed boundary conditions using uniform $M = 7, 14$, and 28 cells, and discuss the accuracy of your numerical estimates.

Section 10.7 Finite Difference Solution of Nonlinear BVPs

E10.36 Use the FDM to solve the following nonlinear two-point BVPs by employing tridiagonal iteration with $M = 10, 20, 40$, and 80 equispaced subintervals. Compare your numerical estimates with the true solution. Use $\varepsilon = 10^{-6}$ for tolerance for convergence.

(a) $y'' = e^y$, $\quad y(0) = \ln 2$, $\quad y(\pi/4) = \ln 4$, $\quad y_{\text{true}}(x) = \ln(2\sec^2 x)$

(b) $yy'' + 3(y')^2 = 0$, $\quad y(0) = 0$, $\quad y(4) = 1$, $\quad y_{\text{true}}(x) = (x/4)^{1/4}$

(c) $y'' = (y' + y(y')^2)/(1 + y^2)$, $\quad y(0) = 1$, $\quad y(2) = e^2$, $\quad y_{\text{true}}(x) = e^x$

(d) $y'' = 2y(y')^2/(1 + y^2)$, $\quad y(0) = 0$, $\quad y(2) = 1$, $\quad y_{\text{true}}(x) = \tan(\pi x/8)$

E10.37 Use the FDM to solve the following nonlinear two-point BVPs by employing Newton's method with $M = 10, 20, 40$, and 80 equispaced subintervals. Compare your numerical estimates with the given true solution. Use $\varepsilon = 10^{-6}$ for tolerance for convergence.

(a) $y'' = 2e^{2y} - (y')^2$, $\quad y(0) = 0$, $\quad y(1) = -\ln 2$, $\quad y_{\text{true}}(x) = -\ln(1 + x)$

(b) $y'' = 12y^3 + 2yy'$, $\quad y(0) = 1$, $\quad y(1/2) = 1/2$, $\quad y_{\text{true}}(x) = 1/(2x + 1)$

(c) $y'' = 2y^5 + (y')^2/y$, $\quad y(0) = 1$, $\quad y(4) = 1/3$, $\quad y_{\text{true}}(x) = 1/\sqrt{2x + 1}$

(d) $y'' = 6e^{-x} - ye^x(y' + y)/3$, $\quad y(0) = 1$, $\quad y(1/3) = 0$, $\quad y_{\text{true}}(x) = (1 - 3x)e^{-x}$

E10.38 Use the FDM to solve the nonlinear two-point BVPs given below using Newton's method with $M = 10, 20, 40$, and 80 equispaced subintervals. Compare your numerical estimates with the given true solution. Take $y_i^{(0)} = 1$ for the initial guess and $\varepsilon = 10^{-6}$ for convergence tolerance.

(a) $2xy'' + 2y' + 3xe^y = 0$, $\quad y'(0) = 0$, $\quad y(1) = 0$, $\quad y_{\text{true}}(x) = 2\ln(4/(3 + x^2))$

(b) $4y'' + (y')^2 + 32y = 16(x^6 + 1)$, $\quad y'(0) = 0$, $\quad y(1) = 0$, $\quad y_{\text{true}}(x) = (x^2 - 1)^2$

(c) $y'' + (y' + y)^2 - y = 6$, $\quad y'(0) = 2$, $\quad y(1) = 3 - 2/e$, $\quad y_{\text{true}}(x) = 3 - 2e^{-x}$

(d) $y'' = 2y^3 - 3yy'$, $\quad y'(0) = -2$, $\quad y(2) = 1/5$, $\quad y_{\text{true}}(x) = 1/(2x+1)$

E10.39 An exothermic reaction takes place in a non-isothermal batch reactor. The conservation of material and energy yields the following second-order nonlinear two-point BVP:

$$y'' + \frac{2y'}{x+1} - \left(1 - \frac{y}{2}\right)\exp\left(\frac{y}{y-2}\right) = -\frac{e^{-x}}{x+1}, \qquad y(0) = 0, \quad y(1) = 1$$

(a) Use the TIM with $M = 5$ and 10 equispaced subintervals to solve the BVP; (b) compare your numerical estimates with the true solution given with $y(x) = 2x/(x+1)$. *Clue*: Obtain initial guess values by linear interpolation using the two BC values and use $\varepsilon = 10^{-6}$ for convergence.

E10.40 Consider the thin-long cantilever beam shown in **Fig. E10.40**. The curvature of the beam is dictated by the following BVP for the deflection angle ϕ:

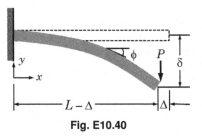

$$\frac{d^2\phi}{ds^2} = -\frac{P}{B}\cos\phi, \qquad \phi(0) = 0, \qquad \frac{d\phi}{ds}(1) = 0$$

where s is the arc length measured from the tip, P is the

Fig. E10.40

concentrated vertical load at the free end, B is the flexural rigidity, L is the beam length, Δ is the horizontal component of the displacement at the loaded end, and δ is the corresponding vertical displacement. Use Newton's method with $M = 5$ and 10 subintervals to estimate the deflection angle throughout the beam for $P/B = 0.025$ and 0.25.

E10.41 A reaction-diffusion process leads to the following nonlinear two-point BVP:

$$\frac{d^2C}{dx^2} = \frac{5C(x)}{10 + 2C(x)}, \qquad C(0) = 1, \qquad \frac{dC}{dx}(1) - \frac{1}{2}C(1) = 0$$

Use the tridiagonal iteration method with $M = 5$ and 10 uniform intervals to estimate the solution of the BVP. *Note*: For the initial guess and convergence tolerance, use $C_i^{(0)} = 1$ and $\varepsilon = 10^{-6}$, respectively.

E10.42 Consider a uniform cable hanging between two unequal poles, as shown in **Fig. E10.42**. As a function of the horizontal tension (T) in the cable and its weight (W) per unit length, the differential equation describing the shape of a hanging cable is given by

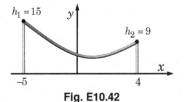

Fig. E10.42

$$y'' = \frac{W}{T}\sqrt{1 + (y')^2}, \qquad y(-5) = h_1, \quad y(4) = h_2$$

Use the tridiagonal iteration method with $M = 5$ and 10 subintervals, estimate the shape of the cable for a case with $W/T = 0.3$, $h_1 = 15$ m and $h_2 = 9$ m, and compare the estimates with the true shape is given by $y(x) = 4.3282723 - 1.12513\sinh 0.3x + 3.51810\cosh 0.3x$ For an initial guess and convergence tolerance, use $y_i^{(0)} = 1$ and $\varepsilon = 10^{-6}$, respectively.

E10.43 Consider heat conduction in a cylindrical nuclear fuel rod, with heat generation proportional to an exponential function of temperature. The dimensionless conduction equation is expressed as

$$\frac{d^2\Theta}{d\eta^2} + \frac{1}{\eta}\frac{d\Theta}{d\eta} + ae^{b\Theta} = 0, \qquad \frac{d\Theta}{d\eta}(0) = 0, \quad \Theta(1) = 1$$

where $\eta = r/R$, $\Theta = T/T_r$, R is the rod radius, and T_r is a reference temperature. Use the finite differences with the TIM method and $M = 5$ and $M = 10$ subintervals to estimate the dimensionless temperature of the rod. *Given*: Use $a = 1.30$, $b = 0.72$, $\Theta^{(0)} = 1$, and $\varepsilon = 10^{-6}$.

Section 10.8 Shooting Method

E10.44 Repeat **E10.36**, using the shooting method along with the RK4 and a step size of $h = 0.1$.

E10.45 Repeat **E10.37**, using the shooting method along with the RK4 and a step size of $h = 0.1$.

E10.46 Repeat **E10.38**(a)-(c), using the shooting method along with the RK4 and a step size of $h = 0.1$.

E10.47 Solve each of the following two-point linear BVPs using the shooting method. Having found the equivalent IVPs, apply the RK4 scheme with step size $h = 0.1$ to obtain the numerical solutions.

(a) $y'' + 2y' + y = 2e^{-x}$, $\quad y(0) = 2$, $\quad y(1) = 0$ $\quad$ (use $c_1 = -1$, $c_2 = 2$)

(b) $y'' + (3 - x)y' + xy = 0$, $\quad y(1) = 3$, $\quad y(4) = 2$, $\quad$ (use $c_1 = -2$, $c_2 = 1$)

(c) $x^4 y'' + x^3 y' - 3xy = -9$, $\quad y(1) = 4$, $\quad y(3) = 2$, $\quad$ (use $c_1 = -2$, $c_2 = 1$)

E10.48 Solve each of the following nonlinear two-point BVPs using the shooting method. Having found the equivalent IVPs, apply the RK4 scheme with step size $h = 0.1$ to obtain the numerical solutions.

(a) $y'' = 2y^3 - 3yy'$, $\quad y(0) = 1$, $\quad y(2) = 1/5$, $\quad y_{\text{true}}(x) = 1/(2x + 1)$, $\quad (c_0 = -2.5$, $c_1 = -1)$

(b) $y'' = 4 - 2x^3 + yy'$, $\quad y(1) = -1$, $\quad y(2) = 3$, $\quad y_{\text{true}}(x) = x^2 - 2/x$, $\quad (c_0 = 1$, $c_1 = 3)$

(c) $y'' - 2y'/x + e^{y/(y-x^4)} = e^{1-x} + 4x^2$, $\quad y(1) = 0$, $\quad y(2) = 8$,
$\quad y_{\text{true}}(x) = x^4 - x^3$ $(c_0 = -1$, $c_1 = 2)$

(d) $y^2 y'' + xy' - y = 0$, $\quad y(0) = 1$, $\quad y(2) = \sqrt{5}$, $\quad y_{\text{true}}(x) = \sqrt{1 + x^2}$ $(c_0 = -1$, $c_1 = 1)$

(e) $y'' + e^{-x^2} y' - 2xe^{-y} = 2$, $\quad y(0) = 0$, $\quad y(1) = 1$, $\quad y_{\text{true}}(x) = x^2$ $(c_0 = -1$, $c_1 = 1)$

(f) $(y - 1)^2 y'' + (x + 1/22)y' - y + 1 = 0$, $\quad y(0) = 2$, $\quad y(2) = 1 + \sqrt{7}$,
$\quad y_{\text{true}}(x) = 1 + \sqrt{1 + x + x^2}$ $(c_0 = 0$, $c_1 = 3)$

E10.49 Solve each of the following two-point linear BVPs defined in a semi-infinite domain using the shooting method. Apply the RK4 scheme with step size $h = 0.1$ to obtain the numerical solutions.

(a) $(1 + x)y'' + xy' + y = 0$, $\quad y(1) = 4$, $\quad y(\infty) = 0$

(b) $(2x^3 + x^2 + 1)y'' + 4x(x^2 - 1)y' - 8xy = 0$, $\quad y'(0) = -2$, $\quad y(\infty) = 0$

E10.50 Solve each of the following nonlinear two-point BVPs defined in a semi-infinite domain using the shooting method. Apply the RK4 scheme with step size $h = 0.1$ to obtain the numerical solutions.

(a) $y'' + yy'/2 = 0$, $\quad y(0) = 0$, $\quad y(\infty) = 2$, $\quad$ (use $c_1 = 0.1$, $c_2 = 1.5$)

(b) $y'' + 5xy' - e^{-xy} = 0$, $\quad y(0) = 1$, $\quad y(\infty) = 3/2$, $\quad$ (use $c_1 = 0.1$, $c_2 = 0.25$)

(c) $y'' + 2xy'/\sqrt{1 - y/4} = 0$, $\quad y(0) = 1/2$, $\quad y(\infty) = -1/2$, $\quad$ (use $c_1 = -0.1$, $c_2 = -0.5$)

(d) $x(1 + x^2)y'' + (1 + x^2)yy' + 2y' = 0$, $\quad y(1) = 0$, $\quad y(\infty) = 3$, $\quad$ (use $c_1 = 1$, $c_2 = 2$)

E10.51 The concentration of species due to diffusion in a planar material, shown in **Fig. E10.51**, is described by the so-called *diffusion equation*. In one-dimensional space, it is expressed as

$$\frac{\partial c}{\partial t} = \frac{\partial}{\partial x}\left(D(c)\frac{\partial c}{\partial x}\right), \quad c(0, t) = 1,$$

$$c(x, 0) = 0, \quad c(\infty, t) = 0$$

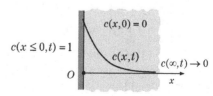

Fig. E10.51

A method called *similarity solution* is used to reduce a partial differential equation into an ODE. Introducing the $\eta = x/\sqrt{t}$ variable, concentration can be expressed as a function of η, i.e., $c(x,t) = f(\eta)$, where $f(\eta)$ is unknown. Upon substituting the similarity variable, the diffusion equation and its BCs result in the following BVP:

$$D\frac{d^2 f}{d\eta^2} + \frac{\eta}{2}\frac{df}{d\eta} = 0, \quad f(0) = 1, \quad f(\infty) = 0$$

Apply the shooting method along with the RK4 scheme to estimate the numerical solution to this problem. Assume a constant diffusion coefficient ($D = 1$) and a step size of $\Delta\eta = 0.025$.

Section 10.9 Fourth-order Linear Differential Equations

E10.52 Consider the following fourth-order linear ODEs with constant coefficients:

(a) $u^{(iv)} - u = -\cos x$, $\quad u(0) = u(\pi) = 0$, $\quad u'(0) = 1/4$, $u'(\pi) = -(1+\pi)/4$,
$u_{true}(x) = (1+x)\sin(x)/4$

(b) $u^{(iv)} + 4u = 0$, $\quad u(0) = u(\pi) = 0$, $\quad u'(0) = 1$, $\quad u'(\pi) = -e^\pi$, $\quad u_{true}(x) = e^x \sin x$

(c) $u^{(iv)} + u'' + 4u = 2e^x \cos x$, $\quad u(0) = u(\pi) = 0$, $\quad u'(0) = 1$, $\quad u'(\pi) = -e^\pi$,
$u_{true}(x) = e^x \sin x$

(d) $u^{(iv)} + 3u'' + 9u = -9x^4$, $\quad u(0) = u(2) = 0$, $\quad u'(0) = 0$, $\quad u'(2) = -16$,
$u_{true}(x) = 4x^2 - x^4$

(e) $u^{(iv)} - 4u'' - 45u = -48\cosh x$, $\quad u(0) = 3$, $\quad u(\ln 2) = 3/2$, $\quad u'(0) = -6$, $\quad u'(\ln 2) = 0$,
$u_{true}(x) = \cosh x + 2e^{-3x}$

To apply the FDM method, (a) divide the solution interval into uniform M subintervals to obtain the general difference equation; (b) implement the BCs and express the difference equations in matrix form; and (c) find the numerical solutions for $M = 8, 16,$ and 32 and compare your solutions with the true solution.

E10.53 Consider the following fourth-order linear ODEs with variable coefficients:

(a) $u^{(iv)} - xu'' + x\,u(x) = 3x - 4$, $\quad u(0) = 0$, $\quad u(1) = e^{-1}$, $\quad u'(0) = 1$, $\quad u'(1) = 0$,
$u_{true}(x) = xe^{-x}$

(b) $u^{(iv)} + (x-1)u(x) = x^5 + 23$, $\quad u(0) = u'(0) = 1$, $\quad u(1) = 5$, $\quad u'(1) = 10$,
$u_{true}(x) = 1 + x + x^2 + x^3 + x^4$

(c) $8(x-2)u^{(iv)} + 2(4-x)u'' + u(x) = 0$, $\quad u(0) = 0$, $\quad u'(0) = 2$, $\quad u(1) = 2/\sqrt{e}$,
$u'(1) = 1/\sqrt{e}$, $\quad u_{true}(x) = 2xe^{-x/2}$

(d) $(x-3)u^{(iv)} + (6-x)u'' - 5u = 0$, $\quad u(0) = 0$, $\quad u(2) = 2$, $\quad u'(0) = 1/e^2$, $\quad u'(2) = 3$,
$u_{true}(x) = xe^{x-2}$

(e) $u^{(iv)} + (x+\pi^2)u'' + x\pi^2 u = 4\pi^2 x^2$, $\quad u(0) = 0$, $\quad u(1) = 4$, $\quad u'(0) = 4 - \pi$,
$u'(1) = 4 + \pi$, $\quad u_{true}(x) = 4x - \sin\pi x$

To employ the FDM method, (a) divide the solution interval into uniform M subintervals to obtain the general difference equation; (b) implement the BCs and express the difference equations in matrix form; and (c) find the numerical solutions for $M = 8, 16,$ and 32 and compare your solutions with the true solution.

E10.54 A beam undergoes deflection when it is subjected to an applied load. The deflection for a beam of length L embedded at both ends with a variable load satisfies the following BVP:

$$\frac{d^4y}{dx^4} = -\frac{W}{EI}\sin\left(\frac{\pi x}{L}\right), \quad y(0) = y'(0) = y(L) = y'(L) = 0$$

where $W/EI = \pi^4$ and $L = 1$ m. To find a numerical solution to this problem, (a) divide $0 \leqslant x \leqslant L$ into uniform M subintervals and derive the general difference equation; (b) implement the BCs and express the difference equations in matrix form; and (c) solve the resulting linear system for $M = 10$, 20, and 40 and compare your estimates with the true solution given by $y_{\text{true}}(x) = \pi x - \pi x^2 - \sin \pi x$

10.12 COMPUTER ASSIGNMENTS

CA10.1 In an isothermal packed-bed reactor, chemical reactions of the form $A + A \to B$ take place. The governing ODE for the fraction of A remaining, y, can be expressed in dimensionless form as

$$\frac{1}{\text{Pe}}\frac{d^2y}{dx^2} + \frac{dy}{dx} - \alpha y = 0$$

where x is the axial coordinate, α is a constant, and Pe is a dimensionless number referred to as the Peclet number. The differential equation is subjected to the following BCs:

$$\text{At } x = 0 \quad y'(0) = 0, \quad \text{At } x = 1 \quad \frac{1}{\text{Pe}}y'(1) + y(1) = 1,$$

Find the numerical solution for $\alpha = 5$ and Pe=0.1, 1, and 10 using the FDM by dividing the solution interval into uniform $M = 10$ intervals.

CA10.2 As shown in **Fig. CA10.2**, a conical-frustum fin is used to dissipate the heat conducted from a base surface through convection across all its surfaces. The dimensionless form of the governing energy equation is given as

$$(1+(\eta-1)\xi)\frac{d^2\Theta}{d\xi^2} + 2(\eta-1)\frac{d\Theta}{d\xi} - 2\text{Bi}\sqrt{a^2+(\eta-1)^2}\,\Theta(\xi) = 0$$

At $\xi = 0$, $\quad \Theta(1) = 1$ $\quad$ (Known base temperature),

At $\xi = 1$, $\quad \Theta'(0) - \text{Bi}\,\Theta(0) = 0$, (Convection at the tip)

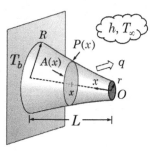

Fig. CA10.2

where $\xi = x/L$ $(0 \leqslant \xi \leqslant 1)$ is the dimensionless axial coordinate, L is the fin length, and $\Theta(\xi) = (T(x) - T_\infty)/(T_b - T_\infty)$ is the dimensionless temperature, T_b and T_∞ are respectively the base and surrounding temperatures, k is the conductivity, h is the convection heat transfer coefficient, η is the ratio of the base (R) to tip (r) radius $(= R/r)$, a is defined as $(= r/L)$, and Bi is called the Biot number, which is a dimensionless parameter defined as Bi$= hL/k$. Write a computer program to solve the given BVP and (a) solve this BVP for Bi=0.5, $\eta = 2$, and $a = 5$ using uniform $M = 5$, 10, 20, and 40 intervals, and comment on the accuracy of your estimates. (b) Investigate the effect of the Biot number (Bi=0.02, 0.2, and 2) on the dimensionless temperature distribution for $\eta = 1.5$

and $a = 10$. (c) Investigate the effect of η ($\eta = 1, 1.25, 1.5, 1.75$, and 2) on the dimensionless temperature distribution for Bi=0.5 and $a = 5$. (d) Investigate the effect of the length-to-tip radius ratio ($a = 2, 5$, and 8) on the dimensionless temperature distribution for Bi=0.2 and $\eta = 1.5$. For (b)-(d), generate comparative plots of Θ.

CA10.3 The steady-state conservation equation for a chemical species subject to reaction and diffusion within a radially symmetric sphere is

$$\frac{d^2C}{dr^2} + \frac{2}{r}\frac{dC}{dr} + \beta C(r) = 0, \qquad 0 \leqslant r \leqslant 1$$

where c is the concentration, r is the normalized radial coordinate, $\beta = kR/D$, D is the diffusion coefficient, R is the radius of the sphere, and k is the chemical reaction rate constant. The boundary conditions are given as

$$\left(\frac{dC}{dr}\right)_{r=0} = 0 \quad \text{and} \quad C(1) = 1$$

Use the FDM to find the numerical solution by first obtaining the difference equations in matrix form. For $\beta = 0.09, 1$, and 4, obtain the numerical estimates for uniform $M = 5$ subintervals, and compare your estimates with those of the true solution given with $C(r) = \sin(\beta r)/(r\sin\beta)$.

CA10.4 The steady-state temperature distribution between two annular cylinders filled with incompressible Newtonian fluid can be expressed as follows:

$$\frac{1}{r}\frac{d}{dr}\left(r\frac{d\theta}{dr}\right) + 4\frac{a}{r^4} = 0, \qquad \kappa \leqslant r \leqslant 1$$

where $\kappa = R_o/R_i$, r is the dimensionless radial coordinate, θ is the dimensionless temperature, a is a dimensionless constant, and R_i and R_o are the radii of the inner and outer cylinders, respectively. The boundary conditions are given as

$$\theta(\kappa) = 0 \quad \text{and} \quad \theta(1) = 1$$

We seek the numerical solution for $a = 0.5$ and $\kappa = 0.25, 0.5$, and 0.75. Obtain solutions with FDM and FVM using $M = 5, 10, 20$, and 40 cells or uniform subintervals. Compare your estimates with those of the true solution given as

$$\theta(r) = \left(a + 1 - \frac{a}{r^2}\right) - \left(a + 1 - \frac{a}{\kappa^2}\right)\frac{\ln r}{\ln \kappa}$$

Eigenvalues and Eigenvalue Problems

I N GERMAN, the word *"eigen"* means *"own"* or *"one's own."* In this sense, the eigenvalue and its associated eigenvector (also referred to as the *eigenpair*) are commonly termed the *characteristic value* and *characteristic vector* of a matrix.

Eigenvalue problems are commonly encountered when solving mathematical models of various physical systems. For example, eigenvalues have applications in physics when studying vibrating strings or the dynamic analysis of structures such as bridges, buildings, or mass-spring systems; in studying elasticity; in electrical networks when studying the behavior of complex electrical circuits; in control systems when determining the response of the system; etc. A physical system can be expressed in terms of *normal modes*, which are the eigenvectors. Each normal mode oscillates at some frequency, which is basically an eigenvalue

DOI: 10.1201/9781003474944-11

associated with the corresponding eigenvector. The eigenvalues and the eigenvectors have a number of useful properties that make them an invaluable tool, and they also yield a number of valuable relationships with the matrices from which they are derived.

Scientists and engineers frequently resort to the numerical computation of the eigenvalues and eigenvectors of matrix equations, especially those resulting from the discretization of differential equations. Differential equations, whose eigenvalues and/or eigenvectors are sought, are called the *Eigenvalue Problems* (*EVPs*) or *Characteristic Value Problems* (*CVPs*). In some cases, only the largest (dominant) or the smallest eigenvalue (in absolute value) and its associated eigenvector may be of primary interest to an analyst.

In this chapter, the methods of computing the eigenvalues and/or associated eigenvectors of a real matrix **A** are discussed. A general overview of the numerical methods that are either fundamental or the basis of more elaborate methods is presented. In practice, the matrices encountered in most eigenvalue problems are symmetric; hence, many of the numerical methods presented here are for symmetric matrices. As in previous chapters, algorithms for the eigenvalues are presented in the form of pseudocodes.

11.1 EIGENVALUE PROBLEM AND PROPERTIES

A matrix-eigenvalue problem is defined as

$$\mathbf{A}\mathbf{x} = \lambda\mathbf{x} \tag{11.1}$$

where **A** is a $n \times n$ square matrix, **x** is a non-zero vector of length n, and λ is a scalar.

Our aim is basically to solve Eq. (11.1), which has a trivial solution of $\mathbf{x} = \mathbf{0}$. However, we are interested in only the non-trivial ($\mathbf{x} \neq \mathbf{0}$) solutions, which may exist for some special λ values of **A**, called *eigenvalues*. The solution of an *eigenvalue problem* involves finding scalar-vector pairs, i.e., a scalar λ and a vector **x**. Note that Eq. (11.1) can also be written as

$$(\mathbf{A} - \lambda\mathbf{I})\mathbf{x} = \mathbf{0} \tag{11.2}$$

where **I** is the identity matrix. This equation poses a homogeneous system of equations, and in order for a system to have a non-trivial solution, it must satisfy the following condition:

$$\det(\mathbf{A} - \lambda\mathbf{I}) = p_n(\lambda) = 0 \tag{11.3}$$

where $p_n(\lambda)$ is an nth-degree polynomial called *characteristic polynomial* or *characteristic equation* and λ denotes the eigenvalues (or roots of the characteristic polynomial). For every real λ, a corresponding real *eigenvector* **x** unique to **A** can be obtained from Eq. (11.2).

Basic Properties: Before moving on to numerical methods, it will be beneficial to refresh our memories of some basic definitions and properties of eigenvalues and eigenvectors. Let **A** be a real $n \times n$ square matrix, then

Property 1. The sum of all eigenvalues of **A** (including the repeated eigenvalues) is equal to the sum of the diagonal elements (trace) of **A**, i.e.,

$$\sum_{i=1}^{n} \lambda_i = \mathrm{tr}(\mathbf{A})$$

Property 2. The product of all eigenvalues of matrix $\mathbf{A}$ is equal to the determinant of matrix $\mathbf{A}$, i.e.,

$$\prod_{i=1}^{n} \lambda_i = \det(\mathbf{A})$$

Property 3. If $\mathbf{A}$ is a singular matrix, then at least one eigenvalue is zero.

Property 4. If $\mathbf{A}$ has imaginary eigenvalues, they exist as complex conjugate pairs.

Property 5. If $\mathbf{A}$ is an upper or lower triangular matrix, then the eigenvalues of $\mathbf{A}$ are equal to the diagonal elements (i.e., $\lambda_i = a_{ii}$).

Property 6. For $\beta \neq 0$, if λ and $\mathbf{x}$ are eigenpairs of a matrix $\mathbf{A}$, then the eigenpairs of $\beta\mathbf{A}$ are $\beta\lambda$ and $\mathbf{x}$.

Property 7. If $\{\lambda, \mathbf{x}\}$ is an eigenpair of a matrix $\mathbf{A}$ and m is a natural number, then the eigenpair of $\mathbf{A}^m$ is $\{\lambda^m, \mathbf{x}\}$.

Property 8. If $\{\lambda, \mathbf{x}\}$ is an eigenpair of an invertible matrix $\mathbf{A}$, then the eigenpair of the inverse matrix, $\mathbf{A}^{-1}$, is $\{\lambda^{-1}, \mathbf{x}\}$.

Property 9. If $\mathbf{A}$ is a diagonal matrix, then its eigenvalues are equal to the diagonal elements (i.e., $\lambda_i = a_{ii}$ for all i), and its eigenvectors are the unit basis vectors for the corresponding ith column.

Property 10. Eigenvectors of a matrix are linearly independent; that is, an eigenvector cannot be expressed as a linear sum of other eigenvectors.

Property 11. If a matrix $\mathbf{A}$ has n independent eigenvectors, there exists a nonsingular matrix $\mathbf{P}$, where $\mathbf{P}^{-1}\mathbf{A}\mathbf{P} = \mathbf{D}$ is diagonal, and the matrix $\mathbf{A}$ here is called a *diagonalizable* matrix. Furthermore, the columns of $\mathbf{P}$ are the eigenvectors, and the diagonal elements are the eigenvalues of $\mathbf{A}$, i.e., $\lambda_i = d_{ii}$.

11.1.1 LOCATION AND BOUNDS OF EIGENVALUES

Some iterative methods for finding the eigenvalues require an estimate of the lower and upper bounds for the interval in which the eigenvalues lie. For this reason, it is important to become familiar with the *Gerschgorin Circle Theorem* before diving into the numerical methods for computing eigenpairs. This theorem, dealing with the location of the eigenvalues, provides a useful tool for estimating the eigenvalues of a matrix solely based on the matrix elements.

In the complex plane, a Gerschgorin circle with the center at a_{ii} and radius equal to $r_i = \Sigma_{i \neq j}|a_{ij}|$ can be defined for each row of $\mathbf{A}$ ($i = 1, 2, \ldots, n$). The theorem states that every eigenvalue of $\mathbf{A}$ is enclosed by one Gerschgorin circle:

$$C_i = \{z \in \mathbf{C} : |z - a_{ii}| \leqslant r_i\} \quad \text{where } r_i = \sum_{i \neq j}|a_{ij}| \text{ for } i = 1, 2, \ldots, n \tag{11.4}$$

where $\mathbf{C}$ denotes the complex plane. If m of these circles form a union set $D = C_i \cup C_j \cdots \cup C_n$ that is disjoint from the remaining $(n - m)$ circles, then D contains exactly m of $\mathbf{A}$'s eigenvalues. Since the eigenvalues of $\mathbf{A}^T$ and $\mathbf{A}$ are the same, we may also construct Gerschgorin circles by columns; that is,

$$\widehat{C}_i : \{z \in \widehat{\mathbf{C}} : |\lambda - a_{jj}| = r_j\} \quad \text{where } r_j = \sum_{i \neq j}|a_{ij}| \text{ for } j = 1, 2, \ldots, n \tag{11.5}$$

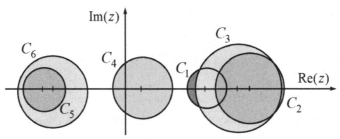

FIGURE 11.1: Depiction of Gerschgorin circles in complex plane.

In Fig. 11.1, the Gerschgorin circles of a real 6×6 square matrix in the complex plane are illustrated: $C_i : |\lambda - a_{ii}| = r_i$ for $i = 1, 2, \ldots, 6$. Since we are dealing with real matrices in this chapter, the centroids of the circles always lie on the real axis, even if some eigenvalues exist in complex conjugate pairs. Circle C4 encloses exactly one eigenvalue, which has to be real. The union of $D_1 = C_5 \cup C_6$ contains two eigenvalues, which may be real eigenvalues or a complex conjugate pair. The union of $D_2 = C_1 \cup C_2 \cup C_3$, on the other hand, contains exactly three eigenvalues, which are either all real or one real and a complex conjugate pair.

 Combining Eqs. (11.4) and Eq. (11.5) provides a way to possibly narrow down the intervals for the eigenvalues. Thus, we may deduce that an eigenvalue of $\mathbf{A}$ is contained in the intersection of $\widehat{C}_i$ and C_i, i.e., $\widehat{C}_i \cap C_i$.

11.1.2 EIGENPAIRS OF REAL SYMMETRIC MATRICES

Many eigenvalue problems resulting from numerical solutions to physical problems generally involve real and symmetric matrices; for this reason, the eigenpairs of symmetric matrices have deserved special attention. Hence, a considerable portion of this chapter is devoted to numerical methods applied to symmetric matrices and their applications.

Let $\mathbf{A}$ be a symmetric matrix (i.e., $\mathbf{A} = \mathbf{A}^T$) whose eigenvalues are all real. There also exists an orthogonal $\mathbf{Q}^T\mathbf{A}\mathbf{Q} = \mathbf{D}$ matrix such that $\mathbf{D}$ is a diagonal matrix with eigenvalues $\lambda_i = d_{ii}$, which correspond to diagonal elements of $\mathbf{D}$.

For an arbitrary column vector $\mathbf{x}$, we may write

$$\mathbf{A}\mathbf{x} = \mathbf{w} \tag{11.6}$$

and

$$\mathbf{x}^T\mathbf{A} = \mathbf{w}^T \tag{11.7}$$

where $\mathbf{w}$ is a column vector of length n, as well. The jth elements of $\mathbf{A}\mathbf{x}$ and $\mathbf{x}^T\mathbf{A}$ are obtained as

$$(\mathbf{A}\mathbf{x})_j = \sum_{k=1}^{n} a_{jk}x_k \tag{11.8}$$

$$(\mathbf{x}^T\mathbf{A})_j = \sum_{k=1}^{n} x_k a_{kj} \tag{11.9}$$

However, since $a_{kj} = a_{jk}$, we can write

$$\mathbf{A}\mathbf{x} = (\mathbf{x}^T\mathbf{A})^T \tag{11.10}$$

EXAMPLE 11.1: Determining the location of eigenvalues

Plot the Gerschgorin circles for matrix $\mathbf{A}$, and determine the approximate locations of the eigenvalues.

$$\mathbf{A} = \begin{bmatrix} -8 & -1 & 1 & 0 & 1 \\ 1 & -1 & 0 & 1 & 1 \\ -1 & 1 & -6 & 1 & -2 \\ 0 & 1 & -1 & 4 & 0 \\ -1 & 0 & 1 & 1 & 10 \end{bmatrix}$$

SOLUTION:

In Section 3.3, an estimate for the spectral radius (i.e., the largest eigenvalue in magnitude) was given as $|\lambda_{\max}| \leqslant \|\mathbf{A}\|_\infty$ according to the Gerschgorin theorem. This result also gives us a rough estimate that, in this case, all eigenvalues are bounded by $|\lambda| \leqslant 13$.

For a more refined determination, we may use a *row by row* approach. The radii of the circles are determined from Eq. (11.4) as

$$r_1 = |-1| + 1 + 0 + 1 = 3,$$
$$r_2 = 1 + 0 + 1 + 1 = 3,$$
$$r_3 = |-1| + 1 + 1 + |-2| = 5,$$
$$r_4 = 0 + 1 + |-1| = 2,$$
$$r_5 = |-1| + 0 + 1 + 1 = 3$$

The circles are then obtained as

$$C_1 : \{z \in \mathbf{C} : |z + 8| \leqslant 3\}, \quad C_2 : \{z \in \mathbf{C} : |z + 1| \leqslant 3\}, \quad C_3 : \{z \in \mathbf{C} : |z + 6| \leqslant 5\}$$
$$C_4 : \{z \in \mathbf{C} : |z - 4| \leqslant 2\}, \quad C_5 : \{z \in \mathbf{C} : |z - 10| \leqslant 3\}$$

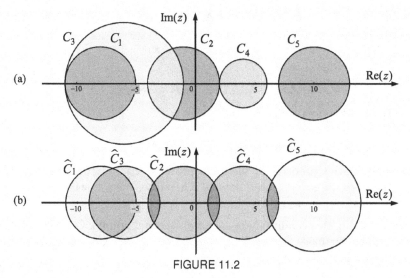

FIGURE 11.2

The circles, along with the true eigenvalues marked by solid dots, are illustrated in Fig. 11.2a. Note that C_5 is disjoint from the others, indicating that one eigenvalue is contained in C_5. Moreover, since it is a real matrix, the eigenvalue contained in this circle must lie in (7,13). The circle C_4 is tangent to C_2, and an eigenvalue contained in C_4 is also real and lies on (2,6), even if it were exactly at the intersection point of the circles C_2 and C_4. On the other hand, the union of the three circles (C_1, C_2, and C_3) contains three eigenvalues, one of which must be real, and the other two eigenvalues may be either real or a complex conjugate pair.

A column-wise approach leads to

$$\widehat{C}_1 : \{z \in \widehat{\mathbf{C}} : |z + 8| \leqslant 3\}, \quad \widehat{C}_2 : \{z \in \widehat{\mathbf{C}} : |z + 1| \leqslant 3\}, \quad \widehat{C}_3 : \{z \in \widehat{\mathbf{C}} : |z + 6| \leqslant 3\}$$
$$\widehat{C}_4 : \{z \in \widehat{\mathbf{C}} : |z - 4| \leqslant 3\}, \quad \widehat{C}_5 : \{z \in \widehat{\mathbf{C}} : |z - 10| \leqslant 4\}$$

where the radii are determined by Eq. (11.5). An optimized case is found by replacing C_3 with $\widehat{C}_3$, as depicted in Fig. 11.2b. The true eigenvalues of $\mathbf{A}$ are -1.285, 4.0011, 9.8736, and $-6.795 \pm 0.4944i$, and they are located approximately near the centroid of the disks.

Discussion: The Gerschgorin circles in rows or columns may result in different bounds. An optimal solution can be deduced by analyzing both approaches.

Consider another arbitrary vector $\mathbf{y}$ and note that

$$(\mathbf{y}, \mathbf{w}) = (\mathbf{w}, \mathbf{y}) \tag{11.11}$$

satisfies the commutative property, and $(\mathbf{w}, \mathbf{y})$ denotes the inner product defined as

$$(\mathbf{w}, \mathbf{y}) = w_1 y_1 + w_2 y_2 + \ldots + w_n y_n$$

Since $\mathbf{w} = \mathbf{A}\mathbf{x}$, Eq. (11.11) can be expressed as

$$(\mathbf{y}, \mathbf{A}\mathbf{x}) = (\mathbf{x}, \mathbf{A}\mathbf{y}) \tag{11.12}$$

Now, consider two different eigenvectors of $\mathbf{A}$ to be $\mathbf{y} = \mathbf{x}_1$ and $\mathbf{y} = \mathbf{x}_2$, substituting these into Eq. (11.12) gives

$$(\mathbf{x}_1, \mathbf{A}\mathbf{x}_2) = (\mathbf{x}_2, \mathbf{A}\mathbf{x}_1) \tag{11.13}$$

But using

$$\mathbf{A}\mathbf{x}_1 = \lambda_1 \mathbf{x}_1 \quad \text{and} \quad \mathbf{A}\mathbf{x}_2 = \lambda_2 \mathbf{x}_2 \tag{11.14}$$

Equation (11.13) may be written as

$$\lambda_1(\mathbf{x}_2, \mathbf{x}_1) = \lambda_2(\mathbf{x}_1, \mathbf{x}_2) \tag{11.15}$$

Due to the commutative property of the inner product, Eq. (11.15) implies $\lambda_1 = \lambda_2$. However, the eigenvalues, in general, are unequal. Therefore, the following conditions must be met to ensure equality: $(\mathbf{x}_1, \mathbf{x}_2) = 0$.

Vectors satisfying this condition are called *orthogonal vectors*. Consequently, the eigenvectors of a symmetric matrix are said to be *orthogonal*. Hence, an ordinary vector may be written as a linear combination of eigenvectors; that is,

$$\mathbf{x} = c_1 \mathbf{x}_1 + c_2 \mathbf{x}_2 + c_3 \mathbf{x}_3 + \ldots + c_n \mathbf{x}_n \tag{11.16}$$

From this point on, we will discuss the methods used to determine the eigenvalues, or eigenpairs.

11.2 POWER METHOD

In many physical problems, computing the dominant eigenvalue is of primary interest. The power method is a widely used iterative method for finding solely the *dominant eigenvalue* of a matrix, i.e., the *largest eigenvalue* in absolute value. The method, which can be applied to any matrix, whether symmetrical or not, simultaneously gives the associated eigenvector as well.

Consider a nonsingular $n \times n$ matrix $\mathbf{A}$ with n eigenvalues, $|\lambda_1| > |\lambda_2| > \ldots > |\lambda_n|$, and linearly independent associated eigenvectors, i.e., $\mathbf{A}\mathbf{x}_k = \lambda_k \mathbf{x}_k$ for $1 \leqslant k \leqslant n$. Using Eq. (11.16), an arbitrary vector $\mathbf{x}^{(0)}$ can be constructed as a linear combination of n eigenvectors. Left-multiplying $\mathbf{x}^{(0)}$ by $\mathbf{A}$ yields

$$\mathbf{A}\mathbf{x}^{(0)} = c_1 \mathbf{A}\mathbf{x}_1 + c_2 \mathbf{A}\mathbf{x}_2 + c_3 \mathbf{A}\mathbf{x}_3 + \ldots + c_n \mathbf{A}\mathbf{x}_n = \mathbf{x}^{(1)} \tag{11.17}$$

Invoking $\mathbf{A}\mathbf{x}_k = \lambda_k \mathbf{x}_k$, Eq. (11.17) can be expressed as

$$\mathbf{x}^{(1)} = c_1 \lambda_1 \mathbf{x}_1 + c_2 \lambda_2 \mathbf{x}_2 + c_3 \lambda_3 \mathbf{x}_3 + \ldots + c_n \lambda_n \mathbf{x}_n \tag{11.18}$$

Left-multiplying both sides of Eq. (11.18) with $\mathbf{A}$ and invoking again the $\mathbf{A}\mathbf{x}_k = \lambda_k \mathbf{x}_k$ relationship yields

$$\mathbf{A}^2 \mathbf{x}^{(0)} = \mathbf{A}\mathbf{x}^{(1)} = \mathbf{x}^{(2)} = c_1 \lambda_1^2 \mathbf{x}_1 + c_2 \lambda_2^2 \mathbf{x}_2 + c_3 \lambda_3^2 \mathbf{x}_3 + \ldots + c_n \lambda_n^2 \mathbf{x}_n \tag{11.19}$$

Each time multiplying by $\mathbf{A}$ and invoking $\mathbf{A}\mathbf{x}_k = \lambda_k \mathbf{x}_k$, the power of the eigenvalues is raised by one. Thus, the sequence of successive multiplications can be expressed as

$$\mathbf{x}^{(p)} = \mathbf{A}^p \mathbf{x}^{(0)} = \mathbf{A}\mathbf{x}^{(p-1)} = c_1 \lambda_1^p \mathbf{x}_1 + c_2 \lambda_2^p \mathbf{x}_2 + c_3 \lambda_3^p \mathbf{x}_3 + \ldots + c_n \lambda_n^p \mathbf{x}_n \tag{11.20}$$

which means that $\mathbf{A}^p \mathbf{x}^{(0)}$ is an eigenvector of $\mathbf{A}$.

Factoring out λ_1^p from the rhs of Eq. (11.20) yields

$$\mathbf{x}^{(p)} = \lambda_1^p \left\{ c_1 \mathbf{x}_1 + c_2 \left(\frac{\lambda_2}{\lambda_1} \right)^p \mathbf{x}_2 + c_3 \left(\frac{\lambda_3}{\lambda_1} \right)^p \mathbf{x}_3 + \ldots + c_n \left(\frac{\lambda_n}{\lambda_1} \right)^p \mathbf{x}_n \right\} \tag{11.21}$$

Given that $|\lambda_1| > |\lambda_k|$ for $k > 1$, Eq. (11.21) converges to $c_1 \lambda_1^p \mathbf{x}_1$ as $p \to \infty$. Thus, it follows that

$$\lim_{p \to \infty} \frac{\mathbf{x}^{(p)}}{\lambda_1^p} = c_1 \mathbf{x}_1 \tag{11.22}$$

If $c_1 = 0$, then the above expression would yield the next eigenvector with a nonzero coefficient. Since any nonzero multiple of an eigenvector is also an associated eigenvector of λ_1, the scaled sequence, $\mathbf{x}^{(p)}/\lambda_1^p$, converges to the eigenvector of the dominant eigenvalue, if $c_1 \neq 0$.

In practice, the dominant eigenvalue can be obtained by comparing the corresponding elements of two consecutive vectors. For a large enough p, we have the following:

$$\lim_{p \to \infty} \frac{(\mathbf{A}^p \mathbf{x})_i}{(\mathbf{A}^{p-1} \mathbf{x})_i} = \lim_{p \to \infty} \frac{(\mathbf{x}^{p+1})_i}{(\mathbf{x}^p)_i} = \lim_{p \to \infty} \frac{x_i^{(p+1)}}{x_i^{(p)}} = \lambda_1 + \mathcal{O}\left(\left(\frac{\lambda_2}{\lambda_1} \right)^p \right) \tag{11.23}$$

where $x_i^{(p)}$ denotes the ith component of the associate vector. The growth (or decay) of λ_1^p does not lead to any difficulty, and the convergence is observed to be at least of first order with the ratio at most $(\lambda_2/\lambda_1)^p$. Then the convergence rate may be defined as $\ln|\lambda_2/\lambda_1|$.

11.2.1 POWER METHOD WITH SCALING

The power method is based on the results of Eqs. (11.22) and (11.23). The procedure, which relies on comparing the corresponding elements of consecutive vector components, results in some problems, especially when one or more of the components of $\mathbf{x}^{(p)}$ is zero. For this reason, making comparisons between corresponding vector components is undesirable. Yet another issue arises as $\mathbf{x}^{(p)}$ grows larger with each iteration. an In order to avoid the *overflow* problem, $\mathbf{x}^{(p)}$ is scaled (i.e., normalized) at each iteration. Accordingly, this procedure is referred to as the *Power method with scaling*. The scaling, or normalizing, of $\mathbf{x}^{(p)}$ is based on the largest component of the estimated eigenvector (i.e., the infinity norm) as follows:

$$\mathbf{x}^{(p+1)} = \frac{1}{\|\mathbf{x}^{(p)}\|_\infty} \mathbf{A}\mathbf{x}^{(p)}$$

Next, comparing the above equation to Eq. (11.1), the largest component of $\mathbf{x}^{(p)}$ converges to the dominant eigenvalue for $p \to \infty$; that is,

$$\mathbf{A}\mathbf{x}^{(p)} = \|\mathbf{x}^{(p)}\|_\infty \mathbf{x}^{(p+1)}, \qquad \lambda = \lim_{p \to \infty} \|\mathbf{x}^{(p)}\|_\infty$$

A convergence criterion can be based on either the relative or absolute error of the largest eigenvalue or the maximum absolute difference of the two consecutive eigenvectors, i.e.,

$$|1 - \lambda^{(p-1)}/\lambda^{(p)}| < \varepsilon \quad \text{or} \quad |\lambda^{(p)} - \lambda^{(p-1)}| < \varepsilon \quad \text{or} \quad \|\mathbf{x}^{(p)} - \mathbf{x}^{(p-1)}\|_\infty < \varepsilon$$

A termination criterion based on the dominant eigenvalue is computationally cheap. However, when $\mathbf{A}$ is ill-conditioned, the eigenvector has yet to converge, even though the dominant eigenvalue generally does. Hence, it is prudent to use the maximum norm or ℓ_2-norm of the difference vector as a termination criterion, that is, $\|\mathbf{x}^{(p)} - \mathbf{x}^{(p-1)}\|_2$.

A pseudomodule, **POWER_METHOD_S**, presented in Pseudocode 11.1 finds the dominant eigenvalue and its associated eigenvector by applying the Power method with scaling. As input, the module requires a matrix $(\mathbf{A})$ and its size (n), an initial guess for the eigenvalue $(\lambda = \lambda^{(0)})$ and its associated eigenvector $(\mathbf{x} = \mathbf{x}^{(0)})$, a convergence tolerance (ε), and an upper bound for the number of iterations $(maxit)$ as input. The module also requires two additional modules: (1) Module **Ax** for carrying out matrix-vector multiplications, and (2) Module **MAX_SIZE**, for finding the element with the largest magnitude (in absolute value sense) among the elements of a vector (i.e., infinity norm, $\max_i |x_i|$). On exit from the module, the estimated eigenvalue (λ), corresponding eigenvector $(\mathbf{x})$, and maximum error $(error)$ are returned.

This process is carried out within a **While**-construct, and the iterative procedure is stopped when the eigenpairs converge within a preset tolerance level or the number of iterations reaches $maxit$. The procedure involves computing $\mathbf{x}^{(p)} = \mathbf{A}\mathbf{x}^{(p-1)}$ at each iteration step. An estimate for the dominant eigenvalue is obtained from $\lambda^{(p)} = \max|x_i^{(p)}|$, i.e., the infinity norm of $|\mathbf{x}^{(p)}|$. Next, the estimated eigenvector is scaled with the estimated eigenvalue $\mathbf{x}^{(p)} \leftarrow \mathbf{x}^{(p)}/\lambda^{(p)}$. The convergence test is based on the maximum of the $\|\mathbf{x}^{(p)} - \mathbf{x}^{(p-1)}\|_\infty$ and $|1 - \lambda^{(p-1)}/\lambda^{(p)}|$. If convergence is not achieved with the maximum number of iterations allowed $(maxit)$, the user is issued a warning along with the error value reached.

Pseudocode 11.1

Module POWER_METHOD_S (n, **A**, **x**, λ, ε, *error*, *maxit*)
\ DESCRIPTION: A pseudomodule to compute the dominant eigenvalue (largest in
\   absolute value) using the Power Method with scaling.
\ USES:
\   **Ax**:: Module computing **Ax** product, Pseudocode 2.5;
\   **ABS**:: Built-in function giving absolute value a real number;
\   **MAX**:: Built-in function giving largest of the argument list;
\   **MAX_SIZE**:: Module returning the largest magnitude element in an array.
Declare: a_{nn}, x_n, xn_n \ Declare array variables
$error \leftarrow$ **MAX_SIZE**($\mathbf{x}$)   \ Initialize error with $\|\mathbf{x}^{(0)}\|_\infty$
$p \leftarrow 0$ \ Initialize iteration counter
$\lambda o \leftarrow \lambda$ \ Initialize eigenvalue
\ Iterate until *maxit* is reached or convergence is achieved
While $\left[error > \varepsilon \ \textbf{And} \ p < maxit \right]$ \ Iteration loop
   $p \leftarrow p + 1$ \ Count iterations
   Ax($n, \mathbf{A}, \mathbf{x}, \mathbf{xn}$)   \ Perform $\mathbf{x}^{(p)} \leftarrow \mathbf{A}\mathbf{x}^{(p-1)}$
   $\lambda \leftarrow$ **MAX_SIZE**($\mathbf{xn}$)   \ Estimate largest eigenvalue
   $\mathbf{xn} \leftarrow \mathbf{xn}/\lambda$ \ Scale eigenvector estimate with λ
   $err1 \leftarrow$ **MAX_SIZE**($|\mathbf{x} - \mathbf{xn}|$)   \ Find infinity norm of $|\mathbf{x}^{(p)} - \mathbf{x}^{(p-1)}|$
   $err2 \leftarrow |1 - \lambda o/\lambda|$ \ Find rel. error for eigenvalue estimate
   $error \leftarrow$ **MAX**($err1, err2$)   \ Find max. error of eigenpair
   Write: p, λ, *error*   \ Print iteration progress
   $\mathbf{x} \leftarrow \mathbf{xn}$ \ Assign current eigenvector estimate as prior
   $\lambda o \leftarrow \lambda$ \ Assign current eigenvalue estimates as prior
End While
\ Issue a warning and error level achieved if *maxit* is reached with no convergence.
If $\left[error > \varepsilon \ \textbf{And} \ p = maxit \right]$ **Then**
   Write: "Not converged with *maxit*, current error is",*error*
End If
End Module POWER_METHOD_S

Function Module MAX_SIZE (n, $\mathbf{x}$)
\ DESCRIPTION: A pseudo Function module finding the largest element
\   (in absolute value) of an array.
\ USES:
\   **ABS**:: Built-in function giving absolute value a real number.
Declare: x_n \ Declare array variables
$xmax \leftarrow x_1$ \ Set $xmax$ to $|x_1|$
For $\left[i = 2, n \right]$
   If $\left[|x_i| > |xmax| \right]$ **Then**   \ If $|x_i| > xmax$, update $xmax$.
      $xmax \leftarrow x_i$ \ Updated max is $xmax = |x_i|$
   End If
End For
MAX_SIZE$\leftarrow xmax$ \ Set $xmax$ to MAX_SIZE
End Function Module MAX_SIZE

Power Method

- The method is simple and easy to implement;
- Each iteration requires only a single matrix-vector product, $\mathbf{Ax}$, which greatly reduces the computational effort, especially when $\mathbf{A}$ is very large;
- It gives, when it converges, the largest magnitude (dominant) eigenvalue and its associated eigenvector simultaneously;
- Rayleigh quotient version, which generally converges faster, is effective for symmetric matrices.

- It only gives the dominant eigenvalue and its associated eigenvector;
- Its convergence rate depends on the dominance ratio $|\lambda_2/\lambda_1|$; its convergence is very slow when $\lambda_2 \approx \lambda_1$;
- It is sensitive to initial guess and matrix condition number; it converges to the wrong eigenpair if an initial guess coincides with an eigenvector other than its dominant eigenvalue of $\mathbf{A}$.

11.2.2 POWER METHOD WITH RAYLEIGH QUOTIENT

Another common approach requires a vector comparison based on a single criterion called the *Rayleigh Quotient (RQ)*. This iterative algorithm, originally developed for real symmetric matrices, is based on minimizing the ℓ_2-norm of $(\mathbf{Ax} - \lambda\mathbf{x})$ at every iteration.

The procedure starts with an initial arbitrary nonzero vector, $\mathbf{x}^{(0)}$, and then the dominant eigenvalue is estimated in each iteration by

$$\lambda^{(p)} = \frac{(\mathbf{x}^{(p-1)}, \mathbf{Ax}^{(p-1)})}{(\mathbf{x}^{(p-1)}, \mathbf{x}^{(p-1)})} = \frac{(\mathbf{x}^{(p-1)}, \mathbf{x}^{(p)})}{(\mathbf{x}^{(p-1)}, \mathbf{x}^{(p-1)})}, \qquad p \geqslant 1 \qquad (11.24)$$

where p denotes the iteration step and "$(\mathbf{u},\mathbf{v})$" the inner product.

The convergence status is checked at the end of each iteration, and the iteration procedure is terminated when the stopping criterion is met.

Convergence Issues: The number of iterations required to converge the dominant eigenvalue, λ_1, within a predetermined tolerance depends on the initial guess $\mathbf{x}^{(0)}$. But note that the power method may not always give the dominant eigenvalue for every matrix. The convergence of the method also depends strongly on how well the two largest eigenvalues (λ_1 and λ_2) are separated; that is, $|\lambda_1| \gg |\lambda_2|$. In this respect, the dominance ratio $|\lambda_2|/|\lambda_1|$ is of fundamental importance in determining how quickly λ_1 overwhelms other terms.

- If $|\lambda_2|/|\lambda_1| < 1$, then λ_1 dominates other terms very quickly, so the convergence rate is fast; otherwise, the convergence rate is slow.

- If the largest two eigenvalues coincide ($\lambda_1 = \lambda_2$), the method may or may not converge after a large number of iterations.

If the two eigenvalues are equal but opposite ($\lambda_1 = -\lambda_2$), the dominant eigenvalue can be estimated by applying the Power method to $\mathbf{A}^2$, where the eigenvalues satisfy $\lambda_k(\mathbf{A}^2) = \lambda_k^2(\mathbf{A})$ for $k = 1, 2, \ldots, n$. However, if the initial guess corresponds to any one of the non-dominant eigenvalues of $\mathbf{A}$, then the method fails due to converging to the wrong eigenvalue.

Pseudocode 11.2

Module POWER_METHOD_RQ (n, $\mathbf{A}$, $\mathbf{x}$, λ, ε, $error$, $maxit$)
\\ DESCRIPTION: A pseudomodule to compute the dominant eigenvalue (largest
\\   in absolute value) using the Power Method with Rayleigh quotient.
\\ USES:
\\   Ax:: Module computing $\mathbf{Ax}$ product, Pseudocode 2.5;
\\   ABS:: A built-in function computing the absolute value;
\\   MAX:: Built-in function giving largest of the argument list;
\\   XdotY:: Module computing the inner product of two vectors, Pseudocode 2.3;
\\   MAX_SIZE:: Module returning the largest magnitude element in an array.
Declare: a_{nn}, x_n, xn_n, w_n
$error \leftarrow$ **MAX_SIZE**($\mathbf{x}$)   \\ Initialize error with $\|\mathbf{x}^{(0)}\|_\infty$
$\lambda o \leftarrow \lambda$ \\ Initialize eigenvalue estimate
$p \leftarrow 0$ \\ Initialize iteration counter
\\ Issue a warning and error level achieved if $maxit$ is reached with no convergence.
While $\left[error > \varepsilon \text{ And } p < maxit \right]$ \\ Iteration loop
   $p \leftarrow p + 1$ \\ Count iterations
   Ax(n, $\mathbf{A}$, $\mathbf{x}$, $\mathbf{w}$)   \\ Perform $\mathbf{w} \leftarrow \mathbf{Ax}^{(p-1)}$
   $\lambda \leftarrow$ **XdotY**(n, $\mathbf{w}$, $\mathbf{x}$)   \\ Find dot product, $(\mathbf{w}, \mathbf{x})$
   $\lambda \leftarrow \lambda /$**XdotY**($n$, $\mathbf{x}$, $\mathbf{x}$)   \\ Find $\lambda \leftarrow (\mathbf{w}, \mathbf{x})/(\mathbf{x}, \mathbf{x})$
   $wmax \leftarrow$ **MAX_SIZE**(n, $\mathbf{w}$)   \\ Find infinity norm of $\mathbf{w}$
   $\mathbf{xn} \leftarrow \mathbf{w}/wmax$ \\ Scale $\mathbf{w}$ with its max norm
   $err1 \leftarrow$ **MAX_SIZE**($|\mathbf{x} - \mathbf{xn}|$)   \\ Find infinity norm of $|\mathbf{x}^{(p)} - \mathbf{x}^{(p-1)}|$
   $err2 \leftarrow |1 - \lambda o/\lambda|$ \\ Find relative error of eigenvalue estimate
   $error \leftarrow$ **MAX**($err1$, $err2$)   \\ Find max. error of eigenpair
   Write: p, λ, $error$ \\ Print iteration progress
   $\mathbf{x} \leftarrow \mathbf{xn}$ \\ Assign current eigenvector estimate as prior
   $\lambda o \leftarrow \lambda$ \\ Assign current eigenvalue estimates as prior
End While
\\ Issue a warning if $maxit$ is reached with no convergence.
If $\left[error > \varepsilon \text{ And } p = maxit \right]$ **Then**
   Write: "Not converged with $maxit$, current error is",$error$
End If
End Module POWER_METHOD_RQ

A pseudomodule, POWER_METHOD_RQ, for finding the dominant eigenvalue and its associated eigenvector by the Power method using the Rayleigh quotient technique is presented in Pseudocode 11.2. The module's input and output lists are the same as those of the one presented in Pseudocode 11.1. This module, besides Ax and MAX_SIZE, requires module XdotY to compute the inner product of vectors, namely $\mathbf{x}$ and $\mathbf{w}$. The iterative procedure requires the computation of a matrix-vector product $\mathbf{w}^{(p)} = \mathbf{Ax}^{(p-1)}$, and the inner products. $(\mathbf{x}^{(p-1)}, \mathbf{w}^{(p)})$ and $(\mathbf{x}^{(p-1)}, \mathbf{x}^{(p-1)})$ to estimate the eigenvalue from Eq. (11.24). The eigenvector is normalized with the maximum norm of $\mathbf{w}$, i.e., $\mathbf{x}^{(p)} = \mathbf{w}^{(p)}/wmax$. This normalization procedure prevents arithmetic overflow (in the case of $|\lambda_1| > 1$) or underflow (in the case of $|\lambda_1| < 1$) during the iterations. The termination procedure is the same as described in Pseudocode 11.1. The user is issued a warning along with the error value reached if convergence is not achieved with the maximum number of iterations allowed ($maxit$).

EXAMPLE 11.2: Finding the dominant eigenvalue with the Power method

Consider the following undamped mass-spring system:

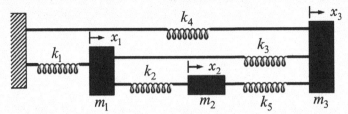

Assuming solutions of the form $x_i = A_i \sin(\omega t + \phi)$, where A_i is the amplitude of x_i; ω and ϕ are the frequency and the phase shift, respectively. The equations of motion for $m_1 = m_2 = m_3 = m$ and $k_3/3 = k_1/2 = k_4/2 = k_2 = k_5 = k$ are reduced to the following eigenvalue problem:

$$\begin{bmatrix} 6 & -1 & -3 \\ -1 & 2 & -1 \\ -3 & -1 & 6 \end{bmatrix} \begin{bmatrix} x_1 \\ x_2 \\ x_3 \end{bmatrix} = \frac{m}{k}\omega^2 \begin{bmatrix} x_1 \\ x_2 \\ x_3 \end{bmatrix}$$

Find the dominant frequency (largest magnitude eigenvalue) of the system, correct to six decimal places, using the Power method with (i) scaling and (ii) the Rayleigh quotient.

SOLUTION:

(i) A nonzero initial guess for the eigenvector is needed to apply the Power method. Let the initial guess be $\mathbf{x} = [1\ 1\ 0]^T$. Setting $\lambda = \omega^2 m/k$ and using the "scaling technique," we find

$$\begin{bmatrix} 6 & -1 & -3 \\ -1 & 2 & -1 \\ -3 & -1 & 6 \end{bmatrix} \begin{bmatrix} 1 \\ 1 \\ 0 \end{bmatrix} = \begin{bmatrix} 5 \\ 1 \\ -4 \end{bmatrix} = 5 \begin{bmatrix} 1 \\ 1/5 \\ -4/5 \end{bmatrix}, \qquad \lambda^{(1)} = 5$$

Note that $\mathbf{Ax}$ is normalized by its largest value (i.e., 5), which is also the first estimate for the dominant eigenvalue.

Using the normalized vector, the second iteration yields

$$\begin{bmatrix} 6 & -1 & -3 \\ -1 & 2 & -1 \\ -3 & -1 & 6 \end{bmatrix} \begin{bmatrix} 1 \\ 1/5 \\ -4/5 \end{bmatrix} = \begin{bmatrix} 41/5 \\ 1/5 \\ -8 \end{bmatrix} = \frac{41}{5} \begin{bmatrix} 1 \\ 1/41 \\ -40/41 \end{bmatrix}, \qquad \lambda^{(2)} = \frac{41}{5} = 8.2$$

and repeating this procedure successively results in

$$\lambda^{(3)} = 8.9024390, \quad \mathbf{x}^{(3)} = [\ 1 \quad 0.002740 \quad -0.997260\ ]^T$$

$$\lambda^{(4)} = 8.9890411, \quad \mathbf{x}^{(4)} = [\ 1 \quad 0.000305 \quad -0.999695\ ]^T$$

$$\lambda^{(5)} = 8.9987809, \quad \mathbf{x}^{(5)} = [\ 1 \quad 0.000034 \quad -0.999966\ ]^T$$

$$\lambda^{(6)} = 8.9998645, \quad \mathbf{x}^{(6)} = [\ 1 \quad 3.76 \times 10^{-6} \quad -0.999996\ ]^T$$

$$\lambda^{(7)} = 8.9999849, \quad \mathbf{x}^{(7)} = [\ 1 \quad 4.18 \times 10^{-7} \quad -1\ ]^T$$

$$\lambda^{(8)} = 8.9999983, \quad \mathbf{x}^{(8)} = [\ 1 \quad 4.65 \times 10^{-8} \quad -1\ ]^T$$

$$\lambda^{(9)} = 8.9999998, \quad \mathbf{x}^{(9)} = [\ 1 \quad 5.16 \times 10^{-9} \quad -1\ ]^T$$

(ii) Using $\mathbf{x}^{(0)} = [\,1\ 1\ 0\,]^T$ with the Rayleigh quotient also yields

$$\mathbf{w} = \mathbf{A}\mathbf{x}^{(0)} = \begin{bmatrix} 6 & -1 & -3 \\ -1 & 2 & -1 \\ -3 & -1 & 6 \end{bmatrix} \begin{bmatrix} 1 \\ 1 \\ 0 \end{bmatrix} = \begin{bmatrix} 5 \\ 1 \\ -4 \end{bmatrix}$$

For the first iteration step for the inner products, we obtain $(\mathbf{w}, \mathbf{x}^{(0)}) = 6$ and $(\mathbf{x}^{(0)}, \mathbf{x}^{(0)}) = 2$, which leads to the first estimate for the eigenvalue:

$$\lambda^{(1)} = \frac{(\mathbf{x}^{(0)}, \mathbf{w})}{(\mathbf{x}^{(0)}, \mathbf{x}^{(0)})} = \frac{6}{2} = 3$$

Having normalized the eigenvector estimate to $\mathbf{x}^{(1)} = \mathbf{w}/\|\mathbf{w}\|_\infty = [\,1\ 1/5\ -4/5\,]^T$, the vector $\mathbf{w}$ at the second iteration becomes

$$\mathbf{w} = \mathbf{A}\mathbf{x}^{(1)} = \begin{bmatrix} 6 & -1 & -3 \\ -1 & 2 & -1 \\ -3 & -1 & 6 \end{bmatrix} \begin{bmatrix} 1 \\ 1/5 \\ -4/5 \end{bmatrix} = \begin{bmatrix} 41/5 \\ 1/5 \\ -8 \end{bmatrix}$$

which leads to $(\mathbf{x}^{(1)}, \mathbf{w}) = 366/25$ and $(\mathbf{x}^{(1)}, \mathbf{x}^{(1)}) = 42/25$. Then the second estimate for the dominant eigenvalue becomes

$$\lambda^{(2)} = \frac{(\mathbf{x}^{(1)}, \mathbf{w})}{(\mathbf{x}^{(1)}, \mathbf{x}^{(1)})} = \frac{61}{5} = 8.7142857$$

Likewise, the dominant eigenvalue estimates of the subsequent iterations, summarized below, are obtained in the same manner.

$$\lambda^{(3)} = 8.9963437, \quad \mathbf{x}^{(3)} = [\ 1 \quad 0.002740 \quad -0.997260\ ]^T$$
$$\lambda^{(4)} = 8.9999548, \quad \mathbf{x}^{(4)} = [\ 1 \quad 0.000305 \quad -0.999695\ ]^T$$
$$\lambda^{(5)} = 8.9999994, \quad \mathbf{x}^{(5)} = [\ 1 \quad 0.000034 \quad -0.999966\ ]^T$$
$$\lambda^{(6)} = 8.9999999, \quad \mathbf{x}^{(6)} = [\ 1 \quad 3.76 \times 10^{-6} \quad -0.999996\ ]^T$$
$$\lambda^{(7)} = 9, \quad \mathbf{x}^{(7)} = [\ 1 \quad 4.18 \times 10^{-7} \quad -1\ ]^T$$

With both methods, the dominant eigenvalue and its associated eigenvector converge to $\lambda = 9$ and $\mathbf{x} = [\,1\ 0\ -1\,]$. The dominant frequency can then be obtained from $\lambda = \omega^2 m/k$ as $\omega = 3\sqrt{k/m}$.

Discussion: The Power method is quite simple and requires very little storage. Normalization of the vector is necessary because the eigenvector iterates will grow exponentially (or decay) if scaling is not applied.

As far as the speed of convergence is concerned, the dominance ratio plays an important role. The dominance ratio $|\lambda_2/\lambda_1|$, where λ_1 and λ_2 denote the largest and next largest eigenvalues (in magnitude), is of fundamental importance in determining how fast the method will converge. In this example, the matrix is symmetrical, and its eigenvalues are 9, 4, and 1. We note that the second-largest eigenvalue ($\lambda_2 = 4$) is considerably smaller than the largest one ($\lambda_1 = 9$). That is why we obtain convergence in both methods very quickly. Since the rate of convergence is $(\lambda_2/\lambda_1)^2 = (4/9)^2$, the procedure with the Rayleigh quotient converged faster.

11.2.3 INVERSE POWER METHOD

Inverse Power Method is designed to determine the smallest eigenvalue of a matrix and its associated eigenvector. To achieve this, both sides of Eq. (11.1) are multiplied with the inverse of $\mathbf{A}$ and rearranged as follows:

$$\frac{1}{\lambda}\mathbf{x} = \mathbf{A}^{-1}\mathbf{x} \quad \text{or} \quad \mathbf{A}^{-1}\mathbf{x} = \mu\mathbf{x} \tag{11.25}$$

where $\mu = 1/\lambda$ and $\mathbf{A}$ is a nonsingular matrix.

Once the problem is cast as $\mathbf{A}^{-1}\mathbf{x} = \mu\mathbf{x}$, the Power method can be applied to find the dominant eigenvalue (μ) corresponding to the smallest value of $1/\lambda$. In practice, of course, the computation of the inverse matrix is avoided by noting that the Power iteration $(\mathbf{x}^{(p+1)} = \mathbf{A}^{-1}\mathbf{x}^{(p)})$ is equivalent to

$$\mathbf{A}\mathbf{x}^{(p+1)} = \mathbf{x}^{(p)} \tag{11.26}$$

The above linear system is then solved for the current eigenvector estimates, $\mathbf{x}^{(p+1)}$. A judicious strategy to tackle this linear system is to employ LU-decomposition on $\mathbf{A}$. Once we have $\mathbf{L}$ and $\mathbf{U}$, we can start with an initial guess $\mathbf{x}^{(0)}$ such that $\|\mathbf{x}^{(0)}\|_\infty = 1$ and $\mathbf{x}^{(0)} \neq 0$. The system is solved by a forward elimination, $\mathbf{L}\mathbf{z} = \mathbf{x}^{(p)}$, followed by a back substitution, $\mathbf{U}\mathbf{x}^{(p+1)} = \mathbf{z}$ algorithm. A new estimate for the minimum eigenvalue (in absolute value) is computed from

$$\mu^{(p+1)} = \frac{(\mathbf{x}^{(p)}, \mathbf{x}^{(p+1)})}{(\mathbf{x}^{(p)}, \mathbf{x}^{(p)})} \tag{11.27}$$

where the prior step eigenvector estimate is obtained as $\mathbf{x}^{(p-1)} \leftarrow \mathbf{x}^{(p)}/\|\mathbf{x}^{(p)}\|$. If $\mu^{(p+1)} < 0$, the eigenvector is updated as $\mathbf{x}^{(p-1)} \leftarrow (-1)\mathbf{x}^{(p)}/\|\mathbf{x}^{(p)}\|$]. The iteration procedure is continued until the stopping criterion is met.

When using the Gauss elimination algorithm to solve Eq. (11.26), the matrix $\mathbf{A}$ is destroyed during the forward elimination step. Hence, a copy of $\mathbf{A}$ should also be created before the Power iteration procedure is applied, and this copy should be retrieved each time $\mathbf{A}$ is needed ($\mathbf{A} \leftarrow \mathbf{B}$). On the other hand, LU-decomposition of $\mathbf{A}$ is a memory-efficient approach that does not require a copy of $\mathbf{A}$ ($\mathbf{L}$ and $\mathbf{U}$ can also be saved on $\mathbf{A}$). Then, $\mathbf{L}\mathbf{U}\mathbf{x}^{(p+1)} = \mathbf{x}^{(p)}$ is solved at each iteration.

EXAMPLE 11.3: Finding the smallest eigenvalue

Find the smallest frequency (eigenvalue) of the spring-mass system given in Example 11.2 (correct to six decimal places) using the power method with Rayleigh quotient.

SOLUTION:

Starting with $\mathbf{x}^{(0)} = \begin{bmatrix} 1 & 1 & 0 \end{bmatrix}^T$, we begin with the following linear system:

$$\mathbf{A}\mathbf{x}^{(1)} = \begin{bmatrix} 6 & -1 & -3 \\ -1 & 2 & -1 \\ -3 & -1 & 6 \end{bmatrix} \begin{bmatrix} x_1 \\ x_2 \\ x_3 \end{bmatrix}^{(1)} = \begin{bmatrix} 1 \\ 1 \\ 0 \end{bmatrix}$$

which leads to $\mathbf{x}^{(1)} = [5/9 \ 1 \ 4/9]^T$.

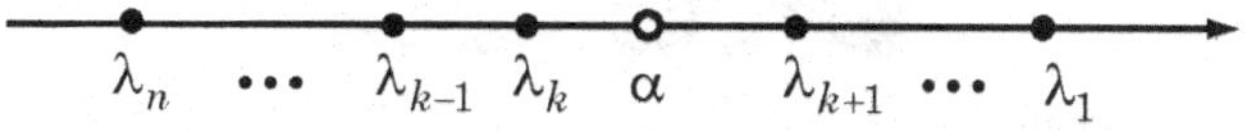

FIGURE 11.3: The distribution of the eigenvalues with respect to the estimate α.

The first estimate for the dominant eigenvalue of $\mathbf{A}^{-1}$ becomes

$$\mu^{(1)} = \frac{(\mathbf{x}^{(1)}, \mathbf{x}^{(0)})}{(\mathbf{x}^{(0)}, \mathbf{x}^{(0)})} = \frac{14/9}{2} = \frac{7}{9}$$

The second iteration leads to

$$\mathbf{A}\mathbf{x}^{(2)} = \begin{bmatrix} 5 & 3 & -1 \\ -2 & 5 & -3 \\ -1 & -3 & 5 \end{bmatrix} \begin{bmatrix} x_1 \\ x_2 \\ x_3 \end{bmatrix}^{(2)} = \begin{bmatrix} 5/9 \\ 1 \\ 4/9 \end{bmatrix} \Rightarrow \begin{bmatrix} x_1 \\ x_2 \\ x_3 \end{bmatrix}^{(2)} = \begin{bmatrix} 41/81 \\ 1 \\ 40/81 \end{bmatrix}$$

which, for the eigenvalue estimate, gives

$$\mu^{(2)} = \frac{(\mathbf{x}^{(2)}, \mathbf{x}^{(1)})}{(\mathbf{x}^{(1)}, \mathbf{x}^{(1)})} = \frac{1094/729}{122/81} = \frac{547}{549}$$

The subsequent iteration steps result in

$$\begin{aligned}
\mu^{(3)} &= 0.9999548, & x^{(3)} &= \begin{bmatrix} 0.500686 & 1 & 0.499314 \end{bmatrix}^T \\
\mu^{(4)} &= 0.9999994, & x^{(4)} &= \begin{bmatrix} 0.500076 & 1 & 0.499924 \end{bmatrix}^T \\
\mu^{(5)} &= 1, & x^{(5)} &= \begin{bmatrix} 0.500008 & 1 & 0.499992 \end{bmatrix}^T \\
\mu^{(6)} &= 1, & x^{(6)} &= \begin{bmatrix} 0.500001 & 1 & 0.499999 \end{bmatrix}^T
\end{aligned}$$

The smallest eigenvalue is then found from $\lambda = 1/\mu = 1$. Using $\lambda = \omega^2 m/k$, the smallest frequency is obtained as $\omega = \sqrt{k/m}$.

Discussion: The smallest eigenvalue is obtained very quickly. The 3×3 system in this example was solved by applying the Gauss elimination method. However, for large matrices, the computational effort with Gaussian elimination will increase for a large number of iterations, so the implementation of the LU-decomposition method should be considered in such cases.

11.2.4   SHIFTED INVERSE POWER METHOD

This method is based on shifting and inverting the eigenvalues. For $i = 1, 2, \ldots, n$, suppose λ_i and $\mathbf{x}_i$ are the eigenpairs of a nonsingular matrix $\mathbf{A}$. For a constant α, the eigenpair of $\mathbf{A} - \alpha\mathbf{I}$ (*shifting*) is $\lambda_i - \alpha$ and $\mathbf{x}_i$. On the other hand, the eigenpair of $\mathbf{A}^{-1}$ (*inverting*) is $1/\lambda_i$ and $\mathbf{x}_i$ for all i. Hence, employing shifting by α ($\alpha \neq \lambda$) and inverting, we find $(\lambda_i - \alpha)^{-1}$ and $\mathbf{x}_i$ as the eigenpair of $(\mathbf{A} - \alpha\mathbf{I})^{-1}$. Further, we will assume that $\mathbf{A}$ has distinct eigenvalues: $|\lambda_1| > |\lambda_2| > |\lambda_3| > \cdots > |\lambda_n|$. If α is chosen for an arbitrary λ_k such that $\alpha \approx \lambda_k$ (*see* Fig. 11.3), then the eigenvalue of $(\mathbf{A} - \alpha\mathbf{I})^{-1}$ closest to α will be the largest one; in other words, $\mu_k \to \infty$ as $\alpha \to \lambda_k$ while the remaining eigenvalues, $\mu_i = (\lambda_i - \lambda_k)^{-1}$, are finite.

Now, starting with $\mathbf{x}^{(0)} \neq 0$ provided that $\|\mathbf{x}^{(0)}\| = 1$, a sequence of eigenpairs can be obtained as follows:

$$(\mathbf{A} - \alpha\mathbf{I})\mathbf{x}^{(p+1)} = \mathbf{x}^{(p)} \tag{11.28}$$

> Inverse Power Method

- The power method with inverse or shift is easy to implement;
- It can provide very fast convergence in cases of symmetric matrices;
- It is also suitable to find complex eigenvectors.

- Only one eigenpair can be determined;
- The approximate value of the desired eigenvalue needs to be known in advance;
- The rate of convergence may be slow if there is a second eigenvalue close to the one sought.

$$\mu^{(p+1)} = \frac{(\mathbf{x}^{(p)}, \mathbf{x}^{(p+1)})}{(\mathbf{x}^{(p)}, \mathbf{x}^{(p)})} \tag{11.29}$$

where, before the next iteration, the current eigenvector estimate is normalized as the prior eigenvector estimate, i.e., $\mathbf{x}^{(p-1)} \leftarrow \mathbf{x}^{(p)}/\|\mathbf{x}^{(p)}\|$. If $\mu^{(p+1)} < 0$, the eigenvector is also updated as $\mathbf{x}^{(p-1)} \leftarrow (-1)\mathbf{x}^{(p)}/\|\mathbf{x}^{(p)}\|$. This sequence converges to the dominant eigenvalue and its associated eigenvector of the matrix $(\mathbf{A} - \alpha\mathbf{I})^{-1}$. Finally, the corresponding eigenvalue of $\mathbf{A}$ is obtained from

$$\lambda_k = \alpha + \frac{1}{\mu_1} \tag{11.30}$$

where μ_1 is the dominant eigenvalue of $(\mathbf{A} - \alpha\mathbf{I})^{-1}$.

The convergence rate of the shifted Inverse Power method depends on how fast the power of the dominance ratio, $|\lambda_2/\lambda_1|^p$, yields zero as $p \to \infty$. The asymptotic error constant, $\mathcal{O}(|\mu_2/\mu_1|)$, for $(\mathbf{A} - \alpha\mathbf{I})^{-1}$ in the vicinity of $\alpha \approx \lambda_k$ can be written as

$$\left|\frac{\mu_{k+1}}{\mu_k}\right| = \left|\frac{1/(\lambda_{k+1} - \alpha)}{1/(\lambda_k - \alpha)}\right| = \left|\frac{\lambda_k - \alpha}{\lambda_{k+1} - \alpha}\right|$$

Despite its linear convergence rate, this approach improves convergence with a judicious choice of α (i.e., $\lambda_k \approx \alpha$) because the asymptotic error constant becomes smaller ($\lambda_k - \alpha \approx 0$). From the preceding arguments, it is clear that λ_k does not necessarily need to be the largest or smallest one; it can be used to determine any desired eigenpair of $\mathbf{A}$. Having cast the eigenvalue problem as $(\mathbf{A} - \alpha\mathbf{I})^{-1}\mathbf{x} = \mu\mathbf{x}$, the inverse Power method is generally employed to determine the dominant eigenvalue and its associated eigenvector. After replacing $\mathbf{A}$ with $(\mathbf{A} - \alpha\mathbf{I})$, the Power method can be employed.

> To achieve fast convergence, a *'smart'* prediction sufficiently close to the α value is required. Otherwise, the benefit of shifting cannot materialize if the prediction is far from the desired eigenvalue.

EXAMPLE 11.4: Finding a specific eigenvalue

Consider the spring-mass system given in Example 11.2. Find the frequency closest to $\alpha = 4.1$ to an accuracy of six decimal places using the shifted inverse power method with the Rayleigh quotient.

SOLUTION:

First, we construct the $(\mathbf{A} - \alpha\mathbf{I})$ matrix, which is symmetrical. To solve the linear system, Eq. (11.28), we may employ Cholesky decomposition, giving

$$\mathbf{A} - \alpha\mathbf{I} = \begin{bmatrix} 1.9 & -1 & -3 \\ -1 & 2.1 & -1 \\ -3 & -1 & 1.9 \end{bmatrix}, \qquad \mathbf{L} = \begin{bmatrix} \sqrt{6} & 0 & 0 \\ -\dfrac{1}{\sqrt{6}} & \sqrt{\dfrac{11}{6}} & 0 \\ -\sqrt{\dfrac{3}{2}} & -3\sqrt{\dfrac{3}{22}} & \dfrac{6}{\sqrt{11}} \end{bmatrix}$$

The starting guess vector, as in previous examples, is set to $\mathbf{x} = [\, 1\ 1\ 0\,]^T$. The first iteration requires the solution of $(\mathbf{A} - \alpha\mathbf{I})\mathbf{x}^{(1)} = \mathbf{x}^{(0)}$, which yields

$$\mathbf{x}^{(1)} = [\ -0.0592495 \quad -0.3225807 \quad -0.2633311\ ]^T$$

Next, the largest eigenvalue is estimated by

$$\mu^{(1)} = \frac{(\mathbf{x}^{(1)}, \mathbf{x}^{(0)})}{(\mathbf{x}^{(0)}, \mathbf{x}^{(0)})} = \frac{-0.3818302}{2} = -0.190915$$

Since the eigenvalue is negative, the solution vector is normalized before the second iteration as

$$\mathbf{x}^{(1)} \leftarrow -\mathbf{x}^{(1)}/\|\mathbf{x}^{(1)}\|_\infty = [\ 0.1836735 \quad 1 \quad 0.8163266\ ]^T$$

Then solving $(\mathbf{A} - \alpha\mathbf{I})\mathbf{x}^{(2)} = \mathbf{x}^{(1)}$ yields

$$\mathbf{x}^{(2)} = [\ -0.2258468 \quad -0.3225806 \quad -0.0967339\ ]^T$$

The second estimate for the dominant eigenvalue yields

$$\mu^{(2)} = \frac{(\mathbf{x}^{(2)}, \mathbf{x}^{(1)})}{(\mathbf{x}^{(1)}, \mathbf{x}^{(1)})} = \frac{-0.4430291}{1.7001249} = -0.2605862$$

and we find the normalized solutions as

$$\mathbf{x}^{(2)} \leftarrow -\mathbf{x}^{(2)}/\|\mathbf{x}^{(2)}\|_\infty = [\ 0.700125 \quad 1 \quad 0.299875\ ]^T$$

In the subsequent iteration steps, the estimates for the dominant eigenvalue and corresponding eigenvector are found as follows:

$$\mu^{(3)} = -0.295883, \quad \mathbf{x}^{(3)} = [\ 0.373390 \quad 1 \quad 0.626610\ ]^T$$
$$\mu^{(4)} = -0.311560, \quad \mathbf{x}^{(4)} = [\ 0.580100 \quad 1 \quad 0.419900\ ]^T$$
$$\mu^{(5)} = -0.318113, \quad \mathbf{x}^{(5)} = [\ 0.449324 \quad 1 \quad 0.550676\ ]^T$$
$$\vdots \qquad\qquad \vdots$$
$$\mu^{(15)} = -10, \qquad \mathbf{x}^{(15)} = [\ 1 \quad -1 \quad 1\ ]^T$$

The largest eigenvalue of $(\mathbf{A} - \alpha\mathbf{I})^{-1}$ (in magnitude) converges to -10. Using this value in Eq. (11.30), the eigenvalue near $\alpha = 4.1$ is obtained as

$$\lambda = \alpha + 1/\mu_1 = 4.1 + (-1/10) = 4$$

Discussion: From $\lambda = \omega^2 m/k$, the corresponding frequency is obtained as $\omega = 2\sqrt{k/m}$, and its associated eigenvector is $[1\ -1\ 1]^T$. For large $\mathbf{A}$, the solution of the linear system $(\mathbf{A} - \alpha\mathbf{I})\mathbf{x}^{(p+1)} = \mathbf{x}^{(p)}$ should be obtained with the Cholesky decomposition, which can result in a significant reduction in the cpu-time.

11.3 SIMILARITY AND ORTHOGONAL TRANSFORMATIONS

The calculation of eigenvalues of diagonal or triangular matrices is much easier. In this regard, it is desirable to transform a dense matrix into a similar matrix with a diagonal or tridiagonal form, as this reduces the complexity of calculating the eigenvalues.

To begin, let $\mathbf{A}$ and $\mathbf{P}$ be square and nonsingular matrices. The matrix $\mathbf{A}$ and $\mathbf{P}^{-1}\mathbf{A}\mathbf{P}$ are said to be *similar matrices*, and the conversion from matrix $\mathbf{A}$ to matrix $\mathbf{P}^{-1}\mathbf{A}\mathbf{P}$ is called *similarity transformation*. The similarity transformation of a matrix $\mathbf{A}$ to a matrix $\mathbf{C}$ is defined as

$$\mathbf{C} = \mathbf{P}^{-1}\mathbf{A}\mathbf{P} \tag{11.31}$$

Note that this relationship is also symmetric since $\mathbf{A} = \mathbf{P}\mathbf{C}\mathbf{P}^{-1}$, and Eq. (11.31) can be expressed as $\mathbf{P}\mathbf{C} = \mathbf{A}\mathbf{P}$.

A similarity relation can be interpreted to say that $\mathbf{A}$ and $\mathbf{C}$ are matrix representations of the same linear transformation with respect to a different basis. Matrix $\mathbf{P}$ may be referred to as the *change of basis matrix*. Now, let us begin with the standard eigenvalue problem:

$$\mathbf{A}\mathbf{x} = \lambda\mathbf{x} \tag{11.32}$$

Left-multiplying Eq. (11.32) with $\mathbf{B}^{-1}$ gives

$$\mathbf{B}^{-1}\mathbf{A}\mathbf{x} = \lambda\mathbf{B}^{-1}\mathbf{x} \tag{11.33}$$

Setting $\mathbf{z} = \mathbf{B}^{-1}\mathbf{x}$ (also $\mathbf{x} = \mathbf{B}\mathbf{z}$) in Eq. (11.33) yields

$$\mathbf{C}\mathbf{z} = \lambda\mathbf{z} \tag{11.34}$$

where $\mathbf{C} = \mathbf{B}^{-1}\mathbf{A}\mathbf{B}$. Note that Eq. (11.34) is similar to Eq. (11.32), and λ remains unaffected by the preceding matrix algebra. In other words, if $\mathbf{A}$ and $\mathbf{C}$ are similar, then the characteristic polynomials and eigenvalues of $\mathbf{A}$ and $\mathbf{C}$ are the same, i.e., $p_{\mathbf{A}}(\lambda) = p_{\mathbf{C}}(\lambda)$ and $\lambda_i(\mathbf{A}) = \lambda_i(\mathbf{C})$. But the eigenvectors are different.

A matrix $\mathbf{A}$ is said to be *diagonalizable* if matrix $\mathbf{D}$ is diagonal and similar to $\mathbf{A}$. Any diagonalizable matrix $\mathbf{A}$ has n distinct eigenvalues, λ_i, and linearly independent associated eigenvectors, $\mathbf{x}_i$. This condition is *sufficient* (but *not necessary*) for diagonalization.

An invertible matrix $\mathbf{P}$ is needed to diagonalize a square matrix. If a suitable matrix can be constructed from its eigenvectors, such as $\mathbf{P} = [\mathbf{x}_1\ \mathbf{x}_2\cdots\mathbf{x}_n]$, then $\mathbf{D} = \mathbf{P}^{-1}\mathbf{A}\mathbf{P}$ is a diagonal matrix whose diagonal elements are eigenvalues:

$$\mathbf{D} = \begin{bmatrix} \lambda_1 & & & \\ & \lambda_2 & & \\ & & \ddots & \\ & & & \lambda_n \end{bmatrix}$$

Note that changing the order of eigenvalues or eigenvectors (by exchanging the column vectors of $\mathbf{P}$) produces a different diagonalization of the same matrix.

In symmetric matrices, the matrix $\mathbf{P}$ can be chosen to have the special property of $\mathbf{P}^{-1} = \mathbf{P}^T$. A matrix $\mathbf{P}$ is called an *orthogonal matrix* if $\mathbf{P}^{-1} = \mathbf{P}^T$, provided that the matrix $\mathbf{P}$ is invertible. A matrix $\mathbf{A}$ is *orthogonally diagonalizable* if an orthogonal matrix $\mathbf{P}$ exists, so that

EXAMPLE 11.5: Application of diagonalization procedure

Determine if matrix $\mathbf{A}$ is *diagonalizable* or not. If so, find $\mathbf{P}$ and $\mathbf{P}^{-1}\mathbf{A}\mathbf{P}$.

$$\mathbf{A} = \begin{bmatrix} 3 & 1 & 1 \\ 1 & 3 & 1 \\ -1 & -1 & 1 \end{bmatrix}$$

SOLUTION:

The characteristic equation of matrix $\mathbf{A}$ is found to be $\lambda^3 - 7\lambda^2 + 16\lambda - 12 = 0$ or $(\lambda - 2)^2(\lambda - 3) = 0$. Note that $\lambda = 2$ is an eigenvalue of the multiplicity of "two." Thus, the diagonalizability of $\mathbf{A}$ depends on having two independent eigenvectors for $\lambda = 2$.

By definition, we seek a nontrivial solution, that is,

$$(\mathbf{A} - \lambda\mathbf{I})\mathbf{x} = \begin{bmatrix} 3-\lambda & 1 & 1 \\ 1 & 3-\lambda & 1 \\ -1 & -1 & 1-\lambda \end{bmatrix} \begin{bmatrix} x_1 \\ x_2 \\ x_3 \end{bmatrix} = \begin{bmatrix} 0 \\ 0 \\ 0 \end{bmatrix}$$

For $\lambda_1 = 3$, the linear system becomes

$$\begin{bmatrix} 0 & 1 & 1 \\ 1 & 0 & 1 \\ -1 & -1 & -2 \end{bmatrix} \begin{bmatrix} x_1 \\ x_2 \\ x_3 \end{bmatrix} = \begin{bmatrix} 0 \\ 0 \\ 0 \end{bmatrix}$$

and the solution of this homogeneous system can be written $\mathbf{x}_1 = \begin{bmatrix} 1 & 1 & -1 \end{bmatrix}^T$.

For $\lambda_{2,3} = 2$, we arrive at

$$\begin{bmatrix} 1 & 1 & 1 \\ 1 & 1 & 1 \\ -1 & -1 & -1 \end{bmatrix} \begin{bmatrix} x_1 \\ x_2 \\ x_3 \end{bmatrix} = \begin{bmatrix} 0 \\ 0 \\ 0 \end{bmatrix}$$

which gives $x_1 + x_2 + x_3 = 0$. With one leading and two free variables, s and t, the general solution is found as $x_1 = s + t$, $x_2 = -s$, and $x_3 = -t$. Two independent eigenvectors can then be obtained by arbitrary choices of $(s, t) = (1, 0)$ and $(s, t) = (0, 1)$, as follows:

$$\mathbf{x}_2 = \begin{bmatrix} 1 \\ -1 \\ 0 \end{bmatrix} \quad \text{and} \quad \mathbf{x}_3 = \begin{bmatrix} 1 \\ 0 \\ -1 \end{bmatrix}$$

Now, constructing matrix $\mathbf{P}$ and calculating $\mathbf{P}^{-1}$ gives

$$\mathbf{P} = \begin{bmatrix} 1 & 1 & 1 \\ 1 & -1 & 0 \\ -1 & 0 & -1 \end{bmatrix} \quad \text{and} \quad \mathbf{P}^{-1} = \begin{bmatrix} 1 & 1 & 1 \\ 1 & 0 & 1 \\ -1 & -1 & -2 \end{bmatrix}$$

Finally, we obtain

$$\mathbf{P}^{-1}\mathbf{A}\mathbf{P} = \begin{bmatrix} 1 & 1 & 1 \\ 1 & 0 & 1 \\ -1 & -1 & -2 \end{bmatrix} \begin{bmatrix} 3 & 1 & 1 \\ 1 & 3 & 1 \\ -1 & -1 & 1 \end{bmatrix} \begin{bmatrix} 1 & 1 & 1 \\ 1 & -1 & 0 \\ -1 & 0 & -1 \end{bmatrix} = \begin{bmatrix} 3 & 0 & 0 \\ 0 & 2 & 0 \\ 0 & 0 & 2 \end{bmatrix}$$

Discussion: Notice that the triple matrix product gives a diagonal matrix whose diagonal elements are the eigenvalues of $\mathbf{A}$.

$$\mathbf{D} = \mathbf{Q}^T \mathbf{A} \mathbf{Q} \tag{11.35}$$

which represents an *orthogonal transformation*. Here, the letter $\mathbf{Q}$ is denoted instead of $\mathbf{P}$ to distinguish this special case. Note that the $\mathbf{Q}\mathbf{A}\mathbf{Q}^T$ matrix product is also an orthogonal transformation, which preserves not only the *eigenvalues* but also the *symmetry* property of the matrix. To diagonalize an orthogonal matrix,

  i. find all eigenvalues of $\mathbf{A}$ and determine the multiplicity of each eigenvalue;

 ii. choose a unit eigenvector for every eigenvalue of multiplicity of $m = 1$, i.e., normalize eigenvectors;

iii. find a set of m eigenvectors for each eigenvalue of multiplicity $m \geqslant 2$.

If an orthonormal set cannot be found, then the Gram-Schmidt orthonormalization process can be applied. Each eigenvector is a column in matrix $\mathbf{Q}$.

11.3.1   DIFFERENTIAL EQUATIONS

The diagonalization property is especially useful to uncouple a linear system of ODEs. Let a set of first-order coupled ODEs be given as

$$\frac{d\mathbf{x}}{dt} = \mathbf{A}\mathbf{x}, \qquad \mathbf{x}(0) = \mathbf{x}_0 \tag{11.36}$$

where $\mathbf{x} = [x_1(t), x_2(t), \cdots, x_n(t)]^T$ and $\mathbf{A}$ is a diagonalizable matrix. Substituting $\mathbf{x} = \mathbf{P}\mathbf{z}$, into Eq. (11.36) and left-multiplying by $\mathbf{P}^{-1}$ on both sides, the system is an uncoupled form:

$$\mathbf{P}^{-1}\mathbf{P}\frac{d\mathbf{z}}{dt} = \mathbf{P}^{-1}\mathbf{A}\mathbf{P}\mathbf{z} \quad \rightarrow \quad \frac{d\mathbf{z}}{dt} = \mathbf{D}\mathbf{z} \tag{11.37}$$

where $\mathbf{D}$ is a diagonal matrix whose diagonal elements are the eigenvalues of $\mathbf{A}$.

EXAMPLE 11.6: Solving coupled ODEs using the diagonalization procedure

Consider the following reversible chemical reactions:

$$X \underset{k_2}{\overset{k_1}{\rightleftharpoons}} Y \underset{k_4}{\overset{k_3}{\rightleftharpoons}} Z$$

The kinetic behavior of the reactions is described by the rate equations and the initial conditions and is given by:

$$\left. \begin{aligned} \frac{dC_X}{dt} &= -k_1 C_X + k_2 C_Y \\[2mm] \frac{dC_Y}{dt} &= k_1 C_X - (k_2 + k_3)C_Y + k_4 C_Z \\[2mm] \frac{dC_Z}{dt} &= k_3 C_Y - k_4 C_Z \end{aligned} \right\} \qquad \begin{aligned} C_X(0) &= 1 \\[2mm] C_Y(0) &= 0 \\[2mm] C_Z(0) &= 0 \end{aligned}$$

where $k_1 = 0.5$, $k_2 = 0.75$, $k_3 = 0.25$, and $k_4 = 1.5$ min^{-1} are the reaction rate constants. (a) Express the given system of coupled ODEs in matrix form, and (b) apply the diagonalization procedure to decouple the system into first-order ODEs.

SOLUTION:

(a) The system can be expressed in matrix form as $d\mathbf{x}/dt = \mathbf{A}\mathbf{x}$, where

$$\mathbf{A} = \frac{1}{4}\begin{bmatrix} -2 & 3 & 0 \\ 2 & -4 & 6 \\ 0 & 1 & -6 \end{bmatrix}, \quad \mathbf{x} = \begin{bmatrix} C_X \\ C_Y \\ C_Z \end{bmatrix}, \quad \text{and} \quad \mathbf{x}(0) = \begin{bmatrix} 1 \\ 0 \\ 0 \end{bmatrix}$$

The matrix has three distinct and real eigenvalues: $\lambda = -2, -1,$ and 0; hence, the given system is diagonalizable.

Using a suitable method, the eigenpairs of $\mathbf{A}$ are determined to construct $\mathbf{D}$, $\mathbf{P}$, and $\mathbf{P}^{-1}$, yielding

$$\mathbf{D} = \begin{bmatrix} -2 & & \\ & -1 & \\ & & 0 \end{bmatrix}, \quad \mathbf{P} = \begin{bmatrix} 1 & -3 & 9 \\ -2 & 2 & 6 \\ 1 & 1 & 1 \end{bmatrix}, \quad \mathbf{P}^{-1} = \frac{1}{16}\begin{bmatrix} 1 & -3 & 9 \\ -2 & 2 & 6 \\ 1 & 1 & 1 \end{bmatrix}$$

As illustrated in Eq. (11.37), by introducing $\mathbf{x} = \mathbf{P}\mathbf{z}$ substitution, followed by left multiplication of both sides with $\mathbf{P}^{-1}$, the system is decoupled as $d\mathbf{z}/dt = \mathbf{D}\mathbf{z}$, leading to

$$\left.\begin{aligned} \frac{dz_1}{dt} &= -2z_1 \\ \frac{dz_2}{dt} &= -z_2 \\ \frac{dz_3}{dt} &= 0 \end{aligned}\right\} \rightarrow \quad \mathbf{z}(0) = \mathbf{P}^{-1}\mathbf{x}(0) = \frac{1}{16}\begin{bmatrix} 1 \\ -2 \\ 1 \end{bmatrix}$$

The analytical solutions of the uncoupled ODEs can then be obtained as

$$z_1(t) = e^{-2t}/16, \qquad z_2(t) = -e^{-t}/8 \qquad z_3(t) = 1/16$$

Finally, using $\mathbf{x} = \mathbf{P}\mathbf{z}$, the solution for the original coupled ODEs is found as

$$x_1(t) = z_1(t) - 3z_2(t) + 9z_3(t) = \frac{1}{16}e^{-2t}(3e^t + 1)^2,$$

$$x_2(t) = -2z_1(t) + 2z_2(t) + 6z_3(t) = \frac{1}{8}e^{-2t}(3e^t + 1)(e^t - 1),$$

$$x_3(t) = z_1(t) + z_2(t) + z_3(t) = \frac{1}{16}e^{-2t}(e^t - 1)^2$$

Discussion: Note that the diagonalization procedure through the decoupling of an ODE system greatly simplifies the mathematical procedure and leads to the easy finding of analytical solutions.

One of the most important consequences of diagonalization is that the very large power of a matrix, $\mathbf{A}$, can easily be evaluated by $\mathbf{A}^k = \mathbf{P}\mathbf{D}^k\mathbf{P}^{-1}$.

11.4 JACOBI METHOD

The *Jacobi method*, based on orthogonal transformations, is a numerical technique frequently used to evaluate the eigenvalues and eigenvectors of a *symmetric matrix*. Nonetheless, this limitation is relatively insignificant since many problems of practical importance in science and engineering involve symmetric matrices [9].

The Jacobi method transforms a symmetric matrix into a diagonal matrix through a series of orthogonal transformations. As indicated earlier, the eigenvalues are unaltered during these transformations. With each transformation (or plane rotation), symmetric pairs of the largest off-diagonal element are zeroed out. However, the zeros established by previous plane rotations are not necessarily preserved in subsequent transformations; that is, new rotations generally undo elements previously zeroed out. Yet, with each transformation, the non-zero off-diagonal elements become smaller and smaller until the matrix is diagonal to machine precision. The diagonal elements of the final transformed matrix are the eigenvalues.

The basic Jacobi orthogonal rotation matrix $\mathbf{U}$ has the following form:

$$
\mathbf{U} =
\begin{bmatrix}
1 & & & & & & & & \\
& \ddots & & & & & & & \\
& & 1 & & & & & & \\
& & & c & & & s & & \\
& & & & 1 & & & & \\
& & & & & \ddots & & & \\
& & & & & & 1 & & \\
& & & -s & & & c & & \\
& & & & & & & 1 & \\
& & & & & & & & \ddots \\
& & & & & & & & & 1
\end{bmatrix}
\begin{matrix} \\ \\ \\ p \\ \\ \vdots \\ \\ q \\ \\ \\ \end{matrix}
\tag{11.38}
$$

where (p, q) denotes the plane of rotation, $u_{ij} = 1$ for $i \neq p$, $i \neq q$, $u_{pp} = u_{qq} = c$ and $u_{pq} = -u_{qp} = s$, where c and s denote the cosine and sine of the rotation angle θ, respectively, so $c^2 + s^2 = 1$. All off-diagonal elements are zero except s and $-s$.

A plane rotation, using matrix $\mathbf{U}$, modifies only the pth and qth rows and columns of matrix $\mathbf{A}$. The objective is to zero the off-diagonal elements $(a_{pq} = a_{qp} = 0)$ by a series of $U^T \mathbf{A} \mathbf{U}$ orthogonal transformations. As a result, with each successive transformation, matrix $\mathbf{A}$ is modified until all off-diagonal elements are zero. Accordingly, this similarity transformation procedure is expressed as

$$
\mathbf{A}_{k+1} = \mathbf{U}^T \mathbf{A}_k \mathbf{U}
\tag{11.39}
$$

where $\mathbf{A}_{k+1}$ and $\mathbf{A}_k$ denote the two successive transformation matrices, and $\mathbf{A}_0$ is the original matrix $\mathbf{A}$.

The matrix $\mathbf{U}$ is orthogonal since $\mathbf{U}$ and $\mathbf{U}^T$ are inverses of each other $(\mathbf{U^T U = I})$. In the triple product $\mathbf{U^T A U}$, only the pth row and the qth column elements of $\mathbf{A}$ are altered, so we may express this operation as

$$
\begin{bmatrix} c & -s \\ s & c \end{bmatrix}
\begin{bmatrix} a_{pp} & a_{pq} \\ a_{qp} & a_{qq} \end{bmatrix}
\begin{bmatrix} c & s \\ -s & c \end{bmatrix}
=
\begin{bmatrix} a'_{pp} & a'_{pq} \\ a'_{qp} & a'_{qq} \end{bmatrix}
\tag{11.40}
$$

Equation (11.40) gives

$$a'_{pp} = c^2 a_{pp} + s^2 a_{qq} - 2cs\, a_{pq}$$
$$a'_{qq} = c^2 a_{qq} + s^2 a_{pp} + 2cs a_{pq} \tag{11.41}$$
$$a'_{pq} = a'_{qp} = (c^2 - s^2)a_{pq} + cs(a_{pp} - a_{qq})$$

The next step involves determining c and s. We already have $c^2 + s^2 = 1$, and we need an additional relationship to find c and s. We enforce $a'_{pq} = a'_{qp} = 0$. Setting $a'_{pq} = 0$ in Eq. (11.41) and noting that $c^2 - s^2 = \cos 2\theta$ and $cs = (1/2)\sin 2\theta$, we obtain

$$\cos 2\theta\, a_{pq} + \frac{1}{2}\sin 2\theta\,(a_{pp} - a_{qq}) = 0 \tag{11.42}$$

or rearranging Eq. (11.42) gives

$$\tan 2\theta = \frac{2a_{pq}}{a_{pp} - a_{qq}} \tag{11.43}$$

For rotations, only $\sin\theta$ and $\cos\theta$ are required. Rather than finding θ from Eq. (11.42) and then computing $\sin\theta$ and $\cos\theta$, it is more convenient and also more accurate to compute c and s directly using trigonometric relationships. After some manipulation, we write

$$c = \cos\theta = \sqrt{\frac{1}{2} + \frac{|\alpha|}{2\beta}} \tag{11.44}$$

$$s = \sin\theta = \frac{\alpha\,(-a_{pq})}{2\beta|\alpha|\cos\theta} \tag{11.45}$$

where $\alpha = (a_{pp} - a_{qq})/2$ and $\beta = \sqrt{a_{pq}^2 + \alpha^2}$.

The pth row and the qth column elements of $\mathbf{A}$ are modified as a result of an orthogonal transformation as follows:

For pth and qth rows $(j \neq p$ or $q)$

$$a'_{pj} = c\,a_{pj} - s\,a_{qj} \quad \text{and} \quad a'_{qj} = s\,a_{pj} + c\,a_{qj} \tag{11.46}$$

For pth and qth columns $(i \neq p$ or $q)$

$$a'_{ip} = c\,a_{ip} - s\,a_{iq} \quad \text{and} \quad a'_{iq} = s\,a_{ip} + c\,a_{iq} \tag{11.47}$$

All other elements of $\mathbf{A}$ remain unchanged.

The procedure now requires choosing the element a_{pq}, which we desire to make zero. Then, having calculated c and s from Eqs. (11.44) and (11.45), the new values of a'_{pp} and a'_{qq} can be determined from Eq. (11.41), and a_{pq} and a_{qp} are set to zero. Finally, Eqs. (11.41) provide the remaining modifications to the matrix $\mathbf{A}$.

The eigenvectors are also computed in the same manner by applying the same orthogonal transformations ($\mathbf{X}_{k+1} = \mathbf{X}_k\mathbf{U}$) to the identity matrix (where the identity matrix and its elements are taken as $\mathbf{X}_0 = \mathbf{I}$), so the transformation yields

$$\begin{aligned} p^{\text{th}} \text{ column}: &\quad x_{ip} = c\,x_{ip} - s\,x_{iq} \\ q^{\text{th}} \text{ column}: &\quad x_{iq} = s\,x_{ip} + c\,x_{iq} \end{aligned} \tag{11.48}$$

All other columns remain unchanged. When the eigenvalue problem converges, the columns of $\mathbf{X}$ become the eigenvectors of the original matrix.

The eigenvalues of matrix $\mathbf{A}$ are the diagonal elements of the modified matrix for sufficiently large k, i.e., $\mathbf{A}_\infty = \lim_{k\to\infty}\mathbf{A}_k$. The jth column of the converged matrix, $\mathbf{X}_\infty$, is the eigenvector corresponding to λ_j, i.e., $\mathbf{X}_\infty^{(j)} = [x_{1j}, \ x_{2j}, \ x_{3j}, \ \dots, \ x_{nj}]^T$.

The algorithm presented in this section is referred to as the classical *Jacobi method*. It is based on searching for the location of the largest off-diagonal element (p and q) to be zeroed out. Nevertheless, a drawback of this algorithm is that searching for the largest element in very large matrices can be time-consuming due to the number of comparisons on the order of $\mathcal{O}(n^2)$. While a'_{pq} and a'_{qp} are zeroed out, the matrix elements are modified with each plane rotation, which also modifies previously annihilated elements, often resulting in nonzero values. But the magnitude of the non-zero off-diagonal elements becomes smaller with each subsequent rotation, and the matrix eventually tends toward diagonal form. The method converges quite quickly once the off-diagonal elements become small.

An alternative, the *cyclic Jacobi* method, has been developed to eliminate excessive (search and compare) operations devoted to finding the location of the largest off-diagonal element. In this variant, the off-diagonal elements are zeroed out in a predetermined order regardless of their magnitudes, i.e., ordering can be devised on a row-by-row or column-by-column basis. Each element of a matrix is zeroed out only once in a sequence of n rotations called a *sweep*. This method also converges quite quickly once off-diagonal elements become small.

The accuracy of the Jacobi method and its variants depends on the criteria adopted for terminating iterations. The simplest criterion that can be used to terminate the iteration process is to check whether the maximum absolute value of the off-diagonal elements is smaller than a preset convergence tolerance; that is, $\max|a_{pq}| < \varepsilon$ for all p and q ($p \neq q$). However, at this stage, the reader must be cautioned to take into account the relative magnitudes of a_{pp} and a_{qq}. However, another criterion involves checking the sum of the squares of the off-diagonal elements at the end of each sweep.

$$T = \sum_{i=1}^{n} \sum_{j=i+1}^{n} a_{ij}^2$$

This procedure is called the *Threshold Method*. The *threshold value* is found by $\mu = \sqrt{T}/n$ which is computed in the subsequent iterations as $\mu^{(k)} = \sqrt{T^{(k)}}/n$, where $T^{(k)}$ is the sum of the squares of the off-diagonal elements of $\mathbf{A}_k$. This process is terminated when $\mu_k \leqslant \varepsilon$.

Jacobi Method

- Jacobi method is simple, reliable, and easy to program;
- It works well for well-conditioned matrices;
- It gives all eigenvalues and eigenvectors of a real symmetric matrix.

- It is limited to symmetric matrices;
- It converges rather slowly and may require a very large number of iterations (the minimum number of rotations is $n(n-1)/2$);
- The zeros created in previous rotations can be undone in subsequent rotations.

Pseudocode 11.3

Module JACOBI (n, **A**, **X**, ε, $maxit$)
\ DESCRIPTION: A pseudomodule to find the eigenvalues and eigenvectors
\   using the Jacobi method.
\ USES:
\   ABS:: A built-in function computing the absolute value;
\   SQRT:: A built-in function computing the square-root of a value;
\   ROTATE:: Module performing plane rotations.
Declare: a_{nn}, x_{nn} \ Declare array variables
X ← **0**   \ Initialize eigenmatrix
For $\big[i = 1, n\big]$ \ Set up Identity matrix
  $x_{ii} \leftarrow 1$ \ Set up unit diagonals, $x_{ii} \leftarrow 1$
End For
$Threshold \leftarrow 1$ \ Initialize the threshold value
$k \leftarrow 0$ \ Initialize iteration count
Repeat   \ Iteration Loop
  $amax \leftarrow 0$ \ Initialize max value
  For $\big[i = 1, n\big]$ \ Search for (p, q) of element with max. magnitude
    For $\big[j = i + 1, n\big]$ \ Search upper diagonal elements
      If $\big[|a_{ij}| > amax\big]$ **Then**
        $amax \leftarrow |a_{ij}|$
        $p \leftarrow i;\ q \leftarrow j$ \ Save (i, j) on (p, q)
      End If
    End For
  End For
  $\alpha \leftarrow (a_{pp} - a_{qq})/2$ \ Compute rotation angles
  $\beta \leftarrow \sqrt{\alpha^2 + a_{pq}^2}$
  $c \leftarrow \sqrt{(1 + |\alpha|/\beta)/2}$ \ Find rotation cosines
  $s \leftarrow \alpha(-a_{pq})/(2 * |\alpha|\, c * \beta)$ \ Find rotation sine
  ROTATE$(n, p, q, c, s, \mathbf{A}, \mathbf{X})$ \ Get rotation done!
  $sums \leftarrow 0$ \ Initialize accumulator, $sums$
  For $\big[i = 1, n\big]$ \ Compute sums of squares of off-diagonals
    For $\big[j = i + 1, n\big]$
      $sums \leftarrow sums + a_{ij}^2$ \ Accumulate a_{ij}^2
    End For
  End For
  $Threshold \leftarrow \sqrt{sums}/n$ \ Compute threshold value
  $k \leftarrow k + 1$ \ Count iterations
  Write: k, $Threshold$ \ Print out the iteration progress
Until $\big[Threshold < \varepsilon$ **Or** $k = maxit\big]$
If $\big[k = maxit\big]$ **Then**   \ Issue info and a warning
  Write: "The $maxit$ reached with no convergence"
  Write: "The threshold value is =", $Threshold$
End If
End Module JACOBI

Module ROTATE (n, p, q, c, s, **A**, **X**)
\ DESCRIPTION: A pseudomodule to perform a plane rotation

Declare: a_{nn}, x_{nn}
For $\left[i = 1, n\right]$
$\quad g1 = a_{pi}$
$\quad a_{pi} = c * g1 - s * a_{qi}$
$\quad a_{qi} = s * g1 + c * a_{qi}$
End For
For $\left[i = 1, n\right]$
$\quad g1 \leftarrow a_{ip};\ g2 \leftarrow x_{ip}$
$\quad a_{ip} \leftarrow c * g1 - s * a_{iq}$
$\quad a_{iq} \leftarrow s * g1 + c * a_{iq}$
$\quad x_{ip} \leftarrow c * g2 - s * x_{iq}$
$\quad x_{iq} \leftarrow s * g2 + c * x_{iq}$
End For
End Module ROTATE

A pseudomodule, **JACOBI**, computing all eigenvalues and eigenvectors of a real symmetric matrix is presented in Pseudocode 11.3. The module requires a symmetric matrix ($\mathbf{A}$), the matrix size (n), a convergence tolerance (ε), and an upper bound for the maximum number of iterations ($maxit$) as input. On exit, the matrix $\mathbf{A}$ is a diagonal matrix whose elements are the eigenvalues ($\lambda_i = a_{ii}$), and the eigenvectors are stored on the square matrix $\mathbf{X}$. At the beginning, $\mathbf{X}$ is initialized with the zero matrix, i.e., $\mathbf{X} = \mathbf{0}$. Then, the diagonals are set as $x_{ii} = 1$ for all i. The location of the matrix element corresponding to the largest off-diagonal value, (p, q), is found with a couple of If-constructs. The rotation angles are determined by Eqs. (11.44) and (11.45). Then the matrix $\mathbf{A}$ is modified according to Eqs. (11.46) and (11.47) and applied to the matrix $\mathbf{X}$ by Eq. (11.48), which zeros the matrix element at (p, q). The procedure is repeated until $\mathbf{A}$ is transformed into a diagonal matrix. The state of the diagonalization process is verified by checking whether the threshold criterion is satisfied or not, i.e., $\mu^{(k)} = \sqrt{T^{(k)}}/n < \varepsilon$. Once the diagonalization is achieved, the diagonal elements of $\mathbf{A}$ are the eigenvalues, and the corresponding columns of $\mathbf{X}$ are the associated eigenvectors $(\lambda_j, \mathbf{x}_j)$ for $j = 1, 2, \ldots, n$.

EXAMPLE 11.7: Application of the Jacobi method

A machine component is subjected to the following stress conditions:

$$\begin{bmatrix} \sigma_{xx} & \sigma_{xy} & \sigma_{xz} \\ \sigma_{yx} & \sigma_{yy} & \sigma_{yz} \\ \sigma_{zx} & \sigma_{zy} & \sigma_{zz} \end{bmatrix} = \begin{bmatrix} 17 & -5 & 41 \\ -5 & 53 & -5 \\ 41 & -5 & 17 \end{bmatrix}$$

where the stresses are in MPa. The principal stresses are obtained from the solution of the following eigenvalue problem:

$$\begin{bmatrix} \sigma_{xx} & \sigma_{xy} & \sigma_{xz} \\ \sigma_{yx} & \sigma_{yy} & \sigma_{yz} \\ \sigma_{zx} & \sigma_{zy} & \sigma_{zz} \end{bmatrix} \begin{bmatrix} \ell_1 \\ \ell_2 \\ \ell_3 \end{bmatrix} = \lambda \begin{bmatrix} \ell_1 \\ \ell_2 \\ \ell_3 \end{bmatrix}$$

where λ denotes the principal stresses (eigenvalues) and ℓ_1, ℓ_2, and ℓ_3 denote the direction cosines defining the principal plane on which σ acts. Use the Jacobi method to find the principal stresses and associated principal planes in the machine component for the given stress condition.

SOLUTION:

Since the stress matrix is symmetric, the Jacobi method can be employed to find its eigenpairs. Setting $\mathbf{A}_0 = \mathbf{A}$, we start searching for the location of the absolute maximum of the off-diagonal elements.

There is only one element with the largest magnitude, namely, $|a_{13}| = 41$. We start the plane rotations with $(p, q) = (3, 1)$, for which we have $a_{33} = a_{11} = 17$ and $a_{13} = 41$.

Next, α and β are determined as follows:

$$\alpha = \frac{a_{33} - a_{11}}{2} = \frac{17 - 17}{2} = 0$$

$$\beta = \sqrt{a_{13}^2 + \alpha^2} = \sqrt{41^2 + 0^2} = 41$$

The rotation sine and cosine are obtained from Eq. (11.44) and (11.45) as

$$c = \sqrt{\frac{1}{2} + \frac{|\alpha|}{2\beta}} = \sqrt{\frac{1}{2} + \frac{0}{2(12.5)}} = \frac{1}{\sqrt{2}}$$

$$s = -\frac{a_{13}}{2\,\beta\,c} = -\frac{41}{2(41)(1/\sqrt{2})} = -\frac{1}{\sqrt{2}}$$

Then, the rotation matrix $\mathbf{U}$ is constructed as

$$\mathbf{U} = \begin{bmatrix} \dfrac{1}{\sqrt{2}} & 0 & \dfrac{1}{\sqrt{2}} \\ 0 & 1 & 0 \\ -\dfrac{1}{\sqrt{2}} & 0 & \dfrac{1}{\sqrt{2}} \end{bmatrix}$$

Subsequently, $\mathbf{A}_1$ is found as follows:

$$\mathbf{A}_1 = \mathbf{U}^T \mathbf{A}_0 \mathbf{U} = \begin{bmatrix} \dfrac{1}{\sqrt{2}} & 0 & -\dfrac{1}{\sqrt{2}} \\ 0 & 1 & 0 \\ \dfrac{1}{\sqrt{2}} & 0 & \dfrac{1}{\sqrt{2}} \end{bmatrix} \begin{bmatrix} 17 & -5 & 41 \\ -5 & 53 & -5 \\ 41 & -5 & 17 \end{bmatrix} \begin{bmatrix} \dfrac{1}{\sqrt{2}} & 0 & \dfrac{1}{\sqrt{2}} \\ 0 & 1 & 0 \\ -\dfrac{1}{\sqrt{2}} & 0 & \dfrac{1}{\sqrt{2}} \end{bmatrix}$$

which leads to

$$\mathbf{A}_1 = \begin{bmatrix} -24 & 0 & 0 \\ 0 & 53 & -5\sqrt{2} \\ 0 & -5\sqrt{2} & 58 \end{bmatrix}$$

Notice that a_{31} and a_{13} in $\mathbf{A}_1$ are zero. This rotation leads coincidentally to $a_{21} = a_{12} = 0$. Now, setting $\mathbf{X}_0 = \mathbf{I}$ and employing the same rotation to $\mathbf{X}$ yields

$$\mathbf{X}_1 = \mathbf{X}_0 \mathbf{U} = \begin{bmatrix} \dfrac{1}{\sqrt{2}} & 0 & \dfrac{1}{\sqrt{2}} \\ 0 & 1 & 0 \\ -\dfrac{1}{\sqrt{2}} & 0 & \dfrac{1}{\sqrt{2}} \end{bmatrix}$$

Having completed the first sweep, we search for the location of the absolute maximum of the off-diagonal element of $\mathbf{A}_1$, which is $|a_{23}| = 5\sqrt{2}$; that is, $p = 2$ and $q = 3$. For this case, the required matrix elements are $a_{22} = 53$, $a_{33} = 58$, and $a_{23} = -5\sqrt{2}$. Then the plane rotation cosine, sine, rotation matrix, and $\mathbf{A}_2$ matrix are obtained as

$$\alpha = \frac{a_{22} - a_{33}}{2} = \frac{53 - 58}{2} = -\frac{5}{2}$$

$$\beta = \sqrt{a_{23}^2 + \alpha^2} = \sqrt{\left(-5\sqrt{2}\right)^2 + \left(-\frac{5}{2}\right)^2} = \frac{15}{2}$$

$$c = \sqrt{\frac{1}{2} + \frac{|\alpha|}{2\beta}} = \sqrt{\frac{1}{2} + \frac{5/2}{2(15/2)}} = \sqrt{\frac{2}{3}}$$

$$s = -\frac{a_{23}}{2\beta c} = -\frac{(-5\sqrt{2})}{2(15/2)(\sqrt{2/3})} = \frac{1}{\sqrt{3}}$$

Substituting these into $\mathbf{U}$ yields

$$\mathbf{U} = \begin{bmatrix} 1 & 0 & 0 \\ 0 & \sqrt{\dfrac{2}{3}} & -\dfrac{1}{\sqrt{3}} \\ 0 & \dfrac{1}{\sqrt{3}} & \sqrt{\dfrac{2}{3}} \end{bmatrix}$$

Then we find

$$\mathbf{A}_2 = \mathbf{U}^T \mathbf{A}_1 \mathbf{U} = \begin{bmatrix} -24 & 0 & 0 \\ 0 & 48 & 0 \\ 0 & 0 & 63 \end{bmatrix}$$

Employing the second rotation to $\mathbf{X}_1$ yields

$$\mathbf{X}_2 = \mathbf{X}_1 \mathbf{U} = \begin{bmatrix} \dfrac{1}{\sqrt{2}} & \dfrac{1}{\sqrt{6}} & \dfrac{1}{\sqrt{3}} \\ 0 & \sqrt{\dfrac{2}{3}} & -\dfrac{1}{\sqrt{3}} \\ -\dfrac{1}{\sqrt{2}} & \dfrac{1}{\sqrt{6}} & \dfrac{1}{\sqrt{3}} \end{bmatrix}$$

Discussion: In this example problem, we observed that after 2 plane rotations, the matrix $\mathbf{A}$ becomes diagonal, containing the eigenvalues of -24, 48, and 63. On the other hand, the matrix $\mathbf{X}$ contains the associated normalized eigenvectors, each in a separate column for $\lambda = -24$, 48, and 63, respectively.

It should be pointed out that non-iterative or direct methods require $\mathcal{O}(n^3)$ operations for square matrices of order n. The direct methods for large n are practically useless. On the other hand, the convergence rate of the Jacobi method is poor. The number of rotations is typically in the order of $\mathcal{O}(n^2)$ and each rotation requires $4n$ operations. The overall number of mathematical operations to zero out the non-diagonal matrix elements is in the order of $\mathcal{O}(n^3)$. In general, other methods are preferred due to the relatively slow convergence properties of the Jacobi method.

11.5 CHOLESKY DECOMPOSITION

The generalized eigenvalue problem of two symmetric $\mathbf{A}$ and $\mathbf{B}$ matrices has the following form:

$$\mathbf{Ax} = \lambda \mathbf{Bx} \tag{11.49}$$

where $\mathbf{A}$ and $\mathbf{B}$ are of the same size, and for convenience, we will assume $\mathbf{B}$ is a *positive definite matrix*, which can be defined mathematically in several ways, but, at this stage, it is sufficient to note that all eigenvalues of a positive definite matrix are positive. It is also worth noting that a diagonally dominant matrix with positive diagonal elements is also positive-definite.

We may attempt to transform Eq. (11.49) simply into the standard eigenvalue problem. For a nonsingular matrix $\mathbf{B}$, obviously the transformation is

$$\mathbf{B}^{-1}\mathbf{Ax} = \lambda \mathbf{x} \tag{11.50}$$

Yet even when both $\mathbf{A}$ and $\mathbf{B}$ matrices are symmetric, the product $\mathbf{C} = \mathbf{B}^{-1}\mathbf{A}$ will not necessarily be symmetric. However, $\mathbf{C}$ is often desired to be symmetrical. For this reason, we adopt a different approach to reach the standard eigenvalue problem.

We propose to decompose matrix $\mathbf{B}$ as follows:

$$\mathbf{B} = \mathbf{LL}^T \tag{11.51}$$

where $\mathbf{L}$ is a lower triangular matrix and $\mathbf{L}^T$ is its transpose.

Now, left-multiplying the rhs of Eq. (11.49) with $\mathbf{L}^{-1}$ and substituting Eq. (11.51) gives

$$\lambda \mathbf{L}^{-1}(\mathbf{LL}^T)\mathbf{x} = \lambda \mathbf{L}^T \mathbf{x} \tag{11.52}$$

Similarly, left-multiplying the lhs of Eq. (11.49) with $\mathbf{L}^{-1}$ and making use of the $(\mathbf{L}^{-1})^T = (\mathbf{L}^T)^{-1}$ identity yields

$$\mathbf{L}^{-1}\mathbf{A}(\mathbf{L}^{-T}\mathbf{L}^T)\mathbf{x} \tag{11.53}$$

where $(\mathbf{L}^{-1})^T$ is denoted by $\mathbf{L}^{-T}$ and $\mathbf{L}^{-T}\mathbf{L}^T = \mathbf{I}$.

Combining Eq. (11.52) and Eq. (11.53) results in

$$(\mathbf{L}^{-1}\mathbf{AL}^{-T})(\mathbf{L}^T\mathbf{x}) = \lambda \mathbf{L}^T \mathbf{x} \tag{11.54}$$

The matrix product $\mathbf{L}^{-1}\mathbf{AL}^{-T}$ is now a symmetric matrix with the same eigenvalues as the original eigenvalue problem. By setting $\mathbf{C} = \mathbf{L}^{-1}\mathbf{AL}^{-T}$ and $\mathbf{z} = \mathbf{L}^T\mathbf{x}$ in Eq. (11.54), the standard eigenvalue problem is obtained, i.e., $\mathbf{Cz} = \lambda \mathbf{z}$. Subsequently, the eigenvectors of the original problem can be determined from $\mathbf{x} = \mathbf{L}^{-T}\mathbf{z}$.

The procedure for determining a lower triangular matrix $\mathbf{L}$ is referred to as *Cholesky Decomposition* and is covered in Chapter 2 (*see* Section 2.8). On the other hand, its inverse, $\mathbf{L}^{-1}$, can be found as follows:

$$\ell_{ii}^{-1} = \frac{1}{\ell_{ii}}, \quad \ell_{ij}^{-1} = -\frac{1}{\ell_{jj}} \sum_{k=j+1}^{i} \ell_{ik}^{-1}\ell_{kj}, \quad \text{for } j = (i-1), \ldots, 3, 2, 1, \ i = 2, 3, \ldots, n \tag{11.55}$$

where $\ell_{11}^{-1} = 1/\ell_{11}$ and ℓ_{ij}^{-1} denotes the ith row and jth column element of $\mathbf{L}^{-1}$, i.e., $\ell_{ij}^{-1} \neq 1/\ell_{ij}$ for $i \neq j$.

EXAMPLE 11.8: Application of Cholesky decomposition

Consider the following electric circuit:

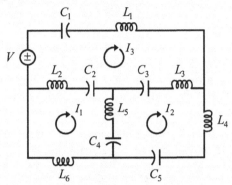

where $2C_2/3 = C_1 = C_3 = C_4 = C_5 = C$ and $L_1 = 2L_2 = L_3 = L_4 = L_5 = 2L_6 = L$. The transient current response of the circuit is described by the following coupled second-order ODEs:

$$2\frac{d^2 I_1}{dt^2} - \frac{d^2 I_2}{dt^2} - \frac{1}{2}\frac{d^2 I_3}{dt^2} + \frac{5}{3LC}I_1 - \frac{1}{LC}I_2 - \frac{2}{3LC}I_3 = 0$$

$$-\frac{d^2 I_1}{dt^2} + 3\frac{d^2 I_2}{dt^2} - \frac{d^2 I_3}{dt^2} - \frac{1}{LC}I_1 + \frac{3}{LC}I_2 - \frac{1}{LC}I_3 = 0$$

$$-\frac{1}{2}\frac{d^2 I_1}{dt^2} - \frac{d^2 I_2}{dt^2} + \frac{5}{2}\frac{d^2 I_3}{dt^2} - \frac{2}{3LC}I_1 - \frac{1}{LC}I_2 + \frac{8}{3LC}I_3 = 0$$

Assuming that the currents have the form $I_i = A_i \sin(\omega t + \phi)$, where A_i is the amplitude, ω is the resonance frequency, and ϕ is the phase shift, the set of ODEs can be reduced to the following eigenvalue problem:

$$\begin{bmatrix} \frac{5}{3} & -1 & -\frac{2}{3} \\ -1 & 3 & -1 \\ -\frac{2}{3} & -1 & \frac{8}{3} \end{bmatrix} \begin{bmatrix} I_1 \\ I_2 \\ I_3 \end{bmatrix} = \lambda \begin{bmatrix} 2 & -1 & -\frac{1}{2} \\ -1 & 3 & -1 \\ -\frac{1}{2} & -1 & \frac{5}{2} \end{bmatrix} \begin{bmatrix} I_1 \\ I_2 \\ I_3 \end{bmatrix}$$

where $\lambda = \omega^2 LC$ denotes the eigenvalues. Apply the Cholesky decomposition to the problem to find the resonance frequencies and associated currents.

SOLUTION:

The problem is set up as a general eigenvalue problem. The first step is to transform it into a standard eigenvalue problem: $\mathbf{Cz} = \lambda\mathbf{z}$. To achieve this, $\mathbf{B}$ is decomposed by using Cholesky's method.

Starting with $\ell_{11} = \sqrt{b_{11}} = \sqrt{2}$ and following the decomposition steps leads to

$$\ell_{21} = \frac{b_{21}}{\ell_{11}} = -\frac{1}{\sqrt{2}} \qquad \ell_{22} = \sqrt{b_{22} - \ell_{21}^2} = \sqrt{3 - (-1/\sqrt{2})^2} = \sqrt{\frac{5}{2}}$$

$$\ell_{31} = \frac{b_{31}}{\ell_{11}} = -\frac{1}{2\sqrt{2}}, \qquad \ell_{32} = \frac{1}{\ell_{22}}(b_{32} - \ell_{31}\ell_{21}) = -\frac{1}{2}\sqrt{\frac{5}{2}}$$

$$\ell_{33} = \sqrt{b_{33} - \ell_{31}^2 - \ell_{32}^2} = \frac{\sqrt{7}}{2}$$

So, the lower triangular matrix $\mathbf{L}$ is found as

$$\mathbf{L} = \begin{bmatrix} \sqrt{2} & 0 & 0 \\ -\dfrac{1}{\sqrt{2}} & \sqrt{\dfrac{5}{2}} & 0 \\ -\dfrac{1}{2\sqrt{2}} & -\dfrac{1}{2}\sqrt{\dfrac{5}{2}} & \dfrac{\sqrt{7}}{2} \end{bmatrix}$$

Now, we employ the algorithm for inverting triangular matrices to find $\mathbf{L}^{-1}$:

$$\ell_{11}^{-1} = \frac{1}{\ell_{11}} = \frac{1}{\sqrt{2}}, \qquad \ell_{22}^{-1} = \frac{1}{\ell_{22}} = \sqrt{\frac{2}{5}}, \qquad \ell_{33}^{-1} = \frac{1}{\ell_{33}} = \frac{2}{\sqrt{7}}$$

$$\ell_{21}^{-1} = -\frac{\ell_{22}^{-1}\ell_{21}}{\ell_{11}} = -\frac{(\sqrt{2/5})(-1/\sqrt{2})}{\sqrt{2}} = \frac{1}{\sqrt{10}}$$

$$\ell_{32}^{-1} = -\frac{\ell_{33}^{-1}\ell_{32}}{\ell_{22}} = -\frac{1}{\sqrt{5/2}}\left(\frac{2}{\sqrt{7}}\right)\left(-\frac{1}{2}\sqrt{\frac{5}{2}}\right) = \frac{1}{\sqrt{7}}$$

$$\ell_{31}^{-1} = -\frac{\ell_{32}^{-1}\ell_{21} + \ell_{33}^{-1}\ell_{31}}{\ell_{11}} = -\frac{1}{\sqrt{2}}\left\{\frac{1}{\sqrt{7}}\left(-\frac{1}{\sqrt{2}}\right) + \frac{2}{\sqrt{7}} - \frac{1}{2\sqrt{2}}\right\} = \frac{1}{\sqrt{7}}$$

In matrix form, we get

$$\mathbf{L}^{-1} = \begin{bmatrix} \dfrac{1}{\sqrt{2}} & 0 & 0 \\ \dfrac{1}{\sqrt{10}} & \sqrt{\dfrac{2}{5}} & 0 \\ \dfrac{1}{\sqrt{7}} & \dfrac{1}{\sqrt{7}} & \dfrac{2}{\sqrt{7}} \end{bmatrix}$$

Then, $\mathbf{C}$ is constructed as follows:

$$\mathbf{C} = \mathbf{L}^{-1}\mathbf{A}\mathbf{L}^{-T} = \begin{bmatrix} \dfrac{1}{\sqrt{2}} & 0 & 0 \\ \dfrac{1}{\sqrt{10}} & \sqrt{\dfrac{2}{5}} & 0 \\ \dfrac{1}{\sqrt{7}} & \dfrac{1}{\sqrt{7}} & \dfrac{2}{\sqrt{7}} \end{bmatrix} \begin{bmatrix} \dfrac{5}{3} & -1 & -\dfrac{2}{3} \\ -1 & 3 & -1 \\ -\dfrac{2}{3} & -1 & \dfrac{8}{3} \end{bmatrix} \begin{bmatrix} \dfrac{1}{\sqrt{2}} & \dfrac{1}{\sqrt{10}} & \dfrac{1}{\sqrt{7}} \\ 0 & \sqrt{\dfrac{2}{5}} & \dfrac{1}{\sqrt{7}} \\ 0 & 0 & \dfrac{2}{\sqrt{7}} \end{bmatrix}$$

$$= \begin{bmatrix} \dfrac{5}{6} & -\dfrac{1}{6\sqrt{5}} & -\dfrac{1}{3}\sqrt{\dfrac{2}{7}} \\ -\dfrac{1}{6\sqrt{5}} & \dfrac{29}{30} & -\dfrac{1}{3}\sqrt{\dfrac{2}{35}} \\ -\dfrac{1}{3}\sqrt{\dfrac{2}{7}} & -\dfrac{1}{3}\sqrt{\dfrac{2}{35}} & \dfrac{20}{21} \end{bmatrix}$$

Notice that matrix $\mathbf{C}$ is symmetric, which now allows us to apply the Jacobi method to the standard $\mathbf{C}\mathbf{z} = \lambda\mathbf{z}$ eigenvalue problem to compute its eigenpairs. (The implementation part of the Jacobi method is left as an exercise for the reader.)

As a result, the eigenvalues of $\mathbf{C}$ are found to be $\lambda_1 = 38/35$, $\lambda_2 = 1$, and $\lambda_3 = 2/3$. For resonance frequencies, this leads to $\omega = \sqrt{38/35LC}$, $\omega = 1/\sqrt{LC}$, and $\omega = \sqrt{2/3LC}$. The associated eigenvectors are found as

$$\mathbf{z}_1 = \begin{bmatrix} -\dfrac{1}{3}\sqrt{\dfrac{7}{2}} \\[2mm] -\dfrac{1}{3}\sqrt{\dfrac{7}{2}} \\[2mm] 1 \end{bmatrix}, \quad \mathbf{z}_2 = \begin{bmatrix} -\dfrac{1}{\sqrt{5}} \\[2mm] 1 \\[2mm] 0 \end{bmatrix}, \quad \mathbf{z}_3 = \begin{bmatrix} \dfrac{5}{\sqrt{14}} \\[2mm] \sqrt{\dfrac{5}{14}} \\[2mm] 1 \end{bmatrix}$$

The eigenvectors of the original system are determined by $\mathbf{x}_i = \mathbf{L}^{-T}\mathbf{z}_i$.

$$\mathbf{x}_1 = \begin{bmatrix} \dfrac{1}{\sqrt{2}} & \dfrac{1}{\sqrt{10}} & \dfrac{1}{\sqrt{7}} \\[2mm] 0 & \sqrt{\dfrac{2}{5}} & \dfrac{1}{\sqrt{7}} \\[2mm] 0 & 0 & \dfrac{2}{\sqrt{7}} \end{bmatrix} \begin{bmatrix} -\dfrac{1}{3}\sqrt{\dfrac{7}{2}} \\[2mm] -\dfrac{1}{3}\sqrt{\dfrac{7}{2}} \\[2mm] 1 \end{bmatrix} = \begin{bmatrix} -\dfrac{2}{5\sqrt{7}} \\[2mm] \dfrac{8}{15\sqrt{7}} \\[2mm] \dfrac{2}{\sqrt{7}} \end{bmatrix}$$

$$\mathbf{x}_2 = \begin{bmatrix} \dfrac{1}{\sqrt{2}} & \dfrac{1}{\sqrt{10}} & \dfrac{1}{\sqrt{7}} \\[2mm] 0 & \sqrt{\dfrac{2}{5}} & \dfrac{1}{\sqrt{7}} \\[2mm] 0 & 0 & \dfrac{2}{\sqrt{7}} \end{bmatrix} \begin{bmatrix} -\dfrac{1}{\sqrt{5}} \\[2mm] 1 \\[2mm] 0 \end{bmatrix} = \begin{bmatrix} 0 \\[2mm] \sqrt{\dfrac{2}{5}} \\[2mm] 0 \end{bmatrix}$$

$$\mathbf{x}_3 = \begin{bmatrix} \dfrac{1}{\sqrt{2}} & \dfrac{1}{\sqrt{10}} & \dfrac{1}{\sqrt{7}} \\[2mm] 0 & \sqrt{\dfrac{2}{5}} & \dfrac{1}{\sqrt{7}} \\[2mm] 0 & 0 & \dfrac{2}{\sqrt{7}} \end{bmatrix} \begin{bmatrix} \dfrac{5}{\sqrt{14}} \\[2mm] \sqrt{\dfrac{5}{14}} \\[2mm] 1 \end{bmatrix} = \begin{bmatrix} \dfrac{4}{\sqrt{7}} \\[2mm] \dfrac{2}{\sqrt{7}} \\[2mm] \dfrac{2}{\sqrt{7}} \end{bmatrix}$$

Note that the computed eigenvalues ($\mathbf{x}$'s) are not normalized. The eigenvectors are normalized as $\mathbf{v}_i = \mathbf{x}_i/\|\mathbf{x}_i\|$ if needed.

Discussion: The transformation of a generalized eigenvalue problem results in a lower triangular matrix, which is easy to obtain and invert. The procedure is also computationally inexpensive.

A pseudomodule, INVERSE_L, computing the inverse of any lower triangular matrix is presented in Pseudocode 11.4. The module requires a lower triangular matrix ($\mathbf{L}$) and its size (n) as input, and returns the inverse matrix, $\mathbf{IL}$ (its elements denoted by $i\ell_{ij}$), on exit. The inversion algorithm presented by Eq. (11.55) is simple and easy to program. Having evaluated $i\ell_{11}$, the rest of the elements are obtained using the two For-loops: (1) the i-loop sweeps rows from top ($i = 2$) all the way down to the last row ($i = n$), while the diagonal elements are inverted according to $i\ell_{ii} = 1/\ell_{ii}$; (2) the loop-j finds the elements of the ith row from right to left using Eq. (11.55). An accumulator, For-loop over index k, is used to perform the summation term that runs from $k = j + 1$ to $k = i$.

```
                          Pseudocode 11.4

  Module INVERSE_L ( n, L, IL)
  \ DESCRIPTION: A pseudomodule to find the inverse of a lower
  \   triangular matrix.
  Declare: ℓnn, iℓnn                          \ Declare array variables
  iℓ11 = 1/ℓ11                                 \ Invert diagonal element
  For [i = 2, n]
      iℓii = 1/ℓii                             \ Invert diagonal element
      For [j = (i − 1), 1, (−1)]
          sums ← 0                             \ Initialize accumulator
          For [k = (j + 1), i]
              sums ← sums + iℓik * ℓkj         \ Accumulate ℓik⁻¹ ℓkj
          End For
          iℓij = −sums/ℓjj                     \ Find Σ(k=j+1 to i) ℓik⁻¹ ℓkj/ℓjj
      End For
  End For
  End Module INVERSE_L
```

Module INVERSE_L (n, $\mathbf{L}$, $\mathbf{IL}$)

Declare: $\ell_{nn}, i\ell_{nn}$

$i\ell_{11} = 1/\ell_{11}$

For $[i = 2, n]$

$\quad i\ell_{ii} = 1/\ell_{ii}$

$\quad$ For $[j = (i-1), 1, (-1)]$

$\quad\quad sums \leftarrow 0$

$\quad\quad$ For $[k = (j+1), i]$

$\quad\quad\quad sums \leftarrow sums + i\ell_{ik} * \ell_{kj}$

$\quad\quad$ End For

$\quad\quad i\ell_{ij} = -sums/\ell_{jj}$ $\quad$ \ Find $\sum_{k=j+1}^{i} \ell_{ik}^{-1}\ell_{kj}/\ell_{jj}$

$\quad$ End For

End For

End Module INVERSE_L

                          Cholesky Decomposition

- Cholesky decomposition is a numerically stable and computationally efficient method;
- Its memory requirements are considerably less;
- It is highly adaptive to parallel architectures; therefore, pivoting is not required;
- It is faster than the LU decomposition as it exploits the symmetry of the matrix;
- It can be used for solving linear equations with symmetric positive and definite matrices; it requires half of the number of operations required by Gaussian elimination.

- The method does not work for any matrix, except for symmetric positive definite matrices.

11.6 HOUSEHOLDER METHOD

Finding the eigenvalues of especially dense matrices can be computationally very expensive. Alternatively, the eigenvalues of a dense matrix are determined much more efficiently by reducing it to a tridiagonal matrix through a series of similarity transformations to overcome the large cpu requirements.

Suppose $\mathbf{A}$ is a well-conditioned symmetric matrix and has n eigenvalues with n linearly independent eigenvectors. The main objective of the *Householder method* is to transform a symmetric matrix into a symmetric tridiagonal matrix having identical eigenpairs. Note that the Householder method does not give the eigenpairs directly; the analyst needs to apply another method for computing the eigenpairs of a symmetric tridiagonal matrix as well.

In this method, orthogonal transformations are sequentially applied to zero out the pertinent row and column elements all at once, except for the tridiagonal elements. Moreover, the elements that are zeroed out in prior steps are not altered in the subsequent operations, unlike the Jacobi method. This algorithm, which is not an iterative one, requires $n - 2$ orthogonal transformations.

The transformations are expressed as

$$\mathbf{A}_1 = \mathbf{A}$$
$$\mathbf{A}_2 = \mathbf{P}_1 \mathbf{A}_1 \mathbf{P}_1$$
$$\vdots$$
$$\mathbf{A}_{n-1} = \mathbf{P}_{n-2} \mathbf{A}_{n-2} \mathbf{P}_{n-2}$$

where n is the size of $\mathbf{A}$, and $\mathbf{P}_k$ is the transformation matrix at the kth step, defined as

$$\mathbf{P} = \mathbf{I} - 2\,\mathbf{v}\mathbf{v}^T$$

where the matrix $\mathbf{P}$, which is symmetric and orthogonal, is called the *Householder matrix*, and $\mathbf{v}$ is a unit vector. Recalling that any vector $\mathbf{u}$ can be normalized to give a unit vector $(\mathbf{v} = \mathbf{u}/|\mathbf{u}|)$ in the direction of $\mathbf{u}$, the Householder matrix can be constructed as

$$\mathbf{P}_m = \mathbf{I} - \frac{\mathbf{u}_m \mathbf{u}_m^T}{(\mathbf{u}_m, \mathbf{u}_m)/2} \tag{11.56}$$

where $(\mathbf{u}, \mathbf{u}) = |\mathbf{u}|^2$.

We construct the vector $\mathbf{u}$, whose first $(m - 1)$ elements as zero, that is,

$$\mathbf{u}_m = \left[0,\ 0,\ \ldots,\ 0,\ u_m^{(m)},\ u_m^{(m+1)},\ \ldots,\ u_m^{(n)} \right]^T$$

or, in short,

$$\mathbf{u} = \mathbf{x} \mp |\mathbf{x}|\,\mathbf{e}_1 \tag{11.57}$$

where $\mathbf{x}$ is the lower $m - 1$ elements of the first column of the Householder matrix and $\mathbf{e}_1$ is defined as the unit vector (i.e., the first column of the identity matrix of size $m - 1$). Then, with Eq. (11.56), we define

$$\mathbf{A}_1 = \mathbf{A}, \qquad \mathbf{A}_m = \mathbf{P}_m^T \mathbf{A}_{m-1} \mathbf{P}_m, \qquad m = 2, 3, \ldots, n - 1 \tag{11.58}$$

where $\mathbf{A}_m$ is a partially tridiagonal up to row m:

$$\mathbf{A}_m = \left[\begin{array}{ccccc:cc}
a_{11} & a'_{12} & & & & & \\
a'_{12} & a'_{2}2 & a'_{23} & & & & \\
& \ddots & \ddots & \ddots & & a'_{m(m+1)} & \\
\hdashline
& & a'_{m(m-1)} & a'_{mm} & & & \\
& & & a'_{m(m+1)} & & & \\
& & & & & & \mathbf{A}_{n-m} \\
\end{array} \right]$$

The matrix $\mathbf{P}_m$ has the form

$$\mathbf{P}_m = \mathbf{I} - 2\mathbf{v}_m \mathbf{v}_m^T = \left[\begin{array}{c:ccc}
\mathbf{I}_m & 0 & \cdots & 0 \\
\hdashline
0 & & & \\
\vdots & & \mathbf{P}_{n-m} & \\
0 & & & \\
\end{array} \right]$$

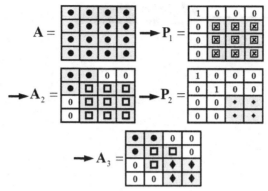

FIGURE 11.4: Depiction of the matrix operations in Householder method.

The objective of this transformation is to obtain $\mathbf{u}$ such that it satisfies $|\mathbf{v}| = 1$. Setting $\mathbf{u} = \mathbf{x} + S\,\mathbf{e}_1$, where $S = \pm|\mathbf{x}|$, we find

$$S = \text{sgn}(a_{m+1,m})\sqrt{\sum_{i=m+1}^{n} a_{im}^2} \tag{11.59}$$

and

$$d = \frac{1}{2}(\mathbf{u}_m, \mathbf{u}_m) = \frac{1}{2}(\mathbf{x} + S\,\mathbf{e}_1, \mathbf{x} + S\,\mathbf{e}_1) = S(S + a_{m+1,m}) \tag{11.60}$$

where sgn is the sign function.

We can now construct the elements of the vector $\mathbf{u}_m$ as follows:

$$u_{im} = \begin{cases} 0, & \text{if } i = 1, 2, \ldots, m \\ a_{im} + S, & \text{if } i = m+1 \\ a_{im}, & \text{if } i = (m+2), \ldots, n \end{cases} \tag{11.61}$$

The orthogonal matrix $\mathbf{P}$ (for $i, j = (k+1), (k+2), \ldots, (n-1)n$, it is embedded in matrix $\mathbf{T}$).

$$p_{ij} = \delta_{ij} - \frac{u_i u_j}{d}, \quad i, j = (m+1), (m+2), \ldots, (n-1)n \tag{11.62}$$

where δ_{ij} is Kronecker delta, i.e., $\delta_{ii} = 1$ and $\delta_{ij} = 0$ if $i \neq j$. It should be noted that the first m rows and m columns remain unaffected by the similarity transformations. Additionally, we can restrict the calculations to the elements in the upper triangle of $\mathbf{A}_m$ since the transformation preserves the symmetry.

Householder Method

- Householder method is fast, accurate, and cpu-time efficient;
- It always works for any nonsingular symmetric matrix;
- It is stable and does not require pivoting.

- Procedure is bandwidth-intensive heavy (diagonally dense) and not parallelizable;
- Every reflection modifies the matrices $\mathbf{P}$ and $\mathbf{A}$.

Pseudocode 11.5

Module HOUSEHOLDER $(n, \mathbf{A})$
\ DESCRIPTION: A pseudomodule transforming a positive definite and symmetric
\   matrix into a symmetric tridiagonal matrix.
\ USES:
\   SIGN:: A built-in Function returning the sign of a real number;
\   MAT_MUL:: Module for computing matrix multiplication, Pseudocode 2.4.
Declare: $a_{nn}, u_n, p_{nn}, c_{nn}$
For $\left[k = 1, (n-2)\right]$ \ Triangularize the first $n-2$ rows
  $m \leftarrow k + 1$ \ Find index of below diagonal element
  $S \leftarrow 0$ \ Initialize accumulator
  For $\left[i = m, n\right]$
   $S \leftarrow S + a_{ik}^2$ \ Accumulate a_{ik}^2 for lower elements
  End For
  $S \leftarrow \mathbf{SIGN}(a_{mk})\sqrt{S}$ \ Find S, Eq. (11.51)
  $d \leftarrow S * (S + a_{mk})$ \ Find d, Eq. (11.52))
  $\mathbf{u} = 0$
  $u_m \leftarrow a_{mk} + S$ \ Set the first element of vector $\mathbf{u}$
  For $\left[i = (m+1), n\right]$
   $u_i \leftarrow a_{ik}$ \ Set rest of the elements of $\mathbf{u}$
  End For
  For $\left[i = m, n\right]$
   $p_{ki} \leftarrow 0; \; p_{ik} \leftarrow 0$ \ Construct transformation matrix $\mathbf{T}$
  End For
  $p_{kk} \leftarrow 1$
  For $\left[i = m, n\right]$
   For $\left[j = m, n\right]$ \ Construct Householder matrix $\mathbf{P}$
    $p_{ij} \leftarrow 0$
    If $\left[i = j\right]$ **Then**
     $p_{ij} \leftarrow 1 - u_i * u_j/d$ \ Construct diagonal elements
    Else
     $p_{ij} \leftarrow -u_i * u_j/d$ \ Construct off-diagonal elements
    End If
   End For
  End For
  MAT_MUL$(n,n,n,\mathbf{P},\mathbf{A},\mathbf{C})$ \ Utilize Pseudocode 2.4 to find $\mathbf{C} = \mathbf{PA}$
  MAT_MUL$(n,n,n,\mathbf{C},\mathbf{P},\mathbf{A})$ \ Utilize Pseudocode 2.4 to find $\mathbf{A} = \mathbf{CP}$
End For
End Module HOUSEHOLDER

A pseudomodule, HOUSEHOLDER, transforming a positive definite symmetric matrix into a symmetric tridiagonal matrix is presented in Pseudocode 11.5. This module requires a square matrix $(\mathbf{A})$ of order n as input. On exit, the transformed (triangularized) matrix is stored on $\mathbf{A}$. The Householder process is depicted in Fig. 11.4. Using Eqs. (11.62), the orthogonal matrix $\mathbf{P}_1$, a full matrix of order $(n-1) \times (n-1)$, is constructed. Then, another orthogonal transformation is applied: $\mathbf{A}_2 = \mathbf{P}_1\mathbf{A}_1\mathbf{P}_1$. This leads to the tridiagonalization of the first row while modifying the rest of the elements in $\mathbf{A}$. Next, an orthogonal matrix $\mathbf{P}_2$, $(n-2) \times (n-2)$, is constructed, and then the transformation matrix $\mathbf{A}_3$ is obtained. This procedure is repeated successively for $n-2$ rows of $\mathbf{A}$. The final transformed matrix $\mathbf{A}_m$ is a tridiagonal matrix. The original matrix $\mathbf{A}$ is destroyed during the process.

EXAMPLE 11.9: Application of Householder method

Use the Householder method to transform $\mathbf{A}$ into a tridiagonal matrix.

$$\mathbf{A} = \begin{bmatrix} 6 & 1 & 2 & 2 \\ 1 & 2 & 3 & 1 \\ 2 & 3 & 11 & 2 \\ 2 & 1 & 2 & 3 \end{bmatrix}$$

SOLUTION:

First, we need to compute S to construct $\mathbf{P}$; thus, we find

$$S = \text{sgn}(a_{21})\sqrt{a_{21}^2 + a_{31}^2 + a_{41}^2} = +\sqrt{1^2 + 2^2 + 2^2} = 3$$

The vector $\mathbf{u}$ is then constructed. Since $a_{21} = 1$, we write

$$\mathbf{u} = \begin{bmatrix} 1+3 \\ 2 \\ 2 \end{bmatrix}$$

Then, $\mathbf{u}\mathbf{u}^T$ is found as

$$\mathbf{u}\mathbf{u}^T = \begin{bmatrix} 4 \\ 2 \\ 2 \end{bmatrix} \begin{bmatrix} 4 & 2 & 2 \end{bmatrix} = \begin{bmatrix} 16 & 8 & 8 \\ 8 & 4 & 4 \\ 8 & 4 & 4 \end{bmatrix}$$

Evaluating $d = S(S + a_{21}) = 3(3 + 1) = 12$ and substituting $\mathbf{u}\mathbf{u}^T$ into Eq. (11.56) we get

$$\mathbf{I} - \frac{1}{d}\mathbf{u}\mathbf{u}^T = \begin{bmatrix} 1 & 0 & 0 \\ 0 & 1 & 0 \\ 0 & 0 & 1 \end{bmatrix} - \frac{1}{12}\begin{bmatrix} 16 & 8 & 8 \\ 8 & 4 & 4 \\ 8 & 4 & 4 \end{bmatrix} = -\frac{1}{3}\begin{bmatrix} 1 & 2 & 2 \\ 2 & -2 & 1 \\ 2 & 1 & -2 \end{bmatrix}$$

Now, the first transformation matrix is found as

$$\mathbf{P}_1 = \begin{bmatrix} 1 & 0 & 0 & 0 \\ 0 & -\dfrac{1}{3} & -\dfrac{2}{3} & -\dfrac{2}{3} \\ 0 & -\dfrac{2}{3} & \dfrac{2}{3} & -\dfrac{1}{3} \\ 0 & -\dfrac{2}{3} & -\dfrac{1}{3} & \dfrac{2}{3} \end{bmatrix}$$

Then the orthogonal transformation (i.e., $\mathbf{P}_1\mathbf{A}\mathbf{P}_1$ product) yields

$$\mathbf{A}_2 = \mathbf{P}_1\mathbf{A}_1\mathbf{P}_1 = \begin{bmatrix} 6 & -3 & 0 & 0 \\ -3 & 10 & -3 & 3 \\ 0 & -3 & 3 & -2 \\ 0 & 3 & -2 & 3 \end{bmatrix}$$

The transformation matrix $\mathbf{P}_2$ is constructed by repeating the procedure above.

Noting that the first element of $\mathbf{x}$ (the first column of $\mathbf{A}_2$, which is a 3×3 matrix) is now a_{32}, S and d are computed as

$$S = \text{sgn}(a_{32})\sqrt{a_{32}^2 + a_{42}^2} = (-1)\sqrt{3^2 + 3^2} = -3\sqrt{2}$$

$$d = S(S + a_{32}) = 3\sqrt{2}(3\sqrt{2} + 3) = 9(2 + \sqrt{2})$$

Next, $\mathbf{u}$ and $\mathbf{uu}^T$ yield

$$\mathbf{u} = \begin{bmatrix} -3 \pm 3\sqrt{2} \\ 3 \end{bmatrix} = \begin{bmatrix} -3 - 3\sqrt{2} \\ 3 \end{bmatrix}$$

$$\mathbf{uu}^T = \begin{bmatrix} -3 - 3\sqrt{2} \\ 3 \end{bmatrix} \begin{bmatrix} -3 - 3\sqrt{2} \\ 3 \end{bmatrix}^T = 9\begin{bmatrix} 3 + 2\sqrt{2} & -(1 + \sqrt{2}) \\ -(1 + \sqrt{2}) & 1 \end{bmatrix}$$

And lastly, matrix $\mathbf{P}$ is formed.

$$\mathbf{I} - \frac{1}{d}\mathbf{uu}^T = \begin{bmatrix} 1 & 0 \\ 0 & 1 \end{bmatrix} - \frac{9}{18 + 9\sqrt{2}}\begin{bmatrix} 3 + 2\sqrt{2} & -(1 + \sqrt{2}) \\ -(1 + \sqrt{2}) & 1 \end{bmatrix} = \frac{1}{\sqrt{2}}\begin{bmatrix} -1 & 1 \\ 1 & 1 \end{bmatrix}$$

Embedding this into matrix $\mathbf{P}_2$ yields

$$\mathbf{P}_2 = \begin{bmatrix} 1 & 0 & 0 & 0 \\ 0 & 1 & 0 & 0 \\ 0 & 0 & -\dfrac{1}{\sqrt{2}} & \dfrac{1}{\sqrt{2}} \\ 0 & 0 & \dfrac{1}{\sqrt{2}} & \dfrac{1}{\sqrt{2}} \end{bmatrix}$$

Finally, the triple product matrix $\mathbf{P}_2\mathbf{A}_2\mathbf{P}_2$ is found as

$$\mathbf{A}_3 = \mathbf{P}_2\mathbf{A}_2\mathbf{P}_2 = \begin{bmatrix} 6 & -3 & 0 & 0 \\ -3 & 10 & 3\sqrt{2} & 0 \\ 0 & 3\sqrt{2} & 5 & 0 \\ 0 & 0 & 0 & 1 \end{bmatrix}$$

In this example, the tridiagonal matrix coincidentally yielded $a_{43} = a_{34} = 0$.

Discussion: The Householder method reduces a full symmetric matrix into a symmetric tridiagonal matrix to be processed later on. This procedure is numerically stable and does not require pivoting, which also makes it a candidate for solving systems of linear equations. In this regard, the method is much more stable than the Gauss elimination method; however, computationally, it is more expensive than the Gaussian elimination method.

11.7 EIGENVALUES OF TRIDIAGONAL MATRICES

Computing the eigenvalues of a large symmetric matrix by conventional methods is not desired due to the large memory requirement and the accumulation of round-off errors as a result of numerous transformations, resulting in a consequently large cpu-time. In practice, to circumvent these adverse effects, the original matrix is first reduced to a symmetric tridiagonal form with *Givens* or *Householder's methods*. The eigenvalues and/or eigenvectors of a symmetric tridiagonal matrix are then determined using a suitable method, some of which will be discussed in this section.

11.7.1 THE STURM SEQUENCE

Consider a tridiagonal symmetric matrix $\mathbf{A}$, defined as

$$\mathbf{A} = \begin{bmatrix} d_1 & e_1 & & & \\ e_1 & d_2 & e_2 & & \\ & \ddots & \ddots & \ddots & \\ & & e_{n-2} & d_{n-1} & e_{n-1} \\ & & & e_{n-1} & d_n \end{bmatrix} \tag{11.63}$$

Furthermore, suppose $e_k \neq 0$ for $k = 1, 2, \ldots, n$, which ensures no repeating eigenvalues.

The eigenvalues of $\mathbf{A}$ are determined by finding the roots of the characteristic polynomial:

$$\det(\mathbf{A} - \lambda\mathbf{I}) = p_n(\lambda) = \begin{vmatrix} d_1 - \lambda & e_1 & & & \\ e_1 & d_2 - \lambda & e_2 & & \\ & \ddots & \ddots & \ddots & \\ & & e_{n-2} & d_{n-1} - \lambda & e_{n-1} \\ & & & e_{n-1} & d_n - \lambda \end{vmatrix} = 0 \tag{11.64}$$

where $p_n(\lambda)$ is a polynomial of degree n.

First, we consider a 2×2 matrix whose characteristic polynomial is

$$p_2(\lambda) = (d_1 - \lambda)(d_2 - \lambda) - e_1^2 = 0 \tag{11.65}$$

Defining $p_0(\lambda) = 1$ and $p_1(\lambda) = d_1 - \lambda$, Eq. (11.65) can also be rewritten as

$$p_2(\lambda) = (d_2 - \lambda)p_1(\lambda) - e_1^2 p_0(\lambda) = 0 \tag{11.66}$$

For a 3×3 matrix, we find

$$p_3(\lambda) = \begin{vmatrix} d_1 - \lambda & e_1 & 0 \\ e_1 & d_2 - \lambda & e_2 \\ 0 & e_2 & d_3 - \lambda \end{vmatrix} = (d_3 - \lambda)p_2(\lambda) - e_3^2 p_1(\lambda) = 0 \tag{11.67}$$

For increasing orders of a matrix, a recurrence relationship between the determinants for $2 \leqslant k \leqslant n$ can be established as follows:

$$p_k(\lambda) = (d_k - \lambda)p_{k-1}(\lambda) - e_{k-1}^2 p_{k-2}(\lambda) = 0 \tag{11.68}$$

which is the characteristic polynomial of a symmetric tridiagonal matrix. It can be shown

---

Sturm Sequence Method

- The method is simple enough to not require elaborate matrix algebra;
- It is relatively easy to implement;
- It can be used to estimate a specified eigenvalue (smallest, largest, or $\lambda = \alpha$ in $\lambda_a < \alpha < \lambda_b$);
- Once an eigenvalue (i.e., root) is bracketed, the bisection method is certain to converge to an approximate value.

- The method is not suitable for very large matrices;
- It is restricted to symmetric tridiagonal matrices only;
- It provides the eigenvalues only (not the eigenvectors);
- It is susceptible to round-off errors even for fairly large matrices;
- Narrowing down the roots with bisections can be expensive for a widely or tightly dispersed roots due to its linear convergence rate.

---

that the sequence, $p_0(\lambda)$, $p_1(\lambda)$, $p_2(\lambda)$, $\ldots, p_n(\lambda)$, also referred to as the *Sturm sequence* (provided that $e_k \neq 0$), can be used to determine the root of $p_n(\lambda)$ in any given interval.

Sturm's method is applied to a symmetric tridiagonal matrix to obtain its eigenvalues from its characteristic equation. The procedure involves selecting trial λ values and determining the number of sign changes to narrow down or update the upper and lower bounds for the roots. For example, let the Sturm-sequence and the corresponding sequence of signs for $\lambda = \alpha$ be given as follows:

$$\{p_0(\alpha), p_1(\alpha), p_2(\alpha), \cdots, p_n(\alpha)\} \quad \rightarrow \quad \{+, -, +, \cdots, +\}$$

We will denote the number of sign changes up to $\lambda = \alpha$ by $N(\alpha)$; for instance, $N(\alpha) = M$ indicates that $p_n(\lambda)$ has M roots on $-\infty < \lambda < \alpha$. This gives the number of distinct eigenvalues of a tridiagonal matrix lying between any two numbers. If we let the number of sign changes corresponding to λ_1 and λ_2 be $N(\lambda_1) = N_1$ and $N(\lambda_2) = N_2$, then the number of roots within (λ_1, λ_2) can be found as $N(\lambda_2) - N(\lambda_1) = N_2 - N_1$. Recalling the Gerschgorin theorem, the lower and upper bounds of the roots of $p_n(\lambda)$ may be found from

$$\lambda_{\min} = \min_{1 \leqslant k \leqslant n} \left(d_k - |e_{k-1}| - |e_k| \right), \qquad \lambda_{\max} = \max_{1 \leqslant k \leqslant n} \left(d_k + |e_{k-1}| + |e_k| \right) \qquad (11.69)$$

where $e_0 = e_n = 0$.

First, the upper and lower bounds covering all eigenvalues are determined, i.e., $(\lambda_{\min}, \lambda_{\max})$. Then, individual roots are isolated by further dividing the $(\lambda_{\min}, \lambda_{\max})$ interval into smaller $(\lambda_{k+1}, \lambda_k)$ subintervals (panels) such that each root is contained in a panel, i.e., $N(\lambda_{k+1}) - N(\lambda_k) = 1$. Then, the bisection method is employed to narrow down the root to the desired accuracy.

As a word of caution, locating the roots (even if they are isolated) with the Newton-Raphson method is too risky in that a root estimate may fall outside of the isolated interval and converge to a root in another interval. However, with the bisection method, the successive estimates are guaranteed to remain within the bracketed interval.

EXAMPLE 11.10: Isolating the eigenvalues using the Sturm sequence

Isolate the eigenvalues of the following matrix:

$$\mathbf{A} = \begin{bmatrix} 2 & -1 & 0 & 0 \\ -1 & 1 & -1 & 0 \\ 0 & -1 & 1 & 2 \\ 0 & 0 & 2 & 2 \end{bmatrix}$$

SOLUTION:

Let us generate the Sturm sequence of matrix $\mathbf{A}$ using Eq. (11.65)–(11.68). Starting with $p_0(\lambda) = 1$, the Sturm sequence yields

$$p_1(\lambda) = (2 - \lambda)$$
$$p_2(\lambda) = (1 - \lambda)p_1(\lambda) - p_0(\lambda) = \lambda^2 - 3\lambda + 1$$
$$p_3(\lambda) = (1 - \lambda)p_2(\lambda) - 4p_1(\lambda) = -\lambda^3 + 4\lambda^2 - 3\lambda - 1$$
$$p_4(\lambda) = (2 - \lambda)p_3(\lambda) - p_2(\lambda) = \lambda^4 - 6\lambda^3 + 7\lambda^2 + 7\lambda - 6$$

Using Eq. (11.69), the minimum and maximum values of each row are determined to be $(1,3)$, $(-1,3)$, $(-2,4)$, and $(0,4)$, respectively. Hence, we find out that all roots of the characteristic polynomial are contained within $(-2,4)$.

The next step is to isolate each root in this interval. For this purpose, we will count the sign changes in the Sturm sequences to isolate individual roots. Table 11.1 depicts the Sturm sequence for various λ's within $-2 \leqslant \lambda \leqslant 4$. Note that for $\lambda = -2$, the sequence does not depict any sign change, i.e., $N(-2) = 0$ indicates no root in $(-\infty, -2]$. For $\lambda = 0$, the sequence changes sign at p_3, resulting in $N(0) = 1$, which implies the presence of a single root in $(-2,0)$ ($N(0) - N(-2) = 1$). For $\lambda = 1$, which yields $N(1) = 2$, the Sturm sequence exhibits two sign changes, one at p_2 and the other at p_4. The number of sign changes between $(-2,1)$ and $(0,1)$ intervals is $N(1) - N(-2) = 2$ and $N(1) - N(0) = 1$, respectively. For $\lambda = 3$, we observe three sign changes on $(-\infty,3]$, i.e., $N(3) = 3$. The number of sign changes in $(-2,3)$ and $(1,3)$ is found by $N(3) - N(-2) = 3$ and $N(3) - N(1) = 1$, respectively. Lastly, for $\lambda = 4$, we obtain $N(4) = 4$, indicating that the last root is in $(3,4)$.

TABLE 11.1

λ	$p_0(\lambda)$	$p_1(\lambda)$	$p_2(\lambda)$	$p_3(\lambda)$	$p_4(\lambda)$	Explanation
-2	1	4	11	29	72	No sign change; No eigenvalue smaller than -2
0	1	2	1	-1	-6	1 sign change in $(-2,0)$; 1 eigenvalue in $(-2,0)$ interval
1	1	1	-1	-1	3	2 sign changes in $(-2,0)$; 2 eigenvalues. Second one is in $(0,1)$.
3	1	-1	1	-1	-3	3 sign changes in $(-2,0)$; 3 eigenvalues. Third one is in $(1,3)$.
4	1	-2	5	-13	6	4 sign changes in $(-2,0)$; the last one is in $(3,4)$.

Discussion: This technique can be applied not only to find the roots (eigenvalues) of the characteristic polynomial but also to isolate and find the real roots of any real polynomial.

11.7.2   THE QR ITERATION

Finding the eigenvalues via the Sturm sequence can be impractical when a tridiagonal matrix is very large. Hence, a tridiagonal symmetric matrix is usually reduced to a diagonal matrix by employing the method of QL or *QR decomposition*. The goal is to decompose the matrix into the product of an orthogonal matrix (denoted by $\mathbf{Q}$) and a lower ($\mathbf{L}$) or upper ($\mathbf{R}$) triangular matrix.

Consider the following symmetric tridiagonal matrix:

$$\mathbf{A} = \begin{bmatrix} d_1 & e_1 & & & & \\ e_1 & d_2 & e_2 & & & \\ & e_2 & d_3 & e_3 & & \\ & & & \ddots & & \\ & & & e_{n-2} & d_{n-1} & e_{n-1} \\ & & & & e_{n-1} & d_n \end{bmatrix} \tag{11.70}$$

If e_1 or e_{n-1} is zero, then one of the eigenvalues of $\mathbf{A}$ is d_1 or d_n, respectively. The *QR iteration method* exploits this property of tridiagonal matrices by successively eliminating the sub-diagonal elements using the following two steps:

$$\left.\begin{array}{l} \text{Decomposition of } \mathbf{A} \text{ as } \mathbf{A}_p = \mathbf{Q}_p \mathbf{R}_p \\ \text{Construction of } \mathbf{A}_{p+1} \text{ as } \mathbf{A}_{p+1} = \mathbf{R}_p \mathbf{Q}_p \end{array}\right\} \text{ for } p = 0, 1, 2, \ldots$$

where $\mathbf{A}_0 = \mathbf{A}$ is the original symmetric tridiagonal matrix, $\mathbf{Q}_p$ is an orthogonal matrix, and $\mathbf{R}_p$ is an upper triangular matrix.

The objective of this numerical method is to apply a series of orthogonal transformations to obtain a sequence of similar matrices $\mathbf{A}_p$ $(p = 1, 2, \ldots)$ that, for sufficiently large p $(p \to \infty)$, eventually converge to a diagonal matrix $(\mathbf{A}_\infty)$ whose diagonal elements are the eigenvalues of $\mathbf{A}$. In this process, after each similarity transformation, not only the tridiagonal structure and the symmetry property but also the eigenvalues of $\mathbf{A}$ are preserved.

A decomposed matrix is expressed by

$$\mathbf{A}_p = \begin{bmatrix} d_1^{(p)} & e_1^{(p)} & & & \\ e_1^{(p)} & d_2^{(p)} & e_2^{(p)} & & \\ & \ddots & \ddots & \ddots & \\ & & e_{n-2}^{(p)} & d_{n-1}^{(p)} & e_{n-1}^{(p)} \\ & & & e_{n-1}^{(p)} & d_n^{(p)} \end{bmatrix} \tag{11.71}$$

where $\mathbf{A}_0$ is the original matrix and the matrices with $p > 1$ denote degenerated or modified matrices, $d_k^{(p)}$ and $e_k^{(p)}$ denote the modified matrix elements at the pth step.

We begin by setting $\mathbf{A}_1 = \mathbf{A}$ as the product of an orthogonal ($\mathbf{Q}_1$) and an upper triangular matrix ($\mathbf{R}_1$): $\mathbf{Q}_1 \mathbf{R}_1$. Then $\mathbf{A}_2 = \mathbf{R}_1 \mathbf{Q}_1$ is constructed. To generalize, we define two matrix products: $\mathbf{A}_p = \mathbf{Q}_p \mathbf{R}_p$ and $\mathbf{A}_{p+1} = \mathbf{R}_p \mathbf{Q}_p$. Since $\mathbf{Q}_p$ is orthogonal (i.e., $\mathbf{Q}_p^{-1} = \mathbf{Q}_p^T$ and $\mathbf{Q}_p^{-1}\mathbf{Q} = \mathbf{Q}_p^T\mathbf{Q} = \mathbf{I}$) and making use of $\mathbf{R}_p = \mathbf{Q}_p^T \mathbf{A}_p$, we find

$$\mathbf{A}_{p+1} = \mathbf{R}_p \mathbf{Q}_p = \mathbf{Q}_p^T \mathbf{A}_p \mathbf{Q}_p \tag{11.72}$$

This relationship clearly demonstrates that $\mathbf{A}_{p+1}$ is symmetric and shares the same eigenvalues as $\mathbf{A}_p$ and $\mathbf{A}$. By defining proper $\mathbf{R}_p$ and $\mathbf{Q}_p$, $\mathbf{A}_{p+1}$ is assured to be tridiagonal.

The essence of the QR iteration method is the computation of the QR decomposition of $\mathbf{A}_p$. Due to the symmetry, we deal with the diagonal (d_k) and sub-diagonal (e_k) elements only. The decomposition is carried out by a so-called rotation matrix. For this purpose, we define an orthogonal matrix, $\mathbf{P}_{(k,k+1)}$, which is basically an identity matrix whose $(k-1)$th and kth rows and columns are modified for an arbitrary rotation angle θ as follows:

$$
\mathbf{P}_{(k,k+1)} =
\begin{bmatrix}
1 & & & & & & \\
& \ddots & & & & & \\
& & 1 & & & & \\
\hline
& & & \cos\theta & \sin\theta & & \\
& & & -\sin\theta & \cos\theta & & \\
\hline
& & & & & 1 & \\
& & & & & & \ddots \\
& & & & & & & 1
\end{bmatrix}
\quad
\begin{matrix} r_{k-1} \\ r_k \end{matrix}
\tag{11.73}
$$

This matrix represents the rotation of the $(k-1)$th and kth axes about the origin of the coordinate system by the angle θ. The right-multiplication of $\mathbf{P}_{(k,k+1)}$ with an arbitrary $\mathbf{M}$ (i.e., $\mathbf{P}_{(k,k+1)}\mathbf{M}$) affects only the $(k-1)$ and kth rows and columns. From this point on, for notational convenience, as we have done so in the Householder method, we will denote the rotation sine and cosines with s and c, respectively.

The decomposition of matrix $\mathbf{A}_k$ is handled in the usual manner. For a tridiagonal matrix of order n, the sub-diagonal elements are zeroed with $n-1$ rotations. In order to illustrate the QR iteration steps and develop a general algorithm, a 4×4 symmetric tridiagonal matrix $\mathbf{A}$ is considered:

$$
\mathbf{A} =
\begin{bmatrix}
d_1 & e_1 & & \\
e_1 & d_2 & e_2 & \\
& e_2 & d_3 & e_3 \\
& & e_3 & d_4
\end{bmatrix}
$$

For the first rotation, $\mathbf{P}_{(1,2)}$ is chosen such that $\mathbf{P}_{(1,2)}\mathbf{A}_p$ yields zero at the second row, first column. The product $\mathbf{P}_{(1,2)}\mathbf{A}_0$ yields

$$
\begin{aligned}
\mathbf{P}_{(1,2)}\mathbf{A}_0 &=
\begin{bmatrix}
c_1 & s_1 & \\
-s_1 & c_1 & \\
& & 1 & \\
& & & 1
\end{bmatrix}
\begin{bmatrix}
d_1 & e_1 & & \\
e_1 & d_2 & e_2 & \\
& e_2 & d_3 & e_4 \\
& & e_3 & d_4
\end{bmatrix} \\[2mm]
&=
\begin{bmatrix}
c_1 d_1 + e_1 s_1 & c_1 e_1 + d_2 s_1 & e_2 s_1 & \\
c_1 e_1 - d_1 s_1 & c_1 d_2 - e_1 s_1 & c_1 e_2 & \\
& e_2 & d_3 & e_4 \\
& & e_3 & d_4
\end{bmatrix}
\end{aligned}
\tag{11.74}
$$

The second row first column element is set to zero, i.e., $c_1 e_1 - d_1 s_1 = 0$. Provided that $c_1^2 + s_1^2 = 1$, the rotation cosine and sine are easily obtained (by solving the two equations with two unknowns) as:

$$
\rho^{(1)} = \sqrt{e_1^2 + d_1^2}, \quad c_1 = d_1/\rho^{(1)}, \quad s_1 = e_1/\rho^{(1)}
\tag{11.75}
$$

Equation (11.74) now takes the following form:

$$
\mathbf{P}_{(1,2)}\mathbf{A}_0 =
\left[
\begin{array}{cc|c}
\rho^{(1)} & (d_1+d_2)e_1/\rho^{(1)} & e_1e_2/\rho^{(1)} \\
0 & (d_1d_2-e_1^2)/\rho^{(1)} & d_1e_2/\rho^{(1)} \\
\hline
e_2 & d_3 & e_3 \\
& e_3 & d_4
\end{array}
\right]
=
\left[
\begin{array}{cc|c}
d_1^{(1)} & e_1^{(1)} & f_1 \\
0 & d_2^{(1)} & e_2^{(1)} \\
\hline
e_2 & d_3 & e_3 \\
& e_3 & d_4
\end{array}
\right]
$$

(11.76)

where $d_1^{(1)} = \rho^{(1)}$, $d_2^{(1)} = (d_1d_2 - e_1^2)/\rho^{(1)}$, $f_1 = e_1e_2/\rho^{(1)}$, and $e_2^{(1)} = d_1e_2/\rho^{(1)}$, where the superscripts denote degenerated elements.

The second rotation $\mathbf{P}_{(2,3)}$ is chosen such that $\mathbf{P}_{(2,3)}\mathbf{P}_{(1,2)}\mathbf{A}_p$ gives zero at the third row, second column. The matrix product $\mathbf{P}_{(2,3)}\mathbf{P}_{(1,2)}\mathbf{A}_p$ can be explicitly written as

$$
\mathbf{P}_{(2,3)}\mathbf{P}_{(1,2)}\mathbf{A}_p =
\left[
\begin{array}{c|cc|c}
1 & & & \\
\hline
& c_2 & s_2 & \\
& -s_2 & c_2 & \\
\hline
& & & 1
\end{array}
\right]
\mathbf{P}_{(1,2)}\mathbf{A}_p
$$

$$
=
\left[
\begin{array}{c|cc|c}
d_1^{(1)} & e_1^{(1)} & f_1 & \\
\hline
0 & e_2s_2 + c_2d_2^{(1)} & d_3s_2 + c_2e_2^{(1)} & s_2e_3 \\
& e_2c_2 - s_2d_2^{(1)} & d_3c_2 - s_2e_2^{(1)} & c_2e_3 \\
\hline
& & e_3 & d_4
\end{array}
\right]
$$

(11.77)

By setting $c_2e_2 - s_2d_2^{(1)} = 0$ and making use of $c_2^2 + s_2^2 = 1$, the cosine and sines for the second rotation can be found as

$$
\rho^{(2)} = \sqrt{e_2^2 + \left(d_2^{(1)}\right)^2}, \quad c_2 = d_2^{(1)}/\rho^{(2)}, \quad s_2 = e_2/\rho^{(2)}
$$

(11.78)

Equation 11.77 now becomes

$$
\mathbf{P}_{(2,3)}\mathbf{P}_{(1,2)}\mathbf{A}_p =
\left[
\begin{array}{c|cc|c}
d_1^{(1)} & e_1^{(1)} & f_1 & \\
\hline
0 & \rho^{(2)} & (d_3e_2 + d_2^{(1)}e_2^{(1)})/\rho^{(2)} & e_3e_2/\rho^{(2)} \\
& 0 & (d_3d_2^{(1)} - e_2e_2^{(1)})/\rho^{(2)} & e_3d_2^{(1)}/\rho^{(2)} \\
\hline
& & e_3 & d_4
\end{array}
\right]
$$

$$
=
\left[
\begin{array}{c|cc|c}
d_1^{(1)} & e_1^{(1)} & f_1 & \\
\hline
0 & d_2^{(2)} & e_2^{(2)} & f_2 \\
& 0 & d_3^{(1)} & e_3^{(1)} \\
\hline
& & e_3 & d_4
\end{array}
\right]
$$

(11.79)

where $d_1^{(2)}$, $e_2^{(2)}$, and f_2 have conformable definitions.

The third rotation, $\mathbf{P}_{(3,4)}$, is chosen as $\mathbf{P}_{(3,4)}\mathbf{P}_{(2,3)}\mathbf{P}_{(1,2)}\mathbf{A}_p$ so that the fourth row, the third column, is zero. The $\mathbf{P}_{(3,4)}\mathbf{P}_{(2,3)}\mathbf{P}_{(1,2)}\mathbf{A}_p$ is obtained as

$$
\mathbf{P}_{(3,4)}\mathbf{P}_{(2,3)}\mathbf{P}_{(1,2)}\mathbf{A}_p =
\left[
\begin{array}{cc|cc}
1 & & & \\
& 1 & & \\
\hline
& & c_3 & s_3 \\
& & -s_3 & c_3
\end{array}
\right]
\mathbf{P}_{(2,3)}\mathbf{P}_{(1,2)}\mathbf{A}_p
$$

(11.80)

$$
=
\left[
\begin{array}{cc|c}
d_1^{(1)} & e_1^{(1)} & f_1 \\
0 & d_2^{(2)} & e_2^{(2)} \\
\hline
0 & & f_2 \\
& e_3s_3 + c_3d_3^{(1)} & d_3s_3 + c_3e_3^{(1)} \\
& e_3c_3 - s_3d_3^{(1)} & d_3c_3 - s_3e_3^{(1)}
\end{array}
\right]
$$

Similarly, setting $c_3 e_3 - s_3 d_3^{(1)} = 0$ and using $c_3^2 + s_3^2 = 1$, the cosine and sines for the third rotation are obtained as

$$\rho^{(3)} = \sqrt{e_3^2 + \left(d_3^{(1)}\right)^2}, \qquad c_3 = d_3^{(1)}/\rho^{(3)}, \qquad s_3 = e_3/\rho^{(3)} \tag{11.81}$$

Finally, the first sweep yields

$$\mathbf{P}_{(3,4)}\mathbf{P}_{(2,3)}\mathbf{P}_{(1,2)}\mathbf{A}_p = \begin{bmatrix} d_1^{(1)} & e_1^{(1)} & f_1 & \\ 0 & d_2^{(2)} & e_2^{(2)} & f_2 \\ & 0 & \rho^{(3)} & (d_3 e_3 + d_3^{(1)} e_3^{(1)})/\rho^{(3)} \\ & & 0 & (d_3 d_3^{(1)} - e_3 e_3^{(1)})/\rho^{(3)} \end{bmatrix}$$

$$= \begin{bmatrix} d_1^{(1)} & e_1^{(1)} & f_1 & \\ 0 & d_2^{(3)} & e_2^{(2)} & f_2 \\ & 0 & d_3^{(2)} & e_3^{(2)} \\ & & 0 & d_4^{(1)} \end{bmatrix} \tag{11.82}$$

where $d_2^{(3)}$, $e_2^{(2)}$, and f_2 are conformably defined.

Note that in the preceding rotations, the matrix elements labeled as $f_1, f_2, \ldots$, and so on were not used in decomposition; hence, these quantities do not need to be evaluated or saved. Moreover, they are not required for the product $\mathbf{A}_{p+1} = \mathbf{R}_p\mathbf{Q}_p$. In fact, by ignoring f_k's, we do not obtain a full QR decomposition of $\mathbf{A}$ but instead retain sufficient information needed to compute $\mathbf{R}_p\mathbf{Q}_p$.

The QR decomposition step of this method can be generalized as follows:

$$\left.\begin{aligned} & t \leftarrow e_1, \\ & \rho \leftarrow \sqrt{t^2 + d_k^2}, \\ & c_k \leftarrow d_k/\rho_k, \quad s_k \leftarrow t/\rho_k, \\ & d_k \leftarrow \rho, \quad t \leftarrow e_k, \\ & e_k \leftarrow t c_k + d_{k+1} s_k, \\ & d_{k+1} \leftarrow -t s_k + d_{k+1} c_k \end{aligned}\right\} \quad k = 1, 2, \ldots, (n-1) \tag{11.83}$$

and

$$\left.t \leftarrow e_k, \qquad e_{k+1} \leftarrow t c_k \right\} \quad k \neq (n-1) \tag{11.84}$$

where the similarity transformation steps are ordered in such a way that memory allocation is kept to a minimum.

Equations (11.83) and (11.84) generate the upper triangular matrix $\mathbf{R}$ of the QR-decomposition of $\mathbf{A}$, which can be defined for a symmetric triangular matrix of order n as

$$\mathbf{R}_p = \mathbf{P}_{(n-1,n)} \cdots \mathbf{P}_{(2,3)}\mathbf{P}_{(1,2)}\mathbf{A}_p$$

Noting that $\mathbf{R}_p = \mathbf{Q}_p^T \mathbf{A}_p$, we find $\mathbf{Q}_p^T = \mathbf{P}_{(n-1,n)} \cdots \mathbf{P}_{(2,3)}\mathbf{P}_{(1,2)}$. Recalling $(\mathbf{AB})^T = \mathbf{B}^T\mathbf{A}^T$ properties of the transpose, this result can be extended by induction to a general case of multiple matrices:

$$\mathbf{Q}_p = \mathbf{P}_{(1,2)}^T \mathbf{P}_{(2,3)}^T \cdots \mathbf{P}_{(n-2,n-1)}^T \mathbf{P}_{(n-1,n)}^T \tag{11.85}$$

At this point, it is necessary to emphasize that there is no need to compute $\mathbf{Q}_p$ explicitly to construct $\mathbf{R}_p\mathbf{Q}_p$. After each rotation, c_k and s_k's can be multiplied.

$$
\begin{bmatrix}
d_1 & e_1 & \vdots & f_1 & \\
0 & d_2 & \vdots & e_2 & f_2 \\
& 0 & \vdots & d_3 & e_3 \\
& & \vdots & 0 & d_4
\end{bmatrix}
\begin{bmatrix}
c_1 & -s_1 & \vdots & \\
s_1 & c_1 & \vdots & \\
& & \vdots & 1 & \\
& & \vdots & & 1
\end{bmatrix}
=
\begin{bmatrix}
d_1^{(1)} & \hat{e}_1 & \vdots & f_1 \\
e_1^{(1)} & d_2^{(1)} & \vdots & e_2 & f_2 \\
& 0 & \vdots & d_3 & e_3 \\
& & \vdots & 0 & d_4
\end{bmatrix}
\tag{11.86}
$$

where $d_1^{(1)} \leftarrow c_1 d_1 + e_1 s_1$, $e_1^{(1)} \leftarrow d_2 s_1$, $e_1^{(1)} \leftarrow c_1 e_1 - d_1 s_1$ and $d_2^{(1)} \leftarrow d_2^{(1)} c_1$.

The next step gives

$$
\begin{bmatrix}
d_1^{(1)} & \vdots & \hat{e}_1 & \hat{f}_1 & \vdots \\
e_1^{(1)} & \vdots & d_2^{(1)} & e_2 & f_2 \\
& \vdots & 0 & d_3 & e_3 \\
& & & & d_4
\end{bmatrix}
\begin{bmatrix}
1 & \vdots & & & \\
& \vdots & c_2 & -s_2 & \\
& \vdots & s_2 & c_2 & \\
& & & & 1
\end{bmatrix}
=
\begin{bmatrix}
d_1^{(1)} & \vdots & c_2\hat{e}_1 + f_1 s_2 & c_2 f_1 - \hat{e}_1 s_2 & \vdots \\
e_1^{(1)} & \vdots & d_2^{(2)} & \hat{e}_2 & \vdots & f_2 \\
& & e_2^{(1)} & d_3^{(1)} & \vdots & e_3 \\
& & & & & d_4
\end{bmatrix}
\tag{11.87}
$$

where similarly $d_2^{(2)} \leftarrow c_2 d_2^{(1)} + e_2 s_2$, $e_2^{(1)} \leftarrow d_3 s_2$ and $d_3^{(1)} \leftarrow c_2 d_3$, and so on.

Now that the tridiagonal matrix is symmetric and $\mathbf{Q}_p$ is orthogonal, $\mathbf{R}_p\mathbf{Q}_p$ is also symmetric. As a result, the elements of the product matrix corresponding to super-diagonal positions do not need to be computed. Finally, we deduce the following steps to construct the $\mathbf{A}_{p+1}$ matrix:

$$
d_k \leftarrow d_k c_k + e_k s_k, \quad e_k \leftarrow d_{k+1} s_k, \quad d_{k+1} \leftarrow d_{k+1} c_k, \quad k = 1, 2, \ldots, (n-1) \tag{11.88}
$$

With this procedure, the eigenvectors can also be simultaneously obtained. Let $\mathbf{V}$ be a matrix of eigenvectors, with the jth column being the associated eigenvector of λ_j. To compute $\mathbf{V}$, it is initialized to the identity matrix ($\mathbf{V}_0 = \mathbf{I}$) before the QR iterations. Once $\mathbf{A}_p$ is decomposed, the matrix $\mathbf{V}$ is improved upon by

$$
\mathbf{V}_{p+1} = \mathbf{V}_p \mathbf{P}_{(1,2)}^T \mathbf{P}_{(2,3)}^T \cdots \mathbf{P}_{(n-2,n-1)}^T \mathbf{P}_{(n-1,n)}^T \tag{11.89}
$$

The eigenvector computations are generalized as

$$
\left.
\begin{aligned}
q &\leftarrow v_{ik}, \\
r &\leftarrow v_{i,k+1}, \\
v_{ik} &\leftarrow c_k q + s_k r, \\
v_{i,k+1} &\leftarrow -s_k q + c_k r
\end{aligned}
\right\}, \quad \text{for } k = 1, 2, \ldots, (n-1) \text{ and } i = 1, 2, \ldots, n \tag{11.90}
$$

where v_{ik} denotes the elements of the matrix of eigenvectors, and q and r are temporary variables used to overwrite the updates on v_{ik} and $v_{i,k+1}$, respectively.

The QR iteration procedure is repeated until the sub-diagonal elements of $\mathbf{A}_p$ (i.e., $e_k^{(p)}$) are sufficiently small. A criterion for convergence is based on either the relative or absolute error of the eigenvalues or ℓ_2-norm of the sub-diagonal elements:

$$
L_2 = \sqrt{\sum_{k=1}^{n-1} e_k^2} < \varepsilon \tag{11.91}
$$

Consider a real symmetric tridiagonal matrix $\mathbf{A}$ satisfying $|\lambda_1| > |\lambda_2| > |\lambda_3| > \cdots >$

QR Iteration Method

- The method presents a simple, elegant algorithm for finding the eigenvectors and eigenvalues of a real, symmetric, full-rank matrix;
- It gives the complete set of all eigenpairs (i.e., the orthogonal matrix Q_p, whose columns approach eigenvectors of $\mathbf{A}$, and a diagonal matrix $\mathbf{A}_p$), whose diagonal elements approach the eigenvalues of $\mathbf{A}$);

- It can be used as a direct solver for the system of linear equations;
- It has been improved by variants having quadratic convergence rates;

- The basic QR method has a linear convergence rate;
- It may fail when no real eigenvalues exist or when the eigenvectors do not form an orthogonal basis;
- It requires excessive memory for large sparse matrices;
- Its convergence can be slow if the eigenvalues are close to each other.

$|\lambda_n| > 0$. The sequence of iterates $\{\mathbf{A}_p\}$ will converge to a diagonal matrix that contains the eigenvalues $\{\lambda_k\}$ in the diagonal position, ordered in descending order. During the QR iterations on matrix $\mathbf{A}_p$, the diagonal elements $d_k^{(p)}$ approach the eigenvalues. The convergence criterion given by Eq. (11.91) is monitored at each step, i.e., $e_k^{(p)} \to 0$ for $p \to \infty$. The rate of convergence is linear, with a convergence ratio of $|\lambda_k/\lambda_{k-1}|$. The basic QR algorithm presented here converges slowly enough to be competitive, especially when the eigenvalues are clustered together. To accelerate the convergence of the QR method, several variants with a second-order convergence rate have been developed. In this section, we will suffice to explain the basic QR algorithm in detail.

The QR algorithm consists of two separate stages. First, the original matrix is transformed through similarity transformations into a tridiagonal form, which is the preparation stage for the second stage. The QR iterations are, in fact, applied to a tridiagonal matrix. The number of floating points in the method is $\mathcal{O}(n^3)$.

EXAMPLE 11.11: Application of QR iteration method

Compute the eigenvalues of the following matrix using the QR iteration method.

$$\mathbf{A} = \begin{bmatrix} 2 & -1 & 0 & 0 \\ -1 & 1 & -1 & 0 \\ 0 & -1 & 1 & 2 \\ 0 & 0 & 2 & 2 \end{bmatrix}$$

SOLUTION:

The diagonal and super-diagonal elements are $d_1 = d_4 = 2$, $d_2 = d_3 = 1$, $e_1 = e_2 = -1$, and $e_3 = 2$. The sub-diagonal elements will be zeroed out by row rotations. Note that the ℓ_2-norm of the sub-diagonal elements is initially $\sqrt{\Sigma e_k^2} = \sqrt{6}$.

We apply Eqs. (11.83) and (11.84) for the row rotations. Setting $t = -1$, the first rotation ($k = 1$) leads to

$$\rho^{(1)} = \sqrt{t^2 + d_1^2} = \sqrt{5}, \quad c_1 = \frac{d_1}{\rho^{(1)}} = \frac{2}{\sqrt{5}} \quad s_1 = \frac{t}{\rho^{(1)}} = -\frac{1}{\sqrt{5}}$$

and we set $d_1 = \rho^{(1)} = \sqrt{5}$.

Next, assigning $t = e_1 = -1$ we calculate

$$e_1 = tc_1 + d_2s_1 = -\frac{3}{\sqrt{5}}, \quad d_2 = -ts_1 + d_2c_1 = \frac{1}{\sqrt{5}}$$

Since $k \neq 3$, we set $t = e_2 = -1$ to find $e_2 = tc_1 = -2/\sqrt{5}$.

The second rotation ($k = 2$) with $t = -1$ results in

$$\rho^{(2)} = \sqrt{t^2 + d_2^2} = \sqrt{\frac{6}{5}}, \quad c_2 = \frac{d_2}{\rho^{(2)}} = \frac{1}{\sqrt{6}}, \quad s_2 = \frac{t}{\rho^{(2)}} = -\sqrt{\frac{5}{6}}$$

and we set $d_2 = \rho^{(2)} = \sqrt{6/5}$.

Setting $t = e_2 = -2/\sqrt{5}$, we find

$$e_2 = tc_2 + d_3s_2 = -\frac{7}{\sqrt{30}}, \quad d_3 = -ts_2 + d_3c_2 = -\frac{1}{\sqrt{6}}$$

Again, since $k \neq 3$, setting $t = e_3 = 2$, we can then compute $e_3 = tc_2 = \sqrt{2/3}$. In the third rotation ($k = 3$) and $t = 2$, we find

$$\rho^{(3)} = \sqrt{t^2 + d_3^2} = \frac{5}{\sqrt{6}}, \quad c_3 = \frac{d_3}{\rho^{(3)}} = -\frac{1}{5}, \quad s_3 = \frac{t}{\rho^{(3)}} = \frac{2\sqrt{6}}{5}$$

and we set $d_3 = \rho^{(3)} = 5/\sqrt{6}$. By $t = e_3 = \sqrt{2/3}$, we find

$$e_3 = tc_3 + d_4s_3 = \frac{11}{5}\sqrt{\frac{2}{3}}, \quad d_4 = -ts_3 + d_4c_3 = -\frac{6}{5}$$

We use Eq. (11.88), for $k = 1$

$$d_1 = d_1c_1 + e_1s_1 = \frac{13}{5}, \quad e_1 = d_2s_1 = \frac{2\sqrt{6}}{5}$$

For $k = 2$, we find

$$d_2 = d_2c_2 + e_2s_2 = \frac{47}{30}, \quad e_2 = d_3s_2 = -\frac{5\sqrt{6}}{6}, \quad d_3 = d_3c_2 = \frac{5}{6}$$

Finally, for $k = 3$, we have

$$d_3 = d_3c_3 + e_3s_3 = \frac{239}{150}, \quad e_3 = d_4s_3 = \frac{6}{25}$$

Using $\varepsilon = 10^{-5}$, the QR method converges to the following eigenvalues after 36 iterations:

$$d_1 = \lambda_1 = 3.747964, \qquad d_2 = \lambda_2 = 2.570181,$$
$$d_3 = \lambda_3 = -0.964160, \qquad d_4 = \lambda_4 = 0.646016$$

Discussion: The method converges to the approximate eigenvalues in 36 iterations. The ℓ_2-norm of $\mathbf{A}_2$ for the 1st to 6th iterations are 2.257127, 1.298705, 0.794264, 0.700586, and 0.625785, respectively.

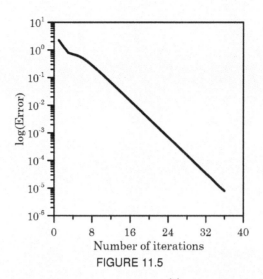

FIGURE 11.5

The 10-based logarithm of the ℓ_2-norm, $\log_{10}(L_2^{(p)})$, is plotted as a function of the number of iterations in Fig. 11.5. After several iteration steps, the variation of the log-ℓ_2-norm settles into a linear form, from which the convergence rate is determined to be 0.164.

Pseudocode 11.6

Module BASIC_QR $(n, \mathbf{d}, \mathbf{e}, \varepsilon, maxit, \mathbf{V})$

\ DESCRIPTION: A module to find eigenvalues of a symmetric tridiagonal matrix.

\ USES:

\   SQRT:: A built-in function computing the square-root of a value.

Declare: $d_n, e_n, c_n, s_n, v_{n,n}$

$err \leftarrow 1$ \ Initialize error

$p \leftarrow 0$ \ Initialize iteration counter

$\mathbf{V} \leftarrow 0$ \ Initialize matrix of eigenvectors

For $\left[i = 1, n \right]$

    $v_{ii} \leftarrow 1$ \ Initialize diagonals, $v_{ii} \leftarrow 1$

End For

While $\left[err > \varepsilon \textbf{ And } p < maxit \right]$ \ Iteration loop

    $t \leftarrow e_1$ \ Start decomposition steps

    For $\left[k = 1, n - 1 \right]$ \ Decomposition steps

       $\rho \leftarrow \sqrt{d_k^2 + t^2}$

       $c_k \leftarrow d_k / \rho$

       $s_k \leftarrow t / \rho$

       $d_k \leftarrow \rho$

       $t \leftarrow e_k$

       $e_k \leftarrow t * c_k + d_{k+1} * s_k$

       $d_{k+1} \leftarrow -t * s_k + d_{k+1} * c_k$

       If $\left[k \neq n - 1 \right]$ **Then**   \ The last step does not contain e_n

          $t \leftarrow e_{k+1}$

          $e_{k+1} \leftarrow t * c_k$

End If
For $\left[i=1,n\right]$
    $q \leftarrow v_{ik}$
    $r \leftarrow v_{i,k+1}$
    $v_{ik} \leftarrow c_k * q + s_k * r$
    $v_{i,k+1} \leftarrow -s_k * q + c_k * r$
End For
End For   \ End of QR-decomposition loop
For $\left[k=1,n-1\right]$ \ Obtain $\mathbf{R}_k\mathbf{Q}_k$ product
    $d_k \leftarrow d_k * c_k + e_k * s_k$
    $t \leftarrow d_{k+1}$
    $e_k \leftarrow t * s_k$
    $d_{k+1} \leftarrow t * c_k$
End For
$err \leftarrow 0$ \ Initialize accumulator for error
For $\left[k=1,n-1\right]$
    $err \leftarrow err + e_k^2$ \ Accumulate e_k^2
End For
$err \leftarrow \sqrt{err}$ \ Error computed as $err = \sqrt{\Sigma_{k=1}^{n-1} e_k^2}$
$p \leftarrow p + 1$ \ Count iterations
Write: "Iteration=", p, "Error=", err \ Print out the iteration progress
End While
If $\left[p=maxit\right]$ **Then**   \ Upper bound reached with no convergence
    Write: "Iteration did not converge, error is", err \ Print a warning
End If
End Module BASIC_QR

A pseudomodule, BASIC_QR, finding all eigenvalues of a symmetric tridiagonal matrix is presented in Pseudocode 11.6. As input, the module requires the order of the tridiagonal matrix (n), its diagonal ($\mathbf{d}$) and sub-diagonal elements ($\mathbf{e}$), a convergence tolerance (ε), and an upper bound for the number of iterations ($maxit$). On exit, the estimated eigenvalues are stored on the diagonal array $\mathbf{d}$. In this procedure, the convergence decision is based on ℓ_2 norm of the sub-diagonal elements (err), which should approach zero after a sufficient number of transformations. The module processes $n-1$ plane rotations to zero out the sub-diagonal elements. The last rotation, Eq. (11.84), is given with a conditional statement since there is no e_n or e_{n+1} to overwrite. If the eigenvector is to be computed, the construction of matrix $\mathbf{V}_p$ is realized by updating the matrix after each rotation using Eq. (11.90). The second loop involves the $\mathbf{RQ}$ product. Finally, the iterative process is terminated when the convergence criterion is satisfied or if the iterations reach the $maxit$.

11.7.3 CLASSIC GRAM-SCHMIDT PROCESS

Gram-Schmidt QR decomposition is used to find an orthogonal basis from a non-orthogonal basis. Recall that an orthogonal matrix has row and column vectors of unit length, and the orthonormal basis simplifies the computations by having $\mathbf{v}_i \cdot \mathbf{v}_i = 1$ and $\mathbf{v}_i \cdot \mathbf{v}_j = 0$. In this regard, an orthogonal basis has many advantageous properties that are desirable for the QR decomposition.

We begin with n independent vectors $\mathbf{t}_1, \mathbf{t}_2, \ldots, \mathbf{t}_n$, i.e., the column vectors of a matrix as

$$\mathbf{T} = [\mathbf{t}_1\ \mathbf{t}_2\ \mathbf{t}_3\ \cdots\ \mathbf{t}_n] = \begin{bmatrix} d_1 & e_1 & 0 & & 0 & 0 \\ e_1 & d_2 & e_2 & & 0 & 0 \\ 0 & e_2 & d_3 & & \vdots & \vdots \\ 0 & 0 & e_3 & \cdots & 0 & 0 \\ \vdots & \vdots & 0 & & e_{n-2} & 0 \\ 0 & 0 & \vdots & & d_{n-1} & e_{n-1} \\ 0 & 0 & 0 & & e_{n-1} & d_n \end{bmatrix} \tag{11.92}$$

The goal of this process is to produce n orthonormal vectors $\mathbf{q}_1, \mathbf{q}_2, \ldots, \mathbf{q}_n$ (i.e., the column vectors of an orthogonal matrix $\mathbf{Q}$). However, matrix $\mathbf{Q}$ has the upper *Hessenberg matrix* form (i.e., consists of an nonzero upper triangular matrix and the first sub-diagonal):

$$\mathbf{Q} = [\mathbf{q}_1\ \mathbf{q}_2\ \mathbf{q}_3\ \cdots\ \mathbf{q}_n] = \begin{bmatrix} q_{11} & q_{12} & q_{13} & & q_{1(n-1)} & q_{1n} \\ q_{21} & q_{22} & q_{23} & & q_{2(n-1)} & q_{2n} \\ 0 & q_{23} & q_{33} & & q_{3(n-1)} & q_{3n} \\ 0 & 0 & q_{43} & \cdots & q_{4(n-1)} & q_{4n} \\ \vdots & \vdots & 0 & & \vdots & \vdots \\ 0 & 0 & \vdots & & q_{(n-1)(n-1)} & q_{(n-1)n} \\ 0 & 0 & 0 & & q_{n(n-1)} & q_{nn} \end{bmatrix} \tag{11.93}$$

In the classic Gram-Schmidt algorithm, we start with $\mathbf{t}_i$ and subtract off its projections onto the previous $\mathbf{q}$'s to find the orthonormal $\mathbf{q}_i$, and then divide by the length of the vector $\mathbf{u}$ to obtain a unit vector as follows:

$$\mathbf{u}_1 = \mathbf{t}_1, \qquad \mathbf{q}_1 = \frac{\mathbf{u}_1}{\|\mathbf{u}_1\|}$$

$$\mathbf{u}_2 = \mathbf{t}_2 - r_{12}\,\mathbf{q}_1, \qquad \mathbf{q}_2 = \frac{\mathbf{u}_2}{\|\mathbf{u}_1\|}$$

$$\mathbf{u}_3 = \mathbf{t}_3 - r_{13}\,\mathbf{q}_1 - r_{23}\,\mathbf{q}_2, \quad \mathbf{q}_3 = \frac{\mathbf{u}_3}{\|\mathbf{u}_1\|} \tag{11.94}$$

$$\cdots \qquad\qquad \cdots$$

$$\mathbf{u}_k = \mathbf{t}_k - \sum_{j=1}^{k-1} r_{jk}\mathbf{q}_j, \qquad \mathbf{q}_k = \frac{\mathbf{u}_k}{\|\mathbf{u}_1\|}$$

where r_{jk} denotes the inner product defined as

$$r_{jk} = (\mathbf{t}_k, \mathbf{q}_j) = \begin{bmatrix} \vdots \\ 0 \\ e_{k-1} \\ d_k \\ e_k \\ 0 \\ \vdots \end{bmatrix}^T \begin{bmatrix} q_{1j} \\ q_{2j} \\ \vdots \\ q_{jj} \\ q_{(j+1)j} \\ 0 \\ 0 \end{bmatrix} \quad \text{and} \quad r_{kk} = (\mathbf{t}_k, \mathbf{q}_k) = \|\mathbf{u}_k\| = \sqrt{\sum_{i=1}^{n} u_{ik}^2} \tag{11.95}$$

Gram-Schmid Method

- The method is simple and easy to implement;
- It is not an iterative method, i.e., $\mathbf{Q}$ and $\mathbf{R}$ are found in n steps;
- Unlike the QR Iteration method, a complete $\mathbf{Q}$ and $\mathbf{R}$ decomposition is performed; hence, the method can be employed to solve a linear system as well.

- The process is inherently numerically unstable;
- It is sensitive to round-off errors; as a result, for very large matrices, the orthogonality can be partially or completely lost, requiring the computation of $\mathbf{q}_i$'s to the machine epsilon level;
- It gives only the eigenvalues; another method should be utilized if eigenvectors are also needed.

The inner product $r_{jk} = (\mathbf{t}_k, \mathbf{q}_j)$ is used in constructing the upper triangular matrix $\mathbf{R}$ (since $(\mathbf{t}_k, \mathbf{q}_j) = 0$ for $j > k$) as follows:

$$
\mathbf{R} = \begin{bmatrix}
\|\mathbf{u}_1\| & (\mathbf{t}_2, \mathbf{q}_1) & (\mathbf{t}_3, \mathbf{q}_1) & \cdots & (\mathbf{t}_n, \mathbf{q}_1) \\
0 & \|\mathbf{u}_2\| & (\mathbf{t}_3, \mathbf{q}_2) & \cdots & (\mathbf{t}_n, \mathbf{q}_2) \\
\vdots & 0 & \|\mathbf{u}_3\| & \cdots & (\mathbf{t}_n, \mathbf{q}_3) \\
\vdots & \vdots & \vdots & \ddots & \vdots \\
0 & 0 & \cdots & 0 & \|\mathbf{u}_n\|
\end{bmatrix}
\tag{11.96}
$$

A pseudomodule, GRAM_SCHMIDT, decomposing a symmetric tridiagonal matrix using the classical Gram-Schmidt process is presented in Pseudocode 11.7. The module requires the order of the matrix (n), a tridiagonal matrix, by its diagonal ($\mathbf{d}$) and sub-diagonal ($\mathbf{e}$) elements as input. On exit, the module returns an orthogonal ($\mathbf{Q}$) and upper triangular matrix ($\mathbf{R}$). The tridiagonal matrix $\mathbf{T}$ is overwritten on $\mathbf{Q}$.

The following algorithm is employed:

(i) Find the unit vector

$$
r_{kk} \leftarrow \|\mathbf{q}_k\| = \left(\sum_{i=1}^{k+1} q_{ik}^2 \right)^{1/2}, \qquad \mathbf{q}_k \leftarrow \mathbf{q}_k / r_{kk} \quad \text{for} \quad k = 1, 2, \ldots, n
$$

(ii) Find projections

$$
\left.
\begin{aligned}
r_{kj} &\leftarrow (\mathbf{q}_i, \mathbf{q}_k) = \sum_{i=1}^{n} q_{ij} q_{ik}, \\
q_{ij} &\leftarrow q_{ij} - r_{kj} * q_{ik}, \quad i = 1, 2, \ldots, n
\end{aligned}
\right\} \quad j = k+1, \ldots, n
$$

In the presented Gram-Schmidt algorithm, $\mathbf{R}$ is constructed row by row, whereas in the modified Gram-Schmidt algorithm, this is done column by column.

Module GRAM_SCHMIDT $(n, \mathbf{d}, \mathbf{e}, \mathbf{Q}, \mathbf{R})$
\ DESCRIPTION: A pseudomodule to apply classic Gram-Schmidt orthogonali-
\   zation process to a symmetric tridiagonal matrix.
\ USES:
\   MAX :: Built-in function to find the max of a set of numbers;
\   SQRT:: A built-in function computing the square-root of a value.
Declare: d_n, e_n, r_{nn}, q_{nn}
$\mathbf{Q} \leftarrow 0; \ \mathbf{R} \leftarrow 0$ \ Initialize matrices, $\mathbf{Q}$ and $\mathbf{R}$
For $\left[i = 1, n \right]$ \ Write tridiagonal matrix on $\mathbf{Q}$
  If $\left[i = 1 \right]$ **Then**   \ Write the first row
    $q_{1i} \leftarrow d_1$
    $q_{2i} \leftarrow e_1$
  Else
    If $\left[i = n \right]$ **Then**   \ Write the last row
      $q_{(n-1)i} \leftarrow e_{n-1}$
      $q_{ni} \leftarrow d_n$
    Else   \ Write 1<row<n
      $q_{(i-1)i} \leftarrow e_{i-1}$
      $q_{ii} \leftarrow d_i$
      $q_{(i+1)i} \leftarrow e_i$
    End If
  End If
End For
For $\left[k = 1, n \right]$ \ Begin GS process
  $imax \leftarrow$ **MIN**$(k + 1, n)$ \ Row no. of sub-diagonal element
  $sums \leftarrow 0$
  For $\left[i = 1, imax \right]$ \ Find r_{kk}
    $sums \leftarrow sums + q_{ik}^2$
  End For
  $r_{kk} \leftarrow \sqrt{sums}$ \ Compute $r_{kk} \leftarrow \sqrt{\Sigma q_{ik}^2}$
  For $\left[i = 1, imax \right]$ \ Normalize kth column
    $q_{ik} \leftarrow q_{ik}/r_{kk}$
  End For
  For $\left[j = k + 1, n \right]$
    $rkj \leftarrow 0$
    For $\left[i = 1, n \right]$ \ Compute r_{kj} projection
      $rkj \leftarrow rkj + q_{ij} * q_{ik}$
    End For
    $r_{kj} \leftarrow rkj$
    For $\left[i = 1, n \right]$ \ Compute q_{ij}
      $q_{ij} \leftarrow q_{ij} - r_{kj} * q_{ik}$
    End For
  End For
End For
End Module GRAM_SCHMIDT

EXAMPLE 11.12: Application of Gram-Schmid method

Apply the Gram-Schmidt method to find the QR-decomposition of the following matrix:

$$\mathbf{T} = \begin{bmatrix} 2 & -1 & 0 & 0 \\ -1 & 1 & -1 & 0 \\ 0 & -1 & 1 & 2 \\ 0 & 0 & 2 & 2 \end{bmatrix}$$

SOLUTION:

Using Eqs. (11.94) and (11.95) and setting $\mathbf{u}_1 = \mathbf{t}_1$ yields

$$\mathbf{q}_1 = \frac{\mathbf{u}_1}{\|\mathbf{u}_1\|} = \frac{1}{\sqrt{5}} \begin{bmatrix} 2 \\ -1 \\ 0 \\ 0 \end{bmatrix},$$

$$(\mathbf{t}_2, \mathbf{q}_1) = \begin{bmatrix} -1 & 1 & -1 & 0 \end{bmatrix} \cdot \frac{1}{\sqrt{5}} \begin{bmatrix} 2 \\ -1 \\ 0 \\ 0 \end{bmatrix} = -\frac{3}{\sqrt{5}}$$

For the second column, we obtain

$$\mathbf{u}_2 = \mathbf{t}_2 - (\mathbf{t}_2, \mathbf{q}_1)\mathbf{q}_1 = \frac{1}{5} \begin{bmatrix} 1 \\ 2 \\ -5 \\ 0 \end{bmatrix},$$

$$\mathbf{q}_2 = \frac{\mathbf{u}_2}{\|\mathbf{u}_2\|} = \frac{1}{\sqrt{30}} \begin{bmatrix} 1 \\ 2 \\ -5 \\ 0 \end{bmatrix}$$

$$r_{22} = \|\mathbf{u}_2\| = \sqrt{\frac{6}{5}}, \quad r_{13} = [\mathbf{t}_3, \mathbf{q}_1] = \frac{1}{\sqrt{5}}, \quad r_{23} = (\mathbf{t}_3, \mathbf{q}_2) = \frac{-7}{\sqrt{30}}$$

For the third column, the Gram-Schmidt process yields

$$\mathbf{u}_3 = \mathbf{t}_3 - (\mathbf{t}_3, \mathbf{q}_1)\mathbf{q}_1 - (\mathbf{t}_3, \mathbf{q}_2)\mathbf{q}_2 = \frac{1}{6} \begin{bmatrix} -1 \\ -2 \\ -1 \\ 12 \end{bmatrix},$$

$$\mathbf{q}_3 = \frac{\mathbf{u}_3}{\|\mathbf{u}_3\|} = \frac{1}{5\sqrt{6}} \begin{bmatrix} -1 \\ -2 \\ -1 \\ 12 \end{bmatrix}$$

$$r_{33} = \|\mathbf{u}_3\| = \frac{5}{\sqrt{6}}, \quad r_{14} = (\mathbf{t}_4, \mathbf{q}_1) = 0,$$

$$r_{24} = (\mathbf{t}_4, \mathbf{q}_2) = -\sqrt{\frac{10}{3}}, \quad r_{34} = (\mathbf{t}_4, \mathbf{q}_3) = \frac{11}{5}\sqrt{\frac{2}{3}}$$

The last step leads to

$$\mathbf{u}_4 = \mathbf{t}_4 - (\mathbf{t}_4, \mathbf{q}_1)\mathbf{q}_1 - (\mathbf{t}_4, \mathbf{q}_2)\mathbf{q}_2 - (\mathbf{t}_4, \mathbf{q}_3)\mathbf{q}_3 = \frac{6}{25}\begin{bmatrix} 2 \\ 4 \\ 2 \\ 1 \end{bmatrix},$$

$$\mathbf{q}_4 = \frac{\mathbf{u}_4}{\|\mathbf{u}_4\|} = \frac{1}{5}\begin{bmatrix} 2 \\ 4 \\ 2 \\ 1 \end{bmatrix}$$

and $r_{44} = \|\mathbf{u}_4\| = 6/5$.

Finally, $\mathbf{Q}$ and $\mathbf{R}$ matrices are found as

$$\mathbf{Q} = \begin{bmatrix} \dfrac{2}{\sqrt{5}} & \dfrac{1}{\sqrt{30}} & -\dfrac{1}{5\sqrt{6}} & \dfrac{2}{5} \\ -\dfrac{1}{\sqrt{5}} & \dfrac{2}{\sqrt{30}} & -\dfrac{2}{5\sqrt{6}} & \dfrac{4}{5} \\ 0 & -\dfrac{5}{\sqrt{30}} & -\dfrac{1}{5\sqrt{6}} & \dfrac{2}{5} \\ 0 & 0 & \dfrac{12}{5\sqrt{6}} & \dfrac{1}{5} \end{bmatrix} \qquad \mathbf{R} = \begin{bmatrix} \sqrt{5} & -\dfrac{3}{\sqrt{5}} & \dfrac{1}{\sqrt{5}} & 0 \\ 0 & \sqrt{\dfrac{6}{5}} & -\dfrac{7}{\sqrt{30}} & -\sqrt{\dfrac{10}{3}} \\ 0 & 0 & \dfrac{5}{\sqrt{6}} & \dfrac{11}{5}\sqrt{\dfrac{2}{3}} \\ 0 & 0 & 0 & \dfrac{6}{5} \end{bmatrix}$$

Discussion: The process, which requires n steps, leads to a complete QR-decomposition that can then be used for any relevant matrix procedures.

EXAMPLE 11.13: Application of QR method to find eigenvalues

Find the eigenvalues of $\mathbf{T}$ given in Example 11.12 using the QR-method.

SOLUTION:

We will make use of $\mathbf{Q}$ and $\mathbf{R}$ matrices obtained in **Example 11.12**. We start with

$$\mathbf{A}_1 = \mathbf{R}_0\mathbf{Q}_0 = \begin{bmatrix} 2.6 & -0.4899 & 0 & 0 \\ -0.4899 & 1.56667 & -1.86339 & 0 \\ 0 & -1.86339 & 1.59334 & 1.17575 \\ 0 & 0. & 1.17575 & 0.24 \end{bmatrix}$$

Next, employing QR decomposition to $\mathbf{A}_1$ yields

$$\mathbf{Q}_1 = \begin{bmatrix} 0.9827 & 0.1137 & -0.0540 & 0.1358 \\ -0.1852 & 0.6032 & -0.2867 & 0.7209 \\ 0 & -0.7894 & -0.2268 & 0.5704 \\ 0 & 0 & 0.9292 & 0.3695 \end{bmatrix}$$

$$\mathbf{R}_1 = \begin{bmatrix} 2.6457 & -0.7715 & 0.3450 & 0 \\ 0 & 2.3604 & -2.3819 & -0.9282 \\ 0 & 0 & 1.2653 & -0.0437 \\ 0 & 0 & 0 & 0.7593 \end{bmatrix}$$

$$\mathbf{A}_2 = \mathbf{R}_1\mathbf{Q}_1 = \begin{bmatrix} 2.7429 & -0.4371 & 0 & 0 \\ -0.4371 & 3.3042 & -0.9989 & 0 \\ 0 & -0.9989 & -0.3276 & 0.7056 \\ 0 & 0 & 0.7056 & 0.2806 \end{bmatrix}$$

Then, the QR decomposition of $\mathbf{A}_2$ gives

$$\mathbf{Q}_2 = \begin{bmatrix} 0.9875 & 0.1502 & -0.0306 & 0.0356 \\ -0.1576 & 0.9425 & -0.1922 & 0.2234 \\ 0 & -0.2985 & -0.6225 & 0.7235 \\ 0 & 0 & 0.7580 & 0.6522 \end{bmatrix}$$

$$\mathbf{R}_2 = \begin{bmatrix} 2.7775 & -0.9515 & 0.1572 & 0 \\ 0 & 3.3467 & -0.8437 & -0.2106 \\ 0 & 0 & 0.9308 & -0.2265 \\ 0 & 0 & 0 & 0.6935 \end{bmatrix}$$

$$\mathbf{A}_3 = \mathbf{R}_2\mathbf{Q}_2 = \begin{bmatrix} 2.8926 & -0.5266 & 0 & 0 \\ -0.5266 & 3.4062 & -0.2778 & 0 \\ 0 & -0.2778 & -0.7511 & 0.5257 \\ 0 & 0 & 0.5257 & 0.4523 \end{bmatrix}$$

This procedure is repeated until the norm of the sub-diagonal elements is sufficiently close to zero. In the 14th and 38th iterations, the iteration matrices are found to be

$$\mathbf{A}_{14} = \begin{bmatrix} 3.7472 & -0.0302 & 0 & 0 \\ -0.0302 & 2.5709 & 0 & 0 \\ 0 & 0 & -0.9641 & 0.0073 \\ 0 & 0 & 0.0073 & 0.6460 \end{bmatrix}$$

$$\mathbf{A}_{38} = \begin{bmatrix} 3.74796 & 0 & 0 & 0 \\ 0 & 2.57018 & 0 & 0 \\ 0 & 0 & -0.96416 & 0 \\ 0 & 0 & 0 & 0.64602 \end{bmatrix}$$

Finally, the eigenvalues are obtained as -0.96416, 0.64602, 2.57018, and 3.74796.

Discussion: The process of finding eigenvalues using the QR-decomposition can then be used for any relevant matrix procedures.

11.7.4 COMPUTING EIGENVECTORS

So far, we have seen that some numerical methods are designed to give only certain or all eigenvalues. However, when associated eigenvectors are also needed, the analyst must resort to another method for computing the eigenvectors. In this section, a method to compute the eigenvectors of a symmetric tridiagonal matrix is presented.

Recalling that the eigenvectors of $\mathbf{A}$ satisfy

$$(\mathbf{A} - \lambda\mathbf{I})\mathbf{x} = \mathbf{0} \tag{11.97}$$

Let λ_k and $\mathbf{x}_k$ be respectively known eigenvalues and their associated eigenvectors. Equation

(11.97) leads to

$$(d_1 - \lambda_k)x_1 + e_1 x_2 = 0$$
$$e_1 x_1 + (d_2 - \lambda_k)x_2 + e_2 x_3 = 0$$
$$e_2 x_2 + (d_3 - \lambda_k)x_3 + e_3 x_4 = 0 \tag{11.98}$$
$$\vdots$$
$$e_{n-1} x_{n-1} + (d_n - \lambda_k)x_n = 0$$

where $x_1, x_2, \ldots, x_n$ are the components of $\mathbf{x}_k$, i.e., the kth associated eigenvector λ_k.

Setting $x_1 = 1$, solving for x_2 from the first equation of Eq. (11.98), we obtain

$$x_2 = -\frac{(d_1 - \lambda_k)}{e_1} = -\frac{p_1(\lambda_k)}{e_1} \tag{11.99}$$

Similarly, substituting Eq. (11.99) into the second equation of Eq. (11.98) and solving for x_3, we get

$$x_3 = \frac{p_2(\lambda_k)}{e_1 e_2} \tag{11.100}$$

Applying the forward elimination procedure in this manner, the remaining components of the associated eigenvector can be obtained from the following relationship:

$$x_m = \frac{(-1)^{m-1} p_{m-1}(\lambda_k)}{e_1 e_2 \cdots e_{m-1}}, \quad m = 2, 3, \ldots, n \tag{11.101}$$

where $p_{m-1}(\lambda_k)$ is the Sturm-sequence for the eigenvalue λ_k.

11.8 FADDEEV-LEVERRIER METHOD

So far, numerical methods for determining the eigenpairs of non-symmetric matrices have not yet been discussed. The methods discussed covered the computation of the eigenvalues of symmetric matrices, which cannot be extended to non-symmetric matrices. Nonetheless, some eigenvalue problems can be reduced to symmetric matrices.

The *Faddeev-Leverrier method* is a method that can be applied to relatively small matrices. It relies on finding the coefficients of the characteristic polynomial, and it can be employed to solve both symmetric and non-symmetric eigenvalue problems.

Suppose the characteristic polynomial of an nth-order square matrix is given as

$$\lambda^n + c_1 \lambda^{n-1} + c_2 \lambda^{n-2} + \ldots + c_{n-1} \lambda + c_n = 0 \tag{11.102}$$

The trace of matrix $\mathbf{A}$ (i.e., the sum of the diagonal elements) is computed as

$$\text{Tr}[\mathbf{A}] = a_{11} + a_{22} + a_{33} + \cdots + a_{nn} = \sum_{i=1}^{n} a_{ii} \tag{11.103}$$

The Faddeev-Leverrier algorithm forms a series of matrices, $\mathbf{B}_k$ ($k = 1, 2, \ldots, n$), and the trace of these matrices is used in the computation of c_k's as follows:

$$
\begin{aligned}
\mathbf{B}_1 &= \mathbf{A} && \text{and} && d_1 = \text{Tr}[\mathbf{B}_1] \\
\mathbf{B}_2 &= \mathbf{A}(\mathbf{B}_1 - d_1 \mathbf{I}) && \text{and} && d_2 = \tfrac{1}{2}\text{Tr}[\mathbf{B}_2] \\
&\vdots && && \vdots \\
\mathbf{B}_n &= \mathbf{A}(\mathbf{B}_{n-1} - d_{n-1}\mathbf{I}) && \text{and} && d_n = \tfrac{1}{n}\text{Tr}[\mathbf{B}_n]
\end{aligned}
\tag{11.104}
$$

Faddeev-Leverrier Method

- The method is simple, efficient, and easy to implement;
- It can be applied to symmetric and non-symmetric matrices;
- Since the eigenvalues are obtained from the roots of the characteristic polynomial, it is possible to determine imaginary eigenvalues as well;
- It does not involve any division by matrix elements, except division by integers at the very end;
- The determinant and/or inverse of a matrix can also be computed simultaneously with no extra computational effort.

- The method is practical only for small matrices; for large matrices, it should be used with caution as it leads to the amplification of round-off errors.

Next, the characteristic polynomial and its coefficients are obtained as

$$\lambda^n - d_1\lambda^{n-1} - d_2\lambda^{n-2} - \ldots - d_{n-1}\lambda - p_n = 0 \qquad (11.105)$$

where $c_k = -d_k$ $(k = 1, 2, 3, \ldots, n)$.

This method also allows the determinant and/or the inverse of the matrix to be computed as follows:

$$\mathbf{A}^{-1} = \frac{1}{d_n}\left(\mathbf{B}_{n-1} - d_{n-1}\mathbf{I}\right) \qquad \text{and} \qquad \det(\mathbf{A}) = (-1)^n d_n \qquad (11.106)$$

EXAMPLE 11.14: Finding characteristic polynomial and inverse of a matrix

Find the characteristic polynomial, the inverse, and the determinant of matrix $\mathbf{A}$.

$$\mathbf{A} = \begin{bmatrix} 3 & 3 & 2 \\ 4 & 2 & 1 \\ -5 & 2 & 2 \end{bmatrix}$$

SOLUTION:

Setting $\mathbf{B}_1 = \mathbf{A}$, the first coefficient is obtained as

$$d_1 = \text{Tr}[\mathbf{B}_1] = \sum_{i=1}^{3} b_{ii} = 3 + 2 + 2 = 7$$

The construction of $\mathbf{B}_2$ yields,

$$\mathbf{B}_1 - d_1\mathbf{I} = \begin{bmatrix} 3 & 3 & 2 \\ 4 & 2 & 1 \\ -5 & 2 & 2 \end{bmatrix} - (7)\begin{bmatrix} 1 & 0 & 0 \\ 0 & 1 & 0 \\ 0 & 0 & 1 \end{bmatrix} = \begin{bmatrix} -4 & 3 & 2 \\ 4 & -5 & 1 \\ -5 & 2 & -5 \end{bmatrix}$$

$$\mathbf{B}_2 = \mathbf{A}(\mathbf{B}_1 - d_1\mathbf{I}) = \begin{bmatrix} 3 & 3 & 2 \\ 4 & 2 & 1 \\ -5 & 2 & 2 \end{bmatrix}\begin{bmatrix} -4 & 3 & 2 \\ 4 & -5 & 1 \\ -5 & 2 & -5 \end{bmatrix} = \begin{bmatrix} -10 & -2 & -1 \\ -13 & 4 & 5 \\ 18 & -21 & -18 \end{bmatrix}$$

From the trace of $\mathbf{B}_2$, we find d_2 as

$$d_2 = \frac{1}{2}\text{Tr}[\mathbf{B}_2] = \frac{1}{2}\sum_{i=1}^{3} b_{ii} = \frac{(-10) + 4 + (-18)}{2} = -12$$

Next, computing $\mathbf{B}_3$ gives

$$\mathbf{B}_2 - p_2\mathbf{I} = \begin{bmatrix} -10 & -2 & -1 \\ -13 & 4 & 5 \\ 18 & -21 & -18 \end{bmatrix} - (-12)\begin{bmatrix} 1 & 0 & 0 \\ 0 & 1 & 0 \\ 0 & 0 & 1 \end{bmatrix} = \begin{bmatrix} 2 & -2 & -1 \\ -13 & 16 & 5 \\ 18 & -21 & -6 \end{bmatrix}$$

$$\mathbf{B}_3 = \mathbf{A}(\mathbf{B}_2 - d_2\mathbf{I}) = \begin{bmatrix} 3 & 3 & 2 \\ 4 & 2 & 1 \\ -5 & 2 & 2 \end{bmatrix}\begin{bmatrix} 2 & -2 & -1 \\ -13 & 16 & 5 \\ 18 & -21 & -6 \end{bmatrix} = \begin{bmatrix} 3 & 0 & 0 \\ 0 & 3 & 0 \\ 0 & 0 & 3 \end{bmatrix}$$

and finally, we obtain

$$d_3 = \frac{1}{3}\text{Tr}[\mathbf{B}_3] = \frac{1}{3}\sum_{i=1}^{3} b_{ii} = \frac{3 + 3 + 3}{3} = 3$$

Now, changing the signs of d_k's, the coefficients of the characteristic polynomial are found as $d_1 = -7$, $d_2 = 12$ and $d_3 = -3$, i.e., the characteristic polynomial is found as

$$P(\lambda) = \lambda^3 - 7\lambda^2 + 12\lambda - 3 = 0$$

Since the roots of the characteristic polynomial are the eigenvalues, the roots in this case are found as $\lambda_1 = 0.300372$, $\lambda_2 = 2.23912$ and $\lambda_3 = 4.4605$. The determinant and the inverse of $\mathbf{A}$ are obtained as

$$\mathbf{A}^{-1} = \frac{1}{d_3}(\mathbf{B}_2 - d_2\mathbf{I}) = \frac{1}{3}\begin{bmatrix} 2 & -2 & -1 \\ -13 & 16 & 5 \\ 18 & -21 & -6 \end{bmatrix} \quad \text{and} \quad \det(\mathbf{A}) = (-1)^3 d_3 = -3$$

Discussions: The Faddeev-Leverrier process generates all the coefficients, c_i's, for the characteristic polynomial, $p_n(\lambda)$. We can then use a root-finding method to obtain all the eigenvalues from $p_n(\lambda) = 0$. Meanwhile, the process also allows the determinant and the inverse to be computed.

11.9 CHARACTERISTIC VALUE PROBLEMS

Characteristic value problems (CVPs) or *Eigenvalue problems (EVPs)* involve the solution of a differential equation with an undetermined parameter and associated boundary conditions. The solution of the differential equation exists only for certain values of a parameter (i.e., the eigenvalue).

To illustrate how eigenvalue problems arise, consider a constant vertical compressive force or a load of P applied to a thin column of uniform cross section (*see* Fig. 11.6). We assume that the column is pinned at both ends. The column deflection, $y(x)$, satisfies the

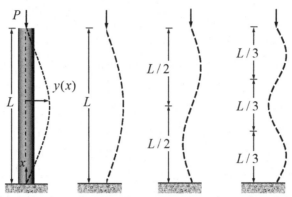

FIGURE 11.6: (a) A slender column under compressive stress, (b,c,d) buckling modes.

following two-point boundary value problem:

$$EI\frac{d^2y}{dx^2} + Py(x) = 0, \qquad y(0) = y(L) = 0 \qquad (11.107)$$

Rearranging and defining $k^2 = P/EI$, Eq. (11.107) becomes $y'' + k^2y = 0$. One of the solutions of this BVP is so-called the trivial solution ($y(x) = 0$) and is not of practical interest due to representing the neutral equilibrium state. However, this BVP has unique (i.e., non-trivial) solutions depending on the value of k denoting an eigenvalue, and the associated eigenvector is the solution of the BVP.

The column deflects or buckles only when the compressive force reaches a critical load (P_n). To determine the critical load, the BVP, a homogeneous second-order ODE, is solved, yielding

$$y(x) = c_1 \cos kx + c_2 \sin kx$$

where c_1 and c_2 are arbitrary constants.

Upon employing the BCs, the left BC yields

$$y(0) = c_1 \cos 0 + c_2 \sin 0 = c_1 = 0$$

Likewise, the right BC results in

$$y(L) = c_2 \sin kL = 0 \qquad (11.108)$$

Setting $c_2 = 0$ leads to the trivial solution; however, recall that we are seeking the non-trivial solutions, i.e., $y \neq 0$ and $c_2 \neq 0$. Nonetheless, for $kL = n\pi$, note that the BVP is satisfied for all $c_2 \neq 0$ and $n = 1, 2, 3$, and so on. In other words, this problem has infinitely many but discrete solutions (called *eigenfunctions*). For a specified value of k, we find

$$y(x) = c_2 \sin(\frac{n\pi x}{L})$$

where k now is replaced with $k_n = n\pi/L$, which leads to $P_{cr,n} = n^2\pi^2 EI/L^2$. But c_2 is indeterminate, and that is why the column displacement cannot be determined. This is to be expected since the column is in equilibrium or a neutral state. The smallest buckling corresponds to the $n = 1$ case; that is, $P_{cr,1} = \pi^2 EI/L^2$. The column buckles into a longitudinal half-sine wave seen in Fig. 11.6a. For $n = 2, 3$, and so on, the buckling loads

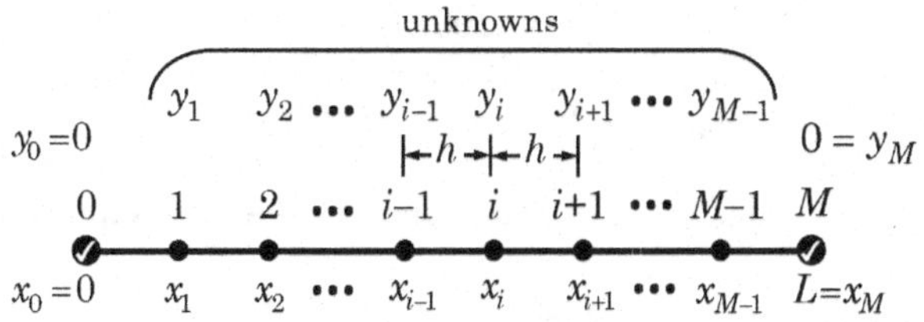

unknowns

$$
\begin{array}{ccccccc}
y_1 & y_2 \cdots & y_{i-1} & y_i & y_{i+1} & \cdots & y_{M-1}
\end{array}
$$

$y_0 = 0$ $|{\leftarrow}h{\rightarrow}|{\leftarrow}h{\rightarrow}|$ $0 = y_M$

FIGURE 11.7: Grid structure.

correspond to $P_{cr,2} = 4\pi^2 EI/L^2$, $P_{cr,3} = 9\pi^2 EI/L^2$, and so on, respectively. The larger values of buckling loads ($P_{cr,n}$, $n > 1$) correspond to more complex buckling loads, as depicted in Fig. 11.6b, 11.6c, and 11.6d. Theoretically, these buckling modes could occur; nevertheless, in reality, the lowest buckling load is never exceeded because the column breaks down due to high stresses.

11.9.1 NUMERICAL SOLUTION OF CVPs

Now let us examine step-by-step how to solve characteristic value problems using the finite difference method.

The procedure for solving CVPs is, for the most part, similar to that for solving BVPs. A CVP differs from the BVPs in that the final matrix equation is in the form of a standard or general eigenvalue problem, i.e., $\mathbf{Ax} = \lambda\mathbf{x}$ or $\mathbf{Ax} = \lambda\mathbf{Bx}$. The numerical solution steps are given as follows:

Step 1. Gridding. The first step is gridding of the physical domain $[0,L]$, which is divided into M uniformly spaced ($h = L/M$) subintervals (*see* Fig. 11.7). The nodes placed at both end points of each subinterval are enumerated starting from 0 to M, as depicted in Fig. 11.7. The corresponding abscissas are determined ($x_i = ih$ for $i = 0, 1, 2, \ldots, M$), and the ordinates are coded with short-hand notation, $y(x_i) = y_i$.

Step 2. Discretizing. The CVP is discretized for any nodal point x_i in $[0, L]$:

$$y_i'' + k^2 y_i = 0$$

The derivatives appearing in the differential equation are approximated by the CDFs. In this case, replacing y_i'' with the CDF gives

$$\frac{y_{i+1} - 2y_i + y_{i-1}}{h^2} + k^2 y_i = 0,$$

or rearranging results in the *general difference equation*:

$$y_{i+1} - 2y_i + y_{i-1} + (kh)^2 y_i = 0, \quad i = 1, 2, \ldots, (M-1) \tag{11.109}$$

Step 3. Implementing BCs. The implementation of the BCs is carried out in the same manner as in the case of BVPs. The solutions at the boundary nodes are known ($y_0 = y(0) = 0$ and $y_M = y(L) = 0$) due to the prescribed Dirichlet BCs. Hence, the difference equations for $i = 0$ and $i = M$ can be omitted, which leads to $M - 1$ unknown nodal points.

To implement the BCs, we concentrate on the first ($i = 1$) and last ($i = M-1$) difference equations. Substituting $i = 1$ and $y_0 = 0$ in Eq. (11.109) gives

$$-2y_1 + y_2 = -(kh)^2 y_1 \tag{11.110}$$

For $i = M-1$ and $y_M = 0$, similarly, Eq. (11.109) is reduced to

$$y_{M-2} - 2y_{M-1} = -(kh)^2 y_{M-1} \qquad (11.111)$$

For the interior nodes, $i = 2, 3, \ldots, M-2$, the general difference equation is valid.

Step 4. Setting up an Eigenvalue Problem. In the Dirichlet BC case, the term containing the known boundary node is moved to the rhs of the difference equation. In the Neumann or Robin BC cases, the resulting fictitious node is eliminated in the same way as in the treatment of the BVPs. Then, putting together all the difference equations (Eqs. (11.109), (11.110), and (11.111)) gives the following matrix form:

$$
\begin{bmatrix}
-2 & 1 & & & & \\
1 & -2 & 1 & & & \\
 & 1 & -2 & 1 & & \\
 & & \ddots & \ddots & \ddots & \\
 & & & 1 & -2 & 1 \\
 & & & & 1 & -2
\end{bmatrix}
\begin{bmatrix}
y_1 \\ y_2 \\ y_3 \\ \vdots \\ y_{M-2} \\ y_{M-1}
\end{bmatrix}
= -(kh)^2
\begin{bmatrix}
y_1 \\ y_2 \\ y_3 \\ \vdots \\ y_{M-2} \\ y_{M-1}
\end{bmatrix}
\qquad (11.112)
$$

By setting $\lambda = -(kh)^2$ and denoting the coefficient matrix and column vector as $\mathbf{A}$ and $\mathbf{y}$, respectively, Eq. (11.112) becomes $\mathbf{Ay} = \lambda \mathbf{y}$, i.e., the standard eigenvalue problem.

Step 5. Solving the Eigenvalue Problem. The solution to an eigenvalue problem can be found with a suitable analytical or numerical method, as discussed earlier in this chapter. Note that the set $\mathbf{x}_k$ is the solution (i.e., *eigenvector*) of the ODE corresponding to the predetermined discrete nodal points for the corresponding λ_k (i.e., eigenvalue).

A standard eigenvalue problem having Toeptliz tridiagonal form, as presented in the critical buckling load problem, has an analytical solution (*see* Section 2.1) and need not be solved numerically. We resort only to numerical methods in cases where analytical solutions are very difficult or impossible to obtain. Returning back to the buckling problem, the eigenpairs of the CVP can easily be found analytically since the coefficient matrix $\mathbf{A}$ is a symmetric Toeplitz tridiagonal matrix (i.e., $a = -2$ and $b = c = 1$). Using the analytical expression, the eigenvalues are found from

$$\lambda_i = -(hk_i)^2 = -2 + 2\cos(i\pi/M), \quad i = 1, 2, \ldots, (M-1)$$

For $M = 10$, the first two eigenvalues (λ_1 and λ_2) are obtained as -0.0978869 and -0.381966, respectively. Then evaluating k^2 yields

$$k_1^2 = -\frac{\lambda_1}{h^2} = 9.78869 \quad \text{and} \quad k_2^2 = -\frac{\lambda_2}{h^2} = 38.1966$$

On the other hand, recall that the analytical solution gave $k_1^2 = \pi^2 = 9.8696$ and $k_2^2 = (2\pi)^2 = 39.47864$. These values of k^2 are very close to those of the analytical solution. In other words, as the number of intervals is increased ($M \to \infty$), not only the eigenvalues but also the eigenvectors, approach the true values.

EXAMPLE 11.15: Finite difference solution of CVPs

Apply the finite difference method to solve the following CVP:

$$\frac{d}{dx}\left[(1+x)^2\frac{dy}{dx}\right] + \lambda y = 0, \qquad y'(0) = y(1) = 0$$

SOLUTION:

We implement the numerical solution step by step as follows:

Step 1. Gridding: The solution domain is divided into uniform M subintervals ($h = 1/M$), and the nodal points are indexed from 0 to M (*see* Fig. 11.8). The corresponding estimates (numerical solutions) are expressed as $y(x_i) = y_i$ where $x_i = ih$.

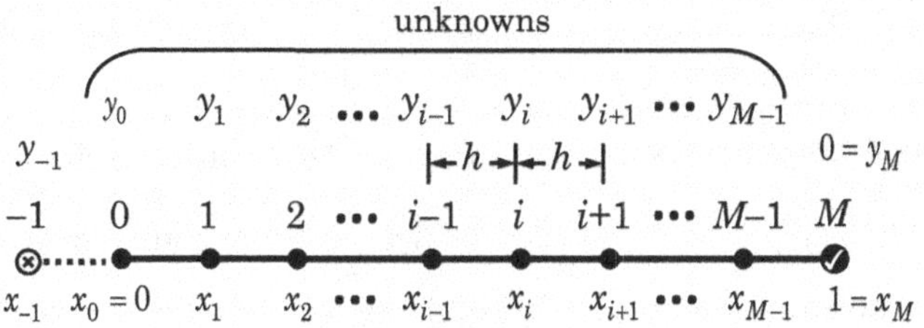

FIGURE 11.8

Step 2. Discretizing: The CVP is valid for an arbitrary x_i in (0,1); thus, we may write

$$(1+x_i)^2 y''_i + 2(1+x_i)y'_i + \lambda y_i = 0, \qquad x_i \in [0,1)$$

Approximating y'_i and y''_i with the CDFs yields

$$-(1+x_i)^2\left(\frac{y_{i+1} - 2y_i + y_{i-1}}{h^2}\right) - 2(1+x_i)\left(\frac{y_{i+1} - y_{i-1}}{2h}\right) = \lambda y_i \qquad (11.113)$$

For simplicity, letting $r_i = (1+x_i)/h$, the general difference equation is expressed as

$$r_i(1 - r_i)y_{i-1} + 2r_i^2 y_i - r_i(1 + r_i)y_{i+1} = \lambda y_i \qquad (11.114)$$

for $i = 1, 2, \ldots, (M-1)$.

Step 3. Implementing the BCs: The Neumann and Dirichlet BCs are imposed on the left and right boundary nodes, respectively, i.e., $y'(0) = y'_0 = 0$ and $y(1) = y_M = 0$. By dropping the right boundary node, the total number of unknowns becomes M.

For the left boundary, setting $i = 1$ in Eq. (11.114) leads to

$$r_0(1 - r_0)y_{-1} + 2r_0^2 y_0 - r_0(1 + r_0)y_1 = \lambda y_0 \qquad (11.115)$$

The fictitious node y_{-1} in Eq. (11.115) is eliminated by discretizing the left BC as $y'_0 \approx (y_1 - y_{-1})/2h = 0$, giving $y_{-1} = y_1$. Substituting this into Eq. (11.115) yields

$$2r_0^2 y_0 - 2r_0^2 y_1 = \lambda y_0 \qquad (11.116)$$

For the right boundary, setting $i = M-1$ and noting $y_M = 0$ results in

$$r_{M-1}(1 - r_{M-1})y_{M-2} + 2r_{M-1}^2 y_{M-1} = \lambda y_{M-1} \qquad (11.117)$$

For the interior nodes ($i = 1, \ldots, M - 2$), Eq. (11.114) is applied.

Step 4. Setting up the CVP: The CVP can be expressed in matrix equation form as

$$\begin{bmatrix} d_0 & a_0 & & & & \\ b_1 & d_1 & a_1 & & & \\ & b_2 & d_2 & a_2 & & \\ & & \ddots & \ddots & \ddots & \\ & & & b_{M-2} & d_{M-2} & a_{M-2} \\ & & & & b_{M-1} & a_{M-1} \end{bmatrix} \begin{bmatrix} y_0 \\ y_1 \\ y_2 \\ \vdots \\ y_{M-2} \\ y_{M-1} \end{bmatrix} = \lambda \begin{bmatrix} y_0 \\ y_1 \\ y_2 \\ \vdots \\ y_{M-2} \\ y_{M-1} \end{bmatrix}$$

where $b_i = r_i(1 - r_i)$, $d_i = 2r_i^2$, and $a_i = -r_i(1 + r_i)$ for $i = 0, 1, \ldots, M - 1$, except $a_0 = -2r_0^2$. The resulting tridiagonal matrix, however, is not symmetrical.

Step 5. Solving the EVP: For $M = 5$, we determine the eigenvalues from the roots of the characteristic polynomial:

$$\lambda^5 + 510\lambda^4 - 88508\lambda^3 + 6.213 \times 10^6 \lambda^2 - 1.567 \times 10^8 \lambda + 7.62 \times 10^8 = 0$$

which gives $\lambda_1 = 6.301$, $\lambda_2 = 40.799$, $\lambda_3 = 90.620$, $\lambda_4 = 142.140$, and $\lambda_5 = 230.137$.

Discussion: Problems for large number of nodes lead to large matrices. Determining the eigenvalues by the characteristic polynomial is not be attempted as the polynomial will become ill-conditioned.

EXAMPLE 11.16: Numerical solution of a CVP

Apply the finite difference method with five uniform intervals to solve the following characteristic value problem:

$$y'' + k^2 x^2 y = 0, \qquad y(0) = 0, \quad 4y'(1) - y(1) = 0$$

SOLUTION:

Step 1. Gridding: The solution interval is $(0,1)$, and the BCs are given as Dirichlet (left BC) and Robin BC (right BC). The interval is divided into uniform M subintervals ($h = (1 - 0)/M$) to create one-dimensional grids. The nodal points placed on both sides of the intervals are numbered from 0 to M, and the corresponding abscissas are found from $x_i = ih$, and the ordinates are set to $y(x_i) = y_i$ (see Fig. 11.9). The total number of unknowns (nodes) is M.

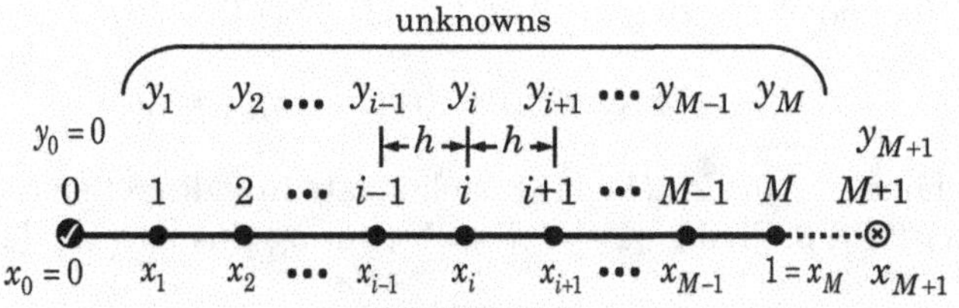

FIGURE 11.9

Step 2. Discretizing: The differential equation for any node within $(0,1)$ is

$$y''_i + k^2 x_i^2 y_i = 0, \quad x_i \in (0,1)$$

Substituting the CDF for y_i'' leads to

$$\frac{y_{i+1} - 2y_i + y_{i-1}}{h^2} + k^2 x_i^2 y_i = 0$$

Rearranging this equation yields the general difference equation:

$$-y_{i-1} + 2y_i - y_{i+1} = (kh)^2 x_i^2 y_i \qquad (11.118)$$

Step 3. Implementing the BCs: The left boundary nodal value is known ($y(0) = y_0 = 0$), and the right boundary nodal value is unknown ($4y_M' - y_M = 0$). Setting $i = 1$ and substituting $y_0 = 0$ (left BC) into Eq. (11.118) gives

$$2y_1 - y_2 = (kh)^2 x_1^2 y_1$$

Setting $i = M$ in Eq. (11.118) results in

$$-y_{M-1} + 2y_M - y_{M+1} = (kh)^2 y_M$$

which creates a fictitious node (y_{M+1}). This fictitious node, as usual, is eliminated by discretizing the right BC with the CDF as follows:

$$4\frac{y_{M+1} - y_{M-1}}{2h} - y_M = 0 \quad \rightarrow \quad y_{M+1} = y_{M-1} + \frac{h}{2}y_M$$

Substituting y_{M+1}, the difference equation for $i = M$ takes the following form:

$$-2y_{M-1} + \left(2 - \frac{h}{2}\right) y_M = (kh)^2 x_i^2 y_i$$

For the interior nodes ($i = 1, 2, \ldots, M-1$), Eq. (11.118) is valid.

Step 4. Setting up the Eigenvalue Problem: Finally, setting $\lambda = (kh)^2$ and combining all difference equations, a generalized eigenvalue problem is obtained, $\mathbf{Ay} = \lambda \mathbf{By}$, or explicitly

$$\mathbf{A} = \begin{bmatrix} 2 & -1 & & & & \\ -1 & 2 & -1 & & & \\ & -1 & 2 & -1 & & \\ & & \ddots & \ddots & \ddots & \\ & & & -1 & 2 & -1 \\ & & & & -1 & 1.9 \end{bmatrix}, \quad \mathbf{B} = \begin{bmatrix} x_1^2 & & & & & \\ & x_2^2 & & & & \\ & & x_3^2 & & & \\ & & & \ddots & & \\ & & & & x_{M-1}^2 & \\ & & & & & x_M^2 \end{bmatrix}$$

Step 5. Solving the EVP: The Cholesky decomposition method is applied to solve the resulting CVP by transforming it to a standard $\mathbf{CZ} = \lambda \mathbf{Z}$ form, where $\mathbf{C} = \mathbf{L}^{-1}\mathbf{AL}^{-T}$. For $M = 5$, the Cholesky decomposition yields

$$\mathbf{L} = \begin{bmatrix} 0.2 & 0 & 0 & 0 & 0 \\ 0 & 0.4 & 0 & 0 & 0 \\ 0 & 0 & 0.6 & 0 & 0 \\ 0 & 0 & 0 & 0.8 & 0 \\ 0 & 0 & 0 & 0 & 1 \end{bmatrix}, \quad \mathbf{L}^{-1} = \begin{bmatrix} 5 & 0 & 0 & 0 & 0 \\ 0 & 5/2 & 0 & 0 & 0 \\ 0 & 0 & 5/3 & 0 & 0 \\ 0 & 0 & 0 & 5/4 & 0 \\ 0 & 0 & 0 & 0 & 1 \end{bmatrix}$$

$$C = \begin{bmatrix} 50 & -12.5 & 0 & 0 & 0 \\ -12.5 & 12.5 & -25/6 & 0 & 0 \\ 0 & -25/6 & 50/9 & -25/12 & 0 \\ 0 & 0 & -25/12 & 3.125 & -1.25 \\ 0 & 0 & 0 & -1.25 & 1.9 \end{bmatrix}$$

Discussion: Notice that C is a symmetric tridiagonal matrix. By implementing the QR algorithm, the eigenvalues are obtained as 53.8152, 11.5853, 4.8848, 2.2513, and 0.5440. Then, k^2 values are computed from $k^2 = \lambda/h^2$, yielding 36.6794, 17.0186, 11.051, 7.5022, and 3.6878 for $k = 1, 2, \ldots, 5$, respectively.

11.10 CLOSURE

Eigenvalues and eigenvectors have a wide and diverse range of applications, some requiring the calculation of a single eigenvalue, others requiring the computation of all eigenvalues. The computation of the associated eigenvector may or may not be required. In this regard, there are a variety of methods in the literature on this subject that serve different purposes.

Many physical problems require either the largest or smallest (in magnitude) eigenvalue and its associated eigenvector. The power method and the inverse power method provide the numerical means to compute the largest or smallest eigenvalue as well as associated eigenvectors, respectively. The convergence rate of the power method is slow when the dominance ratio approaches 1, i.e., $\lambda_1 \approx \lambda_1$. The shift technique, on the other hand, also allows the calculation of an eigenvalue close to a specified value.

The Jacobi method, designed for symmetric matrices, uses similarity transformations based on plane rotations to reduce a matrix into a diagonal form. This method is iterative in nature, and it may require many rotations in order to obtain the diagonal matrix.

Computing all eigenvalues of a symmetric matrix is handled in two steps: (1) transforming the matrix into a symmetric tridiagonal form, and (2) applying an iterative procedure to generate a sequence of matrices that eventually converge to a diagonal matrix whose diagonal elements are the eigenvalues. The Householder method yields the desired symmetrical tridiagonal form by applying a series of similarity transformations to a symmetric matrix. A common method to numerically obtain all eigenvalues of a symmetric tridiagonal matrix is the QR algorithm, which is an iterative method. A sequence of matrices generated as a result of this iterative procedure converges to a diagonal matrix whose elements are the eigenvalues of the original matrix. The Gram-Schmid algorithm, which does not require iterations, performs QR decomposition in n-steps. The method for large matrices is plagued with round-off errors, leading to loss of orthogonality.

The Sturm sequence method is employed on a symmetric tridiagonal matrix (of order $n < 10$) to determine its characteristic polynomial. Another method is needed to find the roots of the characteristic polynomial (eigenvalues). Moreover, a separate method is also required to compute the eigenvectors, if it is needed. The Faddeev-Leverrier method is a general method that gives the characteristic equation, which should be solved by any method calculating the roots of a polynomial.

Since eigenvalue problems can yield very large matrices, the numerical computation of eigenpairs may lead to an accumulation of round-off errors. Thus, numerical computations with high precision should be carried out.

11.11 EXERCISES

Section 11.1 Eigenvalue Problem and Properties

E11.1 Find the characteristic polynomial, eigenvalues, and associated eigenvectors (normalized by ℓ_∞−norm) of the following matrices:

(a) $\begin{bmatrix} 3 & 2 \\ -3 & -4 \end{bmatrix}$,   (b) $\begin{bmatrix} 11 & 4 \\ -15 & -5 \end{bmatrix}$,   (c) $\begin{bmatrix} -5 & -3 \\ 9 & 7 \end{bmatrix}$,

(d) $\begin{bmatrix} 1 & -1 & 3 \\ 1 & 1 & 2 \\ 3 & 1 & 4 \end{bmatrix}$,   (e) $\begin{bmatrix} -5 & -2 & -2 \\ 4 & 4 & 1 \\ 6 & -3 & 4 \end{bmatrix}$,   (f) $\begin{bmatrix} -3 & 1 & 9 \\ -10 & 8 & 1 \\ -3 & 3 & -3 \end{bmatrix}$

E11.2 For matrices given in **E11.1**, show that the sum of the eigenvalues of a matrix is equal to its trace.

E11.3 For matrices given in **E11.1**, show that the product of the eigenvalues of a matrix is equal to its determinant.

E11.4 For matrices given in **E11.1** and $\beta = 5$, show that the eigenpair of $\beta\mathbf{A}$ is $(\beta\lambda, \mathbf{x})$.

E11.5 For matrices given in **E11.1**, show that the eigenpair of $\mathbf{A}^2$ is $(\lambda^2, \mathbf{x})$.

E11.6 For matrices given in **E11.1**, show that the eigenpair of $\mathbf{A}^{-1}$ is $(\lambda^{-1}, \mathbf{x})$.

E11.7 Find the eigenpairs of the following triangular matrices:

(a) $\begin{bmatrix} 3 & 4 \\ 0 & -3 \end{bmatrix}$ (b) $\begin{bmatrix} 4 & 0 \\ 5 & -5 \end{bmatrix}$ (c) $\begin{bmatrix} 2 & 3 & -3 \\ 0 & -5 & -4 \\ 0 & 0 & 4 \end{bmatrix}$ (d) $\begin{bmatrix} 6 & 0 & 0 \\ 2 & -2 & 0 \\ 3 & 2 & 1 \end{bmatrix}$

E11.8 Find the eigenpairs of the following diagonal matrices.

(a) $\begin{bmatrix} 3 & 0 \\ 0 & -5 \end{bmatrix}$ (b) $\begin{bmatrix} 7 & 0 \\ 0 & 3 \end{bmatrix}$ (c) $\begin{bmatrix} -4 & 0 & 0 \\ 0 & 2 & 0 \\ 0 & 0 & 10 \end{bmatrix}$ (d) $\begin{bmatrix} 6 & 0 & 0 \\ 0 & 3 & 0 \\ 0 & 0 & -9 \end{bmatrix}$

E11.9 For the given matrices, plot the Gerschgorin disks in the complex plane and mark the location of the eigenvalues.

(a) $\begin{bmatrix} 2 & 5 & 0 \\ -2 & 0 & 0 \\ 0 & 1 & 4 \end{bmatrix}$ (b) $\begin{bmatrix} -3 & -5 & 0 \\ 4 & -7 & 1 \\ 0 & 0 & 7 \end{bmatrix}$ (c) $\begin{bmatrix} -4 & -1 & 1 \\ 2 & -2 & -1 \\ -1 & 1 & -6 \end{bmatrix}$

(d) $\begin{bmatrix} -8 & -2 & 4 & 0 \\ 4 & -4 & 0 & 2 \\ 0 & 2 & 2 & 21 \\ 0 & 2 & -2 & 12 \end{bmatrix}$ (e) $\begin{bmatrix} -5 & -9 & 0 & 1 \\ 1 & -4 & 1 & 1 \\ -1 & 0 & 5 & -1 \\ 1 & 2 & -1 & 4 \end{bmatrix}$ (f) $\begin{bmatrix} -10 & 1 & 0 & 1 \\ -1 & -4 & -1 & -1 \\ 1 & -2 & 3 & -2 \\ 1 & 1 & -1 & 9 \end{bmatrix}$

E11.10 Consider the symmetric tridiagonal matrix with $a_{ii} = 2$ and $a_{i+1,i} = a_{i,i+1} = 1$ for $i = 1, 2, \ldots, n$. Use the Gerschgorin theorem to determine the effect of the matrix size ($n = 3, 5, 10$) on the location of the eigenvalues.

E11.11 (a) Find the eigenpairs of the following symmetric matrix given below; (b) Determine whether the eigenvectors are orthogonal or not.

$$\begin{bmatrix} 8 & -2 & 4 & -1 & -3 \\ -2 & 10 & 0 & 2 & -2 \\ 4 & 0 & 11 & 4 & 0 \\ -1 & 2 & 4 & 8 & 3 \\ -3 & -2 & 0 & 3 & 12 \end{bmatrix}$$

Section 11.2 Power Method

E11.12 Apply the power method with scaling to determine the dominant eigenvalue of the following matrices. For a starting guess, use (i) $\mathbf{x}^{(0)} = [1, 1, 1]^T$, (ii) $\mathbf{x}^{(0)} = [1, 1, 0]^T$ and (iii) $\mathbf{x}^{(0)} = [1, 0, 0]^T$. Use $\varepsilon = 10^{-6}$.

(a) $\mathbf{A} = \begin{bmatrix} 3 & -8 & 1 \\ -8 & -4 & -8 \\ 1 & -8 & 3 \end{bmatrix}$ (b) $\mathbf{B} = \begin{bmatrix} 2 & -1 & 2 \\ 1 & 0 & -2 \\ 2 & -2 & 3 \end{bmatrix}$ (c) $\mathbf{C} = \begin{bmatrix} 5 & 4 & -2 \\ -1 & 1 & 1 \\ 1 & 2 & -1 \end{bmatrix}$

E11.13 Apply the power method with scaling to find the largest eigenvalue of the following non-symmetric matrices, correct to within $\varepsilon = 10^{-6}$. Use $\mathbf{x}^{(0)} = \mathbf{1}$.

(a) $\begin{bmatrix} 8 & 4 & 4 \\ 6 & -2 & 2 \\ 1 & 2 & 12 \end{bmatrix}$ (b) $\begin{bmatrix} -13 & -15 & -15 \\ -1 & 17 & -1 \\ 31 & 15 & 33 \end{bmatrix}$ (c) $\begin{bmatrix} 5 & 15 & -1 \\ 1 & -9 & 1 \\ -16 & -15 & -10 \end{bmatrix}$

(d) $\begin{bmatrix} -2 & 1 & 9 \\ 1 & 8 & 1 \\ 2 & 3 & -3 \end{bmatrix}$ (e) $\begin{bmatrix} 2 & 3 & 8 \\ 3 & 2 & 8 \\ -9 & 4 & -9 \end{bmatrix}$ (f) $\begin{bmatrix} 1 & -2 & -6 \\ 2 & 2 & 4 \\ -2 & 1 & 0 \end{bmatrix}$

E11.14 Apply the Power method with scaling to find the largest eigenvalue of the following non-symmetric 4×4 matrices, correct to within $\varepsilon = 10^{-6}$. Use $\mathbf{x}^{(0)} = \mathbf{1}$.

(a) $\begin{bmatrix} 6 & 2 & 3 & 2 \\ -6 & 0 & -5 & -4 \\ -4 & -2 & -1 & -2 \\ 4 & 1 & 3 & 5 \end{bmatrix}$ (b) $\begin{bmatrix} 12 & 5 & 9 & 5 \\ -3 & 0 & -5 & -1 \\ -10 & -5 & -7 & -5 \\ -5 & -2 & -3 & -1 \end{bmatrix}$ (c) $\begin{bmatrix} -1 & 1 & 0 & 1 \\ -6 & -4 & -4 & -5 \\ -2 & -1 & -3 & -1 \\ 8 & 2 & 6 & 3 \end{bmatrix}$

(d) $\begin{bmatrix} 4 & -5 & -2 & -7 \\ -1 & -1 & -3 & -2 \\ -2 & 1 & 0 & 3 \\ 3 & -2 & 1 & -3 \end{bmatrix}$ (e) $\begin{bmatrix} 8 & 9 & 10 & 9 \\ -13 & -12 & -12 & -11 \\ -18 & -9 & -20 & -9 \\ 11 & 1 & 10 & 0 \end{bmatrix}$ (f) $\begin{bmatrix} -48 & -19 & -41 & -19 \\ 2 & -1 & 15 & -7 \\ 38 & 19 & 31 & 19 \\ 42 & 13 & 29 & 19 \end{bmatrix}$

E11.15 Repeat **E11.13**, using the power method with the Rayleigh quotient.

E11.16 Repeat **E11.14**, using the power method with the Rayleigh quotient.

E11.17 Apply the power method with the Rayleigh quotient to find the dominant eigenvalue of the following symmetric matrices, correct to within $\varepsilon = 10^{-6}$. Use $\mathbf{x}^{(0)} = \mathbf{1}$. How many iterations are required to find the dominant eigenvalue?

$$\mathbf{A}_1 = \begin{bmatrix} 2 & 1 & 0 & 0 & 0 \\ 1 & 2 & 1 & 0 & 0 \\ 0 & 1 & 2 & 1 & 0 \\ 0 & 0 & 1 & 2 & 1 \\ 0 & 0 & 0 & 1 & 2 \end{bmatrix}, \quad \mathbf{A}_2 = \begin{bmatrix} 4 & 1 & 0 & 0 & 0 \\ 1 & 4 & 1 & 0 & 0 \\ 0 & 1 & 4 & 1 & 0 \\ 0 & 0 & 1 & 4 & 1 \\ 0 & 0 & 0 & 1 & 4 \end{bmatrix}, \quad \mathbf{A}_3 = \begin{bmatrix} 8 & 1 & 0 & 0 & 0 \\ 1 & 8 & 1 & 0 & 0 \\ 0 & 1 & 8 & 1 & 0 \\ 0 & 0 & 1 & 8 & 1 \\ 0 & 0 & 0 & 1 & 8 \end{bmatrix}$$

E11.18 To find the minimum eigenvalue of the following matrices, first find the inverse matrices, and then apply the power method with scaling. Use $\mathbf{x}^{(0)} = \mathbf{1}$ and $\varepsilon = 10^{-6}$.

(a) $\begin{bmatrix} 2 & 3 & 8 \\ 3 & 2 & 8 \\ -9 & 4 & -9 \end{bmatrix}$ (b) $\begin{bmatrix} 1 & -2 & -6 \\ 2 & 2 & 4 \\ -2 & 1 & 0 \end{bmatrix}$ (c) $\begin{bmatrix} -2 & 1 & 1 \\ 3 & 3 & 2 \\ -1 & 4 & 4 \end{bmatrix}$

(d) $\begin{bmatrix} 1 & -1 & 3 \\ 1 & 1 & 2 \\ 3 & 1 & 4 \end{bmatrix}$ (e) $\begin{bmatrix} -4 & 1 & 1 & 1 \\ 1 & 7 & 1 & -1 \\ 2 & -1 & -3 & 2 \\ 1 & 1 & 2 & 4 \end{bmatrix}$ (f) $\begin{bmatrix} 9 & 0 & 1 & -2 \\ 6 & 3 & 1 & 0 \\ 1 & -2 & 5 & 1 \\ 0 & -1 & 1 & 2 \end{bmatrix}$

E11.19 Apply the power method with scaling to find the smallest eigenvalue of the matrices in **E11.13**, correct to within $\varepsilon = 10^{-6}$. Start iterations with $\mathbf{x}^{(0)} = \mathbf{1}$.

E11.20 Apply the power method with scaling to find the smallest eigenvalue of the matrices in **E11.14**, correct to within $\varepsilon = 10^{-6}$. Start iterations with $\mathbf{x}^{(0)} = \mathbf{1}$.

E11.21 Repeat **E11.17** to find the smallest eigenvalue.

E11.22 For a given matrix $\mathbf{A}$, (a) find its eigenpairs; (b) perform two iterations using the Power method with scaling starting and $\mathbf{x} = \mathbf{1}$; (c) does the method appear to converge to the dominant eigenvector?; and (d) repeat Part (b) with four iterations using the Rayleigh Quotient method. What do you observe?

$$\mathbf{A} = \begin{bmatrix} 3 & -1 & 1 \\ -1 & 5 & -1 \\ 1 & -1 & 3 \end{bmatrix}$$

E11.23 Find the dominant eigenvalue of the matrix below using the Power method with (a) scaling and (b) Rayleigh quotient, and determine how many iterations are required for convergence with both methods. Use $\mathbf{x} = \mathbf{1}$ as the starting guess. Apply both convergence criteria: $\|\mathbf{x}\|_\infty < 10^{-6}$ and $|1 - \lambda^{(p-1)}/\lambda^{(p)}| < 10^{-6}$.

$$\begin{bmatrix} 1 & 2 & -1 & 1 & -1 \\ -2 & 8 & 1 & 3 & -2 \\ 1 & -1 & 4 & -2 & 2 \\ 3 & 2 & -2 & -9 & 3 \\ 1 & 2 & -2 & -1 & -3 \end{bmatrix}$$

E11.24 Apply the shifted-inverse power algorithm to find the dominant eigenvalue of the matrix in **E11.23** for $\alpha = -11, -10, -9, -8.5, -8,$ and -7. Use $\varepsilon = 10^{-6}$ and $\mathbf{x} = \mathbf{1}$ as the starting guess.

E11.25 Apply the shifted-inverse power algorithm to find the eigenvalues of the matrix given in **E11.23** in the neighborhood of $\alpha = -3$ and $+3$. How many iterations did it require to reach a converged solution? Use $\varepsilon = 10^{-6}$ and $\mathbf{x} = \mathbf{1}$ as the starting guess.

E11.26 Determine all eigenvalues of the following matrices by applying the shifted inverse power method with the Rayleigh quotient and Gerschgorin theorem. Use $\varepsilon = 10^{-6}$ and $\mathbf{x} = \mathbf{1}$ as the starting guess.

(a) $\begin{bmatrix} -48 & -19 & -41 & -19 \\ 2 & -1 & 15 & -7 \\ 38 & 19 & 31 & 19 \\ 42 & 13 & 29 & 19 \end{bmatrix}$ (b) $\begin{bmatrix} -11 & -5 & -8 & -5 \\ -13 & -5 & -6 & -9 \\ 10 & 5 & 7 & 5 \\ 19 & 7 & 12 & 11 \end{bmatrix}$ (c) $\begin{bmatrix} 10 & 3 & -1 & 1 \\ 3 & 5 & -1 & -1 \\ -1 & -1 & 9 & 1 \\ 1 & -1 & 1 & 6 \end{bmatrix}$

E11.27 Determine the two smallest (in magnitude) eigenvalues of the following matrix by applying the shifted inverse power method with the Rayleigh quotient. Make use of the Gerschgorin theorem to find a good initial guess for the eigenvalue. Use $\varepsilon = 10^{-6}$ and $\mathbf{x} = \mathbf{1}$ as the starting guess.

$$\begin{bmatrix} 30 & -1 & 1 & 2 & -1 \\ -1 & -16 & -3 & -1 & 0 \\ 1 & -3 & 26 & 2 & 1 \\ 2 & -1 & 2 & 8 & -1 \\ -1 & 0 & 1 & -1 & 5 \end{bmatrix}$$

Section 11.3 Similarity and Orthogonal Transformation

E11.28 Are the following 2×2 matrices similar?

(a) $\mathbf{A} = \begin{bmatrix} 1 & 0 \\ 0 & 1 \end{bmatrix}$ $\mathbf{B} = \begin{bmatrix} 1 & 0 \\ -1 & 1 \end{bmatrix}$, (b) $\mathbf{A} = \begin{bmatrix} 1 & 2 \\ 3 & -4 \end{bmatrix}$ $\mathbf{B} = \begin{bmatrix} 7 & 15 \\ -4 & -10 \end{bmatrix}$,

(c) $\mathbf{A} = \begin{bmatrix} 0 & 2 \\ 0 & 0 \end{bmatrix}$ $\mathbf{B} = \begin{bmatrix} 0 & 0 \\ 2 & 0 \end{bmatrix}$

E11.29 Are the following 3×3 matrices similar?

$$\mathbf{A} = \begin{bmatrix} 3 & -6 & 2 \\ 2 & -3 & -2 \\ 1 & -1 & -2 \end{bmatrix}, \qquad \mathbf{B} = \begin{bmatrix} 13 & -32 & 3 \\ 6 & -7 & 3 \\ -22 & 40 & -8 \end{bmatrix}$$

E11.30 For given matrices, obtain the modal matrix $\mathbf{P}$ and find $\mathbf{P}^{-1}\mathbf{AP}$.

(a) $\mathbf{A} = \begin{bmatrix} 1 & 3 \\ 3 & 1 \end{bmatrix}$ (b) $\mathbf{A} = \begin{bmatrix} 5 & 1 \\ 2 & 4 \end{bmatrix}$ (c) $\mathbf{A} = \begin{bmatrix} 4 & -1 & 2 \\ -1 & 5 & -1 \\ 2 & -1 & 4 \end{bmatrix}$ (d) $\mathbf{A} = \begin{bmatrix} 3 & 2 & -1 \\ 1 & 4 & 0 \\ -2 & 2 & 3 \end{bmatrix}$

E11.31 Use the given matrices to evaluate $\mathbf{P}_1^{-1}\mathbf{AP}_1$, $\mathbf{P}_2^{-1}\mathbf{AP}_2$, and $\mathbf{P}_3^{-1}\mathbf{AP}_3$.

	A	**P₁**	**P₂**	**P₃**
(a)	$\begin{bmatrix} 4 & 1 \\ 3 & 2 \end{bmatrix}$	$\begin{bmatrix} 1 & -1 \\ 1 & 3 \end{bmatrix}$	$\begin{bmatrix} -1 & 1 \\ 3 & 1 \end{bmatrix}$	$\begin{bmatrix} 2 & -2 \\ 2 & 6 \end{bmatrix}$
(b)	$\begin{bmatrix} 5 & 3 \\ 6 & -2 \end{bmatrix}$	$\begin{bmatrix} 3 & -1 \\ 2 & 3 \end{bmatrix}$	$\begin{bmatrix} -1 & 3 \\ 3 & 2 \end{bmatrix}$	$\begin{bmatrix} 9 & -3 \\ 6 & 9 \end{bmatrix}$
(c)	$\begin{bmatrix} 23 & -1 & 0 \\ -1 & 22 & -1 \\ 0 & -1 & 23 \end{bmatrix}$	$\begin{bmatrix} 1 & -1 & 1 \\ -1 & 0 & 2 \\ 1 & 1 & 1 \end{bmatrix}$	$\begin{bmatrix} -1 & 1 & 1 \\ 0 & -1 & 2 \\ 1 & 1 & 1 \end{bmatrix}$	$\begin{bmatrix} 1 & -1 & 1 \\ -1 & 0 & 1 \\ 1 & 2 & 1 \end{bmatrix}$
(d)	$\begin{bmatrix} 1 & 6 & -8 \\ -8 & 5 & 8 \\ -3 & 3 & 2 \end{bmatrix}$	$\begin{bmatrix} 1 & 8 & 5 \\ 2 & 8 & 4 \\ 1 & 5 & 3 \end{bmatrix}$	$\begin{bmatrix} 1 & 2 & 1 \\ 8 & 8 & 5 \\ 5 & 4 & 3 \end{bmatrix}$	$\begin{bmatrix} 1 & 8 & 5 \\ -1 & -4 & -2 \\ 1 & 5 & 3 \end{bmatrix}$

E11.32 Apply the similarity identity to evaluate $\mathbf{A}^{10}$.

(a) $\mathbf{A} = \begin{bmatrix} 1 & 1 \\ 2 & 2 \end{bmatrix}$ (b) $\mathbf{A} = \begin{bmatrix} 1 & 1 & 1 \\ 1 & 1 & 0 \\ 0 & 0 & 1 \end{bmatrix}$ (c) $\mathbf{A} = \begin{bmatrix} 1 & 0 & 3 \\ 1 & 1 & 4 \\ 3 & 0 & 1 \end{bmatrix}$

E11.33 (a) Find the eigenpairs of $\mathbf{A}$; (b) determine the matrix inverse by $\mathbf{A}^{-1} = \mathbf{PD}^{-1}\mathbf{P}^{-1}$.

$$\mathbf{A} = \begin{bmatrix} 11 & -1 & 5 & 1 \\ -1 & 11 & 1 & 5 \\ 5 & 1 & 11 & -1 \\ 1 & 5 & -1 & 11 \end{bmatrix}$$

E11.34 If possible, diagonalize the following matrices:

(a) $\begin{bmatrix} 3 & 12 & -16 \\ 1 & 2 & 2 \\ 1 & 2 & 1 \end{bmatrix}$ (b) $\begin{bmatrix} 3 & -1 & 1 \\ -1 & 3 & -1 \\ 1 & -1 & 5 \end{bmatrix}$ (c) $\begin{bmatrix} 2 & 1 & -1 \\ -3 & 3 & 2 \\ 1 & 1 & 0 \end{bmatrix}$

(d) $\begin{bmatrix} 6 & 2 & 3 & 2 \\ -6 & 0 & -5 & -4 \\ -4 & -2 & -1 & -2 \\ 4 & 1 & 3 & 5 \end{bmatrix}$ (e) $\begin{bmatrix} 13 & 1 & 1 & 1 \\ 1 & 13 & 1 & 1 \\ 1 & 1 & 13 & 1 \\ 1 & 1 & 1 & 13 \end{bmatrix}$ (f) $\begin{bmatrix} 4 & 13 & 9 & 8 \\ 7 & -29 & -33 & 3 \\ 1 & 29 & 29 & -1 \\ 2 & 4 & 0 & 12 \end{bmatrix}$

E11.35 The net equilibrium force acting on a free-falling body yields the following first-order coupled ODEs:

$$m\frac{dv}{dt} = mg - Cv, \quad \frac{dh}{dt} = -v \quad \text{with} \quad h(0) = h_0, \quad v(0) = 0$$

where h is altitude, v is free-fall speed, g is acceleration of gravity, and C is the drag coefficient experienced during the fall. (a) Express the ODEs in matrix form, and (b) apply the diagonalization procedure to be able to find an analytical solution for the altitude and velocity.

E11.36 Consider the circuit given in **Fig. E11.36**. Current $I(t)$ through the inductor and voltage $V(t)$ across the capacitor satisfy the following coupled first-order ODEs:

$$\frac{dV}{dt} = -\frac{1}{CR_1}V(t) + \frac{1}{C}I(t), \quad \frac{dI}{dt} = -\frac{1}{L}V(t) - \frac{R_2}{L}I(t)$$

with $V(0) = 10$ V and $I(0) = 2$ A. Apply the diagonalization procedure to uncouple the system of ODEs and find the analytical solution for the current and voltage if $R_1 = 10\Omega$, $R_2 = 5\ \Omega$, $L = 2$ H, and $C = 50$ F.

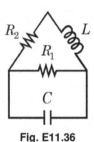

Fig. E11.36

E11.37 Consider the radioactive decay chain, $A \xrightarrow{\lambda_1} B \xrightarrow{\lambda_2} C \xrightarrow{\lambda_3} D$, which satisfies the following coupled first-order linear ODEs:

$$\frac{dN_A}{dt} = -\lambda_1 N_A, \quad \frac{dN_B}{dt} = \lambda_1 N_A - \lambda_2 N_B, \quad \frac{dN_C}{dt} = \lambda_2 N_B - \lambda_3 N_C$$

with $N_A(0) = 1$, $N_B(0) = 0$, and $N_C(0) = 0$. (a) Express the coupled ODEs in matrix form; (b) apply a diagonalization procedure to find an analytical solution for N_A, N_B, and N_C. Given: $\lambda_1 = 1$, $\lambda_2 = 0.75$, and $\lambda_3 = 0.20$ s^{-1}.

E11.38 Two objects (A and B) with initial temperatures $T_A(0) = 500°C$ and $T_B(0) = 300°C$ are placed in a room at $T_r(0) = 10°C$, as shown in **Fig. E11.38**. Newton's law of cooling yields the following first-order coupled ODEs for the temperatures:

$$\frac{dT_r}{dt} = \frac{3}{2000}(T_r - T_A) + \frac{5}{2000}(T_r - T_B)$$

$$\frac{dT_A}{dt} = -\frac{2}{15}(T_A - T_r), \quad \frac{dT_B}{dt} = -\frac{1}{16}(T_B - T_r)$$

(a) Express the set of coupled ODEs in matrix form; (b) apply the diagonalization procedure to find an analytical solution for the temperatures.

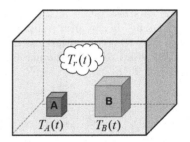

Fig. E11.38

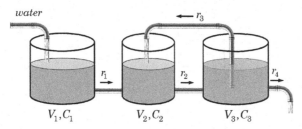

Fig. E11.39

E11.39 Three tanks with volumes V_1, V_2, and V_3 contain brine, as shown in **Fig. E11.39**. Salt concentrations in the tanks are denoted by C_1, C_2, and C_3. The first tank is fed water at a rate of 15 m^3/min. The feed rates between the tanks are $r_1 = 2.5$ m^3/h, $r_2 = 1.8$ m^3/h, and $r_3 = 2$ m^3/h. A mathematical model for the salt content as a function of time is given as

$$\frac{dC_1}{dt} = -r_1 \frac{C_1}{V_1}, \qquad \frac{dC_2}{dt} = r_1 \frac{C_1}{V_1} + r_3 \frac{C_3}{V_3} - r_2 \frac{C_2}{V_2}, \qquad \frac{dC_3}{dt} = r_2 \frac{C_2}{V_2} - r_3 \frac{C_3}{V_3}$$

with the ICs: $C_1(0) = 10$, $C_2(0) = 2$, and $C_3(0) = 1$ kg. (a) express the ODE set in matrix form; (b) apply the diagonalization procedure to find the analytical solutions for the salt contents. Given: $V_1 = 120$, $V_2 = 50$, and $V_3 = 120$ m^3.

Section 11.4 Jacobi Method

E11.40 Use the Jacobi method to find the eigenvalues and associated eigenvectors of the following matrices ($\varepsilon = 10^{-6}$).

(a) $\begin{bmatrix} 2 & 1 & 2 \\ 1 & 4 & -1 \\ 2 & -1 & 5 \end{bmatrix}$
(b) $\begin{bmatrix} 4 & 2 & 1 \\ 2 & 5 & -1 \\ 1 & -1 & 8 \end{bmatrix}$
(c) $\begin{bmatrix} 3 & -2 & 3 \\ -2 & 5 & 4 \\ 3 & 4 & 1 \end{bmatrix}$

(d) $\begin{bmatrix} 3 & -2 & 1 & -1 \\ -2 & 7 & 3 & 5 \\ 1 & 3 & 6 & -3 \\ -1 & 5 & -3 & 8 \end{bmatrix}$
(e) $\begin{bmatrix} 1 & 5 & 2 & -3 \\ 5 & 3 & 2 & -1 \\ 2 & 2 & 7 & 1 \\ -3 & -1 & 1 & 4 \end{bmatrix}$
(f) $\begin{bmatrix} 4 & 1 & 1 & -2 \\ 1 & 4 & -2 & -1 \\ 1 & -2 & 5 & 3 \\ -2 & -1 & 3 & 8 \end{bmatrix}$

E11.41 The stresses in a three-dimensional body are described by the following stress tensor: Find the principle stresses (eigenvalues) and associated direction cosines (eigenvectors). Use $\varepsilon = 10^{-6}$ for convergence tolerance.

$$\begin{bmatrix} \sigma_x & \tau_{xy} & \tau_{xz} \\ \tau_{xy} & \sigma_y & \tau_{yz} \\ \tau_{xz} & \tau_{yz} & \sigma_z \end{bmatrix} = \begin{bmatrix} 50 & 15 & 0 \\ 15 & 40 & 5\sqrt{30} \\ 0 & 5\sqrt{30} & 30 \end{bmatrix}$$

E11.42 A stress tensor at a point in a three-dimensional body is given below. Find the stress invariants before and after transformation (i.e., $\mathbf{T} = \mathbf{U}^T \mathbf{S} \mathbf{U}$) of the axes at 45° about the z-axis. Recall that the stress invariants are defined as $I_1 = \text{Tr}(\mathbf{S})$, $I_2 = (1/2)(\text{Tr}(\mathbf{S})^2 - \text{Tr}(\mathbf{S}^2))$, and $I_3 = |\mathbf{S}|$. *Hint*: After transformation, find the rotation matrix $\mathbf{U}$ first and then compute I_1, I_2, and I_3 for $\mathbf{T}$.

$$\mathbf{S} = \begin{bmatrix} 10 & 8 & 5 \\ 8 & 16 & 0 \\ 5 & 0 & 18 \end{bmatrix}$$

Section 11.5 Cholesky Decomposition

E11.43 Use the algorithm given in Section 11.5 to invert the following lower triangular matrices.

(a) $\begin{bmatrix} 4 & 0 & 0 \\ 3 & 1 & 0 \\ 2 & 3 & 1 \end{bmatrix}$ (b) $\begin{bmatrix} 2 & 0 & 0 \\ -1 & 3 & 0 \\ 3 & -3 & 2 \end{bmatrix}$ (c) $\begin{bmatrix} 5 & 0 & 0 & 0 \\ 2 & -1 & 0 & 0 \\ 1 & -4 & 3 & 0 \\ 1 & 0 & -2 & 2 \end{bmatrix}$ (d) $\begin{bmatrix} 2 & 0 & 0 & 0 \\ 3 & 4 & 0 & 0 \\ 3 & 0 & -1 & 0 \\ 2 & 1 & -1 & 1 \end{bmatrix}$

E11.44 Given $\mathbf{A}$ and $\mathbf{B}$, transform $\mathbf{Ax} = \lambda \mathbf{Bx}$ to the standard eigenvalue problem $\mathbf{Cx} = \lambda \mathbf{x}$ using Cholesky decomposition, and then find all eigenpairs.

(a) $\mathbf{A} = \begin{bmatrix} 2 & 1 & 1 \\ 1 & 6 & 3 \\ 1 & 3 & 9 \end{bmatrix}$ $\mathbf{B} = \begin{bmatrix} 3 & 2 & 1 \\ 2 & 8 & 4 \\ 1 & 4 & 7 \end{bmatrix}$ (b) $\mathbf{A} = \begin{bmatrix} 3 & 1 & 2 \\ 1 & 10 & 4 \\ 2 & 4 & 12 \end{bmatrix}$ $\mathbf{B} = \begin{bmatrix} 5 & 2 & 3 \\ 2 & 4 & 2 \\ 3 & 2 & 8 \end{bmatrix}$

(c) $\mathbf{A} = \begin{bmatrix} 2 & 2 & -1 \\ 2 & 3 & 1 \\ -1 & 1 & 6 \end{bmatrix}$ $\mathbf{B} = \begin{bmatrix} 4 & 3 & -1 \\ 3 & 5 & 2 \\ -1 & 2 & 4 \end{bmatrix}$ (d) $\mathbf{A} = \begin{bmatrix} 8 & -2 & 3 \\ -2 & 5 & 4 \\ 3 & 4 & 3 \end{bmatrix}$ $\mathbf{B} = \begin{bmatrix} 2 & 1 & -2 \\ 1 & 6 & 3 \\ -2 & 3 & 5 \end{bmatrix}$

E11.45 Given $\mathbf{A}$ and $\mathbf{B}$, transform $\mathbf{Ax} = \lambda \mathbf{Bx}$ to the standard eigenvalue problem $\mathbf{Cx} = \lambda \mathbf{x}$ using Cholesky decomposition, and then find all eigenpairs.

(a) $\mathbf{A} = \begin{bmatrix} 3 & 2 & 1 & 1 \\ 2 & 5 & 1 & 4 \\ 1 & 1 & 7 & 1 \\ 1 & 4 & 1 & 8 \end{bmatrix}$ $\mathbf{B} = \begin{bmatrix} 4 & 2 & 2 & 6 \\ 2 & 10 & 7 & 6 \\ 2 & 7 & 6 & 6 \\ 6 & 6 & 6 & 47 \end{bmatrix}$ (b) $\mathbf{A} = \begin{bmatrix} 2 & 1 & 2 & 1 \\ 1 & 10 & 1 & 2 \\ 2 & 1 & 8 & 1 \\ 1 & 2 & 1 & 4 \end{bmatrix}$ $\mathbf{B} = \begin{bmatrix} 1 & 1 & 2 & 1 \\ 1 & 5 & 6 & 3 \\ 2 & 6 & 24 & 16 \\ 1 & 3 & 16 & 12 \end{bmatrix}$

E11.46 Consider the following coupled second-order differential equations:

$$\begin{bmatrix} 4 & 1 \\ 1 & 3 \end{bmatrix} \frac{d^2}{dt^2} \begin{bmatrix} y_1 \\ y_2 \end{bmatrix} = \begin{bmatrix} 2 & -3 \\ -3 & 5 \end{bmatrix} \begin{bmatrix} y_1 \\ y_2 \end{bmatrix}$$

Assuming a harmonic solution in the form of $y_i = K_i \sin \omega t)$, (a) show that the system yields an $\mathbf{Ak} = \lambda \mathbf{Bk}$ eigenvalue problem, where $\lambda = \omega^2$ and $\mathbf{k} = [K_1, K_2]^T$; (b) transform the problem to a standard $\mathbf{Mz} = \lambda \mathbf{z}$ eigenvalue problem, where $\mathbf{z} = \mathbf{L}^T \mathbf{k}$; (c) obtain the eigenpairs of the given system.

E11.47 Consider the spring-mass-pendulum system depicted in **Fig. E11.47**. For small angular displacements (i.e., $\sin \theta \approx \theta$ and $\cos \theta \approx 1$), the equations of motion can be expressed as

$$m_1 \frac{d^2 x}{dt^2} + m_2 \left(\frac{d^2 x}{dt^2} + \ell \frac{d^2 \theta}{dt^2} \right) = -kx,$$

$$m_2 \ell \left(\frac{d^2 x}{dt^2} + \ell \frac{d^2 \theta}{dt^2} \right) = -m_2 \ell g \theta$$

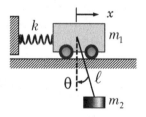

Fig. E11.47

where x and θ are respectively horizontal and angular displacements, ℓ is pendulum length, g is gravitational acceleration, and m_1 and m_2 are the masses of the cart and the pendulum. Letting $\mathbf{x} = [x, \theta]^T$, (a) cast the equations of motion in matrix form; (b) assuming a harmonic solution ($x = K_1 \sin \omega t$, $\theta = K_2 \sin \omega t$), express the given system as an eigenvalue problem $\mathbf{Ak} = \lambda \mathbf{Bk}$, where $\lambda = \omega^2$ and $\mathbf{k} = [K_1, K_2]^T$; and (c) transform the problem to a standard $\mathbf{Mz} = \lambda \mathbf{z}$ eigenvalue problem.

E11.48 The two simple pendulums, shown in **Fig. E11.48**, are connected with a spring (k) with negligible mass. The spring is undeformed when the two pendulums are in the vertical position. Assuming small angular displacements (i.e., $\sin\theta \approx \theta$ and $\cos\theta \approx 1$), the equations of motion can be expressed as

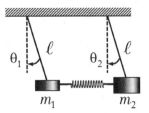

$$m_1\ell\frac{d^2\theta_1}{dt^2} = k\ell(\theta_2 - \theta_1) - m_1g\theta_1,$$

$$m_2\ell\frac{d^2\theta_2}{dt^2} = k\ell(\theta_1 - \theta_2) - m_2g\theta_2,$$

Fig. E11.48

where θ, ℓ, and m denote the angular displacement, length, and mass of the pendulums, and g is gravitational acceleration. Letting $\mathbf{x} = [\theta_1,\ \theta_2]^T$, (a) put the equations of motion in matrix form; (b) assuming a harmonic solution of the form $\theta_1 = c_1 \sin\omega t$ and $\theta_2 = c_2 \sin\omega t$, express the system as $\mathbf{Ac} = \lambda\mathbf{Bc}$, where $\lambda = \omega^2$ and $\mathbf{c} = [c_1,\ c_2]^T$; (c) transform this problem into a standard eigenvalue problem, $\mathbf{Mz} = \lambda\mathbf{z}$, and then obtain the frequencies of vibrations (eigenvalues) along with the relative amplitudes (eigenvectors) for the case of $m_1 = 0.5$ kg, $m_2 = 1$ kg, $k = 25$ N/m, $\ell = 0.5$ m, and $g = 9.81$ m/s^2.

E11.49 Consider the horizontal motion of the spring-mass system, depicted in **Fig. E11.49**. The equations of motion are expressed as

$$m_1\frac{d^2x_1}{dt^2} = -k_1x_1 + k_2\left(x_2 - x_1\right), \qquad m_2\frac{d^2x_2}{dt^2} = -k_2\left(x_2 - x_1\right) + k_3\left(x_3 - x_2\right)$$

$$m_3\frac{d^2x_3}{dt^2} = -k_3\left(x_3 - x_2\right) - k_4x_3$$

where k_i is spring stiffness, x_i is the horizontal deflection of m_i relative to static equilibrium positions.

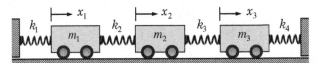

Fig. E11.49

Letting $\mathbf{x} = [x_1,\ x_2,\ x_3]^T$, (a) put the equations of motion in matrix form; (b) show that the general solution has $x_i = c_i \sin\omega t$ form, where ω is the frequency and c_i's amplitudes, leads to $\mathbf{Ac} = \lambda\mathbf{Bc}$ eigenvalue problem, where $\lambda = \omega^2$; (c) transform this problem to the standard $\mathbf{Mc} = \lambda\mathbf{c}$ eigenvalue problem; and (d) for $m_1 = m_2 = m_3 = m$ and $k_1 = k_2 = k_3 = k$, find the eigenpairs of $\mathbf{M}$ and associated natural frequencies.

E11.50 Consider the LC-circuit given in **Fig. E11.50**.

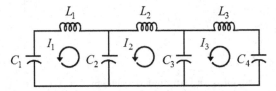

Fig. E11.50

The unsteady behavior of the circuit is described with the following coupled second-order ODEs:

$$L_1\frac{d^2I_1}{dt^2} + \frac{I_1}{C_1} + \frac{I_1 - I_2}{C_2} = 0, \quad L_2\frac{d^2I_2}{dt^2} + \frac{I_2 - I_1}{C_2} + \frac{I_2 - I_3}{C_3} = 0, \quad L_3\frac{d^2I_3}{dt^2} + \frac{I_3 - I_2}{C_3} + \frac{I_3}{C_4} = 0$$

where L, I, and C denote the inductance, current, and capacitance. Letting $\mathbf{i} = [I_1,\ I_2,\ I_3]^T$, (a) put the system in matrix form; (b) assume the general solution is $I_n = c_n \sin\omega t$ form, which leads to the $\mathbf{Ai} = \lambda\mathbf{Bi}$ eigenvalue problem, where $\lambda = \omega^2$; (c) transform the problem to the standard

$\mathbf{Mc} = \lambda\mathbf{c}$ eigenvalue problem; and (d) for $L_1 = 2L_2 = 4L_3 = 1$ and $4C_1 = 2C_2 = C_3 = 1$, find the eigenpairs of $\mathbf{M}$.

E11.51 Consider a simplified representation of the three rotating masses, shown in **Fig. E11.51**. A mathematical model, consisting of coupled second-order ODEs, is given as

$$I_1\frac{d^2\theta_1}{dt^2} = k_1(\theta_2 - \theta_1), \quad I_2\frac{d^2\theta_2}{dt^2} = k_1(\theta_1 - \theta_2) + k_2(\theta_3 - \theta_2), \quad I_3\frac{d^2\theta_3}{dt^2} = k_2(\theta_2 - \theta_3)$$

where I, k, and θ denote the moment of inertia, shaft stiffness, and angular displacement, respectively.

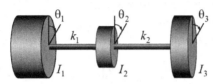

Fig. E11.51

Letting $\mathbf{t} = [\theta_1, \theta_2, \theta_3]^T$, (a) put the system in matrix form; (b) assuming the general solution is in $\theta_n = c_n \sin\omega t$ form, which leads to $\mathbf{Ac} = \lambda\mathbf{Bc}$ eigenvalue problem, where $\lambda = \omega^2$; (c) transform the problem to the standard $\mathbf{Mz} = \lambda\mathbf{z}$ eigenvalue problem; and (d) for $4I_1 = 10I_2 = 5I_3 = 50$ kg·m^2 and $5k_1/4 = k_2 = 10^5$ N/m, find the eigenpairs of $\mathbf{M}$.

Section 11.6 Householder Method

E11.52 Use the Householder method to transform the following matrices into tridiagonal matrices.

$$\text{(a)} \begin{bmatrix} 2 & 3 & 4 \\ 3 & 1 & 1 \\ 4 & 1 & 2 \end{bmatrix} \quad \text{(b)} \begin{bmatrix} 1 & 2 & 1 \\ 2 & \sqrt{5} & \sqrt{5} \\ 1 & \sqrt{5} & 2\sqrt{5} \end{bmatrix} \quad \text{(c)} \begin{bmatrix} 1 & 1 & \sqrt{3} \\ 1 & 3 & 3\sqrt{3} \\ \sqrt{3} & 3\sqrt{3} & 1 \end{bmatrix}$$

E11.53 Use the Householder method to transform the following matrices into tridiagonal matrices.

$$\text{(a)} \begin{bmatrix} 1 & -2 & 1 & 2 \\ -2 & 3 & -2 & 1 \\ 1 & -2 & 4 & 11 \\ 2 & 1 & 11 & 3 \end{bmatrix} \quad \text{(b)} \begin{bmatrix} 1 & 4 & 0 & 3 \\ 4 & 3 & 5 & 4 \\ 0 & 5 & 3 & 1 \\ 3 & 4 & 1 & 8 \end{bmatrix} \quad \text{(c)} \begin{bmatrix} 4 & 1 & -2 & -2 \\ 1 & 3 & 4 & 5 \\ -2 & 4 & -3 & 11 \\ -2 & 5 & 11 & 3 \end{bmatrix}$$

Section 11.7 Eigenvalues of Tridiagonal Matrices

E11.54 Use the Strum sequence to find the coefficients of the characteristic polynomials.

$$\text{(a)} \begin{bmatrix} 2 & 1 & 0 \\ 1 & 2 & 3 \\ 0 & 3 & 4 \end{bmatrix} \quad \text{(b)} \begin{bmatrix} 1 & -2 & 0 \\ -2 & 4 & -3 \\ 0 & -3 & 2 \end{bmatrix} \quad \text{(c)} \begin{bmatrix} 3 & -1 & 0 \\ -1 & 3 & 2 \\ 0 & 2 & 6 \end{bmatrix}$$

E11.55 Use the Strum sequences to find the coefficients of the characteristic polynomials.

$$\text{(a)} \begin{bmatrix} 1 & 4 & 0 & 0 \\ 4 & 3 & -1 & 0 \\ 0 & -1 & 3 & 1 \\ 0 & 0 & 1 & 4 \end{bmatrix} \quad \text{(b)} \begin{bmatrix} 4 & 2 & 0 & 0 \\ 2 & 1 & 1 & 0 \\ 0 & 1 & -1 & -2 \\ 0 & 0 & -2 & 3 \end{bmatrix} \quad \text{(c)} \begin{bmatrix} 3 & -2 & 0 & 0 \\ -2 & -3 & -1 & 0 \\ 0 & -1 & 4 & 1 \\ 0 & 0 & 1 & 5 \end{bmatrix}$$

E11.56 Use the Gram-Schmidt QR decomposition to compute the eigenvalues and the eigenvectors of the following matrices.

$$\text{(a)} \begin{bmatrix} 2 & -\sqrt{5} & 0 \\ -\sqrt{5} & 1 & -2 \\ 0 & -2 & 1 \end{bmatrix} \quad \text{(b)} \begin{bmatrix} 2 & -2 & 0 \\ -2 & 4 & 4 \\ 0 & 4 & 10 \end{bmatrix} \quad \text{(c)} \begin{bmatrix} 12 & -5 & 0 \\ -5 & 1 & -1 \\ 0 & -1 & 3 \end{bmatrix}$$

E11.57 Use the Gram-Schmidt QR decomposition to compute the eigenvalues and the eigenvectors of the following matrices.

$$
\text{(a)}
\begin{bmatrix}
1 & -3 & 0 & 0 \\
-3 & 2 & 3 & 0 \\
0 & 3 & 4 & 2 \\
0 & 0 & 2 & 1
\end{bmatrix}
\quad
\text{(b)}
\begin{bmatrix}
3 & -4 & 0 & 0 \\
-4 & 5 & 1 & 0 \\
0 & 1 & 4 & -2 \\
0 & 0 & -2 & 3
\end{bmatrix}
\quad
\text{(c)}
\begin{bmatrix}
2 & 1 & 0 & 0 \\
1 & 4 & -1 & 0 \\
0 & -1 & 4 & 2 \\
0 & 0 & 2 & 8
\end{bmatrix}
$$

Section 11.8 Leverrier-Faddeev Method

E11.58 Use the Leverrier-Faddeev method to find the characteristic polynomials and the inverse matrices.

$$
\text{(a)}
\begin{bmatrix}
2 & -2 & 3 \\
-2 & 4 & 4 \\
3 & 4 & 10
\end{bmatrix}
\quad
\text{(b)}
\begin{bmatrix}
1 & 2 & -2 \\
3 & 1 & -1 \\
2 & -3 & -2
\end{bmatrix}
\quad
\text{(c)}
\begin{bmatrix}
5 & 3 & 2 \\
1 & 4 & -2 \\
-1 & -3 & 1
\end{bmatrix}
$$

E11.59 Use the Leverrier-Faddeev method to find the characteristic polynomials and the inverse matrices.

$$
\text{(a)}
\begin{bmatrix}
2 & 3 & 1 & 2 \\
1 & 1 & -3 & -1 \\
-1 & 2 & 4 & 1 \\
1 & -2 & 2 & 2
\end{bmatrix}
\quad
\text{(b)}
\begin{bmatrix}
3 & 1 & 1 & -2 \\
1 & -4 & 0 & 1 \\
2 & 6 & 3 & -3 \\
-1 & -2 & 2 & -3
\end{bmatrix}
\quad
\text{(c)}
\begin{bmatrix}
2 & 1 & 2 & 1 \\
0 & 4 & -2 & -1 \\
-2 & -2 & 3 & 2 \\
-3 & -2 & 1 & 4
\end{bmatrix}
$$

E11.60 Find the eigenpairs of the following tridiagonal matrices using the analytical expressions.

$$
\text{(a)}
\begin{bmatrix}
4 & -1 & 0 \\
-2 & 4 & -1 \\
0 & -2 & 4
\end{bmatrix}
\quad
\text{(b)}
\begin{bmatrix}
1 & 2 & 0 \\
3 & 1 & 2 \\
0 & 3 & 1
\end{bmatrix}
\quad
\text{(c)}
\begin{bmatrix}
5 & 3 & 0 \\
1 & 5 & 3 \\
0 & 1 & 5
\end{bmatrix}
$$

E11.61 Find the eigenpairs of the following tridiagonal matrices using the analytical expressions.

$$
\text{(a)}
\begin{bmatrix}
3 & 2 & 0 & 0 \\
1 & 3 & 2 & 0 \\
0 & 1 & 3 & 2 \\
0 & 0 & 1 & 3
\end{bmatrix}
\quad
\text{(b)}
\begin{bmatrix}
3 & -2 & 0 & 0 \\
-2 & 3 & -2 & 0 \\
0 & -2 & 3 & -2 \\
0 & 0 & -2 & 3
\end{bmatrix}
\quad
\text{(c)}
\begin{bmatrix}
4 & 9 & 0 & 0 \\
1 & 4 & 9 & 0 \\
0 & 1 & 4 & 9 \\
0 & 0 & 1 & 4
\end{bmatrix}
$$

Section 11.8 Characteristic Value Problems

E11.62 Solve the following characteristic value problems numerically using the finite difference method: (a) develop the difference equations and express them in matrix form; (b) obtain the numerical solution using uniformly spaced $M = 5$ intervals. Use $\varepsilon = 10^{-5}$ for convergence tolerance.

(a) $y'' + \lambda y = 0$, $\quad y(0) = y(2\pi) = 0$, $\qquad$ (b) $x^2 y'' + 2xy' + \lambda y = 0$, $\quad y(1) = y(2) = 0$

(c) $y'' + 2y' + k^2 e^{-x} y = 0$, $\quad y(0) = y(1) = 0$, $\quad$ (d) $(1 + x^2) y'' + \lambda y = 0$, $\quad y(-1) = y(1) = 0$

E11.63 Solve the following characteristic value problems numerically using the finite difference method: (a) develop the difference equations and express them in matrix form; (b) obtain the numerical solution using uniformly spaced $M = 5$ intervals. Use $\varepsilon = 10^{-5}$ for convergence tolerance.

(a) $y'' + \lambda y = 0$, $\quad y(0) = 2y'(2) - y(2) = 0$, $\quad$ (b) $(x^2 y')' - \lambda x^2 y = 0$, $\quad y'(0) = y(2) = 0$,

(c) $y'' + \lambda y = 0$, $\quad y'(0) = y'(1) + y(1) = 0$, $\quad$ (d) $y'' + 6y' + (9 + \lambda)y = 0$, $\quad y'(0) = y(1) = 0$,

(e) $xy'' + (1 - x)y' + \lambda y = 0$, $\quad y'(0.5) = 5y'(1) - y(1) = 0$

11.12 COMPUTER ASSIGNMENTS

CA11.1 Applying Hückel theory to the Schörinder equation for cyclobutadiene (a) and benzene (b) leads to the following Hamiltonian matrix:

$$
\text{(a)}\quad \mathbf{H} = \begin{bmatrix} \alpha & \beta & & \beta \\ \beta & \alpha & \beta & \\ & \beta & \alpha & \beta \\ \beta & & \beta & \alpha \end{bmatrix}, \qquad \text{(b)}\quad \mathbf{H} = \begin{bmatrix} \alpha & \beta & & & & \beta \\ \beta & \alpha & \beta & & & \\ & \beta & \alpha & \beta & & \\ & & \beta & \alpha & \beta & \\ & & & \beta & \alpha & \beta \\ \beta & & & & \beta & \alpha \end{bmatrix}
$$

The secular equation is the standard eigenvalue problem ($\mathbf{H\Psi} = E\mathbf{\Psi}$), where E (eigenvalues) denote the orbital energies and $\mathbf{\Psi}$ denotes corresponding wave functions. Employing the problem simplifications illustrated in Chapter 2 (Example 2.2), find the molecular orbitals and corresponding wave functions using a suitable numerical method.

CA11.2 Write a pseudomodule to find the smallest (in magnitude) eigenvalue using the inverse power method.

CA11.3 Write a pseudomodule to find any one of the eigenvalues of a matrix using the inverse shift power method.

CA11.4 A four-story building frame, shown in **Fig. CA11.4**, can be modeled as a four-degree-of-freedom mass-spring system. The mass and stiffness of the floors are denoted by m_i and k_i, respectively. The natural frequencies and mode shapes of the frame are given with

$$
\begin{aligned}
m_1\ddot{x}_1 &= -k_1 x_1 + k_2 (x_2 - x_1) \\
m_2\ddot{x}_2 &= -k_2 (x_2 - x_1) + k_3 (x_3 - x_2) \\
m_3\ddot{x}_3 &= -k_3 (x_3 - x_2) + k_4 (x_4 - x_3) \\
m_4\ddot{x}_4 &= -k_4 (x_4 - x_3)
\end{aligned}
$$

where x_i's are the horizontal displacements of the floors, $k_i = 24EI/\ell_i$'s are stiffness constants for columns subjected to shear only, EI is the bending stiffness of the column, and ℓ_i's are the column lengths. Find the natural frequencies and mode shapes of the building frame for $m_1 = m_2 = m_3 = m_4 = m$ and $k_1 = k_2 = k_3 = k_4 = k$. What would the dominant natural frequency be if the number of stories were increased?

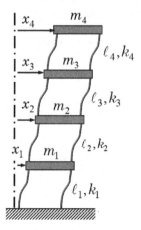

Fig. CA11.4

CA11.5 Find the circular frequencies of the oscillations of the circuit in **Fig. CA11.5**. The Kirchoff equations are:

$$L\frac{d^2i_1}{dt^2} + L\left(\frac{d^2i_1}{dt^2} - \frac{d^2i_2}{dt^2}\right) + \frac{1}{C_1}i_1 = 0,$$

$$L\left(\frac{d^2i_2}{dt^2} - \frac{d^2i_1}{dt^2}\right) + L\left(\frac{d^2i_2}{dt^2} - \frac{d^2i_3}{dt^2}\right) + \frac{1}{C_2}i_2 = 0,$$

$$L\left(\frac{d^2i_3}{dt^2} - \frac{d^2i_2}{dt^2}\right) + L\left(\frac{d^2i_3}{dt^2} - \frac{d^2i_4}{dt^2}\right) + \frac{1}{C_3}i_3 = 0,$$

$$L\left(\frac{d^2i_4}{dt^2} - \frac{d^2i_3}{dt^2}\right) + L\left(\frac{d^2i_4}{dt^2} - \frac{d^2i_5}{dt^2}\right) + \frac{1}{C_4}i_4 = 0,$$

$$L\left(\frac{d^2i_5}{dt^2} - \frac{d^2i_4}{dt^2}\right) + L\frac{d^2i_5}{dt^2} + \frac{1}{C_5}i_5 = 0,$$

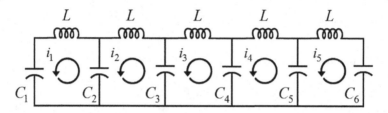

Fig. CA11.5

CA11.6 A mathematical model for the motion of an airfoil (see **Fig. CA11.6**) being tested in a wind-tunnel is given by the following set of coupled second-order differential equations:

$$m\frac{d^2y}{dt^2} + m\ell\frac{d^2\theta}{dt^2} + k_1y = 0,$$

$$m\ell\frac{d^2y}{dt^2} + J\frac{d^2\theta}{dt^2} + k_2\theta = 0$$

where y and θ are horizontal and angular displacement of the airfoil from the horizontal position, m, J, and k_1 are mass, inertia moment about center of gravity (cog) and spring constant of the airfoil, $k = 2$ is the spring constant of the spiral spring. Letting $\mathbf{x} = [y,\ \theta]^T$, (a) cast the equations of motion in matrix form, (b) assuming a harmonic solution ($y = K_1\sin(\omega t)$, $\theta = K_2\sin(\omega t)$), express the given system as an eigenvalue problem $\mathbf{Ac} = \lambda\,\mathbf{Bc}$ where, $\lambda = \omega^2$ and $\mathbf{c} = [C_1,\ C_2]^T$; (c) Transform the problem to standard $\mathbf{Mz} = \lambda\mathbf{z}$ eigenvalue problem; (d) for $k_1 = 1$, $k_2 = 4$, $\ell = 0.20$, $J = 1$, $m = 1$, obtain the natural frequency and modes.

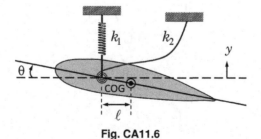

Fig. CA11.6

A Guide on How to Read and Write a Pseudocode

THE algorithms presented in this book are presented in a form that requires very little time and effort to digest. The motivations for using pseudocodes have been stated in the preface and Chapter 1. The aim of this appendix is to present the basic syntax and constructions with simple and concise examples.

A.1 BASICS AND VARIABLES

A.1.1 VARIABLES AND CONSTANTS

A *variable*, in programming languages, is usually represented symbolically with letters or combinations of letters and numbers (**A**, **B**, **AX**, **xy**, **a1**, **tol**, ...). Greek letters (α, β, ...) can also be used as variables in pseudocodes. A variable can be viewed as a named placeholder to which a value (compatible with its type) is assigned and read as needed. By defining, declaring, or initializing, a storage area is allocated for every constant or variable in the computer's memory. The value of a variable changes during the execution of a computer program, while the value of a *constant* does not.

A.1.2 DECLARATION

The **Declare** statement is used to indicate the use of arrays (*see* Section A.5 for details) by assigning subscripts. The *extent* of an array (the range of its index) is denoted by $m : n$ where m and n are integers ($n > m$). If m is omitted, the default value is 1. Arrays a_1, a_2, ..., a_n, or b_0, b_1, b_2, ..., b_n or e_{11}, e_{12}, e_{13}, and so on are declared as follows:

$$\textbf{Declare: } a_n, \ b_{0:n}, \ e_{n,n} \qquad \backslash \text{ Array variables } a_n, \ b_{0:n}, \ e_{n,n}$$
$$\backslash \text{ For } a_k, \ k = 1, 2, ..., n \text{ and } b_k, \ k = 0, 1, 2, ..., n$$
$$\backslash \text{ For } e_{ij}, \ i = j = 1, 2, ..., n$$

Note that array **B** is declared as $b_{0:n}$ indicating the index runs from 0 to n. The memory spaces allocated for **A**, **B**, and **E** are n, $n+1$, and n^2. Arrays declared within a **Module** are not accessible from outside unless they are arguments of the module.

DOI: 10.1201/9781003474944-A

A.1.3 COMMENTING

Commenting is a common practice in actual programs. It is done to allow *"human-readable"* descriptions detailing the purpose of some of the statement(s) and/or to create *in situ* documentation. Everything from the "\" symbol, where it is introduced, to the end of the line is reserved for comments. Some commenting examples are

$$\backslash \text{ Pseudocode for finding the roots of a quadratic equation.}$$
$$\Delta \leftarrow b^2 - 4ac \qquad\qquad\qquad \backslash \text{ Computing the discriminant}$$

A *single-line* comment is applied to describe an expression (or statement) on the corresponding line. A *block* of comments generally at the beginning of each module (*Header Comments*) is used to describe the purpose of the module, its variables, exceptions, other modules used, etc.

A.1.4 ASSIGNMENT

In actual programming languages, the *expression* on the right-hand side of the "=" sign is evaluated first, and its value is placed at the allocated memory location of the *variable* on the left-hand side. Any recently computed value of *variable* replaces its previous value. As a depiction of this process, an *assignment* is denoted by ← (a left-arrow). It has the following general form

$$\textit{\textbf{variable}} \leftarrow \textit{Expression}$$

As an example, consider the pseudocode segment below. By invoking $X \leftarrow 0$, "zero" is placed in the memory location of X. Next, the expression on the right-hand side of the second line is processed first (0+4), and the result is placed in the memory location of the variable X on the left-hand side (i.e., $X \leftarrow 4$). The third expression is processed in the same way $(X \leftarrow 4 + 6)$, which updates the memory location of X with 10.

$$X \leftarrow 0 \qquad\qquad\qquad \backslash \text{ 0 "zero" is assigned to } X\text{-variable}$$
$$X \leftarrow X + 4 \qquad\qquad\qquad \backslash \text{ Increment } X \text{ by 4}$$
$$X \leftarrow X + 6 \qquad\qquad\qquad \backslash \text{ Increment } X \text{ by 6}$$

In this text, the multiplication operator "*" has sometimes been omitted in algebraic expressions where the multiplication is implied. For instance, in the $\Delta \leftarrow \sqrt{b^2 - 4ac}$ expression, $4bc$ is clearly the product of "4", a, and b.

A.1.5 SEQUENTIAL STATEMENTS

Frequently, a sequence of statements (with no imposed conditions) is used to perform a specific task. These statements will be executed in the order they are specified in the program. By using a semicolon (;) as a separator, multiple expressions are placed in a line. For instance, the following pseudocode segment illustrates how the first and second statements are presented on a single line with the use of a semicolon.

$$a \leftarrow b^2 + c^2; \quad d \leftarrow \sqrt{a} \qquad \backslash \text{ Statement-1 and Statement-2}$$
$$x_1 \leftarrow d/(b + c) \qquad \backslash \text{ Statement-3}$$
$$x_2 \leftarrow d/(b - c) \qquad \backslash \text{ Statement-4}$$
$$y \leftarrow x_1 + x_2 \qquad \backslash \text{ Statement-5}$$

A.1.6 INPUT/OUTPUT STATEMENTS

An inevitable element in any program is the communication of the input and output data with the main or sub-programs. This is done through **Read/Write** statements, which are used for reading initial values of variables into the program from a file (or console) or writing out intermediate or final values of variables to a file (or screen).

For instance, the following code segment illustrates supplying numerical values of the variables a, b, and c into the program. After evaluating Δ, its value is printed to the screen or written to a file.

Read: a, b, c \qquad $\backslash$ Read a, b, c variables from console or file
$\Delta \leftarrow \sqrt{b^2 - 4ac}$ \qquad $\backslash$ Compute the discriminant.
Write: Δ \qquad $\backslash$ Write discriminant to screen or file

A.2 LOGICAL VARIABLES AND LOGICAL OPERATORS

Branching in a computer program causes a computer to execute a different block of instructions, deviating from its default behavior of executing instructions sequentially. Branching structures are controlled by *logical variables* and *logical operations*.

Logical calculations are carried out with an assignment statement:

$$Logical\_variable \leftarrow \textbf{Logical\_expression}$$

Logical_expression can be a combination of logical constants, logical variables, and logical operators. A logical operator is defined as an operator on numeric, character, or logical data that yields a logical result. There are two basic types of logical operators: *relational operators* and *combinational operators*.

The relational operators can be expressed in a variety of ways. This pseudocode convention uses the following corresponding formalism:

$$B = C \qquad \backslash \text{ Equal}$$
$$B \neq C \qquad \backslash \text{ Not equal}$$
$$B < C \ (B \leqslant C) \qquad \backslash \text{ Less than or (less than and equal to)}$$
$$B > C \ (B \geqslant C) \qquad \backslash \text{ Greater than or (greater than and equal to)}$$

Logical statements or variables take either **True** or **False** values, and they can be constructed with the aid of logic operators (**Or** , **And**, and **Not**).

Or \qquad $\backslash$ OR operator
And \qquad $\backslash$ AND operator
Not \qquad $\backslash$ NOT operator

Logical operators are used in a program together with relational operators to control the flow of the program. If L_1 and L_2 are two logical prepositions, in order for L_1 **And** L_2 to be True, both L_1 and L_2 must be True. In order for L_1 **Or** L_2 to be True, it is sufficient to have either L_1 or L_2 to be True. When using **Not** in any logical statement, the logic value is changed to True when it is False or changed to False when it is True.

A.3 CONDITIONAL CONSTRUCTIONS (IF-THEN, IF-THEN-ELSE)

The direction of execution flow cannot be changed with sequential statements since they are executed only once. Whereas, there is a need to execute part of a program multiple times or, under some conditions, to change the direction of the execution flow. In high-level programming languages, branching control and conditional structures allow the execution flow to jump to a different part of the program.

Conditional constructions create branches in the execution path based on the evaluation of a condition. When a control statement is reached, the condition is evaluated, and a path is selected according to the result of the condition.

Most programming languages use **If-Then** and **If-Then-Else** constructions that behave in the same way. They allow one to check a condition and execute certain parts of the code if *condition* is True or False. The most common and simplified form (**If-Then** construct) executes the block of statements *if and only if* a logical expression (i.e., *condition*) is True. The **If-Then** construct has the form

> **If** $[condition]$ **Then**   \ Check for the *condition*
>   STATEMENT(s) *if*   \ Execute STATEMENT(s)
>   the *condition*=True   \ in block if *condition* is True
> **End If**

Note that *condition* is enclosed with square brackets. When the condition is tied to only one statement, the **If-Then** construct is further simplified as follows:

> **If** $[condition]$ **Then** STATEMENT

In the above, the STATEMENT reserved to a single statement (algebraic expression, input or output statement, etc.), and it is executed only if *condition* is True.

A more general **If-Then-Else** construct allows the use of another block of statements when the *condition* is False. The **If-Then-Else** construct has the following form:

> **If** $[condition]$ **Then**
>   STATEMENT(s)-1 *if*   \ Execute STATEMENT(s)
>   the *condition*=True   \ in this block if *condition* is True
> **Else**
>   STATEMENT(s)-2 *if*   \ Execute STATEMENT(s)
>   the *condition*=False   \ in this block if *condition* is False
> **End If**

When *condition* is True, STATEMENT(S)-1 is executed; otherwise (*condition*=False), STATEMENT(S)-2 is executed.

A more complicated **If-Then-Else** construct can be devised as follows:

```
If [condition₁] Then
        STATEMENT(s)-1 if              \ Execute STATEMENT(s)-1
        the condition₁=True            \ in this block if condition₁ is True
Else
    If [condition₂] Then
            STATEMENT(s)-2 if          \ Execute STATEMENT(s)-2
            the condition₂=False       \ in this block if condition₁ is False
        Else
        ⋮
        If [conditionₙ] Then
                STATEMENT(s)-n if      \ Execute STATEMENT(s)-n
                the conditionₙ=True    \ in this block if conditionₙ is True
            Else
                STATEMENT(s)) if       \ Execute STATEMENT(s)
                the conditionₙ=False   \ in this block if conditionₙ is False
            End If
        End If
    End If
```

A.4   CONTROL CONSTRUCTIONS

Control (loop) constructions are used when a program needs to execute a block of instructions repeatedly until a *condition* is met, at which time the loop is terminated. There are three control statements in most programming languages that behave in the same way: **While-**, **Repeat-Until**, and **For**-constructs.

A.4.1   **WHILE-** CONSTRUCTS

A **While**-construct has the form

```
While [condition]
        STATEMENT(s)) if              \ Execute STATEMENT(s) in
        condition=True                \ this block so long as condition is True
    End While
```

The flowchart for a **While**-construct is presented in Fig. A.1a. In this construct, the *condition* (enclosed by square brackets) is analyzed first. If the *condition* is True, then the block of statement(s) is executed. Then, the *condition* is analyzed again, and the loop is iterated until the *condition* becomes False; i.e., the only way out of the loop is when *condition*=False. If the *condition*=False at the start, the block of statements will not be executed.

While- construct is used in cases where the number of iterations cannot be determined beforehand. It does not require an index variable, so it is used to execute the STATEMENT(s) until a desired condition is met.

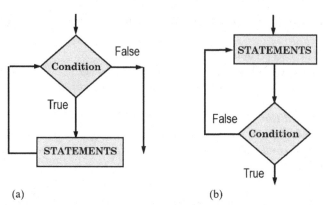

FIGURE A.1: Flowchart for a (a) While- and (b) Repeat-Until constructs.

A.4.2 **REPEAT-UNTIL** CONSTRUCT

A **Repeat-Until** is a loop construct similar to **While**-construct in that a block of STATE-MENT(S) is executed until the *condition* (enclosed with square brackets) becomes **False**. The **Repeat-Until** loop construct has the form

> **Repeat**
> STATEMENT(s)) *if*   \ Execute STATEMENT(s) in this
> *condition*=**False**   \ block so long as *condition* is **False**
> **Until** $\left[condition\right]$

The flowchart for the **Repeat-Until** construct is illustrated in Fig. A.1b. In this construct, however, *condition* is analyzed after the STATEMENT(S) is executed. The **Repeat-Until** construct is convenient in cases where the condition is unknown or undefined until the STATEMENT(S) block is executed at least once. The loop is terminated when the *condition* becomes **True**.

 The condition prior to the **Repeat-Until** construct is important for skipping or executing the entire loop. Also, it is important to make sure that the *condition* is always met to prevent an infinite loop.

A.4.3 **FOR**-CONSTRUCT

For-construct is probably the most common type of loop construct in any programming language. It is used when a block of STATEMENT(S) is to be executed a specified number of times. A **For**-construct has the form

> **For** $\left[index\ variable{=}i_1,\ i_2\ (,\Delta i)\right]$
> STATEMENT(s)   \ Execute STATEMENT(s) in this block
> **End For**

where *index variable* is referred to as the loop index or loop counter, i_1 and i_2 are the initial and terminal values of the index, and an optional parameter Δi denotes increments $(\Delta i > 0)$ or decrements $(\Delta i < 0)$; if it is omitted, it is 1 by default. The total number of

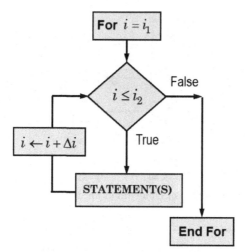

FIGURE A.2: Flowchart for a For-construct.

iterations is $(i_2 - i_1 + \Delta i)/\Delta i$. To break out of a **For**-loop (**Exit**) before it reaches n_2, a *condition* within the STATEMENT(S) block needs to be specified.

A flowchart for a **For**-construct is illustrated in Fig. A.2. The initial step is executed first and only once; the next $i \leq i_2$ condition is evaluated. If this condition is **True**, then the STATEMENT(S) in the iteration loop are executed. Otherwise, (if $i \leq i_2$ is **False**) the STATEMENT(S) are not executed, and the flow of control jumps to the next statement just after **End For**.

 A **For**-construct is used when the number of iterations to be performed is known beforehand. It is easy to use in nested loop settings (with arrays) due to having clearly identified loop indexes.

A.4.4 EXITING A LOOP

The **Exit** statement is used anywhere in the block statements of **While-**, **Repeat-Until**, or **For-** constructs to terminate the loop. When the **Exit** statement is executed, the remaining block of statement(s) in the current iteration loop is skipped.

Consider the pseudocode segment using the **Exit** statement in a **For**-construct. The loop is executed for $k=1, 2, \ldots, 12$, but when *condition* $(k > 12)$ becomes **True** for $k=13$, the loop is terminated before k reaches 30.

> **For** $[k = 1, 30]$
> . . .
> **If** $[k > 12]$ **Then Exit**
> . . .
> **Write:** k
> . . .
> **End For**

Note that any *condition* for exit need not be tied to the index variable.

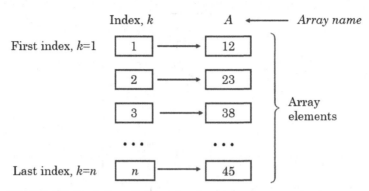

FIGURE A.3: Depiction of elements of an array in computer memory.

 When using **If-Then, If-Then-Else, While-, Repeat-Until,** or **For-** constructs, always indent the block of statements by several spaces to improve the readability of your pseudocode.

A.5 ARRAYS

An array is a special case of a variable representing a set of data (variables) under one group name. In general, for the most part, the data structures only need arrays and placeholders that hold elements of the same type, i.e., a_1, a_2, ..., a_n (or d_{-2}, d_{-1}, ..., d_m, or g_0, g_1, ..., g_9). The values in an array occupy consecutive locations in the computer's memory (*see* Fig. A.3). The subscript of an array is always an *integer*.

Arrays can have two or more dimensions as well. Arrays should be explicitly declared at the beginning of each pseudocode because the range and length of an array are critical in coding. Vectors and matrices are declared as arrays of one and two dimensions, respectively.

This pseudocode convention also adopts *whole array arithmetic*. Under certain circumstances, whole array arithmetic is used in order to keep the pseudocodes as short as possible without sacrificing the intended operation. If two arrays have the *same size*, then they are used in arithmetic operations where the operation is carried out on an element-by-element basis. The whole array arithmetic expressions do not require **For**-loops and operate with the array names as if they were scalars.

A brief summary of whole array operations (using conformable arrays) is presented in the pseudocode segment below:

Declare: a_n, b_n, c_n, $e_{10,10}$, $f_{10,10}$, $g_{10,10}$ \ Declaring array variables
c←a+b   \ Performs **c=a+b**
c← 4a+(−3)b   \ Performs $c_i=4*a_i+(-3)*b_i$ for all i
c← a*b   \ Performs $c_i = a_i * b_i$ for all i
...

g← e+5*f   \ Performs $g_{ij}=e_{ij}+5*f_{ij}$ for all i,j
Write: c   \ Prints c_i sequentially for all i
...

In the pseudocodes in this text, vectors (i.e., one-dimensional arrays) and matrices (i.e., two- or multi-dimensional arrays) are denoted in lowercase and uppercase bold typeface, respectively. For instance, $\mathbf{a}$ denotes $a_1, a_2, \cdots, a_n$ and $\mathbf{E}$ denotes $e_{11}, e_{12}, \cdots, e_{nm}$ and they should be declared as

Declare: a_n, e_{nm}.

A.6   **PROGRAM**, **MODULE**, AND **FUNCTION MODULE** STRUCTURES

A pseudo-program is a collection of instructions designed to perform a task. A pseudo-program is given a *name*, and it has the form

> **Program** EXAMPLE
>   **Declare:** $a_n, e_{n,m}$
>   ...
>
> $$\left\{ \begin{array}{c} ... \\ \text{STATEMENT(S)} \\ ... \end{array} \right\}$$
>
>   ...
>   **Write: a, e**
> **End Program** EXAMPLE

Note that it is delimited by **Program-End Program** statements to indicate where the program begins and where it ends.

In general, it is impractical to write a complete *program* from A to Z that includes everything. Such programs would not only be too long but also too complicated to be practical. When writing a computer program, it is often necessary to repeat the same set of instructions (a task) multiple times within the same program. To avoid this unwanted repetition, we often utilize a sub-program, which is a collection of statements for the purpose of achieving some specific tasks, such as inverting a matrix, finding the real roots of a polynomial, and so on. All high-level languages allow *modular programming*, which is a programming technique that separates the functionality of a program into independent modules containing only the required aspect of the desired functionality. Then these *modules* are accessed and executed from the main or sub-programs whenever needed.

Two types of modules are adopted in this pseudocode convention: **Function Module** for functions and **Module** for procedures or sub-programs. Each module, having its own input/output variables, is a separate program completely isolated from the rest of the program.

An efficient program generally consists of one or more independent modules. The modular programming approach is also easier to conceptualize and write a program as a whole.

A.6.1   SUB-PROGRAM DEFINITIONS: **Module**

A **Module** as a procedure or sub-program requires a module name as well as an argument list (input/output variables). A module is delimited with **Module-End Module** statements.

The general structure of a **Module** is given as

Module NAME$(in_1, in_2, ..., out_1, out_2, ...)$

$$\left\{ \begin{array}{c} ... \\ \text{STATEMENT(S)} \\ ... \end{array} \right\} \quad \text{Body of the module}$$

End Module NAME

where $in_1, in_2, ...$ and $out_1, out_2, ...$ are respectively input and output variable lists, i.e., the arguments of the module. Information in and out of a module is communicated by the argument list. It is invoked by naming it (with its full argument lists) in any pseudocode or pseudomodule. Any array variable is declared in the module by the **Declare** statement.

A module usually has its own internal (*local*) variables that are accessible only internally. Once a specific task is completed, at least one output value is returned to the calling module or program. No restriction is imposed on the number of modules. A module code can be used with any other relevant programs as well. A module can also be accessed many times in the same program. The variables declared in a module argument list are global to the module.

A sample pseudomodule finding the magnitude of the vector $\mathbf{V} = v_1\,\mathbf{i} + v_2\,\mathbf{j} + v_3\,\mathbf{k}$ is given below.

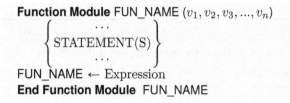

> **Module** VECTOR_MAGNITUDE(v_1, v_2, v_3, mag)
> \ DESCRIPTION: A pseudomodule finding the magnitude of a vector.
> $mag \leftarrow \sqrt{v_1^2 + v_2^2 + v_3^2}$
> **End Module** VECTOR_MAGNITUDE

A.6.2   **FUNCTION MODULE** OR **RECURSIVE FUNCTION MODULE**

A **Function Module** may be considered a smaller version of a **Module** discussed earlier. A module may have more than one output, while **Function Module** is restricted to *single output value*, which can also be used to evaluate an expression. Functions may consist of a single *expression* or be computed as a result of multiple statements.

A function is delimited by **Function Module-End Function Module** statements. The general structure of a **Function Module** is given as

Function Module FUN_NAME $(v_1, v_2, v_3, ..., v_n)$

$$\left\{ \begin{array}{c} ... \\ \text{STATEMENT(S)} \\ ... \end{array} \right\}$$

FUN_NAME $\leftarrow$ Expression
End Function Module FUN_NAME

where $v_1, v_2, v_3, ..., v_n$ are the input variables (*arguments*) of the function, and the result (*output*) is written on the function name (FUN_NAME) on return.

Most common mathematical functions ($\sqrt{x}$, $|x|$, e^x, $\ln(x)$, $\log(x)$, trigonometric, hyperbolic functions and their inverses, remainder or modulo function $\mathbf{MOD}(a, b)$, etc.) are built into most programming languages. Array variables may be used in the argument list of functions; however, they need to be stated in the **Declare** statement of the function module.

A **Function Module** is invoked by naming it in an expression with its full argument list. For instance, a user-defined function $f(x, y, z) = \sqrt{x^2 + y^2 + z^2}$ is coded as shown in Pseudocode A.1. Note that the function $f(x, y, z)$ named FUNC is used in arithmetic expressions just like any other built-in function.

Pseudocode A.1

Program EXAMPLE
\ DESCRIPTION: Purpose of this code is to illustrate the use a function module.
\ USES:
\   FUNC:: A function of three variables.
. . .
$S \leftarrow$ FUNC$(2, 1, 3)$*FUNC$(3, -2, 4)$
. . .
End Program EXAMPLE

Function Module FUNC (x, y, z)
FUNC$\leftarrow \sqrt{x^2 + y^2 + z^2}$
End Function Module FUNC

A program, module, or function module is rarely independent; it frequently requires other built-in or user-defined functions or modules. The modules called by a program or other modules are referred to as *program dependencies*, which are specified in the header section beneath as **USES** statements. That means the program or module will not function without the specified modules. Note that in the pseudocode above, **Function Module** FUNC is identified in the program header. Make sure that the built-in modules given in the pseudocodes are used with compatible names specified in the programming language you are programming in.

If a function is to call itself again, directly or indirectly ($f_n(x) \leftarrow n + x * f_{n-1}(x)$ with $f_0(x) \leftarrow x$), it should be explicitly delimited by **Recursive Function Module-End Recursive Function Module** statements. For instance,

Recursive Function Module FX(n, x)
  If $\left[n = 0 \right]$ **Then**
    FX$\leftarrow x$
  Else
    FX$\leftarrow n + x*$FX$(n - 1, x)$
  End If
  FUN_NAME $\leftarrow$ Expression
End Recursive Function Module FX

Quadratures

B.1 GAUSS-LEGENDRE QUADRATURE

N	$\pm x_k$	w_k	N	$\pm x_k$	w_k
2	0.57735 02691	1.00000 00000		0.90411 72564	0.10693 93260
3	0.00000 00000	0.88888 88888		0.98156 06342	0.04717 53364
	0.77459 66692	0.55555 55556	14	0.10805 49487	0.21526 38535
4	0.33998 10435	0.65214 51548		0.31911 23689	0.20519 84637
	0.86113 63115	0.34785 48451		0.51524 86364	0.18553 83975
5	0.00000 00000	0.56888 88889		0.68729 29048	0.15720 31672
	0.53846 93101	0.47862 86704		0.82720 13151	0.12151 85707
	0.90617 98459	0.23692 68850		0.92843 48837	0.08015 80872
6	0.23861 91860	0.46791 39345		0.98628 38087	0.03511 94603
	0.66120 93864	0.36076 15730	16	0.09501 25098	0.18945 06104
	0.93246 95142	0.17132 44923		0.28160 35507	0.18260 34150
7	0.00000 00000	0.41795 91837		0.45801 67776	0.16915 65193
	0.40584 51514	0.38183 00505		0.61787 62444	0.14959 59888
	0.74153 11856	0.27970 53915		0.75540 44083	0.12462 89712
	0.94910 79123	0.12948 49662		0.86563 12023	0.09515 85116
8	0.18343 46424	0.36268 37833		0.94457 50230	0.06225 35239
	0.52553 24099	0.31370 66458		0.98940 09349	0.02715 24594
	0.79666 64774	0.22238 10344	20	0.07652 65211	0.15275 33871
	0.96028 98564	0.10122 85362		0.22778 58511	0.14917 29864
10	0.14887 43398	0.29552 42247		0.37370 60887	0.14209 61093
	0.43339 53941	0.26926 67193		0.51086 70019	0.13168 86384
	0.67940 95682	0.21908 63625		0.63605 36807	0.11819 45319
	0.86506 33666	0.14945 13491		0.74633 19064	0.10193 01198
	0.97390 65286	0.06667 13443		0.83911 69718	0.08327 67415
12	0.12523 34085	0.24914 70458		0.91223 44282	0.06267 20483
	0.36783 14990	0.23349 25365		0.96397 19272	0.04060 14298
	0.58731 79543	0.20316 74267		0.99312 85991	0.01761 40071
	0.76990 26742	0.16007 83285			

DOI: 10.1201/9781003474944-B

B.2 GAUSS-LAGUERRE QUADRATURE

N	x_k	w_k	$W_k = w_k e^{x_k}$
2	0.58578 64376	8.53553 390593 E-01	1.53332 603312
	3.41421 35623	1.46446 609407 E-01	4.45095 733505
3	0.41577 45567	7.11093 009929 E-01	1.07769 285927
	2.29428 03602	2.78517 733569 E-01	2.76214 296190
	6.28994 50829	1.03892 565016 E-02	5.60109 462543
4	0.32254 76896	6.03154 104342 E-01	0.83273 912384
	1.74576 11011	3.57418 692438 E-01	2.04810 243845
	4.53662 02969	3.88879 085150 E-02	3.63114 630582
	9.39507 09123	5.39294 705561 E-04	6.48714 508441
5	0.26356 03197	5.21755 610583 E-01	0.67909 404221
	1.41340 30591	3.98666 811083 E-01	1.63848 787360
	3.59642 57710	7.59424 496817 E-02	2.76944 324237
	7.08581 00058	3.61175 867992 E-03	4.31565 690092
	12.64080 08442	2.33699 723858 E-05	7.21918 635435
6	0.22284 66042	4.58964 673950 E-01	0.57353 550742
	1.18893 21016	4.17000 830772 E-01	1.36925 259071
	2.99273 63261	1.13373 382074 E-01	2.26068 459338
	5.77514 35691	1.03991 974531 E-02	3.35052 458235
	9.83746 74184	2.61017 202815 E-04	4.88682 680021
	15.9828 73981	8.98547 906430 E-07	7.84901 594559
8	0.17027 96323	3.69188 589342 E-01	0.43772 341049
	0.90370 17767	4.18786 780814 E-01	1.03386 934767
	2.25108 66298	1.75794 986637 E-01	1.66970 976566
	4.26670 01702	3.33434 922612 E-02	2.37692 470176
	7.04590 54023	2.79453 623523 E-03	3.20854 091335
	10.75851 60101	9.07650 877336 E-05	4.26857 551083
	15.74067 86412	8.48574 671627 E-07	5.81808 336867
	22.86313 17368	1.04800 117487 E-09	8.90622 621529
12	0.11572 21173	2.64731 371055 E-01	0.29720 963605
	0.61175 74845	3.77759 275873 E-01	0.69646 298043
	1.51261 02697	2.44082 011320 E-01	1.10778 139462
	2.83375 13377	9.04492 222117 E-02	1.53846 423904
	4.59922 76394	2.01023 811546 E-02	1.99832 760627
	6.84452 54531	2.66397 354187 E-03	2.50074 576910
	9.62131 68424	2.03231 592663 E-04	3.06532 151828
	13.00605 49933	8.36505 585682 E-06	3.72328 911078
	17.11685 51874	1.66849 387654 E-07	4.52981 402998
	22.15109 03793	1.34239 103052 E-09	5.59725 846184
	28.48796 72509	3.06160 163504 E-12	7.21299 546093
	37.09912 10444	8.14807 746743 E-16	10.54383 74619

B.3 GAUSS-HERMITE QUADRATURE

N	$\pm x_k$	w_k	$W_k = w_k e^{x_k^2}$
2	0.70710 67811	8.86226 92545 E-01	1.46114 11826
3	0.00000 00000	1.18163 59006 E+00	1.18163 59006
	1.22474 48713	2.95408 97515 E-01	1.32393 11752
4	0.52464 76232	8.04914 09000 E-01	1.05996 44828
	1.65068 01238	8.13128 35447 E-02	1.24022 58176
5	0.00000 00000	9.45308 72048 E-01	0.94530 87204
	0.95857 24646	3.93619 32315 E-01	0.98658 09967
	2.02018 28704	1.99532 42059 E-02	1.18148 86255
6	0.43607 74119	7.24629 59522 E-01	0.87640 13344
	1.33584 90740	1.57067 32032 E-01	0.93558 05576
	2.35060 49736	4.53000 99055 E-03	1.13690 83326
8	0.38118 69902	6.61147 01255 E-01	0.81026 46175
	1.15719 37124	2.07802 32581 E-01	0.82868 73032
	1.98165 67566	1.70779 83007 E-02	0.89718 46002
	2.93063 74202	1.99604 07221 E-04	1.10133 07296
10	0.34290 13272	6.10862 63373 E-01	0.68708 18539
	1.03661 08297	2.40138 61108 E-01	0.70329 63231
	1.75668 36492	3.38743 94455 E-02	0.74144 19319
	2.53273 16742	1.34364 57467 E-03	0.82066 61264
	3.43615 91188	7.64043 28552 E-06	1.02545 16913
16	0.27348 10461	5.07929 47901 E-01	0.54737 52050
	0.82295 14491	2.80647 45852 E-01	0.55244 19573
	1.38025 85391	8.38100 41398 E-02	0.56321 78290
	1.95178 79909	1.28803 11535 E-02	0.58124 72754
	2.54620 21578	9.32284 00862 E-04	0.60973 69582
	3.17699 91619	2.71186 00925 E-05	0.65575 56728
	3.86944 79048	2.32098 08448 E-07	0.73824 56222
	4.68873 89393	2.65480 74740 E-10	0.93687 44928
20	0.24534 07083	4.62243 66960E-01	0.49092 15006
	0.73747 37285	2.86675 50536E-01	0.49384 33852
	1.23407 62153	1.09017 20602E-01	0.49992 08713
	1.73853 77121	2.48105 20887E-02	0.50967 90271
	2.25497 40020	3.24377 33422E-03	0.52408 03509
	2.78880 60584	2.28338 63601E-04	0.54485 17423
	3.34785 45673	7.80255 64785E-06	0.57526 24428
	3.94476 40401	1.08606 93707E-07	0.62227 86961
	4.60368 24495	4.39934 09922E-10	0.70433 29611
	5.38748 08900	2.22939 36455E-13	0.89859 19614

Bibliography

[1] ABERTH, O., *Introduction to Precise Numerical Methods*. Academic Press, 2007.

[2] ABRAMOWITZ, M., STEGUN, I. A., *Handbook of Mathematical Functions*. U.S Government Printing Office, Washington, D.C., 1964.

[3] ACTON, S. F., *Numerical Methods That Work*. Harper & Row Publishers, New York, 1970.

[4] AKAI, T. J., *Applied Numerical Methods for Engineers*. John Wiley & Sons, Inc., New York, 1994.

[5] ALLEN, M. P., *Understanding Regression Analysis*. Plenum Press, 1997.

[6] AMES, F. W., *Numerical Methods for Partial Differential Equations*. Academic Press, 1977.

[7] ANDERSON, A. D., TANNEHILL, C. J., PLETCHER, H. R., *Computational Fluid Mechanics and Heat Transfer*. Hemisphere Publishing Corporation, New Work, 1984.

[8] ANTON, H., RORRES, C., KAUL, A., *Elementary Linear Algebra: Applications Version*. John Wiley & Sons, 2019.

[9] ASCHER, U. M., GRIEF, C., *A First Course in Numerical Methods (Computational Science and Engineering Edition)*. Society for Industrial and Applied Mathematics (SIAM), 2011.

[10] ATKINSON, K. E., *An Introduction to Numerical Analysis*. John Wiley & Sons, 1989.

[11] ATKINSON, K. E., HAN, W., STEWART, D. E., *Numerical Solution of Ordinary Differential Equations*. John Wiley & Sons, Hoboken, New Jersey, 2009.

[12] AYYUB, B. M., MCCUEN, R. H., *Numerical Analysis for Engineers: Methods and Applications*, Second Edition (Textbooks in Mathematics). Chapman and Hall/CRC, 2015.

[13] AXELSSON, O., *Iterative Solution Methods*. Cambridge University Press, 1996.

[14] BAILY, P. B., SHAMPİNE, L. F., WALTMAN, P. E., *Nonlinear Two Point Boundary Value Problems (Mathematics in Science and Engineering)*. Academic Press, 1968.

[15] BARRETT, R., BERRY, M., CHAN, T. F., DEMMEL, J., DONATO, J., DONGARRA, J., EIJKHOUT, V., POZO, R., ROMINE, C., VAN DER VORST, H., *Templates for the Solution of Linear Systems: Building Blocks for Iterative Methods (Miscellaneous Titles in Applied Mathematics Series No 43)*. Society for Industrial and Applied Mathematics (SIAM), 1987.

[16] BEERS, K. J., *Numerical Methods for Chemical Engineering: Applications in MATLAB*. Cambridge University Press, 2006.

[17] BOBER, W., *Introduction to Numerical and Analytical Methods with MATLAB for Engineers and Scientists*, CRC Press, 2013.

[18] BORZI, A., *Modelling with Ordinary Differential Equations: A Comprehensive Approach*. Chapman & Hall/CRC Numerical Analysis and Scientific Computing Series, CRC Press, 2020.

[19] BOSE, S. K., *Numerical Methods of Mathematics Implemented in Fortran*. Springer, Singapore, 2019.

[20] BOYCE, W. E., DIPRIMA, R. C., MEADE, D. B., *Elementary Differential Equations and Boundary Value Problems*. Wiley, 2017.

[21] BRADIE, B., *A Friendly Introduction to Numerical Analysis*. Pearson Prentice Hall, 2006.

[22] BRONSON, R., COSTA, G. B., *Matrix Methods: Applied Linear Algebra and Sabermetrics*. Academic Press, 2020.

[23] BROOKS, D. R., *Problem Solving with Fortran 90: For Scientists and Engineers*. Springer, New York, 2012.

[24] BURDEN, R. L., FAIRES, J. D., BURDEN, A. M., *Numerical Analysis, 10th ed.* Cengage Learning, 2015.

[25] BUTCHER, J. C., *Numerical Methods for Ordinary Differential Equations*. John Wiley, 2016.

[26] BUTENKO, S., PARDALOS, P.M., *Numerical Methods and Optimization: An Introduction*. Chapman & Hall/CRC Numerical Analysis and Scientific Computing Series, Taylor & Francis, 2014,

[27] CARNAHAN, B., LUTHER, H. A., WILKES, J. O., *Applied Numerical Methods*. John Wiley, 1969.

[28] CHAPRA, S., CANALE, R., *Numerical Methods For Engineers*. 7th ed. McGraw-Hill Education, Boston, 2020.

[29] CHENEY, E. W., KINCAID, D. R., *Numerical Mathematics and Computing*. 7th ed. Cengage Learning, 2012.

[30] CLARC, M. Jr., HANSEN, F. K., *Numerical Methods of Reactor Analysis*. Academic Press, New York, 1964.

[31] CLINE, A. K., MOLER, C. B., STEWART, G. W., WILKINSON, J. H., An Estimate for the Condition Number of a Matrix. *SIAM Journal on Numerical Analysis*, 16(2), 368-375. http://www.jstor.org/stable/2156842

[32] COKER, A. K., *Fortran Programs for Chemical Process Design, Analysis, and Simulation*. Gulf Professional Publishing, 1995.

[33] CONTE, S. D., BOOR, C. D., *Elementary Numerical Analysis: An Algorithmic Approach*. Society for Industrial and Applied Mathematics (SIAM), 2017.

[34] CRANK, J., MARTIN H. G., MELLUIS, D. M., *Numerical Solution of Partial Differential Equations*. Oxford University Press, 1978.

[35] CURTISS, C. F., HIRSCHFELDER, J. O. (1952). Integration of Stiff Equations, *Proceedings of the National Academy of Sciences of the United States of America*, 38(3), 235-243. http://www.jstor.org/stable/88864

[36] ÇENGEL, Y. A., *Heat Transfer: A Practical Approach*. McGraw-Hill Higher Education, 1998.

[37] ÇENGEL, Y. A., *Thermodynamics: An Engineering Approach*. McGraw-Hill Higher Education, 2008.

[38] DAHLQUIST, G., BJÖRCK, A., *Numerical Methods in Scientific Computing: Volume II*. Society for Industrial and Applied Mathematics (SIAM), 2008.

[39] DAVIS, H. T., THOMSON, K. T., *Linear Algebra and Linear Operators in Engineering: With Applications in Mathematica*. Academic Press, 2000.

[40] DAVIS, P. J., RABINOWITZ, P., *Methods of Numerical Integration, 2nd ed.* Dover Publications, 2007.

[41] DAVIS, G. V., *Numerical Methods in Engineering and Science*. Springer, 1986.

[42] DAVIS, M. E., *Numerical Methods and Modeling for Chemical Engineers*. John Wiley & Sons, 1984.

[43] DAVIS, M. E., DAVIS, R. J., *Fundamentals of Chemical Reaction Engineering*. Mac-Graw Hill Inc., 2003.

[44] DE COSTER, C., HABETS, P., *Two-Point Boundary Value Problems: Lower and Upper Solutions*. Elsevier Science, 2006.

[45] DEMMEL, J. W., *Applied Numerical Linear Algebra*. Society for Industrial and Applied Mathematics (SIAM), 1997.

[46] DONGARRA, J. J., DUFF, I. S., SORENSEN, D. C., VAN DER VORST, H. A., *Numerical Linear Algebra on High-Performance Computers (Software, Environments and Tools, Series Number 7)*. Society for Industrial and Applied Mathematics (SIAM), 1998.

[47] DORFMAN, K. D., DAOUTIDIS, P., *Numerical Methods with Chemical Engineering Applications (Cambridge Series in Chemical Engineering)*. Cambridge University Press, 2017.

[48] DRISCOLL, T. A., BRAUN, R. J., *Fundamentals of Numerical Computation*. Society for Industrial & Applied Mathematics (SIAM), 2017.

[49] DUBIN, D., *Numerical and Analytical Methods for Scientists and Engineers, Using Mathematica*. Wiley-Interscience, 2003.

[50] EPPERSON, J. F., *An Introduction to Numerical Methods and Analysis*. Wiley, 2013.

[51] ESFANDIARI, R. S., *Numerical Methods for Engineers and Scientists Using MAT-LAB*. CRC Press, 2017.

[52] FARRASHKHALVAT, M., MILES, J. P., *Basic Structured Grid Generation: With an Introduction to Unstructured Grid Generation*. Butterworth-Heinemann, 2003.

[53] FAUL, A. C., *A Concise Introduction to Numerical Analysis*. CRC Press, 2018.

[54] FORD, W., *Numerical Linear Algebra with Applications: Using MATLAB*. Academic Press, 2014.

[55] FORSYTHE, G. E., MALCOLM, M. A., MOLER, C. B., *Computer Methods for Mathematical Computations (Prentice-Hall series in automatic computation)*. Prentice Hall, 1977.

[56] FROST, J., *Regression Analysis: An Intuitive Guide for Using and Interpreting Linear Models*. Statistics By Jim Publishing, 2020.

[57] FRÖBERG, C., *Introduction to Numerical Analysis*. Second Edition, Addison-Wesley Publishing Company, Massachusetts, 1974.

[58] GANDER, W., GANDER, M. J., KWOK, F., *Scientific Computing—An Introduction using Maple and MATLAB (Texts in Computational Science and Engineering, 11)*. Springer, 2014.

[59] GAUTSCHI, W., *Numerical Analysis*. Birkhäuser, 2011.

[60] GEAR, C. W., *Numerical Initial Value Problems in Ordinary Differential Equations (Automatic Computation)*. Prentice Hall, 1971.

[61] GERALD, C. F., WHEATLEY, P. O., *Applied Numerical Analysis, 6th Ed.* Addison-Wesley Publishing Co., 1998.

[62] GEZERLIS, A., *Numerical Methods in Physics with Python*. Cambridge University Press, 2020.

[63] GIESELA, E. M., UHLIG, F., *Numerical Algorithms with C*. Springer, 2014.

[64] GILAT, A., SUBRAMANIAM, V., *Numerical Methods for Engineers and Scientists: An Introduction with Applications Using MATLAB*. Wiley, 2013.

[65] GOCKENBACH, M. S., *Partial Differential Equations: Analytical and Numerical Methods*. 2nd ed. Society for Industrial and Applied Mathematics (SIAM), 2010.

[66] GOLUB, G. H., ORTEGA, J. M., *Scientific Computing and Differential Equations: An Introduction to Numerical Methods*. Academic Press, San Diego, 1992.

[67] Golub, G. H., Van Loan, C. F., *Matrix Computations*. (Johns Hopkins Studies in the Mathematical Sciences, 3), Johns Hopkins University Press, 2013.

[68] GREENBAUM, A., CHARTIER, T. P., *Numerical Methods: Design, Analysis, and Computer Implementation of Algorithms*. Princeton University Press, 2012.

[69] GREENBAUM, A., *Iterative Methods for Solving Linear Systems*. Society for Industrial and Applied Mathematics (SIAM), 1997.

[70] GREENSPAN, D., *Discrete Numerical Methods in Physics and Engineering*. Academic Press, 1974.

[71] GREWAL, B. S., *Numerical Methods in Engineering and Science: with Programs in C and C++*. Khanna Publishers, 2010.

[72] GRIFFITHS, D. V., SMITH, I. M., *Numerical Methods for Engineers*. Chapman and Hall/CRC, 2006.

[73] HACKBUSCH, W., *Iterative Solution of Large Sparse Systems of Equations*. Applied mathematical sciences, Springer-Verlag, 1994.

[74] HAGEMAN, L. A., YOUNG, D. M., *Applied Iterative Methods*. Dover Publications, 2004.

[75] HAMMING, R. W., *Introduction to Applied Numerical Analysis (Dover Books on Mathematics)*. Dover Publications, 2012.

[76] HEARN, E. J., *Mechanics of Materials Volume 1: An Introduction to the Mechanics of Elastic and Plastic Deformation of Solids and Structural Materials*. Butterworth-Heinemann, 2000.

[77] HEARN, E. J., *Mechanics of Materials Volume 2: The Mechanics of Elastic and Plastic Deformation of Solids and Structural Materials*. Butterworth-Heinemann, 2000.

[78] HAMMING, R. W., *Numerical Methods for Scientists and Engineers*. Dover Publications, 1987.

[79] HENRICI, P., *Elements of Numerical Analysis*. Wiley, 1964.

[80] HESTENES, M. R., STIEFEL, E., Methods of conjugate gradients for solving linear systems. *Journal of Research of the National Bureau of Standards*, 49, 409-435 (1952).

[81] HIGHAM, N. J., *Accuracy and Stability of Numerical Algorithms*. Society for Industrial and Applied Mathematics (SIAM), 2002.

[82] HILDEBRAND, F. B., *Introduction to Numerical Analysis*. 2nd ed. Dover Books on Mathematics), Dover Publications, 1987.

[83] HOFFMAN, J. D., *Numerical Methods for Engineers and Scientists*. 2nd ed., McGraw-Hill Inc., New York, 1992.

[84] HORNBECK, W. R., *Numerical Methods (With Numerous Examples and Solved Illustrative Problems)*. Quantum Publishers, New York, 1975.

[85] HOROWITZ, E., SAHNI, S., RAJASEKARAN, S., *Computer Algorithms*. Computer Science Press, 1997.

[86] ISAACSON, E., KELLER, H. B., *Analysis of Numerical Methods*. Dover Publishing Co., 1994.

[87] ISERLES, A., *A First Course in the Numerical Analysis of Differential Equations*. Cambridge University Press, 1996.

[88] JANA, A. K., *Chemical Process Modelling and Computer Simulation*. PHI Learning Ltd., 2018.

[89] JOHNSON, L., RIESS, D., ARNOLD, J., *Introduction to Linear Algebra (Classic Version) (Pearson Modern Classics for Advanced Mathematics Series)*. Pearson, 2017.

[90] JOHNSTON, R. L., *Numerical Methods: A Software Approach*. John Wiley & Sons, 1982.

[91] KAHANER, David., MOLER, C., NASH, S. *Numerical Methods and Software*. Prentice-Hall, 1989.

[92] KELLY, C. T. *Iterative Methods for Linear and Nonlinear Equations*. (Frontiers in Applied Mathematics, Series Number 18), Society for Industrial and Applied Mathematics (SIAM), 1995.

[93] KHARAB, A., GUENTHER, R., *An Introduction to Numerical Methods*. CRC Numerical Analysis and Scientific Computing Series. CRC Press, 2021.

[94] KHOURY, R., HARDER, D. W., *Numerical Methods and Modelling for Engineering*. Springer Int. Pub., 2016.

[95] KIUSALAAS, J., *Numerical Methods in Engineering with Python 3*. Cambridge University Press, 2013.

[96] KNUTH, D.E., *The Art of Computer Programming: Fundamental algorithms*. Addison-Wesley Publishing Company, 1973.

[97] KOMZSIK, L., *Approximation Techniques for Engineers*. CRC Press, Taylor & Francis Group, 2020.

[98] KUO, S. S., *Computer Applications of Numerical Methods.* Addison-Wesley Publishing Company, Massachusetts, 1972.

[99] LAMARSH, J. R., *Introduction to Nuclear Engineering.* Addison-Wesley, 1983.

[100] LAMBERT, J. D., *Numerical Methods for Ordinary Differential Systems: The Initial Value Problem.* Wiley, 1991.

[101] LANDAU, R. H., MEJI, M. J. P., *Computational Physics: Problem Solving with Computers.* Wiley, 2007.

[102] LARSON, R., FALVO, D. C., *Elementary Linear Algebra.* Brooks Cole Learning, 2008.

[103] LEVEQUE, R. J., *Finite Difference Methods for Ordinary and Partial Differential Equations: Steady-State and Time-dependent Problems.* Society for Industrial and Applied Mathematics (SIAM), 2007.

[104] LINDFIELD, G., PENNY, J., *Numerical Methods: Using MATLAB.* Academic Press, 2018.

[105] LUYBEN W.L., *Process Modeling, Simulation And Control For Chemical Engineers.* 2nd ed. McGraw Hill Exclusive (Cbs), 2014.

[106] MARDEN, M., *Geometry of Polynomials.* American Mathematical Society, 1970,

[107] MARON, M. J., LOPEZ, R. J., *Numerical Analysis: A Practical Approach.* Wadsworth Pub. Co., 1991,

[108] MATTHEIJ, R. M. M., RIENSTRA, S. W., TEN THIJE BOONKKAMP, J. H. M., *Partial Differential Equations: Modeling, Analysis, Computation.* (Monographs on Mathematical Modeling and Computation, Series Number 10). Society for Industrial and Applied Mathematics (SIAM), 1987.

[109] MEURANT, G., *Computer Solution of Large Linear Systems.* Elsevier Science, 1999.

[110] MCNAMEE, J. M., *Numerical Methods for Roots of Polynomials—Part I.* Elsevier Science, 2013,

[111] MCNAMEE, J. M., PAN, V., *Numerical Methods for Roots of Polynomials - Part II.* (Studies in Computational Mathematics, Volume 16), Elsevier Science, 2018,

[112] MESEGUER, A., *Fundamentals of Numerical Mathematics for Physicists and Engineers.* Wiley, 2020,

[113] MIDDLETON, J.A., *Experimental Statistics and Data Analysis for Mechanical and Aerospace Engineers.* Advances in Applied Mathematics, CRC Press, 2021.

[114] MILLER, G., *Numerical Analysis for Engineers and Scientists.* Cambridge University Press, 2014.

[115] MOIN, P., *Fundamentals of Engineering Numerical Analysis.* Cambridge University Press, 2010.

[116] MULLER, J. M., BRUNIE, N., DE DINECHIN, F., JEANNEROD, C. P., JOLDES, M., LEFÈVRE, V., MELQUIOND, G., REVOL, N., TORRES, S., *Handbook of Floating-Point Arithmetic.* Springer International Publishing, 2018.

[117] MYLER, H. R., *Fundamentals of Engineering Programming with C and Fortran.* Cambridge University Press, 1998.

[118] NA, T. Y., *Computational Methods in Engineering Boundary Value Problems.* Academic Press, 2012.

[119] NA, T. Y., SIDHOM, M. M. On Stokes' Problems for Linear Viscoelastic Fluids. *ASME. J. Appl. Mech.* 34(4), 1040–1042 (1967). https://doi.org/10.1115/1.3607818

[120] NASH, J. C., *Compact Numerical Methods for Computers: Linear Algebra and Function Minimisation.* CRC Press, 1990.

[121] NERI, F., *Linear Algebra for Computational Sciences and Engineering.* Springer, 2019.

[122] NEUMAIER, A., *Introduction to Numerical Analysis.* Cambridge University Press, 2001.

[123] NILSSON, J. W., RIEDEL, S. A., *Electric Circuits.* Prentice Hall, 2011.

[124] NORMAN, D., WOLCZUK, D., *Introduction to Linear Algebra for Science and Engineering.* 2nd ed. Pearson Education Canada, 2011.

[125] NYHOFF, L.R., LEESTMA, S., FORTRAN 77 for Engineers and Scientists: With an Introduction to FORTRAN 90. Prentice Hall, 1996.

[126] O'LEARY, D. P. Estimating Matrix Condition Numbers. *SIAM Journal on Scientific and Statistical Computing,* 1(2), 205–209 (1980). DOI: 10.1137/0901014

[127] ORTEGA, J. M., *Numerical Analysis: A Second Course.* Society for Industrial and Applied Mathematics, 1990.

[128] PARLETT, B. N., *The Symmetric Eigenvalue Problem.* (Prentice-Hall series in Computational Mathematics), Prentice-Hall, Englewood Cliffs, NJ, 1980.

[129] PHILLIPS, G. M., TAYLOR, P. J., *Theory and Applications of Numerical Analysis.* Academic Press, 1973.

[130] PINDER, G. F., *Numerical Methods for Solving Partial Differential Equations: A Comprehensive Introduction for Scientists and Engineers.* Wiley, 2018.

[131] POWERS, D. L., *Boundary Value Problems: and Partial Differential Equations.* Academic Press, 2009.

[132] PRESS, W. H., TEUKOLSKY, S. A., VETTERLING, W. T., FLANNERY, B. P., *Numerical Recipes in Fortran 90: Volume 2, Volume 2 of Fortran Numerical Recipes: The Art of Parallel Scientific Computing.* Cambridge University Press, 1996.

[133] QUARTERONI, A., SACCO, R., SALERI, F., *Numerical Mathematics.* Texts in Applied Mathematics, Springer Berlin Heidelberg, 2006.

[134] RALSTON, A., RABINOWITZ, P., *A First Course in Numerical Analysis.* 2nd ed. (Dover Books on Mathematics), Dover Publications, 2001.

[135] RAO, S., *Applied Numerical Methods for Engineers and Scientists.* Pearson, 2001.

[136] RAO, S. G., *Numerical Analysis.* New Age International, 2018.

[137] RASMUSON, A., ANDERSSON, B., OLSSON, L., ANDERSSON, R., *Mathematical Modeling in Chemical Engineering.* Cambridge University Press, 2014.

[138] RAY, S. S., *Numerical Analysis with Algorithms and Programming.* Chapman and Hall/CRC, 2016.

[139] REID, J. K., A Method for Finding the Optimum Successive Over-Relaxation Parameter, *The Computer Journal,* 9(2), 200–204 (1966). https://doi.org/10.1093/comjnl/9.2.200

[140] RHEINBOLDT, W. C., *Methods for Solving Systems of Nonlinear Equations*. (CBMS-NSF Regional Conference Series in Applied Mathematics, Series Number 70), Society for Industrial and Applied Mathematics (SIAM), 1987.

[141] RICE, R. G., DO, D. D., *Applied Mathematics And Modeling For Chemical Engineers*. Wiley-AIChE, 2012.

[142] RICE, R. J., *Numerical Methods, Software, and Analysis*. McGraw-Hill Book Company, New York, 1983.

[143] RIDGWAY, S. L., *Numerical Analysis*. Princeton University Press, 2011.

[144] RIVLIN, T.J., *An Introduction to the Approximation of Functions*. Dover Publications, 1981.

[145] SAAD, Y., *Iterative Methods for Sparse Linear Systems*. (The Pws Series in Computer Science), Pws Pub Co, 1996.

[146] SALVADORI, G. M., BARON, L. M., *Numerical Methods in Engineering*. Prentice-Hall of India LTD., New Delhi, 1966.

[147] SAUER, T., *Numerical Analysis*. Pearson, 2017.

[148] SCHEID, F., *Numerical Analysis*. McGraw-Hill Book Company, New York, 1989.

[149] SCHROEDER, L.D., SJOQUIST, D.L., STEPHAN, P.E., *Understanding Regression Analysis: An Introductory Guide*. Quantitative Applications in the Social Sciences, SAGE Publications, 1986.

[150] SHAMPINE, L. F., ALLEN, R. C., PRUESS, S., *Fundamentals of Numerical Computing*. Wiley, 1997.

[151] SIAUW, T., BAYEN, A., *An Introduction to MATLAB Programming and Numerical Methods for Engineers*. Academic Press, 2014.

[152] SMITH, G. D., *Numerical Solution of Partial Differential Equations: Finite Difference Methods (Oxford Applied Mathematics and Computing Science Series)*. Clarendon Press, 1986.

[153] SPIEGEL, M., *Schaum's Outline of Calculus of Finite Differences and Difference Equations*. McGraw Hill, 1971.

[154] STEPHENS, M. A., D'AGOSTINO, R. B., *Goodness-of-Fit-Techniques*. Statistics: A Series of Textbooks and Monographs, Taylor & Francis, 1986.

[155] STERLING, M. J., *Linear Algebra For Dummies*. Wiley, 2009.

[156] STEWART, G. W., *Matrix Algorithms: Volume 1, Basic Decompositions*. SIAM: Society for Industrial and Applied Mathematics, 1998.

[157] STEWART, G. W., *Matrix Algorithms: Volume 2, Eigensystems*. SIAM: Society for Industrial and Applied Mathematics, 2001.

[158] STOER, J., BULIRSCH, R., *Introduction to Numerical Analysis.*, 2nd ed. Springer Verlag, 1993.

[159] STEWART, G. W., *Matrix Algorithms: Volume 1, Basic Decompositions*. Society for Industrial and Applied Mathematics (SIAM), 1998.

[160] STEWART, G. W., *Matrix Algorithms: Volume 2, Eigensystems*. Society for Industrial and Applied Mathematics (SIAM), 2001.

[161] TOBIA, M.J., *Matrices in Engineering Problems*. Synthesis digital library of engineering and computer science, Morgan & Claypool Publishers, 2011.

[162] THOMPSON, J. F., SONI, B. K., WEATHERILL, N.P. (Eds.) *Handbook of Grid Generation*. 1st ed. CRC Press, 1998.

[163] TREFETHEN, L. N., BAU, D. III., *Numerical Linear Algebra*. Society for Industrial and Applied Mathematics (SIAM), 1997.

[164] TURNER, P. R., *Guide to Numerical Analysis (CRC Mathematical Guides)*. CRC Press, 1990.

[165] YANG, Y. W., CAO, W., KIM, J., PARK, K. W., PARK, H., JOUNG, J., *Applied Numerical Methods Using Matlab*. 2nd Ed., Wiley, 2020.

[166] VAN LOAN, C., COLEMAN, T. F., *Handbook for Matrix Computations (Frontiers in Applied Mathematics)*. Society for Industrial and Applied Mathematics (SIAM), 1987.

[167] VARGA, R. S., *Matrix Iterative Analysis*. Prentice Hall Inc., 1962.

[168] VINOKUR, M., On one-dimensional stretching functions for finite-difference calculations. *Journal of Computational Physics*, 50(2), 215–234 (1980). DOI: 10.1016/0021-9991(83)90065-7

[169] VORST, H. A. van der. *Iterative Krylov Methods for Large Linear Systems*. Cambridge University Press, 2003.

[170] XIE, W.C., *Differential Equations for Engineers*. Cambridge University Press, 2010.

[171] WAIT, R., *The Numerical Solution of Algebraic Equations*. John Wiley & Sons, 1979.

[172] WATKINS, D. S., *The Matrix Eigenvalue Problem: GR and Krylov Subspace Methods*. Society for Industrial and Applied Mathematics (SIAM), 2008.

[173] WATKINS, D. S., *Fundamentals of Matrix Computations*. Wiley, 2010.

[174] WILKINSON, J. H., *The Algebraic Eigenvalue Problem*. Numerical Mathematics and Scientific Computation, Clarendon Press, 1988.

[175] WILKINSON, J. H., REINSCH, C., *Handbook for Automatic Computation, Vol. 2: Linear Algebra (Grundlehren Der Mathematischen Wissenschaften, Vol. 186)*. Editor:Bauer, F. L., Springer Verlag, 1971.

[176] WILKINSON, J. H., *Rounding Errors in Algebraic Processes*. Hassell Street Press, 2021.

[177] WILKINSON, J. H, REINSCH, C., BAUER, F. L., *Handbook for Automatic Computation: Linear Algebra* (Grundlehren Der Mathematischen Wissen-schaften, Vol 186) Springer Verlag, 1986.

[178] YOUNG, D. F., MUNSON, B. R., OKIISHI, T. H., *A Brief Introduction to Fluid Mechanics*. 5th ed. Wiley, 2011.

[179] YOUNG., D. M., *Iterative Solution of Large Linear Systems*. Academic Press, New York, 1971.

[180] ZALIZNIAK, V., *Essentials of Scientific Computing: Numerical Methods for Science and Engineering*. Woodhead Publishing, 2008.

[181] ZAROWSKI, C. J., *An Introduction to Numerical Analysis for Electrical and Computer Engineers*. Wiley-Interscience, 2004.

[182] ZILL, D. G., CULLEN, M. R., *Differential Equations with Boundary-Value Problems.* 7th ed. Cengage Learning, 2009.

[183] ZILL, D. G., *A First Course in Differential Equations with Applications.* Brooks/Cole, Cengage Learning, 2012.

References

[1] ABRAMOWITZ, M., STEGUN, I.A., *Handbook of Mathematical Functions*. U.S Government Printing Office, Washington, D.C., 1964.

[2] AKAI, T. J., *Applied Numerical Methods for Engineers*. John Wiley & Sons, Inc., New York, 1994.

[3] AMES, F. W., *Numerical Methods for Partial Differential Equations*. Academic Press, 1977.

[4] AXELSSON, O., *Iterative Solution Methods*. Cambridge University Press, 1996.

[5] BARRETT, R., BERRY, M., CHAN, T. F., DEMMEL, J., DONATO, J., DONGARRA, J., EIJKHOUT, V., POZO, R., ROMINE, C., VAN DER VORST, H., *Templates for the Solution of Linear Systems: Building Blocks for Iterative Methods (Miscellaneous Titles in Applied Mathematics Series No 43)*. Society for Industrial and Applied Mathematics (SIAM), 1987.

[6] CURTISS, C. F., HIRSCHFELDER, J. O., Integration of Stiff Equations. *Proceedings of the National Academy of Sciences of the United States of America*, 38(3), 235–243 (1952). http://www.jstor.org/stable/88864

[7] FARRASHKHALVAT, M., MILES, J. P., *Basic Structured Grid Generation: With an Introduction to Unstructured Grid Generation*. Butterworth-Heinemann, 2003.

[8] GREENBAUM, A., *Iterative Methods for Solving Linear Systems*. Society for Industrial and Applied Mathematics (SIAM), 1997.

[9] GOLUB, G. H., Van der VORST, H. A., Eigenvalue computation in the 20th century. *Journal of Computational and Applied Mathematics*, 123, 35–65 (2000).

[10] GUGGENHEIMER, H. W., EDELMAN, A. S., JOHNSON, C. R., A Simple Estimate of the Condition Number of a Linear System. *The College Mathematics Journal*, 26(1), 2–5 (1995). DOI: 10.1080/07468342.1995.11973657

[11] HACKBUSCH, W., *Iterative Solution of Large Sparse Systems of Equations*. Applied mathematical sciences, Springer-Verlag, 1994.

[12] HAGEMAN, L. A., YOUNG, D. M., *Applied Iterative Methods*. Dover Publications, 2004.

[13] HART, J. J., A Correction for the Trapezoidal Rule. *The American Mathematical Monthly*, 59(1), 33–37 (1952). https://doi.org/10.2307/2307187

[14] HESTENES, M. R., STIEFEL, E., Methods of conjugate gradients for solving linear systems. *Journal of Research of the National Bureau of Standards*, 49, 409–435 (1952).

[15] *IEEE Standard for Binary Floating-Point Arithmetic*. in ANSI/IEEE Std 754-1985, Oct. (1985). DOI: 10.1109/IEEESTD.1985.82928.

[16] *IEEE Standard for Binary Floating-Point Arithmetic (Revision of IEEE 754-2008).* IEEE Std 754 (2019). DOI: 10.1109/IEEESTD.2019.8766229

[17] ISAACSON, E., KELLER, H. B., *Analysis of Numerical Methods.* Dover Publishing Co., 1994.

[18] KELLY, C. T. *Iterative Methods for Linear and Nonlinear Equations.* (Frontiers in Applied Mathematics, Series Number 18), Society for Industrial and Applied Mathematics (SIAM), 1995.

[19] KIUSALAAS, J., *Numerical Methods in Engineering with Python 3.* Cambridge University Press, 2013.

[20] LAMBERT, J. D., *Numerical Methods for Ordinary Differential Systems: The Initial Value Problem.* Wiley, 1991.

[21] LANCZOS, C.,*Applied Analysis*, Dover Books on Mathematics, Dover Publications, 1988.

[22] LETHER, F. G., WENSTON, P. R., Minimax approximations to the zeros of $P_n(x)$ and Gauss-Legendre quadrature. *Journal of Computational and Applied Mathematics,* 59(2) 245–252 (1995).

[23] MCNAMEE, J. M., *Numerical Methods for Roots of Polynomials—Part I.* Elsevier Science, 2013.

[24] MCNAMEE, J. M., PAN, V., *Numerical Methods for Roots of Polynomials—Part II (Volume 16) (Studies in Computational Mathematics, Volume 16).* Elsevier Science, 2018.

[25] MEURANT, G., *Computer Solution of Large Linear Systems.* Elsevier Science, 1999.

[26] NA, T. Y., SIDHOM, M. M., On Stokes' Problems for Linear Viscoelastic Fluids. *ASME Journal of Applied Mechanics* 34(4), 1040–1042 (1967). https://doi.org/10.1115/1.3607818

[27] ROHSENOW, W. M., HARTNETT, J. P., YOUNG, I. C, eds. *Handbook of Heat Transfer.* 3rd ed. McGraw-Hill Education, New York: 1998.

[28] PRESS, W. H., TEUKOLSKY, S. A., VETTERLING, W. T., FLANNERY, B. P., *Numerical Recipes in Fortran 90: Volume 2, Volume 2 of Fortran Numerical Recipes: The Art of Parallel Scientific Computing.* Cambridge University Press, 1996.

[29] SAAD, Y., *Iterative Methods for Sparse Linear Systems (The Pws Series in Computer Science).* Pws Pub Co, 1996.

[30] TAKEMASA, T., Abscissae and weights for the Gauss-Laguerre quadrature formula. *Computer Physics Communications,* 52(1), 133–140 (1988).

[31] TAKEMASA, T., Abscissae and weights for the Gauss-Hermite quadrature formula. *Computer Physics Communications,* 48, 265–270 (1988).

[32] THOMPSON, J. F., SONI, B. K., WEATHERILL, N. P. (Eds.) *Handbook of Grid Generation.* 1st ed. CRC Press. 1998.

[33] TREFETHEN, L. M., Liquid metal heat transfer in circular tubes and annuli. General discussion on heat transfer, *Journal of IMechE,* 436–438 (1951).

[34] WILKINSON, J. H., *Rounding Errors in Algebraic Processes.* Hassell Street Press, 2021.

[35] VARGA, R. S., *Matrix Iterative Analysis.* Prentice Hall, Inc., 1962.

[36] VINOKUR, M., On one-dimensional stretching functions for finite-difference calculations. *Journal of Computational Physics*, 50(2), 215–234 (1980). DOI: 10.1016/0021-9991(83)90065-7

[37] VORST, H. A. van der. *Iterative Krylov Methods for Large Linear Systems.* Cambridge University Press, 2003.

[38] YOUNG., D. M., *Iterative Solution of Large Linear Systems.* Academic Press, New York, 1971.

Index

Printed in the United States
by Baker & Taylor Publisher Services